Control Systems Engineering

SECOND EDITION

W0235482

Control Systems Engineering

SECOND EDITION

A Nagoor Kani

CBSPD

CBS Publishers & Distributors Pvt Ltd

New Delhi • Bengaluru • Chennai • Kochi • Kolkata • Lucknow • Mumbai

Hyderabad • Jharkhand • Nagpur • Patna • Pune • Uttarakhand

Control Systems Engineering

ISBN: 978-93-89239-00-3

Second Edition: 2020
Reprint 2023
First Edition: 2010

Published by Satish Kumar Jain and produced by Varun Jain for
CBS Publishers & Distributors Pvt Ltd
4819/XI Prahlad Street, 24 Ansari Road, Daryaganj, New Delhi 110 002, India
Ph: 011–23289259, 23266861   e-mail: delhi@cbspd.com   Website: www.cbspd.com

Corporate Office: 204 FIE, Industrial Area, Patparganj, Delhi 110 092, India
Ph: 011–49344934   Fax: 011–49344935   e-mail: publishing@cbspc.com; publicity@cbspd.com

Branches

- **Bengaluru:** Seema House 2975, 17th Cross, K.R. Road, Banasankari 2nd Stage, Bengaluru 560 070, Karnataka, India
  Ph: +91-80-26771678/79   Fax: +91-80-26771680   e-mail: bangalore@cbspd.com
- **Chennai:** 7, Subbaraya Street, Shenoy Nagar, Chennai 600 030, Tamil Nadu, India
  Ph: +91-44-26680620, 26681266   Fax: +91-44-42032115   e-mail: chennai@cbspd.com
- **Kochi:** 42/1325, 1326, Power House Road, Opposite KSEB, Power House, Ernakulum-632018, Kochi, Kerala, India
  Ph: +91-484-4059061–67   Fax: +91-484-4059065   e-mail: kochi@cbspd.com
- **Kolkata:** 147, Hind Ceramics Compound, 1st Floor, Nilgunj Road, Belghoria, Kolkata 700056, West Bengal, India
  Ph: +91-33-25633055/56   e-mail: kolkata@cbspd.com
- **Lucknow:** Basement, Khushuma Complex, 7 Meerabai Marg (Behind Jawahar Bhawan), Lucknow 226001, UP, India
  Ph: +91-522-40000032   e-mail: tiwari.lucknow@cbspd.com
- **Mumbai:** PWD Shed, Gala No. 25/26, Ramchandra Bhatt Marg, Next JJ Hospital Gate No. 2, Opp. Union Bank of India, Noorbaug, Mumbai-400009, Maharashtra, India
  Ph: +91-22-66661880/89   e-mail: mumbai@cbspd.com

Representatives

• Hyderabad	0-9885175004	• Jharkhand	0-9811541605	• Nagpur	0-9421945513
• Patna	0-9334159340	• Pune	0-9923910676	• Uttarakhand	0-9716462459

Printed at Mudrak, Noida, UP, India

to

my brother-in-law Prof NS Abdul Hameed
sister Mrs A Rabia Hameed and
their son Er A Khilji and
daughter Dr A Parveen

PREFACE

Control system plays a significant role in the advance of all engineering disciplines and sciences like space vehicle systems, missile guidance systems, robotic systems and all manufacturing and industrial processes. Hence, it becomes necessary for the engineering students to be familiar with the basic concepts of control systems. This text is designed for the undergraduate course in control systems for engineering students.

This book is organized into 7 chapters and appendices. The fundamental concepts, modeling, design and analysis of control systems are presented in a very easiest and elaborative manner. Analysis and design of both continuous time and discrete time control systems are well structured. Throughout the book, carefully chosen examples are presented so that the reader will have a clear understanding of the concepts discussed. In addition to the fundamentals, MATLAB programs are provided for design and analysis of control systems like block diagram reduction, time response analysis, frequency response analysis and state space analysis with clear explanations.

Chapter 1 presents the basic concepts of control systems and mathematical modeling of mechanical and electrical systems. Development of transfer function of a system, block diagram reduction and signal flow graph methods are dealt with detailed examples and solved problems. The MATLAB based approach to block diagram reduction is also introduced in this chapter aided with neat explanation.

Chapter 2 focuses on time response analysis of first order and second systems. The time domain specifications, error analysis for various types of test input signals, the concepts of steady state error, static and generalized error coefficients are also explained. The rest of this chapter deals with various types of controllers such as P, PI , PD and PID controllers for automatic control systems. The MATLAB programs for step response, impulse response and for any arbitrary input, and estimation of time domain specifications are also presented in this chapter.

Chapter 3 is concerned with frequency response analysis. Various frequency domain specifications, frequency response plots like Bode plot, polar plot, Nichols plot, Nichols chart and closed loop frequency response are presented with clear explanation and more number of solved problems. The MATLAB programs to obtain frequency response plots are provided with neat explanation.

Chapter 4 introduces the concepts of stability analysis and root locus technique. Location of poles on s-plane for stability, Routh–Hurwitz criterion, Nyquist stability criterion, relative stability, and root locus technique are discussed with number of solved numericals. In this chapter, the MATLAB programs are included to plot Nyquist and root locus plots.

Chapter 5 is devoted to linear system design using compensators. The various types of compensators such as, lag, lead and lag-lead are designed using both Bode diagram and root locus approach. The design of P, PI, PID controllers are also presented in this chapter. The concept of feedback compensation is also discussed briefly.

Chapter 6 deals with state space analysis of both continuous time and discrete time systems. State space representation, concepts of controllability and observability are given with proper explanation and examples. Finally this chapter concludes with MATLAB programs to obtain state model, transfer function and to check the controllability and observability of the system.

Chapter 7 introduces the concepts of sampled data control systems. The concepts of sampling, reconstruction of sampled signals, Z-transform, impulse sampling, pulse transfer function and stability of sampled data systems are discussed in this chapter.

Several concepts and procedures are illustrated through simple examples rather than generalized formulation. Each chapter provides the foundations and practical implications of their own topic with a large number of solved problems for better understanding.

Special care has been taken in presenting the concepts, choosing examples and solved problems. I hope that the teaching and student community will welcome the book. The readers can feel free to convey their suggestions or criticism for further improvement of the book.

A Nagoor Kani

ACKNOWLEDGEMENTS

I express my heartfelt thanks to my wife Mrs C Gnanaparanjothi Nagoor Kani, and my sons N Bharath Raj alias Chandrakani Allaudeen and N Vikram Raj for their support, encouragement and cooperation they have extended to me throughout my career. I thank Ms T A Benazir, Manager, RBA Group, and all my office staff for their cooperation in carrying out my day-to-day activities.

I sincerely acknowledge the contributions of our technical editors, Ms C Mohana Priya and Ms S Saranya and Ms P Kanimozhi, for editing, proof-reading and type-setting of the manuscript and preparing the layout of the book.

My sincere thanks to all reviewers for their valuable suggestions and comments which helped me to explore the subject to a greater depth.

I am also grateful to Mr Satish K Jain, CMD, CBS Publishers & Distributors, for his keen interest in publishing this work in CBS banner. My sincere thanks to all team members of CBS Publishers & Distributors, for their concern and care in publishing this work.

Finally, a special note of appreciation is due to my sisters, brothers, relatives, friends, students and the entire teaching community, for their overwhelming support and encouragement to my writing.

A Nagoor Kani

CONTENTS

CHAPTER 4: Concepts of Stability and Root Locus   4.1–4.106

CHAPTER 5: Linear System Design   5.1–5.102

CHAPTER 6: State Space Analysis   6.1–6.106

CHAPTER 7: Sampled Data Control Systems   7.1–7.66

Appendices   A.1–A.11

Index   1.1– 1.5

LIST OF SYMBOLS AND ABBREVIATIONS

Symbols

B — Viscous friction coefficient, N-sec/m

e_b — Back emf, v

e_{ss} — Steady state error

f — Applied force, N

f_m — Opposing force offered by the mass of the body, N

f_k — Opposing force offered by the elasticity of the body, N

f_b — Opposing force offered by the friction of the body, N

F_s — Sampling frequency, Hz

G — Conductance, mho

H — Transformation or operator

i_a — Armature current, A

i_f — Field current, A

J — Moment of inertia, kg-m^2/rad

j — Complex operator

K — Stiffness of the spring, N-m / rad

K_a — Acceleration error constant

K_b — Back emf constant, V / (rad/sec)

K_d — Derivative constant or gain

K_i — Integral constant or gain

K_g — Gain Margin

K_m — Motor gain constant

K_p — Proportional gain

K_t — Torque constant, N-m/A

K_{tf} — Torque constant, Nm / A

K_v — Velocity error constant

L_a — Armature inductance, H

L_f — Field inductance, H

M — Mass, kg

M_p — Maximum overshoot

M_r — Resonant peak

n — Order of the system

N — Type number

p — Pole of a system

p_c — Pole of compensator

P_k — Forward path gain of Kth forward path

q — Charge

R_a — Armature resistance, Ω

R_f — Field resistance, Ω

s — Complex variable

s_d — Dominant pole

T — Applied torque, N-m

T_a — Electrical time constant

T_b — Opposing torque due to friction, N-m

t_d — Delay time

T_d — Derivative time

T_f — Field time constant

T_i — Integral time

T_j — Opposing torque due to moment of inertia, N-m

T_k — Opposing torque due to elasticity, N-m

T_m — Mechanical time constant

t_p — Peak time

t_r — Rise time

t_s — Settling time

u_b — Normalized bandwidth

u_r — Normalized resonant frequency

V_a — Armature voltage, V

V_f — Field voltage, V

x — Displacement, m

z — Zero of a system

z_c — Zero of compensator

θ — Angular displacement, rad

$\dfrac{d\theta}{dt}$ — Angular velocity, rad/sec

$\dfrac{d^2\theta}{dt}$ — Angular acceleration, rad/sec^2

ω_n — Undamped natural frequency, rad/sec

ζ — Damping ratio

ω_r — Resonant frequency

ω_b	Bandwidth
γ	Phase margin
ω_{pc}	Phase crossover frequency
ω_{gc}	Gain crossover frequency
ϕ	Flux, weber
ω_c	Corner frequency
ω_d	Damped frequency of oscillation
α	Phase angle
ω_m	Frequency of maximum phase lag/lead
ϕ_m	Maximum lag/lead angle
ε	Additional phase lead
ϕ_a	Angle of asymptotes
ϕ_p	Angle of departure
ϕ_z	Angle of arrival
λ	Eigen value
δ_T	Impulse train

Standard Input/Output signals

$c(t)$	Response in time domain
$c(k)$	Response of discrete signal
$e(t)$	Error signal
$f(kT)$	Digital error signal
$g(kT)$	Digital control signal
$r(t)$	Input in time domain
$r(k)$	Discrete time input signal
$u(t)$	Control signal (Analog)
$\delta(t)$	Impulse signal

Matrices and Vectors

$\mathbf{A}$	System matrix
A^κ	State transition matrix of discrete system
$\mathbf{B}$	Input matrix
$\mathbf{C}$	Output matrix
$\mathbf{D}$	Transmission matrix
e^{At}	State transition matrix

$\mathbf{I}$	Identity matrix
$\mathbf{J}$	Jordan matrix
$\mathbf{M}$	Modal matrix or diagonalization matrix
$\mathbf{P}_0 / \mathbf{P}_c$	Transformation matrix
$\mathbf{Q}_c$	Composite matrix for controllability
$\mathbf{Q}_0$	Composite matrix for observability
$\mathbf{U}(t)$	Input vector
$\mathbf{U}(k)$	Input vector of discrete time system
$\mathbf{X}(t)$	State variable vector
$\mathbf{X}_o$	Initial condition vector
$\mathbf{X}(k)$	State vector of discrete time system
$\mathbf{Y}(t)$	Output vector
$\mathbf{Y}(k)$	Output vector of discrete time system
$\wedge$	Grammian matrix

Transform Operators and Functions

$A(s)$	Auxiliary polynomial
$E(s)$	Error signal in s-domain
$G(s)$	Open loop transfer function
$G(s)H(s)$	Loop transfer function
$H(s)$	Feedback transfer function
$\mathcal{L}$	Laplace transform
$\mathcal{L}^{-1}$	Inverse Laplace transform
$M(s)$	Closed loop transfer function
$T(s)$	Transfer function of the system
Z	Z-transform
Z^{-1}	Inverse Z-transform

Abbreviations

BIBO	Bounded Input Bounded Output
LDS	Linear Discrete Time System
LTI	Linear Time Invariant System
ROC	Region of convergence
ZOH	Zero Order Hold

CHAPTER 1

MATHEMATICAL MODELS OF CONTROL SYSTEM

1.1 CONTROL SYSTEM

Control system theory evolved as an engineering discipline and due to universality of the principles involved, it is extended to various fields like economy, sociology, biology, medicine, etc. Control theory has played a vital role in the advance of engineering and science. The automatic control has become an integral part of modern manufacturing and industrial processes. For example, numerical control of machine tools in manufacturing industries, controlling pressure, temperature, humidity, viscosity and flow in process industry.

When a number of elements or components are connected in a sequence to perform a specific function, the group thus formed is called a *system.* In a system when the output quantity is controlled by varying the input quantity, the system is called *control system.* The output quantity is called controlled variable or response and input quantity is called command signal or excitation.

OPEN LOOP SYSTEM

Any physical system which does not automatically correct the variation in its output, is called an *open loop system*, or control system in which the output quantity has no effect upon the input quantity are called open-loop control system. This means that the output is not fedback to the input for correction.

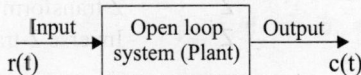

Fig 1.1 : *Open loop system.*

In open loop system the output can be varied by varying the input. But due to external disturbances the system output may change. When the output changes due to disturbances, it is not followed by changes in input to correct the output. In open loop systems the changes in output are corrected by changing the input manually.

CLOSED LOOP SYSTEM

Control systems in which the output has an effect upon the input quantity in order to maintain the desired output value are called *closed loop systems.*

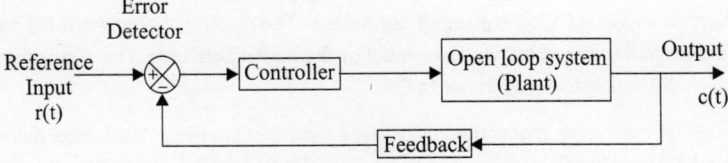

Fig 1.2 : *Closed loop system.*

The open loop system can be modified as closed loop system by providing a feedback. The provision of feedback automatically corrects the changes in output due to disturbances. Hence the closed loop system is also called *automatic control system*. The general block diagram of an automatic control system is shown in fig 1.2. It consists of an error detector, a controller, plant (open loop system) and feedback path elements.

The reference signal (or input signal) corresponds to desired output. The feedback path elements samples the output and converts it to a signal of same type as that of reference signal. The feedback signal is proportional to output signal and it is fed to the error detector. The error signal generated by the error detector is the difference between reference signal and feedback signal. The controller modifies and amplifies the error signal to produce better control action. The modified error signal is fed to the plant to correct its output.

Advantages of open loop systems

1. The open loop systems are simple and economical.
2. The open loop systems are easier to construct.
3. Generally the open loop systems are stable.

Disadvantages of open loop systems

1. The open loop systems are inaccurate and unreliable.
2. The changes in the output due to external disturbances are not corrected automatically.

Advantages of closed loop systems

1. The closed loop systems are accurate.
2. The closed loop systems are accurate even in the presence of non-linearities.
3. The sensitivity of the systems may be made small to make the system more stable.
4. The closed loop systems are less affected by noise.

Disadvantages of closed loop systems

1. The closed loop systems are complex and costly.
2. The feedback in closed loop system may lead to oscillatory response.
3. The feedback reduces the overall gain of the system.
4. Stability is a major problem in closed loop system and more care is needed to design a stable closed loop system.

1.2 EXAMPLES OF CONTROL SYSTEMS

EXAMPLE 1 : TEMPERATURE CONTROL SYSTEM

OPEN LOOP SYSTEM

The electric furnace shown in fig 1.3. is an open loop system. The output in the system is the desired temperature. The temperature of the system is raised by heat generated by the heating element. The output temperature depends on the time during which the supply to heater remains ON.

The ON and OFF of the supply is governed by the time setting of the relay. The temperature is measured by a sensor, which gives an analog voltage corresponding to the temperature of the furnace. The analog signal is converted to digital signal by an Analog - to - Digital converter (A/D converter).

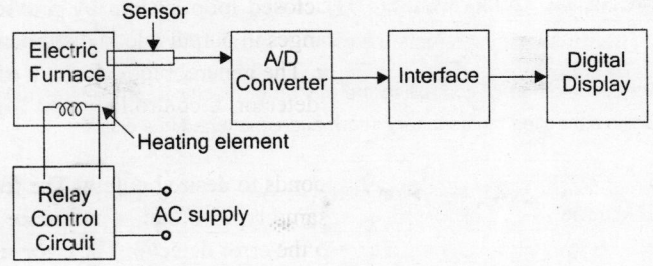

Fig 1.3 : *Open loop temperature control system.*

The digital signal is given to the digital display device to display the temperature. In this system if there is any change in output temperature then the time setting of the relay is not altered automatically.

CLOSED LOOP SYSTEM

The electric furnace shown in fig 1.4 is a closed loop system. The output of the system is the desired temperature and it depends on the time during which the supply to heater remains ON.

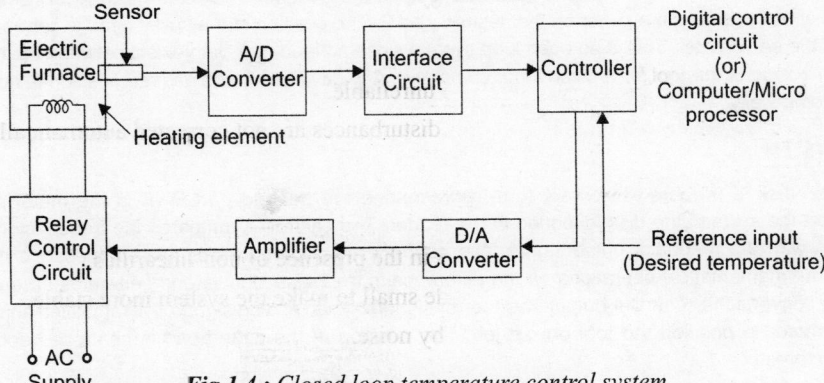

Fig 1.4 : *Closed loop temperature control system.*

The switching ON and OFF of the relay is controlled by a controller which is a digital system or computer. The desired temperature is input to the system through keyboard or as a signal corresponding to desired temperature via ports. The actual temperature is sensed by sensor and converted to digital signal by the A/D converter. The computer reads the actual temperature and compares with desired temperature. If it finds any difference then it sends signal to switch ON or OFF the relay through D/A converter and amplifier. Thus the system automatically corrects any changes in output. Hence it is a closed loop system.

EXAMPLE 2 : TRAFFIC CONTROL SYSTEM

OPEN LOOP SYSTEM

Traffic control by means of traffic signals operated on a time basis constitutes an open-loop control system. The sequence of control signals are based on a time slot given for each signal. The time slots are decided based on a traffic study. The system will not measure the density of the traffic before giving the signals. Since the time slot does not changes according to traffic density, the system is open loop system.

CLOSED LOOP SYSTEM

Traffic control system can be made as a closed loop system if the time slots of the signals are decided based on the density of traffic. In closed loop traffic control system, the density of the traffic is measured on all the sides and the information is fed to a computer. The timings of the control signals are decided by the computer based on the density of traffic. Since the closed loop system dynamically changes the timings, the flow of vehicles will be better than open loop system.

EXAMPLE 3 : NUMERICAL CONTROL SYSTEM

OPEN LOOP SYSTEM

Numerical control is a method of controlling the motion of machine components using numbers. Here, the position of work head tool is controlled by the binary information contained in a disk.

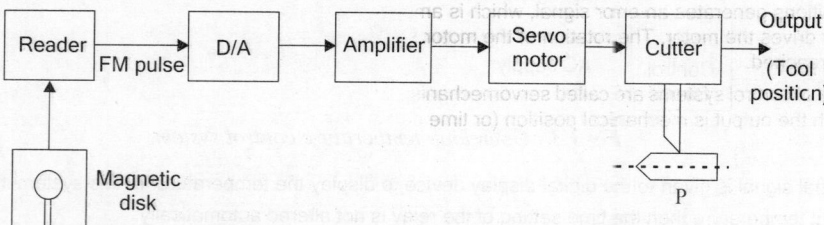

Fig 1.5 : *Open loop numerical control system.*

A magnetic disk is prepared in binary form representing the desired part P (P is the metal part to be machined). The tool will operate on the desired part P. To start the system, the disk is fed through the reader to the D/A converter. The D/A converter converts the FM(frequency modulated) output of the reader to a analog signal. It is amplified and fed to servometer which positions the cutter on the desired part P. The position of the cutter head is controlled by the angular motion of the servometer. This is an open loop system since no feedback path exists between the output and input. The system positions the tool for a given input command. Any deviation in the desired position is not checked and corrected automatically.

CLOSED LOOP SYSTEM

A magnetic disk is prepared in binary form representing the desired part P (P is the metal part to be machined). To start the system, the disk is loaded in the reader. The controller compares the frequency modulated input pulse signal with the feedback pulse signal. The controller is a computer or microprocessor system. The controller carries out mathematical operations on the difference in the pulse signals and generates an error signal. The D/A converter converts the controller output pulse (error signal) into an analog signal . The amplified analog signal rotates the servomotor to position the tool on the job. The position of the cutterhead is controlled according to the input of the servomotor.

The transducer attached to the cutterhead converts the motion into an electrical signal. The analog electrical signal is converted to the digital pulse signal by the A/D converter. Then this signal is compared with the input pulse signal. If there is any difference between these two, the controller sends a signal to the servomotor to reduce it. Thus the system automatically corrects any deviation in the desired output tool position. An advantage of numerical control is that complex parts can be produced with uniform tolerances at the maximum milling speed.

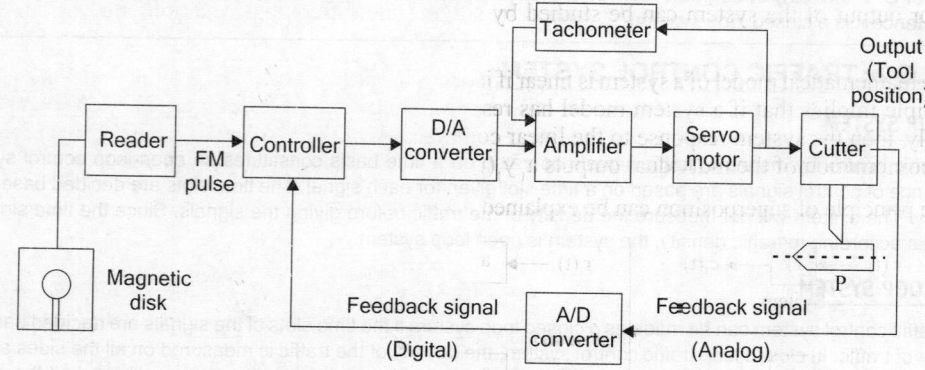

Fig 1.6 : *Closed loop numerical control system.*

EXAMPLE 4 : POSITION CONTROL SYSTEM USING SERVOMOTOR

The position control system shown in fig 1.7 is a closed loop system. The system consists of a servomotor powered by a generator. The load whose position has to be controlled is connected to motor shaft through gear wheels. Potentiometers are used to convert the mechanical motion to electrical signals. The desired load position (θ_R) is set on the input potentiometer and the actual load position (θ_c) is fed to feedback potentiometer. The difference between the two angular positions generates an error signal, which is amplified and fed to generator field circuit. The induced emf of the generator drives the motor. The rotation of the motor stops when the error signal is zero, i.e. when the desired load position is reached.

This type of control systems are called servomechanisms .The servo or servomechanisms are feedback control systems in which the output is mechanical position (or time derivatives of position e.g. velocity and acceleration).

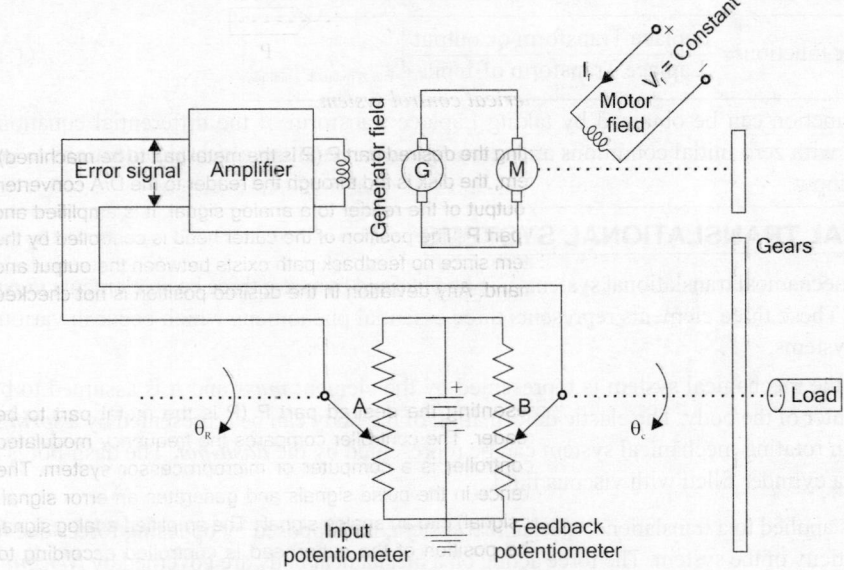

Fig 1.7 : *A position control system (servomechanism).*

1.3 MATHEMATICAL MODELS OF CONTROL SYSTEMS

A *control system* is a collection of physical objects (components) connected together to serve an objective. The input output relations of various physical components of a system are governed by *differential equations.* The mathematical model of a control system constitutes a set of differential equations. The response or output of the system can be studied by solving the differential equations for various input conditions.

The mathematical model of a system is linear if it obeys the principle of superposition and homogenity. This principle implies that if a system model has responses $y_1(t)$ and $y_2(t)$ to any inputs $x_1(t)$ and $x_2(t)$ respectively, then the system response to the linear combination of these inputs $a_1 x_1(t) + a_2 x_2(t)$ is given by linear combination of the individual outputs $a_1 y_1(t) + a_2 y_2(t)$, where a_1 and a_2 are constants.

The principle of superposition can be explained diagrammatically as shown in fig. 1.8.

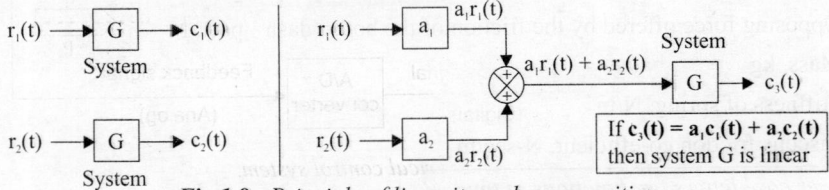

Fig 1.8 : *Principle of linearity and superposition.*

A mathematical model will be linear if the differential equations describing the system has constant coefficients (or the coefficients may be functions of independent variables). If the coefficients of the differential equation describing the system are constants then the model is ***linear time invariant.*** If the coefficients of differential equations governing the system are functions of time then the model is ***linear time varying.***

The differential equations of a linear time invariant system can be reshaped into different form for the convenience of analysis. One such model for single input and single output system analysis is transfer function of the system. The ***transfer function*** of a system is defined as the ratio of Laplace transform of output to the Laplace transform of input with zero initial conditions.

$$\text{Transfer function} = \frac{\text{Laplace Transform of output}}{\text{Laplace Transform of input}} \Bigg|_{\text{with zero initial condition}} \quad \dots (1.1)$$

The transfer function can be obtained by taking Laplace transform of the differential equations governing the system with zero initial conditions and rearranging the resulting algebraic equations to get the ratio of output to input.

1.4 MECHANICAL TRANSLATIONAL SYSTEMS

The model of mechanical translational systems can be obtained by using three basic elements ***mass, spring and dash-pot.*** These three elements represents three essential phenomena which occur in various ways in mechanical systems.

The weight of the mechanical system is represented by the element ***mass*** and it is assumed to be concentrated at the center of the body. The elastic deformation of the body can be represented by a ***spring.*** The friction existing in rotating mechanical system can be represented by the ***dash-pot.*** The dash-pot is a piston moving inside a cylinder filled with viscous fluid.

When a force is applied to a translational mechanical system, it is opposed by opposing forces due to mass, friction and elasticity of the system. The force acting on a mechanical body are governed by ***Newton's second law of motion.*** For translational systems it states that the sum of forces acting on a body is zero. (or Newton's second law states that the sum of applied forces is equal to the sum of opposing forces on a body).

LIST OF SYMBOLS USED IN MECHANICAL TRANSLATIONAL SYSTEM

x = Displacement, m

$v = \dfrac{dx}{dt}$ = Velocity, m/sec

$a = \dfrac{dv}{dt} = \dfrac{d^2x}{dt^2}$ = Acceleration, m/sec^2

f = Applied force, N (Newtons)

f_m = Opposing force offered by mass of the body, N

f_k = Opposing force offered by the elasticity of the body (spring), N

f_b = Opposing force offered by the friction of the body (dash - pot), N

M = Mass, kg

K = Stiffness of spring, N/m

B = Viscous friction co-efficient, N-sec/m

Note : *Lower case letters are functions of time*

FORCE BALANCE EQUATIONS OF IDEALIZED ELEMENTS

Consider an ideal mass element shown in fig 1.9 which has negligible friction and elasticity. Let a force be applied on it. The mass will offer an opposing force which is proportional to acceleration of the body.

Let, f = Applied force

f_m = Opposing force due to mass

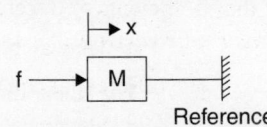

Fig 1.9 : Ideal mass element.

Here, $f_m \propto \dfrac{d^2x}{dt^2}$ or $f_m = M\dfrac{d^2x}{dt^2}$

By Newton's second law, $\boxed{f = f_m = M\dfrac{d^2x}{dt^2}}$ .....(1.2)

Consider an ideal frictional element dashpot shown in fig 1.10 which has negligible mass and elasticity. Let a force be applied on it. The dash-pot will offer an opposing force which is proportional to velocity of the body.

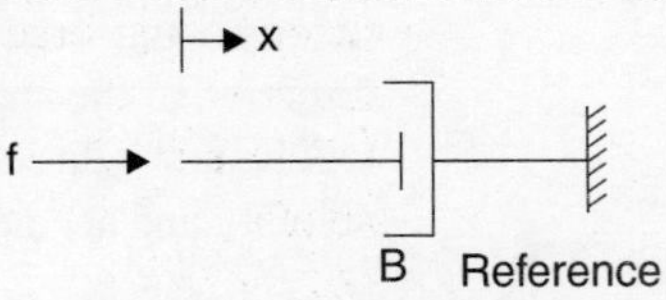

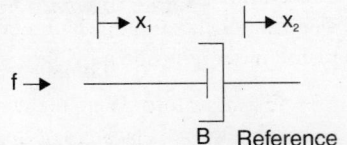

Fig 1.10 : Ideal dashpot with one end fixed to reference.

Let, f = Applied force

f_b = Opposing force due to friction

Here, $f_b \propto \dfrac{dx}{dt}$ or $f_b = B\dfrac{dx}{dt}$

By Newton's second law, $\boxed{f = f_b = B\dfrac{dx}{dt}}$ .....(1.3)

When the dashpot has displacement at both ends as shown in fig 1.11, the opposing force is proportional to difference between velocity at both ends.

$f_b \propto \dfrac{d}{dt}(x_1 - x_2)$ or $f_b = B\dfrac{d}{dt}(x_1 - x_2)$

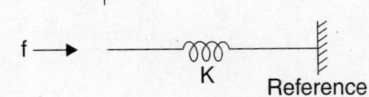

Fig 1.11 : Ideal dashpot with displacement at both ends.

$\therefore \boxed{f = f_b = B\dfrac{d}{dt}(x_1 - x_2)}$ .....(1.4)

Consider an ideal elastic element spring shown in fig 1.12, which has negligible mass and friction. Let a force be applied on it. The spring will offer an opposing force which is proportional to displacement of the body.

Let, f = Applied force

f_k = Opposing force due to elasticity

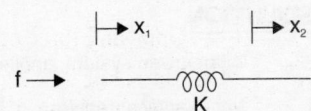

Fig 1.12 : Ideal spring with one end fixed to reference.

Here $f_k \propto x$ or $f_k = K x$

By Newton's second law, $\boxed{f = f_k = Kx}$ .....(1.5)

When the spring has displacement at both ends as shown in fig 1.13 the opposing force is proportional to difference between displacement at both ends.

$f_k \propto (x_1 - x_2)$ or $f_k = K(x_1 - x_2)$

$\therefore$ $f = f_k = K(x_1 - x_2)$ .....(1.6)

Fig 1.13 : Ideal spring with displacement at both ends.

Guidelines to determine the Transfer Function of Mechanical Translational System

1. In mechanical translational system, the differential equations governing the system are obtained by writing force balance equations at nodes in the system. The nodes are meeting point of elements. Generally the nodes are mass elements in the system. In some cases the nodes may be without mass element.

2. The linear displacement of the masses (nodes) are assumed as x_1, x_2, x_3, etc., and assign a displacement to each mass(node). The first derivative of the displacement is velocity and the second derivative of the displacement is acceleration.

3. Draw the free body diagrams of the system. The free body diagram is obtained by drawing each mass separately and then marking all the forces acting on that mass (node). Always the opposing force acts in a direction opposite to applied force. The mass has to move in the direction of the applied force. Hence the displacement, velocity and acceleration of the mass will be in the direction of the applied force. If there is no applied force then the displacement, velocity and acceleration of the mass will be in a direction opposite to that of opposing force.

4. For each free body diagram, write one differential equation by equating the sum of applied forces to the sum of opposing forces.

5. Take Laplace transform of differential equations to convert them to algebraic equations. Then rearrange the s-domain equations to eliminate the unwanted variables and obtain the ratio between output variable and input variable. This ratio is the transfer function of the system.

Note : *Laplace transform of $x(t) = \mathcal{L}\{x(t)\} = X(s)$*

Laplace transform of $\dfrac{dx(t)}{dt} = \mathcal{L}\left\{ \dfrac{d}{dt} x(t) \right\} = s\, X(s)$ (with zero initial conditions)

Laplace transform of $\dfrac{d^2 x(t)}{dt^2} = \mathcal{L}\left\{ \dfrac{d^2}{dt^2} x(t) \right\} = s^2\, X(s)$ (with zero initial conditions)

EXAMPLE 1.1

Write the differential equations governing the mechanical system shown in fig 1. and determine the transfer function.

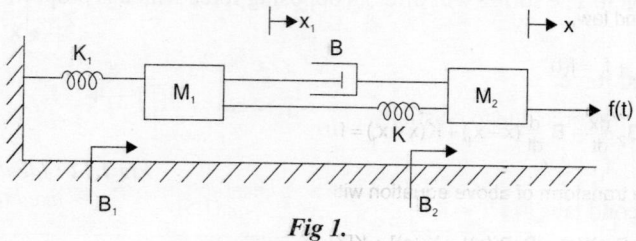

Fig 1.

SOLUTION

In the given system, applied force 'f(t)' is the input and displacement 'x' is the output.

Let, Laplace transform of f(t) $= \mathcal{L}\{f(t)\} = F(s)$

Laplace transform of x $= \mathcal{L}\{x\} = X(s)$

Laplace transform of $x_1 = \mathcal{L}\{x_1\} = X_1(s)$

Hence the required transfer function is $\dfrac{X(s)}{F(s)}$

The system has two nodes and they are mass M_1 and M_2. The differential equations governing the system are given by force balance equations at these nodes.

Let the displacement of mass M_1 be x_1. The free body diagram of mass M_1 is shown in fig 2. The opposing forces acting on mass M_1 are marked as f_{m1}, f_{b1}, f_b, f_{k1} and f_k.

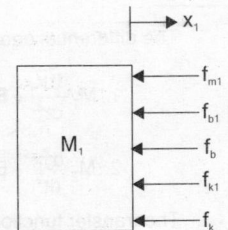

$$f_{m1} = M_1 \frac{d^2x_1}{dt^2} \quad ; \quad f_{b1} = B_1 \frac{dx_1}{dt} \quad ; \quad f_{k1} = K_1 x_1$$

$$f_b = B \frac{d}{dt}(x_1 - x) \quad ; \quad f_k = K(x_1 - x)$$

By Newton's second law,

$$f_{m1} + f_{b1} + f_b + f_{k1} - f_k = 0$$

$$\therefore \; M_1 \frac{d^2x_1}{dt^2} + B_1 \frac{dx_1}{dt} + B \frac{d}{dt}(x_1 - x) + K_1 x_1 + K(x_1 - x) = 0$$

Fig 2 : *Free body diagram of mass M_1 (node 1).*

On taking Laplace transform of above equation with zero initial conditions we get,

$$M_1 s^2 X_1(s) + B_1 s X_1(s) + Bs[X_1(s) - X(s)] + K_1 X_1(s) + K[X_1(s) - X(s)] = 0$$

$$X_1(s) [M_1 s^2 + (B_1 + B)s + (K_1 + K)] - X(s) [Bs + K] = 0$$

$$X_1(s) [M_1 s^2 + (B_1 + B)s + (K_1 + K)] = X(s) [Bs + K]$$

$$\boxed{\therefore \; X_1(s) = X(s) \frac{Bs + K}{M_1 s^2 + (B_1 + B)s + (K_1 + K)}} \qquad \qquad ......(1)$$

The free body diagram of mass M_2 is shown in fig 3. The opposing forces acting on M_2 are marked as f_{m2}, f_{b2}, f_b and f_k.

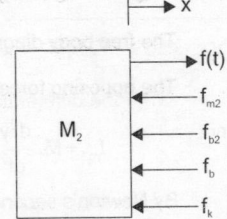

$$f_{m2} = M_2 \frac{d^2x}{dt^2} \quad ; \quad f_{b2} = B_2 \frac{dx}{dt}$$

$$f_{m2} = B \frac{dx}{dt}(x - x_1) \quad ; \quad f_k = K(x - x_1)$$

By Newton's second law,

$$f_{m2} + f_{b2} + f_b + f_k = f(t)$$

$$M_2 \frac{d^2x}{dt^2} + B_2 \frac{dx}{dt} + B \frac{d}{dt}(x - x_1) + K(x - x_1) = f(t)$$

Fig 3 : *Free body diagram of mass M_2 (node 2).*

On taking Laplace transform of above equation with zero initial conditions we get,

$$M_2 s^2 X(s) + B_2 s X(s) + Bs[X(s) - X_1(s)] + K[X(s) - X_1(s)] = F(s)$$

$$X(s)[M_2 s^2 + (B_2 + B)s + K] - X_1(s)[Bs + K] = F(s) \qquad \qquad .....(2)$$

Substituting for $X_1(s)$ from equation (1) in equation (2) we get,

$$X(s)[M_2 s^2 + (B_2 + B)s + K] - X(s) \frac{(Bs + K)^2}{M_1 s^2 + (B_1 + B)s + (K_1 + K)} = F(s)$$

$$X(s) \left[\frac{[M_2s^2 + (B_2 + B)s + K][M_1s^2 + (B_1 + B)s + (K_1 + K)] - (Bs + K)^2}{M_1s^2 + (B_1 + B)s + (K_1 + K)} \right] = F(s)$$

$$\therefore \frac{X(s)}{F(s)} = \frac{M_1s^2 + (B_1 + B)s + (K_1 + K)}{[M_1s^2 + (B_1 + B)s + (K_1 + K)][M_2s^2 + (B_2 + B)s + K] - (Bs + K)^2}$$

RESULT

The differential equations governing the system are,

1. $M_1 \dfrac{d^2x_1}{dt^2} + B_1 \dfrac{dx_1}{dt} + B \dfrac{d}{dt}(x_1 - x) + K_1x_1 + K(x_1 - x) = 0$

2. $M_2 \dfrac{d^2x}{dt^2} + B_2 \dfrac{dx}{dt} + B \dfrac{d}{dt}(x - x_1) + K(x - x_1) = f(t)$

The transfer function of the system is,

$$\frac{X(s)}{F(s)} = \frac{M_1s^2 + (B_1 + B)s + (K_1 + K)}{[M_1s^2 + (B_1 + B)s + (K_1 + K)][M_2s^2 + (B_2 + B)s + K] - (Bs + K)^2}$$

EXAMPLE 1.2

Determine the transfer function $\dfrac{Y_2(s)}{F(s)}$ of the system shown in fig 1.

SOLUTION

Let,   Laplace transform of $f(t) = \mathcal{L}\{f(t)\} = F(s)$

Laplace transform of $y_1 = \mathcal{L}\{y_1\} = Y_1(s)$

Laplace transform of $y_2 = \mathcal{L}\{y_2\} = Y_2(s)$

The system has two nodes and they are mass M_1 and M_2. The differential equations governing the system are the force balance equations at these nodes.

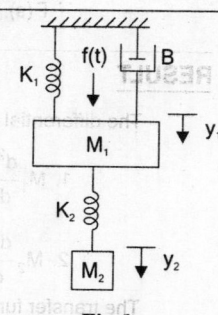

Fig 1.

The free body diagram of mass M_1 is shown in fig 2.

The opposing forces are marked as f_{m1}, f_b, f_{k1} and f_{k2}

$$f_{m1} = M_1 \frac{d^2y_1}{dt^2} \quad ; \quad f_b = B \frac{dy_1}{dt} \quad ; \quad f_{k1} = K_1y_1 \quad ; \quad f_{k2} = K_2(y_1 - y_2)$$

By Newton's second law, $f_{m1} + f_b + f_{k1} + f_{k2} = f(t)$

$$\therefore M_1 \frac{d^2y_1}{dt^2} + B \frac{dy_1}{dt} + K_1y_1 + K_2(y_1 - y_2) = f(t) \quad .....(1)$$

On taking Laplace transform of equation (1) with zero initial condition we get,

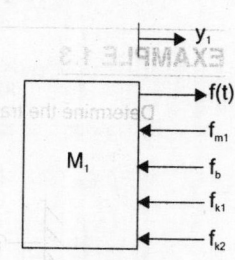

Fig 2.

$$M_1s^2Y_1(s) + BsY_1(s) + K_1Y_1(s) + K_2[Y_1(s) - Y_2(s)] = F(s)$$

$$Y_1(s)[M_1s^2 + Bs + (K_1 + K_2)] - Y_2(s)K_2 = F(s) \quad .....(2)$$

The free body diagram of mass M_2 is shown in fig 3. The opposing forces acting on M_2 are f_{m2} and f_{k2}.

$$f_{m2} = M_2 \frac{d^2y_2}{dt^2} \quad ; \quad f_{k2} = K_2(y_2 - y_1)$$

By Newton's second law, $f_{m2} + f_{k2} = 0$

$$\therefore M_2 \frac{d^2 y_2}{dt^2} + K_2(y_2 - y_1) = 0$$

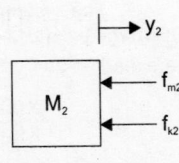

Fig 3.

On taking Laplace transform of above equation we get,

$$M_2 s^2 Y_2(s) + K_2[Y_2(s) - Y_1(s)] = 0$$

$$Y_2(s)[M_2 s^2 + K_2] - Y_1(s) K_2 = 0$$

$$\therefore Y_1(s) = Y_2(s) \frac{M_2 s^2 + K_2}{K_2} \qquad \qquad .....(3)$$

Substituting for $Y_1(s)$ from equation (3) in equation (2) we get,

$$Y_2(s) \left[\frac{M_2 s^2 + K_2}{K_2} \right] [M_1 s^2 + Bs + (K_1 + K_2)] - Y_2(s) K_2 = F(s)$$

$$Y_2(s) \left[\frac{(M_2 s^2 + K_2)[M_1 s^2 + Bs + (K_1 + K_2)] - K_2^2}{K_2} \right] = F(s)$$

$$\therefore \frac{Y_2(s)}{F(s)} = \left[\frac{K_2}{[M_1 s^2 + Bs + (K_1 + K_2)](M_2 s^2 + K_2) - K_2^2} \right]$$

RESULT

The differential equations governing the system are,

$$1. \ M_1 \frac{d^2 y_1}{dt^2} + B \frac{dy_1}{dt} + K_1(y_1 - y_2) = f(t)$$

$$2. \ M_2 \frac{d^2 y_2}{dt^2} + K_2(y_2 - y_1) = 0$$

The transfer function of the system is,

$$\frac{Y_2(s)}{F(s)} = \frac{K_2}{[M_1 s^2 + Bs - (K_1 + K_2)][M_2 s^2 + K_2] - K_2^2}$$

EXAMPLE 1.3

Determine the transfer function, $\dfrac{X_1(s)}{F(s)}$ and $\dfrac{X_2(s)}{F(s)}$ for the system shown in fig 1.

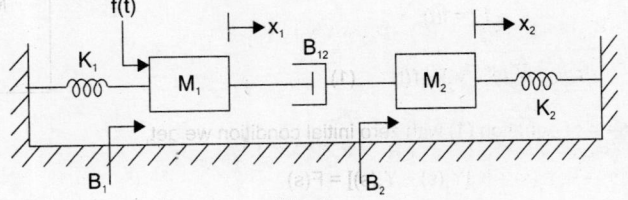

Fig 1.

SOLUTION

Let,   Laplace transform of $f(t)$ = $\mathcal{L}\{f(t)\}$ = $F(s)$

       Laplace transform of x_1 = $\mathcal{L}\{x_1\}$ = $X_1(s)$

       Laplace transform of x_2 = $\mathcal{L}\{x_2\}$ = $X_2(s)$

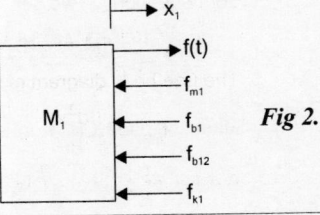

Fig 2.

The system has two nodes and they are mass M_1 and M_2. The differential equations governing the system are the force balance equations at these nodes. The free body diagram of mass M_1 is shown in fig 2. The opposing forces are marked as f_{m1}, f_{b1}, f_{b12} and f_{k1}.

$$f_{m1} = M_1 \frac{d^2x_1}{dt^2} \quad ; \quad f_{b1} = B_1 \frac{dx_1}{dt} \quad ; \quad f_{b12} = B_{12} \frac{d}{dt}(x_1 - x_2) \quad ; \quad f_{k1} = K_1 x_1$$

By Newton's second law, $f_{m1} + f_{b1} + f_{b12} + f_{k1} = f(t)$

$$M_1 \frac{d^2x_1}{dt^2} + B_1 \frac{dx_1}{dt} + B_{12} \frac{d(x_1 - x_2)}{dt} + K_1 x_1 = f(t)$$

On taking Laplace transform of above equation with zero initial conditions we get,

$$M_1 s^2 X_1(s) + B_1 s X_1(s) + B_{12} s [X_1(s) - X_2(s)] + K_1 X_1(s) = F(s)$$

$$X_1(s)[M_1 s^2 + (B_1 + B_{12})s + K_1] - B_{12} s X_2(s) = F(s) \qquad \ldots\ldots(1)$$

The free body diagram of mass M_2 is shown in fig 3. The opposing forces are marked as f_{m2}, f_{b2}, f_{b12} and f_{k2}.

$$f_{m2} = M_2 \frac{d^2x_2}{dt^2} \qquad ; \qquad f_{b2} = B_2 \frac{dx_2}{dt}$$

$$f_{b12} = B_{12} \frac{d}{dt}(x_2 - x_1) \quad ; \quad f_{k2} = K_2 x_2$$

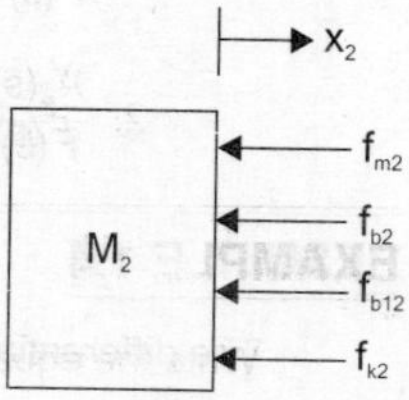

By Newton's second law, $f_{m2} + f_{b2} + f_{b12} + f_{k2} = 0$

$$M_2 \frac{d^2x_2}{dt^2} + B_2 \frac{dx_2}{dt} + B_{12} \frac{d(x_2 - x_1)}{dt} + K_2 x_2 = f(t) \qquad \ldots\ldots(2)$$

Fig 3.

On taking Laplace transform of equation (2) with zero initial conditions we get,

$$M_2 s^2 X_2(s) + B_2 s X_2(s) + B_{12} s [X_2(s) - X_1(s)] + K_2 X_2(s) = 0$$

$$X_2(s)[M_2 s^2 + (B_2 + B_{12})s + K_2] - B_{12} s X_1(s) = 0$$

$$X_2(s)[M_2 s^2 + (B_2 + B_{12})s + K_2] = B_{12} s X_1(s)$$

$$X_2(s) = \frac{B_{12} s X_1(s)}{[M_2 s^2 + (B_2 + B_{12})s + K_2]} \qquad \ldots\ldots(3)$$

Substituting for $X_2(s)$ from equation (3) in equation (1) we get,

$$X_1(s)[M_1 s^2 + (B_1 + B_{12})s + K_1] - \frac{(B_{12} s)^2 X_1(s)}{M_2 s^2 + (B_2 + B_{12})s + K_2} = F(s)$$

$$X_1(s) \frac{\left[[M_1 s^2 + (B_1 + B_{12})s + K_1][M_2 s^2 + (B_2 + B_{12})s + K_2] - (B_{12} s)^2 \right]}{M_2 s^2 + (B_2 + B_{12})s + K_2} = F(s)$$

$$\therefore \frac{X_1(s)}{F(s)} = \frac{M_2 s^2 + (B_2 + B_{12})s + K_2}{[M_1 s^2 + (B_1 + B_{12})s + K_1][M_2 s^2 + (B_2 + B_{12})s + K_2] - (B_{12} s)^2}$$

From equation (3) we get,

$$X_1(s) = \frac{[M_2 s^2 + (B_2 + B_{12})s + K_2] X_2(s)}{B_{12} s} \qquad \ldots\ldots(4)$$

Substituting for $X_1(s)$ from equation (4) in equation (1) we get,

$$\frac{X_2(s)[M_2 s^2 + (B_2 + B_{12})s + K_2]}{B_{12} s} [M_1 s^2 + (B_1 + B_{12})s + K_1] - B_{12} s X_2(s) = F(s)$$

$$X_2(s)\left[\frac{[M_2s^2+(B_2+B_{12})s+K_2][M_1s^2+(B_1+B_{12})s+K_1]-(B_{12}s)^2}{B_{12}s}\right]=F(s)$$

$$\therefore \frac{X_2(s)}{F(s)}=\frac{B_{12}s}{[M_2s^2+(B_2+B_{12})s+K_2][M_1s^2+(B_1+B_{12})s+K_1]-(B_{12}s)^2}$$

RESULT

The differential equations governing the system are,

1. $M_1\dfrac{d^2x_1}{dt^2}+B_1\dfrac{dx_1}{dt}+B_{12}\dfrac{d(x_1-x_2)}{dt}+K_1x_1=f(t)$

2. $M_2\dfrac{d^2x_2}{dt^2}+B_2\dfrac{dx_2}{dt}+B_{12}\dfrac{d(x_2-x_1)}{dt}+K_2x_2=0$

The transfer functions of the system are,

1. $\dfrac{X_1(s)}{F(s)}=\dfrac{M_2s^2+(B_2+B_{12})s+K_2}{[M_1s^2+(B_1+B_{12})s+K_1][M_2s^2+(B_2+B_{12})s+K_2]-(B_{12}s)^2}$

2. $\dfrac{X_2(s)}{F(s)}=\dfrac{B_{12}s}{[M_2s^2+(B_2+B_{12})s+K_2][M_1s^2+(B_1+B_{12})s+K_1]-(B_{12}s)^2}$

EXAMPLE 1.4

Write the equations of motion in s-domain for the system shown in fig 1. Determine the transfer function of the system.

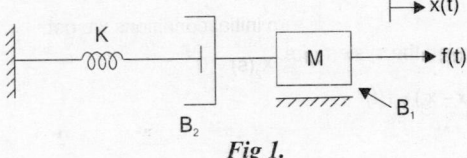

Fig 1.

SOLUTION

Let, Laplace transform of $x(t)=\mathcal{L}\{x(t)\}=X(s)$

Laplace transform of $f(t)=\mathcal{L}\{f(t)\}=F(s)$

Let x_1 be the displacement at the meeting point of spring and dashpot. Laplace transform of x_1 is $X_1(s)$.

The system has two nodes and they are mass M and the meeting point of spring and dashpot. The differential equations governing the system are the force balance equations at these nodes. The equations of motion in the s-domain are obtained by taking Laplace transform of the differential equations.

The free body diagram of mass M is shown in fig 2. The opposing forces are marked as f_m, f_{b1} and f_{b2}.

$$f_m=M\frac{d^2x}{dt^2}\quad;\quad f_{b1}=B_1\frac{dx}{dt}\quad;\quad f_{b2}=B_2\frac{d}{dt}(x-x_1)$$

By Newton's second law the force balance equation is,

$$f_m+f_{b1}+f_{b2}=f(t)$$

$$\therefore M\frac{d^2x}{dt^2}+B_1\frac{dx}{dt}+B_2\frac{d}{dt}(x-x_1)=f(t)$$

On taking Laplace transform of the above equation we get,

$$Ms^2X(s)+B_1(s)X(s)+B_2s[X(s)-X_1(s)]=F(s)$$

$$[Ms^2+(B_1+B_2)s]X(s)-B_2sX_1(s)=F(s) \qquad \qquad .....(1)$$

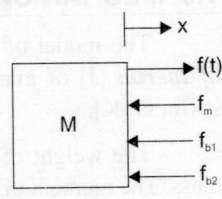

Fig 2.

The free body diagram at the meeting point of spring and dashpot is shown in fig 3. The opposing forces are marked as f_k and f_{b2}.

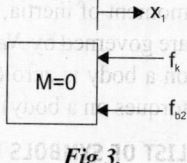

Fig 3.

$$f_{b2} = B_2 \frac{d}{dt}(x_1 - x) \quad ; \quad f_k = Kx_1$$

By Newton's second law, $f_{b2} + f_k = 0$

$$\therefore B_2 \frac{d}{dt}(x_1 - x) + Kx_1 = 0$$

On taking Laplace transform of the above equation we get,

$$B_2 s[X_1(s) - X(s)] + K X_1(s) = 0$$

$$(B_2 s + K) X_1(s) - B_2 sX(s) = 0$$

$$\therefore X_1(s) = \frac{B_2 s}{B_2 s + K} X(s) \qquad \qquad .....(2)$$

Substituting for $X_1(s)$ from equation (2) in equation (1) we get,

$$\left[Ms^2 + (B_1 + B_2)s \right] X(s) - B_2 s \left[\frac{B_2 s}{B_2 s + K} \right] X(s) = F(s)$$

$$X(s) \frac{\left[Ms^2 + (B_1 + B_2)s \right](B_2 s + K) - (B_2 s)^2}{B_2 s + K} = F(s)$$

$$\therefore \frac{X(s)}{F(s)} = \frac{B_2 s + K}{\left[Ms^2 + (B_1 + B_2)s \right](B_2 s + K) - (B_2 s)^2}$$

RESULT

The differential equations governing the system are,

1. $M \dfrac{d^2x}{dt^2} + B_1 \dfrac{dx}{dt} + B_2 \dfrac{d}{dt}(x - x_1) = f(t)$

2. $B_2 \dfrac{d}{dt}(x_1 - x) + K x_1 = 0$

The equations of motion in s-domain are,

1. $[M s^2 + (B_1 + B_2)s] X(s) - B_2 sX_1(s) = F(s)$

2. $(B_2 s + K) X_1(s) - B_2 sX(s) = 0$

The transfer function of the system is,

$$\frac{X(s)}{F(s)} = \frac{B_2 s + K}{[M s^2 + (B_1 + B_2)s](B_2 s + K) - (B_2 s)^2}$$

1.5 MECHANICAL ROTATIONAL SYSTEMS

The model of rotational mechanical systems can be obtained by using three elements, ***moment of inertia*** [J] of mass, ***dash-pot*** with rotational frictional coefficient [B] and ***torsional spring*** with stiffness [K].

The weight of the rotational mechanical system is represented by the moment of inertia of the mass. The moment of inertia of the system or body is considered to be concentrated at the centre of gravity of the body. The elastic deformation of the body can be represented by a spring (torsional spring). The friction existing in rotational mechanical system can be represented by the dash-pot. The dash-pot is a piston rotating inside a cylinder filled with viscous fluid.

When a torque is applied to a rotational mechanical system, it is opposed by opposing torques due to moment of inertia, friction and elasticity of the system. The torques acting on a rotational mechanical body are governed by *Newton's second law of motion* for rotational systems. It states that the sum of torques acting on a body is zero (or Newton's law states that the sum of applied torques is equal to the sum of opposing torques on a body).

LIST OF SYMBOLS USED IN MECHANICAL ROTATIONAL SYSTEM

θ = Angular displacement, rad

$\dfrac{d\theta}{dt}$ = Angular velocity, rad/sec

$\dfrac{d^2\theta}{dt}$ = Angular acceleration, rad/sec^2

T = Applied torque, N-m

J = Moment of inertia, Kg-m^2/rad

B = Rotational frictional coefficient, N-m/(rad/sec)

K = Stiffness of the spring, N-m/rad

TORQUE BALANCE EQUATIONS OF IDEALISED ELEMENTS

Consider an ideal mass element shown in fig 1.14 which has negligible friction and elasticity. The opposing torque due to moment of inertia is proportional to the angular acceleration.

Let, T = Applied torque.

T_j = Opposing torque due to moment of inertia of the body.

Here $T_j \propto \dfrac{d^2\theta}{dt^2}$ or $T_j = J\dfrac{d^2\theta}{dt^2}$

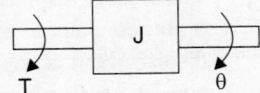

By Newton's second law,

Fig 1.14 : Ideal rotational mass element.

$$\boxed{T = T_j = J\dfrac{d^2\theta}{dt^2}}$$ .....(1.7)

Consider an ideal frictional element dash pot shown in fig 1.15 which has negligible moment of inertia and elasticity. Let a torque be applied on it. The dash pot will offer an opposing torque which is proportional to the angular velocity of the body.

Let, T = Applied torque.

T_b = Opposing torque due to friction.

$T_b \propto \dfrac{d\theta}{dt^2}$ or $T_b = B\dfrac{d\theta}{dt}$

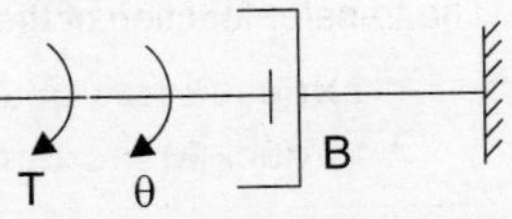

Fig 1.15 : Ideal rotational dash-pot with one end fixed to reference.

By Newton's second law, $\boxed{T = T_b = B\dfrac{d\theta}{dt}}$ .....(1.8)

When the dash pot has angular displacement at both ends as shown in fig 1.16, the opposing torque is proportional to the difference between angular velocity at both ends.

$T_b \propto \dfrac{d}{dt}(\theta_1 - \theta_2)$ or $T_b = B\dfrac{d}{dt}(\theta_1 - \theta_2)$

$\therefore\ T = T_b = B\dfrac{d}{dt}(\theta_1 - \theta_2)$ .....(1.9)

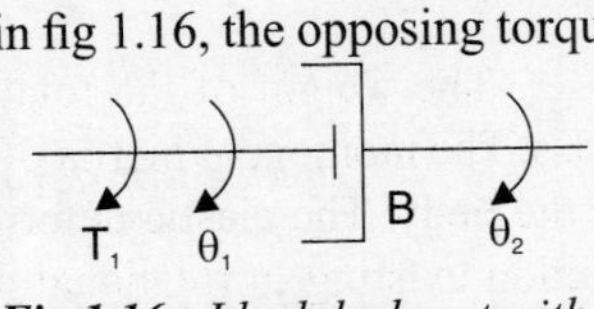

Fig 1.16 : Ideal dash-pot with angular displacement at both ends.

Consider an ideal elastic element, torsional spring as shown in fig 1.17, which has negligible moment of inertia and friction. Let a torque be applied on it. The torsional spring will offer an opposing torque which is proportional to angular displacement of the body.

Let, T = Applied torque.

T_k = Opposing torque due to elasticity.

$T_k \propto \theta$ or $T_k = K\theta$

By Newton's second law, $\boxed{T = T_k = K\theta}$ .....(1.10)

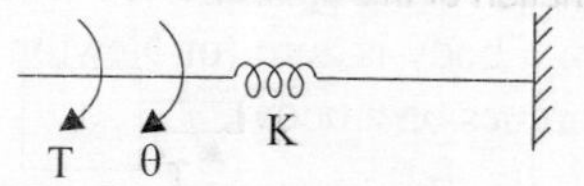

Fig 1.17 : Ideal spring with one end fixed to reference.

When the spring has angular displacement at both ends as shown in fig 1.18 the opposing torque is proportional to difference between angular displacement at both ends.

$T_k \propto (\theta_1 - \theta_2)$ or $T_k = K(\theta_1 - \theta_2)$

$\boxed{\therefore T = T_k = K(\theta_1 - \theta_2)}$ .....(1.11)

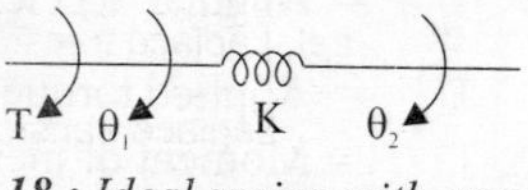

Fig 1.18 : Ideal spring with angular displacement at both ends.

Guidelines to determine the Transfer Function of Mechanical Rotational System

1. In mechanical rotational system, the differential equations governing the system are obtained by writing torque balance equations at nodes in the system. The nodes are meeting point of elements. Generally the nodes are mass elements with moment of inertia in the system. In some cases the nodes may be without mass element.

2. The angular displacement of the moment of inertia of the masses (nodes) are assumed as θ_1, θ_2, θ_3, etc., and assign a displacement to each mass (node). The first derivative of angular displacement is angular velocity and the second derivative of the angular displacement is angular acceleration.

3. Draw the free body diagrams of the system. The free body diagram is obtained by drawing each moment of inertia of mass separately and then marking all the torques acting on that body. Always the opposing torques acts in a direction opposite to applied torque.

4. The mass has to rotate in the direction of the applied torque. Hence the angular displacement, velocity and acceleration of the mass will be in the direction of the applied torque. If there is no applied torque then the angular displacement, velocity and acceleration of the mass is in a direction opposite to that of opposing torque.

5. For each free body diagram write one differential equation by equating the sum of applied torques to the sum of opposing torques.

6. Take Laplace transform of differential equation to convert them to algebraic equations. Then rearrange the s-domain equations to eliminate the unwanted variables and obtain the relation between output variable and input variable. This ratio is the transfer function of the system.

> **Note :**
>
> *Laplace transform of $\theta = \mathcal{L}\{\theta\} = \theta(s)$*
>
> *Laplace transform of $\dfrac{d\theta}{dt} = \mathcal{L}\left\{\dfrac{d\theta}{dt}\right\} = s\,\theta(s)$ (with zero initial conditions)*
>
> *Laplace transform of $\dfrac{d^2\theta}{dt^2} = \mathcal{L}\left\{\dfrac{d^2\theta}{dt^2}\right\} = s\,\theta(s)$ (with zero initial conditions)*

EXAMPLE 1.5

Write the differential equations governing the mechanical rotational system shown in fig 1. Obtain the transfer function of the system.

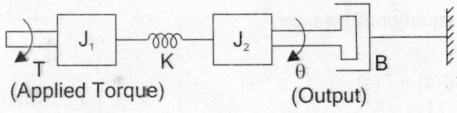

(Applied Torque)   (Output)

Fig 1.

SOLUTION

In the given system, applied torque T is the input and angular displacement θ is the output.

Let, Laplace transform of $T = \mathcal{L}\{T\} = T(s)$

Laplace transform of $\theta = \mathcal{L}\{\theta\} = \theta(s)$

Laplace transform of $\theta_1 = \mathcal{L}\{\theta_1\} = \theta_1(s)$

Hence the required transfer function is $\dfrac{\theta(s)}{T(s)}$

The system has two nodes and they are masses with moment of inertia J_1 and J_2. The differential equations governing the system are given by torque balance equations at these nodes.

Let the angular displacement of mass with moment of inertia J_1 be θ_1. The free body diagram of J_1 is shown in fig 2. The opposing torques acting on J_1 are marked as T_{j1} and T_k.

$$T_{j1} = J_1\frac{d^2\theta_1}{dt^2} \quad ; \quad T_k = K(\theta_1 - \theta)$$

By Newton's second law, $T_{j1} + T_k = T$

$$J_1\frac{d^2\theta_1}{dt^2} + K(\theta_1 - \theta) = T$$

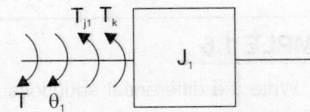

Fig 2 : Free body diagram of mass with moment of inertia J_1.

$$J_1\frac{d^2\theta_1}{dt^2} + K\theta_1 - K\theta = T \qquad .....(1)$$

On taking Laplace transform of equation (1) with zero initial conditions we get,

$$J_1 s^2\theta_1(s) + K\theta_1(s) - K\theta(s) = T(s)$$

$$(J_1 s^2 + K)\theta_1(s) - K\theta(s) = T(s) \qquad .....(2)$$

The free body diagram of mass with moment of inertia J_2 is shown in fig 3. The opposing torques acting on J_2 are marked as T_{j2}, T_b and T_k.

$$T_{j2} = J_2\frac{d^2\theta}{dt^2} \quad ; \quad T_b = B\frac{d\theta}{dt} \quad ; \quad T_k = K(\theta - \theta_1)$$

By Newton's second law, $T_{j2} + T_b + T_k = 0$

$$\therefore \ J_2\frac{d^2\theta}{dt^2} + B\frac{d\theta}{dt} + K(\theta - \theta_1) = 0$$

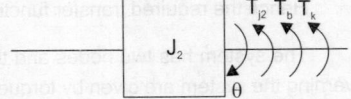

Fig 3 : Free body diagram of mass with moment of inertia J_2.

$$J_2\frac{d^2\theta}{dt^2} + B\frac{d\theta}{dt} + K\theta - K\theta_1 = 0$$

On taking Laplace transform of above equation with zero initial conditions we get,

$$J_2 s^2\theta(s) + B s\,\theta(s) + K\theta(s) - K\theta_1(s) = 0$$

$$(J_2 s^2 + Bs + K)\, \theta(s) - K\theta_1(s) = 0$$

$$\theta_1(s) = \frac{(J_2 s^2 + Bs + K)}{K}\, \theta(s) \qquad \qquad .....(3)$$

Substituting for $\theta_1(s)$ from equation (3) in equation (2) we get,

$$(J_1 s^2 + K)\, \frac{(J_2 s^2 + Bs + K)}{K}\, \theta(s) - K\theta(s) = T(s)$$

$$\left[\frac{(J_1 s^2 + K)(J_2 s^2 + Bs + K) - K^2}{K} \right] \theta(s) = T(s)$$

$$\therefore \frac{\theta(s)}{T(s)} = \frac{K}{(J_1 s^2 + K)(J_2 s^2 + Bs + K) - K^2}$$

RESULT

The differential equations governing the system are,

1. $J_1 \dfrac{d^2\theta_1}{dt^2} + K\theta_1 - K\theta = T$

2. $J_2 \dfrac{d^2\theta}{dt^2} + B\dfrac{d\theta}{dt} + K\theta - K\theta_1 = 0$

The transfer function of the system is,

$$\frac{\theta(s)}{T(s)} = \frac{K}{(J_1 s^2 + K)(J_2 s^2 + Bs + K) - K^2}$$

EXAMPLE 1.6

Write the differential equations governing the mechanical rotational system shown in fig 1. and determine the transfer function $\theta(s)/T(s)$.

SOLUTION

In the given system, the torque T is the input and the angular displacement θ is the output.

Let, Laplace transform of $T = \mathcal{L}\{T\}\ = T(s)$

Laplace transform of $\theta = \mathcal{L}\{\theta\}\ = \theta(s)$

Laplace transform of $\theta_1 = \mathcal{L}\{\theta_1\} = \theta_1(s)$

Hence the required transfer function is $\dfrac{\theta(s)}{T(s)}$

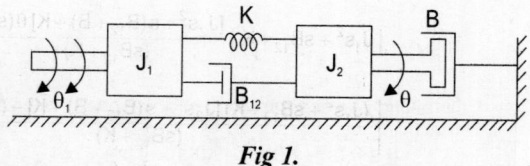

Fig 1.

The system has two nodes and they are masses with moment of inertia J_1 and J_2. The differential equations governing the system are given by torque balance equations at these nodes.

Let the angular displacement of mass with moment of inertia J_1 be θ_1. The free body diagram of J_1 is shown in fig 2. The opposing torques acting on J_1 are marked as T_{j1}, T_{b12} and T_k.

$$T_{j1} = J_1 \frac{d^2\theta_1}{dt^2} \quad ; \quad T_{b12} = B_{12}\frac{d}{dt}(\theta_1 - \theta) \quad ; \quad T_k = K(\theta_1 - \theta)$$

By Newton's second law, $T_{j1} + T_{b12} + T_k = T$

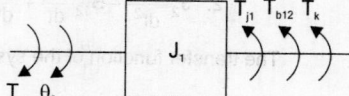

Fig 2 : *Free body diagram of mass with moment of inertia J_1.*

$$J_1 \frac{d^2\theta_1}{dt^2} + B_{12}\frac{d}{dt}(\theta_1 - \theta) + K(\theta_1 - \theta) = T$$

On taking Laplace transform of above equation with zero initial conditions we get,

$$J_1 s^2\theta_1(s) + s\,B_{12}\,[\theta_1(s) - \theta(s)] + K\theta_1(s) - K\theta(s) = T(s)$$

$$\theta_1(s)\,[J_1 s^2 + s\,B_{12} + K\,] - \theta(s)\,[s\,B_{12} + K] = T(s) \qquad \qquad .....(1)$$

The free body diagram of mass with moment of inertia J_2 is shown in fig 3. The opposing torques are marked as T_{j2}, T_{b12}, T_b and T_k.

$$T_{j2} = J_2\frac{d^2\theta}{dt^2} \quad ; \quad T_{b12} = B_{12}\frac{d}{dt}(\theta - \theta_1)$$

$$T_b = B\frac{d\theta}{dt} \quad ; \quad T_k = K(\theta - \theta_1)$$

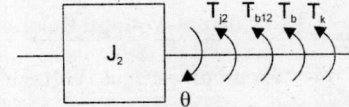

Fig 3 : Free body diagram of mass with moment of inertia J_2.

By Newton's second law, $T_{j2} + T_{b12} + T_b + T_k = 0$

$$J_2\frac{d^2\theta}{dt^2} + B_{12}\frac{d}{dt}(\theta - \theta_1) + B\frac{d\theta}{dt} + K(\theta - \theta_1) = 0$$

$$J_2\frac{d^2\theta}{dt^2} - B_{12}\frac{d\theta_1}{dt} + \frac{d\theta}{dt}(B_{12} + B) + K\theta - K\theta_1 = 0$$

On taking Laplace transform of above equation with zero initial conditions we get,

$$J_2 s^2\theta(s) - B_{12}\,s\theta_1(s) + s\theta(s)\,[B_{12} + B] + K\theta(s) - K\theta_1(s) = 0$$

$$\theta(s)\,[s^2 J_2 + s(B_{12} + B) + K] - \theta_1(s)\,[sB_{12} + K] = 0$$

$$\theta_1(s) = \frac{[s^2 J_2 + s(B_{12} + B) + K]}{[sB_{12} + K]}\,\theta(s) \qquad \qquad .....(2)$$

Substituting for $\theta_1(s)$ from equation (2) in equation (1) we get,

$$\Big[J_1 s^2 + sB_{12} + K\,\Big]\frac{[J_2 s^2 + s(B_{12} + B) + K]\,\theta(s)}{(sB_{12} + K)} - (sB_{12} + K)\theta(s) = T(s)$$

$$\left[\frac{(J_1 s^2 + sB_{12} + K)\,[J_2 s^2 + s(B_{12} + B) + K] - (sB_{12} + K)^2}{(sB_{12} + K)}\right]\theta(s) = T(s)$$

$$\therefore \frac{\theta(s)}{T(s)} = \frac{(sB_{12} + K)}{(J_1 s^2 + sB_{12} + K)\,[J_2 s^2 + s(B_{12} + B) + K] - (sB_{12} + K)^2}$$

RESULT

The differential equations governing the system are,

1. $\displaystyle J_1\frac{d^2\theta_1}{dt^2} + B_{12}\frac{d}{dt}(\theta_1 - \theta) + K(\theta_1 - \theta) = T$

2. $\displaystyle J_2\frac{d^2\theta_1}{dt^2} - B_{12}\frac{d\theta_1}{dt} + \frac{d\theta}{dt}(B_{12} + B) + K(\theta - \theta_1) = 0$

The transfer function of the system is,

$$\frac{\theta(s)}{T(s)} = \frac{(sB_{12} + K)}{(J_1 s^2 + sB_{12} + K)\,[J_2 s^2 + s(B_{12} + B) + K] - (sB_{12} + K)^2}$$

1.6 ELECTRICAL SYSTEMS

The models of electrical systems can be obtained by using resistor, capacitor and inductor. The current-voltage relation of resistor, inductor and capacitor are given in table-1. For modelling electrical systems, the electrical network or equivalent circuit is formed by using R, L and C and voltage or current source.

The differential equations governing the electrical systems can be formed by writing Kirchoff 's current law equations by choosing various nodes in the network or Kirchoff 's voltage law equations by choosing various closed paths in the network. The transfer function can be obtained by taking Laplace transform of the differential equations and rearranging them as a ratio of output to input.

Table-1.1 : Current-Voltage Relation of R, L and C

Element	Voltage across the element	Current through the element
i(t) R v(t)	$v(t) = Ri(t)$	$i(t) = \dfrac{v(t)}{R}$
i(t) L v(t)	$v(t) = L\dfrac{d}{dt}i(t)$	$i(t) = \dfrac{1}{L}\int v(t)\,dt$
i(t) C v(t)	$v(t) = \dfrac{1}{C}\int i(t)\,dt$	$i(t) = C\dfrac{dv(t)}{dt}$

EXAMPLE 1.7

Obtain the transfer function of the electrical network shown in fig 1.

SOLUTION

In the given network, input is e(t) and output is $v_2(t)$.

Let, Laplace transform of $e(t) = \mathcal{L}\{e(t)\} = E(s)$

Laplace transform of $v_2(t) = \mathcal{L}\{v_2(t)\} = V_2(s)$

The transfer function of the network is $\dfrac{V_2(s)}{E(s)}$

Transform the voltage source in series with resistance R_1 into equivalent current source as shown in figure 2. The network has two nodes. Let the node voltages be v_1 and v_2. The Laplace transform of node voltages v_1 and v_2 are $V_1(s)$ and $V_2(s)$ respectively. The differential equations governing the network are given by the Kirchoff 's current law equations at these nodes.

At node-1, by Kirchoff's current law (refer fig 3)

$$\frac{v_1}{R_1} + C_1\frac{dv_1}{dt} + \frac{v_1 - v_2}{R_2} = \frac{e}{R_1}$$

On taking Laplace transform of above equation with zero initial conditions we get,

$$\frac{V_1(s)}{R_1} + C_1 s V_1(s) + \frac{V_1(s)}{R_2} - \frac{V_2(s)}{R_2} = \frac{E(s)}{R_1}$$

$$V_1(s)\left[\frac{1}{R_1} + sC_1 + \frac{1}{R_2}\right] - \frac{V_2(s)}{R_2} = \frac{E(s)}{R_1} \qquad .....(1)$$

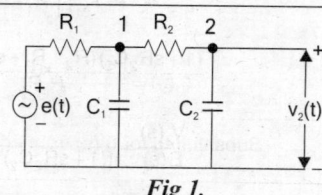

Fig 1.

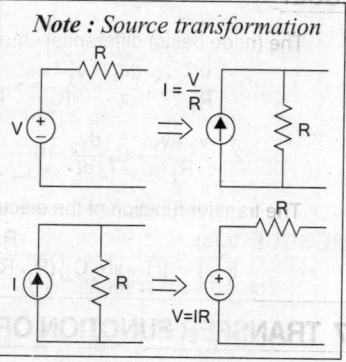

Note : Source transformation

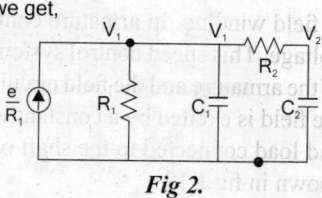

Fig 2.

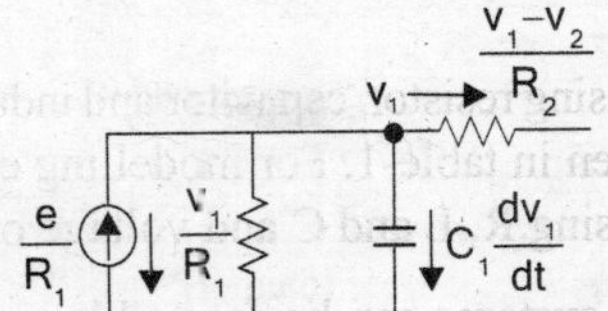

$$\frac{v_1 - v_2}{R_2}$$

Fig 3. 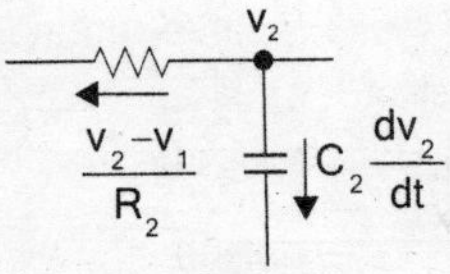   *Fig 4.*

At node-2, by Kirchoff's current law (refer fig 4)

$$\frac{v_2 - v_1}{R_2} + C_2 \frac{dv_2}{dt} = 0$$

On taking Laplace transform of above equation with zero initial conditions we get,

$$\frac{V_2(s)}{R_2} - \frac{V_1(s)}{R_2} + C_2 s V_2(s) = 0$$

$$\frac{V_1(s)}{R_2} = \frac{V_2(s)}{R_2} + C_2 s V_2(s) = \left[\frac{1}{R_2} + sC_2\right] V_2(s)$$

$$\therefore \ V_1(s) = [1 + sC_2 R_2] \, V_2(s) \qquad \qquad .....(2)$$

Substituting for $V_1(s)$ from equation (2) in equation (1) we get,

$$(1 + sR_2 C_2) V_2(s) \left[\frac{1}{R_1} + sC_1 + \frac{1}{R_2}\right] - \frac{V_2(s)}{R_2} = \frac{E(s)}{R_1}$$

$$\left[\frac{(1 + sR_2 C_2)(R_2 - R_1 + sC_1 R_1 R_2) - R_1}{R_1 R_2}\right] V_2(s) = \frac{E(s)}{R_1}$$

$$\therefore \ \frac{V_2(s)}{E(s)} = \frac{R_2}{[(1 + sR_2 C_2)(R_1 + R_2 + sC_1 R_1 R_2) - R_1]}$$

RESULT

The (node basis) differential equations governing the electrical network are,

1. $\dfrac{v_1}{R_1} + C_1 \dfrac{dv_1}{dt} + \dfrac{v_1 - v_2}{R_2} = \dfrac{e}{R_1}$

2. $\dfrac{v_2 - v_1}{R_2} + C_2 \dfrac{dv_2}{dt} = 0$

The transfer function of the electrical network is,

$$\frac{V_2(s)}{E(s)} = \frac{R_2}{[(1 + sR_2 C_2)(R_1 + R_2 + sC_1 R_1 R_2) - R_1]}$$

1.7 TRANSFER FUNCTION OF ARMATURE CONTROLLED DC MOTOR

The speed of DC motor is directly proportional to armature voltage and inversely proportional to flux in field winding. In armature controlled DC motor the desired speed is obtained by varying the armature voltage. This speed control system is an electro-mechanical control system. The electrical system consists of the armature and the field circuit but for analysis purpose, only the armature circuit is considered because the field is excited by a constant voltage. The mechanical system consists of the rotating part of the motor and load connected to the shaft of the motor. The armature controlled DC motor speed control system is shown in fig 1.19.

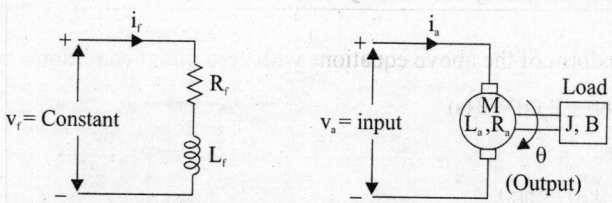

Fig 1.19 : Armature controlled DC motor.

Let, R_a = Armature resistance, Ω

       L_a = Armature inductance, H

       i_a = Armature current, A

       v_a = Armature voltage, V

       e_b = Back emf, V

       K_t = Torque constant, N-m/A

       T = Torque developed by motor, N-m

       θ = Angular displacement of shaft, rad

       ω = Angular velocity, rad/sec

       J = Moment of inertia of motor and load, Kg-m²/rad

       B = Frictional coefficient of motor and load, N-m/(rad/sec)

       K_b = Back emf constant, V/(rad/sec)

The equivalent circuit of armature is shown in fig 1.20.

By Kirchoff's voltage law, we can write,

$$i_a R_a + L_a \frac{di_a}{dt} + e_b = v_a \qquad \ldots\ldots(1.12)$$

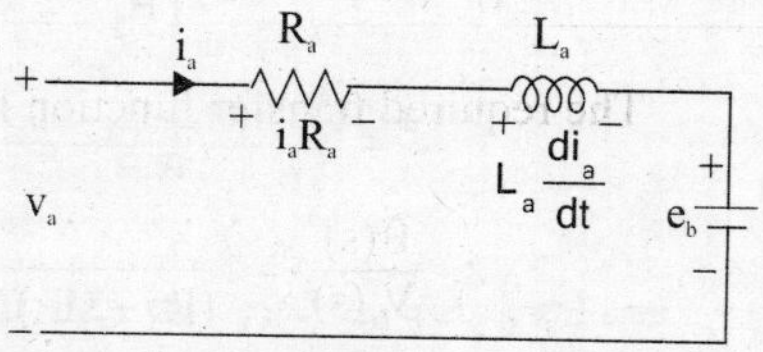

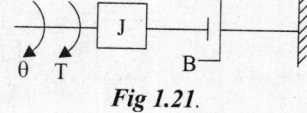

Torque of DC motor is proportional to the product of flux and current. Since flux is constant in this system, the torque is proportional to i_a alone.

Fig 1.20 : Equivalent circuit of armature.

$$T \propto i_a$$

$$\therefore \text{ Torque, } T = K_t i_a \qquad \ldots\ldots(1.13)$$

The mechanical system of the motor is shown in fig 1.21.

The differential equation governing the mechanical system of motor is given by,

Fig 1.21.

$$J\frac{d^2\theta}{dt^2} + B\frac{d\theta}{dt} = T \qquad \ldots\ldots(1.14)$$

The back emf of DC machine is proportional to speed (angular velocity) of shaft.

$$e_b \propto \omega \quad \text{and} \quad \omega = \frac{d\theta}{dt} \ ; \quad \therefore e_b \propto \frac{d\theta}{dt} \quad \text{or} \quad \text{Back emf, } e_b = K_b \frac{d\theta}{dt} \qquad \ldots\ldots(1.15)$$

The Laplace transform of various time domain signals involved in this system are shown below.

$$\mathcal{L}\{v_a\} = V_a(s) \ ; \ \mathcal{L}\{e_b\} = E_b(s) \ ; \ \mathcal{L}\{T\} = T(s) \ ; \ \mathcal{L}\{i_a\} = Ia(s) \ ; \ \mathcal{L}\{\theta\} = \theta(s)$$

The differential equations governing the armature controlled DC motor speed control system are,

$$i_a R_a + L_a \frac{di_a}{dt} + e_b = v_a \ ; \quad T = K_t i_a \ ; \quad J\frac{d^2\theta}{dt^2} + B\frac{d\theta}{dt} = T \ ; \quad e_b = K_b \frac{d\theta}{dt}$$

Taking Laplace transform of the above equations with zero initial conditions we get,

$$I_a(s) R_a + L_a sI_a(s) + E_b(s) = V_a(s) \qquad .....(1.16)$$

$$T(s) = K_t I_a(s) \qquad .....(1.17)$$

$$Js^2\theta(s) + B s\,\theta(s) = T(s) \qquad .....(1.18)$$

$$E_b(s) = K_b s\theta(s) \qquad .....(1.19)$$

On equating equations (1.17) and (1.18) we get,

$$K_t I_a(s) = (Js^2 + Bs)\,\theta(s)$$

$$I_a(s) = \frac{(Js^2 + Bs)}{K_t}\,\theta(s) \qquad .....(1.20)$$

Equation (1.16) can be written as,

$$(R_a + sL_a)\,I_a(s) + E_b(s) = V_a(s) \qquad .....(1.21)$$

Substituting for $E_b(s)$ and $I_a(s)$ from equation (1.19) and (1.20) respectively in equation (1.21),

$$(R_a + sL_a)\,\frac{(Js^2 + Bs)}{K_t}\,\theta(s) + K_b s\theta(s) = V_a(s)$$

$$\left[\frac{(R_a + sL_a)(Js^2 + Bs) + K_b K_t s}{K_t}\right]\theta(s) = V_a(s)$$

The required transfer function is $\dfrac{\theta(s)}{V_a(s)}$

$$\therefore \frac{\theta(s)}{V_a(s)} = \frac{K_t}{(R_a + sL_a)(Js^2 + Bs) + K_b K_t s} \qquad ....(1.22)$$

$$= \frac{K_t}{R_a Js^2 + R_a Bs + L_a Js^3 + L_a Bs^2 + K_b K_t s}$$

$$= \frac{K_t}{s\left[JL_a s^2 + (JR_a + BL_a)s + (BR_a + K_b K_t)\right]}$$

$$= \frac{K_t/JL_a}{s\left[s^2 + \left(\dfrac{JR_a + BL_a}{JL_a}\right)s + \left(\dfrac{BR_a + K_b K_t}{JL_a}\right)\right]} \qquad .....(1.23)$$

The transfer function of armature controlled dc motor can be expressed in another standard form as shown below. From equation (1.22) we get,

$$\frac{\theta(s)}{V_a(s)} = \frac{K_t}{(R_a + sL_a)(Js^2 + Bs) + K_b K_t s} = \frac{K_t}{R_a\left(\dfrac{sL_a}{R_a} + 1\right)Bs\left(1 + \dfrac{Js^2}{Bs}\right) + K_b K_t s}$$

$$= \frac{K_t/R_a B}{s\left[(1 + sT_a)(1 + sT_m) + \dfrac{K_b K_t}{R_a B}\right]} \qquad .....(1.24)$$

where, $\dfrac{L_a}{R_a} = T_a$ = Electrical time constant

$\dfrac{J}{B} = T_m$ = Mechanical time constant

1.8 TRANSFER FUNCTION OF FIELD CONTROLLED DC MOTOR

The speed of a DC motor is directly proportional to armature voltage and inversely proportional to flux. In field controlled DC motor the armature voltage is kept constant and the speed is varied by varying the flux of the machine. Since flux is directly proportional to field current, the flux is varied by varying field current. The speed control system is an electromechanical control system. The electrical system consists of armature and field circuit but for analysis purpose, only field circuit is considered because the armature is excited by a constant voltage. The mechanical system consists of the rotating part of the motor and the load connected to the shaft of the motor. The field controlled DC motor speed control system is shown in fig 1.22.

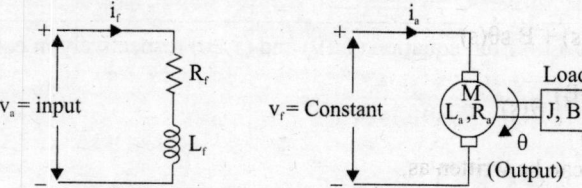

Fig 1.22 : Field controlled DC motor.

Let, R_f = Field resistance, Ω

L_f = Field inductance, H

i_f = Field current, A

v_f = Field voltage, V

T = Torque developed by motor, N-m

K_{tf} = Torque constant, N-m/A

J = Moment of inertia of rotor and load, Kg-m²/rad

B = Frictional coefficient of rotor and load, N-m/(rad/sec)

The equivalent circuit of field is shown in fig 1.23.

By Kirchoff's voltage law, we can write

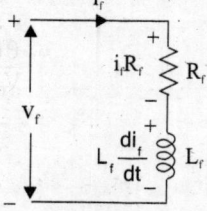

Fig 1.23 : Equivalent circuit of field.

$$R_f i_f + L_f \dfrac{di_f}{dt} = v_f \qquad \qquad \ldots\ldots(1.25)$$

The torque of DC motor is proportional to product of flux and armature current. Since armature current is constant in this system, the torque is proportional to flux alone, but flux is proportional to field current.

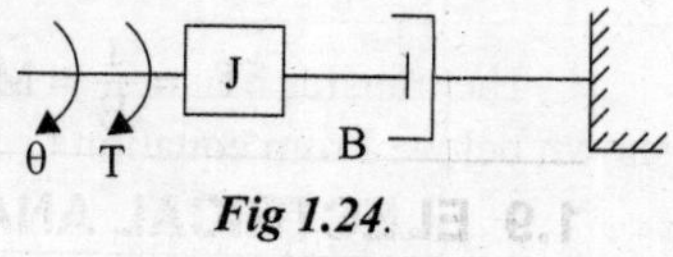

Fig 1.24.

$$T \propto i_f, \quad \therefore \text{Torque, } T = K_{tf} i_f \qquad \qquad \ldots\ldots(1.26)$$

The mechanical system of the motor is shown in fig 1.24. The differential equation governing the mechanical system of the motor is given by,

$$J \dfrac{d^2\theta}{dt^2} + B \dfrac{d\theta}{dt} = T \qquad \qquad \ldots\ldots(1.27)$$

The Laplace transform of various time domain signals involved in this system are shown below.

$$\mathcal{L}\{i_f\} = I_f(s) \quad ; \quad \mathcal{L}\{T\} = T(s) \quad ; \quad \mathcal{L}\{v_f\} = V_f(s) \quad ; \quad \mathcal{L}\{\theta\} = \theta(s)$$

The differential equations governing the field controlled DC motor are,

$$R_f i_f + L_f \frac{di_f}{dt} = v_f \quad ; \quad T = K_{tf} i_f \quad ; \quad J \frac{d^2\theta}{dt^2} + B \frac{d\theta}{dt} = T$$

On taking Laplace transform of the above equations with zero initial condition we get,

$$R_f I_f(s) + L_f s I_f(s) = V_f(s) \hspace{5cm} .....(1.28)$$

$$T(s) = K_{tf} I_f(s) \hspace{5cm} .....(1.29)$$

$$Js^2\theta(s) + B\, s\theta(s) = T(s) \hspace{5cm} .....(1.30)$$

Equating equations (1.29) and (1.30) we get,

$$K_{tf} I_f(s) = Js^2\theta(s) + B\, s\theta(s)$$

$$I_f(s) = s \frac{(Js + B)}{K_{tf}} \theta(s) \hspace{4cm} .....(1.31)$$

The equation (1.28) can be written as,

$$(R_f + sL_f)\, I_f(s) = V_f(s) \hspace{4cm} .....(1.32)$$

On substituting for $I_f(s)$ from equation (1.31) in equation (1.32) we get,

$$(R_f + sL_f)\, s \frac{(Js + B)}{K_{tf}} \theta(s) = V_f(s)$$

$$\frac{\theta(s)}{V_f(s)} = \frac{K_{tf}}{s(R_f + sL_f)(B + sJ)}$$

$$= \frac{K_{tf}}{sR_f\left(1 + \frac{sL_f}{R_f}\right)B\left(1 + \frac{sJ}{B}\right)} = \frac{K_m}{s(1 + sT_f)(1 + sT_m)} \hspace{1cm} .....(1.33)$$

where, $K_m = \dfrac{K_{tf}}{R_f B}$ = Motor gain constant

$T_f = \dfrac{L_f}{R_f}$ = Field time constant

$T_n = \dfrac{J}{B}$ = Mechanical time constant

1.9 ELECTRICAL ANALOGOUS OF MECHANICAL TRANSLATIONAL SYSTEMS

Systems remain *analogous* as long as the differential equations governing the systems or transfer functions are in identical form. The electric analogue of any other kind of system is of greater importance since it is easier to construct electrical models and analyse them.

The three basic elements mass, dash-pot and spring that are used in modelling mechanical translational systems are analogous to resistance, inductance and capacitance of electrical systems.

The input force in mechanical system is analogous to either voltage source or current source in electrical systems. The output velocity (first derivative of displacement) in mechanical system is analogous to either current or voltage in an element in electrical system.

Since the electrical systems has two types of inputs either voltage or current source, there are two types of analogies : *force-voltage analogy* and *force-current analogy*.

FORCE-VOLTAGE ANALOGY

The force balance equations of mechanical elements and their analogous electrical elements in force-voltage analogy are shown in table-1.2. The table-1.3 shows the list of analogous quantities in force-voltage analogy.

The following points serve as guidelines to obtain electrical analogous of mechanical systems based on force-voltage analogy.

1. In electrical systems the elements in series will have same current, likewise in mechanical systems, the elements having same velocity are said to be in series.

2. The elements having same velocity in mechanical system should have the same analogous current in electrical analogous system.

3. Each node (meeting point of elements) in the mechanical system corresponds to a closed loop in electrical system. A mass is considered as a node.

4. The number of meshes in electrical analogous is same as that of the number of nodes (masses) in mechanical system. Hence the number of mesh currents and system equations will be same as that of the number of velocities of nodes (masses) in mechanical system.

Table- 1.2 : Analogous Elements in Force-Voltage Analogy

Mechanical system	Electrical system
Input : Force Output : Velocity	Input : Voltage source Output : Current through the element
$v = \dfrac{dx}{dt}$ $f = B\dfrac{dx}{dt} = Bv$	$e = v$ and $v = Ri$ $\therefore e = Ri$
$a = \dfrac{d^2x}{dt^2} = \dfrac{dv}{dt}$ $f = M\dfrac{d^2x}{dt^2} = M\dfrac{dv}{dt}$	$e = v$ and $v = L\dfrac{di}{dt}$ $\therefore e = L\dfrac{di}{dt}$
$x = \int v\,dt$ $f = Kx = K\int v\,dt$	$e = v$ and $v = \dfrac{1}{C}\int i\,dt$ $\therefore e = \dfrac{1}{C}\int i\,dt$

Table -1.3 : Analogous Quantities in Force-Voltage Analogy

Item	Mechanical system	Electrical system (mesh basis system)
Independent variable (input)	Force, f	Voltage, e, v
Dependent variable (output)	Velocity, v	Current, i
	Displacement, x	Charge, q
Dissipative element	Frictional coefficient of dashpot, B	Resistance, R
Storage element	Mass, M	Inductance, L
	Stiffness of spring, K	Inverse of capacitance, 1/C
Physical law	Newton's second law $\sum f = 0$	Kirchoff's voltage law $\sum v = 0$
Changing the level of independent variable	Lever $\dfrac{f_1}{f_2} = \dfrac{l_1}{l_2}$	Transformer $\dfrac{e_1}{e_2} = \dfrac{N_1}{N_2}$

Table-1.4 : Analogous Elements in Force-Current Analogy

Mechanical system	Electrical system
Input : Force Output : Velocity	Input : Current source Output : Voltage across the element

Table-1.5 : Analogous Quantities in Force-Current Analogy

Item	Mechanical system	Electrical system (node basis system)
Independent variable (input)	Force, f	Current, i
Dependent variable (output)	Velocity, v	Voltage, v
	Displacement, x	Flux, φ
Dissipative element	Frictional coefficient of dashpot, B	Conductance G=1/R
Storage element	Mass, M	Capacitance, C
	Stiffness of spring, K	Inverse of inductance, 1/L
Physical law	Newton's second law $\sum f = 0$	Kirchoff's current law $\sum i = 0$
Changing the level of independent variable	Lever $\dfrac{f_1}{f_2} = \dfrac{l_1}{l_2}$	Transformer $\dfrac{e_1}{e_2} = \dfrac{N_1}{N_2}$

5.　The mechanical driving sources (force) and passive elements connected to the node (mass) in mechanical system should be represented by analogous elements in a closed loop in analogous electrical system.

6.　The element connected between two (nodes) masses in mechanical system is represented as a common element between two meshes in electrical analogous system.

FORCE-CURRENT ANALOGY

The force balance equations of mechanical elements and their analogous electrical elements in force-current analogy are shown in table-1.4. The table-1.5 shows the list of analogous quantities in force-current analogy.

The following points serve as guidelines to obtain electrical analogous of mechanical systems based on force-current analogy.

1.　In electrical systems elements in parallel will have same voltage, likewise in mechanical systems, the elements having same force are said to be in parallel.

2.　The elements having same velocity in mechanical system should have the same analogous voltage in electrical analogous system.

3.　Each node (meeting point of elements) in the mechanical system corresponds to a node in electrical system. A mass is considered as a node.

4.　The number of nodes in electrical analogous is same as that of the number of nodes (masses) in mechanical system. Hence the number of node voltages and system equations will be same as that of the number of velocities of (nodes) masses in mechanical system.

5.　The mechanical driving sources (forces) and passive elements connected to the node (mass) in mechanical system should be represented by analogous elements connected to a node in electrical system.

6.　The element connected between two nodes (masses) in mechanical system is represented as a common element between two nodes in electrical analogous system.

EXAMPLE 1.8

Write the differential equations governing the mechanical system shown in fig 1. Draw the force-voltage and force-current electrical analogous circuits and verify by writing mesh and node equations.

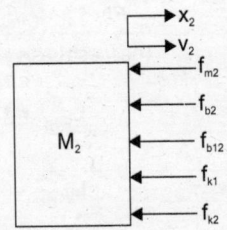

Fig 1.

SOLUTION

The given mechanical system has two nodes (masses). The differential equations governing the mechanical system are given by force balance equations at these nodes. Let the displacements of masses M_1 and M_2 be x_1 and x_2 respectively. The corresponding velocities be v_1 and v_2.

The free body diagram of M_1 is shown in fig 2. The opposing forces are marked as f_{m1}, f_{b1}, f_{b12} and f_{k1}.

$$f_{m1} = M_1 \frac{d^2 x_1}{dt^2} \quad ; \quad f_{b1} = B_1 \frac{dx_1}{dt}$$

$$f_{b12} = B_{12} \frac{d}{dt}(x_1 - x_2) \quad ; \quad f_{k1} = K_1(x_1 - x_2)$$

By Newton's second law, $f_{m1} + f_{b1} + f_{b12} + f_{k1} = f(t)$

$$\therefore M_1 \frac{d^2 x_1}{dt^2} + B_1 \frac{dx_1}{dt} + B_{12} \frac{d}{dt}(x_1 - x_2) + K_1(x_1 - x_2) = f(t) \quad \text{......(1)}$$

Fig 2.

The free body diagram of M_2 is shown in fig 3. The opposing forces are marked as f_{m2}, f_{b2}, f_{b12}, f_{k1} and f_{k2}.

$$f_{m2} = M_2 \frac{d^2 x_2}{dt^2} \quad ; \quad f_{b2} = B_2 \frac{dx_2}{dt} \quad ; \quad f_{b12} = B_{12} \frac{d}{dt}(x_2 - x_1)$$

$$f_{k1} = K_1(x_2 - x_1) \quad ; \quad f_{k2} = K_2 x_2$$

By Newton's second law, $f_{m2} + f_{b2} + f_{b12} + f_{k1} = 0$

$$M_2 \frac{d^2 x_2}{dt^2} + B_2 \frac{dx_2}{dt} + K_2 x_2 + B_{12} \frac{d}{dt}(x_2 - x_1) + K_1(x_2 - x_1) = 0 \quad \text{.....(2)}$$

On replacing the displacements by velocity in the differential equations (1) and (2) of the mechanical system we get,

Fig 3.

$$\left(\text{i.e., } \frac{d^2 x}{dt^2} = \frac{dv}{dt} \quad ; \quad \frac{dx}{dt} = v \quad \text{and} \quad x = \int v \, dt \right)$$

$$M_1 \frac{dv_1}{dt} + B_1 v_1 + B_{12}(v_1 - v_2) + K_1 \int (v_1 - v_2) \, dt = f(t) \quad \text{.....(3)}$$

$$M_2 \frac{dv_2}{dt} + B_2 v_2 + K_2 \int v_2 dt + B_{12}(v_2 - v_1) + K_1 \int (v_2 - v_1) \, dt = 0 \quad \text{.....(4)}$$

FORCE-VOLTAGE ANALOGOUS CIRCUIT

The given mechanical system has two nodes (masses). Hence the force-voltage analogous electrical circuit will have two meshes.

The force applied to mass, M_1 is represented by a voltage source in first mesh. The elements M_1, B_1, K_1 and B_{12} are connected to first node. Hence they are represented by analogous element in mesh-1 forming a closed path. The elements K_1, B_{12}, M_2, K_2, and B_2 are connected to second node. Hence they are represented by analogous element in mesh-2 forming a closed path.

The elements K_1 and B_{12} are common between node-1 and 2 and so they are represented by analogous element as common elements between two meshes. The force-voltage electrical analogous circuit is shown in fig 4.

The electrical analogous elements for the elements of mechanical system are given below.

$$f(t) \to e(t) \qquad M_1 \to L_1 \qquad B_1 \to R_1 \qquad K_1 \to 1/C_1$$

$$v_1 \to i_1 \qquad M_2 \to L_2 \qquad B_2 \to R_2 \qquad K_2 \to 1/C_2$$

$$v_2 \to i_2 \qquad\qquad\qquad B_{12} \to R_{12}$$

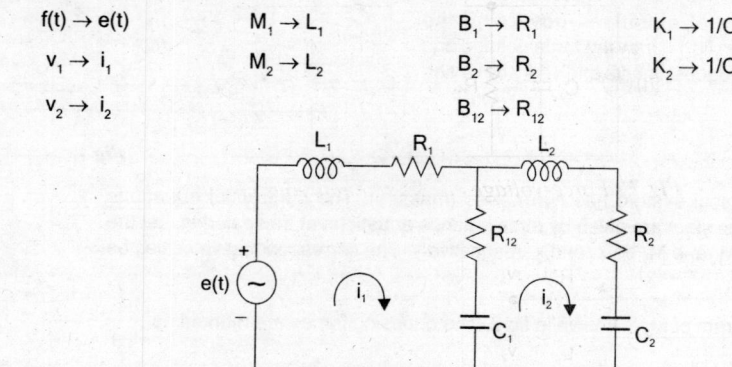

Fig 4 : Force-voltage electrical analogous circuit.

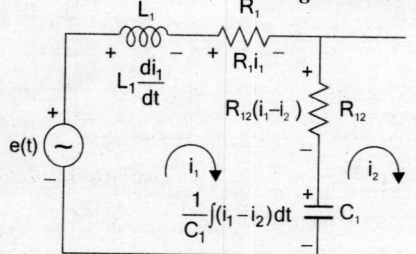

Fig 5 : Mesh-1 of analogous circuit.   *Fig 6 : Mesh-2 of analogous circuit.*

The mesh basis equations using Kirchoff's voltage law for the circuit shown in fig 4 are given below (Refer fig 5 and 6).

$$L_1 \frac{di_1}{dt} + R_1 i_1 + R_{12}(i_1 - i_2) + \frac{1}{C_1} \int (i_1 - i_2) \, dt = e(t) \qquad .....(5)$$

$$L_2 \frac{di_2}{dt} + R_2 i_2 + \frac{1}{C_2} \int i_2 \, dt + R_{12}(i_2 - i_1) + \frac{1}{C_1} \int (i_2 - i_1) \, dt = 0 \qquad .....(6)$$

It is observed that the mesh basis equations (5) and (6) are similar to the differential equations (3) and (4) governing the mechanical system.

FORCE-CURRENT ANALOGOUS CIRCUIT

The given mechanical system has two nodes (masses). Hence the force-current analogous electrical circuit will have two nodes.

The force applied to mass M_1 is represented as a current source connected to node-1 in analogous electrical circuit. The elements M_1, B_1, K_1 and B_{12} are connected to first node. Hence they are represented by analogous elements connected to node-1 in analogous electrical circuit. The elements K_1, B_{12}, M_2, K_2, and B_2 are connected to second node. Hence they are represented by analogous elements as elements connected to node-2 in analogous electrical circuit.

The elements K_1 and B_{12} are common between node-1 and 2 and so they are represented by analogous elements as common element between two nodes in analogous circuit. The force-current electrical analogous circuit is shown in fig 7.

The electrical analogous elements for the elements of mechanical system are given below.

$$v(t) \to i(t) \qquad M_1 \to C_1 \qquad B_1 \to 1/R_1 \qquad K_1 \to 1/L_1$$

$$v_1 \to v_1 \qquad M_2 \to C_2 \qquad B_2 \to 1/R_2 \qquad K_2 \to 1/L_2$$

$$v_2 \to v_2 \qquad B_{12} \to 1/R_{12}$$

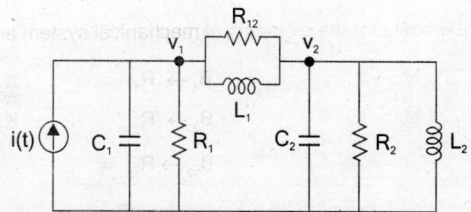

Fig 7 : *Force-voltage electrical analogous circuit.*

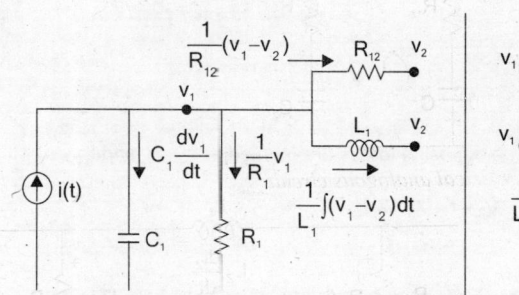

Fig 8 : *Node-1 of analogous circuit.*   **Fig 9 :** *Node-2 of analogous circuit.*

The node basis equations using Kirchoff's current law for the circuit shown in fig 7 are given below (Refer fig 8 and 9).

$$C_1 \frac{dv_1}{dt} + \frac{1}{R_1} v_1 + \frac{1}{R_{12}}(v_1 - v_2) dt = i(t) \qquad .....(7)$$

$$C_2 \frac{dv_2}{dt} + \frac{1}{R_2} v_2 + \frac{1}{L_2} \int v_2 dt + \frac{1}{R_{12}}(v_2 - v_1) + \frac{1}{L_1} \int (v_2 - v_1) dt = 0 \qquad .....(8)$$

It is observed that the node basis equations (7) and (8) are similar to the differential equations (3) and (4) governing the mechanical system.

EXAMPLE 1.9

Write the differential equations governing the mechanical system shown in fig 1. Draw the force -voltage and force-current electrical analogous circuits and verify by writing mesh and node equations.

SOLUTION

The given mechanical system has three nodes masses. The differential equations governing the mechanical system are given by force balance equations at these nodes. Let the displacements of masses M_1, M_2 and M_3 be x_1, x_2 and x_3 respectively. The corresponding velocities be v_1, v_2 and v_3.

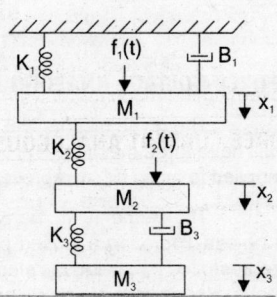

Fig 1.

The free body diagram of M_1 is shown in fig 2. The opposing forces are marked as f_{m1}, f_{b1}, f_{k2} and f_{k1}.

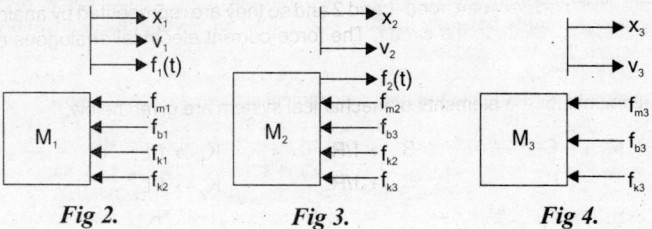

Fig 2.   *Fig 3.*   *Fig 4.*

$$f_{m1} = \frac{M_1 d^2 x_1}{dt^2} \quad ; \quad f_{b1} = B_1 \frac{dx_1}{dt} \quad ; \quad f_{k2} = K_2(x_1 - x_2) \quad ; \quad f_{k1} = K_1 x_1$$

By Newton's second law, $f_{m1} + f_{b1} + f_{k2} + f_{k1} = f_1(t)$

$$M_1 \frac{d^2 x_1}{dt^2} + B_1 \frac{dx_1}{dt} + K_2(x_1 - x_2) + K_1 x_1 = f_1(t) \qquad \text{.....(1)}$$

Free body diagram of M_2 is shown in fig 3. The opposing forces are marked as f_{m2}, f_{b3}, f_{k2} & f_{k3}.

$$f_{m2} = M_2 \frac{d^2 x_2}{dt^2} \quad ; \quad f_{b3} = B_3 \frac{d}{dt}(x_2 - x_3) \quad ; \quad f_{k2} = K_2(x_2 - x_1) \quad ; \quad f_{k3} = K_3(x_2 - x_3)$$

By Newton's second law,

$$M_2 \frac{d^2 x_2}{dt^2} + B_3 \frac{d}{dt}(x_2 - x_3) + K_2(x_2 - x_1) + K_3(x_2 - x_3) = f_2(t) \qquad \text{.....(2)}$$

The free body diagram of M_3 is shown in fig 4. The opposing forces are marked as f_{m3}, f_{b3} and f_{k3}.

$$f_{m3} = M_3 \frac{d^2 x_3}{dt^2} \quad ; \quad f_{b3} = B_3 \frac{d}{dt}(x_3 - x_2) \quad ; \quad f_{k3} = K_3(x_3 - x_2)$$

By Newton's second law,

$$M_3 \frac{d^2 x_3}{dt^2} + B_3 \frac{d}{dt}(x_3 - x_2) + K_3(x_3 - x_2) = 0 \qquad \text{.....(3)}$$

On replacing the displacements by velocity in the differential equations (1), (2) and (3) governing the mechanical system we get,

$$\left(\text{i.e., } \frac{d^2 x}{dt^2} = \frac{dv}{dt} \quad ; \quad \frac{dx}{dt} = v \text{ and } x = \int v \, dt \right)$$

$$M_1 \frac{dv_1}{dt} + B_1 v_1 + K_1 \int v_1 dt + K_2 \int (v_1 - v_2) \, dt = f_1(t) \qquad \text{.....(4)}$$

$$M_2 \frac{dv_2}{dt} + B_3(v_2 - v_3) + K_2 \int (v_2 - v_1) \, dt + K_3 \int (v_2 - v_3) \, dt = f_2(t) \qquad \text{.....(5)}$$

$$M_3 \frac{dv_3}{dt} + B_3(v_3 - v_2) + K_3 \int (v_3 - v_2) \, dt = 0 \qquad \text{.....(6)}$$

FORCE-VOLTAGE ANALOGOUS CIRCUIT

The given mechanical system has three nodes (masses). Hence the force-voltage analogous electrical circuit will have three meshes. The force applied to mass, M_1 is represented by a voltage source in first mesh and the force applied to mass, M_2 is represented by a voltage source in second mesh.

The elements M_1, B_1, K_1 and K_2 are connected to first node. Hence they are represented by analogous element in mesh-1 forming a closed path. The elements M_2, B_3, K_2 and K_3 are connected to second node. Hence they are represented by analogous element in mesh-2 forming a closed path. The elements M_3, K_3 and B_3 are connected to third node. Hence they are represented by analogous element in mesh-3 forming a closed path.

The element K_2 is common between node-1 and 2 and so it is represented by analogous element as common element between mesh 1 and 2. The elements K_3 and B_3 are common between node-2 and 3 and so they are represented by analogous elements as common elements between mesh-2 and 3. The force-voltage electrical analogous circuit is shown in fig 5.

The electrical analogous elements for the elements of mechanical system are given below.

$f_1(t) \rightarrow e_1(t)$	$v_1 \rightarrow i_1$	$M_1 \rightarrow L_1$	$B_1 \rightarrow R_1$	$K_1 \rightarrow 1/C_1$
$f_2(t) \rightarrow e_2(t)$	$v_2 \rightarrow i_2$	$M_2 \rightarrow L_2$	$B_3 \rightarrow R_3$	$K_2 \rightarrow 1/C_2$
	$v_3 \rightarrow i_3$	$M_3 \rightarrow L_3$		$K_3 \rightarrow 1/C_3$

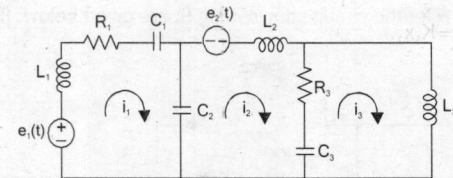

Fig 5 : Force-voltage electrical analogous circuit.

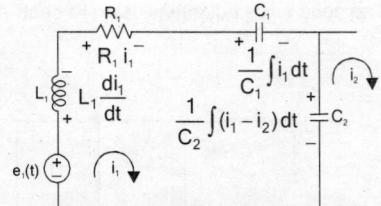

Fig 6 : Mesh-1 analogous circuit.

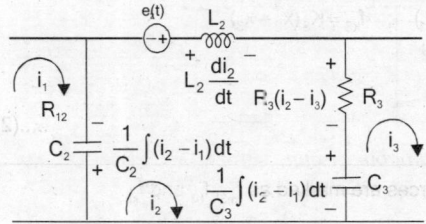

Fig 7 : Mesh-2 of analogous circuit.

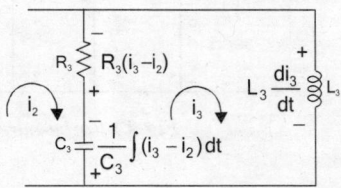

Fig 8 : Mesh-3 of analogous circuit.

The mesh basis equations using Kirchoff's voltage law for the circuit shown in fig 5 are given below (Refer fig 6, 7, 8).

$$L_1\frac{di_1}{dt} + R_1i_1 + \frac{1}{C_1}\int i_1 dt + \frac{1}{C_2}\int (i_1 - i_2)dt = e_1(t) \qquad .....(7)$$

$$L_2\frac{di_2}{dt} + R_3(i_2 - i_3) + \frac{1}{C_3}\int (i_2 - i_3)dt + \frac{1}{C_2}\int (i_2 - i_1)dt = e_2(t) \qquad .....(8)$$

$$L_3\frac{di_3}{dt} + R_3(i_3 - i_2) + \frac{1}{C_3}\int (i_3 - i_2)dt = 0 \qquad .....(9)$$

It is observed that the mesh equations (7), (8) and (9) are similar to the differential equations (4), (5) and (6) governing the mechanical system.

FORCE-CURRENT ANALOGOUS CIRCUIT

The given mechanical system has three nodes (masses). Hence the force-current analogous electrical circuit will have three nodes.

The force applied to mass M_1 is represented as a current source connected to node-1 in analogous electrical circuit. The force applied to mass M_2 is represented as a current source connected to node-2 in analogous electrical circuit.

The elements M_1, B_1, K_1 and K_2 are connected to first node. Hence they are represented by analogous elements as elements connected to node-1 in analogous electrical circuit. The elements M_2, B_3, K_2 and K_3 are connected to second node. Hence they are represented by analogous elements as elements connected to node-2 in analogous electrical circuit. The elements M_3, B_3 and K_3 are connected to third node. Hence they are represented by analogous elements as elements connected to node-3 in analogous electrical circuit.

The element K_2 is common between node-1 and 2 and so it is represented by analogous element as common element between node-1 and 2 in analogous circuit. The elements B_3 and K_3 are common between node-2 and 3 and so they are represented by analogous elements as common elements between node-2 and 3. The force-current electrical analogous circuit is shown in fig 9.

The electrical analogous elements for the elements of mechanical system are given below.

$f_1(t) \rightarrow i_1(t)$	$v_1 \rightarrow v_1$	$M_1 \rightarrow C_1$	$B_1 \rightarrow 1/R_1$	$K_1 \rightarrow 1/L_1$
$f_2(t) \rightarrow i_2(t)$	$v_2 \rightarrow v_2$	$M_2 \rightarrow C_2$	$B_3 \rightarrow 1/R_3$	$K_2 \rightarrow 1/L_2$
	$v_3 \rightarrow v_3$	$M_3 \rightarrow C_3$		$K_3 \rightarrow 1/L_3$

The node basis equations using Kirchoff's current law for the circuit shown in fig 9. are given below. (Refer fig 10, 11,12).

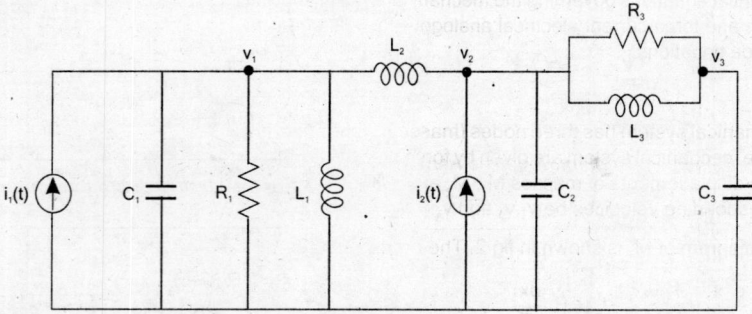

Fig 9 : *Force-current electrical analogous circuit.*

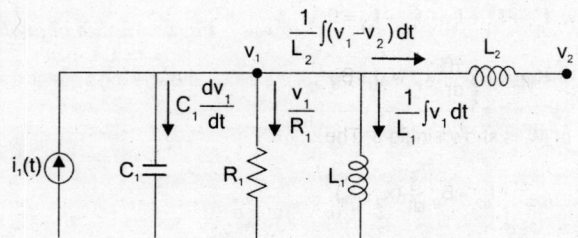

Fig 10 : *Node-1 of analogous circuit.*

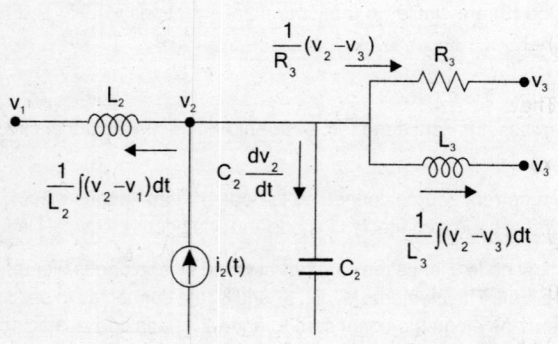

Fig 11 : *Node-2 of analogous circuit.*   **Fig 12 :** *Node-3 of analogous circuit.*

$$C_1 \frac{dv_1}{dt} + \frac{1}{R_1} v_1 + \frac{1}{L_1} \int (v_1 - v_2) dt = i_1(t) \qquad \qquad .....(10)$$

$$C_2 \frac{dv_2}{dt} + \frac{1}{R_3} (v_2 - v_3) + \frac{1}{L_3} \int (v_2 - v_3) dt + \frac{1}{L_2} \int (v_2 - v_1) dt = i_2(t) \qquad \qquad .....(11)$$

$$C_3 \frac{dv_3}{dt} + \frac{1}{R_3} (v_3 - v_2) + \frac{1}{L_3} \int (v_3 - v_2) dt = 0 \qquad \qquad .....(12)$$

It is observed that node basis equations (10), (11) and (12) are similar to the differential equations (4), (5) and (6) governing the mechanical system.

EXAMPLE 1.10

Write the differential equations governing the mechanical system shown in fig 1.Draw force-voltage and force-current electrical analogous circuits and verify by writing mesh and node equations.

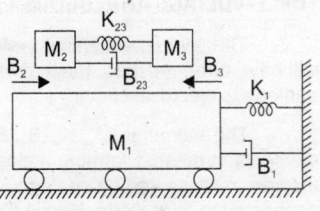

Fig 1.

SOLUTION

The given mechanical system has three nodes (masses). The differential equations governing the mechanical system are given by force balance equations at these nodes. Let the displacements of masses M_1, M_2 and M_3 be x_1, x_2 and x_3 respectively. The corresponding velocities be v_1, v_2 and v_3.

The free body diagram of M_1 is shown in fig 2. The opposing forces are marked as f_{b1}, f_{k1}, f_{b2}, f_{b3}, and f_{m1}.

$$f_{m1} = M_1\frac{d^2x_1}{dt^2} \quad ; \quad f_{b1} = B_1\frac{dx_1}{dt} \quad ; \quad f_{k1} = K_1x_1$$

$$f_{b2} = B_2\frac{d}{dt}(x_1 - x_2) \quad ; \quad f_{b3} = B_3\frac{d}{dt}(x_1 - x_3)$$

By Newton's second law, $f_{m1} + f_{b1} + f_{k1} + f_{b2} + f_{b3} = 0$

$$M_1\frac{d^2x_1}{dt^2} + B_1\frac{dx_1}{dt} + K_1x_1 + B_2\frac{d}{dt}(x_1 - x_2) + B_3\frac{d}{dt}(x_1 - x_3) = 0 \qquad .....(1)$$

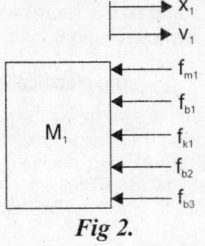

Fig 2.

The free body diagram of M_2 is shown in fig 3. The opposing forces are marked as f_{m2}, f_{b2}, f_{b23} and f_{k23}.

$$f_{m2} = M_2\frac{d^2x_2}{dt^2} \quad\quad ; \quad f_{b2} = B_2\frac{d}{dt}(x_2 - x_1)$$

$$f_{b23} = B_{23}\frac{d}{dt}(x_2 - x_3) \quad ; \quad f_{k23} = K_{23}(x_2 - x_3)$$

By Newton's second law, $f_{m2} + f_{b2} + f_{b23} + f_{k23} = 0$

$$M_2\frac{d^2x_2}{dt^2} + B_2\frac{d}{dt}(x_2 - x_1) + B_{23}\frac{d}{dt}(x_2 - x_3) + K_{23}(x_2 - x_3) = 0 \qquad ......(2)$$

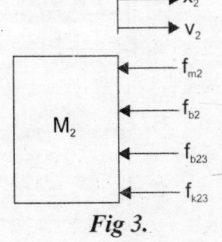

Fig 3.

The free body diagram of M_3 is shown in fig 4. The opposing forces are marked as f_{m3}, f_{b3}, f_{b23}, and f_{k23}.

$$f_{m3} = M_3\frac{d^2x_3}{dt^2} \quad\quad ; \quad f_{b3} = B_3\frac{d}{dt}(x_3 - x_1)$$

$$f_{b23} = B_{23}\frac{d}{dt}(x_3 - x_2) \quad ; \quad f_{k23} = K_{23}(x_3 - x_2)$$

By Newton's second law, $f_{m3} + f_{b3} + f_{b23} + f_{k23} = 0$

$$M_3\frac{d^2x_3}{dt^2} + B_3\frac{d}{dt}(x_3 - x_1) + B_{23}\frac{d}{dt}(x_3 - x_2) + K_{23}(x_3 - x_2) = 0 \qquad .....(3)$$

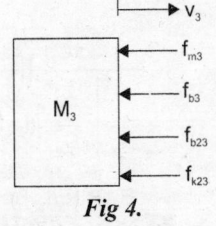

Fig 4.

On replacing the displacements by velocity in the differential equations (1), (2) and (3) governing the mechanical system we get,

$$\left(\text{i.e., } \frac{d^2x}{dt^2} = \frac{dv}{dt}, \quad \frac{dx}{dt} = v \quad \text{and} \quad x = \int v\,dt\right)$$

$$M_1\frac{dv_1}{dt} + B_1v_1 + K_1\int v_1 dt + B_2(v_1 - v_2) + B_3(v_1 - v_3) = 0 \qquad\qquad .....(4)$$

$$M_2\frac{dv_2}{dt} + B_2(v_2 - v_1) + B_{23}(v_2 - v_3) + K_{23}\int(v_2 - v_3)\,dt = 0 \qquad .....(5)$$

$$M_3\frac{dv_3}{dt} + B_3(v_3 - v_1) + B_{23}(v_3 - v_2) + K_{23}\int(v_3 - v_2)\,dt = 0 \qquad .....(6)$$

FORCE-VOLTAGE ANALOGOUS CIRCUIT

The given mechanical system has three nodes (masses). Hence the force-voltage analogous electrical circuit will have three meshes. Since there is no applied force in mechanical system there will not be any voltage source in analogous electrical circuit.

The elements M_1, K_1, B_1, B_3 and B_2 are connected to first node. Hence they are represented by analogous elements in mesh-1 forming a closed path. The elements M_2, K_{23}, B_{23} and B_2 are connected to second node. Hence they are represented by analogous elements in mesh-2 forming a closed path. The elements M_3, K_{23}, B_{23} and B_3 are connected to third node. Hence they are represented by analogous elements in mesh-3 forming a closed path.

The elements K_{23} and B_{23} are common between node-2 and 3 and so they are represented by analogous element as common elements between mesh-2 and 3. The element B_2 is common between node-1 and 2 and so it is represented by analogous element as common element between mesh-1 and 2. The element B_3 is common between node-1 and 3 and so it is represented by analogous element between mesh-1 and 3. The force-voltage electrical analogous circuit is shown in fig 5.

The electrical analogous elements for the elements of mechanical system are given below.

$v_1 \to i_1$	$M_1 \to L_1$	$K_1 \to 1/C_1$	$B_2 \to R_2$
$v_2 \to i_2$	$M_2 \to L_2$	$K_{23} \to 1/C_{23}$	$B_3 \to R_3$
$v_3 \to i_3$	$M_3 \to L_3$	$B_1 \to R_1$	$B_{23} \to R_{23}$

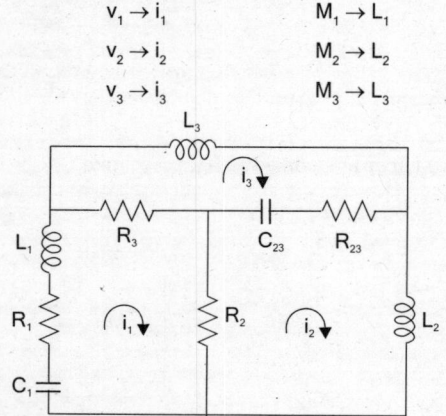

Fig 5 : *Force-voltage electrical analogous circuit.*

Fig 6 : *Mesh-1 of analogous circuit.*

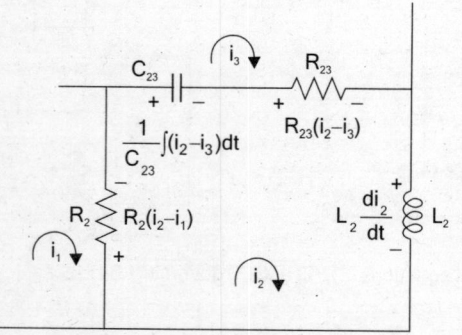

Fig 7 : *Mesh-2 of analogous circuit.*

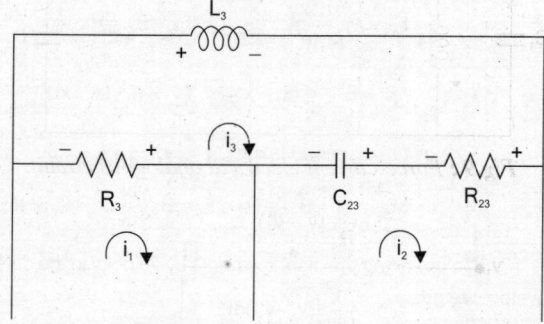

Fig 8 : *Mesh-3 of analogous circuit..*

The mesh basis equations using Kirchoff's voltage law for the circuit shown in fig 5 are given below. (Refer fig 6, 7 and 8).

$$L_1 \frac{di_1}{dt} + R_1 i_1 + \frac{1}{C_1} \int i_1 dt + R_2(i_1 - i_2) + R_3(i_1 - i_2) = 0 \qquad .....(7)$$

$$L_2 \frac{di_2}{dt} + R_2(i_2 - i_1) + \frac{1}{C_{23}} \int (i_2 - i_3) dt + R_{23}(i_2 - i_3) = 0 \qquad .....(8)$$

$$L_3 \frac{di_3}{dt} + R_3(i_3 - i_1) + \frac{1}{C_{23}} \int (i_3 - i_2)\,dt + R_{23}(i_3 - i_2) = 0 \qquad \qquad .....(9)$$

It is observed that the mesh basis equations (7), (8) and (9) are similar to the differential equations (4), (5) and (6) governing the mechanical system.

FORCE-CURRENT ANALOGOUS CIRCUIT

The given mechanical system has three nodes (masses). Hence the force-current analogous electrical circuit will have three nodes. Since there is no applied force in mechanical system there will not be any current source in analogous electrical circuit.

The elements M_1, K_1, B_1, B_2 and B_3 are connected to first node. Hence they are represented by analogous elements as elements connected to node-1 in analogous electrical circuit. The elements M_2, K_{23}, B_{23} and B_2 are connected to second node. Hence they are represented by analogous elements as elements connected to node-2 in analogous electrical circuit. The elements M_3, K_{23}, B_{23} and B_3 are connected to third node. Hence they are represented by analogous elements as elements connected to node-3 in analogous electrical circuit.

The elements K_{23} and B_{23} are common between node-2 and 3 and so they are represented by analogous element as common elements between node-2 and 3 in electrical analogous circuit. The element B_2 is common between node-1 and 2 and so it is represented by analogous element as common element between node-1 and 2 in electrical analogous circuit. The element B_3 is common between node-1 and 3 and so it is represented by analogous element as common element between node-1 and 3 in electrical analogous circuit. The force-current electrical analogous circuit is shown in fig 9.

The electrical analogous elements for the elements of mechanical system are given below.

$v_1 \to v_1$	$M_1 \to C_1$	$K_1 \to 1/L_1$	$B_2 \to 1/R_2$
$v_2 \to v_2$	$M_2 \to C_2$	$K_{23} \to 1/L_{23}$	$B_3 \to 1/R_3$
$v_3 \to v_3$	$M_3 \to C_3$	$B_1 \to 1/R_1$	$B_{23} \to 1/R_{23}$

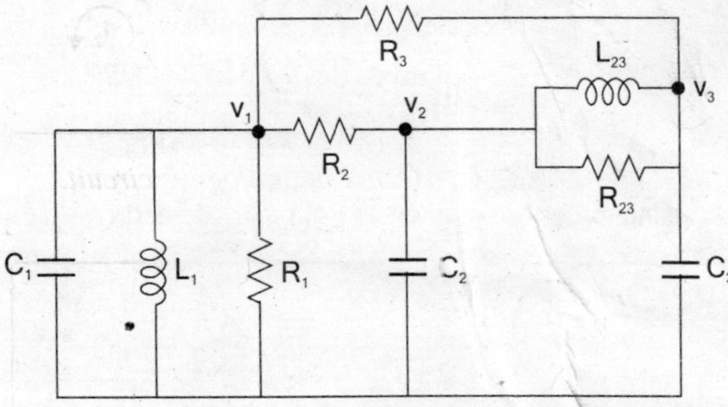

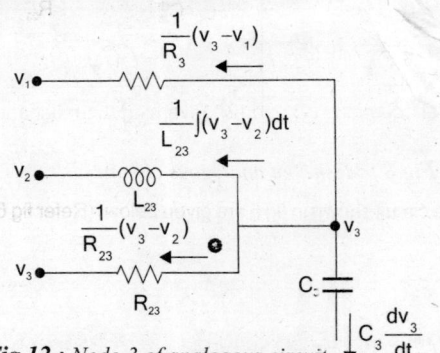

Fig 9 : Force-current electrical analogous circuit.

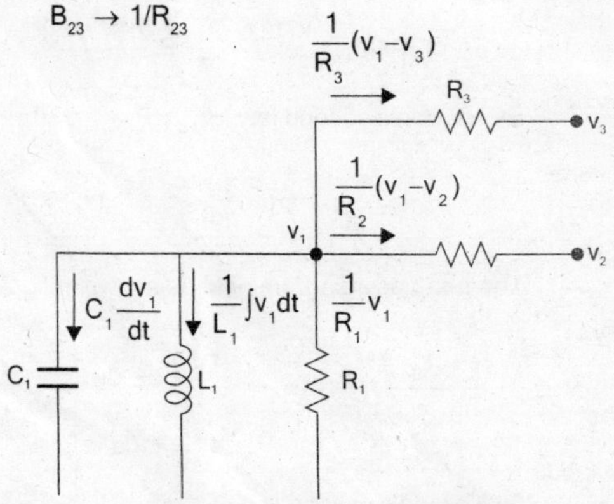

Fig 10 : Node-1 of analogous circuit.

Fig 12 : Node-3 of analogous circuit.

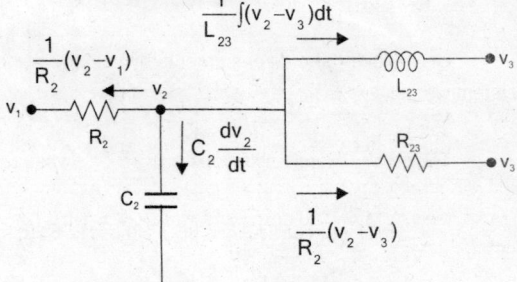

Fig 11 : Node-2 of analogous circuit.

The node basis equations using Kirchoff's current law for the circuit shown in fig 9 are given below. (Refer fig 10, 11 and 12).

$$C_1 \frac{dv_1}{dt} + \frac{1}{R_1} v_1 + \frac{1}{L_1} \int v_1 dt + \frac{1}{R_2}(v_1 - v_3) = 0 \qquad .....(10)$$

$$C_2 \frac{dv_2}{dt} + \frac{1}{R_2}(v_2 - v_1) + \frac{1}{L_{23}} \int (v_2 - v_3) dt + \frac{1}{R_{23}}(v_2 - v_3) = 0 \qquad .....(11)$$

$$C_3 \frac{dv_3}{dt} + \frac{1}{R_3}(v_3 - v_1) + \frac{1}{L_{23}} \int (v_3 - v_2) dt + \frac{1}{R_{23}}(v_3 - v_2) = 0 \qquad .....(12)$$

It is observed that the node basis equations (10), (11) and (12) are similar to the differential equations (4), (5) and (6) governing the mechanical system.

EXAMPLE 1.11

Write the differential equations governing the mechanical system shown in fig 1. Draw the force-voltage and force-current electrical analogous circuits and verify by writing mesh and node equations.

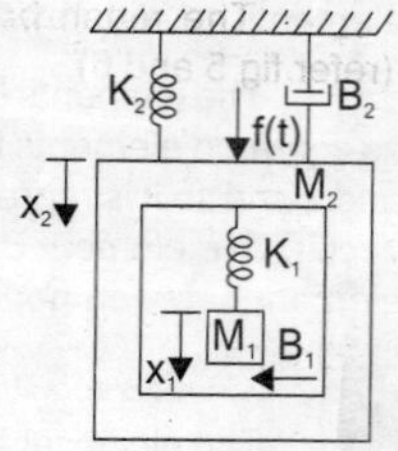

SOLUTION

The given mechanical system has two nodes (masses). The differential equations governing the mechanical system are given by force balance equations at these nodes. Let the displacement of masses M_1 and M_2 be x_1 and x_2 respectively. The corresponding velocities be v_1 and v_2.

Fig 1.

The free body diagram of M_1 is shown in fig 2. The opposing forces are marked as f_{m1}, f_{b1} and f_{k1}.

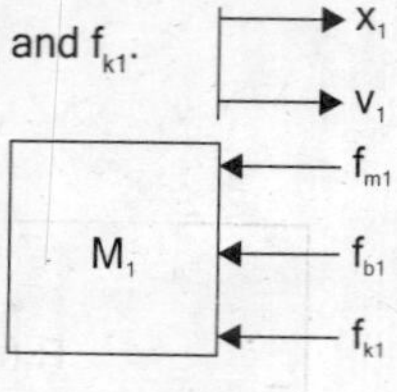

$$f_{m1} = M_1 \frac{d^2 x_1}{dt^2} \quad ; \quad f_{b1} = B_1 \frac{d(x_1 - x_2)}{dt} \quad ; \quad f_{k1} = K_1(x_1 - x_2)$$

By Newton's second law, $f_{m1} + f_{b1} + f_{k1} = 0$

$$M_1 \frac{d^2 x_1}{dt^2} + B_1 \frac{d(x_1 - x_2)}{dt} + K_1(x_1 - x_2) = 0 \qquad .....(1)$$

Fig 2.

The free body diagram of M_2 is shown in fig 3. The opposing forces are marked as f_{m2}, f_{b2}, f_{b1}, f_{k2} and f_{k1}.

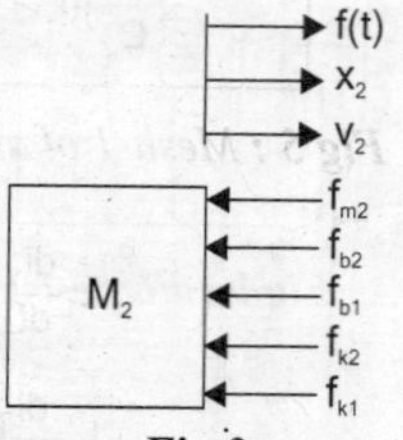

$$f_{m2} = M_2 \frac{d^2 x_2}{dt^2} \quad ; \quad f_{b2} = B_2 \frac{dx_2}{dt} \quad ; \quad f_{b1} = B_1 \frac{d}{dt}(x_2 - x_1)$$

$$f_{k2} = K_2 x_2 \quad ; \quad f_{k1} = K_1(x_2 - x_1)$$

By Newton's second law, $f_{m2} + f_{b2} + f_{k2} + f_{b1} + f_{k1} = f(t)$

$$M_2 \frac{d^2 x_2}{dt^2} + B_2 \frac{dx_2}{dt} + K_2 x_2 + B_1 \frac{d}{dt}(x_2 - x_1) + K_1(x_2 - x_1) = f(t) \qquad .....(2)$$

Fig 3.

On replacing the displacements by velocity in the differential equations (1) and (2) governing the mechanical system we get,

$$\left(\text{i.e., } \frac{d^2 x}{dt^2} = \frac{dv}{dt}, \quad \frac{dx}{dt} = v \text{ and } x = \int v dt \right)$$

$$M_1 \frac{dv_1}{dt} + B_1(v_1 - v_2) + K_1 \int (v_1 - v_2) dt = 0 \qquad .....(3)$$

$$M_2 \frac{dv_2}{dt} + B_2 v_2 + K_2 \int v_2 dt + B_1(v_2 - v_1) + K_1 \int (v_2 - v_1) dt = f(t) \qquad .....(4)$$

FORCE-VOLTAGE ANALOGOUS CIRCUIT

The given mechanical system has two nodes (masses). Hence the force voltage analogous electrical circuit will have two meshes. The force applied to mass, M_2 is represented by a voltage source in second mesh.

The elements M_1, K_1 and B_1 are connected to first node. Hence they are represented by analogous element in mesh 1 forming a closed path. The elements M_2, K_2, B_2, B_1 and K_1 are connected to second node. Hence they are represented by analogous element in mesh 2 forming a closed path.

The elements B_1 and K_1 are common between node 1 and 2 and so they are represented as common elements between mesh 1 and 2. The force-voltage electrical analogous circuit is shown in fig 4.

The electrical analogous elements for the elements of mechanical system are given below.

$$f(t) \rightarrow e(t) \quad v_1 \rightarrow i_1 \qquad M_1 \rightarrow L_1 \qquad K_1 \rightarrow 1/C_1 \qquad B_1 \rightarrow R_1$$
$$v_2 \rightarrow i_2 \qquad M_2 \rightarrow L_2 \qquad K_2 \rightarrow 1/C_2 \qquad B_2 \rightarrow R_2$$

The mesh basis equations using Kirchoff's voltage law for the circuit shown in fig 4. are given below, (refer fig 5 and 6).

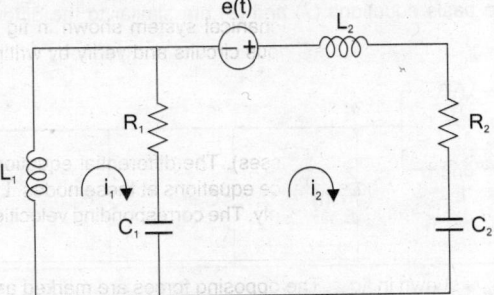

Fig 4 : *Force-voltage electrical analogous circuit.*

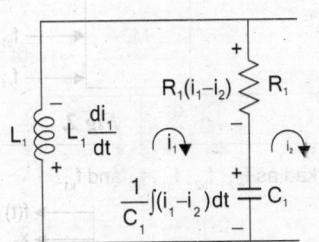

Fig 5 : *Mesh-1 of analogous circuit.*

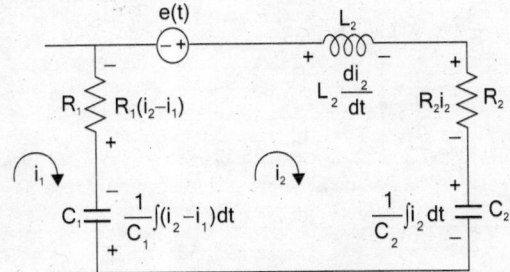

Fig 6 : *Mesh-2 of analogous circuit.*

$$L_1 \frac{di_1}{dt} + R_1(i_1 - i_2) + \frac{1}{C_1} \int (i_1 - i_2) dt = 0 \qquad \qquad .....(5)$$

$$L_2 \frac{di_2}{dt} + R_2 i_2 + \frac{1}{C_2} \int i_2 dt + \frac{1}{C_1} \int (i_2 - i_1) dt + R_1(i_2 - i_1) = e(t) \qquad .....(6)$$

It is observed that the mesh basis equations (5) and (6) are similar to the differential equations (3) and (4) governing the mechanical system.

FORCE-CURRENT ANALOGOUS CIRCUIT

The given mechanical system has two nodes (masses). Hence the force-current analogous electrical circuit will have two nodes. The force applied to mass M_2 is represented as a current source connected to node-2 in analogous electrical circuit.

The elements M_1, K_1 and B_1 are connected to first node. Hence they are represented by analogous elements as elements connected to node-1 in analogous electrical circuit. The elements M_2, K_2, B_2, B_1 and K_1 are connected to second node. Hence they are represented by analogous elements as elements connected to node-1 in analogous electrical circuit.

The elements K_1 and B_1 is common to node-1 and 2 and so they are represented by analogous element as common elements between two nodes in analogous circuit. The force-current electrical analogous circuit is shown in fig 7.

The electrical analogous elements for the elements of mechanical system are given below.

$$f(t) \rightarrow i(t) \quad v_1 \rightarrow v_1 \qquad M_1 \rightarrow C_1 \qquad B_1 \rightarrow 1/R_1 \qquad K_1 \rightarrow 1/L_1$$
$$v_2 \rightarrow v_2 \qquad M_2 \rightarrow C_2 \qquad B_2 \rightarrow 1/R_2 \qquad K_2 \rightarrow 1/L_2$$

The node basis equations using Kirchoff's current law for the circuit shown in fig.7, are given below, (Refer fig 8 and 9).

$$C_1 \frac{dv_1}{dt} + \frac{1}{R_1}(v_1 - v_2) + \frac{1}{L_1}\int (v_1 - v_2)\,dt = 0 \qquad \qquad .....(7)$$

$$C_2 \frac{dv_2}{dt} + \frac{1}{R_2}v_2 + \frac{1}{L_2}\int v_2\,dt + \frac{1}{R_1}(v_2 - v_1) + \frac{1}{L_1}\int (v_2 - v_1) = i(t) \qquad .....(8)$$

It is observed that the node basis equations (7) and (8) are similar to the differential equations (3) and (4) governing the mechanical system.

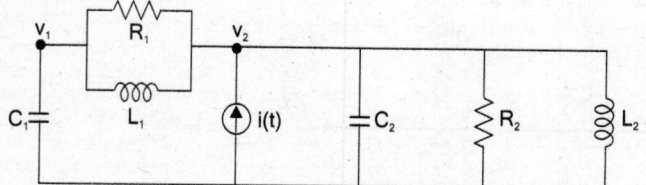

Fig 7 : *Force-current electrical analogous circuit.*

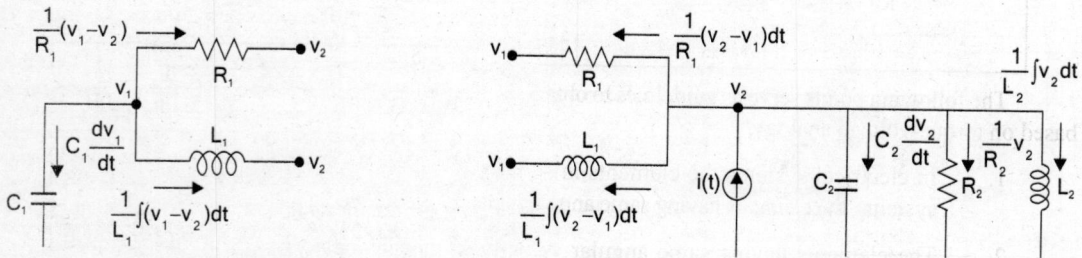

Fig 8 : *Node-1 of analogous circuit.*   **Fig 9** : *Node-2 of analogous circuit.*

1.10 ELECTRICAL ANALOGOUS OF MECHANICAL ROTATIONAL SYSTEMS

The three basic elements moment of inertia, rotational dashpot and torsional spring that are used in modelling mechanical rotational systems are analogous to resistance, inductance and capacitance of electrical systems. The input torque in mechanical system is analogous to either voltage source or current source in electrical systems. The output angular velocity (first derivative of angular displacement) in mechanical rotational system is analogous to either current or voltage in an element in electrical system. Since the electrical systems has two types of inputs either voltage source or current source, there are two types of analogies: *torque-voltage analogy and torque-current analogy*.

TORQUE-VOLTAGE ANALOGY

The torque balance equations of mechanical rotational elements and their analogous electrical elements in torque-voltage analogy are shown in table-1.6. The table-1.7 shows the list of analogous quantities in torque-voltage analogy.

TABLE-1.6 : Analogous Element of Torque-Voltage Analogy

Mechanical rotational system	Electrical system
Input : Torque Output : Angular velocity	Input : Voltage source Output : Current through the element
$T = B\dfrac{d\theta}{dt} = B\omega$ $\omega = \dfrac{d\theta}{dt}$	$e = v$; $v = Ri$ $\therefore e = Ri$
$T = J\dfrac{d^2\theta}{dt^2} = j\dfrac{d\omega}{dt}$ $\propto = \dfrac{d^2\theta}{dt^2} = \dfrac{d\omega}{dt}$	$e = v$; $v = L\dfrac{di}{dt}$ $\therefore e = L\dfrac{di}{dt}$
$T = K\theta = K\int\omega\,dt$ $\theta = \int\omega\,dt$	$e = v$; $v = \dfrac{1}{C}\int i\,dt$ $\therefore e = \dfrac{1}{C}\int i\,dt$

The following points serve as guidelines to obtain electrical analogous of mechanical rotational systems based on torque-voltage analogy.

1.   In electrical systems the elements in series will have same current, likewise in mechanical systems, the elements having same angular velocity are said to be in series.

2.   The elements having same angular velocity in mechanical system should have analogous same current in electrical analogous system.

3.   Each node (meeting point of elements) in the mechanical system corresponds to a closed loop in electrical system. The moment of inertia of mass is considered as a node.

4.   The number of meshes in electrical analogous is same as that of the number of nodes (moment of inertia of mass) in mechanical system. Hence the number of mesh currents and system equations will be same as that of the number of angular velocities of nodes (moment of inertia of mass) in mechanical system.

5.   The mechanical driving sources (Torque) and passive elements connected to the node (moment of inertia of mass) in mechanical system should be represented by analogous element in a closed loop in analogous electrical system.

6.   The element connected between two nodes (moment of inertia) in mechanical system is represented as a common element between two meshes in electrical analogous system.

Table-1.7 : Analogous Quantities in Torque-Voltage Analogy

Item	Mechanical rotational system	Electrical system (mesh basis system)
Independent variable (input)	Torque, T	Voltage, e, v
Dependent variable (output)	Angular Velocity, ω	Current, i
	Angular displacement, θ	Charge, q
Dissipative element	Rotational coefficient of dashpot, B	Resistance, R
Storage element	Moment of inertia, J	Inductance, L
	Stiffness of spring, K	Inverse of capacitance, 1/C
Physical law	Newton's second law $\sum T = 0$	Kirchoff's voltage law $\sum v = 0$
Changing the level of independent variable	Gear $\dfrac{T_1}{T_2} = \dfrac{n_1}{n_2}$	Transformer $\dfrac{e_1}{e_2} = \dfrac{N_1}{N_2}$

TORQUE-CURRENT ANALOGY

The torque balance equations of mechanical elements and their analogous electrical elements in torque-current analogy are shown in table-1.8. The table-1.9 shows the list of analogous quantities in torque-current analogy.

The following points serve as guidelines to obtain electrical analogous of mechanical rotational systems based on Torque-current analogy.

1. In electrical systems the elements in parallel will have same voltage, likewise in mechanical systems, the elements having same torque are said to be in parallel.

2. The elements having same angular velocity in mechanical system should have analogous same voltage in electrical analogous system.

3. Each node (meeting point of elements) in the mechanical system corresponds to a node in electrical system. The moment of inertia of mass is considered as a node.

4. The number of nodes in electrical analogous is same as that of the number of nodes (moment of inertia of mass) in mechanical system. Hence the number of node voltages and system equations will be same as that of the number of angular velocities of nodes (moment of inertia of mass) in mechanical system.

5. The mechanical driving sources (Torque) and passive elements connected to the node in mechanical system should be represented by analogous element connected to a node in analogous electrical system.

6. The element connected between two nodes (moment of inertia of mass) in mechanical system is represented as a common element between two nodes in electrical analogous system.

TABLE-1.8 : Analogous Elements in Torque-Current Analogy

Mechanical rotational system	Electrical system
Input : Torque Output : Angular velocity	Input : Current source Output : Voltage across the element
$$T = B\frac{d\theta}{dt} = B\omega \qquad \omega = \frac{d\epsilon}{dt}$$	$$i = \frac{1}{R}v$$
$$T = K\theta = K\int \omega \, dt \qquad \theta = \int \omega \, dt$$	$$i = \frac{1}{L}\int v \, dt$$
$$T = J\frac{d^2\theta}{dt^2} = J\frac{d\omega}{dt} \qquad \alpha = \frac{d^2\theta}{dt^2} = J\frac{d\omega}{dt}$$	$$i = C\frac{dv}{dt}$$

Table-1.9 : Analogous Quantities in Torque-Current Analogy

Item	Mechanical rotational system	Electrical system (node basis system)
Independent variable (input)	Torque, T	Current, i
Dependent variable (output)	Angular Velocity, ω	Voltage, v
	Angular displacement, θ	Flux, ϕ
Dissipative element	Rotational frictional coefficient of dashpot, B	Conductance, G = 1/R
Storage element	Moment of inertia, J	Capacitance, C
	Stiffness of spring, K	Inverse of inductance, 1/L
Physical law	Newton's second law $\sum T = 0$	Kirchoff's current law $\sum i = 0$
Changing the level of independent variable	Gear $\dfrac{T_1}{T_2} = \dfrac{n_1}{n_2}$	Transformer $\dfrac{i_1}{i_2} = \dfrac{N_2}{N_1}$

EXAMPLE 1.12

Write the differential equations governing the mechanical rotational system shown in fig 1. Draw the torque-voltage and torque-current electrical analogous circuits and verify by writing mesh and node equations.

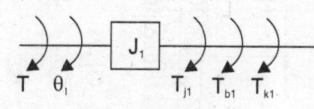

Fig 1.

SOLUTION

The given mechanical rotational system has two nodes (moment of inertia of masses). The differential equations governing the mechanical rotational system are given by torque balance equations at these nodes.

Let the angular displacements of J_1 and J_2 be θ_1 and θ_2 respectively. The corresponding angular velocities be ω_1 and ω_2.

The free body diagram of J_1 is shown in fig 2. The opposing torques are marked as T_{j1}, T_{b1} and T_{k1}.

$$T_{j1} = J_1 \frac{d^2\theta_1}{dt^2} \quad ; \quad T_{b1} = B_1 \frac{d\theta_1}{dt} \quad ; \quad T_{k1} = K_1(\theta_1 - \theta_2)$$

By Newton's second law, $\quad T_{j1} + T_{b1} + T_{k1} = T$

$$J_1 \frac{d^2\theta_1}{dt^2} + B_1 \frac{d\theta_1}{dt} + K_1(\theta_1 - \theta_2) \qquad \qquad .....(1)$$

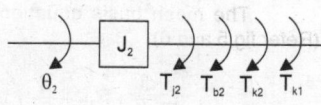

Fig 2.

The free body diagram of J_2 is shown in fig 3. The opposing torques are marked as T_{j2}, T_{b2}, T_{k2} and T_{k1}.

$$T_{j2} = J_2 \frac{d^2\theta_2}{dt^2} \quad ; \quad T_{b2} = B_2 \frac{d\theta_2}{dt}$$

$$T_{k2} = K_2\theta_2 \quad ; \quad T_{k1} = K_1(\theta_2 - \theta_1)$$

By Newton's second law, $\quad T_{j2} + T_{b2} + T_{k2} + T_{k1} = 0$

$$J_2 \frac{d^2\theta_2}{dt^2} + B_2 \frac{d\theta_2}{dt} + K_2\theta_2 + K_1(\theta_2 - \theta_1) \qquad \qquad .....(2)$$

Fig 3.

On replacing the angular displacements by angular velocity in the differential equations (1) and (2) governing the mechanical rotational system we get,

$$\left(\text{i.e., } \frac{d^2\theta}{dt^2} = \frac{d\omega}{dt} \quad ; \quad \frac{d\theta}{dt} = \omega \text{ and } \theta = \int \omega \, dt \right)$$

$$J_1 \frac{d\omega_1}{dt} + B_1\omega_1 + K_1 \int (\omega_1 - \omega_2) \, dt = T \qquad \qquad .....(3)$$

$$J_2 \frac{d\omega_2}{dt} + B_2\omega_2 + K_2 \int \omega_2 dt + K_1 \int (\omega_2 - \omega_1) \, dt = 0 \qquad \qquad .....(4)$$

TORQUE-VOLTAGE ANALOGOUS CIRCUIT

The given mechanical system has two nodes (J_1 and J_2). Hence the torque-voltage analogous electrical circuit will have two meshes. The torque applied to J_1 is represented by a voltage source in first mesh. The elements J_1, B_1 and K_1 are connected to first node. Hence they are represented by analogous element in mesh-1 forming a closed path. The elements J_2, B_2, K_2 and K_1 are connected to second node. Hence they are represented by analogous elements in mesh-2 forming a closed path.

The element K_1 is common between node-1 and 2 and so it is represented by analogous element as common element between two meshes. The torque-voltage electrical analogous circuit is shown in fig 4.

The electrical analogous elements for the elements of mechanical rotational system are given below.

$$
\begin{array}{llll}
T \rightarrow e(t) & J_1 \rightarrow L_1 & B_1 \rightarrow R_1 & K_1 \rightarrow 1/C_1 \\
\omega_1 \rightarrow i_1 & J_2 \rightarrow L_2 & B_2 \rightarrow R_2 & K_2 \rightarrow 1/C_2 \\
\omega_2 \rightarrow i_2 & & &
\end{array}
$$

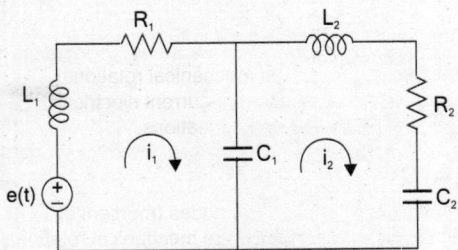

Fig 4 : *Torque-voltage electrical analogous circuit.*

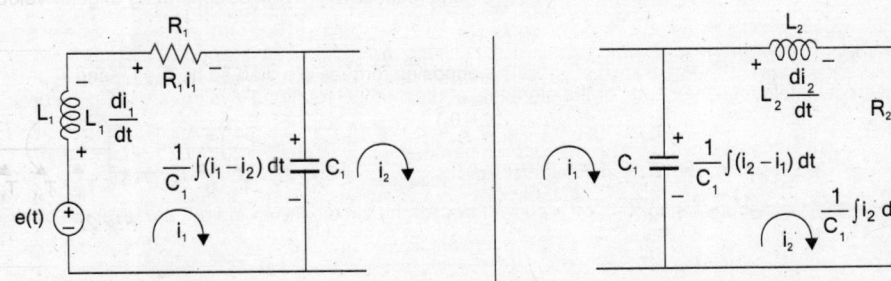

Fig 5 : *Mesh-1 of analogous circuit.*   **Fig 6 :** *Mesh-2 of analogous circuit..*

The mesh basis equations using Kirchoff 's voltage law for the circuit shown in fig 4 are given below (Refer fig 5 and 6).

$$L_1 \frac{di_1}{dt} + R_1 i_1 + \frac{1}{C_1} \int (i_1 - i_2) = e(t) \quad \text{.....(5)}$$

$$L_2 \frac{di_2}{dt} + R_2 i_2 + \frac{1}{C_2} \int i_2 dt + \frac{1}{C_2} \int (i_2 - i_1) dt = 0 \quad \text{.....(6)}$$

It is observed that the mesh basis equations (5) and (6) are similar to the differential equations (3) and (4) governing the mechanical system.

TORQUE-CURRENT ANALOGOUS CIRCUIT

The given mechanical system has two nodes (J_1 and J_2). Hence the torque-current analogous electrical circuit will have two nodes. The torque applied to J_1 is represented as a current source connected to node-1 in analogous electrical circuit.

The elements J_1, B_1 and K_1 are connected to first node. Hence they are represented by analogous elements as elements connected to node-1 in analogous electrical circuit. The elements J_2, B_2, K_2 and K_1 are connected to second node. Hence they are represented by analogous elements as elements connected to node-2 in analogous electrical circuit.

The element K_1 is common between node-1 and 2. So it is represented by analogous element as common element between node-1 and 2. The torque-current electrical analogous circuit is shown in fig 7.

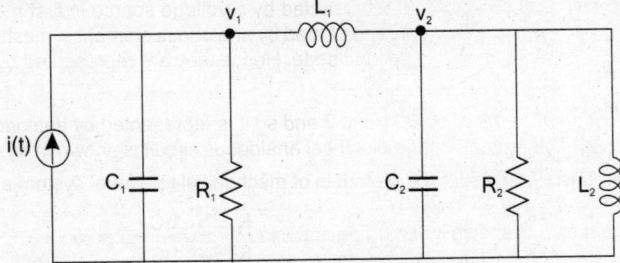

Fig 7 : *Torque-current electrical analogous circuit.*

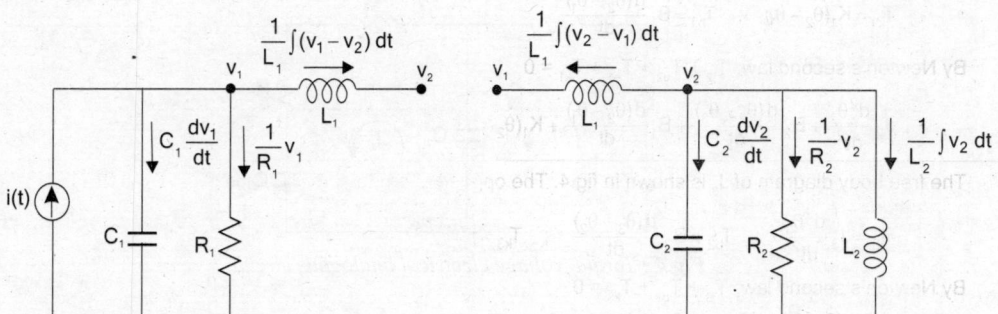

Fig 8 : *Node-1 of analogous circuit.*   **Fig 9** : *Node-2 of analogous circuit.*

The electrical analogous elements for the elements of mechanical rotational system are given below.

$$T \to i(t) \qquad B_1 \to 1/R_1 \qquad \omega_1 \to v_1 \qquad J_1 \to C_1 \qquad K_1 \to 1/L_1$$
$$B_2 \to 1/R_2 \qquad \omega_2 \to v_2 \qquad J_2 \to C_2 \qquad K_2 \to 1/L_2$$

The node basis equations using Kirchoff's current law for the circuit shown in fig 7 are given below (Refer fig 8 and 9).

$$C_1 \frac{dv_1}{dt} + \frac{1}{R_1} v_1 + \frac{1}{L_1} \int (v_1 - v_2) \, dt = i(t) \qquad \qquad .....(7)$$

$$C_1 \frac{dv_2}{dt} + \frac{1}{R_2} v_2 + \frac{1}{L_2} \int v_2 \, dt + \frac{1}{L_1} \int (v_2 - v_1) \, dt = i(t) \qquad .....(8)$$

It is observed that the mesh basis equations (5) and (6) are similar to the differential equations (3) and (4) governing the mechanical system.

EXAMPLE 1.13

Write the differential equations governing the mechanical rotational system shown in fig 1. Draw the torque-voltage and torque-current electrical analogous circuits and verify by writing mesh and node equations.

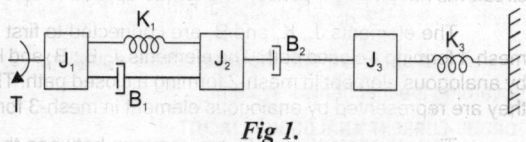

Fig 1.

SOLUTION

The given mechanical rotational system has three nodes (moment of inertia of masses). The differential equations governing the mechanical rotational system are given by torque balance equations at these nodes.

Let the angular displacements of J_1, J_2 and J_3 be θ_1, θ_2 and θ_3 respectively. The corresponding angular velocities be ω_1, ω_2 and ω_3.

The free body diagram of J_1 is shown in fig 2. The opposing torques are marked as T_{j1}, T_{b1} and T_{k1}.

$$T_{j1} = J_1 \frac{d^2\theta_1}{dt^2} \quad ; \quad T_{b1} = B_1 \frac{d(\theta_1 - \theta_2)}{dt}$$

$$T_{k1} = K_1(\theta_1 - \theta_2)$$

By Newton's second law, $T_{j1} + T_{b1} + T_{k1} = T$

$$J_1 \frac{d^2\theta_1}{dt^2} + B_1 \frac{d(\theta_1 - \theta_2)}{dt} + K_1(\theta_1 - \theta_2) = T$$

Fig 2.

$$.....(1)$$

The free body diagram of J_2 is shown in fig 3. The opposing torques are marked as T_{j2}, T_{b2}, T_{b1} and T_{k1}.

$$T_{j2} = J_2 \frac{d^2\theta_2}{dt^2} \quad ; \quad T_{b2} = B_2 \frac{d(\theta_2 - \theta_3)}{dt}$$

$$T_{k1} = K_1(\theta_2 - \theta_1) \quad ; \quad T_{b1} = B_1 \frac{d(\theta_2 - \theta_1)}{dt}$$

By Newton's second law, $T_{j2} + T_{b2} + T_{b1} + T_{k1} = 0$

Fig 3.

$$J_2 \frac{d^2\theta_2}{dt^2} + B_2 \frac{d(\theta_2 - \theta_3)}{dt} + B_1 \frac{d(\theta_2 - \theta_1)}{dt} + K_1(\theta_2 - \theta_1) = 0 \qquad \text{.....(2)}$$

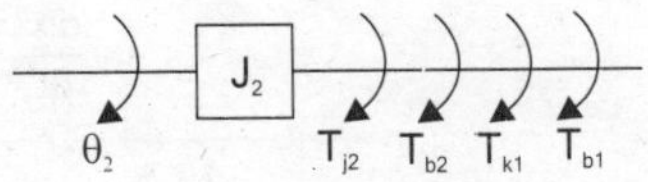

The free body diagram of J_3 s shown in fig 4. The opposing torques are marked as T_{j3}, T_{b2}, and T_{k3}.

$$T_{j3} = J_3 \frac{d^2\theta_3}{dt^2} \quad ; \quad T_{b2} = B_2 \frac{d(\theta_3 - \theta_2)}{dt} \quad ; \quad T_{k3} = K_3\theta_3$$

By Newton's second law, $T_{j3} + T_{b2} + T_{k3} = 0$

$$\therefore \ J_3 \frac{d^2\theta_3}{dt^2} + B_2 \frac{d(\theta_3 - \theta_2)}{dt} + K_3\theta_3 = 0 \qquad \qquad \textit{Fig 4.} \qquad \text{.....(3)}$$

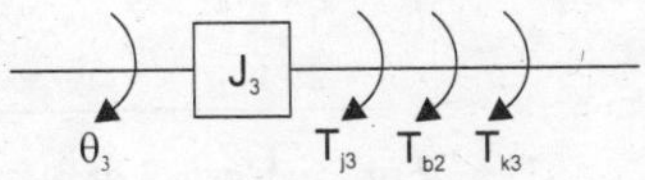

On replacing the angular displacements by angular velocity in the differential equations (1) and (2) governing the mechanical rotational system we get,

$$\left(\text{i.e.,} \ \frac{d^2\theta}{dt^2} = \frac{d\omega}{dt} \quad ; \quad \frac{d\theta}{dt} = \omega \ \text{ and } \ \theta = \int \omega \, dt \right)$$

$$J_1 \frac{d\omega_1}{dt} + B_1(\omega_1 - \omega_2) + K_1 \int (\omega_1 - \omega_2) \, dt = T \qquad \text{.....(4)}$$

$$J_2 \frac{d\omega_2}{dt} + B_1(\omega_2 - \omega_1) + B_2(\omega_2 - \omega_3) + K_1 \int (\omega_2 - \omega_1) \, dt = 0 \qquad \text{.....(5)}$$

$$J_3 \frac{d\omega_3}{dt} + B_2(\omega_3 - \omega_2) + K_3 \int \omega_3 \, dt = 0 \qquad \text{.....(6)}$$

TORQUE-VOLTAGE ANALOGOUS CIRCUIT

The given mechanical system has three nodes (J_1, J_2 and J_3). Hence the torque-voltage analogous electrical circuit will have three meshes. The torque applied to J_1 is represented by a voltage source in first mesh.

The elements J_1, K_1 and B_1 are connected to first node. Hence they are represented by analogous element in mesh-1 forming a closed path. The elements J_2, B_2, B_1 and K_1 are connected to second node. Hence they are represented by analogous element in mesh-2 forming a closed path. The element J_3, B_2 and K_3 are connected to third node. Hence they are represented by analogous element in mesh-3 forming a closed path.

The elements K_1 and B_1 are common between the nodes-1 and 2 and so they are represented by analogous element as common between mesh-1 and 2. The element B_2 is common between the nodes-2 and 3 and so it is represented by analogous element as common element between the mesh-2 and 3. The torque-voltage electrical analogous circuit is shown in fig 5.

The electrical analogous elements for the elements of mechanical rotational system are given below.

$$T \to e(t) \qquad \omega_1 \to i_1 \qquad J_1 \to L_1 \qquad B_1 \to R_1 \qquad K_1 \to 1/C_1$$
$$\omega_2 \to i_2 \qquad J_2 \to L_2 \qquad B_2 \to R_2 \qquad K_3 \to 1/C_3$$
$$\omega_3 \to i_3 \qquad J_3 \to L_3 \qquad \qquad L_1 \frac{di_1}{dt}$$

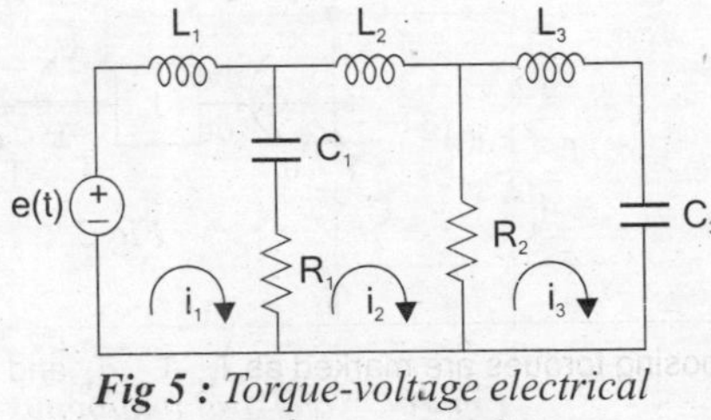

Fig 5 : Torque-voltage electrical analogous circuit.

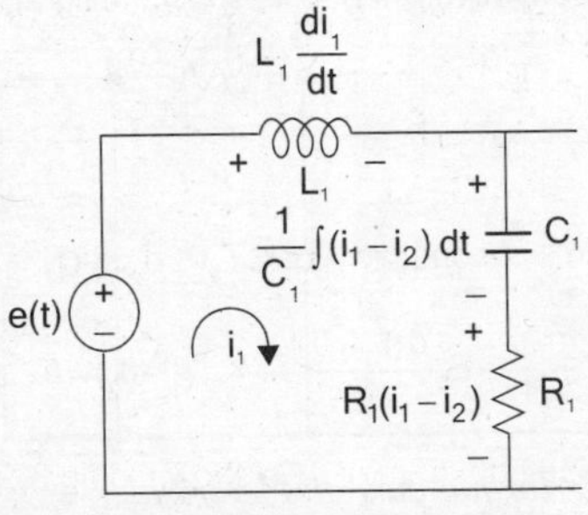

Fig 6 : Mesh-1 of analogous circuit.

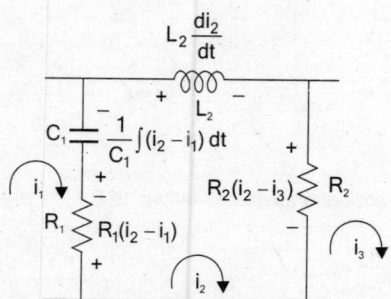

Fig 7 : *Mesh-2 of analogous circuit.*

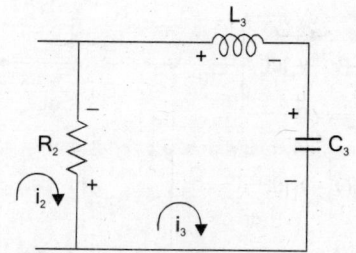

Fig 8 : *Mesh-3 of analogous circuit.*

The mesh basis equations using Kirchoff's voltage law for the circuit shown in fig 5 are given below (Refer fig 6, 7 and 8).

$$L_1 \frac{di_1}{dt} + R_1(i_1 - i_2) + \frac{1}{C_1} \int (i_1 - i_2) dt = e(t) \qquad .....(7)$$

$$L_2 \frac{di_2}{dt} + R_1(i_2 - i_1) + R_2(i_2 - i_3) + \frac{1}{C_1} \int (i_2 - i_1) dt = 0 \qquad .....(8)$$

$$L_3 \frac{di_3}{dt} + R_2(i_3 - i_2) + \frac{1}{C_3} \int i_3 dt = 0 \qquad .....(9)$$

It is observed that the mesh basis equations (7), (8) and (9) are similar to the differential equations (4), (5) and (6) governing the mechanical system.

TORQUE-CURRENT ANALOGOUS CIRCUIT

The given mechanical system has three nodes (J_1, J_2 and J_3). Hence the torque-current analogous electrical circuit will have three nodes. The torque applied to J_1 is represented as a current source connected to node-1 in analogous electrical circuit.

The elements K_1, J_1 and B_1 are connected to first node. Hence they are represented by analogous elements as elements connected to node-1 in analogous electrical circuit. The elements J_2, B_2, B_1 and K_1 are connected to second node. Hence they are represented by analogous elements as elements connected to node-2 in analogous electrical circuit. The elements J_3, B_2, and K_3 are connected to third node. Hence they are represented by analogous elements as elements connected to node-3 in analogous electrical circuit.

The elements K_1 and B_1 are common between node-1 and 2 and so they are represented by analogous element as common elements between node-1 and 2. The element B_2 is common between node-2 and 3 and so it is represented as common element between node-2 and 3 in analogous circuit. The torque-current electrical analogous circuit is shown in fig 9.

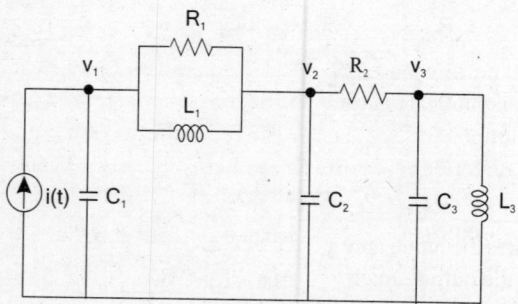

Fig 9 : *Torque-current electrical analogous circuit.*

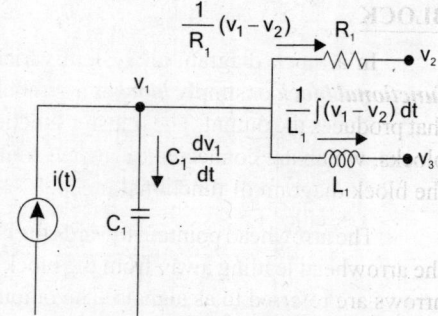

Fig 10 : *Node-1 of analogous circuit.*

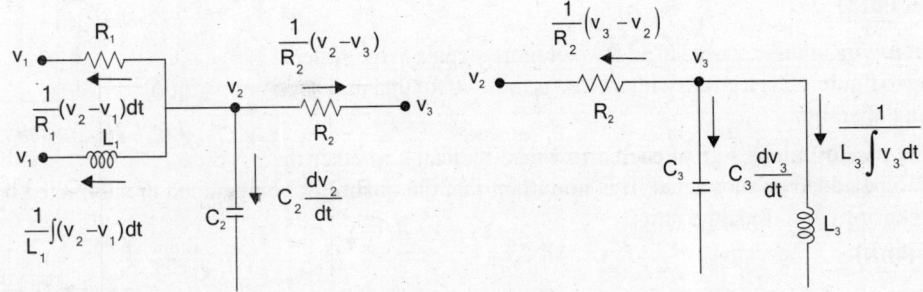

Fig 11 : *Node-2 of analogous circuit*   **Fig 12 :** *Node-3 of analogous circuit*

The electrical analogous elements for the elements of mechanical rotational system are given below.

$$T \rightarrow i(t) \quad \omega_1 \rightarrow v_1 \quad J_1 \rightarrow C_1 \quad B_1 \rightarrow 1/R_1 \quad K_1 \rightarrow 1/L_1$$
$$\omega_2 \rightarrow v_2 \quad J_2 \rightarrow C_2 \quad B_2 \rightarrow 1/R_2 \quad K_3 \rightarrow 1/L_3$$
$$\omega_3 \rightarrow v_3 \quad J_3 \rightarrow C_3$$

The node basis equations using Kirchoff's current law for the circuit shown in fig 9 are given below (Refer fig 10, 11 and 12).

$$C_1 \frac{dv_1}{dt} + \frac{1}{R_1}(v_1 - v_2) + \frac{1}{L_1}\int (v_1 - v_2)\, dt = i(t) \qquad \text{.....(10)}$$

$$C_2 \frac{dv_2}{dt} + \frac{1}{R_1}(v_2 - v_1) + \frac{1}{R_2}(v_2 - v_3) + \frac{1}{L_1}\int (v_2 - v_1)\, dt = 0 \qquad \text{.....(11)}$$

$$C_3 \frac{dv_3}{dt} + \frac{1}{R_2}(v_3 - v_2) + \frac{1}{L_3}\int v_3\, dt = 0 \qquad \text{.....(12)}$$

It is observed that the node basis equations (10), (11) and (12) are similar to the differential equations (4), (5) and (6) governing the mechanical system.

1.11 BLOCK DIAGRAMS

A control system may consist of a number of components. In control engineering to show the functions performed by each component, we commonly use a diagram called the block diagram. A *block diagram* of a system is a pictorial representation of the functions performed by each component and of the flow of signals. Such a diagram depicts the interrelationships that exist among the various components. The elements of a block diagram are *block*, *branch point* and *summing point.*

BLOCK

In a block diagram all system variables are linked to each other through functional blocks. The *functional block* or simply *block* is a symbol for the mathematical operation on the input signal to the block that produces the output. The transfer functions of the components are usually entered in the corresponding blocks, which are connected by arrows to indicate the direction of the flow of signals. Figure 1.25 shows the block diagram of functional block.

The arrowhead pointing towards the block indicates the input, and the arrowhead leading away from the block represents the output. Such arrows are referred to as signals. The output signal from the block is given by the product of input signal and transfer function in the block.

Input, A →	Transfer function G(s)	Output, B → B = A G(s)

Fig 1.25 : *Functional block.*

SUMMING POINT

Summing points are used to add two or more signals in the system. Referring to figure 1.26, a circle with a cross is the symbol that indicates a summing operation.

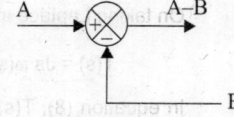

Fig 1.26 : Summing point.

The plus or minus sign at each arrowhead indicates whether the signal is to be added or subtracted. It is important that the quantities being added or subtracted have the same dimensions and the same units.

BRANCH POINT

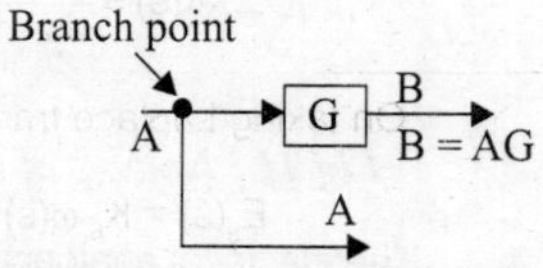

A **branch point** is a point from which the signal from a block goes concurrently to other blocks or summing points.

Fig 1.27 : Branch point.

CONSTRUCTING BLOCK DIAGRAM FOR CONTROL SYSTEMS

A control system can be represented diagramatically by block diagram. The differential equations governing the system are used to construct the block diagram. By taking Laplace transform the differential equations are converted to algebraic equations. The equations will have variables and constants. From the working knowledge of the system the input and output variables are identified and the block diagram for each equation can be drawn. Each equation gives one section of block diagram. The output of one section will be input for another section. The various sections are interconnected to obtain the overall block diagram of the system.

EXAMPLE 1.14

Construct the block diagram of armature controlled dc motor.

SOLUTION

The differential equations governing the armature controlled dc motor are (refer section 1.7),

$$V_a = i_a R_a + L_a \frac{di_a}{dt} + e_b \qquad .....(1)$$

$$T = K_t i_a \qquad .....(2)$$

$$T = J \frac{d\omega}{dt} + B\omega \qquad .....(3)$$

$$e_b = K_b \omega \qquad .....(4)$$

$$\omega = \frac{d\theta}{dt} \qquad .....(5)$$

On taking Laplace transform of equation (1) we get,

$$V_a(s) = I_a(s)R_a + L_a s\, I_a(s) + E_b(s) \qquad .....(6)$$

In equation (6), $V_a(s)$ and $E_b(s)$ are inputs and $I_a(s)$ is the output. Hence the equation (6) is rearranged and the block diagram for this equation is shown in fig 1.

$$V_a(s) - E_b(s) = I_a(s)\,[R_a + s\, L_a]$$

$$\therefore\; I_a(s) = \frac{1}{R_a + s\, L_a}\,[V_a(s) - E_b(s)]$$

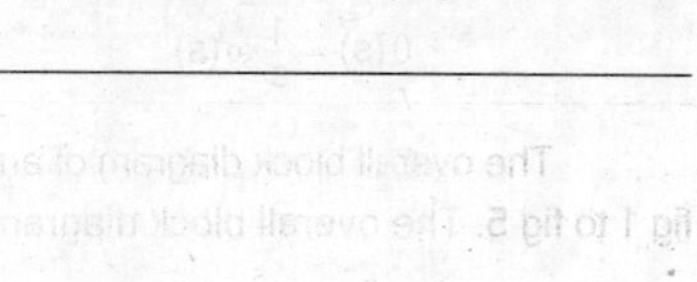

Fig 1.

On taking Laplace transform of equation (2) we get,

$$T(s) = K_t\, I_a(s) \qquad .....(7)$$

Fig 2.

In equation (7), $I_a(s)$ is the input and $T(s)$ is the output. The block diagram for this equation is shown in fig 2.

On taking Laplace transform of equation (3) we get,

$$T(s) = Js\,\omega(s) + B\,\omega(s) \qquad .....(8)$$

In equation (8), $T(s)$ is the input and $\omega(s)$ is the output. Hence the equation (8) is rearranged and the block diagram for this equation is shown in fig (3).

$$T(s) = (Js + B)\,\omega(s)$$

$$\therefore\ \omega(s) = \frac{1}{Js + B}\,T(s)$$

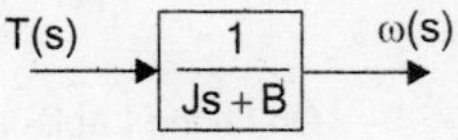

Fig 3.

On taking Laplace transform of equation (4) we get,

$$E_b(s) = K_b\,\omega(s) \qquad .....(9)$$

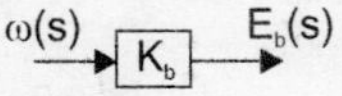

In equation (9), $\omega(s)$ is the input and $E_b(s)$ is the output. The block diagram for this equation is shown in fig 4.

Fig 4.

On taking Laplace transform of equation (5) we get,

$$\omega(s) = s\,\theta(s) \qquad .....(10)$$

In equation (10), $\omega(s)$ is the input and $\theta(s)$ is the output. Hence equation (10) is rearranged and the block diagram for this equation is shown in fig 5.

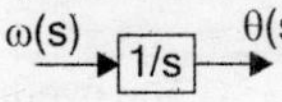

$$\theta(s) = \frac{1}{s}\,\omega(s)$$

Fig 5.

The overall block diagram of armature controlled dc motor is obtained by connecting the various sections shown in fig 1 to fig 5. The overall block diagram is shown in fig 6.

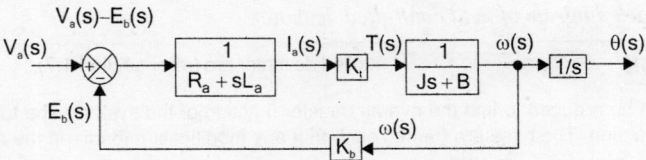

Fig 6 : Block diagram of armature controlled dc motor.

EXAMPLE 1.15

Construct the block diagram of field controlled dc motor.

SOLUTION

The differential equations governing the field controlled dc motor are (refer section 1.8),

$$v_t = R_f i_f + L_f \frac{di_f}{dt} \qquad .....(1)$$

$$T = K_{tf} i_f \qquad .....(2)$$

$$T = J\frac{d^2\theta}{dt^2} + B\frac{d\theta}{dt} \qquad .....(3)$$

On taking Laplace transform of equation (1) we get,

$$V_f(s) = R_f I_f(s) + L_f s\, I_f(s) \qquad .....(4)$$

In equation (4), $V_f(s)$ is the input and $I_f(s)$ is the output. Hence the equation (4) is rearranged and the block diagram for this equation is shown in fig 1.

$$V_f(s) = I_f(s)\left[R_f + sL_f\right]$$

$$\therefore\ I_f(s) = \frac{1}{R_f + sL_f}\ V_f(s)$$

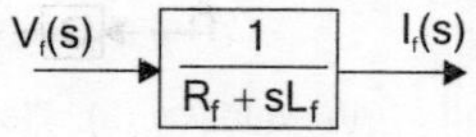

Fig 1.

On taking Laplace transform of equation (2) we get,

$$T(s) = K_{tf} I_f(s) \hspace{3cm} .....(5)$$

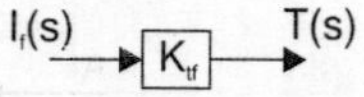

In equation (5), $I_f(s)$ is the input and $T(s)$ is the output. The block diagram for this equation is shown in fig 2.

Fig 2.

On taking Laplace transform of equation (3) we get,

$$T(s) = J s^2\,\theta(s) + B s\,\theta(s) \hspace{3cm} .....(6)$$

In equation (6), $T(s)$ is input and $\theta(s)$ is the output. Hence equation (6) is rearranged and the block diagram for this equation is shown in fig 3.

$$T(s) = (Js^2 + Bs)\,\theta(s)$$

$$\therefore\ \theta(s) = \frac{1}{Js^2 + Bs}\ T(s)$$

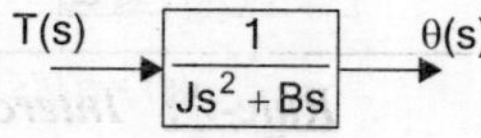

Fig 3.

The overall block diagram of field controlled dc motor is obtained by connecting the various section shown in fig 1 to fig 3. The overall block diagram is shown in fig 4.

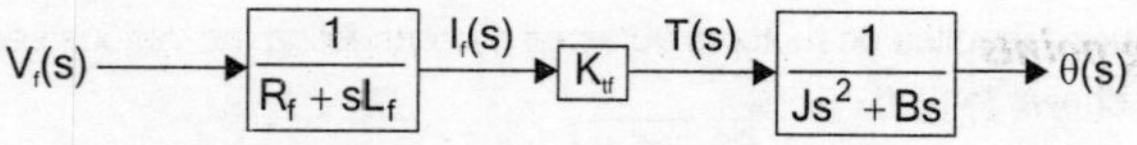

Fig 4 : Block diagram of field controlled dc motor.

BLOCK DIAGRAM REDUCTION

The block diagram can be reduced to find the overall transfer function of the system. The following rules can be used for block diagram reduction. The rules are framed such that any modification made on the diagram does not alter the input-output relation.

RULES OF BLOCK DIAGRAM ALGEBRA
Rule-1 : **Combining the blocks in cascade**
Rule-2 : **Combining Parallel blocks (or combining feed forward paths)**
Rule-3 : **Moving the branch point ahead of the block**

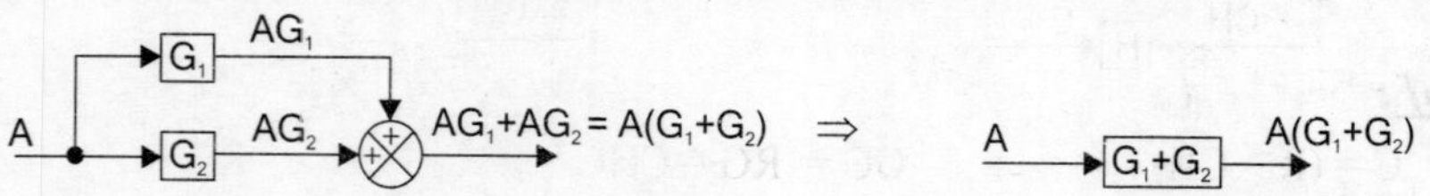

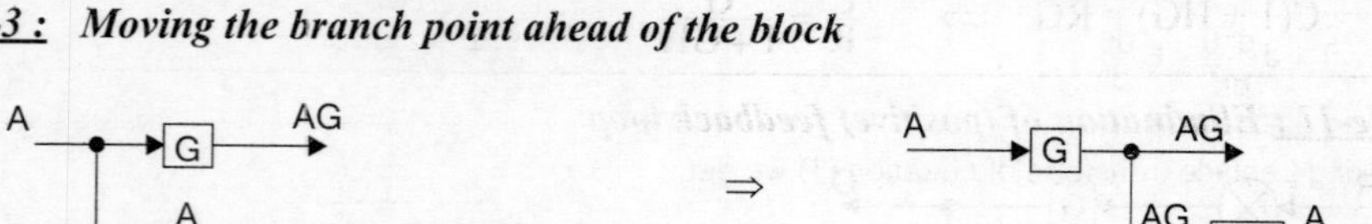

Rule-4 : *Moving the branch point before the block*

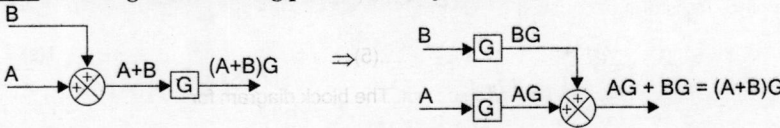

Rule-5 : *Moving the summing point ahead of the block*

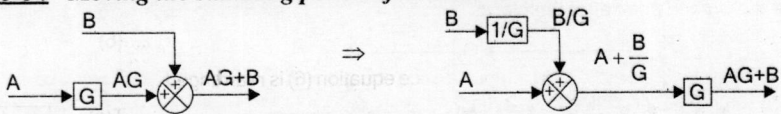

Rule-6 : *Moving the summing point before the block*

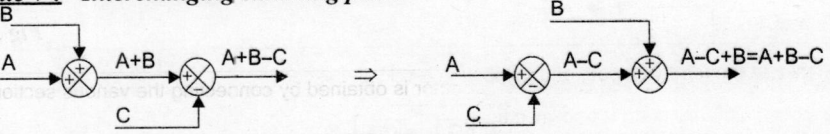

Rule-7 : *Interchanging summing point*

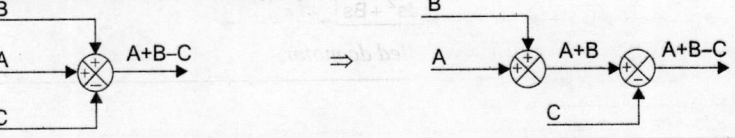

Rule-8 : *Splitting summing points*

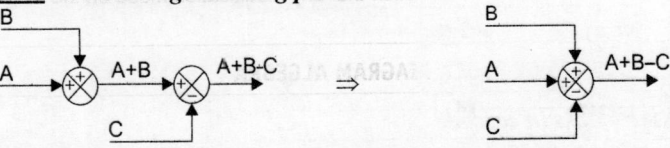

Rule-9 : *Combining summing points*

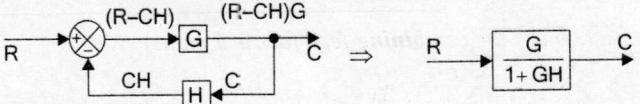

Rule-10 : *Elimination of (negative) feedback loop*

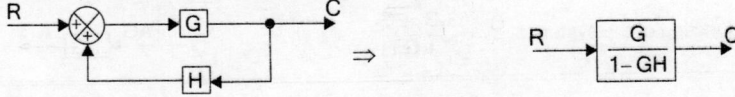

Proof :

$$C = (R - CH) \implies GC = RG - CHG \implies C + CHG = RG$$

$$C(1 + HG) = RG \implies \frac{C}{R} = \frac{G}{1 + GH}$$

Rule-11 : *Elimination of (positive) feedback loop*

EXAMPLE 1.16

Reduce the block diagram shown in fig 1 and find C/R.

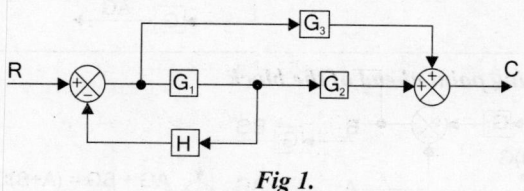

Fig 1.

SOLUTION

Step 1 : Move the branch point after the block.

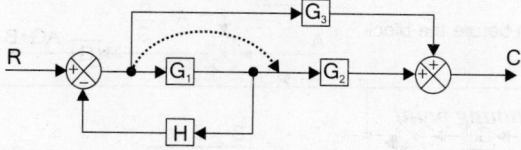

Step 2 : Eliminate the feedback path and combining blocks in cascade.

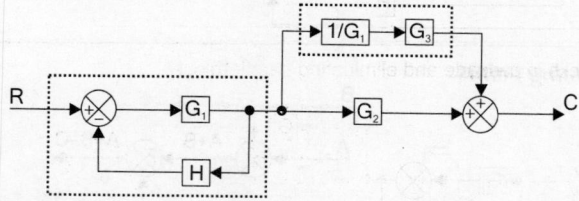

Step 3 : Combining parallel blocks

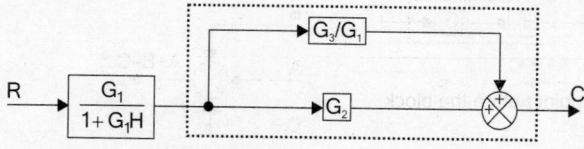

Step 4 : Combining blocks in cascade

$$\frac{C}{R} = \left(\frac{G_1}{1+G_1H}\right)\left(G_2+\frac{G_3}{G_1}\right) = \left(\frac{G_1}{1+G_1H}\right)\left(\frac{G_1G_2+G_3}{G_1}\right) = \frac{G_1G_2+G_3}{1+G_1H}$$

RESULT

The overall transfer function of the system, $\dfrac{C}{R} = \dfrac{G_1G_2+G_3}{1+G_1H}$

EXAMPLE 1.17

Using block diagram reduction technique find closed loop transfer function of the system whose block diagram is shown in fig 1.

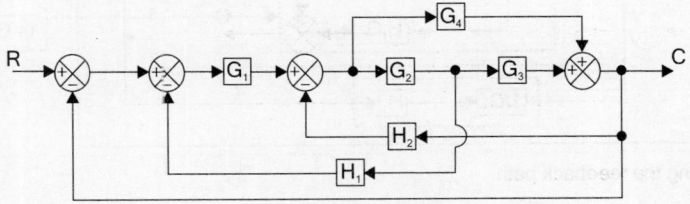

SOLUTION   *Fig 1.*

Step 1 : Moving the branch point before the block

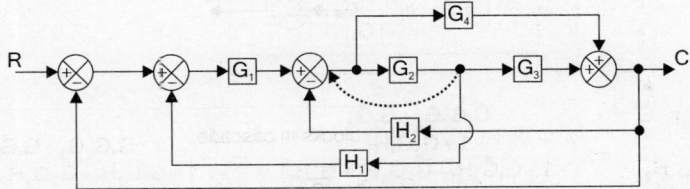

Step 2 : Combining the blocks in cascade and eliminating parallel blocks

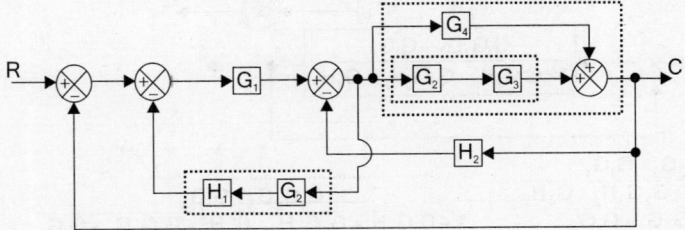

Step 3 : Moving summing point before the block.

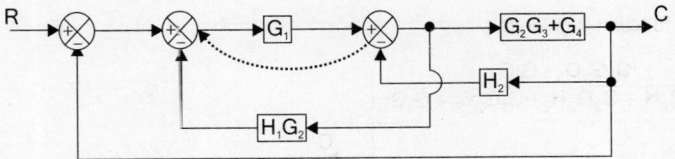

Step 4 : Interchanging summing points and modifying branch points.

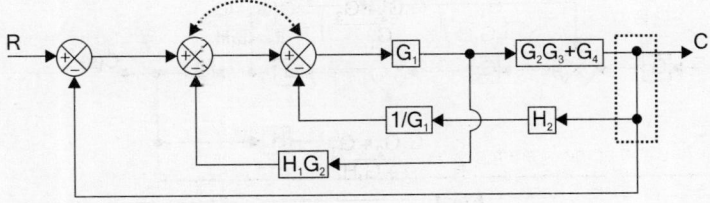

Step 5 : Eliminating the feedback path and combining blocks in cascade

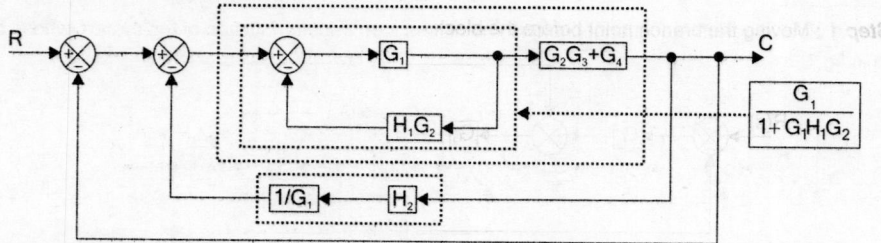

Step 6 : Eliminating the feedback path

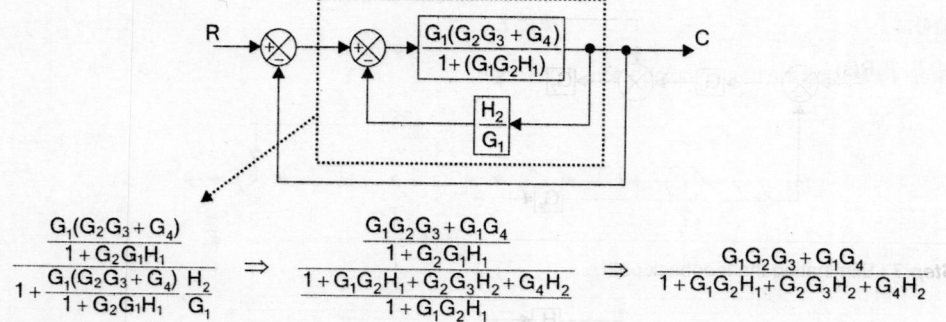

$$\frac{\dfrac{G_1(G_2G_3 + G_4)}{1 + G_2G_1H_1}}{1 + \dfrac{G_1(G_2G_3 + G_4)}{1 + G_2G_1H_1}\dfrac{H_2}{G_1}} \quad \Rightarrow \quad \frac{\dfrac{G_1G_2G_3 + G_1G_4}{1 + G_2G_1H_1}}{\dfrac{1 + G_1G_2H_1 + G_2G_3H_2 + G_4H_2}{1 + G_1G_2H_1}} \quad \Rightarrow \quad \frac{G_1G_2G_3 + G_1G_4}{1 + G_1G_2H_1 + G_2G_3H_2 + G_4H_2}$$

Step 7 : Eliminating the feedback path

$$R \longrightarrow \bigotimes \longrightarrow \boxed{\frac{G_1G_2G_3 + G_1G_4}{1 + G_1G_2H_1 + G_2G_3H_2 + G_4H_2}} \longrightarrow C$$

$$\frac{C}{R} = \frac{\dfrac{G_1G_2G_3 + G_1G_4}{1 + G_1G_2H_1 + G_2G_3H_2 + G_4H_2}}{1 + \dfrac{G_1G_2G_3 + G_1G_4}{1 + G_1G_2H_1 + G_2G_3H_2 + G_4H_2}} = \frac{G_1G_2G_3 + G_1G_4}{1 + G_1G_2H_1 + G_2G_3H_2 + G_4H_2 + G_1G_2G_3 + G_1G_4}$$

RESULT

The overall transfer function is given by,

$$\frac{C}{R} = \frac{G_1G_2G_3 + G_1G_4}{1 + G_1G_2H_1 + G_2G_3H_2 + G_1G_2G_3 + G_1G_4}$$

EXAMPLE 1.18

Determine the overall transfer function $\dfrac{C(s)}{R(s)}$ for the system shown in fig 1.

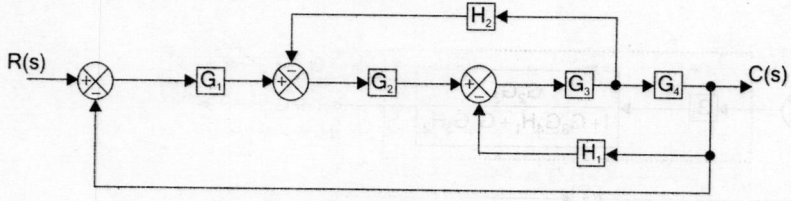

Fig 1.

SOLUTION

Step 1 : Moving the branch point before the block

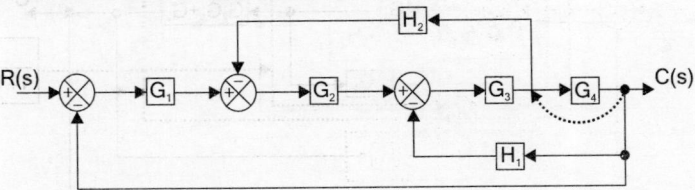

Step 2 : Combining the blocks in cascade and rearranging the branch points

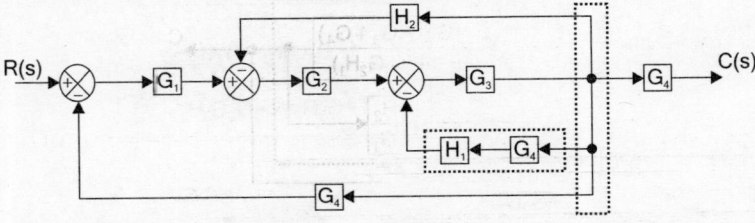

Step 3 : Eliminating the feedback path

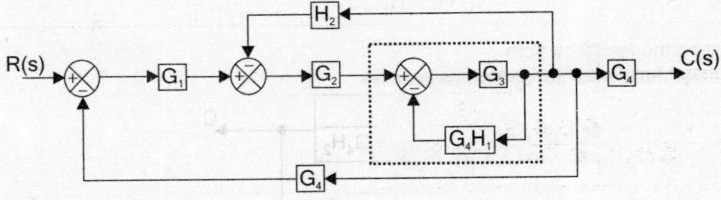

Step 4 : Combining the blocks in cascade and eliminating feedback path

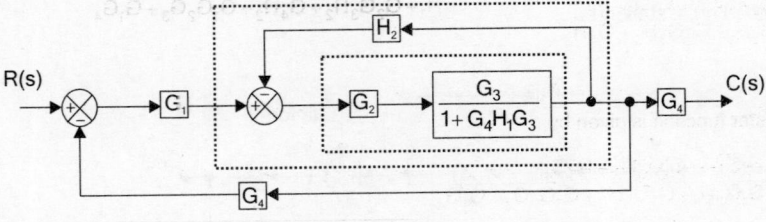

Step 5 : Combining the blocks in cascade

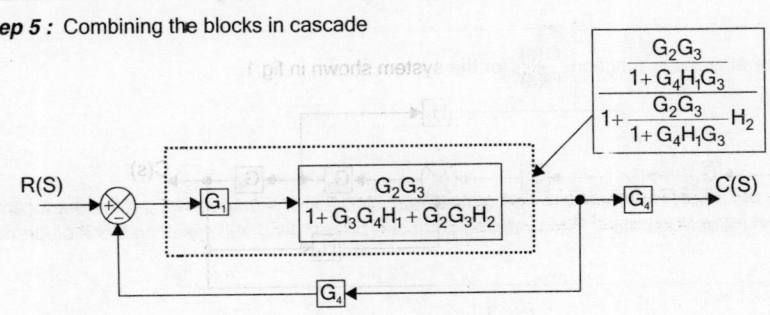

Step 6 : Eliminating the feedback path

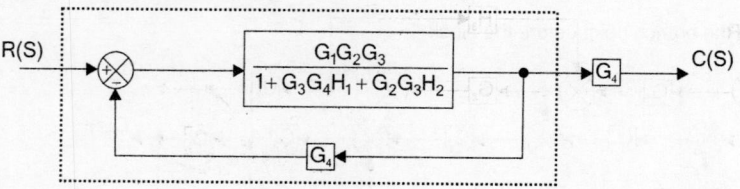

Step 7 : Combining the blocks in cascade

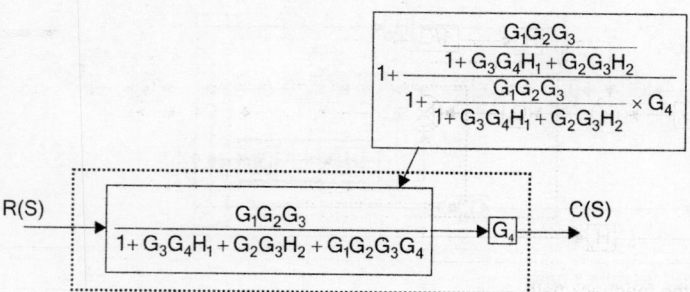

RESULT

The overall transfer function of the system is given by,

$$\frac{C(s)}{R(s)} = \frac{G_1G_2G_3G_4}{1 + G_3G_4H_1 + G_2G_3H_2 + G_1G_2G_3G_4}$$

EXAMPLE 1.19

For the system represented by the block diagram shown in fig 1. Evaluate the closed loop transfer function when the input R is (i) at station-I (ii) at station-II.

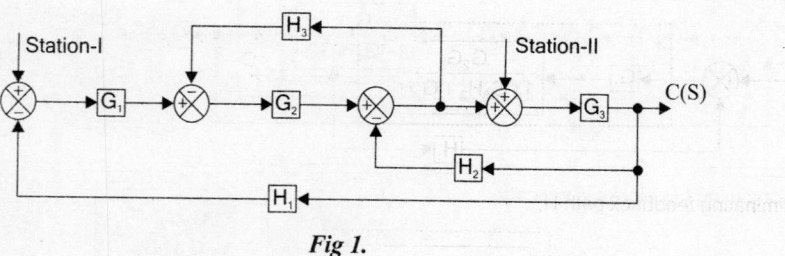

Fig 1.

SOLUTION

(i)   Consider the input R is at station-I and so the input at station-II is made zero. Let the output be C^1. Since there is no input at station-II that summing point can be removed and resulting block diagram is shown in fig 2.

Step 1 : Shift the take off point of feedback H_3 beyond G_3 and rearrange the branch points

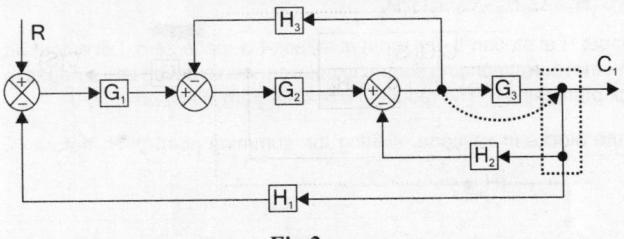

Fig 2.

Step 2 : Eliminating the feedback H_2 and combining blocks in cascade

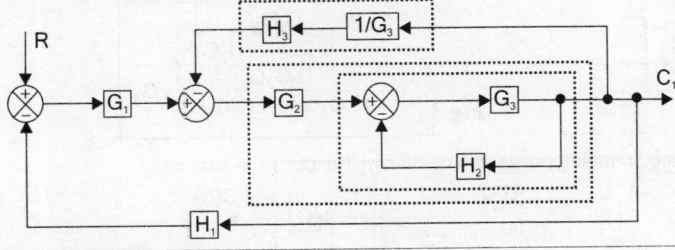

Step 3 : Eliminating the feecback path

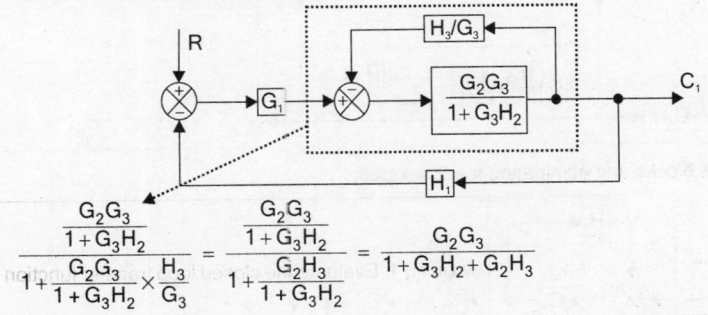

$$\frac{\dfrac{G_2G_3}{1+G_3H_2}}{1+\dfrac{G_2G_3}{1+G_3H_2}\times\dfrac{H_3}{G_3}}=\frac{\dfrac{G_2G_3}{1+G_3H_2}}{1+\dfrac{G_2H_3}{1+G_3H_2}}=\frac{G_2G_3}{1+G_3H_2+G_2H_3}$$

Step 4 : Combining the blocks in cascade

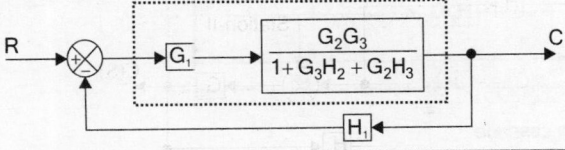

Step 5 : Eliminating feedback path H_1

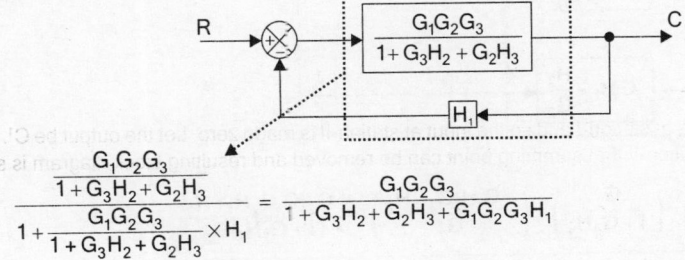

$$\frac{\dfrac{G_1G_2G_3}{1+G_3H_2+G_2H_3}}{1+\dfrac{G_1G_2G_3}{1+G_3H_2+G_2H_3}\times H_1}=\frac{G_1G_2G_3}{1+G_3H_2+G_2H_3+G_1G_2G_3H_1}$$

$$\therefore \frac{C_1(s)}{R(s)} = \frac{G_1G_2G_3}{1 + G_3H_2 + G_2H_3 + G_1G_2G_3H_1}$$

(ii) Consider the input R at station-II, the input at station-I is made zero. Let output be C_2. Since there is no input in station-I that corresponding summing point can be removed and a negative sign can be attached to the feedback path gain H_1. The resulting block diagram is shown in fig 3.

Step 1 : Combining the blocks in cascade, shifting the summing point of H_2 before G_2 and rearranging the branch points.

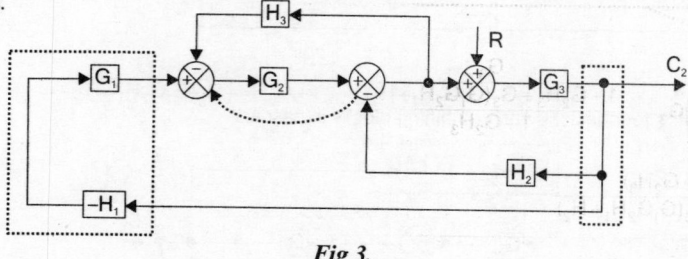

Fig 3.

Step 2 : Interchanging summing points and combining the blocks in cascade.

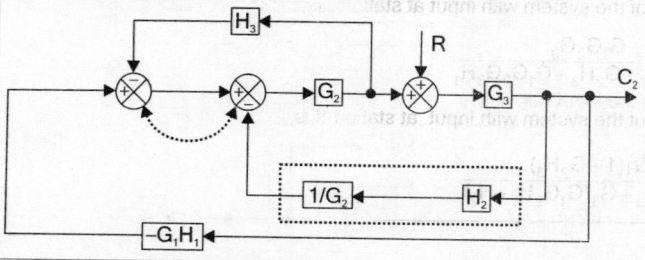

Step 3 : Combining parallel blocks and eliminating feedback path

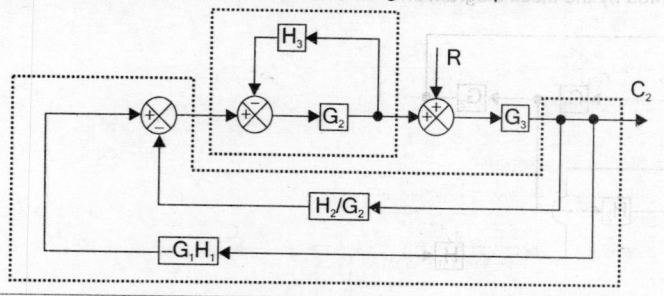

Step 4 : Combining the blocks in cascade

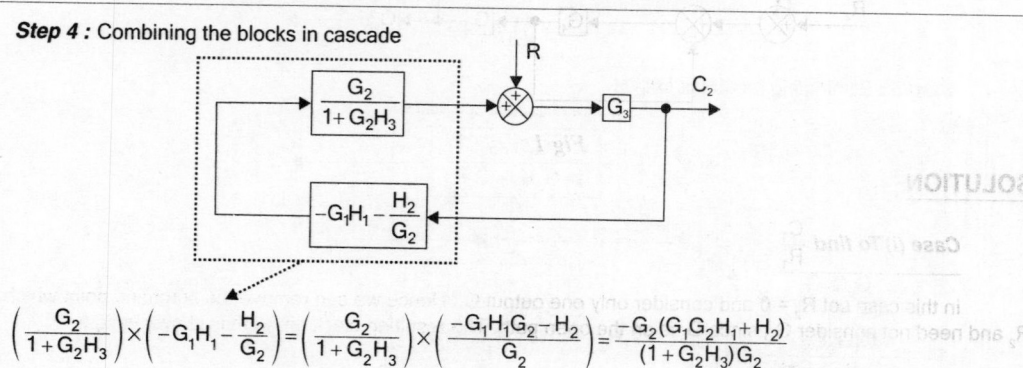

$$\left(\frac{G_2}{1+G_2H_3}\right) \times \left(-G_1H_1 - \frac{H_2}{G_2}\right) = \left(\frac{G_2}{1+G_2H_3}\right) \times \left(\frac{-G_1H_1G_2 - H_2}{G_2}\right) = \frac{-G_2(G_1G_2H_1 + H_2)}{(1+G_2H_3)G_2}$$

Step 5 : Eliminating the feedback path

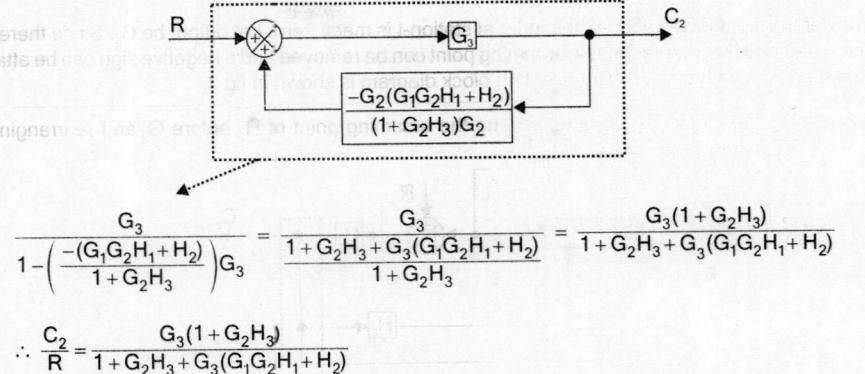

$$\cfrac{G_3}{1-\left(\cfrac{-(G_1G_2H_1+H_2)}{1+G_2H_3}\right)G_3} = \cfrac{G_3}{\cfrac{1+G_2H_3+G_3(G_1G_2H_1+H_2)}{1+G_2H_3}} = \cfrac{G_3(1+G_2H_3)}{1+G_2H_3+G_3(G_1G_2H_1+H_2)}$$

$$\therefore \frac{C_2}{R} = \frac{G_3(1+G_2H_3)}{1+G_2H_3+G_3(G_1G_2H_1+H_2)}$$

RESULT

The transfer function of the system with input at station-I is,

$$\frac{C_1}{R} = \frac{G_1G_2G_3}{1+G_3H_2+G_2H_3+G_1G_2G_3H_1}$$

The transfer function of the system with input at station-II is,

$$\frac{C_2}{R} = \frac{G_3(1+G_2H_3)}{1+G_2H_3+G_3(G_1G_2H_1+H_2)}$$

EXAMPLE 1.20

For the system represented by the block diagram shown in the fig 1, determine and .

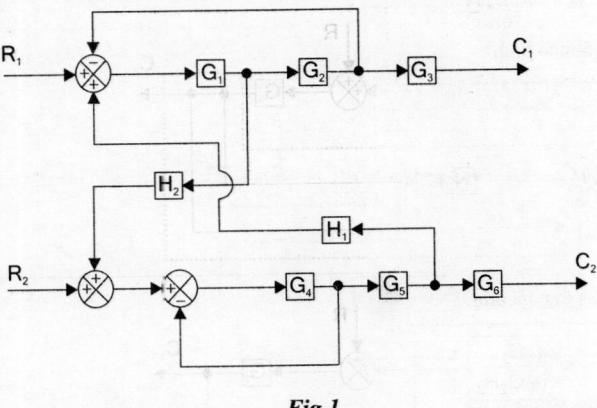

Fig 1

SOLUTION

Case (i) To find $\dfrac{C_1}{R_1}$

In this case set $R_2 = 0$ and consider only one output C_1. Hence we can remove the summing point which adds R_2 and need not consider G_6, since G_6 is on the open path. The resulting block diagram is shown in fig 2.

Step 1 : Eliminating the feedback path

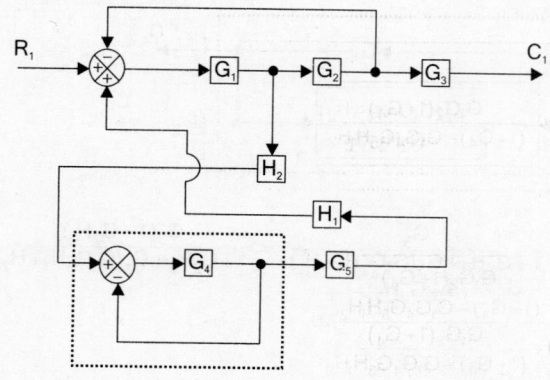

Fig 2.

Step 2 : Combining the blocks in cascade and splitting the summing point

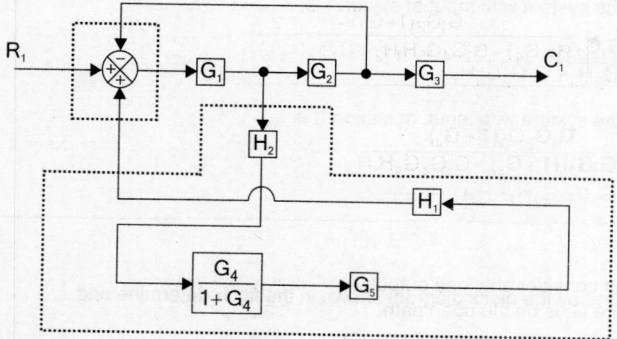

Step 3 : Eliminating the feedback path

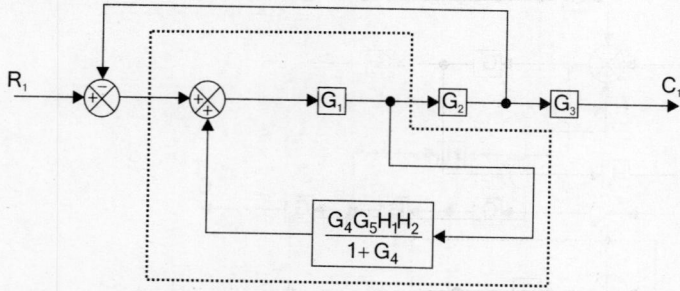

Step 4 : Combining the blocks in cascade

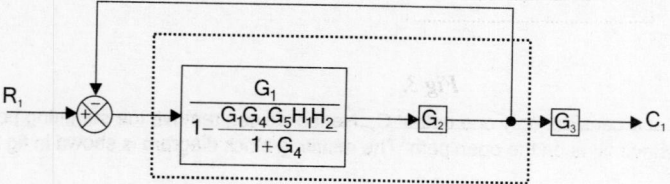

Step 5 : Eliminating the feedback path

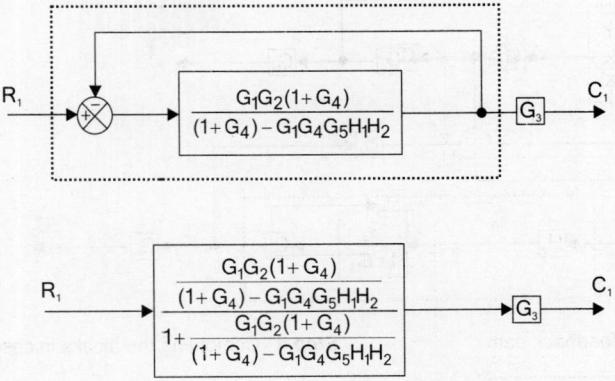

Step 6 : Combining the blocks in cascade

$$R_1 \longrightarrow \boxed{\dfrac{G_1G_2(1+G_4)}{(1+G_4)-G_1G_4G_5H_1H_2+G_1G_2(1+G_4)}} \longrightarrow C_1$$

$$\frac{C_1}{R_1} = \frac{G_1G_2G_3(1+G_4)}{(1+G_1G_2)(1+G_4)-G_1G_4G_5H_1H_2}$$

Case 2 : To find $\dfrac{C_2}{R_1}$

In this case set $R_2 = 0$ and consider only one output C_2. Hence we can remove the summing point which adds R_2 and need not consider G_3, since G_3 is on the open path. The resulting block diagram is shown in fig 3.

Step 1 : Eliminate the feedback path.

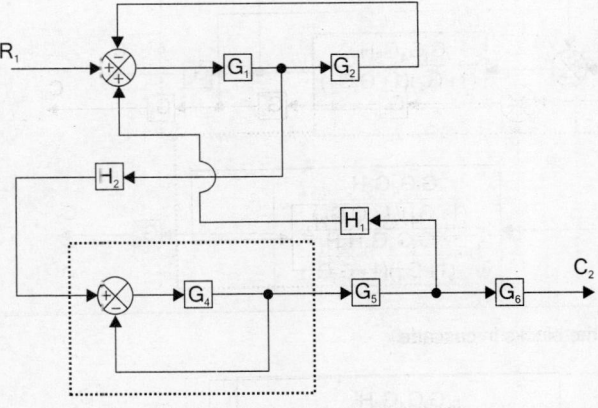

Fig 3.

Step 2 : Combining blocks in cascade and splitting the summing point

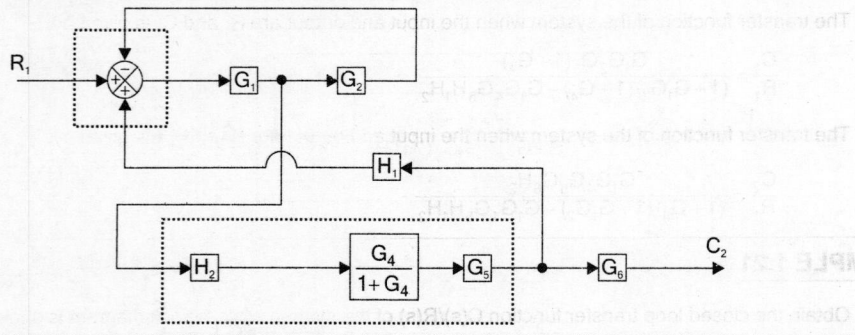

Step 3 : Eliminating the feedback path   **Step 4** : Combining the blocks in cascade

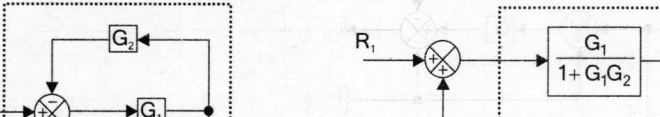

Step 5 : Eliminating the feedback path

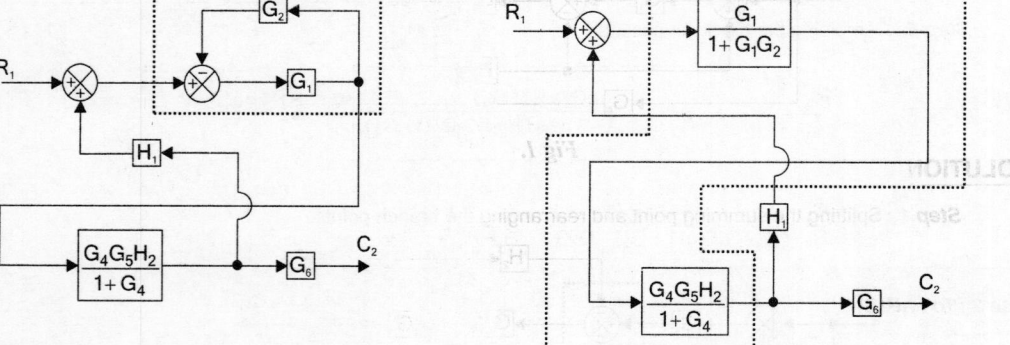

Step 6 : Combining the blocks in cascade

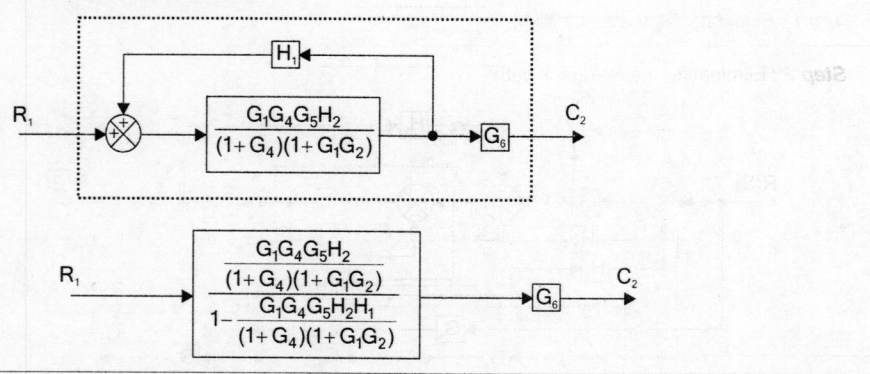

$$\frac{C_2}{R_1} = \frac{G_1 G_4 G_5 G_6 H_2}{(1+G_4)(1+G_1 G_2) - G_1 G_4 G_5 H_1 H_2}$$

RESULT

The transfer function of the system when the input and output are R_1 and C_1 is given by,

$$\frac{C_1}{R_1} = \frac{G_1 G_2 G_3 (1 + G_4)}{(1 + G_1 G_2)(1 + G_4) - G_1 G_4 G_5 H_1 H_2}$$

The transfer function of the system when the input and output are R_1 and C_2 is given by,

$$\frac{C_2}{R_1} = \frac{G_1 G_4 G_5 G_6 H_2}{(1 + G_4)(1 + G_1 G_2) - G_1 G_4 G_5 H_1 H_2}$$

EXAMPLE 1.21

Obtain the closed loop transfer function C(s)/R(s) of the system whose block diagram is shown in fig 1.

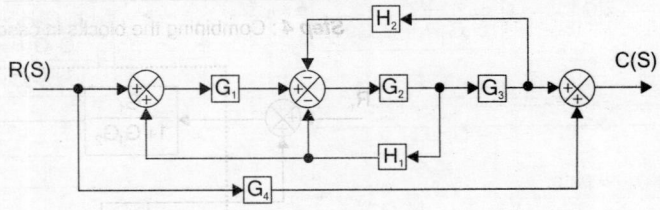

Fig 1.

SOLUTION

Step 1 : Splitting the summing point and rearranging the branch points

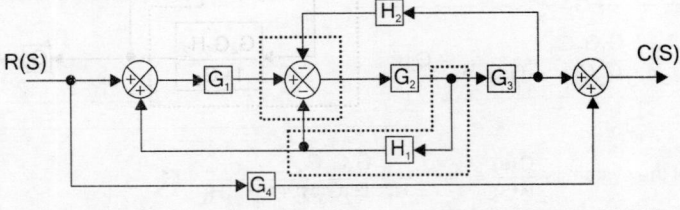

Step 2 : Eliminating the feedback path

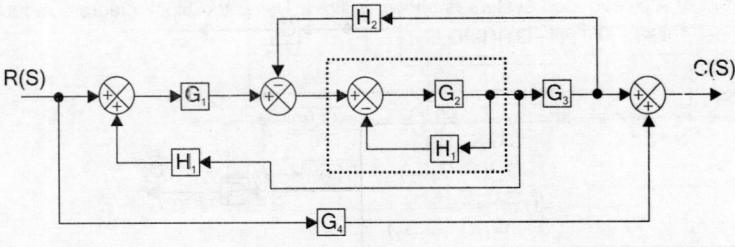

Step 3 : Shifting the branch point after the block.

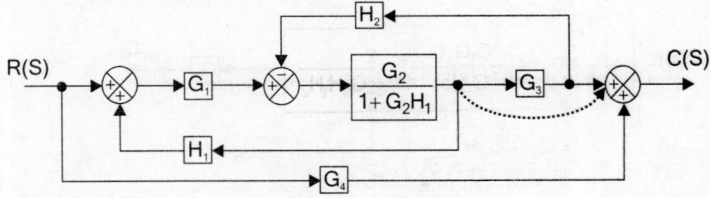

Step 4 : Combining the blocks in cascade and eliminating feedback path

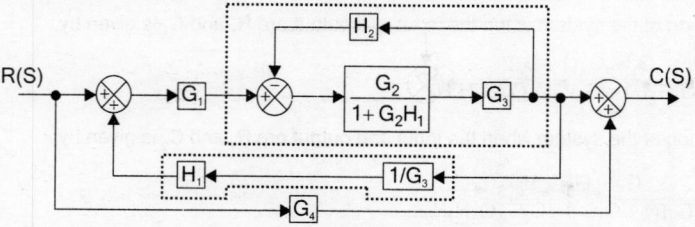

Step 5 : Combining the blocks in cascade and eliminating feedback path

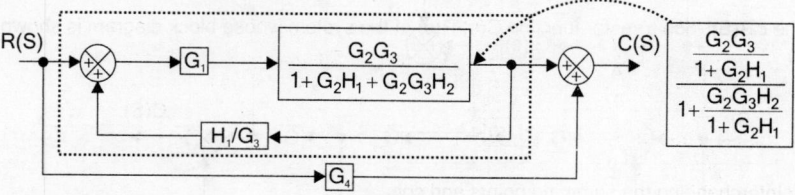

Step 6 : Eliminating forward path

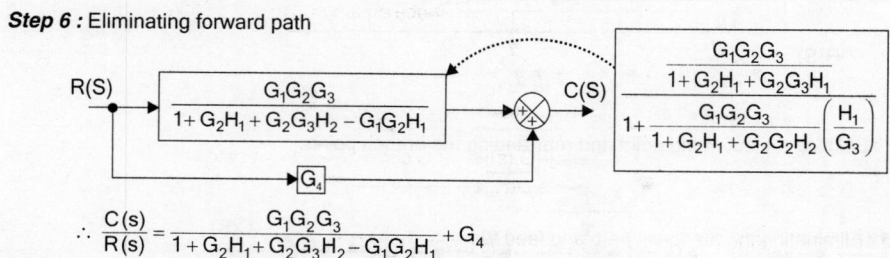

$$\therefore \frac{C(s)}{R(s)} = \frac{G_1 G_2 G_3}{1 + G_2 H_1 + G_2 G_3 H_2 - G_1 G_2 H_1} + G_4$$

RESULT

The transfer function of the system is $\dfrac{C(s)}{R(s)} = \dfrac{G_1 G_2 G_3}{1 + G_2 H_1 + G_2 G_3 H_2 - G_1 G_2 H_1} + G_4$

EXAMPLE 1.22

The block diagram of a closed loop system is shown in fig 1. Using the block diagram reduction technique determine the closed loop transfer function C(s)/R(s).

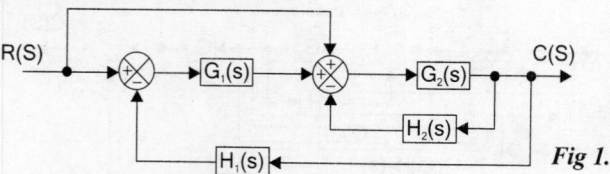

Fig 1.

SOLUTION

Step 1 : Splitting the summing point.

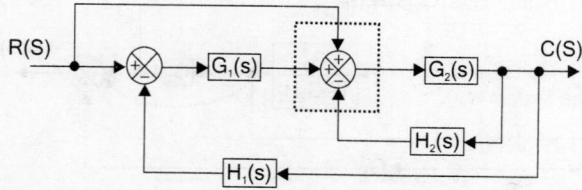

Step 2 : Eliminating the feedback path.

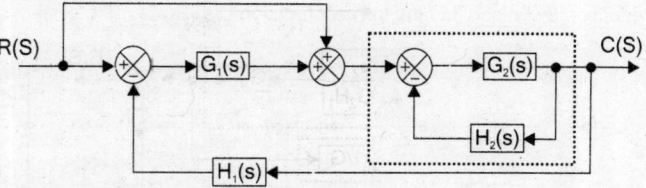

Step 3 : Moving the summing point after the block.

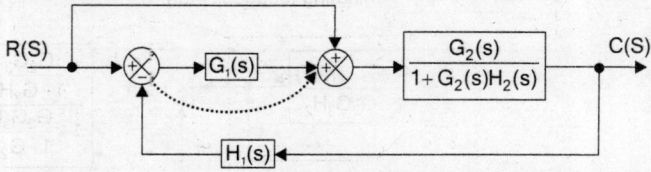

Step 4 : Interchanging the summing points and combining the blocks in cascade

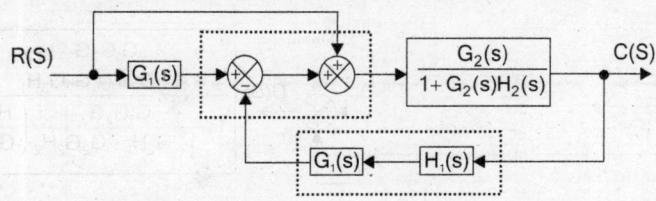

Step 5 : Eliminating the feedback path and feed forward path

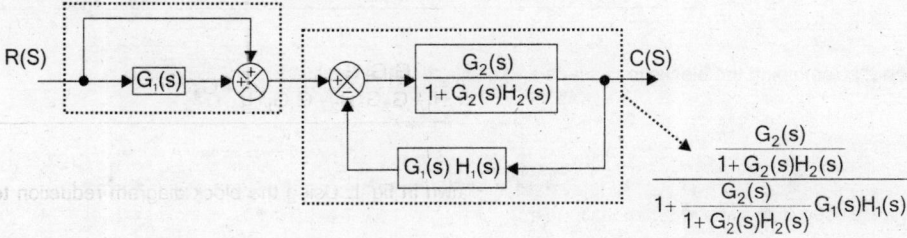

Step 6 : Combining the blocks in cascade

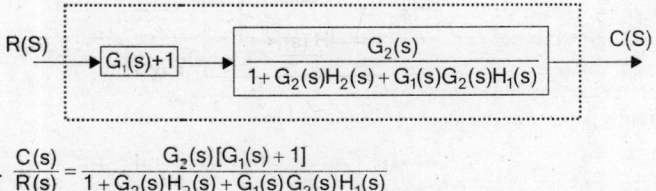

$$\therefore \frac{C(s)}{R(s)} = \frac{G_2(s)[G_1(s)+1]}{1+G_2(s)H_2(s)+G_1(s)G_2(s)H_1(s)}$$

RESULT

The transfer function of the system is,

$$\frac{C(s)}{R(s)} = \frac{G_2(s)[G_1(s)+1]}{1+G_2(s)H_2(s)+G_1(s)G_2(s)H_1(s)}$$

EXAMPLE 1.23

Using block diagram reduction technique find the transfer function C(s)/R(s) for the system shown in fig 1.

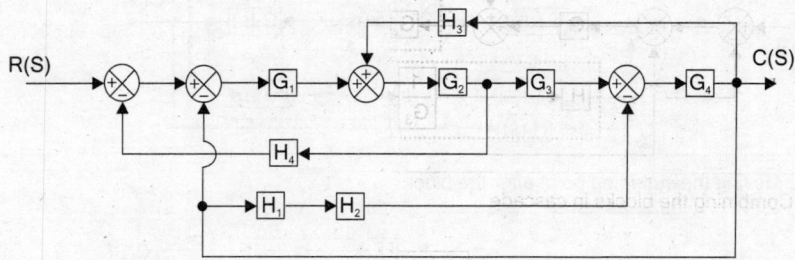

Fig 1.

SOLUTION

Step 1 : Rearranging the branch points

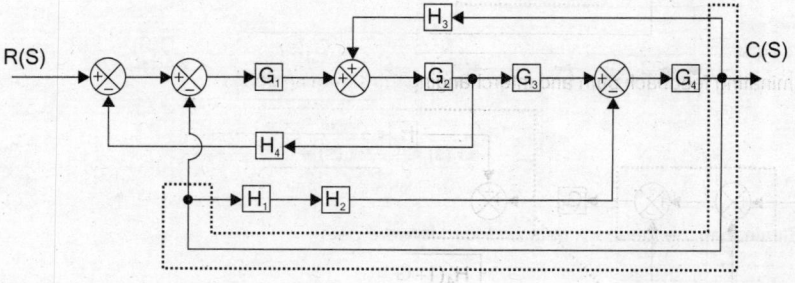

Step 2 : Combining the blocks in cascade and eliminating the feedback path.

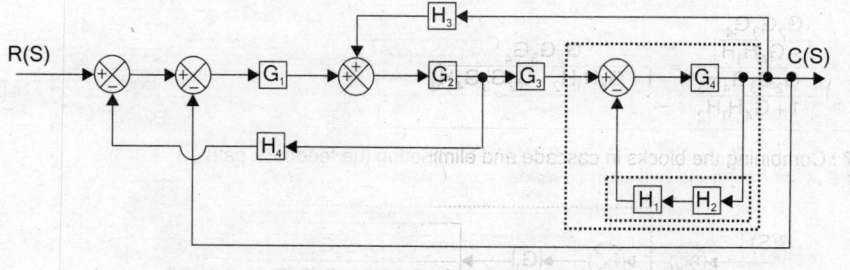

Step 3 : Moving the branch point after the block.

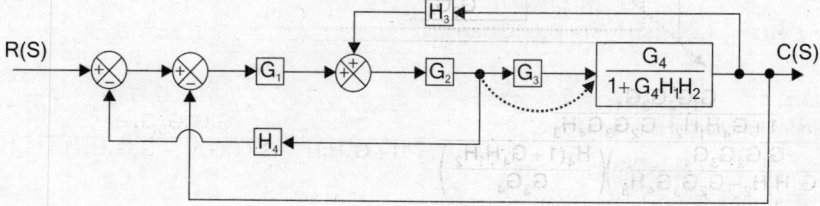

Step 4 : Moving the branch point and combining the blocks in cascade.

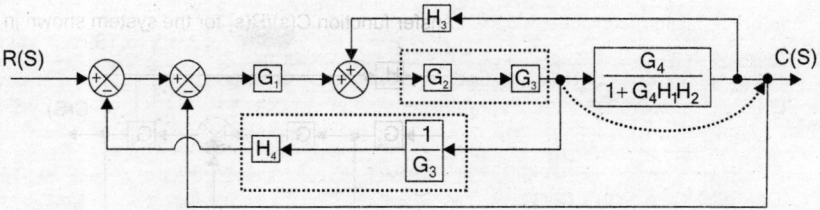

Step 5 : Combining the blocks in cascade

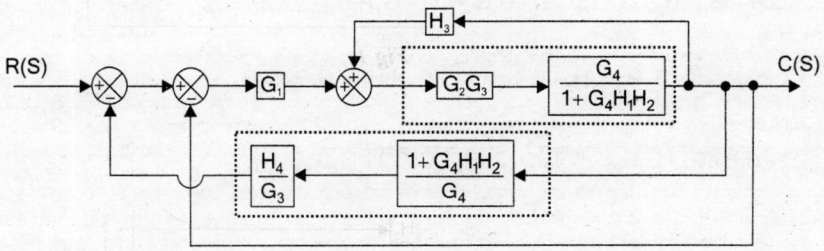

Step 6 : Eliminating feedback path and interchanging the summing points.

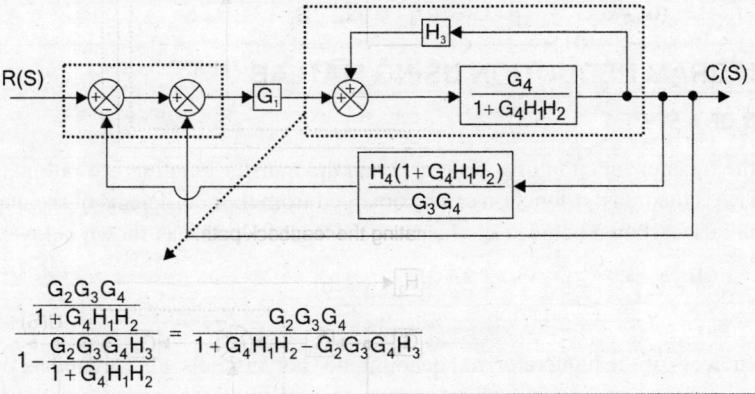

$$\frac{\dfrac{G_2G_3G_4}{1+G_4H_1H_2}}{1-\dfrac{G_2G_3G_4H_3}{1+G_4H_1H_2}} = \frac{G_2G_3G_4}{1+G_4H_1H_2-G_2G_3G_4H_3}$$

Step 7 : Combining the blocks in cascade and eliminating the feedback path

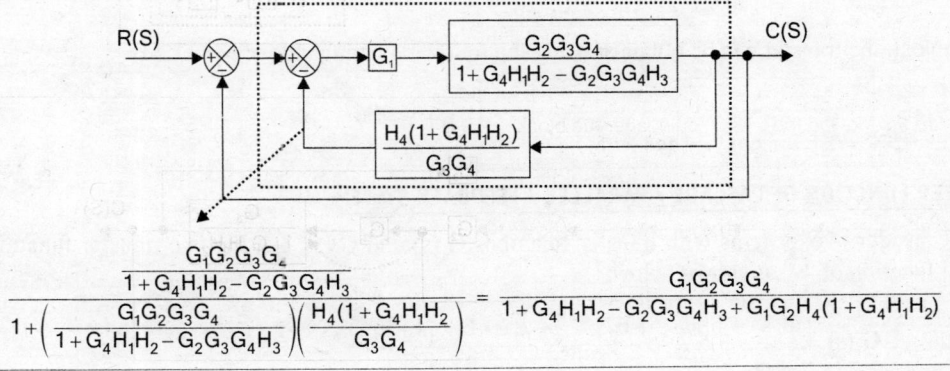

$$\frac{\dfrac{G_1G_2G_3G_4}{1+G_4H_1H_2-G_2G_3G_4H_3}}{1+\left(\dfrac{G_1G_2G_3G_4}{1+G_4H_1H_2-G_2G_3G_4H_3}\right)\left(\dfrac{H_4(1+G_4H_1H_2)}{G_3G_4}\right)} = \frac{G_1G_2G_3G_4}{1+G_4H_1H_2-G_2G_3G_4H_3+G_1G_2H_4(1+G_4H_1H_2)}$$

Step 8 : Eliminating the unity feedback path.

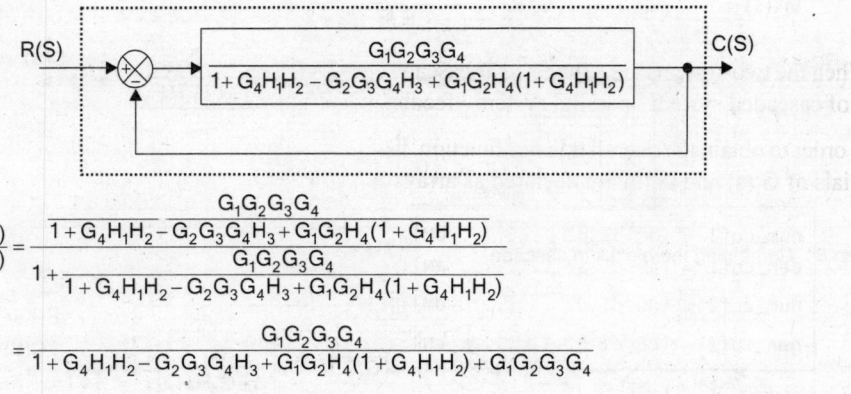

$$\therefore \frac{C(s)}{R(s)} = \frac{\dfrac{G_1 G_2 G_3 G_4}{1 + G_4 H_1 H_2 - G_2 G_3 G_4 H_3 + G_1 G_2 H_4 (1 + G_4 H_1 H_2)}}{1 + \dfrac{G_1 G_2 G_3 G_4}{1 + G_4 H_1 H_2 - G_2 G_3 G_4 H_3 + G_1 G_2 H_4 (1 + G_4 H_1 H_2)}}$$

$$= \frac{G_1 G_2 G_3 G_4}{1 + G_4 H_1 H_2 - G_2 G_3 G_4 H_3 + G_1 G_2 H_4 (1 + G_4 H_1 H_2) + G_1 G_2 G_3 G_4}$$

$$= \frac{G_1 G_2 G_3 G_4}{1 + H_1 H_2 (G_4 + G_1 G_2 G_4 H_4) + G_1 G_2 (H_4 + G_3 G_4) - G_2 G_3 G_4 H_3}$$

RESULT

The transfer function of the system is,

$$\frac{C(s)}{R(s)} = \frac{G_1 G_2 G_3 G_4}{1 + H_1 H_2 (G_4 + G_1 G_2 G_4 H_4) + G_1 G_2 (H_4 + G_3 G_4) - G_2 G_3 G_4 H_3}$$

1.12 BLOCK DIAGRAM REDUCTION USING MATLAB

TRANSFER FUNCTION OF A SYSTEM

Let, $G(s)$ be the transfer function of a system. When the transfer function is a rational function of "s", then using MATLAB the transfer function can be obtained from the coefficients of the numerator and denominator polynomials as shown below. Let, the general form of $G(s)$ be as shown below.

$$G(s) = \frac{b_0 s^M + b_1 s^{M-1} + b_2 s^{M-2} + \ldots + b_{M-1} s + b_M}{a_0 s^N + a_1 s^{N-1} + a_2 s^{N-2} + \ldots + a_{N-1} s + a_N}$$

First, the coefficients of the numerator and denominator polynomials are declared as two arrays as shown below.

```
num_cof = [b0 b1 b2 ........ bM];
den_cof = [a0 a1 a2 ........ aN];
```

Next, the transfer can be obtained using the following commands of MATLAB.

```
G = tf('s');
G = ([num_cof], [den_cof])
```

TRANSFER FUNCTION OF CASCADE / PARALLEL / FEEDBACK SYSTEM

Consider two systems with transfer functions $G_1(s)$ and $G_2(s)$. Let the two transfer functions be rational function of "s" as shown below.

$$G_1(s) = \frac{b_0 s^M + b_1 s^{M-1} + b_2 s^{M-2} + \ldots + b_{M-1} s + b_M}{a_0 s^N + a_1 s^{N-1} + a_2 s^{N-2} + \ldots + a_{N-1} s + a_N}$$

$$G_2(s) = \frac{d_0 s^M + d_1 s^{M-1} + d_2 s^{M-2} + \ldots + d_{M-1} s + d_M}{c_0 s^N + c_1 s^{N-1} + c_2 s^{N-2} + \ldots + c_{N-1} s + c_N}$$

When the two systems are connected as cascade / parallel / feedback system, then the overall transfer function of cascaded system / parallel system / feedback system can be obtained using MATLAB.

In order to obtain the overall transfer function, first the coefficients of the numerator and denominator polynomials of $G_1(s)$ and $G_2(s)$ are declared as arrays as shown below.

```
num_cof1 = [b0 b1 b2 ....... bM];
den_cof1 = [a0 a1 a2 ....... aN];
num_cof2 = [d0 d1 d2 ....... dM];
den_cof2 = [c0 c1 c2 ....... cN];
```

When the two systems are connected in cascade as shown below, then the overall transfer function $G_c(s)$ of the cascaded system can be obtained using the following commands of MATLAB.

```
GC = tf('s');
[num_cofC, den_cofC] = series(num_cof1, den_cof1, num_cof2, den_cof2);
GC = ([num_cofC], [den_cofC])
```

When the two systems are connected in parallel as shown below, then the overall transfer function $G_p(s)$ of parallel system can be obtained using the following commands of MATLAB.

```
GP = tf('s');
[num_cofP, den_cofP]=parallel(num_cof1, den_cof1, num_cof2, den_cof2);
GP = ([num_cofP], [den_cofP])
```

When the two systems are connected in feedback as shown below, then the overall transfer function $G_F(s)$ of feedback system can be obtained using the following commands of MATLAB.

$$G_F = \frac{G_1}{1 + G_1 G_2}$$

```
GF = tf('s');
[num_cofF, den_cofF] = feedback(num_cof1, den_cof1, num_cof2, den_cof2);
GF = ([num_cofF], [den_cofF])
```

PROGRAM 1.1

Consider the transfer functions of the two systems given below.

$$G_1(s)=8/(s^2+2s+9) \quad \text{and} \quad G_2(s)=4/(s+6)$$

Write a MATLAB program to find the overall transfer function if the two systems are connected as cascade system, parallel system and feedback system.

```
clc
clear all
G1=tf('s'); G2=tf('s'); GC=tf('s');GP=tf('s');GF=tf('s');
num_cof1=[0 0 8];
den_cof1=[1 2 9];
disp('System1');
G1=tf([num_cof1], [den_cof1])
num_cof2=[0 4];
den_cof2=[1 6];
disp('System2');
G2=tf([num_cof2], [den_cof2])
[num_cofC,den_cofC]=series(num_cof1,den_cof1,num_cof2,den_cof2);
disp('Cascade system');
GC=tf([num_cofC], [den_cofC])
[num_cofP,den_cofP]=parallel(num_cof1,den_cof1,num_cof2,den_cof2);
disp('Parallel system');
GP=tf([num_cofP], [den_cofP])
[num_cofF,den_cofF]=feedback(num_cof1,den_cof1,num_cof2,den_cof2);
disp('Feedback system');
GF=tf([num_cofF], [den_cofF])
```

OUTPUT

```
System1
Transfer function:
        8
   ---------------
    s^2 + 2 s + 9
System2
Transfer function:
      4
    ------
    s + 6
Cascade system
Transfer function:
            32
   -----------------------
    s^3 + 8 s^2 + 21 s + 54

Parallel system
Transfer function:
      4 s^2 + 16 s + 84
   -----------------------
    s^3 + 8 s^2 + 21 s + 54

Feedback system
Transfer function:
          8 s + 48
   -----------------------
    s^3 + 8 s^2 + 21 s + 86
```

1.13 Signal flow graph

The signal flow graph is used to represent the control system graphically and it was developed by **S.J. Mason.**

A signal flow graph is a diagram that represents a set of simultaneous linear algebraic equations. By taking Laplace transform, the time domain differential equations governing a control system can be transferred to a set of algebraic equations in s-domain. The signal flow graph of the system can be constructed using these equations.

It should be noted that the signal flow graph approach and the block diagram approach yield the same information. The advantage in signal flow graph method is that, using Mason's gain formula the overall gain of the system can be computed easily. This method is simpler than the tedious block diagram reduction techniques.

The signal flow graph depicts the flow of signals from one point of a system to another and gives the relationships among the signals. A signal flow graph consists of a network in which nodes are connected by directed branches. Each node represents a system variable and each branch connected between two nodes acts as a signal multiplier. Each branch has a gain or transmittance. When the signal pass through a branch, it gets multiplied by the gain of the branch.

In a signal flow graph, the signal flows in only one direction. The direction of signal flow is indicated by an arrow placed on the branch and the gain (multiplication factor) is indicated along the branch.

EXPLANATION OF TERMS USED IN SIGNAL FLOW GRAPH

Node	:	A node is a point representing a variable or signal.
Branch	:	A branch is directed line segment joining two nodes. The arrow on the branch indicates the direction of signal flow and the gain of a branch is the transmittance.
Transmittance	:	The gain acquired by the signal when it travels from one node to another is called transmittance. The transmittance can be real or complex.
Input node (Source)	:	It is a node that has only outgoing branches.
Output node (Sink)	:	It is a node that has only incoming branches.
Mixed node	:	It is a node that has both incoming and outgoing branches.
Path	:	A path is a traversal of connected branches in the direction of the branch arrows. The path should not cross a node more than once.
Open path	:	A open path starts at a node and ends at another node.
Closed path	:	Closed path starts and ends at same node.
Forward path	:	It is a path from an input node to an output node that does not cross any node more than once.
Forward path **gain**	:	It is the product of the branch transmittances (gains) of a forward path.
Individual loop	:	It is a closed path starting from a node and after passing through a certain part of a graph arrives at same node without crossing any node more than once.
Loop gain	:	It is the product of the branch transmittances (gains) of a loop.
Non-touching Loops	:	If the loops does not have a common node then they are said to be non- touching loops.

PROPERTIES OF SIGNAL FLOW GRAPH

The basic properties of signal flow graph are the following:

(i) The algebraic equations which are used to construct signal flow graph must be in the form of cause and effect relationship.

(ii) Signal flow graph is applicable to linear systems only.

(iii) A node in the signal flow graph represents the variable or signal.

(iv) A node adds the signals of all incoming branches and transmits the sum to all outgoing branches.

(v) A mixed node which has both incoming and outgoing signals can be treated as an output node by adding an outgoing branch of unity transmittance.

(vi) A branch indicates functional dependence of one signal on the other.

(vii) The signals travel along branches only in the marked direction and when it travels it gets multiplied by the gain or transmittance of the branch.

(viii) The signal flow graph of system is not unique. By rearranging the system equations different types of signal flow graphs can be drawn for a given system.

SIGNAL FLOW GRAPH ALGEBRA

Signal flow graph for a system can be reduced to obtain the transfer function of the system using the following rules. The guideline in developing the rules for signal flow graph algebra is that the signal at a node is given by sum of all incoming signals.

Rule 1 : Incoming signal to a node through a branch is given by the product of a signal at previous node and the gain of the branch.

Example:

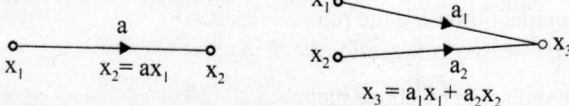

Rule 2 : Cascaded branches can be combined to give a single branch whose transmittance is equal to the product of individual branch transmittance.

Example:

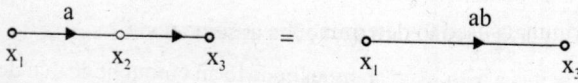

Rule 3 : Parallel branches may be represented by single branch whose transmittance is the sum of individual branch transmittances.

Example:

Rule 4 : A mixed node can be eliminated by multiplying the transmittance of outgoing branch (from the mixed node) to the transmittance of all incoming branches to the mixed node.

Example

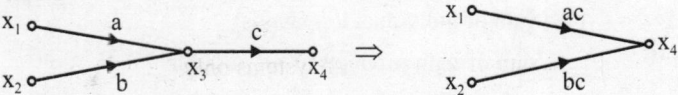

Rule 5 : A loop may be eliminated by writing equations at the input and output node and rearranging the equations to find the ratio of output to input. This ratio gives the gain of resultant branch.

Example

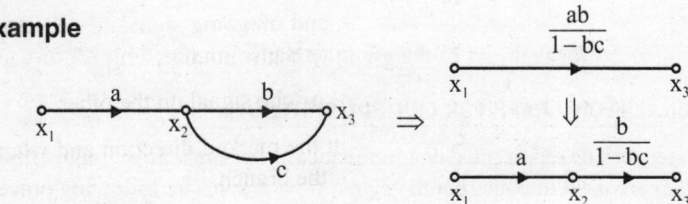

Proof:

$$x_2 = ax_1 + cx_3 \quad ; \quad x_3 = bx_2$$

Put, $x_2 = ax_1 + cx_3$ in the equation for x_3.

$$\therefore x_3 = b(ax_1 + cx_3) \quad \Rightarrow \quad x_3 = abx_1 + bcx_3 \quad \Rightarrow \quad x_3 - bcx_3 = abx_1 \quad \Rightarrow \quad x_3(1 - bc) = abx_1$$

$$\therefore \frac{x_3}{x_1} = \frac{ab}{1 - bc}$$

SIGNAL FLOW GRAPH REDUCTION

The signal flow graph of a system can be reduced either by using the rules of a signal flow graph algebra or by using Mason's gain formula.

For signal flow graph reduction using the rules of signal flow graph, write equations at every node and then rearrange these equations to get the ratio of output and input (transfer function).

The signal flow graph reduction by above method will be time consuming and tedious. **S.J.Mason** has developed a simple procedure to determine the transfer function of the system represented as a signal flow graph. He has developed a formula called by his name **mason's gain formula** which can be directly used to find the transfer function of the system.

MASON'S GAIN FORMULA

The Mason's gain formula is used to determine the transfer function of the system from the signal flow graph of the system.

Let,   R(s) = Input to the system

C(s) = Output of the system

Now, Transfer function of the system, $T(s) = \dfrac{C(s)}{R(s)}$ .....(1.34)

Mason's gain formula states the overall gain of the system [transfer function] as follows,

$$\text{Overall gain, } T = \frac{1}{\Delta} \sum_k P_k \Delta_k$$ .....(1.35)

where,   T   = $T(s)$ = Transfer function of the system

          P_K =   Forward path gain of Kth forward path

          K   =   Number of forward paths in the signal flow graph

          Δ =   1 − (Sum of individual loop gains)

$$+ \left(\begin{array}{l} \text{Sum of gain products of all possible} \\ \text{combinations of two non − touching loops} \end{array} \right)$$

$$- \left(\begin{array}{l} \text{Sum of gain products of all possible} \\ \text{combinations of three non − touching loops} \end{array} \right)$$

$$+ ....$$

          Δ_K = Δ for that part of the graph which is not touching K^{th} forward path

CONSTRUCTING SIGNAL FLOW GRAPH FOR CONTROL SYSTEMS

A control system can be represented diagrammatically by signal flow graph. The differential equations governing the system are used to construct the signal flow graph. The following procedure can be used to construct the signal flow graph of a system.

1.   Take Laplace transform of the differential equations governing the system in order to convert them to algebraic equations in s-domain.

2.   The constants and variables of the s-domain equations are identified.

3.   From the working knowledge of the system, the variables are identified as input, output and intermediate variables.

4.   For each variable a node is assigned in signal flow graph and constants are assigned as the gain or transmittance of the branches connecting the nodes.

5.   For each equation a signal flow graph is drawn and then they are interconnected to give overall signal flow graph of the system.

PROCEDURE FOR CONVERTING BLOCK DIAGRAM TO SIGNAL FLOW GRAPH

The signal flow graph and block diagram of a system provides the same information but there is no standard procedure for reducing the block diagram to find the transfer function of the system. Also the block diagram reduction technique will be tedious and it is difficult to choose the rule to be applied for simplification. Hence it will be easier if the block diagram is converted to signal flow graph and **Mason's gain formula** is applied to find the transfer function. The following procedure can be used to convert block diagram to signal flow graph.

1.   Assume nodes at input, output, at every summing point, at every branch point and in between cascaded blocks.

2.   Draw the nodes separately as small circles and number the circles in the order 1, 2, 3, 4, ... etc.

3.   From the block diagram find the gain between each node in the main forward path and connect all the corresponding circles by straight line and mark the gain between the nodes.

4.   Draw the feed forward paths between various nodes and mark the gain of feed forward path along with sign.

5.   Draw the feedback paths between various nodes and mark the gain of feedback paths along with sign.

EXAMPLE 1.24

Construct a signal flow graph for armature controlled dc motor.

SOLUTION

The differential equations governing the armature controlled dc motor are (refer section 1.7).

$$v_a = i_a R_a + L_a \frac{di_a}{dt} + e_b \quad ; \quad T = K_f i_a \quad ; \quad T = J\frac{d\omega}{dt} + B\omega \quad ; \quad e_b = K_b \omega \quad ; \quad \omega = d\theta/dt$$

On taking Laplace transform of above equations we get,

$$V_a(s) = I_a(s)\,R_a + L_a\,s\,I_a(s) + E_b(s) \qquad \qquad .....(1)$$

$$T(s) = K_f I_a(s) \qquad \qquad .....(2)$$

$$T(s) = J\,s\,\omega(s) + B\,\omega(s) \qquad \qquad .....(3)$$

$$E_b(s) = K_b\,\omega(s) \qquad \qquad .....(4)$$

$$\omega(s) = s\,\theta(s) \qquad \qquad .....(5)$$

The input and output variables of armature controlled dc motor are armature voltage $V_a(s)$ and angular displacement $\theta(s)$ respectively. The variables $I_a(s)$, $T(s)$, $E_b(s)$ and $\omega(s)$ are intermediate variables.

The equations (1) to (5) are rearranged & individual signal flow graph are shown in fig 1 to fig 5.

$$V_a(s) - E_b(s) = I_a(s)\,[R_a + s\,L_a]$$

$$\therefore\ I_a(s) = \frac{1}{R_a + sL_a}\,[V_a(s) - E_b(s)]$$

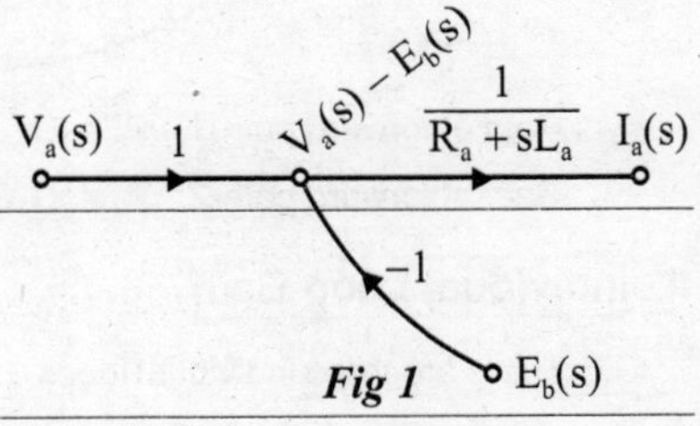

$$T(s) = K_t I_a(s)$$

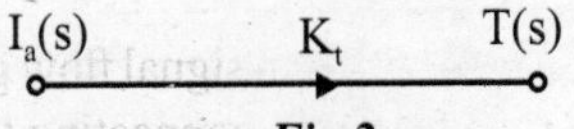

Fig 2   *Fig 1*

$$T(s) = \omega(s)\,[Js + B]$$

$$\therefore\ \omega(s) = \frac{1}{Js + B}\,T(s)$$

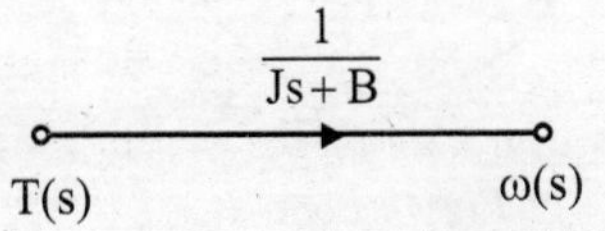

Fig 3

$$E_b(s) = K_b\,\omega(s)$$

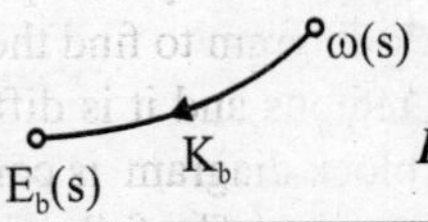

Fig 4

$$\omega(s) = s\theta(s)$$

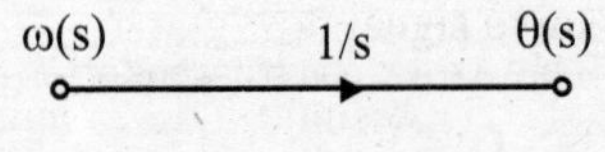

$$\therefore\ \theta(s) = \frac{1}{s}\,\omega(s)$$

Fig 5

The overall signal flow graph of armature controlled dc motor is obtained by interconnecting the individual signal flow graphs shown in fig 1 to fig 5. The overall signal flow graph is shown in fig 6.

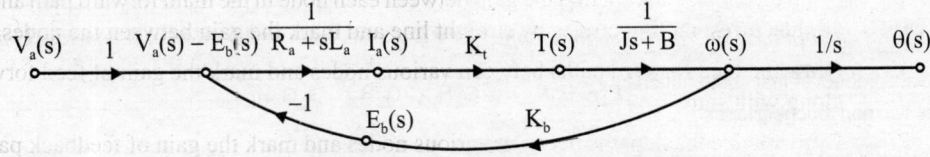

Fig 6 : Signal flow graph of armature controlled dc motor.

EXAMPLE 1.25

Find the overall transfer function of the system whose signal flow graph is shown in fig 1.

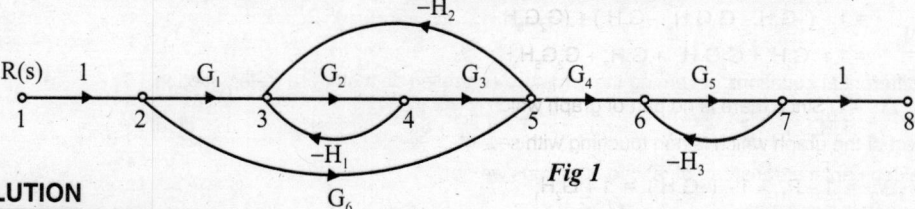

Fig 1

SOLUTION

I. Forward Path Gains

There are two forward paths. $\therefore K = 2$

Let forward path gains be P_1 and P_2.

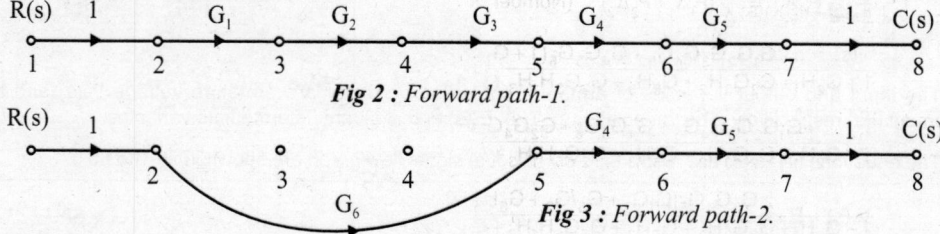

Fig 2 : Forward path-1.

Fig 3 : Forward path-2.

Gain of forward path-1, $P_1 = G_1 \ G_2 \ G_3 \ G_4 \ G_5$

Gain of forward path-2, $P_2 = G_4 G_5 G_6$

II. Individual Loop Gain

There are three individual loops. Let individual loop gains be P_{11}, P_{21} and P_{31}.

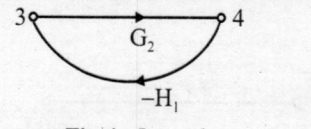

Fig 4 : Loop-1.

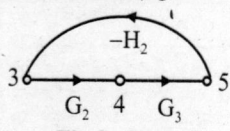

Fig 5 : Loop-2.

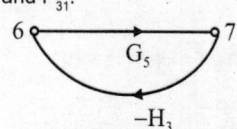

Fig 6 : Loop-3.

Loop gain of individual loop-1, $P_{11} = -G_2H_1$

Loop gain of individual loop-2, $P_{21} = -G_2G_3H_2$

Loop gain of individual loop-3, $P_{31} = -G_5H_3$

III. Gain Products of Two Non-touching Loops

There are two combinations of two non-touching loops. Let the gain products of two non touching loops be P_{12} and P_{22}.

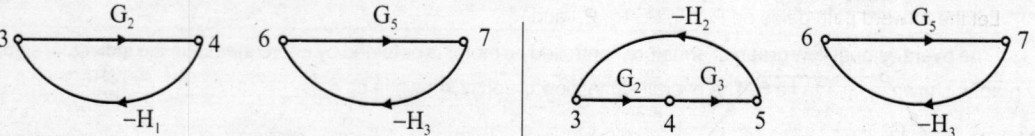

Fig 7 : First combination of 2 non-touching loops. | *Fig 8 : Second combination of 2 non-touching loops.*

Gain product of first combination of two non touching loops $\Big\}$ $P_{12} = P_{11}P_{31} = (-G_2H_1)(-G_5H_3) = G_2G_5H_1H_3$

Gain product of second combination of two non touching loops $\Big\}$ $P_{22} = P_{21}P_{31} = (-G_2G_3H_2)(-G_5H_3) = G_2G_3G_5H_2H_3$

IV. Calculation of Δ and Δ_K

$$\Delta = 1 - (P_{11} + P_{21} + P_{31}) + (P_{12} + P_{22})$$

$$= 1 - (-G_2H_1 - G_2G_3H_2 - G_5H_3) + (G_2G_5H_1H_3 + G_2G_3G_5H_2H_3)$$

$$= 1 + G_2H_1 + G_2G_3H_2 + G_5H_3 + G_2G_5H_1H_3 + G_2G_3G_5H_2H_3$$

$\Delta_1 = 1$, Since there is no part of graph which is not touching with first forward path.

The part of the graph which is non touching with second forward path is shown in fig 9.

$$\Delta_2 = 1 - P_{11} = 1 - (-G_2H_1) = 1 + G_2H_1$$

V. Transfer Function, T

By Mason's gain formula the transfer function, T is given by,

$$T = \frac{1}{\Delta} \sum_K P_K \Delta_K = \frac{1}{\Delta}(P_1\Delta_1 + P_2\Delta_2) \quad \text{(Number of forward paths is 2 and so K = 2)}$$

$$= \frac{G_1G_2G_3G_4G_5 + G_4G_5G_6(1 + G_2H_1)}{1 + G_2H_1 + G_2G_3H_2 + G_5H_3 + G_2G_5H_1H_3 + G_2G_3G_5H_2H_3}$$

$$= \frac{G_1G_2G_3G_4G_5 + G_4G_5G_6 + G_2G_4G_5G_6H_1}{1 + G_2H_1 + G_2G_3H_2 + G_5H_3 + G_2G_5H_1H_3 + G_2G_3G_5H_2H_3}$$

$$= \frac{G_2G_4G_5[G_1G_3 + G_6/G_2 + G_6H_1]}{1 + G_2H_1 + G_2G_3H_2 + G_5H_3 + G_2G_5H_1H_3 + G_2G_3G_5H_2H_3}$$

EXAMPLE 1.26

Find the overall gain of the system whose signal flow graph is shown in fig 1.

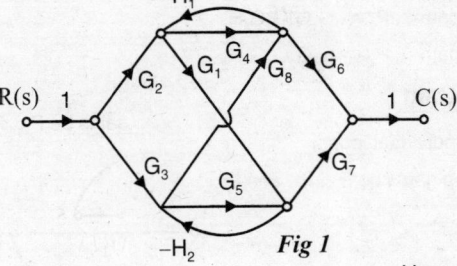

Fig 1

SOLUTION

Let us number the nodes as shown in fig 2.

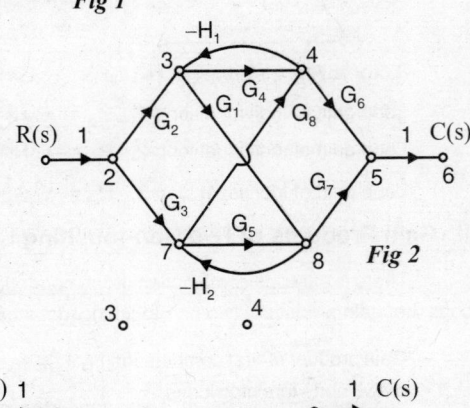

Fig 2

I. Forward Path Gains

There are six forward paths. $\therefore$ K = 6

Let the forward path gains be P_1, P_2, P_3, P_4, P_5 and P_6.

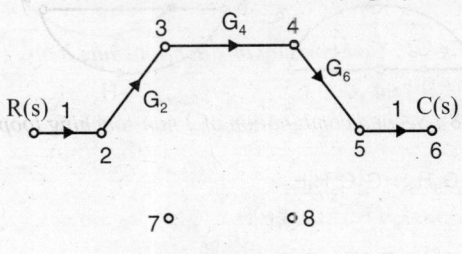

Fig 3 : Forward path-1.

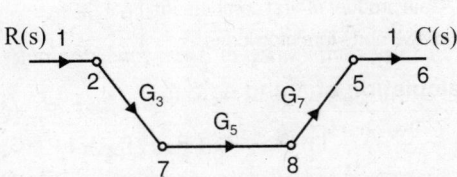

Fig 4 : Forward path-2.

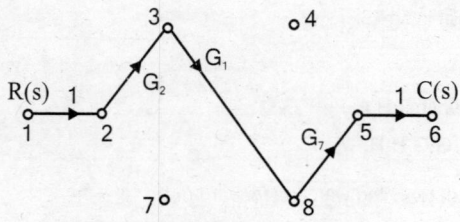

Fig 5 : Forward path-3

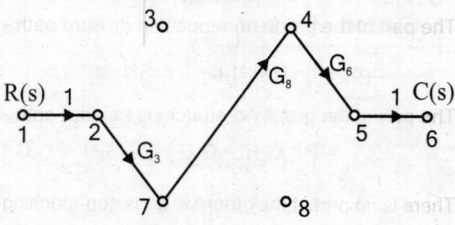

Fig 6 : Forward path-4

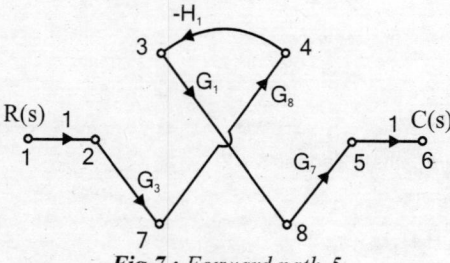

Fig 7 : Forward path-5

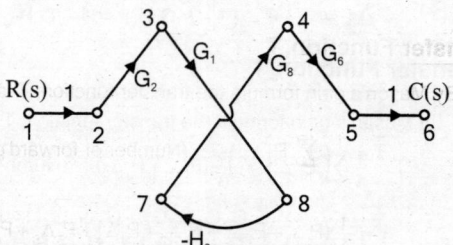

Fig 8 : Forward path-6

Gain of forward path-1, $P_1 = G_2G_4G_6$

Gain of forward path-2, $P_2 = G_3G_5G_7$

Gain of forward path-3, $P_3 = G_1G_2G_7$

Gain of forward path-4, $P_4 = G_3G_8G_6$

Gain of forward path-5, $P_5 = -G_1G_3G_7G_8H_1$

Gain of forward path-6, $P_6 = -G_1G_2G_6G_8H_2$

II. Individual Loop Gain

There are three individual loops.

Let individual loop gains be P_{11}, P_{21} and P_{31}.

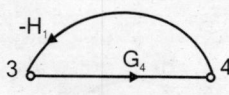

Fig 9 : Loop-1

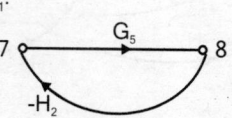

Fig 10 : Loop-2

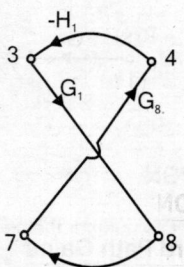

Fig 11 : Loop-3

Loop gain of individual loop-1, $P_{11} = -G_4H_1$

Loop gain of individual loop-2, $P_{21} = -G_5H_2$

Loop gain of individual loop-3, $P_{31} = G_1G_8H_1H_2$

III. Gain Products of Two Non-touching Loops

There is only one combination of two non-touching loops. Let gain product of two non-touching loops be P^{12}.

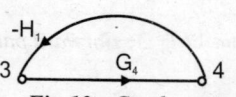
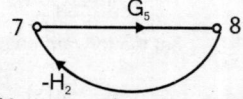

Fig 12 : Combination of 2 non-touching loops

Gain product of first combination of two non - touching loops $\Bigg\}$ $P_{12} = P_{11}P_{21} = (-G_4H_1)(-G_5H_2) = G_4G_5H_1H_2$

IV. Calculation of Δ and Δ_K

$$\Delta = 1 - (P_{11} + P_{21} + P_{31}) + P_{12} = 1 - (-G_4H_1 - G_5H_2 + G_1G_8H_1H_2) + G_4G_5H_1H_2$$

$$= 1 + G_4H_1 + G_5H_2 - G_1G_8H_1H_2 + G_4G_5H_1H_2$$

The part of the graph non-touching forward path -1 is shown in fig 13.

$$\therefore \Delta_1 = 1 - (-G_5 H_2) = 1 + G_5 H_2$$

The part of the graph non-touching forward path -2 is shown in fig 14.

$$\therefore \Delta_2 = 1 - (-G_4 H_1) = 1 + G_4 H_1$$

There is no part of the graph which is non-touching with forward paths 3, 4, 5 and 6.

$$\therefore \Delta_3 = \Delta_4 = \Delta_5 = \Delta_6 = 1$$

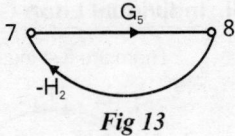

Fig 13

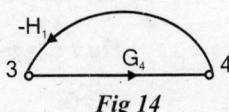

Fig 14

V. Transfer Function, T

By Mason's gain formula the transfer function, T is given by,

$$T = \frac{1}{\Delta}\left(\sum_K P_K \Delta_K\right) \quad \text{(Number of forward paths is six and so K = 6)}$$

$$= \frac{1}{\Delta}(P_1\Delta_1 + P_2\Delta_2 + P_3\Delta_3 + P_4\Delta_4 + P_5\Delta_5 + P_6\Delta_6)$$

$$= \frac{\begin{aligned}G_2 G_4 G_6(1 + G_5 H_2) + G_3 G_5 G_7(1 + G_4 H_1) + G_1 G_2 G_7 + G_3 G_6 G_8\\ - G_1 G_3 G_7 G_8 H_1 - G_1 G_2 G_6 G_8 H_2\end{aligned}}{1 + G_4 H_1 + G_5 H_2 - G_1 G_8 H_1 H_2 + G_4 G_5 H_1 H_2}$$

EXAMPLE 1.27

Find the overall gain C(s)/R(s) for the signal flow graph shown in fig 1.

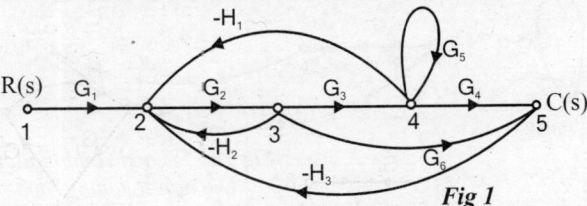

Fig 1

SOLUTION

I. Forward Path Gains

There are two forward paths. $\therefore$ K = 2.   Let the forward path gains be P_1 and P_2.

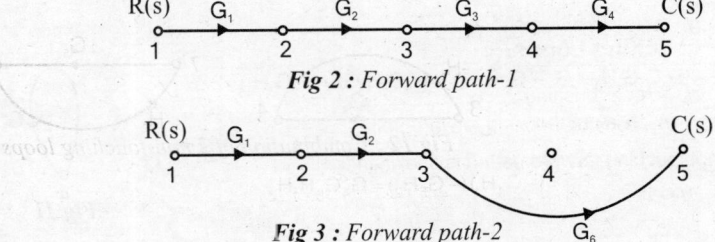

Fig 2 : Forward path-1

Fig 3 : Forward path-2

Gain of forward path-1, $P_1 = G_1 G_2 G_3 G_4$

Gain of forward path-2, $P_2 = G_1 G_2 G_6$

II. Individual Loop Gain

There are five individual loops. Let the individual loop gains be p_{11}, p_{21}, p_{31}, p_{41} and p_{51}.

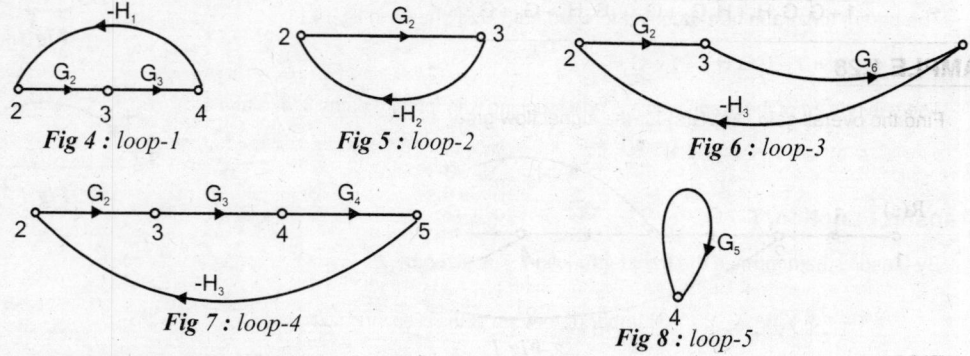

Fig 4 : *loop-1*   **Fig 5 :** *loop-2*   **Fig 6 :** *loop-3*

Fig 7 : *loop-4*

Fig 8 : *loop-5*

Loop gain of individual loop-1, $P_{11} = -G_2G_3H_1$

Loop gain of individual loop-2, $P_{21} = -H_2G_2$

Loop gain of individual loop-3, $P_{31} = -G_2G_6H_3$

Loop gain of individual loop-4, $P_{41} = -G_2G_3G_4H_3$

Loop gain of individual loop-5, $P_{51} = G_5$

III. Gain Products of Two Non-touching Loops

There are two combinations of two non-touching loops.

Let the gain products of two non-touching loops be P_{12} and P_{22}.

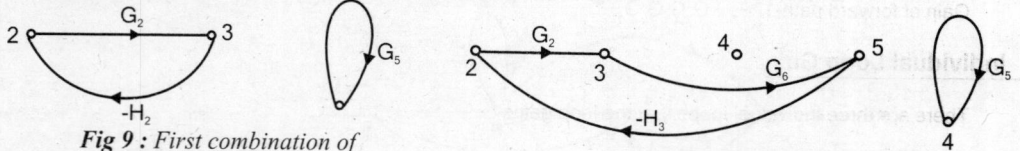

Fig 9 : *First combination of*
two non-touching loops

Fig 10 : *Second combination of*
two non-touching loops

Gain product of first combination
of two non touching loops $\Big\}$ $P_{12} = P_{21}P_{51} = (-G_2H_2)(G_5) = -G_2G_5H_2$

Gain product of second combination
of two non touching loops $\Big\}$ $P_{22} = P_{31}P_{51} = (-G_2G_6H_3)(G_5) = -G_2G_5G_6H_3$

IV. Calculation of Δ and Δ_K

$\Delta = 1 - (P_{11} + P_{21} + P_{31} + P_{41} + P_{51}) + (P_{12} + P_{22})$

$= 1 - (-G_2G_3H_1 - H_2G_2 - G_2G_3G_4H_3 + G_5 - G_2G_6H_3) + (-G_2H_2G_5 - G_2G_5H_3)$

Since there is no part of graph which is not touching forward path-1, $\Delta_1 = 1$.

The part of graph which is not touching forward path-2 is shown in fig 11.

$\therefore \Delta_2 = 1 - G_5$

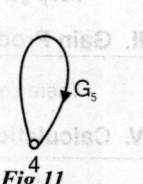

Fig 11

V. Transfer Function, T

By Mason's gain formula the transfer function, T is given by,

$T = \dfrac{1}{\Delta} \displaystyle\sum_K P_K \Delta_K$ (Number of forward path is 2 and so K = 2)

$$= \frac{1}{\Delta}[P_1\Delta_1 + P_2\Delta_2] = \frac{1}{\Delta}[G_1G_2G_3G_4 \times 1 + G_1G_2G_6(1-G_5)]$$

$$= \frac{G_1G_2G_3G_4 + G_1G_2G_6 - G_1G_2G_5G_6}{1 + G_2G_3H_1 + H_2G_2 + G_2G_3G_4H_3 - G_5 + G_2G_6H_3 - G_2H_2G_5 - G_2G_5G_6H_3}$$

EXAMPLE 1.28

Find the overall gain C(s)/R(s) for the signal flow graph shown in fig 1.

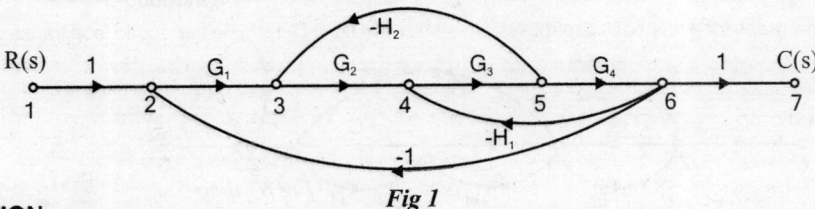

Fig 1

SOLUTION

I. Forward Path Gains

There is only one forward path. $\therefore$ K = 1.

Let the forward path gain be P_1.

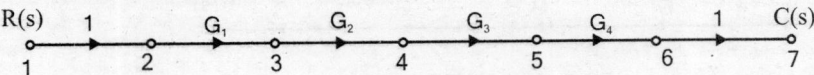

Fig 2 : Forward path-1

Gain of forward path-1, $P_1 = G_1G_2G_3G_4$

II. Individual Loop Gain

There are three individual loops. Let the loop gains be P_{11}, P_{21}, P_{31}.

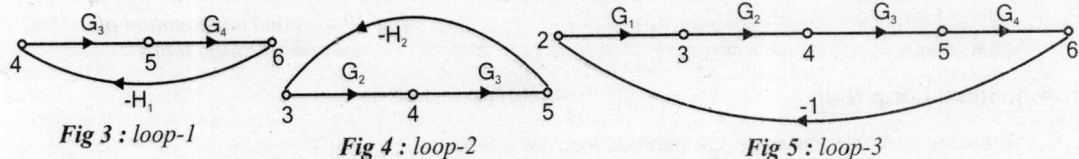

Fig 3 : loop-1   **Fig 4 : loop-2**   **Fig 5 : loop-3**

Loop gain of individual loop-1, $P_{11} = -G_3G_4H_1$

Loop gain of individual loop-2, $P_{21} = -G_2G_3H_2$

Loop gain of individual loop-3, $P_{31} = -G_1G_2G_3G_4$

III. Gain Products of Two Non-touching Loops

There are no possible combinations of two non-touching loops, three non-touching loops, etc.

IV. Calculation of Δ and Δ_K

$$\Delta = 1 - (P_{11} + P_{21} + P_{31})$$

$$= 1 - (-G_3G_4H_1 - G_2G_3H_2 - G_1G_2G_3G_4)$$

$$= 1 + G_3G_4H_1 + G_2G_3H_2 + G_1G_2G_3G_4$$

Since no part of the graph is non-touching with forward path-1, $\Delta_1 = 1$.

V. Transfer Function, T

By Mason's gain formula the transfer function, T is given by,

$$T = \frac{C(s)}{R(s)} = \frac{1}{\Delta} \sum_{K} P_K \Delta_K = \frac{1}{\Delta} P_1 \Delta_1 \text{ (Number of forward path is 1 and so K = 1)}$$

$$= \frac{G_1 G_2 G_3 G_4}{1 + G_3 G_4 H_1 + G_2 G_3 H_2 + G_1 G_2 G_3 G_4}$$

EXAMPLE 1.29

The signal flow graph for a feedback control system is shown in fig 1. Determine the closed loop transfer function C(s)/R(s).

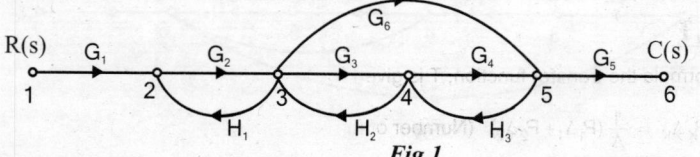

Fig 1

SOLUTION

I. Forward Path Gains

There are two forward paths. ∴ K = 2.
Let forward path gains be P_1 and P_2.

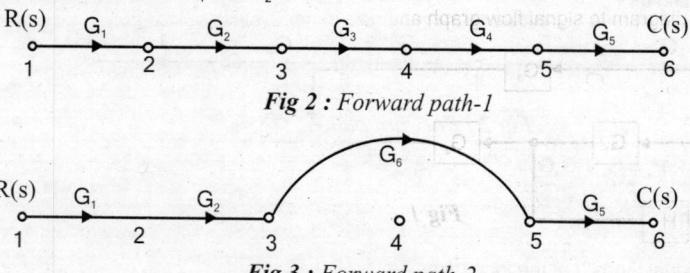

Fig 2 : *Forward path-1*

Fig 3 : *Forward path-2*

Gain of forward path-1, $P_1 = G_1 G_2 G_3 G_4 G_5$

Gain of forward path-2, $P_2 = G_1 G_2 G_6 G_5$

II. Individual Loop Gain

There are four individual loops. Let individual loop gains be P_{11}, P_{21}, P_{31} and P_{41}.

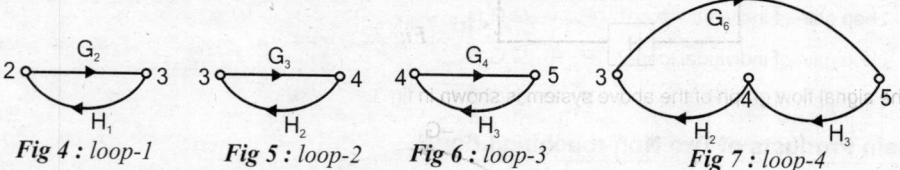

Fig 4 : *loop-1*   **Fig 5 :** *loop-2*   **Fig 6 :** *loop-3*   **Fig 7 :** *loop-4*

Loop gain of individual loop-1, $P_{11} = G_2 H_1$

Loop gain of individual loop-2, $P_{21} = G_3 H_2$

Loop gain of individual loop-3, $P_{31} = G_4 H_3$

Loop gain of individual loop-4, $P_{41} = G_6 H_2 H_3$

III. Gain Products of Two Non-touching Loops

There is only one combination of two non-touching loops. Let the gain products of two non-touching loops be P_{12}.

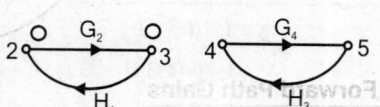

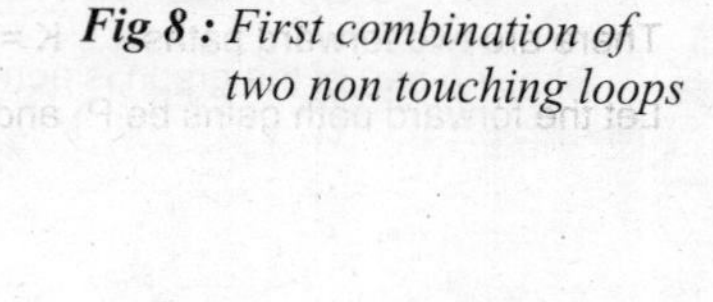

Fig 8 : *First combination of two non touching loops*

Gain product of first combination $\left.\begin{matrix}\end{matrix}\right\}$ $P_{12} = (G_2 H_1)(G_4 H_3)$
of two non - touching loops

IV. Calculation of Δ and Δ_K

$$\Delta = 1 - (P_{11} + P_{21} + P_{31} + P_{41}) + P_{12}$$

$$= 1 - (G_2 H_1 + G_3 H_2 + G_4 H_3 + G_6 H_2 H_3) + G_2 G_4 H_1 H_3$$

$$= 1 - G_2 H_1 - G_3 H_2 - G_4 H_3 - G_6 H_2 H_3 + G_2 G_4 H_1 H_3$$

Since there is no part of graph which is non-touching with forward path-1 and 2, $\Delta_1 = \Delta_2 = 1$

V. Transfer Function, T

By Mason's gain formula the transfer function, T is given by,

$$T = \frac{1}{\Delta} \sum_K P_K \Delta_K = \frac{1}{\Delta}(P_1 \Delta_1 + P_2 \Delta_2) \quad \text{(Number of forward paths is two and so K = 2)}$$

$$= \frac{G_1 G_2 G_3 G_4 G_5 + G_1 G_2 G_5 G_6}{1 - G_2 H_1 - G_3 H_2 - G_4 H_3 - G_6 H_2 H_3 + G_2 G_4 H_1 H_3}$$

EXAMPLE 1.30

Convert the given block diagram to signal flow graph and determine C(s)/R(s).

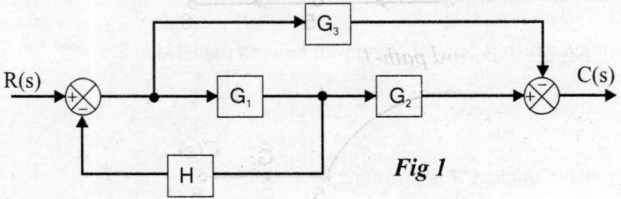

Fig 1

SOLUTION

The nodes are assigned at input, output, at every summing point & branch point as shown in fig 2.

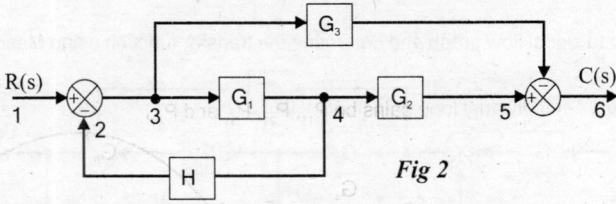

Fig 2

The signal flow graph of the above system is shown in fig 3.

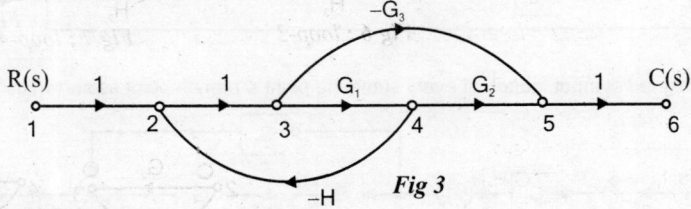

Fig 3

I. Forward Path Gains

There are two forward paths. $\therefore$ K = 2

Let the forward path gains be P_1 and P_2.

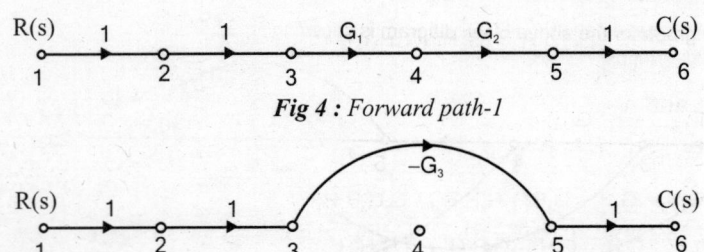

Fig 4 : Forward path-1

Fig 5 : Forward path-2

Gain of forward path-1, $P_1 = G_1 G_2$

Gain of forward path-2, $P_2 = -G_3$

II. Individual Loop Gain

There is only one individual loop. Let the individual loop gain be P_{11}.

Loop gain of individual loop-I, $P_{11} = -G_1 H$.

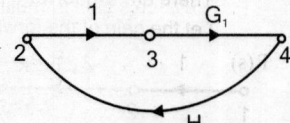

Fig 6 : loop-1

III. Gain Products of Two Non-touching Loops

There are no combinations of non-touching loops.

IV. Calculation of Δ and Δ_K

$\Delta = 1 - [P_{11}] = 1 + G_1 H$

Since there are no part of the graph which is non-touching with forward path-1 and 2,

$\Delta_1 = \Delta_2 = 1$

V. Transfer Function, T

By Mason's gain formula the transfer function, T is given by,

$$T = \frac{1}{\Delta}\sum_K P_K \Delta_K = \frac{1}{\Delta}[P_1 \Delta_1 + P_2 \Delta_2] = \frac{G_1 G_2 - G_3}{1 + G_1 H}$$

EXAMPLE 1.31

Convert the block diagram to signal flow graph and determine the transfer function using Mason's gain formula.

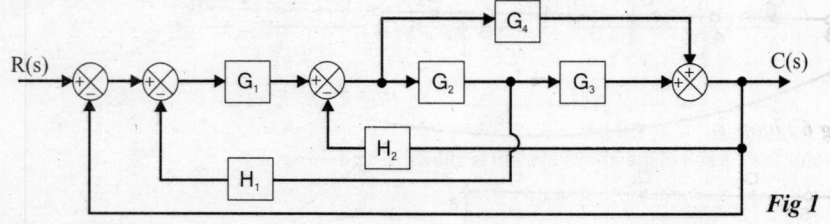

Fig 1

SOLUTION

The nodes are assigned at input, ouput, at every summing point & branch point as shown in fig 2.

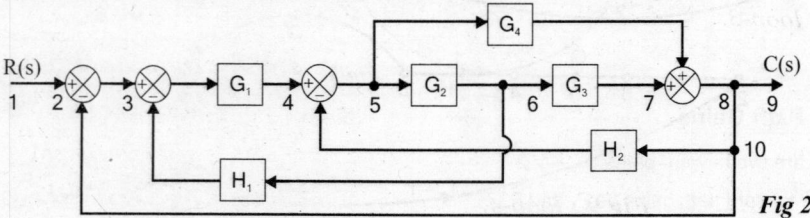

Fig 2

The signal flow graph for the above block diagram is shown in fig 3.

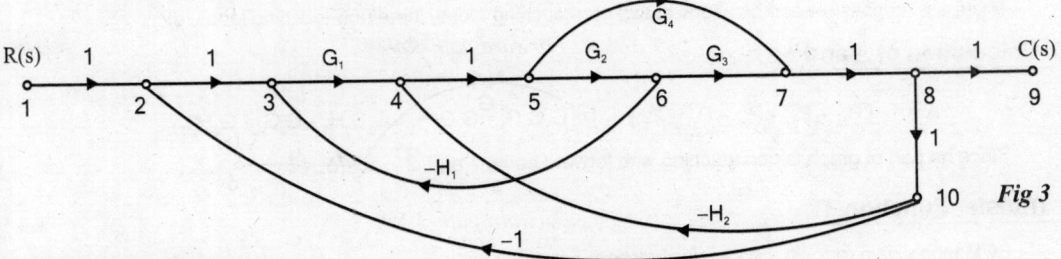

Fig 3

I. Forward Path Gains

There are two forward paths.   ∴ K=2.

Let the gain of the forward paths be P_1 and P_2.

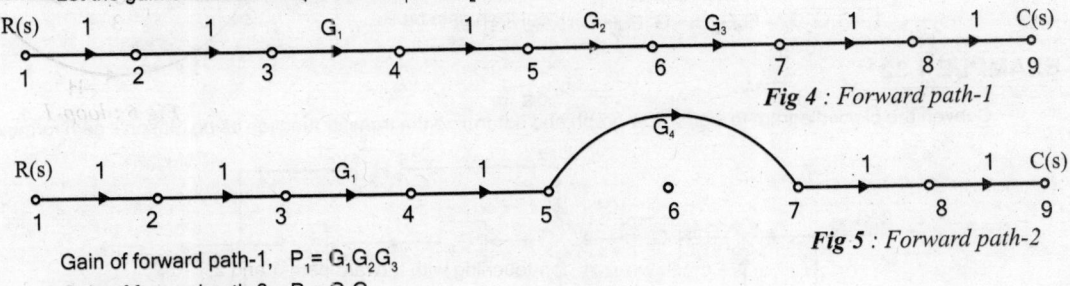

Fig 4 : Forward path-1

Fig 5 : Forward path-2

Gain of forward path-1, $P_1 = G_1 G_2 G_3$

Gain of forward path-2, $P_2 = G_1 G_4$

II. Individual Loop Gain

There are five individual loops. Let the individual loop gain be P_{11}, P_{21}, P_{31}, P_{41} and P_{51}.

Loop gain of individual loop-1, $P_{11} = -G_1 G_2 G_3$

Loop gain of individual loop-2, $P_{21} = -G_2 G_1 H_1$

Loop gain of individual loop-3, $P_{31} = -G_2 G_3 H_2$

Loop gain of individual loop-4, $P_{41} = -G_1 G_4$

Loop gain of individual loop-5, $P_{51} = -G_4 H_2$

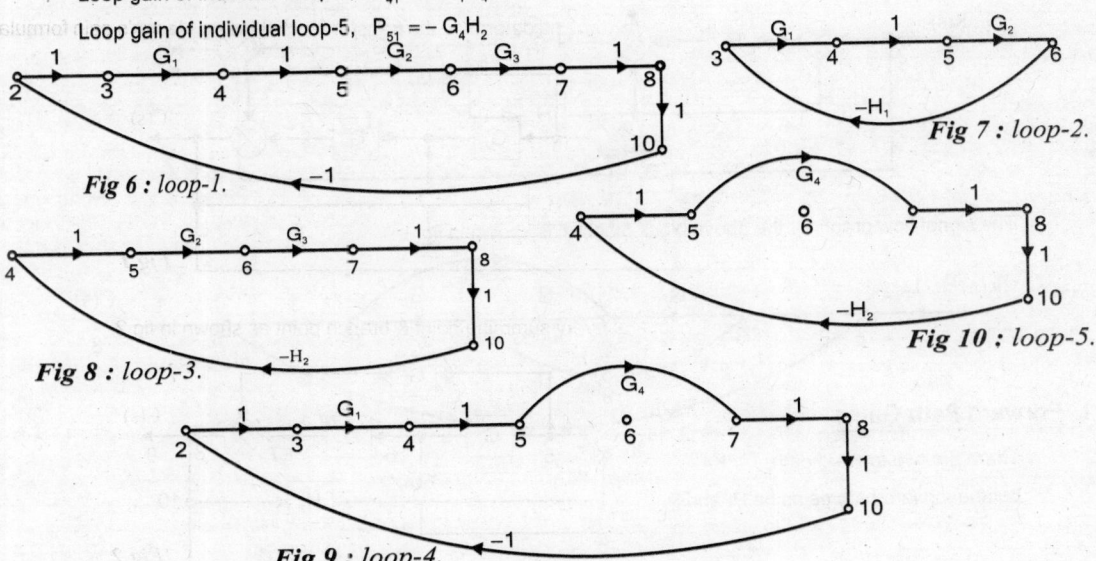

Fig 6 : loop-1.

Fig 7 : loop-2.

Fig 8 : loop-3.

Fig 10 : loop-5.

Fig 9 : loop-4.

III. Gain Products of Two Non-touching Loops

There are no possible combinations of two non-touching loops, three non-touching loops, etc.,.

IV. Calculation of Δ and Δ_K

$$\Delta = 1 - [P_{11} + P_{21} + P_{31} + P_{41} + P_{51}] = 1 + G_1 G_2 G_3 + G_1 G_2 H_1 + G_2 G_3 H_2 + G_1 G_4 + G_4 H_2$$

Since no part of graph is non touching with forward paths-1 and 2, $\Delta_1 = \Delta_2 = 1$.

V. Transfer Function, T

By Mason's gain formula the transfer function, T is given by,

$$T = \frac{1}{\Delta} \sum_K P_K \Delta_K = \frac{1}{\Delta} [P_1 \Delta_1 + P_2 \Delta_2]$$

$$= \frac{G_1 G_2 G_3 + G_1 G_4}{1 + G_1 G_2 G_3 + G_1 G_2 H_1 + G_2 G_3 H_2 + G_1 G_4 + G_4 H_2}$$

EXAMPLE 1.32

Convert the block diagram to signal flow graph and determine the transfer function using Mason's gain formula.

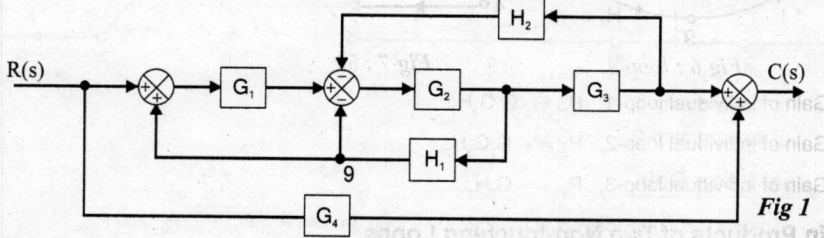

Fig 1

SOLUTION

The nodes are assigned at input, output, at every summing point & branch point as shown in fig 2.

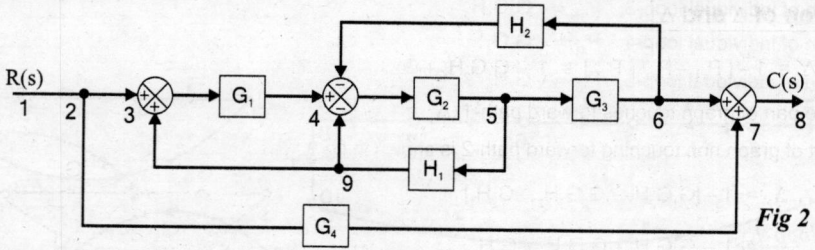

Fig 2

The signal flow graph for the above block diagram is shown in fig 3.

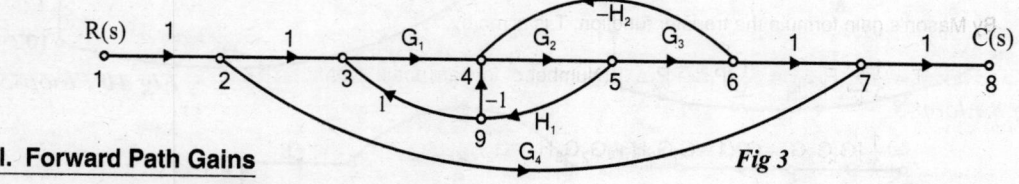

Fig 3

I. Forward Path Gains

There are two forward path, $\therefore$ K=2.

Let the forward path gains be P_1 and P_2.

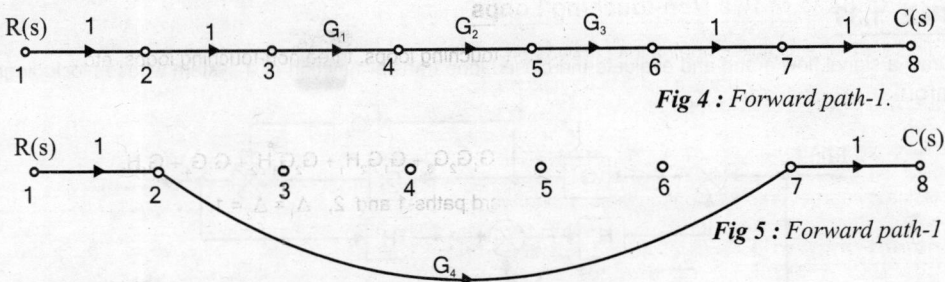

Fig 4 : *Forward path-1.*

Fig 5 : *Forward path-1*

Gain of forward path-1, $P_1 = G_1 G_2 G_3$

Gain of forward path-2, $P_2 = G_4$

II. Individual Loop Gain

There are three individual loops with gains P_{11}, P_{21} and P_{31}.

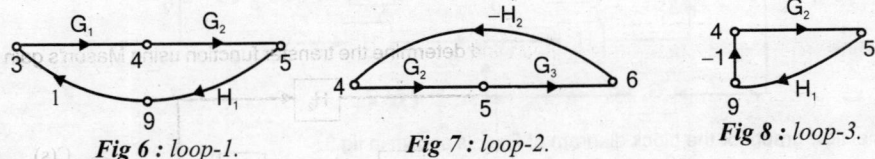

Fig 6 : *loop-1.*   **Fig 7 :** *loop-2.*   **Fig 8 :** *loop-3.*

Gain of individual loop-1, $P_{11} = G_1 G_2 H_1$

Gain of individual loop-2, $P_{21} = -G_2 G_3 H_2$

Gain of individual loop-3, $P_{31} = -G_2 H_1$

III. Gain Products of Two Non-touching Loops

There are no possible combinations of two-non touching loops, three non-touching loops, etc.,.

IV. Calculation of Δ and Δ_K

$$\Delta = 1 - [P_{11} + P_{21} + P_{31}] = 1 - G_1 G_2 H_1 + G_2 G_3 H_2 + G_2 H_1$$

Since no part of graph touches forward path-1, $\Delta_1 = 1$.

The part of graph non touching forward path-2 is shown in fig 9.

$$\therefore \Delta_2 = 1 - [G_1 G_2 H_1 - G_2 G_3 H_2 - G_2 H_1]$$

$$= 1 - G_1 G_2 H_1 + G_2 G_3 H_2 + G_2 H_1$$

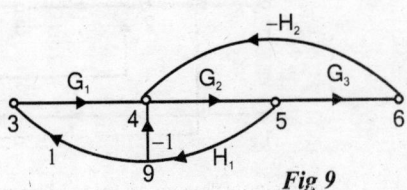

Fig 9

V. Transfer Function, T

By Mason's gain formula the transfer function, T is given by,

$$T = \frac{1}{\Delta} \sum_K P_K \Delta_K = \frac{1}{\Delta} [P_1 \Delta_1 + P_2 \Delta_2] \quad \text{(Number of forward paths is 2 and so K = 2)}$$

$$= \frac{1}{\Delta} [G_1 G_2 G_3 + G_4 (1 - G_1 G_2 H_1 + G_2 G_3 H_2 + G_2 H_1)]$$

$$= \frac{1}{\Delta} [G_1 G_2 G_3 + G_4 - G_4 G_1 G_2 H_1 + G_2 G_3 G_4 H_2 + G_2 G_4 H_1]$$

$$= \frac{G_1 G_2 G_3 + G_4 - G_1 G_2 G_4 H_1 + G_2 G_3 G_4 H_2 + G_2 G_4 H_1}{1 - G_1 G_2 H_1 + G_2 G_3 H_2 + G_2 H_1}$$

EXAMPLE 1.33

Draw a signal flow graph and evaluate the closed loop transfer function of a system whose block diagram is shown in fig 1.

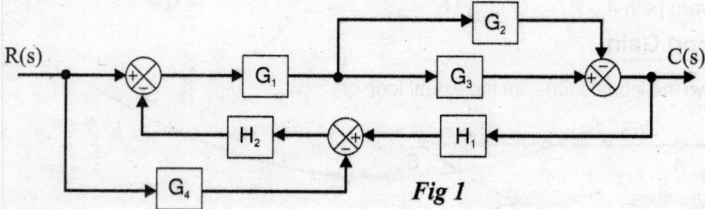

Fig 1

SOLUTION

The nodes are assigned at input, output, at every summing point & branch point as shown in fig 2.

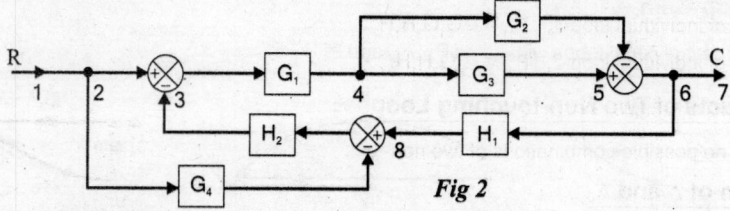

Fig 2

The signal flow graph for the block diagram of fig 2, is shown in fig 3.

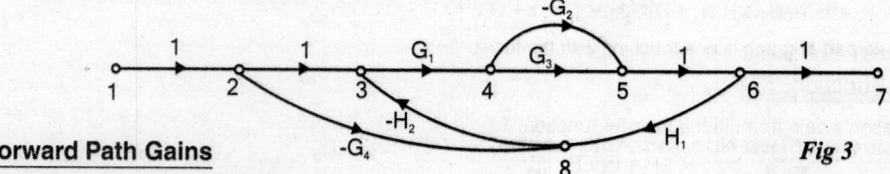

Fig 3

I. Forward Path Gains

There are four forward paths, $\therefore$ K = 4

Let the forward path gains be P_1, P_2, P_3 and P_4.

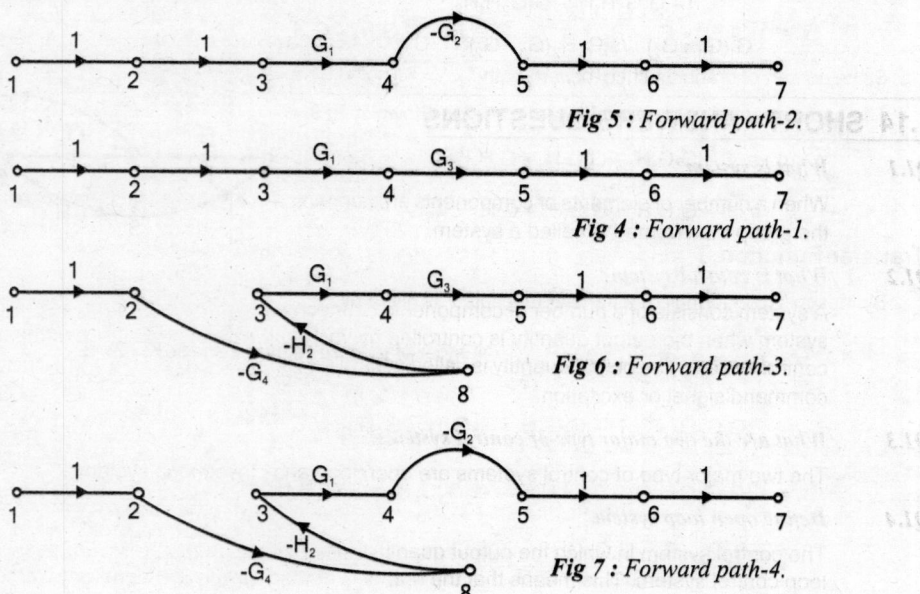

Fig 5 : Forward path-2.

Fig 4 : Forward path-1.

Fig 6 : Forward path-3.

Fig 7 : Forward path-4.

Gain of forward path-1, $P_1 = G_1G_3$

Gain of forward path-2, $P_2 = -G_1G_2$

Gain of forward path-3, $P_3 = G_1G_3G_4H_2$

Gain of forward path-4, $P_4 = -G_1G_2G_4H_2$

II. Individual Loop Gain

There are two individual loops, let individual loop gains be P_{11} and P_{21}.

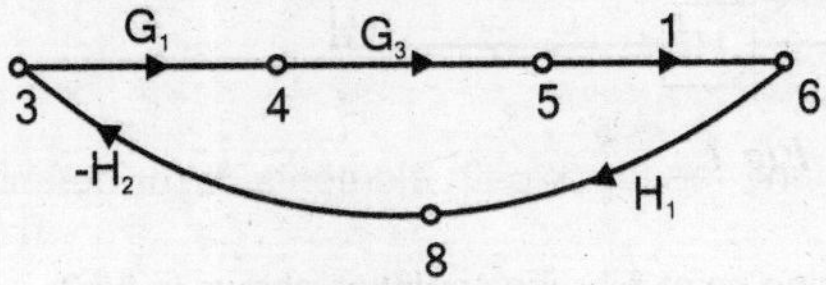

Fig 8 : *loop-1*

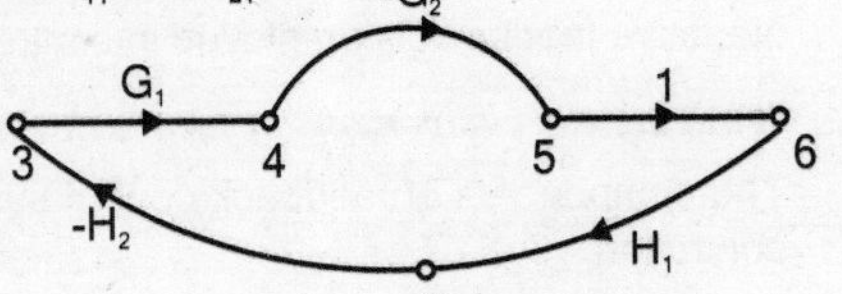

Fig 9 : *loop-2*

Loop gain of individual loop-1, $P_{11} = -G_1G_3H_1H_2$

Loop gain of individual loop-2, $P_{21} = G_1G_2H_1H_2$

III. Gain Products of Two Non-touching Loops

There are no possible combinations of two non-touching loops, three non-touching loops, etc.,.

IV. Calculation of Δ and Δ_K

$$\Delta = 1- [\text{ sum of individual loop gain }] = 1 - (P_{11} + P_{21})$$

$$= 1- [-G_1G_3H_1H_2 + G_1G_2H_1H_2] = 1 + G_1G_3H_1H_2 - G_1G_2H_1H_2$$

Since no part of graph is non touching with the forward paths, $\Delta_1 = \Delta_2 = \Delta_3 = \Delta_4 = 1$.

V. Transfer Function, T

By Mason's gain formula the transfer function, T is given by,

$$T = \frac{1}{\Delta} \sum_K P_K \Delta_K = \frac{P_1 + P_2 + P_3 + P_4}{\Delta} \text{ (Number of forward paths is 4 and so K = 4)}$$

$$= \frac{G_1G_3 - G_1G_2 - G_1G_3G_4H_2 - G_1G_2G_4H_2}{1 + G_1G_3H_1H_2 - G_1G_2H_1H_2}$$

$$= \frac{G_1(G_3 - G_2) + G_1G_4H_2(G_3 - G_2)}{1 + G_1H_1H_2(G_3 - G_2)} = \frac{G_1(G_3 - G_2)(1 + G_4H_2)}{1 + G_1H_1H_2(G_3 - G_2)}$$

1.14 SHORT - ANSWERS QUESTIONS

Q1.1   *What is system?*

When a number of elements or components are connected in a sequence to perform a specific function, the group thus formed is called a system.

Q1.2   *What is control system?*

A system consists of a number of components connected together to perform a specific function. In a system when the output quantity is controlled by varying the input quantity, then the system is called control system. The output quantity is called controlled variable or response and input quantity is called command signal or excitation.

Q1.3   *What are the two major type of control systems?*

The two major type of control systems are open loop and closed loop systems.

Q1.4   *Define open loop system.*

The control system in which the output quantity has no effect upon the input quantity are called open loop control system. This means that the output is not fedback to the input for correction.

Q1.5   *Define closed loop system.*

The control systems in which the output has an effect upon the input quantity in order to maintain the desired output value are called closed loop control systems.

Q1.6   *What is feedback? What type of feedback is employed in control system?*

The feedback is a control action in which the output is sampled and a proportional signal is given to input for automatic correction of any changes in desired output.

Negative feedback is employed in control system.

Q1.7   *What are the components of feedback control system?*

The components of feedback control system are plant, feedback path elements, error detector and controller.

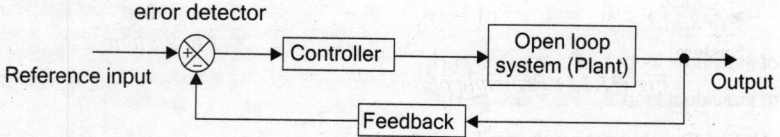

Q1.8   *Why negative feedback is invariably preferred in a closed loop system?*

The negative feedback results in better stability in steady state and rejects any disturbance signals. It also has low sensitivity to parameter variations. Hence negative feedback is preferred in closed loop systems.

Q1.9   *What are the characteristics of negative feedback?*

The characteristics of negative feedback are as follows :

    (i)   accuracy in tracking steady state value.

    (ii)   rejection of disturbance signals.

    (iii)   low sensitivity to parameter variations.

    (iv)   reduction in gain at the expense of better stability.

Q1.10.   *What is the effect of positive feedback on stability?*

The positive feedback increases the error signal and drives the output to instability. But sometimes the positive feedback is used in minor loops in control systems to amplify certain internal signals or parameters.

Q1.11.   *Distinguish between open loop and closed loop system.*

Open loop	Closed loop
1. Inaccurate & unreliable.	1. Accurate & reliable.
2. Simple and economical.	2. Complex and costly.
3. Changes in output due to external disturbances are not corrected automatically.	3. Changes in output due to external disturbances are corrected automatically.
4. They are generally stable.	4. Great efforts are needed to design a stable system.

Q1.12   *What is servomechanism?*

The servomechanism is a feedback control system in which the output is mechanical position (or time derivatives of position e.g. velocity and acceleration).

Q1.13   *State the principle of homogenity (or) State the principle of superposition.*

The principle of superposition and homogenity states that if the system has responses $c_1(t)$ and $c_2(t)$ for the inputs $r_1(t)$ and $r_2(t)$ respectively then the system response to the linear combination of these input $a_1 r_1(t) + a_2 r_2(t)$ is given by linear combination of the individual outputs $a_1 c_1(t) + a_2 c_2(t)$, where a_1 and a_2 are constants.

Q1.14   *Define linear system.*

A system is said to be linear, if it obeys the principle of superposition and homogenity, which states that the response of a system to a weighed sum of signals is equal to the corresponding weighed sum of the responses of the system to each of the individual input signals. The concept of linear system is diagrammatically shown below.

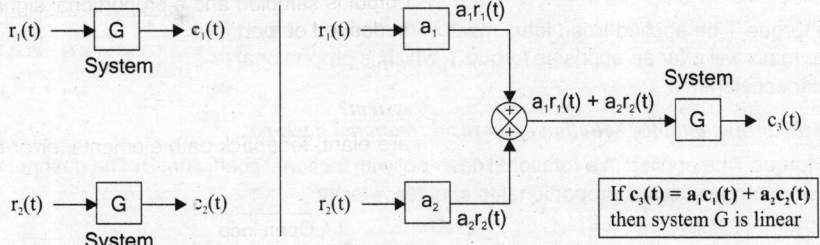

Fig Q1.14 : Principle of linearity and superposition.

Q1.15   *What is time invariant system?*

A system is said to be time invariant if its input-output characteristics do not change with time. A linear time invariant system can be represented by constant coefficient differential equations. (In linear time varying systems the coefficients of the differential equation governing the system are function of time).

Q1.16   *Define transfer function.*

The transfer function of a system is defined as the ratio of Laplace transform of output to Laplace transform of input with zero initial conditions. (It is also defined as the Laplace transform of the impulse response of system with zero initial conditions).

Q1.17   *What are the basic elements used for modelling mechanical translational system?*

The model of mechanical translational system can be obtained by using three basic elements mass, spring and dashpot.

Q1.18   *Write the force balance equation of ideal mass element.*

Let a force f be applied to an ideal mass M. The mass will offer an opposing force, f_m which is proportional to acceleration.

$$\therefore \ f = f_m = M \frac{d^2x}{dt^2}$$

Q1.19   *Write the force balance equation of ideal dashpot.*

Let a force f be applied to an ideal dashpot, with viscous frictional coefficient B. The dashpot will offer an opposing force, f_b which is proportional to velocity.

$$f = f_b = B \frac{dx}{dt}$$

Q1.20   *Write the force balance equation of ideal spring.*

Let a force f be applied to an ideal spring with spring constant K. The spring will offer an opposing force, f_k which is proportional to displacement.

$$f = f_k = Kx$$

$$f = f_k = K(X_1 - X_2)$$

Q1.21   *What are the basic elements used for modelling mechanical rotational system?*

The model of mechanical rotational system can be obtained using three basic elements mass with moment of inertia, J, dash-pot with rotational frictional coefficient, B and torsional spring with stiffness, K.

Q1.22   *Write the torque balance equation of an ideal rotational mass element.*

Let a torque T be applied to an ideal mass with moment of inertia, J. The mass will offer an opposing torque T_j which is proportional to angular acceleration.

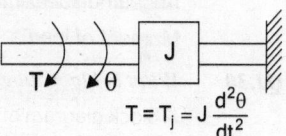

$$T = T_j = J \frac{d^2\theta}{dt^2}$$

Q1.23   *Write the torque balance equation of an ideal rotational dash-pot.*

Let a torque T be applied to a rotational dash-pot with frictional coefficient B. The dashpot will offer an opposing torque which is proportional to angular velocity.

$$T = T_b = B \frac{d\theta}{dt}$$

$$T = T_b = B \frac{d}{dt} (\theta_1 - \theta_2)$$

Q1.24   *Write the torque balance equation of ideal rotational spring.*

Let a torque T be applied to an ideal rotational spring with spring constant K. The spring will offer an opposing torque T_k which is proportional to angular displacement.

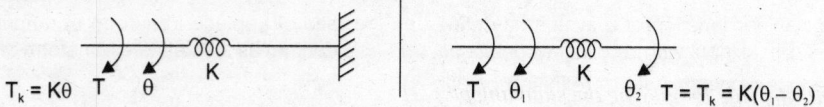

$$T_k = K\theta$$

$$T = T_k = K(\theta_1 - \theta_2)$$

Q1.25   *Name the two types of electrical analogous for mechanical system.*

The two types of analogies for the mechanical system are force-voltage and force-current analogy.

Q1.26   *Write the analogous electrical elements in force-voltage analogy for the elements of mechanical translational system.*

Force, f	→ Voltage, e	Frictional coefficient , B	→ Resistance, R
Velocity, v	→ Current, i	Stiffness, K	→ Inverse of capacitance, 1/C
Displacement, x	→ Charge, q	Newton's second law, $\Sigma f = 0$	→ Kirchoff's voltage law, $\Sigma v = 0$
Mass, M	→ Inductance, L		

Q1.27   *Write the analogous electrical elements in force-current analogy for the elements of mechanical translational system.*

Force, f	→ Current, i	Frictional coefficient, B	→ Conductance, G =1/ R
Velocity, v	→ Voltage, v	Stiffness, K	→ Inverse of Inductance, 1/L
Displacement, x	→ Flux, ϕ	Newton's second law, $\Sigma f = 0$	→ Kirchoff's current law, $\Sigma i = 0$
Mass, M	→ Capacitance, C		

Q1.28   *Write the analogous electrical elements in torque-voltage analogy for the elements of mechanical rotational system.*

Torque, T	→Voltage, e	Stiffness of spring, K	→ Inverse of capacitance, 1/C
Angular velocity, ω	→Current, i	Frictional coefficient, B	→ Resistance, R
Moment of inertia, J	→Inductance, L	Newton's second law, $\Sigma T = 0$	→ Kirchoff's voltage law, $\Sigma v = 0$.
Angular displacement, θ	→ Charge, q		

Q1.29 *Write the analogous electrical elements in torque-current analogy for the elements of mechanical rotational system.*

Torque, T   → Current, i | Frictional coefficient , B   → Conductance, G = 1/R

Angular velocity, ω   → Voltage, v | Stiffness of spring, K   → Inverse of inductance, 1/L

Angular displacement, θ   → Flux, φ | Newton's second law, ΣT = 0 → Kirchoff 's current law, Σi = 0

Moment of inertia, J   → Capacitance, C

Q1.30 *What is block diagram? What are the basic components of block diagram?*

A block diagram of a system is a pictorial representation of the functions performed by each component of the system and shows the flow of signals. The basic elements of block diagram are block, branch point and summing point.

Q1.31 *What is the basis for framing the rules of block diagram reduction technique?*

The rules for block diagram reduction technique are framed such that any modification made on the diagram does not alter the input output relation.

Q1.32 *Write the rule for eliminating negative feedback loop.*

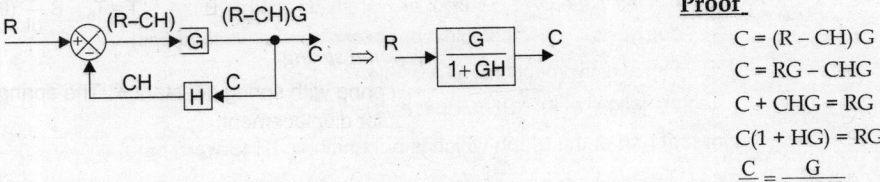

Proof

$C = (R - CH)\, G$

$C = RG - CHG$

$C + CHG = RG$

$C(1 + HG) = RG$

$\dfrac{C}{R} = \dfrac{G}{1 + GH}$

Q1.33 *Write the rule for moving the summing point ahead of a block.*

$AG + BG = (A+B)G$

Q1.34 *What is a signal flow graph?*

A signal flow graph is a diagram that represents a set of simultaneous linear algebraic equations. By taking Laplace transform, the time domain differential equations governing a control system can be transferred to a set of algebraic equations in s-domain. The signal flow graph of the system can be constructed using these equations.

Q1.35 *What is transmittance?*

The transmittance is the gain acquired by the signal when it travels from one node to another node in signal flow graph.

Q1.36 *What is sink and source?*

Source is the input node in the signal flow graph and it has only outgoing branches. Sink is a output node in the signal flow graph and it has only incoming branches.

Q1.37 *Define non-touching loop.*

The loops are said to be non-touching if they do not have common nodes.

Q1.38 *What are the basic properties of signal flow graph?*

The basic properties of signal flow graph are,

(i)   Signal flow graph is applicable to linear systems.

(ii)   It consists of nodes and branches. A node is a point representing a variable or signal. A branch indicates functional dependence of one signal on the other.

(iii)   A node adds the signals of all incoming branches and transmits this sum to all outgoing branches.

(iv)   Signals travel along branches only in the marked direction and when it travels it gets multiplied by the gain or transmittance of the branch.

(v)   The algebraic equations must be in the form of cause and effect relationship.

Q1.39   *Write the Mason's gain formula.*

Mason's gain formula states that the overall gain of the system [transfer function] as follows,

Overall gain, $T = \dfrac{1}{\Delta} \displaystyle\sum_{K} P_K \Delta_K$

T = $T(s)$ = Transfer function of the system

K = Number of forward paths in the signal flow graph

P_K = Forward path gain of Kth forward path

$$\Delta = 1 - \left[\begin{array}{c}\text{sum of individual}\\ \text{loop gains}\end{array}\right] + \left[\begin{array}{c}\text{sum of gain products of all possible}\\ \text{combinations of two non – touching loops}\end{array}\right]$$
$$- \left[\begin{array}{c}\text{sum of gain products of all possible}\\ \text{combinations of three non – touching loops}\end{array}\right] + \ldots\ldots$$

$\Delta_K = \Delta$ for that part of the graph which is not touching K^{th} forward path

Q1.40   *For the given signal flow graph, identify the number of forward path and number of individual loop.*

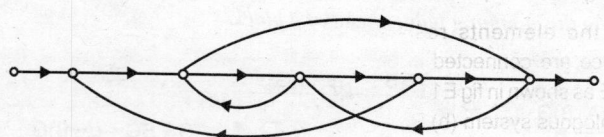

Number of forward paths = 2
Number of individual loops = 4

1.15 EXERCISES

E1.1   For the mechanical system shown in fig E1.1 derive the transfer function. Also draw the force-voltage and force-current analogous circuits.

E1.2   For the mechanical system shown in fig E1.2 draw the force-voltage and force-current analogous circuits.

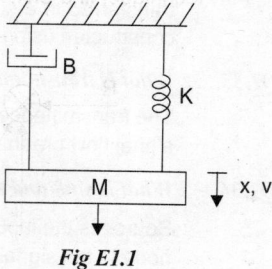

Fig E1.1

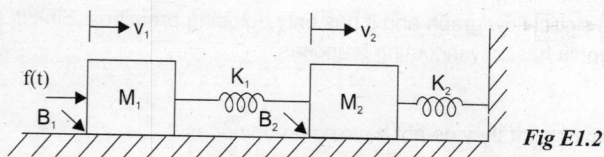

Fig E1.2

E1.3   Write the differential equations governing the mechanical system shown in fig E1.3(a) & (b). Also draw the force-voltage and force-current analogous circuit.

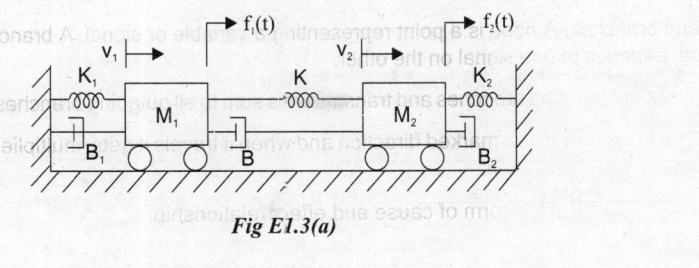

Fig E1.3(a)

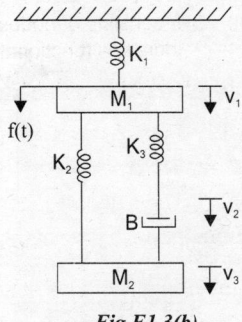

Fig E1.3(b)

E1.4   Consider the mechanical translational system shown in fig E1.4, Draw(a) force-voltage and (b) force-current analogous circuits.

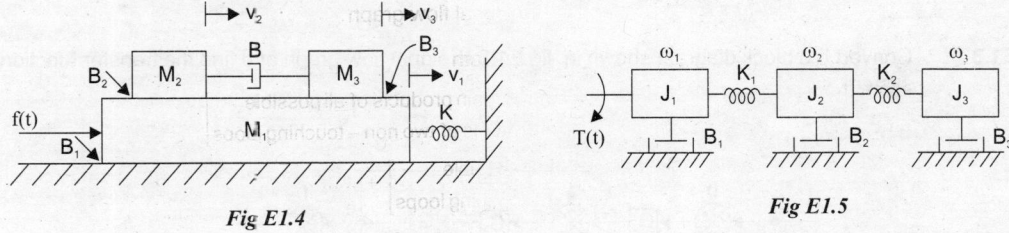

Fig E1.4   *Fig E1.5*

E1.5   Write the differential equations governing the rotational mechanical system shown in fig E1.5. Also draw the torque-voltage and torque-current analogous circuits.

E1.6   In an electrical circuit the elements resistance, capacitance and inductance are connected in parallel across the voltage source E as shown in fig E1.6, Draw(a) Translation mechnical analogous system (b) Rotational mechanical analogous system.

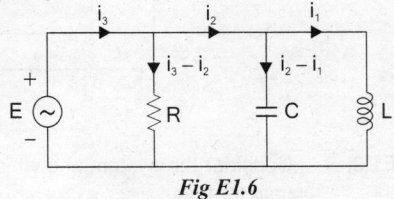

Fig E1.6

E1.7   Consider the block diagram shown in fig E.1.7(a), (b) (c) & (d). Using the block diagram reduction technique, find C/R.

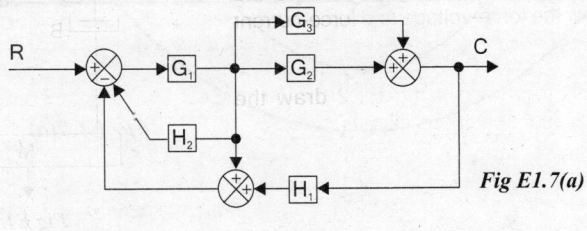

Fig E1.7(a)

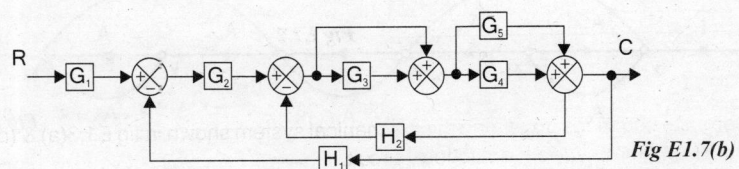

Fig E1.7(b)

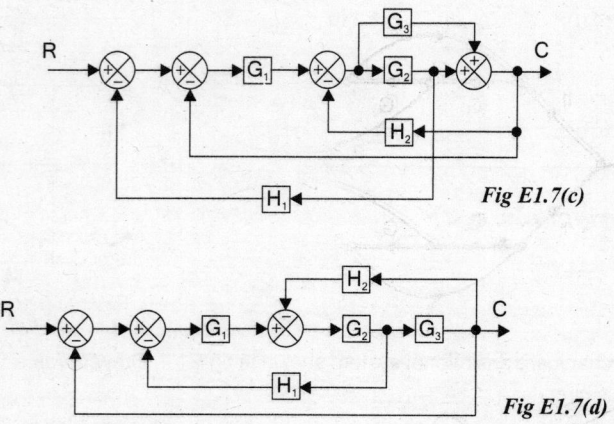

Fig E1.7(c)

Fig E1.7(d)

E1.8   Convert the block diagram shown in fig E1.8 to signal flow graph and find the transfer function of the system.

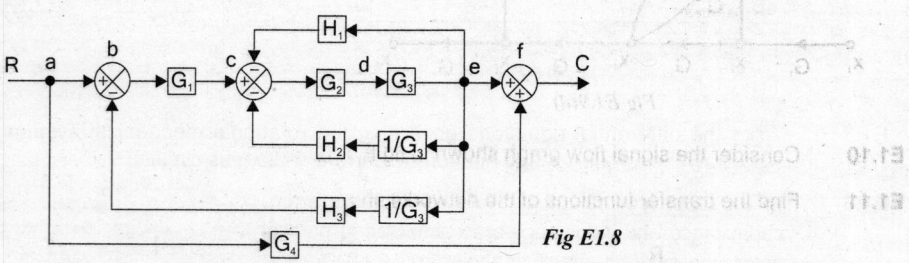

Fig E1.8

E1.9   Consider the system shown in fig E1.9(a), (b), (c) & (d). obtain the transfer function using Mason's gain formula.

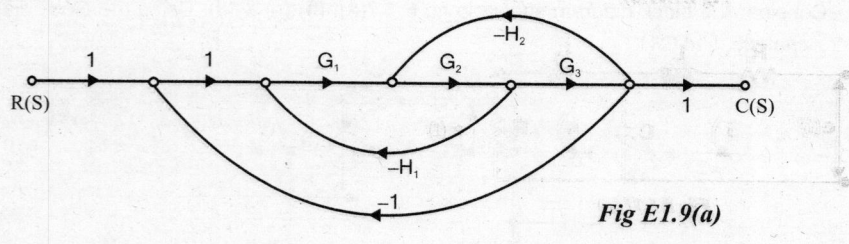

Fig E1.9(a)

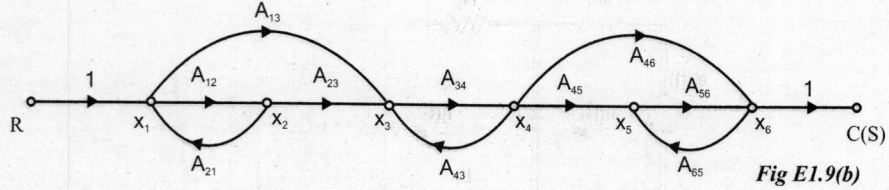

Fig E1.9(b)

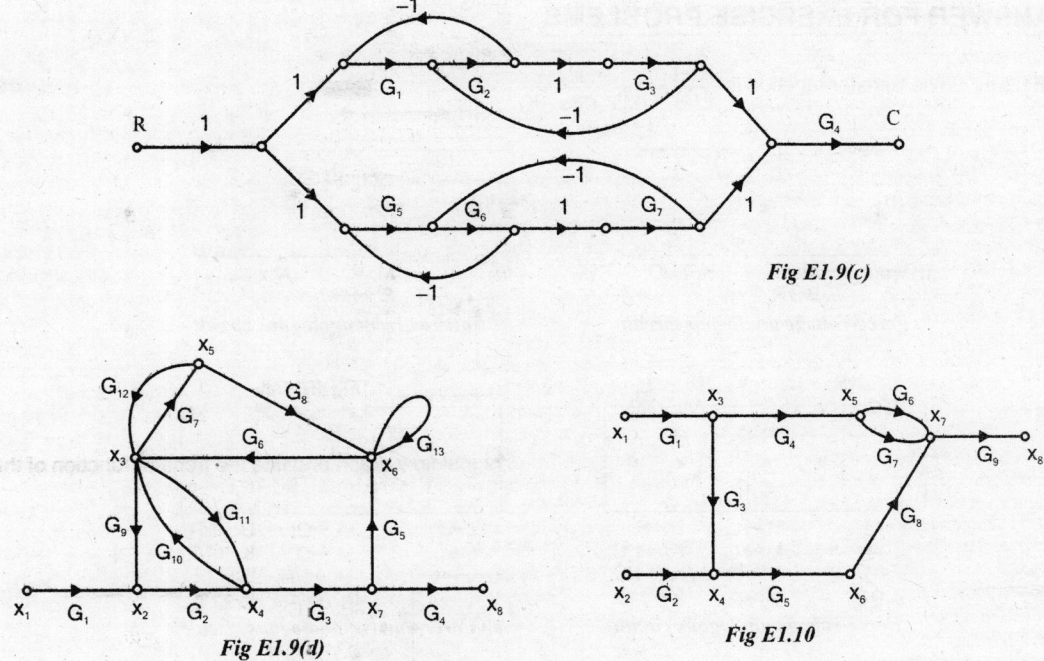

Fig E1.9(c)

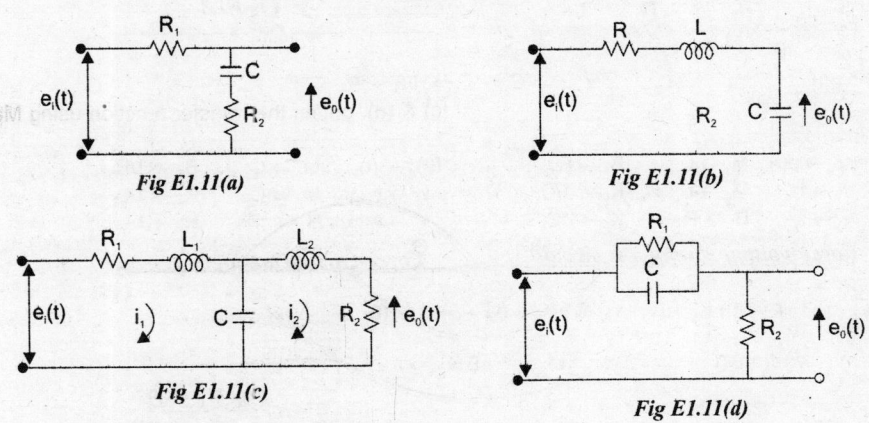

Fig E1.9(d)

Fig E1.10

E1.10   Consider the signal flow graph shown in fig E.1.10 obtain

E1.11   Find the transfer functions of the networks shown in fig E1.11(a), (b), (c) & (d).

Fig E1.11(a)

Fig E1.11(b)

Fig E1.11(c)

Fig E1.11(d)

E1.12   Find the transfer function of the circuit shown in fig E1.12.

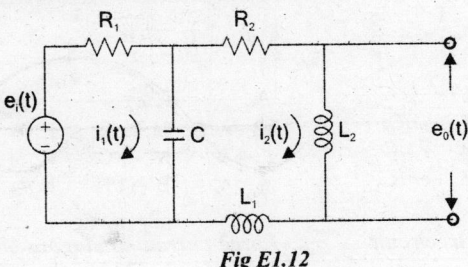

Fig E1.12

ANSWER FOR EXERCISE PROBLEMS

E1.1   The transfer function is $\dfrac{X(s)}{F(s)} = \dfrac{1}{(Ms^2 + Bs + K)}$

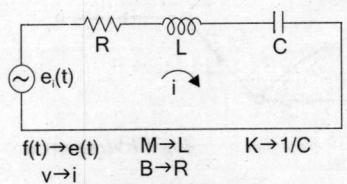

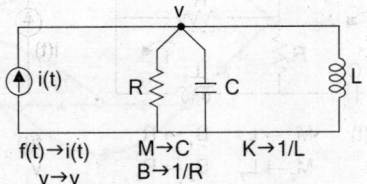

$f(t) \rightarrow e_i(t)$ $M \rightarrow L$ $K \rightarrow 1/C$
$v \rightarrow i$ $B \rightarrow R$

Force-voltage analogous circuit

$f(t) \rightarrow i(t)$ $M \rightarrow C$ $K \rightarrow 1/L$
$v \rightarrow v$ $B \rightarrow 1/R$

Force-current analogous circuit

E1.2

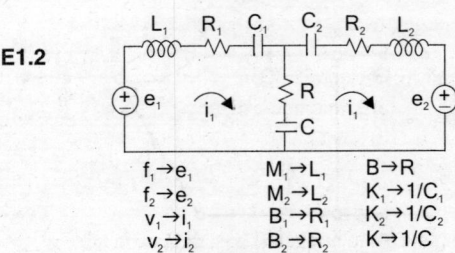

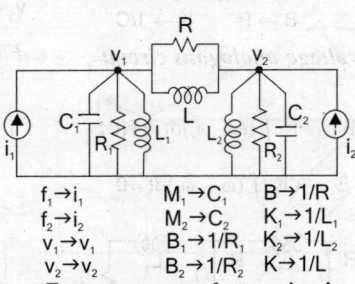

$f_1 \rightarrow e_1$ $M_1 \rightarrow L_1$ $B \rightarrow R$
$f_2 \rightarrow e_2$ $M_2 \rightarrow L_2$ $K_1 \rightarrow 1/C_1$
$v_1 \rightarrow i_1$ $B_1 \rightarrow R_1$ $K_2 \rightarrow 1/C_2$
$v_2 \rightarrow i_2$ $B_2 \rightarrow R_2$ $K \rightarrow 1/C$

Force-voltage analogous circuit

$f_1 \rightarrow i_1$ $M_1 \rightarrow C_1$ $B \rightarrow 1/R$
$f_2 \rightarrow i_2$ $M_2 \rightarrow C_2$ $K_1 \rightarrow 1/L_1$
$v_1 \rightarrow v_1$ $B_1 \rightarrow 1/R_1$ $K_2 \rightarrow 1/L_2$
$v_2 \rightarrow v_2$ $B_2 \rightarrow 1/R_2$ $K \rightarrow 1/L$

Force-current analogous circuit

E1.3(a) $M_1 \dfrac{dv_1}{dt} + B_1 v_1 + B(v_1 - v_2) + K_1 \int v_1 dt + K \int (v_1 - v_2)\, dt = f_1(t)$

$M_2 \dfrac{dv_2}{dt} + B_2 v_2 + B(v_2 - v_1) + K_2 \int v_2 dt + K \int (v_2 - v_1)\, dt = f_2(t)$

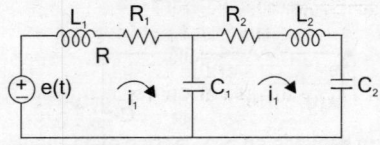

$f(t) \rightarrow e(t)$ $M_1 \rightarrow L_1$ $B_2 \rightarrow R_2$
$v_1 \rightarrow i_1$ $M_2 \rightarrow L_2$ $K_1 \rightarrow 1/C_1$
$v_2 \rightarrow i_2$ $B_1 \rightarrow R_1$ $K_2 \rightarrow 1/C_2$

Force-voltage analogous circuit

$f(t) \rightarrow i(t)$ $M_1 \rightarrow C_1$ $B_2 \rightarrow 1/R_2$
$v_1 \rightarrow v_1$ $M_2 \rightarrow C_2$ $K_1 \rightarrow 1/L_1$
$v_2 \rightarrow v_2$ $B_1 \rightarrow 1/R_1$ $K_2 \rightarrow 1/L_2$

Force-current analogous circuit

E1.3(b) $M_1 \dfrac{dv_1}{dt} + .K_1 \int v_1 dt + K_2 \int (v_1 - v_3)\, dt + K_3 \int (v_1 - v_2)\, dt = f(t)$

$K_3 \int (v_2 - v_1)\, dt + B(v_2 - v_3) = 0$; $M_2 \dfrac{dv_3}{dt} + B(v_3 - v_2) + K_2 \int (v_3 - v_1)\, dt = 0$

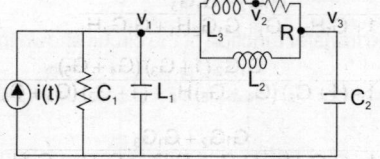

$f(t) \rightarrow e(t)$ $M_1 \rightarrow L_1$ $K_1 \rightarrow 1/C_1$
$v_1 \rightarrow i_1$ $M_2 \rightarrow L_2$ $K_2 \rightarrow 1/C_2$
$v_2 \rightarrow i_2$ $B \rightarrow R$ $K_3 \rightarrow 1/C_3$
$v_3 \rightarrow i_3$

Force-voltage analogous circuit

$f(t) \rightarrow i(t)$ $M_1 \rightarrow C_1$ $K_1 \rightarrow 1/L_1$
$v_1 \rightarrow v_1$ $M_2 \rightarrow C_2$ $K_2 \rightarrow 1/L_2$
$v_2 \rightarrow v_2$ $B \rightarrow 1/R$ $K_3 \rightarrow 1/L_3$
$v_3 \rightarrow v_3$

Force-current analogous circuit

E1.4

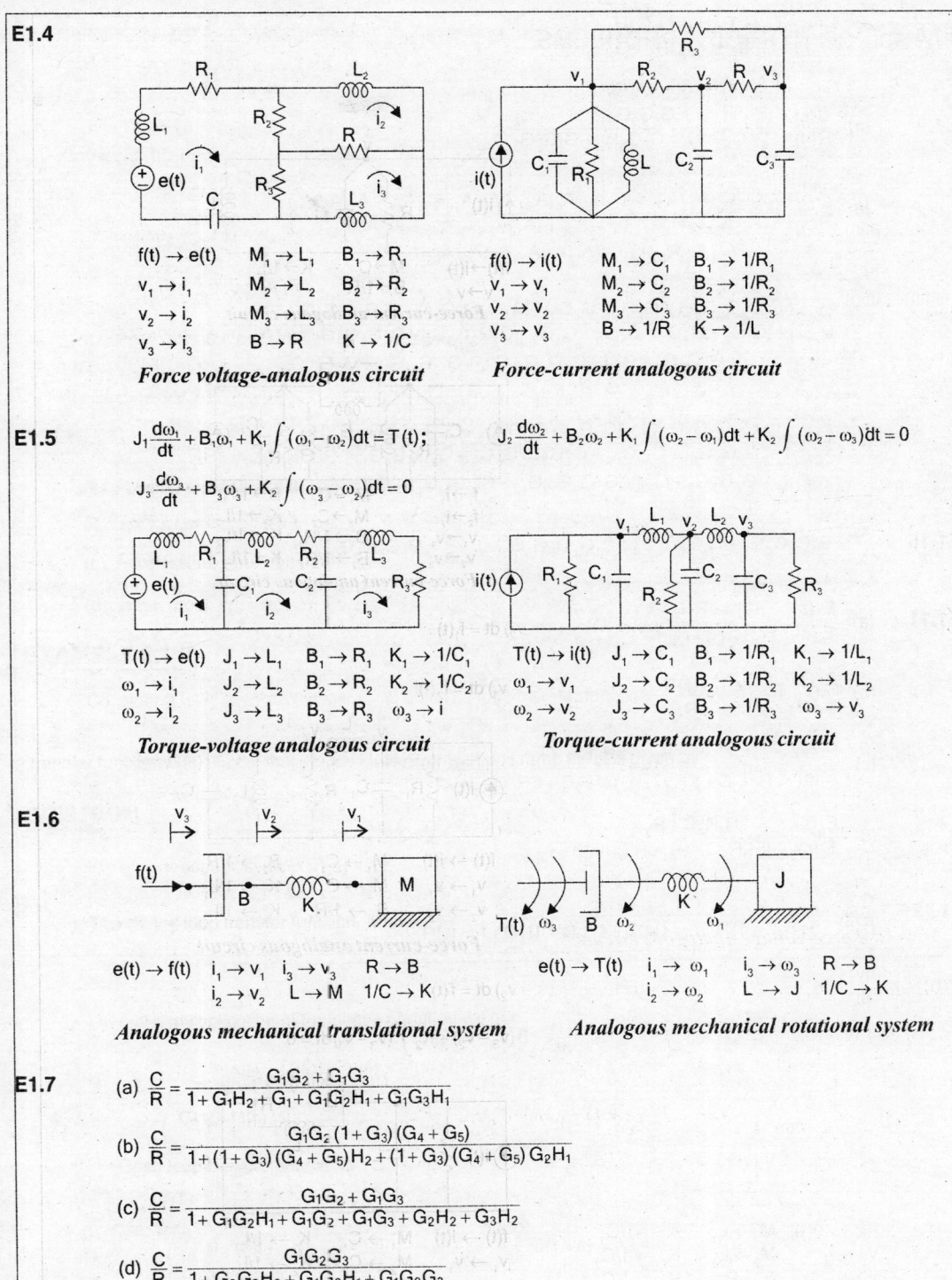

$f(t) \rightarrow e(t)$ $M_1 \rightarrow L_1$ $B_1 \rightarrow R_1$
$v_1 \rightarrow i_1$ $M_2 \rightarrow L_2$ $B_2 \rightarrow R_2$
$v_2 \rightarrow i_2$ $M_3 \rightarrow L_3$ $B_3 \rightarrow R_3$
$v_3 \rightarrow i_3$ $B \rightarrow R$ $K \rightarrow 1/C$

Force voltage-analogous circuit

$f(t) \rightarrow i(t)$ $M_1 \rightarrow C_1$ $B_1 \rightarrow 1/R_1$
$v_1 \rightarrow v_1$ $M_2 \rightarrow C_2$ $B_2 \rightarrow 1/R_2$
$v_2 \rightarrow v_2$ $M_3 \rightarrow C_3$ $B_3 \rightarrow 1/R_3$
$v_3 \rightarrow v_3$ $B \rightarrow 1/R$ $K \rightarrow 1/L$

Force-current analogous circuit

E1.5

$$J_1 \frac{d\omega_1}{dt} + B_1\omega_1 + K_1 \int (\omega_1 - \omega_2)dt = T(t);$$

$$J_2 \frac{d\omega_2}{dt} + B_2\omega_2 + K_1 \int (\omega_2 - \omega_1)dt + K_2 \int (\omega_2 - \omega_3)dt = 0$$

$$J_3 \frac{d\omega_3}{dt} + B_3\omega_3 + K_2 \int (\omega_3 - \omega_2)dt = 0$$

$T(t) \rightarrow e(t)$ $J_1 \rightarrow L_1$ $B_1 \rightarrow R_1$ $K_1 \rightarrow 1/C_1$
$\omega_1 \rightarrow i_1$ $J_2 \rightarrow L_2$ $B_2 \rightarrow R_2$ $K_2 \rightarrow 1/C_2$
$\omega_2 \rightarrow i_2$ $J_3 \rightarrow L_3$ $B_3 \rightarrow R_3$ $\omega_3 \rightarrow i$

Torque-voltage analogous circuit

$T(t) \rightarrow i(t)$ $J_1 \rightarrow C_1$ $B_1 \rightarrow 1/R_1$ $K_1 \rightarrow 1/L_1$
$\omega_1 \rightarrow v_1$ $J_2 \rightarrow C_2$ $B_2 \rightarrow 1/R_2$ $K_2 \rightarrow 1/L_2$
$\omega_2 \rightarrow v_2$ $J_3 \rightarrow C_3$ $B_3 \rightarrow 1/R_3$ $\omega_3 \rightarrow v_3$

Torque-current analogous circuit

E1.6

$e(t) \rightarrow f(t)$ $i_1 \rightarrow v_1$ $i_3 \rightarrow v_3$ $R \rightarrow B$
                 $i_2 \rightarrow v_2$ $L \rightarrow M$ $1/C \rightarrow K$

Analogous mechanical translational system

$e(t) \rightarrow T(t)$ $i_1 \rightarrow \omega_1$ $i_3 \rightarrow \omega_3$ $R \rightarrow B$
                 $i_2 \rightarrow \omega_2$ $L \rightarrow J$ $1/C \rightarrow K$

Analogous mechanical rotational system

E1.7

(a) $\dfrac{C}{R} = \dfrac{G_1G_2 + G_1G_3}{1 + G_1H_2 + G_1 + G_1G_2H_1 + G_1G_3H_1}$

(b) $\dfrac{C}{R} = \dfrac{G_1G_2(1 + G_3)(G_4 + G_5)}{1 + (1 + G_3)(G_4 + G_5)H_2 + (1 + G_3)(G_4 + G_5)G_2H_1}$

(c) $\dfrac{C}{R} = \dfrac{G_1G_2 + G_1G_3}{1 + G_1G_2H_1 + G_1G_2 + G_1G_3 + G_2H_2 + G_3H_2}$

(d) $\dfrac{C}{R} = \dfrac{G_1G_2G_3}{1 + G_2G_3H_2 + G_1G_2H_1 + G_1G_2G_3}$

E1.8 $\dfrac{G_1G_2G_3 + G_2G_3G_4H_1 + G_2G_4H_2 + G_1G_2G_4H_3}{1 + G_2G_3H_1 + G_2H_2 + G_1G_2H_3}$

E1.9 (a) $\dfrac{C(s)}{R(s)} = \dfrac{G_1G_2G_3}{1 + G_1G_2H_1 + G_2G_3H_2 + G_1G_2G_3}$

(b) $\dfrac{C}{R} = \dfrac{A_{12}A_{23}A_{34}A_{45}A_{50} + A_{13}A_{34}A_{45}A_{56} + A_{12}A_{23}A_{34}A_{46} + A_{13}A_{34}A_{46}}{1 - (A_{12}A_{21} + A_{34}A_{43} + A_{56}A_{65}) + (A_{12}A_{21}A_{34}A_{43} + A_{12}A_{21}A_{56}A_{65})}$
$\qquad\qquad\qquad\qquad + A_{34}A_{43}A_{56}A_{65}) - (A_{12}A_{21}A_{34}A_{43}A_{56}A_{65})$

(c) $\dfrac{C}{R} = \dfrac{G_1G_2G_3G_4(1 + G_5G_6 + G_6G_7) + G_4G_5G_6G_7(1 + G_1G_2 + G_2G_3)}{1 + [G_1G_2 + G_2G_3 + G_5G_6 + G_6G_7 + G_1G_2G_5G_6 + G_5G_6G_2G_3}$
$\qquad\qquad\qquad\qquad + G_6G_7G_8 + G_{13}] + G_2G_9G_{10}G_{13} + G_{10}G_{11}G_{13} + G_7G_{12}G_{13}$

(d) $\dfrac{x_8}{x_1} = \dfrac{[G_1G_2G_3G_4][1 - (G_7G_{12} + G_6G_7G_8 + G_{13}) + G_7G_{12}G_{13}]}{1 - [G_2G_9G_{10} + G_{10}G_{11}] + G_2G_3G_5G_6G_9 + G_3G_5G_6G_{11} + G_7G_{12}}$
$\qquad\qquad\qquad + G_6G_7G_8 + G_{13}] + G_2G_9G_{10}G_{13} + G_{10}G_{11}G_{13} + G_7G_{12}G_{13}$

E1.10 $\dfrac{x_8}{x_1} = G_1G_4G_6G_9 + G_1G_4G_7G_9 + G_1G_3G_5G_8G_9$; $\dfrac{x_8}{x_1} = G_2G_5G_8G_9$

E1.11 (a) $\dfrac{E_o(s)}{E_i(s)} = \dfrac{1 + sR_2C}{1 + s(R_1 + R_2)C}$

(b) $\dfrac{E_o(s)}{E_i(s)} = \dfrac{1}{s^2LC + sRC + 1}$

(c) $\dfrac{E_o(s)}{E_i(s)} = \dfrac{sR_2C}{(s^2L_1C + sR_1C + 1)(s^2L_2C + sR_2C + 1) - 1}$

(d) $\dfrac{E_o(s)}{E_i(s)} = \dfrac{sR_1R_2C + R_2}{sR_1R_2C + (R_1 + R_2)}$

E1.12 $\dfrac{C(s)}{R(s)} = \dfrac{s^2L_2C}{[sR_1C + 1][s^2(L_1 + L_2)C + sR_2C + 1] - 1}$

CHAPTER 2

TIME RESPONSE ANALYSIS

2.1 TIME RESPONSE

The time response of the system is the output of the closed loop system as a function of time. It is denoted by c(t). The time response can be obtained by solving the differential equation governing the system. Alternatively, the response c(t) can be obtained from the transfer function of the system and the input to the system.

The closed loop transfer function, $\dfrac{C(s)}{R(s)} = \dfrac{G(s)}{1 + G(s)H(s)} = M(s)$ .....(2.1)

The Output or Response in s-domain, C(s) is given by the product of the transfer function and the input, R(s). On taking inverse Laplace transform of this product the time domain response, c(t) can be obtained.

Response in s-domain, $C(s) = R(s)\,M(s)$ .....(2.2)

Response in time domain, $c(t) = \mathcal{L}^{-1}\{C(s)\} = \mathcal{L}^{-1}\{R(s) \times M(s)\}$ .....(2.3)

where, $M(s) = \dfrac{G(s)}{1 + G(s)H(s)}$

The time response of a control system consists of two parts : *the transient and the steady state response*. The transient response is the response of the system when the input changes from one state to another. The steady state response is the response as time, **t** approaches infinity.

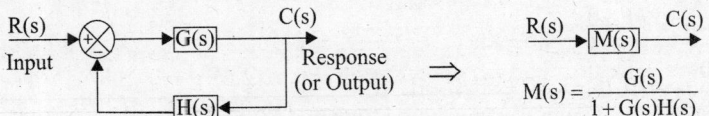

Fig 2.1 : Closed loop system.

2.2 TEST SIGNALS

The knowledge of input signal is required to predict the response of a system. In most of the systems the input signals are not known ahead of time and also it is difficult to express the input signals mathematically by simple equations. The characteristics of actual input signals are a sudden shock, a sudden change, a constant velocity and a constant acceleration. Hence test signals which resembles these characteristics are used as input signals to predict the performance of the system. The commonly used test input signals are impulse, step, ramp, acceleration and sinusoidal signals.

The standard test signals are,

1. a) Step signal
   b) Unit step signal
4. Impulse signal

2. a) Ramp signal
   b) Unit ramp signal
5. Sinusoidal signal.

3. a) Parabolic signal
   b) Unit parabolic signal

Since the test signals are simple functions for time, they can be easily generated in laboratories. The mathematical and experimental analysis of control systems using these signals can be carried out easily. The use of the test signals can be justified because of a correlation existing between the response characteristics of a system to a test input signal and capability of the system to cope with actual input signals.

STEP SIGNAL

The step signal is a signal whose value changes from zero to A at t = 0 and remains constant at A for t > 0. The step signal resembles an actual steady input to a system. A special case of step signal is unit step in which **A** is unity.

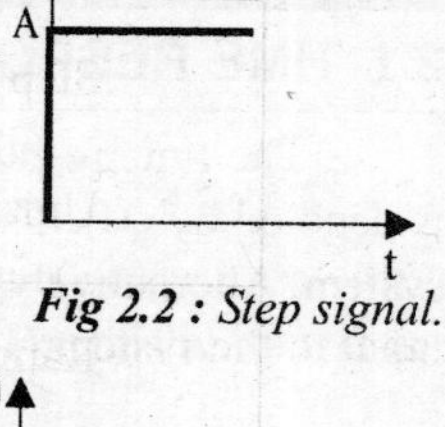

The mathematical representation of the step signal is,

$$r(t) = 1 \; ; \; t \geq 0$$
$$= 0 \; ; \; t < 0 \qquad\qquad .....(2.4)$$

Fig 2.2 : Step signal.

RAMP SIGNAL

The ramp signal is a signal whose value increases linearly with time from an initial value of zero at t = 0. The ramp signal resembles a constant velocity input to the system. A special case of ramp signal is unit ramp signal in which the value of **A** is unity.

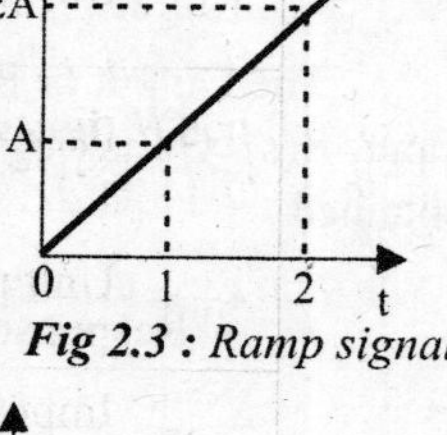

The mathematical representation of the ramp signal is,

$$r(t) = A t \; ; \; t \geq 0$$
$$= 0 \quad ; \; t < 0 \qquad\qquad .....(2.5)$$

Fig 2.3 : Ramp signal.

PARABOLIC SIGNAL

In parabolic signal, the instantaneous value varies as square of the time from an initial value of zero at t = 0. The sketch of the signal with respect to time resembles a parabola. The parabolic signal resembles a constant acceleration input to the system. A special case of parabolic signal is unit parabolic signal in which A is unity.

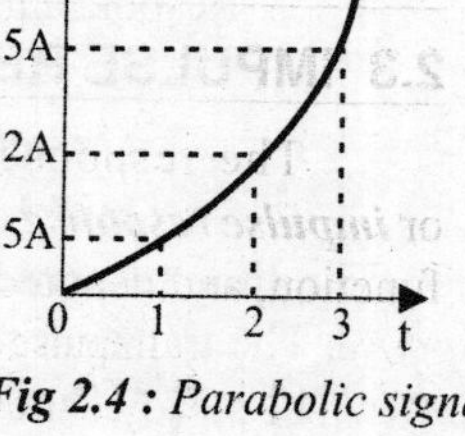

The mathematical representation of the parabolic signal is,

$$r(t) = \frac{At^2}{2} \; ; \; t \geq 0$$
$$= 0 \quad ; \; t < 0 \qquad\qquad .....(2.6)$$

Fig 2.4 : Parabolic signal.

> **Note :** *Integral of step signal is ramp signal. Integral of ramp signal is parabolic signal.*

IMPULSE SIGNAL

A signal of very large magnitude which is available for very short duration is called *impulse signal.* Ideal impulse signal is a signal with infinite magnitude and zero duration but with an area of A. The unit impulse signal is a special case, in which A is unity.

The impulse signal is denoted by $\delta(t)$ and mathematically it is expressed as,

$$\delta(t) = \infty \; ; \; t = 0 \quad \text{and} \quad \int_{-\infty}^{+\infty} \delta(t) \, dt = A$$
$$= 0 \; ; \; t \neq 0 \qquad\qquad .....(2.7)$$

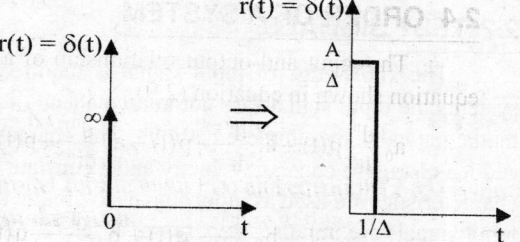

Fig 2.5 : Impulse signal.

Since a perfect impulse cannot be achieved in practice it is usually approximated by a pulse of small width but with area, A. Mathematically an impulse signal is the derivative of a step signal. Laplace transform of the impulse function is unity.

TABLE 2.1 : Standard Test Signals

Name of the signal	Time domain equation of signal, r(t)	Laplace transform of the signal, R(s)
Step	A	$\dfrac{A}{s}$
Unit step	1	$\dfrac{1}{s}$
Ramp	At	$\dfrac{A}{s^2}$
Unit ramp	t	$\dfrac{1}{s^2}$
Parabolic	$\dfrac{At^2}{2}$	$\dfrac{A}{s^3}$
Unit parabolic	$\dfrac{t^2}{2}$	$\dfrac{1}{s^3}$
Impulse	$\delta(t)$	1

2.3 IMPULSE RESPONSE

The response of the system, with input as impulse signal is called *weighing function* or *impulse response* of the system. It is also given by the inverse Laplace transform of the system transfer function, and denoted by m(t).

$$\text{Impulse response, } m(t) = \mathcal{L}^{-1}\{R(s)\,M(s)\} = \mathcal{L}^{-1}\{M(s)\} \qquad \qquad .....(2.8)$$

$$\text{where, } M(s) = \frac{G(s)}{1 + G(s)H(s)} \qquad \boxed{R(s) = 1, \text{ for impulse}}$$

Since impulse response (or weighing function) is obtained from the transfer function of the system, it shows the characteristics of the system. Also the response for any input can be obtained by convolution of input with impulse response.

2.4 ORDER OF A SYSTEM

The input and output relationship of a control system can be expressed by n^{th} order differential equation shown in equation (2.9).

$$a_0 \frac{d^n}{dt^n} p(t) + a_1 \frac{d^{n-1}}{dt^{n-1}} p(t) + a_2 \frac{d^{n-2}}{dt^{n-2}} p(t) + ... + a_{n-1} \frac{d}{dt} p(t) + a_n p(t) = b_0 \frac{d^m}{dt^m} q(t)$$

$$+ b_1 \frac{d^{m-1}}{dt^{m-1}} q(t) + b_2 \frac{d^{m-2}}{dt^{m-2}} q(t) + ... + b_{m-1} \frac{d}{dt} q(t) + b_m q(t) \qquad .....(2.9)$$

where, p(t) = Output / Response ; q(t) = Input / Excitation.

The order of the system is given by the order of the differential equation governing the system. If the system is governed by n^{th} order differential equation, then the system is called *n^{th} order system.*

Alternatively, the order can be determined from the transfer function of the system. The transfer function of the system can be obtained by taking Laplace transform of the differential equation governing the system and rearranging them as a ratio of two polynomials in **s**, as shown in equation (2.10).

$$\text{Transfer function, } T(s) = \frac{P(s)}{Q(s)} = \frac{b_0 s^m + b_1 s^{m-1} + b_2 s^{m-2} + \ldots + b_{m-1} s + b_m}{a_0 s^n + a_1 s^{n-1} + a_2 s^{n-2} + \ldots + a_{n-1} s + a_n} \quad \ldots(2.10)$$

where, $P(s)$ = Numerator polynomial

$Q(s)$ = Denominator polynomial

The order of the system is given by the maximum power of **s** in the denominator polynomial, $Q(s)$.

Here, $Q(s) = a_0 s^n + a_1 s^{n-1} + a_2 s^{n-2} + \ldots + a_{n-1} s + a_n$.

Now, **n** is the order of the system

When n = 0, the system is zero order system.

When n = 1, the system is first order system.

When n = 2, the system is second order system and so on.

> **Note :** *The order can be specified for both open loop system and closed loop system.*

The numerator and denominator polynomial of equation (2.10) can be expressed in the factorized form as shown in equation (2.11).

$$T(s) = \frac{P(s)}{Q(s)} = \frac{(s + z_1)(s + z_2) \ldots (s + z_m)}{(s + p_1)(s + p_2) \ldots (s + p_n)} \quad \ldots(2.11)$$

where, $z_1, z_2, \ldots z_m$ are zeros of the system.

$p_1, p_2, \ldots, p_n$ are poles of the system.

Now, the value of **n** gives the number of poles in the transfer function. Hence the order is also given by the number of poles of the transfer function.

> **Note :** *The zeros and poles are critical value, of s, at which the function T(s) attains extreme values 0 or ∞. When s takes the value of a zero, the function T(s) will be zero. When s takes the value of a pole, the function T(s) will be infinite.*

2.5 REVIEW OF PARTIAL FRACTION EXPANSION

The time response of the system is obtained by taking the inverse Laplace transform of the product of input signal and transfer function of the system. Taking inverse Laplace transform requires the knowledge of partial fraction expansion. In control systems three different types of transfer function are encountered. They are,

Case 1 : Functions with separate poles.

Case 2 : Functions with multiple poles.

Case 3 : Functions with complex conjugate poles.

The partial fraction of all the three cases are explained with an example.

Case 1 : *When the transfer function has distinct poles*

$$\text{Let, } T(s) = \frac{K}{s(s + p_1)(s + p_2)}$$

By partial fraction expansion, T(s) can be expressed as,

$$T(s) = \frac{K}{s(s+p_1)(s+p_2)} = \frac{A}{s} + \frac{B}{s+p_1} + \frac{C}{s+p_2}$$

The residues A, B and C are given by,

$$A = T(s) \times s\big|_{s=0} \quad B = T(s) \times (s+p_1)\big|_{s=-p_1} \quad C = T(s) \times (s+p_2)\big|_{s=-p_2}$$

Example

Let, $T(s) = \dfrac{2}{s(s+1)(s+2)}$

By partial fraction expansion, T(s) can be expressed as,

$$T(s) = \frac{2}{s(s+1)(s+2)} = \frac{A}{s} + \frac{B}{s+1} + \frac{C}{s+2}$$

A is obtained by multiplying T(s) by s and letting s = 0.

$$A = T(s) \times s\big|_{s=0} = \frac{2}{s(s+1)(s+2)} \times s\bigg|_{s=0} = \frac{2}{(s+1)(s+2)}\bigg|_{s=0} = \frac{2}{1 \times 2} = 1$$

B is obtained by multiplying T(s) by (s +1) and letting s = −1.

$$B = T(s) \times (s+1)\big|_{s=-1} = \frac{2}{s(s+1)(s+2)} \times (s+1)\bigg|_{s=-1} = \frac{2}{s(s+2)}\bigg|_{s=-1} = \frac{2}{-1(-1+2)} = -2$$

C is obtained by multiplying T(s) by (s +2) and letting s = −2.

$$C = T(s) \times (s+2)\big|_{s=-2} = \frac{2}{s(s+1)(s+2)} \times (s+2)\bigg|_{s=-2} = \frac{2}{s(s+1)}\bigg|_{s=-2} = \frac{2}{-2(-2+1)} = +1$$

$$\therefore T(s) = \frac{2}{s(s+1)(s+2)} = \frac{1}{s} - \frac{2}{s+1} + \frac{1}{s+2}$$

Case 2 : When the transfer function has multiple poles

Let, $T(s) = \dfrac{K}{s(s+p_1)(s+p_2)^2}$

By partial fraction expansion, T(s) can be expressed as,

$$T(s) = \frac{K}{s(s+p_1)(s+p_2)^2} = \frac{A}{s} + \frac{B}{s+p_1} + \frac{C}{(s+p_2)^2} + \frac{D}{(s+p_2)}$$

The residues A, B, C and D are given by,

$$A = T(s) \times s\big|_{s=0} \qquad\qquad B = T(s) \times (s+p_1)\big|_{s=-p_1}$$

$$C = T(s) \times (s+p_2)^2\big|_{s=-p_2} \qquad D = \frac{d}{ds}\big[T(s) \times (s+p_2)^2\big]\big|_{s=-p_2}$$

Example

Let, $T(s) = \dfrac{2}{s(s+1)(s+2)^2}$

By partial fraction expansion, T(s) can be expressed as,

$$T(s) = \frac{K}{s(s+1)(s+2)^2} = \frac{A}{s} + \frac{B}{(s+1)} + \frac{C}{(s+2)^2} + \frac{D}{(s+2)}$$

A is obtained by multiplying T(s) by s and letting s = 0.

$$A = T(s) \times s \Big|_{s=0} = \frac{2}{s(s+1)(s+2)^2} \times s \Big|_{s=0} = \frac{2}{(s+1)(s+2)^2} \Big|_{s=0} = \frac{2}{1 \times 2^2} = 0.5$$

B is obtained by multiplying T(s) by (s +1) and letting s = −1.

$$B = T(s) \times (s+1) \Big|_{s=-1} = \frac{2}{s(s+1)(s+2)^2} \times (s+1) \Big|_{s=-1} = \frac{2}{s(s+2)^2} \Big|_{s=-1} = \frac{2}{-1(-1+2)^2} = -2$$

C is obtained by multiplying T(s) by (s + 2)² and letting s = −2.

$$C = T(s) \times (s+2)^2 \Big|_{s=-2} = \frac{2}{s(s+1)(s+2)^2} \times (s+2)^2 \Big|_{s=-2} = \frac{2}{s(s+1)} \Big|_{s=-2} = \frac{2}{-2(-2+1)} = 1$$

D is obtained by differentiating the product T(s) (s + 2)² with respect to s and then letting s = −2.

$$D = \frac{d}{ds} \Big[T(s) \times (s+2)^2 \Big] \Big|_{s=-2} = \frac{d}{ds} \Big[\frac{2}{s(s+1)} \Big] \Big|_{s=-2} = \frac{-2(2s+1)}{s^2(s+1)^2} \Big|_{s=-2} = \frac{-2(-2)+1)}{(-2)^2(-2+1)^2} = 1.5$$

$$\therefore \; T(s) = \frac{2}{s(s+1)(s+2)^2} = \frac{0.5}{s} - \frac{2}{s+1} + \frac{1}{(s+2)^2} + \frac{1.5}{s+2}$$

Case 3 : *When the transfer function has complex conjugate poles*

$$\text{Let, } T(s) = \frac{K}{(s+p_1)(s^2+bs+c)}$$

By partial fraction expansion, T(s) can be expressed as,

$$T(s) = \frac{K}{(s+p_1)(s^2+bs+c)} = \frac{A}{s+p_1} + \frac{Bs+C}{s^2+bs+c} \qquad \qquad .....(2.12)$$

The residue A is given by, $A = T(s) \times (s+p_1) \big|_{s=-p_1}$

The residues B and C are solved by cross multiplying the equation (2.12) and then equating the coefficient of like power of **s**.

Finally express T(s) as shown below,

$$T(s) = \frac{A}{s+p_1} + \frac{Bs+C}{s^2+bs+c} \qquad \qquad \boxed{(x+y)^2 = x^2 + 2xy + y^2}$$

Let us express, $s^2 + bs$, in the form of $(x + y)^2$. This will require addition and subtraction of an extra term $(b/2)^2$.

$$\therefore \; T(s) = \frac{A}{s+p_1} + \frac{Bs+C}{s^2 + 2 \times \frac{b}{2}s + \left(\frac{b}{2}\right)^2 + c - \left(\frac{b}{2}\right)^2} = \frac{A}{s+p_1} + \frac{Bs+C}{\left(s+\frac{b}{2}\right)^2 + \left(c - \frac{b^2}{4}\right)}$$

$$= \frac{A}{s+p_1} + \frac{Bs}{\left(s+\frac{b}{2}\right)^2 + \left(c - \frac{b^2}{4}\right)} + \frac{C}{\left(s+\frac{b}{2}\right)^2 + \left(c - \frac{b^2}{4}\right)}$$

Example

Let, $T(s) = \dfrac{1}{(s+2)(s^2+s+1)}$

By partial fraction expansion,

$$T(s) = \dfrac{1}{(s+2)(s^2+s+1)} = \dfrac{A}{s+2} + \dfrac{Bs+C}{s^2+s+1}$$

A is obtained by multiplying T(s) by (s + 2) and letting s = −2.

$$\therefore \ A = T(s) \times (s+2)\Big|_{s=-2} = \dfrac{1}{(s+2)(s^2+s+1)} \times (s+2)\Big|_{s=-2} = \dfrac{1}{(-2)^2-2+1} = \dfrac{1}{3}$$

To solve B and C, cross multiply the following equation and substitute the value of A. Then equate the like power of s.

$$\dfrac{1}{(s+2)(s^2+s+1)} = \dfrac{A}{s+2} + \dfrac{Bs+C}{s^2+s+1}$$

$$1 = A(s^2+s+1) + (Bs+C)(s+2)$$

$$1 = \dfrac{1}{3}(s^2+s+1) + Bs^2 + 2Bs + Cs + 2C$$

$$1 = \dfrac{s^2}{3} + \dfrac{s}{3} + \dfrac{1}{3} + Bs^2 + 2Bs + Cs + 2C$$

$$\boxed{\begin{aligned} &s^2 + s + 1 \\ &= s^2 + 2 \times \dfrac{s}{2} + \left(\dfrac{1}{2}\right)^2 + 1 - \left(\dfrac{1}{2}\right)^2 \\ &= \left(s + \dfrac{1}{2}\right)^2 + \left(1 - \dfrac{1}{4}\right) \\ &= (s+0.5)^2 + 0.75 \end{aligned}}$$

On equating the coefficient of s² terms, $0 = \dfrac{1}{3} + B \ ; \ \therefore \ B = -\dfrac{1}{3}$

On equating the coefficient of s terms, $0 = \dfrac{1}{3} + 2B + C \ ; \ \therefore \ C = -\dfrac{1}{3} - 2B = -\dfrac{1}{3} + \dfrac{2}{3} = \dfrac{1}{3}$

$$T(s) = \dfrac{\frac{1}{3}}{s} + \dfrac{-\frac{1}{3}s + \frac{1}{3}}{s^2+s+1} = \dfrac{1}{3s} - \dfrac{1}{3}\dfrac{s}{(s^2+s+1)} + \dfrac{1}{3}\dfrac{1}{(s^2+s+1)}$$

$$= \dfrac{1}{3s} - \dfrac{1}{3}\dfrac{s}{(s+0.5)^2+0.75} + \dfrac{1}{3}\dfrac{1}{(s+0.5)^2+0.75}$$

2.6 RESPONSE OF FIRST ORDER SYSTEM FOR UNIT STEP INPUT

The closed loop order system with unity feedback is shown in fig 2.6.

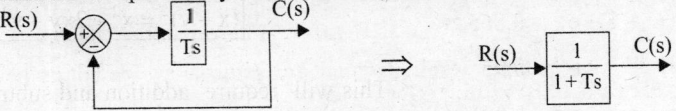

Fig 2.6 : *Closed loop for first order system.*

The closed loop transfer function of first order system, $\dfrac{C(s)}{R(s)} = \dfrac{1}{1+Ts}$

If the input is unit step then, $r(t) = 1$ and $R(s) = \dfrac{1}{s}$.

The response in s-domain, $C(s) = R(s)\dfrac{1}{(1+Ts)} = \dfrac{1}{s}\dfrac{1}{(1+Ts)} = \dfrac{1}{sT\left(\frac{1}{T}+s\right)} = \dfrac{\frac{1}{T}}{s\left(s+\frac{1}{T}\right)}$

By partial fraction expansion,

$$C(s) = \frac{\frac{1}{T}}{s\left(s + \frac{1}{T}\right)} = \frac{A}{s} + \frac{B}{\left(s + \frac{1}{T}\right)}$$

A is obtained by multiplying C(s) by s and letting s = 0.

$$A = C(s) \times s \Big|_{s=0} = \frac{\frac{1}{T}}{s\left(s + \frac{1}{T}\right)} \times s \Bigg|_{s=0} = \frac{\frac{1}{T}}{s + \frac{1}{T}} \Bigg|_{s=0} = \frac{\frac{1}{T}}{\frac{1}{T}} = 1$$

B is obtained by multiplying C(s) by (s+1/T) and letting s = −1/T.

$$B = C(s) \times \left(s + \frac{1}{T}\right) \Bigg|_{s=-\frac{1}{T}} = \frac{\frac{1}{T}}{s\left(s + \frac{1}{T}\right)} \times \left(s + \frac{1}{T}\right) \Bigg|_{s=-\frac{1}{T}} = \frac{\frac{1}{T}}{s} \Bigg|_{s=-\frac{1}{T}} = \frac{\frac{1}{T}}{\frac{-1}{T}} = -1$$

$$\therefore \ C(s) = \frac{1}{s} - \frac{1}{s + \frac{1}{T}}$$

$$\boxed{\mathcal{L}\{e^{-at}\} = \frac{1}{s+a}}$$

The response in time domain is given by,

$$c(t) = \mathcal{L}^{-1}\{C(s)\} = \mathcal{L}^{-1}\left\{\frac{1}{s} - \frac{1}{s + \frac{1}{T}}\right\} = 1 - e^{-\frac{t}{T}} \qquad \qquad \dots\dots(2.13)$$

The equation (2.13) is the response of the closed loop first order system for unit step input. For step input of step value, A, the equation (2.13) is multiplied by A.

$$\boxed{\begin{array}{l} \therefore \ \text{For closed loop first order system, } \ \text{Unit step response} = 1 - e^{-\frac{t}{T}} \\[2mm] \qquad\qquad\qquad\qquad \text{Step response} \ = \ A\left(1 - e^{-\frac{t}{T}}\right) \end{array}}$$

When,   t = 0,   c(t) = $1 - e^0 = 0$

When,   t = 1T,   c(t) = $1 - e^{-1} = 0.632$

When,   t = 2T,   c(t) = $1 - e^{-2} = 0.865$

When,   t = 3T,   c(t) = $1 - e^{-3} = 0.95$

When,   t = 4T,   c(t) = $1 - e^{-4} = 0.9817$

When,   t = 5T,   c(t) = $1 - e^{-5} = 0.993$

When,   t = ∞,   c(t) = $1 - e^{-\infty} = 1$

Here T is called Time constant of the system. In a time of 5T, the system is assumed to have attained steady state. The input and output signal of the first order system is shown in fig 2.7.

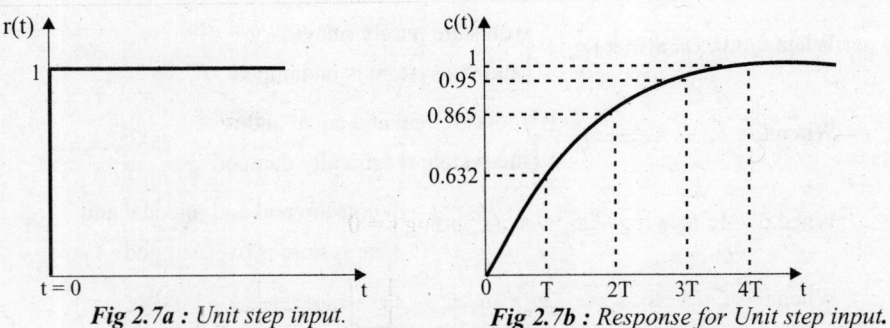

Fig 2.7a : *Unit step input.*   **Fig 2.7b** : *Response for Unit step input.*

Fig 2.7 : *Response of first order system to unit step input.*

2.7 SECOND ORDER SYSTEM

The closed loop second order system is shown in fig 2.8

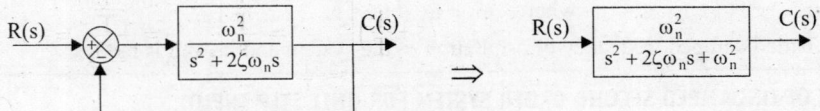

Fig 2.8 : *Closed loop for second order system.*

The standard form of closed loop transfer function of second order system is given by,

$$\frac{C(s)}{R(s)} = \frac{\omega_n^2}{s^2 + 2\zeta\omega_n s + \omega_n^2} \qquad \qquad .....(2.14)$$

where, ω_n = Undamped natural frequency, rad/sec.

ζ = Damping ratio.

*The **damping ratio** is defined as the ratio of the actual damping to the critical damping. The response c(t) of second order system depends on the value of damping ratio. Depending on the value of ζ, the system can be classified into the following four cases,

Case 1 : Undamped system, $\zeta = 0$

Case 2 : Under damped system, $0 < \zeta < 1$

Case 3 : Critically damped system, $\zeta = 1$

Case 4 : Over damped system, $\zeta > 1$

The characteristics equation of the second order system is,

$$s^2 + 2\zeta\omega_n s + \omega_n^2 = 0 \qquad \qquad .....(2.15)$$

It is a quadratic equation and the roots of this equation is given by,

$$s_1, s_2 = \frac{-2\zeta\omega_n \pm \sqrt{4\zeta^2\omega_n^2 - 4\omega_n^2}}{2} = \frac{-2\zeta\omega_n \pm \sqrt{4\omega_n^2(\zeta^2 - 1)}}{2}$$

$$= -\zeta\omega_n \pm \omega_n\sqrt{\zeta^2 - 1} \qquad \qquad .....(2.16)$$

When $\zeta = 0$, $s_1, s_2 = \pm j\omega_n$; $\begin{cases} \text{roots are purely imaginary} \\ \text{and the system is undamped} \end{cases}$ .....(2.17)

When $\zeta = 1$, $s_1, s_2 = -\omega_n$; $\begin{cases} \text{roots are real and equal and} \\ \text{the system is critically damped} \end{cases}$ .....(2.18)

When $\zeta > 1$, $s_1, s_2 = -\zeta\omega_n + \omega_n\sqrt{\zeta^2 - 1}$; $\begin{cases} \text{roots are real and unequal and} \\ \text{the system is overdamped} \end{cases}$ .....(2.19)

When $0 < \zeta < 1$, $s_1, s_2 = -\zeta\omega_n \pm \omega_n\sqrt{\zeta^2 - 1} = -\zeta\omega_n \pm \omega_n\sqrt{(-1)(1-\zeta^2)}$

$$= -\zeta\omega_n \pm \omega_n\sqrt{-1}\sqrt{1-\zeta^2} = -\zeta\omega_n \pm j\omega_n\sqrt{1-\zeta^2}$$

$$= -\zeta\omega_n \pm j\omega_d \; ; \begin{cases} \text{roots are complex conjugate} \\ \text{the system is underdamped} \end{cases}$$ .....(2.20)

where, $\omega_d = \omega_n\sqrt{1-\zeta^2}$ .....(2.21)

Here ω_d is called damped frequency of oscillation of the system and its unit is rad/sec.

2.7.1 RESPONSE OF UNDAMPED SECOND ORDER SYSTEM FOR UNIT STEP INPUT

The standard form of closed loop transfer function of second order system is,

$$\frac{C(s)}{R(s)} = \frac{\omega_n^2}{s^2 + 2\zeta\omega_n s + \omega_n^2}$$

For undamped system, $\zeta = 0$.

$$\therefore \frac{C(s)}{R(s)} = \frac{\omega_n^2}{s^2 + \omega_n^2}$$ .....(2.22)

When the input is unit step, $r(t) = 1$ and $R(s) = \frac{1}{s}$.

$\therefore$ The response in s-domain, $C(s) = R(s)\dfrac{\omega_n^2}{s^2 + \omega_n^2} = \dfrac{1}{s}\dfrac{\omega_n^2}{s^2 + \omega_n^2}$ .....(2.23)

By partial fraction expansion,

$$C(s) = \frac{\omega_n^2}{s(s^2 + \omega_n^2)} = \frac{A}{s} + \frac{B}{s^2 + \omega_n^2}$$

A is obtained by multiplying C(s) by s and letting s = 0.

$$A = C(s) \times s\big|_{s=0} = \frac{\omega_n^2}{s(s^2 + \omega_n^2)} \times s\bigg|_{s=0} = \frac{\omega_n^2}{s^2 + \omega_n^2}\bigg|_{s=0} = \frac{\omega_n^2}{\omega_n^2} = 1$$

B is obtained by multiplying C(s) by $(s^2 + \omega_n^2)$ and letting $s^2 = -\omega_n^2$ or $s = j\omega_n$.

$$B = C(s) \times (s^2 + \omega_n^2)\big|_{s=j\omega} = \frac{\omega_n^2}{s(s^2 + \omega_n^2)} \times (s^2 + \omega_n^2)\bigg|_{s=j\omega} = \frac{\omega_n^2}{s}\bigg|_{s=j\omega_n} = \frac{\omega_n^2}{j\omega_n} = -j\omega_n = -s$$

$$\therefore C(s) = \frac{A}{s} + \frac{B}{s^2 + \omega_n^2} = \frac{1}{s^2} - \frac{s}{s^2 + \omega_n^2} \qquad \boxed{\mathcal{L}\{1\} = \frac{1}{s} \quad \bigg| \quad \mathcal{L}\{\cos\omega t\} = \frac{s}{s^2 + \omega^2}}$$

Time domain response, $c(t) = \mathcal{L}^{-1}\{C(s)\} = \mathcal{L}^{-1}\left\{\dfrac{1}{s} - \dfrac{s}{s^2 + \omega_n^2}\right\} = 1 - \cos\omega_n t$ .....(2.24)

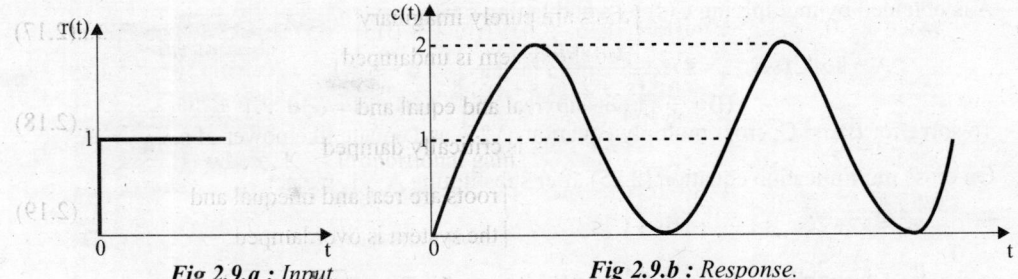

Fig 2.9.a : Input.   **Fig 2.9.b : Response.**

Fig 2.9 : Response of undamped second order system for unit step input.

Using equation (2.24), the response of undamped second order system for unit step input is sketched in fig 2.9, and observed that the response is completely oscillatory.

Note : *Every practical system has some amount of damping. Hence undamped system does not exist in practice.*

The equation (2.24) is the response of undamped closed loop second order system for unit step input. For step input of step value A, the equation (2.24) should be multiplied by A.

$\therefore$ For closed loop undamped second order system,

Unit step response $= 1 - \cos \omega_n t$

Step response $\quad = A(1 - \cos \omega_n t)$

2.7.2 RESPONSE OF UNDERDAMPED SECOND ORDER SYSTEM FOR UNIT STEP INPUT

The standard form of closed loop transfer function of second order system is,

$$\frac{C(s)}{R(s)} = \frac{\omega_n^2}{s^2 + 2\zeta\omega_n s + \omega_n^2}$$

For underdamped system, $0 < \zeta < 1$ and roots of the denominator (characteristic equation) are complex conjugate.

The roots of the denominator are, $s = -\zeta\omega_n \pm \omega_n \sqrt{\zeta^2 - 1}$

Since $\zeta < 1$, ζ^2 is also less then 1, and so $1 - \zeta^2$ is always positive.

$$\therefore \ s = -\zeta\omega_n \pm \omega_n \sqrt{(-1)(1 - \zeta^2)} = -\zeta\omega_n \pm j\omega_n \sqrt{1 - \zeta^2}$$

The damped frequency of oscillation, $\omega_d = \omega_n \sqrt{1 - \zeta^2}$

$$\therefore \ s = -\zeta\omega_n \pm j\omega_d$$

The response in s-domain, $\quad C(s) = R(s) \dfrac{\omega_n^2}{s^2 + 2\zeta\omega_n s + \omega_n^2}$

For unit step input, r(t) = 1 and R(s) = 1/s.

$$\therefore \ C(s) = \frac{\omega_n^2}{s(s^2 + 2\zeta\omega_n s + \omega_n^2)}$$

By partial fraction expansion, $\quad C(s) = \dfrac{\omega_n^2}{s(s^2 + 2\zeta\omega_n s + \omega_n^2)} = \dfrac{A}{s} + \dfrac{Bs + C}{s^2 + 2\zeta\omega_n s + \omega_n^2}$ .....(2.25)

A is obtained by multiplying C(s) by s and letting s = 0.

$$\therefore \ A = s \times C(s)\big|_{s=0} = s \times \frac{\omega_n^2}{s(s^2 + 2\zeta\omega_n s + \omega_n^2)}\bigg|_{s=0} = \frac{\omega_n^2}{\omega_n^2} = 1$$

To solve for B and C, cross multiply equation (2.25) and equate like power of s.

On cross multiplication equation (2.25) after substituting A = 1, we get,

$$\omega_n^2 = s^2 + 2\zeta\omega_n s + \omega_n^2 + (Bs + C)s$$

$$\omega_n^2 = s^2 + 2\zeta\omega_n s + \omega_n^2 + Bs^2 + Cs$$

Equating coefficients of s^2 we get, $0 = 1 + B$ $\therefore B = -1$

Equating coefficient of s we get, $0 = 2\zeta\omega_n + C$ $\therefore C = -2\zeta\omega_n$

$$\therefore \ C(s) = \frac{1}{s} - \frac{s + 2\zeta\omega_n}{s^2 + 2\zeta\omega_n s + \omega_n^2}$$

.....(2.26)

Let us add and subtract $\zeta^2\omega_n^2$ to the denominator of second term in the equation (2.26).

$$\therefore \ C(s) = \frac{1}{s} - \frac{s + 2\zeta\omega_n}{s^2 + 2\zeta\omega_n s + \omega_n^2 + \zeta^2\omega_n^2 - \zeta^2\omega_n^2} = \frac{1}{s} - \frac{s + 2\zeta\omega_n}{(s^2 + 2\zeta\omega_n s + \zeta^2\omega_n^2) + (\omega_n^2 - \zeta^2\omega_n^2)}$$

$$= \frac{1}{s} - \frac{s + 2\zeta\omega_n}{(s + \zeta\omega_n)^2 + \omega_n^2(1 - \zeta^2)} = \frac{1}{s} - \frac{s + 2\zeta\omega_n}{(s + \zeta\omega_n)^2 + \omega_d^2}$$

$$\boxed{\omega_d = \omega_n\sqrt{1 - \zeta^2}}$$

$$= \frac{1}{s} - \frac{s + \zeta\omega_n}{(s + \zeta\omega_n)^2 + \omega_d^2} - \frac{\zeta\omega_n}{(s + \zeta\omega_n)^2 + \omega_d^2}$$

.....(2.27)

Let us multiply and divide by ω_d in the third term of the equation (2.27).

$$\therefore \ C(s) = \frac{1}{s} - \frac{s + \zeta\omega_n}{(s + \zeta\omega_n)^2 + \omega_d^2} - \frac{\zeta\omega_n}{\omega_d}\frac{\zeta\omega_n}{(s + \zeta\omega_n)^2 + \omega_d^2}$$

$$\boxed{\mathcal{L}\{1\} = \frac{1}{s}}$$

$$\boxed{\mathcal{L}\{e^{-at}\sin\omega t\} = \frac{\omega}{(s + a)^2 + \omega^2}}$$

The response in time domain is given by,

$$c(t) = \mathcal{L}^{-1}\{C(s)\} = \mathcal{L}^{-1}\left\{\frac{1}{s} - \frac{s + \zeta\omega_n}{(s + \zeta\omega_n)^2 + \omega_d^2} - \frac{\zeta\omega_n}{\omega_d}\frac{\omega_d}{(s + \zeta\omega_n)^2 + \omega_d^2}\right\}$$

$$\boxed{\mathcal{L}\{e^{-at}\cos\omega t\} = \frac{s + a}{(s + a)^2 + \omega^2}}$$

$$\boxed{\omega_d = \omega_n\sqrt{1 - \zeta^2}}$$

$$= 1 - e^{-\zeta\omega_n t}\cos\omega_d t - \frac{\zeta\omega_n}{\omega_d}e^{-\zeta\omega_n t}\sin\omega_d t = 1 - e^{-\zeta\omega_n t}\left(\cos\omega_d t + \frac{\zeta\omega_n}{\omega_n\sqrt{1 - \zeta^2}}\sin\omega_d t\right)$$

$$= 1 - \frac{e^{-\zeta\omega_n t}}{\sqrt{1 - \zeta^2}}\left(\sqrt{1 - \zeta^2}\cos\omega_d t + \zeta\sin\omega_d t\right) = 1 - \frac{e^{-\zeta\omega_n t}}{\sqrt{1 - \zeta^2}}\left(\sin\omega_d t \times \zeta + \cos\omega_d t \times \sqrt{1 - \zeta^2}\right)$$

Let us express c(t) in a standard form as shown below.

$$c(t) = 1 - \frac{e^{-\zeta\omega_n t}}{\sqrt{1 - \zeta^2}}(\sin\omega_d t \times \cos\theta + \cos\omega_d t \times \sin\theta)$$

$$\boxed{\begin{array}{l}\sin(A+B) = \sin A\cos B \\ \qquad\qquad + \cos A\sin B\end{array}}$$

Note : *On constructing right angle triangle with ζ and $\sqrt{1 - \zeta^2}$, we get*

$$= 1 - \frac{e^{-\zeta\omega_n t}}{\sqrt{1 - \zeta^2}}\sin(\omega_d t + \theta)$$

.....(2.28)

$$\sin\theta = \sqrt{1 - \zeta^2}$$

$$\cos\theta = \zeta$$

where, $\left(\theta = \tan^{-1}\dfrac{\sqrt{1 - \zeta^2}}{\zeta}\right)$

$$\tan\theta = \frac{\sqrt{1 - \zeta^2}}{\zeta}$$

The equation (2.28) is the response of under damped closed loop second order system for unit step input. For step input of step value, A, the equation (2.28) should be multiplied by A.

∴ For closed loop under damped second order system,

$$\text{Unit step response } = 1 - \frac{e^{-\zeta\omega_n t}}{\sqrt{1-\zeta^2}}\sin(\omega_d t + \theta) \ ; \ \theta = \tan^{-1}\frac{\sqrt{1-\zeta^2}}{\zeta}$$

$$\text{Step response } = A\left[1 - \frac{e^{-\zeta\omega_n t}}{\sqrt{1-\zeta^2}}\sin(\omega_d t + \theta) \ ; \ \theta = \tan^{-1}\frac{\sqrt{1-\zeta^2}}{\zeta}\right]$$

Using equation (2.28) the response of underdamped second order system for unit step input is sketched and observed that the response oscillates before settling to a final value. The oscillations depends on the value of damping ratio.

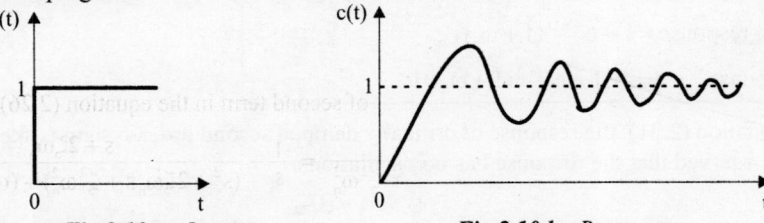

Fig 2.10.a : Input.   *Fig 2.10.b : Response.*

Fig 2.10 : Response of under damped second order system for unit step input.

2.7.3 RESPONSE OF CRITICALLY DAMPED SECOND ORDER SYSTEM FOR UNIT STEP INPUT

The standard form of closed loop transfer function of second order system is

$$\frac{C(s)}{R(s)} = \frac{\omega_n^2}{s^2 + 2\zeta\omega_n s + \omega_n^2}$$

For critical damping $\zeta = 1$.

$$\therefore \frac{C(s)}{R(s)} = \frac{\omega_n^2}{s^2 + 2\omega_n s + \omega_n^2} = \frac{\omega_n^2}{(s + \omega_n)^2} \qquad \qquad .....(2.29)$$

When input is unit step, $r(t) = 1$ and $R(s) = 1/s$.

∴ The response in s-domain,

$$C(s) = R(s)\frac{\omega_n^2}{(s + \omega_n)^2} = \frac{1}{s}\frac{\omega_n^2}{(s + \omega_n)^2} = \frac{\omega_n^2}{s(s + \omega_n)^2} \qquad .....(2.30)$$

By partial fraction expansion, we can write,

$$C(s) = \frac{\omega_n^2}{s(s + \omega_n)^2} = \frac{A}{s} + \frac{B}{(s + \omega_n)^2} + \frac{C}{s + \omega_n}$$

$$A = s \times C(s)\Big|_{s=0} = \frac{\omega_n^2}{(s + \omega_n)^2}\Big|_{s=0} = \frac{\omega_n^2}{\omega_n^2} = 1$$

$$B = (s + \omega_n)^2 \times C(s)\Big|_{s=-\omega_n} = \frac{\omega_n^2}{s}\Big|_{s=-\omega_n} = -\omega_n$$

$$C = \frac{d}{ds}\Big[(s + \omega_n)^2 \times C(s)\Big]\Big|_{s=-\omega_n} = \frac{d}{ds}\left(\frac{\omega_n^2}{s}\right)\Big|_{s=-\omega_n} = \frac{-\omega_n^2}{s^2}\Big|_{s=-\omega_n} = -1$$

$$\therefore C(s) = \frac{A}{s} + \frac{B}{(s+\omega_n)^2} + \frac{C}{s+\omega_n} = \frac{1}{s} - \frac{\omega_n}{(s+\omega_n)^2} - \frac{1}{s+\omega_n}$$

The response in time domain,

$$c(t) = \mathcal{L}^{-1}\{C(s)\} = \mathcal{L}^{-1}\left\{\frac{1}{s} - \frac{\omega_n}{(s+\omega_n)^2} - \frac{1}{s+\omega_n}\right\}$$

$$c(t) = 1 - \omega_n t\, e^{-\omega_n t} - e^{-\omega_n t}$$

$$c(t) = 1 - e^{-\omega_n t}(1 + \omega_n t) \qquad\qquad .....(2.31)$$

$$\boxed{\mathcal{L}\{1\} = \frac{1}{s}}$$

$$\boxed{\mathcal{L}\{te^{-at}\} = \frac{1}{(s+a)^2}}$$

$$\boxed{\mathcal{L}\{e^{-at}\} = \frac{1}{s+a}}$$

The equation (2.31) is the response of critically damped closed loop second order system for unit step input. For step input of step value, A, the equation (2.31) should be multiplied by A.

$\therefore$ For closed loop critically damped second order system,

Unit step response $= 1 - e^{-\omega_n t}(1 + \omega_n t)$

Step response $\quad = A[1 - e^{-\omega_n t}(1 + \omega_n t)]$

Using equation (2.31), the response of critically damped second order system is sketched as shown in fig 2.11 and observed that the response has no oscillations.

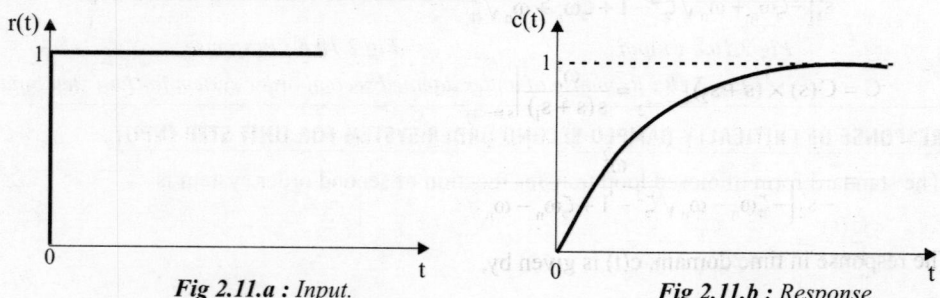

Fig 2.11.a : Input.   Fig 2.11.b : Response.

***Fig 2.11** : Response of critically damped second order system for unit step input.*

2.7.4 RESPONSE OF OVER DAMPED SECOND ORDER SYSTEM FOR UNIT STEP INPUT

The standard form of closed loop transfer function of second order system is,

$$\frac{C(s)}{R(s)} = \frac{\omega_n^2}{s^2 + 2\zeta\omega_n s + \omega_n^2}$$

For overdamped system $\zeta > 1$. The roots of the denominator of transfer function are real and distinct. Let the roots of the denominator be s_a, s_b.

$$s_a, s_b = -\zeta\omega_n \pm \omega_n\sqrt{\zeta^2 - 1} = -\left[\zeta\omega_n \pm \omega_n\sqrt{\zeta^2 - 1}\right] \qquad .....(2.32)$$

Let $s_1 = -s_a$ and $s_2 = -s_b$ $\qquad \therefore s_1 = -\zeta\omega_n - \omega_n\sqrt{\zeta^2 - 1}$ $\qquad .....(2.33)$

$$s_2 = \zeta\omega_n + \omega_n\sqrt{\zeta^2 - 1} \qquad\qquad .....(2.34)$$

The closed loop transfer function can be written in terms of s_1 and s_2 as shown below.

$$\frac{C(s)}{R(s)} = \frac{\omega_n^2}{s^2 + 2\zeta\omega_n s + \omega_n^2} = \frac{\omega_n^2}{(s+s_1)(s+s_2)} \qquad .....(2.35)$$

For unit step input r(t) = 1 and R(s) = 1/s.

$$\therefore \; C(s) = R(s)\, \frac{\omega_n^2}{(s+s_1)(s+s_2)} = \frac{\omega_n^2}{s(s+s_1)(s+s_2)}$$

By partial fraction expansion we can write,

$$C(s) = \frac{\omega_n^2}{s(s+s_1)(s+s_2)} = \frac{A}{s} + \frac{B}{s+s_1} + \frac{C}{s+s_2}$$

$$A = s \times C(s)\big|_{s=0} = s \times \frac{\omega_n^2}{s(s+s_1)(s+s_2)}\bigg|_{s=0} = \frac{\omega_n^2}{s_1 s_2}$$

$$= \frac{\omega_n^2}{\left[\zeta\omega_n - \omega_n\sqrt{\zeta^2-1}\right]\left[\zeta\omega_n + \omega_n\sqrt{\zeta^2-1}\right]} = \frac{\omega_n^2}{\zeta^2\omega_n^2 - \omega_n^2(\zeta^2-1)} = \frac{\omega_n^2}{\omega_n^2} = 1$$

$$B = (s+s_1) \times C(s)\big|_{s=-s_1} = \frac{\omega_n^2}{s(s+s_2)}\bigg|_{s=-s_1} = \frac{\omega_n^2}{-s_1(-s_1+s_2)}$$

$$= \frac{-\omega_n^2}{s_1\left[-\zeta\omega_n + \omega_n\sqrt{\zeta^2-1} + \zeta\omega_n + \omega_n\sqrt{\zeta^2-1}\right]} = \frac{-\omega_n^2}{\left[2\omega_n\sqrt{\zeta^2-1}\right]s_1} = \frac{-\omega_n}{2\sqrt{\zeta^2-1}}\frac{1}{s_1}$$

$$C = C(s) \times (s+s_2)\big|_{s=-s_2} = \frac{\omega_n^2}{s(s+s_1)}\bigg|_{s=-s_2} = \frac{\omega_n^2}{-s_2(-s_2+s_1)}$$

$$= \frac{\omega_n^2}{-s_2\left[-\zeta\omega_n - \omega_n\sqrt{\zeta^2-1} + \zeta\omega_n - \omega_n\sqrt{\zeta^2-1}\right]} = \frac{\omega_n^2}{\left[2\omega_n\sqrt{\zeta^2-1}\right]s_2} = \frac{\omega_n}{2\sqrt{\zeta^2-1}}\frac{1}{s_2}$$

The response in time domain, c(t) is given by,

$$c(t) = \mathcal{L}^{-1}\left\{\frac{1}{s} - \frac{\omega_n}{2\sqrt{\zeta^2-1}}\frac{1}{s_1}\frac{1}{(s+s_1)} + \frac{\omega_n}{2\sqrt{\zeta^2-1}}\frac{1}{s_2}\frac{1}{(s+s_2)}\right\}$$

$$c(t) = 1 - \frac{\omega_n}{2\sqrt{\zeta^2-1}}\frac{1}{s_1}e^{-s_1 t} + \frac{\omega_n}{2\sqrt{\zeta^2-1}}\frac{1}{s_2}e^{-s_2 t}$$

$$c(t) = 1 - \frac{\omega_n}{2\sqrt{\zeta^2-1}}\left(\frac{e^{-s_1 t}}{s_1} - \frac{e^{-s_2 t}}{s_2}\right) \qquad \qquad \ldots(2.36)$$

where, $s_1 = \zeta\omega_n - \omega_n\sqrt{\zeta^2-1}$

$$s_2 = \zeta\omega_n + \omega_n\sqrt{\zeta^2-1}$$

The equation (2.36) is the response of overdamped closed loop system for unit step input. For step input of value, A, the equation (2.36) is multiplied by A.

∴ For closed loop over damped second order system,

Unit step response $= 1 - \dfrac{\omega_n}{2\sqrt{\zeta^2-1}}\dfrac{1}{s_1}\left(\dfrac{e^{-s_1 t}}{s_1} - \dfrac{e^{-s_2 t}}{s_2}\right)$ where, $s_1 = \zeta\omega_n - \omega_n\sqrt{\zeta^2-1}$

Step response $= A\left[1 - \dfrac{\omega_n}{2\sqrt{\zeta^2-1}}\dfrac{1}{s_1}\left(\dfrac{e^{-s_1 t}}{s_1} - \dfrac{e^{-s_2 t}}{s_2}\right)\right]$ $s_2 = \zeta\omega_n + \omega_n\sqrt{\zeta^2-1}$

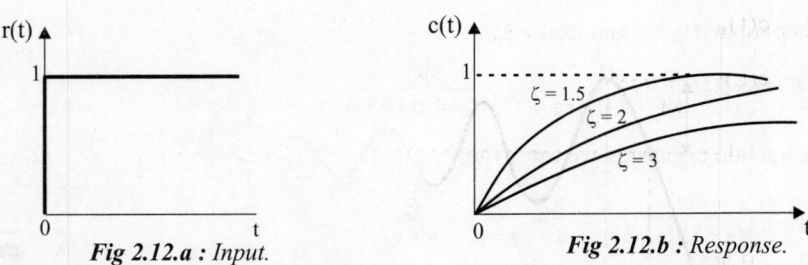

Fig 2.12.a : Input.   *Fig 2.12.b : Response.*

Fig 2.12 : Response of over damped second order system for unit step input.

Using equation (2.36), the response of overdamped second order system is sketched as shown in fig 2.12 and observed that the response has no oscillations but it takes longer time for the response to reach the final steady value.

2.8 TIME DOMAIN SPECIFICATIONS

The desired performance characteristics of control systems are specified in terms of time domain specifications. Systems with energy storage elements cannot respond instantaneously and will exhibit transient responses, whenever they are subjected to inputs or disturbances.

The desired performance characteristics of a system of any order may be specified in terms of the transient response to a unit step input signal. The response of a second order system for unit-step input with various values of damping ratio is shown in fig 2.13.

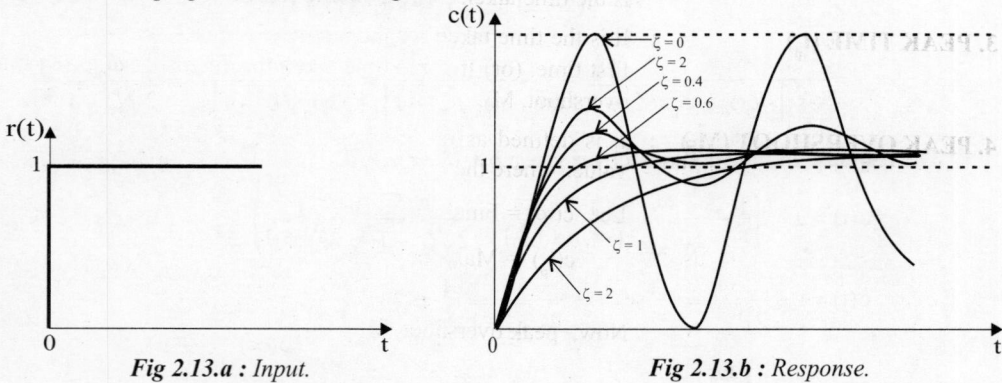

Fig 2.13.a : Input.   *Fig 2.13.b : Response.*

Fig 2.13 : Unit step response of second order system.

The transient response of a system to a unit step input depends on the initial conditions. Therefore to compare the time response of various systems it is necessary to start with standard initial conditions. The most practical standard is to start with the system at rest and so output and all time derivatives before t = 0 will be zero. The transient response of a practical control system often exhibits damped oscillation before reaching steady state. A typical damped oscillatory response of a system is shown in fig 2.14.

The transient response characteristics of a control system to a unit step input is specified in terms of the following time domain specifications.

     1. Delay time, t_d
     2. Rise time, t_r
     3. Peak time, t_p
     4. Maximum overshoot, M_p
     5. Settling time, t_s

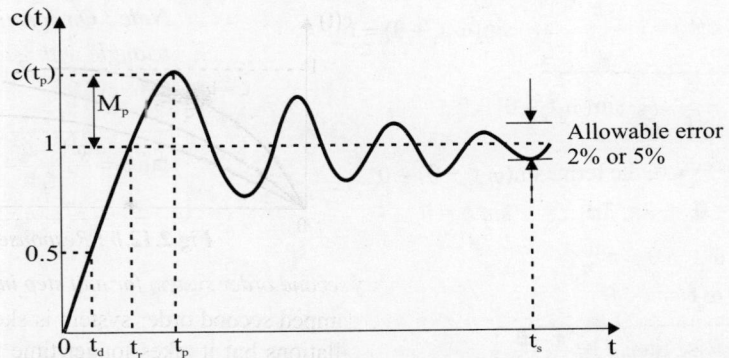

Fig 2.14 : Damped oscillatory response of second order system for unit step input.

The time domain specifications are defined as follows.

1. DELAY TIME (t_d) : It is the time taken for response to reach 50% of the final value, for the very first time.

2. RISE TIME (t_r) : It is the time taken for response to raise from 0 to 100% for the very first time. For underdamped system, the rise time is calculated from 0 to 100%. But for overdamped system it is the time taken by the response to raise from 10% to 90%. For critically damped system, it is the time taken for response to raise from 5% to 95%.

3. PEAK TIME (t_p) : It is the time taken for the response to reach the peak value the very first time. (or) It is the time taken for the response to reach the peak overshoot, M_p.

4. PEAK OVERSHOOT (M_p) : It is defined as the ratio of the maximum peak value to the final value, where the maximum peak value is measured from final value.

Let, $c(\infty)$ = Final value of c(t)

$c(t_p)$ = Maximum value of c(t).

Now, peak overshoot, $M_p = \dfrac{c(t_p) - c(\infty)}{c(\infty)}$ .....(2.37)

% Peak overshoot, % $M_p = \dfrac{c(t_p) - c(\infty)}{c(\infty)} \times 100$ .....(2.38)

5. SETTLING TIME (t_s) : It is defined as the time taken by the response to reach and stay within a specified error. It is usually expressed as % of final value. The usual tolerable error is 2 % or 5% of the final value.

EXPRESSIONS FOR TIME DOMAIN SPECIFICATIONS

Rise time (t_r)

The unit step response of second order system for underdamped case is given by,

$$c(t) = 1 - \frac{e^{-\zeta\omega_n t_r}}{\sqrt{1-\zeta^2}} \sin(\omega_d t + \theta)$$

At $t = t_r$, $c(t) = c(t_r) = 1$ **(Refer fig 2.14).**

$$\therefore \ c(t_r) = 1 - \frac{e^{-\zeta\omega_n t_r}}{\sqrt{1-\zeta^2}} \ \sin(\omega_d t_r + \theta) = 1$$

$$\therefore \ \frac{-e^{-\zeta\omega_n t_r}}{\sqrt{1-\zeta^2}} \ \sin(\omega_d t_r + \theta) = 0$$

> **Note :** On constructing right angle triangle with ζ and $\sqrt{1-\zeta^2}$, we get
>
> $$\tan\theta = \frac{\sqrt{1-\zeta^2}}{\zeta}$$

Since $-e^{-\zeta\omega_n t_r} \neq 0$, the term, $\sin(\omega_d t_r + \theta) = 0$.

When, $\phi = 0,\ \pi,\ 2\pi,\ 3\pi \ ....,\qquad \sin\phi = 0$

$$\therefore \ \omega_d t_r + \theta = \pi$$

$$\omega_d t_r = \pi - \theta$$

$$\boxed{\therefore \ \text{Rise time, } t_r = \frac{\pi-\theta}{\omega_d}}$$

.....(2.39)

Here, $\theta = \tan^{-1}\dfrac{\sqrt{1-\zeta^2}}{\zeta}$;   Damped frequency of oscillation, $\omega_d = \omega_n\sqrt{1-\zeta^2}$ (Refer note)

$$\boxed{\therefore \ \text{Rise time, } t_r = \frac{\pi - \tan^{-1}\dfrac{\sqrt{1-\zeta^2}}{\zeta}}{\omega_n\sqrt{1-\zeta^2}} \ \text{in sec}}$$

.....(2.40)

> **Note :** θ or $\tan^{-1}\dfrac{\sqrt{1-\zeta^2}}{\zeta}$ should be measured in radians.

Peak time (t_p)

To find the expression for peak time, t_p, differentiate $c(t)$ with respect to t and equate to 0.

i.e., $\dfrac{d}{dt}\,c(t)\Big|_{t=t_p} = 0$

> **Note :** For any signal $f(t)$, maximum value is obtained by $\dfrac{df(t)}{dt} = 0$

The unit step response of under damped second order system is given by,

$$c(t) = 1 - \frac{e^{-\zeta\omega_n t}}{\sqrt{1-\zeta^2}} \ \sin(\omega_d t + \theta)$$

Differentiating $c(t)$ with respect to **t**.

$$\frac{d}{dt}c(t) = \frac{-e^{-\zeta\omega_n t}}{\sqrt{1-\zeta^2}}(-\zeta\omega_n)\sin(\omega_d t + \theta) + \left(\frac{-e^{-\zeta\omega_n t}}{\sqrt{1-\zeta^2}}\right)\cos(\omega_d t + \theta)\omega_d$$

Put, $\omega_d = \omega_n\sqrt{1-\zeta^2}$

$$\therefore \ \frac{d}{dt}c(t) = \frac{e^{-\zeta\omega_n t}}{\sqrt{1-\zeta^2}}(\zeta\omega_n)\sin(\omega_d t + \theta) - \frac{\omega_n\sqrt{1-\zeta^2}}{\sqrt{1-\zeta^2}}e^{-\zeta\omega_n t}\cos(\omega_d t + \theta)$$

$$= \frac{\omega_n e^{-\zeta\omega_n t}}{\sqrt{1-\zeta^2}}\left[\zeta\sin(\omega_d t + \theta) - \sqrt{1-\zeta^2}\cos(\omega_d t + \theta)\right]$$

$$= \frac{\omega_n}{\sqrt{1-\zeta^2}}e^{-\zeta\omega_n t}\left[\cos\theta\sin(\omega_d t + \theta) - \sin\theta\cos(\omega_d t + \theta)\right] \ \text{(refer note)}$$

$$= \frac{\omega_n}{\sqrt{1-\zeta^2}}e^{-\zeta\omega_n t}\left[\sin(\omega_d t + \theta)\cos\theta - \cos(\omega_d t + \theta)\sin\theta\right]$$

$$= \frac{\omega_n}{\sqrt{1-\zeta^2}} \, e^{-\zeta\omega_n t} [\sin((\omega_d t + \theta) - \theta)] = \frac{\omega_n}{\sqrt{1-\zeta^2}} \, e^{-\zeta\omega_n t} \sin(\omega_d t)$$

at $t = t_p$, $\frac{d}{dt} c(t) = 0$

$$\therefore \frac{\omega_n}{\sqrt{1-\zeta^2}} \, e^{-\zeta\omega_n t_p} \sin(\omega_d t_p) = 0$$

Since, $e^{-\zeta\omega_n t_p} \neq 0$, the term, $\sin(\omega_d t_p) = 0$.

When $\phi = 0, \pi, 2\pi, 3\pi, \sin\phi = 0$

$$\therefore \omega_d t_p = \pi$$

$$\boxed{\therefore \text{ Peak time, } t_p = \frac{\pi}{\omega_d}} \qquad \qquad .....(2.41)$$

The damped frequency of oscillation, $\omega_d = \omega_n \sqrt{1-\zeta^2}$

$$\boxed{\therefore \text{ Peak time, } t_p = \frac{\pi}{\omega_n \sqrt{1-\zeta^2}}} \qquad .....(2.42)$$

> **Note :** *On constructing right angle triangle with ζ and $\sqrt{1-\zeta^2}$, we get*
> $$\sin\theta = \sqrt{1-\zeta^2}$$
> $$\cos\theta = \zeta$$

Peak overshoot (M$_p$)

$$\boxed{\text{\% Peak overshoot, } \%M_p = \frac{c(t_p) - c(\infty)}{c(\infty)} \times 100} \qquad .....(2.43)$$

where, $c(t_p) = $ Peak response at $t = t_p$.

$c(\infty) = $ Final steady state value.

The unit step response of second order system is given by,

$$c(t) = 1 - \frac{e^{-\zeta\omega_n t}}{\sqrt{1-\zeta^2}} \sin(\omega_d t + \theta)$$

At $t = \infty$, $c(t) = c(\infty) = 1 - \frac{e^{-\infty}}{\sqrt{1-\zeta^2}} \sin(\omega_d t + \theta) = 1 - 0 = 1$

$$\boxed{t_p = \frac{\pi}{\omega_d}}$$

At $t = t_p$, $c(t) = c(t_p) = 1 - \frac{e^{-\zeta\omega_n t_p}}{\sqrt{1-\zeta^2}} \sin(\omega_d t_p + \theta)$

$$= 1 - \frac{e^{-\zeta\omega_n \frac{\pi}{\omega_d}}}{\sqrt{1-\zeta^2}} \sin\left(\omega_d \frac{\pi}{\omega_d} + \theta\right)$$

$$\boxed{\omega_d = \omega_n \sqrt{1-\zeta^2}}$$
$$\boxed{\sin(\pi + \theta) = -\sin\theta}$$

$$= 1 - \frac{e^{-\zeta\omega_n \frac{\pi}{\omega_n\sqrt{1-\zeta^2}}} \sin(\pi + \theta)}{\sqrt{1-\zeta^2}}$$

> **Note :** *On constructing right angle triangle with ζ and $\sqrt{1-\zeta^2}$, we get*
> $$\sin\theta = \sqrt{1-\zeta^2}$$

$$= 1 + \frac{e^{-\frac{\zeta\pi}{\sqrt{1-\zeta^2}}}}{\sqrt{1-\zeta^2}} \sin\theta$$

$$= 1 + \frac{e^{-\frac{\zeta\pi}{\sqrt{1-\zeta^2}}}}{\sqrt{1-\zeta^2}} \sqrt{1-\zeta^2} = 1 + e^{-\frac{\zeta\pi}{\sqrt{1-\zeta^2}}} \qquad .....(2.44)$$

$$\text{Percentage Peak Overshoot, } \% \, M_p = \frac{c(t_p) - c(\infty)}{c(\infty)} \times 100 = \frac{1 + e^{-\frac{\zeta\pi}{\sqrt{1-\zeta^2}}} - 1}{1} \times 100$$

$$= e^{-\frac{\zeta\pi}{\sqrt{1-\zeta^2}}} \times 100$$

$$\therefore \text{ Percentage Peak Overshoot, } \% \, M_p = = e^{-\frac{\zeta\pi}{\sqrt{1-\zeta^2}}} \times 100 \qquad \text{.....(2.45)}$$

Settling time (t_s)

The response of second order system has two components. They are,

1. Decaying exponential component, $\dfrac{e^{-\zeta\omega_n t}}{\sqrt{1-\zeta^2}}$.

1. Decaying exponential component, $\sin(\omega_d t + \theta)$.

2. Sinusoidal component, $\sin(\omega_d t + \theta)$.

In this the decaying exponential term dampens (or) reduces the oscillations produced by sinusoidal component. Hence the settling time is decided by the exponential component. The settling time can be found out by equating exponential component to percentage tolerance errors.

For 2 % tolerance error band, at $t = t_s$, $\dfrac{e^{-\zeta\omega_n t_s}}{\sqrt{1-\zeta^2}} = 0.02$

For least values of ζ, $e^{-\zeta\omega_n t_s} = 0.02$.

On taking natural logarithm we get,

$$-\zeta\omega_n t_s = ln(0.02) \qquad \Rightarrow \qquad -\zeta\omega_n t_s = -4 \qquad \Rightarrow \qquad t_s = \frac{4}{\zeta\omega_n}$$

For the second order system, the time constant, $T = \dfrac{1}{\zeta\omega_n}$

$$\therefore \text{ Settling time, } t_s = \frac{1}{\zeta\omega_n} = 4T \qquad \text{(for 2\% error)} \qquad \text{.....(2.46)}$$

For 5% error, $e^{-\zeta\omega_n t_s} = 0.05$

On taking natural logarithm we get,

$$-\zeta\omega_n t_s = ln(0.05) \qquad \Rightarrow \qquad -\zeta\omega_n t_s = -3 \qquad \Rightarrow \qquad t_s = \frac{3}{\zeta\omega_n}$$

$$\therefore \text{ Settling time, } t_s = \frac{3}{\zeta\omega_n} = 3T \qquad \text{(for 5\% error)} \qquad \text{.....(2.47)}$$

In general for a specified percentage error, Settling time can be evaluted using equation (2.48).

$$\therefore \text{ Settling time, } t_s = \frac{ln(\% \text{ error})}{\zeta\omega_n} = \frac{ln(\% \text{ error})}{T} \qquad \text{.....(2.48)}$$

EXAMPLE 2.1

Obtain the response of unity feedback system whose open loop transfer function is $G(s) = \dfrac{4}{s(s+5)}$ and when the input is unit step.

SOLUTION

The closed loop system is shown in fig 1.

The closed loop transfer function, $\dfrac{C(s)}{R(s)} = \dfrac{G(s)}{1+G(s)}$

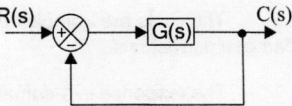

Fig 1 : Closed loop system.

$$\therefore \frac{C(s)}{R(s)} = \frac{\dfrac{4}{s(s+5)}}{1+\dfrac{4}{s(s+5)}} = \frac{\dfrac{4}{s(s+5)}}{\dfrac{s(s+5)+4}{s(s+5)}} = \frac{4}{s(s+5)+4} = \frac{4}{s^2+5s+4} = \frac{4}{(s+4)(s+1)}$$

The response in s-domain, $C(s) = R(s)\dfrac{4}{(s+1)(s+4)}$

Since the input is unit step, $R(s) = \dfrac{1}{s}$; $\therefore C(s) = \dfrac{4}{(s+1)(s+4)}$

By partial fraction expansion, we can write,

$$C(s) = \frac{4}{s(s+1)(s+4)} = \frac{A}{s} + \frac{B}{s+1} + \frac{C}{s+4}$$

$$A = C(s) \times s\Big|_{s=0} = \frac{4}{(s+1)(s+4)}\Big|_{s=0} = \frac{4}{1\times4} = 1$$

$$B = C(s) \times (s+1)\Big|_{s=-1} = \frac{4}{s(s+4)}\Big|_{s=-1} = \frac{4}{-1(-1+4)} = \frac{-4}{3}$$

$$C = C(s) \times (s+4)\Big|_{s=-4} = \frac{4}{s(s+1)}\Big|_{s=-4} = \frac{4}{-4(-4+1)} = \frac{1}{3}$$

The time domain response c(t) is obtained by taking inverse Laplace transform of C(s).

Response in time domain, $c(t) = \mathcal{L}^{-1}\{C(s)\} = \mathcal{L}^{-1}\left\{\dfrac{1}{s} - \dfrac{4}{3}\dfrac{1}{s+1} + \dfrac{1}{3}\dfrac{1}{s+4}\right\}$

$$= 1 - \frac{4}{3}e^{-t} + \frac{1}{3}e^{-4t} = 1 - \frac{1}{3}\left[4e^{-t} - e^{-4t}\right]$$

RESULT

Response of unity feedback system, $c(t) = 1 - \dfrac{1}{3}\left[4e^{-t} - e^{-4t}\right]$

EXAMPLE 2.2

A positional control system with velocity feedback is shown in fig 1. What is the response of the system for unit step input.

SOLUTION

The closed loop transfer function,

$$\frac{C(s)}{R(s)} = \frac{G(s)}{1+G(s)H(s)}$$

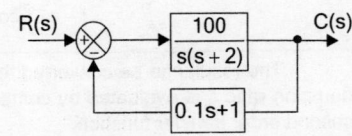

Fig 1 : Positional control system.

Given that, $G(s) = \dfrac{100}{s(s+2)}$ and $H(s) = 0.1s + 1$

$$\therefore \frac{C(s)}{R(s)} = \frac{\dfrac{100}{s(s+2)}}{1+\left(\dfrac{100}{s(s+2)}\right)(0.1s+1)} = \frac{\dfrac{100}{s(s+2)}}{\dfrac{s(s+2)+100(0.1s+1)}{s(s+2)}} = \frac{100}{s^2+2s+10s+100} = \frac{100}{s^2+12s+100}$$

Here $(s^2 + 12s + 100)$ is characteristic polynomial. The roots of the characteristic polynomial are,

$$s_1, s_2 = \frac{-12 \pm \sqrt{144 - 400}}{2} = \frac{-12 \pm j16}{2} = -6 \pm j8$$

The roots are complex conjugate. The system is underdamped and so the response of the system will have damped oscillations.

The response in s-domain, $C(s) = R(s) \dfrac{100}{s^2 + 12s + 100}$

Since input is unit step, $R(s) = \dfrac{1}{s}$

$$\therefore \ C(s) = \frac{1}{s} \frac{100}{s^2 + 12s + 100} = \frac{100}{s(s^2 + 12s + 100)}$$

By partial fraction expansion we can write,

$$C(s) = \frac{100}{s(s^2 + 12s + 100)} = \frac{A}{s} + \frac{Bs + C}{s^2 + 12s + 100}$$

The residue A is obtained by multiplying C(s) by s and letting s = 0.

$$A = C(s) \times s \Big|_{s=0} = \frac{100}{s^2 + 12s + 100} \Big|_{s=0} = \frac{100}{100} = 1$$

The residue B and C are evaluated by cross multiplying the following equation and equating the coefficients of like power of s.

$$\frac{100}{s(s^2 + 12s + 100)} = \frac{A}{s} + \frac{Bs + C}{s^2 + 12s + 100}$$

$$100 = A(s^2 + 12s + 100) + (Bs + C)s$$

$$100 = As^2 + 12As + 100A + Bs^2 + Cs$$

On equating the coefficients of s^2 we get, $0 = A + B$ $\therefore \ B = -A = -1$

On equating coefficients of s we get, $0 = 12A + C$ $\therefore \ C = -12A = 12$

$$\therefore \ C(s) = \frac{1}{s} + \frac{-s - 12}{s^2 + 12s + 100} = \frac{1}{s} - \frac{s + 12}{s^2 + 12s + 36 + 64} = \frac{1}{s} - \frac{s + 6 + 6}{(s + 6)^2 + 8^2}$$

$$= \frac{1}{s} - \frac{s + 6}{(s + 6)^2 + 8^2} - \frac{6}{(s + 6)^2 + 8^2} = \frac{1}{s} - \frac{s + 6}{(s + 6)^2 + 8^2} - \frac{6}{8} \frac{8}{(s + 6)^2 + 8^2}$$

The time domain response is obtained by taking inverse Laplace transform of C(s).

Time response, $c(t) = \mathcal{L}^{-1}\{C(s)\} = \mathcal{L}^{-1}\left\{ \dfrac{1}{s} - \dfrac{s + 6}{(s + 6)^2 + 8^2} - \dfrac{6}{8} \dfrac{8}{(s + 6)^2 + 8^2} \right\}$

$$= 1 - e^{-6t} \cos 8t - \frac{6}{8} e^{-6t} \sin 8t = 1 - e^{-6t} \left[\frac{6}{8} \sin 8t + \cos 8t \right]$$

The result can be converted to another standard form by constructing right angle triangle with ζ and. The damping ratio ζ is evaluated by comparing the closed loop transfer function of the system with standard form of second order transfer function.

$$\therefore \ \frac{C(s)}{R(s)} = \frac{\omega_n^2}{s^2 + 2\zeta\omega_n + \omega_n^2} = \frac{100}{s^2 + 12s + 100}$$

On comparing we get, $\omega_n^2 = 100$ $\Big|$ $2\zeta\omega_n = 12$

$$\therefore \ \omega_n = 10 \qquad \Big| \qquad \therefore \ \zeta = \frac{12}{2\omega_n} = \frac{12}{2 \times 10} = 0.6$$

Constructing right angled triangle with ζ and $\sqrt{1-\zeta^2}$ we get,

$$\sin\theta = 0.8 \quad ; \quad \cos\theta = 0.6 \quad ; \quad \tan\theta = \frac{0.8}{0.6}$$

$$\therefore \theta = \tan^{-1}\frac{0.8}{0.6} = 53° = 53° \times \frac{\pi}{180°} \text{ rad} = 0.925 \text{ rad}.$$

$$\therefore \text{ Time response, } c(t) = 1 - \epsilon^{-6t}\left[\frac{6}{8}\sin 8t + \cos 8t\right] = 1 - e^{-6t}\frac{10}{8}\left[\frac{6}{10}\sin 8t + \frac{8}{10}\cos 8t\right]$$

$$= 1 - \frac{10}{3}e^{-6t}[\sin 8t \times 0.6 + \cos 8t \times 0.8] = 1 - 1.25\,e^{-6t}[\sin 8t\cos\theta + \cos 8t\sin\theta]$$

$$= 1 - 1.25e^{-6t}[\sin(8t+\theta)] = 1 - 1.25e^{-6t}\sin(8t+0.925)$$

Note : θ is expressed in radians

$$\cos\theta = 0.6$$
$$\sin\theta = 0.8$$

RESULT

The response in time domain,

$$c(t) = 1 - e^{-6t}\left[\frac{6}{8}\sin 8t + \cos 8t\right] \quad \text{or} \quad c(t) = 1 - 1.25\,e^{-6t}\sin(8t+0.925)$$

EXAMPLE 2.3

The response of a servomechanism is, $c(t) = 1 + 0.2\,e^{-60t} - 1.2\,e^{-10t}$ when subject to a unit step input. Obtain an expression for closed loop transfer function. Determine the undamped natural frequency and damping ratio.

SOLUTION

Given that, $c(t) = 1 + 0.2\,e^{-60t} - 1.2\,e^{-10t}$

On taking Laplace transform of c(t) we get,

$$C(s) = \frac{1}{s} + 0.2\frac{1}{(s+60)} - 1.2\frac{1}{(s+10)} = \frac{(s+60)(s+10) + 0.2s(s+10) - 1.2s(s+60)}{s(s+60)(s+10)}$$

$$= \frac{s^2 + 70s + 600 + 0.2s^2 + 2s - 1.2s^2 - 72s}{s(s+60)(s+10)} = \frac{600}{s(s+60)(s+10)} = \frac{1}{s}\frac{600}{(s+60)(s+10)}$$

Since input is unit step, R(s) = 1/s.

$$\therefore C(s) = R(s)\frac{600}{(s+60)(s+10)} = R(s)\frac{600}{s^2 + 70s + 600}$$

The damping ratio and natural frequency of oscillation can be estimated by comparing the system transfer function with standard form of second order transfer function.

$$\therefore \frac{C(s)}{R(s)} = \frac{\omega_n^2}{s^2 + 2\zeta\omega_n s + \omega_n^2} = \frac{600}{s^2 + 70s + 60}$$

On comparing we get,

$$\omega_n^2 = 600 \qquad\qquad 2\zeta\omega_n = 70$$

$$\therefore \omega_n = \sqrt{600} = 24.49 \text{ rad/ sec} \qquad \therefore \zeta = \frac{70}{2\omega_n} = \frac{70}{2 \times 24.49} = 1.43$$

RESULT

The closed loop transfer function of the system, $\dfrac{C(s)}{R(s)} = \dfrac{600}{s^2 + 70s + 600}$

Natural frequency of oscillation, $\omega_n = 24.49$ rad/sec

Damping ratio, $\zeta = 1.43$

EXAMPLE 2.4

The unity feedback system is characterized by an open loop transfer function G(s) =K/s (s +10). Determine the gain K, so that the system will have a damping ratio of 0.5 for this value of K. Determine peak overshoot and time at peak overshoot for a unit step input.

SOLUTION

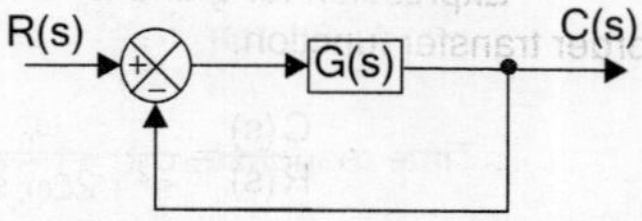

The unity feedback system is shown in fig 1.

The closed loop transfer function $\dfrac{C(s)}{R(s)} = \dfrac{G(s)}{1+G(s)}$

Fig 1 : *Unity feedback system.*

Given that, G(s) = K/s (s +10)

$$\therefore \frac{C(s)}{R(s)} = \frac{\dfrac{K}{s(s+10)}}{1+\dfrac{K}{s(s+10)}} = \frac{K}{s(s+10)+K} = \frac{K}{s^2+10s+K}$$

The value of K can be evaluated by comparing the system transfer function with standard form of second order transfer function.

$$\therefore \frac{C(s)}{R(s)} = \frac{\omega_n^2}{s^2+2\zeta\omega_n s+\omega_n^2} = \frac{K}{s^2+10s+K}$$

On comparing we get,

$\omega_n^2 = K$	$2\zeta\omega_n = 10$	K=100
$\therefore\ \omega_n = \sqrt{K}$	Put $\zeta = 0.5$ and $\omega_n = \sqrt{K}$	$\omega_n = 10$ rad/sec
	$\therefore\ 2\times0.5\times\sqrt{K} = 10$	
	$\sqrt{K} = 10$	

The value of gain, K =100.

Percentage peak overshoot, $\% \ M_p = e^{-\zeta\pi/\sqrt{1-\zeta^2}} \times 100$

$$= e^{-0.5\pi/\sqrt{1-0.5^2}} \times 100 = 0.163 \times 100 = 16.3\%$$

Peak time, $t_p = \dfrac{\pi}{\omega_d} = \dfrac{\pi}{\omega_n\sqrt{1-\zeta^2}} = \dfrac{\pi}{10\sqrt{1-0.5^2}} = 0.363$ sec

RESULT

The value of gain,	K =	100
Percentage peak overshoot,	%M_p =	16.3%
Peak time,	t_p =	0.363 sec.

EXAMPLE 2.5

The open loop transfer function of a unity feedback system is given by G(s) = K/s (sT + 1), where K and T are positive constant. By what factor should the amplifier gain K be reduced, so that the peak overshoot of unit step response of the system is reduced from 75% to 25%.

SOLUTION

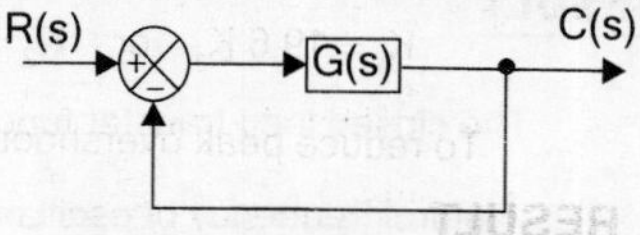

The unity feedback system is shown in fig 1.

The closed loop transfer function, $\dfrac{C(s)}{R(s)} = \dfrac{G(s)}{1+G(s)}$

Fig 1 : *Unity feedback system.*

Given that, G(s) = K/s (sT +1)

$$\therefore \frac{C(s)}{R(s)} = \frac{K/s\,(sT+1)}{1+K/s\,(sT+1)} = \frac{K}{s(sT+1)+K} = \frac{K}{s^2T+s+K} = \frac{K/T}{s^2+\frac{1}{T}s+\frac{K}{T}}$$

Expression for ζ and ω_n can be obtained by comparing the transfer function with the standard form of second order transfer function.

$$\therefore \frac{C(s)}{R(s)} = \frac{\omega_n^2}{s^2+2\zeta\omega_n s+\omega_n^2} = \frac{K/T}{s^2+\frac{1}{T}s+\frac{K}{T}}$$

On comparing we get,

$$\omega_n^2 = K/T \qquad\qquad 2\zeta\omega_n = 1/T$$

$$\therefore \omega_n = \sqrt{K/T} \qquad\qquad \zeta = \frac{1}{2\omega_n T} = \frac{1}{2\sqrt{\frac{K}{T}}\,T} = \frac{1}{2\sqrt{KT}}$$

The peak overshoot, M_p is reduced by increasing the damping ratio ζ. The damping ratio ζ is increased by reducing the gain K.

When $M_p = 0.75$, Let $\zeta = \zeta_1$ and $K = K_1$
When $M_p = 0.25$, Let $\zeta = \zeta_2$ and $K = K_2$

Peak overshoot, $M_p = e^{-\zeta\pi\sqrt{1-\zeta^2}}$

Taking natural logarithm on both sides, $lnM_p = \dfrac{-\zeta\pi}{\sqrt{1-\zeta^2}}$

On squarring we get, $(lnM_p)^2 = \dfrac{\zeta^2\pi^2}{1-\zeta^2}$

On crossing multiplication we get,

$$(1-\zeta^2)(lnM_p) = \zeta^2\pi^2$$

$$(lnM_p)^2 - \zeta^2(lnM_p) = \zeta^2\pi^2$$

$$(lnM_p)^2 = \zeta^2\pi^2 + \zeta^2(lnM_p)^2$$

$$(lnM_p)^2 = \zeta^2[\pi^2+(lnM_p)^2]$$

$$\therefore \zeta^2 = \frac{(lnM_p)^2}{\pi^2+(lnM_p)^2} \qquad .....(1)$$

But $\zeta = \dfrac{1}{2\sqrt{KT}}$, $\quad \therefore \zeta^2 = \dfrac{1}{4KT} \qquad .....(2)$

On equating, equation (1) & (2) we get,

$$\frac{1}{4KT} = \frac{(lnM_p)^2}{\pi^2+(lnM_p)^2}$$

$$\frac{1}{K} = \frac{4T\,(lnM_p)^2}{\pi^2+(lnM_p)^2}$$

$$K = \frac{\pi^2+(lnM_p)^2}{4T\,(lnM_p)^2}$$

When, $K = K_1$, $M_p = 0.75$. $\therefore K_1 = \dfrac{\pi^2+(ln0.75)^2}{4T(ln0.75)^2} = \dfrac{9.952}{0.331T} = \dfrac{30.06}{T}$

When, $K = K_2$, $M_p = 0.25$, $\therefore K_2 = \dfrac{\pi^2+(ln0.25)^2}{4T(ln0.25)^2} = \dfrac{11.79}{7.68T} = \dfrac{1.53}{T}$

$$\therefore \frac{K_1}{K_2} = \frac{(1/T)30.06}{(1/T)1.53} = 19.6$$

$$K_1 = 19.6\,K_2 \quad \text{or} \quad K_2 = \frac{1}{19.6}K_1$$

To reduce peak overshoot from 0.75 to 0.25, K should be reduced by 19.6 times (approximately 20 times).

RESULT

The value of gain, K should be reduced approximately 20 times to reduce peak overshoot from 0.75 to 0.25.

EXAMPLE 2.6

A positional control system with velocity feedback is shown in fig 1. What is the response c(t) to the unit step input. Given that $\zeta = 0.5$. Also calculate rise time, peak time, maximum overshoot and settling time.

SOLUTION

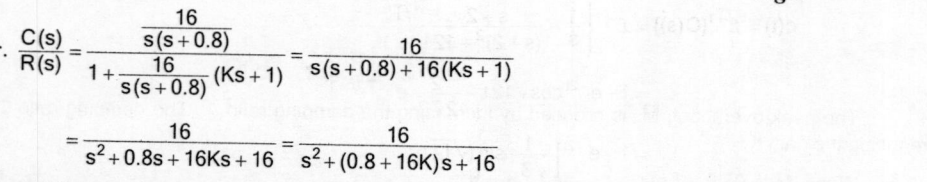

Fig 1

The closed loop transfer function, $\dfrac{C(s)}{R(s)} = \dfrac{G(s)}{1 + G(s)H(s)}$

Given that $G(s) = 16/s(s + 0.8)$ and $H(s) = Ks + 1$

$$\therefore \frac{C(s)}{R(s)} = \frac{\dfrac{16}{s(s+0.8)}}{1 + \dfrac{16}{s(s+0.8)}(Ks+1)} = \frac{16}{s(s+0.8) + 16(Ks+1)}$$

$$= \frac{16}{s^2 + 0.8s + 16Ks + 16} = \frac{16}{s^2 + (0.8 + 16K)s + 16}$$

The values of K and ω_n are obtained by comparing the system transfer function with standard form of second order transfer function.

$$\therefore \frac{C(s)}{R(s)} = \frac{\omega_n^2}{s^2 + 2\zeta\omega_n + \omega_n^2} = \frac{16}{s^2 + (0.8 + 16K)s + 16}$$

On comparing we get.

$$\omega_n^2 = 16 \qquad \qquad 0.8 + 16K = 2\zeta\omega_n$$

$$\therefore \omega_n = 4 \text{ rad/sec} \qquad \therefore K = \frac{2\zeta\omega_n - 0.8}{16} = \frac{2 \times 0.5 \times 4 - 0.8}{16} = 0.2$$

$$\therefore \frac{C(s)}{R(s)} = \frac{16}{s^2 + (0.8 + 16 \times 0.2)s + 16} = \frac{16}{s^2 + 4s + 16}$$

Given that the damping ratio, $\zeta = 0.5$. Hence the system is underdamped and so the response of the system will have damped oscillations. The roots of characteristic polynomial will be complex conjugate.

The response in s-domain, $C(s) = R(s) \dfrac{16}{s^2 + 4s + 16}$

For unit step input, $R(s) = 1/s$.

$$\therefore C(s) = \frac{1}{s} \frac{16}{s^2 + 4s + 16} = \frac{16}{s(s^2 + 4s + 16)}$$

By partial fraction expansion we can write,

$$C(s) = \frac{16}{s(s^2 + 4s + 16)} = \frac{A}{s} + \frac{Bs + C}{s^2 + 4s + 16}$$

The residue A is obtained by multiplying C(s) by s and letting s = 0.

$$A = C(s) \times s\Big|_{s=0} = \frac{16}{s^2 + 4s + 16}\Big|_{s=0} = \frac{16}{16} = 1$$

The residues B and C are evaluated by cross multiplying the following equation and equating the coefficients of like powers of s.

$$\frac{16}{s(s^2 + 4s + 16)} = \frac{A}{s} + \frac{Bs + C}{s^2 + 4s + 16}$$

On cross multiplication we get, $16 = A(s^2 + 4s + 16) + (Bs + C)s$

$$16 = As^2 + 4As + 16A + Bs^2 + Cs$$

On equating the coefficients of s^2 we get, $0 = A + B$ $\therefore B = -A = -1$

On equating the coefficients of s we get, $0 = 4A + C$ $\therefore C = -4A = -4$

$$\therefore C(s) = \frac{1}{s} + \frac{-s-4}{s^2+4s+16} = \frac{1}{s} - \frac{s+4}{s^2+4s+4+12}$$

$$= \frac{1}{s} - \frac{s+2+2}{(s+2)^2+12} = \frac{1}{s} - \frac{s+2}{(s+2)^2+12} - \frac{2}{\sqrt{12}}\frac{\sqrt{12}}{(s+2)^2+12}$$

The time domain response is obtained by taking inverse Laplace transform of C(s).

The response in time domain,

$$c(t) = \mathcal{L}^{-1}\{C(s)\} = \mathcal{L}^{-1}\left\{\frac{1}{s} - \frac{s+2}{(s+2)^2+12} - \frac{2}{\sqrt{12}}\frac{\sqrt{12}}{(s+2)^2+12}\right\}$$

$$= 1 - e^{-2t}\cos\sqrt{12}t - \frac{2}{2\sqrt{3}}e^{-2t}\sin\sqrt{12}t$$

$$= 1 - e^{-2t}\left[\frac{1}{\sqrt{3}}\sin(\sqrt{12}t) + \cos(\sqrt{12}t)\right]$$

The result can be converted to another standard form by constructing right angle triangle with ζ and $\sqrt{1-\zeta^2}$.

On constructing right angle triangle with ζ and $\sqrt{1-\zeta^2}$ we get,

$\sin\theta = 0.866 = \sqrt{3}/2$; $\cos\theta = 0.5 = 1/2$; $\tan\theta = 1.732$

$\therefore \theta = \tan^{-1}1.732 = 60° = 1.047$ rad

$\therefore$ The response in time domain,

$$c(t) = 1 - e^{-2t}\left[\frac{1}{\sqrt{3}} \times 2 \times \sin\sqrt{12}t \times \frac{1}{2} + \frac{2}{\sqrt{3}} \times \cos\sqrt{12}t \times \frac{\sqrt{3}}{2}\right]$$

$$= 1 - e^{-2t}\frac{2}{\sqrt{3}}\left[\sin\sqrt{12}t\cos\theta + \cos\sqrt{12}t\sin\theta\right]$$

$$= 1 - \frac{2}{\sqrt{3}}e^{-2t}\left[\sin\sqrt{12}t + \theta\right] = 1 - \frac{2}{\sqrt{3}}e^{-2t}\left[\sin\sqrt{12}t + 1.047\right]$$

Note : θ is expressed in radians

Damped frequency of oscillation $\left.\right\}$ $\omega_d = \omega_n\sqrt{1-\zeta^2} = 4\sqrt{1-0.5^2} = 3.464$ rad/sec

$\therefore$ Rise time, $t_r = \dfrac{\pi - \theta}{\omega_d} = \dfrac{\pi - 1.047}{3.464} = 0.6046$ sec

Peak time, $t_p = \dfrac{\pi}{\omega_d} = \dfrac{\pi}{3.464} = 0.907$ sec

% Maximum overshoot $\left.\right\}$ $\% M_p = e^{\frac{-\zeta\pi}{\sqrt{1-\zeta^2}}} \times 100 = e^{\frac{-0.5 \times \pi}{\sqrt{1-0.5^2}}} \times 100 = 0.163 \times 100 = 16.3\%$

Time constant, $T = \dfrac{1}{\zeta\omega_n} = \dfrac{1}{0.5 \times 4} = 0.5$ sec

For 5% error, Settling time, $t_s = 3T = 3 \times 0.5 = 1.5$ sec

For 2% error, Settling time, $t_s = 4T = 4 \times 0.5 = 2$ sec

RESULT

The time domain response, $c(t) = 1 - e^{-2t}\left[\dfrac{1}{\sqrt{3}}\sin(\sqrt{12}t) + \cos(\sqrt{12}t)\right]$

$$\text{(or)} \quad c(t) = 1 - \frac{2}{\sqrt{3}}e^{-2t}\left[\sin(\sqrt{12}t + 1.047)\right]$$

Rise time,	$t_r = 0.6046$ sec
Peak time,	$t_p = 0.907$ sec
% Maximum overshoot,	$\%M_p = 16.3\%$
Settling time,	$t_s = 1.5$ sec, for 5% error
	$= 2$ sec, for 2% error

EXAMPLE 2.7

A unity feedback control system is characterized by the following open loop transfer function G(s) = (0.4 s +1)/s(s +0.6). Determine its transient response for unit step input and sketch the response. Evaluate the maximum overshoot and the corresponding peak time.

SOLUTION

The closed loop transfer function, $\dfrac{C(s)}{R(s)} = \dfrac{G(s)}{1 + G(s)H(s)}$

Given that, G(s) = (0.4 s +1)/s(s + 0.6)

For unity feedback system, H(s) = 1.

$$\therefore \frac{C(s)}{R(s)} = \frac{G(s)}{1+G(s)} = \frac{\dfrac{0.4s+1}{s(s+0.6)}}{1+\dfrac{0.4s+1}{s(s+0.6)}} = \frac{0.4s+1}{s(s+0.6)+0.4s+1}$$

$$= \frac{0.4s+1}{s^2+0.6s+0.4s+1} = \frac{0.4s+1}{s^2+s+1}$$

The s-domain response, $C(s) = R(s) \times \dfrac{0.4s+1}{s^2+s+1}$

For step input, R(s) = 1/s.

$$\therefore C(s) = \frac{1}{s}\frac{0.4s+1}{(s^2+s+1)} = \frac{0.4s+1}{s(s^2+s+1)}$$

By partial fraction expansion C(s) can be expressed as,

$$C(s) = \frac{0.4s+1}{s(s^2+s+1)} = \frac{A}{s} + \frac{Bs+C}{s^2+s+1}$$

The residue A is solved by multiplying C(s) by s and letting s = 0.

$$\therefore A = C(s) \times s \Big|_{s=0} = \frac{0.4s+1}{s^2+s+1}\Big|_{s=0} = 1$$

The residues B and C are solved by cross multiplying the following equation and equating the coefficients of like powers of s.

$$\frac{0.4s+1}{s(s^2+s+1)} = \frac{A}{s} + \frac{Bs+C}{s^2+s+1}$$

On cross multiplication we get,

$$0.4\,s + 1 = A(s^2 + s + 1) + (Bs + C)\,s$$
$$0.4\,s + 1 = As^2 + As + A + Bs^2 + Cs$$

On equating coefficients of s^2 we get, 0 = A + B $\therefore$ B = – A = –1

On equating coefficients of s we get,   0.4 = A + C $\therefore$ C = 0.4 – A = – 0.6

$$\therefore C(s) = \frac{1}{s} + \frac{-s-0.6}{s^2+s+1} = \frac{1}{s} - \frac{s+0.6}{s^2+s+0.25+0.75} = \frac{1}{s} - \frac{s+0.6}{(s^2+2\times0.5s+0.5^2)+0.75}$$

$$= \frac{1}{s} - \frac{s+0.5+0.1}{(s+0.5)^2+0.75} = \frac{1}{s} - \frac{s+0.5}{(s+0.5)^2+0.75} - \frac{0.1}{\sqrt{0.75}}\frac{\sqrt{0.75}}{(s+0.5)^2+0.75}$$

The time domain response is obtained by taking inverse Laplace transform of C(s).

$\therefore$ The response in time domain,

$$c(t) = \mathcal{L}^{-1}\{C(s)\} = \mathcal{L}^{-1}\left\{ \frac{1}{s} - \frac{s+0.5}{(s+0.5)^2+0.75} - \frac{0.1}{\sqrt{0.75}}\,\frac{\sqrt{0.75}}{(s+0.5)^2+0.75} \right\}$$

$$= 1 - e^{-0.5t}\cos\sqrt{0.75}\,t - \frac{0.1}{\sqrt{0.75}}\,e^{-0.5t}\sin\sqrt{0.75}\,t$$

$$= 1 - e^{-0.5t}\left[\, 0.1155\sin(\sqrt{0.75}\,t) + \cos(\sqrt{0.75}\,t) \,\right]$$

The transient response is the part of the output which vanishes as t tends to infinity. Here as t tends to infinity the exponential component $e^{-0.5t}$ tends to zero. Hence the transient response is given by the damped sinusoidal component.

The transient response of $c(t) = e^{-0.5t}\left[\, 0.1155\sin(\sqrt{0.75}\,t) + \cos(\sqrt{0.75}\,t) \,\right]$

The value of ζ and ω_n can be estimated by comparing the characteristic equation of the system with standard form of second order characteristic equation.

$$\therefore\; s^2 + 2\zeta\omega_n s + \omega_n^2 = s^2 + s + 1$$

On comparing we get,

$$\omega_n^2 = 1 \qquad\qquad 2\zeta\omega_n = 1$$

$$\omega_n = 1 \text{ rad/sec} \qquad \therefore\; \zeta = \frac{1}{2\omega_n} = \frac{1}{2} = 0.5$$

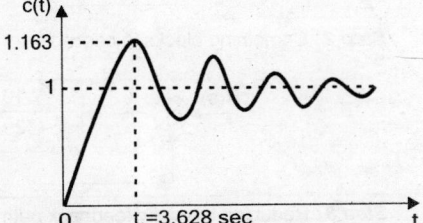

Fig 1 : Response of under damped system.

Maximum overshoot, $M_p = e^{\frac{-\zeta\pi}{\sqrt{1-\zeta^2}}} = e^{\frac{-0.5\pi}{\sqrt{1-0.5^2}}} = 0.163$

% Maximum overshoot, $\% M_p = M_p \times 100 = 0.163 \times 100 = 16.3\%$

Peak time, $t_p = \dfrac{\pi}{\omega_d} = \dfrac{\pi}{\omega_n\sqrt{1-\zeta^2}} = \dfrac{\pi}{1\times\sqrt{1-0.5^2}} = 3.628$ sec

The response of the system is underdamped and it is shown in fig 1.

RESULT

Transient response of the system, $c(t)$ = $e^{-0.5t}\left[\, 0.1155\sin(\sqrt{0.75}\,t) + \cos(\sqrt{0.75}\,t) \,\right]$

% Maximum peak overshoot, $\%M_p$ = 16.3 %

Peak time, t_p = 3.628 sec

EXAMPLE 2.8

A unity feedback control system has an amplifier with gain K_A = 10 and gain ratio, G(s) = 1/s(s +2) in the feed forward path. A derivative feedback, H(s) = sK_o is introduced as a minor loop around G(s). Determine the derivative feedback constant, K_o so that the system damping factor is 0.6.

SOLUTION

The given system can be represented by the block diagram shown in fig 1.

Here, K_A = 10 ; $G(s) = \dfrac{1}{s(s+2)}$ and H(s) = sK_o

The closed loop transfer function of the system can be obtained by block diagram reduction techniques.

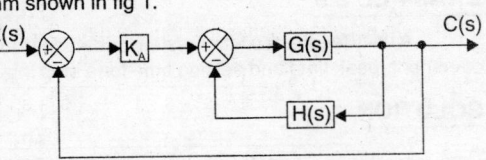

Fig 1.

Step 1: Reducing the inner feedback loop.

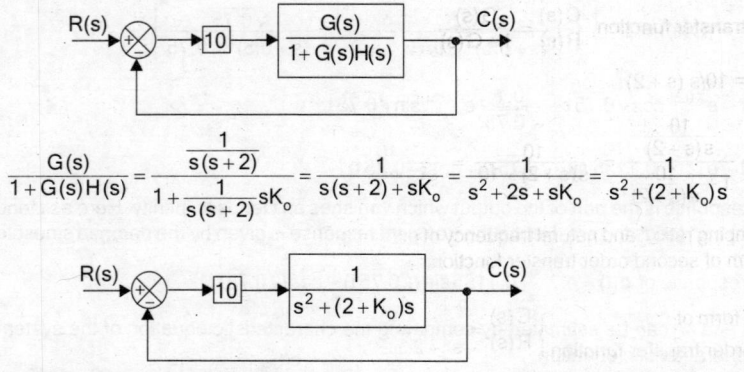

$$\frac{G(s)}{1+G(s)H(s)} = \frac{\dfrac{1}{s(s+2)}}{1+\dfrac{1}{s(s+2)}sK_o} = \frac{1}{s(s+2)+sK_o} = \frac{1}{s^2+2s+sK_o} = \frac{1}{s^2+(2+K_o)s}$$

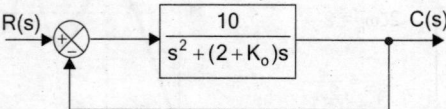

Step 2 : Combining blocks in cascade

Step 3 : Reducing the unity feedback path

The closed loop transfer function, $\dfrac{C(s)}{R(s)} = \dfrac{10}{s^2+(2+K_o)s+10}$ .....(1)

The given system is a second order system. The value of K_o can be determined by comparing the system transfer function with standard form of second order transfer function given below.

Standard form of second order transfer function $\Big\}$ $\dfrac{C(s)}{R(s)} = \dfrac{\omega_n^2}{s^2+2\zeta\omega_n s+\omega_n^2}$ .....(2)

On comparing equation (1) & (2) we get,

$\omega_n^2 = 10$

$\therefore \omega_n = \sqrt{10} = 3.162 \text{ rad/sec}$

$2 + K_o = 2\zeta\omega_n$

$\therefore K_o = 2\zeta\omega_n - 2$

$= 2 \times 0.6 \times 3.162 - 2 = 1.7944$

RESULT

The value of constant, $K_o = 1.7944$

EXAMPLE 2.9

A unity feedback control system has an open loop transfer function, $G(s) = 10/s(s+2)$. Find the rise time, percentage overshoot, peak time and settling time for a step input of 12 units.

SOLUTION

Note : The formulae for rise time, percentage overshoot and peak time remains same for unit step and step input.

The unity feedback system is shown in fig 1.

The closed loop transfer function, $\dfrac{C(s)}{R(s)} = \dfrac{G(s)}{1+G(s)}$

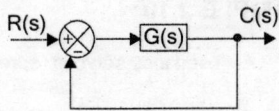

Given that, $G(s) = 10/s\,(s+2)$

Fig 1 : Unity feedback system.

$$\therefore \frac{C(s)}{R(s)} = \frac{\dfrac{10}{s(s+2)}}{1+\dfrac{10}{s(s+2)}} = \frac{10}{s(s+2)+10} = \frac{10}{s^2+2s+10} \qquad \dots\dots(1)$$

The values of damping ratio ζ and natural frequency of oscillation ω_n are obtained by comparing the system transfer function with standard form of second order transfer function.

$$\left.\begin{array}{l}\text{Standard form of}\\ \text{Second order transfer function}\end{array}\right\} \frac{C(s)}{R(s)} = \frac{\omega_n^2}{s^2+2\zeta\omega_n s+\omega_n^2} \qquad \dots\dots(2)$$

On comparing equation (1) & (2) we get,

$$\omega_n^2 = 10 \qquad\qquad\qquad 2\zeta\omega_n = 2$$

$$\therefore\ \omega_n = \sqrt{10} = 3.162\,\text{rad/sec} \qquad \therefore\ \zeta = \frac{2}{2\omega_n} = \frac{1}{3.162} = 0.316$$

$$\theta = \tan^{-1}\frac{\sqrt{1-\zeta^2}}{\zeta} = \tan^{-1}\frac{\sqrt{1-0.316^2}}{0.316} = 1.249\,\text{rad}$$

$$\omega_d = \omega_n\sqrt{1-\zeta^2} = 3.162\sqrt{1-0.316^2} = 3\,\text{rad/sec}$$

Rise time, $t_r = \dfrac{\pi-\theta}{\omega_d} = \dfrac{\pi-1.249}{3} = 0.63\,\text{sec}$

Percentage overshoot, $\% M_p = e^{\frac{-\zeta\pi}{\sqrt{1-\zeta^2}}} \times 100 = e^{\frac{-0.316\pi}{\sqrt{1-0.316^2}}} \times 100$

$$= 0.3512 \times 100 = 35.12\%$$

Peak overshoot $= \dfrac{35.12}{100} \times 12\,\text{units} = 4.2144\,\text{units}$

Peak time, $t_p = \dfrac{\pi}{\omega_d} = \dfrac{\pi}{3} = 1.047\,\text{sec}$

Time constant, $T = \dfrac{1}{\zeta\omega_n} = \dfrac{1}{0.316 \times 3.162} = 1\,\text{sec}$

$\therefore$ For 5% error, Settling time, $t_s = 3T = 3\,\text{sec}$

For 2% error, Settling time, $t_s = 4T = 4\,\text{sec}$

RESULT

Rise time, t_r	=	0.63 sec
Percentage overshoot, $\% M_p$	=	35.12%
Peak overshoot	=	4.2144 units, (for a input of 12 units)
Peak time, t_p	=	1.047 sec
Settling time, t_s	=	3 sec for 5% error
	=	4 sec for 2% error

EXAMPLE 2.10

A closed loop servo is represented by the differential equation $\dfrac{d^2c}{dt^2} + 8\dfrac{dc}{dt} = 64\,e$

Where c is the displacement of the output shaft, r is the displacement of the input shaft and e = r – c. Determine undamped natural frequency, damping ratio and percentage maximum overshoot for unit step input.

SOLUTION

The mathematical equations governing the system are,

$$\frac{d^2c}{dt^2} + 8\frac{dc}{dt} = 64\,e \qquad\qquad .....(1)$$

$$e = r - c \qquad\qquad .....(2)$$

Put e = r – c in equation (1),

$$\therefore \frac{d^2c}{dt^2} + 8\frac{dc}{dt} = 64\,(r - c) \qquad\qquad .....(3)$$

Let $\mathcal{L}\{c\}$ = C(s) and $\mathcal{L}\{r\}$ = R(s)

On taking Laplace transform of equation (3) we get,

$$s^2\,C(s) + 8s\,C(s) = 64\,[R(s) - C(s)]$$

$$\therefore\ s^2\,C(s) + 8s\,C(s) + 64\,C(s) = 64\,R(s)$$

$$(s^2 + 8s + 64)\,C(s) = 64\,R(s)$$

$$\therefore\ \frac{C(s)}{R(s)} = \frac{64}{s^2 + 8s + 64} \qquad\qquad .....(4)$$

The ratio C(s)/R(s) is the closed loop transfer function of the system. On comparing the system transfer function with standard form of second order transfer function, we can estimate the values of ζ and ω_n.

$$\left.\begin{array}{l}\text{Standard form of}\\ \text{Second order transfer function}\end{array}\right\}\ \frac{C(s)}{R(s)} = \frac{\omega_n^2}{s^2 + 2\zeta\omega_n s + \omega_n^2} \qquad\qquad .....(5)$$

On comparing equation (1) & (2) we get,

$$\omega_n^2 = 64 \qquad\qquad 2\zeta\omega n = 8$$

$$\therefore\ \omega_n = 8\ \text{rad/sec} \qquad\qquad \zeta = \frac{8}{2\omega_n} = \frac{8}{2\times 8} = 0.5$$

Percentage peak overshoot, $\%\,M_p = e^{\frac{-\zeta\pi}{\sqrt{1-\zeta^2}}} \times 100 = e^{\frac{-0.5\pi}{\sqrt{1-0.5^2}}} \times 100 = 16.3\%$

RESULT

Undamped natural frequency of oscillation, ω_n = 8 rad/sec

Damping ratio, ζ = 0.5

Percentage peak overshoot, $\%M_p$ = 16.3%

2.9 TYPE NUMBER OF CONTROL SYSTEMS

The type number is specified for loop transfer function G(s) H(s). The number of poles of the loop transfer function lying at the origin decides the type number of the system. In general, if N is the number of poles at the origin then the type number is **N.**

The loop transfer function can be expressed as a ratio of two polynomials in **s.**

$$G(s)H(s) = K\,\frac{P(s)}{Q(s)} = K\,\frac{(s+z_1)(s+z_2)(s+z_3)\dots\dots}{s^N(s+p_1)(s+p_2)(s+p_3)\dots\dots} \qquad \dots\dots(2.49)$$

where, $z_1, z_2, z_3, \dots\dots\dots\dots$ are zeros of transfer function

$p_1, p_2, p_3, \dots\dots\dots\dots$ are poles of transfer function

K = Constant

N = Number of poles at the origin

The value of **N** in the denominator polynomial of loop transfer function shown in equation (2.49) decides the type number of the system.

If N = 0, then the system is type – 0 system

If N = 1, then the system is type – 1 system

If N = 2, then the system is type – 2 system

If N = 3, then the system is type – 3 system and so on.

2.10 STEADY STATE ERROR

The steady state error is the value of error signal e(t), when **t** tends to infinity. The steady state error is a measure of system accuracy. These errors arise from the nature of inputs, type of system and from non linearity of system components. The steady state performance of a stable control system is generally judged by its steady state error to step, ramp and parabolic inputs.

Consider a closed loop system shown in fig 2.15.

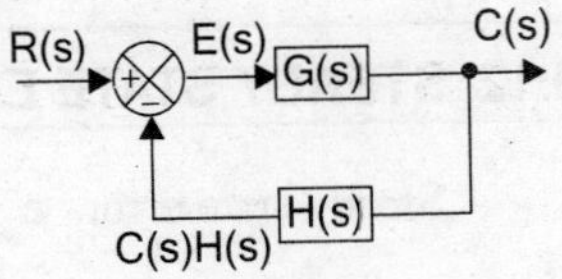

Let,   R(s)   =   Input signal

E(s)   =   Error signal

C(s) H(s)   =   Feedback signal

C(s)   =   Output signal or response

Fig 2.15 .

The error signal, E(s) = R(s) – C(s) H(s)   .....(2.50)

The output signal, C(s) = E(s) G(s)   .....(2.51)

On substituting for C(s) from equation (2.51) in equation (2.50) we get,

E(s) = R(s) – [E(s) G(s)] H(s)

E(s) + E(s) G(s) H(s) = R(s)

E(s) [1 + G(s) H(s)] = R(s)

$$\therefore\; E(s) = \frac{R(s)}{1 + G(s)H(s)} \qquad \dots\dots(2.52)$$

Let, e(t) = error signal in time domain.

$$\therefore \; e(t) = \mathcal{L}^{-1}\{E(s)\} = \mathcal{L}^{-1}\left\{\frac{R(s)}{1+G(s)H(s)}\right\} \qquad .....(2.53)$$

Let, e_{ss} = steady state error.

The steady state error is defined as the value of e(t) when **t** tends to infinity.

$$\therefore \; e_{ss} = \underset{t\to\infty}{Lt}\; e(t) \qquad\qquad ......(2.54)$$

The final value theorem of Laplace transform states that,

$$\textbf{If,}\;\; \textbf{F(s)} = \mathcal{L}\{\textbf{f(t)}\} \;\textbf{then,}\;\; \underset{t\to\infty}{\textbf{Lt}}\; \textbf{f(t)} = \underset{s\to 0}{\textbf{Lt}}\; \textbf{s}\, \textbf{F(s)} \qquad .....(2.55)$$

Using final value theorem,

The steady state error, $\quad e_{ss} = \underset{t\to\infty}{Lt}\; e(t) = \underset{s\to 0}{Lt}\; sE(s) = \underset{s\to 0}{Lt}\; \dfrac{sR(s)}{1+G(s)H(s)} \qquad .....(2.56)$

2.11 STATIC ERROR CONSTANTS

When a control system is excited with standard input signal, the steady state error may be zero, constant or infinity. The value of steady state error depends on the type number and the input signal. Type-0 system will have a constant steady state error when the input is step signal. Type-1 system will have a constant steady state error when the input is ramp signal or velocity signal. Type-2 system will have a constant steady state error when the input is parabolic signal or acceleration signal. For the three cases mentioned above the steady state error is associated with one of the constants defined as follows,

Positional error constant, $\qquad K_p = \underset{s\to 0}{Lt}\; G(s)H(s) \qquad\qquad .....(2.57)$

Velocity error constant, $\qquad K_v = \underset{s\to 0}{Lt}\; s\,G(s)H(s) \qquad\qquad .....(2.58)$

Acceleration error constant, $\quad K_a = \underset{s\to 0}{Lt}\; s^2 G(s)H(s) \qquad\qquad .....(2.59)$

The K_p, K_v and K_a are in general called static error constants.

2.12 STEADY STATE ERROR WHEN THE INPUT IS UNIT STEP SIGNAL

Steady state error, $e_{ss} = \underset{s\to 0}{Lt}\; \dfrac{sR(s)}{1+G(s)H(s)}$

When the input is unit step, $R(s) = 1/s$

$$\therefore \; e_{ss} = \underset{s\to 0}{Lt}\; \frac{s\,\dfrac{1}{s}}{1+G(s)H(s)} = \underset{s\to 0}{Lt}\; \frac{1}{1+G(s)H(s)} = \frac{1}{1+\underset{s\to 0}{Lt}G(s)H(s)} = \frac{1}{1+K_p} \qquad .....(2.60)$$

where, $K_p = \underset{s\to 0}{Lt}\; G(s)H(s)$

The constant K_p is called *positional error constant*.

Type-0 system

$$K_p = \underset{s\to 0}{Lt}\; G(s)H(s) = \underset{s\to 0}{Lt}\; K\frac{(s+z_1)(s+z_2)(s+z_3)......}{(s+p_1)(s+p_2)(s+p_3)......} = K\frac{z_1 z_2 z_3....}{p_1 p_2 p_3....} = \text{constant}$$

$$\therefore \; e_{ss} = \frac{1}{1+K_p} = \text{constant}$$

Hence in type-0 systems when the input is unit step there will be a constant steady state error.

Type-1 system

$$K_p = \underset{s \to 0}{Lt} \; G(s)H(s) = \underset{s \to 0}{Lt} \; K \frac{(s+z_1)(s+z_2)(s+z_3)......}{(s+p_1)(s+p_2)(s+p_3)......} = \infty$$

$$\therefore \; e_{ss} = \frac{1}{1+K_p} = \frac{1}{1+\infty} = 0$$

In systems with type number 1 and above, for unit step input the value of K_p is infinity and so the steady state error is zero.

2.13 STEADY STATE ERROR WHEN THE INPUT IS UNIT RAMP SIGNAL

Steady state error, $e_{ss} = \underset{s \to 0}{Lt} \dfrac{sR(s)}{1+G(s)H(s)}$

When the input is unit ramp, $R(s) = \dfrac{1}{s^2}$

$$\therefore \; e_{ss} = \underset{s \to 0}{Lt} \frac{s\frac{1}{s^2}}{1+G(s)H(s)} = \underset{s \to 0}{Lt} \frac{1}{s+sG(s)H(s)} = \frac{1}{\underset{s \to 0}{Lt}\,sG(s)H(s)} = \frac{1}{K_v} \qquad .....(2.61)$$

where, $K_v = \underset{s \to 0}{Lt} \; sG(s)H(s)$

The constant K_v is called **velocity error constant.**

Type-0 system

$$K_v = \underset{s \to 0}{Lt} \; sG(s)H(s) = \underset{s \to 0}{Lt} \; sK \frac{(s+z_1)(s+z_2)(s+z_3).......}{(s+p_1)(s+p_2)(s+p_3).......} = 0$$

$$\therefore \; e_{ss} = 1/K_v = 1/0 = \infty$$

Hence in type-0 systems when the input is unit ramp, the steady state error is infinity.

Type-1 system

$$K_v = \underset{s \to 0}{Lt} \; sG(s)H(s) = \underset{s \to 0}{Lt} \; sK \frac{(s+z_1)(s+z_2)(s+z_3).......}{(s+p_1)(s+p_2)(s+p_3).......} = K \frac{z_1 z_2 z_3 .....}{p_1 p_2 p_3 .....} = constant$$

$$\therefore \; e_{ss} = 1/K_v = constant$$

Hence in type-1 systems when the input is unit ramp there will be a constant steady state error.

Type-2 system

$$K_v = \underset{s \to 0}{Lt} \; sG(s)H(s) = \underset{s \to 0}{Lt} \; sK \frac{(s+z_1)(s+z_2)(s+z_3).......}{s^2(s+p_1)(s+p_2)(s+p_3).......} = \infty$$

$$\therefore \; e_{ss} = 1/K_v = 1/\infty = 0$$

In systems with type number 2 and above, for unit ramp input, the value of K_v is infinity so the steady state error is zero.

2.14 STEADY STATE ERROR WHEN THE INPUT IS UNIT PARABOLIC SIGNAL

Steady state error, $e_{ss} = \underset{s \to 0}{Lt} \dfrac{sR(s)}{1+G(s)H(s)}$

When the input is unit parabola, $R(s) = \dfrac{1}{s^3}$

$$\therefore e_{ss} = \underset{s \to 0}{Lt} \dfrac{s \dfrac{1}{s^3}}{1+G(s)H(s)} = \underset{s \to 0}{Lt} \dfrac{1}{s^2 + s^2 G(s)H(s)} = \dfrac{1}{\underset{s \to 0}{Lt} s^2 G(s)H(s)} = \dfrac{1}{K_a} \qquad .....(2.62)$$

where, $K_a = \underset{s \to 0}{Lt} s^2 G(s)H(s)$

The constant K_a is called *acceleration error constant.*

Type-0 system

$$K_a = \underset{s \to 0}{Lt} s^2 G(s)H(s) = \underset{s \to 0}{Lt} s^2 K \dfrac{(s+z_1)(s+z_2)(s+z_3).......}{(s+p_1)(s+p_2)(s+p_3).......} = 0$$

$$\therefore e_{ss} = \dfrac{1}{K_a} = \dfrac{1}{0} = \infty$$

Hence in type-0 systems for unit parabolic input, the steady state error is infinity.

Type-1 system

$$K_a = \underset{s \to 0}{Lt} s^2 G(s)H(s) = \underset{s \to 0}{Lt} s^2 K \dfrac{(s+z_1)(s+z_2)(s+z_3).......}{(s+p_1)(s+p_2)(s+p_3).......} = 0$$

$$\therefore e_{ss} = \dfrac{1}{K_a} = \dfrac{1}{0} = \infty$$

Hence in type-1 systems for unit parabolic input, the steady state error is infinity.

Type-2 system

$$K_a = \underset{s \to 0}{Lt} s^2 G(s)H(s) = \underset{s \to 0}{Lt} s^2 K \dfrac{(s+z_1)(s+z_2)(s+z_3).......}{(s+p_1)(s+p_2)(s+p_3).......} = K \dfrac{z_1.z_2.z_3.....}{p_1.p_2.p_3.....} = \text{constant}$$

$$\therefore e_{ss} = \dfrac{1}{K_a} = \text{constant}$$

Hence in type-2 system when the input is unit parabolic signal there will be a constant steady state error.

Type-3 system

$$K_a = \underset{s \to 0}{Lt} s^2 G(s)H(s) = \underset{s \to 0}{Lt} s^2 K \dfrac{(s+z_1)(s+z_2)(s+z_3).......}{(s+p_1)(s+p_2)(s+p_3).......} = \infty$$

$$\therefore e_{ss} = \dfrac{1}{K_a} = \dfrac{1}{\infty} = 0$$

In systems with type number 3 and above for unit parabolic input the value of K_a is infinity and so the steady state error is zero.

TABLE-2.2 : Static Error Constant for Various Type Number of Systems

Error Constant	Type number of system			
	0	1	2	3
K_p	constant	∞	∞	∞
K_v	0	constant	∞	∞
K_a	0	0	constant	∞

TABLE-2.3 : Steady State Error for Various Types of Inputs

Input Signal	Type number of system			
	0	1	2	3
Unit Step	$\dfrac{1}{1+K_p}$	0	0	0
Unit Ramp	∞	$\dfrac{1}{K_v}$	0	0
Unit Parabolic	∞	∞	$\dfrac{1}{K_a}$	0

2.15 GENERALIZED ERROR COEFFICIENT

The drawback in static error coefficients is that it does not show the variation of error with time and input should be a standard input. The generalized error coefficients gives the steady state error as a function of time. Also using the generalized error coefficients, the steady state error can be found for any type of input.

The error signal in s-domain, E(s) can be expressed as a product of two s-domain functions.

$$E(s) = \frac{R(s)}{1+G(s)H(s)} = \frac{1}{1+G(s)H(s)} R(s) = F(s) R(s) \qquad .....(2.63)$$

$$\text{where, } F(s) = \frac{1}{1+G(s)H(s)}$$

Let, $e(t)$ = $\mathcal{L}^{-1}\{E(s)\}$ (error signal in time domain)

$f(t)$ = $\mathcal{L}^{-1}\{F(s)\}$

$r(t)$ = $\mathcal{L}^{-1}\{R(s)\}$ (input signal in time domain)

The convolution theorem of Laplace transform states that the Laplace transform of the convolution of two time domain signals is equal to the product of their individual Laplace transform.

i.e., $\mathcal{L}\{f(t) * r(t)\} = F(s) R(s)$

where $*$ is the symbol for convolution operation

$$\mathcal{L}^{-1}\{F(s) R(s)\} = f(t) * r(t) \qquad .....(2.64)$$

From equation (2.63) & (2.64) we can write,

$e(t) = f(t) * r(t)$

Mathematically the convolution of f(t) and r(t) is defined as,

$$f(t) * r(t) = \int_{-\infty}^{+\infty} f(T) \, r(t-T) \, dT \quad ; \quad \text{where T is a dummy variable}$$

$$\therefore e(t) = \int_{-\infty}^{+\infty} f(T) \, r(t-T) \, dT$$

It is assumed that the input signal starts only at t = 0 and does not exist before t = 0. Also we are interested in finding error signal at any time t after t = 0 (i.e., for t > 0). Hence in the above equation the limit of integral can be changed as 0 to t.

$$\therefore e(t) = \int_{0}^{t} f(T) \, r(t-T) \, dT$$

Using Taylor's series expansion the signal r(t–T) can be expressed as,

$$r(t-T) = r(t) - T\,\dot{r}(t) + \frac{T^2}{2!}\ddot{r}(t) - \frac{T^3}{3!}\dddot{r}(t) + \dots\dots + (-1)^n\frac{T^n}{n!}\overset{n.}{r}(t)\dots$$

where, $\dot{r}(t) = 1^{st}$ derivative of r(t)

$\ddot{r}(t) = 2^{nd}$ derivative of r(t)

$$\vdots$$

$\overset{n.}{r}(t) = n^{th}$ derivative of r(t)

On substituting the Taylor's series expansion of r(t – T), the error e(t) can be written as,

$$e(t) = \int_0^t f(t)\left[r(t) - T\,\dot{r}(t) + \frac{T^2}{2!}\ddot{r}(t) - \frac{T^3}{3!}\dddot{r}(t) + \dots + (-1)^n\frac{T^n}{n!}\overset{n.}{r}(t)\dots\right]dT$$

$$e(t) = \int_0^t f(T)\,r(t)\,dT - \int_0^t f(T)\,T\,\dot{r}(t)\,dT + \int_0^t f(T)\,\frac{T^2}{2!}\ddot{r}(t)\,dT$$

$$- \int_0^t f(T)\,\frac{T^3}{3!}T\,\dddot{r}(t) + \dots\dots + \int_0^t f(T)(-1)^n\frac{T^n}{n!}\overset{n.}{r}(t)\,dT\dots\infty$$

Since r(t), $\dot{r}$(t), $\ddot{r}$(t).... $\overset{n.}{r}$(t) are constants when the integration is done with respect to T, the error signal can be written as,

$$e(t) = r(t)\int_0^t f(T)\,dT - \dot{r}(t)\int_0^t Tf(T)\,dT + \frac{\ddot{r}(t)}{2!}\int_0^t T^2 f(T)\,dt$$

$$- \frac{\dddot{r}(t)}{3!}\int_0^t T^3 f(T)\,dt + \dots + (-1)^n\frac{\overset{n.}{r}(t)}{n!}\int_0^t T^n f(T)\,dt\dots$$

Let, $C_0 = + \int_0^t f(T)\,dT$ $C_3 = - \int_0^t T^3 f(T)\,dT$

$C_1 = - \int_0^t Tf(T)\,dT$ $\vdots$

$C_2 = + \int_0^t T^2 f(T)\,dT$ $C_n = (-1)^n\int_0^t T^n f(T)\,dT$

$$e(t) = r(t)C_0 + \dot{r}(t)C_1 + \ddot{r}(t)\frac{C_2}{2!} + \dddot{r}(t)\frac{C_3}{3!} + \dots + \overset{n.}{r}(t)\frac{C_n}{n!} + \dots$$

$$= C_0 r(t) + C_1\dot{r}(t) + \frac{C_2}{2!}\ddot{r}(t) + \frac{C_3}{3!}\dddot{r}(t) + \dots\dots + \frac{C_n}{n!}\overset{n.}{r}(t)\dots\dots \quad\dots(2.65)$$

The equation (2.65) is the general equation for error signal, e(t).

The coefficients C_0, C_1, C_2 ,......C_n are called the generalized error coefficients or dynamic error coefficients.

The steady state error e_{ss} is obtained by taking limit $t \to \infty$ on e(t).

$\therefore$ Steady state error, $e_{ss} = \underset{t\to\infty}{Lt}\left[r(t)C_0 + \dot{r}(t)C_1 + \ddot{r}(t)\frac{C_2}{2!} + \dddot{r}(t)\frac{C_3}{3!} + \dots\dots + \overset{n.}{r}(t)\frac{C_n}{n!} + \dots\right]$

$$= C_0 r(t) + C_1\dot{r}(t) + \frac{C_2}{2!}\ddot{r}(t) + \frac{C_3}{3!}\dddot{r}(t) + \dots + \frac{C_n}{n!}\overset{n.}{r}(t)\dots \qquad\dots(2.66)$$

2.16 EVALUATION OF GENERALIZED ERROR COEFFICIENTS

The generalized error coefficient is given by,

$$C_n = (-1)^n \int_0^t T^r f(T) \, dT \; ; \quad \text{where } F(s) = \frac{1}{1 + G(s)H(s)}$$

We know that $\mathcal{L}\{f(T)\} = F(s)$, hence by the definition of Laplace transform,

$$F(s) = \int_0^t f(T) \, e^{-sT} \, dT \qquad \qquad .....(2.67)$$

On taking $\underset{s \to 0}{\text{Lt}}$ on both sides of equation (2.67) we get,

$$\underset{s \to 0}{\text{Lt}} \, F(s) = \underset{s \to 0}{\text{Lt}} \int_0^t f(T) \, e^{-sT} \, dT$$

$$= \int_0^t f(T) \underset{s \to 0}{\text{Lt}} \, e^{-sT} \, dT = \int_0^t f(T) \, dT = C_0$$

$$\boxed{\therefore \; C_0 = \underset{s \to 0}{\text{Lt}} \, F(s)} \qquad \qquad .....(2.68)$$

On differentiating equation (2.68) with respect to s we get,

$$\frac{d}{ds} F(s) = \frac{d}{ds} \int_0^t f(T) \, e^{-sT} \, dT$$

$$= \int_0^t f(T) \frac{d}{ds} (e^{-sT}) \, dT = \int_0^t f(T) \, (-T) e^{-sT} \, dT$$

$$= - \int_0^t T f(t) \, e^{-sT} \, dT \qquad \qquad .....(2.69)$$

On taking on both sides of equation (2.69) we get,

$$\underset{s \to 0}{\text{Lt}} \, \frac{d}{ds} F(s) = \underset{s \to 0}{\text{Lt}} - \int_0^t T f(t) \, e^{-sT} \, dT$$

$$= - \int_0^t T f(T) \underset{s \to 0}{\text{Lt}} \, e^{-sT} \, dT = - \int_0^t T f(T) \, dT = C_1$$

$$\boxed{\therefore \; C_1 = \underset{s \to 0}{\text{Lt}} \frac{d}{ds} F(s)} \qquad \qquad .....(2.70)$$

On differentiating equation (2.68) on both sides with respect to s we get,

$$\frac{d}{ds}\left[\frac{d}{ds}(F(s)) \right] = \frac{d}{ds}\left[-\int_0^t T f(T) e^{-sT} dT \right]$$

$$\frac{d^2}{ds^2} F(s) = \left[-\int_0^t T f(T) \frac{d}{ds}(e^{-sT}) \, dT \right] = -\int_0^t T f(T) (-T) e^{-sT} \, dT$$

$$\frac{d^2(F(s))}{ds^2} = \int_0^t T^2 f(T) e^{-sT} \, dT \qquad \qquad .....(2.71)$$

Applying the limit $s \to 0$ on both sides of the equation (2.71) we get,

$$\underset{s \to 0}{Lt} \frac{d^2}{ds^2} F(s) = \underset{s \to 0}{Lt} \int_0^t T^2 f(T) e^{-st} dT$$

$$= \int_0^t T^2 f(T) \underset{s \to 0}{Lt} e^{-st} dT = \int_0^t T^2 f(T) dT = C_2$$

$$\therefore \boxed{C_2 = \underset{s \to 0}{Lt} \frac{d^2}{ds^2} F(s)} \qquad\qquad .....(2.72)$$

Similarly it can be shown that,

$$\boxed{C_n = \underset{s \to 0}{Lt} \frac{d^n}{ds^n} F(s)} \qquad\qquad .....(2.73)$$

2.17 CORRELATION BETWEEN STATIC AND DYNAMIC ERROR COEFFICIENTS

The values of dynamic error coefficients can be used to calculate static error coefficients. The following expressions shows the relationship between them.

$$C_0 = \frac{1}{1 + K_p} \qquad\qquad .....(2.74)$$

$$C_1 = \frac{1}{K_v} \qquad\qquad .....(2.75)$$

$$C_2 = \frac{1}{K_a} \qquad\qquad .....(2.76)$$

Proof

$$C_0 = \underset{s \to 0}{Lt} F(s) = \underset{s \to 0}{Lt} \frac{1}{1 + G(s)H(s)} = \frac{1}{1 + \underset{s \to 0}{Lt} G(s)H(s)} = \frac{1}{1 + K_p}$$

2.18 ALTERNATE METHOD FOR GENERALIZED ERROR COEFFICIENTS

The error signal in s-domain, $E(s) = \dfrac{R(s)}{1 + G(s)H(s)}$

$$\therefore \frac{E(s)}{R(s)} = \frac{1}{1 + G(s)H(s)} \qquad\qquad .....(2.77)$$

The equation (2.77) can be expressed as a power series of **s** as shown in equation (2.78).

$$\frac{E(s)}{R(s)} = \frac{1}{1 + G(s)H(s)} = C_0 + C_1 s + \frac{C_2}{2!} s^2 + \frac{C_3}{3!} s^3 + ... \qquad\qquad .....(2.78)$$

$$\therefore E(s) = C_0 R(s) + C_1 s R(s) + \frac{C_2}{2!} s^2 R(s) + \frac{C_3}{3!} s^3 R(s) + ... \qquad\qquad .....(2.79)$$

On taking inverse Laplace transform of equation (2.79) we get,

$$e(t) = C_0 r(t) + C_1 s R(s) + \frac{C_2}{2!} s^2 r(t) + \frac{C_3}{3!} s^3 r(t) + ... \qquad\qquad .....(2.80)$$

The equation (2.80) is same as that of equation (2.65) in section 2.14. This method will be useful to find the generalized error coefficients without using differentiation, but using laplace transform.

EXAMPLE 2.11

For a unity feedback control system the open loop transfer function, $G(s) = \dfrac{10(s+2)}{s^2(s+1)}$. Find

a) the position, velocity and acceleration error constants,

b) the steady state error when the input is R(s), where $R(s) = \dfrac{3}{s} - \dfrac{2}{s^2} + \dfrac{1}{3s^3}$

SOLUTION

a) To find static error constants

For a unity feedback system, H(s)=1

Position error constant, $K_p = \underset{s \to 0}{Lt}\ G(s)H(s) = \underset{s \to 0}{Lt}\ G(s) = \underset{s \to 0}{Lt}\ \dfrac{10(s+2)}{s^2(s+1)} = \infty$

Velocity error constant, $K_v = \underset{s \to 0}{Lt}\ s\,G(s)H(s) = \underset{s \to 0}{Lt}\ s\,G(s) = \underset{s \to 0}{Lt}\ s\,\dfrac{10(s+2)}{s^2(s+1)} = \infty$

Acceleration error constant, $K_a = \underset{s \to 0}{Lt}\ s^2 G(s)H(s) = \underset{s \to 0}{Lt}\ s^2 G(s)$

$$= \underset{s \to 0}{Lt}\ s^2\,\dfrac{10(s+2)}{s^2(s+1)} = \dfrac{10 \times 2}{1} = 20$$

b) To find steady state error

Method-I

Steady state error for non-standard input is obtained using generalized error series, given below.

The error signal, $e(t) = r(t)C_0 + \dot r(t)C_1 + \ddot r(t)\dfrac{C_2}{2!} + ... + \overset{n}{r}(t)\dfrac{C_n}{2!} + ...$

Given that, $R(s) = \dfrac{3}{s} - \dfrac{2}{s^2} + \dfrac{1}{3s^3}$

Input signal in time domain, $r(t) = \mathcal{L}^{-1}\{R(s)\} = \mathcal{L}^{-1}\left\{ \dfrac{3}{s} - \dfrac{2}{s^2} + \dfrac{1}{3s^3} \right\}$

$$= 3 - 2t + \dfrac{1}{3}\dfrac{t^2}{2!} = 3 - 2t + \dfrac{t^2}{6}$$

$$\therefore\ \dot r(t) = \dfrac{d}{dt} r(t) = -2 + \dfrac{1}{6} 2t = -2 + \dfrac{t}{3}$$

$$\ddot r(t) = \dfrac{d^2}{dt^2} r(t) = \dfrac{d}{dt} \dot r(t) = \dfrac{1}{3}$$

$$\dddot r(t) = \dfrac{d^3}{dt^3} r(t) = \dfrac{d}{dt} \ddot r(t) = 0$$

The derivatives of r(t) is zero after second derivative. Hence we have to evaluate only three constants C_0, C_1 and C_2.

The generalized error constants are given by,

$$C_0 = \underset{s \to 0}{Lt}\ F(s)\ ; \qquad C_1 = \underset{s \to 0}{Lt}\ \dfrac{d}{ds} F(s)\ ; \qquad C_2 = \underset{s \to 0}{Lt}\ \dfrac{d^2}{ds^2} F(s)$$

$$F(s) = \dfrac{1}{1+G(s)H(s)} = \dfrac{1}{1+G(s)} = \dfrac{1}{1 + \dfrac{10(s+2)}{s^2(s+1)}} = \dfrac{s^2(s+1)}{s^2(s+1) + 10(s+2)} = \dfrac{s^3 + s^2}{s^3 + s^2 + 10s + 20}$$

$$C_0 = \underset{s \to 0}{Lt}\ F(s) = \underset{s \to 0}{Lt}\left[\dfrac{s^3 + s^2}{s^3 + s^2 + 10s + 20} \right] = 0$$

$$C_1 = \underset{s \to 0}{Lt} \frac{d}{ds} F(s) = \underset{s \to 0}{Lt} \frac{d}{ds} \left[\frac{s^3 + s^2}{s^3 + s^2 + 10s + 20} \right]$$

$$= \underset{s \to 0}{Lt} \left[\frac{(s^3 + s^2 + 10s + 20)(3s^2 + 2s) - (s^3 + s^2)(3s^2 + 2s + 10)}{(s^3 + s^2 + 10s + 20)^2} \right]$$

$$= \underset{s \to 0}{Lt} \left[\frac{3s^5 + 2s^4 + 3s^4 + 2s^3 + 30s^3 + 20s^2 + 60s^2 + 40s - 3s^5 - 2s^4 - 10s^3 - 3s^4 - 2s^3 - 10s^2}{(s^3 + s^2 + 10s + 20)^2} \right]$$

$$= \underset{s \to 0}{Lt} \left[\frac{20s^3 + 70s^2 + 40s}{(s^3 + s^2 + 10s + 20)^2} \right] = 0$$

$$C_2 = \underset{s \to 0}{Lt} \frac{d^2}{ds^2} F(s) = \underset{s \to 0}{Lt} \frac{d}{ds} \left[\frac{d}{ds} F(s) \right] = \underset{s \to 0}{Lt} \frac{d}{ds} \left[\frac{20s^3 + 70s^2 + 40s}{(s^3 + s^2 + 10s + 20)^2} \right]$$

$$= \underset{s \to 0}{Lt} \left[\frac{(s^3 + s^2 + 10s + 20)^2 (60s^2 + 140s + 40)}{-(20s^3 + 70s^2 + 40s) 2 \times (s^3 + s^2 + 10s + 20)(3s^2 + 2s + 10)}{(s^3 + s^2 + 10s + 20)^4} \right] = \frac{20^2 \times 40}{20^4} = \frac{1}{10}$$

Error signal, $e(t) = r(t) C_0 + \dot{r}(t) C_1 + \ddot{r}(t) \dfrac{C_2}{2!} = \left(3 - 2t + \dfrac{t^2}{6} \right) \times 0 + \left(-2 + \dfrac{t}{3} \right) \times 0 + \dfrac{1}{3} \times \dfrac{1}{10} \times \dfrac{1}{2!} = \dfrac{1}{60}$

Steady state error, $e_{ss} = \underset{t \to \infty}{Lt} e(t) = \underset{t \to \infty}{Lt} \dfrac{1}{60} = \dfrac{1}{60}$

Method - II

The error signal in s-domain, $E(s) = \dfrac{R(s)}{1 + G(s)H(s)}$

Given that, $R(s) = \dfrac{3}{s} - \dfrac{2}{s^2} + \dfrac{1}{3s^3}$; $G(s) = \dfrac{10(s+2)}{s^2(s+1)}$; $H(s) = 1$

$$\therefore E(s) = \frac{\dfrac{3}{s} - \dfrac{2}{s^2} + \dfrac{1}{3s^3}}{1 + \dfrac{10(s+2)}{s^2(s+1)}} = \frac{\dfrac{3}{s} - \dfrac{2}{s^2} + \dfrac{1}{3s^3}}{\dfrac{s^2(s+1) + 10(s+2)}{s^2(s+1)}}$$

$$= \frac{3}{s} \left[\frac{s^2(s+1)}{s^2(s+1) + 10(s+2)} \right] - \frac{2}{s^2} \left[\frac{s^2(s+1)}{s^2(s+1) + 10(s+2)} \right] + \frac{1}{3s^3} \left[\frac{s^2(s+1)}{s^2(s+1) + 10(s+2)} \right]$$

The steady state error e_{ss} can be obtained from final value theorem.

Steady state error, $e_{ss} = \underset{t \to \infty}{Lt} e(t) = \underset{s \to 0}{Lt} s E(s)$

$$\therefore e_{ss} = \underset{s \to 0}{Lt} s \left\{ \frac{3}{s} \left[\frac{s^2(s+1)}{s^2(s+1) + 10(s+2)} \right] - \frac{2}{s^2} \left[\frac{s^2(s+1)}{s^2(s+1) + 10(s+2)} \right] + \frac{1}{3s^3} \left[\frac{s^2(s+1)}{s^2(s+1) + 10(s+2)} \right] \right\}$$

$$= \underset{s \to 0}{Lt} s \left\{ \frac{3s^2(s+1)}{s^2(s+1) + 10(s+2)} - \frac{2s(s+1)}{s^2(s+1) + 10(s+2)} + \frac{(s+1)}{3s^2(s+1) + 30(s+2)} \right\} = 0 - 0 + \frac{1}{60}$$

$$= \frac{1}{60}$$

Method - III

Error signal in s-domain, $E(s) = \dfrac{R(s)}{1 + G(s)H(s)}$

$\therefore \dfrac{E(s)}{R(s)} = \dfrac{1}{1 + G(s)H(s)}$

Given that, $G(s) = \dfrac{10(s+2)}{s^2(s+1)}$; $H(s) = 1$

$\therefore \dfrac{E(s)}{R(s)} = \dfrac{1}{1 + \dfrac{10(s+2)}{s^2(s+1)}} = \dfrac{s^2(s+1)}{s^2(s+1) + 10(s+2)}$

$= \dfrac{s^3 + s^2}{s^3 + s^2 + 10s + 20} = \dfrac{s^2 + s^3}{20 + 10s + s^2 + s^3} = \dfrac{s^2}{20} + \dfrac{s^3}{40} + \dots$

$$
\begin{array}{r}
\dfrac{s^2}{20} + \dfrac{s^3}{40} \, \dots \\[4pt]
\hline
20 + 10s + s^2 + s^3 \left.\right) \; s^2 + s^3 \\
s^2 + \dfrac{s^3}{2} + \dfrac{s^4}{20} + \dfrac{s^5}{20} \\
(-)\;\;(-)\;\;(-)\;\;(-) \\
\hline
\dfrac{s^3}{2} - \dfrac{s^4}{20} - \dfrac{s^5}{20} \\
\dfrac{s^3}{2} + \dfrac{s^4}{4} + \dfrac{s^5}{40} + \dfrac{s^6}{40} \\
(-)\;\;(-)\;\;(-)\;\;(-) \\
\hline
-\dfrac{3s^4}{10} - \dfrac{3s^5}{40} - \dfrac{s^6}{40} \\
\vdots
\end{array}
$$

Dividing numerator polynomial by denominator polynomial.

$E(s) = R(s)\left[\dfrac{s^2}{20} + \dfrac{s^3}{40} + \dots\right] = \dfrac{1}{20}s^2 R(s) + \dfrac{1}{40}s^2 R(s) + \dots$

On taking inverse Laplace transform of the above equation we get,

$e(t) = \dfrac{1}{20}\ddot{r}(t) + \dfrac{1}{40}\dddot{r}(t) + \dots$

Given that, $R(s) = \dfrac{3}{s} - \dfrac{2}{s^2} + \dfrac{1}{3s^3}$

$\therefore r(T) = \mathcal{L}^{-1}\{R(s)\} = \mathcal{L}^{-1}\left\{\dfrac{3}{s} - \dfrac{2}{s^2} + \dfrac{1}{3s^3}\right\} = 3 - 2t + \dfrac{1}{3}\dfrac{t^2}{2!} = 3 - 2t + \dfrac{t^2}{6}$

$\dot{r}(t) = \dfrac{d}{dt}r(t) = -2 + \dfrac{1}{6}2t = -2 + \dfrac{t}{3}$

$\ddot{r}(t) = \dfrac{d^2}{dt^2}r(t) = \dfrac{d}{dt}\dot{r}(t) = \dfrac{1}{3}$

$\dddot{r}(t) = \dfrac{d^3}{dt^3}r(t) = \dfrac{d}{dt}\ddot{r}(t) = 0$

Error signal in time domain, $e(t) = \dfrac{1}{20}\ddot{r}(t) = \dfrac{1}{20}\left(\dfrac{1}{3}\right) = \dfrac{1}{60}$

Steady state error, $e_{ss} = \underset{t \to \infty}{Lt} \, e(t) = \underset{t \to \infty}{Lt} \dfrac{1}{60} = \dfrac{1}{60}$

RESULT

(a)   Position error constant, $K_p = \infty$

Velocity error constant, $K_v = \infty$

Acceleration error constant, $K_a = 20$

(b)   When, $R(s) = \dfrac{3}{s} - \dfrac{2}{s^2} + \dfrac{1}{3s^3}$,   Steady state error, $e_{ss} = \dfrac{1}{60}$

EXAMPLE 2.12

For servomechanisms with open loop transfer function given below explain what type of input signal give rise to a constant steady state error and calculate their values.

a) $G(s) = \dfrac{20(s+2)}{s(s+1)(s+3)}$;   b) $G(s) = \dfrac{10}{(s+2)(s+3)}$;   c) $G(s) = \dfrac{10}{s^2(s+1)(s+2)}$

SOLUTION

a) $G(s) = \dfrac{20(s+2)}{s(s+1)(s+3)}$

Let us assume unity feedback system, $\therefore H(s)=1$

The open loop system has a pole at origin. Hence it is a type-1 system. In systems with type number-1, the velocity (ramp) input will give a constant steady state error.

The steady state error with unit velocity input, $e_{ss} = \dfrac{1}{K_v}$

Velocity error constant, $K_v = \underset{s \to 0}{Lt}\ sG(s)H(s) = \underset{s \to 0}{Lt}\ sG(s)$

$\qquad = \underset{s \to 0}{Lt}\ s\dfrac{20(s+2)}{s(s+1)(s+3)} = \dfrac{20 \times 2}{1 \times 3} = \dfrac{40}{3}$

Steady state error, $e_{ss} = \dfrac{1}{K_v} = \dfrac{3}{40} = 0.075$

b) $G(s) = \dfrac{10}{(s+2)(s+3)}$

Let us assume unity feedback system, $\therefore H(s)=1$.

The open loop system has no pole at origin. Hence it is a type-0 system. In systems with type number-0, the step input will give a constant steady state error.

The steady state error with unit step input, $e_{ss} = \dfrac{1}{1+K_p}$

Position error constant, $K_p = \underset{s \to 0}{Lt}\ G(s)H(s) = \underset{s \to 0}{Lt}\ G(s) = \underset{s \to 0}{Lt}\ \dfrac{10}{(s+2)(s+3)} = \dfrac{10}{2 \times 3} = \dfrac{5}{3}$

Steady state error, $e_{ss} = \dfrac{1}{1+K_p} = \dfrac{1}{1+\dfrac{5}{3}} = \dfrac{3}{3+5} = \dfrac{3}{8} = 0.375$

c) $G(s) = \dfrac{10}{s^2(s+1)(s+2)}$

Let us assume unity feedback system, $\therefore H(s)=1$.

The open loop system has two poles at origin. Hence it is a type-2 system. In systems with type number-2, the acceleration (parabolic) input will give a constant steady state error.

The steady state error with unit step input, $e_{ss} = \dfrac{1}{K_a}$

Position error constant, $K_p = \underset{s \to 0}{Lt}\ s^2 G(s)H(s) = \underset{s \to 0}{Lt}\ s^2 G(s) = \underset{s \to 0}{Lt}\ s^2 \dfrac{10}{s^2(s+1)(s+2)} = \dfrac{10}{1 \times 2} = 5$

Steady state error, $e_{ss} = \dfrac{1}{K_a} = \dfrac{1}{5} = 0.2$

RESULT

1. In system (a) with unit velocity input,	Steady state error = 0.075
2. In system (b) with unit step input,	Steady state error = 0.375
3. In system (c) with unit acceleration input,	Steady state error = 0.2

EXAMPLE 2.13

The open loop transfer function of a servo system with unity feedback is G(s) = 10/s(0.1s+1). Evaluate the static error constants of the system. Obtain the steady state error of the system, when subjected to an input given by the polynominal, $r(t) = a_0 + a_1 t + \dfrac{a_2}{2} t^2$.

SOLUTION

To find static error constant

For unity feedback system, H(s) = 1.

$\therefore$ Loop transfer function, G(s) H(s) = G(s)

The static error constants are K_p, K_v and K_a.

Position error constant, $K_p = \underset{s \to 0}{Lt}\ G(s) = \underset{s \to 0}{Lt}\ \dfrac{10}{s(0.1s+1)} = \infty$

Velocity error constant, $K_v = \underset{s \to 0}{Lt}\ s\,G(s) = \underset{s \to 0}{Lt}\ s\,\dfrac{10}{s(0.1s+1)} = 10$

Acceleration error constant, $K_a = \underset{s \to 0}{Lt}\ s^2 G(s) = \underset{s \to 0}{Lt}\ s^2\,\dfrac{10}{s(0.1s+1)} = 0$

To find steady state error

Method - I

Steady state error for non-standard input is obtained using generalized error series, given below.

The error signal, $e(t) = r(t)\,C_0 + \dot{r}(t)\,C_1 + \ddot{r}(t)\,\dfrac{C_2}{2!} + \dots + \overset{n}{r}(t)\,\dfrac{C_n}{n!} + \dots$

Given that, $r(t) = a_0 + a_1 t + \dfrac{a_2}{2} t^2$

$\therefore\ \dot{r}(t) = \dfrac{d}{dt} r(t) = \dfrac{d}{dt}\left(a_0 + a_1 t + \dfrac{a_2}{2} t^2 \right) = a_1 + a_2 t$

$\ddot{r}(t) = \dfrac{d^2}{dt^2} r(t) = \dfrac{d}{dt}\left(\dfrac{d}{dt} r(t) \right) = \dfrac{d}{dt}(a_1 + a_2 t) = a_2$

$\dddot{r}(t) = \dfrac{d^3}{dt^3} r(t) = \dfrac{d}{dt}\left(\dfrac{d^2}{dt^2} r(t) \right) = \dfrac{d}{dt}(a_2) = 0$

Derivatives of r(t) is zero after 2nd derivative. Hence, let us evaluate three constants C_0, C_1 & C_2.

The generalized error constants are given by,

$C_0 = \underset{s \to 0}{Lt}\ F(s)$; $\qquad C_1 = \underset{s \to 0}{Lt}\ \dfrac{d}{ds} F(s)$; $\qquad C_2 = \underset{s \to 0}{Lt}\ \dfrac{d^2}{ds^2} F(s)$

$F(s) = \dfrac{1}{G(s)H(s)} = \dfrac{1}{1 + G(s)} = \dfrac{1}{1 + \dfrac{10}{s(0.1s+1)}} = \dfrac{s(0.1s+1)}{s(0.1s+1)+10} = \dfrac{0.1s^2 + s}{0.1s^2 + s + 10}$

$C_0 = \underset{s \to 0}{Lt}\ F(s) = \underset{s \to 0}{Lt}\ \dfrac{0.1s^2 + s}{0.1s^2 + s + 10} = 0$

$C_1 = \underset{s \to 0}{Lt}\ \dfrac{d}{ds} F(s) = \underset{s \to 0}{Lt}\ \dfrac{d}{ds}\left[\dfrac{0.1s^2 + s}{0.1s^2 + s + 10} \right]$

$= \underset{s \to 0}{Lt}\left[\dfrac{(0.1s^2 + s + 10)(0.2s+1) - (0.1s^2 + s)(0.2s+1)}{(0.1s^2 + s + 10)^2} \right] = \underset{s \to 0}{Lt}\ \dfrac{2s + 10}{(0.1s^2 + s + 10)^2} = \dfrac{10}{10^2} = 0.1$

$$C_2 = \underset{s \to 0}{Lt} \frac{d^2}{ds^2} F(s) = \underset{s \to 0}{Lt} \frac{d}{ds}\left[\frac{d}{ds}F(s)\right] = \underset{s \to 0}{Lt} \frac{d}{ds}\left[\frac{2s+10}{(0.1s^2+s+10)^2}\right]$$

$$= \underset{s \to 0}{Lt} \frac{d}{ds}\left[\frac{(0.1s^2+s+10)^2 \times 2 - (2s+10) \times 2(0.1s^2+s+10)(0.2s+1)}{(0.1s^2+s+10)^4}\right]$$

$$\therefore C_2 = \frac{10^2 \times 2 - 10 \times 2 \times 10 \times 1}{10^4} = 0$$

Error signal, $e(t) = r(t)C_0 + \dot{r}(t)C_1 + \ddot{r}(t)\dfrac{C_2}{2!} = \dot{r}(t)C_1 + 0 + 0 = (a_1 + a_2 t)0.1$

$\therefore$ Steady state error, $e_{ss} = \underset{t \to \infty}{Lt}\, e(t) = \underset{t \to \infty}{Lt}\,[(a_1 + a_2 t)0.1] = \infty$

Method - II

The error signal in s-domain, $E(s) = \dfrac{R(s)}{1+G(s)H(s)}$

Given that, $r(t) = a_0 + a_1 t + \dfrac{a_2}{2}t^2$; $G(s) = \dfrac{10}{s(0.1s+1)}$; $H(s) = 1$

On taking Laplace transform of r(t) we get R(s),

$$\therefore R(s) = \frac{a_0}{s} + \frac{a_1}{s} + \frac{a_2}{s}\frac{2!}{s^3} = \frac{a_0}{s} + \frac{a_1}{s^2} + \frac{a_2}{s^3}$$

$$\therefore E(s) = \frac{R(s)}{1+G(s)H(s)} = \frac{\dfrac{a_0}{s}+\dfrac{a_1}{s^2}+\dfrac{a_2}{s^3}}{1+\dfrac{10}{s(0.1s+1)}} = \frac{\dfrac{a_0}{s}+\dfrac{a_1}{s^2}+\dfrac{a_2}{s^3}}{\dfrac{s(0.1s+1)+10}{s(0.1s+1)}}$$

$$= \frac{a_0}{s}\left[\frac{s(0.1s+1)}{s(0.1s+1)+10}\right] + \frac{a_1}{s^2}\left[\frac{s(0.1s+1)}{s(0.1s+1)+10}\right] + \frac{a_2}{s^3}\left[\frac{s(0.1s+1)}{s(0.1s+1)+10}\right]$$

The steady state error e_{ss} can be obtained from final value theorem.

Steady state error, $e_{ss} = \underset{t \to \infty}{Lt}\, e(t) = \underset{s \to 0}{Lt}\, sE(s)$

$$\therefore e_{ss} = \underset{s \to 0}{Lt}\, s\left\{\frac{a_0}{s}\left[\frac{s(0.1s+1)}{s(0.1s+1)+10}\right] + \frac{a_1}{s^2}\left[\frac{s(0.1s+1)}{s(0.1s+1)+10}\right] + \frac{a_2}{s^3}\left[\frac{s(0.1s+1)}{s(0.1s+1)+10}\right]\right\}$$

$$= \underset{s \to 0}{Lt}\left\{\frac{a_0 s(0.1s+1)}{s(0.1s+1)+10} + \frac{a_1(0.1s+1)}{s(0.1s+1)+10} + \frac{a_2(0.1s+1)}{s(0.1s+1)+10}\right\} = 0 + \frac{a_1}{10} + \infty = \infty$$

Method - III

Error signal in s-domain, $E(s) = \dfrac{R(s)}{1+G(s)H(s)}$; $\therefore \dfrac{E(s)}{R(s)} = \dfrac{1}{1+G(s)H(s)}$

Given that, $G(s) = \dfrac{10}{s(0.1s+1)}$ and $H(s) = 1$.

$$\therefore \frac{E(s)}{R(s)} = \frac{1}{1+\dfrac{10}{s(0.1s+1)}} = \frac{s(0.1s+1)}{s(0.1s+1)+10} = \frac{0.1s^2+s}{0.1s^2+s+10} = \frac{s+0.1s^2}{10+s+0.1s^2} = \frac{s}{10} - \frac{s^3}{1000} + \dots$$

$$\therefore E(s) = \frac{s}{10}R(s) - \frac{s^3}{1000}R(s) + \dots$$

| Dividing numerator polynomial by denominator polynomial. |

On taking inverse Laplace transform,

$$e(t) = \frac{1}{10}\dot{r} - \frac{1}{1000}\ddot{r}(t) + \dots$$

Given that, $r(t) = a_0 + a_1 t + \frac{a_2}{2} t^2$

$$\therefore \dot{r} = \frac{d}{dt} r(t) = a_1 + a_2 t$$

$$\ddot{r}(t) = \frac{d}{dt} \dot{r}(t) = a_2$$

$$\dddot{r}(t) = \frac{d}{dt} \ddot{r}(t) = 0$$

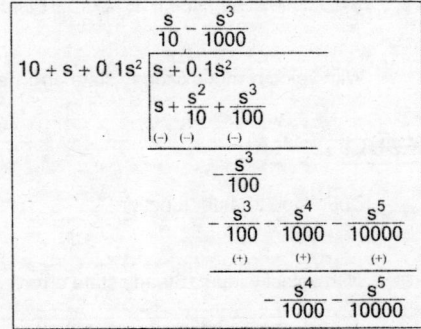

$\therefore$ Error signal in time domain, $e(t) = \frac{1}{10}\dot{r}(t) = \frac{1}{10}(a_1 + a_2 t)$

Steady state error, $e_{ss} = \underset{t \to \infty}{Lt}\ e(t) = \underset{t \to \infty}{Lt}\ \frac{1}{10}(a_1 + a_2 t) = \infty$

RESULT

(a) Position error constant, $K_p = \infty$

(b) Velocity error constant, $K_v = 10$

(c) Acceleration error constant, $K_a = 0$

(d) When input, $r(t) = a_0 + a_1 t + \frac{a_2 t^2}{2}$,   Steady state error, $e_{ss} = \infty$

EXAMPLE 2.14

Consider a unity feedback system with a closed loop transfer function $\dfrac{C(s)}{R(s)} = \dfrac{Ks + b}{s^2 + as + b}$ Determine open loop

transfer function G(s). Show that steady state error with unit ramp input is given by $\dfrac{(a - K)}{b}$

SOLUTION

For unity feedback system, H(s) = 1

The closed loop transfer function, $M(s) = \dfrac{C(s)}{R(s)} = \dfrac{G(s)}{1 + G(s)H(s)} = \dfrac{G(s)}{1 + G(s)}$

$$\therefore \frac{G(s)}{1 + G(s)} = M(s)$$

On cross multiplication of the above equation we get,

$$G(s) = M(s)[1 + G(s)] = M(s) + M(s)\,G(s)$$

$\therefore\ G(s) - M(s)\,G(s) = M(s) \quad \Rightarrow \quad G(s)[1 - M(s)] = M(s) \quad \Rightarrow \quad M(s) = \dfrac{Ks + b}{s^2 + as + b}$

$\therefore$ Open loop transfer function,

$$G(s) = \frac{M(s)}{1 - M(s)} = \frac{\dfrac{Ks + b}{s^2 - as + b}}{1 - \dfrac{Ks + b}{s^2 + as + b}} = \frac{Ks + b}{(s^2 + as + b) - (Ks + b)}$$

$$= \frac{Ks + b}{s^2 + as + b - Ks - b} = \frac{Ks + b}{s^2 + (a - K)s} = \frac{Ks + b}{s[s + (a - K)]}$$

Velocity error constant, $K_v = \underset{s \to 0}{Lt}\, s\, G(s)H(s) = \underset{s \to 0}{Lt}\, sG(s) = \underset{s \to 0}{Lt}\, s\dfrac{Ks+b}{s[s+a(-K)]} = \dfrac{b}{a-K}$

With velocity input, Steady state error, $e_{ss} = \dfrac{1}{K_v} = \dfrac{a-K}{b}$

RESULT

Open loop transfer function, $G(s) = \dfrac{Ks+b}{s[s+(a-K)]}$

With velocity input, Steady state error, $e_{ss} = \dfrac{a-K}{b}$

EXAMPLE 2.15

A unity feedback system has the forward transfer function $G(s) = \dfrac{K_1(2s+1)}{s(5s+1)(1+s)^2}$. When the input r(t) = 1+6t,

determine the minimum value of K_1 so that the steady error is less than 0.1.

SOLUTION

Given that, input r(t) = 1 + 6t

On taking laplace transform of r(t) we get R(s).

$$\therefore R(s) = \mathcal{L}\{r(t)\} = \mathcal{L}\{1 + 6t\} = \frac{1}{s} + \frac{6}{s^2}$$

The error signal in s-domain E(s) is given by,

$$\therefore E(s) = \frac{R(s)}{1+G(s)H(s)} = \frac{\dfrac{1}{s}+\dfrac{6}{s^2}}{1+\dfrac{K_1(2s+1)}{s(5s+1)(1+s)^2}} = \frac{\dfrac{1}{s}+\dfrac{6}{s^2}}{\dfrac{s(5s+1)(1+s)^2+K_1(2s+1)}{s(5s+1)(1+s)^2}}$$

$\boxed{\text{Here H(s) = 1}}$

$$= \frac{1}{s}\left[\frac{s(5s+1)(1+s)^2}{s(5s+1)(1+s)^2+K_1(2s+1)}\right] + \frac{6}{s^2}\left[\frac{s(5s+1)(1+s)^2}{s(5s+1)(1+s)^2+K_1(2s+1)}\right]$$

The steady state error ess can be obtained from final value theorem.

$$e_{ss} = \underset{t \to \infty}{Lt}\, e(t) = \underset{s \to 0}{Lt}\, sE(s)$$

$$= \underset{t \to \infty}{Lt}\, s\left\{\frac{1}{s}\left[\frac{s(5s+1)(1+s)^2}{s(5s+1)(1+s)^2+K_1(2s+1)}\right] + \frac{6}{s^2}\left[\frac{s(5s+1)(1+s)^2}{s(5s+1)(1+s)^2+K_1(2s+1)}\right]\right\}$$

$$= \underset{s \to 0}{Lt}\,\left\{\frac{s(5s+1)(1+s)^2}{s(5s+1)(1+s)^2+K_1(2s+1)} + \frac{6(5s+1)(1+s)^2}{s(5s+1)(1+s)^2+K_1(2s+1)}\right\} = 0 + \frac{6}{K_1} = \frac{6}{K_1}$$

Given that, $e_{ss} < 0.1$, $\therefore 0.1 = \dfrac{6}{K_1}$ or $K_1 = \dfrac{6}{0.1} = 60$

RESULT

For steady state error, $e_{ss} < 0.1$, the value of K_1 should be greater than 60.

2 .19 COMPONENTS OF AUTOMATIC CONTROL SYSTEM

The basic components of an automatic control system are Error detector, Amplifier and Controller, Actuator (Power actuator), Plant and Sensor or Feedback system. The block diagram of an automatic control system is shown in fig 2.16.

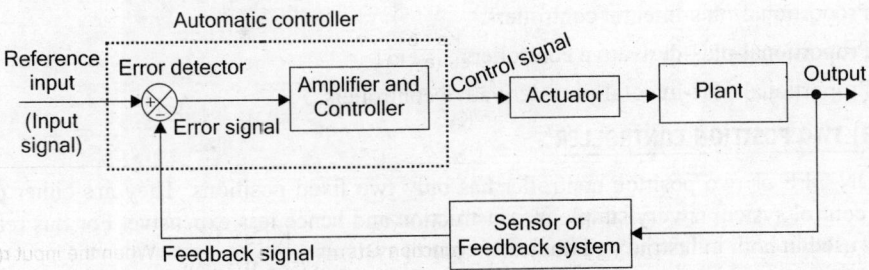

Fig 2.16 : *Block diagram of automatic control system.*

The plant is the open loop system whose output is automatically controlled by closed loop system. The combined unit of error detector, amplifier and controller is called ***automatic controller***, because without this unit the system becomes open loop system.

In automatic control systems the reference input will be an input signal proportional to desired output. The feedback signal is a signal proportional to current output of the system. The error detector compares the reference input and feedback signal and if there is a difference it produces an error signal. An amplifier can be used to amplify the error signal and the controller modifies the error signal for better control action.

The actuator amplifies the controller output and converts to the required form of energy that is acceptable for the plant. Depending on the input to the plant, the output will change. This process continues as long as there is a difference between reference input and feedback signal. If the difference is zero, then there is no error signal and the output settles at the desired value.

Generally, the error signal will be a weak signal and so it has to be amplified and then modified for better control action. In most of the system the controller itself amplifies the error signal and integrates or differentiates to produce a control signal (i.e., modified error signal). The different types of controllers are P, PI, PD and PID controllers.

2.20 CONTROLLERS

A controller is a device introduced in the system to modify the error signal and to produce a control signal. The manner in which the controller produces the control signal is called the ***control action***. The controller modifies the transient response of the system. The electronic controllers using operational amplifiers are presented in this section.

The following six basic control actions are very common among industrial analog controllers.

1. Two-position or ON-OFF control action.
2. Proportional control action.
3. Integral control action.
4. Proportional- plus- integral control action.
5. Proportional-plus-derivative control action.
6. Proportional-plus-integral-plus-derivative control action.

Depending on the control actions provided the controllers can be classified as follows.

1. Two position or ON-OFF controllers.
2. Proportional controllers.
3. Integral controllers.
4. Proportional-plus-integral controllers.
5. Proportional-plus-derivative controllers.
6. Proportional-plus-integral-plus-derivative controllers.

ON-OFF (OR) TWO POSITION CONTROLLER

The ON-OFF or two position controller has only two fixed positions. They are either on or off. The on-off control system is very simple in construction and hence less expensive. For this reason, it is very widely used in both industrial and domestic control systems.

The ON-OFF control action may be provided by a relay. There are different types of relay. The most popular one is electromagnetic relay. It is a device which has NO (Normally Open) and NC (Normally Closed) contacts, whose opening and closing are controlled by the relay coil. When the relay coil is excited, the relay operates and the contacts change their positions (i.e., NO → NC and NC → NO).

Let the output signal from the controller be u(t) and the actuating error signal be e(t). In this controller, u(t) remains at either a maximum or minimum value.

$$u(t) = u_1 \; ; \; \text{for } e(t) < 0$$

$$= u_2 \; ; \; \text{for } e(t) > 0$$

$$\boxed{E(s) = \mathcal{L}\{e(t)\} \; ; \; U(s) = \mathcal{L}\{u(t)\}}$$

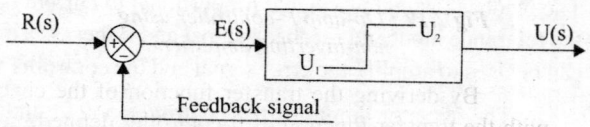

Fig 2.17 : Block diagram of on-off controller.

PROPORTIONAL CONTROLLER (P - CONTROLLER)

The proportional controller is a device that produces a control signal, u(t) proportional to the input error signal, e(t).

In P-controller, $u(t) \propto e(t)$

$$\therefore \quad u(t) = K_p \, e(t) \qquad \qquad \qquad \text{.....(2.81)}$$

where, K_p = Proportional gain or constant

On taking Laplace transform of equation (2.81) we get,

$$U(s) = K_p E(s) \qquad \qquad \qquad \text{.....(2.82)}$$

$$\boxed{\therefore \text{ Transfer function of P-controller, } \frac{U(s)}{E(s)} = K_p} \qquad \qquad \text{.....(2.83)}$$

The equation (2.82) gives the output of the P-controller for the input E(s) and equation (2.83) is the transfer function of the P-controller. The block diagram of the P-controller is shown in fig 2.18.

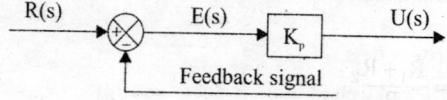

Fig 2.18 : Block diagram of proportional controller.

From the equation (2.82), we can conclude that the proportional controller amplifies the error signal by an amount K_p. Also the introduction of the controller on the system increases the loop gain by an amount K_p. The increase in loop gain improves the steady state tracking accuracy, disturbance signal rejection and the relative stability and also makes the system less sensitive to parameter variations. But increasing the gain to very large values may lead to instability of the system. The drawback in P-controller is that it leads to a constant steady state error.

EXAMPLE OF ELECTRONIC P-CONTROLLER

The proportional controller can be realized by an amplifier with adjustable gain. Either the non-inverting operational amplifier or the inverting operational amplifier followed by sign changer will work as a proportional controller. The op-amp proportional controller is shown in fig 2.19 and 2.20.

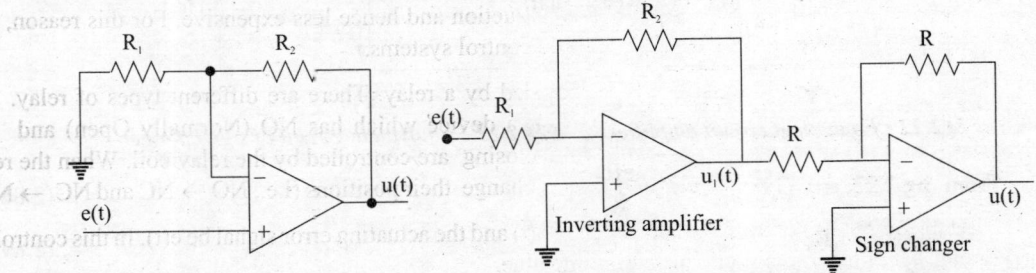

Fig 2.19 : *Op-amp P-controller using non-inverting amplifier.*

Fig 2.20 : *Op-amp P-controller using inverting amplifier.*

By deriving the transfer function of the controllers shown in fig 2.11 and 2.12 and comparing with the transfer function of P-controller defined by equation (2.83), it can be shown that they work as P-controllers.

ANALYSIS OF P-CONTROLLER SHOWN IN FIG 2.19

In fig 2.19, the input e(t) is applied to positive input. By symmetry of op-amp the voltage of negative input is also e(t). Also we assume an ideal op-amp so that input current is zero. Based on the above assumptions the equivalent circuit of the controller is shown in fig 2.21.

By voltage division rule,

$$e(t) = \frac{R_1}{R_1 + R_2} u(t) \quad ; \quad \therefore \ u(t) = \frac{R_1 + R_2}{R_1} e(t) \quad .....(2.84)$$

On taking Laplace transform of equation (2.84) we get,

$$U(s) = \frac{R_1 + R_2}{R_1} E(s) \quad .....(2.85)$$

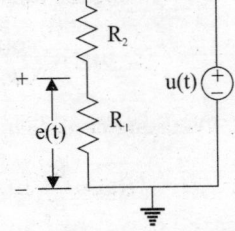

$$\therefore \ \frac{U(s)}{E(s)} = \frac{R_1 + R_2}{R_1} \quad .....(2.86)$$

Fig 2.21 : *Equivalent circuit of P-controller shown in fig 2.19.*

The equation (2.86) is the transfer function of op-amp P-controller. On comparing equation (2.86) with equation (2.83) we get,

Proportional gain, $\quad K_p = \dfrac{R_1 + R_2}{R_1} \quad .....(2.87)$

Therefore by adjusting the values of R_1 and R_2 the value of gain, K_p can be varied.

ANALYSIS OF P-CONTROLLER SHOWN IN FIG 2.20

The assumption made in op-amp circuit analysis are,

1. The voltages at both inputs are equal

2. The input current is zero.

Based on the above assumptions, the equivalent circuit of op-amp amplifier and sign changer are shown in fig 2.22 and 2.23.

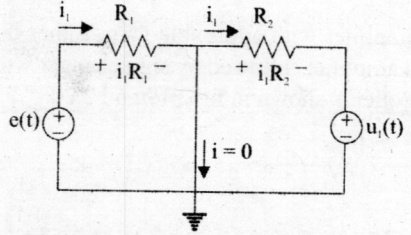   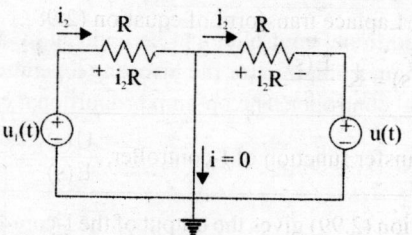

Fig 2.22 : *Equivalent circuit amplifier.*   **Fig 2.23** : *Equivalent circuit of sign changer.*

From fig 2.22, $e(t) = i_1 R_1$; $\therefore i_1 = \dfrac{e(t)}{R_1}$ .....(2.88)

$u_1(t) = -i_1 R_2$ .....(2.89)

Substitute for i_1 from equation (2.88) in equation (2.89).

$\therefore u_1(t) = -\dfrac{e(t)}{R_1} R_2$ .....(2.90)

From fig 2.23, $u(t) = -i_2 R$; $\therefore i_2 = -\dfrac{u(t)}{R}$ .....(2.91)

$u_1(t) = i_2 R$ .....(2.92)

Substitute for i_2 from equation (2.91) in equation (2.92).

$\therefore u_1(t) = -\dfrac{u(t)}{R} R = -u(t)$ .....(2.93)

On equating the equations (2.90) and (2.93) we get,

$\therefore u(t) = -\dfrac{e(t)}{R_1} R_2$; $u(t) = \dfrac{R_2}{R_1} e(t)$ .....(2.94)

On taking Laplace transform of equation (2.94) we get,

$U(s) = \dfrac{R_2}{R_1} E(s)$ .....(2.95)

$\therefore \dfrac{U(s)}{E(s)} = \dfrac{R_2}{R_1}$ .....(2.96)

The equation (2.96) is the transfer function of op-amp P-controller. On comparing equation (2.96) with equation (2.83) we get,

Proportional gain, $K_p = \dfrac{R_2}{R_1}$ .....(2.97)

Therefore by adjusting the values of R_1 and R_2 the value of gain K_p can be varied.

INTEGRAL CONTROLLER (I-CONTROLLER)

The integral controller is a device that produces a control signal u(t) which is proportional to integral of the input error signal, e(t).

In I-controller, $u(t) \alpha \int e(t)\, dt$; $\therefore u(t) = K_i \int e(t)\, dt$

.....(2.98)

where, K_i = Integral gain or constant.

On taking Laplace transform of equation (2.98) with zero initial conditions we get,

$$U(s) = K_i \frac{E(s)}{s}$$

.....(2.99)

$$\boxed{\text{Transfer function of I-controller, } \frac{U(s)}{E(s)} = \frac{K_i}{s}}$$

.....(2.100)

The equation (2.99) gives the output of the I-controller for the input E(s) and equation (2.101) is the transfer function of the I-controller. The block diagram of I-controller is shown in fig 2.24.

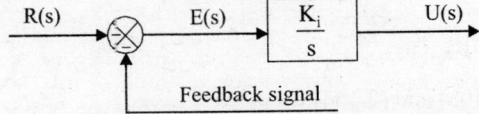

Fig 2.24 : *Block diagram of an integral controller.*

The integral controller removes or reduces the steady error without the need for manual reset. Hence the I-controller is sometimes called ***automatic reset***. The drawback in integral controller is that it may lead to oscillatory response of increasing or decreasing amplitude which is undesirable and the system may become unstable.

EXAMPLE OF ELECTRONIC I-CONTROLLER

The integral controller can be realized by an integrator using op-amp followed by a sign changer as shown in fig 2.25.

By deriving the transfer function of the controller shown in fig 2.25 and comparing with the transfer function of I-controller defined by equation(2.101), it can be shown that it work as I-controller.

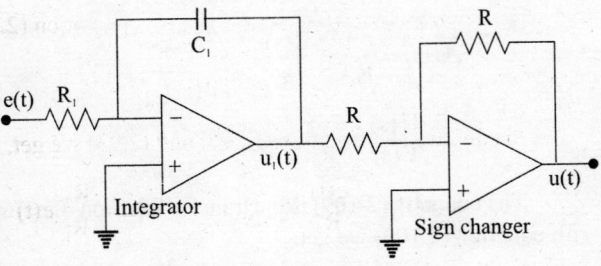

Fig 2.25 : *I-controller using op-amp.*

ANALYSIS OF I-CONTROLLER SHOWN IN FIG 2.25

The assumptions made in op-amp circuit analysis are,

    1. The voltages of both inputs are equal.

    2. The input current is zero.

Based on the above assumptions the equivalent circuit of op-amp integrator and sign changer are shown in fig 2.26 and 2.27.

From fig 2.26, $e(t) = i_1 R_1$; $\therefore i_1 = \dfrac{e(t)}{R_1}$

.....(2.101)

$$u_1(t) = -\frac{1}{C_1} \int i_1\, dt$$

.....(2.102)

Substitute for i_1 from equation (2.101) in equation (2.102).

$$\therefore \ u_1(t) = -\frac{1}{C_1} \int \frac{e(t)}{R_1} dt = -\frac{1}{R_1 C_1} \int e(t)\, dt \qquad \qquad .....(2.103)$$

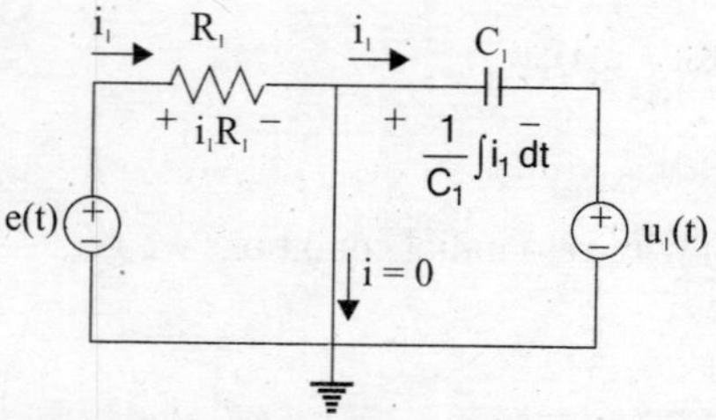

Fig 2.26 : *Equivalent circuit of integrator.* 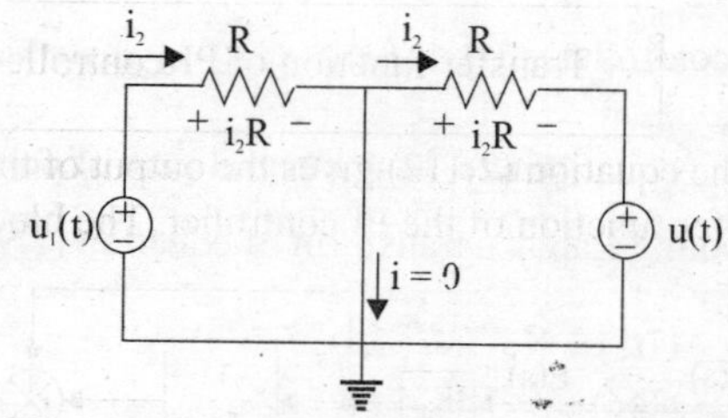   Fig 2.27 : *Equivalent circuit of sign changer.*

From fig 2.27, $u(t) = -i_2 R, \therefore \ i_2 = \dfrac{-u(t)}{R}$ .....(2.104)

$$u_1(t) = i_2 R \qquad \qquad .....(2.105)$$

Substitute for i_2 from equation (2.106) in equation (2.107),

$$\therefore \ u_1(t) = -\frac{u(t)}{R} R = -u(t) \qquad \qquad .....(2.106)$$

On equating the equations (2.103) and (2.106) we get,

$$-u(t) = -\frac{1}{R_1 C_1} \int e(t)\, dt$$

$$\therefore \ u(t) = \frac{1}{R_1 C_1} \int e(t)\, dt \qquad \qquad .....(2.107)$$

On taking Laplace transform of equation (2.107) with zero initial conditions we get,

$$U(s) = \frac{1}{R_1 C_1} \frac{E(s)}{s} \qquad \qquad .....(2.108)$$

$$\therefore \ \frac{U(s)}{E(s)} = \frac{1}{R_1 C_1} \frac{1}{s} \qquad \qquad .....(2.109)$$

The equation (2.109) is the transfer function of op-amp I-controller. On comparing equation (2.109) with equation (2.100) we get,

$$\text{Integral gain, } K_i = \frac{1}{R_1 C_1} \qquad \qquad .....(2.110)$$

Therefore by adjusting the values of R_1 and C_1 the value of gain K_i can be varied.

PROPORTIONAL PLUS INTEGRAL CONTROLLER (PI-CONTROLLER)

The proportional plus integral controller (PI-controller) produces an output signal consisting of two terms : *one proportional to error signal and the other proportional to the integral of error signal.*

In PI-controller, $u(t) \, \alpha \left[e(t) + \int e(t)\, dt \right]$; $\therefore \ u(t) = K_p e(t) + \dfrac{K_p}{T_i} \int e(t)\, dt$ .....(2.111)

where, K_p = Proportional gain

T_i = Integral time.

On taking Laplace transform of equation (2.111) with zero initial conditions we get,

$$U(s) = K_p E(s) + \frac{K_p}{T_i} \frac{E(s)}{s} \qquad .....(2.112)$$

$$\therefore \text{Transfer function of PI-controller, } \frac{U(s)}{E(s)} = K_p \left(1 + \frac{1}{T_i s}\right) \qquad .....(2.113)$$

The equation (2.112) gives the output of the PI-controller for the input E(s) and equation (2.113) is the transfer function of the PI controller. The block diagram of PI-controller is shown in fig 2.28.

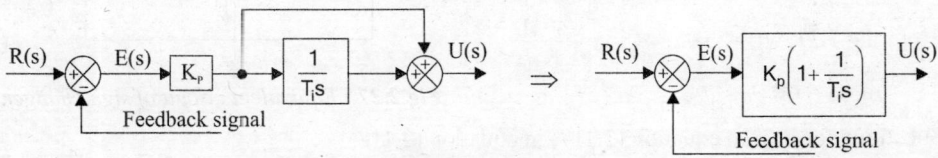

Fig 2.28 : *Block diagram of PI-controller.*

The advantages of both P-controller and I-controller are combined in PI-controller. The proportional action increases the loop gain and makes the system less sensitive to variations of system parameters. The integral action eliminates or reduces the steady state error.

The integral control action is adjusted by varying the integral time. The change in value of K_p affects both the proportional and integral parts of control action. The inverse of the integral time T_i is called the *reset rate.*

EXAMPLE OF ELECTRONIC PI-CONTROLLER

The PI-controller can be realized by an op-amp integrator with gain followed by a sign changer as shown in fig 2.29.

By deriving the transfer function of the controller shown in fig (2.29) and comparing with the transfer function of PI-controller defined by equation (2.114), it can be proved that the circuit shown in fig 2.29, work as PI-controller.

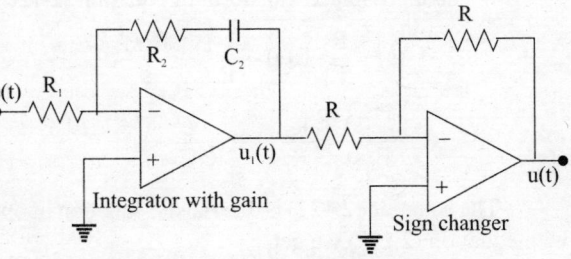

Fig 2.29 : *PI-controller using op-amp.*

ANALYSIS OF PI-CONTROLLER SHOWN IN FIG 2.29

The assumptions made in op-amp circuit analysis are,

    1. The voltages at both inputs are equal.

    2. The input current is zero.

Based on the above assumptions the equivalent circuit of op-amp integrator and sign changer are shown in fig 2.30. and 2.31.

From fig 2.30, $e(t) = i_1 R_1$; $\therefore i_1 = \dfrac{e(t)}{R_1}$ .....(2.114)

$$u_1(t) = -i_1 R_2 - \frac{1}{C_2} \int i_1 dt \qquad .....(2.115)$$

Substitute for i_1 from equation (2.114) in equation (2.115).

$$\therefore u_1(t) = -\frac{e(t)}{R_1} R_2 - \frac{1}{C_2} \int \frac{e(t)}{R_1} dt \qquad .....(2.116)$$

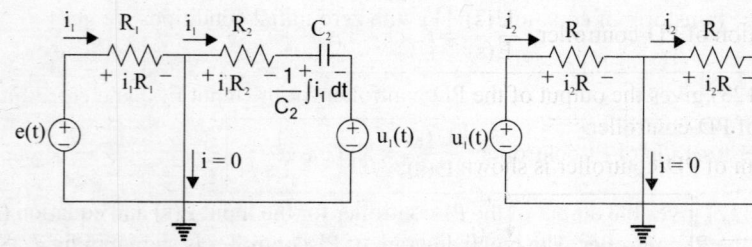

Fig 2.30 : *Equivalent circuit of integrator.*   **Fig 2.31** : *Equivalent circuit of sign changer.*

From fig 2.31, $u(t) = -i_2 R$, $\therefore i_2 = \dfrac{-u(t)}{R}$ .....(2.117)

$$u_1(t) = i_2 R \qquad\qquad\qquad .....(2.118)$$

Substitute for i_2 from equation (2.117) in equation (2.118),

$$\therefore u_1(t) = -\frac{u(t)}{R} R = -u(t) \qquad\qquad .....(2.119)$$

On equating the equations (2.116) and (2.119) we get,

$$-u(t) = -\frac{e(t)}{R_1} R_2 - \frac{1}{C_2} \int \frac{e(t)}{R_1} dt$$

$$\therefore u(t) = \frac{R_2}{R_1} e(t) + \frac{1}{R_1 C_2} \int e(t) dt \qquad .....(2.120)$$

On taking Laplace transform of equation (2.120) with zero initial conditions we get,

$$U(s) = \frac{R_2}{R_1} E(s) + \frac{1}{R_1 C_2} \frac{E(s)}{s}$$

$$\therefore \frac{U(s)}{E(s)} = \frac{R_2}{R_1}\left(1 + \frac{1}{R_2 C_2 s}\right) \qquad .....(2.121)$$

The equation (2.121) is the transfer function of op-amp PI-controller. On comparing equation(2.121) with equation (2.113) we get,

Proportional gain, $K_p = \dfrac{R_2}{R_1}$;   Integral time, $T_i = R_2 C_2$.

By varying the values of R_1 and R_2, the value of gain K_p and T_i can be adjusted.

PROPORTIONAL PLUS DERIVATIVE CONTROLLER (PD-CONTROLLER)

The proportional plus derivative controller produces an output signal consisting of two terms : *one proportional to error signal and the other proportional to the derivative of error signal.*

In PD-controller, $u(t) \alpha \left[e(t) + \dfrac{d}{dt} e(t)\right]$; $\therefore u(t) = K_p e(t) + K_p T_d \dfrac{d}{dt} e(t)$ .....(2.122)

where, K_p = Proportional gain

T_d = Derivative time

On taking Laplace transform of equation (2.123) with zero initial conditions we get,

$$U(s) = K_p E(s) + K_p T_d sE(s) \qquad\qquad .....(2.123)$$

$$\therefore \text{ Transfer function of PD-controller, } \frac{U(s)}{E(s)} = K_p(1 + T_d s) \quad\quad .....(2.124)$$

The equation (2.123) gives the output of the PD-controller for the input E(s) and equation (2.124) is the transfer function of PD-controller.

The block diagram of PD-controller is shown in fig 2.32.

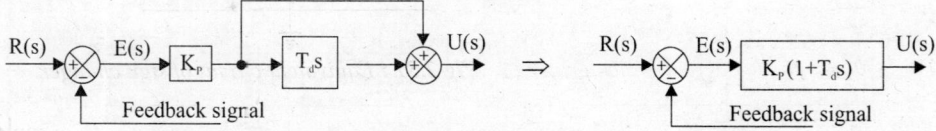

Fig 2.32 : *Block diagram of PD-controller.*

The derivative control acts on rate of change of error and not on the actual error signal. The derivative control action is effective only during transient periods and so it does not produce corrective measures for any constant error. Hence the derivative controller is never used alone, but it is employed in association with proportional and integral controllers. The derivative controller does not affect the steady-state error directly but anticipates the error, initiates an early corrective action and tends to increase the stability of the system. While derivative control action has an advantage of being anticipatory it has the disadvantage that it amplifies noise signals and may cause a saturation effect in the actuator.

The derivative control action is adjusted by varying the derivative time. The change in the value of K_p affects both the proportional and derivative parts of control action. The derivative control is also called *rate control.*

EXAMPLE OF ELECTRONIC PD-CONTROLLER

The PD-controller can be realized by an op-amp differentiator with gain followed by a sign changer as shown in fig 2.33.

By deriving the transfer function of the controller shown in fig 2.33 and comparing with the transfer function of PD-controller defined by equation (2.124) it can be proved that the circuit shown in fig 2.33 will work as PD-controller.

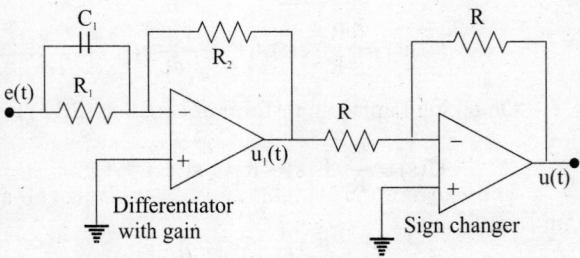

Fig 2.33 : *PD controller using op-amp.*

ANALYSIS OF PD-CONTROLLER SHOWN IN FIG 2.33

The assumptions made in op-amp circuit analysis are,

      1. The voltages at both inputs are equal.

      2. The input current is zero.

Based on the above assumptions the equivalent circuit of op-amp differentiator and sign changer are shown in fig 2.34 and 2.35.

$$\text{From fig 2.34, } \therefore \ i_1 = \frac{e(t)}{R_1} + C_1 \frac{de(t)}{dt} \quad\quad .....(2.125)$$

$$i_1 R_2 = -u_1(t), \quad\quad \therefore \ i_1 = \frac{-u_1(t)}{R_2} \quad\quad .....(2.126)$$

On equating the equations (2.125) and (2.126) we get,

$$-\frac{u_1 t}{R_2} = \frac{e(t)}{R_1} + C_1 \frac{d}{dt} e(t) \ ; \quad\quad \therefore \ u_1(t) = -\left(\frac{R_2}{R_1} e(t) + R_2 C_1 \frac{d}{dt} e(t) \right) \quad\quad .....(2.127)$$

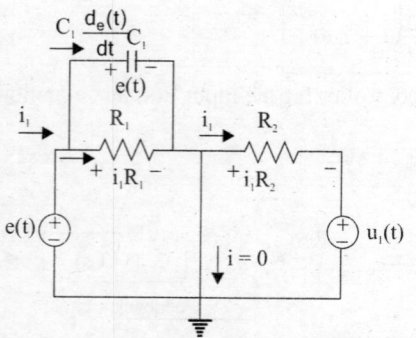

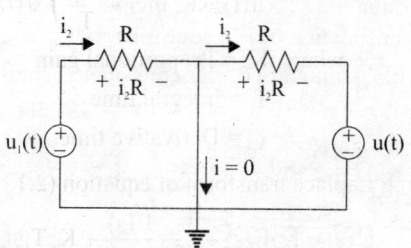

Fig 2.34 : *Equivalent circuit of differentiator.*   **Fig 2.35 :** *Equivalent circuit of sign changer.*

From fig 2.35, $u(t) = -i_2 R$; $\therefore i_2 = \dfrac{-u(t)}{R}$ .....(2.128)

$u_1(t) = i_2 R$ .....(2.129)

Substitute for i_2 from equation (2.128) in equation (2.129).

$\therefore u_1(t) = -\dfrac{u(t)}{R} R = -u(t)$ .....(2.130)

On equating the equations (2.127) and (2.130) we get,

$$-u(t) = -\left(\dfrac{R_2}{R_1} e(t) + R_2 C_1 \dfrac{d}{dt} e(t) \right)$$

$\therefore u(t) = \dfrac{R_2}{R_1} e(t) + R_2 C_1 \dfrac{d}{dt} e(t)$ .....(2.131)

On taking Laplace transform of equation (2.131) with zero initial conditions we get,

$U(s) = \dfrac{R_2}{R_1} E(s) + R_2 C_1 sE(s)$ .....(2.132)

$\therefore \dfrac{U(s)}{E(s)} = \dfrac{R_2}{R_1} (1 + R_1 C_1 s)$ .....(2.133)

The equation (2.133) is the transfer function of op-amp PD-controller. On comparing equation (2.133) with equation (2.124) we get,

Proportional gain, $K_p = \dfrac{R_2}{R_1}$

Derivative time, $T_d = R_1 C_1$

By varying the values of R_1 and R_2, the value of gain K_p and T_d can be adjusted.

PROPORTIONAL PLUS INTEGRAL PLUS DERIVATIVE CONTROLLER (PID-CONTROLLER)

The PID-controller produces an output signal consisting of three terms : *one proportional to error signal, another one proportional to integral of error signal and the third one proportional to derivative of error signal.*

In PID-controller, $u(t) \quad \alpha \quad \left[e(t) + \int e(t)\, dt + \dfrac{d}{dt}\, e(t) \right]$

$$\therefore \; u(t) = K_p\, e(t) + \frac{K_p}{T_i} \int e(t)\, dt + K_p T_d \frac{d}{dt}\, e(t) \qquad \dots \dots (2.134)$$

where, K_p = Proportional gain

$\quad \quad T_i$ = Integral time

$\quad \quad T_d$ = Derivative time

On taking Laplace transform of equation (2.134) with zero initial conditions we get,

$$U(s) = K_p E(s) + \frac{K_p}{T_i} \frac{E(s)}{s} + K_p T_d s E(s) \qquad \dots \dots (2.135)$$

$$\boxed{\therefore \; \text{Transfer function of PID-controller}, \quad \frac{U(s)}{E(s)} = K_p \left(1 + \frac{1}{T_i s} + T_d s \right)} \qquad \dots \dots (2.136)$$

The equation (2.135) gives the output of the PID-controller for the input E(s) and equation (2.136) is the transfer function of the PID-controller. The block diagram of PID-contoller is shown in fig 2.36.

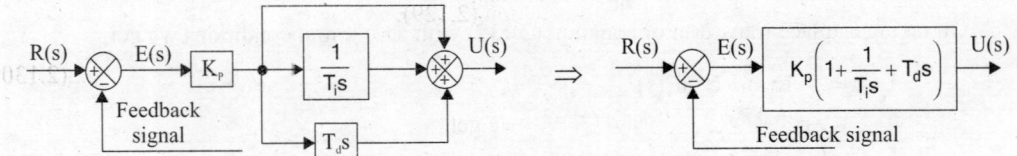

Fig 2.36 : *Block diagram of PID-controller.*

The combination of proportional control action, integral control action and derivative control action is called PID-control action. This combined action has the advantages of the each of the three individual control actions.

The proportional controller stabilizes the gain but produces a steady state error. The integral controller reduces or eliminates the steady state error. The derivative controller reduces the rate of change of error.

EXAMPLE OF ELECTRONIC PID-CONTROLLER

The PID-controller can be realized by op-amp amplifier with integral and derivative action followed by sign changer as shown in fig 2.37.

By deriving the transfer function of the controller shown in fig (2.37) and comparing with the transfer function of PID-controller defined by equation (2.136) it can be proved that the circuit shown in fig 2.37 work as PID-controller.

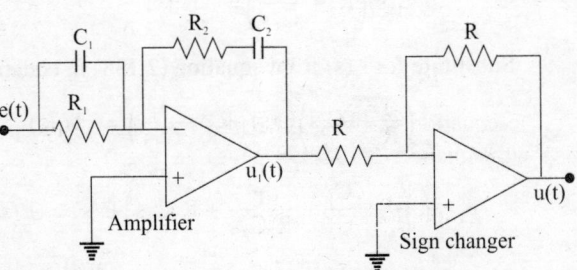

Fig 2.37 : *PID-controller using op-amp.*

ANALYSIS OF PID-CONTROLLER SHOWN IN FIG 2.37

The assumptions made in op-amp circuit analysis are.

1. The voltages of both inputs are equal.

2. The input current is zero.

Based on the above assumptions the equivalent circuit of op-amp amplifier and sign changer are shown in fig 2.38 and 2.39.

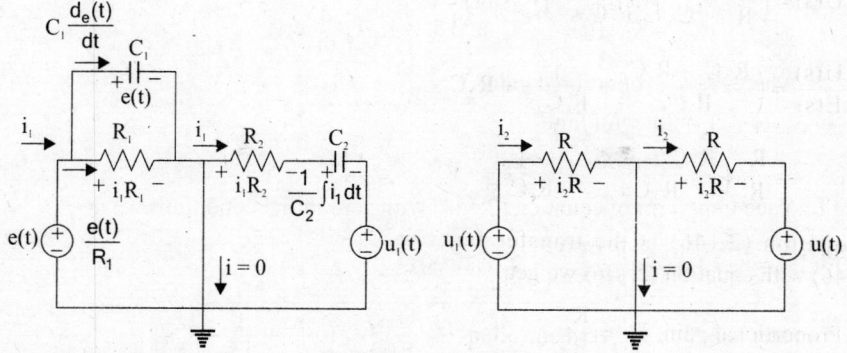

Fig 2.38 : *Equivalent circuit of amplifier.*   Fig 2.39 : *Equivalent circuit of sign changer.*

From fig 2.38, $i_1 = \dfrac{e(t)}{R_1} + C_1 \dfrac{de(t)}{dt}$ .....(2.137)

On taking Laplace transform of equation (2.137) with zero initial conditions we get,

$$I_1(s) = \frac{1}{R_1} E(s) + C_1\, sE(s)$$

$$I_1(s) = \left(\frac{1}{R_1} + C_1 s \right) E(s)$$ .....(2.138)

From fig 2.38, $i_1 R_2 + \dfrac{1}{C_2} \displaystyle\int i_1 \, dt = -u_1(t)$ .....(2.139)

On taking Laplace transform of equation (2.138) with zero initial conditions we get,

$$I_1(s) R_2 + \frac{1}{C_2} \frac{I_1(s)}{s} = -U_1(s)$$

$$\therefore \; I_1(s)\left(R_2 + \frac{1}{C_2 s} \right) = -U_1(s)$$ .....(2.140)

Substitute for $I_1(s)$ from equation (2.138) in equation (2.140).

$$\therefore \; \left(\frac{1}{R_1} + C_1 s \right) E(s) \left(R_2 + \frac{1}{C_2 s} \right) = -U_1(s)$$

$$-\left(\frac{R_2}{R_1} + \frac{C_1}{C_2} + \frac{1}{R_1 C_2 s} + R_2 C_1 s \right) E(s) = U_1(s)$$ .....(2.141)

From fig 2.39, $u(t) = -i_2 R$; $\therefore \; i_2 = -\dfrac{u(t)}{R}$ .....(2.142)

$$u_1(t) = i_2 R$$ .....(2.143)

Substitute for i_2 from equation (2.142) in equation (2.143).

$$\therefore \; u_1(t) = -\frac{u(t)}{R} R = -u(t)$$ .....(2.144)

On taking Laplace transform of equation (2.144) we get,

$$U_1(s) = -U(s)$$ .....(2.145)

From equations (2.142) and (2.146) we get,

$$U(s) = \left(\frac{R_2}{R_1} + \frac{C_1}{C_2} + \frac{1}{R_1 C_2 s} + R_2 C_1 s \right) E(s)$$

$$\therefore \frac{U(s)}{E(s)} = \left(\frac{R_2 C_2 + R_1 C_1}{R_1 C_2} + \frac{1}{R_1 C_2 s} + R_2 C_1 s \right)$$

$$= \frac{R_2}{R_1} \left(\frac{R_2 C_2 + R_1 C_1}{R_2 C_2} + \frac{1}{R_1 C_2 s} + R_1 C_1 s \right)$$

The equation (2.146) is the transfer function of op-amp PID-controller. On comparing equation (2.146) with equation (2.136) we get,

Proportional gain, $K_p = \dfrac{R_2}{R_1}$

Derivative time, $T_d = R_1 C_1$;   Integral time, $T_i = R_2 C_2$

Also, $\dfrac{R_1 C_1 + R_2 C_2}{R_2 C_2} = 1$

By varying the values of R_1 and R_2 the values of K_p, T_d and T_i are adjusted.

2.21 RESPONSE WITH P, PI, PD AND PID CONTROLLERS

In feedback control systems a controller may be introduced to modify the error signal and to achieve better control action. The introduction of controllers will modify the transient response and the steady state error of the system. The effects due to introduction of P, PI, PD and PID controllers are discussed in this section.

EFFECT OF PROPORTIONAL CONTROLLER (P-CONTROLLER)

The proportional controller produces an output signal which is proportional to error signal. The transfer function of proportional controller is given below. (Refer equation 2.83).

Transfer function of P-controller, $\dfrac{U(s)}{E(s)} = K_p$

The term K_p in the transfer function of proportional controller is called the gain of the controller. Hence the proportional controller amplifies the error signal and increases the loop gain of the system. The following aspects of system behaviour are improved by increasing loop gain.

    ❋   Steady state tracking accuracy.
    ❋   Disturbance signal rejection.
    ❋   Relative stability.

In addition to increase in loop gain it decreases the sensitivity of the system to parameter variations. The drawback in proportional control action is that it produces a constant steady state error.

EFFECT OF PI-CONTROLLER

The proportional plus integral controller (PI-controller) produces an output signal consisting of two terms : *one proportional to error signal and the other proportional to the integral of error signal.*

Transfer function of PI-controller, $G_c(s) = K_p \left(1 + \dfrac{1}{T_i s} \right) = K_p \left(\dfrac{T_i s + 1}{T_i s} \right)$ (Refer equation 2.113)

where, K_p is proportional gain and, T_i is integral time.

The block diagram of unity feedback system with PI-controller is shown in fig 2.40.

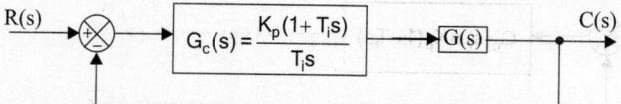

Fig 2.40 : *Block diagram of feedback system with PI-controller.*

Let the open loop transfer function G(s) be a second order system with transfer function, as shown in equation (2.148).

Open loop transfer function, $G(s) = \dfrac{\omega_n^2}{s(s + 2\zeta\omega_n)}$ ......(2.147)

Now, loop transfer function = $G_c(s)G(s)H(s) = G_c(s)\,G(s)$

$$= K_p\left(\frac{1 + T_i s}{T_i s}\right) \times \frac{\omega_n^2}{s(s + 2\zeta\omega_n)} = \frac{K_p\omega_n^2(1 + T_i s)}{s^2 T_i(s + 2\zeta\omega_n)} \qquad .....(2.148)$$

Now the closed loop transfer function is given by,

$$\frac{C(s)}{R(s)} = \frac{G_c(s)G(s)}{1 + G_c(s)G(s)} = \frac{\dfrac{K_p\omega_n^2(1 + T_i s)}{s^2 T_i(s + 2\zeta\omega_n)}}{1 + \dfrac{K_p\omega_n^2(1 + T_i s)}{s^2 T_i(s + 2\zeta\omega_n)}} = \frac{K_p\omega_n^2(1 + T_i s)}{s^2 T_i(s + 2\zeta\omega_n) + K_p\omega_n^2(1 + T_i s)}$$

$$= \frac{K_p\omega_n^2(1 + T_i s)}{T_i s^3 + 2\zeta\omega_n T_i s^2 + K_p\omega_n^2 T_i s + K_p\omega_n^2}$$

$$= \frac{(K_p/T_i)\omega_n^2(1 + T_i s)}{s^2 + 2\zeta\omega_n s^2 + K_p\omega_n^2 s + \dfrac{K_p}{T_i}\omega_n^2}$$

$$= \frac{K_i\omega_n^2(1 + T_i s)}{s^3 + 2\zeta\omega_n s^2 + K_p\omega_n^2 s + K_i\omega_n^2} \qquad .....(2.149)$$

From the closed loop transfer function (equation (2.149)) it is observed that the PI-controller introduces a zero in the system and increases the order by one. The increase in the order of the system results in a less stable system than the original one because higher order systems are less stable than lower order systems.

From the loop transfer function (equation (2.148)) it is observed that the PI-controller increase the type number by one. The increase in type number results in reducing the steady state error. For example if the steady state error of the original system is constant, then the integral controller will reduce the error to zero.

EFFECT OF PD-CONTROLLER

The proportional plus derivative controller produces an output signal consisting of two terms : *one proportional to error signal and the other proportional to the derivative of error signal.*

The transfer function of PD - controller, $G_c(s) = K_p\,(1 + T_d s)$ (Refer equation 2.124)

where K_p is Proportional gain, T_d is Derivative time.

The block diagram of unity feedback system with PD-controller is shown in fig 2.41.

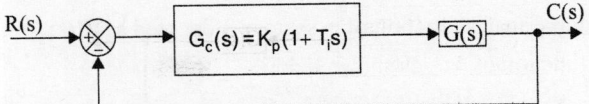

$$\text{Fig 2.41 : Block diagram of feedback system with PD-controller.}$$

Let the open loop transfer function $G(s)$ be a second order system with transfer function as shown in equation (2.150).

Open loop transfer function, $G(s) = \dfrac{\omega_n^2}{s(s + 2\zeta\omega_n)}$ .....(2.150)

Now, loop transfer function $= G_c(s)G(s)H(s) = G_c(s)G(s)$ $\boxed{H(s) = 1}$

$$= K_p(1 + T_d(s)) \times \dfrac{\omega_n^2}{s(s + 2\zeta\omega_n)} = \dfrac{K_p\omega_n^2(1 + T_d s)}{s(s + 2\zeta\omega_n)}$$ .....(2.151)

Now the closed loop transfer function is given by,

$$\dfrac{C(s)}{R(s)} = \dfrac{G_c(s)G(s)}{1 + G(s)G_c(s)} = \dfrac{\dfrac{K_p\omega_n^2(1 + T_d s)}{s(s + 2\zeta\omega_n)}}{1 + \dfrac{K_p\omega_n^2(1 + T_d s)}{s(s + 2\zeta\omega_n)}}$$

$$= \dfrac{K_p\omega_n^2(1 + T_d s)}{s(s + 2\zeta\omega_n) + K_p\omega_n^2(1 + T_d s)}$$

$$= \dfrac{K_p\omega_n^2(1 + T_d s)}{s^2 + 2\zeta\omega_n s + K_p\omega_n^2 + K_p\omega_n^2 T_d s}$$

$$= \dfrac{K_p\omega_n^2(1 + T_d s)}{s^2 + (2\zeta\omega_n + K_p\omega_n^2 T_d)s + K_p\omega_n^2}$$ $\boxed{K_d = K_p T_d}$

$$= \dfrac{\omega_n^2(K_p + K_d s)}{s^2 + (2\zeta\omega_n + K_p\omega_n^2)s + K_p\omega_n^2}$$ .....(2.152)

From the closed loop transfer function (equation (2.152)) it is observed that the PD-controller introduces a zero in the system and increases the damping ratio. The addition of the zero may increase the peak overshoot and reduce the rise time. But the effect of increased damping ultimately reduces the peak overshoot.

From the loop transfer function (equation (2.151)) it is observed that the PD-controller does not modify the type number of the system. Hence PD-controller will not act modify steady state error.

EFFECT OF PID-CONTROLLER

A suitable combination of the three basic modes : *proportional, integral and derivative* (PID) can improve all aspects of the system performance.

The proportional controller stabilizes the gain but produces a steady state error. The integral controller reduces or eliminates the steady state error. The derivative controller reduces the rate of change of error. The combined effect of all the three cannot be judged from the parameters K_p, K_i and K_d.

2.22 TIME RESPONSE ANALYSIS USING MATLAB

In general, the closed loop transfer function of a system is denoted as M(s).

Let, M(s) be a rational function of "s", as shown below.

$$M(s) = \frac{b_0 s^M + b_1 s^{M-1} + b_2 s^{M-2} + \ldots + b_{M-1} s + b_M}{a_0 s^N + a_1 s^{N-1} + a_2 s^{N-2} + \ldots + a_{N-1} s + a_N}$$

For time response analysis, the coefficients of the numerator and denominator polynomials are declared as two arrays as shown below.

```
num_cof = [b0 b1 b2 ........ bM];
den_cof = [a0 a1 a2 ........ aN];
```

UNIT STEP RESPONSE

To compute step response

The unit step response can be computed and displayed using following commands.

```
syms s complex;
R = 1/s;
M = (b0*s^M+b1*s^(M-1)+...+bM)/(a0*s^N+a1*s^(N-1)+...+aN);
S = R*M;
disp('Unit step response of the system is,');
step_res = ilaplace(s)
```

To plot step response

Method 1 :

The unit step response can be plotted using the following command.

```
step(num_cof, den_cof);
```

Method 2 :

The unit step response of the system can be plotted using the following commands.

```
t = t_start : t_step : t_end ;
c = step(num_cof, den_cof,t);
plot(t,c,'k');
      where, c is an array where the values of response are stored.
```

The unit step response can be computed "n" times by varying some parameter of the system (coefficient / damping ratio / natural frequency of oscillation) using the following commands.

```
t = t_start : t_step : t_end ;
for i = 1 : n
    .
    .
    .
    c(1:k, i) = step(num_cof, den_cof,t);
    .
    .
end
plot(t,c,'k');
      where, c is an array where the values of response are stored.
            k is the number of samples of response to be computed.
```

Method 3 :

The unit step response of the system can be plotted using the following commands.

```
s = tf('s');
M = (b0*s^M+b1*s^(M-1)+...+bM)/(a0*s^N+a1*s^(N-1)+...+aN);
t = t_start : t_step : t_end ;
sr = step(M,t);
plot(t,sr,'k');
```

IMPULSE RESPONSE

To compute impulse response

The impulse response can be computed and displayed using following commands.

```
syms s complex;
M = (b0*s^M+b1*s^(M-1)+...+bM)/(a0*s^N+a1*s^(N-1)+...+aN);
disp('Unit step response of the system is,');
imp_res = ilaplace(s)
```

To plot impulse response

Method 1 :

The impulse response can be plotted using the following command.

```
impulse(num_cof, den_cof);
```

Method 2 :

The impulse response of the system can be plotted using the following commands.

```
t = t_start : t_step : t_end ;
m = impulse(num_cof, den_cof,t);
plot(t,m,'k');
        where, m is an array where the values of impulse response are stored.
```

Method 3 :

The impulse response of the system can be plotted using the following commands.

```
s = tf('s');
M = (b0*s^M+b1*s^(M-1)+...+bM)/(a0*s^N+a1*s^(N-1)+...+aN);
t = t_start : t_step : t_end ;
imp = impulse(M,t);
plot(t,imp,'k');
```

Response for arbitrary input

The response of a system for an arbitrary input, r(t) can be plotted using the following commands.

```
t = t_start : t_step : t_end ;
c = Lsim(num_cof, den_cof, r, t);
plot(t,c,'k');
        where, c is an array where the values of response are stored.
```

PROGRAM 2.1

Consider the standard closed loop transfer function of the second order system given below.

$$M(s) = \omega_n^2 / (s^2 + 2\zeta\omega_n s + \omega_n^2)$$

Write a MATLAB program to find the unit step response for various values of damping ratio, ζ. Take, natural frequency of oscillation, $\omega_n = 1$ rad/sec.

```
%Unit step response for various values of damping ratio, zeta.
%The natural frequency of oscillation, wn=1.
clc
t=0:0.2:12;                    %specify a time vector
c=zeros(61,6);                 %initialize response array as zero
zeta=[0 0.2 0.4 0.6 0.8 1];    %store zeta as an array
    for n=1:6;                 %for loop to compute c(t) 6 times
        num_cof=[0 0 1];
        den_cof=[1 2*zeta(n) 1];
        c(1:61,n)=step(num_cof,den_cof,t);
    end

plot(t,c,'k'); grid
xlabel('time,t in sec'); ylabel('Unit step response,c(t)');

text(2.8,1.86,'\zeta=0')
text(2.8,1.58,'\zeta=0.2')
text(2.8,1.30,'\zeta=0.4')
text(2.8,1.12,'\zeta=0.6')
text(2.8,0.95,'\zeta=0.8')
text(2.8,0.72,'\zeta=1.0')
```

OUTPUT

The output waveforms are shown in fig p2.1.

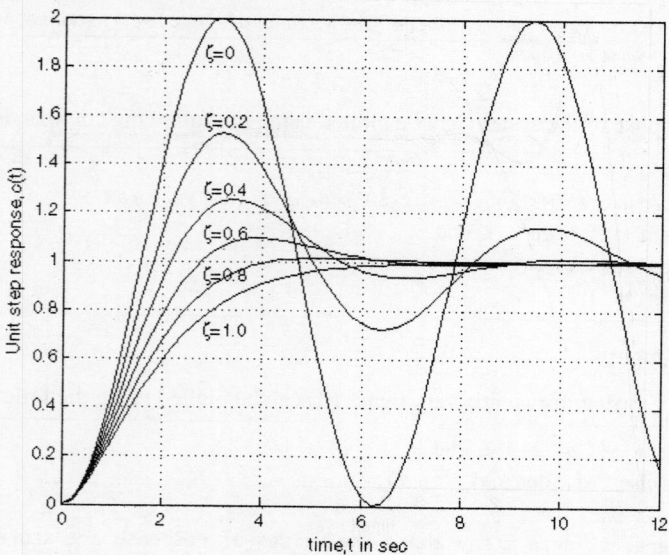

Fig P2.1 : *Unit step response of second order system for various values of damping ratio.*

PROGRAM 2.2

Consider the standard closed loop transfer function of the second order system given below.

$$M(s) = \omega_n^2/(s^2 + 2\zeta\omega_n s + \omega_n^2)$$

Write a MATLAB program to find the unit step response for various values of natural frequency of oscillation, ω_n. Take, damping ratio, $\zeta = 0.4$.

```
%Unit step response for various natural frequency of oscillation,wn.
%The damping ratio, zeta=0.4.
clc
t=0:0.1:8;              %specify a time vector
wn=[1 2 4 6];          %store wn as an array
zeta=0.4;
c=zeros(81,4);         %initialize the response array as zeros

  for i=1:4;           %for loop to compute c(t) 4 times
  b2=wn(i)*wn(i);
  a1=2*zeta*wn(i);
  num_cof=[0 0 b2];
  den_cof=[1 a1 b2];
  c(1:81,i)=step(num_cof,den_cof,t);
  end

plot(t,c(:,1),'--k',t,c(:,2),'xk',t,c(:,3),'-k',t,c(:,4),'-.k');
grid; xlabel('time,t in sec'); ylabel('Unit step response,c(t)');
text(4.25,1.25,'wn=1')
text(1.5,1.30,'wn=2')
text(0.7,1.30,'wn=3')
text(0.1,1.25,'wn=4')
```

OUTPUT

The output waveforms are shown in fig p2.2.

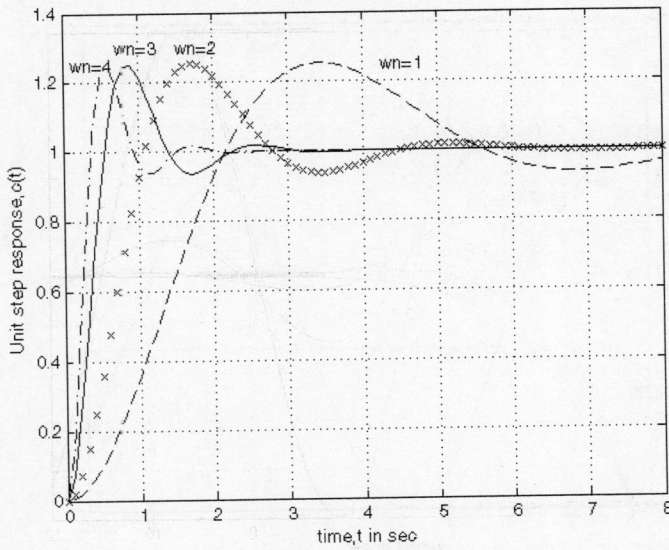

Fig P2.2 : *Unit step response of second order system for various values of natural frequency of oscillation.*

PROGRAM 2.3

Write a MATLAB program to find impulse response of the following systems.

a) $M_1(s) = (2s+1)/(s+1)^2$ b) $M_2(s) = s/(s+1)$ c) $M_3(s) = 1/(s^2+1)$

```
%Program to find impulse response
clc
syms s complex;
M1=(2*s+1)/((s+1)^2);
disp('Impulse response of the system1 is,');
m1=ilaplace(M1)

M2=s/(s+1);
disp('Impulse response of the system2 is,');
m2=ilaplace(M2)

M3=1/(s^2+1);
disp('Impulse response of the system3 is,');
m3=ilaplace(M3)

s=tf('s');
M1=(2*s+1)/((s+1)^2);
M2=s/(s+1);
M3=1/(s^2+1);

t=0:.005:10;
m1=impulse(M1,t);
m2=impulse(M2,t);
m3=impulse(M3,t);

plot(t,m1,'--k',t,m2,'-.k',t,m3,'-k');grid
xlabel('time,t in sec');
ylabel('Impulse responses,m1(t),m2(t),m3(t)');
text(0.4,1.30,'m1(t)')
text(0.3,-0.30,'m2(t)')
text(2.2,0.90,'m3(t)')
```

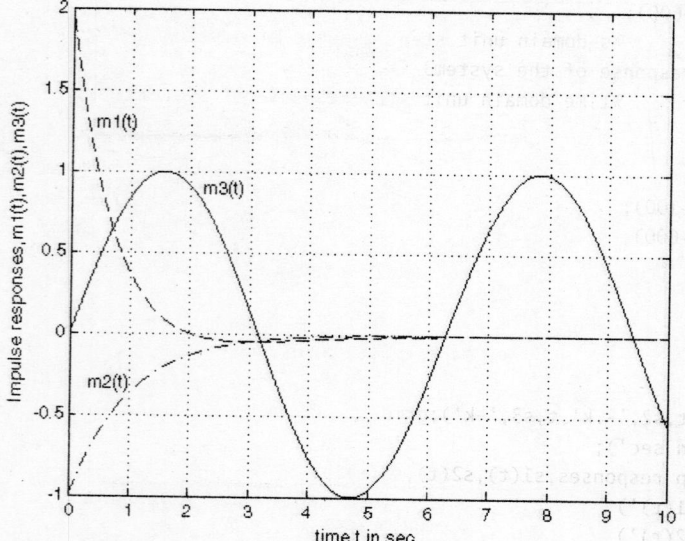

Fig P2.3 *: Impulse response of systems given in program 2.3.*

OUTPUT

Impulse response of the system1 is,
m1 =
(2-t)*exp(-t)

Impulse response of the system2 is,
m2 =
dirac(t)-exp(-t)

Impulse response of the system3 is,
m3 =
sin(t)

The output waveforms are shown in fig p2.3.

PROGRAM 2.4

Write a MATLAB program to find unit step response of the following systems.

a) $M_1(s)=4/(s^2+5s+4)$ b) $M_2(s)=100/(s^2+12s+100)$ c) $M_3(s)=600/(s^2+70s+600)$

```
%program to find unit step response
clc
syms s complex;
R=1/s;                  %Laplace of unit step input
M1=4/(s^2+5*s+4);
S1=R*M1;                %s-domain unit step response of system1
disp('Unit step response of the system1 is,');
s1=ilaplace(S1)         %time domain unit step response of system1

M2=100/(s^2+12*s+100);
S2=R*M2;                %s-domain unit step response of system2
disp('Unit step response of the system2 is,');
s2=ilaplace(S2)         %time domain unit step response of system2

M3=600/(s^2+70*s+600);
S3=R*M3;                %s-domain unit step response of system3
disp('Unit step response of the system3 is,');
s3=ilaplace(S3)         %time domain unit step response of system3

s=tf('s');
M1=4/(s^2+5*s+4);
M2=100/(s^2+12*s+100);
M3=600/(s^2+70*s+600);

t=0:.005:10;
s1=step(M1,t);
s2=step(M2,t);
s3=step(M3,t);

plot(t,s1,'--k',t,s2,'-.k',t,s3,'-k');grid
xlabel('time,t in sec');
ylabel('Unit step responses,s1(t),s2(t),s3(t)');
text(2.2,0.85,'s1(t)')
text(0.2,1.15,'s2(t)')
text(0.5,0.95,'s3(t)')
```

OUTPUT

Unit step response of the system1 is,

s1 =

1/3*exp(-4*t)+1-4/3*exp(-t)

Unit step response of the system2 is,

s2 =

1-exp(-6*t)*cos(8*t)-3/4*exp(-6*t)*sin(8*t)

Unit step response of the system3 is,

s3 =

1+1/5*exp(-60*t)-6/5*exp(-10*t)

The output waveform is shown in fig p2.4.

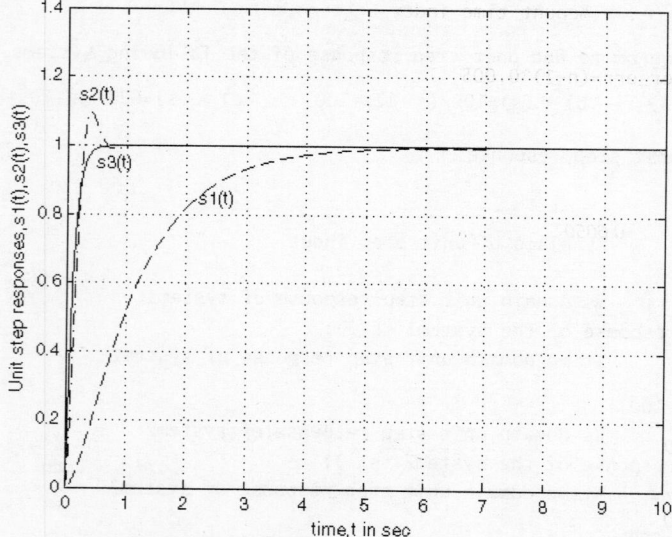

Fig P2.4 : Unit step response of systems given in program 2.4.

PROGRAM 2.5

Consider the closed loop transfer function of the following second order system,

M(s)=16/(s²+4s+16)

Write a MATLAB program to find the rise time, peak time, maximum peak overshoot, and settling time from the unit step response of the system.

```
clc
t=0:0.005:5;          %set time vector

num_cof=[0 0 16];     %store the numerator coefficients as an array
den_cof=[1 4 16];     %store denominator coefficients as an array
[c,x,t]=step(num_cof,den_cof,t);

n=1;                  %initialize count as 1
while c(n)<1.0001;    %count the time index as along as c(t)<1
    n=n+1;
end;
```

```
rise_time=(n-1)*0.005   %rise time=(count-1)*time interval
[cmax,tp]=max(c);       %determine maximum value of c(t) &
                                        %corresponding time

peak_time=(tp-1)*0.005  %peak time=(tp-1)*time interval
max_overshoot=cmax-1    %compute peak overshoot
n=1001;                 %initialize count as (5/.005)+1=1001
while c(n)>0.95&c(n)<1.05;
    n=n-1;              %count time index between c(t)>0.95&c(t)<1.05
end;
settling_time_5per_err=(n-1)*0.005
n=1001;                 %initialize count as (5/.005)+1=1001
while c(n)>0.98 & c(n)<1.02;
    n=n-1;              %count time index between c(t)>0.98&c(t)<1.02
end;
settling_time_2per_err=(n-1)*0.005
```

OUTPUT

rise_time =
                            0.6050

peak_time =
                0.9050

max_overshoot =
                0.1630

settling_time_5per_err =
                1.3200

settling_time_2per_err =
                2.0150

PROGRAM 2.6

Consider the closed loop transfer function of the following second order system,

$$M(s)=64/(s^2+8s+64)$$

Write a MATLAB program to find the response for unit step, unit ramp and unit parabolic input signals.

```
%unit step/ramp/parabolic response
clc
num_cof=[0 0 64];
den_cof=[1 8 64];

t=0:0.005:2;

r1=t;               %unit ramp input signal
r2=0.5*t.^2;        %unit parabolic input signal

c1=step(num_cof, den_cof,t);
c2=Lsim(num_cof, den_cof,r1,t);
c3=Lsim(num_cof, den_cof,r2,t);
```

```
plot(t,c1,'--k',t,c2,'-.k',t,c3,'-k'); grid
xlabel('time,t in sec');
ylabel('Responses,c1(t),c2(t),c3(t)');
text(0.25,1.15,'c1(t)')
text(1.45,1.5,'c2(t)')
text(1.35,0.7,'c3(t)')
```

OUTPUT

The output waveform is shown in fig p2.6.

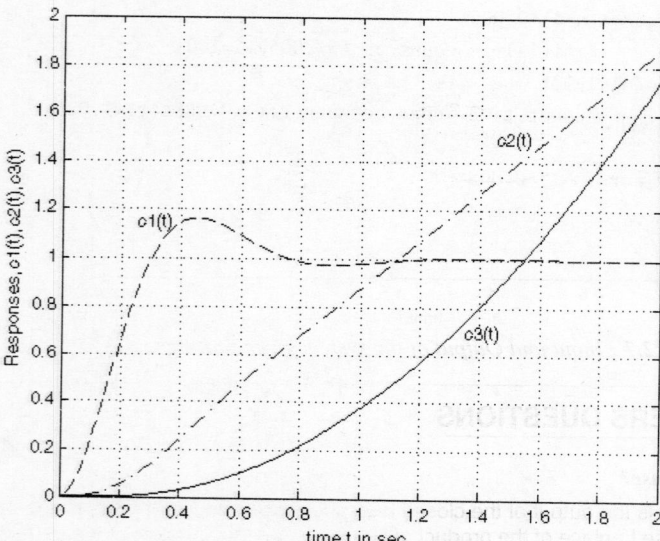

Fig P2.6 : *Step, ramp and parabolic response of system given in program 2.6.*

PROGRAM 2.7

Consider the closed loop transfer function of the following second order system,

$$M(s)=5/(s^2+s+5)$$

Write a MATLAB program to find the response for the input signal, $r(t)=2-2t+t^2$.

```
%program to find response for given input
clc
num_cof=[0 0 5];
den_cof=[1 1 5];
t=0:0.005:3;                    %specify a time vector
r=2-2*t+t.^2;                   %input signal
c=Lsim(num_cof,den_cof,r,t);    %compute response using Lsim function
plot(t,r,'--k',t,c,'-.k'); grid
xlabel('time,t in sec');
ylabel('Input,r(t) and output,c(t)');
text(0.25,1.65,'r(t)')
text(0.25,0.6,'c(t)')
```

OUTPUT

The output waveform is shown in fig p2.7.

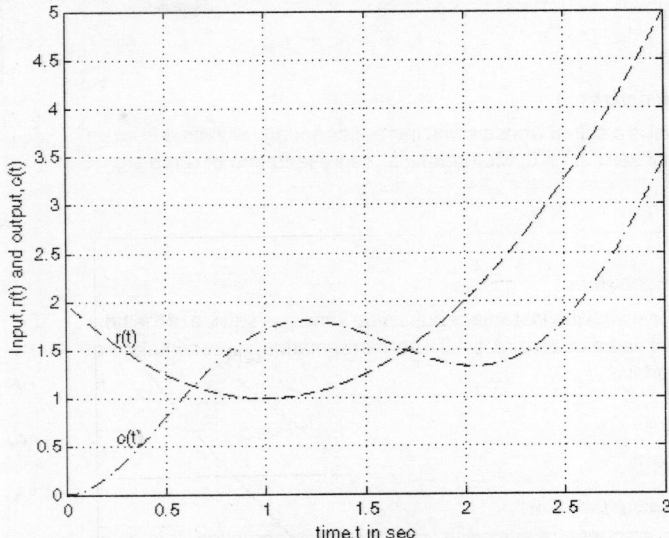

Fig P2.7 : Input and Output of the system given in program 2.7.

2.23 SHORT - ANSWERS QUESTIONS

Q2.1   *What is time response?*

The time response is the output of the closed loop system as a function of time. It is denoted by c(t). It is given by inverse Laplace of the product of input and transfer function of the system.

The closed loop transfer function, $\dfrac{C(s)}{R(s)} = \dfrac{G(s)}{1+G(s)H(s)}$

Response in s-domain, $C(s) = \dfrac{R(s)G(s)}{1+G(s)H(s)}$

Response in time domain, $C(t) = \mathcal{L}^{-1}\{C(s)\} = \mathcal{L}^{-1}\left\{\dfrac{R(s)G(s)}{1+G(s)H(s)}\right\}$

Q2.2   *What is transient and steady state response?*

The transient response is the response of the system when the input changes from one state to another. The response of the system as $t \to \infty$ is called steady state response.

Q2.3   *What is the importance of test signals?*

The test signals can be easily generated in test laboratories and the characteristics of test signals resembles, the characteristics of actual input signals. The test signals are used to predetermine the performance of the system. If the response of a system is satisfactory for a test signal, then the system will be suitable for practical applications.

Q2.4   *Name the test signals used in control system.*

The commonly used test input signals in control system are Impulse, Step, Ramp, Acceleration and Sinusoidal signals.

Q2.5 **Define step signal.**

The step signal is a signal whose value changes from 0 to A and remains constant at A for t > 0. The mathematical representation of step signal is,

r(t) = A, t ≥ 0

= 0, t < 0

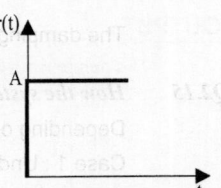

Fig Q2.5 : Step signal.

Q2.6 **Define ramp signal.**

A ramp signal is a signal whose value increases linearly with time from an initial value of zero at t = 0. Mathematical representation of ramp signal is,

r(t) = At , t ≥ 0

= 0 , t < 0

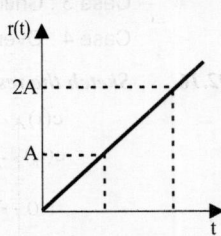

Fig Q2.6 : Ramp signal.

Q2.7 **Define parabolic signal.**

It is a signal in which the instantaneous value varies as square of the time from an initial value of zero at t = 0. The mathematical representation of parabolic signal is,

$r(t) = \dfrac{At^2}{2}$, t ≥ 0

= 0,   t < 0

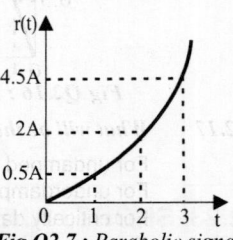

Fig Q2.7 : Parabolic signal.

Q2.8 **What is weighing function?**

The impulse response of system is called weighing function. It is given by inverse Laplace transform of system transfer function.

Q2.9 **What is an impulse signal?**

A signal which is available for very short duration is called impulse signal. Ideal impulse signal is a unit impulse signal which is defined as a signal having zero values at all time except at t = 0. At t = 0 the magnitude becomes infinite. It is denoted by δ(t) and mathematically expressed as,

$$\delta(t) = \infty \; ; \; t = 0 \quad \text{and} \quad \int_{-\infty}^{+\infty} \delta(t)\, dt = 1$$
$$= 0 \; ; \; t = 0$$

Q2.10 **Define pole.**

The pole of a function, F(s) is the value at which the function, F(s) becomes infinite, where F(s) is a function of complex variable s.

Q2.11 **Define zero.**

The zero of a function, F(s) is the value at which the function, F(s) becomes zero, where F(s) is a function of complex variable s.

Q2.12 **What is the order of a system?**

The order of the system is given by the order of the differential equation governing the system. It is also given by the maximum power of s in the denominator polynomial of transfer function. The maximum power of s also gives the number of poles of the system and so the order of the system is also given by number of poles of the transfer function.

Q2.13 **Define damping ratio.**

The damping ratio is defined as the ratio of actual damping to critical damping.

Q2.14 **Give the expression for damping ratio of mechanical and electrical system.**

The damping ratio of second order mechanical translational system, $\zeta = \dfrac{B}{2\sqrt{MK}}$

The damping ratio of second order mechanical rotational system, $\zeta = \dfrac{B}{2\sqrt{JK}}$

The damping ratio of second order electrical system, $\zeta = \dfrac{R}{\sqrt{L/C}}$

Q2.15   *How the system is classified depending on the value of damping?*

Depending on the value of damping, the system can be classified into the following four cases.

Case 1 : Undamped system, $\zeta = 0$

Case 2 : Underdamped system, $0 < \zeta < 1$

Case 3 : Critically damped system, $\zeta = 1$

Case 4 : Over damped system, $\zeta > 1$

Q2.16.   *Sketch the response of a second order under damped system.*

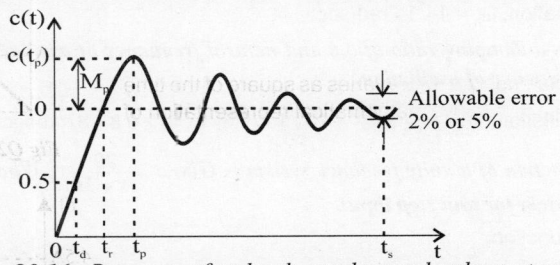

Fig Q2.16 : *Response of under damped second order system.*

Q2.17   *What will be the nature of response of a second order system with different types of damping?*

For undamped system the response is oscillatory.

For underdamped system the response is damped oscillatory.

For critically damped system the response is exponentially rising.

For overdamped system the response is exponentially rising but the rise time will be very large.

Q2.18.   *What is damped frequency of oscillation?*

In underdamped system the response is damped oscillatory. The frequency of damped oscillation is given by, $\omega_d = \omega_n \sqrt{1 - \zeta^2}$.

Q2.19.   *Give the expression for natural frequency of oscillations of electrical and mechanical system.*

The natural frequency of oscillation of second order mechanical translational system $\Big\}$ $\omega_n = \sqrt{\dfrac{K}{M}}$

The natural frequency of oscillation of second order mechanical rotational system $\Big\}$ $\omega_n = \sqrt{\dfrac{K}{J}}$

The natural frequency of oscillation of second order electrical system $\Big\}$ $\omega_n = \dfrac{1}{\sqrt{LC}}$

Q2.20.   *The closed loop transfer function of second order system is $\dfrac{C(s)}{R(s)} = \dfrac{10}{s^2 + 6s + 10}$. What is the type of damping in the system?.*

Let us compare the given transfer function with the standard form of second order transfer function

$$\therefore \; \frac{C(s)}{R(s)} = \frac{\omega_n^2}{s^2 + 2\zeta\omega_n + \omega_n^2} = \frac{10}{s^2 + 6s + 10}$$

$\omega_n^2 = 10$ | $2\zeta\omega_n = 6$

$\therefore \; \omega_n = \sqrt{10} = 3.1622$ rad./ sec | $\therefore \; \zeta = \dfrac{6}{2 \times \omega_n} = \dfrac{6}{2 \times \sqrt{10}} = 0.95$

Since $\zeta < 1$, the system is underdamped.

Q2.21 *The closed loop transfer function of a second order system is given by $\dfrac{200}{s^2+20s+200}$. Determine the damping ratio and natural frequency of oscillation.*

Let us compare the given transfer function with the standard form of second order transfer function

$$\therefore \frac{C(s)}{R(s)} = \frac{\omega_n^2}{s^2+2\zeta\omega_n s+\omega_n^2} = \frac{200}{s^2+20s+200}$$

$$\therefore \omega_n^2 = 200 \qquad\qquad 2\zeta\omega_n = 20$$

$$\omega_n = \sqrt{200} = 14.14\,\text{rad/sec} \qquad \zeta = \frac{20}{2\times\omega_n} = \frac{20}{2\times14.14} = 0.707$$

Damping ratio, $\zeta = 0.707$

Natural frequency of oscillation, $\omega_n = 14.14$ rad/sec.

Q2.22 *A second order system has a damping ratio of 0.6 and natural frequency of oscillation is 10 rad/sec. Determine the damped frequency of oscillation.*

Damped frequency of oscillation, $\omega_d = \omega_n\sqrt{1-\zeta^2} = 10\sqrt{1-(0.6)^2} = 10\times0.8 = 8\,\text{rad/sec}$

Q2.23 *The open loop transfer function of a unity feedback system is $G(s) = \dfrac{20}{s(s+10)}$. What is the nature of response of closed loop system for unit step input.*

The closed loop transfer function,

$$\frac{C(s)}{R(s)} = \frac{G(s)}{1+G(s)} = \frac{20/s(s+10)}{1+\dfrac{20}{s(s+10)}} = \frac{20}{s(s+10)+20} = \frac{20}{s^2+10s+20}$$

The standard form of second order transfer function is, $\dfrac{C(s)}{R(s)} = \dfrac{\omega_n^2}{s^2+2\zeta\omega_n s+\omega_n^2}$

On comparing system transfer function with standard form of second order transfer function we get,

$$\omega_n^2 = 20 \qquad\qquad 2\zeta\omega_n = 10$$

$$\therefore \omega_n = \sqrt{20} = 4.47\,\text{rad/sec} \qquad \zeta = \frac{10}{2\times\omega_n} = \frac{10}{2\times4.47} = 1.12$$

Since damping ratio, $\zeta > 1$, the system is overdamped and the response will be exponentially rising.

Q2.24 *List the time domain specifications.*

The time domain specifications are,

(i) Delay time   (ii) Rise time   (iii) Peak time
(iv) Maximum overshoot   (v) Settling time.

Q2.25 *Define delay time.*

It is the time taken for response to reach 50% of the final value, the very first time.

Q2.26 *Define rise time.*

It is the time taken for response to raise from 0 to 100%, the very first time. For underdamped system, the rise time is calculated from 0 to 100%. But for overdamped system it is the time taken by the response to raise from 10% to 90%. For critically damped system, it is the time taken for response to raise from 5% to 95%.

Q2.27 *Define peak time.*

It is the time taken for the response to reach the peak value, the very first time (or) It is the time taken for the response to reach peak overshoot, M_p.

Q2.28 *Define peak overshoot.*

It is defined as the ratio of the maximum peak value to final value, where maximum peak value is measured from final value.

Let final value = $c(\infty)$,   Maximum value = $c(t_p)$ $\therefore$ Peak overshoot, $M_p = \dfrac{c(t_p)-c(\infty)}{c(\infty)}$

Q2.29   *Define settling time.*

It is defined as the time taken by the response to reach and stay within a specified error and the error is usually specified as % of final value. The usual tolerable error is 2% or 5% of the final value.

Q2.30   *The damping ratio of a system is 0.75 and the natural frequency of oscillation is 12 rad/sec. Determine the peak overshoot and the peak time.*

Peak overshoot, $M_p = e^{\frac{-\zeta\pi}{\sqrt{1-\zeta^2}}} = e^{-\frac{0.75\times\pi}{\sqrt{1-(0.75)^2}}} = 0.028$;   % $M_p = 0.028 \times 100 = 2.8\%$

Damped frequency of oscillation, $\omega_d = \omega_n\sqrt{1-\zeta^2} = 12\sqrt{1-(0.75)^2} = 7.94 \, \text{rad/sec}$

Peak time, $t_p = \frac{\pi}{\omega_d} = \frac{\pi}{7.94} = 0.396 \, \text{sec}$

Q2.31   *The damping ratio of system is 0.6 and the natural frequency of oscillation is 8 rad/sec. Determine the rise time.*

Rise time, $t_r = \frac{\pi - \theta}{\omega_d}$

$\theta = \tan^{-1}\frac{\sqrt{1-\zeta^2}}{\zeta} = \tan^{-1}\frac{\sqrt{1-(0.6)^2}}{0.6} = 53.13^\circ = \frac{53.13}{180}\times\pi \, \text{rad} = 0.927 \, \text{rad}$

$\omega_d = \omega_n\sqrt{1-\zeta^2} = 8\sqrt{1-(0.6)^2} = 6.4 \, \text{rad/sec}$

$\therefore$ Rise time, $t_r = \frac{\pi - 0.927}{6.4} = 0.34 \, \text{sec}$

Q2.32   *What is type number of a system? What is its significance?*

The type number is given by number of poles of loop transfer function at the origin. The type number of the system decides the steady state error.

Q2.33   *Distinguish between type and order of a system.*

(i) Type number is specified for loop transfer function but order can be specified for any transfer function. (open loop or closed loop transfer function).

(ii) The type number is given by number of poles of loop transfer function lying at origin of s-plane but the order is given by the number of poles of transfer function.

Q2.34   *For the system with following transfer function, determine type and order of the system.*

(i) $G(s) H(s) = \dfrac{K}{s(s+1)(s^2+6s+8)}$

(ii) $G(s) H(s) = \dfrac{20(s+2)}{s^2(s+3)(s+0.5)}$

(iii) $G(s) H(s) = \dfrac{(s+4)}{(s-2)(s+0.252)}$

(iv) $G(s) H(s) = \dfrac{10}{s^3(s^2+2s+1)}$

Ans: (i)   Type - 1, order - 4   (ii)   Type - 2, order - 4

(iii)   Type - 0, order - 2   (iv)   Type - 3, order - 5.

Q2.35   *What is steady state error?*

The steady state error is the value of error signal e(t), when t tends to infinity . The steady state error is a measure of system accuracy. These errors arise from the nature of inputs, type of system and from non-linearity of system components.

Q2.36   *What are static error constants?*

The K_p, K_v and K_a are called static error constants. These constants are associated with steady state error in a particular type of system and for a standard input.

Q2.37   *Define positional error constant.*

The positional error constant .The steady state error in type-0 system when the input is unit step is given by $1/1+K_p$.

Q2.38   *Define velocity error constant.*

The velocity error constant $K_v = \underset{s \to 0}{\text{Lt}}\, sG(s)H(s)$. The steady state error in type-1 system for unit ramp input is given by $1/K_v$.

Q2.39   *Define acceleration error constant.*

The acceleration error constant $K_a = \underset{s \to 0}{\text{Lt}}\, s^2 G(s)H(s)$. The steady state error in type-2 system for unit parabolic input is given by $1/K_a$.

Q2.40   *A unity feedback system has a open loop transfer function of* $G(s) = \dfrac{10}{(s+1)(s+2)}$. *Determine the steady state error for unit step input.*

The steady state error for unit step input, $e_{ss} = \dfrac{1}{1+K_p}$, where, $K_p = \underset{s \to 0}{\text{Lt}}\, G(s)H(s)$

For unity feedback system $H(s) = 1$.

$$\therefore K_p = \underset{s \to 0}{\text{Lt}}\, G(s) = \underset{s \to 0}{\text{Lt}}\, \dfrac{10}{(s+1)(s+2)} = 5 \quad \text{and} \quad e_{ss} = \dfrac{1}{1+K_p} = \dfrac{1}{1+5} = \dfrac{1}{6}$$

Q2.41   *A unity feedback system has a open loop transfer function of* $G(s) = \dfrac{25(s+4)}{s(s+0.5)(s+2)}$. *Determine the steady state error for unit ramp input.*

The steady state error for unit ramp input is, $e_{ss} = \dfrac{1}{K_v}$, where, $K_v = \underset{s \to 0}{\text{Lt}}\, sG(s)H(s)$. For unity feedback system $H(s) = 1$.

$$\therefore K_v = \underset{s \to 0}{\text{Lt}}\, sG(s) = \underset{s \to 0}{\text{Lt}}\, s\left[\dfrac{25(s+4)}{s(s+0.5)(s+2)}\right] = \dfrac{25 \times 4}{0.5 \times 2} = 100 \quad \text{and} \quad e_{ss} = \dfrac{1}{K_v} = \dfrac{1}{100} = 0.01$$

Q2.42   *A unity feedback system has a open loop transfer function of* $G(s) = \dfrac{20(s+5)}{s(s+0.1)(s+3)}$. *Determine the steady state error for parabolic input.*

The steady state error for unit ramp input is, $e_{ss} = \dfrac{1}{K_a}$, where, $K_a = \underset{s \to 0}{\text{Lt}}\, s^2 G(s)H(s)$. For unity feedback system $H(s) = 1$.

$$\therefore K_a = \underset{s \to 0}{\text{Lt}}\, s^2\left[\dfrac{20(s+5)}{s^2(s+0.1)(s+3)}\right] = \dfrac{20 \times 5}{0.1 \times 3} = \dfrac{100}{0.3} = 333.33 \quad \text{and} \quad e_{ss} = \dfrac{1}{K_a} = \dfrac{1}{333.33} = 0.003$$

Q2.43   *What are generalized error coefficients?*

They are the coefficients of generalized error series. The generalized error series is given by,

$$e(t) = C_0 r(t) + C_1 \dot{r}(t) + \dfrac{C_2}{2!} \ddot{r}(t) + \dfrac{C_3}{3!} \dddot{r}(t) + \dots + \dfrac{C_n}{n!} \overset{n}{r}(t) \dots$$

The coefficients $C_0, C_1, C_2 \dots C_n$ are called generalized error coefficients or dynamic error coefficients.

The nth coefficient, $C_n = \underset{s \to 0}{\text{Lt}}\, \dfrac{d^n}{ds^n} F(s)$, where $F(s) = \dfrac{1}{1+G(s)H(s)}$

Q2.44   *Give the relation between generalized and static error coefficients.*

The following expression shows the relation between generalized and static error coefficient.

$$C_0 = \dfrac{1}{1+K_p} \; ; \qquad C_1 = \dfrac{1}{K_v} \; ; \qquad C_2 = \dfrac{1}{K_a}$$

Q2.45   *Mention two advantages of generalized error constants over static error constants.*

(i) Generalized error series gives error signal as a function of time.

(ii) Using generalized error constants the steady state error can be determined for any type of input but static error constants are used to determine steady state error when the input is anyone of the standard input.

Q2.46. **What are the basic components of an automatic control system ?**

The basic components of an automatic control system are,

1. Error detector
2. Amplifier and controller
3. Actuator (Power actuator)

4. Plant
5. Sensor or feedback system

Q2.47 **What is automatic controller ?**

The combined unit of error detector, amplifier and controller is called automatic controller.

Q2.48 **What is the need for a controller?**

The controller is provided to modify the error signal for better control action.

Q2.49 **What are the different types of controllers?**

The different types of controller used in control system are P, PI, PD and PID controllers.

Q2.50 **What is Proportional controller and what are its advantages?**

The Proportional controller is a device that produces a control signal which is proportional to the input error signal.

The advantages in the proportional controller are improvement in steady-state tracking accuracy, disturbance signal rejection and the relative stability. It also makes a system less sensitive to parameter variations.

Q2.51 **What is the drawback in P-controller?**

The drawback in P-controller is that it develop a constant steady-state error.

Q2.52 **What is integral control action?**

In integral control action, the control signal is proportional to integral of error signal.

Q2.53 **What is the advantage and disadvantage in integral controller?**

The advantage in Integral controller is that it eliminates or reduces the steady-state error. The disadvantage is that it can make a system unstable.

Q2.54 **Write the transfer function of P, PI, PD and PID controllers.**

The transfer function of P-controller, $\dfrac{U(s)}{E(s)} = K_p$; where, K_p = Proportional gain.

The transfer function of PI-controller, $\dfrac{U(s)}{E(s)} = K_p \left(1 + \dfrac{1}{T_i s}\right)$; where, T_i = Integral time constant.

The transfer function of PD-controller, $\dfrac{U(s)}{E(s)} = K_p (1 + T_d s)$; where, T_d = Derivative time constant.

The transfer function of PID-controller, $\dfrac{U(s)}{E(s)} = K_p \left(1 + \dfrac{1}{T_i s} + T_d s\right)$

Q2.55 **What is Reset rate?**

The Reset rate is the reciprocal of integral time or reset time. The reset rate is the number of times per minute that the proportional part of the control action is duplicated and it is measured in terms of repeats/minute.

Q2.56 **Why derivative control is not employed in isolation?**

A derivative control mode in isolation produces no corrective efforts for any constant errors. Because, it acts only on rate of change of error.

Q2.57 **What is PI-controller?**

The PI-controller is a device which produces a control signal consisting of two terms : one proportional to error signal and the other proportional to the integral of error signal.

Q2.58   What is PD-controller?

The PD-controller is a device which produces a control signal consisting of two terms : one proportional to error signal and the other proportional to the derivative of error signal.

Q2.59   What is PID-Controller?

The PID-controller is a device which produces a control signal consisting of three terms : one proportional to error signal, another one proportional to integral of error signal and the third one proportional to derivative of error signal.

Q2.60   Give an example of electronic PID-controller

The electronic PID-controller can be realized by an op-amp amplifier with integral and derivative action followed by sign changer, as shown in figure Q2.60.

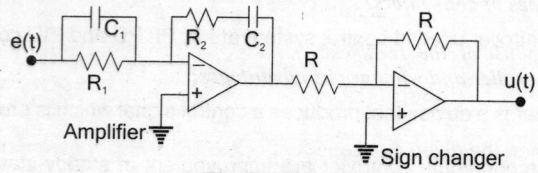

Fig Q2.60.

Q2.61   Sketch the step response of a P and PI-controller ?

Let e(t) be the input signal to the controller and u(t) be the output signal to the controller. The input and output signals are shown in the figure Q2.61.

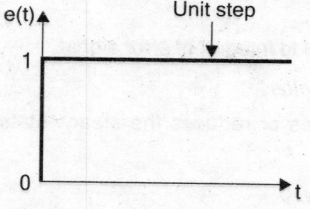
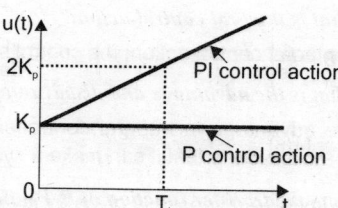

Fig Q2.61

Q2.62   Sketch the ramp response of P, PD and PID-controller?

Let e(t) be the input signal to the controller and u(t) be the output signal to the controller. The input and output signals are shown in the figure Q2.62.

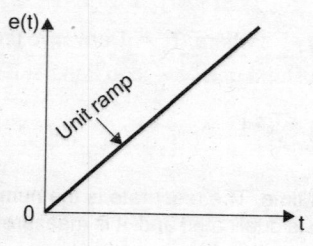
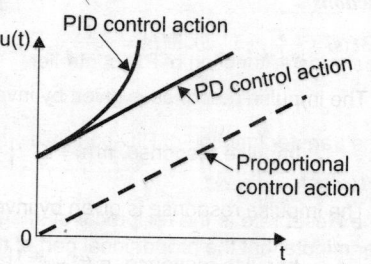

Fig Q2.62

Q2.63   What is the effect on system performance when a proportional controller is introduced in a system?

The proportional controller improves the steady-state tracking accuracy, disturbance signal rejection and relative stability of the system. It also increases the loop gain of the system which results in reducing the sensitivity of the system to parameter variations.

Q2.64   What is the disadvantage in proportional controller?

The disadvantage in proportional controller is that it produces a constant steady state error.

Q2.65   *What is the effect of PI-controller on the system performance?*

The PI - controller increases the order of the system by one, which results in reducing, the steady state error. But the system becomes less stable than the original system.

Q2.66   *What is the effect of PD-controller on the system performance?*

The effect of PD - controller is to increase the damping ratio of the system and so the peak overshoot is reduced.

Q2.67   *Why derivative controller is not used in control systems?*

The derivative controller produces a control action based on rate of change of error signal and it does not produce corrective measures for any constant error. Hence derivative controller is not used in control systems.

Q2.68   *Determine the impulse response of the feedback system governed by the closed loop transfer function,*

$$M(s) = \frac{2s+1}{(s+1)^2}.$$

By partial fraction expansion the given closed loop transfer function can be expressed as,

$$\therefore \; M(s) = \frac{2s+1}{(s+1)^2} = \frac{A}{(s+1)^2} + \frac{B}{s+1}$$

$$A = \frac{2s+1}{(s+1)^2} \times (s+1)^2 \bigg|_{s=-1} = 2s+1 \big|_{s=-1} = 2(-1) + 1 = -1$$

$$B = \frac{d}{ds} \left[\frac{2s+1}{(s+1)^2} \times (s+1)^2 \right] \bigg|_{s=-1} = \frac{d}{ds} [2s+1] \big|_{s=-1} = 2$$

$$\therefore \; M(s) = \frac{-1}{(s+1)^2} + \frac{2}{(s+1)}$$

The impulse response is given by inverse Laplace transform of closed loop transfer function.

$$\therefore \; \text{Impulse response, } m(t) = \mathcal{L}^{-1} \left\{ \frac{-1}{(s+1)^2} + \frac{2}{(s+1)} \right\} = -te^{-t} + 2e^{-t}$$

Q2.69   *Determine the impulse response of the feedback systems governed by the following closed loop transfer functions,*

a) $M(s) = \dfrac{s}{s+1}$;   *b)* $M(s) = \dfrac{1}{s^2+1}$

a) The impulse response is given by inverse Laplace transform of closed loop transfer function.

$$\therefore \; \text{Impulse response, } m(t) = \mathcal{L}^{-1} \left\{ \frac{s}{s+1} \right\} = \mathcal{L}^{-1} \left\{ 1 - \frac{1}{s+1} \right\} = \delta(t) - e^{-t}$$

b) The impulse response is given by inverse Laplace transform of closed loop transfer function.

$$\therefore \; \text{Impulse response, } m(t) = \mathcal{L}^{-1} \left\{ \frac{1}{s^2+1} \right\} = \sin t$$

Q2.70   *Determine the impulse response of the feedback system governed by the closed loop transfer function,*

$$M(s) = \frac{2(s+3)}{(s+3)^2+1}.$$

The impulse response is given by inverse Laplace transform of closed loop transfer function.

$$\therefore \; \text{Impulse response, } m(t) = \mathcal{L}^{-1} \left\{ \frac{2(s+3)}{(s+3)^2+1} \right\} = 2e^{-3t} \cos t$$

2.24 EXERCISES

E2.1 What is the unit-step response of the system shown in fig E2.1

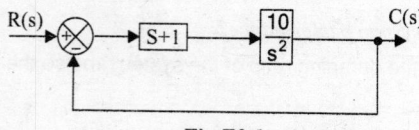

Fig E2.1.

E2.2 Obtain the unit-step response of a unity-feedback system

whose open-loop transfer function is $G(s) = \dfrac{5(s+20)}{s(s+4.59)(s^2+3.41s+16.35)}$

E2.3 The open loop transfer function of an unity feedback control system is given by $G(s) = \dfrac{100}{s(s+2)(s+5)}$

For unit step input, find the time response of the closed loop system and determine % over shoot and the rise time.

E2.4 A Servomechanism has its moment of inertia $J = 10\times10^{-6}$ Kg-m², retarding friction, $B = 400\times10^{-6}$ N-m/(rad/sec) and elasticity coefficient, $K = 0.004$ N-m/rad. Find the natural frequency and damping factor of the system.

E2.5 For a second order system whose open loop transfer function $G(s) = \dfrac{4}{s(s+2)}$, determine the maximum over

shoot, the time to reach the maximum overshoot when a step displacement of 18° is given to the system. Find the rise time, time constant and the settling time for an error of 7% .

E2.6 Consider the unity feedback closed loop system where the forward transfer function is $G(s) = \dfrac{25}{s(s+5)}$. Obtain

the rise time, Peak time, Maximum overshoot and the settling time when the system is subjected to a unit-step input.

E2.7 Consider the system shown in fig E2.7, where $\zeta = 0.6$ and $\omega_n = 0.5$ rad/sec. Determine the rise time, peak time, maximum overshoot and Settling time, when the system is subjected to a unit-step input.

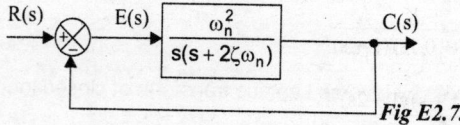

Fig E2.7.

E2.8 For the system shown in fig E2.8, determine the values of K and K_h so that the maximum overshoot in the unit step response is 0.2 and the peak time is 1 sec. With these values of K and K_h, obtain rise time and settling time.

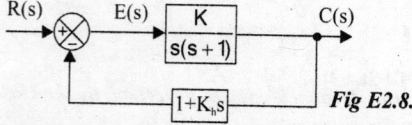

Fig E2.8.

E2.9 The system shown in fig E2.9 subjected to a unit-step input. Determine the values of K and T, where the Maximum overshoot of the system is 25.4% corresponding to $\zeta = 0.4$.

E2.10 Determine the values of K and T of the closed-loop system shown in Fig E2.10, so that the maximum overshoot in unit-step response is 25% and the peak time is 2 sec. Assume that J=1 Kg-m².

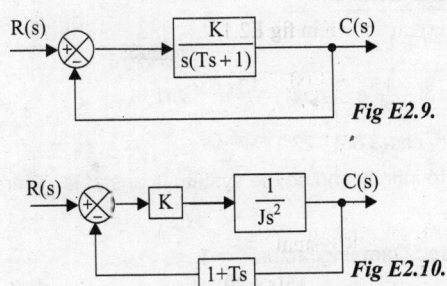

Fig E2.9.

Fig E2.10.

E2.11 A unity-feedback system is characterized by the open-loop transfer function $G(s) = \dfrac{1}{s(0.5s+1)(0.2s+1)}$

a) Determine the steady-state errors to unit-step, unit-ramp and unIt parabolic inputs.

b) Determine rise time, peak time, peak overshoot and settling time of the unit-step respone of the system.

E2.12 For a system whose $G(s) = \dfrac{10}{s(s+1)(s+2)}$,

find the steady state error when it is subjected to the input, $r(t) = 1+2t+1.5\ t^2$.

E2.13 A unity feedback system has $G(s) = \dfrac{1}{s(1+s)}$.

The input to the system is described by $r(t) = 4+6t+2t^3$. Find the generalized error coefficients and steady state error.

E2.14 A unity feedback system has the forward path transfer function $G(s) = \dfrac{10}{(s+1)}$.

Find the steady state error and the generalised error coefficient for $r(t) = t$.

E2.15 Find out the position, velocity and acceleration error coefficients for the following unity feedback systems having forward loop transfer function $G(s)$ as,

(a) $\dfrac{100}{(1+0.5s)(1+2s)}$ (b) $\dfrac{K}{s(1+0.1s)(1+s)}$

(c) $\dfrac{K}{s^2(s^2+8s+100)}$ (d) $\dfrac{K(1+s)(1+2s)}{s^2(s^2+4s+20)}$

E2.16 The open loop transfer function of a unity feedback control system is $G(s) = 9/(s+1)$, using the generalized error series determine the error signal and steady state error of the system when the system is excited by,

(i) $r(t) = 2$ (ii) $r(t) = t$

(iii) $r(t) = 3t^2/2$ (iv) $r(t) = 1+2t+3t^2/2$

E2.17 For unity feedback system having open loop transfer function as $G(s) = \dfrac{K(s+2)}{s^2(s^2+7s+12)}$. Determine,

(i) type of system,

(ii) error constants K_p, K_v and K_a

(iii) steady state error for parabolic input.

ANSWER FOR EXERCISE PROBLEMS

E2.1 $\quad c(t) = -1.1455e^{-8.87t} + 0.1455e^{-1.13t} + 1$

E2.2 $\quad c(t) = 1 + \dfrac{3}{8}e^{-t}\cos 3t - \dfrac{17}{24}e^{-t}\sin 3t - \dfrac{11}{8}e^{-3t}\cos t - \dfrac{13}{8}e^{-3t}\sin t$

E2.3 $\quad c(t) = \left[1 - 0.816e^{-7.45t} - 0.88e^{0.225t}\cos(3.65t - 22°)\right]$

As t tends to infinity, c(t) tends to infinity and so the system is unstable. Therefore % over shoot and rise time are not defined.

E2.4 $\quad$ Natural frequency, ω_n = 20 rad/sec, Damping factor, ζ = 1.

E2.5 $\quad$ Maximum overshoot = 0.16, when input is 18%, M_p = 2.88%

Peak time, t_p = 1.81sec. $\quad$ Rise time, t_r = 1.21 sec

Time constant, T = 1 sec, $\quad$ Settling time for 7% error = 2.66 sec.

E2.6 $\quad$ Rise time, t_r = 0.55sec, $\quad$ %Peak overshoot, M_p = 9.5%

Peak time, t_p = 0.785sec, $\quad$ Settling time, t_s = 1.33sec (for 2% error); $\quad$ t_s =1 sec(for 5% error)

E2.7 $\quad$ Rise time, t_r = 0.55sec, $\quad$ Maximum overshoot, $\quad$ M_p = 0.095

Peak time, t_p = 0.785sec, $\quad$ Settling time, t_s = 1 sec $\quad$ (for 5% criterion)

E2.8 $\quad$ K=12.5, $\quad$ Rise time, $\quad$ t_r = 0.65sec

K_h = 0.178 ; $\quad$ Settling time, t_s = 2.48 sec (for 2% error) ; $\quad$ t_s = 1.86 sec (for 5% error)

E2.9 $\quad$ K=1.42, $\quad$ T=1.09 $\qquad$ **E2.10** $\quad$ K=2.95 N-m $\quad$ T=0.471 sec

E2.11 $\quad$ (a) $\quad e_{ss}\big|_{unit\ step} = 0$ $\qquad\qquad$ (b) $\quad$ Rise time, $\quad$ t_r = 1.91 sec

$\qquad\qquad\quad e_{ss}\big|_{unit\ ramp} = 1$ $\qquad\qquad\qquad\qquad$ Peak time, $\quad$ t_p = 2.79 sec

$\qquad\qquad\quad e_{ss}\big|_{unit\ parabola} = \infty$ $\qquad\qquad\qquad\quad$ Peak overshoot, $\quad$ M_p = 0.1265

$\qquad\qquad\qquad\qquad\qquad\qquad\qquad\qquad\qquad\qquad$ Settling time, $\quad$ t_s = 5.4 sec

E2.12 $\quad$ The total steady state error is ∞.

E2.13 $\quad$ C_0 = 0; $\qquad\qquad$ C_1 = 1; $\qquad\qquad$ C_2 = 0; $\quad$ C_3 = -6; $\qquad$ $e_{ss} = \infty$

E2.14 $\quad$ C_o =1/11; $\qquad$ C_1 =10/121; $\qquad$ $e_{ss} = \infty$.

E2.15 $\quad$

Question	K_p	K_v	K_a
(a)	100	0	0
(b)	∞	K	0
(c)	∞	∞	K/100
(d)	∞	∞	K/20

E2.16 $\quad$ (i) e(t) = 0.2 $\qquad\qquad\qquad$; e_{ss} = 0.2 $\qquad$ **E2.17** $\quad$ (i) It is type-2 system

$\qquad\qquad$ (ii) e(t) = 0.1t+0.09 $\qquad\qquad$; e_{ss} = ∞. $\qquad\qquad\qquad$ (ii) K_p = ∞; $\quad$ K_v = ∞ ; $\quad$ K_a = K/6

$\qquad\qquad$ (iii) e(t) = 0.15t² + 0.27 t - 0.054 $\quad$; e_{ss} = ∞. $\qquad\qquad$ (iii) e_{ss} = 6/K

$\qquad\qquad$ (iv) e(t) = 0.15t² + 0.77 t + 0.226 $\quad$; e_{ss} = ∞.

CHAPTER 3

FREQUENCY RESPONSE ANALYSIS

3.1 SINUSOIDAL TRANSFER FUNCTION AND FREQUENCY RESPONSE

The response of a system for the sinusoidal input is called sinusoidal response. The ratio of sinusoidal response and sinusoidal input is called **sinusoidal transfer function** of the system and in general, it is denoted by $T(j\omega)$. The sinusoidal transfer function is the frequency domain representation of the system, and so it is also called **frequency domain transfer function.**

The sinusoidal transfer, function $T(j\omega)$ can be obtained as shown below.

1. Contruct a physical model of a system using basic elements/parameters.
2. Determine the differential equations governing the system from the physical model of the system.
3. Take Laplace transform of differential equations in order to convert them to s-domain equation.
4. Determine s-domain transfer function, $T(s)$, which is ratio of s-domain output and input.
5. Determine the frequency domain transfer function, $T(j\omega)$ by replacing s by $j\omega$ in the s-domain transfer function, $T(s)$.

> **Note :** If the s-domain transfer function, $T(s)$ is known, then frequency domain transfer function, $T(j\omega)$, can be obtained directly from $T(s)$ by replacing s by $j\omega$.
>
> i.e., $T(s) \xrightarrow{\ s=j\omega\ } T(j\omega)$

Consider a linear time invariant system with frequency domain transfer function, $T(j\omega)$ shown in fig 3.1. Let the system be excited by a sinusoidal signal frequency ω, amplitude A, and phase θ. Now the response or output will also be a sinusoidal signal of same frequency ω, but the amplitude and phase of response will be modified by amplitude and phase of the transfer function respectively.

Now, the amplitude of the response is given by the product of the amplitude of the input and transfer function. The phase of the response is given by the sum of the phase of the input and transfer function.

Let, $\quad T(j\omega) = |T(j\omega)| \angle T(j\omega)$

where, $|T(j\omega)|$ = Magnitude of $T(j\omega)$, and, $\angle T(j\omega)$ = Phase of $T(j\omega)$.

Let, $\quad$ Input, $r(t) = A \sin(\omega t + \theta) = A\angle\theta$

where, A = Amplitude of input , ω = Frequency of input, and θ = Phase of input.

Now, Response, $c(t) = r(t) \times T(j\omega) = A\angle\theta \times |T(j\omega)|\angle T(j\omega) = A \times |T(j\omega)| \angle(\theta + \angle T(j\omega)) = B\angle\phi$

where, $B = A \times |T(j\omega)| =$ Magnitude of response, and, $\phi = \theta + \angle T(j\omega)$ = Phase of response.

$$\begin{array}{ccc}
& r(t) & \\
\xrightarrow{\hspace{2cm}} & \boxed{T(j\omega) = |T(j\omega)| \angle T(j\omega)} & \xrightarrow{\hspace{1.5cm}} \\
r(t) = A\sin(\omega t + \theta) = A \angle\theta & & c(t)
\end{array}$$

$c(t) = B\angle\phi$
where, $B = A \times |T(j\omega)|$
$\phi = \theta + \angle |T(j\omega)|$

Fig 3.1 : *System with sinusoidal transfer function $T(j\omega)$.*

FREQUENCY RESPONSE

The frequency domain transfer function T(jω) is a complex function of ω. Hence it can be seperated into magnitude function and phase function. Now, the magnitude and phase functions will be real functions of ω, and they are called *frequency response.*

The frequency response can be evaluated for open loop system and closed loop system. The frequency domain transfer function of open loop and closed loop systems can be obtained from the s-domain transfer function by replacing s by jω shown below.

Open loop transfer function : $G(s) \xrightarrow{s=j\omega} G(j\omega) = |G(j\omega)| \angle G(j\omega)$ .....(3.1)

Loop transfer function : $G(s)H(s) \xrightarrow{s=j\omega} G(j\omega)H(j\omega) = |G(j\omega)H(j\omega)| \angle G(j\omega)H(j\omega)$ .....(3.2)

Closed loop transfer function : $M(s) \xrightarrow{s=j\omega} M(j\omega) = |M(j\omega)| \angle M(j\omega)$ .....(3.3)

where, $|G(j\omega)|, |M(j\omega)|, |G(j\omega)H(j\omega)|$ are Magnitude functions

$\angle G(j\omega), \angle M(j\omega), \angle G(j\omega)H(j\omega)$ are Phase functions.

> *Note : For unity feedback system, H(s) = 1 and open loop and loop transfer functions are same.*

The advantages of frequency response analysis are the following.

1. The absolute and relative stability of the closed loop system can be estimated from the knowledge of their open loop frequency response.

2. The practical testing of systems can be easily carried with available sinusoidal signal generators and precise measurement equipments .

3. The transfer function of complicated systems can be determined experimentally by frequency response tests.

4. The design and parameter adjustment of the open loop transfer function of a system for specified closed loop performance is carried out more easily in frequency domain.

5. When the system is designed by use of the frequency response analysis, the effects of noise disturbance and parameters variations are relatively easy to visualize and incorporate corrective measures.

6. The frequency response analysis and designs can be extended to certain nonlinear control systems.

3.2 FREQUENCY DOMAIN SPECIFICATIONS

The performance and characteristics of a system in frequency domain are measured in terms of frequency domain specifications. The requirements of a system to be designed are usually specified in terms of these specifications.

The frequency domain specifications are,

1. Resonant peak , M_r
2. Resonant Frequency , ω_r
3. Bandwidth, ω_b
4. Cut-off rate
5. Gain margin, K_g
6. Phase margin, γ

Resonant Peak (M$_r$)

The maximum value of the magnitude of closed loop transfer function is called the resonant peak, M_r. A large resonant peak corresponds to a large overshoot in transient response.

Resonant Frequency (ω_r)

The frequency at which the resonant peak occurs is called resonant frequency, ω_r. This is related to the frequency of oscillation in the step response and thus it is indicative of the speed of transient response.

Bandwidth (ω_b)

The Bandwidth is the range of frequencies for which normalized gain of the system is more than -3 db. The frequency at which the gain is -3 db is called cut-off frequency. Bandwidth is usually defined for closed loop system and it transmits the signals whose frequencies are less than the cut-off frequency. The Bandwidth is a measure of the ability of a feedback system to reproduce the input signal, noise rejection characteristics and rise time. A large bandwidth corresponds to a small rise time or fast response.

Cut-off Rate

The slope of the log-magnitude curve near the cut off frequency is called cut-off rate. The cut -off rate indicates the ability of the system to distinguish the signal from noise.

Gain Margin , K_g

The gain margin, K_g is defined as the value of gain, to be added to system, in order to bring the system to the verge of instability.

The gain margin, K_g is given by the reciprocal of the magnitude of open loop transfer function at phase cross over frequency. The frequency at which the phase of open loop transfer function is $180°$ is called the phase cross-over frequency, ω_{pc}.

$$\text{Gain margin, } K_g = \frac{1}{|G(j\omega_{pc})|} \qquad\qquad .....(3.4)$$

The gain margin in db can be expressed as,

$$K_g \text{ in db} = 20 \log K_g = 20 \log \frac{1}{|G(j\omega_{pc})|} \qquad\qquad .....(3.5)$$

> *Note :* $|G(j\omega_{pc})|$ is the magnitude of $G(j\omega)$ at $\omega = \omega_{pc}$

The Gain margin in db is given by the negative of the db magnitude of $G(j\omega)$ at phase cross-over frequency. The gain margin indicates the additional gain that can be provided to system without affecting the stability of the system.

Phase Margin (γ)

The phase margin γ, is defined as the additional phase lag to be added at the gain cross over frequency in order to bring the system to the verge of instability. The gain cross over frequency ω_{gc} is the frequency at which the magnitude of the open loop transfer function is unity (or it is the frequency at which the db magnitude is zero).

The phase margin γ, is obtained by adding $180°$ to the phase angle ϕ of the open loop transfer function at the gain cross over frequency

$$\text{Phase margin, } \gamma = 180° + \phi_{gc}, \qquad\qquad .....(3.6)$$

$$\text{where, } \phi_{gc} = \angle G(j\omega_{gc})$$

> *Note :* $\angle G(j\omega_{gc})$ is the phase angle of $G(j\omega)$ at $\omega = \omega_{gc}$

The phase margin indicates the additional phase lag that can be provided to the system without affecting stability.

3.3 FREQUENCY DOMAIN SPECIFICATIONS OF SECOND ORDER SYSTEM

RESONANT PEAK (M_R)

Consider the closed loop transfer function of second order system,

$$\frac{C(s)}{R(s)} = M(s) = \frac{\omega_n^2}{s^2 + 2\zeta\omega_n s + \omega_n^2} \qquad .....(3.7)$$

The sinusoidal transfer function $M(j\omega)$ is obtained by letting $s = j\omega$.

$$\therefore\ M(j\omega) = \frac{\omega_n^2}{(j\omega)^2 + 2\zeta\omega_n(j\omega) + \omega_n^2} \qquad .....(3.8)$$

$$= \frac{\omega_n^2}{-\omega^2 + j2\zeta\omega_n\omega + \omega_n^2} = \frac{\omega_n^2}{\omega_n^2\left(-\dfrac{\omega^2}{\omega_n^2} + j2\zeta\dfrac{\omega}{\omega_n} + 1\right)} = \frac{1}{1 - \left(\dfrac{\omega}{\omega_n}\right)^2 + j2\zeta\dfrac{\omega}{\omega_n}}$$

Let, Normalized frequency, $u = \left(\dfrac{\omega}{\omega_n}\right)$

$$\therefore\ M(j\omega) = \frac{1}{(1 - u^2) + j2\zeta u}$$

Let, M = Magnitude of closed loop transfer function

α = Phase of closed loop transfer function.

$$M = |M(j\omega)| = \left[\frac{1}{(1 - u^2)^2 + (2\zeta u)^2}\right]^{\frac{1}{2}} = [(1 - u^2)^2 + 4\zeta^2 u^2]^{-\frac{1}{2}} \qquad .....(3.9)$$

$$\alpha = \angle M(j\omega) = -\tan^{-1}\frac{2\zeta u}{1 - u^2} \qquad .....(3.10)$$

The resonant peak is the maximum value of M. The condition for maximum value of M can be obtained by differentiating the equation of M with respect to u and letting dM/du = 0 when $u = u_r$,

where, $u_r = \dfrac{\omega_r}{\omega_n}$ = Normalized resonant frequency.

On differentiating equation (3.9) with respect to u we get,

$$\frac{dM}{du} = \frac{d}{du}[(1 - u^2)^2 + 4\zeta^2 u^2]^{-\frac{1}{2}} = -\frac{1}{2}[(1 - u^2)^2 + 4\zeta^2 u^2]^{-\frac{3}{2}}[2(1 - u^2)(-2u) + 8\zeta^2 u]$$

$$= \frac{-[-4u(1 - u^2) + 8\zeta^2 u]}{2[(1 - u^2)^2 + 4\zeta^2 u^2]^{\frac{3}{2}}} = \frac{4u(1 - u)^2 - 8\zeta^2 u}{2[(1 - u^2)^2 + 4\zeta^2 u^2]^{\frac{3}{2}}} \qquad .....(3.11)$$

Replace u by u_r in equation (3.11) and equate to zero.

$$\frac{4u_r(1 - u_r^2) - 8\zeta^2 u_r}{2[(1 - u_r^2)^2 + 4\zeta^2 u_r^2]^{\frac{3}{2}}} = 0 \qquad ......(3.12)$$

The equation (3.12) will be zero if numerator is zero. Hence, on equating numerator to zero we get,

$$4u_r(1 - u_r^2) - 8\zeta^2 u_r = 0 \qquad \Rightarrow \qquad 4u_r - 4u_r^3 - 8\zeta^2 u_r = 0$$

$$\therefore \; 4u_r^3 = 4u_r - 8\zeta^2 u_r \qquad \Rightarrow \qquad u_r^2 = 1 - 2\zeta^2 \qquad \Rightarrow \qquad u_r = \sqrt{1 - 2\zeta^2} \qquad ......(3.13)$$

Therefore, the resonant peak occurs when $u_r = \sqrt{1 - 2\zeta^2}$

Put this condition in the equation for M and solve for M_r.

$$\therefore \; M_r = \frac{1}{[(1 - u^2)^2 + 4\zeta^2 u^2]^{\frac{1}{2}}} \bigg|_{u = u_r} = \frac{1}{[(1 - u_r^2)^2 + 4\zeta^2 u_r^2]^{\frac{1}{2}}} = \frac{1}{[(1 - (1 - 2\zeta^2))^2 + 4\zeta^2(1 - 2\zeta^2)]^{\frac{1}{2}}}$$

$$= \frac{1}{[4\zeta^4 + 4\zeta^2 - 8\zeta^4]^{\frac{1}{2}}} = \frac{1}{[4\zeta^2 - 4\zeta^4]^{\frac{1}{2}}} = \frac{1}{[4\zeta^2(1 - \zeta^2)]^{\frac{1}{2}}} = \frac{1}{2\zeta\sqrt{1 - \zeta^2}}$$

$$\boxed{\therefore \quad \text{Resonant peak, } M_r = \frac{1}{2\zeta\sqrt{1 - \zeta^2}}} \qquad .....(3.14)$$

RESONANT FREQUENCY (ω_r)

Normalized resonant frequency, $u_r = \dfrac{\omega_r}{\omega_n} = \sqrt{1 - 2\zeta^2}$ $\qquad .....(3.15)$

The resonant frequency, $\omega_r = \omega_n \sqrt{1 - 2\zeta^2}$ $\qquad .....(3.16)$

BANDWIDTH (ω_b)

Let, Normalized bandwidth, $u_b = \dfrac{\omega_b}{\omega_n}$

When $u = u_b$, the magnitude M, of the closed loop system is $1/\sqrt{2}$ (or –3db).

Hence in the equation for M (equation 3.9), put $u = u_b$ and equate to $1/\sqrt{2}$.

$$\therefore \; M = \frac{1}{[1 - u_b^2)^2 + 4\zeta^2 u_b^2]^{\frac{1}{2}}} = \frac{1}{\sqrt{2}} \qquad .....(3.17)$$

On squaring and cross multiplying we get,

$$(1 - u_b^2)^2 + 4\zeta^2 u_b^2 = 2 \quad \Rightarrow \quad 1 + u_b^4 - 2u_b^2 + 4\zeta^2 u_b^2 = 2 \quad \Rightarrow \quad u_b^4 - 2u_b^2(1 - \zeta^2) - 1 = 0$$

Let, $x = u_b^2$; $\therefore \; x^2 - 2(1 - 2\zeta^2)x - 1 = 0$

$$\therefore \; x = \frac{2(1 - 2\zeta^2) \pm \sqrt{4(1 - 2\zeta^2)^2 + 4}}{2} = \frac{2(1 - 2\zeta^2) \pm 2\sqrt{(1 + 4\zeta^4 - 4\zeta^2) + 1}}{2}$$

Let us take only the positive sign,

$$\therefore \; x = 1 - 2\zeta^2 + \sqrt{2 - 4\zeta^2 + 4\zeta^4}$$

But, $u_b = \sqrt{x}$; $\therefore \; u_b = \sqrt{x} = [1 - 2\zeta^2 + \sqrt{2 - 4\zeta^2 + 4\zeta^4}]^{\frac{1}{2}}$; Also, $u_b = \dfrac{\omega_b}{\omega_n}$

$$\therefore \text{Bandwidth, } \omega_b = \omega_n u_b = \omega_n \left[1 - 2\zeta^2 + \sqrt{2 - 4\zeta^2 + 4\zeta^4}\right]^{\frac{1}{2}} \qquad .....(3.18)$$

PHASE MARGIN (γ)

The open loop transfer function of second order system,

$$G(s) = \frac{\omega_n^2}{s(s + 2\zeta\omega_n)} \qquad .....(3.19)$$

The sinusoidal transfer function $G(j\omega)$ is obtained by letting $s = j\omega$.

$$G(j\omega) = \frac{\omega_n^2}{j\omega(j\omega + 2\zeta\omega_n)} = \frac{\omega_n^2}{\omega_n\left(\dfrac{j\omega}{\omega_n}\right)\omega_n\left(2\zeta + j\dfrac{\omega}{\omega_n}\right)} = \frac{1}{j\dfrac{\omega}{\omega_n}\left(2\zeta + j\dfrac{\omega}{\omega_n}\right)} \qquad .....(3.20)$$

Let normalized frequency, $u = \omega/\omega_n$

On substituting $u = \omega/\omega_n$ in equation (3.20) we get,

$$G(j\omega) = \frac{1}{ju(2\zeta + ju)} \qquad .....(3.21)$$

Magnitude of $G(j\omega) = \left|G(j\omega)\right| = \dfrac{1}{u\sqrt{4\zeta^2 + u^2}} = \dfrac{1}{\sqrt{u^4 + 4\zeta^2 u^2}} \qquad .....(3.22)$

Phase of $G(j\omega) = -90° - \tan^{-1}\dfrac{u}{2\zeta} \qquad .....(3.23)$

At the gain cross-over frequency ω_{gc}, the magnitude of $G(j\omega)$ is unity.

Let normalized gain cross over frequency, $u_{gc} = \omega_{gc}/\omega_n$

On substituting u by u_{gc} in the equation (3.22) and equating to unity, we get,

$$\therefore \text{ At } u = u_{gc}, \left|G(j\omega)\right| = \frac{1}{\sqrt{u_{gc}^4 + 4\zeta^2 u_{gc}^2}} = 1 \quad \Rightarrow \quad u_{gc}^4 + 4\zeta^2 u_{gc}^2 = 1 \quad \Rightarrow \quad u_{gc}^4 + 4\zeta^2 u_{gc}^2 - 1 = 0$$

Let, $x = u_{gc}^2$; $\therefore x^2 + 4\zeta^2 x - 1 = 0$

$$\therefore x = \frac{-4\zeta^2 \pm \sqrt{16\zeta^4 + 4}}{2} = -2\zeta^2 \pm \sqrt{4\zeta^4 + 1}$$

Let us take only the positive sign,

$$\therefore x = -2\zeta^2 + \sqrt{4\zeta^4 + 1}$$

But, $u_{gc} = \sqrt{x}$; $\therefore u_{gc} = \sqrt{x} = \left[-2\zeta^2 + \sqrt{4\zeta^4 + 1}\right]^{\frac{1}{2}} \qquad .....(3.24)$

The phase margin, $\gamma = 180 + \angle G(j\omega)\big|_{\omega = \omega_{gc}, \, u = u_{gc}} \qquad .....(3.25)$

Substituting for $\angle G(j\omega)$ from equation (3.23) in equation (3.25) we get,

$$\gamma = 180 + \left(-90° - \tan^{-1}\frac{u_{gc}}{2\zeta}\right) = 90 - \tan^{-1}\left[\frac{\left[-2\zeta^2 + \sqrt{4\zeta^4 + 1}\right]^{\frac{1}{2}}}{2\zeta}\right] \qquad .....(3.26)$$

Note : *The gain margin of second order system is infinite.*

3.4 CORRELATION BETWEEN TIME AND FREQUENCY RESPONSE

The correlation between time and frequency response has an explicit form only for first and second order systems. The correlation for second-order system is discussed here.

Consider the magnitude and phase of a closed loop second order system as a function of normalized frequency, as given by equations (3.9) and (3.10).

Magnitude of closed loop system, $M = |M(j\omega)| = \dfrac{1}{\sqrt{(1 - u^2)^2 + (2\zeta u)^2}}$

Phase of closed loop system, $\alpha = \angle M(j\omega) = -\tan^{-1} \dfrac{2\zeta u}{1 - u^2}$

The magnitude and phase angle characteristics for normalized frequency u, for certain values of ζ are shown in fig 3.2 and 3.3. The frequency at which M has a peak value is known as the resonant frequency. The peak value of the magnitude is the resonant peak M_r. At this frequency the slope of the magnitude curve is zero. The frequency corresponding to M_r is u_r, which is the normalized resonant frequency.

From equations (3.14) and (3.15) we get,

Resonant peak, $M_r = \dfrac{1}{2\zeta\sqrt{1 - \zeta^2}}$

Resonant frequency, $\omega_r = \omega_n\sqrt{1 - 2\zeta^2}$ .....(3.27)

When $\zeta = 0$, $\omega_r = \omega_n\sqrt{1 - 2\zeta^2} = \omega_n$

When $\zeta = 0$, $M_r = \dfrac{1}{2\zeta\sqrt{1 - \zeta^2}} = \infty$ .....(3.28)

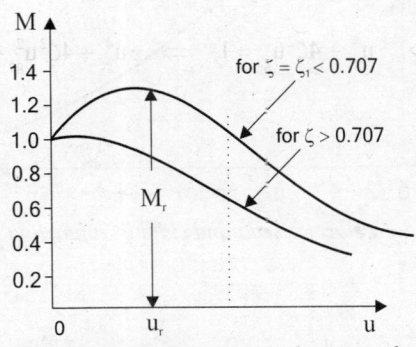

Fig 3.2 : *Magnitude, M as a function of u.*

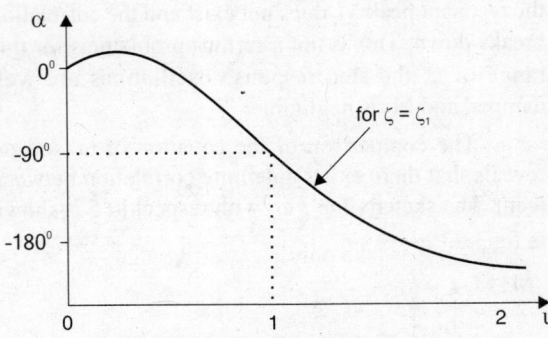

Fig 3.3 : *Phase, α as a function of u.*

From equations (3.27) and (3.28), it is clear that as ζ tends to zero, ω_r approaches ω_n, and M_r approaches infinity.

When $1 - 2\zeta^2 = 0$, $\omega_r = 0$, which means there is no resonant peak at this condition.

Let, $1 - 2\zeta^2 = 0$; $\therefore \zeta^2 = \dfrac{1}{2}$ $\Rightarrow$ $\zeta = \dfrac{1}{\sqrt{2}}$

For $0 < \zeta \le 1/\sqrt{2}$, the resonant frequency always has a value less than ω_n, and the resonant peak has a value greater than one.

For $\zeta > 1/\sqrt{2}$, the condition (dM/du) = 0, will not be satisfied for any real value of ω.

Hence when $\zeta > 1/\sqrt{2}$ the magnitude M decreases monotonically from M = 1 at u = 0 with increasing u. It follows that for $\zeta > 1/\sqrt{2}$ there is no resonant peak and the greatest value of M equals one.

The frequency at which M has a value of $\zeta > 1/\sqrt{2}$ is of special significance and is called the cut-off frequency ω_c. The signal frequencies above cut-off are greatly attenuated on passing through a system.

For feedback control system, the range of frequencies over which $M \geq 1/\sqrt{2}$ is defined as bandwidth ω_b. Control system being low-pass filters (at zero frequency M = 1), the bandwidth ω_b is equal to cut-off frequency ω_c.

In general the bandwidth of a control system indicates the noise-filtering characteristics of the system. Also, bandwidth gives a measure of the transient response.

The normalized bandwidth, $u_b = \dfrac{\omega_b}{\omega_n} = \left[1 - 2\zeta^2 + \sqrt{2 - 4\zeta^2 + \zeta^4}\right]^{\frac{1}{2}}$

From the equation of u_b it is clear that u_b is a function of ζ alone. The graph between u_b and ζ is shown in fig 3.4.

The expression for the damped frequency of oscillation ω_d and peak overshoot M_p of the step response, for $0 \leq \zeta \leq 1$ are,

Damped frequency, $\omega_d = \omega_n \sqrt{1 - \zeta^2}$ and Peak overshoot, $M_p = e^{\frac{-\zeta\pi}{\sqrt{1-\zeta^2}}}$

Comparison of the equation of M_r and M_p reveals that both are functions of only ζ.

The sketch of M_r and M_p for various value of ζ are shown in fig 3.5. The sketches reveals that a system with a given value of M_r must exhibit a corresponding value of M_p if subjected to a step input. For $\zeta > 1/\sqrt{2}$, the resonant peak M_r does not exist and the correlation breaks down. This is not a serious problem as for this range of ζ, the step response oscillations are well damped and M_p is negligible.

The comparison of the equation of ω_r and ω_d reveals that there exists a definite correlation between them. The sketch of ω_r / ω_d with respect to ζ is shown in fig 3.6.

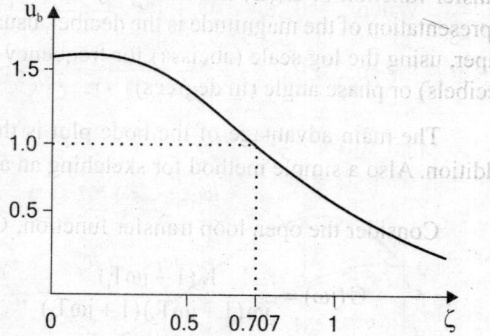

Fig 3.4 : *Normalised bandwidth as a function of ζ.*

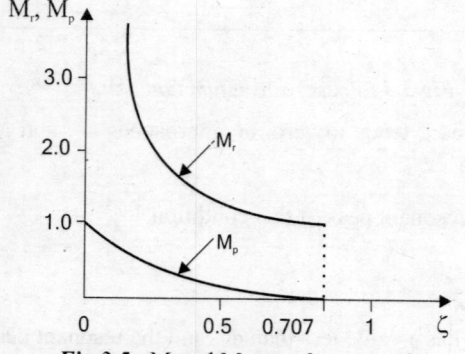

Fig 3.5 : M_r *and* M_p *as a function of ζ.*

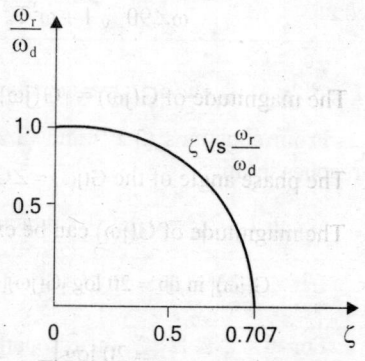

Fig 3.6 : ω_r / ω_d *as a function of ζ.*

3.5 FREQUENCY RESPONSE PLOTS

Frequency response analysis of control systems can be carried either analytically or graphically. The various graphical techniques available for frequency response analysis are,

1. Bode plot
2. Polar plot (or Nyquist plot)
3. Nichols plot
4. M and N circles
5. Nichols chart

The Bode plot, Polar plot and Nichols plot are usually drawn for open loop systems. From the open loop response plot, the performance and stability of closed loop system are estimated. The M and N circles and Nichols chart are used to graphically determine the frequency response of unity feedback closed loop system from the knowledge of open loop response.

The frequency response plots are used to determine the frequency domain specifications, to study the stability of the systems and to adjust the gain of the system to satisfy the desired specifications.

3.6 BODE PLOT

The Bode plot is a frequency response plot of the sinusoidal transfer function of a system. A Bode plot consists of two graphs. One is a plot of the magnitude of a sinusoidal transfer function versus $\log \omega$. The other is a plot of the phase angle of a sinusoidal transfer function versus $\log \omega$.

The Bode plot can be drawn for both open loop and closed loop system. Usually the bode plot is drawn for open loop system. The standard representation of the logarithmic magnitude of open loop transfer function of $G(j\omega)$ is $20 \log |G(j\omega)|$ where the base of the logarithm is 10. The unit used in this representation of the magnitude is the decibel, usually abbreviated as db. The curves are drawn on semilog paper, using the log scale (abcissa) for frequency and the linear scale (ordinate) for either magnitude (in decibels) or phase angle (in degrees).

The main advantage of the bode plot is that multiplication of magnitudes can be converted into addition. Also a simple method for sketching an approximate log-magnitude curve is available.

Consider the open loop transfer function, $G(s) = \dfrac{K(1+sT_1)}{s(1+sT_2)(1+sT_3)}$

$$G(j\omega) = \frac{K(1+j\omega T_1)}{j\omega(1+j\omega T_2)(1+j\omega T_3)}$$

$$= \frac{K\angle 0° \sqrt{1+\omega^2 T_1^2} \angle \tan^{-1}\omega T_1}{\omega\angle 90° \sqrt{1+\omega^2 T_2^2} \angle \tan^{-1}\omega T_2 \sqrt{1+\omega^2 T_3^2} \angle \tan^{-1}\omega T_3}$$

The magnitude of $G(j\omega) = |G(j\omega)| = \dfrac{K\sqrt{1+\omega^2 T_1^2}}{\omega\sqrt{1+\omega^2 T_2^2}\sqrt{1+\omega^2 T_3^2}}$

The phase angle of the $G(j\omega) = \angle G(j\omega) = \tan^{-1}\omega T_1 - 90° - \tan^{-1}\omega T_2 - \tan^{-1}\omega T_3$

The magnitude of $G(j\omega)$ can be expressed in decibels as shown below.

$|G(j\omega)|$ in db $= 20 \log |G(j\omega)|$

$$= 20 \log\left[\frac{K\sqrt{1+\omega^2 T_1^2}}{\omega\sqrt{1+\omega^2 T_2^2}\sqrt{1+\omega^2 T_3^2}}\right]$$

$$= 20 \log \left[\frac{K}{\omega} \times \sqrt{1 + \omega^2 T_1^2} \times \frac{1}{\sqrt{1 + \omega^2 T_2^2}} \times \frac{1}{\sqrt{1 + \omega^2 T_3^2}} \right]$$

$$= 20 \log \frac{K}{\omega} + 20 \log \sqrt{1 + \omega^2 T_1^2} + 20 \log \frac{1}{\sqrt{1 + \omega^2 T_2^2}} + 20 \log \frac{1}{\sqrt{1 + \omega^2 T_3^2}}$$

$$= 20 \log \frac{K}{\omega} + 20 \log \sqrt{1 + \omega^2 T_1^2} - 20 \log \sqrt{1 + \omega^2 T_2^2} - 20 \log \sqrt{1 + \omega^2 T_3^2} \qquad .....(3.29)$$

From the equation (3.29) it is clear that, when the magnitude is expressed in db, the multiplication is converted to addition. Hence in magnitude plot, the db magnitudes of individual factors of $G(j\omega)$ can be added.

Therefore to sketch the magnitude plot, a knowledge of the magnitude variations of individual factor is essential. The magnitude plot and phase plot of various factors of $G(j\omega)$ are explained in the following section.

BASIC FACTORS OF G(jω)

The basic factors that very frequently occur in a typical transfer function $G(j\omega)$ are,

1. Constant gain, K

2. Integral factor, $\dfrac{K}{j\omega}$ or $\dfrac{K}{(j\omega)^n}$

3. Derivative factor, $K \times j\omega$ or $K \times (j\omega)^n$

4. First order factor in denominator, $\dfrac{1}{1 + j\omega T}$ or $\dfrac{1}{(1 + j\omega T)^m}$

5. First order factor in numerator, $(1 + j\omega T)$ or $(1 + j\omega T)^m$

6. Quadratic factor in denominator, $\left[\dfrac{1}{1 + 2\zeta(j\omega/\omega_n) + (j\omega/\omega_n)^2} \right]$

7. Quadratic factor in numerator, $\left[1 + 2\zeta\left(\dfrac{j\omega}{\omega_n}\right) + \left(\dfrac{j\omega}{\omega_n}\right)^2 \right]$

CONSTANT GAIN, K

Let, $G(s) = K$

$$\therefore \ G(j\omega) = K = K \angle 0°$$

$$A = |G(j\omega)| \text{ in db} = 20 \log K$$

$$\phi = \angle G(j\omega) = 0°$$

The magnitude plot for a constant gain K is a horizontal straight line at the magnitude of 20 log K db. The phase plot is straight line at 0°.

When K > 1,   20 log K is positive.

When 0 < K < 1,   20 log K is negative.

When K = 1,   20 log K is zero.

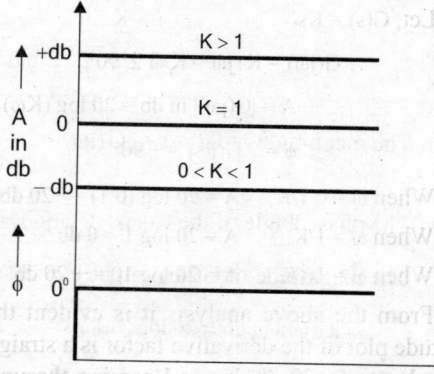

Fig 3.7 : Bode plot of constant gain, K.

Integral Factor

Let, $G(s) = \dfrac{K}{s}$

$\therefore\ G(j\omega) = \dfrac{K}{j\omega} = \dfrac{K}{\omega} \angle -90°$

$A = |G(j\omega)|$ in db $= 20 \log (K/\omega)$

$\phi = \angle G(j\omega) = -90°$

When $\omega = 0.1\ K$, $A = 20 \log (1/0.1) = 20$ db

When $\omega = K$, $A = 20 \log 1 = 0$ db

When $\omega = 10\ K$, $A = 20 \log (1/10) = -20$ db

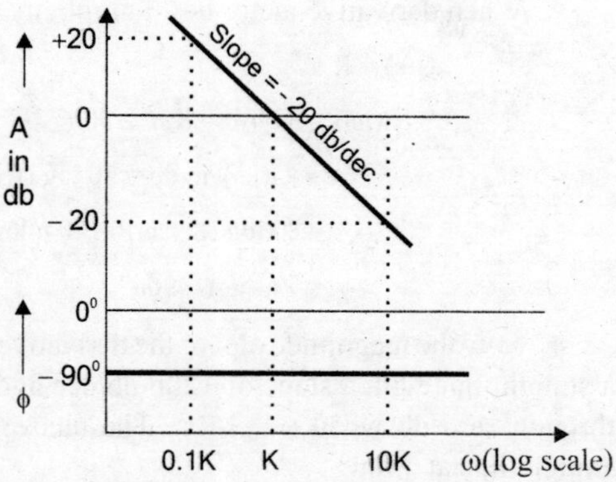

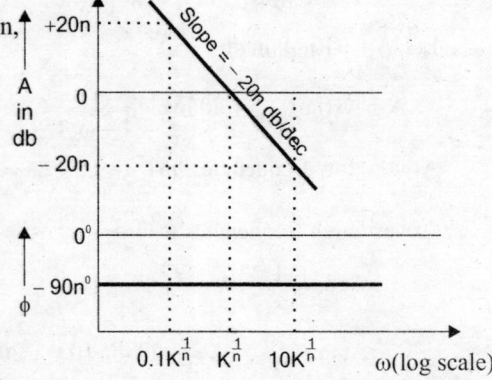

Fig 3.8 : Bode plot of integral factor, $\dfrac{K}{j\omega}$

From the above analysis it is evident that the magnitude plot of the integral factor is a straight line with a slope of −20 db/dec and passing through zero db, when $\omega = K$. Since the $\angle G(j\omega)$ is a constant and independent of ω the phase plot is a straight line at −90°.

When an integral factor has multiplicity of n, then,

$G(s) = K/s^n$

$G(j\omega) = K/(j\omega)^n = K/\omega^n \angle -90n°$

$A = |G(j\omega)|$ in db $= 20 \log \dfrac{K}{\omega^n}$

$= 20 \log \left(\dfrac{K^{\frac{1}{n}}}{\omega}\right)^n = 20\ n \log\left(\dfrac{K^{\frac{1}{n}}}{\omega}\right)$

$\phi = \angle G(j\omega) = -90\ n°$

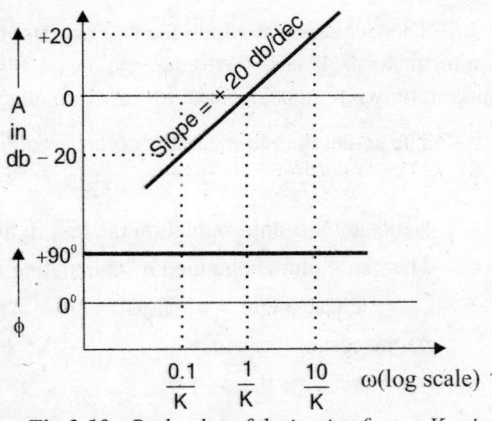

Fig 3.9 : Bode plot of integral factor, $K/(j\omega)^n$.

Now the magnitude plot of the integral factor is a straight line with a slope of −20n db/dec and passing through zero db when $\omega = K^{1/n}$. The phase plot is a straight line at −90n°.

Derivative Factor

Let, $G(s) = Ks$

$\therefore\ G(j\omega) = K\ j\omega = K\ \omega \angle 90°$

$A = |G(j\omega)|$ in db $= 20 \log (K\omega)$

$\phi = \angle G(j\omega) = +90°$

When $\omega = 0.1/K$, $A = 20 \log (0.1) = -20$ db

When $\omega = 1/K$, $A = 20 \log 1 = 0$ db

When $\omega = 10/K$, $A = 20 \log 10 = +20$ db

From the above analysis it is evident that the magnitude plot of the derivative factor is a straight line with a slope of +20 db/dec and passing through zero db when $\omega = 1/K$. Since the $\angle G(j\omega)$ is a constant and independent of ω, the phase plot is a straight line at +90°.

Fig 3.10 : Bode plot of derivative factor, $K \times j\omega$.

When derivative factor has multiplicity of n then,

$$G(s) = K\,s^n$$

$$\therefore\ G(j\omega) = K(j\omega)^n = K\omega^n\ \angle 90n°$$

$$A = |G(j\omega)|\ \text{in db} = 20\log(K\omega^n)$$

$$= 20\log(K^{1/n}\,\omega)^n = 20\,n\log(K^{1/n}\,\omega)$$

$$\phi = \angle G(j\omega) = 90n°$$

Now the magnitude plot of the derivative factor is a straight line with a slope of +20n db/dec and passing through zero db when $\omega = 1/K^{1/n}$. The phase plot is a straight line at +90n°.

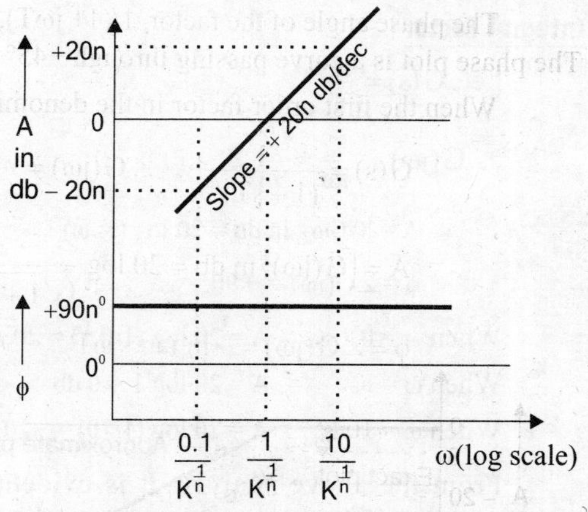

Fig 3.11 : Bode plot of derivative factor, $K(j\omega)^n$.

First order factor in denominator

$$G(s) = \frac{1}{1+sT}$$

$$\therefore\ G(j\omega) = \frac{1}{1+j\omega T} = \frac{1}{\sqrt{1+\omega^2 T^2}}\ \angle -\tan^{-1}\omega T$$

Let, $A = |G(j\omega)|$ in db.

$$\therefore\ A = \big|G(j\omega)\big|_{\text{in db}} = 20\log\frac{1}{\sqrt{1+\omega^2 T^2}} = -20\log\sqrt{1+\omega^2 T^2}$$

At very low frequencies, $\omega T \ll 1$; $\therefore A = -20\log\sqrt{1+\omega^2 T^2} \approx -20\log 1 = 0$

At very high frequencies, $\omega T \gg 1$; $\therefore A = -20\log\sqrt{1+\omega^2 T^2} \approx -20\log\sqrt{1+\omega^2 T^2} = -20\log\omega T$

At $\omega = \dfrac{1}{T}$, $\quad A = -20\log 1 = 0$

At $\omega = \dfrac{10}{T}$, $\quad A = -20\log 10 = -20\text{db}$

The above analysis shows that the magnitude plot of the factor $1/(1+j\omega T)$ can be approximated by two straight lines, one is a straight line at 0 db for the frequency range, **$0 < \omega < 1/\,T$**, and the other is a straight line with slope −20 db/dec for the frequency range, **$1/T < \omega < \infty$**. The two straight lines are asymptotes of the exact curve.

The frequency at which the two asymptotes meet is called *corner frequency* or *break frequency*. For the factor $1/(1+j\omega T)$ the frequency, $\omega = 1/T$ is the corner frequency, ω_c. It divides the frequency response curve into two regions, a curve for low frequency region and a curve for high frequency region.

The actual magnitude at the corner frequency, $\omega_c = \dfrac{1}{T}$ is,

$$A = -20\log\sqrt{1+1} = -3\text{ db}$$

Hence by this approximation the loss in db at the corner frequency is −3 db.

The phase plot is obtained by calculating the phase angle of $G(j\omega)$ for various values of ω

Phase angle, $\phi = \angle G(j\omega) = -\tan^{-1}\omega T$

At the corner frequency, $\omega = \omega_c = \dfrac{1}{T}$, $\quad \phi = -\tan^{-1}\omega T = -\tan^{-1}1 = -45°$

As $\omega \to 0$, $\ \phi \to 0°$

As $\omega \to \infty$, $\ \phi \to -90°$

The phase angle of the factor, $1/(1+j\omega T)$, varies from $0°$ to $-90°$ as ω is varied from zero to infinity. The phase plot is a curve passing through $-45°$ at ω_c.

When the first order factor in the denominator has a multiplicity of m, then,

$$G(s) = \frac{1}{(1+sT)^m} \; ; \; \therefore G(j\omega) = \frac{1}{(1+j\omega T)^m} = \frac{1}{(\sqrt{1+\omega^2 T^2})^m \angle m \tan^{-1}\omega T}$$

$$A = |G(j\omega)| \text{ in db} = 20 \log \frac{1}{(\sqrt{1+\omega^2 T^2})^m} = -20m \log \sqrt{1+\omega^2 T^2}$$

$$\phi = \angle G(j\omega) = -m \tan^{-1}\omega T$$

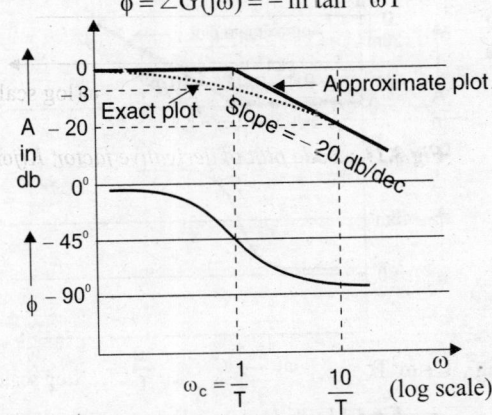

Fig 3.12 : *Bode plot of the factor* $\frac{1}{1+j\omega T}$.   **Fig 3.13 :** *Bode plot of the factor* $1/(1+j\omega T)^m$.

Now the magnitude plot of the factor $1/(1+j\omega T)^m$ can be approximated by two straight lines, one is a straight line at zero db for the frequency range, $0 < \omega < 1/T$, and the other is a straight line with slope -20 m db/dec for the frequency range, $1/T < \omega < \infty$. The corner frequency, $\omega_c = 1/T$ and the loss in db at the corner frequency is $-3m$ db.

The phase angle of the factor $1/(1+j\omega T)^m$ varies from $0°$ to $-90m°$ as ω is varied from zero to infinity. The phase plot is a curve passing through $-45m°$ at .

FIRST ORDER FACTOR IN THE NUMERATOR

$$G(s) = 1 + sT$$

$$G(j\omega) = 1 + j\omega T = \sqrt{1+\omega^2 T^2} \angle \tan^{-1}\omega T$$

$$A = |G(j\omega)| \text{ in db} = 20 \log \sqrt{1+\omega^2 T^2}$$

$$\phi = \angle G(j\omega) = \tan^{-1}\omega T$$

By an analysis similar to that of previous section it can be shown that the magnitude plot of the factor $(1+j\omega T)$ can be approximated by two straight lines, one is a straight line at zero db for the frequency range $0 < \omega < 1/T$ and the other is a straight line with slope $+20$ db/dec for the frequency range $1/T < \omega < \infty$. The two straight lines are asymptotes of the exact curve.

The frequency at which the two asymptotes meet is called the corner frequency or break frequency. For the factor $(1+j\omega T)$, the frequency, $(\omega = 1/T)$ is the corner frequency, ω_c. By this approximation the loss in db at the corner frequency is $+3$ db. The phase angle of the factor $(1+j\omega T)$ varies from zero to $+90°$ as ω is varied from 0 to ∞. The phase plot is a curve passing through $+45°$ at ω_c.

When the first order factor in the numerator has a multiplicity of m, then,

$$G(s) = (1 + sT)^m$$

$$G(j\omega) = (1 + j\omega T)^m = \left(\sqrt{1 + \omega^2 T^2}\right)^m \angle\, m \tan^{-1} \omega T$$

$$A = |G(j\omega)| \text{ in db} = 20 \log\left(\sqrt{1 + \omega^2 T^2}\right)^m = 20m \log \sqrt{1 + \omega^2 T^2}$$

$$\phi = \angle G(j\omega) = m \tan^{-1} \omega T$$

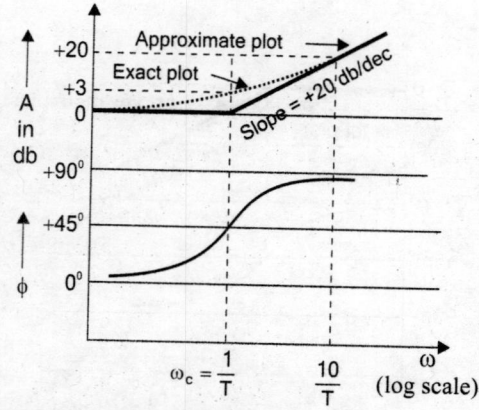

Fig 3.14 : Bode plot of the factor (1+jωT).

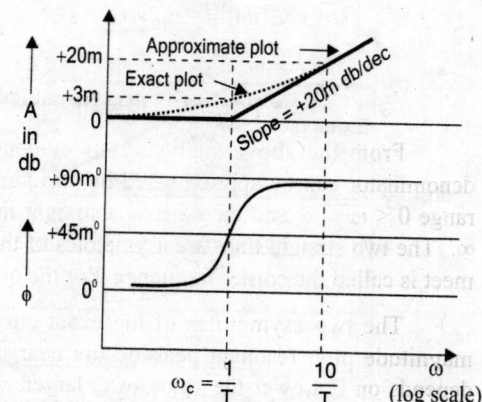

Fig 3.15 : Bode plot of the factor (1+jωT)^m.

Now the magnitude plot of the factor $(1 + j\omega T)^m$ can be approximated by two straight lines, one is a straight line at zero db for the frequency range $0 < \omega < 1/T$ and the other is a straight line with a slope of +20m db/dec for the frequency range $1/T < \omega < \infty$. The corner frequency, $\omega_c = 1/T$ and the loss in db at this corner frequency is +3m db.

The phase angle of the factor $(1 + j\omega T)^m$ varies from zero to +90m° as ω is varied from zero to infinity. The phase plot is a curve passing through +45 m° at ω_c.

QUADRATIC FACTOR IN THE DENOMINATOR

$$G(s) = \frac{\omega_n^2}{s^2 + 2\zeta\omega_n s + \omega_n^2} = \frac{1}{1 + 2\zeta \dfrac{s}{\omega_n} + \left(\dfrac{s}{\omega_n}\right)^2}$$

$$\therefore\; G(j\omega) = \frac{1}{1 + j\dfrac{2\zeta\omega}{\omega_n} + \left(\dfrac{j\omega}{\omega_n}\right)^2} = \frac{1}{1 - \left(\dfrac{\omega}{\omega_n}\right)^2 + j2\zeta\dfrac{\omega}{\omega_n}} = \frac{1}{\sqrt{\left(1 - \dfrac{\omega^2}{\omega_n^2}\right)^2 + 4\zeta^2\dfrac{\omega^2}{\omega_n^2}}} \angle \tan^{-1} \frac{2\zeta\dfrac{\omega}{\omega_n}}{1 - \dfrac{\omega^2}{\omega_n^2}}$$

Let, $A = |G(j\omega)|$ in db.

$$A = 20 \log \frac{1}{\sqrt{\left(1 - \dfrac{\omega^2}{\omega_n^2}\right)^2 + 4\zeta^2\dfrac{\omega^2}{\omega_n^2}}} = -20 \log \sqrt{\left(1 - \dfrac{\omega^2}{\omega_n^2}\right)^2 + 4\zeta^2\dfrac{\omega^2}{\omega_n^2}}$$

$$= -20 \log \sqrt{1 + \dfrac{\omega^4}{\omega_n^4} - 2\dfrac{\omega^2}{\omega_n^2} + 4\zeta^2\dfrac{\omega^2}{\omega_n^2}} = -20 \log \sqrt{1 - \dfrac{\omega^2}{\omega_n^2}(2 - 4\zeta^2) + \dfrac{\omega^4}{\omega_n^4}}$$

At very low frequencies when $\omega << \omega_n$, the magnitude is,

$$A = -20 \log \sqrt{1 - \frac{\omega^2}{\omega_n^2}(2 - 4\zeta^2) + \frac{\omega^4}{\omega_n^4}} \approx -20 \log 1 = 0$$

At very high frequencies when $\omega >> \omega_n$, the magnitude is,

$$A = -20 \log \sqrt{1 - \frac{\omega^2}{\omega_n^2}(2 - 4\zeta^2) + \frac{\omega^4}{\omega_n^4}} \approx -20 \log \sqrt{\frac{\omega^4}{\omega_n^4}} = -20 \log \frac{\omega^2}{\omega_n^2} = -20 \log \left(\frac{\omega}{\omega_n}\right)^2$$

$$\therefore \; A = -40 \log \frac{\omega}{\omega_n}$$

At $\omega = \omega_n$, $A = -40 \log 1 = 0$ db

At $\omega = 10\omega_n$, $A = -40 \log 10 = -40$ db

From the above analysis it is evident that the magnitude plot of the quadratic factor in the denominator can be approximated by two straight lines, one is a straight line at 0 db for the frequency range $0 < \omega < \omega_n$ and the other is a straight line with slope -40 db/dec for the frequency range $\omega_n < \omega < \infty$. The two straight lines are asymptotes of the exact curve. The frequency at which the two asymptotes meet is called the corner frequency. For the quadratic factor, the frequency ω_n is the corner frequency, ω_c.

The two asymptotes of the exact curve are independent of the damping ratio, ζ. In the exact magnitude plot, resonant peak occurs near the corner frequency and the magnitude of resonant peak depends on ζ. Lower the value of ζ, larger will be the resonant peak. Hence by this approximation the error at the corner frequency depends on damping ratio ζ. The phase plot is obtained by calculating the phase angle of $G(j\omega)$ for various values of ω.

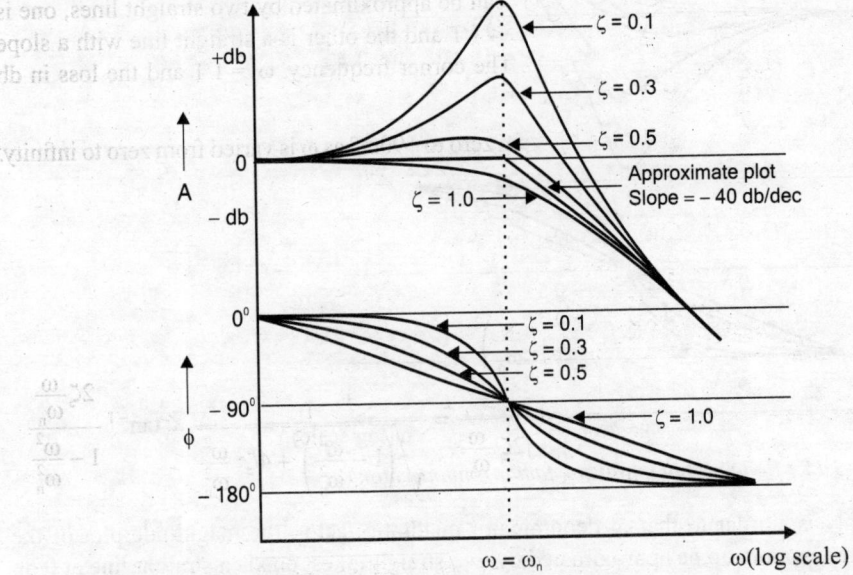

Fig 3.16 : Bode plot of quadratic factor in denominator.

$$\phi = \angle G(j\omega) = -\tan^{-1}\left(\frac{2\zeta \frac{\omega}{\omega_n}}{1 - \frac{\omega^2}{\omega_n^2}}\right)$$

As $\omega = \omega_n$, $\phi = -\tan^{-1}\dfrac{2\zeta}{0} = -\tan^{-1}\infty = -90°$

As $\omega \to 0$, $\phi \to 0$

As $\omega \to \infty$, $\phi \to -180°$

The phase angle of the quadratic factor varies from 0 to $-180°$ as ω is varied from 0 to ∞. The phase plot is a curve passing through $-90°$ at ω_c. At the corner frequency phase angle is $-90°$ and independent of ζ, but at all other frequency it depends on ζ.

QUADRATIC FACTOR IN THE NUMERATOR

$$G(s) = \frac{s^2 + 2\zeta\omega_n s + \omega_n^2}{\omega_n^2} = 1 + 2\zeta\left(\frac{s}{\omega_n}\right) + \left(\frac{s}{\omega_n}\right)^2$$

$$G(j\omega) = 1 + j2\zeta\frac{\omega}{\omega_n} + \left(\frac{j\omega}{\omega_n}\right)^2 = \sqrt{\left(1 - \frac{\omega^2}{\omega_n^2}\right)^2 + 4\zeta^2\frac{\omega^2}{\omega_n^2}} \angle \tan^{-1}\frac{2\zeta\dfrac{\omega}{\omega_n}}{1 - \dfrac{\omega^2}{\omega_n^2}}$$

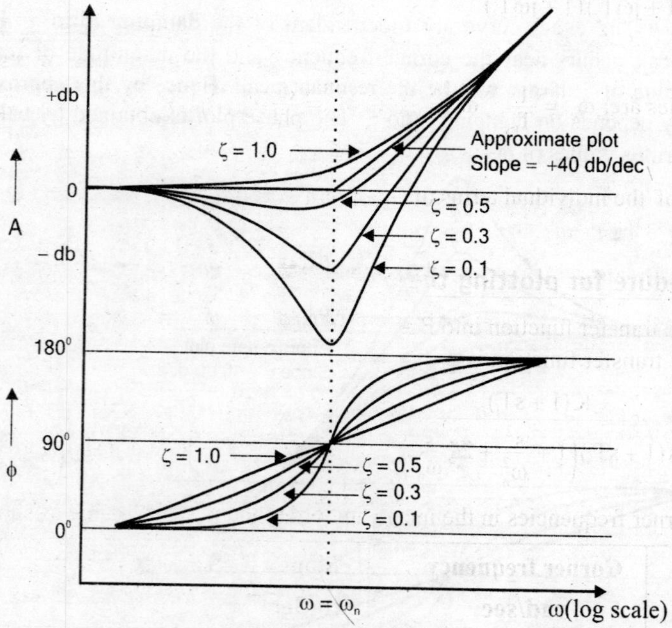

Fig 3.17 : Bode plot of quadratic factor in numerator.

Based on an analysis similar to that of denominator quadratic factor, the magnitude plot of the quadratic factor in the numerator can be approximated by two straight lines, one is a straight line at 0 db for the frequency range $0 < \omega < \omega_n$ and the other is a straight line with slope +40 db/dec for the frequency range $\omega_n < \omega < \infty$. The corner frequency is ω_n. Due to this approximation the error at the corner frequency depends on ζ.

The phase angle varies from 0 to $+180°$, as ω is varied from 0 to ∞. At the corner frequency the phase angle is $+90°$ and independent of ζ, but at all other frequency it depends on ζ.

PROCEDURE FOR MAGNITUDE PLOT OF BODE PLOT

From the analysis of previous sections the following conclusions can be obtained.

1. The constant gain K, integral and derivative factors contribute gain (magnitude) at all frequencies.

2. In approximate plot the first, quadratic and higher order factors contribute gain (magnitude) only when the frequency is greater than the corner frequency.

Hence the low frequency response upto the lowest corner frequency is decided by K or $K / (j\omega)^n$ or $K(j\omega)^n$ term. Then at every corner frequency the slope of the magnitude plot is altered by the first, quadratic and higher order terms. Therefore the magnitude plot can be started with K or $K/(j\omega)^n$ or $K(j\omega)^n$ term and then the db magnitude of every first and higher order terms are added one by one in the increasing order of the corner frequency.

This is illustrated in the following example.

Let, $G(s) = \dfrac{K(1 + sT_1)^2}{s^2(1 + sT_2)(1 + sT_3)}$

$\therefore G(j\omega) = \dfrac{K(1 + j\omega T_1)^2}{(j\omega)^2(1 + j\omega T_2)(1 + j\omega T_3)}$

Let, $T_2 < T_3 < T_1$.

The corner frequencies are, $\omega_{c1} = \dfrac{1}{T_1}$, $\omega_{c2} = \dfrac{1}{T_2}$, $\omega_{c3} = \dfrac{1}{T_3}$.

Let, $\omega_{c1} < \omega_{c3} < \omega_{c2}$.

The magnitude plot of the individual terms of $G(j\omega)$, and their combined magnitude plot are shown in fig 3.18.

The step by step procedure for plotting the magnitude plot is given below

Step1 : Convert the transfer function into Bode form or time constant form. An exampleof Bode form of the transfer function is

$$G(s) = \dfrac{K(1 + sT_1)}{s(1 + sT_2)\left(1 + \dfrac{s^2}{\omega_n^2} + 2\zeta\dfrac{s}{\omega_n}\right)} \xrightarrow{s = j\omega} G(j\omega) = \dfrac{K(1 + j\omega T_1)}{j\omega(1 + j\omega T_2)\left(1 + \dfrac{\omega^2}{\omega_n^2} + j2\zeta\dfrac{\omega}{\omega_n}\right)}$$

Step 2 : List the corner frequencies in the increasing order and prepare a table as shown below.

Term	Corner frequency rad/sec	Slope db/dec	Change in slope db/dec

In the above table enter K or $K/(j\omega)^n$ or $K(j\omega)^n$ as the first term and the other terms in the increasing order of corner frequencies. Then enter the corner frequency, slope contributed by each term and change in slope at every corner frequency.

Step 3 : Choose an arbitrary frequency ω_1 which is lesser than the lowest corner frequency. Calculate the db magnitude of K or $K/(j\omega)^n$ or $K(j\omega)^n$ at ω_1 and at the lowest corner frequency.

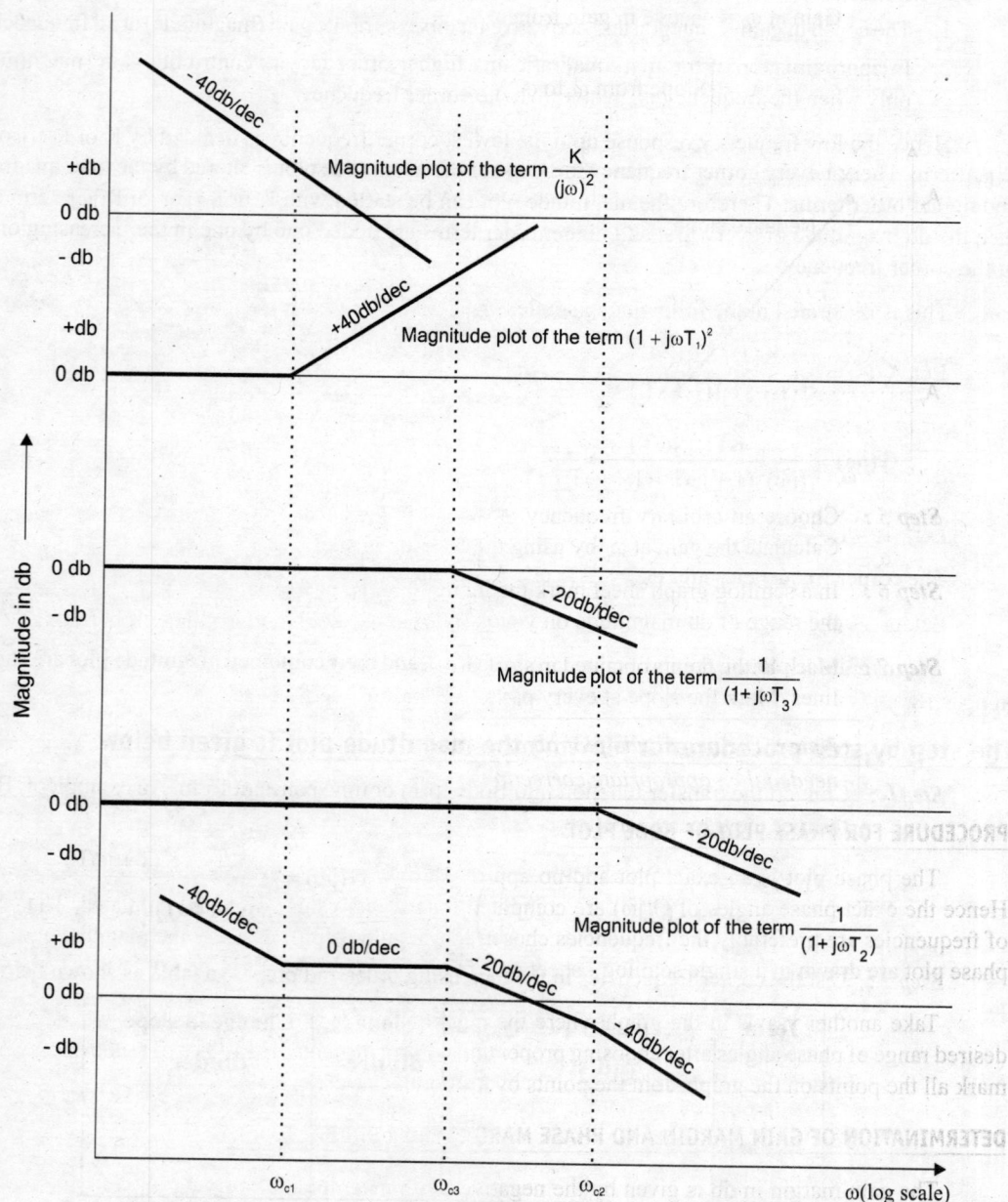

$$ \textbf{\textit{Fig 3.18 :}} \textit{ Magnitude plot of bode plot of, } G(j\omega) = \frac{K(1+j\omega T_1)^2}{(j\omega)^2(1+j\omega T_2)(1+j\omega T_3)} $$

Step 4 :   Then calculate the gain (db magnitude) at every corner frequency one by one by using the formula,

Gain at ω_y = change in gain from ω_x to ω_y + Gain at ω_x

$$A_y = \left[\text{Slope from } \omega_x \text{ to } \omega_y \times \log \frac{\omega_y}{\omega_x} \right] + \text{Gain at } \omega_x$$

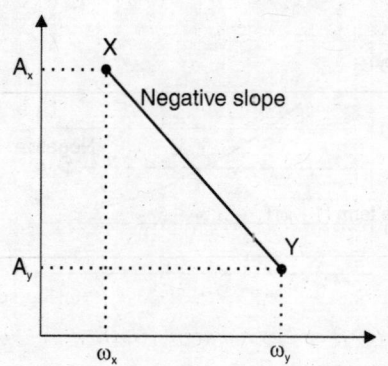 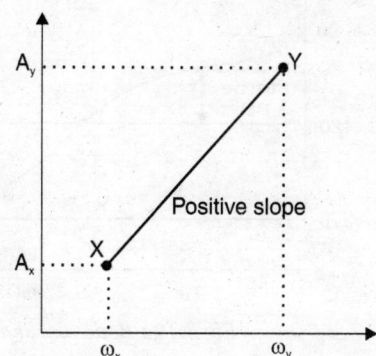

Step 5 :   Choose an arbitrary frequency ω_h which is greater than the highest corner frequency. Calculate the gain at ω_h by using the formula in step 4.

Step 6 :   In a semilog graph sheet mark the required range of frequency on x-axis (log scale) and the range of db magnitude on y-axis (ordinary scale) after choosing proper units.

Step 7 :   Mark all the points obtained in steps 3, 4, and 5 on the graph and join the points by straight lines. Mark the slope at every part of the graph.

> ***Note :*** *The magnitude plot obtained above is an approximate plot. If an exact plot is needed then appropriate corrections should be made at every corner frequencies.*

PROCEDURE FOR PHASE PLOT OF BODE PLOT

The phase plot is an exact plot and no approximations are made while drawing the phase plot. Hence the exact phase angles of $G(j\omega)$ are computed for various values of ω and tabulated. The choice of frequencies are preferably the frequencies chosen for magnitude plot. Usually the magnitude plot and phase plot are drawn in a single semilog - sheet on a common frequency scale.

Take another y-axis in the graph where the magnitude plot is drawn and in this y-axis mark the desired range of phase angles after choosing proper units. From the tabulated values of ω and phase angles, mark all the points on the graph. Join the points by a smooth curve.

DETERMINATION OF GAIN MARGIN AND PHASE MARGIN FROM BODE PLOT

The gain margin in db is given by the negative of db magnitude of $G(j\omega)$ at the phase cross-over frequency, ω_{pc}. The ω_{pc} is the frequency at which phase of $G(j\omega)$ is $-180°$. If the db magnitude of $G(j\omega)$ at ω_{pc} is negative then gain margin is positive and vice versa.

Let ϕ_{gc} be the phase angle of $G(j\omega)$ at gain cross over frequency ω_{gc}. The ω_{gc} is the frequency at which the db magnitude of $G(j\omega)$ is zero. Now the phase margin, γ is given by, $\gamma = 180° + \phi_{gc}$. If ϕ_{gc} is less negative than $-180°$ then phase margin is positive and vice versa.

The positive and negative gain margins and phase margins are illustrated in fig 3.19.

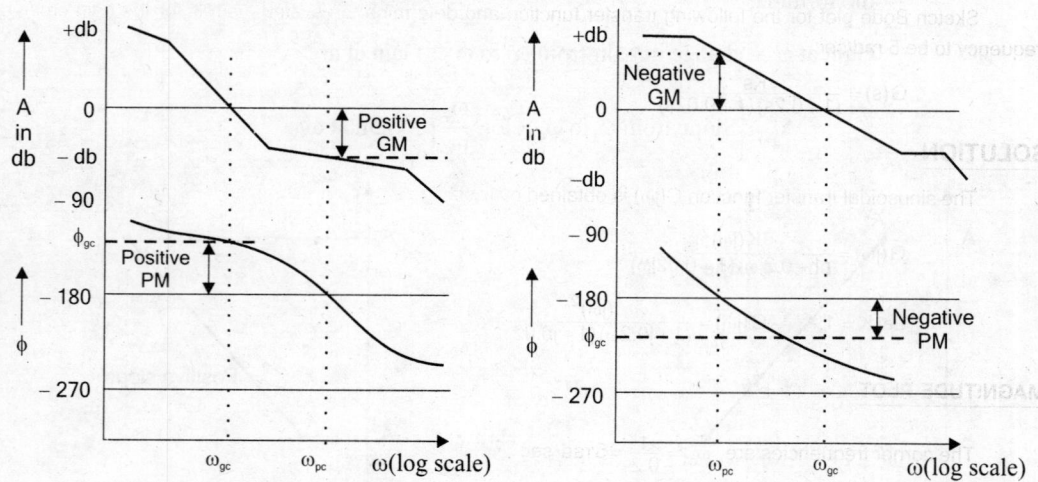

Fig 3.19 : Bode plot showing phase margin (PM) and gain margin (GM).

GAIN ADJUSTMENT IN BODE PLOT

In the open loop transfer function $G(j\omega)$ the constant K contributes only magnitude. Hence by changing the value of K the system gain can be adjusted to meet the desired specifications. The desired specifications are gain margin, phase margin, ω_{pc} and ω_{gc}. In a system transfer function if the value of K required to be estimated to satisfy a desired specification then draw the bode plot of the system with K = 1. The constant K can add 20 log K to every point of the magnitude plot and due to this addition the magnitude plot will shift vertically up or down. Hence shift the magnitude plot vertically up or down to meet the desired specification. Equate the vertical distance by which the magnitude plot is shifted to 20 log K and solve for K.

Let, x = change in db   (x is positive if the plot is shifted up and vice versa).

Now, $20 \log K = x$; $\log K = x/20$; $\therefore K = 10^{x/20}$

Note : *A point in complex plane can be represented by rectangular coordinates or by polar coordinates. Consider a point, z = a+jb in complex plane.*

Now, $|z| = \sqrt{a^2 + b^2}$ *and* $\angle z = tan^{-1} b/a.$

If the point lies in first or fourth quadrant then the argument as calculated by $tan^{-1} b/a$ will be the correct values. But if it lies either in second or third quadrant then a correction should be made in the calculated values of argument, because the calculator will always give the values of $tan^{-1} b/a$ either from 0 to + 90° or from 0 to −90°. The corrections to be made while converting from rectangular to polar coordinates is shown below.

A point in Ist quadrant, $a + jb = \sqrt{a^2 + b^2} \angle tan^{-1} b/a$

A point in IInd quadrant, $-a + jb = \sqrt{a^2 + b^2} \angle (\pi - tan^{-1} b/a)$

A point in IIIrd quadrant, $-a - jb = \sqrt{a^2 + b^2} \angle (\pi + tan^{-1} b/a)$

A point in IVth quadrant, $a - jb = \sqrt{a^2 + b^2} \angle - tan^{-1} b/a$

EXAMPLE 3.1

Sketch Bode plot for the following transfer function and determine the system gain K for the gain cross over frequency to be 5 rad/sec.

$$G(s) = \frac{Ks^2}{(1+0.2s)(1+0.02s)}$$

SOLUTION

The sinusoidal transfer function $G(j\omega)$ is obtained by replacing s by $j\omega$ in the given s-domain transfer function.

$$G(j\omega) = \frac{K(j\omega)^2}{(1+0.2j\omega)(1+0.02j\omega)}$$

Let K = 1, $\quad \therefore \ G(j\omega) = \frac{(j\omega)^2}{(1+j0.2\omega)(1+j0.02\omega)}$

MAGNITUDE PLOT

The corner frequencies are, $\omega_{c1} = \dfrac{1}{0.2} = 5 \text{ rad/sec}$ and $\omega_{c2} = \dfrac{1}{0.02} = 50 \text{ rad/sec}$

The various terms of $G(j\omega)$ are listed in Table-1 in the increasing order of their corner frequency. Also the table shows the slope contributed by each term and the change in slope at the corner frequency.

Table-1

Term	Corner frequency rad/sec	Slope db/dec	Change in slope db/dec
$(j\omega)^2$	–	+ 40	
$\dfrac{1}{1+j0.2\omega}$	$\omega_{c1} = \dfrac{1}{0.2} = 5$	– 20	40 – 20 = 20
$\dfrac{1}{1+j0.02\omega}$	$\omega_{c2} = \dfrac{1}{0.02} = 50$	– 20	20 – 20 = 0

Choose a low frequency ω_l such that $\omega_l < \omega_{c1}$ and choose a high frequency ω_h such that $\omega_h > \omega_{c2}$.

Let, ω_l = 0.5 rad/sec and ω_h = 100 rad/sec.

Let, $A = |G(j\omega)|$ in db

Let us calculate A at $\omega_l, \omega_{c1}, \omega_{c2}$ and ω_h.

At $\omega = \omega_l,\quad A = 20\log\left|(j\omega)^2\right| = 20\log(\omega)^2 = 20\log(0.5)^2 = -12 \text{ db}$

At $\omega = \omega_{c1},\quad A = 20\log\left|(j\omega)^2\right| = 20\log(\omega)^2 = 20\log(5)^2 = 28 \text{ db}$

At $\omega = \omega_{c2},\quad A = \left[\text{slope from } \omega_{c1} \text{ to } \omega_{c2} \times \log \dfrac{\omega_{c2}}{\omega_{c1}}\right] + A_{(at\ \omega=\omega_{c1})} = 20 \times \log \dfrac{50}{5} + 28 = 48 \text{ db}$

At $\omega = \omega_h,\quad A = \left[\text{slope from } \omega_{c2} \text{ to } \omega_h \times \log \dfrac{\omega_h}{\omega_{c2}}\right] + A_{(at\ \omega=\omega_{c2})} = 0 \times \log \dfrac{100}{50} + 48 = 48 \text{ db}$

Let the points a, b, c and d be the points corresponding to frequencies ω_1, ω_{c1}, ω_{c2} and ω_h respectively on the magnitude plot. In a semilog graph sheet choose a scale of 1unit = 10db on y-axis. The frequencies are marked in decades from 0.1 to 100 rad/sec on, logarithmic scales in x-axis. Fix the points a, b, c and d on the graph. Join the points by straight lines and mark the slope on the respective region.

PHASE PLOT

The phase angle of G(jω) as a function of ω is given by,

$$\phi = \angle G(j\omega) = 180^\circ - \tan^{-1}0.2\omega - \tan^{-1}0.02\omega$$

The phase angle of G(jω) are calculated for various values of ω and listed in table-2.

TABLE-2

ω rad/sec	$\tan^{-1}0.2\omega$ deg	$\tan^{-1}0.02\omega$ deg	$\phi = \angle G(j\omega)$ deg	Point in phase plot
0.5	5.7	0.6	173.7 ≈ 174	e
1	11.3	1.1	167.6 ≈ 168	f
5	45	5.7	129.3 ≈ 130	g
10	63.4	11.3	105.3 ≈ 106	h
50	84.3	45	50.7 ≈ 50	i
100	87.1	63.4	29.5 ≈ 30	j

On the same semilog graph sheet choose a scale of 1unit = 20°, on the y-axis on the right side of semilog graph sheet. Mark the calculated phase angle on the graph sheet. Join the points by a smooth curve.

CALCULATION OF K

Given that the gain crossover frequency is 5 rad/sec. At ω = 5 rad/sec the gain is 28 db. If gain crossover frequency is 5 rad/sec then at that frequency the db gain should be zero. Hence to every point of magnitude plot a db gain of –28db should be added. The addition of –28db shifts the plot downwards. The corrected magnitude plot is obtained by shifting the plot with K = 1 by –28db downwards. The magnitude correction is independent of frequency. Hence the magnitude of –28db is contributed by the term K. The value of K is calculated by equating 20 logK to –28 db.

$$\therefore \ 20\log K = -28 \ db$$

$$\log K = \frac{-28}{20} \ ; \ K = 10^{-\left(\frac{28}{20}\right)} = 0.0398$$

The magnitude plot with K = 1 and 0.0398 and the phase plot are shown in fig 3.1.1

> *Note : The frequency ω = 5 rad/sec is a corner frequency. Hence in the exact plot the db gain at ω = 5 rad/sec will be 3db less than the actual plot. Therefore for exact plot the 20 log K will contribute a gain of –25db.*
>
> $$\therefore \ 20\log K = -25 \ db$$
>
> $$\log K = \frac{-25}{20} \ ; \ K = 10^{-\left(\frac{25}{20}\right)} = 0.0562$$

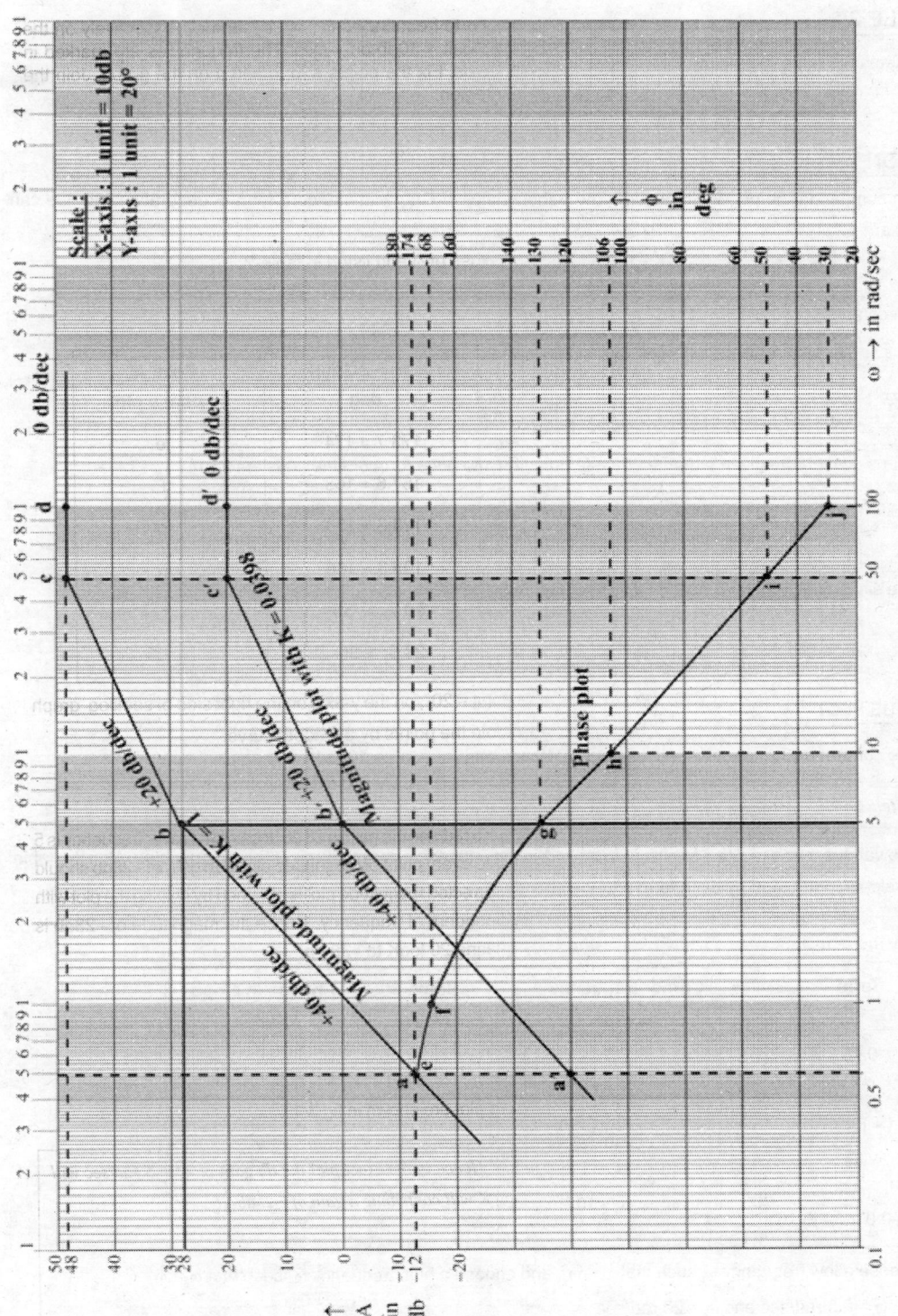

Fig 3.1.1 : Bode plot of transfer function, $G(j\omega) = \dfrac{K(j\omega)^2}{(1 + j0.2\omega)(1 + j0.02\omega)}$

EXAMPLE 3.2

Sketch the bode plot for the following transfer function and determine phase margin and gain margin.

$$G(s) = \frac{75(1+0.2s)}{s(s^2+16s+100)}$$

SOLUTION

On comparing the quadratic factor in the denominator of G(s) with standard form of quadratic factor we can estimate ζ and ω_n.

$$\therefore\ s^2+16s+100 = s^2+2\zeta\omega_n s + \omega_n^2$$

On comparing we get,

$$\omega_n^2 = 100 \quad\Rightarrow\quad \omega_n = 10$$

$$2\zeta\omega_n = 16 \quad\Rightarrow\quad \zeta = \frac{16}{2\omega_n} = \frac{16}{2\times 10} = 0.8$$

Let us convert the given s-domain transfer function into bode form or time constant form.

$$\therefore\ G(s) = \frac{75(1+0.2s)}{s(s^2+16s+100)} = \frac{75(1+0.2s)}{s\times 100\left(\frac{s^2}{100}+\frac{16s}{100}+1\right)} = \frac{0.75(1+0.2s)}{s(1+0.01s^2+0.16s)}$$

The sinusoidal transfer function G(jω) is obtained by replacing s by jω in G(s).

$$\therefore\ G(j\omega) = \frac{75(1+0.2j\omega)}{j\omega(1+0.01(j\omega)^2+0.16j\omega)} = \frac{0.75(1+j0.2\omega)}{j\omega(1-0.01\omega^2+j0.16\omega)}$$

MAGNITUDE PLOT

The corner frequencies are, $\omega_{c1} = \frac{1}{0.2} = 5$ rad/sec and $\omega_{c2} = \omega_n = 10$ rad/sec

> *Note : For the quadratic factor the corner frequency is ω_n.*

The various terms of G(jω) are listed in table-1 in the increasing order of their corner frequencies. Also the table shows the slope contributed by each term and the change in slope at the corner frequency.

TABLE-1

Term	Corner frequency rad/sec	Slope db/dec	Change in slope db/dec
$\frac{0.75}{j\omega}$	–	-20	
$1+j0.2\omega$	$\omega_{c1} = \frac{1}{0.2} = 5$	20	$-20+20 = 0$
$\frac{1}{1-0.01\omega^2+j0.16\omega}$	$\omega_{c2} = \omega_n = 10$	-40	$0-40 = -40$

Choose a low frequency ω_l such that $\omega_l < \omega_{c1}$ and choose a high frequency ω_h such that $\omega_h > \omega_{c2}$.

Let, $\omega_l = 0.5$ rad/sec and $\omega_h = 20$ rad/sec.

Let, $A = \left|G(j\omega)\right|$ in db

Let us calculate A at ω_l, ω_{c1}, ω_{c2} and ω_h.

At, $\omega = \omega_1$, $A = 20 \log \left| \dfrac{0.75}{j\omega} \right| = 20 \log \dfrac{0.75}{0.5} = 3.5 \, db$

At, $\omega = \omega_{c1}$, $A = 20 \log \left| \dfrac{0.75}{j\omega} \right| = 20 \log \dfrac{0.75}{5} = -16.5 \, db$

At, $\omega = \omega_{c2}$, $A = \left[\text{slope from } \omega_{c1} \text{ to } \omega_{c2} \times \log \dfrac{\omega_{c2}}{\omega_{c1}} \right] + A_{(at \, \omega = \omega_{c1})}$

$$= 0 \times \log \dfrac{10}{5} + (-16.5) = -16.5 \, db$$

At, $\omega = \omega_h$, $A = \left[\text{slope from } \omega_{c2} \text{ to } \omega_h \times \log \dfrac{\omega_h}{\omega_{c2}} \right] + A_{(at \, \omega = \omega_{c2})}$

$$= -40 \times \log \dfrac{20}{10} + (-16.5) = -28.5 \, db$$

Let the points a, b, c and d be the points corresponding to frequencies ω_l, ω_{c1}, ω_{c2} and ω_h respectively on the magnitude plot. In a semilog graph sheet choose a scale of 1unit = 5 db on y-axis. The frequencies are marked in decades from 0.1 to 100 rad/sec on logarithmic scales in x-axis. Fix the points a,b,c and d on the graph. Join the points by straight lines and mark the slope on the respective region.

PHASE PLOT

The phase angle of $G(j\omega)$ as a function of ω is given by,

$$\phi = \angle G(j\omega) = \tan^{-1} 0.2\omega - 90° - \tan^{-1} \dfrac{0.16\omega}{1 - 0.01\omega^2} \quad \text{for } \omega \le \omega_n$$

$$\phi = \angle G(j\omega) = \tan^{-1} 0.2\omega - 90° - \left(\tan^{-1} \dfrac{0.16\omega}{1 - 0.01\omega^2} + 180° \right) \quad \text{for } \omega > \omega_n$$

> Note : In quadratic factors the phase varies from 0° to −180°. But calculator calculates $\tan^{-1}$ only between 0° to 90°. Hence a correction of 180° should be added to phase after ω_n.

The phase angle of $G(j\omega)$ are calculated for various values of ω and listed in Table-2.

TABLE-2

ω rad/sec	$\tan^{-1} 0.2\omega$ deg	$\tan^{-1} \dfrac{0.16\omega}{1 - 0.01\omega^2}$ deg	$\phi = \angle G(j\omega)$ deg	Points in phase plot
0.5	5.7	4.6	$-88.9 \approx -88$	e
1	11.3	9.2	$-87.9 \approx -88$	f
5	45	46.8	$-91.8 \approx -92$	g
10	63.4	90	$-116.6 \approx -116$	h
20	75.9	$-46.8+180 = 133.2$	$-147.3 \approx -148$	i
50	84.3	$-18.4+180 = 161.6$	$-167.3 \approx -168$	j
100	87.1	$-92+180 = 170.8$	$-173.7 \approx -174$	k

On the same semilog graph sheet choose a scale of 1unit = $20°$ on the y-axis on the right side of semilog graph sheet. Mark the calculated phase angle on the graph sheet. Join the points by a smooth curve.

The magnitude plot and the phase plot are shown in fig 3.2.1.

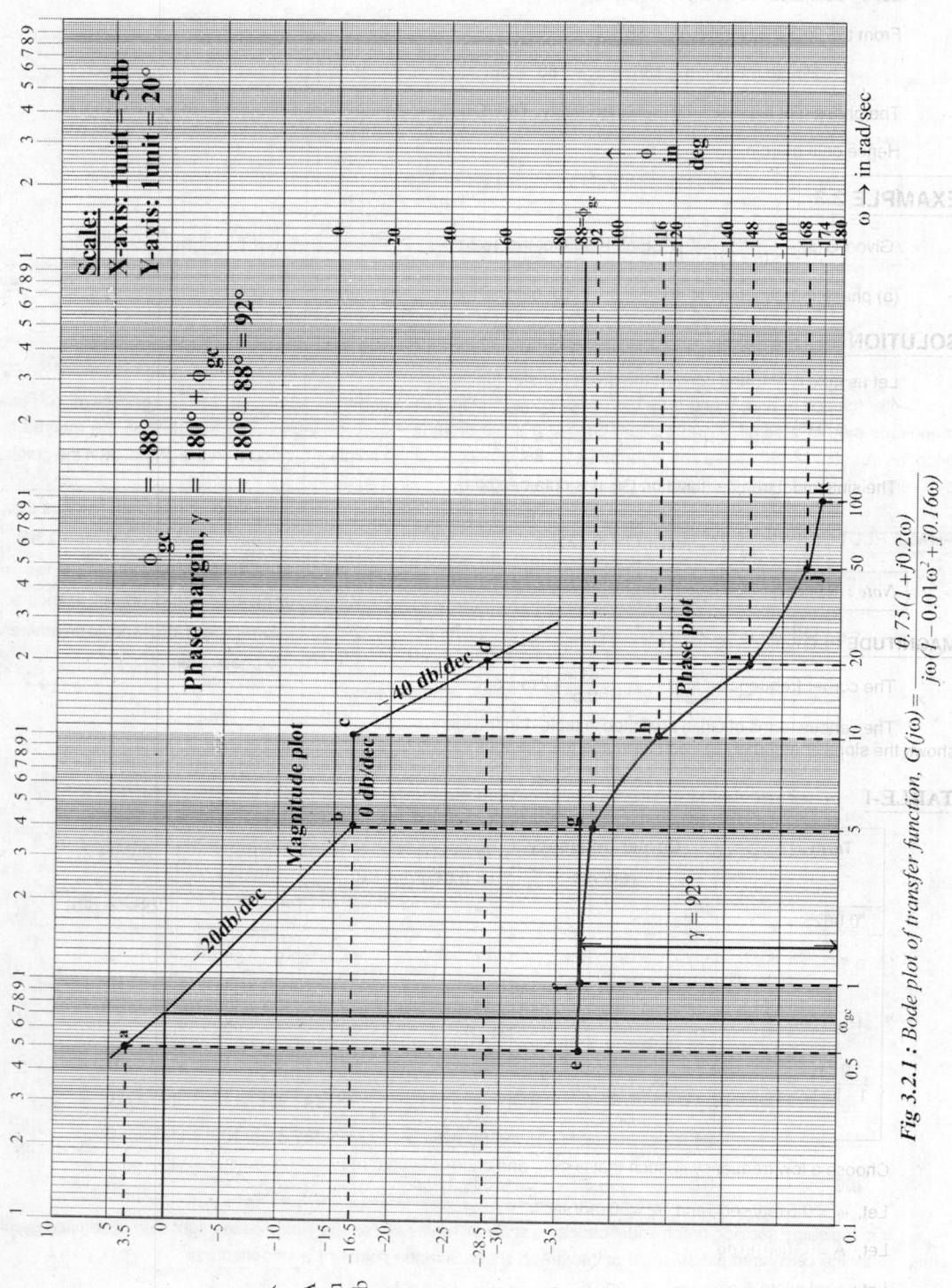

Fig 3.2.1 : Bode plot of transfer function, $G(j\omega) = \dfrac{0.75(1+j0.2\omega)}{j\omega(1-0.01\omega^2+j0.16\omega)}$

Let ϕ_{gc} be the phase of $G(j\omega)$ at gain cross-over frequency, ω_{gc}.

From the fig 3.2.1, we get, $\phi_{gc} = 88°$

$\qquad \therefore$ Phase margin, $g = 180° + \phi_{gc} = 180° - 88° = 92°$

The phase plot crosses $-180°$ only at infinity. The $|G(j\omega)|$ at infinity is $-\infty$ db.

Hence gain margin is $+\infty$.

EXAMPLE 3.3

Given $G(s) = \dfrac{Ke^{-0.2s}}{s(s+2)(s+8)}$. Find K so that the system is stable with,

(a) phase margin equal to $45°$ (b) gain margin equal to 2db,

SOLUTION

Let us take K = 1, and convert the given transfer function to time constant form or bode form.

$$\therefore G(s) = \frac{e^{-0.2s}}{s(s+2)(s+8)} = \frac{e^{-0.2s}}{s \times 2\left(1+\frac{s}{2}\right) \times 8\left(1+\frac{s}{8}\right)} = \frac{0.0625e^{-0.2s}}{s(1+0.5s)(1+0.125s)}$$

The sinusoidal transfer function $G(j\omega)$ is obtained by replacing s by $j\omega$ in $G(s)$.

$$\therefore G(j\omega) = \frac{0.0625e^{-j0.2\omega}}{j\omega(1+j0.5\omega)(1+j0.125\omega)}$$

> *Note :* $|0.0625e^{-j0.2\omega}| = 0.0625$ *and* $\angle (0.0625e^{-j0.2\omega}) = -0.2\omega$ *radians.*

MAGNITUDE PLOT

The corner frequencies are, $\omega_{c1} = \dfrac{1}{0.5} = 2\,rad/sec$ and $\omega_{c2} = \dfrac{1}{0.125} = 8\,rad/sec$

The various terms of $G(j\omega)$ are listed in table-1 in the increasing order of their corner frequencies. Also the table shows the slope contributed by each term and the change in slope at the corner frequency.

TABLE-1

Term	Corner frequency rad/sec	Slope db/dec	Change in slope db/dec
$\dfrac{0.0625}{j\omega}$	—	-20	
$\dfrac{1}{1+j0.5\omega}$	$\omega_{c1} = \dfrac{1}{0.5} = 2$	-20	$-20-20 = -40$
$\dfrac{1}{1+j0.125\omega}$	$\omega_{c2} = \dfrac{1}{0.125} = 8$	-20	$-40-20 = -60$

Choose a low frequency ω_l such that $\omega_l < \omega_{c1}$ and choose a high frequency ω_h such that $\omega_h > \omega_{c2}$.

Let, $\omega_l = 0.5\,rad/sec$ and $\omega_h = 50\,rad/sec$

Let, $A = |G(j\omega)|$ in db

Let us calculate A at $\omega_l, \omega_{c1}, \omega_{c2}$ and ω_h.

At $\omega = \omega_1$, $A = 20\log\left|\dfrac{0.0625}{j\omega}\right| = 20\log\dfrac{0.0625}{0.5} = -18$ db

At $\omega = \omega_{c1}$, $A = 20\log\left|\dfrac{0.0625}{j\omega}\right| = 20\log\dfrac{0.0625}{2} = -30$ db

At $\omega = \omega_{c2}$, $A = \left[\text{Slope from } \omega_{c1} \text{ to } \omega_{c2} \times \log\dfrac{\omega_{c2}}{\omega_{c1}}\right] + A_{(at\ \omega=\omega_{c1})} = -40 \times \log\dfrac{8}{2} + (-30) = -54$ db

At $\omega = \omega_h$, $A = \left[\text{Slope from } \omega_{c2} \text{ to } \omega_h \times \log\dfrac{\omega_h}{\omega_{c2}}\right] + A_{(at\ \omega=\omega_{c2})} = -60 \times \log\dfrac{50}{8} + (-54) = -102$ db

Let the points a,b,c,d be the points corresponding to frequencies ω_1, ω_{c1}, ω_{c2} and ω_h respectively on the magnitude plot. In a semilog graph sheet choose a scale of 1unit = 10db on y-axis. The frequencies are marked in decades from 0.01 to 100 rad/sec on logarithmic scale in x-axis. Fix the points a, b, c, and d on the graph. Join the points by straight line and mark the slope on the respective region.

PHASE PLOT

The phase angle of $G(j\omega)$ as a function of ω is given by,

$$\phi = -0.2\omega \times \dfrac{180^\circ}{\pi} - 90^\circ - \tan^{-1}0.125\omega$$

The phase angle of $G(j\omega)$ are calculated for various values of ω and listed in table-2.

TABLE-2

ω rad/sec	$-0.2\,\omega\,(180^\circ/\pi)$ deg	$\tan^{-1} 0.5\,\omega$ deg	$\tan^{-1} 0.125\,\omega$ deg	$\phi = \angle G(j\omega)$ deg	Point in phase plot
0.01	− 0.1145	0.2864	0.0716	−90.4 ≈ −90	e
0.1	−1.145	2.862	0.716	−94.7 ≈ −94	f
0.5	−5.7	14	3.6	−113.3 ≈ −114	g
1	−11.4	26	7.12	−134.4 ≈ −134	h
2	− 22.9	45	14	−171.9 ≈ −172	i
3	− 34.37	56.30	20.56	−201.2 ≈ −202	j
4	− 45.84	63.43	26.57	−225.8 ≈ − 226	k

On the same semilog graph sheet choose a scale of 1unit = 20° on the y-axis on the right side of the semilog graph sheet. Mark the calculated phase angle on the graph sheet. Join the points by smooth curve.

The magnitude and phase plot are shown in fig 3.3.1.

CALCULATION OF K

Phase margin, $\gamma = 180^0 + \phi_{gc}$, where ϕ_{gc} is the phase of $G(j\omega)$ at $\omega = \omega_{gc}$.

When $\gamma = 45^\circ$, $\phi_{gc} = \gamma - 180^0 = 45^0 - 180^0 = -135^\circ$.

With K = 1, the db gain at $\phi = -135^\circ$ is −24 db. This gain should be made zero to have to PM of 45°. Hence to every point of magnitude plot a db gain of 24 db should be added. The corrected magnitude plot is obtained by shifting the plot with K = 1 by 24 db upwards. The magnitude correction is independent of frequency. Hence the magnitude of 24 db is contributed by the term K. The value of K is calculated by equating 20logK to 24 db.

$\therefore$ 20 log K = 24 ; K = $10^{24/20}$; K = 15.84

With K = 1, the gain margin = −(−32) = 32 db. But the required gain margin is 2 db. Hence to every point of magnitude plot a db gain of 30 db should be added. This addition of 30 db shifts the plot upwards. The magnitude correction is independent of frequency. Hence the magnitude of 30 db is contributed by the term K. The value of K is calculated by equating 20logK to 30 db.

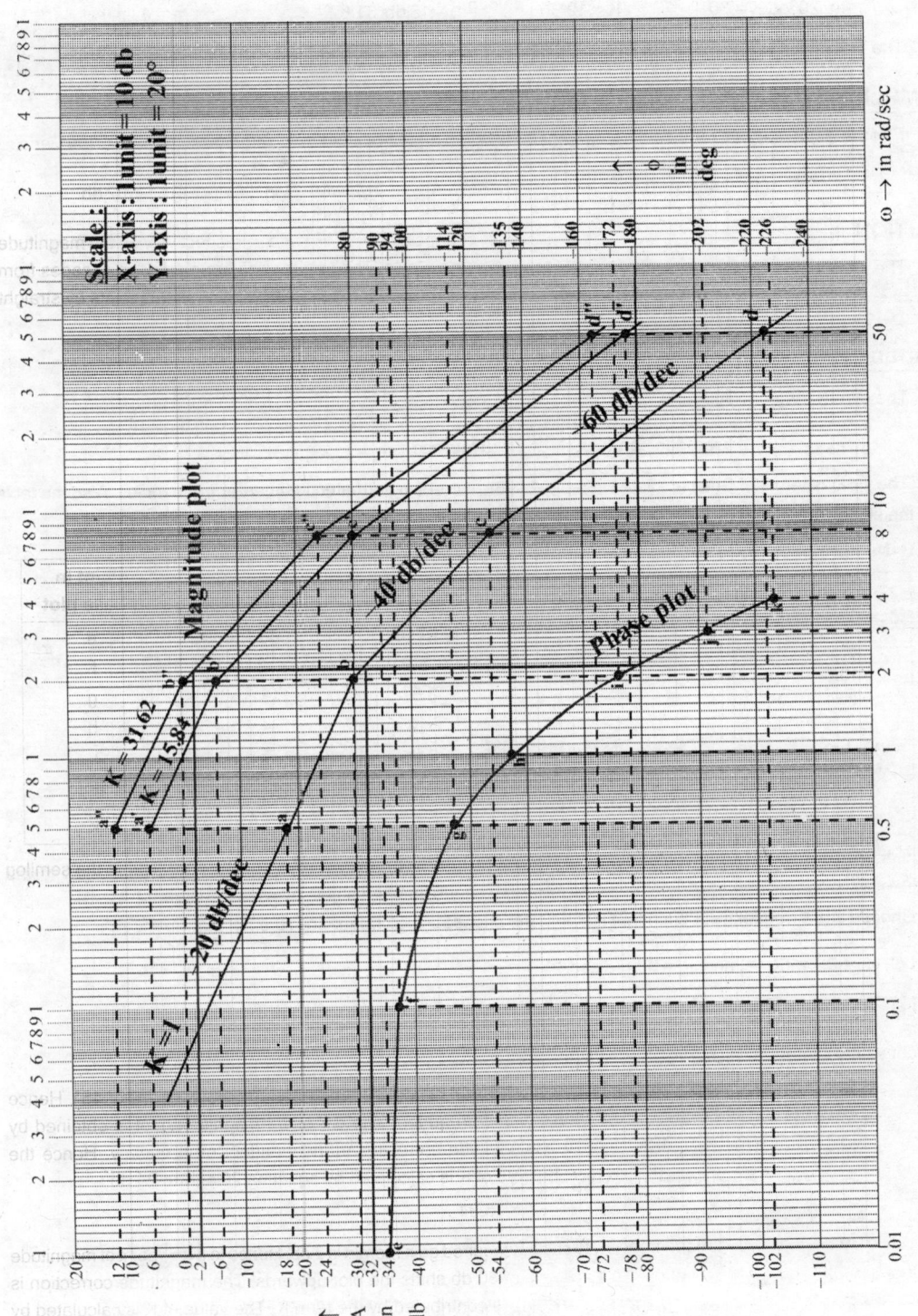

Fig 3.3.1 : Bode plot of transfer function, $G(j\omega) = \dfrac{0.0625 K e^{-j0.2\sim}}{j\sim (1 + j0.5\omega)(1 + j0.125\omega)}$

$$\therefore \ 20 \log K = 30 \qquad ; \qquad K = 10^{30/20} \qquad ; \qquad K = 31.62$$

The magnitude plot with K = 15.84 and 31.62 are shown in fig 3.3.1.

EXAMPLE 3.4

Plot the Bode diagram for the following transfer function and obtain the gain and phase cross over frequencies.

$$G(s) = \frac{10}{s(1 + 0.4s)(1 + 0.1s)}$$

SOLUTION

The sinusoidal transfer function of $G(j\omega)$ is obtained by replacing s by $j\omega$ in the given transfer function.

$$\therefore \ G(j\omega) = \frac{10}{j\omega(1 + j0.4\omega)(1 + j0.1\omega)}$$

MAGNITUDE PLOT

The corner frequencies are,

$$\omega_{c1} = \frac{1}{0.4} = 2.5 \, \text{rad/sec} \quad \text{and} \quad \omega_{c2} = \frac{1}{0.1} = 10 \, \text{rad/sec}$$

The various terms of $G(j\omega)$ are listed in table-1 in the increasing order of their corner frequencies. Also the table shows the slope contributed by each term and the change in slope at the corner frequency.

TABLE-1

Term	Corner frequency rad/sec	Slope db/dec	Change in slope db/dec
$\dfrac{10}{j\omega}$	–	– 20	
$\dfrac{1}{1 + j0.4\omega}$	$\omega_{c1} = \dfrac{1}{0.4} = 2.5$	– 20	$-20 -20 = - 40$
$\dfrac{1}{1 + j0.1\omega}$	$\omega_{c2} = \dfrac{1}{0.1} = 10$	– 20	$- 40 - 20 = - 60$

Choose a low frequency ω_l such that $\omega_l < \omega_{c1}$ and choose a high frequency ω_h such that $\omega_h > \omega_{c2}$.

Let, ω_l = 0.1 rad/sec, and ω_h = 50 rad/sec.

Let, $A = |G(j\omega)|$ in db.

Let us calculate A at ω_l, ω_{c1}, ω_{c2} and ω_h.

At $\omega = \omega_1$, $A = 20 \log \left| \dfrac{10}{j\omega} \right| = 20 \log \dfrac{10}{0.1} = 40 \, \text{db}$

At $\omega = \omega_2$, $A = 20 \log \left| \dfrac{10}{j\omega} \right| = 20 \log \dfrac{10}{2.5} = 12 \, \text{db}$

At $\omega = \omega_{c2}$, $A = \left[\text{Slope from } \omega_{c1} \text{ to } \omega_{c2} \times \log \dfrac{\omega_{c2}}{\omega_{c1}} \right] + A_{(\text{at } \omega = \omega_{c1})} = -40 \times \log \dfrac{10}{2.5} + 12 = - 12 \, \text{db}$

At $\omega = \omega_h$, $A = \left[\text{Slope from } \omega_{c2} \text{ to } \omega_h \times \log \dfrac{\omega_h}{\omega_{c2}} \right] + A_{(\text{at } \omega = \omega_{c2})} = - 60 \times \log \dfrac{50}{10} + (- 12) = - 54 \, \text{db}$

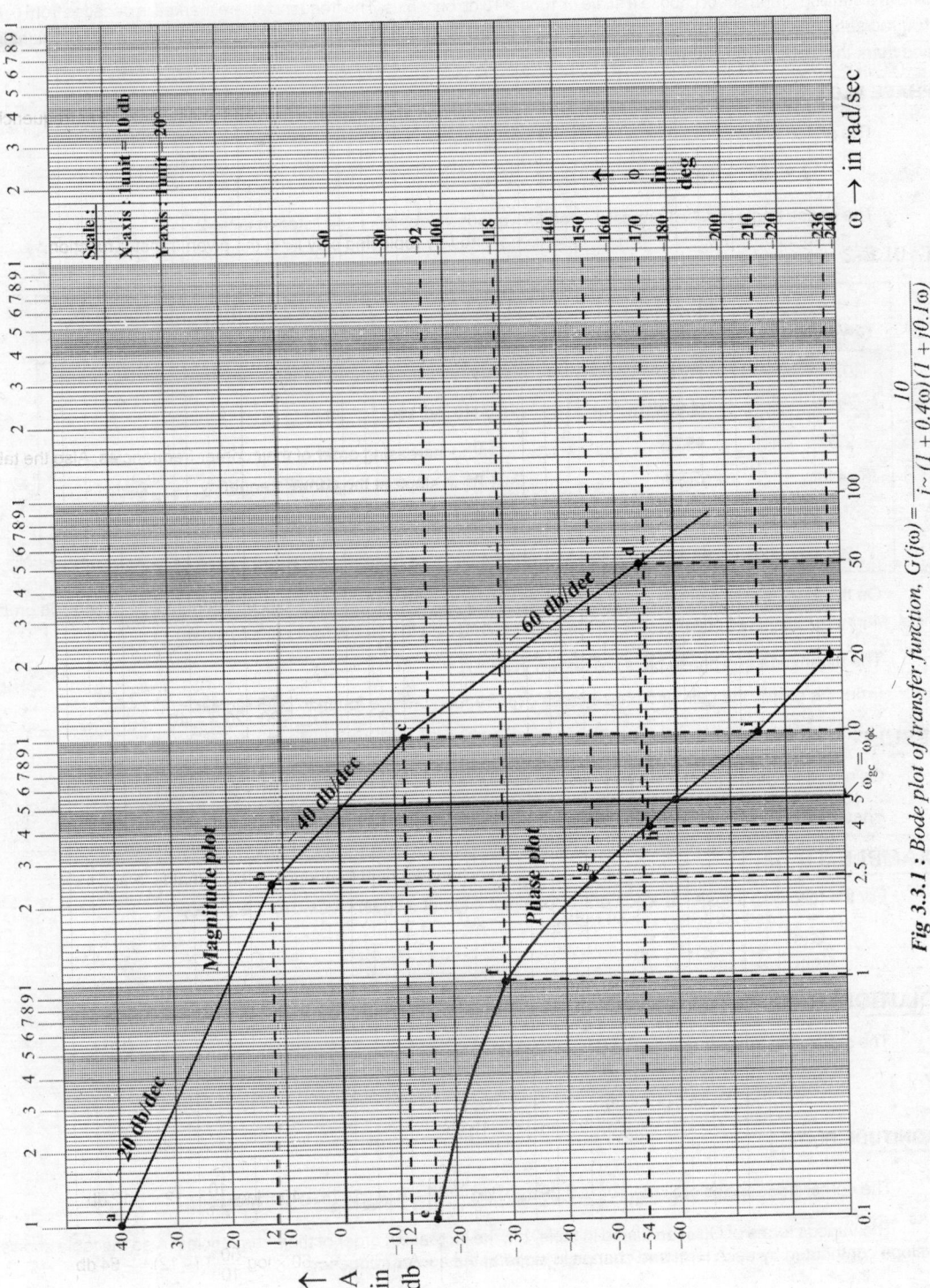

Fig 3.3.1 : Bode plot of transfer function, $G(j\omega) = \dfrac{10}{j\omega\,(1 + 0.4\omega)\,(1 + j0.1\omega)}$

Let the points a, b, c and d be the points corresponding to frequencies ω_l, ω_{c1}, ω_{c2} and ω_h respectively on the magnitude plot. In a semilog graph sheet choose a scale of 1unit = 10 db on y-axis. The frequencies are marked in decades from 0.1 to 100 rad/sec on logarithmic scales in x-axis. Fix the points a, b, c and d on the graph. Join the points by a straight line and mark the slope in the respective region.

PHASE PLOT

The phase angle of $G(j\omega)$ as a function of ω is given by,

$$\phi = -90° - \tan^{-1} 0.4\omega - \tan^{-1} 0.1\omega$$

The phase angle of $G(j\omega)$ are calculated for various values of ω and listed in table-2.

TABLE-2

ω rad/sec	$\tan^{-1} 0.4\,\omega$ deg	$\tan^{-1} 0.1\,\omega$ deg	$\phi = \angle G(j\omega)$ deg	Points in phase plot
0.1	2.29	0.57	$-92.86 \approx -92$	e
1	21.80	5.71	$-117.5 \approx -118$	f
2.5	45.0	14.0	$-149 \approx -150$	g
4	57.99	21.8	$-169.79 \approx -170$	h
10	75.96	45.0	$-210.96 \approx -210$	i
20	82.87	63.43	$-236.3 \approx -236$	j

On the same semilog graph sheet choose a scale of 1unit = 20° on the y-axis on the right side of semilog graph sheet. Mark the calculated phase angle on the graph sheet. Join the points by a smooth curve.

The magnitude and phase plots are shown in fig 3.4.1.

From the graph, the gain and phase cross over frequencies are found to be 5 rad/sec.

RESULT

Gain cross-over frequency = 5 rad/sec.

Phase cross-over frequency = 5 rad/sec.

EXAMPLE 3.5

For the following transfer function draw bode plot and obtain gain cross-over frequency.

$$G(s) = \frac{20}{s(1+3s)(1+4s)}$$

SOLUTION

The sinusoidal transfer function of $G(j\omega)$ is obtained by replacing s by $j\omega$ in the given transfer function.

$$G(j\omega) = \frac{20}{j\omega(1+3j\omega)(1+4j\omega)}$$

MAGNITUDE PLOT

The corner frequencies are, $\omega_{c1} = \frac{1}{4} = 0.25 \text{ rad/sec}$, $\omega_{c2} = \frac{1}{3} = 0.333 \text{ rad/sec}$,

The various terms of $G(j\omega)$ are listed in table-1 in the increasing order of their frequencies. Also the table shows the slope contributed by each term and change in slope at the corner frequency.

TABLE-1

Term	Corner frequency rad/sec	Slope db/dec	Change in slope db/dec
$\dfrac{20}{j\omega}$	–	– 20	
$\dfrac{1}{1+j4\omega}$	$\omega_{c1} = \dfrac{1}{4} = 0.25$	– 20	$-20 -20 = -40$
$\dfrac{1}{1+j3\omega}$	$\omega_{c2} = \dfrac{1}{3} = 0.33$	– 20	$-40 -20 = -60$

Choose a frequency ω_l such that $\omega_l < \omega_{c1}$ and choose a high frequency ω_h such that $\omega_h > \omega_{c2}$.

Let, $\omega_l = 0.15$ rad/sec and $\omega_h = 1$ rad/sec.

Let, $A = |G(j\omega)|$ in db.

Let us calculate A at ω_l, ω_{c1}, ω_{c2} and ω_h.

At $\omega = \omega_1$, $A = \left| G(j\omega) \right| = 20 \log \left| \dfrac{20}{0.15} \right| = 42.5$ db

At $\omega = \omega_{c1}$, $A = \left| G(j\omega) \right| = 20 \log \left| \dfrac{20}{0.25} \right| = 38$ db

At $\omega = \omega_{c2}$, $A = \left| \text{Slope from } \omega_{c1} \text{ to } \omega_{c2} \times \log \dfrac{\omega_{c2}}{\omega_{c1}} \right| + A_{(at\ \omega=\omega_{c1})}$

$= -40 \times \log \dfrac{0.33}{0.25} + 38 = 33$ db

At $\omega = \omega_h$, $A = \left| \text{Slope from } \omega_{c2} \text{ to } \omega_h \times \log \dfrac{\omega_h}{\omega_{c2}} \right| + A_{(at\ \omega=\omega_{c2})}$

$= -60 \times \log \dfrac{1}{0.33} + 33 = 4$ db

Let the points a, b, c and c be the points corresponding to frequencies ω_l, ω_{c1}, ω_{c2} and ω_h respectively on the magnitude plot. In a semilog graph sheet choose a scale of 1unit = 10 db on y-axis. The frequencies are marked in decades from 0.01 to 10 rad/sec on logarithmic scales on x-axis. Fix the points a, b, c and d on the graph sheet. Join the points by a straight line and mark the slope in the respective region.

PHASE PLOT

The phase angle of $G(j\omega)$, $\phi = -90^0 - \tan^{-1} 3\omega - \tan^{-1} 4\omega$

The phase angle of $G(j\omega)$ are calculated for various values of ω and listed in table-2.

TABLE-2

ω, rad/sec	$\tan^{-1} 3\omega$, deg	$\tan^{-1} 4\omega$, deg	$\omega = \angle G(j\omega)$, deg	Points in phase plot
0.15	24.22	30.96	$-145.18 \approx -146$	e
0.2	30.96	38.66	$-159.61 \approx -160$	f
0.25	36.86	45.0	$-171.86 \approx -172$	g
0.33	44.7	52.8	$-187.5 \approx -188$	h
0.6	60.14	67.38	$-218.32 \approx -218$	i
1	71.56	75.96	$-237.56 \approx -238$	j

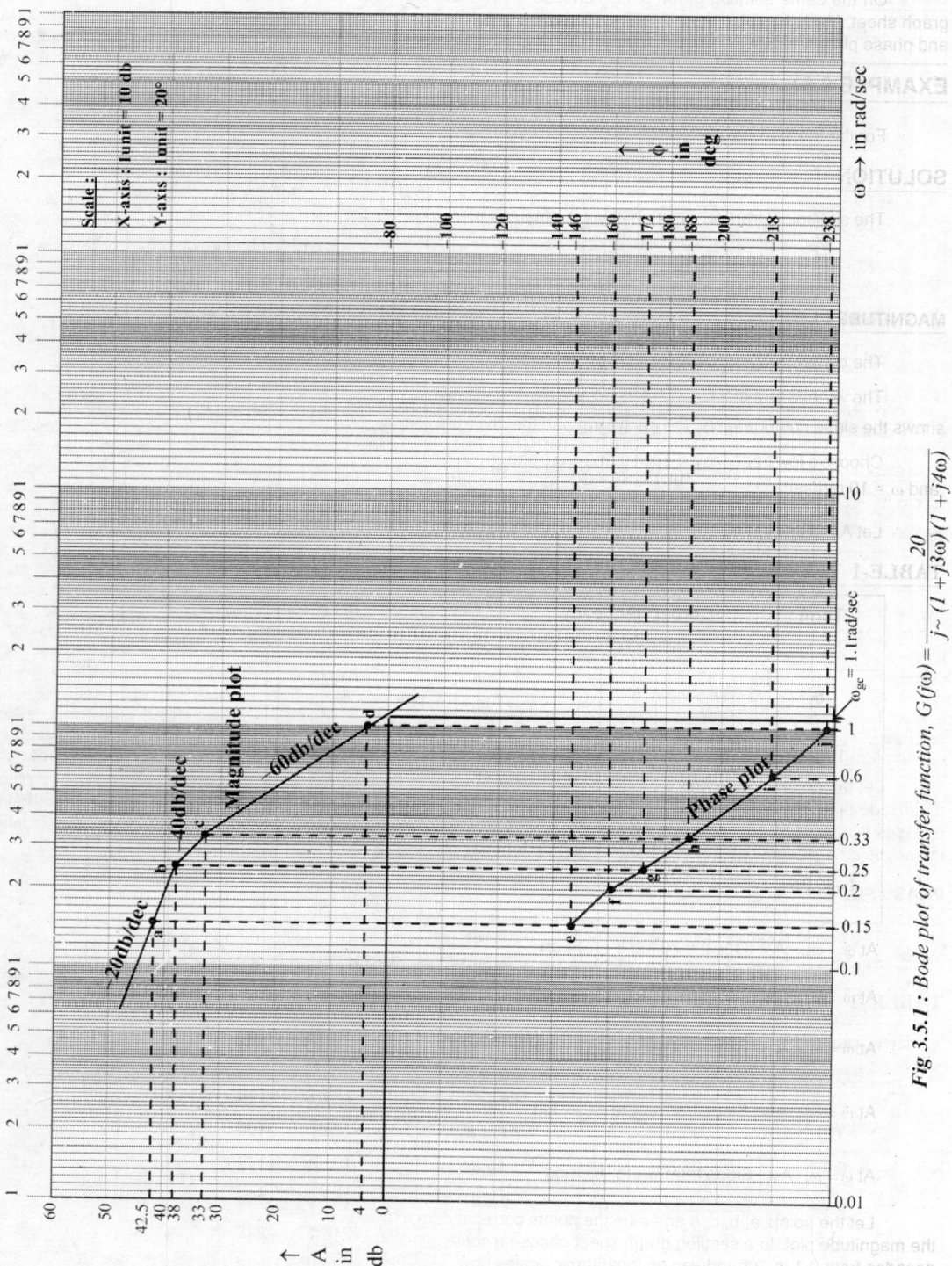

$$\textit{Fig 3.5.1 : Bode plot of transfer function, } G(j\omega) = \frac{20}{j\sim (1 + j3\omega)(1 + j4\omega)}$$

On the same semilog graph sheet choose a scale of 1unit = 20° on the y-axis on the right side of semilog graph sheet. Mark the calculated phase angle on the graph sheet. Join the points by a smooth curve. The magnitude and phase plots are shown in fig 3 5.1. From the graph the gain cross-over frequency is found to be ω_{gc} = 1.1 rad/sec.

EXAMPLE 3.6

For the function, $G(s) = \dfrac{5(1+2s)}{(1+4s)(1+0.25s)}$ draw the bode plot.

SOLUTION

The sinusoidal transfer function G(jω) is obtained by replacing s by jω in G(s).

$$\therefore G(j\omega) = \frac{5(1+2j\omega)}{(1+4j\omega)(1+0.25j\omega)}$$

MAGNITUDE PLOT

The corner frequencies are, $\omega_{c1} = \dfrac{1}{4} = 0.25$ rad/sec, $\omega_{c2} = \dfrac{1}{2} = 0.5$ rad/sec, $\omega_{c3} = \dfrac{1}{0.25} = 4$ rad/sec

The various terms of G(jω) are listed in table-1 in the increasing order of their corner frequencies. Also the table shows the slope contributed by the each term and the change in slope at the corner frequency.

Choose a low frequency ω_l such that $\omega_l < \omega_{c1}$ and choose a high frequency ω_h such that $\omega_h > \omega_{c2}$. Let $\omega_l = 0.1$ rad/sec and $\omega_h = 10$ rad/sec.

Let A = |G(jω)| in db and let us calculate A at ω_l, ω_{c1}, ω_{c2}, ω_{c3} and ω_h.

TABLE-1

Term	Corner frequency rad/sec	Slope db/dec	Change in slope db/deg
5	–	0	–
$\dfrac{1}{1+j4\omega}$	$\omega_{c1} = \dfrac{1}{4} = 0.25$	–20	0 – 20 = –20
$1 + j2\omega$	$\omega_{c2} = \dfrac{1}{2} = 0.5$	20	–20 + 20 = 0
$\dfrac{1}{1+j0.25\omega}$	$\omega_{c3} = \dfrac{1}{0.25} = 4$	–20	0 – 20 = –20

At $\omega = \omega_l$, A = |G(jω)| = 20 log 5 = +14 db

At $\omega = \omega_{c1}$, A = |G(jω)| = 20 log 5 = +14 db

At $\omega = \omega_{c2}$, A = $\left[\text{Slope from } \omega_{c1} \text{ to } \omega_{c2} \times \log \dfrac{\omega_{c2}}{\omega_{c1}} \right] + A_{(at\ \omega=\omega_{c1})} = -20 \times \log \dfrac{0.5}{0.25} + 14 = +8 db$

At $\omega = \omega_{c3}$, A = $\left[\text{Slope from } \omega_{c2} \text{ to } \omega_{c3} \times \log \dfrac{\omega_{c3}}{\omega_{c2}} \right] + A_{(at\ \omega=\omega_{c2})} = 0 \times \log \dfrac{4}{0.5} + 8 = +8 db$

At $\omega = \omega_h$, A = $\left[\text{Slope from } \omega_{c3} \text{ to } \omega_h \times \log \dfrac{\omega_h}{\omega_{c3}} \right] + A_{(at\ \omega=\omega_{c3})} = -20 \log \dfrac{10}{4} + 8 = 0 db$

Let the points a, b, c, d and e be the points correponding to frequencies ω_l, ω_{c1}, ω_{c2}, ω_{c3} and ω_h respectively on the magnitude plot. In a semilog graph sheet choose a scale of 1unit = 5 db on y axis. The frequencies are marked in decades from 0.1 to 100 rad/sec on logarithmic scales on x-axis. Fix the points a, b, c, d and e on the graph. Join the points by a straight line and mark the slope in the respective region.

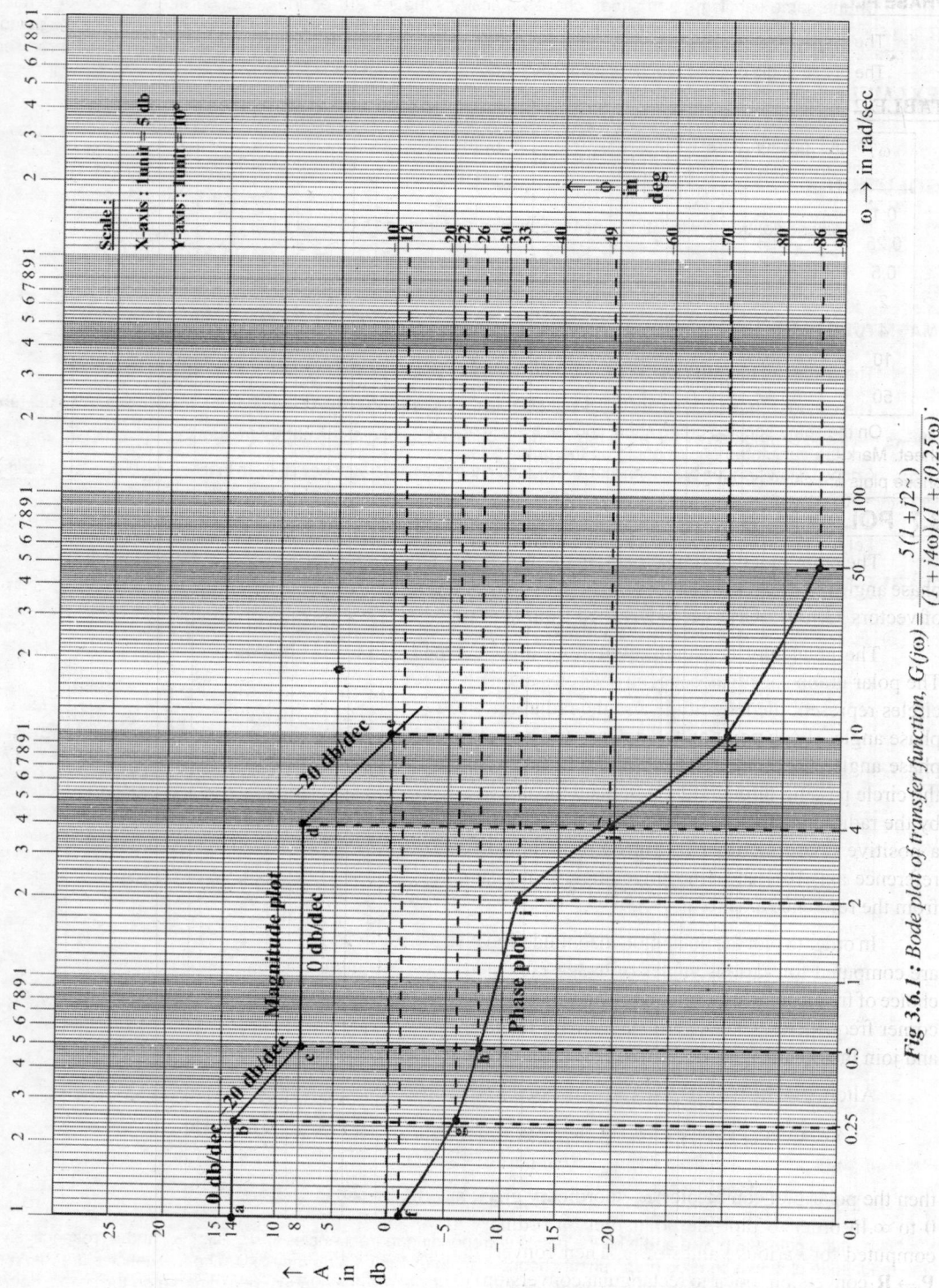

Fig 3.6.1 : Bode plot of transfer function, $G(j\omega) = \dfrac{5(1+j2\omega)}{(1+j4\omega)(1+j0.25\omega)}$.

PHASE PLOT

The phase angle of $G(j\omega)$, $\phi = \tan^{-1}(2\omega) - \tan^{-1}(4\omega) - \tan^{-1}(0.25\omega)$

The phase angle of $G(j\omega)$ are calculated for various values of ω and listed in the table-2.

TABLE-2

ω	$\tan^{-1} 2\omega$ deg	$\tan^{-1} 4\omega$ deg	$\tan^{-1} 0.25\omega$ deg	$\phi = \angle G(j\omega)$	Points in phase plot
0.1	11.3	21.8	1.43	$-11.93 \approx -12$	f
0.25	26.56	45.0	3.5	$-21.94 \approx -22$	g
0.5	45.0	63.43	7.1	$-25.53 \approx -26$	h
2	75.96	82.87	26.56	$-33.47 \approx -33$	i
4	82.87	86.42	45.0	$-48.55 \approx -49$	j
10	87.13	88.56	68.19	$-69.62 \approx -70$	k
50	89.42	89.71	85.42	$-85.71 \approx -86$	l

On the same semilog graph sheet choose a scale of 1unit = 10° on y-axis on the right side of the semilog graph sheet. Mark the calculated phase angle on the graph sheet, Join the points by a smooth curve. The magnitude and phase plots are shown in fig 3.6.1.

3.7 POLAR PLOT

The polar plot of a sinusoidal transfer function $G(j\omega)$ is a plot of the magnitude of $G(j\omega)$ versus the phase angle of $G(j\omega)$ on polar coordinates as ω is varied from zero to infinity. Thus the polar plot is the locus of vectors $|G(j\omega)| \angle G(j\omega)$ as ω is varied from zero to infinity. The polar plot is also called *Nyquist plot.*

The polar plot is usually plotted on a polar graph sheet. The polar graph sheet has concentric circles and radial lines. The circles represent the magnitude and the radial lines represent the phase angles. Each point on the polar graph has a magnitude and phase angle. The magnitude of a point is given by the value of the circle passing through that point and the phase angle is given by the radial line passing through that point. In polar graph sheet a positive phase angle is measured in anticlockwise from the reference axis (0°) and a negative angle is measured clockwise from the reference axis (0°).

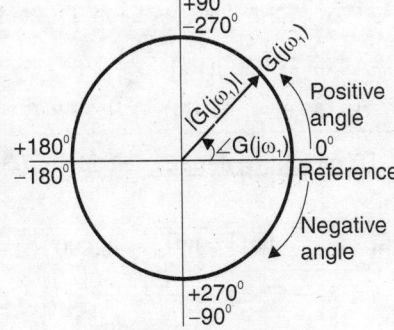

Fig 3.20 : *Polar graph.*

In order to plot the polar plot, magnitude and phase of $G(j\omega)$ are computed for various values of ω and tabulated. Usually the choice of frequencies are corner frequencies and frequencies around corner frequencies. Choose proper scale for the magnitude circles. Fix all the points on polar graph sheet and join the points by smooth curve. Write the frequency corresponding to each point of the plot.

Alternatively, if $G(j\omega)$ can be expressed in rectangular coordinates as,

$$G(j\omega) = G_R(j\omega) + jG_I(j\omega)$$

where, $G_R(j\omega)$ = Real part of $G(j\omega)$; $G_I(j\omega)$ = Imaginary part of $G(j\omega)$.

then the polar plot can be plotted in ordinary graph sheet between $G_R(j\omega)$ and $G_I(j\omega)$ by varying ω from 0 to ∞. In order to plot the polar plot on ordinary graph sheet, the magnitude and phase of $G(j\omega)$ are computed for various values of ω. Then convert the polar coordinates to rectangular coordinates using $P \rightarrow R$ conversion (polar to rectangular conversion) in the calculator. Sketch the polar plot using rectangular coordinates.

For minimum phase transfer function with only poles, type number of the system determines the quadrant at which the polar plot starts and the order of the system determines the quadrant at which the polar plot ends. The minimum phase systems are systems with all poles and zeros on left half of s-plane. The start and end of polar plot of all pole minimum phase system are shown in fig 3.21 & 3.22 respectively. Some typical sketches of polar plot are shown in table-3.1.

The change in shape of polar plot can be predicted due to addition of a pole or zero.

1.   When a pole is added to a system, the polar plot end point will shift by -90°.
2.   When a zero is added to a system the polar plot end point will shift by $+90^\circ$.

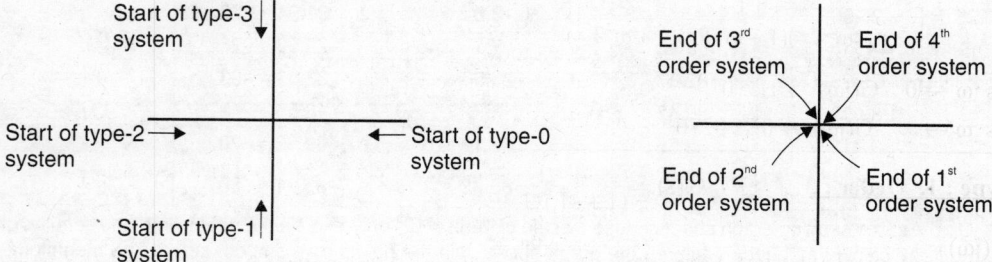

Fig 3.21 : *Start of polar plot of all pole minimum phase system.*　　　**Fig 3.22** : *Start of polar plot of all pole minimum phase system.*

TABLE-3.1 : Typical Sketches of Polar Plot

Type : 0, Order : 1 $G(s) = \dfrac{1}{1+sT}$	

$G(j\omega) = \dfrac{1}{1+j\omega T} = \dfrac{1}{\sqrt{1+\omega^2 T^2} \angle \tan^{-1}\omega T} = \dfrac{1}{\sqrt{1+\omega^2 T^2}} \angle -\tan^{-1}\omega T$

As $\omega \to 0$, $G(j\omega) \to 1\angle 0^\circ$

As $\omega \to \infty$, $G(j\omega) \to 0\angle -90^\circ$

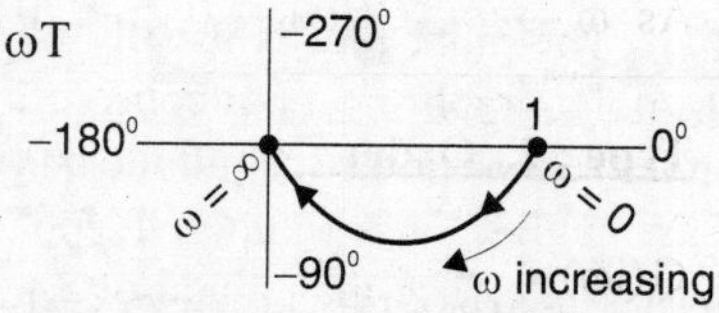

Type : 1, Order : 2 $G(s) = \dfrac{1}{s(1+sT)}$

$G(j\omega) = \dfrac{1}{j\omega(1+j\omega T)} = \dfrac{1}{\omega\angle 90^\circ \sqrt{1+\omega^2 T^2} \angle \tan^{-1}\omega T}$

$= \dfrac{1}{\omega\sqrt{1+\omega^2 T^2}} \angle (-90^\circ - \tan^{-1}\omega T$

As $\omega \to 0$, $G(j\omega) \to \infty\angle -90^\circ$

As $\omega \to \infty$, $G(j\omega) \to 0\angle -180^\circ$

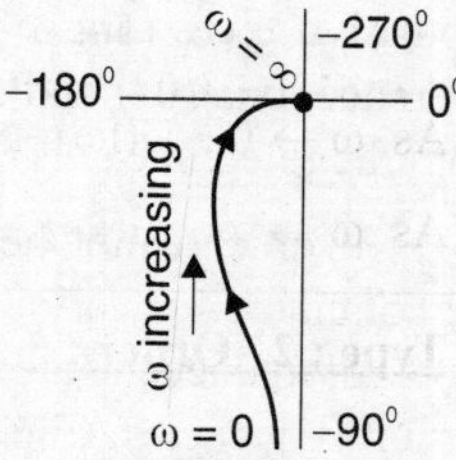

Type : 0, Order : 2 $G(s) = \dfrac{1}{(1+sT_1)(1+sT_2)}$

$G(j\omega) = \dfrac{1}{(1+j\omega T_1)(1+j\omega T_2)} = \dfrac{1}{\sqrt{1+\omega^2 T_1^2} \angle \tan^{-1}\omega T_1 \sqrt{1+\omega^2 T_2^2} \angle \tan^{-1}\omega T_2}$

$= \dfrac{1}{\sqrt{(1+\omega^2 T_1^2)(1+\omega^2 T_2^2)}} \angle (-\tan^{-1}\omega T_1 - \tan^{-1}\omega T_2)$

As $\omega \to 0$, $G(j\omega) \to 1\angle 0^\circ$

As $\omega \to \infty$, $G(j\omega) \to 0\angle -180^\circ$

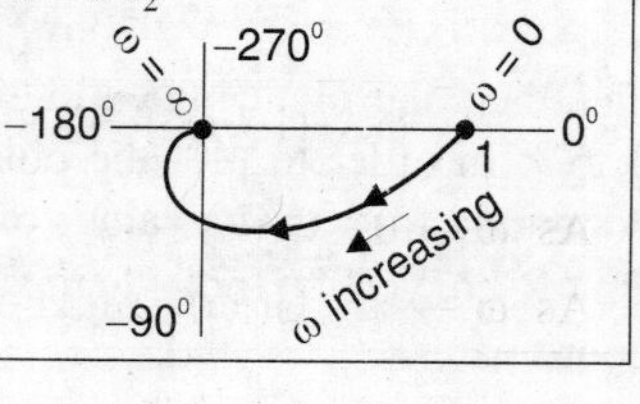

TABLE-3.1 : Typical Sketches of Polar Plot

Type : 0, Order : 3 $G(s) = \dfrac{1}{(1+sT_1)(1+sT_2)(1+sT_3)}$

$G(j\omega) = \dfrac{1}{(1+j\omega T_1)(1+j\omega T_2)(1+j\omega T_3)}$

$\quad = \dfrac{1}{\sqrt{1+\omega^2 T_1^2}\angle \tan^{-1}\omega T_1\ \sqrt{1+\omega^2 T_2^2}\angle \tan^{-1}\omega T_2\ \sqrt{1+\omega^2 T_3^2}\angle \tan^{-1}\omega T_3}$

$\quad = \dfrac{1}{\sqrt{(1+\omega^2 T_1^2)(1+\omega^2 T_2^2)(1+\omega^2 T_3^2)}} \angle (-\tan^{-1}\omega T_1 - \tan^{-1}\omega T_2 - \tan^{-1}\omega T_3)$

As $\omega \to 0$, $G(j\omega) \to 1 \angle 0^\circ$

As $\omega \to \infty$, $G(j\omega) \to 0 \angle -270^\circ$

Type : 1, Order : 3 $G(s) = \dfrac{1}{(1+sT_1)(1+sT_2)}$

$G(j\omega) = \dfrac{1}{(1+j\omega T_1)(1+j\omega T_2)} = \dfrac{1}{\omega\angle 90^\circ\ \sqrt{1+\omega^2 T_1^2}\angle \tan^{-1}\omega T_1\ \sqrt{1+\omega^2 T_2^2}\angle \tan^{-1}\omega T_2}$

$\quad = \dfrac{1}{\omega\sqrt{(1+\omega^2 T_1^2)(1+\omega^2 T_2^2)}}\angle(-90^\circ - \tan^{-1}\omega T_1 - \tan^{-1}\omega T_2)$

As $\omega \to 0$, $G(j\omega) \to \infty \angle -90^\circ$

As $\omega \to \infty$, $G(j\omega) \to 0 \angle -270^\circ$

Type : 2, Order : 4 $G(s) = \dfrac{1}{s^2(1+sT_1)(1+sT_2)}$

$G(j\omega) = \dfrac{1}{(j\omega)^2(1+j\omega T_1)(1+j\omega T_2)} = \dfrac{1}{\omega^2\angle -180^\circ\ \sqrt{1+\omega^2 T_1^2}\angle \tan^{-1}\omega T_1\ \sqrt{1+\omega^2 T_2^2}\angle \tan^{-1}\omega T_2}$

$\quad = \dfrac{1}{\omega^2\sqrt{(1+\omega^2 T_1^2)(1-\omega^2 T_2^2)}}\angle(-180^\circ - \tan^{-1}\omega T_1 - \tan^{-1}\omega T_2)$

As $\omega \to 0$, $G(j\omega) \to \infty \angle -180^\circ$

As $\omega \to \infty$, $G(j\omega) \to 0 \angle -360^\circ$

Type : 2, Order : 5 $G(s) = \dfrac{1}{s^2(1+sT_1)(1+sT_2)(1+sT_3)}$

$G(j\omega) = \dfrac{1}{(j\omega)^2(1+j\omega T_1)(1+j\omega T_2)(1+j\omega T_3)}$

$\quad = \dfrac{1}{\omega^2\angle -180^\circ\ \sqrt{1+\omega^2 T_1^2}\angle \tan^{-1}\omega T_1\ \sqrt{1+\omega^2 T_2^2}\angle \tan^{-1}\omega T_2\ \sqrt{1+\omega^2 T_3^2}\angle \tan^{-1}\omega T_3}$

$\quad = \dfrac{1}{\omega^2\sqrt{(1+\omega^2 T_1^2)(1+\omega^2 T_2^2)(1+\omega^2 T_3^2)}}\angle(-180^\circ - \tan^{-1}\omega T_1 - \tan^{-1}\omega T_2 - \tan^{-1}\omega T_3)$

As $\omega \to 0$, $G(j\omega) \to \infty \angle -180^\circ$

As $\omega \to \infty$, $G(j\omega) \to 0 \angle -450^\circ = 0 \angle -90^\circ$

TABLE-3.1 : Typical Sketches of Polar Plot

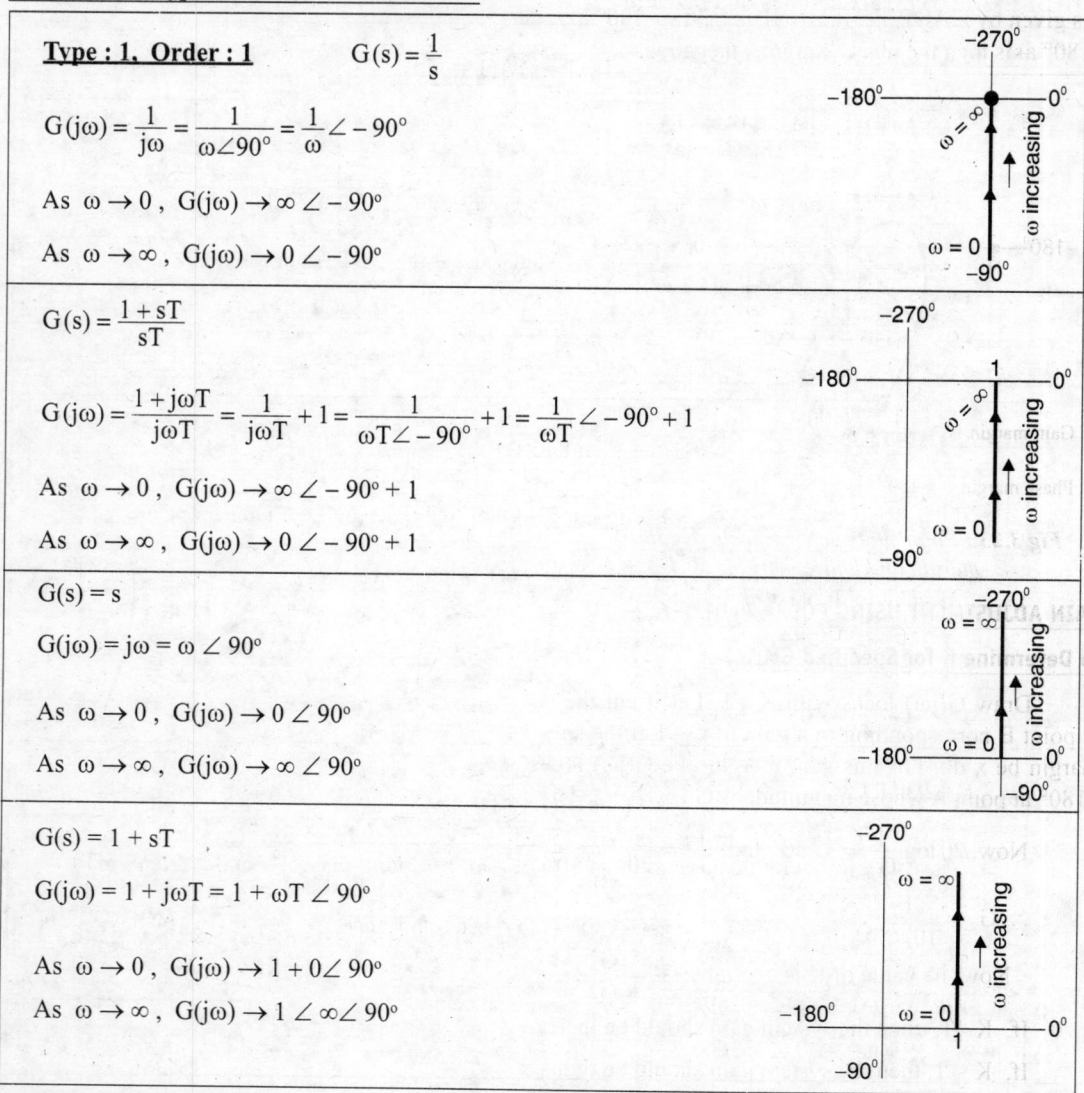

Type : 1, Order : 1 $G(s) = \dfrac{1}{s}$	
$G(j\omega) = \dfrac{1}{j\omega} = \dfrac{1}{\omega \angle 90°} = \dfrac{1}{\omega} \angle -90°$	
As $\omega \to 0$, $G(j\omega) \to \infty \angle -90°$	
As $\omega \to \infty$, $G(j\omega) \to 0 \angle -90°$	
$G(s) = \dfrac{1+sT}{sT}$	
$G(j\omega) = \dfrac{1+j\omega T}{j\omega T} = \dfrac{1}{j\omega T} + 1 = \dfrac{1}{\omega T \angle -90°} + 1 = \dfrac{1}{\omega T} \angle -90° + 1$	
As $\omega \to 0$, $G(j\omega) \to \infty \angle -90° + 1$	
As $\omega \to \infty$, $G(j\omega) \to 0 \angle -90° + 1$	
$G(s) = s$	
$G(j\omega) = j\omega = \omega \angle 90°$	
As $\omega \to 0$, $G(j\omega) \to 0 \angle 90°$	
As $\omega \to \infty$, $G(j\omega) \to \infty \angle 90°$	
$G(s) = 1 + sT$	
$G(j\omega) = 1 + j\omega T = 1 + \omega T \angle 90°$	
As $\omega \to 0$, $G(j\omega) \to 1 + 0 \angle 90°$	
As $\omega \to \infty$, $G(j\omega) \to 1 \angle \infty \angle 90°$	

DETERMINATION OF GAIN MARGIN AND PHASE MARGIN FROM POLAR PLOT

The **gain margin** is defined as the inverse of the magnitude of $G(j\omega)$ at phase crossover frequency. The **phase crossover frequency** is the frequency at which the phase of $G(j\omega)$ is 180°.

Let the polar plot cut the 180° axis at point **B** and the magnitude circle passing through the point B be G_B. Now the Gain margin, $K_g = 1/G_B$. If the point B lies within unity circle, then the Gain margin is positive otherwise negative. (If the polar plot is drawn in ordinary graph sheet using rectangular coordinates then the point B is the cutting point of $G(j\omega)$ locus with negative real axis and $K_g = 1/|G_B|$ where G_B is the magnitude corresponding to point B).

The **phase margin** is defined as, phase margin, $\gamma = 180° + \phi_{gc}$ where ϕ_{gc} is the phase angle of $G(j\omega)$ at gain crossover frequency. The **gain crossover frequency** is the frequency at which the magnitude of $G(j\omega)$ is unity.

Let the polar plot cut the unity circle at point A as shown in fig 3.23 and 3.24. Now the phase margin, γ is given by $\angle AOP$, i.e. if $\angle AOP$ is below $-180°$ axis then the phase margin is positive and if it is above $-180°$ axis then the phase margin is negative.

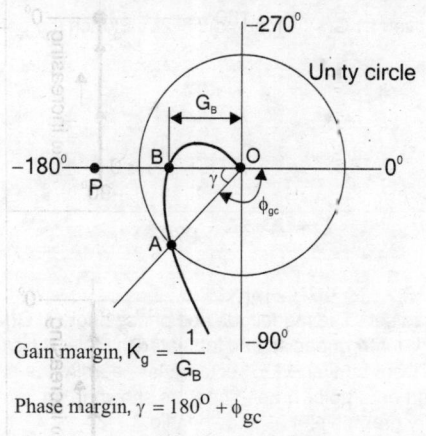

Gain margin, $K_g = \dfrac{1}{G_B}$

Phase margin, $\gamma = 180° + \phi_{gc}$

Fig 3.23 : *Polar plot showing positive gain margin and phase margin.*

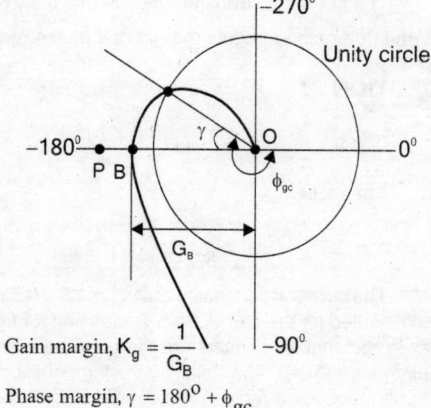

Gain margin, $K_g = \dfrac{1}{G_B}$

Phase margin, $\gamma = 180° + \phi_{gc}$

Fig 3.24 : *Polar plot showing negative gain margin and phase margin.*

GAIN ADJUSTMENT USING POLAR PLOT

To Determine K for Specified GM

Draw $G(j\omega)$ locus with K =1. Let it cut the $-180°$ axis at point B corresponding to a gain of G_B. Let the specified gain margin be x db. For this gain margin, the $G(j\omega)$ locus will cut $-180°$ at point A whose magnitude is G_A.

Now, $20 \log \dfrac{1}{G_A} = x \implies \log \dfrac{1}{G_A} = \dfrac{x}{20} \implies \dfrac{1}{G_A} = 10^{x/20}$

$\therefore G_A = \dfrac{1}{10^{x/20}}$

Now the value of K is given by, $K = \dfrac{G_A}{G_B}$.

If, K >1, then the system gain should be increased.

If, K < 1, then the system gain should be reduced.

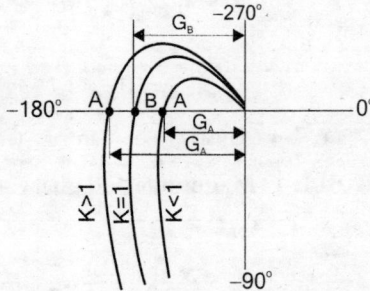

Fig 3.25 : *Polar plot for different values of K*

To Determine K for Specified PM

Draw $G(j\omega)$ locus with K = 1. Let it cut the unity circle at point B. (The gain at point B is G_B and equal to unity). Let the specified phase margin be x°

For a phase margin of x°, let ϕ_{gcx} be the phase angle of $G(j\omega)$ at gain crossover frequency.

$$\therefore x° = 180° + \phi_{gcx} \implies \phi_{gcx} = x° - 180°$$

In the polar plot, the radial line corresponding to ϕ_{gcx} will cut the locus of $G(j\omega)$ with K = 1 at point A and the magnitude corresponding to that point be G_A

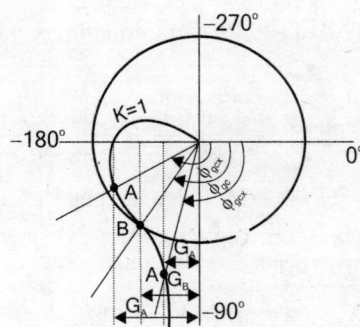

Fig 3.26 : *Gain adjustment for required phase margin.*

Now, $K = \dfrac{G_B}{G_A} = \dfrac{1}{G_A}$ $(\because G_B = 1)$

EXAMPLE 3.7

The open loop transfer function of a unity feedback system is given by G(s) = 1/s(1+s) (1+2s). Sketch the polar plot and determine the gain margin and phase margin.

SOLUTION

Given that, G(s) = 1/s(1+s) (1 +2s)

Put s = jω.

$\therefore\ G(j\omega) = \dfrac{1}{j\omega(1+j\omega)(1+j2\omega)}$

The corner frequencies are $\omega_{c1} = 1/2 = 0.5$ rad/sec and $\omega_{c2} = 1$ rad/sec. The magnitude and phase angle of G(jω) are calculated for the corner frequencies and for frequencies around corner frequencies and tabulated in table-1. Using polar to rectangular conversion, the polar coordinates listed in table-1 are converted to rectangular coordinates and tabulated in table-2. The polar plot using polar coordinates is sketched on a polar graph sheet as shown in fig 3.7.1. The polar plot using rectangular coordinates is sketched on an ordinary graph sheet as shown in fig 3.7.2.

$$G(j\omega) = \dfrac{1}{(j\omega)(1+j\omega)(1+j2\omega)} = \dfrac{1}{\omega\angle 90^\circ\ \sqrt{1+\omega^2}\ \angle\tan^{-1}\omega\ \sqrt{1+4\omega^2}\ \angle\tan^{-1}2\omega}$$

$$= \dfrac{1}{\omega\sqrt{(1+\omega^2)(1+4\omega^2)}}\ \angle -90^\circ - \tan^{-1}\omega - \tan^{-1}2\omega$$

$$\therefore\ |G(j\omega)| = \dfrac{1}{\omega\sqrt{(1+\omega^2)(1+4\omega^2)}} = \dfrac{1}{\omega\sqrt{1+4\omega^2+\omega^2+4\omega^4}} = \dfrac{1}{\omega\sqrt{1+5\omega^2+4\omega^4}}$$

$$\angle G(j\omega) = -90^\circ - \tan^{-1}\omega - \tan^{-1}2\omega$$

TABLE-1 : Magnitude and phase of G(jω) at various frequencies

ω rad/sec	0.35	0.4	0.45	0.5	0.6	0.7	1.0
|G(jω)|	2.2	1.8	1.5	1.2	0.9	0.7	0.3
∠G(jω) deg	-144	-150	-156	-162	-171	-179.5 ≈ -180	-198

TABLE-2 : Real and imaginary part of G(jω) at various frequencies

ω rad/sec	0.35	0.4	0.45	0.5	0.6	0.7	1.0
$G_R(j\omega)$	-1.78	-1.56	-1.37	-1.14	-0.89	-0.7	-0.29
$G_I(j\omega)$	-1.29	-0.9	-0.61	-0.37	-0.14	0	0.09

RESULT

Gain margin, K_g = 1.4286

Phase margin, γ = +12°

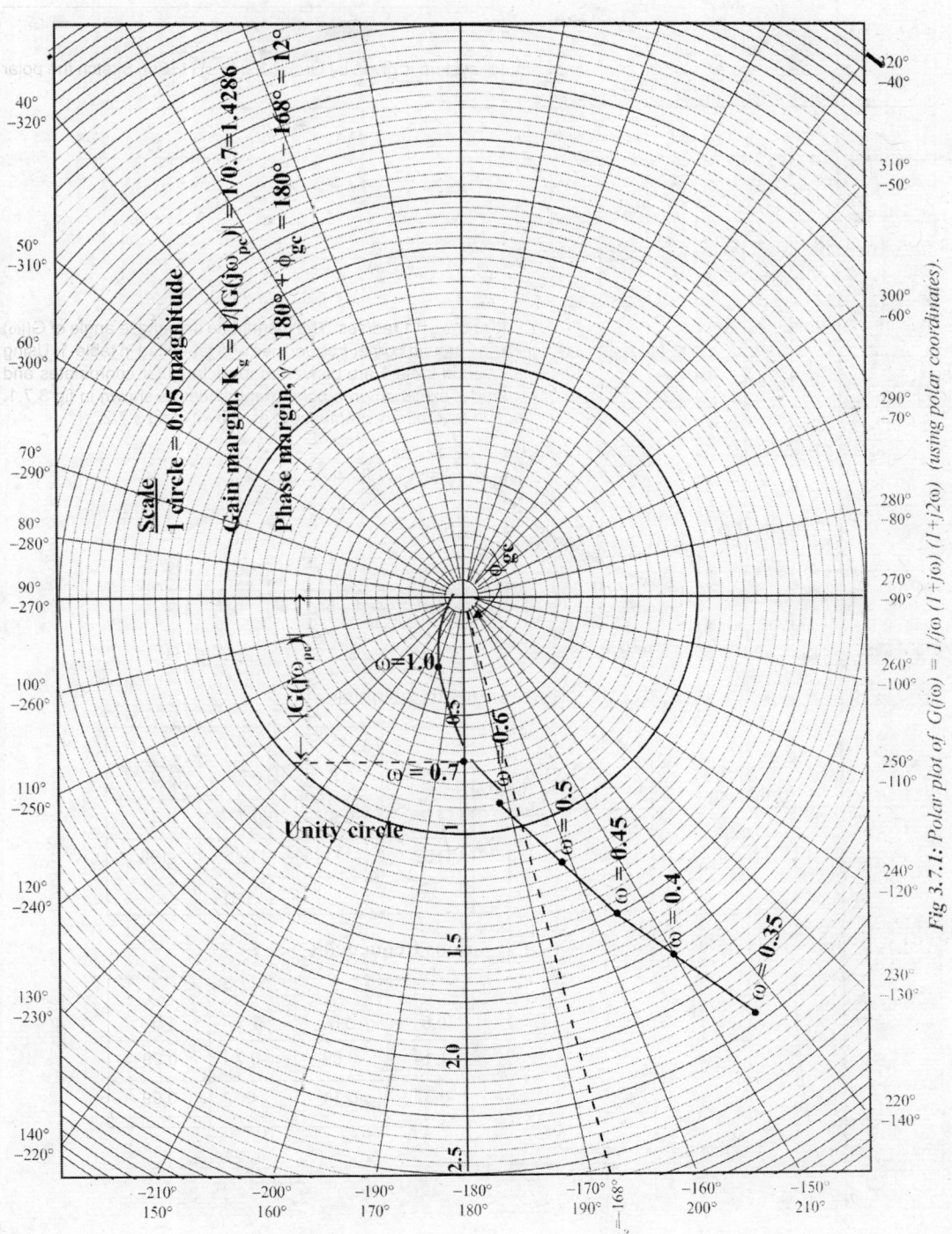

Fig 3.7.1: Polar plot of $G(j\omega) = 1/j\omega (1+j\omega) (1+j2\omega)$ (using polar coordinates).

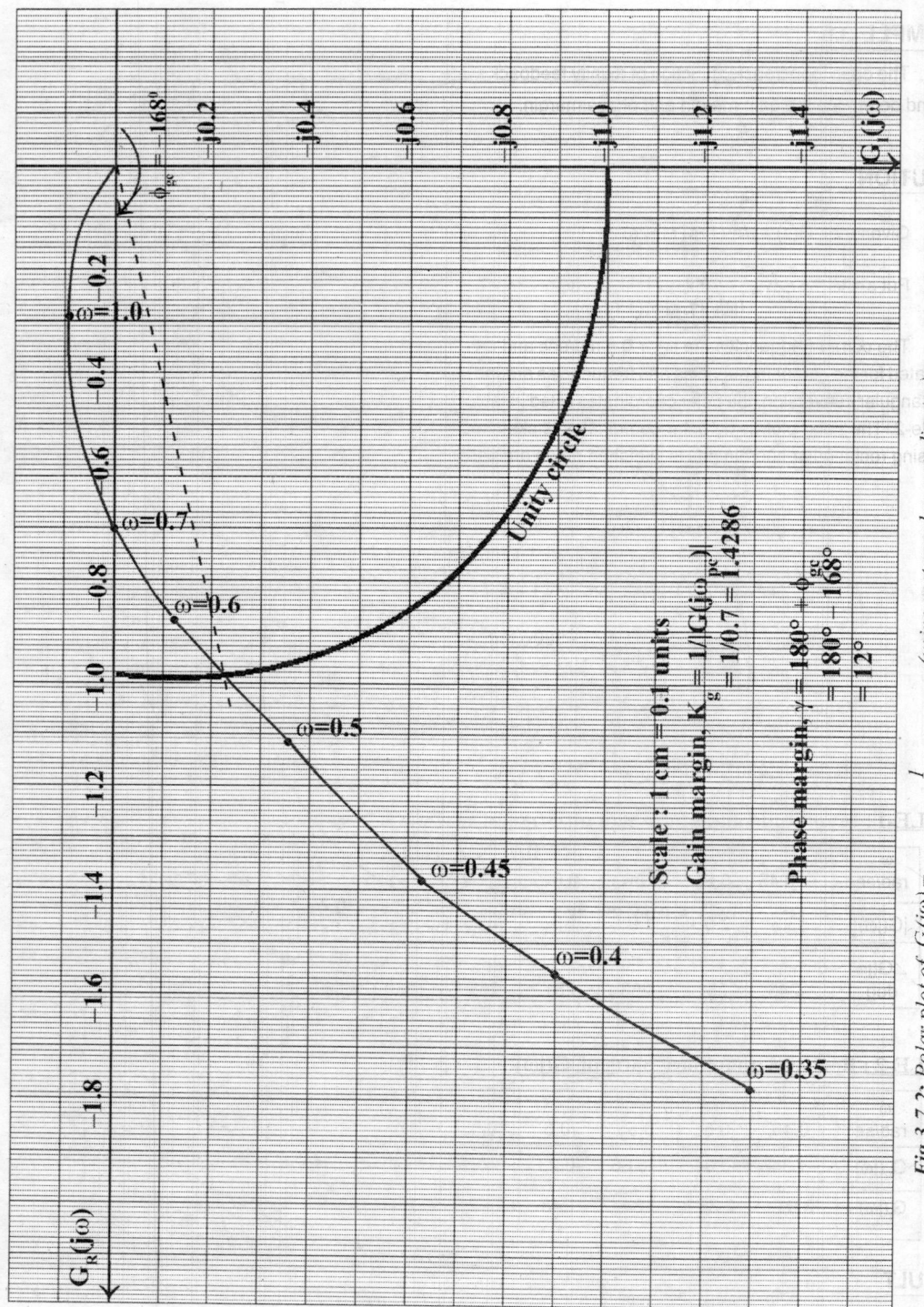

Scale : 1 cm = 0.1 units

Gain margin, $K_g = 1/|G(j\omega)|^5$

$= 1/0.7 = 1.4286$

Phase margin, $\gamma = 180° + \phi_{gc}$

$= 180° - 168°$

$= 12°$

$$G(j\omega) = \frac{1}{j\omega(1 + j\omega)(1 + j2\omega)} \quad \text{(using rectangular coordinates)}.$$

Fig 3.7.2: Polar plot of, $G(j\omega)$

EXAMPLE 3.8

The open loop transfer function of a unity feedback system is given by $G(s) = 1/s^2(1+s)(1+2s)$. Sketch the polar plot and determine the gain margin and phase margin.

SOLUTION

Given that, $G(s) = 1/s^2(1+s)(1+2s)$

Put $s = j\omega$, $\quad G(j\omega) = \dfrac{1}{(j\omega)^2(1+j\omega)(1+j2\omega)}$

The corner frequencies are $\omega_{c1} = 0.5$ rad/sec and $\omega_{c2} = 1$ rad/sec. The magnitude and phase angle of $G(j\omega)$ are calculated for the corner frequencies and frequencies around corner frequencies are tabulated in table-1. Using the polar to rectangular conversion, the polar coordinates listed in table-1 are converted to rectangular coordinates and tabulated in table-2. The polar plot using polar coordinates is sketched on a polar graph sheet as shown in fig 3.8.1. The polar plot using rectangular coordinates is sketched on an ordinary graph sheet as shown in fig 3.8.2.

$$G(j\omega) = \frac{1}{(j\omega)^2(1+j\omega)(1+j2\omega)}$$

$$= \frac{1}{\omega^2\angle 180° \sqrt{1+\omega^2}\angle\tan^{-1}\omega \sqrt{1+4\omega^2}\angle\tan^{-1}2\omega}$$

$$G(j\omega) = \frac{1}{\omega^2\sqrt{1+\omega^2}\sqrt{1+4\omega^2}}\angle(-180-\tan^{-1}\omega-\tan^{-1}2\omega)$$

$$|G(j\omega)| = \frac{1}{\omega^2\sqrt{1+\omega^2}\sqrt{1+4\omega^2}} = \frac{1}{\omega^2\sqrt{(1+\omega^2)(1+4\omega^2)}}$$

$$\angle G(j\omega) = -180° - \tan^{-1}\omega - \tan^{-1}2\omega$$

TABLE-1 : Magnitude and phase plot of $G(j\omega)$ at various frequencies

ω rad/sec	0.45	0.5	0.55	0.6	0.65	0.7	0.75	1.0
$\lvert G(j\omega)\rvert$	3.3	2.5	1.9	1.5	1.2	$0.97 \approx 1$	0.8	0.3
$\angle G(j\omega)$ deg	−246	−251	−256	−261	−265	−269	−273	−288

TABLE-2 : Real and imaginary parts of $G(j\omega)$

ω rad/sec	0.45	0.5	0.55	0.6	0.65	0.7	0.75	1.0
$G_R(j\omega)$	−1.34	−0.81	−0.46	−0.23	−0.1	−0.02	0.04	0.09
$G_I(j\omega)$	3.01	2.36	1.84	1.48	1.2	1.0	0.8	0.29

RESULT

Gain margin, $K_g = 0$

Phase margin, $\gamma = -90°$

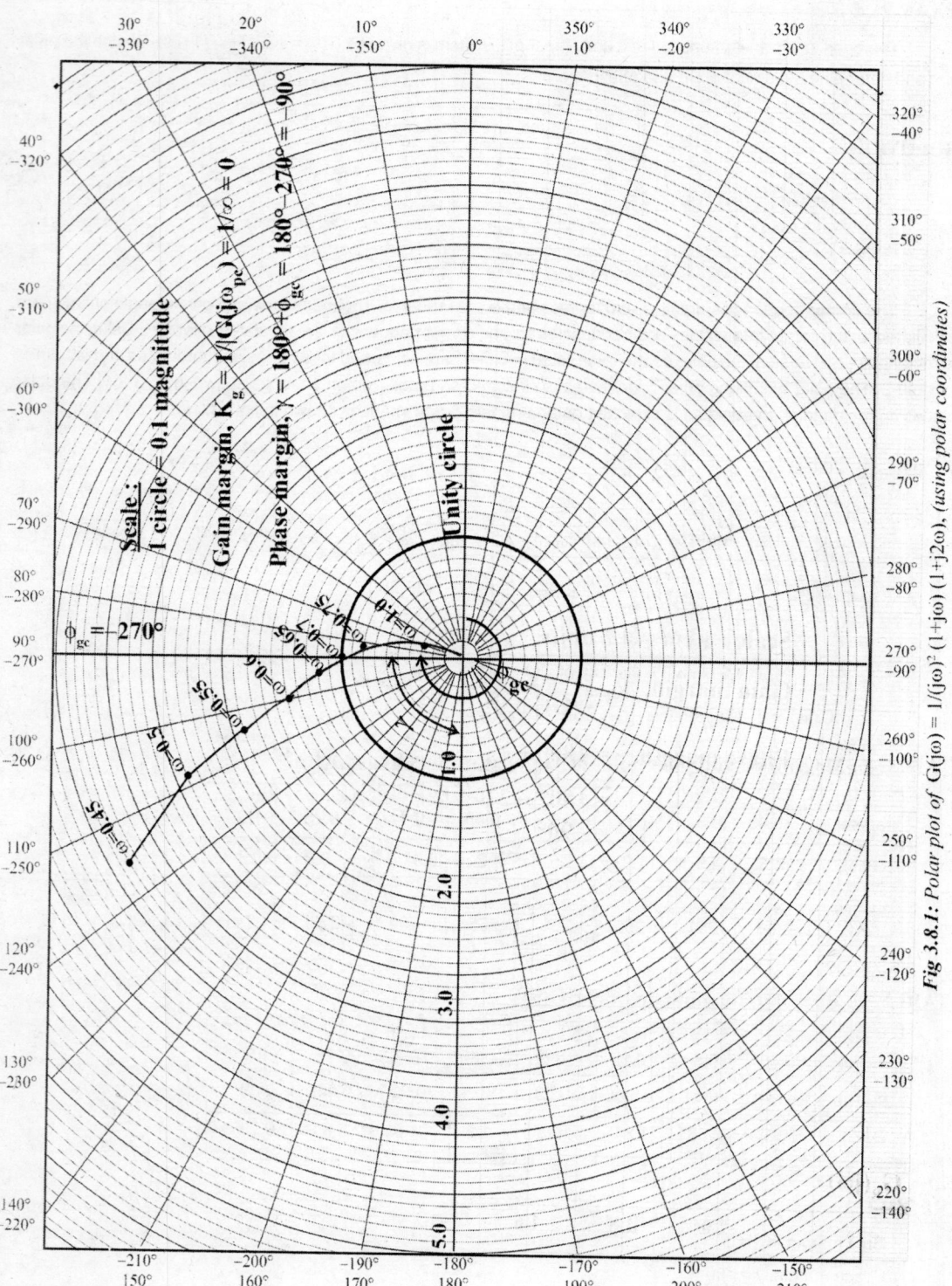

Fig 3.8.1: Polar plot of $G(j\omega) = 1/(j\omega)^2 (1+j\omega) (1+j2\omega)$, (using polar coordinates)

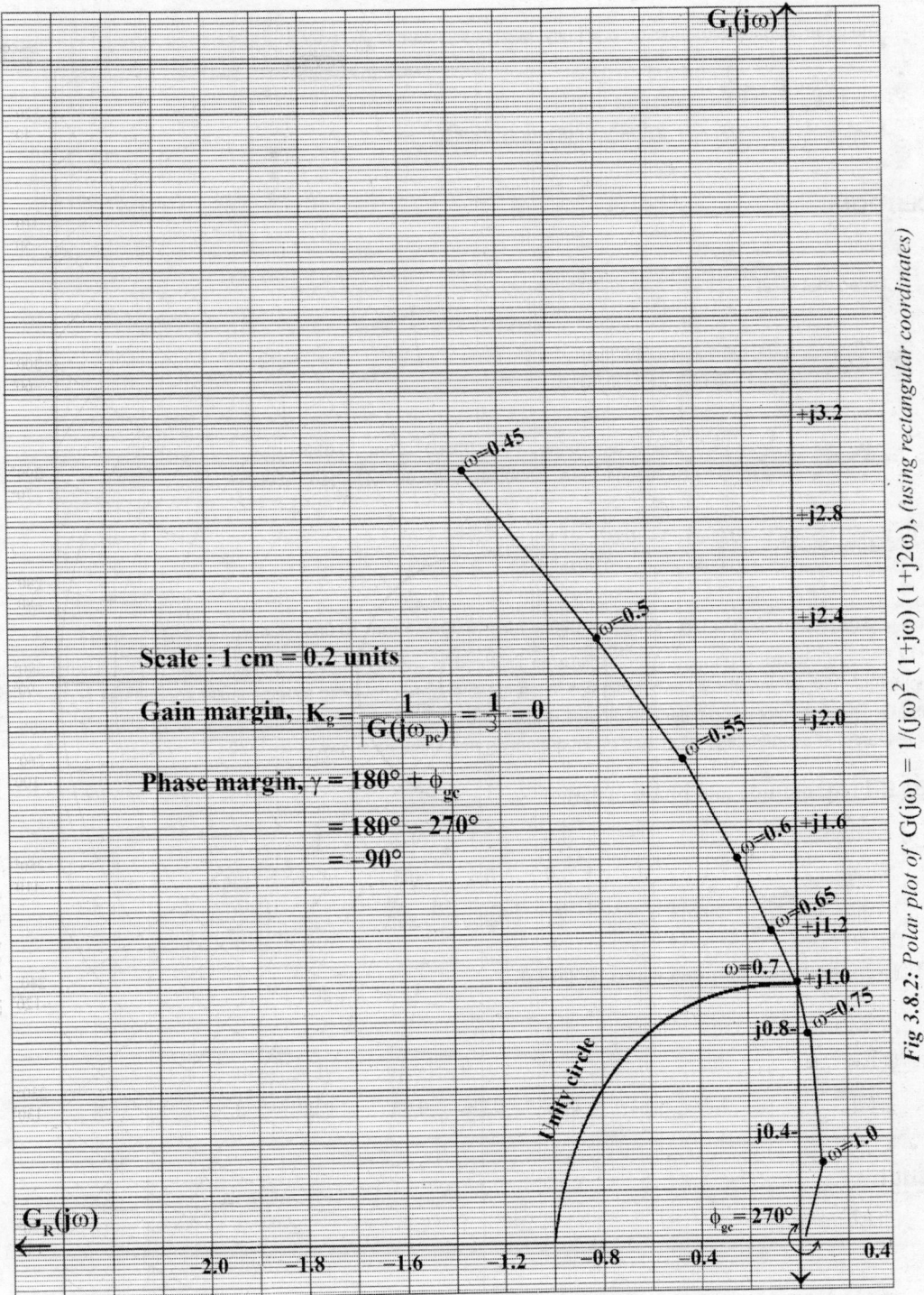

Scale : 1 cm = 0.2 units

Gain margin, $K_g = \dfrac{1}{|G(j\omega_{pc})|} = \dfrac{1}{3} = 0$

Phase margin, $\gamma = 180° + \phi_{gc}$

$= 180° - 270°$

$= -90°$

Fig 3.8.2: Polar plot of $G(j\omega) = 1/(j\omega)^2 (1+j\omega) (1+j2\omega)$, (using rectangular coordinates)

EXAMPLE 3.9

The open loop transfer function of a unity feedback system is given by,

$$G(s) = \frac{(1+0.2s)(1+0.025s)}{s^3(1+0.005s)(1+0.001s)}$$

Sketch the polar plot and determine the phase margin.

SOLUTION

Given that $G(s) = \dfrac{(1+0.2s)(1+0.025s)}{s^3(1+0.005s)(1+0.001s)}$

$\therefore G(j\omega) = \dfrac{(1+j0.2\omega)(1+j0.025\omega)}{(j\omega)^3(1+j0.005\omega)(1+j0.001\omega)}$

$$= \frac{\sqrt{1+(0.2\omega)^2}\angle\tan^{-1}0.2\omega\sqrt{1+(0.025\omega)^2}\angle\tan^{-1}0.025\omega}{\omega^3\angle270°\sqrt{1+(0.005\omega)^2}\angle\tan^{-1}0.005\omega\sqrt{1+(0.001\omega)^2}\angle\tan^{-1}0.001\omega}$$

$$\left|G(j\omega)\right| = \frac{\sqrt{1+(0.2\omega)^2}\sqrt{1+(0.025\omega)^2}}{\omega^3\sqrt{1+(0.005\omega)^2}\sqrt{1+(0.001\omega)^2}}$$

$$\angle G(j\omega) = \tan^{-1}0.2\omega + \tan^{-1}0.025\omega - 270° - \tan^{-1}0.005\omega - \tan^{-1}0.001\omega$$

The magnitude and phase angle of $G(j\omega)$ are calculated for various frequencies and listed in table-1. Using the polar to rectangular conversion, the polar coordinates listed in table-1 are converted to rectangular coordinates and tabulated in table-2. The polar plot using polar coordinates is sketched on a polar graph sheet as shown in fig 3.9.1. The polar plot using rectangular coordinates is sketched on an ordinary graph sheet as shown in fig 3.9.2.

TABLE-1 : Magnitude and phase of $G(j\omega)$

ω,rad/sec	0.9	0.95	1.0	1.1	1.2	1.4	1.7
$\left|G(j\omega)\right|$	1.4	1.2	1.0	0.8	0.6	0.4	0.2
$\angle G(j\omega)$, deg	−259	−258	−257	−256	−255	−253	−249

TABLE-2 : Real and imaginary part of $G(j\omega)$

ω, rad/sec	0.9	0.95	1.0	1.1	1.2	1.4	1.7
$G_R(j\omega)$	−0.27	−0.25	−0.22	−0.19	−0.16	−0.12	−0.07
$G_i(j\omega)$	1.37	1.17	0.97	0.78	0.58	0.38	0.19

RESULT

Phase margin, $\gamma = -77°$

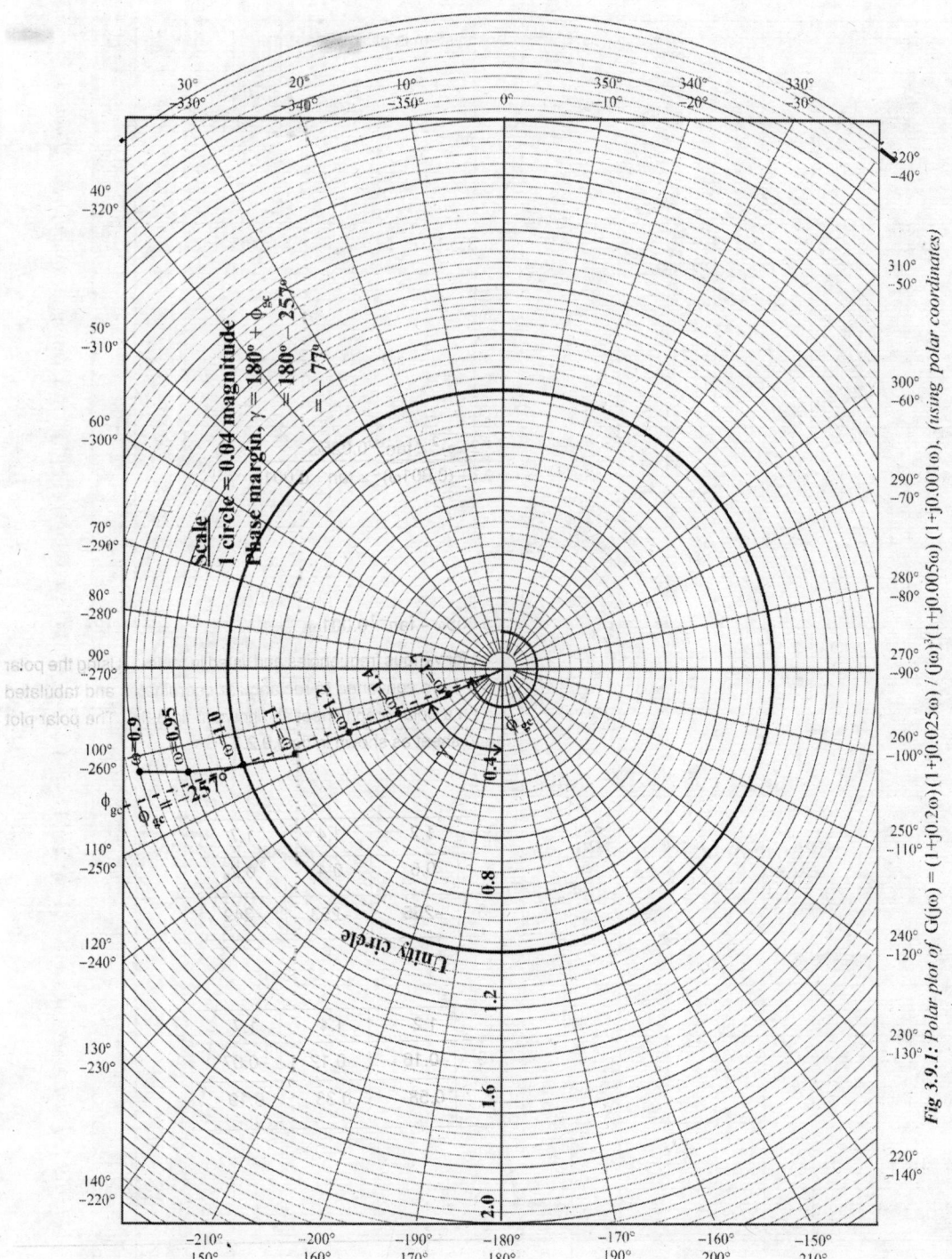

Fig 3.9.1: Polar plot of $G(j\omega) = (1+j0.2\omega)(1+j0.025\omega) / (j\omega)^3(1+j0.005\omega) (1+j0.001\omega)$, *(using polar coordinates)*

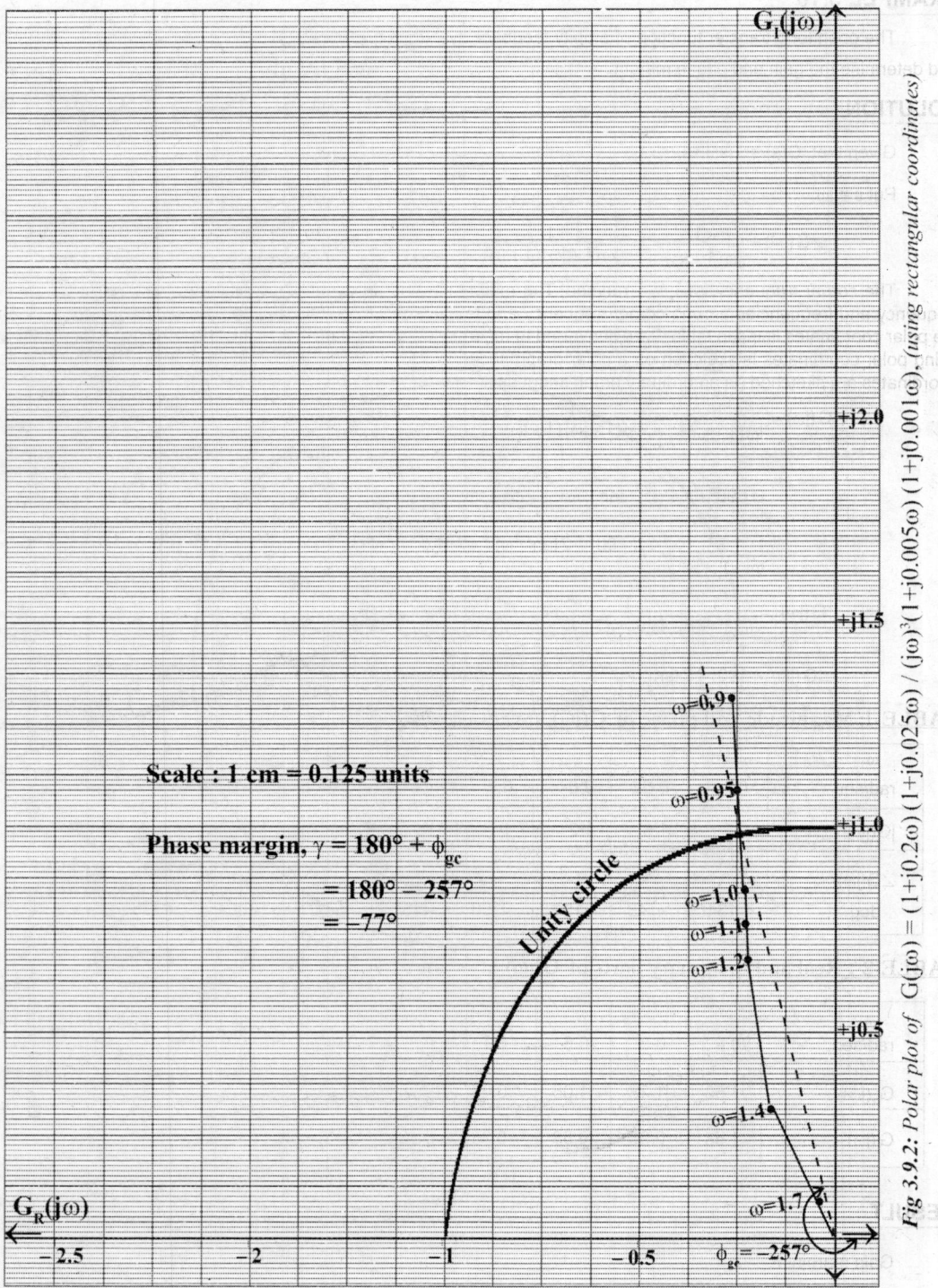

$G_i(j\omega)$

+j2.0

+j1.5

$\omega=0.9$

$\omega=0.95$

+j1.0

Scale : 1 cm = 0.125 units

Phase margin, $\gamma = 180° + \phi_{gc}$

$= 180° - 257°$

$= -77°$

Unity circle

$\omega=1.0$

$\omega=1.1$

$\omega=1.2$

+j0.5

$\omega=1.4$

$G_R(j\omega)$

$\omega=1.7$

−2.5   −2   −1   −0.5 $\phi_g = -257°$

Fig 3.9.2: Polar plot of $G(j\omega) = (1+j0.2\omega)(1+j0.025\omega) / (j\omega)^3(1+j0.005\omega)(1+j0.001\omega)$, *(using rectangular coordinates)*

EXAMPLE 3.10

The open loop transfer function of a unity feedback system is given by $G(s) = 1/s(1+s)^2$. Sketch the polar plot and determine the gain and phase margin.

SOLUTION

Given that, $G(s) = 1/s(1+s)^2$.

Put $s = j\omega$,

$$\therefore G(j\omega) = \frac{1}{j\omega(1+j\omega)^2} = \frac{1}{j\omega(1+j\omega)(1+j\omega)}$$

The corner frequency is $\omega_{c1} = 1$ rad/sec. The magnitude and phase angle of $G(j\omega)$ are calculated for corner frequency and frequencies around corner frequency and tabulated in table-1. Using polar to rectangular conversion the polar coordinates listed in table-1 are converted to rectangular coordinates and tabulated in table-2. The polar plot using polar coordinates is sketched on a polar graph sheet as shown in fig 3.10.1. The polar plot using rectangular coordinates are sketched on an ordinary graph sheet as shown in fig 3.10.2.

$$G(j\omega) = \frac{1}{j\omega(1+j\omega)^2} = \frac{1}{j\omega(1+j\omega)(1+j\omega)}$$

$$= \frac{1}{\omega\angle 90° \sqrt{1+\omega^2} \angle\tan^{-1}\omega \sqrt{1+\omega^2} \angle\tan^{-1}\omega}$$

$$= \frac{1}{\omega(\sqrt{1+\omega^2})^2} \angle(-90° - 2\tan^{-1}\omega)$$

$$|G(j\omega)| = \frac{1}{\omega(1+\omega^2)} = \frac{1}{\omega + \omega^3}$$

$$\angle G(j\omega) = -90° - 2\tan^{-1}\omega$$

TABLE-1: Magnitude and phase of $G(j\omega)$ at various frequencies

ω rad/sec	0.4	0.5	0.6	0.7	0.8	0.9	1.0	1.1
$|G(j\omega)|$	2.2	1.6	1.2	1	0.8	0.6	0.5	0.4
$\angle G(j\omega)$ deg	−134	−143	−151	−159	−167	−174	−180	−185

TABLE-2 : Real and imaginary parts of $G(j\omega)$ at various frequencies

ω rad/sec	0.4	0.5	0.6	0.7	0.8	0.9	1.0	1.1
$G_R(j\omega)$	−1.53	−1.28	−1.05	−0.93	−0.78	−0.6	−0.5	−0.4
$G_i(j\omega)$	−1.58	−0.96	−0.58	−0.36	−0.18	0.06	0	0.03

RESULT

Gain margin, $K_g = 2$

Phase margin, $\gamma = 21°$

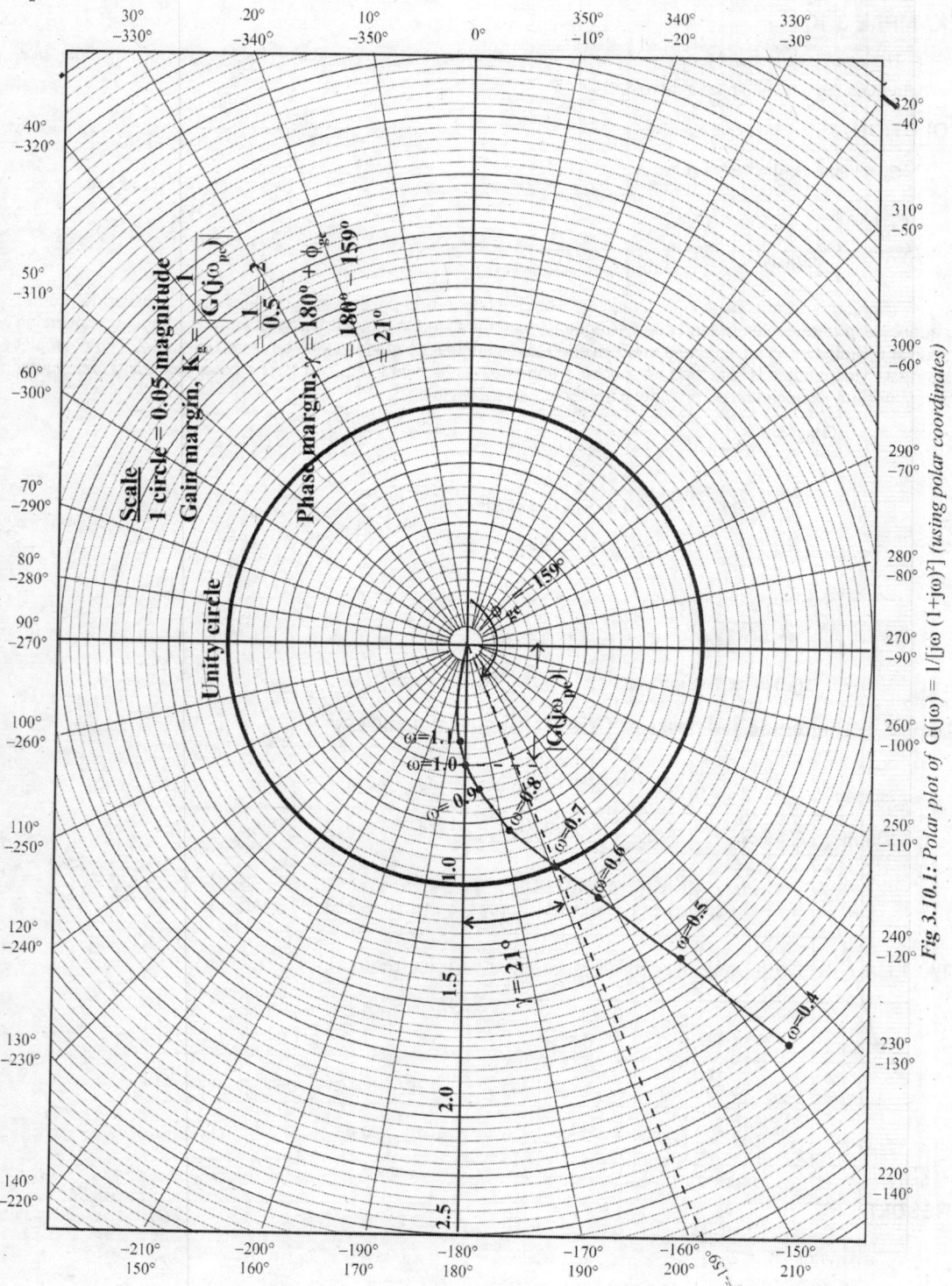

Scale
1 circle = 0.05 magnitude

Gain margin, $K_g = \dfrac{1}{|G(j\omega_{pc})|}$

$= \dfrac{1}{0.5} = 2$

Phase margin, $\gamma = 180^\circ + \phi_{gc}$

$= 180^\circ - 159^\circ$

$= 21^\circ$

Unity circle

$\phi_{pc} = 159^\circ$

$\omega = 1.1$

$\omega = 1.0$

$\omega = 0.9$

$|G(j\omega_{pc})|$

$\omega = 0.8$

$\omega = 0.7$

$\omega = 0.6$

$\omega = 0.5$

$\omega = 0.4$

$\gamma = 21^\circ$

$\phi_{gc} = -159^\circ$

Fig 3.10.1: Polar plot of $G(j\omega) = 1/[j\omega\,(1+j\omega)^2]$ (using polar coordinates)

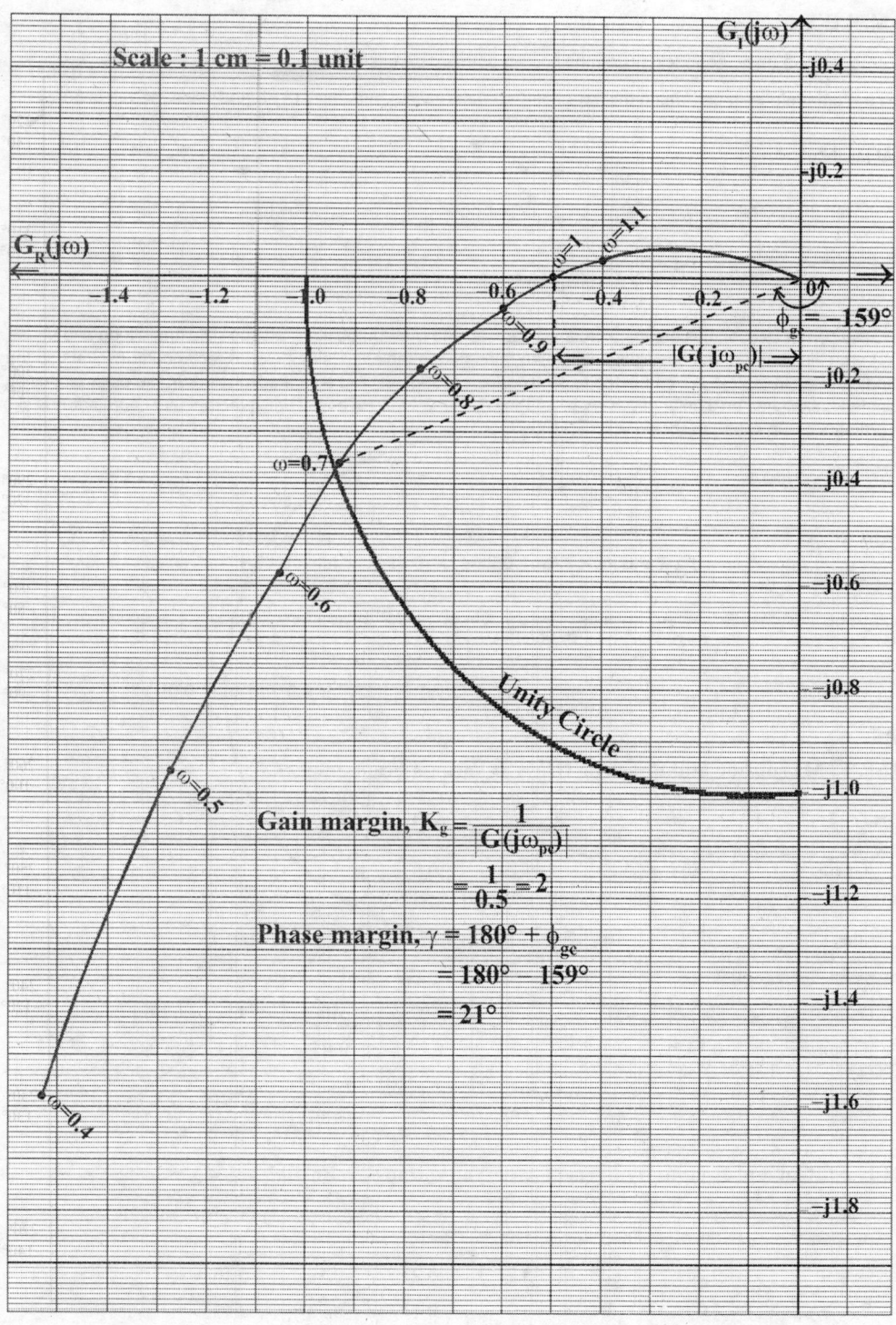

Fig 3.10.2: *Polar plot of* $G(j\omega) = 1/[j\omega\,(1+j\omega)^2]$ *(using rectangular coordinates)*

EXAMPLE 3.11

Consider a unity feedback system having an open loop transfer function $G(s) = \dfrac{K}{s(1+0.2s)(1+0.05s)}$
Sketch the polar plot and determine the value of K so that (i) Gain margin is 18 db (ii) Phase margin is 60°.

SOLUTION

Given that, $G(s) = \dfrac{K}{s(1+0.2s)(1+0.05s)}$. The polar plot is sketched by taking K = 1.

∴ Put K = 1 and s = jω in G(s). ∴ $G(j\omega) = \dfrac{K}{j\omega(1+0.2j\omega)(1+0.05j\omega)}$

The corner frequencies are ω_{c1} = 1/0.2 = 5 rad/sec and ω_{c2} =1/0.05 = 20 rad/sec. The magnitude and phase angle of G(jω) are calculated for various frequencies and tabulated in table-1. Using polar to rectangular conversion the polar coordinates listed in table-1 are converted to rectangular coordinates and tabulated in table-2. The polar plot using polar coordinates is sketched on a polar graph sheet as shown in fig 3.11.1. Polar plot using rectangular coordinates is sketched on an ordinary graph sheet as shown in fig 3.11.2.

$$G(j\omega) = \dfrac{K}{j\omega(1+j0.2\omega)(1+j0.05\omega)}$$

$$= \dfrac{1}{\omega\angle 90° \sqrt{1+(0.2\omega)^2} \angle \tan^{-1}0.2\omega \sqrt{1+(0.05\omega)^2} \angle \tan^{-1}0.05\omega}$$

$$= \dfrac{1}{\omega\sqrt{1+(0.2\omega)^2}\sqrt{1+(0.05\omega)^2}} \angle(-90° - \tan^{-1}0.2\omega - \tan^{-1}0.05\omega)$$

∴ $|G(j\omega)| = \dfrac{1}{\omega\sqrt{1+(0.2\omega)^2}\sqrt{1+(0.05\omega)^2}}$ and $\angle G(j\omega) = -90° - \tan^{-1}0.2\omega - \tan^{-1}0.05\omega$

TABLE-1 : Magnitude and Phase of G(jω) at Various Frequencies

ω rad/sec	0.6	0.8	1	2	3	4
|G(jω)|	1.65	1.23	1.0	0.5	0.3	0.2
∠G(jω) deg	−98	−101	−104	−117.5	−129.4	−140

ω rad/sec	5	6	7	9	10	11	14
|G(jω)|	0.14	0.1	0.07	0.05	0.04	0.03	0.02
∠G(jω) deg	−149	−157	−164	−176	−180	−184	−195

TABLE-2 : Real and Imaginary Parts of G(jω) at Various Frequencies

ω rad/sec	0.6	0.8	1	2	3	4
$G_R(j\omega)$	−0.23	−0.23	−0.24	−0.23	−0.19	−0.15
$G_I(j\omega)$	−1.63	−1.21	−0.97	−0.44	−0.23	−0.13

ω rad/sec	5	6	7	9	10	11	14
$G_R(j\omega)$	−0.120	−0.092	−0.067	−0.050	−0.04	−0.030	−0.019
$G_I(j\omega)$	−0.072	−0.039	−0.019	−0.0034	0	0.002	0.005

In the polar plot shown in fig 3.11.1 and 3.11.2 there are two plots, marked as curve-I and curve-II. These two loci are sketched with different scales to clearly determine the gain margin and phase margin.

From the polar plot, with K = 1,

Gain margin, K_g = 1/0.04 = 25.

Gain margin in db = 20 log 25 = 28 db.

Phase margin, γ = 76°.

Case (i)

With reference to curve I, when K = 1, let G(jω) cut the −180° axis at point B and gain corresponding to that point be G_B. From the polar plot G_B = 0.04. The gain margin of 28 db with K = 1 has to be reduced to 18 db and so K has to be increased to a value greater than one.

Let G_A be the gain at −180° for a gain margin of 18 db.

Now, $20\log\dfrac{1}{G_A} = 18$ $\Rightarrow$ $\log\dfrac{1}{G_A} = \dfrac{18}{20}$ $\Rightarrow$ $\dfrac{1}{G_A} = 10^{18/20}$

$\therefore G_A = \dfrac{1}{10^{18/20}} = 0.125$

The value of K is given by, $K = \dfrac{G_A}{G_B} = \dfrac{0.125}{0.04} = 3.125$

Case (ii)

With reference to curve II, when K = 1, the phase margin is 76°. This has to be reduced to 60°. Hence gain has to be increased.

Let ϕ_{gc2} be the phase of G(jω) for a phase margin of 60°

$\therefore 60° = 180° + \phi_{gc2}$

$\phi_{gc2} = 60° - 180° = -120°$

In the polar plot the −120° line cut the locus of G(jω) at point C and cut the unity circle at point D.

Let, G_C = Magnitude of G(jω) at point C.

G_D = Magnitude of G(jω) at point D.

From the polar plot, G_C = 0.425 and G_D = 1.

Now, $K = \dfrac{G_D}{G_C} = \dfrac{1}{0.425} = 2.353$

RESULT

(a)   When K = 1, Gain margin, K_g = 25

                      Gain margin in db = 28db

(b)   When K = 1, Phase margin, γ = 76°

(c)   For a gain margin of 18 db, K   = 3.125

(d)   For a phase margin of 60°, K   = 2.353

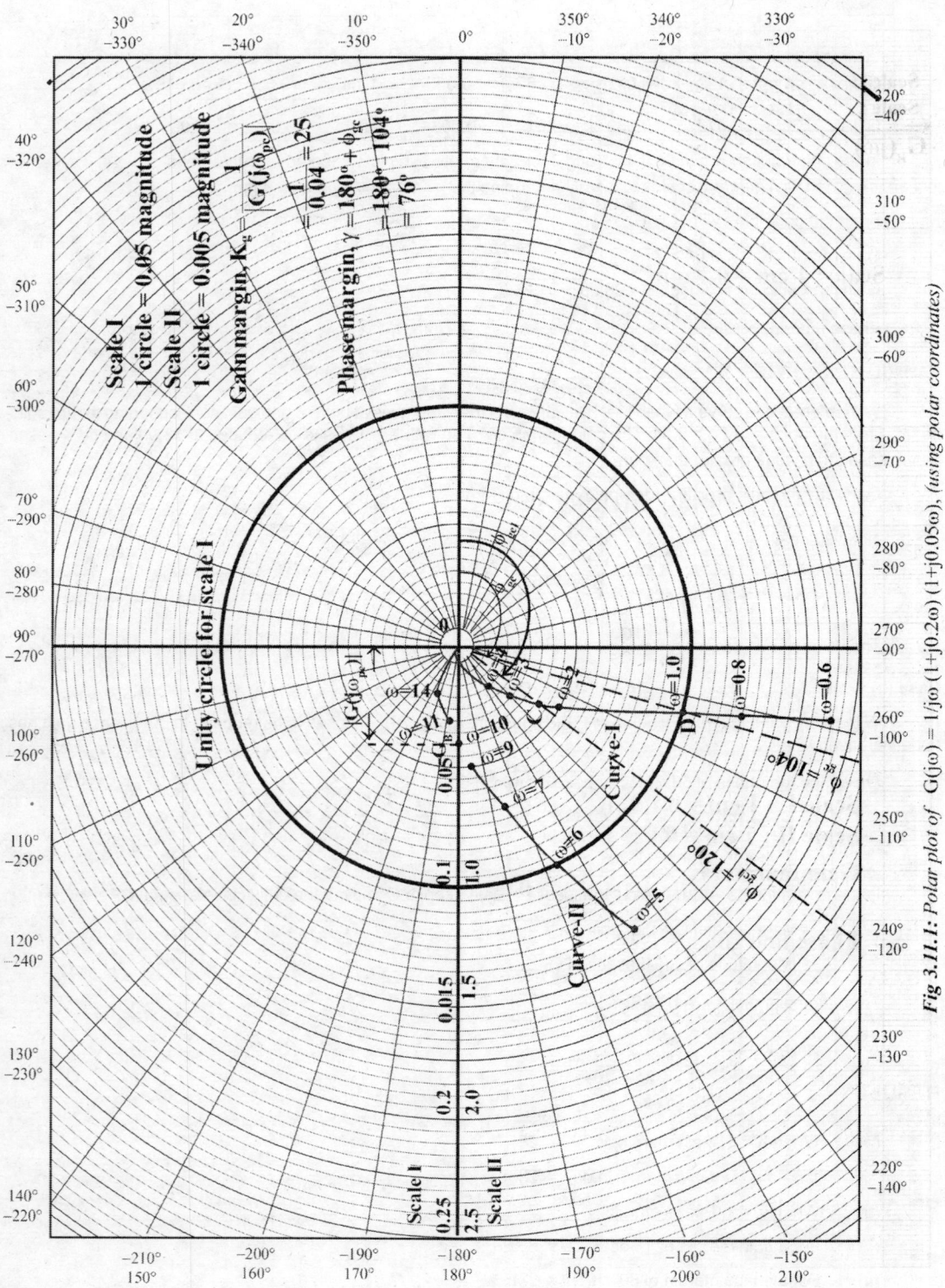

Fig 3.11.1: Polar plot of G(jω) = 1/jω (1+j0.2ω) (1+j0.05ω), (using polar coordinates)

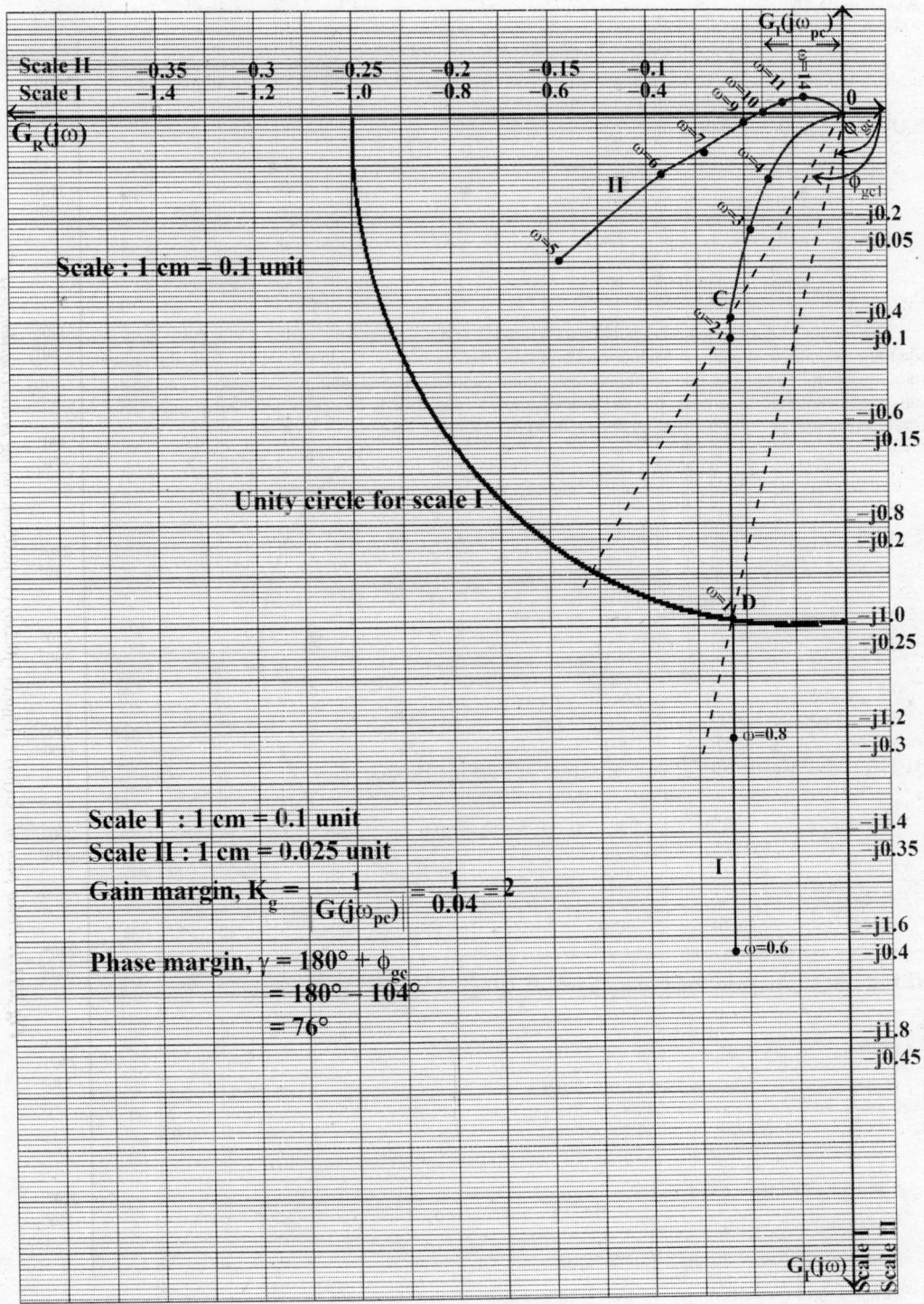

Fig 3.11.2: *Polar plot of* $G(j\omega) = 1/j\omega\,(1+j0.2\omega)\,(1+j0.05\omega)$, *(using rectangular coordinates)*

EXAMPLE 3.12

Consider a unity feedback system having an open loop transfer function, $G(s) = \dfrac{K}{s(1+0.5s)(1+4s)}$ .. Sketch the polar plot and determine the value of K so that (i) Gain margin is 20 db and (ii) Phase margin is 30°.

SOLUTION

Given that, $G(s) = K/s(1+0.5s)(1+4s)$

The polar plot is sketched by taking K = 1.

Put K = 1 and s = jω in G(s).

$$\therefore\ G(j\omega) = \frac{1}{j\omega(1+j0.5\omega)(1+j4\omega)}$$

The corner frequencies are $\omega_{c1} = 1/4 = 0.25$ rad/sec and $\omega_{c2} = 1/0.5 = 2$ rad/sec. The magnitude and phase angle of G(jω) are calculated for various frequencies and tabulated in table-1. Using polar to rectangular conversion the polar coordinates listed in table-1 are converted to rectangular coordinates and tabulated in table-2. The polar plot using polar coordinates is sketched on a polar graph sheet as shown in fig 3.12.1. The polar plot using rectangular coordinates is sketched on an ordinary graph sheet as shown in fig 3.12.2.

$$G(j\omega) = \frac{1}{j\omega(1+j0.5\omega)(1+j4\omega)}$$

$$= \frac{1}{\omega\angle 90°\sqrt{1+(0.5\omega)^2}\angle\tan^{-1}0.5\omega\sqrt{1+(4\omega)^2}\angle\tan^{-1}4\omega}$$

$$= \frac{1}{\omega\sqrt{1+0.25\omega^2}\sqrt{1+16\omega^2}}\angle(-90°-\tan^{-1}0.5\omega-\tan^{-1}4\omega)$$

$$\therefore\ \left|G(j\omega)\right| = \frac{1}{\omega\sqrt{1+0.25\omega^2}\sqrt{1+16\omega^2}}$$

$$\angle G(j\omega) = -90° - \tan^{-1}0.5\omega - \tan^{-1}4\omega$$

TABLE-1 : Magnitude and Phase of G(jω) at Various Frequencies

ω rad/sec	0.3	0.4	0.5	0.6	0.8	1.0	1.2
|G(jω)|	2.11	1.3	0.87	0.61	0.35	0.22	0.15
∠G(jω) deg	−149	−159	−167	−174	−184	−193	−199

TABLE-2 : Real part and Imaginary parts of G(jω) at Various Frequencies

ω rad/sec	0.3	0.4	0.5	0.6	0.8	1.0	1.2
$G_R(j\omega)$	−1.8	−1.21	−0.85	−0.61	−0.35	−0.21	−0.14
$G_i(j\omega)$	−1.09	−0.47	−0.2	−0.06	0.02	0.05	0.05

From the polar plot, with K = 1,

Gain margin, $K_g = 1/0.44 = 2.27$

Gain margin in db = 20 log 2.27 = 7.12 db

Phase margin, $\gamma = 180° + \phi_{gc} = 180° - 165° = 15°$

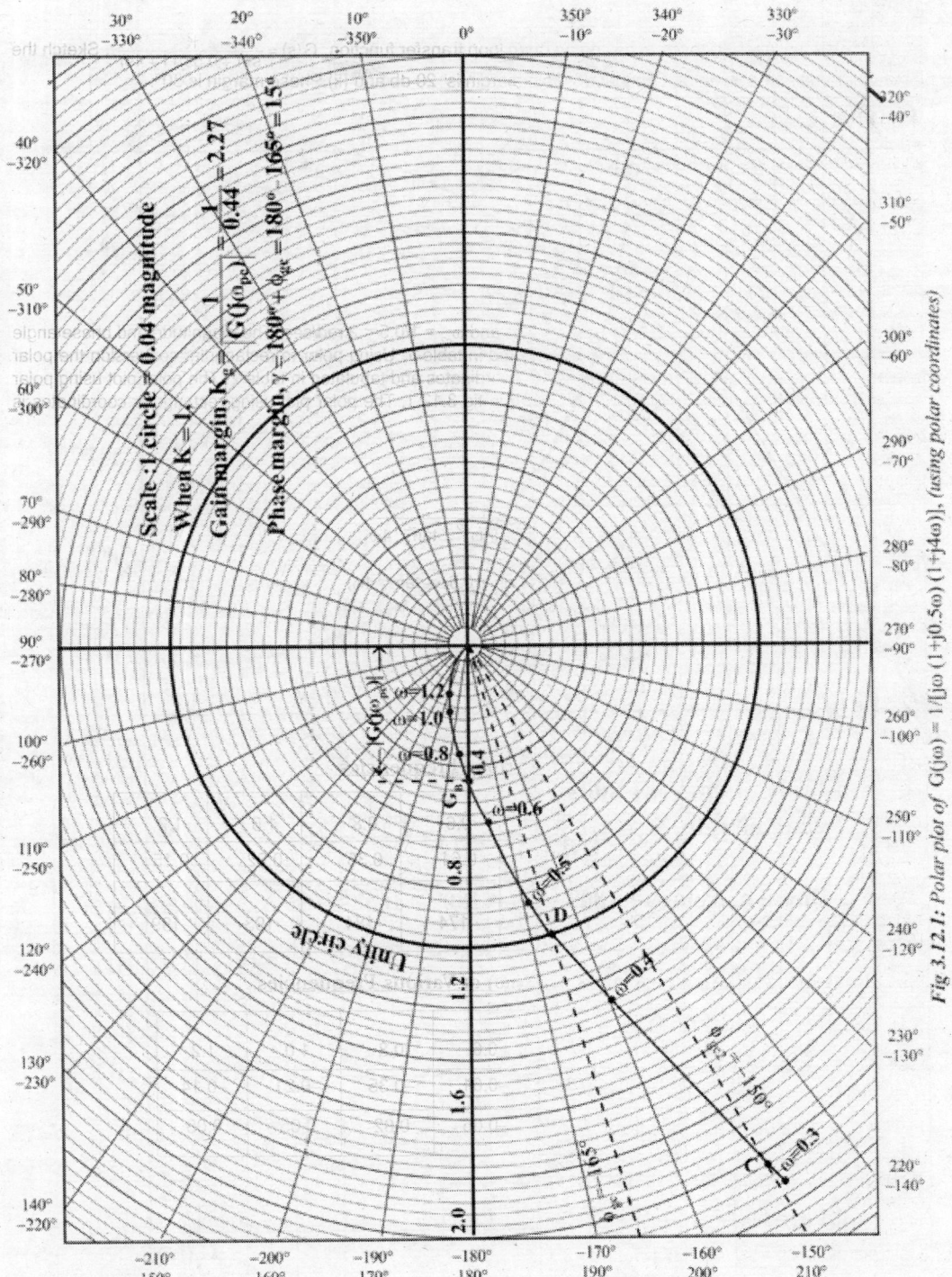

Fig 3.12.1: Polar plot of $G(j\omega) = 1/[j\omega (1+j0.5\omega) (1+j4\omega)]$, (using polar coordinates)

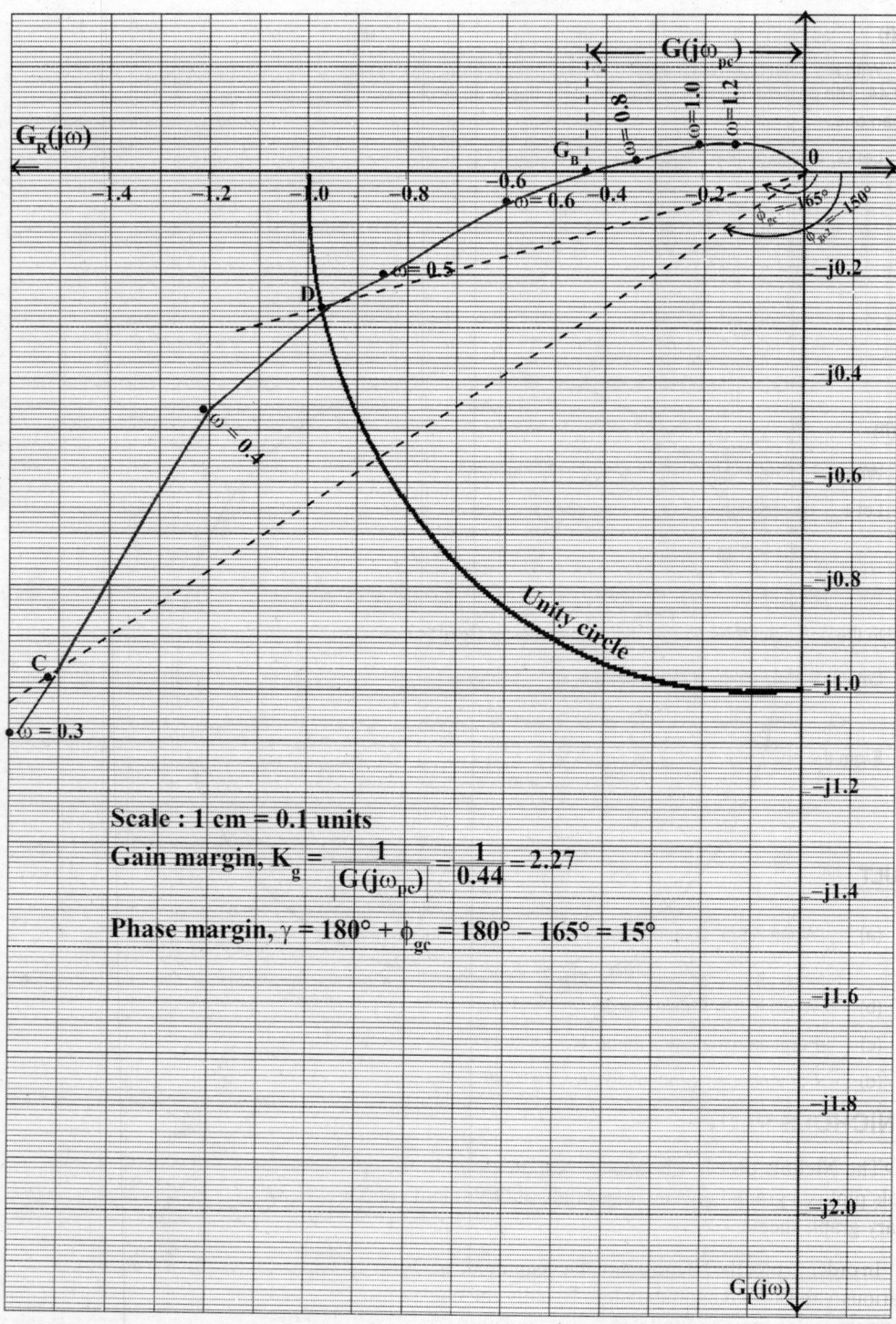

Scale : 1 cm = 0.1 units

Gain margin, $K_g = \dfrac{1}{G(j\omega_{pc})} = \dfrac{1}{0.44} = 2.27$

Phase margin, $\gamma = 180° + \phi_{gc} = 180° - 165° = 15°$

Fig 3.12.2: *Polar plot of* $G(j\omega) = 1/[j\omega\,(1+j0.5\omega)\,(1+j4\omega)]$, *(using rectangular coordinates).*

Case (i)

With K = 1, let $G(j\omega)$ cut the $-180°$ axis at point B and gain corresponding to that point be G_B. From the polar plot, $G_B = 0.44$. The gain margin of 7.12 db with K = 1 has to be increased to 20 db and so K has to be decreased to a value less than one.

Let G_A be the gain at $-180°$ for a gain margin of 20 db.

$$\text{Now, } 20\log\frac{1}{G_A} = 20$$

$$\log\frac{1}{G_A} = \frac{20}{20} = 1$$

$$\frac{1}{G_A} = 10^1 = 10$$

$$\therefore \; G_A = \frac{1}{10} = 0.1$$

Case (ii)

With K = 1, the phase margin is 15°. This has to be increased to 30°. Hence the gain has to be decreased.

Let ϕ_{gc2} be the phase of $G(j\omega)$ for a phase margin of 30°.

$$\therefore \; 30° = 180° + \phi_{gc2}$$

$$\phi_{gc2} = 30° - 180° = -150°$$

In the polar plot the $-150°$ line cuts the locus of $G(j\omega)$ at point C and cut the unity circle at point D.

Let, G_C = Magnitude of $G(j\omega)$ at point C.

G_D = Magnitude of $G(j\omega)$ at point D.

From the polar plot, $G_C = 2.04$ and $G_D = 1$

$$\text{Now, } K = \frac{G_D}{G_C} = \frac{1}{2.04} = 0.49$$

RESULT

(a)	When K = 1, Gain margin, K_g	=	2.27
	Gain margin in db	=	7.12 db
(b)	When K = 1, Phase margin, γ	=	15°
(c)	For a gain margin of 20 db, K	=	0.227
(d)	For a phase margin of 30°, K	=	0.49

3.8 NICHOLS PLOT

The **Nichols plot** is a frequency response plot of the open loop transfer function of a system. The Nichols plot is a graph between magnitude of $G(j\omega)$ in db and the phase of $G(j\omega)$ in degree, plotted on a ordinary graph sheet.

In order to plot the Nichols plot, the magnitude of $G(j\omega)$ in db and phase of $G(j\omega)$ in deg are computed for various values of ω and tabulated. Usually the choice of frequencies are corner frequencies. Choose appropriate scales for magnitude on y–axis and phase on x–axis. Fix all the points on ordinary graph sheet and join the points by smooth curve, and mark frequencies corresponding to each point.

In another method, first the Bode plot of G(jω) is sketched. From the Bode plot the magnitude and phase for various values of frequency, ω are noted and tabulated. Using these values the Nichols plot is sketched as explained earlier.

DETERMINATION OF GAIN MARGIN AND PHASE MARGIN FROM NICHOLS PLOT

The gain margin in db is given by the negative of db magnitude of G(jω) at the phase crossover frequency, ω_{pc}. The ω_{pc} is the frequency at which phase of G(jω) is $-180°$. If the db magnitude of G(jω) at ω_{pc} is negative then gain margin is positive and vice versa.

Let ϕ_{gc} be the phase angle of G(jω) at gain cross over frequency ω_{gc}. The ω_{gc} is the frequency at which the db magnitude of G(jω) is zero. Now the phase margin, γ is given by $\gamma = 180° + \phi_{gc}$. If ϕ_{gc} is less negative than $-180°$ then phase margin is positive and vice versa. The positive and negative gain margins are illustrated in fig 3.27.

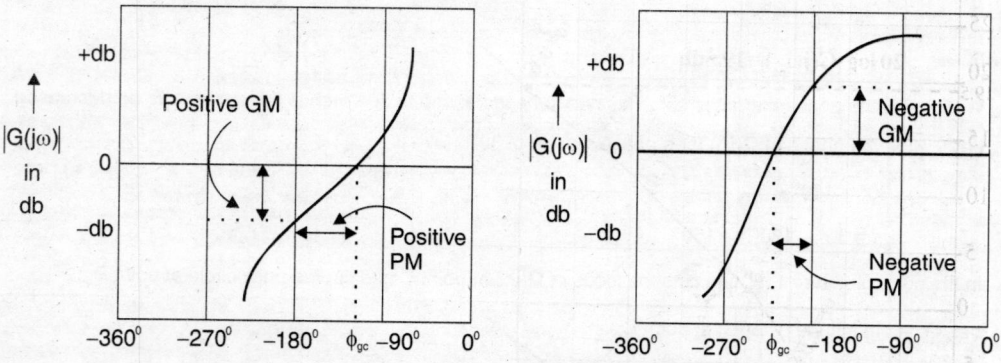

Fig 3.27 : *Nichols plot showing phase margin(PM) and gain margin(GM).*

GAIN ADJUSTMENT IN NICHOLS PLOT

In the open loop transfer function, G(jω) the constant K contributes only magnitude. Hence by changing the value of K the system gain can be adjusted to meet the desired specifications. The desired specifications are gain margin and phase margin.

In a system transfer function, if the value of K required to be estimated, in order to satisfy a desired specification, then draw the Nichols plot of the system with K=1. The constant K can add 20logK to every point of the plot. Due to this addition, the Nichols plot will shift vertically up or down. Hence shift the plot vertically up or down to meet the desired specification. Equate the vertical distance by which the Nichols plot is shifted to 20logK and solve for K.

Let, x = change in db (x is positive if the plot is shifted up and vice versa).

Now, $20 \log K = x$ $\Rightarrow$ $\log K = \dfrac{x}{20}$ $\Rightarrow$ $\therefore K = 10^{\frac{x}{20}}$

EXAMPLE 3.13

Consider a unity feedback system having an open loop transfer function $G(s) = \dfrac{K(1+10s)}{s^2(1+s)(1+2s)}$. Sketch the Nichols plot and determine the value of K so that (i) Gain margin is 10db, (ii) Phase margin is 10°.

SOLUTION

Given that, $G(s) = \dfrac{K(1+10s)}{s^2(1+s)(1+2s)}$

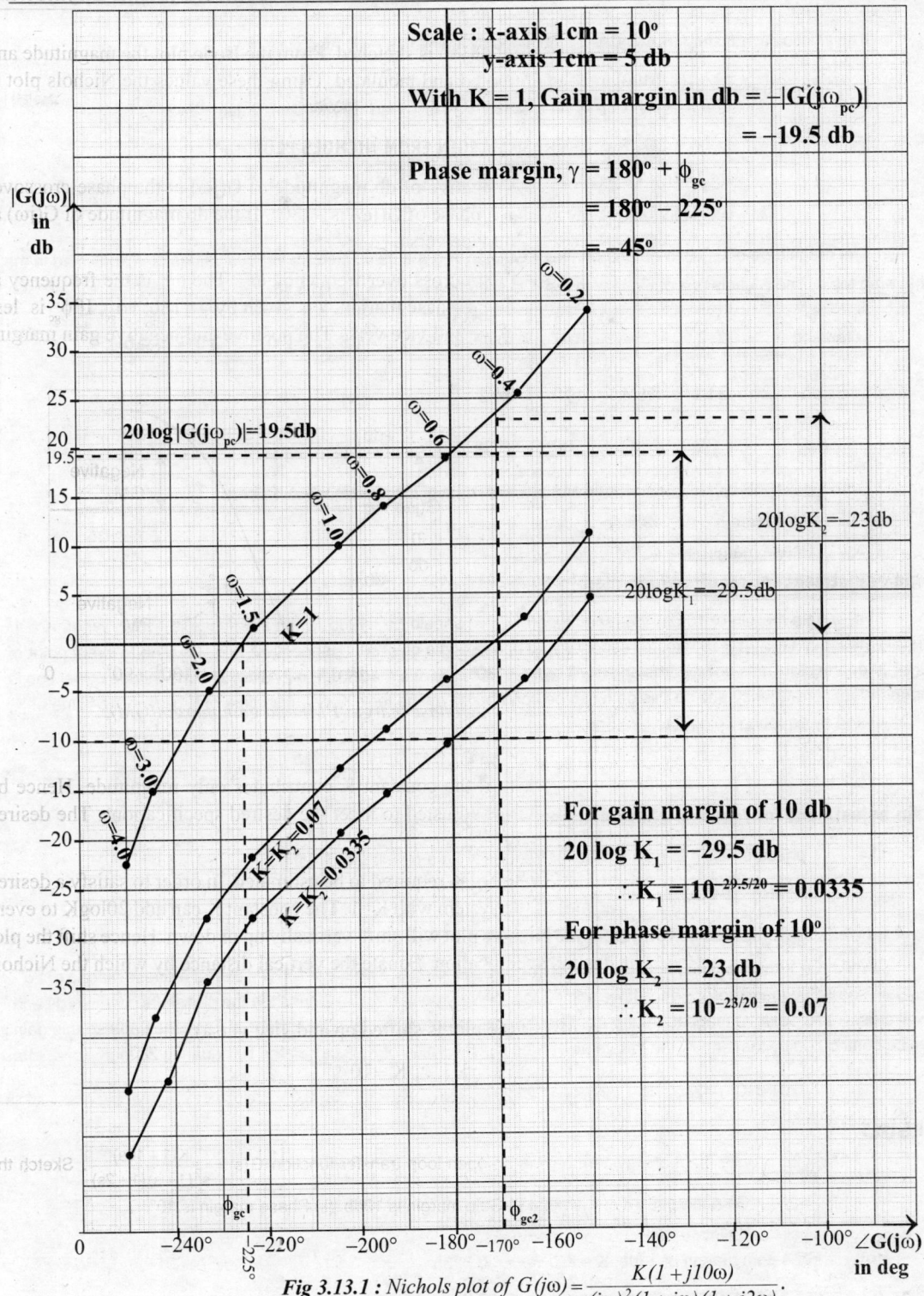

Scale : x-axis 1cm = 10°
y-axis 1cm = 5 db

With K = 1, Gain margin in db = $-|G(j\omega_{pc})|$

= −19.5 db

Phase margin, $\gamma = 180° + \phi_{gc}$

= 180° − 225°

= −45°

$|G(j\omega)|$ in db

$\omega = 0.2$

$\omega = 0.4$

$\omega = 0.6$

20 log $|G(j\omega_{pc})|$=19.5db

$\omega = 0.8$

$\omega = 1.0$

20logK$_2$=−23db

$\omega = 1.5$

K=1

20logK$_1$=−29.5db

$\omega = 2.0$

$\omega = 3.0$

$\omega = 4.0$

K=K$_2$=0.07

K=K$_2$=0.0335

For gain margin of 10 db

20 log K$_1$ = −29.5 db

$\therefore$ K$_1$ = $10^{-29.5/20}$ = 0.0335

For phase margin of 10°

20 log K$_2$ = −23 db

$\therefore$ K$_2$ = $10^{-23/20}$ = 0.07

ϕ_{gc}

ϕ_{gc2}

0   −240° −225° −220°   −200°   −180°−170° −160°   −140°   −120°   −100° $\angle$G(jω)

in deg

***Fig 3.13.1** : Nichols plot of $G(j\omega) = \dfrac{K(1+j10\omega)}{(j\omega)^2(1+j\omega)(1+j2\omega)}$.*

The sinusoidal transfer function $G(j\omega)$ is obtained by letting to $s = j\omega$. Also put $K = 1$.

$$\therefore G(j\omega) = \frac{(1+j10\omega)}{(j\omega)^2(1+j\omega)(1+j2\omega)} = \frac{\sqrt{1+(10\omega)^2} \angle \tan^{-1}10\omega}{\omega^2\angle 180° \sqrt{1+\omega^2} \angle \tan^{-1}\omega \sqrt{1+(2\omega)^2} \angle \tan^{-1}2\omega}$$

$$|G(j\omega)| = \frac{\sqrt{1+100\omega^2}}{\omega^2\sqrt{1+\omega^2}\sqrt{1+4\omega^2}} \; ; \quad \therefore |G(j\omega)|_{in\,db} = 20\log\left[\frac{\sqrt{1+100\omega^2}}{\omega^2\sqrt{1+\omega^2}\sqrt{1+4\omega^2}}\right]$$

$$\angle G(j\omega) = \tan^{-1}10\omega - 180° - \tan^{-1}\omega - \tan^{-1}2\omega$$

The magnitude of $G(j\omega)$ in db and phase of $G(j\omega)$ in deg are calculated for various values of ω and listed in the following table. The Nichols plot of $G(j\omega)$ with $K = 1$ is sketched as shown in fig 3.13.1

ω rad/sec	0.2	0.4	0.6	0.8	1.0	1.5	2.0	3.0	4.0
$|G(j\omega)|$ db	34.1	25.4	19.3	14.3	10	1.4	−5.3	−15.2	−22.5
$\angle G(j\omega)$ deg	−150	−164	−181	−194	−204	−222	−232	−244	−250

From the Nichols plot the gain margin and phase margin of the system when K=1 are,

Gain margin   = − 19.5 db

Phase margin = − 45°

Gain adjustment for required gain margin

For a gain margin of 10 db, the magnitude of $G(j\omega)$ should be −10db, when the phase is −180°. When $K = 1$, the magnitude of $G(j\omega)$ is +19.5db corresponding to phase angle of −180°. Hence if we add −29.5 db to every point of $G(j\omega)$,then the plot shifts downwards and it will cross −180° axis at a magnitude of −10db. The magnitude correction is independent of frequency and so this gain can be contributed by the term K. Let this value of K be K_1. The value of K_1 is calculated by equating $20\log K_1$ to −29.5db.

$$\therefore 20\log K_1 = -29.5\,db \quad \Rightarrow \quad \log K_1 = -\frac{29.5}{20} \quad \Rightarrow \quad K_1 = 10^{\frac{-29.5}{20}} = 0.0335$$

Gain adjustment for required phase margin

Let ϕ_{gc2}= phase of $G(j\omega)$ at gain crossover frequency for a phase margin of 10°

$$\therefore \text{Phase margin}, \gamma_2 = 180° + \phi_{gc2}$$

$$\therefore \phi_{gc2} = \gamma_2 - 180° = 10° - 180° = -170°$$

When $K = 1$, the magnitude of $G(j\omega)$ is +23 db corresponding to a phase of −170°. But for a phase margin of 10°, this gain should be made zero. Hence if we add −23db to every point of $G(j\omega)$ locus then the plot shifts downwards and it will cross −170° axis at magnitude of 0 db. The magnitude correction is independent of frequency and so this gain can be contributed by the term K. Let this value of K be K_2. The value of K_2 is calculated by equating $20\log K_2$ to −23db.

$$\therefore 20\log K_2 = -23 \quad \Rightarrow \quad \log K_2 = -23/20 \quad \Rightarrow \quad K_2 = 10^{\frac{-23}{20}} = 0.07$$

RESULT

(a)   When $K = 1$,

Gain margin   = −19.5 db

Phase margin   = − 45°

(b)   For a gain margin of 10db, $K = K_1$ =   0.0335

(c)   For a phase margin of 10°, $K = K_2$ =   0.07

3.9 CLOSED LOOP RESPONSE FROM OPEN LOOP RESPONSE

The closed loop transfer function of the system is given by,

$$\frac{C(s)}{R(s)} = \frac{G(s)}{1 + G(s)H(s)} = M(s)$$

The sinusoidal transfer function is obtained by replacing s by jω.

$$M(j\omega) = \frac{G(j\omega)}{1 + G(j\omega)H(j\omega)}$$

Let, $M(j\omega) = M\angle\alpha$

where, M = Magnitude of closed loop transfer function

α = Phase of closed loop transfer function.

The magnitude and phase of closed loop system are functions of frequency, ω. The sketch of magnitude and phase of closed loop system with respect to ω is closed loop frequency response plot. The magnitude and phase of closed loop system for various values of frequency can be evaluated analytically or graphically. The analytical method of determining the frequency response involves tedious calculations. Two graphical methods are available to determine the closed loop frequency response from open loop frequency response. They are,

1. M and N circles

2. Nichols chart.

3.10 M AND N CIRCLES

The magnitude of closed loop transfer function with unity feedback can be shown to be in the form of circle for every value of M. These circles are called **M-circles.**

If the phase of closed loop transfer function with unity feedback is α, then it can be shown that tan α will be in the form of circle for every value of α. These circles are called **N-circles.**

The M and N circles are used to find the closed loop frequency response graphically from the open loop frequency response G(jω) without calculating the magnitude and phase of the closed loop transfer function at each frequency.

The M and N circles are available as standard chart. The chart consists of M and N circles superimposed on ordinary graph sheet. Using ordinary graph the locus of G(jω) (Polar Plot) is sketched. The locus of G(jω) will cut the M-circles and N-circles at various points. The intersection of G(jω) locus with M and N circles gives the magnitude and phase of the closed loop system at frequencies corresponding to the cutting point of G(jω).

The M and α for various values of ω are tabulated. The magnitude and phase response of closed loop system are sketched on semilog graph sheet by taking ω on the logarithmic scale on x-axis. [The closed loop frequency response has two plots. They are M Vs ω and α Vs ω].

M-CIRCLES

Consider the closed loop transfer function of unity feedback system, $\quad M(s) = \dfrac{G(s)}{1+G(s)}$

Put $s = j\omega$, $\therefore$ $M(j\omega) = \dfrac{G(j\omega)}{1+G(j\omega)}$

Let, $G(j\omega) = X + jY$

where, X = Real part of $G(j\omega)$.

Y = Imaginary part of $G(j\omega)$.

$\therefore$ $M(j\omega) = \dfrac{X+jY}{1+X+jY} = \dfrac{\sqrt{X^2+Y^2}\angle\tan^{-1}\dfrac{Y}{X}}{\sqrt{(1+X^2)}\angle\tan^{-1}\dfrac{Y}{1+X}} = \dfrac{\sqrt{X^2+Y^2}}{\sqrt{(1+X^2)+Y^2}}\angle\left(\tan^{-1}\dfrac{Y}{X} - \tan^{-1}\dfrac{Y}{1+X}\right)$

Let, M = Magnitude of $M(j\omega)$

$\therefore$ $M = \dfrac{\sqrt{X^2+Y^2}}{\sqrt{(1+X^2)+Y^2}}$

On squaring the above equation we get,

$M^2 = \dfrac{X^2+Y^2}{(1+X^2)+Y^2}$ $\quad\Rightarrow\quad$ $M^2((1+X)^2+Y^2) = X^2+Y^2$ $\quad\Rightarrow\quad$ $M^2(1+X^2+2X+Y^2) = X^2+Y^2$

$M^2 + M^2X^2 + M^2 2X + M^2Y^2 - X^2 - Y^2 = 0$

$X^2(M^2-1) + M^2 2X + M^2 + Y^2(M^2-1) = 0$ $\quad\quad\quad\quad\quad\quad\quad\quad$ .....(3.30)

When $M = 1$, the equation (3.30) represents a straight line.

When $M = 1$, the equation (3.30) is,

$X^2(1-1) + 2X + 1 + Y^2(1-1) = 0$ $\quad\Rightarrow\quad$ $2X + 1 = 0$ $\quad\Rightarrow\quad$ $X = -1/2$

Hence when $M = 1$, equation (3.30) represents a straight line passing through $X = -1/2$ & $Y = 0$.

When $M \neq 1$, the equation (3.30) represents a family of circles.

When $M \neq 1$, equation (3.30) can be rearranged in the form of equation of a circle as shown below.

$X^2(M^2 - 1) + M^2 2X + M^2 + Y^2(M^2 - 1) = 0$

Divide the above equation throughout by $(M^2 - 1)$.

$\therefore$ $X^2 + \dfrac{M^2}{M^2-1}2X + \dfrac{M^2}{M^2-1} + Y^2 = 0$

Add $\dfrac{M^2}{(M^2-1)^2}$ on both sides of the above equation.

$$X^2 + \frac{M^2}{M^2-1}\,2X + \frac{M^2}{M^2-1} + \frac{M^2}{(M^2-1)^2} + Y^2 = \frac{M^2}{(M^2-1)^2}$$

$$X^2 + \frac{M^2}{M^2-1}\,2X + \frac{M^2(M^2-1)+M^2}{(M^2-1)^2} + Y^2 = \frac{M^2}{(M^2-1)^2}$$

$$X^2 + \frac{M^2}{M^2-1}\,2X + \frac{M^4}{(M^2-1)^2} + Y^2 = \frac{M^2}{(M^2-1)^2} \qquad \boxed{a^2 + 2ab + b^2 = (a+b)^2}$$

$$\left(X + \frac{M^2}{M^2-1}\right)^2 + Y^2 = \frac{M^2}{(M^2-1)^2} \qquad\qquad .....(3.31)$$

The equation of circle with centre at (X_1, Y_1) and radius r is given by,

$$(X - X_1)^2 + (Y - Y_1)^2 = r^2 \qquad\qquad .....(3.32)$$

On comparing equation (3.31) and equation (3.32), it can be concluded that the equation (3.31) represents a family circles with centre at $(-M^2/M^2 - 1)$, 0) and with radius, $r = M/(M^2 - 1)$ for various values of M. The circles given by equation (3.31) are called M-circles.

When M = 0	**When M = ∞**
Centre $= (X_1, Y_1)$	Centre $= (X_1, Y_1)$
$X_1 = -\dfrac{M^2}{M^2-1} = 0$	$X_1 = \dfrac{-M^2}{M^2-1} \approx \dfrac{-M^2}{M^2} = -1$
$Y_1 = 0$	$Y_1 = 0$
Radius, $r = \dfrac{M^2}{M^2-1} = 0$	Radius, $r = \dfrac{M}{M^2-1} \approx \dfrac{M}{M^2} = \dfrac{1}{M} = \dfrac{1}{\infty} = 0$
Hence when M = 0, the magnitude circle becomes a point at (0,0).	Hence when M = ∞, the magnitude circle becomes a point at (-1,0).

From the above analysis it is clear that the magnitude of closed loop transfer function will be in the form of circles when M $\neq$ 1 and when M = 1, the magnitude is a straight line passing through $(-1/2,0)$.

For values of M less than 1, the magnitude is a circle to the right of the straight line corresponding to M = 1. It is observed that the circles for M<1 passes through $(-1/2,0)$ and $(0,0)$ on the negative real axis. For decreasing values of M, the radius decreases and the circle, becomes a point at $(0,0)$ when M = 0.

For values of M greater than 1, the magnitude is a circle to the left of the straight line corresponding to M = 1. It is observed that circle passes between the points $(-1,0)$ and $(-1/2,0)$ on the negative real axis. For increasing values of M the radius decreases and the circle becomes a point at $(-1,0)$ when M = ∞. The family of M-circles are shown in fig 3.28.

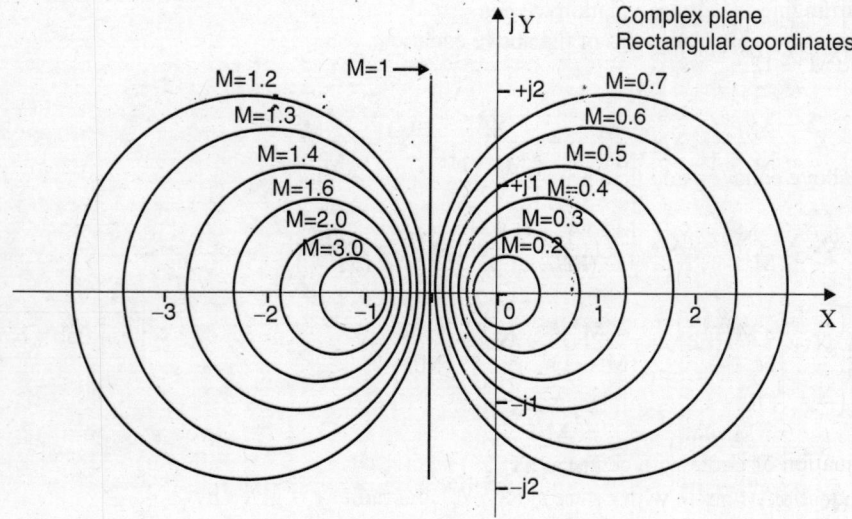

Fig 3.28 : *The family of constant M-circles.*

N-CIRCLES

Consider the closed loop transfer function of unity feedback system.

$$\frac{C(s)}{R(s)} = \frac{G(s)}{1 + G(s)} = M(s)$$

Put $s = j\omega$, $M(j\omega) = \dfrac{G(j\omega)}{1 + G(j\omega)}$

Let $G(j\omega) = X + jY$,   where, X = Real part of $G(j\omega)$.

Y = Imaginary part of $G(j\omega)$.

$$\therefore M(j\omega) = \frac{X + jY}{1 + X + jY} = \frac{\sqrt{X^2 + Y^2} \angle \tan^{-1} \dfrac{Y}{X}}{\sqrt{(1+X)^2 + Y^2} \angle \tan^{-1} \dfrac{Y}{1+X}}$$

$$= \frac{\sqrt{X^2 + Y^2}}{\sqrt{(1+X)^2 + Y^2}} \angle \left(\tan^{-1} \frac{Y}{X} - \tan^{-1} \frac{Y}{1+X} \right)$$

Let, α = Phase of $M(j\omega)$; $\therefore \alpha = \tan^{-1} \dfrac{Y}{X} - \tan^{-1} \dfrac{Y}{1+X}$

Let, $N = \tan \alpha$

$$\therefore N = \tan \left(\tan^{-1} \frac{Y}{X} - \tan^{-1} \frac{Y}{1+X} \right)$$

$$\therefore N = \frac{\tan \left(\tan^{-1} \dfrac{Y}{X} - \tan^{-1} \dfrac{Y}{1+X} \right)}{1 + \tan \left(\tan^{-1} \dfrac{Y}{X} - \tan^{-1} \dfrac{Y}{1+X} \right)} = \frac{\dfrac{Y}{X} - \dfrac{Y}{1+X}}{1 + \dfrac{Y}{X} \times \dfrac{Y}{1+X}} = \frac{\dfrac{Y(1+X) - XY}{X(1+X)}}{\dfrac{X(1+X) + Y^2}{X(1+X)}}$$

$$= \frac{Y + XY - XY}{X + X^2 + Y^2} = \frac{Y}{X + X^2 + Y^2}$$

$$\therefore N = \frac{Y}{X + X^2 + Y^2}$$

> **Note :** $\tan(A - B) = \dfrac{\tan A - \tan B}{1 + \tan A \times \tan B}$

On rearranging the above equation we get.

$$X + X^2 + Y^2 = \frac{Y}{N}$$

$$\therefore X + X^2 + Y^2 - \frac{Y}{N} = 0$$

In the above equation add the term $\frac{1}{4} + \left(\frac{1}{2N}\right)^2$ on both sides,

$$X + X^2 + Y^2 - \frac{Y}{N} + \frac{1}{4} + \left(\frac{1}{2N}\right)^2 = \frac{1}{4} + \left(\frac{1}{2N}\right)^2$$

$$\left(X^2 + \frac{1}{4} + X\right) + \left(Y^2 + \frac{1}{(2N)^2} - \frac{Y}{N}\right) = \frac{1}{4} + \frac{1}{(2N)^2}$$

$$\left(X + \frac{1}{2}\right)^2 + \left(Y - \frac{1}{2N}\right)^2 = \frac{1}{4} + \frac{1}{(2N)^2} \qquad\qquad .....(3.33)$$

The equation of circle with centre at (X_1, Y_1) and radius r is,

$$(X - X_1)^2 + (Y - Y_1)^2 = r^2 \qquad\qquad .....(3.34)$$

On comparing equation (3.33) and (3.34), it can be concluded that the equation (3.33) represents a family of circle with centre at $(-1/2, 1/2N)$ and with radius $\sqrt{\frac{1}{4} + \frac{1}{(2N)^2}}$ for various values of N. The circles given by the equation (3.33) are called N-circles.

For any value of N, the equation of N-circles is satisfied at two points $(0,0)$ and $(-1,0)$. Hence the N-circles passes through these two points for all values of α. ($N = \tan \alpha$).

Consider the equation of N-circle,

When $X = 0$ and $Y = 0$,

$$\left(X + \frac{1}{2}\right)^2 + \left(Y - \frac{1}{2N}\right)^2 = \frac{1}{4} + \frac{1}{(2N)^2}$$

$$\left(\frac{1}{2}\right)^2 + \left(-\frac{1}{2N}\right)^2 = \frac{1}{4} + \frac{1}{(2N)^2}$$

$$\frac{1}{4} + \frac{1}{4N^2} = \frac{1}{4} + \frac{1}{4N^2}$$

Consider the equation of N-circle,

When $X = -1$ and $Y = 0$,

$$\left(X + \frac{1}{2}\right)^2 + \left(Y - \frac{1}{2N}\right)^2 = \frac{1}{4} + \frac{1}{(2N)^2}$$

$$\left(-1 + \frac{1}{2}\right)^2 + \left(-\frac{1}{2N}\right)^2 = \frac{1}{4} + \frac{1}{(4N)^2}$$

$$\frac{1}{4} + \frac{1}{4N^2} = \frac{1}{4} + \frac{1}{4N^2}$$

The above analysis shows that the equation of N-circle is satisfied at points $(0,0)$ and $(-1, 0)$.

When $\alpha = 180°$ the circle becomes a straight line passing through real axis. It is also observed that the circle for $\alpha = \theta°-180°$ above the real axis will be a part of circle for $\alpha = \theta°$ below the real axis, as shown in fig 3.29. The family of N circles are shown in fig 3.30.

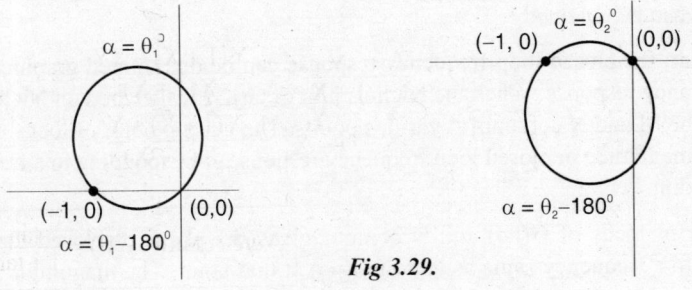

Fig 3.29.

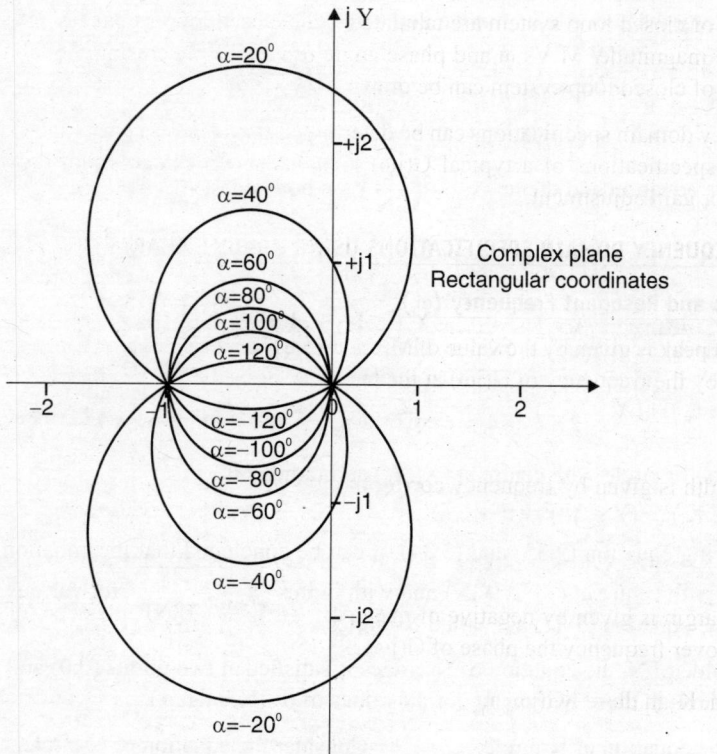

Fig 3.30 : *The family of constant N-circles.*

3.11 NICHOLS CHART

N.B Nichols transformed the constant M and N circles to log-magnitude and phase angle coordinates and the resulting chart is known as *Nichols chart*.

Nichols chart consist of M and N contours, superimposed on ordinary graph. The M contours are the magnitude of closed loop system in decibels and the N contours are the corresponding phase angle locus of closed loop system. The ordinary graph consist of magnitude in db marked on the Y-axis and the phase in degrees marked on the X-axis.

The Nichols plot of open loop system can be plotted on the ordinary graph. The Nichols plot is a graph between magnitude of $G(j\omega)$ in db and the phase of $G(j\omega)$ in degree,plotted on an ordinary graph sheet. To draw the Nichols plot the magnitude and phase angle of $G(j\omega)$ are calculated for various values of ω. Alternatively the Bode plot of $G(j\omega)$ is sketched and from Bode plot, the magnitude and phase of $G(j\omega)$ for any frequency can be obtained.

Using Nichols chart the closed loop frequency response can be determined graphically from the locus of open loop frequency response. When the Nichols plot of $G(j\omega)$ is sketched on Nichols chart, the locus of $G(j\omega)$ will cut the M and N contours at various points. The cutting point of locus of $G(j\omega)$ with the M-contour gives the magnitude of closed loop frequency response corresponding to a frequency same as that of $G(j\omega)$ at that point.

The cutting point of locus of $G(j\omega)$ and N contour gives the phase of closed loop frequency response corresponding to a frequency same as that of $G(j\omega)$ at that point. The magnitude M and phase angle α (N = tanα) of closed loop system are tabulated . The closed loop frequency response consists of two plots. They are magnitude M Vs ω and phase angle α Vs ω. Hence using the tabulated values the bode plot of closed of closed loop system can be drawn.

The frequency domain specifications can be determined from Nicols chart. Fig 3.31, shows various frequency domain specifications of a typical G(jω) locus. Also the Nichols plot drawn on a Nichols chart can be used for gain adjustment.

ESTIMATION OF FREQUENCY DOMAIN SPECIFICATIONS USING NICHOLS CHART

Resonant Peak (M$_r$) and Resonant Frequency (ω_r)

The resonant peak is given by the value of M-contour which is tangent to G(jω) locus. The resonant frequency is given by the frequency of G(jω) at the tangency point.

Bandwidth

The Bandwidth is given by frequency corresponding to the intersection point of G(jω) and −3 db M -contour.

Gain Margin

The gain margin is given by negative of magnitude of G(jω) in db at phase crossover frequency, ω_{pc}. At phase crossover frequency the phase of G(jω) is −180°

Gain Margin, K$_g$ in db = $-|G(j\omega_{pc})|_{in\ db}$

Phase Margin

The phase margin,γ is given by $\gamma = 180^0 + \phi_{gc}$ where ϕ_{gc} is the phase of G(jω) at gain crossover frequency. At gain crossover frequency the magnitude of G(jω) is zero db.

GAIN ADJUSTMENT USING NICHOLS CHART

Determination of K for Specified Gain Margin

Draw the G(jω) locus with K=1. Determine the amount of gain to be added at $\phi = -180^0$, so that db magnitude of G(jω) locus at −180^0 is negative of the specified gain margin. Let the db gain to be added be x db. The gain contribution is independent of frequency and so it can be achieved by choosing proper value of K. The value of K is obtained by equating 20logK to x db.

Now, 20logK = x

$\therefore$ K = $10^{\frac{x}{20}}$

Determination of K for Specified Phase Margin

Draw the G(jω) locus with K= 1. The phase margin, $\gamma = 180° + \phi_{gc}$ where ϕ_{gc}, is phase of G(jω) at gain crossover frequency. $\therefore \phi_{gc} = \gamma - 180°$. For specified phase margin, calculate ϕ_{gc} and from the Nichols plot determine the db gain at o$_{gc}$. Let this gain be y db. For the specified phase margin, this gain should be made zero. Hence −y db should be added to every point of G(jω). This is achieved by choosing proper value of K. The value of K is obtained by equating 20logK to −y db.

Now, 20logK = −y $\quad \Rightarrow \quad$ log K = $-\dfrac{y}{20}$ $\quad \Rightarrow \quad$ K = $10^{\frac{-y}{20}}$

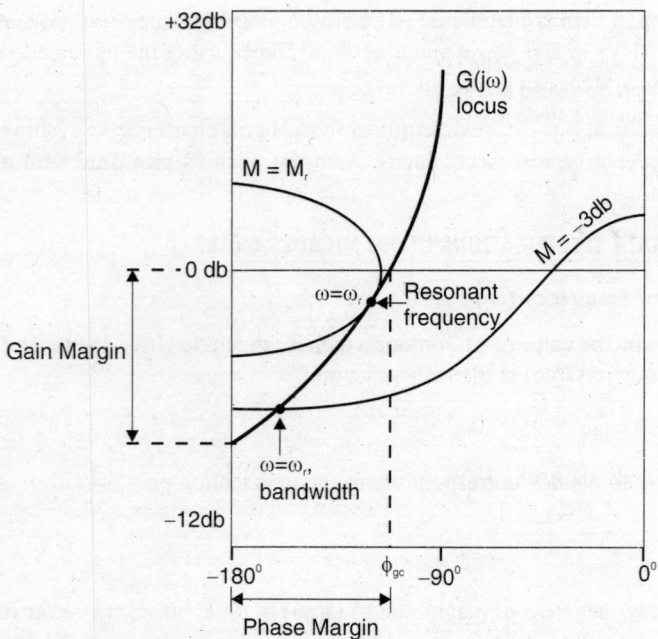

Fig 3.31 : Determination of frequency domain specification from Nichols Chart.

Determination of K for Specified Resonant Peak, M_r

Draw the $G(j\omega)$ locus with K = 1. Using a tracing paper, trace the locus of $G(j\omega)$. A standard tracing paper, called Nichols overlay is available). Then shift the locus vertically up or down, so that $M = M_r$ contour is tangent to $G(j\omega)$ locus. Measure the vertical shift in db. Let the shift be ± x db. (+ for up and –for down).

$$\text{Now,}\quad 20\log K = \pm x \quad\Rightarrow\quad \log K = \pm \frac{x}{20} \quad\Rightarrow\quad K = 10^{\pm \frac{x}{20}}$$

Determination of K for a Specified Bandwidth

Draw the $G(j\omega)$ locus with K=1. Determine the open loop gain $G(j\omega)$ at $\omega = \omega_b$ where, ω_b is the specified bandwith. Determine the point of intersection of -3db M-contour and this open loop gain on the Nichols chart. Let this point be point A. Trace the $G(j\omega)$ locus. Shift the $G(j\omega)$ locus vertically up or down, so that it passes through point A. Measure the vertical shift in db. Let the shift be ± x db (+for up and –for down).

$$\text{Now,}\quad 20\log K = \pm x \quad\Rightarrow\quad \log K = \pm \frac{x}{20} \quad\Rightarrow\quad K = 10^{\pm \frac{x}{20}}$$

EXAMPLE 3.14

The open loop transfer function of unity feedback system is, $G(s) = Ke^{-0.2s}/s(1+0.25s)(1+0.1s)$. Using Nichols chart, determine the following.

(a) The value of K so that the gain margin of the system is 4 db.

(b) The value of K so that the phase margin of the system is 40°

(c) The value of K so that resonant peak M_r of the system is 1 db. What are the corresponding values of ω_r and ω_b?

(d) The value of K so that the bandwidth ω_b of the system is 1.5 rad/sec.

SOLUTION

First the actual bode plot of $G(j\omega)$ with K=1 is plotted on semilog graph sheet. The magnitude of $G(j\omega)$ in db and phase of $G(j\omega)$ for various frequencies are calculated and listed in Table-1. The choice of frequencies are chosen such that the magnitude plot extends in the range of 40 db to −14db and the phase plot extends in the range of 0° to −180°.

Given that, $G(s) = \dfrac{Ke^{-0.2s}}{s(1+0.25s)(1+0.1s)}$

Let, K = 1 and put, s = jω

$\therefore G(j\omega) = \dfrac{e^{-j0.2\omega}}{j\omega(1+0.25s)(1+0.1s)} = \dfrac{1\angle -0.2\omega \times \dfrac{180°}{\pi}}{\omega\angle 90° \sqrt{1+0.0625\ \omega^2}\angle\tan^{-1}0.25\omega\sqrt{1+0.01\omega^2}\angle\tan^{-1}0.1\omega}$

$\therefore |G(j\omega)| = \dfrac{1}{\omega\sqrt{1+0.0625\ \omega^2}\sqrt{1+0.01\omega^2}}$; $\therefore |G(j\omega)|_{in\,db} = 20\log\left[\dfrac{1}{\omega\sqrt{1+0.0625\ \omega^2}\sqrt{1+0.01\omega^2}}\right]$

$\angle G(j\omega) = -0.2\omega \times \dfrac{180°}{\pi} - 90° - \tan^{-1}0.25\omega - \tan - 10.1\omega$

TABLE-1: Calculated values of $|G(j\omega)|$ and $\angle G(j\omega)$

ω rad/sec	0.01	0.02	0.05	0.1	0.2	0.5	1.0	2.0	4.0
|G(jω)| db	40	34	26	20	14	6	0	−7	−16
∠G(jω) deg	−90	−91	−91	−93	−96	−106	−121	−151	−203

The magnitude and phase plot of Bode plot of $G(j\omega)$ are shown in fig 3.14.1 From the bode plot the phase and frequency for various values of magnitudes are noted and tabulated in table-2. (The choice of magnitudes are 20,16,12,.... i.e, in steps of 4 db, which is convenient for Nichols plot on Nichols chart). Using the values listed in table-2 the locus of $G(j\omega)$ on Nichols chart is sketched as shown in fig 3.14.2.

TABLE-2: Values of $|G(j\omega)|$ and $\angle G(j\omega)$ Noted from Bode Plot

ω rad/sec	0.1	0.16	0.25	0.4	0.64	1.0	1.5	2.2	3.0
|G(jω)| db	20	16	12	8	4	0	−4	−8	−12
∠G(jω) deg	−90	−92	−96	−102	−110	−120	−136	−156	−180

Gain Margin and Phase Margin when K = 1

When K = 1, the $G(j\omega)$ locus cuts the −180° axis at −12db. Hence the magnitude at phase crossover frequency is −12db.

$\therefore$ Gain Margin, $K_g = -|G(j\omega_{pc})|_{indb} = -(-12) = +12$db.

When K = 1, the phase of $G(j\omega)$ is −120° corresponding to magnitude of 0 db. Hence the phase at gain crossover frequency is −120°.

$\therefore$ Phase margin, $\gamma = 180° + \phi_{gc} = 180° - 120° = 60°$

To find K for a gain margin of 4db

When gain margin is 4 db, the locus of $G(j\omega)$ should cross the 180° axis at −4 db. When K = 1, the magnitude of $G(j\omega)$ is −12db corresponding to a phase of 180°. Hence if we add, $-4 - (-12) = 8$ db to every point of $G(j\omega)$ then the plot shifts upwards and crosses −180° axis at −4 db. This magnitude correction is achieved by choosing appropriate value of K. The value of K is obtained by equating $20\log K$ to 8 db.

$$\therefore 20\log K = 20\,db \quad \Rightarrow \quad \log K = \frac{8}{20} \quad \Rightarrow \quad K = 10^{\frac{8}{20}} = 2.5$$

The $G(j\omega)$, when $K = 2.5$ is shown in fig 3.14.2.

To find K for phase margin of 40°

Let ϕ_{gc2} be the phase of $G(j\omega)$ at gain crossover frequency when the phase margin is 40°.

$$\therefore \text{Phase margin, } \gamma_2 = 180° + \phi_{gc2}$$

$$\therefore \phi_{gc2} = \gamma_2 - 180° = 40° - 180° = -140°$$

From the above calculation it is evident that for a phase margin of 40°, the magnitude of $G(j\omega)$ should be 0 db corresponding to a phase of $-140°$. When K = 1, the magnitude of $G(j\omega)$ is $-5db$ corresponding to a phase of $-140°$. Hence if we add +5 db to every point of $G(j\omega)$ locus then the plot shifts upwards and crosses $-140°$ axis at 0 db. This magnitude correction is achieved by choosing appropriate values of K. The value of K is obtained by equating 20 logK to 5 db.

$$\therefore 20\log K = 5 \quad \Rightarrow \quad \log K = \frac{5}{20} \quad \Rightarrow \quad K = 10^{\frac{5}{20}} = 1.78$$

The $G(j\omega)$ locus, when $K = 1.78$ is shown in fig 3.14.2

To find K for a resonant peak of 1 db

The resonant peak, M_r is given by M-contour which is tangent to $G(j\omega)$ locus. When K = 1, the $G(j\omega)$ locus is tangent to M = 0.25 db contour. Hence when, K = 1, resonant peak is 0.25 db.

For a resonant peak of 1 db, the M = 1 db contour should be made tangent to $G(j\omega)$ locus. For this, $G(j\omega)$ locus can be shifted vertically up or down so that it becomes tangent to M = 1 db contour. In this problem the $G(j\omega)$ locus is shifted vertically up to make it tangent to M = 1 db contour. The shifted $G(j\omega)$ locus is shown in fig 3.14.3.

> **Note :** *Trace the $G(j\omega)$ locus when K = 1 on a tracing paper and shift the traced locus over the Nichols chart vertically so that it is tangent to required M-contour. By keeping the tracing paper at the shifted position darken the traced locus, so that it makes an impression on nichols chart.*

The vertical shift is equivalent to adding a magnitude of 20 log K to every point of $G(j\omega)$ locus. From the shifted locus of $G(j\omega)$ it is observed that +2db is added to every point of $G(j\omega)$ locus. Hence the value of K is obtained by equating 20 log K to +2 db.

$$20\log K = 2\,db \quad \Rightarrow \quad \log K = \frac{2}{20} \quad \Rightarrow \quad K = 10^{\frac{2}{20}} = 1.26$$

The resonant frequency, ω_r is given by the frequency of $G(j\omega)$ at the tangency point. The magnitude of $G(j\omega)$ is 0 db at the tangency point of M = 1 db contour. The corresponding frequency is noted from the bode plot of $G(j\omega)$. From the bode plot the frequency at 0 db is 1.0 rad/sec. Hence the resonant frequency, $\omega_r = 1.0$ rad/sec.

To find K so that ω_b = 1.5 rad/sec

The bandwidth, ω_b is given by the frequency of $G(j\omega)$ corresponding to the meeting point of $G(j\omega)$ locus and M = $-3db$ contour. From the bode plot find the magnitude of $G(j\omega)$ when $\omega = 1.5$ rad/sec. From fig 3.14.1 it is observed that magnitude of $G(j\omega)$ is $-4db$ when $\omega = 1.5$ rad/sec.

In the Nichols chart, find the point where the M = $-3db$ contour passes through $-4db$ line. Let this point be P. Now the $G(j\omega)$ locus with K = 1 is shifted vertically down so that is passes through point P. The shifted $G(j\omega)$ locus is shown in fig 3.1.3.

> **Note :** *Trace the $G(j\omega)$ locus when K = 1 on a tracing paper and shift the raced locus over Nichols chart so that is passes through point P. By keeping the tracing paper at the shifted position, darken the traced locus, so that is makes an impression on Nichols chart.*

The vertical shift is equivalent to adding a magnitude of 20logK to every point of $G(j\omega)$ locus. From the shifted locusof $G(j\omega)$ it is observed that -6db is added to every point of $G(j\omega)$ locus. Hence the value of K is obtained by equating 20logK to -6 db.

$$20\log K = -6\,db \quad \Rightarrow \quad \log K = -\frac{6}{20} \quad \Rightarrow \quad K = 10^{-6/20} = 0.5$$

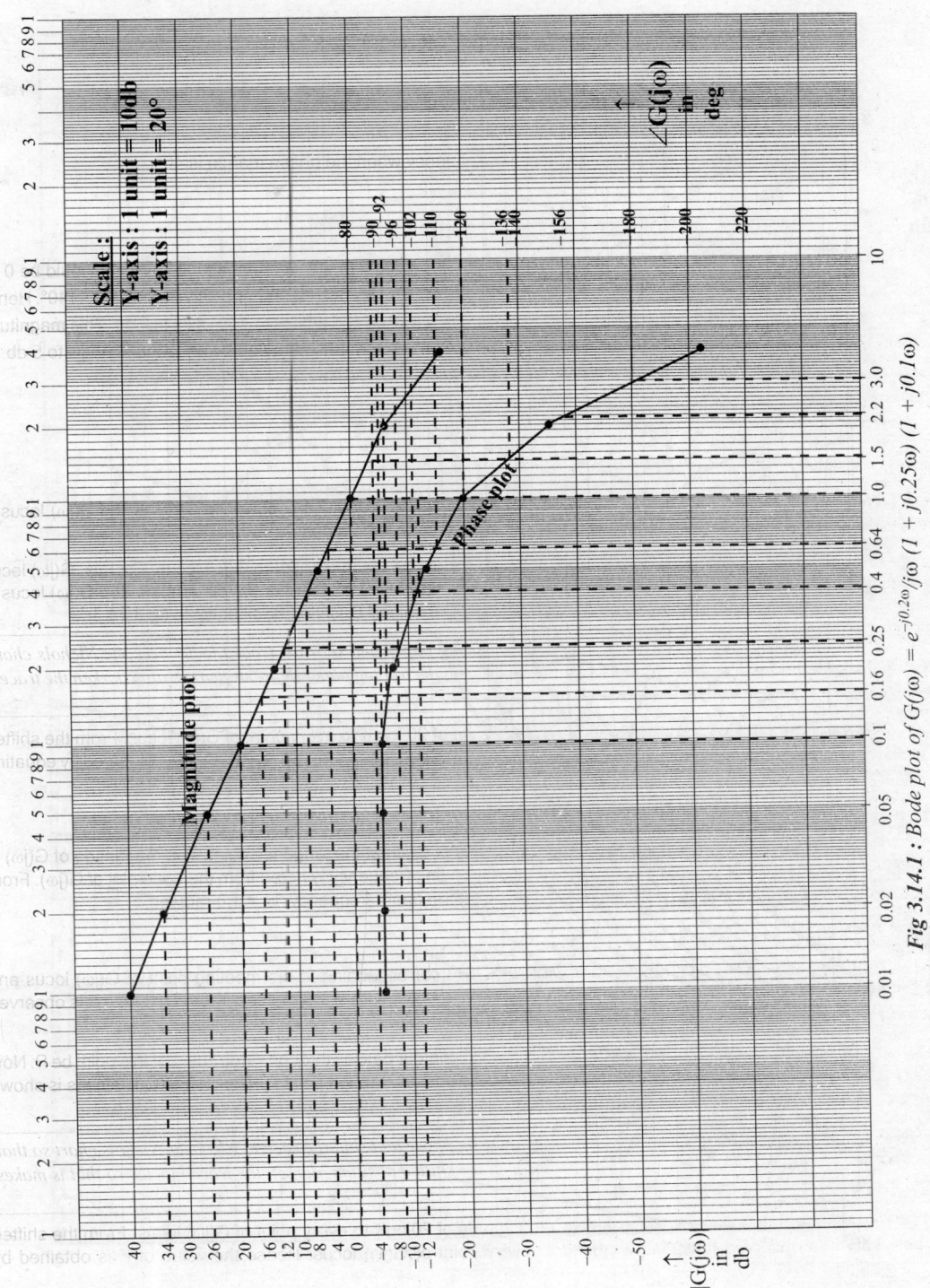

Fig 3.14.1 : Bode plot of $G(j\omega) = e^{-j0.2\omega}/j\omega \, (1 + j0.25\omega) \, (1 + j0.1\omega)$

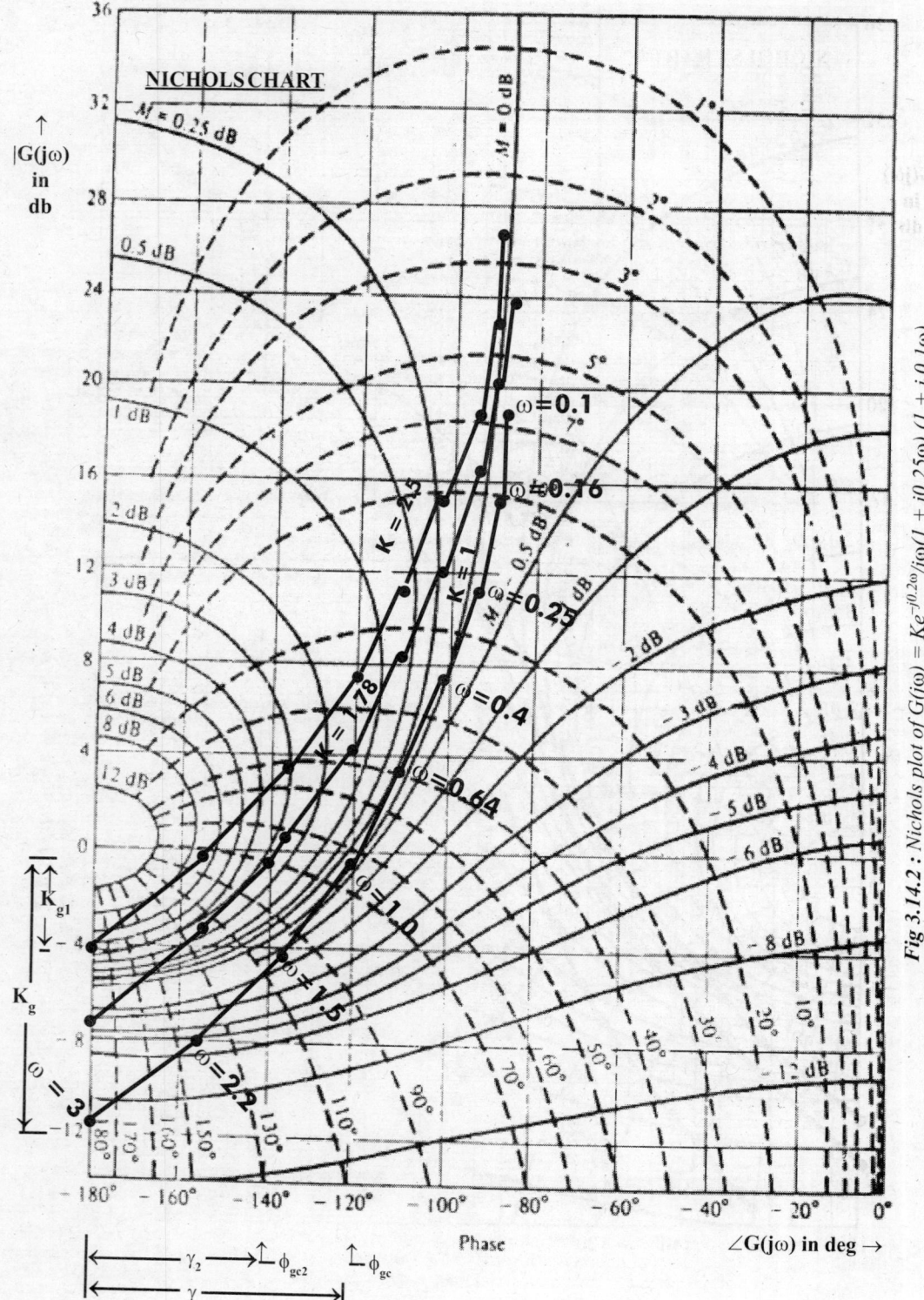

Fig 3.14.2 : Nichols plot of G(jω) = Ke^{-j0.2ω}/jω(1 + j0.25ω) (1 + j 0.1ω)

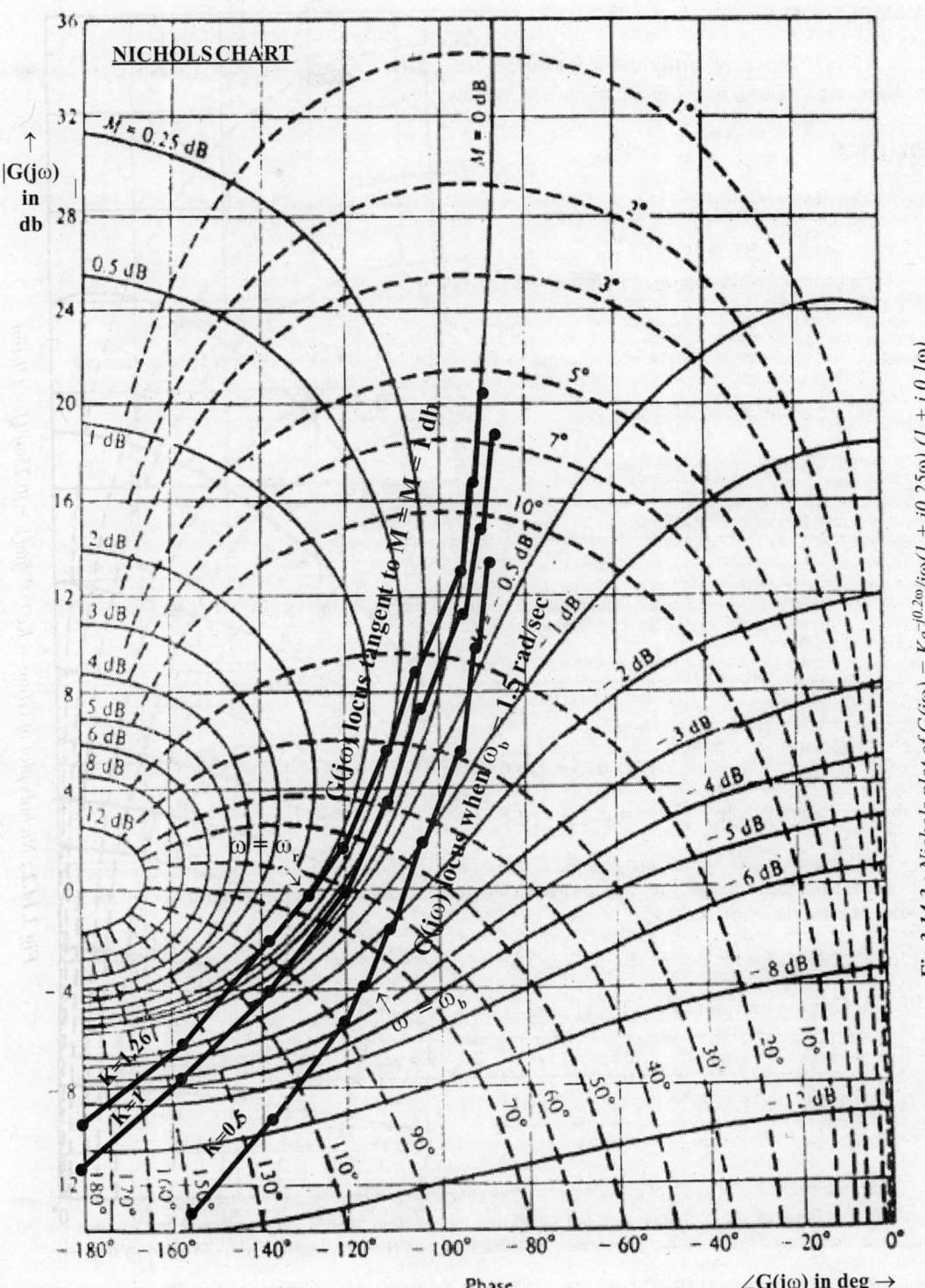

Fig 3.14.3 : Nichols plot of $G(j\omega) = Ke^{-j0.2\omega}/j\omega(1 + j0.25\omega)(1 + j\,0.1\omega)$

EXAMPLE 3.15

A unity feedback system has open loop transfer function, $G(s) = \dfrac{20}{s(s+2)(s+5)}$. Using Nichols chart, determine the closed loop frequency response and estimate M_r, ω_r and ω_b.

SOLUTION

Given that,

$$G(s) = \frac{20}{s(s+2)(s+5)}$$

The transfer function G(s) is converted to time constant or bode form.

$$G(s) = \frac{20}{s \times 2\left(\frac{s}{2}+1\right) \times 5\left(\frac{s}{5}+1\right)} = \frac{20/(2 \times 5)}{s\left(1+\frac{s}{2}\right)\left(1+\frac{s}{5}\right)} = \frac{2}{s(1+0.5s)(1+0.2s)}$$

Put, $s = j\omega$, in G(s) to get G(jω).

$$\therefore G(j\omega) = \frac{2}{j\omega(1+j0.5\omega)(1+j0.2\omega)}$$

$$= \frac{2}{\omega\angle 90° \sqrt{1+0.25\omega^2}\angle\tan^{-1}0.5\omega \sqrt{1+0.04\omega^2}\angle\tan^{-1}0.2\omega}$$

$$= \frac{2}{\omega\sqrt{1+0.25\omega^2}\sqrt{1+0.04\omega^2}} \angle(-90° - \tan^{-1}0.5\omega - \tan^{-1}0.2\omega)$$

$$\therefore \left|G(j\omega)\right| = \frac{2}{\omega\sqrt{1+0.25\omega^2}\sqrt{1+0.04\omega^2}}$$

$$\left|G(j\omega)\right|_{\text{in db}} = 20\log\left[\frac{2}{\omega\sqrt{1+0.25\omega^2}\sqrt{1+0.04\omega^2}}\right]$$

$$\angle G(j\omega) = -90° - \tan^{-1}0.5\omega - \tan^{-1}0.2\omega$$

The magnitude of G(jω) in db and phase of G(jω)for various frequencies are calculated and listed in table-1. The choice of frequencies are chosen such that the magnitude plot extends in the range of 40 db to –14 db and the phase plot extends in the range of 0° to –180°.

Using table-1, the actual bode plot of G(jω) is plotted on semilog graph sheet, as shown in fig 3.15.1.

TABLE-1: Calculated values of $|G(j\omega)|$ and $\angle G(j\omega)$

ω, rad/sec	0.2	0.5	1.0	2.0	3.0	4.0
$|G(j\omega)|$, db	20	12	5	–4	–10	–15
$\angle G(j\omega)$, deg	–98	–110	–128	–157	–177	–192

The magnitude and phase plot of bode plot of G(jω) are shown in fig 3.15.1 From the bode plot, the phase and frequency for various values of magnitudes are noted and tabulated in table-2. (The choice of magnitudes are 20, 16,12,..., i.e., in steps of 4 db, which is convenient for Nichols plot on Nichols chart).

Using the values listed in table-2, the locus of G(jω) is sketched on the Nichols chart as shown in fig 3.15.2.

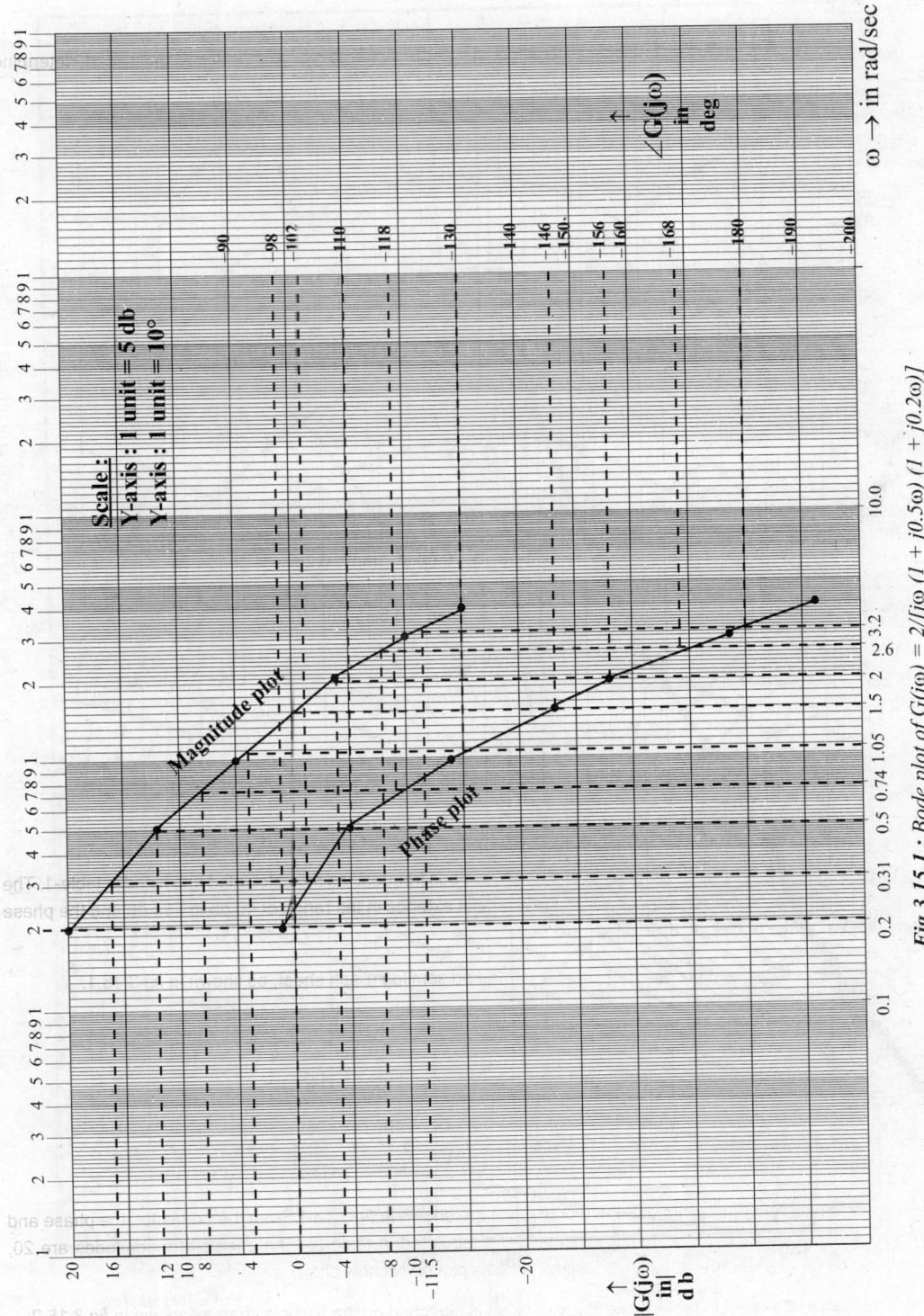

Fig 3.15.1 : Bode plot of G(jω) = 2/[jω (1 + j0.5ω) (1 + j0.2ω)]

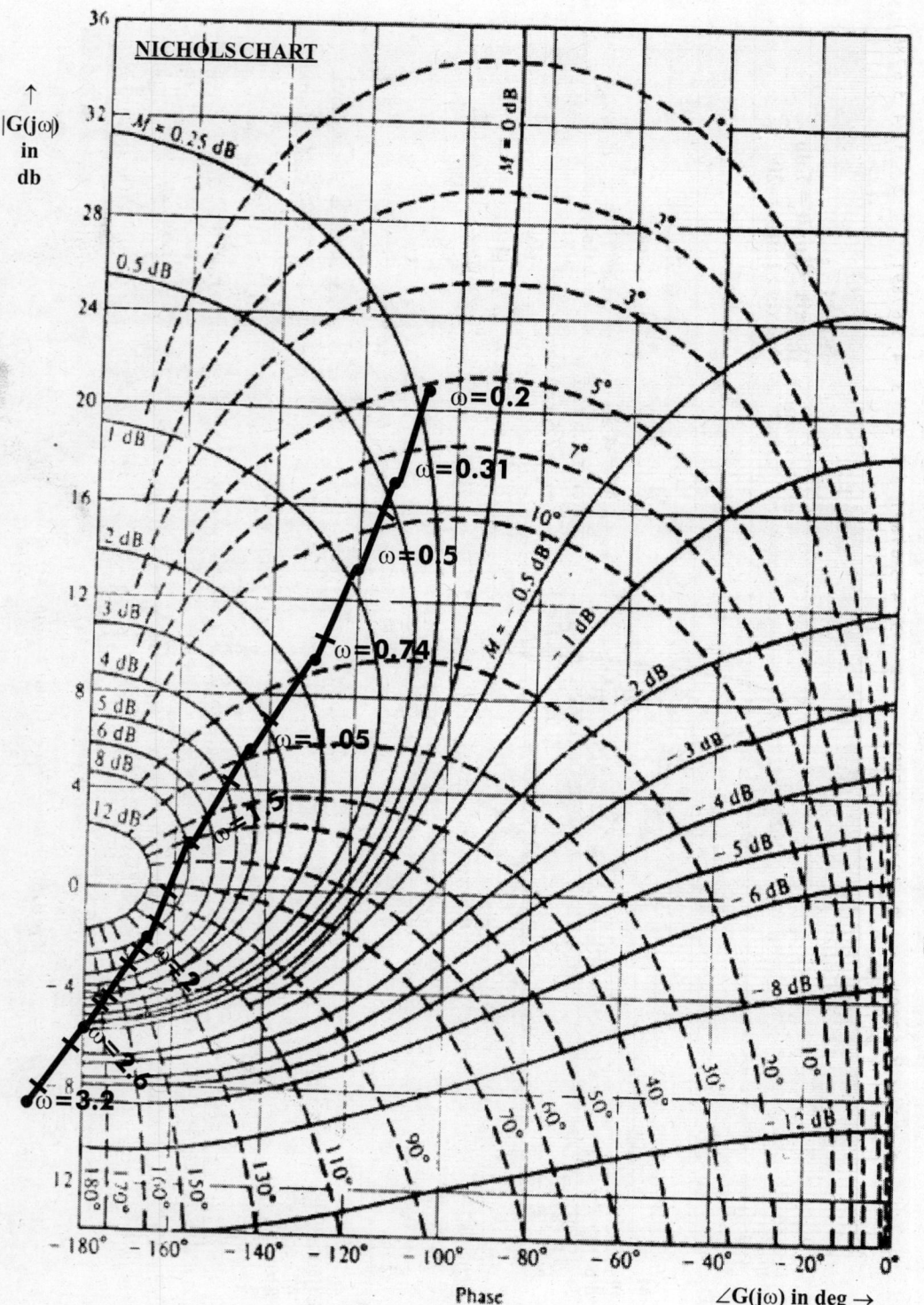

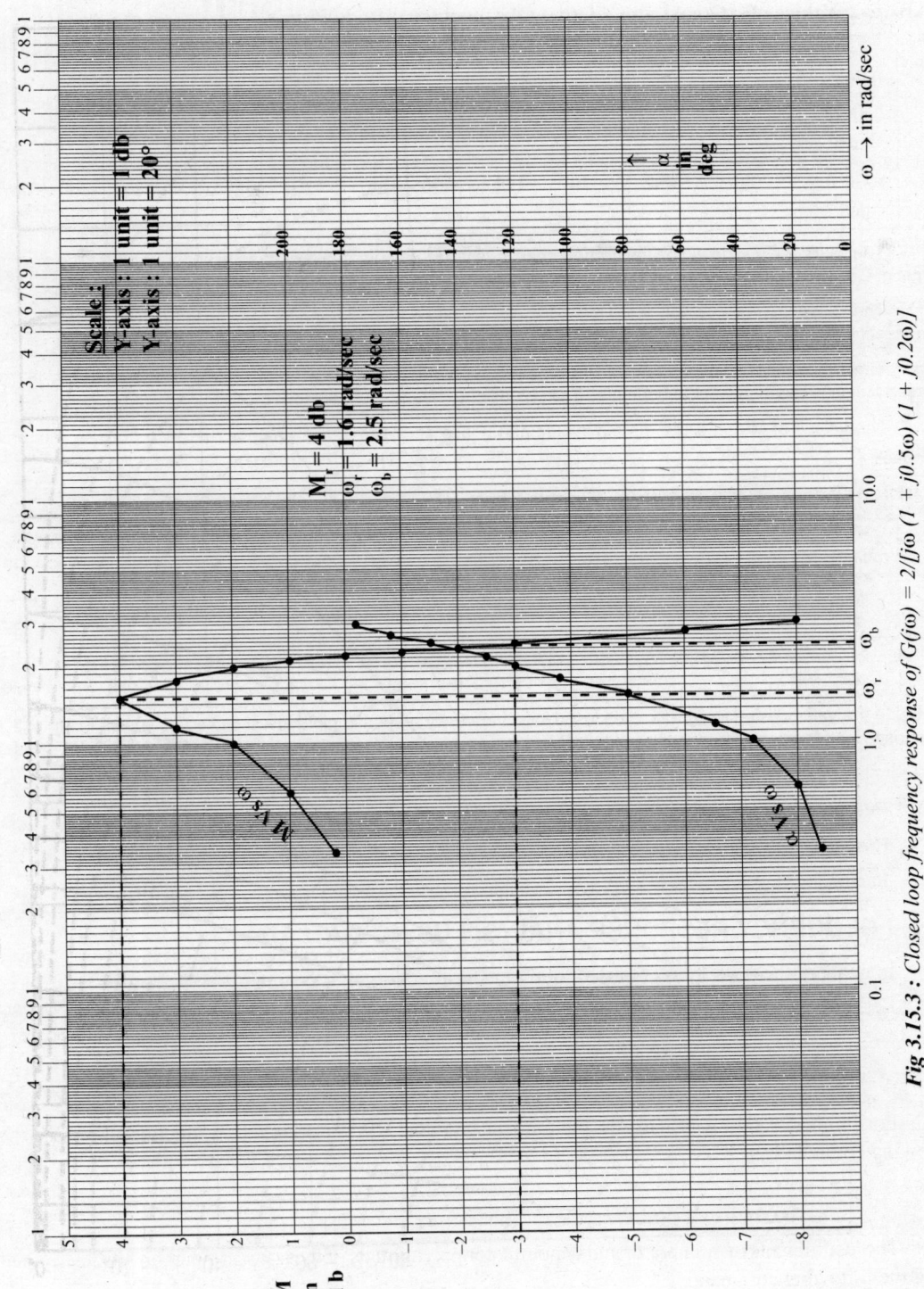

Fig 3.15.3 : Closed loop frequency response of G(jω) = 2/[jω (1 + j0.5ω) (1 + j0.2ω)]

TABLE-2: Values of $|G(j\omega)|$ and $\angle G(j\omega)$ Obtained from Bode Plot

ω rad/sec	0.2	0.31	0.5	0.74	1.05	1.5	2.0	2.6	3.2
$\|G(j\omega)\|$ db	20	16	12	8	4	0	− 4	− 8	−11.5
$\angle G(j\omega)$ deg	−98	−102	−110	−118	−130	−146	−156	−168	−180

The locus of $G(j\omega)$ drawn on the Nichols chart cuts the M-contour and N-contour at various points. The meeting points of $G(j\omega)$ locus and various M-contours are noted.

The phase α corresponding to the meeting point are noted from N-contours passing through the meeting point. (If a meeting point lies between two N-contours, then choose an approximate value of α).

The frequency corresponding to the meeting point are noted from bode plot by transferring the $|G(j\omega)|$ corresponding to meeting point to bode plot. The values of ω, M and α are listed in table-3.

The values of M and α are the magnitude and phase of closed loop frequency response of $G(j\omega)$ with unity feedback.

TABLE-3: Values of M and α from Nichols Chart

ω rad/sec	0.36	0.62	1.0	1.2	1.6	1.8	2.0	2.1	2.3	2.4	2.5	2.8	3.0
M db	0.25	1	2	3	4	3	2	1	0	− 2	− 3	− 6	− 8
α deg	12	21	35	50	78	102	120	130	140	151	155	165	175

Using the values listed in table-3, the closed loop frequency response plots are sketched as shown in fig 3.15.3. The closed loop frequency response consists of two plots and they are magnitude plot, M Vs ω and phase plot, α Vs ω.

From the closed loop frequency response the values of M_r, ω_r and ω_b are noted.

Resonant peak, M_r = +4 db

Resonant Frequency ω_r = 1.6 rad/sec

Bandwidth, ω_b = 2.5 rad/sec

3.12 FREQUENCY RESPONSE ANALYSIS USING MATLAB

In general, the open loop or closed loop transfer function of a system is denoted as T(s).

Let, T(s) be a rational function of "s", as shown below.

$$T(s) = \frac{b_0 s^M + b_1 s^{M-1} + b_2 s^{M-2} + ... + b_{M-1} s + b_M}{a_0 s^N + a_1 s^{N-1} + a_2 s^{N-2} + ... + a_{N-1} s + a_N}$$

For frequency response analysis, the transfer function T(s) is declared as a function of s using the following commands.

```
s=tf('s');
Ts=(b0*s^M+b1*s^(M-1)+...+bM)/(a0*s^N+a1*s^(N-1)+...+aN);
```

The coefficients of numerator and denominator polynomials of the transfer function are determined using the following command.

```
[num_cof den_cof]=tfdata(Ts);
```

The gain margin, phase margin, gain crossover frequency and phase crossover frequency can be determined using the following command.

```
[GM PM wgc wpc] = margin(Ts);
```

BODE PLOT

In order to draw Bode plot the frequecy range can be specified using following commands.

```
w = logspace(ds, de, n);
where, ds represents Start decade as 10^ds
       de represents end decade as 10^de
       n represents number of points to be calculated between 10^ds & 10^de
```

Method 1 :

The Bode plot can be plotted using any one of the following command.

```
bode(Ts,'k');
bode(Ts, w);
bode(num_cof, den_cof);
bode(num_cof, den_cof,w);
```

Method 2 :

The Bode plot can also be plotted using semilog plot command as shown below.

```
[Mag Phase w] = bode(Ts,w);
MagdB = 20*log10(Mag);
subplot(2,1,1);semilogx(w,MagdB,'k');
subplot(2,1,2);semilogx(w,Phase,'k');
```

In this method the magnitude and phase can be scaled by drawing two lines at two specified upper and lower values. For drawing these lines, one dimensional arrays consisting of same values has to be created by multiplying the specified value with one. The length of the array should be same as number of frequency points for which the magnitude and phase are computed. (Refer program 3.3).

POLAR PLOT

The polar plot can be plotted using the following commands.

```
w = w_start : w_step : w_end ;
[re, im, w] = ryquist(num_cof,den_cof,w);
z = re + i*im; r = abs(z); theta = angle(z);
polar(theta,r,'w')
```

NICHOLS PLOT

The Nichols plot of open loop transfer function, G(s) can be plotted using the following commands.

```
[num_cof den_cof ] = tfdata(Gs);
nichols(Gs);
axis([ph_start, ph_end, mag_start, mag_end]);
```

PROGRAM 3.1

Write a MATLAB program to draw the Bode plot for the open loop system governed by the following transfer function.

$$G(s)=s^2/(1+0.2s)(1+0.02s)$$

```
%program to plot Bode plot

clear all
clc
s=tf('s');
disp('The given transfer function is,');
Gs=(s^2)/((1+0.2*s)*(1+0.02*s))

w=logspace(-1,2,200);        %specify the frequency range
bode(Gs,w)
grid
```

OUTPUT

The given transfer function is,

Transfer function:
```
            s^2
    ---------------------
    0.004 s^2 + 0.22 s + 1
```

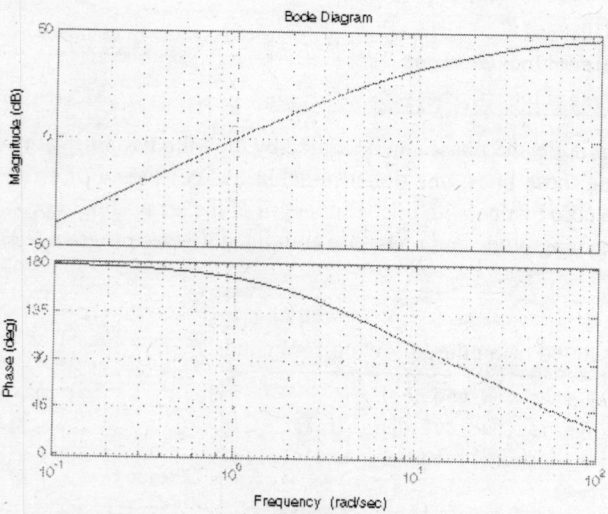

Fig P3.1 : *Bode plot of the open loop system given in problem 3.1.*

The Bode plot of program 3.1 is shown in fig p3.1.

PROGRAM 3.2

Write a MATLAB program to draw the Bode plot and to calculate gain margin, phase margin, gain crossover frequency & phase crossover frequency for the open loop system governed by the following transfer function.

$$G(s)=10/(0.04s^3+0.5s^2+s)$$

```
%program to find gain & phase margins using bode plot
clear all
clc
s=tf('s');
disp('The given transfer function is,');
Gs=10/((0.04*s^3)+(0.5*s^2)+s)

bode(Gs,'k')
grid

[GM,PM,wgc,wpc]=margir(Gs);
GMdB=20*log10(GM);
disp('Gain margin in dB,'); GMdB
disp('Phase margin in deg,');PM
disp('Gain cross over frequency in rad/sec,');wgc
disp('Phase cross over frequency in rad/sec,');wpc
```

OUTPUT

The given transfer function is,
Transfer function:

$$\frac{10}{0.04\ s^3 + 0.5\ s^2 + s}$$

Gain margin in dB,
         GMdB =
             1.9382

Phase margin in deg,
         PM =
             5.2057

Gain cross over frequency in rad/sec,
             wgc =
                 5.0000

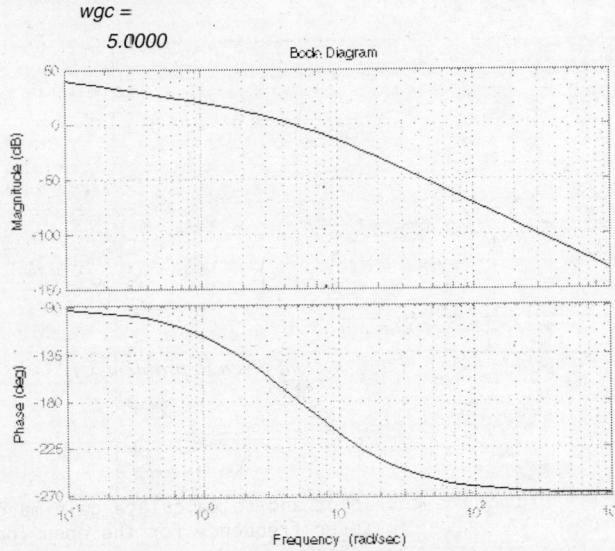

Fig P3.2 : *Bode plot of the open loop system given in problem 3.2.*

Phase cross over frequency in rad/sec,

wpc =

4.4629

The Bode plot of program 3.2 is shown in fig p3.2.

PROGRAM 3.3

Write a MATLAB program to draw the Bode plot for the open loop system governed by the following transfer function. The program should take care of drawing magnitude plot in the range +10 to -50 dB, and phase plot in the range -60 to -180 deg.

$$G(s)=(75(1+0.2s))/(s(s^2+16s+100)).$$

```
%Bode plot with magnitude and phase scaling

clear all
clc
s=tf('s');
disp('The given transfer function is,');
Gs=(75*(1+0.2*s))/(s*(s^2+16*s+100))
[num_cof den_cof]=tfdata(Gs);        %determine numerator and denominator
                                     %coefficients of G(s)

w=logspace(-1,2,200);
[Mag,Phase,w]=bode(num_cof,den_cof,w);
MagdB=20*log10(Mag);

Mscale1=10*ones(1,200);
Mscale2=-50*ones(1,200);
subplot(2,1,1);semilogx(w,MagdB,'-k',w,Mscale1,'k',w,Mscale2,'k')
grid;
xlabel('Frequency in rad/sec'); ylabel('Magnitude in dB');
Pscale1=-60*ones(1,200);
Pscale2=-180*ones(1,200);
subplot(2,1,2);semilogx(w,Phase,'k',w,Pscale1,'-k',w,Pscale2,'-k')
grid;
xlabel('Frequency in rad/sec'); ylabel('Phase in deg');
```

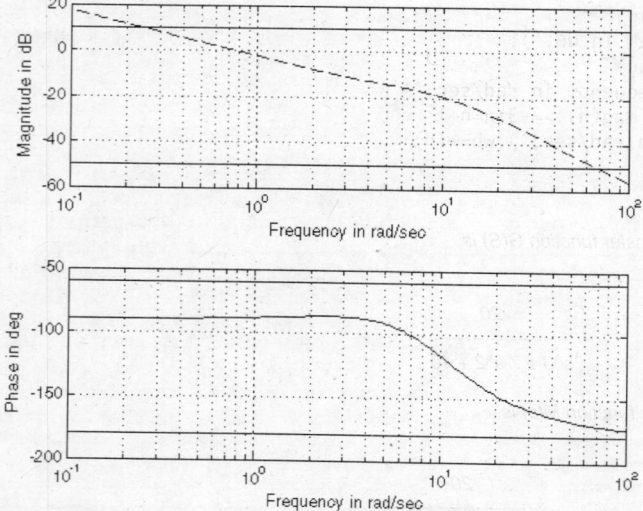

Fig P3.3 : Bode plot of the open loop system given in problem 3.3.

OUTPUT

The given transfer function is,

Transfer function:

$$\frac{15\ s + 75}{s^3 + 16\ s^2 + 100\ s}$$

The Bode plot of program 3.3 is shown in fig p3.3.

PROGRAM 3.4

Consider the transfer function of open loop system given below.

$$G(s)=20/(s^3+7s^2+10s)$$

Write a MATLAB program to determine the transfer function of closed loop system with unity feedback, to plot the Bode plot of closed loop system, and to calculate resonant peak, resonant frequency and bandwidth.

```
%Bode plot of unity feedback closed loop system
clear all
clc
s=tf('s');
disp('The given open loop transfer function G(S) is,');
Gs=20/(s^3+(7*s^2)+10*s)



disp('The closed loop transfer function M(s) is,');
Ms=feedback(Gs,1)
[num_cof den_cof]=tfdata(Ms); %determine numerator & denominator
                              %coefficients of M(s)

w=logspace(-1,1);            %specify frequency range
bode(Ms,w)
grid

[Mag,Phase,w]=bode(Ms,w);
[PeakMag,k]=max(Mag);
disp('Resonant peak in dB,');
Mp=20*log10(PeakMag)
disp('Resonant frequency in rad/sec,');Wr=w(k)
n=1; while 20*log(Mag(n))>=-3;n=n+1; end
disp('Bandwidth in rad/sec,');Wb=w(n)
```

OUTPUT

The given open loop transfer function G(S) is,

Transfer function:

$$\frac{20}{s^3 + 7\ s^2 + 10\ s}$$

The closed loop transfer function M(s) is,

Transfer function:

$$\frac{20}{s^3 + 7\ s^2 + 10\ s + 20}$$

Resonant peak in dB,
          Mp =
                                4.3953

Resonant frequency in rad/sec,
                    Wr =
                              1.6768

Bandwidth in rad/sec,
              Wb =
                              2.4421
The Bode plot of program 3.4 is shown in fig p3.4.

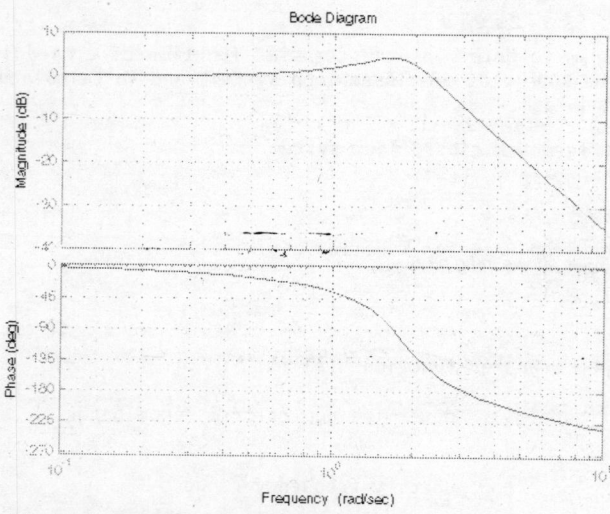

Fig P3.4 : *Bode plot of the closed loop system of program 3.4.*

PROGRAM 3.5

Write a MATLAB program to draw the polar plot and to calculate gain margin and phase margin for the open loop system governed by the following transfer function.

$$G(s)=1/(S(1+s)^2)$$

```
%Program to draw polar plot and compute gain & phase margins

clear all
clc
s=tf('s');
disp('The given transfer function is,');
Gs=1/(s*(1+s)*(1+s))

[num_cof den_cof]=tfdata(Gs);          %determine numerator and
                                       %denominator coeff. of G(s)
w=0.4 : 0.01 : 4;                      %specify frequency range
[re,im,w]=nyquist(num_cof,den_cof,w);  %determine the real and
                                       %imaginary parts of G(jw)

[GM PM]=margin(num_cof, den_cof);      %compute gain & phase margins
disp('Gain margin,');GM
disp('Phase margin in deg,');PM
```

```
z=re+i*im;                          %convert rectangular
                                    %coordinates to polar
r=abs(z);
theta=angle(z);
polar(theta,r,'k')                  %draw polar plot
```

OUTPUT

The given transfer function is,

Transfer function:

$$\frac{1}{s^3 + 2 s^2 + s}$$

Gain margin,
       GM =
         2

Phase margin in deg,
       PM =
         21.3877

The Polar plot of program 3.5 is shown in fig p3.5.

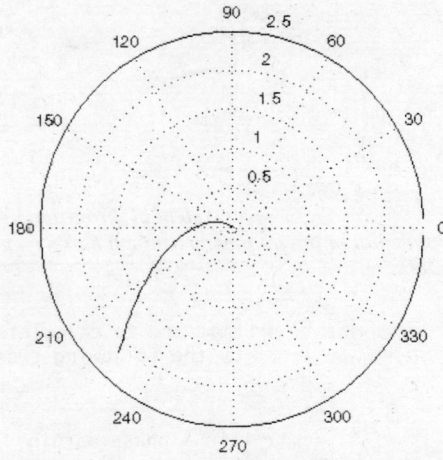

Fig P3.5 *: Polar plot of the open loop system given in problem 3.5.*

PROGRAM 3.6

Write a MATLAB program to draw the polar plot for the open loop system governed by the following transfer function. a) In the frequency range 0.6 to 8 rad/sec b)In the frequency range 5 to 18 rad/sec.

        G(s)=1/(S(1+0.2s)(1+0.05s))

```
%program to draw polar plot for two different frequency ranges

clear all
clc
```

```
s=tf('s');
disp('The given transfer function is,');
Gs=1/(s*(1+0.2*s)*(1+0.05*s))

[num_cof den_cof]=tfdata(Gs);    %determine numerator & denominator
                                 %coefficients of G(s)

w1=0.6 :0.001: 8;                           %specify frequency range1
[re1,im1,w1]=nyquist(num_cof,den_cof,w1);  %determine the real and
                                            %imaginary part of G(jw)
z1=re1+i*im1;                               %convert rectangular
r1=abs(z1);                                 %coordinates to polar
theta1=angle(z1);

subplot(2,1,1);polar(theta1,r1,'k');       %draw polar plot for
                                            %frequency range1


w2= 5 :0.001: 18;                           %specify frequency range2
[re2,im2,w2]=nyquist(num_cof,den_cof,w2);  %determine the real and
                                            %imaginary part of G(jw)
z2=re2+i*im2;                               %convert rectangular
r2=abs(z2);                                 %coordinates to polar
theta2=angle(z2);

subplot(2,1,2);polar(theta2,r2,'k');       %draw polar plot for
                                            %frequency range2
```

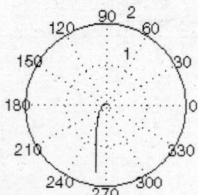

Fig P3.6a : *Polar plot in the frequency range 0.6 to 8 rad/sec.*

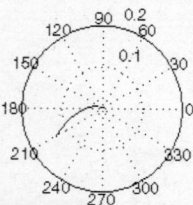

Fig P3.6b : *Polar plot in the frequency range 5 to 18 rad/sec.*

Fig P3.6 : *Polar plots of the open loop system given in problem 3.6.*

OUTPUT

The given transfer function is,

Transfer function:

$$\frac{1}{0.01\ s^3 + 0.25\ s^2 + s}$$

The Polar plot of program 3.6 is shown in fig p3.6.

PROGRAM 3.7

Write a MATLAB program to draw the polar plot using rectangular to polar coordinates for various values of K for the open loop system governed by the following transfer function.

$$G(s)=K/(S(1+s)(1+2S))$$

```
%polar plot for various values of gain,K
clc
s=tf('s');
K=1;
disp('when K=1,the given transfer function is,');
Gs=K/(s*(1+s)*(1+2*s))
[num_cof den_cof]=tfdata(Gs);

for i=1:3;
    if i==1;K=1;[re1,im1]=nyquist([num_cof den_cof],K); end;
    if i==2;K=2;[re2,im2]=nyquist([num_cof den_cof],K);end;
    if i==3;K=25;[re3,im3]=nyquist([num_cof den_cof],K);end;
end

plot(re1,im1,'-k',re2,im2,'-.k',re3,im3,'-k')
axis([-5 1 -5 1]); grid
xlabel('Real axis'); ylabel('Imaginary axis');
text(-3.8,-1.3,'K=1')
text(-3.8,-2.7,'K=2')
text(-1.8,-2.7,'K=25')
```

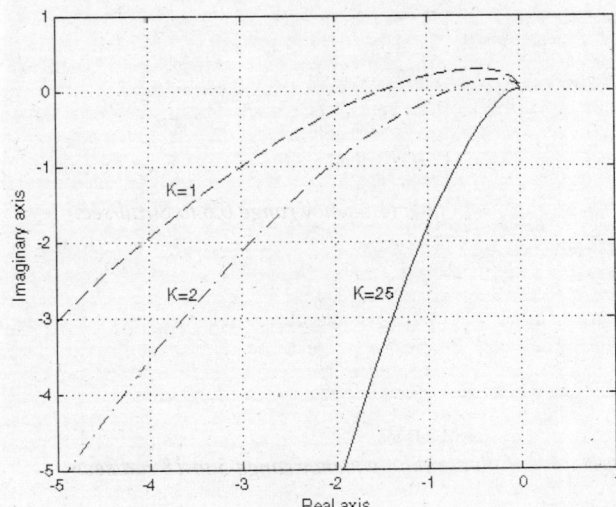

***Fig P3.7 :** Polar plot of the open loop system for various values of K.*

OUTPUT

When K=1, the given transfer function is,

Transfer function:

$$\frac{1}{2\ s^3 + 3\ s^2 + s}$$

PROGRAM 3.8

Write a MATLAB program to draw the polar plot and calculate gain margin and phase margin for the open loop system governed by the following transfer function.

$$G(s) = 1/(s(1+s)^2)$$

```
%program to draw nichols plot on nichols chart

clear all
clc

s=tf('s');
disp('The given transfer function is,');
Gs=20/(s^3+(7*s^2)+10*s)

[num_cof den_cof]=tfdata(Gs);    %determine numerator & denominator
                                 %coefficients of G(s)

nichols(Gs);

axis([-180 0 -15 20]);           %specify the range of horizontal and
                                 %vertical axis

ngrid
```

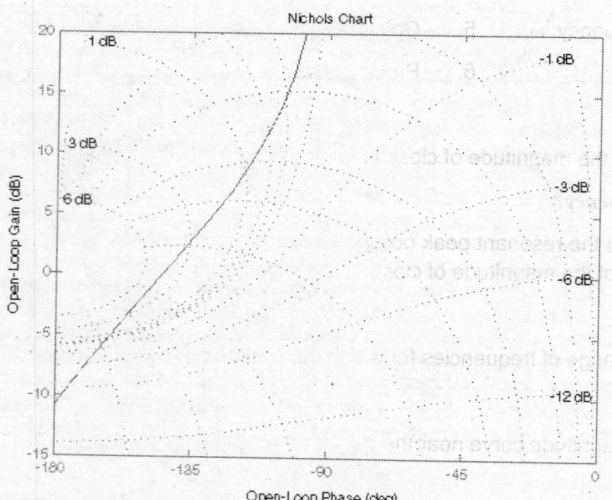

Fig P3.8 : *Nichols plot of the open loop system given in problem 3.8.*

OUTPUT

```
The given transfer function is,

Transfer function:
          20
    -------------------
    s^3 + 7 s^2 + 10 s
```

3.13 SHORT - ANSWERS QUESTIONS

Q3.1   *What is frequency response ?*

The magnitude and phase function of sinusoidal transfer function of a system are real function of frequency ω, and so they are called frequency response.

Q3.2   *What are advantages of frequency response analysis ?*

1. The absolute and relative stability of the closed loop system can be estimated from the knowledge of the open loop frequency response.
2. The practical testing of system can be easily carried with available sinusoidal signal generators and precise measurement equipments.
3. The transfer function of complicated functions can be determined experimentally by frequency response tests.
4. The design and parameter adjustment can be carried more easily.
5. The corrective measure for noise disturbance and parameter variation can be easily carried.
6. It can be extended to certain non-linear systems.

Q3.3   *What are frequency domain specifications?*

The frequency domain specifications indicates the performance of the system in frequency domain, and they are,

1.   Resonant peak, M_r 4.   Cut-off rate

2.   Resonant frequency, ω_r 5.   Gain margin, K_g

3.   Bandwidth, ω_b 6.   Phase margin, γ

Q3.4   *Define Resonant Peak?*

The maximum value of the magnitude of closed loop transfer function is called Resonant Peak.

Q3.5   *What is Resonant frequency?*

The frequency at which the resonant peak occurs is called Resonant frequency. The resonant peak is the maximum value of the magnitude of closed loop transfer function.

Q3.6   *Define Bandwidth?*

The Bandwidth is the range of frequencies for which the system gain is more than -3db.

Q3.7   *What is cut-off rate?*

The slope of the log-magnitude curve near the cut-off frequency is called cut-off rate.

Q3.8   *Define gain margin?*

The gain margin, K_g is defined as the value by which gain of the system has to be increased to drive system to be verge of instability. It is given by the reciprocal of the magnitude of open loop transfer function, at phase cross-over frequency, ω_{pc}. When expressed in decibels, it is given by, the negative of db magnitude of $G(j\omega)$ at phase cross-over frequency,

$$\text{Gain margin, } K_g = \frac{-}{\left| G(j\omega) \right|_{\omega = \omega_{pc}}} \quad \text{and} \quad K_g \text{ in db} = 20 \log \frac{1}{\left| G(j\omega) \right|_{\omega = \omega_{pc}}} = -20 \log \left| G(j\omega) \right|_{\omega = \omega_{pc}}$$

Q3.9   *Define phase margin?*

The phase margin, γ is that amount of additional phase lag at the gain cross-over frequency, ω_{gc} required to bring the system to the verge of instability. It is given by, $180° + \phi_{gc}$, where ϕ_{gc} is the phase of $G(j\omega)$ at the gain cross over frequency.

Phase margin, $\gamma = 180° + \phi_{gc}$;   where, $\phi_{gc} = \angle G(j\omega)\big|_{\omega = \omega_{gc}}$

Q3.10   *What is phase and Gain cross-over frequency?*

The gain cross over frequency is the frequency at which the magnitude of the open loop transfer function is unity. The phase cross over frequency is the frequency at which the phase of the open loop transfer function is 180°.

Q3.11   *Write the expression for resonant peak and resonant frequency.*

$$\text{Resonant peak, } M_r = \frac{1}{2\zeta\sqrt{1-\zeta^2}} \qquad ; \qquad \text{Resonant frequency, } \omega_r = \omega_n\sqrt{1-2\zeta^2}$$

Q3.12   *Write a short note on the correlation between the time and frequency response?*

Correlation exists between time and frequency response of first or second order systems. The frequency domain specifications can be expressed in terms of the time domain parameters ζ and ω_n. For a peak overshoot in time domain there is a corresponding resonant peak in frequency domain.

For higher order systems, there is no explicit correlation between time and frequency response. But if there is a pair of dominant complex conjugate poles, then the system can be approximated to second order system and the correlation between time and frequency response can be estimated.

Q3.13   *The damping ratio and natural frequency of oscillation of a second order system is 0.5 and 8 rad/sec respectively. Calculate the resonant peak and resonant frequency?*

$$\text{Resonant peak, } M_r = \frac{1}{2\zeta\sqrt{1-\zeta^2}} = \frac{1}{2\times(0.5)\sqrt{1-(0.5)^2}} = 1.154$$

$$\text{Resonant frequency, } \omega_r = \omega_n\sqrt{1-2\zeta^2} = 8\times\sqrt{1-2\times0.5^2} = 5.657 \text{ rad/sec}$$

Q3.14   *What is Bode plot?*

The bode plot is a frequency response plot of the transfer function of a system. It consists of two plots : Magnitude plot and Phase plot.

The magnitude plot is a graph between magnitude of a system transfer function in db and frequency, ω. The phase plot is a graph between the phase or argument of a system transfer function in degrees and the frequency, ω. Usually, both the plots are plotted on a common x-axis in which the frequencies are expressed in logarithmic scale.

Q3.15   *What is approximate bode plot?*

In approximate bode plot, the magnitude plot of first and second order factors are approximated by two straight lines, which are asymptotes to exact plot. One straight line is at 0db, for the frequency range 0 to ω_c and the other straight line is drawn with a slope of $\pm 20n$ db/dec for frequency range ω_c to ∞. Here ω_c is the corner frequency.

Q3.16   *Define corner frequency?*

The magnitude plot can be approximated by asymptotic straight lines. The frequencies corresponding to the meeting point of asymptotes are called corner frequency. The slope of the magnitude plot changes at every corner frequency.

Q3.17   *What are the advantages of Bode Plot?*

1. The magnitudes are expressed in db, and so, a simple procedure is available to add magnitude of each term one by one.

2. The approximate bode plot can be quickly sketched, and the corrections can be made at corner frequencies to get the exact plot.

3. The frequency domain specifications can be easily determined.

4. The bode plot can be used to analyse both open loop and closed loop system.

Q3.18   *What is the value of error in the approximate magnitude plot of a first order factor at the corner frequency?*

The error in the approximate magnitude plot of a first order factor at the corner frequency is ± 3m db, where m is multiplicity factor. Positive error for numerator factor and negative error for denominator factor.

Q3.19   *What is the value of error in the approximate magnitude plot of a quadratic factor with $\zeta=1$ at the corner frequency?*

The error is ± 6db, for the quadratic factor with $\zeta=1$. Positive error for numerator factor and negative error for denominator factor.

Q3.20   *Draw the bode plot of,* $G(s) = \dfrac{K}{s^n}$

Let s = jω,

$$\therefore G(j\omega) = \frac{K}{(j\omega)^n}$$

The magnitude of G(jω) is unity when $\omega = K^{1/n}$.

The magnitude plot is a straight line with slope of –20n db/dec and passing through $\omega = K^{1/n}$. The Phase plot is straight line parallel to x-axis at –90n°.

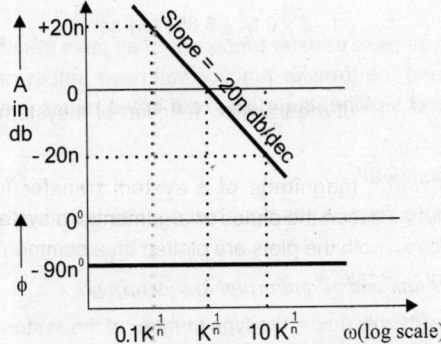

Fig Q3.20 : *Bode plot of integral factor, $K/(j\omega)^n$.*

Q3.21   *Sketch the bode plot of G(s) = 1/(1+sT).*

Let s = jω, $\therefore G(j\omega) = \dfrac{1}{1+j\omega T}$

The corner frequency, $\omega_c = \dfrac{1}{T}$

The magnitude plot is approximated by two straight lines : one straight line at 0db in the frequency range 0 to ω_c and the other straight line with the slope of -20db/dec in the frequency range ω_c to ∞. The phase of G(jω) varies from 0 to -90° as ω is varied from 0 to ∞. Hence, the phase plot is a curve passing through -45° at the corner frequency.

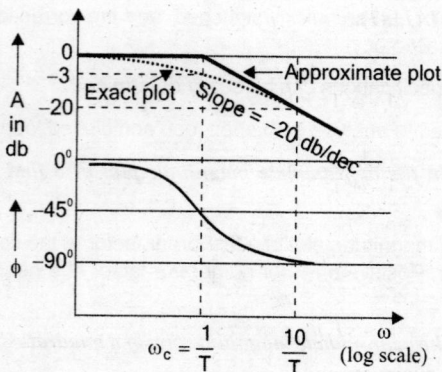

Fig Q3.21 : Bode plot of the factor $\dfrac{1}{1+j\omega T}$

Q3.22   *What is polar plot?*

The polar plot of a sinusoidal transfer function $G(j\omega)$ is a plot of the magnitude of $G(j\omega)$ versus the phase angle/argument of $G(j\omega)$ on polar or rectangular co-ordinates as ω is varied from zero to infinity.

Q3.23   *What is minimum phase system?*

The minimum phase systems are systems with minimum phase transfer functions. In minimum phase transfer functions, all poles and zeros will lie on the left half of s-plane.

Q3.24   *What is All-Pass systems?*

The all pass systems are systems with all pass transfer functions. In all pass transfer functions, the magnitude is unity at all frequencies and the transfer function will have anti-symmetric pole zero pattern (i.e., for every pole in the left half s-plane, there is a zero in the mirror image position with respect to imaginary axis).

Q3.25   *What is non-minimum phase transfer function?*

A transfer function which has one or more zeros in the right half s-plane is known as non-minimum phase transfer function.

Q3.26   *In minimum phase system, how the start and end of polar plot are identified?*

For minimum phase transfer functions, with only poles, the type number of the system determines the quadrant in which the polar plot starts, and the order of a system determines the quadrant in which the polar plot ends.

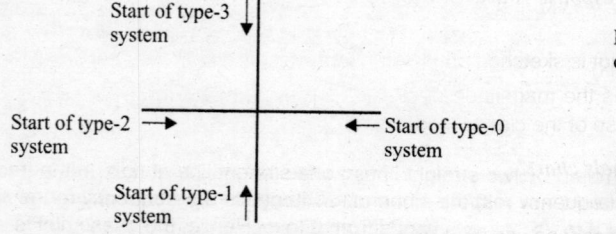

Fig Q3.26a : Start of polar plot of all pole minimum phase system.

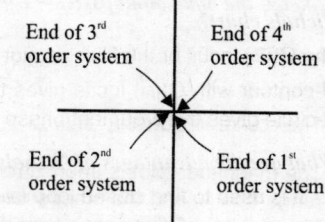

Fig Q3.26b : End of polar plot of all pole minimum phase system.

Q3.27   *Draw the polar plot of $G(s) = 1/(1+sT)$.*

Let $s = j\omega$, $\therefore$ $G(j\omega) = \dfrac{1}{1+j\omega T} = \dfrac{1}{\sqrt{1+\omega^2 T^2}} \angle \tan^{-1}\omega T$

$$= \dfrac{1}{\sqrt{1+\omega^2 T^2}} \angle \tan^{-1}\omega T$$

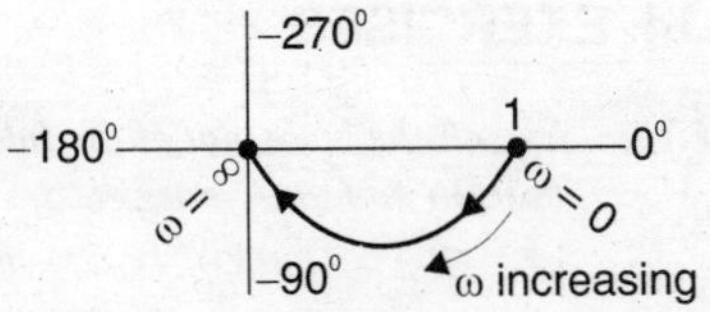

As $\omega \to 0$, $G(j\omega) \to 1\angle 0°$

As $\omega \to \infty$, $G(j\omega) \to 0 \angle -90°$

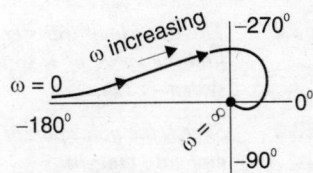

Fig Q3.27 : Polar plot of $G(s) = 1/(1+sT)$.

Q3.28   *Sketch the polar plot of, $G(s) = \dfrac{1}{s^2(1+sT_1)(1+sT_2)(1+sT_3)}$*

The given system is all pole minimum phase system. The type number of the system is 2 and the order is 5. Hence, the polar plot starts in second quadrant and ends in fourth quadrant.

Fig Q3.28 : Polar plot of type-2, 5th order system.

Q3.29   *What is Nichols plot?*

The Nichols plot is a frequency response plot of the open loop transfer function of a system. It is a graph between magnitude of $G(j\omega)$ in db and the phase of $G(j\omega)$ in degree, plotted on a ordinary graph sheet.

Q3.30   *What are M and N circles?*

The magnitude, M of closed loop transfer function with unity feedback will be in the form of circle in complex plane for each constant value of M. The family of these circles are called M-circles.

Let N = tan α, where a is the phase of closed loop transfer function with unity feedback. For each constant value of N, a circle can be drawn in the complex plane. The family of these circles are called N-circles.

Q3.31   *How closed loop frequency response is determined from open loop frequency response using M and N circles?*

The $G(j\omega)$ locus or polar plot of open loop system is sketched on the standard M and N circles chart. The meeting point of M circle with $G(j\omega)$ locus gives the magnitude of closed loop system. (the frequency being same as that of open loop system). The meeting point of $G(j\omega)$ locus with N-circle gives the value of phase of closed loop system, (frequency being same as that of open loop system).

Q3.32   *What is Nichols chart?*

The Nichols chart consists of M and N contours superimposed on ordinary graph. Along each M-contour the magnitude of closed loop system, M will be a constant. Along each N-contour, the phase α of closed loop system will be constant. The ordinary graph consists of magnitude in db, marked on the y-axis and the phase in degrees marked on x-axis. The Nichols chart is used to find the closed loop frequency response from the open loop frequency response.

Q3.33   *How the closed loop frequency response is determined from the open loop frequency response using Nichols chart?*

The $G(j\omega)$ locus or the Nichols plot is sketched on the standard Nichols chart. The meeting point of M-contour with $G(j\omega)$ locus gives the magnitude of closed loop system and the meeting point with N-circle gives the argument/phase of the closed loop system.

Q3.34   *What are the advantages of Nichols chart?*

1. It is used to find closed loop frequency response from open loop frequency response.

2. The frequency domain specifications can be determined from Nichols chart.

3. The gain of the system can be adjusted to satisfy the given specification.

3.14 EXERCISES

E3.1   *Sketch the bode plot of the following open loop transfer functions and from the plot determine the phase margin and gain margin.*

a) $G(s) = 100(1+0.1s)/s(1+0.2s)(1+0.5s)$

b) $G(s) = 50(1+0.1s)/(1+0.01s)(1+s)$

c) $G(s) = 30(1+0.1s)/s(1+0.01s)(1+s)$

d) $G(s) = s^2(s+10)/(s+5)^2(s+0.1)$

e) $G(s) = 40(1+s)/(1+5s)(s^2+2s+4)$

f) $G(s) = 10(1+s)\,e^{-0.1s}/s(1+0.2s)$

E3.2   *The open loop transfer function of a system is given by $G(s) = K/s(1+0.5s)(1+0.2s)$. Using bode plot find the value of K so that (i) The gain margin of the system is 6db and (ii) The phase margin of the system is 25°.*

E3.3   *Sketch the polar plot of the following transfer functions and from the plot, determine the phase margin and gain margin.*

a) $G(s) = 10(s+1)/(s+10)^2$

b) $G(s) = 200(s+2)/s(s^2+10s+100)$

c) $e^{-0.1s}/s(s+1)(s+5)$

d) $1/s(s+4)(s+8)$

E3.4   *The open loop transfer function of a system is given by $G(s) = K/s(s^2+s+4)$. Using polar plot, determine the value of K, so that phase margin is 50°. What is the corresponding value of gain margin?*

E3.5   *A unity feedback system has $G(s) = K/s(1+0.1s)$. Using Nichols chart find the value of K so that resonant peak, $M_r = 1.4$. Find the corresponding value of ω_r.*

E3.6   *The open loop transfer function of unity feedback system is, $G(s) = K/(1+0.05s)(1+0.1s)(1+0.3s)$.*

Using Nichols chart find the value of K so that gain margin of the system is 10db. What is the corresponding value of phase margin.

E3.7   *Using Nichols chart determine the closed loop frequency response of the unity feedback system, whose open loop transfer function is, $G(s) = 200(s+1)/s(s+10)^2$.*

E3.8   *A unity feedback system has open loop transfer function $G(s) = 54/(1+0.1s)(s^2+8s+25)$.*

Using Nichols chart determine the closed loop frequency response. From the closed loop response determine, the resonant peak, resonant frequency and bandwidth.

CHAPTER 4

CONCEPTS OF STABILITY AND ROOT LOCUS

4.1 IMPULSE RESPONSE AND STABILITY

DEFINITIONS OF STABILITY

The term stability refers to the stable working condition of a control system. Every working system is designed to be stable. In a stable system, the response or output is predictable, finite and stable for a given input (or for any changes in input or for any changes in system parameters).

The different definitions of the stability are the following

1. A system is stable, if its output is bounded (finite) for any bounded (finite) input.

2. A system is asymptotically stable, if in the absence of the input, the output tends towards zero (or to the equilibrium state) irrespective of initial conditions.

3. A system is stable if for a bounded disturbing input signal the output vanishes ultimately as **t** approaches infinity.

4. A system is unstable if for a bounded disturbing input signal the output is of infinite amplitude or oscillatory.

5. For a bounded input signal, if the output has constant amplitude oscillations then the system may be stable or unstable under some limited constraints. Such a system is called *limitedly stable.*

6. If a system output is stable for all variations of its parameters, then the system is called *absolutely stable system.*

7. If a system output is stable for a limited range of variations of its parameters, then the system is called *conditionally stable system.*

IMPULSE RESPONSE OF A SYSTEM

Let, M(s) = Closed loop transfer function of a system.

C(s) = Output / Response in s-domain.

R(s) = Input in s-domain

Now, $M(s) = \dfrac{C(s)}{R(s)}$

∴ Response or Output in s-domain, C(s) = M(s) R(s)

Now, Response in time domain, $c(t) = \mathcal{L}^{-1}\{C(s)\}$

Input in time domain, $r(t) = \mathcal{L}^{-1}\{R(s)\}$

For an impulse input, $r(t) = \delta(t)$; ∴ $R(s) = \mathcal{L}[\delta(t)] = 1$

$\therefore$ Impulse response $= \mathcal{L}^{-1}\{C(s)\} = \mathcal{L}^{-1}\{M(s)\,R(s)\} = \mathcal{L}^{-1}\{M(s)\} = m(t)$ .....(4.1)

Hence, impulse response of a system is the inverse Laplace transform of system transfer function.

The importance of impulse response is that, the output of a system for any arbitrary input can be obtained by convolution of input and impulse response.

i.e., Response, $c(t) = m(t) * r(t)$

where $*$ is the symbol for convolution.

Mathematically the convolution operation is defined as,

$$c(t) = \int\limits_{-\infty}^{+\infty} m(\tau)\, r(t-\tau)\, d\tau \qquad\qquad .....(4.2)$$

where t is the dummy variable used for integration.

BOUNDED - INPUT BOUNDED - OUTPUT (BIBO) STABILITY

A linear relaxed system is said to have BIBO stability if every bounded (finite) input results in a bounded (finite) output. A condition for BIBO stability can be obtained from convolution operation defined by equation (4.2).

For a relaxed system the equation (4.2) can be written as,

$$\text{Response, } c(t) = \int\limits_{0}^{\infty} m(\tau)\, r(t-\tau)\, d\tau \qquad\qquad .....(4.3)$$

> **Note :** *A relaxed system is one in which the initial conditions are zero. Hence the limits of integration is from 0 to ∞.*

If the input $r(t)$ is bounded then there exists a constant A_1, such that $|r(t)| \le A_1 < \infty$. The condition for bounded output for this bounded input condition can be derived as follows.

On taking the absolute value on both sides of equation (4.3), we get,

$$|c(t)| = \left| \int\limits_{0}^{\infty} m(\tau)\, r(t-\tau)\, d\tau \right| \qquad |c(t)| \le \int\limits_{0}^{\infty} |m(\tau)\, r(t-\tau)\, d\tau| \qquad .....(4.4)$$

Since the absolute value of an integral is not greater than the integral of the absolute value of the integrand the equation (4.4) can be written as,

$$|c(t)| \le \int\limits_{0}^{\infty} |m(\tau)\, r(t-\tau)|\, d\tau \;\Rightarrow\; |c(t)| \le \int\limits_{0}^{\infty} |m(\tau)| |r(t-\tau)|d\tau \;\Rightarrow\; |c(t)| \le \int\limits_{0}^{\infty} |m(\tau)| A_1\, d\tau$$

$$\therefore |c(t)| \le A_1 \int\limits_{0}^{\infty} |m(\tau)|\, d\tau$$

> For bounded input, a constant exists such that, $|r(t-\tau)| \le A_1$.

If the output $c(t)$ is bounded then there exists a constant A_2 such that $|c(t)| \le A_2 < \infty$.

$$\therefore A_1 \int\limits_{0}^{\infty} |m(\tau)|\, d\tau \le A_2 < \infty \qquad\qquad .....(4.5)$$

The above condition is satisfied if, $\int\limits_{0}^{\infty} |m(\tau)|\, d\tau < \infty$

τ is a dummy variable and so can be replaced by t

Hence for bounded output, $\int\limits_{0}^{\infty} |m(t)|\, dt < \infty$ .....(4.6)

Therefore we can conclude that a system with impulse response $m(t)$ is BIBO stable if and only if the impulse response is absolutely integrable (i.e., $\int_{0}^{\infty} |m(t)|\, dt$ is finite. This means that area under the absolute value curve of the impulse response $m(t)$ evaluated from $t = 0$ to $t = \infty$ must be finite).

4.2 LOCATION OF POLES ON s-PLANE FOR STABILITY

The closed loop transfer function, $M(s)$ can be expressed as a ratio of two polynomials in s. The denominator polynomial of closed loop transfer function is called characteristic equation. The roots of characteristic equation are poles of closed loop transfer function.

For BIBO stability the integral of impulse response should be finite, which implies that the impulse response should be finite as t tends to infinity. *[The impulse response is the inverse Laplace transform of the transfer function]*. This requirement for stability can be linked to the location of roots of characteristic equation in the s-plane.

The closed loop transfer function $M(s)$ can be expressed as a ratio of two polynomials in s as shown in equation (4.7).

$$M(s) = \frac{b_0 s^m + b\, s^{m-1} + b_2 s^{m-2} + ... + b_{m-1} s + b_m}{a_0 s^n + \varepsilon_1 s^{n-1} + a_2 s^{n-2} + ... + a_{n-1} s + a_n} \qquad .....(4.7)$$

$$= \frac{(s+z_1)(s+z_2)(s+z_3)......(s+z_m)}{(s+p_1)(s+p_2)(s+p_3)......(s+p_n)} \qquad .....(4.8)$$

The roots of numerator polynomial $z_1, z_2, ...z_n$ are zeros. The roots of denominator polynomial $p_1, p_2, ...p_n$ are poles. The denominator polynomial is the characteristic equation and so the poles are roots of characteristic equation.

By partial fraction expansion we can write,

$$M(s) = \frac{A_1}{s+p_1} + \frac{A_2}{s+p_2} + \frac{A_3}{s+p_3} + ... + \frac{A_n}{s+p_n} \qquad .....(4.9)$$

The roots (or poles) $p_1, p_2, p_3, ... p_n$ may be at origin or lying on imaginary axis or lying on right or left half of s-plane. The impulse response is given by inverse Laplace transform of $M(s)$. The inverse Laplace transform of each term of $M(s)$ depends on the location of roots (or poles) in s-plane. The impulse response of various types of $M(s)$ are shown in table-4.1.

From table 4.1, the following conclusions are drawn based on the location of roots of characteristic equation.

1. If all the roots of characteristic equation have negative real parts (i.e., lying on left half s-plane) then the impulse response is bounded (i.e., it decreases to zero as t tends to ∞). Hence $\int_{0}^{\infty} |m(t)|$ is finite and the system is bounded-input bounded-output stable.

TABLE 4.1

Transfer function, M(s) and location of roots on s-plane	Impulse response, m(t)
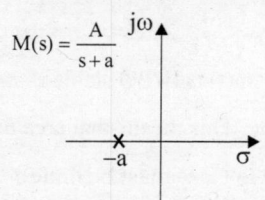 $$M(s) = \frac{A}{s+a}$$ *Root on negative real axis*	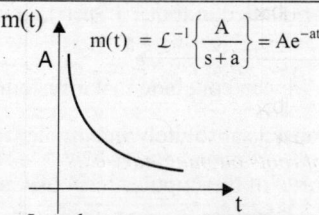 $$m(t) = \mathcal{L}^{-1}\left\{\frac{A}{s+a}\right\} = Ae^{-at}$$ *Impulse response is exponentially decaying. Stable system.*
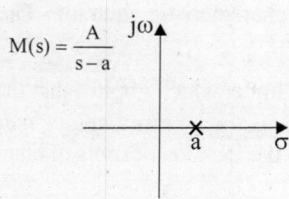 $$M(s) = \frac{A}{s-a}$$ *Root on positive real axis*	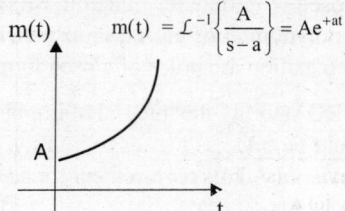 $$m(t) = \mathcal{L}^{-1}\left\{\frac{A}{s-a}\right\} = Ae^{+at}$$ *Impulse response is exponentially increasing. Unstable system.*
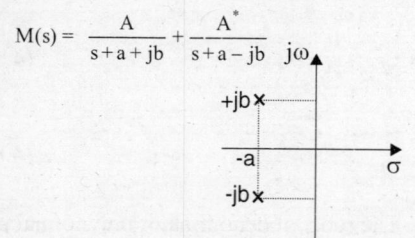 $$M(s) = \frac{A}{s+a+jb} + \frac{A^*}{s+a-jb}$$ *Complex conjugate roots on left half of s-plane*	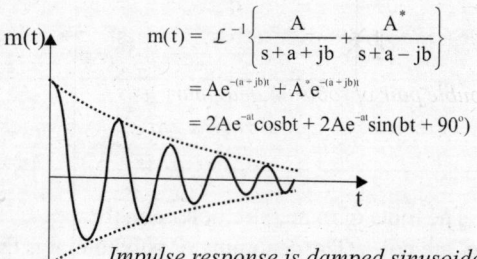 $$m(t) = \mathcal{L}^{-1}\left\{\frac{A}{s+a+jb} + \frac{A^*}{s+a-jb}\right\}$$ $$= Ae^{-(a-jb)t} + A^*e^{-(a+jb)t}$$ $$= 2Ae^{-at}\cos bt + 2Ae^{-at}\sin(bt + 90°)$$ *Impulse response is damped sinusoidal (i.e., Damped oscillatory). Stable system*
$$M(s) = \frac{A}{s-a+jb} + \frac{A^*}{s-a-jb}$$ 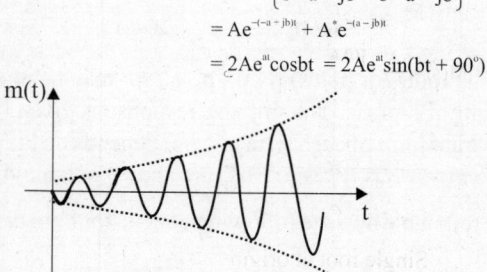 *Complex conjugate roots on right half of s-plane*	$$m(t) = \mathcal{L}^{-1}\left\{\frac{A}{s-a+jb} + \frac{A^*}{s-a-jb}\right\}$$ $$= Ae^{-(-a+jb)t} + A^*e^{-(-a-jb)t}$$ $$= 2Ae^{at}\cos bt = 2Ae^{at}\sin(bt + 90°)$$ *Impulse response is exponentially increasing sinusoidal (i.e., Amplitude of oscillations exponentially increases with time). Unstable system.*

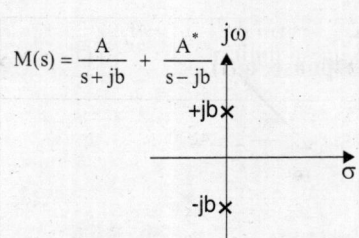

$$M(s) = \frac{A}{s+jb} + \frac{A^*}{s-jb}$$

Single pair of roots on imaginary axis

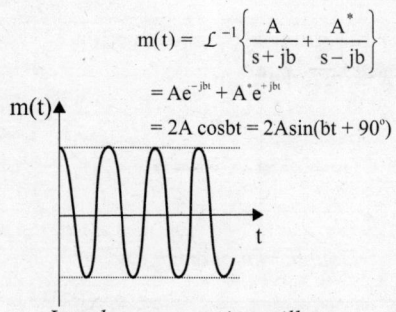

$$m(t) = \mathcal{L}^{-1}\left\{\frac{A}{s+jb} + \frac{A^*}{s-jb}\right\}$$

$$= Ae^{-jbt} + A^*e^{+jbt}$$

$$= 2A\cos bt = 2A\sin(bt + 90°)$$

Impulse response is oscillatory
Marginally stable

$$M(s) = \frac{A}{(s+jb)^2} + \frac{A^*}{(s-jb)^2}$$

Double pair of roots on imaginary axis

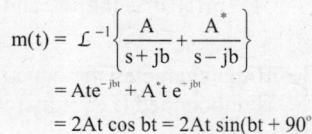

$$m(t) = \mathcal{L}^{-1}\left\{\frac{A}{s+jb} + \frac{A^*}{s-jb}\right\}$$

$$= Ate^{-jbt} + A^*t\,e^{-jbt}$$

$$= 2At\cos bt = 2At\sin(bt + 90°)$$

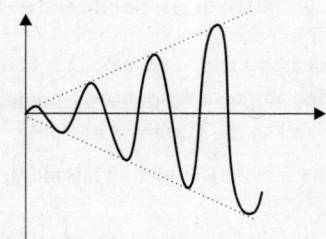

Impulse response is linearly increasing sinusoidal
(i.e., amplitude of oscillations linearly increases
with time). Unstable system.

$$M(s) = \frac{A}{s}$$

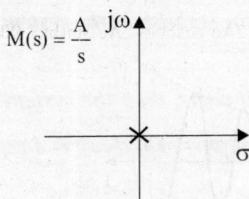

Single root at origin

$$m(t) = \mathcal{L}^{-1}\left\{\frac{A}{s}\right\} = A$$

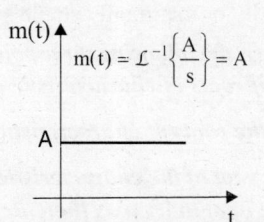

Impulse response is constant.
Marginally stable system.

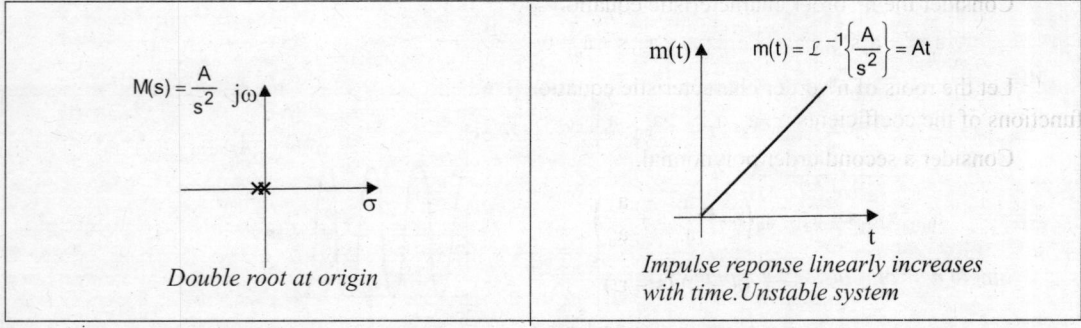

Double root at origin

Impulse reponse linearly increases with time. Unstable system

2. If any root of the characteristic equation has a positive real part (i.e., lying on right half s-plane) then impulse response is unbounded, (i.e., it increases to ∞ as t tends to ∞). Hence $\int_0^\infty |m(t)| \, dt$ is infinite and so system is unstable.

3. If the characteristic equation has repeated roots on the imaginary axis then impulse response is unbounded (i.e., it increases to ∞ as t tends to ∞).

   Hence $\int_0^\infty |m(t)| \, dt$ is infinite and so the system is unstable.

4. If one or more non - repeated roots of the characteristic equation are lying on the imaginary axis, then impulse response is bounded (i.e., it has constant amplitude oscillations) but is infinite and so the system is unstable.

5. If the characteristic equation has single root at origin then the impulse response is bounded (i.e, it has constant amplitude) but $\int_0^\infty |m(t)| \, dt$ is infinite and so the system is unstable.

6. If the characteristic equation has repeated roots at origin then the impulse response is unbounded (i.e., it linearly increases to infinity as t tends to ∞) and so the system is unstable.

7. In system with one or more non-repeated roots on imaginary axis or with single root at origin, the output is bounded for bounded inputs except for the inputs having poles matching the system poles. These cases may be treated as acceptable or non-acceptable. Hence when the system has non repeated poles on imaginary axis or single pole at origin, it is referred as limitedly or marginally stable system.

In summary, the following three points may be stated regarding the stability of the system depending on the location of roots of characteristic equation.

1. *If all the roots of characteristic equation has negative real parts, then the system is stable.*

2. *If any root of the characteristic equation has a positive real part or if there is a repeated root on the imaginary axis then the system is unstable.*

3. *If the condition (i) is satisfied except for the presence of one or more non repeated roots on the imaginary axis, then the system is limitedly or marginally stable.*

In order to ascertain the stability of a system, it is necessary to determine if any of the roots of the characteristic equation lie in the right half s-plane. The characteristic equation is given by the denominator polynomial of closed loop transfer function, [equation (4.7)].

Consider the nth order characteristic equation shown below.

$$a_0 s^n + a_1 s^{n-1} + a_2 s^{n-2} + \ldots\ldots\ldots a_{n-1} s + a_n = 0 \qquad \ldots\ldots(4.10)$$

Let the roots of nth order characteristic equation [equation (4.10)] be $s = r_1, r_2, \ldots r_n$. These roots are functions of the coefficients $a_0, a_1, a_2, \ldots a_{n-1}, a_n$.

Consider a second order polynomial,

$$a_0 s^2 + a_1 s + a_2 = a_0 \left(s^2 + \frac{a_1}{a_0} s + \frac{a_2}{a_0} \right)$$

$$= a_0 (s - r_1)(s - r_2)$$

$$= a_0 s^2 - a_0 (r_1 + r_2) s + a_0 r_1 r_2 \qquad \ldots\ldots(4.11)$$

Consider a third order polynomial

$$a_0 s^3 + a_1 s^2 + a_2 s + a_3 = a_0 \left(s^3 + \frac{a_1}{a_0} s^2 + \frac{a_2}{a_0} s + \frac{a_3}{a_0} \right)$$

$$= a_0 (s - r_1)(s - r_2)(s - r_3)$$

$$= a_0 s^3 - a_0 (r_1 + r_2 + r_3)s - a_0 r_1 r_2 r_3$$

$$+ a_0 (r_1 r_2 + r_1 r_3 + r_2 r_3)s - a_0 r_1 r_2 r_3 \qquad \ldots\ldots(4.12)$$

On extending this expansion to the nth order polynomial, we get.

$$a_0 s^n + a_1 s^{n-1} + \ldots + a_{n-1} s + a_n = a_0 s^n - a_0 (\text{sum of all the roots}) s^{n-1}$$

$$+ a_0 \left(\begin{array}{c} \text{sum of the products of the roots} \\ \text{taken 2 at a time} \end{array} \right) s^{n-2}$$

$$- a_0 \left(\begin{array}{c} \text{sum of the products of the roots} \\ \text{taken 3 at a time} \end{array} \right) s^{n-3}$$

$$+ \ldots\ldots + a_0 (-1)^n (\text{Product of all the n roots}) \qquad \ldots\ldots(4.13)$$

If all the roots of a polynomial are real and in the left half of s-plane, then all r_i in equations (4.11) and (4.12) are real and negative. Therefore all polynomial coefficients are positive. This characteristic also applies to the general case of equation (4.13). If atleast one root is in the right half of s-plane then some of the coefficients will be negative. Also, it can be observed that if all the roots are in the left half of s-plane, no coefficient can be zero.

Since the characteristic polynomial coefficients are real, the complex roots should occur as conjugate pairs. From equation (4.13) it can be inferred that when polynomial coefficients are formed, the imaginary parts of roots/products of roots will cancel. Therefore, if all roots occur in the left half plane, (whether it is complex or real) then all coefficients of the general polynomial of equation (4.13) will be positive. Presence of a negative coefficient implies that there is atleast one root in the right half of s- plane.

A zero coefficient indicates presence of complex-conjugate roots on the imaginary axis and/or one or more roots in the right half of s- plane.

In summary, following conclusions can be made about coefficients of characteristic polynomial.

1. ***If all the coefficients are positive and if no coefficient is zero, then all the roots are in the left half of s- plane.***

2. ***If any coefficient a_i is equal to zero then, some of the roots may be on the imaginary axis or on the right half of s- plane.***

3. *If any coefficient a$_i$ is negative then atleast one root is in the right half of s- plane.*

It can be concluded that the absence or negativeness of any of the coefficients of a characteristic polynomial indicates that the system is either unstable or at most marginally stable. Thus *the necessary condition for stability of the system is that all the coefficients of its characteristic polynomial be positive.* If any coefficient is zero/negative, we can immediately say that the system is unstable.

In order for all the roots to have negative real parts, it is necessary that all of the coefficients of characteristic equation be positive, but it is not sufficient, because there may be roots in the right half plane and/or on the imaginary axis, even when coefficients are positive.(i.e., when roots have negative real part, then all the coefficients of characteristic polynomial will be positive, but the reverse condition is not true always).

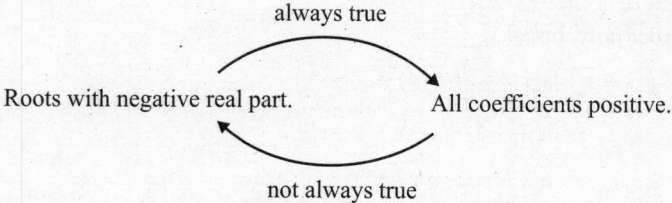

Hence, when all the coefficients are positive, the system may or may not be stable, because there may be roots in the right half plane and/or on the imaginary axis.

For example, consider the characteristic polynomial with all positive coefficients,

$$s^3 + s^2 + 2s + 8 = 0.$$

The characteristic polynomial can be written as,

$$(s^3 + s^2 + 2s + 8) = (s+2)\left(s - \frac{1}{2} - j\frac{\sqrt{15}}{2}\right)\left(s - \frac{1}{2} + j\frac{\sqrt{15}}{2}\right) = 0$$

Now the roots are,

$$s = -2, \ +\frac{1}{2} + j\frac{\sqrt{15}}{2}, \ +\frac{1}{2} - j\frac{\sqrt{15}}{2}$$

The coefficients of the polynomial are all positive, but two roots have positive real part and so will lie on on right half of s-plane, therefore the system is unstable.

4.3 ROUTH HURWITZ CRITERION

The Routh-Hurwitz stability criterion is an analytical procedure for determining whether all the roots of a polynomial have negative real part or not.

The first step in analysing the stability of a system is to examine its characteristic equation. The necessary condition for stability is that all the coefficients of the polynomial be positive. If some of the coefficients are zero or negative it can be concluded that the system is not stable.

When all the coefficients are positive, the system is not necessarily stable. Eventhough the coefficients are positive, some of the roots may lie on the right half of s-plane or on the imaginary axis. In order for all the roots to have negative real parts, it is necessary but not sufficient that all coefficients of the characteristic equation be positive. If all the coefficients of the characteristic equation are positive, then the system may be stable and one should proceed further to examine the sufficient conditions of stability.

A. Hurwitz and E.J. Routh independently published the method of investigating the sufficient conditions of stability of a system. The Hurwitz criterion is in terms of determinants and Routh criterion is in terms of array formulation. The Routh stability criterion is presented here.

The Routh stability criterion is based on ordering the coefficients of the characteristic equation, into a schedule, called the Routh array as shown below.

$$a_0 s^n + a_1 s^{n-1} + a_2 s^{n-2} + ... + a_{n-1} s + a_n = 0, \quad \text{where } a_0 > 0,$$

s^n	:	a_0	a_2	a_4	a_6	a_8	......
s^{n-1}	:	a_1	a_3	a_5	a_7	a_9	......
s^{n-2}	:	b_0	b_1	b_2	b_3	b_4	......
s^{n-3}	:	c_0	c_1	c_2	c_3	c_4	......
s^1	:	g_0					
s_0	:	h_0					

The Routh stability criterion can be stated as follows.

"The necessary and sufficient condition for stability is that all of the elements in the first column of the Routh array be positive. If this condition is not met, the system is unstable and the number of sign changes in the elements of the first column of the Routh array corresponds to the number of roots of the characteristic equation in the right half of the s-plane".

> *Note : If the order of sign of first column element is +, +, −, + and +. Then + to − is considered as one sign change and − to + as another sign change.*

CONSTRUCTION OF ROUTH ARRAY

Let the characteristic polynomial be,

$$a_0 s^n + a_1 s^{n-1} + a_2 s^{n-2} + a_3 s^{n-3} + ... + a_{n-1} s^1 + a_n s^0$$

The coefficients of the polynomial are arranged in two rows as shown below.

s^n	:	a_0	a_2	a_4	a_6
s^{n-1}	:	a_1	a_3	a_5	$a_7 s$

When n is even, the s^n row is formed by coefficients of even order terms (i.e., coefficients of even powers of s) and s^{n-1} row is formed by coefficients of odd order terms (i.e., coefficients of odd powers of s).

When n is odd, the s^n row is formed by coefficients of odd order terms (i.e., coefficients of odd powers of s) and s^{n-1} row is formed by coefficients of even order terms(i.e., coefficients of even powers of s) .

The other rows of routh array upto s^0 row can be formed by the following procedure. Each row of Routh array is constructed by using the elements of previous two rows.

Consider two consecutive rows of Routh array as shown below.

$$s^{n-x} \quad : \quad x_0 \quad x_1 \quad x_2 \quad x_3 \quad x_4 \quad x_5 \dots$$

$$s^{n-x-1} \quad : \quad y_0 \quad y_1 \quad y_2 \quad y_3 \quad y_4 \quad y_5 \dots$$

Let the next row be,

$$s^{n-x-2} \quad : \quad z_0 \quad z_1 \quad z_2 \quad z_3 \quad z_4 \dots$$

The elements of s^{n-x-2} row are given by,

$$z_0 = \frac{(-1)\begin{vmatrix} x_0 & x_1 \\ y_0 & y_1 \end{vmatrix}}{y_0} = \frac{y_0 x_1 - y_1 x_0}{y_0}$$

$$z_1 = \frac{(-1)\begin{vmatrix} x_0 & x_2 \\ y_0 & y_2 \end{vmatrix}}{y_0} = \frac{y_0 x_2 - y_2 x_0}{y_0}$$

$$z_2 = \frac{(-1)\begin{vmatrix} x_0 & x_3 \\ y_0 & y_3 \end{vmatrix}}{y_0} = \frac{y_0 x_3 - y_3 x_0}{y_0}$$

$$z_3 = \frac{(-1)\begin{vmatrix} x_0 & x_4 \\ y_0 & y_4 \end{vmatrix}}{y_0} = \frac{y_0 x_4 - y_4 x_0}{y_0}$$

$$z_4 = \frac{(-1)\begin{vmatrix} x_0 & x_5 \\ y_0 & y_5 \end{vmatrix}}{y_0} = \frac{y_0 x_5 - y_5 x_0}{y_0} \quad \text{and so on.}$$

The elements $z_0, z_1, z_2, z_3, \dots$ are computed for all possible computations as shown above.

In the process of constructing Routh array the missing terms are considered as zeros. Also, all the elements of any row can be multiplied or divided by a positive constant to simplify the computational work.

In the construction of Routh array one may come across the following three cases.

Case-I   : Normal Routh array (Non-zero elements in the first column of routh array).

Case-II : A row of all zeros.

Case-III : First element of a row is zero but some or other elements are not zero.

Case-I : Normal routh array

In this case, there is no difficulty in forming routh array. The routh array can be constructed as explained above. The sign changes are noted to find the number of roots lying on the right half of s-plane and the stability of the system can be estimated.

In this case,

1. If there is no sign change in the first column of Routh array then all the roots are lying on left half of s-plane and the system is stable.

2. If there is sign change in the first column of routh array, then the system is unstable and the number of roots lying on the right half of s-plane is equal to number of sign changes. The remaining roots are lying on the left half of s-plane.

Case-II : A row of all zeros

An all zero row indicates the existence of an even polynomial as a factor of the given characteristic equation. In an even polynomial the exponents of s are even integers or zero only. This even polynomial factor is also called *auxiliary polynomial.* The coefficients of the auxiliary polynomial will always be the elements of the row directly above the row of zeros in the array.

The roots of an even polynomial occur in pairs that are equal in magnitude and opposite in sign. Hence, these roots can be purely imaginary, purely real or complex. The purely imaginary and purely real roots occur in pairs. The complex roots occur in groups of four and the complex roots have quadrantal symmetry, that is the roots are symmetrical with respect to both the real and imaginary axes. The fig 4.1 shows the roots of an even polynomial.

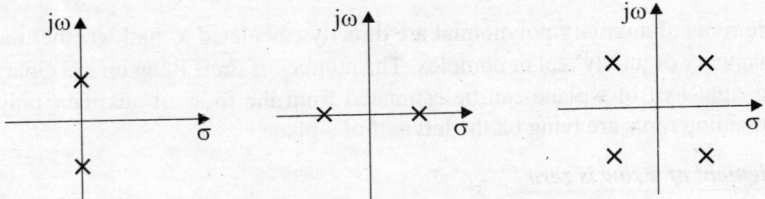

Fig 4.1 : *The roots of an even polynomial.*

The case-II polynomial can be analyzed by any one of the following two methods.

METHOD-1

1. Determine the auxiliary polynomial, A(s)

2. Differentiate the auxiliary polynomial with respect to s, to get d A(s)/ds

3. The row of zeros is replaced with coefficients of dA(s)/ds.

4. Continue the construction of the array in the usual manner (as that of case-I) and the array is interpreted as follows.

   a. If there are sign changes in the first column of routh array then the system is unstable. The number of roots lying on right half of s-plane is equal to number of sign changes. The number of roots on imaginary axis can be estimated from the roots of auxiliary polynomial. The remaining roots are lying on the left half of s-plane.

   b. If there are no sign changes in the first column of routh array then the all zeros row indicate the existence of purely imaginary roots and so the system is limitedly or marginally stable. The roots of auxiliary equation lies on imaginary axis and the remaining roots lies on left half of s-plane.

METHOD-2

1. Determine the auxiliary polynomial, A(s).

2. Divide the characteristic equation by auxiliary polynomial.

3.   Construct Routh array using the coefficients of quotient polynomial.

4.   The array is interpretted as follows.

   a.   If there are sign changes in the first column of routh array of quotient polynomial then the system is unstable. The number of roots of quotient polynomial lying on right half of s-plane is given by number of sign changes in first column of routh array.

   The roots of auxiliary polynomial are directly calculated to find whether they are purely imaginary or purely real or complex.

   The total number of roots on right half of s-plane is given by the sum of number of sign changes and the number of roots of auxiliary polynomial with positive real part. The number of roots on imaginary axis can be estimated from the roots of auxiliary polynomial. The remaining roots are lying on the left half of s-plane.

   b.   If there is no sign change in the first column of routh array of quotient polynomial then the system is limitedly or marginally stable. Since there is no sign change all the roots of quotient polynomial are lying on the left half of s-plane.

   The roots of auxiliary polynomial are directly calculated to find whether they are purely imaginary or purely real or complex. The number of roots lying on imaginary axis and on the right half of s-plane can be estimated from the roots of auxiliary polynomial. The remaining roots are lying on the left half of s-plane.

Case-III : First element of a row is zero

   While constructing routh array, if a zero is encountered as first element of a row then all the elements of the next row will be infinite. To overcome this problem let $0 \rightarrow \in$ and complete the construction of array in the usual way (as that of case-I)

   Finally let $\in \rightarrow 0$ and determine the values of the elements of the array which are functions of $\in$. The resultant array is interpreted as follows.

> **Note :** *If all the elements of a row are zeros then the solution is attempted by considering the polynomial as case-II polynomial. Even if there is a single element zero on s^1 row, it is considered as a row of all zeros.*

   a.   If there is no sign change in first column of routh array and if there is no row with all zeros, then all the roots are lying on left half of s-plane and the system is stable.

   b.   If there are sign changes in first column of routh array and there is no row with all zeros, then some of the roots are lying on the right half of s-plane and the system is unstable. The number of roots lying on the right half of s-plane is equal to number of sign changes and the remaining roots are lying on the left half of s-plane.

   c.   If there is a row of all zeros after letting $\in \rightarrow 0$, then there is a possibility of roots on imaginary axis. Determine the auxiliary polynomial and divide the characteristic equation by auxiliary polynomial to eliminate the imaginary roots. The routh array is constructed using the coefficients of quotient polynomial and the characteristic equation is interpreted as explained in method-2 of case-II polynomial.

EXAMPLE 4.1

Using Routh criterion, determine the stability of the system represented by the characteristic equation, $s^4 + 8s^3 + 18s^2 + 16s + 5 = 0$. Comment on the location of the roots of characteristic equation.

SOLUTION

The characteristic equation of the system is, $s^4 + 8s^3 + 18s^2 + 16s + 5 = 0$.

The given characteristic equation is 4th order equation and so it has 4 roots. Since the highest power of s is even number, form the first row of routh array using the coefficients of even powers of s and form the second row using the coefficients of odd powers of s.

$$s^4 \quad : \quad 1 \qquad 18 \qquad 5 \qquad \text{.... Row-1}$$

$$s^3 \quad : \quad 8 \qquad 16 \qquad\qquad \text{.... Row-2}$$

The elements of s^3 row can be divided by 8 to simplify the computations.

$$s^4 \quad : \quad 1 \qquad 18 \qquad 5 \qquad \text{.... Row-1}$$
$$s^3 \quad : \quad 1 \qquad 2 \qquad\qquad \text{.... Row-2}$$
$$s^2 \quad : \quad 16 \qquad 5 \qquad\qquad \text{.... Row-3}$$
$$s^1 \quad : \quad 1.7 \qquad\qquad\qquad \text{.... Row-4}$$
$$s^0 \quad : \quad 5 \qquad\qquad\qquad \text{.... Row-5}$$

Column-1

$$s^2 : \frac{1 \times 18 - 2 \times 1}{1} \quad \frac{1 \times 5 - 0 \times 1}{1}$$

$$s^2 : \qquad 16 \qquad\qquad 5$$

$$s^1 : \frac{16 \times 2 - 5 \times 1}{16}$$

$$s^1 : 1.6875 \approx 1.7$$

$$s^0 : \frac{1.7 \times 5 - 0 \times 16}{1.7}$$

$$s^0 : 5$$

On examining the elements of first column of routh array it is observed that all the elements are positive and there is no sign change. Hence all the roots are lying on the left half of s-plane and the system is stable.

RESULT

1. Stable system

2. All the four roots are lying on the left half of s-plane.

EXAMPLE 4.2

Construct Routh array and determine the stability of the system whose characterisitc equation is $s^6 + 2s^5 + 8s^4 + 12s^3 + 20s^2 + 16s + 16 = 0$. Also determine the number of roots lying on right half of s-plane, left half of s-plane and on imaginary axis.

SOLUTION

The characteristic equation of the system is, $s^6 + 2s^5 + 8s^4 + 12s^3 + 20s^2 + 16s + 16 = 0$.

The given characteristic polynomial is 6th order equation and so it has 6 roots. Since the highest power of s is even number, form the first row of routh array using the coefficients of even powers of s and form the second row using the coefficients of odd powers of s.

$$s^6 \quad : \quad 1 \qquad 8 \qquad 20 \qquad 16 \qquad \text{.... Row-1}$$

$$s^5 \quad : \quad 2 \qquad 12 \qquad 16 \qquad\qquad \text{.... Row-2}$$

The elements of s^5 row can be divided by 2 to simplify the calculations.

s^6 :   1   8   20   16   .... Row-1

s^5 :   1   6   8   ....Row-2

s^4 :   1   6   8   ....Row-4

s^3 :   0   0   ....Row-4

s^3 :   1   3   ....Row-4

s^2 :   3   8   ....Row-5

s^1 :   0.33   ....Row-6

s^0 :   8   ....Row-7

└─── Column-1

s^4 : $\dfrac{1 \times 8 - 6 \times 1}{1}$ $\dfrac{1 \times 20 - 8 \times 1}{1}$ $\dfrac{1 \times 16 - 0 \times 1}{1}$

s^4 :   2   12   16

divide by 2

s^4 :   1   6   8

s^3 : $\dfrac{1 \times 6 - 6 \times 1}{1}$ $\dfrac{1 \times 8 - 8 \times 1}{1}$

s^3 :   0   0

The auxiliary equation is, $A = s^4 + 6s^2 + 8$. On differentiating A with respect to s we get,

$$\frac{dA}{ds} = 4s^3 + 12s$$

The coefficients of $\dfrac{dA}{ds}$ are used to form s3 row.

s^3 :   4   12

divide by 4

s^3 :   1   3

s^2 : $\dfrac{1 \times 6 - 3 \times 1}{1}$ $\dfrac{1 \times 8 - 0 \times 1}{1}$

s^2 :   3   8

s^1 : $\dfrac{3 \times 3 - 8 \times 1}{3}$

s^1 :   0.33

s^0 : $\dfrac{0.33 \times 8 - 0 \times 3}{0.33}$

s^0 :   8

On examining the elements of 1st column of routh array it is observed that there is no sign change. The row with all zeros indicate the possibility of roots on imaginary axis. Hence the system is limitedly or marginally stable.

The auxiliary polynomial is,

$$s^4 + 6s^2 + 8 = 0$$

Let, $s^2 = x$

$$\therefore x^2 + 6x + 8 = 0$$

The roots of quadratic are, $x = \dfrac{-6 \pm \sqrt{6^2 - 4 \times 8}}{2}$

$$= -3 \pm 1 = -2 \text{ or } -4$$

The roots of auxiliary polynomial is,

$$s = \pm\sqrt{x} = \pm\sqrt{-2} \text{ and } \pm\sqrt{-4}$$

$$= +j\sqrt{2}, -j\sqrt{2}, +j2 \text{ and } -j2$$

The roots of auxiliary polynomial are also roots of characteristic equation. Hence 4 roots are lying on imaginary axis and the remaining two roots are lying on the left half of s-plane.

RESULT

1. The system is limitedly or marginally stable.

2. Four roots are lying on imaginary axis and remaining two roots are lying on left half of s-plane.

EXAMPLE 4.3

Construct Routh array and determine the stability of the system represented by the characteristic equation, $s^5 + s^4 + 2s^3 + 2s^2 + 3s + 5 = 0$. Comment on the location of the roots of characteristic equation.

SOLUTION

The characteristic equation of the system is, $s^5 + s^4 + 2s^3 + 2s^2 + 3s + 5 = 0$.

The given characteristic polynomial is 5th order equation and so it has 5 roots. Since the highest power of s is odd number, form the first row of routh array using the coefficients of odd powers of s and form the second row using the coefficients of even powers of s.

s^5 :   1   2   3   .... Row-1

s^4 :   1   2   5   .... Row-2

s^3	:	$\in$		-2	.... Row-3
s^2	:	$\dfrac{2\in+2}{\in}$		5	.... Row-4
s^1	:	$\dfrac{-(5\in^2+4\in+4)}{2\in+2}$			.... Row-5
s^0	:	5			.... Row-6

On letting $\in \to 0$, we get

	s^5	:	1	2	3	.... Row-1
	s^4	:	1	2	5	.... Row-2
	s^3	:	0	-2		.... Row-3
	s^2	:	∞	5		.... Row-4
	s^1	:	-2			.... Row-5
	s^0	:	5			.... Row-6

— to — / — to + / + to — (arrows indicating sign changes)

Column-1

s^3 : $\dfrac{1\times2-2\times1}{1}$ $\dfrac{1\times3-5\times1}{1}$

s^3 : $\qquad 0 \qquad\qquad -2$

Replace 0 by $\in$

s^2 : $\qquad \in \qquad\qquad -2$

s^2 : $\dfrac{\in\times2-(-2\times1)}{\in}$ $\dfrac{\in\times5-0\times1}{\in}$

s^2 : $\dfrac{2\in+2}{\in}$

s^1 : $\dfrac{\dfrac{2\in+2}{\in}\times(-2)-(5\times\in)}{\dfrac{2\in+2}{\in}}$

s^1 : $\dfrac{-(5\in^2+4\in+4)}{2\in+2}$

s^0 : $\dfrac{\dfrac{-(5\in^2+4\in+4)}{2\in+2}\times5-0\times\dfrac{2\in+2}{\in}}{\dfrac{-(5\in^2+4\in+4)}{2\in+2}}$

s^0 : 5

On observing the elements of first column of routh array, it is found that there are two sign changes. Hence two roots are lying on the right half of s-plane and the system is unstable. The remaining three roots are lying on the left half of s-plane.

RESULT

(a). The system is unstable.

(b). Two roots are lying on right half of s-plane and three roots are lying on left half of s-plane.

EXAMPLE 4.4

By routh stability criterion determine the stability of the system represented by the characteristic equation, $9s^5 - 20s^4 + 10s^3 - s^2 - 9s - 10 = 0$. Comment on the location of roots of characteristic equation.

SOLUTION

The characteristic polynomial of the system is, $9s^5 - 20s^4 + 10s^3 - s^2 - 9s - 10 = 0$

On examining the coefficients of the characteristic polynomial, it is found that some of the coefficients are negative and so some roots will lie on the right half of s-plane. Hence the system is unstable. The routh array can be constructed to find the number of roots lying on right half of s-plane.

The given characteristic polynomial is 5th order equation and so it has 5 roots. Since the highest power of s is odd number, form the first row of routh array using the coefficients of odd powers of s and form the second row using the coefficients of even powers of s.

	s^5	:	9	10	-9	.... Row-1
	s^4	:	-20	-1	-10	.... Row-2
	s^3	:	9.55	-13.5		.... Row-3
	s^2	:	-29.3	-10		.... Row-4
	s^1	:	-16.8			.... Row-5
	s^0	:	-10			.... Row-6

— to — / + to + / — to — / + to — (arrows indicating sign changes)

Column-1

s^3 : $\dfrac{-20\times10-(-1)\times9}{-20}$ $\dfrac{-20\times(-9)-(-10)\times9}{-20}$

s^3 : $\qquad 9.55 \qquad\qquad -13.5$

s^2 : $\dfrac{9.55\times(-1)-(-13.5)\times(-20)}{9.55}$ $\dfrac{9.55\times(-10)}{9.55}$

s^2 : $\qquad -29.3 \qquad\qquad -10$

By examining the elements of Ist column of routh array it is observed that there are three sign changes and so three roots are lying on the right half of s-plane and the remaining two roots are lying on the left half of s-plane.

RESULT

(a). The system is unstable.

(b). Three roots are lying on right half of s-plane and two roots are lying on left half of s-plane.

s^1 :	$\dfrac{-29.3\times(-13.5)-(-10)\times 9.55}{-29.3}$
s^1 :	-16.8
s^0 :	$\dfrac{-16.8\times(-10)}{-16.8}$
s^0 :	-10

EXAMPLE 4.5

The characteristic polynomial of a system is, $s^7+9s^6+24s^5+24s^4+24s^3+24s^2+23s+15=0$. Determine the location of roots on s-plane and hence the stability of the system.

SOLUTION

METHOD-I

The characteristic equation is, $s^7 + 9s^6 + 24s^5 + 24s^4 + 24s^3 + 24s^2 + 23s +15 = 0$.

The given characteristic polynomial is 7th order equation and so it has 7 roots. Since the highest power of s is odd number, form the first row of array using the coefficients of odd powers of s and form the second row using the coefficients of even powers of s as shown below.

s^7	:	1	24	24	23	.... Row-1
s^6	:	9	24	24	15	.... Row-2

Divide s^6 row by 3 to simplify the computations.

s^7	:	1	24	24	23	.... Row-1
s^6	:	3	8	8	5	.... Row-2
s^5	:	1	1	1		.... Row-3
s^4	:	1	1	1		.... Row-4
s^3	:	0	0			.... Row-5
s^3	:	2	1			.... Row-5
s^2	:	0.5	1			.... Row-6
s^1	:	-3				.... Row-7
s^0	:	1				.... Row-8

└── Column-1

On examining the first column elements of routh array it is found that there are two sign changes. Hence two roots are lying on the right half of s-plane and so the system is unstable.

The row of all zeros indicates the possibility of roots on imaginary axis. This can be tested by evaluating the roots of auxiliary polynomial.

The auxiliary equation is, $s^4 + s^2 + 1 = 0$

Put, $s^2 = x$ in the auxiliary equation,

$$s^4 + s^2 + 1 = x^2 + x + 1 = 0$$

s^5 :	$\dfrac{3\times24-8\times1}{3}\quad\dfrac{3\times24-8\times1}{3}\quad\dfrac{3\times23-5\times1}{3}$
s^5 :	$21.33\qquad\qquad 21.33\qquad\qquad 21.33$
Divide by 21.33	
s^5 :	$1\qquad\qquad 1\qquad\qquad 1$
s^4 :	$\dfrac{1\times8-1\times3}{1}\quad\dfrac{1\times8-1\times3}{1}\quad\dfrac{1\times5-0\times3}{1}$
s^5 :	$5\qquad\qquad 5\qquad\qquad 5$
Divide by 5	
s^4 :	$1\qquad\qquad 1\qquad\qquad 1$
s^3 :	$\dfrac{1\times1-1\times1}{1}\quad\dfrac{1\times1-1\times1}{1}$
s^3 :	$0\qquad\qquad 0$

The auxiliary polynomial is,

$$A = s^4 + s^2 + 1$$

Differentiate A with respect to s.

$$\frac{dA}{ds} = 4s^2 + 2s$$

s^3 :	4	2

Divide by 2

s^3 :	2	1

The roots of quadratic are, $x = \dfrac{-1 \pm \sqrt{1-4}}{2} = -\dfrac{1}{2} \pm j\dfrac{\sqrt{3}}{2}$

$$= 1\angle 120° \quad \text{or} \quad 1\angle -120°$$

But $s^2 = x$, $\quad \therefore \quad s = \pm\sqrt{x} = \pm\sqrt{1\angle 120°} \quad \text{or} \quad \pm\sqrt{1\angle 120°}$

$$= \pm\sqrt{1}\angle 120°/2 \quad \text{or} \quad \pm\sqrt{1}\angle -120°/2$$

$$= \pm 1\angle 60° \quad \text{or} \quad \pm 1\angle -60°$$

$$= \pm(0.5 + j0.866) \quad \text{or} \quad \pm(0.5 - j0.866)$$

s^2:	$\dfrac{2\times 1 - 1\times 1}{2}$	$\dfrac{2\times 1 - 0\times 1}{2}$
s^2:	0.5	1
s^1:	$\dfrac{0.5\times 1 - 1\times 2}{0.5}$	
s^1:	-3	
s^0:	$\dfrac{-3\times 1}{-3}$	
s^0:	1	

Two roots of auxiliary polynomial are lying on the right half of s-plane and the remaining two on the left half of s-plane. The roots of auxiliary equation are also the roots of characteristic polynomial. The two roots lying on the right half of s-plane are indicated by two sign changes in the first column of routh array. The remaining five roots are lying on the left half of s-plane. No roots are lying on imaginary axis.

RESULT

1. The system is unstable.

2. Two roots are lying on right half of s-plane and five roots are lying on left half of s-plane.

METHOD-II

The characteristic equation is,

The given characteristic polynomial is 7th order equation and so it has 7 roots. Since the highest power of s is odd number, form the first row of array using the coefficients of odd powers of s and form the second row using the coefficients of even powers of s as shown below.

s^7	:	1	24	24	23	.... Row-1
s^6	:	9	24	24	15	.... Row-2

Divide s^6 row by 3 to simplify the computations.

s^7	:	1	24	24	23	.... Row-1
s^6	:	3	8	8	5	.... Row-2
s^5	:	1	1	1		.... Row-3
s^4	:	1	1	1		.... Row-4
s^3	:	0	0			.... Row-5

s^4:	$\dfrac{3\times 24 - 8\times 1}{3}$	$\dfrac{3\times 24 - 8\times 1}{3}$	$\dfrac{3\times 23 - 5\times 1}{3}$
s^4:	21.33	21.33	21.33
Divide by 21.33			
s^4:	1	1	1
s^4:	$\dfrac{1\times 8 - 1\times 3}{1}$	$\dfrac{1\times 8 - 1\times 3}{1}$	$\dfrac{1\times 5 - 0\times 3}{1}$
s^4:	5	5	5
Divide by 5			
s^4:	1	1	1

s^3:	$\dfrac{1\times 1 - 1\times 1}{1}$	$\dfrac{1\times 1 - 1\times 1}{1}$	
s^3:	0	0	

Since we get a row of zeros, there exists an even polynomial, the even polynomial is nothing but, the auxiliary polynomial.

The auxiliary polynomial is,

$$s^4 + s^2 + 1 = 0$$

Divide the characteristic equation by auxiliary polynomial to get the quotient polynomial.

The characteristic polynomial can be expressed as a product of quotient polynomial and auxiliary polynomial.

$s^7 + 9s^6 + 24s^5 + 24s^4 + 24s^3 + 24\,s^2 + 23\,s + 15 = 0$

$$\Downarrow$$

$(s^4 + s^2 + 1)\,(s^3 + 9s^2 + 23s + 15) = 0$

Even   Quotient
polynomial   polynomial

The routh array is constructed for quotient polynomial as shown below.

s^3 :   1   23

s^2 :   9   15

Divide s^2 row by 3,

s^3 :   1   23

s^2 :   3   5

s^1 :   21.33

s^0 :   5

└── Column-1

$s^1 : \dfrac{3 \times 23 - 5 \times 1}{3}$

$s^1 : 21.33$

$s^1 : \dfrac{21.33 \times 5 - 0 \times 3}{21.33}$

$s^0 : 5$

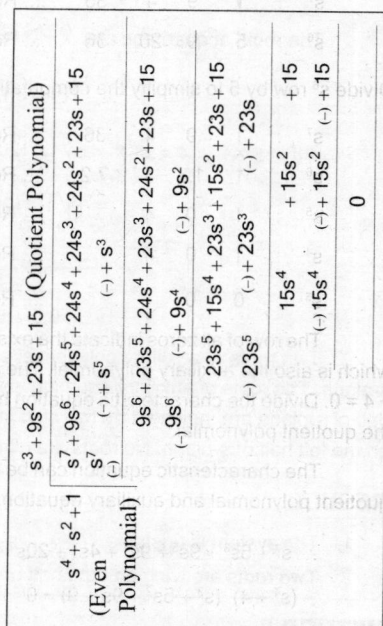

The elements of column-1 of quotient polynomial are all positive and there is no sign change. Hence all the roots of quotient polynomial are lying on the left half of s-plane. To determine the stability, the roots of auxiliary polynomial should be evaluated.

The auxiliary equation is, $s^4 + s^2 + 1 = 0$.

Put, $s^2 = x$ in the auxiliary equation. $s^4 + s^2 + 1 = x^2 + x + 1 = 0$

The roots of quadratic are, $x = \dfrac{-1 \pm \sqrt{1-4}}{2} = -\dfrac{1}{2} \pm j\dfrac{\sqrt{3}}{2} = 1\angle 120°$ or $1\angle -120°$

But $s^2 = x$, $\therefore s = \pm\sqrt{x} = \pm\sqrt{1\angle 120°}$ or $\pm\sqrt{1\angle 120°}$

$= \pm\sqrt{1}\,\angle 120°/2$ or $\pm\sqrt{1}\,\angle -120°/2$

$= \pm 1\angle 60°$ or $\pm 1\angle -60°$

$= \pm(0.5 + j0.866)$ or $\pm(0.5 - j0.866)$

The roots of auxiliary equation are complex and has quadrantal symmetry. Two roots of auxiliary equation are lying on the right half of s-plane and the other two on the left half of s-plane.

The roots of characteristic equation are given by the roots of auxiliary polynomial and the roots of quotient polynomial. Hence we can conclude that two roots of characteristic equation are lying on the right half of s-plane and so system is unstable. The remaining five roots are lying on left half of s-plane.

EXAMPLE 4.6

The characteristic polynomial of a system is $s^7 + 5s^6 + 9s^5 + 9s^4 + 4s^3 + 20s^2 + 36s + 36 = 0$. Determine the location of roots on the s-plane and hence the stability of the system.

SOLUTION

The characteristic equation is, $s^7 + 5s^6 + 9s^5 + 9s^4 + 4s^3 + 20s^2 + 36s + 36 = 0$.

The given characteristic polynomial is 7th order equation and so it has 7 roots. Since the highest power of s is odd number, form the first row of array using the coefficients of odd powers of s and form the second row of array using the coefficients of even powers of s as shown below.

s^7 : 1 9 4 36 .... Row-1

s^6 : 5 9 20 36 .... Row-2

Divide s^6 row by 5 to simplify the computations.

s^7 : 1 9 4 36 .... Row-1

s^6 : 1 1.8 4 7.2 .... Row-2

s^5 : 1 0 4 .... Row-3

s^4 : 1 0 4 .... Row-4

s^3 : 0 0 .... Row-5

The row of all zeros indicate the existence of even polynomial, which is also the auxiliary polynomial. The auxiliary polynomial is, $s^4 + 4 = 0$. Divide the characteristic equation by auxiliary equation to get the quotient polynomial.

The characteristic equation can be expressed as a product of quotient polynomial and auxiliary equation.

$\therefore\ s^7 + 5s^6 + 9s^5 + 9s^4 + 4s^2 + 20s^2 + 36s + 36 = 0$

$(s^4 + 4)\ (s^3 + 5s^2 + 9s + 9) = 0$

Even   Quotient
polynomial polynomial

The routh array is constructed for quotient polynomial as shown below.

s^3 : 1 9

s^2 : 5 9

s^1 : 7.2

s^0 : 9

Column-1

s^5 : $\dfrac{1\times9 - 1.8\times1}{1}$ $\dfrac{1\times4 - 4\times1}{1}$ $\dfrac{1\times36 - 7.2\times1}{1}$

s^5 : 7.2 0 28.8

Divide by 7.2

s^5 : 1 0 4

s^4 : $\dfrac{1\times1.8 - 0\times1}{1}$ $\dfrac{1\times4 - 4\times1}{1}$ $\dfrac{1\times7.2 - 0\times1}{1}$

s^5 : 1.8 0 7.2

Divide by 1.8

s^4 : 1 0 4

s^3 : $\dfrac{1\times0 - 0\times1}{1}$ $\dfrac{1\times4 - 4\times1}{1}$

s^3 : 0 0

s^1 : $\dfrac{5\times9 - 9\times1}{5}$

s^1 : 7.2

s^0 : $\dfrac{7.2\times9 - 0\times5}{7.2}$

s^0 : 9

There is no sign change in the elements of first column of routh array of quotient polynomial. Hence all the roots of quotient polynomial are lying on the left half of s-plane.

To determine the stability, the roots of auxiliary polynomial should be evaluated.

The auxiliary polynomial is, $s^4 + 4 = 0$.

Put, $s^2 = x$ in the auxiliary equation, $\therefore\ s^4 + 4 = x^2 + 4 = 0$

$\therefore x^2 = -4$ $x = \pm\sqrt{-4} = \pm j2 = 2\angle90°$ or $2\angle-90°$

But, $s = \pm\sqrt{x} = \pm\sqrt{2\angle90°}$ or $\pm\sqrt{2\angle-90°} = \pm\sqrt{2}\angle90°/2$ or $\pm\sqrt{2}\angle-90°/2$

$= \pm\sqrt{2}\angle45°$ or $\pm\sqrt{2}\angle-45° = \pm(1+j1)$ or $\pm(1-j1)$

The roots of auxiliary equation are complex and has quadrantal symmetry. Two roots of auxiliary equation are lying on the right half of s-plane and the other two on the left half of s-plane.

The roots of characteristic equation are given by roots of quotient polynomial and auxiliary polynomial. Hence we can conclude that two roots of characteristic equation are lying on the right half of s-plane and so the system is unstable. The remaining five roots are lying on the left half of s-plane.

RESULT

(a) The system is unstable.

(b) Two roots are lying on the right half of s-plane and five roots are lying on the left half of s-plane.

EXAMPLE 4.7

Use the routh stability criterion to determine the location of roots on the s-plane and hence the stability for the system represented by the characteristic equation $s^5 + 4s^4 + 8s^3 + 8s^2 + 7s + 4 = 0$.

SOLUTION

The characteristic equation of the system is, $s^5 + 4s^4 + 8s^3 + 8s^2 + 7s + 4 = 0$.

The given characteristic polynomial is 5th order equation and so it has 5 roots. Since the highest power of s is odd number, form the first row of routh array using the coefficients of odd powers of s and form the second row using the coefficients of even powers of s.

$$s^5 \quad : \quad 1 \quad 8 \quad 7 \qquad .... \text{Row-1}$$
$$s^4 \quad : \quad 4 \quad 8 \quad 4 \qquad .... \text{Row-2}$$

Divide s^4 row by 4 to simplify the calculations.

$$s^5 \quad : \quad 1 \quad 8 \quad 7 \qquad .... \text{Row-1}$$
$$s^4 \quad : \quad 1 \quad 2 \quad 1 \qquad .... \text{Row-2}$$
$$s^3 \quad : \quad 1 \quad 1 \qquad\qquad .... \text{Row-3}$$
$$s^2 \quad : \quad 1 \quad 1 \qquad\qquad .... \text{Row-4}$$
$$s^1 \quad : \quad \in \qquad\qquad\qquad .... \text{Row-5}$$
$$s^0 \quad : \quad 1 \qquad\qquad\qquad .... \text{Row-6}$$

$s^3 : \dfrac{1\times 8 - 2\times 1}{1} \quad \dfrac{1\times 7 - 1\times 1}{1}$

$s^3 : \qquad 6 \qquad\qquad 6$

Divide by 6

$s^3 : \qquad 1 \qquad\qquad 1$

$s^2 : \dfrac{1\times 2 - 1\times 1}{1} \quad \dfrac{1\times 1 - 0\times 1}{1}$

$s^2 : \qquad 1 \qquad\qquad 1$

$s^1 : \dfrac{1\times 1 - 1\times 1}{1}$

$s^1 : \qquad 0$

Let $0 \to \in$

$s^1 : \qquad \in$

$s^0 : \dfrac{\in \times 1 - 0\times 1}{\in}$

$s^0 : \quad 1$

When $\in \to 0$, there is no sign change in the first column of routh array. But we have a row of all zeros (s^1 row or row-5) and so there is a possibility of roots on imaginary axis. This can be found from the roots of auxiliary polynomial. Here the auxiliary polynomial is given by s^2 row.

The auxiliary polynomial is, $s^2 + 1 = 0$; $\qquad \therefore s^2 = -1$ or $s = \pm\sqrt{-1} = \pm j1$

The roots of auxiliary polynomial are $+j1$ and $-j1$, lying on imaginary axis. The roots of auxiliary polynomial are also roots of characteristic equation. Hence two roots of characteristic equation are lying on imaginary axis and so the system is limitedly or marginally stable. The remaining three roots of characteristic equation are lying on the left half of s-plane.

RESULT

(a) The system is limitedly or marginally stable.

(b) Two roots are lying on imaginary axis and three roots are lying on left half of s-plane.

EXAMPLE 4.8

Use the routh stability criterion to determine the location of roots on the s-plane and hence the stability for the system represented by the characteristic equation, $s^6 + s^5 + 3s^4 + 3s^3 + 3s^2 + 2s + 1 = 0$.

SOLUTION

The characteristic polynomial of the system is, $s^6 + s^5 + 3s^4 + 3s^3 + 3s^2 + 2s + 1 = 0$.

The given characteristic polynomial is 6th order equation and so it has 6 roots. Since the highest power of s is even number, form the first row of routh array using the coefficients of even powers of s and form the second row using the coefficients of odd powers of s as shown below.

$$s^6 \quad : \quad 1 \qquad 3 \qquad 3 \qquad 1 \quad .... \text{Row-1}$$

$$s^5 \quad : \quad 1 \qquad 3 \qquad 2 \qquad\quad .... \text{Row-2}$$

$$s^4 \quad : \quad \epsilon \qquad 1 \qquad 1 \qquad\quad .... \text{Row-3}$$

$$s^3 \quad : \quad \dfrac{3\epsilon-1}{\epsilon} \quad \dfrac{2\epsilon-1}{\epsilon} \qquad .... \text{Row-4}$$

$$s^2 \quad : \dfrac{-2\epsilon^2+4\epsilon-1}{3\epsilon-1} \quad 1 \qquad .... \text{Row-5}$$

$$s^1 \quad : \dfrac{4\epsilon^2-\epsilon}{2\epsilon^2-4\epsilon+1} \qquad\qquad .... \text{Row-6}$$

$$s^0 \quad : \quad 1 \qquad\qquad\qquad\quad .... \text{Row-7}$$

On letting $\epsilon \to 0$, we get,

$$s^6 \quad : \quad 1 \qquad 3 \qquad 3 \qquad 1 \quad .... \text{Row-1}$$

$$s^5 \quad : \quad 1 \qquad 3 \qquad 2 \qquad\quad .... \text{Row-2}$$

$$s^4 \quad : \quad 0 \qquad 1 \qquad 1 \qquad\quad .... \text{Row-3}$$

$$s^3 \quad : \quad -\infty \qquad -\infty \qquad\quad .... \text{Row-4}$$

$$s^2 \quad : \quad 1 \qquad 1 \qquad\qquad .... \text{Row-5}$$

$$s^1 \quad : \quad 0 \qquad\qquad\qquad .... \text{Row-6}$$

$$s^0 \quad : \quad 1 \qquad\qquad\qquad .... \text{Row-7}$$

$\left. \begin{array}{c} -\text{to }- \\ +\text{to }+ \\ -\text{to }+ \end{array} \right.$

Since there is a row of all zeros (s^1 row) there is a possibility of roots on imaginary axis. The auxiliary polynomial is $s^2 + 1 = 0$.

The roots of auxiliary polynomial are, $s = \pm\sqrt{-1} = \pm j1$

The roots of auxiliary polynomial are also roots of characteristic equation. Hence two roots are lying on imaginary axis. Therefore divide the characteristic polynomial by auxiliary equation and construct the routh array for quotient polynomial to find the roots lying on right half of s-plane.

The characteristic polynomial can be expressed as a product of auxiliary polynomial and quotient polynomial.

$$s^6 + s^5 + 3s^4 + 3s^3 + 3s^2 + 2s + 1 = 0 \quad \Rightarrow \quad \underset{\substack{\text{Even} \\ \text{polynomial}}}{(s^2 + 1)} \; \underset{\substack{\text{Quotient} \\ \text{polynomial}}}{(s^4 + s^3 + 2s^2 + 2s + 1)} = 0$$

The routh array for quotient polynomial is constructed as shown below.

$$s^4 \quad : \quad 1 \qquad 2 \qquad 1 \qquad .... \text{Row-1}$$

$$s^3 \quad : \quad 1 \qquad 2 \qquad\qquad .... \text{Row-2}$$

$$s^2 \quad : \quad \epsilon \qquad 1 \qquad\qquad .... \text{Row-3}$$

$$s^1 \quad : \quad \dfrac{2\epsilon-1}{\epsilon} \qquad\qquad .... \text{Row-4}$$

$$s^0 \quad : \quad 1 \qquad\qquad\qquad .... \text{Row-5}$$

Side boxes (working):

$$s^4 : \dfrac{1\times3-3\times1}{1} \quad \dfrac{1\times3-2\times1}{1} \quad \dfrac{1\times1-0\times1}{1}$$

$$s^4 : \quad 0 \qquad\qquad 1 \qquad\qquad 1$$

$$\text{let } 0 \to \epsilon$$

$$s^4 : \quad \epsilon \qquad\qquad 1 \qquad\qquad 1$$

$$s^3 : \dfrac{\epsilon\times3-1\times1}{\epsilon} \quad \dfrac{\epsilon\times2-1\times1}{\epsilon}$$

$$s^3 : \quad \dfrac{3\epsilon-1}{\epsilon} \qquad \dfrac{2\epsilon-1}{\epsilon}$$

$$s^2 : \dfrac{\dfrac{3\epsilon-1}{\epsilon}-\dfrac{2\epsilon-1}{\epsilon}\times\epsilon}{\dfrac{3\epsilon-1}{\epsilon}} \qquad \dfrac{\dfrac{3\epsilon-1}{\epsilon}\times1-0\times\epsilon}{\dfrac{3\epsilon-1}{\epsilon}}$$

$$s^2 : \quad \dfrac{-2\epsilon^2+4\epsilon-1}{3\epsilon-1} \qquad\qquad 1$$

$$s^1 : \dfrac{\dfrac{-2\epsilon^2+4\epsilon-1}{3\epsilon-1}\times\dfrac{2\epsilon-1}{\epsilon}-\dfrac{3\epsilon-1}{\epsilon}\times1}{\dfrac{-2\epsilon^2+4\epsilon-1}{3\epsilon-1}}$$

$$s^1 : \dfrac{(-2\epsilon^2+4\epsilon-1)(2\epsilon-1)-(3\epsilon-1)(3\epsilon-1)}{\epsilon(-2\epsilon^2+4\epsilon-1)}$$

$$s^1 : \dfrac{-4\epsilon^3+\epsilon^2}{\epsilon(-2\epsilon^2+4\epsilon-1)} = \dfrac{4\epsilon^2-\epsilon}{2\epsilon^2-4\epsilon+1}$$

$$s^0 : \dfrac{\dfrac{4\epsilon^2-\epsilon}{4\epsilon^2-4\epsilon+1}\times1-0\times\dfrac{-2\epsilon^2+4\epsilon-1}{3\epsilon-1}}{(4\epsilon^2-\epsilon)/(4\epsilon^2-4\epsilon+1)}$$

$$s^0 : \quad 1$$

$$s^2 : \dfrac{1\times2-2\times1}{1} \quad \dfrac{1\times1-0\times1}{1}$$

$$s^2 : \quad 0$$

$$\text{let } 0 \to \epsilon$$

$$s^2 : \quad \epsilon \qquad\qquad 1$$

$$s^1 : \dfrac{\epsilon\times2-1\times1}{\epsilon}$$

$$s^1 : \dfrac{2\epsilon-1}{\epsilon}$$

$$s^0 : \dfrac{\dfrac{2\epsilon-1}{\epsilon}\times1-0\times\epsilon}{(2\epsilon-1)/\epsilon}$$

On letting $\epsilon \to 0$, we get

$$
\begin{array}{lll}
s^4 & : & \boxed{1 \quad 2 \quad 1} \quad \dots \text{Row-1} \\
s^3 & : & 1 \quad 2 \qquad \dots \text{Row-2} \\
s^2 & : & 0 \quad 1 \qquad \dots \text{Row-3} \\
s^1 & : & -\infty \qquad \dots \text{Row-4} \\
s^0 & : & 1 \qquad \dots \text{Row-5}
\end{array}
$$

Column-1

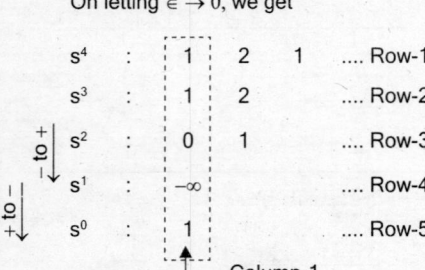

(Quotient polynomial)

$$
s^2 + 1 \overline{\big)\ s^6 + s^5 + 3s^4 + 3s^3 + 3s^2 + 2s + 1}
$$
$$
\begin{aligned}
& s^6_{(-)} \quad {}^+_{(-)}s^4 \\
\hline
& s^5 + 2s^4 + 3s^3 + 3s^2 + 2s + 1 \\
& {}_{(-)}s^5 \qquad {}^+_{(-)}s^3 \\
\hline
& 2s^4 + 2s^3 + 3s^2 + 2s + 1 \\
& {}_{(-)}2s^4 \qquad {}_{(-)}{+}\,2s^2 \\
\hline
& 2s^3 + s^2 + 2s + 1 \\
& {}_{(-)}2s^3 \qquad {}_{(-)}{+}\,2s \\
\hline
& s^2 \qquad\quad + 1 \\
& {}_{(-)}s^2 \qquad {}_{(-)}{+}\,1 \\
\hline
& \qquad\qquad 0
\end{aligned}
$$

(Even polynomial)

On examining the first column of the routh array of quotient polynomial, we found that there are two sign changes. Hence two roots are lying on the right half of s-plane and other two roots of quotient polynomial are lying on the left half of s-plane.

The roots of characteristic equation are given by roots of auxiliary polynomial and quotient polynomial. Hence two roots are lying on imaginary axis, two roots are lying on right half of s-plane and the remaining two roots are lying on left half of s-plane. Hence the system is unstable.

RESULT

(a) The system is unstable.

(b) Two roots are lying on imaginary axis, two roots are lying on right half of s-plane and two roots are lying on left half of s-plane.

EXAMPLE 4.9

Determine the range of K for stability of unity feedback system whose open loop transfer function is $G(s) = \dfrac{K}{s(s+1)(s+2)}$.

SOLUTION

The closed transfer function, $\dfrac{C(s)}{R(s)} = \dfrac{G(s)}{1+G(s)} = \dfrac{\dfrac{K}{s(s+1)(s+2)}}{1 + \dfrac{K}{s(s+1)(s+2)}} = \dfrac{K}{s(s+1)(s+2)+K}$

The characteristic equation is, $s(s+1)(s+2) + K = 0$

$s(s^2 + 3s + 2) + K = 0 \qquad \Rightarrow \qquad s^3 + 3s^2 + 2s + K = 0$

The routh array is constructed as shown below.

The highest power of s in the characteristic polynomial is odd number. Hence form the first row using the coefficients of odd powers of s and form the second row using the coefficients of even powers of s.

$$
\begin{array}{lll}
s^3 & : & \boxed{1 \qquad 2} \\
s^2 & : & 3 \qquad K \\
s^1 & : & \dfrac{6-K}{3} \\
s^0 & : & K
\end{array}
$$

Column-1

$$
\begin{aligned}
s^1 &: \frac{3 \times 2 - K \times 1}{3} \\
s^1 &: \frac{6-K}{3} \\
\hline
s^0 &: \frac{\frac{6-K}{3} \times K - 0 \times 3}{(6-K)/3} \\
s^0 &: K
\end{aligned}
$$

For the system to be stable there should not be any sign change in the elements of first column. Hence choose the value of K so that the first column elements are positive.

From s^0 row, for the system to be stable, K > 0

From s^1 row, for the system to be stable, $\dfrac{6-K}{3} > 0$

For $\dfrac{6-K}{3} > 0$, the value of K should be less than 6.

∴ The range of K for the system to be stable is 0 < K < 6.

RESULT

The value of K is in the range 0<K<6 for the system to be stable.

EXAMPLE 4.10

The open loop transfer function of a unity feedback control system is given by,

$$G(s) = \frac{K}{(s+2)(s+4)(s^2+6s+25)}$$

By applying the routh criterion, discuss the stability of the closed-loop system as a function of K. Determine the value of K which will cause sustained oscillations in the closed-loop system. What are the corresponding oscillating frequencies?

SOLUTION

The closed loop transfer function

$$\frac{C(s)}{R(s)} = \frac{G(s)}{1+G(s)} = \frac{\dfrac{K}{(s+2)(s+4)(s^2+6s+25)}}{1+\dfrac{K}{(s+2)(s+4)(s^2+6s+25)}} = \frac{K}{(s+2)(s+4)(s^2+6s+25)+K}$$

The characteristic equation is given by the denominator polynomial of closed loop transfer function.

The characteristic equation is, $(s+2)(s+4)(s^2+6s+25)+K=0$.

∴ $(s^2+6s+8)(s^2+6s+25)+K=0$ ⇒ $s^4+12s^3+69s^2+198s+200+K=0$

The routh array is constructed as shown below. The highest power of s in the characteristic equation is even number. Hence form the first row using the coefficients of even powers of s and form the second row using the coefficients of odd powers of s.

s^4 :	1	69	200+K	..... Row-1
s^3 :	12	198		..... Row-2

Divide s^3 row by 12 to simplify the calculations

s^4 :	1	69	200+K	..... Row-1
s^3 :	1	16.5		..... Row-2
s^2 :	52.5	200+K		..... Row-3
s^1 :	$\dfrac{666.25-K}{52.5}$			..... Row-4
s^0 :	200+K			..... Row-5

Column-1

s^2 : $\dfrac{1\times69-16.5\times1}{1}$ $\dfrac{1\times(200+K)}{1}$

s^2 :   52.5   200+K

s^1 : $\dfrac{52.5\times16.5-(200+K)\times1}{52.5}$

s^1 : $\dfrac{666.25-K}{52.5}$

s^0 : $\dfrac{\dfrac{666.25-K}{52.5}\times(200+K)}{(666.25-K)/52.5}$

s^0 :   200+K

For the system to be stable there should not be any sign change in the elements of first column. Hence choose the value of K so that the first column elements are positive.

From s^1 row, for the system to be stable, $(666.25 - K) > 0$.

Since $(666.25 - K) > 0$, K should be less than 666.25.

From s^0 row, for the system to be stable, $(200 + K) > 0$

Since $(200 + K) > 0$, K should be greater than –200, but practical values of K starts from 0. Hence K should be greater than zero.

$\therefore$ The range of K for the system to be stable is $0 < K < 666.25$.

When K = 666.25 the s^1 row becomes zero, which indicates the possibility of roots on imaginary axis. A system will oscillate if it has roots on imaginary axis and no roots on right half of s-plane.

When K = 666.25, the coefficients of auxiliary equation are given by the s^2 row.

$\therefore$ The auxiliary equation is, $52.5s^2 + 200 + K = 0$

$$52.5s^2 + 200 + 666.25 = 0$$

$$s^2 = \frac{-200 - 666.25}{52.5} = -16.5$$

$$s = \pm\sqrt{-16.5} = \pm j\sqrt{16.5} = \pm j4.06$$

When K = 666.25, the system has roots on imaginary axis and so it oscillates. The frequency of oscillation is given by the value of root on imaginary axis.

$\therefore$ The frequency of oscillation, $\omega = 4.06$ rad/sec.

RESULT

(a) The range of K for stability is $0 < K < 666.25$
(b) The system oscillates when K = 666.25
(c) The frequency of oscillation, $\omega = 4.06$ rad/sec. (When K = 666.25).

EXAMPLE 4.11

The open loop transfer function of a unity feedback system is given by, $G(s) = \dfrac{K(s+1)}{s^3 + as^2 + 2s + 1}$. Determine the value of K and a so that the system oscillates at a frequency of 2 rad/sec.

SOLUTION

The closed loop transfer function $\left. \right\}$ $\dfrac{C(s)}{R(s)} = \dfrac{G(s)}{1 + G(s)} = \dfrac{\dfrac{K(s+1)}{s^3 + as^2 + 2s + 1}}{1 + \dfrac{K(s+1)}{s^3 + as^2 + 2s + 1}} = \dfrac{K(s+1)}{s^3 + as^2 + 2s + 1 + K(s+1)}$

The characteristic equation is, $s^3 + as^2 + 2s + 1 + K(s+1) = 0$.

$$s^3 + as^2 + 2s + 1 + Ks + K = 0 \quad \Rightarrow \quad s^3 + as^2 + (2 + K)s + 1 + K = 0$$

The routh array of characteristic polynomial is constructed as shown below. The maximum power of s is odd, hence the first row of routh array is formed using coefficients of odd powers of s and the second row of routh array is formed using coefficients of even powers of s.

If the elements of s^1 row are all zeros then there exists an even polynomial (or auxiliary polynomial). If the roots of the auxiliary polynomial are purely imaginary then the roots are lying on imaginary axis and the system oscillates. The frequency of oscillation is the root of auxiliary polynomial.

Routh array

s^3 :   1 $2 + K$

s^2 :   a $1 + K$

s^1 : $\dfrac{a(2+K)-(1+K)}{a}$

s^0 : $1 + K$

From s^2 row, the auxiliary polynomial is,

$$as^2 + (1 + K) = 0 \quad \Rightarrow \quad as^2 = -(1+K) \quad \Rightarrow \quad s = \pm j\sqrt{\dfrac{1+K}{a}}$$

Given that, $s = \pm j2$, $\therefore \sqrt{\dfrac{1+K}{a}} = 2 \quad \Rightarrow \quad \dfrac{1+K}{a} = 4 \quad \Rightarrow \quad K = 4a - 1$

From s^1 row, $\dfrac{a(2+K)-(1+K)}{a} = 0 \quad \Rightarrow \quad a(2+K)-(1+K)=0 \quad \Rightarrow \quad 2a + Ka - 1 - K = 0$

$$\therefore \quad 2a - 1 + K(a-1) = 0$$

Put, $K = 4a - 1$

$$\therefore 2a - 1 + (4a-1)(a-1) = 0 \quad \Rightarrow \quad 2a - 1 + 4a^2 - 4a - a + 1 = 0 \quad \Rightarrow \quad 4a^2 - 3a = 0 \text{ (or) } a(4a-3)=0$$

Since $a \neq 0$, $4a - 3 = 0$, $\therefore a = 3/4$

When $a = (3/4)$, $K = 4a - 1 = 4 \times (3/4) - 1 = 2$

RESULT

When the system oscillates at a frequency of 2 rad/sec, $K = 2$ and $a = 3/4$.

EXAMPLE 4.12

A feedback system has open loop transfer function of $G(s) = \dfrac{Ke^{-s}}{s(s^2+5s+9)}$. Determine the maximum value of K for stability of closed loop system.

SOLUTION

Generally control systems have very low bandwidth which implies that it has very low frequency range of operation. Hence for low frequency ranges the term e^{-s} can be replaced by, $1-s$, (i.e., $e^{s-T} \approx 1 - sT$).

$$\therefore \ G(s) = \dfrac{Ke^{-s}}{s(s^2+5s+9)} \approx \dfrac{K(1-s))}{s(s^2+5s+9)}$$

The close loop transfer function $\Big\}$ $\dfrac{C(s)}{R(s)} = \dfrac{G(s)}{1+G(s)} = \dfrac{\dfrac{K(1-s)}{s(s^2+5s+9)}}{1+\dfrac{K(1-s)}{s(s^2+5s+9)}} = \dfrac{K(1-s)}{s(s^2+5s+9)+K(1-s)}$

The characteristic equation is given by the denominator polynomial of closed loop transfer function.

$\therefore$ The characteristic equation is, $s(s^2+5s+9) + K(1-s) = 0$

$\therefore \ s(s^2+5s+9) + K(1-s) = s^3 + 5s^2 + 9s + K - Ks = 0 \quad \Rightarrow \quad s^3 + 5s^2 + (9-K)s + K = 0$

The routh array of characteristic polynomial is constructed as shown below.

The maximum power of s in the characteristic polynomial is odd, hence form the first row of routh array using coefficients of odd powers of s and second row of routh array using coefficients of even powers of s.

s^3 :   1 $9 - K$

s^2 :   5 K

s^1 : $9 - 1.2K$

s^0 : K

$s^1 : \dfrac{5 \times (9 - K) - K \times 1}{5}$
$s^1 : \dfrac{45 - 5K - K}{5}$
$s^1 : \dfrac{45 - 6K}{5} \approx 9 - 1.2K$
$s^0 : \dfrac{(9 - 1.2K) \times K}{(9 - 1.2K)}$
$s^0 : K$

From s^1 row, for stability of the system, $(9 - 1.2K) > 0$

If $(9 - 1.2K) > 0$ then $1.2K < 9$; $\therefore K < \dfrac{9}{12} = 7.5$

From s^0 row, for stability of the system, $K > 0$

Finally we can conclude that for stability of the system K should be in the range of $0 < K < 7.5$

RESULT

For stability of the system K should be in the range of, $0 < K < 7.5$.

4.4 MATHEMATICAL PRELIMINARIES FOR NYQUIST STABILITY CRITERION

Let F(s) be a function of s, which is expressed as a ratio of two polynomials in s, as shown in equation (4.14), (the polynomials are expressed in the factored form).

$$F(s) = \frac{(s - z_1)(s - z_2)......(s - z_m)}{(s - p_1)(s - p_2)......(s - p_n)} \qquad .....(4.14)$$

The roots of numerator polynomial are zeros and the roots of denominator polynomial are poles. The function has m number of zeros and n number of poles.

Here, s is a complex variable expressed as, $s = \sigma + j\omega$, where s is real part of σ and ω is imaginary part of s. (The s is also called complex frequency). For a particular value of σ and ω, the s will represent a point in the s-plane.

Since s is a complex variable, the function F(s) will also be a complex quantity for any value of s. Hence, F(s) can also be expressed as, $F(s) = u + jv$, where u is real part of F(s) and v is imaginary part of F(s). Let us define another complex plane called F(s)-plane, with coordinates u and v. For a particular value of s, the F(s) will represent a point in F(s)-plane.

Therefore, for every point s in the s-plane at which F(s) is analytic, there exists a corresponding point F(s) in the F(s)-plane. Hence it can be concluded that the function F(s) maps the points in the s-plane into the F(s)-plane.

> **Note :** *A function is analytic in the s-plane provided the function and all its derivatives exist. The points in the s-plane where the function (or it derivatives) does not exist are called singular points.*

Since any number of points of analycity in the s-plane can be mapped into the F(s)- plane it can be concluded that for a contour in the s-plane which does not go through any singular point, there exists a corresponding contour in the F(s)-plane as shown in fig 4.2.

The table 4.2 shows examples of arbitrary s-plane contours and their corresponding F(s)-plane contours (exact shape is not shown).

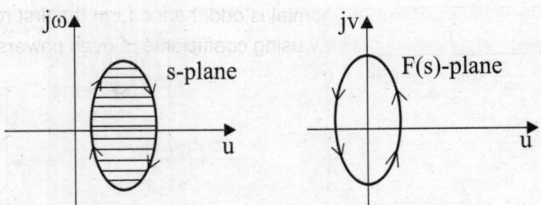

Fig 4.2 : *An arbitrary contour in s-plane and its corresponding contour in F(s)-plane*

Normally the direction of arbitrary contour in s-plane is chosen as clockwise. Here zeros are marked by small circles (o) and poles by (X).

On observing the s-plane contours and the corresponding F(s)-plane contours shown in table-4.2, it can be proved that there exists a relationship between the enclosure of poles and zeros by the s-plane closed contour and number of encirclements of the origin of F(s)- plane by the corresponding F(s)-plane contour.

Note : *For the development of Nyquist criterion, the exact shape of the contour is not required but only the number of encirclements of the origin of the F(s) - plane is essential.*

Concept of encircled and enclosed

It is important to distinguish between the concept of encircled and enclosed which are frequently used to apply Nyquist criterion.

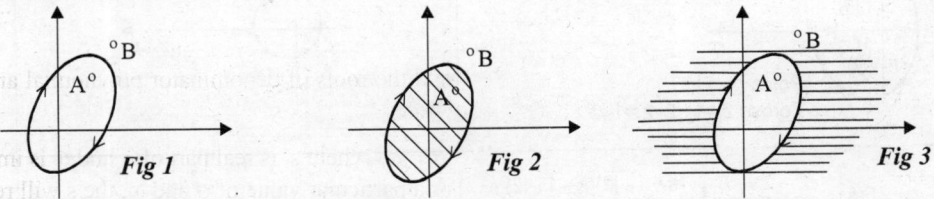

Encircled : *A point is said to be encircled by a closed path if it is found inside the path. With reference to fig 1, the point A is encircled in the clockwise direction and the point B is not encircled.*

Enclosed : *Any point or region is said to be enclosed by a closed path, if it is found to lie to the right of the path when the path is traversed in the prescribed direction. The shaded regions in fig 2 and 3 are the regions enclosed by the closed path. With reference to fig 2, the point A is enclosed by closed path and the point B is not enclosed. With reference to fig 3 the point A is not enclosed by closed path but point B is enclosed.*

TABLE-4.2

The function, F(s) and s-plane contour	F(s)-plane contour
$F(s) = s^2 - 2s + 6 = (s-1-j2)(s-1+j2)$ s-plane Zeros : $z_1 = 1+j2$, $z_2 = 1-j2$	F(s)-plane

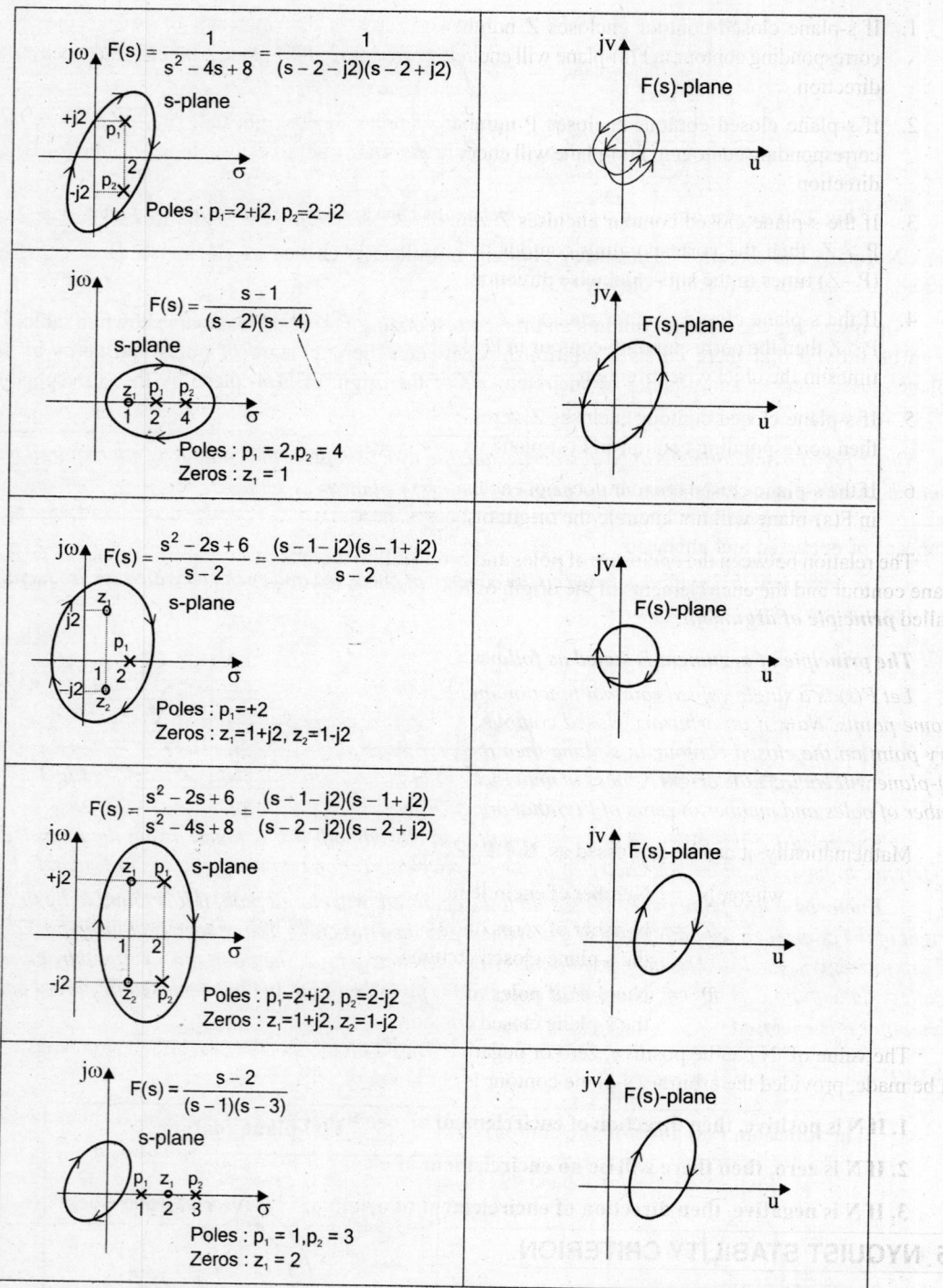

The summary of relationship between the enclosure of poles and zeros by the s-plane closed contour and number of encirclements of the origin of F(s)-plane by the corresponding F(s)-plane contour, are given below.

1. If s-plane closed contour encloses Z number of zeros in the right half of s-plane then the corresponding contour in F(s)-plane will encircle, the origin of F(s)-plane Z times in the clockwise direction.

2. If s-plane closed contour encloses P number of poles in the right half of s-plane then the corresponding contour in F(s)-plane will encircle the origin of F(s)- plane P times in anticlockwise direction.

3. If the s-plane closed contour encloses Z zeros and P poles in the right half of s-plane and if P > Z, then the corresponding contour in F(s)-plane will encircle the origin of F(s)-plane (P−Z) times in the anti-clockwise direction.

4. If the s-plane closed contour encloses Z zeros and P poles in the right half of s-plane and if P < Z then the corresponding contour in F(s)-plane will encircle the origin of F(s)-plane (Z−P) times in the clockwise direction.

5. If s-plane closed contour encloses Z zeros and P poles in right half of s-plane and if P = Z, then corresponding contour in F(s)-plane will not encircle the origin of F(s)-plane.

6. If the s-plane closed contour does not enclose any pole or zero, then the corresponding contour in F(s)-plane will not encircle the origin of F(s)-plane.

The relation between the enclosure of poles and zeros of F(s) lying on the right half of s-plane by the s-plane contour and the encirclements of the origin of F(s)-plane by the corresponding F(s)-plane contour is called ***principle of argument***.

The principle of argument is stated as follows.

Let F(s) is a single valued rational function and is analytic in a given region in the s-plane except at some points. Now, if an arbitrary closed contour is chosen in the s-plane, so that F(s) is analytic at every point on the closed contour in s-plane then the corresponding F(s)-plane contour mapped in the F(s)-plane will encircle the origin N times in anticlockwise direction where N is the difference between the number of poles and number of zeros of F(s) that are encircled by the chosen closed contour in s-plane.

Mathematically, it can be expressed as, N = P−Z.

where, N = Number of encirclement of origin of F(s)-plane, made by F(s)-contour.

Z = Number of zeros of F(s) lying on right half of s-plane and enclosed by the s-plane closed contour.

P = Number of poles of F(s) lying on right half of s-plane and enclosed by the s-plane closed contour.

The value of N can be positive, zero or negative. Based on the sign of N, following conclusions can be made, provided the arbitrary s-plane contour is chosen in the clockwise direction.

1. If N is positive, then direction of encirclement of origin of F(s)-plane will be anticlockwise.

2. If N is zero, then there will be no encirclement of origin of F(s)-plane.

3. If N is negative, then direction of encirclement of origin of F(s)-plane will be clockwise.

4.5 NYQUIST STABILITY CRITERION

Consider the closed loop transfer function, $\dfrac{C(s)}{R(s)} = \dfrac{G(s)}{1 + G(s)H(s)}$

The characteristic equation of the system is given by the condition, $1 + G(s)\,H(s) = 0$.

Let, $F(s) = 1 + G(s)H(s)$.

The loop transfer function $G(s)H(s)$ can be expressed as,

$$G(s)H(s) = \frac{K(s+z_1)(s+z_2)......(s+z_m)}{(s+p_1)(s+p_2).......(s+p_n)}, \text{ where } m \le n \quad .....(4.15)$$

$$\therefore F(s) = 1 + G(s)H(s) = 1 + \frac{K(s+z_1)(s+z_2)......(s+z_m)}{(s+p_1)(s+p_2).......(s+p_n)}$$

$$= \frac{(s+p_1)(s+p_2).......(s+p_n)K(s+z_1)(s+z_2)......(s+z_m)}{(s+p_1)(s+p_2).......(s+p_n)}$$

$$= \frac{(s+z_1')(s+z_2')......(s+z_n')}{(s+p_1)(s+p_2).......(s+p_n)} \quad .....(4.16)$$

In equation (4.16), $z_1', z_2' ..... z_n'$, are zeros of $F(s)$, which are obtained by combining the numerator and denominator polynomial of $G(s)H(s)$.

For the condition $F(s) = 0$, the numerator of $F(s)$ should be equal to zero.

$$\therefore (s+z_1')(s+z_2').......(s+z_n') = 0 \quad .....(4.17)$$

We can say that equation (4.17) is the characteristic equation of the system. For the stability of the system the roots of the characteristic equation should not lie on the right half s-plane. The roots of characteristic equation are zeros of $F(s)$ and also they are poles of closed loop transfer function.

Hence we can conclude that for the stability of closed loop system the zeros of $F(s)$ should not lie on the right half s-plane.

Note : *For a unity feedback system.*

$$G(s)H(s) = G(s) = \frac{K(s+z_1)(s+z_2)........(s+z_m)}{(s+p_1)(s+p_2).......(s+p_n)}$$

$$\therefore \frac{C(s)}{R(s)} = \frac{G(s)}{1+G(s)H(s)} = \frac{\dfrac{K(s+z_1)(s+z_2)........(s+z_m)}{(s+p_1)(s+p_2).......(s+p_n)}}{1 + \dfrac{K(s+z_1)(s+z_2)........(s+z_m)}{(s+p_1)(s+p_2).......(s+p_n)}}$$

$$= \frac{K(s+z_1)(s+z_2)........(s+z_m)}{(s+p_1)(s+p_2).......(s+p_n)+K(s+z_1)(s+z_2)........(s+z_m)}$$

$$= \frac{K(s+z_1)(s+z_2)........(s+z_m)}{(s+z_1')(s+z_2').......(s+z_n')}$$

From the above equation we can say that the poles of closed loop transfer function are $z_1', z_2', ... z_n'$.

Let us choose an arbitrary contour in the s-plane which encircles the right half zeros and poles of $F(s)$ (equation (4.16)). The principle of argument (explained in section 4.4) states that the corresponding contour in $F(s)$-plane will encircle the origin of $F(s)$-plane, N times in the anticlockwise direction.

Let, N = Number of anticlockwise encirclement

Now, $N = P - Z$ \quad .....(4.18)

where, P = Number of poles of $F(s)$ (or poles of loop transfer function) lying on right half s-plane

Z = Number of zeros of F(s) (or poles of closed loop transfer function) lying on right half s-plane

> **Note :** *The stability is related to poles lying on right half s-plane and so, while applying principle of argument only poles and zeros lying on right half s-plane alone are considered.*

For the stability of the system the roots of characteristic equation and so the zeros of F(s) should not lie on the right half of s-plane. Hence for a stable system Z = 0. Hence from equation (4.18) we get,

When Z = 0,   N = P   .....(4.19)

When Z ≠ 0,   N ≠ P   .....(4.20)

From equation (4.15) and (4.16) we can say that the poles of F(s) are also poles of loop transfer function. Hence for the stability of the system, (with reference to equation (4.19) and equation (4.20)) number of poles of loop transfer function lying on right of s-plane should be equal to anticlockwise encirclement of the origin of F(s)-plane. If this condition is not met the system is unstable.

> The principle of argument can also be used to find the number of poles of closed loop transfer function lying on right half of s-plane.
>
> Let,   M = Number of clockwise encirclement
>
> Now, M = Z – P
>
> When P = 0, M = Z
>
> Therefore, when there is no right half open loop poles, number of clockwise encirclement of origin of F(s)-plane gives number of poles of closed-loop transfer function lying on right half s-plane.

The loop transfer function, G(s)H(s) can be expressed as,

$$G(s)H(s) = [1+G(s) H(s)]-1 = F(s) - 1 \qquad .....(4.21)$$

From equation (4.21) it can be concluded that the contour of F(s) drawn with respect to origin of F(s)-plane is same as the contour of F(s)–1 drawn with respect to –1+j0 of F(s)-plane as shown in fig 4.3.

Thus the encirclement of the origin of F(s)-plane by the contour of F(s) is equivalent to the encirclement of the point –1 + j0 by the contour of F(s) – 1.

From equation (4.21), we can say that F(s) – 1, represents loop transfer function G(s)H(s). Hence contour of F(s) – 1 is same as contour of G(s)H(s), and F(s)-plane is G(s)H(s)-plane.

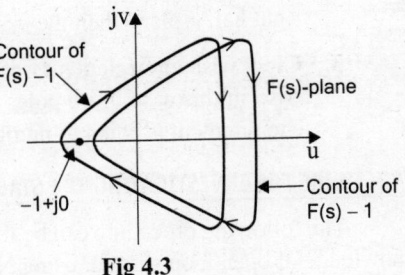

Fig 4.3

Therefore, the encirclement of –1 + j0 point of G(s) H(s)-contour in the G(s)H(s)-plane can be used to determine the stability of closed loop system. The Nyquist stability criterion have been proposed based on this concept.

In order to investigate the presence of poles of G(s)H(s) on the right half s-plane a contour, C is chosen such that it encloses the entire right half s-plane as shown in fig 4.4, such a contour C is called *Nyquist contour*.

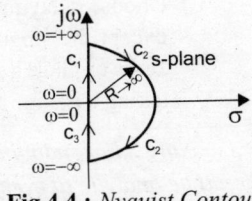

Fig 4.4 : *Nyquist Contour*

The Nyquist contour is directed clockwise and comprises of three segments,

1. An infinite line segment C_1 along the positive imaginary axis.
2. An arc, C_2 of infinite radius, enclosing the entire right half of s-plane.
3. An infinite line segment C_3 along the negative imaginary axis.

Along C_1, $s = j\omega$, with ω varying from 0 to $+\infty$.

Along C_2, $s = \underset{R \to \infty}{\text{Lt}} \ Re^{j\theta}$, with θ varying from $+\dfrac{\pi}{2}$ to $-\dfrac{\pi}{2}$.

Along C_3, $s = j\omega$, with ω varying from $-\infty$ to 0.

Using the loop transfer function G(s)H(s), the Nyquist contour-C of s-plane, is mapped to G(s)H(s)-plane. The mapped contour in G(s)H(s)-plane is called G(s)H(s)-contour.

> *Note : The s-plane is a complex plane. Any point on a complex plane can be expressed by the complex number in polar form, $Re^{j\theta}$, where R is the magnitude and θ is the argument (or phase).*

Now the Nyquist stability criterion can be stated as follows.

"If the G(s)H(s) contour in the G(s)H(s)-plane corresponding to Nyquist contour in the s-plane encircles the point -1 + j0 in the anticlockwise direction as many times as the number of right half s-plane poles of G(s)H(s), then the closed loop system is stable".

In examining the stability of linear control systems using the Nyquist stability criterion, we come across the following three situations.

1. **No encirclement of −1 + j0 point :** This implies that the system is stable if there are no poles of G(s)H(s) in the right half s-plane. If there are poles on right half s-plane then the system is unstable.

2. **Anticlockwise encirclements of −1 + j0 point :** In this case the system is stable if the number of anticlockwise encirclements is same as the number of poles of G(s)H(s) in the right half s-plane. If the number of anticlockwise encirclements is not equal to number of poles on right half s-plane then the system is unstable.

3. **Clockwise encirclements of the −1 + j0 point :** In this case the system is always unstable. Also in this case, if no poles of G(s)H(s) in right half s-plane, then the number of clockwise encirclement is equal to number of poles of closed loop system on right half s-plane.

PROCEDURE FOR INVESTIGATING THE STABILITY USING NYQUIST CRITERION

The following procedure can be followed to investigate the stability of closed loop system from the knowledge of open loop system, using Nyquist stability criterion.

1. Choose a Nyquist contour as shown in fig 4.5, which encloses the entire right half s-plane except the singular points. The Nyquist contour encloses all the right half s-plane poles and zeros of G(s)H(s). [The poles on imaginary axis are singular points and so they are avoided by taking a detour around it as shown in fig 4.5 b and c].

> *Note : For mapping a contour from s-plane to G(s)H(s) plane the Nyquist contour in s-plane should be analytic at every point. At singular points it is not analytic.*

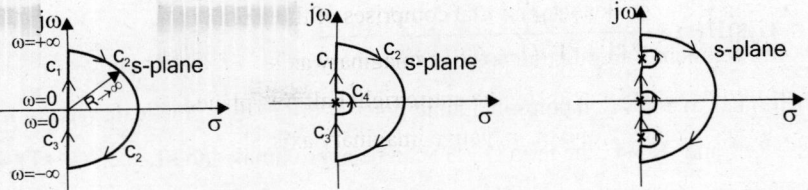

a. *Nyquist Contour when there is no pole on imaginary axis*   **b.** *Nyquist Contour when there are poles at origin*   **c.** *Nyquist Contour when there are poles on imaginary axis and at origin*

Fig 4.5 : *Nyquist Contour*

2. The Nyquist contour should be mapped in the G(s)H(s)-plane using the function G(s)H(s) to determine the encirclement $-1+ j0$ point in the G(s)H(s)-plane. The Nyquist contour of fig 4.5b can be divided into four sections C_1, C_2, C_3 and C_4. The mapping of the four sections in the G(s)H(s)-plane can be carried sectionwise and then combined together to get entire G(s)H(s)-contour.

3. In section C_1 the value of ω varies from 0 to $+ \infty$. The mapping of section C_1 is obtained by letting $s = j\omega$ in G(s)H(s) and varying ω from 0 to $+ \infty$,

$$\text{i.e. } G(s)H(s)\Big|_{\substack{s=j\omega \\ \omega=0 \text{ to } \infty}} = G(j\omega)H(j\omega)\Big|_{\omega=0 \text{ to } \infty}$$

The locus of $G(j\omega)H(j\omega)$ as ω is varied from 0 to $+\infty$ will be the G(s)H(s)-contour in G(s)H(s)-plane corresponding to section C_1 in s-plane. This locus is the polar plot of $G(j\omega)H(j\omega)$. There are three ways of mapping this section of G(s)H(s)-contour, they are,

   (i)   Calculate the values of $G(j\omega)H(j\omega)$ for various values of ω and sketch the actual locus of $G(j\omega)H(j\omega)$.

(or)

   (ii)   Separate the real part and imaginary part of $G(j\omega)H(j\omega)$. Equate the imaginary part to zero, to find the frequency at which the $G(j\omega)H(j\omega)$ locus crosses real axis (to find phase crossover frequency). Substitute this frequency on real part and find the crossing point of the locus on real axis. Sketch the approximate locus of $G(j\omega)H(j\omega)$ from the knowledge of type number and order of the system (or from the value of $G(j\omega)H(j\omega)$ at $\omega = 0$ and $\omega = \infty$).

   (iii)   Separate the magnitude and phase of $G(j\omega)H(j\omega)$. Equate the phase of $G(j\omega)H(j\omega)$ to -180^0 and solve for ω. This value of ω is the phase crossover frequency and the magnitude at this frequency is the crossing point on real axis. Sketch the approximate root locus as mentioned in method (ii).

4. The section C_2 of Nyquist contour has a semicircle of infinite radius. Therefore, every point on section C_2 has infinite magnitude but the argument varies from $+\pi/2$ to $-\pi/2$. Hence the mapping of section C_2 from s-plane to G(s)H(s) plane can be obtained by letting $s = \underset{R \to \infty}{Lt} Re^{j\theta}$ in G(s)H(s) and varying θ from $+\pi/2$ to $-\pi/2$.

Consider the loop transfer function in time constant form and with y number of poles at origin, as shown below.

$$G(s)H(s) = \frac{K(1+sT_1)(1+sT_2)(1+sT_3)\dots\dots}{s^y(1+sT_a)(1+sT_b)(1+sT_c)\dots\dots}$$

Let $G(s)H(s)$ has m zeros & n poles including poles at origin. For practical systems, $n > m$.

Since, $s \to Re^{j\theta}$ and $R \to \infty$, the term $(1+sT)$ can be approximated to sT, [i.e., $(1+sT) \approx sT$].

$$\therefore G(s)H(s) \approx K\frac{sT_1 \times sT_2 \times sT_3 \dots\dots}{s^y \times sT_a \times sT_b \times sT_c} = K_1\frac{s^m}{s^n} = \frac{K_1}{s^{n-m}}$$

On letting, $s = \underset{R \to \infty}{Lt} Re^{j\theta}$ we get,

$$G(s)H(s) \Bigg|_{s = \underset{R \to \infty}{Lt} Re^{j\theta}} = \frac{K_1}{\underset{R \to \infty}{Lt}(Re^{j\theta})^{n-m}} = 0e^{-j\theta(n-m)}$$

When $\theta = \frac{\pi}{2}$, $G(s)H(s) = 0e^{-j\frac{\pi}{2}(n-m)}$

When $\theta = -\frac{\pi}{2}$, $G(s)H(s) = 0e^{+j\frac{\pi}{2}(n-m)}$

From the above two equations we can conclude that the section C_2 of Nyquist contour in s-plane is mapped as circles/circular arc around origin with radius tending to zero in the $G(s)H(s)$-plane.

5. In section C_3, the value of ω varies from $-\infty$ to 0. The mapping of section C_3 is obtained by letting $s = +j\omega$ in $G(s)H(s)$ and varying ω from $-\infty$ to 0.

$$\text{i.e., } G(s)H(s) \Bigg|_{\substack{s = +j\omega \\ \omega = -\infty \text{ to } 0}} = G(j\omega)H(j\omega) \Bigg|_{\omega = -\infty \text{ to } 0.}$$

The locus of $G(j\omega)H(j\omega)$ as ω is varied from $-\infty$ to 0 will be the $G(s)H(s)$-contour in $G(s)H(s)$-plane corresponding to section C_3 in s-plane. This locus is the inverse polar plot of $G(j\omega)H(j\omega)$. The inverse polar plot is given by the mirror image of polar plot with respect to real axis.

6. The section C_4 of Nyquist contour has a semicircle of zero radius. Therefore every point on semicircle has zero magnitude but the argument varies from $-\pi/2$ to $+\pi/2$. Hence the mapping of section C_4 from s-plane to $G(s)H(s)$-plane can be obtained by letting in $G(s)H(s)$ and varying θ from $-\pi/2$ to $+\pi/2$.

Consider the loop transfer function in time constant form and with y number of poles at origin as shown below.

$$G(s)H(s) = \frac{K(1+sT_1)(1+sT_2)(1+sT_3)\dots\dots}{s^y(1+sT_a)(1+sT_b)(1+sT_c)\dots\dots}$$

Let $G(s)H(s)$ has m zeros & n poles including poles at origin. For practical systems, $n > m$.

Since, $s \to Re^{j\theta}$ and $R \to 0$, the term $1+sT$ can be approximated to 1, [i.e. $(1+sT) \approx 1$].

$$\therefore \ G(s)H(s) \approx K \frac{1}{s^y}$$

On letting, $s = \underset{R \to 0}{Lt} \, Re^{j\theta}$ we get

$$G(s)H(s)\Big|_{s = \underset{R \to 0}{Lt} Re^{j\theta}} = \frac{K}{\underset{R \to 0}{Lt}\left(Re^{j\theta}\right)^y} = \infty e^{-j\theta y}$$

When $\theta = -\dfrac{\pi}{2}$, $G(s)H(s) = \infty e^{j\frac{\pi}{2}y}$

When $\theta = \dfrac{\pi}{2}$, $G(s)H(s) = \infty e^{-j\frac{\pi}{2}y}$

From the above two equations we can conclude that the section C_4 of Nyquist contour in s-plane is mapped as circles/circular arc in G(s)H(s)-plane with origin as centre and infinite radius.

Note :

1. *If there are no poles on the origin then the section C_4 of Nyquist contour will be absent.*

2. *If there are poles on imaginary axis as shown below then the Nyquist contour is divided into the following 8 sections and the mapping is performed sectionwise.*

    Section C_1 : $s = j\omega$; $\omega = 0^+$ to $+ \omega_1^-$

    Section C_2 : $s = \underset{R \to 0}{Lt} \, Re^{j\theta}$; $\theta = -\dfrac{\pi}{2}$ to $+\dfrac{\pi}{2}$

    Section C_3 : $s = j\omega$; $\omega = +\omega_1^+$ to $+ \infty$

    Section C_4 : $s = \underset{R \to \infty}{Lt} \, Re^{j\theta}$; $\theta = +\dfrac{\pi}{2}$ to $-\dfrac{\pi}{2}$

    Section C_5 : $s = j\omega$; $\omega = -\infty$ to $- \omega_1^-$

    Section C_6 : $s = \underset{R \to 0}{Lt} \, Re^{j\theta}$; $\theta = -\dfrac{\pi}{2}$ to $+\dfrac{\pi}{2}$

    Section C_7 : $s = j\omega$; $\omega = -\omega_1^+$ to 0^-

    Section C_8 : $s = \underset{R \to 0}{Lt} \, Re^{j\theta}$; $\theta = -\dfrac{\pi}{2}$ to $+\dfrac{\pi}{2}$

EXAMPLE 4.13

Draw the Nyquist plot for the system whose open loop transfer function is, $G(s)H(s) = \dfrac{K}{s(s+2)(s+10)}$. Determine the range of K for which closed loop system is stable.

SOLUTION

Given that, $G(s)H(s) = \dfrac{K}{s(s+2)(s+10)} = \dfrac{K}{s \times 2\left(\frac{s}{2}+1\right) \times 10\left(\frac{s}{10}+1\right)} = \dfrac{0.05K}{s(1+0.5s)(1+0.1s)}$

The open loop transfer function has a pole at origin. Hence choose the Nyquist contour on s-plane enclosing the entire right half plane except the origin as shown in fig 4.13.1.

The Nyquist contour has four sections C_1, C_2, C_3 and C_4. The mapping of each section is performed separately and the overall Nyquist plot is obtained by combining the individual sections.

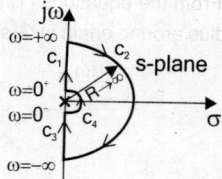

MAPPING OF SECTION C_1

In section C_1, ω varies from 0 to $+\infty$. The mapping of section C_1 is given by the locus of $G(j\omega)H(j\omega)$ as ω is varied from 0 to ∞. This locus is the polar plot of $G(j\omega)H(j\omega)$.

Fig 4.13.1 : *Nyquist Contour in s-plane.*

$$G(s)H(s) = \frac{0.05K}{s(1+0.5s)(1+0.1s)}$$

Let $s = j\omega$

$$\therefore \ G(j\omega)H(j\omega) = \frac{0.05K}{j\omega(1+j0.5\omega)(1+j0.1\omega)} = \frac{0.05K}{j\omega(1+j0.6\omega - 0.05\omega^2)} = \frac{0.05K}{-0.6\omega^2 + j\omega(1 - 0.05\omega^2)}$$

When the locus of $G(j\omega)H(j\omega)$ crosses real axis the imaginary term will be zero and the corresponding frequency is the phase crossover frequency, ω_{pc}.

$$\therefore \ \text{At } \omega = \omega_{pc}, \ \ \omega_{pc}(1 - 0.05\omega_{pc}^2) = 0 \ \ \Rightarrow \ \ 1 - 0.05\omega_{pc}^2 = 0 \ \ \Rightarrow \ \ \omega_{pc} = \sqrt{\frac{1}{0.05}} = 4.472 \text{ rad/sec}$$

At $\omega = \omega_{pc} = 4.472$ rad/sec, $\ \ G(j\omega)H(j\omega) = \frac{0.05K}{-0.6\omega^2} = -\frac{0.05K}{0.6 \times (4.472)^2} = -0.00417K$

The open loop system is type-1 and third order system. Also it is a minimum phase system with all poles. Hence the polar plot of $G(j\omega)H(j\omega)$ starts at $-90°$ axis at infinity, crosses real axis at $-0.00417K$ and ends at origin in second quadrant. The section C_1 and its mapping are shown in fig 4.13.2. and 4.13.3.

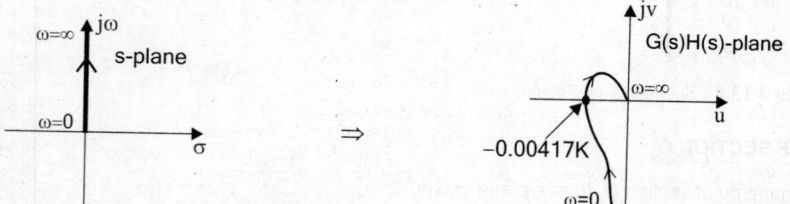

Fig 4.13.2 : Section C_1 in s-plane.

Fig 4.13.3 : *Mapping of section C_1 in G(s)H(s)-plane.*

MAPPING OF SECTION C_2

The mapping of section C_2 from s-plane to G(s)H(s)-plane is obtained by letting $s = \underset{R \to \infty}{\text{Lt}} \ Re^{j\theta}$ in G(s)H(s) and varying θ from $+\pi/2$ to $-\pi/2$. Since $s \to R\,e^{j\theta}$ and $R \to \infty$, the G(s)H(s) can be approximated as shown below, [i.e., $(1+sT) \approx sT$].

$$G(s)H(s) = \frac{0.05K}{s(1+0.5s)(1+0.1s)} \approx \frac{0.05K}{s \times 0.5s \times 0.1s} = \frac{K}{s^3}$$

Let, $s = \underset{R \to \infty}{\text{Lt}} \ Re^{j\theta}$

$$\therefore G(s)H(s) \bigg|_{s = \underset{R \to \infty}{\text{Lt}} Re^{j\theta}} = \frac{K}{s^3} \bigg|_{s = \underset{R \to \infty}{\text{Lt}} Re^{j\theta}} = \frac{K}{\underset{R \to \infty}{\text{Lt}} (Re^{j\theta})^3} = 0e^{-j3\theta}$$

When $\theta = \frac{\pi}{2}$, $\ \ G(s)H(s) = 0e^{-j3\frac{\pi}{2}}$ .....(1)

When $\theta = -\frac{\pi}{2}$, $\ \ G(s)H(s) = \infty e^{+j3\frac{\pi}{2}}$ .....(2)

From the equations (1) and (2) we can say that section C_2 in s-plane (fig 4.13.4.) is mapped as circular arc of zero radius around origin in G(s)H(s)-plane with argument (phase) varying from $-3\pi/2$ to $+3\pi/2$ as shown in fig 4.13.5.

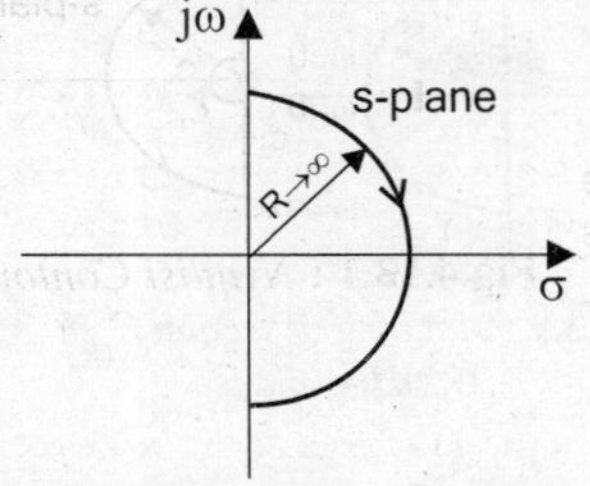

Fig 4.13.4 : Section C_2 in s-plane.

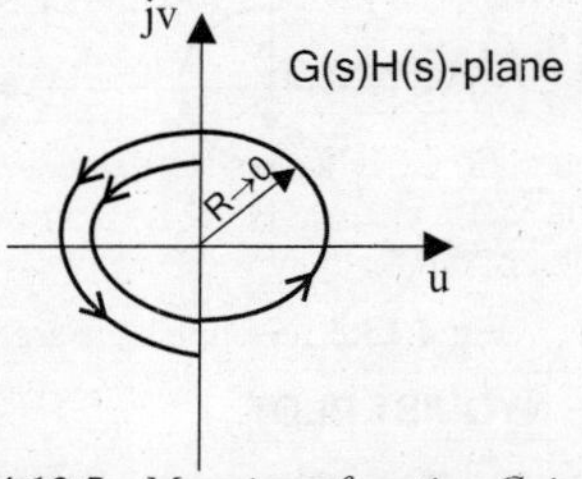

Fig 4.13.5 : *Mapping of section C_2 in G(s)H(s)-plane.*

MAPPING OF SECTION C_3

In section C_3, ω varies from $-\infty$ to 0.The mapping of section C_3 is given by the locus of $G(j\omega)H(j\omega)$ as ω is varied from $-\infty$ to 0. This locus is the inverse polar plot of $G(j\omega)H(j\omega)$.

The inverse polar plot is given by the mirror image of polar plot with respect to real axis. The section C_3 in s-plane and its corresponding contour in G(s)H(s) plane are shown in fig 4.13.6 and fig 4.13.7.

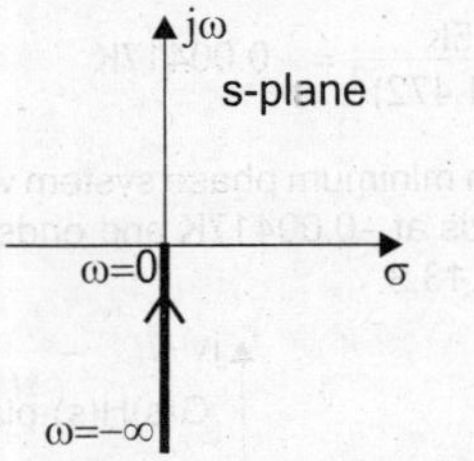

Fig 4.13.6 : Section C_3 in s-plane.

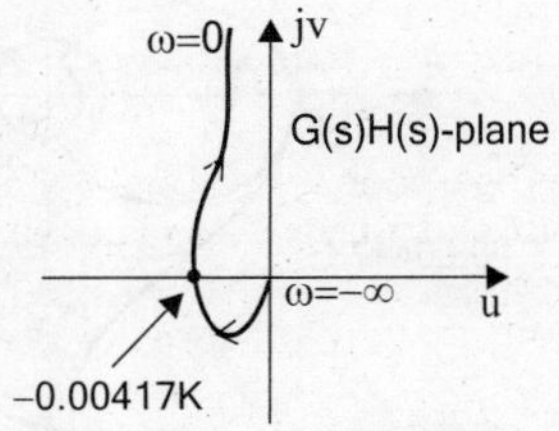

Fig 4.13.7 : *Mapping of section C_3 in G(s)H(s)-plane.*

MAPPING OF SECTION C_4

The mapping of section C_4 from s-plane to G(s)H(s)-plane is obtained by letting in G(s)H(s) and varying θ from $-\pi/2$ to $+\pi/2$. Since $s \to R\,e^{j\theta}$ and $R \to 0$, the G(s) H(s) can be approximated as shown below, [i.e., $(1+sT) \approx 1$].

$$G(s)H(s) = \frac{0.05K}{s(1+0.5s)(1+0.1s)} \approx \frac{0.05K}{s \times 1 \times 1} = \frac{0.05K}{s}$$

Let $s = \underset{R \to 0}{Lt}\ Re^{j\theta}$

$$\therefore\ G(s)H(s)\Big|_{s=\underset{R \to 0}{Lt}Re^{j\theta}} = \frac{0.05K}{s}\Big|_{s=\underset{R \to 0}{Lt}Re^{j\theta}} = \frac{0.05K}{\underset{R \to 0}{Lt}(Re^{j\theta})} = \infty e^{-j\theta}$$

When $\theta = -\dfrac{\pi}{2}$, $\qquad G(s)H(s) = \infty e^{+j\frac{\pi}{2}}$ .....(3)

When $\theta = \dfrac{\pi}{2}$, $\qquad G(s)H(s) = \infty e^{-j\frac{\pi}{2}}$ .....(4)

From the equations (3) and (4) we can say that section C_4 in s-plane (fig 4.13.8.) is mapped as a circular arc of infinite radius with argument (phase) varying from $+\pi/2$ to $-\pi/2$ as shown in fig 4.13.9.

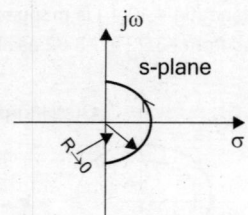

Fig 4.13.8 : Section C_4 in s-plane.

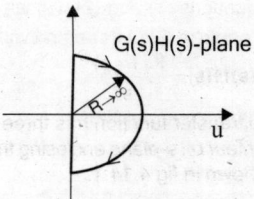

Fig 4.13.9 : *Mapping of section C_4 in G(s)H(s)-plane.*

COMPLETE NYQUIST PLOT

The entire Nyquist plot in G(s)H(s)-plane can be obtained by combining the mappings of individual sections, as shown in fig 4.13.10.

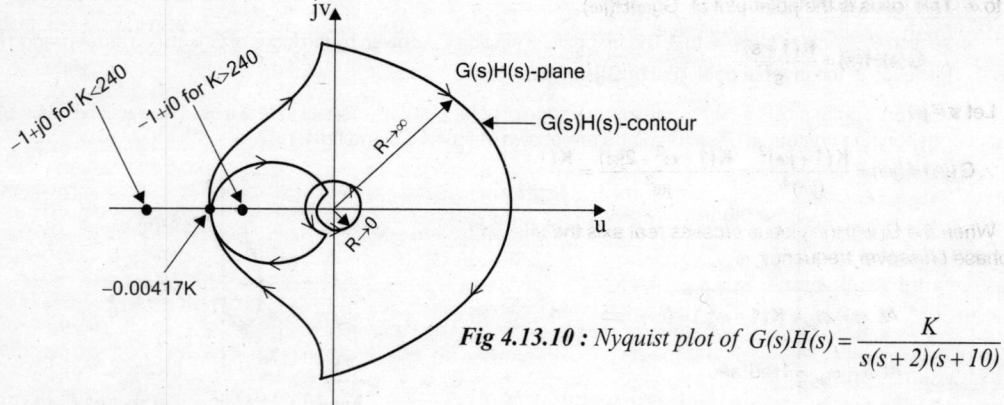

Fig 4.13.10 : *Nyquist plot of* $G(s)H(s) = \dfrac{K}{s(s+2)(s+10)}$

STABILITY ANALYSIS

When, $-0.00417K = -1$, the contour passes through $(-1+j0)$ point and corresponding value of K is the limiting value of K for stability.

$$\therefore \text{ Limiting value of } K = \frac{1}{0.00417} = 240$$

When K < 240

When K is less than 240, the contour crosses real axis at a point between 0 and $-1+j0$. On travelling through Nyquist plot along the indicated direction it is found that the point $-1+j0$ is not encircled. Also the open loop transfer function has no poles on the right half of s-plane. Therefore the closed loop system is stable.

When K > 240

When K is greater than 240, the contour crosses real axis at a point between $-1+j0$ and $-\infty$. On travelling through Nyquist plot along the indicated direction it is found that the point $-1+j0$ is encircled in clockwise direction two times. [Since there are two clockwise encirclement and no right half open loop poles, the closed loop system has two poles on right half of s-plane]. Therefore the closed loop system is unstable.

RESULT

The value of K for stability is $0 < K < 240$

EXAMPLE 4.14

Construct the Nyquist plot for a system whose open loop transfer function is given by $G(s)H(s) = \dfrac{K(1+s)^2}{s^3}$. Find the range of K for stability.

SOLUTION

Given that, $G(s)H(s) = \dfrac{K(1+s)^2}{s^3}$

The open loop transfer function has three poles at origin. Hence choose the Nyquist contour on s-plane enclosing the entire right half plane except the origin as shown in fig 4.14.1.

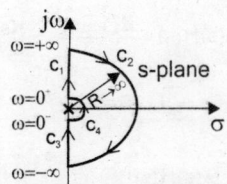

The Nyquist contour has four sections C_1, C_2, C_3 and C_4. The mapping of each section is performed separately and the overall Nyquist plot is obtained by combining the individual sections.

Fig 4.14.1 : *Nyquist Contour in s-plane.*

MAPPING OF SECTION C_1

In section C_1, ω varies from 0 to $+\infty$. The mapping of section C_1 is given by the locus of $G(j\omega)H(j\omega)$ as ω is varied from 0 to ∞. This locus is the polar plot of $G(j\omega)H(j\omega)$.

$$G(s)H(s) = \frac{K(1+s)^2}{s^3}$$

Let $s = j\omega$

$$\therefore G(j\omega)H(j\omega) = \frac{K(1+j\omega)^2}{(j\omega)^3} = \frac{K(1-\omega^2+2j\omega)}{-j\omega^3} = \frac{K(1-\omega^2)}{-j\omega^3} + \frac{K2j\omega}{-j\omega^3} = -\frac{2K}{\omega^2} + j\frac{K(1-\omega^2)}{\omega^3}$$

When the $G(j\omega)H(j\omega)$ locus crosses real axis the imaginary term will be zero and the corresponding frequency is the phase crossover frequency, ω_{pc}.

$$\therefore \text{At } \omega = \omega_{pc}, \; K(1-\omega_{pc}^2) = 0 \quad \Rightarrow \quad 1-\omega_{pc}^2 = 0 \quad \Rightarrow \quad \omega_{pc} = 1\,\text{rad/sec}$$

At $\omega = \omega_{pc} = 1\,\text{rad/sec}$

$$G(j\omega)H(j\omega) = -\frac{2K}{\omega^2} = -\frac{2K}{1^2} = -2K \qquad\qquad .....(1)$$

$$G(j\omega)H(j\omega) = \frac{K(1+j\omega)^2}{(j\omega)^3} = \frac{K\sqrt{1+\omega^2}\angle\tan^{-1}\omega\sqrt{1+\omega^2}\angle\tan^{-1}\omega}{\omega^3\angle270°} = \frac{K(1+\omega^2)}{\omega^3}\angle(2\tan^{-1}\omega - 270°)$$

As $\omega \to 0$, $G(j\omega)H(j\omega) \to \infty \angle -270°$ ......(2)

As $\omega \to \infty$, $G(j\omega)H(j\omega) \to 0 \angle -90°$ .....(3)

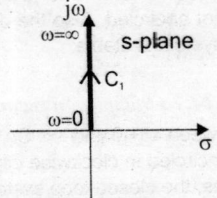
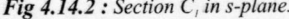

Fig 4.14.2 : *Section C_1 in s-plane.*

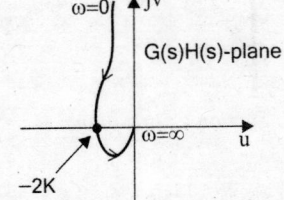

Fig 4.14.3 : *Mapping of section C_1 in G(s)H(s)-plane.*

From equations (1), (2) and (3) we can say that the polar plot starts at $-270°$ axis at infinity, crosses real axis at $-2K$ and ends at origin in third quadrant. The section C_1 and its mapping are shown in fig 2 and 3.

MAPPING OF SECTION C_2

The mapping of section C_2 from s-plane to G(s) H(s)-plane is obtained by letting $s = \underset{R\to\infty}{Lt}\; Re^{j\theta}$ in $G(s)H(s)$ and varying θ from $+\pi/2$ to $-\pi/2$. Since $s \to R e^{j\theta}$ and $R\to\infty$, the G(s) H(s) can be approximated as shown below, [i.e., $(1+sT) \approx sT$].

$$G(s)H(s) = \frac{K(1+s)^2}{s^3} \approx \frac{Ks^2}{s^3} = \frac{K}{s}$$

Let $s = \underset{R \to \infty}{Lt} Re^{j\theta}$

$$\therefore G(s)H(s)\Big|_{s = \underset{R \to \infty}{Lt} Re^{j\theta}} = \frac{K}{s}\Big|_{s = \underset{R \to \infty}{Lt} Re^{j\theta}} = \frac{K}{\underset{R \to \infty}{Lt} Re^{j\theta}} = 0e^{-j\theta}$$

When $\theta = \dfrac{\pi}{2}$, $G(s)H(s) = 0e^{-j\frac{\pi}{2}}$ .....(4)

When $\theta = \dfrac{\pi}{2}$, $G(s)H(s) = 0e^{j\frac{\pi}{2}}$ .....(5)

From the equations (4) and (5) we can say that section C_2 in s-plane (fig 4.14.4.) is mapped as circular arc of zero radius around origin in G(s)H(s)-plane with argument (phase) varying from $-\pi/2$ to $+\pi/2$ as shown in fig 4.14.5.

Fig 4.14.4 : *Section C_2 in s-plane.*

Fig 4.14.5 : *Mapping of section C_2 in G(s)H(s)-plane.*

MAPPING OF SECTION C_3

In section C_3, ω varies from $-\infty$ to 0. The mapping of section C_3 is given by locus of $G(j\omega) H(j\omega)$ as ω is varied from $-\infty$ to 0. This locus is the inverse polar plot of $G(j\omega) H(j\omega)$.

The inverse polar plot is given by the mirror image of polar plot with respect to real axis. The section C_3 in s-plane and its corresponding contour in G(s)H(s) plane are shown in fig 4.14.6 and fig 4.14.7.

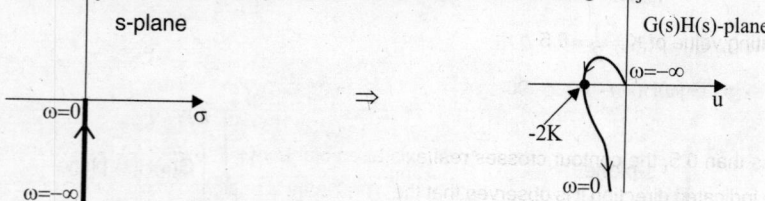

Fig 4.14.6 : *Section C_3 in s-plane.*

Fig 4.14.7 : *Mapping of section C_3 in G(s)H(s)-plane.*

MAPPING OF SECTION C_4

The mapping of section C_4 from s-plane to G(s) H(s)-plane is obtained by letting in G(s)H(s) and varying θ from $-\pi/2$ to $+\pi/2$. Since $s \to R e^{j\theta}$ and $R \to 0$, the G(s)H(s) can be approximated as shown below, [i.e., $(1+sT) \approx 1$].

$$G(s)H(s) = \frac{K(1+s)^2}{s^3} \approx \frac{K \times 1}{s^3} = \frac{K}{s^3}$$

Let $s = \underset{R \to 0}{Lt} Re^{j\theta}$

$$\therefore G(s)H(s)\Big|_{s = \underset{R \to 0}{Lt} Re^{+j\theta}} = \frac{K}{s^3}\Big|_{s = \underset{R \to 0}{Lt} Re^{+j\theta}} = \frac{K}{\underset{R \to 0}{Lt} (Re^{+j\theta})^3} = \infty e^{-j3\theta}$$

When $\theta = -\dfrac{\pi}{2}$, $G(s)H(s) = \infty e^{+j3\frac{\pi}{2}}$ ......(6)

When $\theta = -\dfrac{\pi}{2}$, $G(s)H(s) = \infty e^{-j3\frac{\pi}{2}}$ ......(7)

From the equations (6) and (7) we can say that section C_4 in s-plane (fig 4.14.8.) is mapped as a circular arc of infinite radius with argument (phase) varying from $+3\pi/2$ to $-3\pi/2$ as shown in fig 4.14.9.

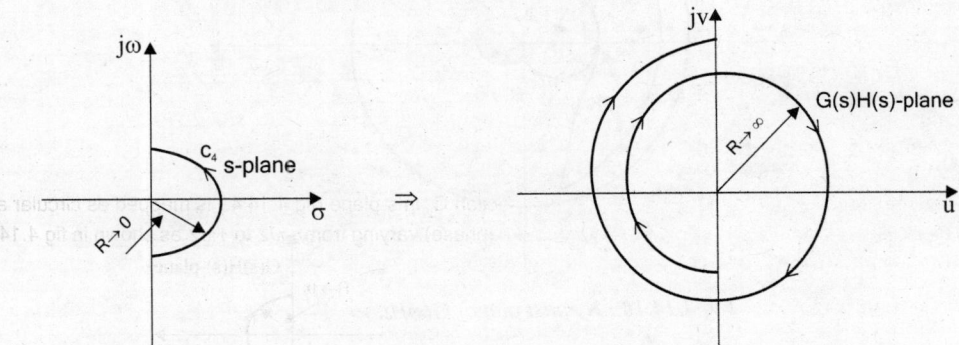

Fig 4.14.8 : *Section C_4 in s-plane.*   **Fig 4.14.9 :** *Mapping of section C_4 in*
                                                                     G(s)H(s)-plane.

COMPLETE NYQUIST PLOT

The entire Nyquist plot in G(s)H(s)-plane can be obtained by combining the mappings of individual sections, as shown in fig 4.14.10.

STABILITY ANALYSIS

When, $-2K = -1$, the contour passes through $-1+j0$ point and corresponding value of K is the limiting value of K for stability.

$$\therefore \text{ Limiting value of } K = \frac{1}{2} = 0.5$$

When K < 0.5

When K is less than 0.5, the contour crosses real axis at a point between 0 and $-1+j0$. On travelling through Nyquist plot along the indicated direction it is observed that the $-1+j0$ point is encircled in clockwise direction two times. Therefore the system is unstable. [Since there are two clockwise encirclement and no right half open loop poles, the closed loop system will have two poles on right half of s-plane]

When K > 0.5

When K is greater than 0.5, the contour crosses real axis at a point between $-1+j0$ and $-\infty$. On travelling through Nyquist plot along the indicated direction it is observed that $(-1+j0)$ point is encircled in both clockwise and anticlockwise direction one time. Hence net encirclement is zero. Also the open loop system has no poles at the right half of s-plane. Therefore the closed loop system is stable.

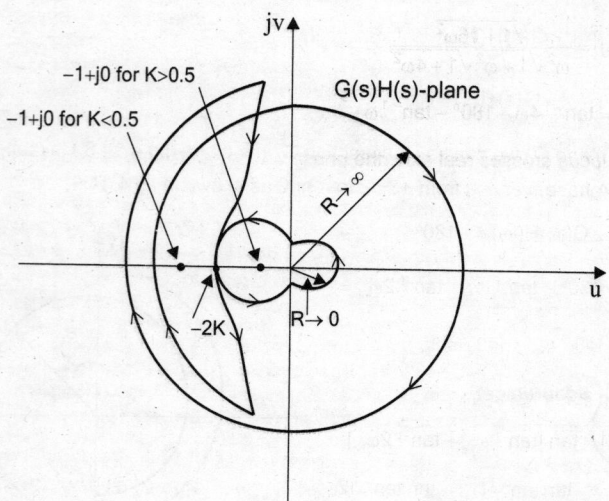

$$\text{Fig 4.14.10 : } \textit{Nyquist plot of } G(s)H(s) = \frac{K(1+s)^2}{s^3}.$$

RESULT

The system is stable when K > 0.5.

EXAMPLE 4.15

The open loop transfer function of a system is $G(s)H(s) = \dfrac{(1+4s)}{s^2(1+s)(1+2s)}$. Determine the stability of closed loop

system. If the closed loop system is not stable then find the number of closed-loop poles lying on the right half of s-plane.

SOLUTION

Given that, $G(s)H(s) = \dfrac{(1+4s)}{s^2(1+s)(1+2s)}$

The open loop transfer function has two poles at origin. Hence choose the Nyquist contour on s-plane enclosing the entire right half plane except the origin as shown in fig 4.15.1.

The Nyquist contour has four sections C_1, C_2, C_3 and C_4. The mapping of each section is performed separately and the overall Nyquist plot is obtained by combining the individual sections.

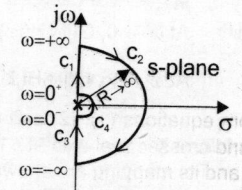

Fig 4.15.1 : Nyquist Contour in s-plane.

MAPPING OF SECTION C_1

In section C_1, ω varies from 0 to $+\infty$. The mapping of section C_1 is given by the locus of $G(j\omega)H(j\omega)$ as ω is varied from 0 to ∞. This locus is the polar plot of $G(j\omega)H(j\omega)$.

$$G(s)H(s) = \frac{(1+4s)}{s^2(1+s)(1+2s)}$$

Let $s = j\omega$,

$$\therefore \ G(j\omega)H(j\omega) = \frac{(1+j4\omega)}{(j\omega)^2(1+j\omega)(1+j2\omega)} = \frac{\sqrt{1+16\omega^2}\ \angle \tan^{-1}4\omega}{\omega^2 \angle 180° \sqrt{1+\omega^2}\ \angle \tan^{-1}\omega \sqrt{1+4\omega^2}\ \angle \tan^{-1}2\omega}$$

$$= \frac{\sqrt{1+16\omega^2}}{\omega^2 \sqrt{1+\omega^2}\sqrt{1+4\omega^2}} \ \angle (\tan^{-1}4\omega - 180° - \tan^{-1}\omega - \tan^{-1}2\omega)$$

$$\therefore \ |G(j\omega)\,H(j\omega)| = \frac{\sqrt{1+16\omega^2}}{\omega^2\sqrt{1+\omega^2}\sqrt{1+4\omega^2}}.$$

$$\angle G(j\omega)\,H(j\omega) = \tan^{-1}4\omega - 180° - \tan^{-1}\omega - \tan^{-1}2\omega$$

When the $G(j\omega)H(j\omega)$ locus crosses real axis, the phase will be $-180°$ and the corresponding frequency is the phase crossover frequency, ω_{pc}.

$$\therefore \ \text{At } \omega = \omega_{pc}, \ \angle G(j\omega)H(j\omega) = -180°$$

$$\therefore \ \tan^{-1}4\omega_{pc} - 180° - \tan^{-1}\omega_{pc} - \tan^{-1}2\omega_{pc} = -180°$$

$$\tan^{-1}4\omega_{pc} = \tan^{-1}\omega_{pc} + \tan^{-1}2\omega_{pc}$$

On taking tan on both sides we get,

$$\tan[\tan^{-1}4\omega_{pc}] = \tan[\tan^{-1}\omega_{pc} + \tan^{-1}2\omega_{pc}]$$

$$4\omega_{pc} = \frac{\tan\tan^{-1}\omega_{pc} + \tan\tan^{-1}2\omega_{pc}}{1 - \tan\tan^{-1}\omega_{pc} \times \tan\tan^{-1}2\omega_{pc}}$$

> **Note :** $\tan(A+B) = \dfrac{\tan A + \tan B}{1 - \tan A \times \tan B}$

$$4\omega_{pc} = \frac{\omega_{pc} + 2\omega_{pc}}{1 - 2\omega_{pc}^2} \ \Rightarrow \ 1 - 2\omega_{pc}^2 = \frac{3\omega_{pc}}{4\omega_{pc}} \ \Rightarrow \ -2\omega_{pc}^2 = \frac{3}{4} - 1$$

$$\therefore \ \omega_{pc} = \sqrt{\frac{-0.25}{-2}} = 0.354 \ \text{rad/sec}$$

At $\omega = \omega_{pc} = 0.354$ rad/sec

$$|G(j\omega)\,H(j\omega)| = \frac{\sqrt{1+16\omega_{pc}^2}}{\omega^2\sqrt{1+\omega_{pc}^2}\sqrt{1+4\omega_{pc}^2}} = \frac{\sqrt{1+16\times0.354^2}}{(0.354)^2\sqrt{1+0.354^2}\sqrt{1+4\times0.354^2}} = 10.64 \qquad \text{.....(1)}$$

Hence $G(j\omega)H(j\omega)$ locus crosses the real axis at -10.64.

$$\text{At } \omega \to 0, \ G(j\omega)H(j\omega) \to \infty \angle -180° \qquad \text{.....(2)}$$

$$\text{At } \omega \to \infty, \ G(j\omega)H(j\omega) \to 0 \angle -270° \qquad \text{.....(3)}$$

From equations (1), (2) and (3) we can say that the polar plot starts at $-180°$ axis at infinity, travels in third quadrant and crosses real axis at -10.64 to enter second quadrant and then ends at origin in second quadrant. The section C_1 and its mapping are shown in fig 4.15.2 and 4.15.3.

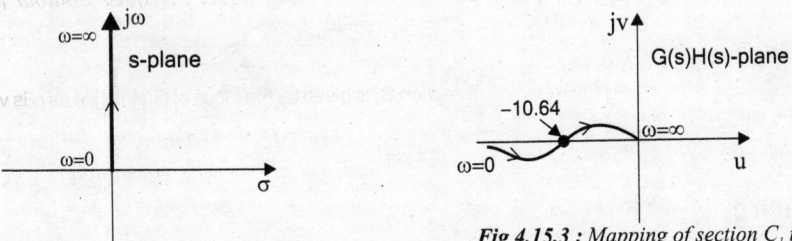

Fig 4.15.2 : *Section C_1 in s-plane.*

Fig 4.15.3 : *Mapping of section C_1 in $G(s)H(s)$-plane.*

MAPPING OF SECTION C_2

The mapping of section C_2 from s-plane to $G(s)H(s)$-plane is obtained by letting in $G(s)H(s)$ and varying θ from $+\pi/2$ to $-\pi/2$. Since $s \to R\,e^{j\theta}$ and $R \to \infty$, the $G(s)H(s)$ can be approximated as shown below, [i.e., $(1+sT) \approx sT$].

$$G(s)H(s) = \frac{(1+4s)}{s^2(1+s)(1+2s)} \approx \frac{4s}{s^2 \times s \times 2s} = \frac{2}{s^3}$$

Let, $s = \underset{R \to \infty}{Lt} Re^{j\theta}$

$$\therefore \ G(s)H(s)\Big|_{s = \underset{R \to \infty}{Lt} Re^{j\theta}} = \frac{2}{s^3}\Big|_{s = \underset{R \to \infty}{Lt} Re^{j\theta}} = \frac{2}{\underset{R \to \infty}{Lt}(Re^{j\theta})^3} = 0e^{-j3\theta}$$

When $\theta = \dfrac{\pi}{2}$, $G(s)H(s) = 0e^{-j3\frac{\pi}{2}}$ .....(4)

When $\theta = -\dfrac{\pi}{2}$, $G(s)H(s) = 0e^{j3\frac{\pi}{2}}$ .....(5)

From the equations (4) and (5) we can say that section C_2 in s-plane (fig 4.15.4.) is mapped as circular arc of zero radius around origin in G(s)H(s)-plane with argument (phase) varying from $-3\pi/2$ to $+3\pi/2$ as shown in fig 4.15.5.

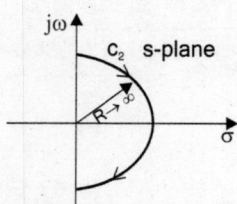

Fig 4.15.4 : Section C_2 in s-plane.

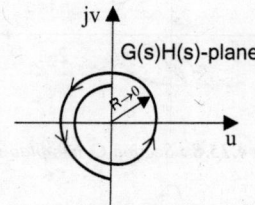

Fig 4.15.5 : Mapping of section C_2 in G(s)H(s)-plane.

MAPPING OF SECTION C₃

In section C_3, ω varies from $-\infty$ to 0. The mapping of section C_3 is given by the locus of G(jω) H(jω) as ω is varied from $-\infty$ to 0. This locus is the inverse polar plot of G(jω) H(jω).

The inverse polar plot is given by the mirror image of polar plot with respect to real axis. The section C_3 in s-plane and its corresponding contour in G(s)H(s) plane are shown in fig 4.15.6 and fig 4.15.7.

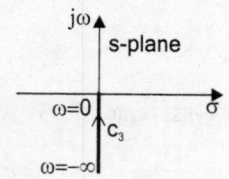

Fig 4.15.6 : Section C_3 in s-plane.

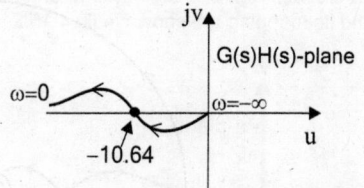

Fig 4.15.7 : Mapping of section C_3 in G(s)H(s)-plane.

MAPPING OF SECTION C₄

The mapping of section C_4 from s-plane to G(s)H(s)-plane is obtained by letting $s = \underset{R \to 0}{Lt} Re^{j\theta}$ in G(s)H(s) and varying θ from $-\pi/2$ to $+\pi/2$. Since $s \to R\,e^{j\theta}$ and $R \to 0$, the G(s)H(s) can be approximated as shown below [i.e., $(1+sT) \approx 1$].

$$G(s)H(s) = \frac{(1+4s)}{s^2(1+s)(1+2s)} \approx \frac{1}{s^2 \times 1 \times 1} = \frac{1}{s^2}$$

Let, $s = \underset{R \to 0}{Lt} \, Re^{j\theta}$

$$\therefore \; G(s)H(s) \Big|_{s = \underset{R \to 0}{Lt} Re^{j\theta}} = \frac{1}{s^2} \Big|_{s = \underset{R \to 0}{Lt} Re^{j\theta}} = \frac{1}{\underset{R \to 0}{Lt}(Re^{j\theta})^2} = \infty e^{-j2\theta}$$

When $\theta = -\dfrac{\pi}{2}$, $\quad G(s)H(s) = \infty e^{j\pi}$ .....(6)

When $\theta = \dfrac{\pi}{2}$, $\quad G(s)H(s) = \infty e^{-j\pi}$ .....(7)

From the equations (6) and (7) we can say that section C_4 in s-plane (fig 4.15.8.) is mapped as a circle of infinite radius with argument (phase) varying from $+\pi$ to $-\pi$ as shown in fig 4.15.9.

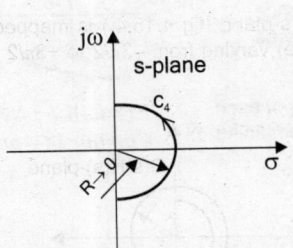

Fig 4.15.8 : *Section C_4 in s-plane.*

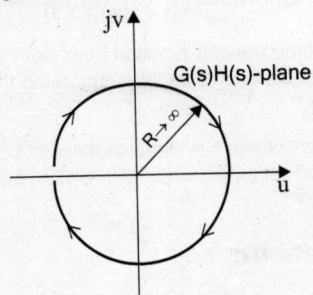

Fig 4.15.9 : *Mapping of section C_4 in G(s)H(s)-plane.*

COMPLETE NYQUIST PLOT

The entire Nyquist plot in G(s)H(s)-plane can be obtained by combining the mappings of individual sections, as shown in fig 4.15.10.

STABILITY ANALYSIS

On travelling through Nyquist contour in G(s)H(s)-plane it is observed that (−1+j0) point is encircled in clockwise direction two times. Therefore the closed loop system is unstable.

Since the −1+j0 is encircled two times in clockwise and no right half open loop poles, two poles of closed loop system are lying on the right half s-plane.

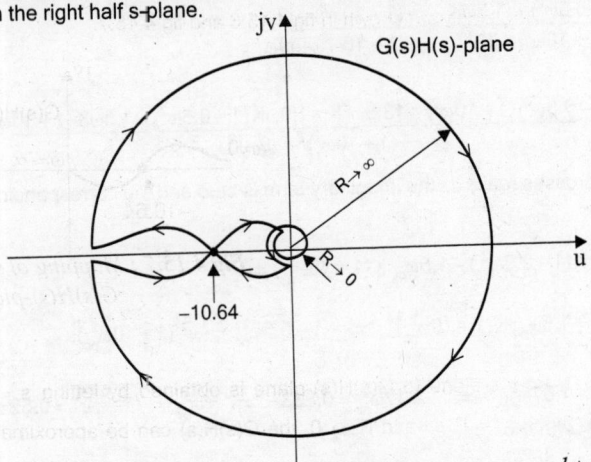

Fig 4.15.10 : *Nyquist plot of $G(s)H(s) = \dfrac{1+4s}{s^2(1+s)(1+2s)}$.*

RESULT

(a)   Closed loop system is unstable.

(b)   Two poles of closed loop system are lying on the right half s-plane.

EXAMPLE 4.16

Sketch the Nyquist plot for a system with the open loop transfer function $G(s)H(s) = \dfrac{K(1+0.5s)(1+s)}{(1+10s)(s-1)}$. Determine the range of values of K for which the system is stable.

SOLUTION

Given that, $G(s)H(s) = \dfrac{K(1+0.5s)(1+s)}{(1+10s)(s-1)}$

The open loop transfer function does not have a pole at origin. Hence choose the Nyquist contour on s-plane enclosing the entire right half plane as shown in fig 4.16.1.

The Nyquist contour has three sections C_1, C_2 and C_3. The mapping of each section is performed separately and the overall Nyquist plot is obtained by combining the individual sections.

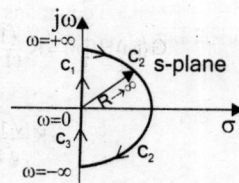

Fig 4.16.1 : *Nyquist Contour in s-plane.*

MAPPING OF SECTION C_1

In section C_1, ω varies from 0 to $+\infty$. The mapping of section C_1 is given by the locus of $G(j\omega)H(j\omega)$ as ω is varied from 0 to ∞. This locus is the polar plot of $G(j\omega)H(j\omega)$.

$$G(s)H(s) = \frac{K(1+0.5s)(1+s)}{(1+10s)(s-1)}$$

Let $s = j\omega$.

$$\therefore\ G(j\omega)H(j\omega) = \frac{K(1+j0.5\omega)(1+j\omega)}{(1+j10\omega)(-1+j\omega)} = \frac{K(1+j1.5\omega-0.5\omega^2)}{-1-j9\omega-10\omega^2} = \frac{K(1-0.5\omega^2)+j1.5\omega K}{-(1+10\omega^2)-j9\omega}$$

On multiplying the numerator and denominator by the complex conjugate of denominator we get,

$$G(j\omega)H(j\omega) = \frac{K(1-0.5\omega^2)+j1.5\omega K}{-(1+10\omega^2)-j9\omega} \times \frac{-(1+10\omega^2)+j9\omega}{-(1+10\omega^2)+j9\omega}$$

$$= \frac{-K(1-0.5\omega^2)(1+10\omega^2)-13.5\omega^2 K + j[9\omega K(1-0.5\omega^2)-1.5\omega K(1+10\omega^2)]}{(1+10\omega^2)^2+(9\omega)^2}$$

When the $G(j\omega)H(j\omega)$ locus crosses real axis the imaginary term is zero and the corresponding frequency is the phase crossover frequency.

$$\therefore\ \text{At } \omega = \omega_{pc}, \quad 9\omega_{pc}K(1-0.5\omega_{pc}^2)-1.5\omega_{pc}K(1+10\omega_{pc}^2) = 0$$

$$\therefore\ 9\omega_{pc}K(1-0.5\omega_{pc}^2) = 1.5\omega_{pc}K(1+10\omega_{pc}^2) \quad \Rightarrow \quad 1-0.5\omega_{pc}^2 = \frac{1.5}{9}(1+10\omega_{pc}^2)$$

$$\therefore\ 1-0.5\omega_{pc}^2 = 0.167+1.67\omega_{pc}^2 \quad \Rightarrow \quad 2.17\omega_{pc}^2 = 0.833 \quad \Rightarrow \quad \omega_{pc} = \sqrt{\frac{0.833}{2.17}} = 0.62 \text{ rad/sec}$$

At $\omega = \omega_{pc} = 0.62$ rad/sec

$$G(j\omega)H(j\omega) = \frac{-K\left(1 - 0.5\omega_{pc}^2\right)\left(1 - 10\omega_{pc}^2\right) - 13.5\omega_{pc}^2\,K}{\left(1 + 10\omega_{pc}^2\right)^2 + (9\omega_{pc})^2}$$

$$= -K\left[\frac{(1 - 0.5 \times 0.62^2)(1 + 10 \times 0.62^2) + 13.5 \times 0.62^2}{(1 + 10 \times 0.62^2)^2 + (9 \times 0.62)^2}\right] = -K\left[\frac{3.913 + 5.189}{23.464 + 31.136}\right] = -0.16667K$$

Therefore, $G(j\omega)H(j\omega)$ locus crosses real axis at a point $-0.1667K$.

The exact shape of $G(j\omega)H(j\omega)$ locus is determined by calculating the magnitude and phase of $G(j\omega)H(j\omega)$ for various values of ω.

$$G(j\omega)H(j\omega) = K\frac{(1 + j0.5\omega)(1 + j\omega)}{(1 + j10\omega)(-1 + j\omega)} = K\frac{\sqrt{1 + (0.5\omega)^2}\angle\tan^{-1}0.5\sqrt{1 + \omega^2}\angle\tan^{-1}\omega}{\sqrt{1 + (10\omega)^2}\angle\tan^{-1}10\omega\sqrt{1 + \omega^2}\angle(180° - \tan^{-1}\omega)}$$

$$= K\frac{\sqrt{1 + 0.25\omega^2}}{\sqrt{1 + 100\omega^2}}\angle(\tan^{-1}0.5\omega + 2\tan^{-1}\omega - \tan^{-1}10\omega - 180°)$$

$$\therefore |G(j\omega)H(j\omega)| = K\frac{\sqrt{1 + 0.25\omega^2}}{\sqrt{1 + 100\omega^2}}$$

$$\angle G(j\omega)H(j\omega) = \tan^{-1}0.5\omega + 2\tan^{-1}\omega - \tan^{-1}10\omega - 180°$$

As $\omega \to 0$, $|G(j\omega)H(j\omega)| = K$

As $\omega \to 0$, $\angle G(j\omega)H(j\omega) = -180°$

As $\omega \to \infty$, $|G(j\omega)H(j\omega)| = \underset{\omega\to\infty}{Lt}\ K\frac{\sqrt{1 + 0.25\omega^2}}{\sqrt{1 + 100\omega^2}} = K\underset{\omega\to\infty}{Lt}\sqrt{\frac{\omega^2\left(\frac{1}{\omega^2} + 0.25\right)}{\omega^2\left(\frac{1}{\omega^2} + 100\right)}} = K\underset{\omega\to\infty}{Lt}\sqrt{\frac{\left(\frac{1}{\omega^2} + 0.25\right)}{\left(\frac{1}{\omega^2} + 100\right)}}$

$$= K\sqrt{\frac{0 + 0.25}{0 + 100}} = 0.05K$$

As $\omega \to \infty$, $\angle G(j\omega) = \tan^{-1}\infty + 2\tan^{-1}\infty - \tan^{-1}\infty - 180° = 90° + 180° - 90° - 180° = 0$

ω rad/sec	0	0.1	0.5	1.5	2.0	5.0	∞
$\lvert G(j\omega)H(j\omega)\rvert$	K	0.707K	0.202K	0.083K	0.07K	0.054K	0.05K
$\angle G(j\omega)H(j\omega)$ deg	−180	−210	−191	−116	−95	−43	0

From the above analysis, the following conclusions are made,

1. The locus of $G(j\omega)H(j\omega)$ starts at $K\angle-180°$ when $\omega = 0$ and travels in second quadrant.

2. The locus crosses real axis at $-0.1667K$ and enters third quadrant.

3. Then the locus crosses negative imaginary axis and enters fourth quadrant.

4. Finally the locus ends at $0.05K \angle0°$ when $\omega = \infty$.

Note : The exact plot can also be sketched on polar graph sheet.

The section C_1 in s-plane and its corresponding mapping in $G(s)H(s)$ plane are shown in fig 4.16.2. and 4.16.3.

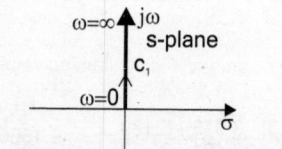

Fig 4.16.2 : *Section C₁ in s-plane.*

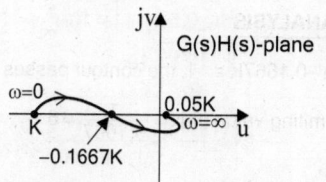

Fig 4.16.3 : *Mapping of section C₁ in G(s)H(s)-plane.*

MAPPING OF SECTION C_2

The mapping of section C_2 from s-plane to G(s)H(s)-plane is obtained by letting $s = \underset{R \to \infty}{\text{Lt}} \, Re^{j\theta}$ in G(s)H(s) and varying θ from $-\pi/2$ to $+\pi/2$. Since $s \to R\,e^{j\theta}$ and $R \to \infty$, G(s)H(s) can be approximated as shown below [i.e., $(1+sT) \approx sT$; Here $(s-1) \approx s$].

$$G(s)H(s) = \frac{K(1+0.5s)(1+s)}{(1+10s)(s-1)}$$

$$= \frac{K\,0.5s \times s}{10s \times s} = 0.05K$$

The approximate G(s)H(s) is independent of s and so the contour of section C_2 in s-plane is mapped as a point at 0.05K in G(s)H(s)-plane.

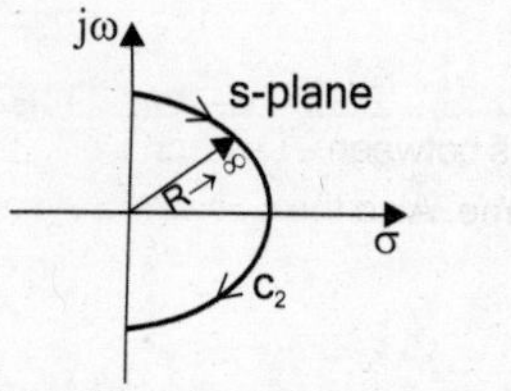

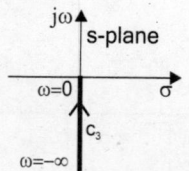

Fig 4.16.4 : *Section C₂ in s-plane.*

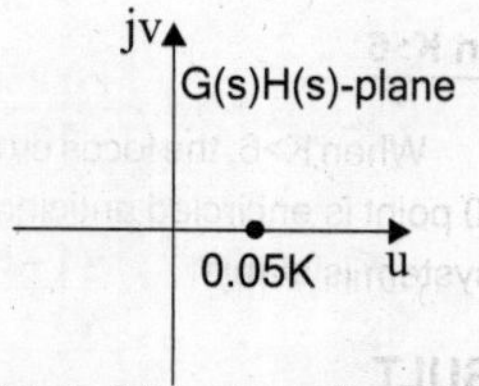

Fig 4.16.5 : *Mapping of section C₂ in G(s)H(s)-plane.*

MAPPING OF SECTION C_3

In section C_3, ω varies from $-\infty$ to 0. The mapping of section C_3 is given by the locus of G(jω) H(jω) as ω is varied from $-\infty$ to 0. This locus is the inverse polar plot of G(jω) H(jω).

The inverse polar plot is given by the mirror image of polar plot with respect to real axis. The section C_3 in s-plane and its corresponding contour in G(s)H(s) plane are shown in fig 4.16.6 and fig 4.16.7.

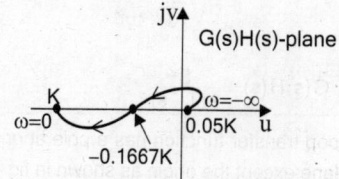

Fig 4.16.6 : *Section C₃ in s-plane.*

Fig 4.16.7 : *Mapping of section C₃ in G(s)H(s)-plane.*

COMPLETE NYQUIST PLOT

The entire Nyquist plot in G(s)H(s)-plane can be obtained by combining the mappings of individual sections, as shown in fig 4.16.8.

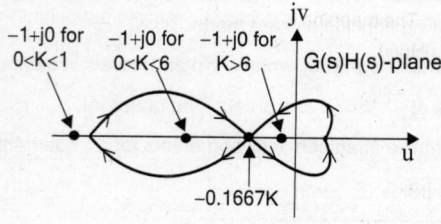

Fig 4.16.8 : *Nyquist plot of* $G(s)H(s) = \dfrac{K(1+0.5s)(1+s)}{(1+10s)(s-1)}.$

STABILITY ANALYSIS

When $-0.1667K = -1$, the contour passes through $-1+j0$ point and this value of K is the limiting value for stability.

The limiting value of $K = \dfrac{1}{0.1667} = 6$

When 0<K<1

When $0<K<1$, the $-1+j0$ point is not encircled, but there is one open loop right half pole and so system is unstable.

When 1<K<6

When $1<K<6$, the locus crosses real axis between 0 and $-1+j0$. On travelling through the locus it is observed that the $-1+j0$ point is encircled clockwise and so the closed loop system is unstable.

When K>6

When $K>6$, the locus crosses real axis between $-1+j0$ and $-\infty$. On travelling through the locus it is observed that the $-1+j0$ point is encircled anticlockwise one time. Also the open loop system has one pole at the right half s-plane. Hence the system is stable.

RESULT

(a)   The open loop system is unstable.

(b)   For stability of the closed loop system, K>6.

EXAMPLE 4.17

Construct Nyquist plot for a feedback control system whose open loop transfer function is given by, $G(s)H(s) = \dfrac{5}{s(1-s)}$. Comment on the stability of open-loop and closed loop system.

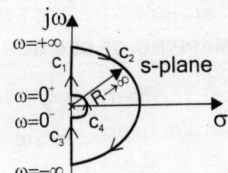

Fig 4.17.1 : *Nyquist Contour in s-plane.*

SOLUTION

Given that, $G(s)H(s) = \dfrac{5}{s(1-s)}$

The open loop transfer function has a pole at origin. Hence choose the Nyquist contour on s-plane enclosing the entire right half s-plane except the origin as shown in fig 4.17.1.

The Nyquist contour has four sections C_1, C_2, C_3 and C_4. The mapping of each section is performed separately and the overall Nyquist plot is obtained by combining the individual sections.

MAPPING OF SECTION C_1

In section C_1, ω varies from 0 to $+\infty$. The mapping of section C_1 is given by the locus of $G(j\omega)H(j\omega)$ as ω is varied from 0 to ∞. This locus is the polar plot of $G(j\omega)H(j\omega)$.

$$G(s)H(s) = \frac{5}{s(1-s)}$$

Let $s = j\omega$, $\therefore G(j\omega)H(j\omega) = \dfrac{5}{j\omega(1-j\omega)}$

$$= \frac{5}{\omega\angle 90^\circ \sqrt{1+\omega^2}\angle\tan^{-1}\omega}$$

$$= \frac{5}{\omega\sqrt{1+\omega^2}}\angle(-90+\tan^{-1}\omega)$$

$\therefore |G(j\omega)H(j\omega)| = \frac{5}{\omega\sqrt{1+\omega^2}}$

$\angle G(j\omega)H(j\omega) = -90^\circ + \tan^{-1}\omega$

> ***Note :*** (1–jω) represents a point in fourth quadrant.

The exact shape of G(jω)H(jω) locus is determined by calculating the magnitude and phase of G(jω)H(jω) for various values of ω.

ω rad/sec	0	0.6	1.0	2.0	10.0	∞
|G(jω)H(jω)|	∞	7.15	3.53	1.12	0.05	0
∠G(jω)H(jω) deg	–90	–59	–45	–26	–5	0

From the above analysis, we can conclude that G(jω)H(jω) locus starts at –90° axis at infinity for ω = 0 and meets the origin along 0° axis when ω = ∞.

The section C_1 in s-plane and its corresponding mapping in G(s)H(s)-plane are shown in fig 4.17.2 and 4.17.3.

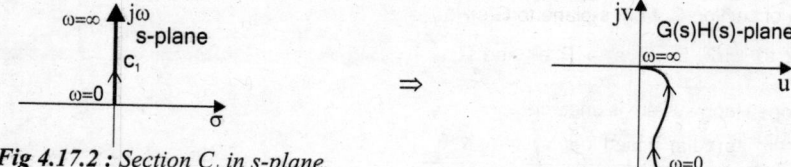

Fig 4.17.2 : Section C_1 in s-plane.

⇒

Fig 4.17.3 : Mapping of section C_1 in G(s)H(s)-plane.

MAPPING OF SECTION C_2

The mapping of section C_2 from s-plane to G(s)H(s)-plane is obtained by letting $s = \underset{t\to\infty}{\text{Lt}}\ R e^{j\theta}$ in G(s)H(s) and varying θ from +π/2 to –π/2. Since $s \to R\,e^{j\theta}$ and R→ ∞, the G(s)H(s) can be approximated as shown below, [i.e., (1–s) ≈ –s)]

$$G(s)H(s) = \frac{5}{s(1-s)} \approx \frac{5}{s(-s)} = \frac{5}{s^2 e^{j\pi}}$$

Let, $s = \underset{R\to\infty}{\text{Lt}}\ Re^{j\theta}$

$$\therefore G(s)H(s)\Big|_{s=\underset{R\to\infty}{\text{Lt}} Re^{j\theta}} = \frac{5}{\underset{R\to\infty}{\text{Lt}}\ (Re^{j\theta})^2\, e^{j\pi}} = 0\,e^{-j(2\theta+\pi)}$$

> ***Note :*** $-1 = e^{j\pi}$

When $\theta = \frac{\pi}{2}$, $\quad G(s)H(s) = 0e^{-j2\pi}$ .....(1)

When $\theta = -\frac{\pi}{2}$, $\quad G(s)H(s) = 0e^{-j0}$ .....(2)

From the equations (1) and (2) we can say that section C_2 in s-plane (fig 4.17.4) is mapped as circular arc of zero radius around origin in G(s)H(s) plane with argument varying from –2π to + 0 as shown in fig 4.17.5.

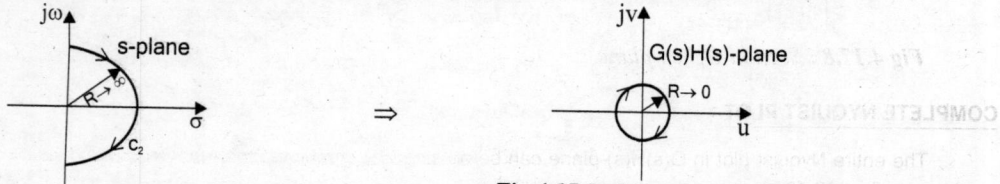

Fig 4.17.4 : Section C_2 in s-plane.

⇒

Fig 4.17.5 : Mapping of section C_2 in G(s)H(s)-plane.

MAPPING OF SECTION C₃

In section C_3, ω varies from $-\infty$ to 0. The mapping of section C_3 is given by the locus of $G(j\omega)H(j\omega)$ as ω is varied from $-\infty$ to 0. This locus is the inverse polar plot of $G(j\omega)H(j\omega)$.

The inverse polar plot is given by the mirror image of polar plot with respect to real axis. The section C_3 in s-plane and its corresponding contour in $G(s)H(s)$ plane are shown in fig 4.17.6 and fig 4.17.7.

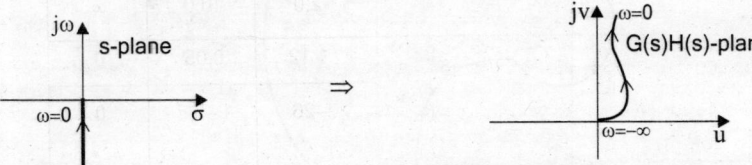

Fig 4.17.6 : *Section C_3 in s-plane.*   **Fig 4.17.7 :** *Mapping of section C_3 in $G(s)H(s)$-plane.*

MAPPING OF SECTION C₄

The mapping of section C_4 from s-plane to $G(s)H(s)$-plane is obtained by letting $s = \underset{R \to 0}{Lt}\, R\, e^{j\theta}$ in $G(s)H(s)$ and varying θ from $-\pi/2$ to $+\pi/2$. Since $s \to R\, e^{j\theta}$ and $R \to 0$, $G(s)H(s)$ can be approximated as shown below, [i.e., $(1-s) \approx 1$].

$$G(s)H(s) = \frac{5}{s(1-s)} \approx \frac{5}{s \times 1} = \frac{5}{s}$$

Let $s = \underset{R \to 0}{Lt}\, R e^{j\theta}$

$$\therefore\ G(s)H(s)\Big|_{s = \underset{R \to 0}{Lt}\, R e^{j\theta}} = \frac{5}{\underset{R \to 0}{Lt}\, R e^{j\theta}} = \infty e^{-j\theta}$$

When $\theta = -\dfrac{\pi}{2}$, $G(s)H(s) = \infty e^{j\frac{\pi}{2}}$ .....(3)

When $\theta = \dfrac{\pi}{2}$, $G(s)H(s) = \infty e^{j\frac{\pi}{2}}$ .....(4)

From the equations (3) and (4) we can say that section C_4 in s-plane (fig 4.17.8.) is mapped as a circular arc of infinite radius with argument varying from $\pi/2$ to $-\pi/2$ as shown in fig 4.17.9.

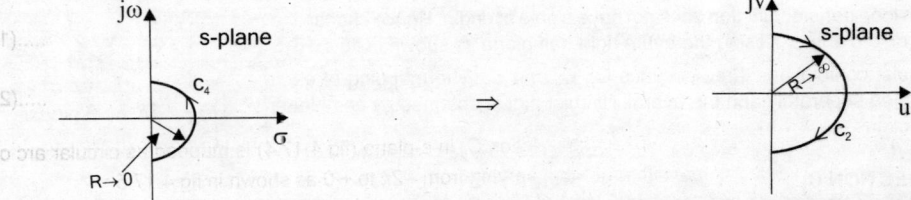

Fig 4.17.8 : *Section C_4 in s-plane.*   **Fig 4.17.9 :** *Mapping of section C_4 in $G(s)H(s)$-plane.*

COMPLETE NYQUIST PLOT

The entire Nyquist plot in $G(s)H(s)$-plane can be obtained by combining the mappings of individual sections, as shown in fig 4.17.10.

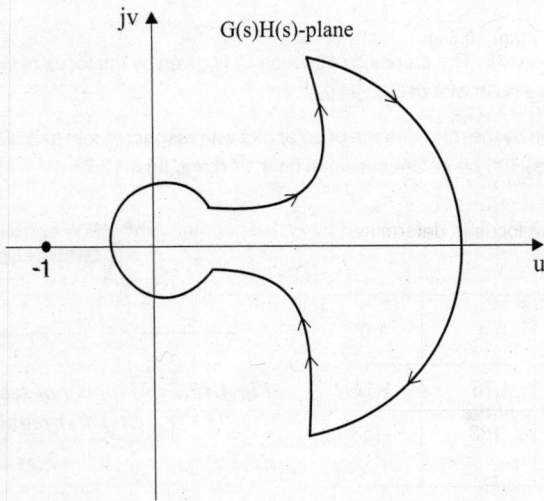

$$\textit{Fig 4.17.10 : Nyquist plot of } G(s)H(s) = \frac{5}{s(1-s)}.$$

STABILITY ANALYSIS

The Nyquist contour in G(s)H(s)-plane does not encircle the point (−1+j0) but the open loop transfer function has one pole on the right half s-plane. Therefore the system is unstable.

RESULT

Both open loop and closed loop systems are unstable.

EXAMPLE 4.18

By Nyquist stability criterion determine the stability of closed loop system, whose open loop transfer function is given by, $G(s)H(s) = \dfrac{(s+2)}{(s+1)(s-1)}$. Comment on the stability of open-loop and closed loop system.

SOLUTION

Given that, $G(s)H(s) = \dfrac{(s+2)}{(s+1)(s-1)}$

The open loop transfer function does not have a pole at origin. Hence choose the Nyquist contour on s-plane enclosing the entire right half plane as shown in fig 4.18.1.

The Nyquist contour has three sections C_1, C_2 and C_3. The mapping of each section is performed separately and the overall Nyquist plot is obtained by combining the individual sections.

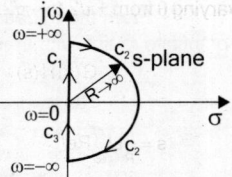

Fig 4.18.1 : Nyquist Contour in s-plane.

MAPPING OF SECTION C_1

In section C_1, ω varies from 0 to +∞. The mapping of section C_1 is given by the locus of G(jω)H(jω) as ω is varied from 0 to ∞. This locus is the polar plot of G(jω)H(jω).

$$G(s)H(s) = \frac{(s+2)}{(s+1)(s-1)} = \frac{2(1+0.5s)}{(1+s)(-1+s)}$$

> **Note :** (−1+j) represents a point in second quadrant.

Let s = jω, ∴ $G(j\omega)H(j\omega) = \dfrac{2(1+j0.5\omega)}{(1+j\omega)(-1+j\omega)} = \dfrac{2\sqrt{1+0.25\omega^2}\angle\tan^{-1}0.5\omega}{\sqrt{1+\omega^2}\angle\tan^{-1}\omega\sqrt{1+\omega^2}\angle(180°-\tan^{-1}\omega)}$

$$= \frac{2\sqrt{1+0.25\omega^2}}{1+\omega^2} \angle(-80 + \tan^{-1} 0.5\omega)$$

$$\therefore |G(j\omega)H(j\omega)| = \frac{2\sqrt{1+0.25\omega^2}}{1+\omega^2}$$

$$\angle G(j\omega)H(j\omega) = -180° + \tan^{-1} 0.5\omega$$

The exact shape of $G(j\omega)H(j\omega)$ locus is determined by calculating the magnitude and phase of $G(j\omega)H(j\omega)$ for various values of ω.

ω rad/sec	0	0.4	1.0	2.0	10.0	∞
$\|G(j\omega)H(j\omega)\|$	2	1.76	1.12	0.57	0.1	0
$\angle G(j\omega)H(j\omega)$ deg	-180	-168	-153	-135	-101	-90

From the above analysis, we can conclude that $G(j\omega)H(j\omega)$ locus starts at $-180°$ axis at a magnitude of -2 for $\omega = 0$ and meets the origin along $-90°$ axis when $\omega = +\infty$.

The section C_1 in s-plane and its corresponding mapping in $G(s)H(s)$-plane are shown in fig 4.18.2. and 4.18.3.

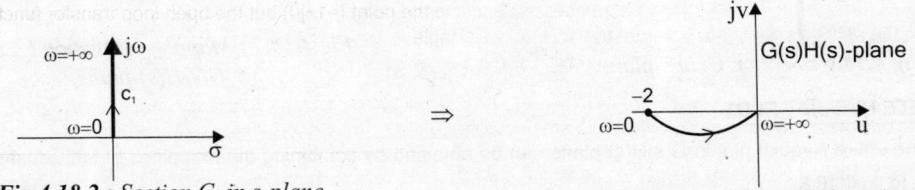

Fig 4.18.2 : *Section C_1 in s-plane.*

Fig 4.18.3 : *Mapping of section C_1 in $G(s)H(s)$-plane.*

MAPPING OF SECTION C_2

The mapping of section C_2 from s-plane to $G(s)H(s)$-plane is obtained by letting $s = \underset{R \to \infty}{Lt} Re^{j\theta}$ in $G(s)H(s)$ and varying θ from $+\pi/2$ to $-\pi/2$. Since $s \to Re^{j\theta}$ and $R \to \infty$, $G(s)H(s)$ can be approximated as shown below, [i.e., $(1+sT) \approx sT$].

$$G(s)H(s) = \frac{2(1+0.5s)}{(1+s)(-1+s)} \approx \frac{2 \times 0.5s}{s \times s} = \frac{1}{s}$$

$$s = \underset{R \to \infty}{Lt} Re^{j\theta}$$

$$\therefore G(s)H(s)\Big|_{s = \underset{R \to \infty}{Lt} Re^{j\theta}} = \frac{1}{\underset{R \to \infty}{Lt} Re^{j\theta}} = 0e^{-j\theta}$$

When $\theta = \frac{\pi}{2}$, $G(s)H(s) = 0e^{-j\frac{\pi}{2}}$ .....(1)

When $\theta = -\frac{\pi}{2}$, $G(s)H(s) = 0e^{j\frac{\pi}{2}}$ .....(2)

From the equations (1) and (2) we can say that section C_2 in s-plane (fig 4.18.4) is mapped as circular arc of zero radius around origin in $G(s)H(s)$-plane with argument varying from $-\pi/2$ to $+\pi/2$ as shown in fig 4.18.5.

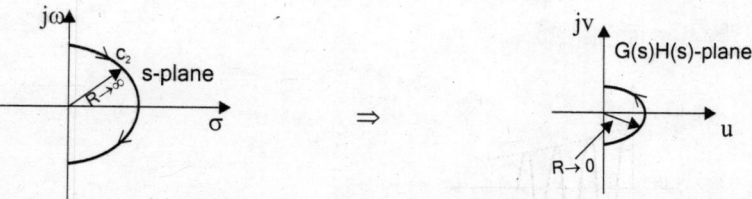

Fig 4.18.4 : *Section C₁ in s-plane.*

Fig 4.18.5 : *Mapping of section C₂ in G(s)H(s)-plane.*

MAPPING OF SECTION C₃

In section C_3, ω varies from $-\infty$ to 0. The mapping of section C_3 is given by the locus of $G(j\omega)H(j\omega)$ as ω is varied from $-\infty$ to 0. This locus is the inverse polar plot of $G(j\omega)H(j\omega)$.

The inverse polar plot is given by the mirror image of polar plot with respect to real axis. The section C_3 in s-plane and its corresponding contour in G(s)H(s) plane are shown in fig 4.18.6 and fig 4.18.7.

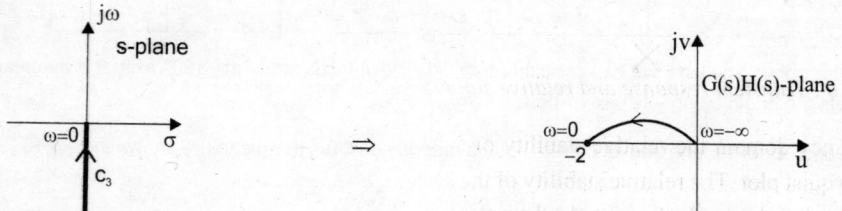

Fig 4.18.6 : *Section C₃ in s-plane.*

Fig 4.18.7 : *Mapping of section C₃ in G(s)H(s)-plane.*

COMPLETE NYQUIST PLOT

The entire Nyquist plot in G(s)H(s)-plane can be obtained by combining the mappings of individual sections, as shown in fig 4.18.8.

STABILITY ANALYSIS

On travelling through Nyquist contour it is observed that $-1+j0$ point is encircled in anticlockwise direction one time. Also the open loop transfer function has one pole at right half s-plane. Since the number of anticlockwise encirclement is equal to number of open loop poles on right half s-plane, the closed loop system is stable.

RESULT

(a)   Open loop system is unstable

(b)   Closed loop system is stable.

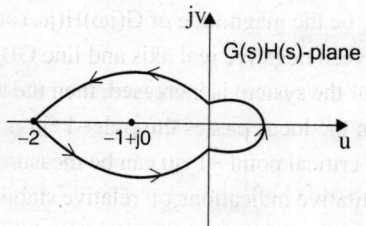

Fig 4.18.8 : *Nyquist plot of $G(s)H(s) = \dfrac{(s+2)}{(s+1)(s-1)}$.*

4.6 RELATIVE STABILITY

The ***Relative stability*** indicates the closeness of the system to stable region. It is an indication of the strength or degree of stability.

In time domain, the relative stability may be measured by relative settling times of each root or pair of roots. The settling time is inversely proportional to the location of roots of characteristic equation. If the root is located far away from the imaginary axis, then the transients dies out faster and so the relative stability of system will improve. The transient response and so the relative stability for various location of roots in s-plane are shown in fig 4.6.

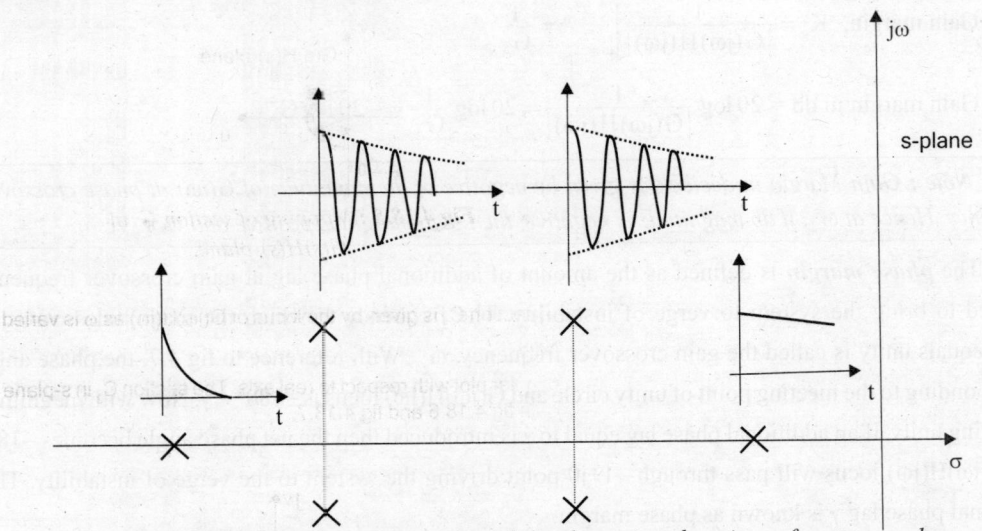

Fig 4.6 : *Transient response and relative stability for various locations of roots on s-plane.*

In frequency domain the relative stability of a system can be studied from Nyquist plot. The relative stability of the system is given by closeness of polar plot to -1+j0 point. As the polar plot gets closer to -1+j0 point the system moves towards instability.

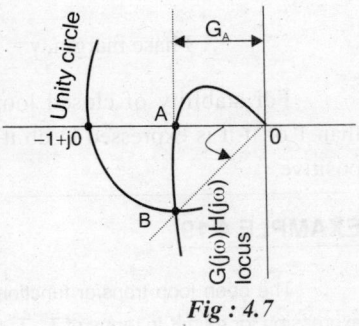

Fig : 4.7

The relative stability in frequency domain are quantitatively measured in terms of phase margin and gain margin. Consider a $G(j\omega)$ $H(j\omega)$ locus as shown in fig 4.7. Let this locus cross the real axis at point-**A** and a unit circle drawn with origin as centre cuts this locus at point-**B**. Let G_A be the magnitude of $G(j\omega)H(j\omega)$ at point-A , and γ be the angle between negative real axis and line OB.

If the gain of the system is increased, then the locus will shift upwards and it may cross real axis at $-1+j0$ point. When the locus passes through $-1+j0$ point, $G_A \to 1$ and $\gamma \to 0$. Hence the closeness of $G(j\omega)$ $H(j\omega)$ locus to the critical point $-1+j0$ can be measured in terms of intercept G_A and angle γ. The value of G_A and γ are quantitative indications of relative stability. These values are used to define gain margin and phase margin as practical measures of relative stability.

The concepts of gain margin and phase margin are defined for open loop systems but from the values of gain margin and phase margin the stability of closed loop system can be judged.

4.7 GAIN MARGIN AND PHASE MARGIN

Gain margin is a factor by which the system gain can be increased to drive the system to the verge of instability. With reference to fig 4.7 the magnitude of $G(j\omega)H(j\omega)$ is G_A when it crosses real axis and the phase corresponding to that point is $-180°$. The frequency corresponding to that point be ω_{pc}. If the gain of the system is increased by a factor $\dfrac{1}{G_A}$ then the magnitude at the frequency ω_{pc} will be, $G_A \times \dfrac{1}{G_A} = -1$.

Now the $G(j\omega)H(j\omega)$ locus will pass through $-1+j0$ point driving the system to the verge of instability. Hence the gain margin, K_g of the system may be defined as the reciprocal of the gain at which the phase angle is 180°. The frequency at which the phase angle is 180° is called phase crossover frequency.

Gain margin, $K_g = \dfrac{1}{|G(j\omega)H(j\omega)|}\bigg|_{\omega = \omega_{pc}} = \dfrac{1}{G_A}$

Gain margin in db $= 20\log\dfrac{1}{|G(j\omega)H(j\omega)|} = 20\log\dfrac{1}{G_A} = -20\log G_A$

Note : Gain Margin in decibels is given by negative of db magnitude of G(jω) at phase crossover frequency. Hence at ω_{pc}, if db magnitude is negative, then gain margin is positive and vice versa.

The **phase margin** is defined as the amount of additional phase lag at gain crossover frequency required to bring the system to verge of instability. The frequency at which the magnitudes of G(jω) H(jω) equals unity is called the gain crossover frequency, ω_{gc}. With reference to fig 4.7, the phase angle corresponding to the meeting point of unity circle and G(jω)H(jω) locus is $-180° + \gamma$. Now with magnitude remaining unity, if an additional phase lag equal to γ is introduced then the net phase angle becomes $-180°$ and G(jω)H(jω) locus will pass through $-1+j0$ point driving the system to the verge of instability. This additional phase lag γ is known as phase margin.

Let, $\phi_{gc} = \angle G(j\omega)H(j\omega)\big|_{\omega = \omega_{gc}}$; Now $-180° + \gamma = \phi_{gc}$; Now $-180°+\gamma = \phi_{gc}$

$\therefore$ Phase margin, $\gamma = 180° + \phi_{gc}$

For stability of closed loop system the gain margin of open loop system should be greater than 1 or if it is expressed in db it should be positive and phase margin of open loop system should be positive.

EXAMPLE 4.19

The open loop transfer function of a unity feedback system is given by, $G(s) = \dfrac{K}{s(1+sT_1)(1+sT_2)}$. Derive an expression for gain K in terms of T_1, T_2 and specified gain margin, K_g.

SOLUTION

Given that, $G(s) = \dfrac{K}{s(1+sT_1)(1+sT_2)}$

Let $s = j\omega$.

$G(j\omega) = \dfrac{K}{j\omega(1+j\omega T_1)(1+j\omega T_2)} = \dfrac{K}{j\omega(1+j\omega T_2+j\omega T_1 - \omega^2 T_1 T_2)}$

$= \dfrac{K}{j\omega\left[1+j\omega(T_1+T_2)-\omega^2 T_1 T_2\right]} = \dfrac{K}{j\omega - \omega^2(T_1+T_2)-j\omega^3 T_1 T_2}$

$= \dfrac{K}{-\omega^2(T_1+T_2)+j\omega(1-\omega^2 T_1 T_2)}$

The gain margin, K_g is defined as the reciprocal of the magnitude of $G(j\omega)$ at phase crossover frequency. At phase crossover frequency the magnitude is purely real. Hence at phase crossover frequency, ω_{pc}, the imaginary part of $G(j\omega)$ is zero.

At $\omega = \omega_{pc}$, $\quad \omega_{pc}\left(1 - \omega_{pc}^2 T_1 T_2\right) = 0 \quad \Rightarrow \quad 1 - \omega_{pc}^2 T_1 T_2 = 0 \quad \Rightarrow \quad -\omega_{pc}^2 T_1 T_2 = -1$

$$\therefore \ \omega_{pc} = \frac{1}{\sqrt{T_1 T_2}}$$

At $\omega = \omega_{pc}$, the imaginary part is zero,

$$\therefore \ |G(j\omega)| = \left| \frac{K}{-\omega^2(T_1 + T_2)} \right| = \frac{K}{\omega^2(T_1 + T_2)}$$

$$K_g = \frac{1}{|G(j\omega)|_{\omega = \omega_{pc}}} = \frac{1}{K/\omega_{pc}^2(T_1 + T_2)} = \frac{\omega_{pc}^2(T_1 + T_2)}{K}$$

Put, $\omega_{pc}^2 = \dfrac{1}{T_1 T_2}$ in the above equation.

$$\therefore \ K_g = \frac{\left(\frac{1}{T_1 T_2}\right)(T_1 + T_2)}{K} \quad \Rightarrow \quad K = \frac{1}{K_g} \frac{T_1 - T_2}{T_1 T_2} = \frac{1}{K_g}\left(\frac{T_1}{T_1 T_2} + \frac{T_2}{T_1 T_2}\right) \quad \Rightarrow \quad K = \frac{1}{K_g}\left(\frac{1}{T_1} + \frac{1}{T_2}\right)$$

RESULT

The expression for gain K in terms of K_g, T_1 and T_2 is, $K = \dfrac{1}{K_g}\left(\dfrac{1}{T_1} + \dfrac{1}{T_2}\right)$

EXAMPLE 4.20

Determine the Gain crossover frequency, Phase crossover frequency, Gain margin and Phase margin of a system with open loop transfer function, $G(s) = \dfrac{1}{s(1 + 2s)(1 + s)}$.

SOLUTION

(i) To find phase crossover frequency and gain margin

Given that, $G(s) = \dfrac{1}{s(1 + 2s)(1 + s)}$

Let $s = j\omega$.

$$\therefore \ G(j\omega) = \frac{1}{j\omega(1 + j2\omega)(1 + j\omega)} = \frac{1}{j\omega(1 + j\omega + j2\omega - 2\omega^2)} = \frac{1}{j\omega(j3\omega + 1 - 2\omega^2)} = \frac{1}{-3\omega^2 + j\omega(1 - 2\omega^2)}$$

At phase crossover frequency the imaginary part of $G(j\omega)$ is zero. Hence put $\omega = \omega_{pc}$ in imaginary part and equate to zero to solve for ω_{pc}.

$$\therefore \ \omega_{pc}\left(1 - 2\omega_{pc}^2\right) = 0$$

since $\omega_{pc} \neq 0$, $\quad 1 - 2\omega_{pc}^2 = 0 \quad \Rightarrow \quad -2\omega_{pc}^2 = -1 \quad \Rightarrow \quad \omega_{pc}^2 = \frac{1}{2} \quad \Rightarrow \quad \omega_{pc} = \frac{1}{\sqrt{2}} = 0.707 \, \text{rad/sec}$

The gain margin, K_g is defined as reciprocal of magnitude of $G(j\omega)$ at phase cross over frequency.

Gain margin, $K_g = \dfrac{1}{|G(j\omega)|_{\omega = \omega_{pc}}} = \dfrac{1}{|1/-3\omega^2|_{\omega = \omega_{pc}}} = 3\omega_{pc}^2 = 3 \times 0.707^2 = 1.5$

Gain margin in db = 20 log K_g = 20 log 1.5 = 3.5 db

(ii) To find gain crossover frequency and phase margin

Given that, $G(s) = \dfrac{1}{s(1+2s)(1+s)}$

Let $s = j\omega$.

$$\therefore G(j\omega) = \frac{1}{j\omega(1+j2\omega)(1+j\omega)} = \frac{1}{\omega\angle 90° \sqrt{1+4\omega^2} \angle \tan^{-1}2\omega \sqrt{1+\omega^2} \angle \tan^{-1}\omega}$$

$$\therefore |G(j\omega)| = \frac{1}{\omega\sqrt{1+4\omega^2}\sqrt{1+\omega^2}}$$

$$\angle G(j\omega) = -90° - \tan^{-1}2\omega - \tan^{-1}\omega$$

At gain crossover frequency, ω_{gc} the magnitude of $G(j\omega)$ is unity.

$$\therefore \text{At } \omega = \omega_{gc}, \qquad |G(j\omega)| = \frac{1}{\omega_{gc}\sqrt{1+4\omega_{gc}^2}\sqrt{1+\omega_{gc}^2}} = 1$$

Solving the above equation for ω_{gc} will be tedious. Hence by trial and error find the root of the above equation.

When $\omega = 1$, $|G(j\omega)| = \dfrac{1}{\omega\sqrt{1+4\omega^2}\sqrt{1+\omega^2}} = \dfrac{1}{1\sqrt{1+4}\sqrt{1+1}} = 0.3$

When $\omega = 0.5$, $|G(j\omega)| = \dfrac{1}{\omega\sqrt{1+4\omega^2}\sqrt{1+\omega^2}} = \dfrac{1}{0.5\sqrt{1+4\times0.5^2}\sqrt{1+0.5^2}} = 1.26$

From the two calculations shown above, we can conclude that the unity magnitude will occur for a frequency between 0.5 and 1.0.

When $\omega = 0.6$, $|G(j\omega)| = \dfrac{1}{0.6\sqrt{1+4\times0.6^2}\sqrt{1+0.6^2}} = 0.915$

When $\omega = 0.57$, $|G(j\omega)| = \dfrac{1}{0.57\sqrt{1+4\times0.57^2}\sqrt{1+0.57^2}} = 1.005$

Let $\omega = 0.57$ be the gain crossover frequency, since for this value of ω the magnitude of $G(j\omega)$ is approximately equal to one.

$$\therefore \text{Gain crossover frequency, } \omega_{gc} = 0.57 \text{ rad/sec.}$$

Let the phase of $G(j\omega)$ at ω_{gc} be ϕ_{gc}.

$$\text{At } \omega = \omega_{gc} = 0.57, \quad \phi_{gc} = -90° - \tan^{-1}2\omega - \tan^{-1}\omega$$

$$= -90° - \tan^{-1}(2\times0.57) - \tan^{-1}0.57 = -168°$$

$$\therefore \text{Phase margin, } \gamma = 180° + \phi_{gc} = 180° - 168° = 12°$$

RESULT

(a)   The phase crossover frequency, $\omega_{pc} = 0.707$ rad/sec

(b)   The gain crossover frequency, $\omega_{gc} = 0.57$ rad/sec

(c)   The gain margin, $K_g = 1.5$

        The gain margin in db = 3.5 db

(d)   The phase margin, $\gamma = 12°$

EXAMPLE 4.21

The open loop transfer function of a system is $G(s) = \dfrac{K}{s(1+0.1s)(1+s)}$.

(i)   Determine the value of K so that gain margin is 6 db.

(ii)   Determine the value of K so that phase margin is 40°.

SOLUTION

(i) To find K for specified gain margin

Given that, $G(s) = \dfrac{K}{s(1+0.1s)(1+s)}$

Let $s = j\omega$.

$\therefore\ G(j\omega) = \dfrac{K}{j\omega(1+j0.1\omega)(1+j\omega)} = \dfrac{K}{j\omega(1+j1.1\omega-0.1\omega^2)} = \dfrac{K}{-1.1\omega^2 + j\omega(1-0.1\omega^2)}$

At phase crossover frequency ω_{pc}, the $G(j\omega)$ is real and so equate the imaginary part to zero to solve for ω_{pc}.

At $\omega = \omega_{pc}$, $\omega_{pc}\left(1-0.1\omega_{pc}^2\right) = 0$ $\Rightarrow$ $1-0.1\omega_{pc}^2$ $\Rightarrow$ $-0.1\omega_{pc}^2 = -1$

$$\therefore\ \omega_{pc} = \frac{1}{\sqrt{0.1}} = 3.162 \text{ rad/sec}$$

$$\therefore\ |G(j\omega)|_{\omega = \omega_{pc}} = \left|\frac{K}{-1.1\omega^2}\right|_{\omega = \omega_{pc}} = \frac{K}{1.1 \times 3.162^2} = 0.0909K$$

Given that gain margin = 6 db, $\therefore\ 20\log K_g = 6$ $\Rightarrow$ $\log K_g = \dfrac{6}{20}$

$$\therefore\ \text{Gain margin, } K_g = 10^{\frac{6}{20}} = 1.9953$$

By definition of gain margin,

$$\text{Gain margin, } K_g = \frac{1}{|G(j\omega)|_{\omega = \omega_{pc}}}$$

$$\therefore\ 1.9953 = \frac{1}{0.0909K}$$

$$\therefore\ K = \frac{1}{0.0909 \times 1.9953} = 5.5135$$

(ii) To find K for specified phase margin

Given that, $G(s) = \dfrac{K}{s(1+0.1s)(1+s)}$.   Let $s = j\omega$.

$\therefore\ G(j\omega) = \dfrac{K}{j\omega(1+j0.1\omega)(1+j\omega)} = \dfrac{K}{\omega\angle 90°\sqrt{1+(0.1\omega)^2}\angle\tan^{-1}0.1\omega\sqrt{1+\omega^2}\angle\tan^{-1}\omega}$

$|G(j\omega)| = \dfrac{K}{\omega\sqrt{1+0.01\omega^2}\sqrt{1+\omega^2}}$

$\angle G(j\omega) = -90° - \tan^{-1}0.1\omega - \tan^{-1}\omega$

Let, ω_{gc} = Gain crossover frequency

$\phi_{gc} = \angle G(j\omega)$ at $\omega = \omega_{gc}$.

At $\omega = \omega_{gc}$, $\quad \phi_{gc} = \angle G(j\omega)|_{\omega = \omega_{gc}} = -90^\circ - \tan^{-1} 0.1\omega_{gc} - \tan^{-1}\omega_{gc}$

By definition of phase margin,

Phase margin, $\gamma = 180^\circ + \phi_{gc}$

The required phase margin is 40°, $\quad \therefore \gamma = 40^\circ$

$\therefore 40^\circ = 180^\circ - 90^\circ - \tan^{-1} 0.1\omega_{gc} - \tan^{-1}\omega_{gc} \quad \Rightarrow \quad \tan^{-1} 0.1\omega_{gc} + \tan^{-1}\omega_{gc} = 180^\circ - 90^\circ - 40^\circ$

$\therefore \tan^{-1} 0.1\omega_{gc} + \tan^{-1}\omega_{gc} = 50^\circ$

On taking tan on either side we get,

$$\tan\left[\tan^{-1} 0.1\omega_{gc} + \tan^{-1}\omega_{gc}\right] = \tan 50^\circ$$

$$\boxed{\tan(A+B) = \frac{\tan A + \tan B}{1 - \tan A \ \tan B}}$$

$$\frac{\tan \tan^{-1} 0.1\omega_{gc} + \tan \tan^{-1}\omega_{gc}}{1 - \tan \tan^{-1} 0.1\omega_{gc} \times \tan \tan^{-1}\omega_{gc}} = \tan 50^\circ \quad \Rightarrow \quad \frac{0.1\omega_{gc} + \omega_{gc}}{1 - 0.1\omega_{gc} \times \omega_{gc}} = 1.192 \quad \Rightarrow \quad \frac{1.1\omega_{gc}}{1 - 0.1\omega_{gc}^2} = 1.192$$

On cross multiplying the above equation we get,

$1.1\omega_{gc} = 1.192(1 - 0.1\omega_{gc}^2) \quad \Rightarrow \quad 0.1192\,\omega_{gc}^2 + 1.1\omega_{gc} - 1.192 = 0$

$\therefore \omega_{gc}^2 + \dfrac{1.1}{0.1192}\,\omega_{gc} - \dfrac{1.192}{0.1192} = 0 \quad \Rightarrow \quad \omega_{gc}^2 + 9.228\,\omega_{gc} - 10 = 0$

$\therefore \omega_{gc} = \dfrac{-9.228 \pm \sqrt{9.228^2 + 4 \times 10}}{2} = \dfrac{-9.228 \pm 11.1873}{2}$

On taking positive value we get,

$\therefore \omega_{gc} = \dfrac{-9.228 + 11.1873}{2} = 0.98 \text{ rad/sec}$

At $\omega = \omega_{gc}$, $\quad |G(j\omega)| = 1$; $\quad |G(j\omega)|_{\omega = \omega_{gc}} = \dfrac{K}{\omega_{gc}\sqrt{1 + 0.01\omega_{gc}^2}\sqrt{1 + \omega_{gc}^2}} = 1$

$\therefore K = \omega_{gc}\sqrt{1 + 0.01\omega_{gc}^2}\sqrt{1 + \omega_{gc}^2} = 0.98\sqrt{1 + 0.01 \times 0.98^2}\sqrt{1 + 0.98^2} = 1.3787$

RESULT

For a gain margin of 6 db, $\quad$ K = 5.5135

For a phase margin of 40°, $\quad$ K = 1.3787

4.8 ROOT LOCUS

The root locus technique was introduced by **W.R.Evans** in 1948 for the analysis of control systems. The root locus technique is a powerful tool for adjusting the location of closed loop poles to achieve the desired system performance by varying one or more system parameters.

Consider the open loop transfer function of system, $G(s) = \dfrac{K}{s(s + p_1)(s + p_2)}$

The closed loop transfer function of the system with unity feedback is given by,

$$\frac{C(s)}{R(s)} = \frac{G(s)}{1 + G(s)} = \frac{\dfrac{K}{s(s + p_1)(s + p_2)}}{1 + \dfrac{K}{s(s + p_1)(s + p_2)}} = \frac{K}{s(s + p_1)(s + p_2) + K}$$

The denominator polynomial of C(s)/R(s) is the characteristic equation of the system. The characteristic equation is given by,

$$s(s + p_1)(s + p_2) + K = 0.$$

The roots of characteristic equation is a function of open loop gain K. [In other words the roots of characteristic equation depend on open loop gain K]. When the gain K is varied from 0 to ∞, the roots of characteristic equation will take different values. When K = 0, the roots are given by open loop poles. When K→ ∞, the roots will take the value of open loop zeros.

The path taken by the roots of characteristic equation when open loop gain K is varied from 0 to ∞ are called *root loci* (or the path taken by a root of characteristic equation when open loop gain K is varied from 0 to ∞ is called root locus).

> **Note :** *In general the roots of characteristic equation can be varied by varying any other system parameter other than gain.*

In general the closed loop transfer function of system with multiple loops is obtained from the signal flow graph of the system using Mason's gain formula.

$$\frac{C(s)}{R(s)} = T(s) = \frac{1}{\Delta} \sum_k P_k \Delta_k \quad \text{(Refer chapter 1 section 1.12)}$$

The determinant, Δ is the denominator polynominal of C(s)/R(s). The characteristic equation of the system is given by , $\Delta = 0$

For the single loop system shown in fig 4.8

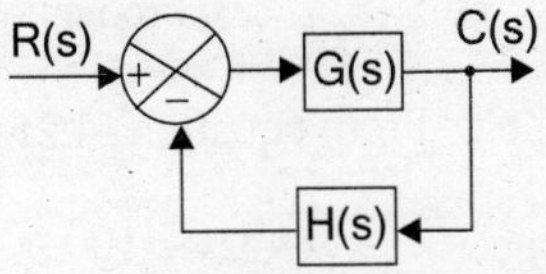

$$\frac{C(s)}{R(s)} = \frac{G(s)}{1 + G(s)H(s)}$$

Fig 4.8

The Characteristic equation is,

$$1 + G(s) H(s) = 0$$

$$\therefore G(s) H(s) = -1 \qquad\qquad .....(4.22)$$

From equation (4.22) it can be concluded that the roots of the characteristic equation occur only for those values of s for which, $G(s)H(s) = -1$.

The equation (4.22) can be converted to two Evans conditions given below,

$$|G(s)H(s)| = 1 \qquad\qquad .....(4.23)$$

$$\angle G(s)H(s) = \pm 180^\circ (2q+1), \qquad \text{where } q = 0, 1, 2, 3 ...... \qquad .....(4.24)$$

The equation (4.23) is called magnitude criterion and equation (4.24) is called angle criterion.

The magnitude criterion states that s = s_a will be a point on root locus if for that value of s,

$|G(s) H(s)| = 1.$

The angle criterion states that $s = s_a$ will be a point on root locus if for that value of s,

$\angle G(s) H(s)$ is equal to an odd multiple of 180°.

The function G(s)H(s) can be expressed as a ratio of two polynomials in s as shown below.

$$G(s)H(s) = K \frac{(s+z_1)(s+z_2)(s+z_3)......}{(s+p_1)(s+p_2)(s+p_3)......} \qquad ....(4.25)$$

$$\therefore |G(s)H(s)| = K \frac{|s+z_1| \times |s+z_2| \times |s+z_3|.....}{|s+z_3| \times |s+p_2| \times |s+p_3|.....} = K \frac{\prod_{i=1}^{m} |s+z_i|}{\prod_{i=1}^{n} |s+p_i|}$$

where, m = Number of zeros of loop transfer function.

n = Number of poles of loop transfer function.

The magnitude criterion states that $|G(s)H(s)| = 1$.

$$\therefore K \frac{\prod_{i=1}^{m} |s+z_i|}{\prod_{i=1}^{n} |s+p_i|} = 1 \quad \text{or} \quad K = \frac{\prod_{i=1}^{n} |s+p_i|}{\prod_{i=1}^{m} |s+z_i|} \qquad .....(4.26)$$

The open-loop gain K corresponding to a point $s = s_a$ on root locus can be calculated using equation (4.26). It can be shown that $|s + p_i|$ is equal to the length of vector drawn from $s = p_i$ to $s = s_a$ and $|s + z_i|$ is equal to the length of vector drawn from $s = z_i$ to $s = s_a$. Hence the equation K can be written as,

$$K = \frac{\text{Product of length of vector from open loop poles to the point } s = s_a}{\text{Product of length of vector from open loop zeros to the point } s = s_a}$$

From equation (4.25),

$$\angle G(s)H(s) = \angle(s+z_1) + \angle(s+z_2) + \angle(s+z_3) + ... - \angle(s+p_1) - \angle(s+p_2) - \angle(s+p_3)....$$

$$= \sum_{i=1}^{m} \angle(s+z_i) - \sum_{i=1}^{n} \angle(s+p_i)$$

where, m = Number of zeros of loop transfer function.

n = Number of poles of loop transfer function.

The angle criterion states that $\angle G(s)H(s) = \pm 180° (2q + 1)$

$$\therefore \sum_{i=1}^{m} \angle(s+z_i) - \sum_{i=1}^{n} \angle(s+p_i) = \pm 180°(2q+1) \qquad .....(4.27)$$

The equations (4.27) can be used to check whether a point $s = s_a$ is a point on root locus or not. It can be shown that $\angle(s+p_i)$ is equal to the angle of vector drawn from $s = p_i$ to $s = s_a$ and $\angle(s+z_i)$ is equal to the angle of vector drawn from $s = z_i$ to $s = s_a$. Hence equation (5.27) can be written as

$$\begin{pmatrix} \text{Sum of angles of vector} \\ \text{from open loop zeros} \\ \text{to the point } s = s_a \end{pmatrix} - \begin{pmatrix} \text{Sum of angles of vector} \\ \text{from open loop poles} \\ \text{to the point } s = s_a \end{pmatrix} = \pm 180°(2q+1)$$

CONSTRUCTION OF ROOT LOCUS

The exact root locus is sketched by trial and error procedure. In this method, the poles and zeros of G(s)H(s) are located on the s-plane on a graph sheet and a trial point $s = s_a$ is selected. Determine the angles of vectors drawn from poles and zeros to the trial point. From the angle criterion, determine the angle to be contributed by these vectors to make the trial point as a point on root locus. Shift the trial point suitably so that the angle criterion is satisfied.

A number of points are determined using the above procedure. Join the points by a smooth curve which is the root locus. The value of K for a particular root can be obtained from the magnitude criterion.

The trial and error procedure for sketching root locus is tedious. A set of rules have been developed to reduce the task involved in sketching root locus and to develop a quick approximate sketch. From the approximate sketch, a more accurate root locus can be obtained by a few trials.

RULES FOR CONSTRUCTION OF ROOT LOCUS

Rule 1 : The root locus is symmetrical about the real axis.

Rule 2 : Each branch of the root locus originates from an open-loop pole corresponding to K = 0 and terminates at either on a finite open loop zero (or open loop zero at infinity) corresponding to K = ∞. The number of branches of the root locus terminating on infinity is equal to n−m, (i.e., the number of open loop poles minus the number of finite zeros)

Rule 3 : Segments of the real axis having an odd number of real axis open-loop poles plus zeros to their right are parts of the root locus.

Rule 4 : The n−m root locus branches that tend to infinity, do so along straight line asymptotes making angles with the real axis given by,

$$\phi_A = \frac{180°(2q+1)}{n-m} \quad ; \quad q = 0, 1, 2, ...n-m$$

Rule 5 : The point of intersection of the asymptotes with the real axis is at $s = \sigma_A$ where,

$$\sigma_A = \frac{\text{Sum of poles} - \text{Sum of zeros}}{n-m}$$

Rule 6 : The breakaway and breakin points of the root locus are determined from the roots of the equation dK/ds = 0. If r numbers of branches of root locus meet at a point, then they break away at an angle of ±180°/r.

Rule 7 : The angle of departure from a complex open-loop pole is given by,

$$\phi_p = ±180° (2q + 1) + \phi ; \qquad q = 0, 1, 2, ...$$

where ϕ is the net angle contribution at the pole by all other open loop poles and zeros. Similarly the angle of arrival at a complex open loop zero is given by,

$$\phi_z = ±180° (2q + 1) + \phi ; \qquad q = 0, 1, 2, ...$$

where ϕ is the net angle contribution at the zero by all other open-loop poles and zeros.

Rule 8 : The points of intersection of root locus branches with the imaginary axis can be determined by use of the Routh criterion. Alternatively they can be evaluated by letting $s = j\omega$ in the characteristic equation and equating the real part and imaginary part to zero, to solve for ω and K. The values of ω are the intersection points on imaginary axis and K is the value of gain at the intersection points.

Rule 9 : The open-loop gain K at any point $s = s_a$ on the root locus is given by,

$$K = \frac{\prod\limits_{i=1}^{n} |s_a + p_i|}{\prod\limits_{i=1}^{m} |s_a + z_i|} = \frac{\text{Product of vector lengths from open loop poles to the point } s_a}{\text{Product of vector lengths from open loop zeros to the point } s_a}$$

> ***Note :*** *The length of vector should be measured to scale. If there is no finite zero then the product of vector lengths from zeros is equal to 1.*

TYPICAL SKETCHES OF ROOT LOCUS PLOTS

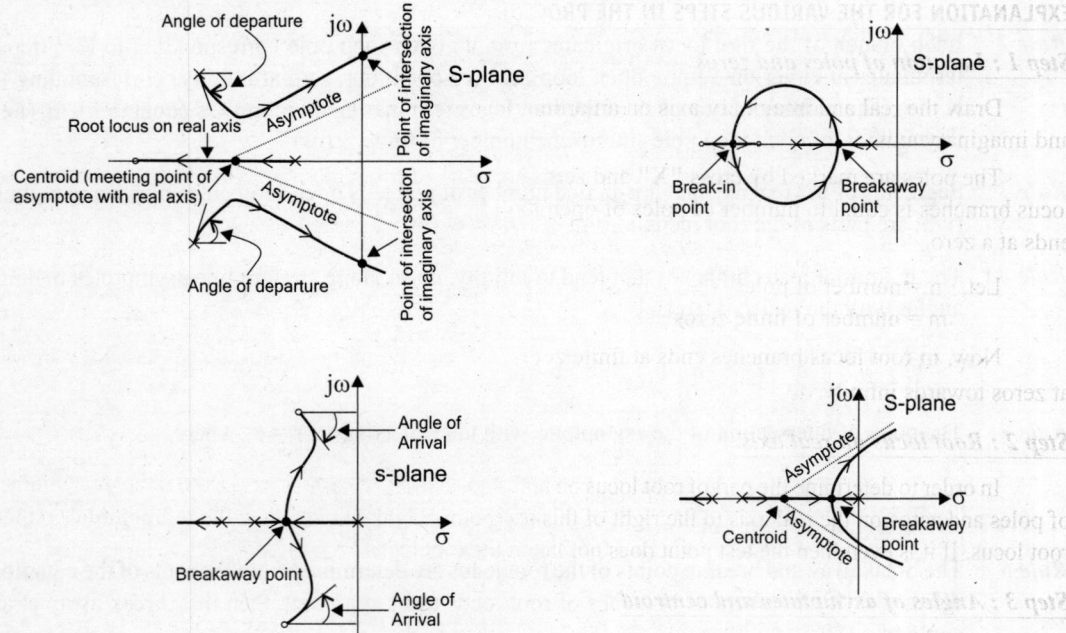

PROCEDURE FOR CONSTRUCTING ROOT LOCUS

Step 1 : Locate the poles and zeros of G(s)H(s) on the s-plane. The root locus branch starts from open loop poles and terminates at zeros.

Step 2 : Determine the root locus on real axis.

Step 3 : Determine the asymptotes of root locus branches and meeting point of asymptotes with real axis.

Step 4 : Find the breakaway and breakin points.

Step 5 :   If there is a complex pole then determine the angle of departure from the complex pole. If there is a complex zero then determine the angle of arrival at the complex zero.

Step 6 :   Find the points where the root loci may cross the imaginary axis.

Step 7 :   Take a series of test points in the broad neighbourhood of the origin of the s-plane and adjust the test point to satisfy angle criterion. Sketch the root locus by joining the test points by smooth curve.

Step 8 :   The value of gain K at any point on the locus can be determined from magnitude condition.

The value of K at a point $s = s_a$, is given by,

$$K = \frac{\text{Product of length of vectors from poles to the point, } s = s_a}{\text{Product of length of vectors from finite zeros to the point, } s = s_a}$$

Note : When there is no finite zero, the denominator is taken as unity. The length of vectors should be measured to scale.

EXPLANATION FOR THE VARIOUS STEPS IN THE PROCEDURE FOR CONSTRUCTING ROOT LOCUS

Step 1 : Location of poles and zeros

Draw the real and imaginary axis on an ordinary graph sheet and choose same scales both on real and imaginary axis.

The poles are marked by cross **"X"** and zeros are marked by small circle **"o"**. The number of root locus branches is equal to number of poles of open loop transfer function. Root locus starts at a pole and ends at a zero.

Let,   n = number of poles

m = number of finite zeros

Now, **m** root locus branches ends at finite zeros. The remaining **n–m** root locus branches will end at zeros towards infinity.

Step 2 : Root locus on real axis

In order to determine the part of root locus on real axis, take a test point on real axis. If the total number of poles and zeros on the real axis to the right of this test point is odd number, then the test point lies on the root locus. If it is even then the test point does not lie on the root locus.

Step 3 : Angles of asymptotes and centroid

If n is number of poles and m is number of finite zeros, then n–m root locus branches will terminate at zeros towards infinity.

These n–m root locus branches will go along an asymptotic path and meets the asymptotes at infinity. Hence number of asymptotes is equal to number of root locus branches going to infinity. The angles of asymptotes and the centroid are given by the following formulae.

$$\text{Angle of asymptotes} = \frac{\pm 180(2q+1)}{n - m}$$

where, q = 0, 1, 2, 3, ...(n–m)

$$\text{Centroid (meeting point of asymptote with real axis)} = \frac{\text{Sum of poles} - \text{Sum of zeros}}{n - m}$$

Step 4 : Breakaway and Breakin points

The breakaway or breakin points either lie on real axis or exist as complex conjugate pairs. If there is a root locus on real axis between 2 poles then there exist a breakaway point. If there is a root locus on real axis between 2 zeros then there exist a breakin point. If there is a root locus on real axis between pole and zero then there may be or may not be breakaway or breakin point.

Let the characteristic equation be in the form,

$$B(s) + K\, A(s) = 0$$

$$\therefore K = \frac{-B(s)}{A(s)}$$

The breakaway and breakin point is given by roots of the equation $dK/ds = 0$. The roots of $dK/ds = 0$ are actual breakaway or breakin point provided for this value of root, the gain K should be positive and real.

Step 5 : Angle of Departure and angle of arrival

$$\left.\begin{array}{l}\text{Angle of Departure}\\ \text{(from a complex pole A)}\end{array}\right\} = 180^\circ - \left(\begin{array}{l}\text{Sum of angles of vector to the}\\ \text{complex pole A from other poles}\end{array}\right) + \left(\begin{array}{l}\text{Sum of angles of vectors to the}\\ \text{complex pole A from zeros}\end{array}\right)$$

> **Note :** *The angles can be calculated as shown in fig 4.9 or they can be measured using protractor.*

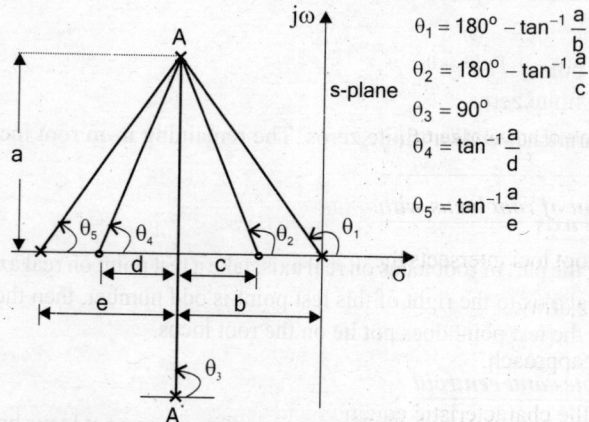

$$\theta_1 = 180^\circ - \tan^{-1}\frac{a}{b}$$
$$\theta_2 = 180^\circ - \tan^{-1}\frac{a}{c}$$
$$\theta_3 = 90^\circ$$
$$\theta_4 = \tan^{-1}\frac{a}{d}$$
$$\theta_5 = \tan^{-1}\frac{a}{e}$$

Fig 4.9 : *Calculation of angle of departure.*

Angle of departure and angle of arrival do not exist for real poles and zeros

Example :

Consider the two complex conjugate poles A and A* shown in fig 4.9.(If poles are complex then they exist only as conjugate pairs)

$$\left.\begin{array}{l}\text{Angle of departure}\\ \text{at pole A}\end{array}\right\} = 180^\circ - (\theta_1 + \theta_3 + \theta_5) + (\theta_2 + \theta_4)$$

$$\left.\begin{array}{l}\text{Angle of departure}\\ \text{at pole A}^*\end{array}\right\} = -\left[\text{Angle of departure at pole A}\right]$$

$$\left.\begin{array}{l}\text{Angle of arrival at a}\\ \text{complex zero A}\end{array}\right\} = 180^\circ - \left(\begin{array}{c}\text{Sum of angles of vectors to the}\\ \text{complex zero A from all the other zeros}\end{array}\right) + \left(\begin{array}{c}\text{Sum of angles of vectors to the}\\ \text{complex zero A from poles}\end{array}\right)$$

> **Note :** *The angles can be calculated as shown in fig 4.10 or they can be measured using protractor.*

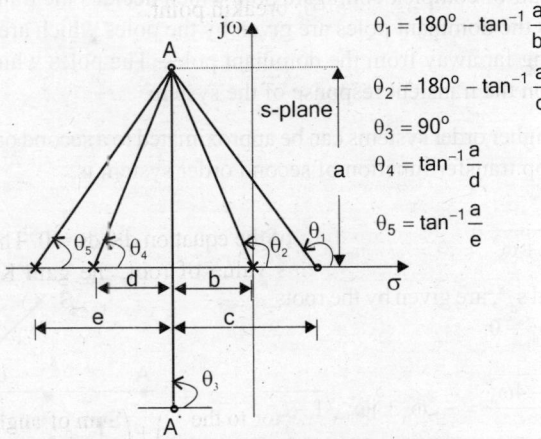

$$\theta_1 = 180^\circ - \tan^{-1}\frac{a}{b}$$

$$\theta_2 = 180^\circ - \tan^{-1}\frac{a}{c}$$

$$\theta_3 = 90^\circ$$

$$\theta_4 = \tan^{-1}\frac{a}{d}$$

$$\theta_5 = \tan^{-1}\frac{a}{e}$$

Fig 4.10 : *Calculation of angle of arrival.*

Example:

Consider the two complex conjugate zeros B and B* as shown in fig 4.10.(If zeros are complex then they exist only as conjugate pairs)

$$\left.\begin{array}{l}\text{Angle of arrival}\\ \text{at zero B}\end{array}\right\} = 180^\circ - (\theta_1 + \theta_3) + (\theta_2 + \theta_4 + \theta_5)$$

$$\left.\begin{array}{l}\text{Angle of arrival}\\ \text{at zero B}^*\end{array}\right\} = -\left[\text{Angle of arrival at zero B}\right]$$

Step 6 : Point of intersection of root locus with imaginary axis

The point where the root loci intersects the imaginary axis can be found by following three methods.

1. By Routh Hurwitz array.

2. By trial and error approach.

3. Letting s = jω in the characteristic equation and separate the real part and imaginary part. Two equations are obtained : *one by equating real part to zero and the other by equating imaginary part to zero.* Solve the two equations for ω and **K**. The values of ω gives the points where the root locus crosses imaginary axis. The value of K gives the value of gain K at there crossing points. Also this value of K is the limiting value of K for stability of the system.

Step 7 : Test points and root locus

Choose a test point. Using a protractor roughly estimate the angles of vectors drawn to this point and adjust the point to satisfy angle criterion. Repeat the procedure for few more test points. Sketch the root locus from the knowledge of typical sketches and the informations obtained in steps 1 through 6.

> **Note :** *In practice the approximate root locus can be sketched from the informations obtained in steps 1 through 6 and from the knowledge of typical sketches of root locus.*

DETERMINATION OF OPEN LOOP GAIN FOR A SPECIFIED DAMPING OF THE DOMINANT ROOTS

The dominant pole is a pair of complex conjugate pole which decides the transient response of the system. In higher order systems the dominant poles are given by the poles which are very close to origin, provided all other poles are lying far away from the dominant poles. The poles which are far away from the origin will have less effect on the transient response of the system.

The transfer function of higher order systems can be approximated to a second order transfer function. The standard form of closed loop transfer function of second order system is,

$$\frac{C(s)}{R(s)} = \frac{\omega_n^2}{s^2 + 2\zeta\omega_n s + \omega_n^2}$$

The dominant poles, s_d and s_d^*, are given by the roots of quadratic factor, $s^2 + 2\zeta\omega_n s + \omega_n^2 = 0$.

$$\therefore s_d = \frac{-2\zeta\omega_n \pm \sqrt{4\zeta^2\omega_n^2 - 4\omega_n^2}}{2} = -\zeta\omega_n \pm j\omega_n\sqrt{1-\zeta^2}$$

The dominant pole can be plotted on the s-plane as shown in fig 4.11.

In fig 4.11, the right angle triangle OAP,

$$\cos\alpha = \frac{\zeta\omega_n}{\omega_n} = \zeta, \qquad \therefore \alpha = \cos^{-1}\zeta$$

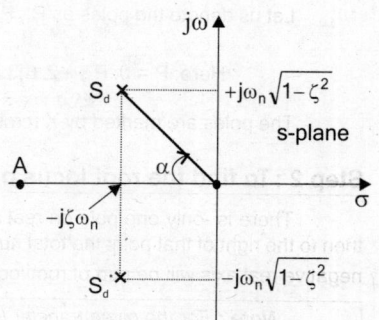

Fig 4.11 : Dominant pole, s_d.

To fix a dominant pole on root locus, draw a line at an angle of $\cos^{-1}\zeta$ with respect to negative real axis. The meeting point of this line with root locus will give the location of dominant pole. The value of K corresponding to dominant pole can be obtained from magnitude condition.

Let, K_{sd} be the value of gain at dominant pole s_d,

Now, $K_{sd} = \dfrac{\text{Product of length of vectors from open loop poles to dominant pole}}{\text{Product of length of vectors from open loop zeros to dominant pole}}$

Importance of root locus

The root locus technique is an important tool in designing control systems with desired performance characteristics. The desired performance of the system can be achieved by adjusting the location of its closed loop poles in the s-plane by varying one or more system parameters.

The root locus can be plotted in the s-plane by varying a system parameter (usually gain, K) over the complete range of values. The roots corresponding to a particular value of the system parameter can then be located on the locus or the value of the parameter for a desired root location can be determined from the locus.

The root locus technique is also used for stability analysis. Using root locus the range of values of K, for a stable system can be determined. It is also easier to study the relative stability of the system from the knowledge of location of closed loop poles. The dominant roots are used to estimate the damping ratio and natural frequency of oscillation of the system. From ζ and ω_n the time domain specifications can be calculated.

EXAMPLE 4.22

A unity feedback control system has an open loop transfer function, $G(s) = \dfrac{K}{s(s^2 + 4s + 13)}$. Sketch the root locus.

SOLUTION

Step 1 : To locate poles and zeros

The poles of open loop transfer function are the roots of the equation, $s(s^2 + 4s + 13) = 0$.

The roots of the quadratic are, $s = \dfrac{-4 \pm \sqrt{4^2 - 4 \times 13}}{2} = -2 \pm j3$

∴ The poles are lying at s = 0, –2 + j3 and –2 –j3.

Let us denote the poles as P_1, P_2, and P_3.

Here, $P_1 = 0$, $P_2 = -2 + j3$ and $P_3 = -2 -j3$.

The poles are marked by X (cross) as shown in fig 4.22.1.

Step 2 : To find the root locus on real axis

There is -only one pole on real axis at the origin. Hence if we choose any test point on the negative real axis then to the right of that point the total number of real poles and zeros is one, which is an odd number. Hence the entire negative real axis will be part of root·locus. The root locus on real axis is shown as a bold line in fig 4.22.1.

> **Note :** *For the given transfer function one root locus branch will start at the pole at the origin and meet the zero at infinity through the negative real axis.*

Step 3 : To find angles of asymptotes and centroid

Since there are 3 poles, the number of root locus branches are three. There is no finite zero. Hence all the three root locus branches ends at zeros at infinity. The number of asymptotes required are three.

Angle of asymptotes = $\dfrac{\pm 180°(2q + 1)}{n - m}$;   q = 0, 1, ...n – m

Here n = 3, and m = 0.   ∴ q = 0, 1, 2, 3 .

When q = 0,   Angles $= \pm \dfrac{180°}{3} = \pm 60°$

When q = 1,   Angles $= \pm \dfrac{180° \times 3}{3} = \pm 180°$

When q = 2,   Angles $= \pm \dfrac{180° \times 5}{3} = \pm 300° = \mp 60°$

When q = 3,   Angles $= \pm \dfrac{180° \times 7}{3} = \pm 420° = \pm 60°$

> **Note :** It is enough if you calculate the required number of angles. Here it is given by first three values of angles. The remaining values will be repetitions of the previous values.

Centroid = $\dfrac{\text{Sum of poles} - \text{Sum of zeros}}{n - m} = \dfrac{0 - 2 + j3 - 2 - j3 - 0}{3} = \dfrac{-4}{3} = -1.33$

The centroid is marked on real axis and from the centroid the angles of asymptotes are marked using a protractor. The asymptotes are drawn as dotted lines as shown in fig 4.22.1.

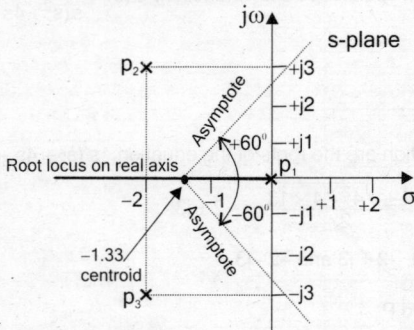

Fig 4.22.1 : *Figure showing the asymptote, root locus on real axis and location of poles and centroid.*

Step 4 : To find the breakaway and breakin points

The closed loop transfer function
$$\frac{C(s)}{R(s)} = \frac{G(s)}{1 + G(s)} = \frac{\dfrac{K}{s(s^2 + 4s + 13)}}{1 + \dfrac{K}{s(s^2 + 4s + 13)}} = \frac{K}{s(s^2 + 4s + 13) + K}$$

The characteristic equation is, $s(s^2 + 4s + 13) + K = 0$

$$\therefore \quad s^3 + 4s^2 + 13s + K = 0 \qquad \Rightarrow \qquad K = -s^3 - 4s^2 - 13s$$

On differentiating the equation of K with respect to s we get,

$$\frac{dK}{ds} = -(3s^2 + 8s + 13)$$

Put $\dfrac{dK}{ds} = 0$

$$\therefore -(3s^2 + 8s + 13) = 0 \quad \Rightarrow \quad (3s^2 + 8s + 13) = 0$$

$$\therefore \quad s = \frac{-8 \pm \sqrt{8^2 - 4 \times 13 \times 3}}{2 \times 3} = -1.33 \pm j1.6$$

Check for K : When, s = −1.33 + j1.6, the value of K is given by,

$$K = -(s^3 + 4s^2 + 13s) = -[(-1.33 + j1.6)^3 + 4(-1.33 + j1.6)^2 + 13(-1.33 + j1.6)]$$

$$\neq \text{ positive and real.}$$

Also it can be shown that when s = −1.33 − j1.6 the value of K is not equal to real and positive.

Since the values of K for, s = −1.33 ± j1.6, are not real and positive, these points are not an actual breakaway or breakin points. The root locus has neither breakaway nor breakin point.

Step 5 : To find the angle of departure

Let us consider the complex pole p_2 shown in fig 4.22.2. Draw vectors from all other poles to the pole p_2 as shown in fig 4.22.2. Let the angles of these vectors be θ_1 and θ_2.

Here, $\theta_1 = 180° - \tan^{-1}(3/2) = 123.7°$; $\theta_2 = 90°$

Angle of departure from the complex pole p_2 = $180° - (\theta_1 + \theta_2)$

$$= 180° - (123.7° + 90°)$$

$$= -33.7°$$

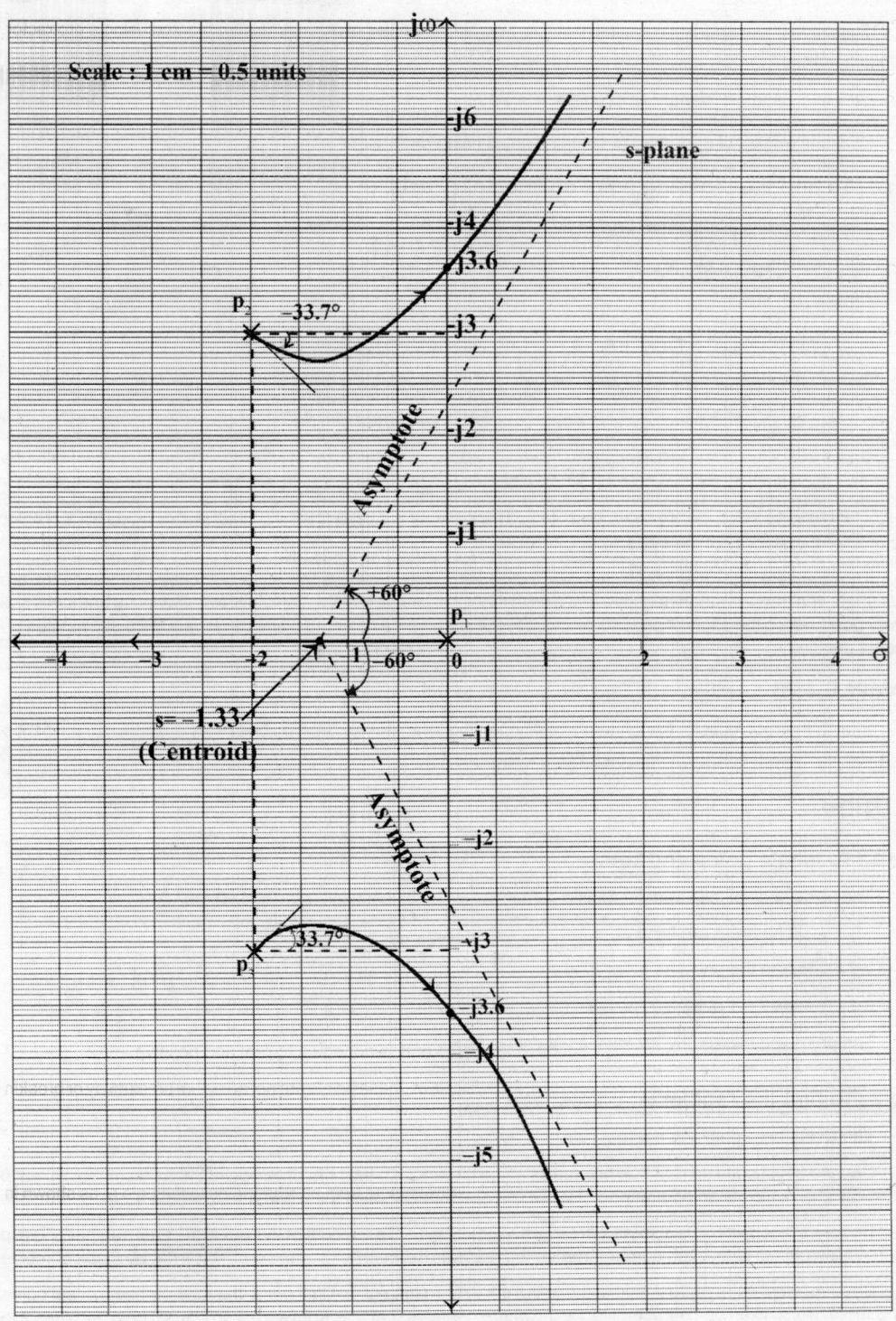

Fig 4.22.3 : *Root locus sketch of,* $1 + G(s) = 1 + \dfrac{K}{s(s^2 + 4s + 13)}$.

The angle of departure at complex pole p_3 is negative of the angle of departure at complex pole A.

$$\therefore \text{ Angle of departure at pole } p_3 = + 33.7°$$

Mark the angles of departure at complex poles using protractor.

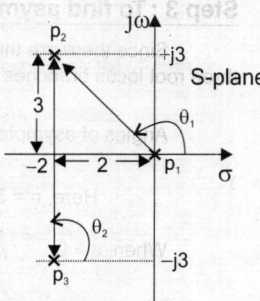

Fig 4.22.2

Step 6 : To find the crossing point on imaginary axis

The characteristic equation is given by,

$$s^3 + 4s^2 + 13s + K = 0$$

Put $s = j\omega$

$$(j\omega)^3 + 4(j\omega)^2 + 13(j\omega) + K = 0 \quad \Rightarrow \quad -j\omega^3 - 4\omega^2 + 13j\omega + K = 0$$

On equating imaginary part to zero, we get, | On equating real part to zero, we get,

$$-\omega^3 + 13\omega = 0$$ $$-4\omega^2 + K = 0$$

$$-\omega^3 = -13\omega$$ $$K = 4\omega^2$$

$$\omega^2 = 13 \quad \Rightarrow \quad \omega = \pm\sqrt{13} = \pm 3.6$$ $$= 4 \times 13 = 52$$

The crossing point of root locus is $\pm\, j3.6$. The value of K at this crossing point is K = 52. (This is the limiting value of K for the stability of the system).

The complete root locus sketch is shown in fig 4.22.3. The root locus has three branches one branch starts at the pole at origin and travel through negative real axis to meet the zero at infinity. The other two root locus branches starts at complex poles (along the angle of departure), crosses the imaginary axis at ± j3.6 and travel parallel to asymptotes to meet the zeros at infinity.

EXAMPLE 4.23

Sketch the root locus of the system whose open loop transfer function is, $G(s) = \dfrac{K}{s(s+2)(s+4)}$. Find the value of K so that the damping ratio of the closed loop system is 0.5.

SOLUTION

Step 1 : To locate poles and zeros

The poles of open loop transfer function are the roots of the equation, $s(s+2)(s+4) = 0$.

$$\therefore \text{The poles are lying at, } s = 0, -2, -4.$$

Let us denote the poles as p_1, p_2, and p_3.

Here, $p_1 = 0$, $p_2 = -2$, $p_3 = -4$.

The poles are marked by X(cross) as shown in fig 4.23.1.

Step 2 : To find the root locus on real axis

There are three poles on the real axis.

Choose a test point on real axis between s = 0 and s = –2. To the right of this point the total number of real poles and zeros is one, which is an odd number. Hence the real axis between s = 0 and s = –2 will be a part of root locus.

Choose a test point on real axis between s = –2 and s = – 4. To the right of this point, the total number of real poles and zeros is two which is an even number. Hence the real axis between s = –2 and s = – 4 will not be a part of root locus.

Choose a test point on real axis to the left of s = – 4. To the right of this point, the total number of real poles and zeros is three, which is an odd number. Hence the entire negative real axis from s = – 4 to – ∞ will be a part of root locus.

The root locus on real axis are shown as bold lines in fig 4.23.1.

Step 3 : To find asymptotes and centroid

Since there are three poles the number of root locus branches are three. There is no finite zero. Hence all the three root locus branches ends at zeros at infinity. The number of asymptotes required are three.

$$\text{Angles of asymptotes} = \frac{\pm 180°(2q+1)}{n-m} \quad ; \quad q = 0, 1, 2, \ldots n-m.$$

Here, n = 3 and m = 0. ∴ q = 0, 1, 2, 3 .

When q = 0, Angles $= \pm \dfrac{180°}{3} = \pm 60°$

When q = 1, Angles $= \pm \dfrac{180° \times 3}{3} = \pm 180°$

Note : *It is enough if you calculate the required number of angles. Here it is given by first three values of angles. The remaining values will be repetitions of the previous values.*

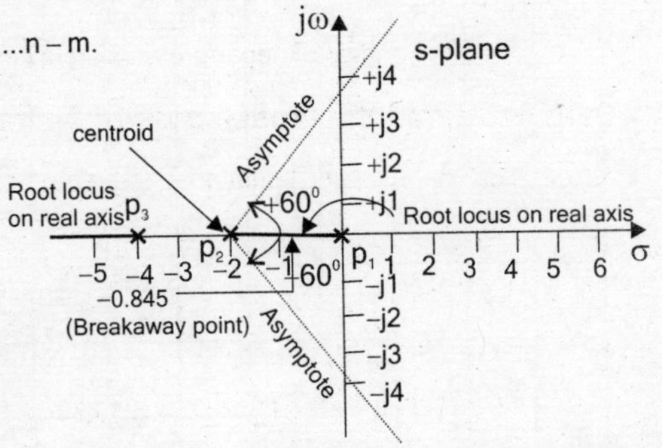

$$\text{Centroid} = \frac{\text{Sum of poles} - \text{Sum of zeros}}{n-m} = \frac{0-2-4-0}{3} = -2$$

Fig 4.23.1 : *Figure showing the asymptotes, root locus on real axis and location of poles, centroid, and breakaway points.*

The centroid is marked on real axis and from the centroid the angles of asymptotes are marked using a protractor. The asymptotes are drawn as dotted lines as shown in fig 4.23.1.

Step 4 : To find the breakaway and breakin points

The closed loop transfer function $\left.\begin{array}{c}\\\\\end{array}\right\}$ $\dfrac{C(s)}{R(s)} = \dfrac{G(s)}{1+G(s)} = \dfrac{\dfrac{K}{s(s+2)(s+4)}}{1+\dfrac{K}{s(s+2)(s+4)}} = \dfrac{K}{s(s+2)(s+4)+K}$

The characteristic equation is given by,

s (s + 2) (s + 4) + K = 0 $\Rightarrow$ s (s² + 6s + 8) + K = 0 $\Rightarrow$ $s^3 + 6s^2 + 8s + K = 0$

∴ $K = -s^3 - 6s^2 - 8s$

On differentiating the equation of K with respect to s we get,

$$\frac{dK}{ds} = -(3s^2 + 12s + 8)$$

Put $\dfrac{dK}{ds} = 0$

∴ $-(3s^2 + 12s + 8) = 0$ $\Rightarrow$ $(3s^2 + 12s + 8) = 0$

$$s = \frac{-12 \pm \sqrt{12^2 - 4 \times 3 \times 8}}{2 \times 3} = -0.845 \quad \text{or} \quad -3.154$$

Check for K : When s = -0.845, the value of K is given by,

$$K = -[(-0.845)^3 + 6(-0.845)^2 + 8(-0.845)] = 3.08$$

Since K, is positive and real for, s = -0.845, this point is actual breakaway point.

When s = -3.154, the value of K is given by,

$$K = -[(-3.154)^3 + 6(-3.154)^2 + 8(-3.154)] = -3.08$$

Since K, is negative for, s = -3.154, this is not a actual breakaway point.

The breakaway point is marked on the negative real axis as shown in fig 4.23.1.

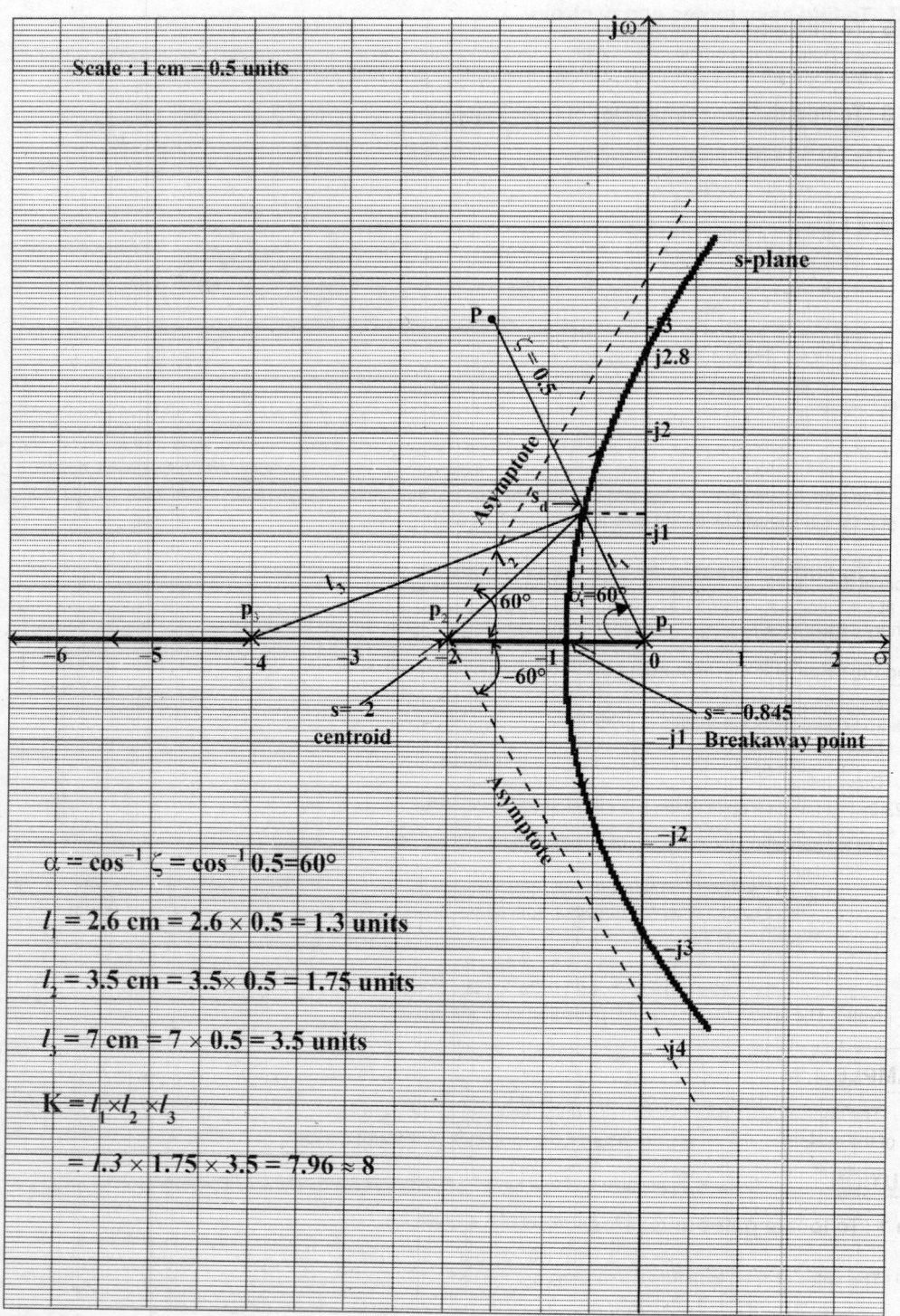

Scale : 1 cm = 0.5 units

s-plane

j3
j2.8
j2
j1

P

$\zeta = 0.5$

Asymptote

$-s_d$

60°

$\zeta = 60$

P_3 l_3 P_2 l_2 60° l_1 p_1
-6   -5   -4   -3   -2   -1   0   1   2

-60°

s= -2   s= -0.845
centroid   j1   Breakaway point

Asymptote

-j2

-j3

-j4

$\alpha = \cos^{-1} \zeta = \cos^{-1} 0.5 = 60°$

$l_1 = 2.6 \text{ cm} = 2.6 \times 0.5 = 1.3 \text{ units}$

$l_2 = 3.5 \text{ cm} = 3.5 \times 0.5 = 1.75 \text{ units}$

$l_3 = 7 \text{ cm} = 7 \times 0.5 = 3.5 \text{ units}$

$K = l_1 \times l_2 \times l_3$

$= 1.3 \times 1.75 \times 3.5 = 7.96 \approx 8$

Fig 4.23.2 : *Root locus sketch of*, $1 + G(s) = 1 + \dfrac{1}{s(s+2)(s+4)}$.

Step 5 : To find angle of departure

Since there are no complex pole or zero, we need not find angle of departure or arrival.

Step 6 : To find the crossing point of imaginary axis

The characteristic equation is given by,

$$s^3 + 6s^2 + 8s + K = 0$$

Put $s = j\omega$

$$(j\omega)^3 + 6(j\omega)^2 + 8(j\omega) + K = 0$$

$$-j\omega^3 - 6\omega^2 + j8\omega + K = 0$$

Equating imaginary part to zero

$$-j\omega^3 + j8\omega = 0$$

$$-j\omega^3 = -j8\omega$$

$$\omega^2 = 8 \qquad \omega = \pm\sqrt{8} = \pm 2.8$$

Equating real part to zero

$$-6\omega^2 + K = 0$$

$$K = 6\omega^2 = 6 \times 8 = 48$$

The crossing point of root locus is ± j2.8. The value of K corresponding to this point is K = 48. (This is the limiting value of K for the stability of the system).

The complete root locus sketch is shown in fig 4.23.2. The root locus has three branches. One branch starts at the pole at s = – 4 and travel through negative real axis to meet the zero at infinity. The other two root locus branches starts at s = 0 and s = – 2 and travel through negative real axis, breakaway from real axis at s = –0.845, then crosses imaginary axis at s= ± j2.8 and travel parallel to asymptotes to meet the zeros at infinity

To find the value of K corresponding to ζ = 0.5

Given that $\zeta = 0.5$

Let $\alpha = \cos^{-1}\zeta = \cos^{-1} 0.5 = 60°$

Draw a line OP, such that the angle between line OP and negative real axis is 60° ($\alpha = 60°$) as shown in fig 4.23.2. The meeting point of the line OP and root locus gives the dominant pole, s_d.

Let K_{sd} be value of K corresponding to the point $s = s_d$

$$K_{sd} = \frac{\text{Product of length of vector from all poles to the point, } s = s_d}{\text{Product of length of vector from all zeros to the point, } s = s_d}$$

$$= \frac{l_1 \times l_2 \times l_3}{1} = 1.3 \times 1.75 \times 3.5 = 7.96 \approx 8$$

Note : The length of vectors are measured to scale.

EXAMPLE 4.24

The open loop transfer function of a unity feedback system is given by, $G(s) = \dfrac{K(s+9)}{s(s^2+4s+11)}$. Sketch the root locus of the system.

SOLUTION

Step 1 : To locate poles and zeros

The poles of open loop transfer function are the roots of the equations, $(s^2 + 4s + 11) = 0$.

The roots of the quadratic are, $s = \dfrac{-4 \pm \sqrt{4^2 - 4 \times 11}}{2} = -2 \pm j2.64$

∴ The poles are lying at, s = 0, –2 + j2.64, –2 –j2.64

The zeros are lying at, s = –9 and infinity.

Let us denote the poles as p_1, p_2, p_3 finite zero by z_1.

Here, $p_1 = 0$, $p_2 = –2 + j2.64$, $p_3 = –2 – j2.64$ and $z_1 = –9$.

The poles are marked by X(cross) and zeros by "o" (circle) as shown in fig 4.24.1.

Step 2 : To find the root locus on real axis.

One pole and one zero lie on real axis.

Choose a test point to the left of s = 0, then to the right of this point, the total number of poles and zeros is one which is an odd number. Hence the portion of real axis from s = 0 to s = –9 will be a part of root locus.

If we choose a test point to the left of s = –9 then to the right of this point, the total number of poles and zeros is two, which is an even number. Hence the real axis from s = –9 to –∞ will not be a part of root locus.

The root locus on real axis is shown as a bold line in fig 4.24.1.

Step 3 : To find angles of asymptotes and centroid

Since there are 3 poles, the number of root locus branches are three. One root locus branch starts at the pole at origin and travel along negative real axis to meet the zero at s = –9. The other two root locus branches meet the zeros at infinity. The number of asymptotes required are two.

$$\text{Angle of asymptotes} = \frac{\pm 180^\circ (2q+1)}{n-m} \quad ; \quad q = 0, 1, 2, ... n-m.$$

Here, n = 3 and m = 1.   ∴ q = 0, 1, 2, 3 .

When q = 0, $\text{Angles} = \pm \dfrac{180^\circ}{2} = \pm 90^\circ$

When q = 1, $\text{Angles} = \pm \dfrac{180^\circ \times 3}{2} = \pm 270^\circ = \mp 90^\circ$

When q = 2, $\text{Angles} = \pm \dfrac{180^\circ \times 5}{2} = \pm 450^\circ = \mp 90^\circ$

> **Note :** It is enough if you calculate the required number of angles. Here it is given by first two values of angles. The remaining values will be repetitions of the previous values.

$$\text{Centroid} = \frac{\text{Sum of poles} - \text{Sum of zeros}}{n-m} = \frac{0-2+j2.64-2-j2.64-(-9)}{2} = 2.5$$

The centroid is marked and from the centroid, the angles of asymptotes are marked using a protractor. The asymptotes are drawn as dotted lines as shown 4.24.1.

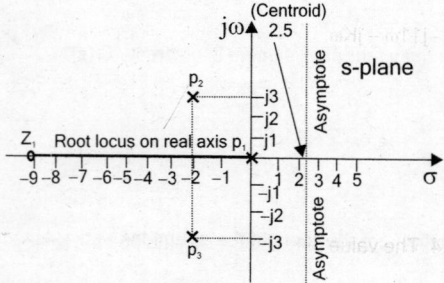

Fig 4.24.1 : *Figure showing the asymptotes, root locus on real axis and location of poles, zero and centroid.*

Step 4 : To find the breakaway and breakin points

From the location of poles and zero and from the knowledge of typical sketches of root locus, it can be concluded that there is no possibility of breakaway or breakin points.

Step 5 : To find the angle of departure

Let us consider the complex pole p_2 as shown in fig 4.24.2. Draw vectors from all other poles and zero to the pole p_2 as shown in fig 4.24.2. Let the angles of these vectors be θ_1, θ_2 and θ_3.

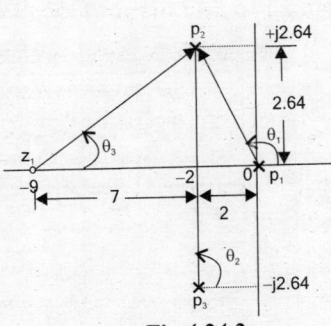

Here, $\theta_1 = 180° - \tan^{-1}\dfrac{2.64}{2} = 127.1°$

$\theta_2 = 90°$

$\theta_3 = \tan^{-1}\dfrac{2.64}{7} = 20.7°$

$$\left.\begin{array}{c}\text{Angle of departure from}\\\text{the complex pole } p_2\end{array}\right\} = 180° - (\theta_1 + \theta_2) + \theta_3$$

$$= 180° - (127.1° + 90°) + 20.7° = -16.4°$$

Fig 4.24.2

The angle of departure at the complex pole p_3 is negative of the angle of departure at complex pole p_3.

$\therefore$ Angle of departure at pole p_3 = $-(-16.4) = +16.4°$

Mark the angles of departure at complex poles using protractor.

Step 6 : To find the crossing point of imaginary axis

$$\left.\begin{array}{c}\text{The closed loop}\\\text{transfer function}\end{array}\right\} \frac{C(s)}{R(s)} = \frac{G(s)}{1+G(s)} = \frac{\dfrac{K(s+9)}{s(s^2+4s+11)}}{1 + \dfrac{K(s+9)}{s(s^2+4s+11)}} = \frac{K(s+9)}{s(s^2+4s+11) + K(s+9)}$$

The characteristic equation is the denominator polynomial of $C(s)/R(s)$.

$\therefore$ The characteristic equation is,

$$s(s^2+4s+11) + K(s+9) = 0 \implies (s^3 + 4s^2 + 11s) + Ks + 9K = 0$$

put $s = j\omega$

$$(j\omega)^3 + 4(j\omega)^2 + 11(j\omega) + K(j\omega) + 9K = 0 \implies -j\omega^3 - 4\omega^2 + j11\omega + jK\omega + 9K = 0$$

On equating imaginary part to zero,

$-j\omega^3 + j11\omega + jK\omega = 0 \implies -j\omega^3 = -j11\omega - jK\omega$

$\omega^2 = 11 + K$

Put $K = 8.8$, $\omega^2 = 11 + 8.8 = 19.8$

$\omega = \pm\sqrt{19.8} = \pm 4.4$

On equating real part to zero,

$-4\omega^2 + 9K = 0 \quad 9K = 4\omega^2$

Put, $\omega^2 = 11 + K \quad \therefore 9K = 4(11 + K) = 44 + 4K$

$\therefore 9K - 4K = 44$

$\therefore 5K = 44 \implies K = \dfrac{44}{5} = 8.8$

The crossing point of root locus is $\pm j4.4$. The value of K at this crossing point is K = 8.8 (This is the limiting value of K for the stability of the system).

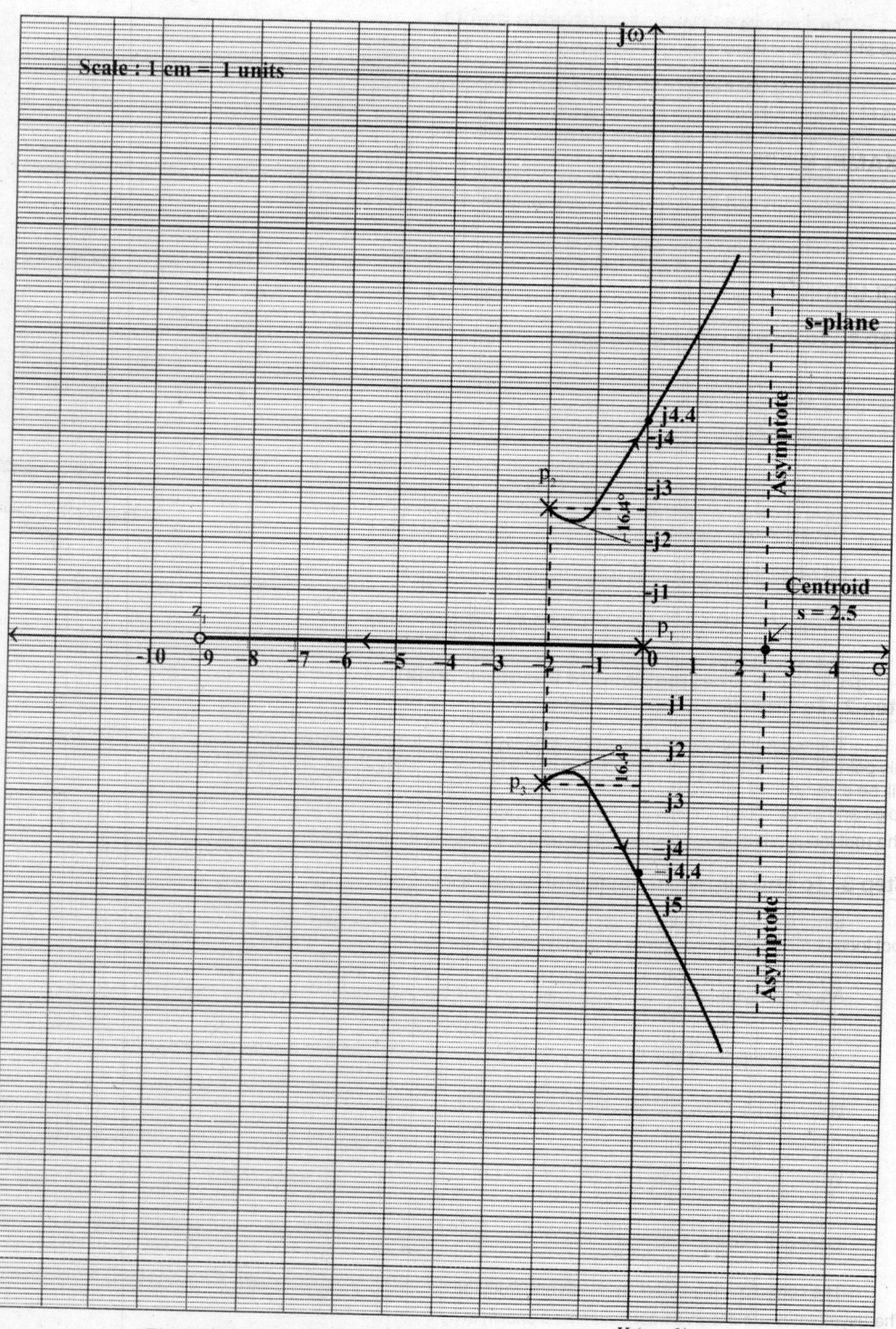

Fig 4.24.3. : *Root locus sketch for,* $1 + G(s) = 1 + \dfrac{K(s+9)}{s(s^2+4s+11)}$.

The complete root locus sketch is shown in fig 4.24.3. The root locus has three branches. One branch starts at pole at origin and travel through negative real axis to meet the zero at s = –9.

The other two root locus branches starts at complex poles (along the angle of departure) crosses the imaginary axis at ±j4.4 and travel parallel to asymptotes to meet the zeros at infinity.

EXAMPLE 4.25

Sketch the root locus for the unity feedback system whose open loop transfer function is,

$$G(s)\,H(s) = \frac{K}{s\,(s+4)\,(s^2+4s+20)}.$$

SOLUTION

Step 1 : To locate poles and zeros

The poles of open loop transfer function are the roots of the equation, s (s + 4) (s² + 4s + 20) = 0.

The roots of the quadratic are, $s = \dfrac{-4 \pm \sqrt{4^2 - 4 \times 1 \times 20}}{2} = -2 \pm j4$

∴ The poles are lying at, s = 0, –4, –2 + j4 and –2 –j4.

The zeros are lying at infinity.

Let us denote the poles as p_1, p_2, p_3 and p_4.

Here, p_1= 0, p_2= –4, p_3= –2 + j4, and p_4= –2 – j4.

The poles are marked by X (cross) as shown in fig 4.25.1.

Step 2 : To find root locus on real axis

There are two poles on the real axis. Choose a test point on real axis between s = 0 and s = –4. To the right of this point, the total number of real poles is one which is an odd number. Hence the real axis between s = 0 and s = –4 will be a part of root locus. Choose a test point to the left of s = –4, now to the right of this test point the total number of poles and zeros is two which is even number. Hence the real axis from s = –4 to s = –∞ will not be a part of root locus. The root locus on real axis is shown as a bold line in fig 4.25.1.

Step 3 : To find angles of asymptotes and centroid

Since there are four poles, the number of root locus branches are four. There is no finite zero. Hence all the four root locus branches ends at zeros at infinity. Hence the number of asymptotes required is four.

Angle of asymptotes = $\dfrac{\pm 180^\circ(2q+1)}{n-m}$; $\quad a = 0, 1, 2, \ldots .n-m$

Here, n = 4 and m = 0. ∴ q = 0, 1, 2, 3 .

When q = 0, Angles = $\pm \dfrac{180^\circ}{4} = \pm 45^\circ$

When q = 1, Angles = $\pm \dfrac{180^\circ \times 3}{4} = \pm 135^\circ$

> **Note :** *It is enough if you calculate the required number of angles. Here it is given by first four values of angles. The remaining will be repetitions of the previous values.*

Centroid = $\dfrac{\text{Sum of poles} - \text{Sum of zeros}}{n-m} = \dfrac{0-4-2+j4-2-j4-0}{4-0} = \dfrac{-8}{4} = -2$

The centroid is marked on real axis and from the centroid the angles of asymptotes are marked using a protractor. The asymptotes are drawn as dotted lines as shown in fig 4.25.1.

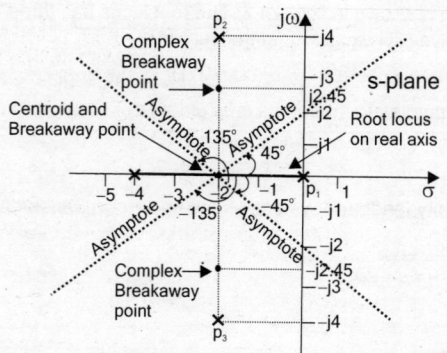

Fig 4.25.1 : *Figure showing the asymptotes, root locus on real axis and location of poles, centroid and breakaway points.*

Step 4 : To find the breakaway and breakin point

The closed loop $\left.\begin{array}{l}\text{The closed loop}\\\text{transfer function}\end{array}\right\}$ $\dfrac{C(s)}{R(s)} = \dfrac{G(s)}{1+G(s)} = \dfrac{\dfrac{K}{s(s+4)(s^2+4s+20)}}{1+\dfrac{K}{s(s+4)(s^2+4s+20)}} = \dfrac{K}{s(s+4)(s^2+4s+20)+K}$

The characteristic equation is, $s(s+4)(s^2+4s+20)+K = 0$.

$\therefore$ $K = -s(s+4)(s^2+4s+20) = -(s^2+4s)(s^2+4s+20)$

$\therefore$ $K = -(s^4+8s^3+36s^2+80s)$

On differentiating the equation of K with respect to s we get, $\dfrac{dK}{ds} = -(4s^3+24s^2+72s+80)$

To find the real root of = 0 by Lin's method.

The first trial divisor is chosen as the last two terms of the polynomial

Ist trial	IInd trial

Ist trial

Trial divisor $= 18s+20 = s+\dfrac{20}{18} = s+1.11$

$$
\begin{array}{r}
s^2+4.89s+12.57 \\
s+1.11\overline{\smash{\big)}\ s^3+6s^2+18s+20} \\
\underline{s^3+1.11s^2} \\
4.89s^2+18s \\
\underline{4.89s^2+5.43s} \\
\end{array}
$$

Next trial divisor $\rightarrow$ $12.57s+20$
                                              $\underline{12.57s+13.95}$
                                                         6.05

IInd trial

Trial divisor $= 12.57s+20 = s+\dfrac{20}{12.57} = s+1.59$

$$
\begin{array}{r}
s^2+4.41s+11 \\
s+1.59\overline{\smash{\big)}\ s^3+6s^2+18s+20} \\
\underline{s^3+1.59s^2} \\
4.41s^2+18s \\
\underline{4.41s^2+7s} \\
\end{array}
$$

Next trial divisor $\rightarrow$ $11s+20$
                                              $\underline{11s+17.49}$
                                                      2.51

IIIrd trial

Trial divisor = 11s + 20

$$= s + \frac{20}{11} = s + 1.82$$

$$
\begin{array}{r}
s^2 + 4.18s + 10.4 \\
s + 1.82\overline{\smash{\big)}\ s^3 + 6s^2 + 18s + 20} \\
\underline{s^3 + 1.82s^2} \\
4.18s^2 + 18s + 20 \\
\underline{4.18s^2 + 7.6s}
\end{array}
$$

Next trial divisor → 10.4s + 20

$$
\begin{array}{r}
10.4s + 18.9 \\
\hline
1.1
\end{array}
$$

Since the remainder converge for every trial, let us approximate the root to s = −2. On dividing the polynomial by s + 2, we found that (s+2) is a divisor of the polynomial.

$$
\begin{array}{r}
s^2 + 4s + 10 \\
s + 2\overline{\smash{\big)}\ s^3 + 6s^2 + 18s + 20} \\
\underline{s^3 + 2s^2} \\
4s^2 + 18s \\
\underline{4s^2 + 8s} \\
10s + 20 \\
\underline{10s + 20} \\
0
\end{array}
$$

Next trial divisor → 10s + 20

Put, $\dfrac{dK}{ds} = 0,$ $\therefore (-4s^3 + 24s^2 + 72s + 80) = 0$ $\Rightarrow$ $4s^3 + 24s^2 + 72s + 80 = 0$

On dividing by 4 we get, $s^3 - 6s^2 + 18s + 20 = 0$.

The equation $s^3 + 6s^2 + 18s + 20 = 0$ will have atleast one real root. By trial and error, the real root is found to be s = −2. (Refer Appendix II for Lin's method.)

The polynomial, $(s^3 + 6s^2 + 18s + 20) = 0$, can be expressed as,

$$s^3 + 6s^2 + 18s + 20 = (s + 2)(s^2 + 4s + 10) = 0$$

The root of the quadratic, $s^2 + 4s + 10 = 0,$ are given by,

$$s = \frac{-4 \pm \sqrt{4^2 - 4 \times 10}}{2} = -2 \pm j2.45$$

Check for K : When, $s = -2$, $K = -(s^4 + 8s^3 + 36s^2 + 80s) = -\left[(-2)^4 + 8 \times (-2)^3 + 36 \times (-2)^2 + 80 \times (-2)\right]$

$$= -[-64] = 64$$

When, $s = -2 \pm j2.45 = 3.16 \angle \pm 129°$

$$K = -(s^4 + 8s^3 + 36s^2 + 80s)$$

$$= -(3.16 \angle \pm 129°)^4 + 8 \times (3.16\angle \pm 129°)^3 + 36 (3.16\angle \pm 129°)^2 + 80 \times 3.16\angle \pm 129°$$

$$= - [99.7\angle \pm 156° + 252.4\angle \pm 27° + 359.5\angle \pm 258° + 252.8\angle \pm 129]$$

For positive values of angles,

$$K = -[-91 + j40 + 225 + j115 - 75 - j351 - 159 + j196] = -[-100] = 100$$

For negative values of angles,

$$K = -[-91 - j40 + 225 + j115 - 75 + j351 - 159 - j196] = -[-100] = 100$$

For all the roots of the equation dK/ds = 0, the value of K is positive and real. Hence all the three roots are actual breakaway points. The breakaway points are shown in fig 4.25.1.

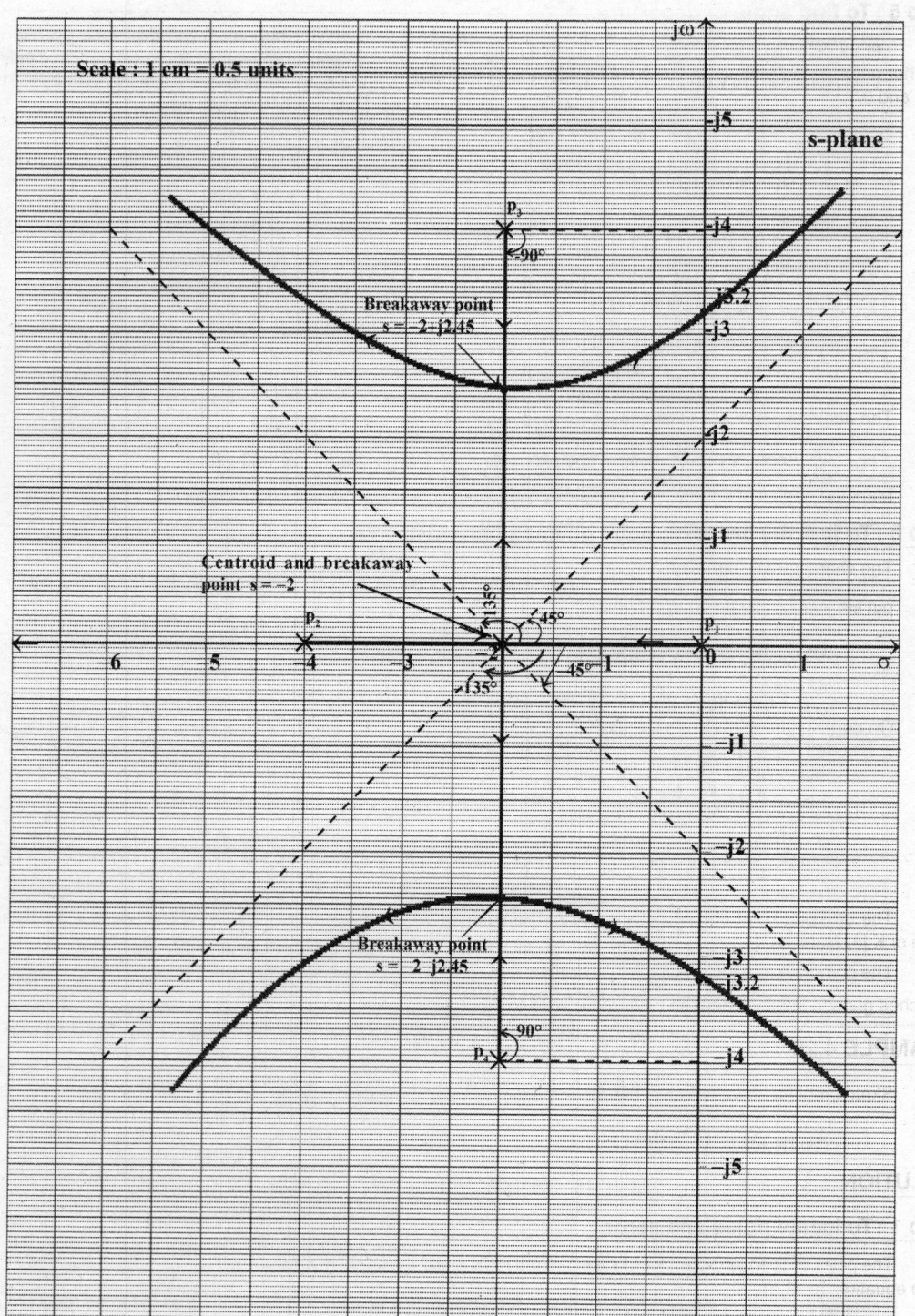

Fig 4.25.3 : *Root locus sketch of,* $1 + G(s) = 1 + \dfrac{K}{s(s+4)(s^2+4s+20)}$.

Step 5 : To find angle of departure

Let us consider the complex pole p_3 shown in fig 4.25.2. Draw vectors from all other poles to the pole p_3 as shown in fig 4.25.2. Let angles of these vectors be θ_1, θ_2 and θ_3.

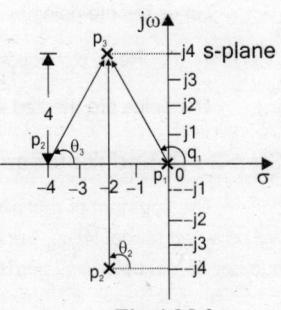

Fig 4.25.2

Here,

$$\theta_1 = 180° - \tan^{-1}\frac{4}{2} = 117°$$

$$\theta_2 = 90°$$

$$\theta_3 = \tan^{-1}\frac{4}{2} = 63°$$

$$\left.\begin{array}{l}\text{Angle of departure} \\ \text{from complex pole } p_3\end{array}\right\} = 180° - (\theta_1 + \theta_2 + \theta_3)$$

$$= 180° - (117° + 90° + 63°) = -90°$$

The angle of departure at complex pole p_4 is negative of the angle of departure at complex pole p_3.

∴ Angle of departure from complex pole p_4 = +90°

Mark the angles of departure at complex poles using protractor.

Step 6 : To find the crossing point on imaginary axis

The characteristic equation is given by, $s^4 + 8s^3 + 36s^2 + 80s + K = 0$.

put $s = j\omega$,

$$(j\omega)^4 + 8(j\omega)^3 + 36(j\omega)^2 + 80(j\omega) + K = 0.$$

$$\omega^4 - j8\omega^3 - 36\omega^2 + j80\omega + K = 0.$$

On equating imaginary part to zero,

$$-j8\omega^3 + j80\omega = 0$$

$$-j8\omega^3 = -j80\omega$$

$$\omega^2 = 10$$

$$\omega = \pm\sqrt{10} = \pm 3.2$$

On equating real part to zero,

$$\omega^4 - 36\omega^2 + K = 0$$

$$K = -\omega^4 + 36\omega^2$$

Put $\omega^2 = 10$

$$\therefore K = -(10)^2 + (36 \times 10) = 260$$

The crossing point of root locus is ±j3.2. The value of K at this crossing point is K = 260. (This is the limiting value of K for stability).

The complete root locus is sketched as shown in fig 4.25.3. The root locus has four branches. All the root locus branches goes to infinity along the asymptotic lines to meet the zeros at infinity.

EXAMPLE 4.26

Sketch root locus for the unity feedback system whose open loop transfer function is,

$$G(s)H(s) = \frac{K(s+1.5)}{s(s+1)(s+5)}.$$

SOLUTION

Step 1 : To locate poles and zeros

The poles of open loop transfer function are the roots of the equation, $s(s+1)(s+5) = 0$ and the zeros are the roots of the equation, $(s+1.5) = 0$.

The poles are lying at, $s = 0, -1, -5$.

The zeros are lying at, $s = -1.5$ and infinity.

Let us denote poles by, p_1, p_2, p_3 and finite zero by z_1.

Here, $p_1 = 0$, $p_2 = -1$, $p_3 = -5$ and $z_1 = -1.5$.

The poles are marked by X(cross) and zeros by "o" (circle) as shown in fig 4.26.1.

Step 2 : To find root locus on real axis

The segment of real axis between s = 0 and s = -1 and the segment of real axis between s = -1.5 and s = -5 will be a part of root locus. Because if we choose a test point in this segment then to the right of this point we have odd number of real poles and zeros. The root locus on real axis are shown as bold lines in fig 4.26.1.

Step 3 : To find angles of asymptotes and centroid

Since there are three poles, the number of root locus branches are three. There is one finite zero, so one root locus branch will end at finite zero. The other two branches will meet the zeros at infinity. Hence the number of asymptotes required is two.

$$\text{Angle of asymptotes} = \frac{\pm 180^\circ (2q + 1)}{n - m} \ ; \quad q = 0, 1, 2, \ldots\ldots n - m$$

Here, n = 3 and m = 1. $\therefore$ q = 0, 1, 2 .

When q = 0, $\text{Angles} = \pm \dfrac{180^\circ}{2} = \pm 90^\circ$

$$\text{Centroid} = \frac{\text{Sum of poles} - \text{Sum of zeros}}{n - m} = \frac{0 - 1 - 5 - (-1.5)}{2} = -2.25$$

The centroid is marked on real axis and from the centroid the angles of asymptotes are marked using a protractor. The asymptotes are drawn as dotted lines as shown in fig 4.26.1.

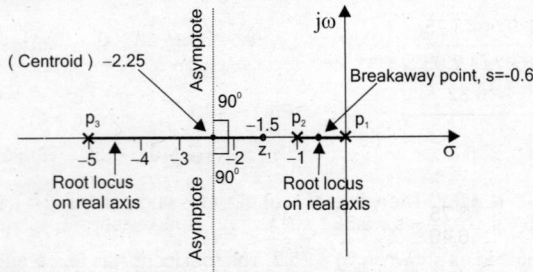

Fig 4.26.1 : *Figure showing the asymptotes, root locus on real axis and location of poles, zeros, centroid and breakaway points.*

Step 4 : To find the breakaway and breakin points

The closed loop transfer function $\left.\right\}$ $\dfrac{C(s)}{R(s)} = \dfrac{G(s)}{1 + G(s)} = \dfrac{\dfrac{K(s + 1.5)}{s(s + 1)(s + 5)}}{1 + \dfrac{K(s + 1.5)}{s(s + 1)(s + 5)}} = \dfrac{K(s + 1.5)}{s(s + 1)(s + 5) + K(s + 1.5)}$

The characteristic equation is, $s(s + 1)(s + 5) + K(s + 1.5) = 0$

$$\therefore K = \frac{-s(s + 1)(s + 5)}{s + 1.5} = \frac{-s(s^2 + 6s + 5)}{s + 1.5} = \frac{-(s^3 + 6s^2 + 5s)}{s + 1.5}$$

On differentiating K with respect to s we get,

$$\frac{dK}{ds} = \frac{-(3s^2 + 12s + 5)(s + 1.5) - [-(s^3 + 6s^2 + 5s)](1)}{(s + 1.5)^2}$$

$$= \frac{-3s^3 - 4.5s^2 - 12s^2 - 18s - 5s - 7.5 + s^3 + 6s^2 + 5s}{(s + 1.5)^2}$$

$$= \frac{-2s^3 - 10.5s^2 - 18s - 7.5}{(s + 1.5)^2} = \frac{-2(s^3 + 5.25s^2 + 9s + 3.75)}{(s + 1.5)^2}$$

For $\frac{dK}{ds} = 0$, the numerator should be zero.

$$\therefore s^3 + 5.25s^2 + 9s + 3.75 = 0$$

The third order polynomia will have one real root. The real root of the above polynomial can be determined by Lin's method. (Refer Appendix II).

To find the real root of, $s^3 + 5.25s^2 + 9s + 3.75 = 0$, by Lin's method

The last two terms of the polynomial are chosen as Ist trial divisor.

Ist trial

Ist Trial divisor = $9s + 3.75 = s + \frac{3.75}{9} = s + 0.42$

$$\begin{array}{r} s^2 + 4.83s + 6.97 \\ s + 0.42 \overline{\smash{\big)}\ s^3 + 5.25s^2 - 9s + 3.75} \\ s^3 + 0.42s^2 \\ \hline 4.83s^2 + 9s \\ 4.83s^2 + 2.03s \\ \hline \end{array}$$

Ist trial divisor → $\boxed{6.97s + 3.75}$
$$6.97s + 2.93$$
$$\overline{\quad 0.82 \quad}$$

IInd trial

IInd Trial divisor = $6.97s + 3.75 = s + \frac{3.75}{6.97} = s + 0.54$

$$\begin{array}{r} s^2 + 4.71s + 6.46 \\ s + 0.54 \overline{\smash{\big)}\ s^3 + 5.25s^2 + 9s + 3.75} \\ s^3 + 0.54s^2 \\ \hline 4.71s^2 + 9s \\ 4.71s^2 + 2.54s \\ \hline \end{array}$$

IInd trial divisor → $\boxed{6.46s + 3.75}$
$$6.46s + 3.49$$
$$\overline{\quad 0.26 \quad}$$

IIIrd trial

IIIrd Trial divisor = $6.46s + 3.75 = s + \frac{3.75}{6.46} = s + 0.58$

$$\begin{array}{r} s^2 + 4.67s + 6.3 \\ s + 0.58 \overline{\smash{\big)}\ s^3 + 5.25s^2 + 9s + 3.75} \\ s^3 + 0.58s^2 \\ \hline 4.67s^2 + 9s \\ 4.67s^2 + 2.7s \\ \hline \end{array}$$

IVth trial divisor → $\boxed{6.3s + 3.75}$
$$6.3s + 3.65$$
$$\overline{\quad 0.1 \quad}$$

IVth trial

IVth Trial divisor = $6.3s + 3.75 = s + \frac{3.75}{6.3} = s + 0.6$

$$\begin{array}{r} s^2 + 4.65s + 6.2 \\ s + 0.6 \overline{\smash{\big)}\ s^3 + 5.25s^2 + 9s + 3.75} \\ s^3 + 0.6s^2 \\ \hline 4.65s^2 + 9s \\ 4.65s^2 + 2.8s \\ \hline \end{array}$$

$$\boxed{6.2s + 3.75}$$
$$6.2s + 3.72$$
$$\overline{\quad 0.03 \quad}$$

On neglecting the small value of 0.03, one of the root of the polynomial is, s = -0.6.

The polynomial, $s^3 + 5.25s^2 + 9s + 3.75 = 0$, can be expressed,

$$s^3 + 5.25s^2 + 9s + 3.75 = (s + 0.6)(s^2 + 4.65s + 6.2) = 0.$$

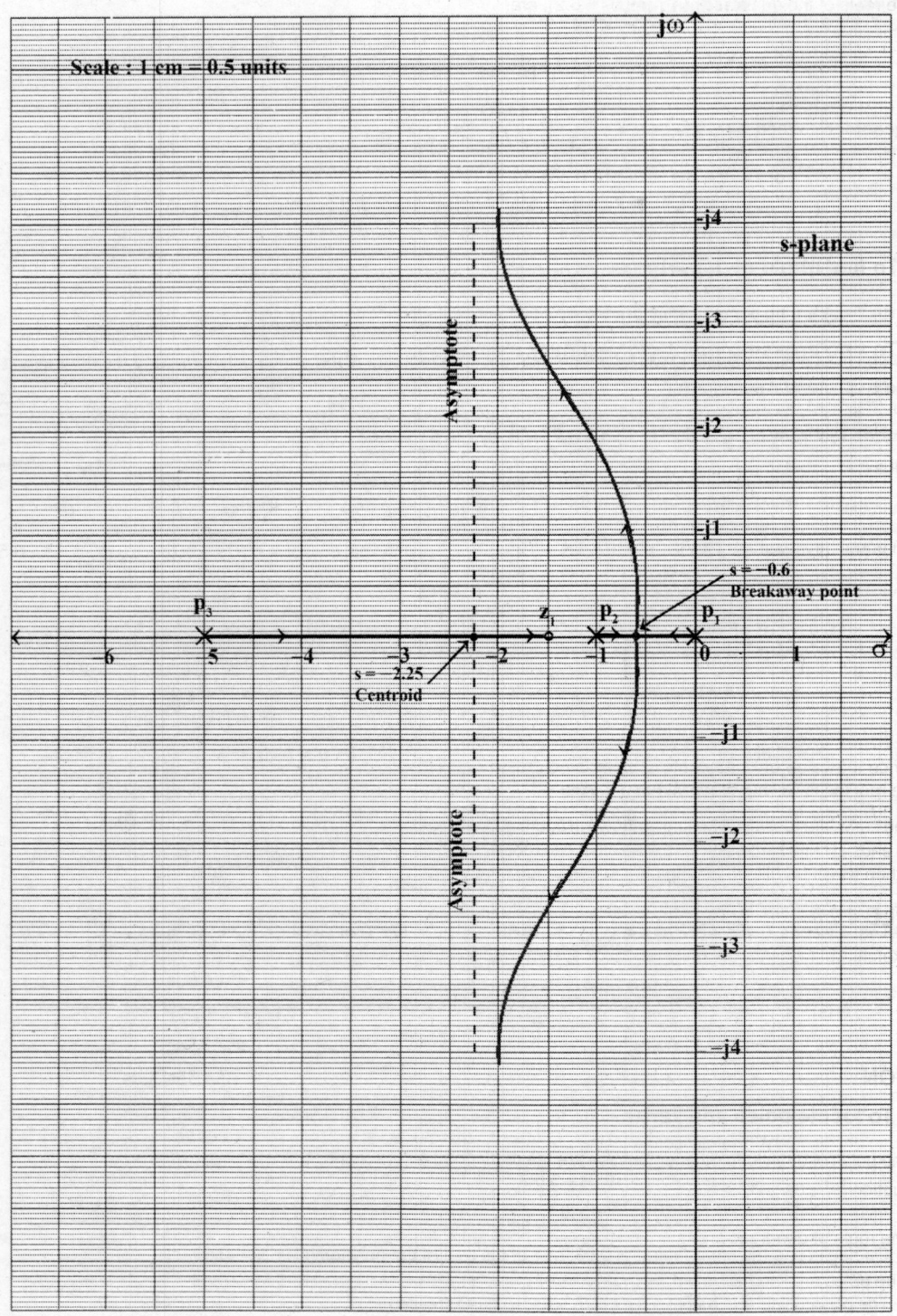

Fig 4.26.2 : *Root locus sketch of* $1 + G(s) = 1 + \dfrac{K(s + 1.5)}{s(s + 1)(s + 5)}$.

The roots of the quadratic, $(s^2 + 4.65s + 6.2)$, are,

$$s = \frac{-4.65 \pm \sqrt{4.65^2 - 4 \times 6.2}}{2} = -2.3 \pm j0.89$$

Check for K : When $s = -0.6$, $K = \frac{-(s^3 + 6s^2 + 5s)}{s + 1.5} = \frac{-[(-0.6)^3 + 6(-0.6)^2 + 5(-0.6)]}{-0.6 + 1.5} = 1.17$

For $s = -0.6$, the value of K is positive and real and so it is actual breakaway point. It can be shown that for $s = -2.3 \pm j0.36$ the value of K is not positive and real and so they cannot be breakaway points. The actual breakaway point is shown in fig 4.26.1.

Step 5 : To find angle of departure

Since there are no complex pole or zero we need not find angle of departure or arrival.

Step 6 : To find crossing point of imaginary axis.

The characteristic equation is,

$$s(s + 1)(s + 5) + K(s + 1.5) = 0 \implies s(s^2 + 6s + 5) + Ks + 1.5K = 0 \implies s^3 + 6s^2 + 5s + Ks + 1.5K = 0$$

Put $s = j\omega$

$$(j\omega)^3 + 6(j\omega)^2 + 5(j\omega) + K(j\omega) + 1.5K = 0 \implies -j\omega^3 - 6\omega^2 + j5\omega + jK\omega + 1.5K = 0$$

On equating imaginary part to zero, we get,

$$-j\omega^3 + j5\omega + jK\omega = 0$$

$$-j\omega^3 = -j5\omega - jK\omega$$

$$\omega^2 = 5 + K$$

On equating real part to zero we get,

$$-6\omega^2 + 1.5K = 0$$

Put $\omega^2 = 5 + K$

$$-6(5 + K) + 1.5K = 0$$

$$-30 - 4.5K = 0$$

$$-4.5K = 30 \implies K = -\frac{30}{4.5} = -6.67$$

Since the value of K is negative, there is no crossing point on imaginary axis, or for any positive values of K, and so the root locus will not cross imaginary axis.

The complete root locus sketch is shown in figure 4.26.2. The root locus has three branches. One branch starts at $s = -5$ and ends at finite zero at $s = -1.5$. The other two root locus starts at $s = 0$ and $s = -1$ and breakaway from real axis at $s = -0.6$, then travel parallel to asymptotes to meet the zeros at infinity.

EXAMPLE 4.27

Sketch root locus for the unity feedback system whose open loop transfer function is,

$$G(s) = \frac{K(s^2 + 6s + 25)}{s(s + 1)(s + 2)}$$

SOLUTION

Step 1 : To locate poles and zeros

The poles of open loop transfer function are the roots of the equation $s(s+1)(s+2) = 0$ and the zeros are the roots of the equation $(s^2 + 6s + 25) = 0$.

The roots of quadratic are, $s = \frac{-6 \pm \sqrt{6^2 - 4 \times 25}}{2} = -3 \pm j4$

The poles are lying at, $s = 0, -1, -2$

The zeros are lying at, $s = -3 + j4, -3 - j4$.

Let us denote poles by p_1, p_2, p_3 and zeros by z_1, z_2.

Here, $p_1 = 0$, $p_2 = -1$, $p_3 = -2$, $z_1 = -3 + j4$, $z_2 = -3 - j4$.

The poles are marked by X (cross) and zeros by "o" (circle) as shown in fig 4.27.1.

Step 2 : To find root locus on real axis

The segment of real axis between s = 0 and s = -1 and the entire negative real axis from s = -2 will be part of root locus. Because if we choose a test point in this segment then to the right of this point we have odd number of real poles and zeros. The root locus on real axis are shown as a bold line in fig 4.27.1.

Step 3 : To find angles of asymptotes and centroid

Since there are three poles the number of root locus branches are three. There are two finite zeros, so two root locus branch will end at finite zeros. The third root locus will meet the zero at infinity by travelling through negative real axis. Here the number of asymptote is one and the angle of asymptote is ±180°.

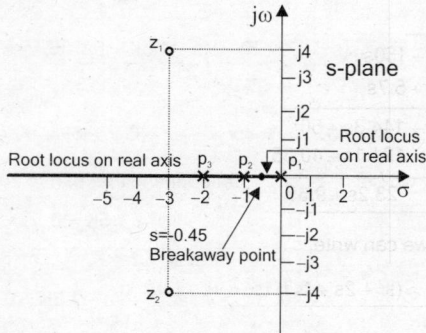

*Fig 4.27.1 : Figure showing the root locus on real axis
location of poles, zeros and breakaway points.*

Step 4 : To find the breakaway and breakin points

The closed loop transfer function $\left.\right\}$
$$\frac{C(s)}{R(s)} = \frac{G(s)}{1 + G(s)} = \frac{\dfrac{K(s^2 + 6s + 25)}{s(s+1)(s+2)}}{1 + \dfrac{K(s^2 + 6s + 25)}{s(s+1)(s+2)}} = \frac{K(s^2 + 6s + 25)}{s(s+1)(s+2) + K(s^2 + 6s + 25)}$$

The characteristic equation is, $s(s+1)(s+2) + K(s^2 + 6s + 25) = 0$.

$$\therefore K = \frac{-s(s+1)(s+2)}{s^2 + 6s + 25} = \frac{-s(s^2 + 3s + 2)}{s^2 + 6s + 25} = \frac{-s^3 - 3s^2 - 2s}{s^2 + 6s + 25}$$

On differentiating K with respect to s we get,

$$\frac{dK}{ds} = \frac{(-3s^2 - 6s - 2)(s^2 + 6s + 25) - (-s^3 - 3s^2 - 2s)(2s + 6)}{(s^2 + 6s + 25)^2}$$

$$= \frac{\begin{array}{c} -3s^4 - 18s^3 - 75s^2 - 6s^3 - 36s^2 - 150s - 2s^2 - 12s - 50 \\ + 2s^4 + 6s^3 + 6s^3 + 18s^2 + 4s^2 + 12s \end{array}}{(s^2 + 6s + 25)^2} = \frac{-(s^4 + 12s^3 + 91s^2 + 150s + 50)}{(s^2 + 6s + 25)^2}$$

For $\dfrac{dK}{ds} = 0$, the numerator should be zero.

$$\therefore s^4 + 12s^3 + 91s^2 + 150s + 50 = 0$$

The fourth order polynomial can be split into two quadratic equations. The two quadratic factors can be obtained by Lin's method. (Refer Appendix-II).

To find quadratic factors by Lin's Method.

The first trial divisor be the last three terms

Ist trial

IInd trial

Ist Trial divisor = $91s^2 + 150s + 50$

$$= s^2 + \frac{150}{91}s + \frac{50}{91} = s^2 + 1.65s + 0.55$$

IInd Trial divisor = $73.37s^2 + 144.3s + 50$

$$= s^2 + \frac{144.3}{73.37}s + \frac{50}{73.37} = s^2 + 2s + 0.7$$

$$
\begin{array}{r}
s^2 + 10.35s + 73.37 \\
s^2 + 1.65s + 0.55 \overline{\big)\, s^4 + 12s^3 + 91s^2 + 150s + 50} \\
s^4 + 1.65s^3 + 0.55s^2 \\
\hline
10.35s^3 + 90.45s^2 + 150s \\
10.35s^3 + 17.08s^2 + 5.7s \\
\hline
\end{array}
$$

IInd trial divisor $\rightarrow$ $73.37s^2 + 144.3s + 50$
$73.37s^2 + 121.1s + 40.35$

$\overline{23.2s + 9.65}$

$$
\begin{array}{r}
s^2 + 10s + 70.3 \\
s^2 + 2s + 0.7 \overline{\big)\, s^4 + 12s^3 + 91s^2 + 150s + 50} \\
s^4 + 2s^3 + 0.7s^2 \\
\hline
10s^3 + 90.3s^2 + 150s \\
10s^3 + 20s^2 + 7s \\
\hline
70.3s^2 + 143s + 50 \\
70.3s^2 + 140.6s + 49.2 \\
\hline
2.4s + 0.8
\end{array}
$$

On neglecting the small remainder we can write,

$$s^4 + 12s^3 + 91s^2 + 150s + 50 \approx (s^2 + 2s + 0.7)(s^2 + 10s + 70.3)$$

The roots of the quadratic, $s^2 + 2s + 0.7 = 0$, are,

$$s = \frac{-2 \pm \sqrt{2^2 - 4 \times 0.7}}{2} = -0.45, -1.55$$

The roots of the quadratic, $s^2 + 10s + 70.3 = 0$, are,

$$s = \frac{-10 \pm \sqrt{10^2 - 4 \times 70.3}}{2} = -5 \pm j6.73$$

Here, $s = -1.55$ is not a point on root locus, hence it cannot be a breakaway point.

Check the other three values for actual breakaway point.

When $s = -0.45$, $\quad K = \dfrac{-s^3 - 3s^2 - 2s}{s^2 + 6s + 25} = \dfrac{-(-0.45)^3 - 3(-0.45)^2 - 2(-0.45)}{(-0.45)^2 + 6(-0.45) + 25} = 0.017$

For $s = -0.45$, the value of K is positive and real and so it is actual breakaway point. It can be shown that for $s = -5 \pm j6.73$ the value of K is not positive and real and so they cannot be breakaway points. The actual breakaway point is shown in fig 4.27.1.

Step 5 : To find angle of arrival

Let us consider the complex zero z_1 shown in fig 4.27.2. Draw vectors from all other poles and zero to the zero z_1 as shown in fig 4.27.2. Let the angles of these vectors be θ_1, θ_2, θ_3 and θ_4.

Here, $\theta_1 = 180° - \tan^{-1}\dfrac{4}{3} = 126.9°$

$\theta_2 = 180° - \tan^{-1}\dfrac{4}{2} = 116.6°$

$\theta_3 = 180° - \tan^{-1}\dfrac{4}{1} = 104°$

$\theta_4 = 90°$

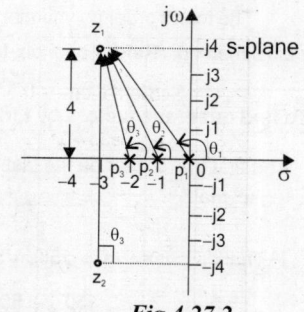

Fig 4.27.2

$\left.\begin{array}{l}\text{Angle of arrival at} \\ \text{complex zero } z_1\end{array}\right\} = 180° - (\theta_4) + (\theta_1 + \theta_2 + \theta_3)$

$= 180° - 90° + 126.9° + 116.6° + 104°$

$= 437.5° = 77.5°$

Angle of arrival at complex zero z_2 is negative of the angle of arrival at complex zero z_1.

$\left.\begin{array}{l}\text{Angle of arrival at} \\ \text{complex zero } z_2\end{array}\right\} = -77.5°$

Mark the angles of arrival at complex zeros using protractor.

Step 6 : To find the crossing point on imaginary axis

The characteristic equation is,

$s(s+1)(s+2) + K(s^2 + 6s + 25) = 0$

$s(s^2 + 3s + 2) + Ks^2 + 6Ks + 25K = 0$

$s^3 + 3s^2 + 2s + Ks^2 + 6Ks + 25K = 0$

$s^3 + (3+K)s^2 + (2+6K)s + 25K = 0$

Put $s = j\omega$.

$(j\omega)^3 + (3+K)(j\omega)^2 + (2+6K)(j\omega) + 25K = 0 \quad \Rightarrow \quad -j\omega^3 - (3+K)\omega^2 + j(2+6K)\omega + 25K = 0$

On equating imaginary part to zero

$-j\omega^3 + j(2+6K)\omega = 0$

$-j\omega^3 = -j(2+6K)\omega$

$\omega^2 = (2+6K)$

On equating real part to zero

$-(3+K)\omega^2 + 25K = 0$

Put $\omega^2 = 2 + 6K$

$-(3+K)(2+6K) + 25K = 0$

$-(6 + 12K + 2K + 6K^2) + 25K = 0$

$-6K^2 + 5K - 6 = 0$

$K = \dfrac{-5 \pm \sqrt{5^2 - 4 \times (-6)(-6)}}{2 \times (-6)} = 0.4 \pm j0.9$

Since the value of K is not real and positive, there is no crossing point on imaginary axis, or for any positive values of K the root locus will not cross imaginary axis.

Step 7 : To find points on root locus

Choose test points a, b, c, d on the s-plane and adjust the test points to satisfy angle criterion. The test points are shown in fig 4.27.3. On the upper half of s-plane the root locus is sketched through the test points a, b, c and d. The root locus on the lower half of s-plane is the mirror image of the root locus on the upper half of s-plane.

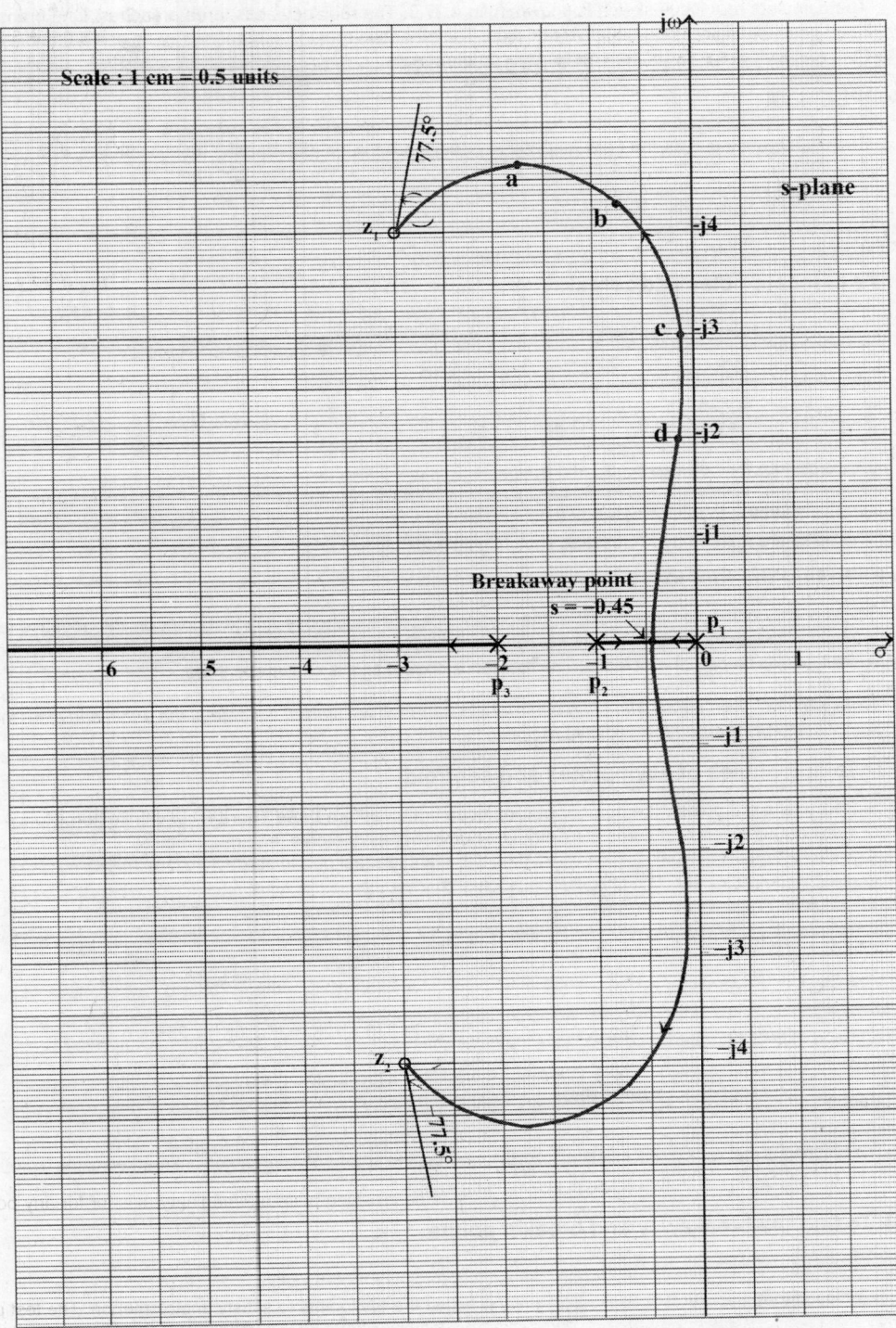

Fig 4.27.3 : *Root locus sketch of,* $1 + G(s) = 1 + \dfrac{K(s^2 + 6s + 25)}{s(s+1)(s+2)}$.

The complete root locus sketch is shown in fig 4.27.3. The root locus has three branches. One branch starts at s = -2 and goes to infinity along negative real axis. The other two root locus branches starts at s = 0 and s = -1 and breaks from real axis at s = -0.45, then meets the complex zeros.

EXAMPLE 4.28

Sketch the root locus for the unity feedback system whose open loop transfer function is,

$$G(s) = \frac{K}{s(s^2 + 6s + 10)}.$$

SOLUTION

Step 1 : To locate poles and zeros

The poles of open loop transfer function are the roots of the equation, $s(s^2 + 6s + 10) = 0$.

The roots of the quadratic are, $s = \dfrac{-6 \pm \sqrt{6^2 - 4 \times 10}}{2} = -3 \pm j1$

The poles are lying at, s= 0, -3 + j1 and -3 - j1

Let us denote the poles as p_1, p_2, and p_3.

Here, $p_1 = 0$, $p_2 = -3 + j1$, and $p_3 = -3 - j1$.

The poles are marked by X(cross) as shown in fig 4.28.1.

Step 2 : To find the root locus on real axis

There is only one pole on real axis at the origin. Hence if we choose any test point on the negative real axis then to the right of that point the total number of real poles and zeros is one, which is an odd number. Hence the entire negative real axis will be part of root locus. The root locus on real axis is shown are three.

> **Note :** For the given transfer function one root locus branch will start at the pole at the origin and meet the zero at infinity through the negative real axis.

Step 3 : To find angles of asymptotes and centroid

Since there are 3 poles , the number of root locus branches are three. There is no infinite zero. Hence all the three root locus branches ends at zeros at infinity. The number of asymptotes required are three

Angle of asymptotes = $\dfrac{\pm 180^\circ (2q + 1)}{n - m}$;   q = 0, 1, 2, ...n - m.

Here, n = 3 and m = 0.   ∴ q = 0, 1, 2, 3,......

When q = 0,   Angles $= \pm \dfrac{180^\circ}{3} = \pm 60^\circ$

When q = 1,   Angles $= \pm \dfrac{180^\circ \times 3}{3} = \pm 180^\circ$

Centroid = $\dfrac{\text{sum of poles} - \text{sum of zeros}}{n - m} = \dfrac{-3 + j1 - 3 - j1}{3} = -2$

The centroid is marked on real axis and from the centroid the angles of asymptotes are marked using a protractor. The asymptotes are drawn as dotted lines as shown in fig 4.28.1.

Step 4: To find the breakaway and breakin points

The closed loop transfer function, $\dfrac{C(s)}{R(s)} = \dfrac{G(s)}{1 + G(s)} = \dfrac{\dfrac{K}{s(s^2 + 6s + 10)}}{1 + \dfrac{K}{s(s^2 + 6s + 10)}} = \dfrac{K}{s(s^2 + 6s + 10) + K}$

The characteristic equation is, $s(s^2 + 6s + 10) + K = 0$

On differentiating the equation of K with respect to s we get,

$$\frac{dK}{ds} = -3s^2 - 12s - 10$$

Put $\frac{dK}{ds} = 0$

$$-3s^2 - 12s - 10 = 0 \qquad \Rightarrow \qquad 3s^2 + 12s + 10 = 0$$

$$\therefore \ s = \frac{-12 \pm \sqrt{12^2 - 4 \times 3 \times 10}}{2 \times 3} = -1.18 \ \text{ or } \ -2.82$$

Check for K : When, s = –1.18, $K = -s^3 - 6s^2 - 10s = -(-1.18)^3 - 6(-1.18)^2 - 10(-1.18) = 5.09$

When, s = –2.82, $K = -s^3 - 6s^2 - 10s = -(-2.82)^3 - 6(-2.82)^2 - 10(-2.82) = 2.91$

Since the values of K for s = –1.18 and –2.82 are positive and real, both the points are actual breakaway or breakin points. It can be proved that s = –2.82 is a breakin point and s = –1.18 is a breakaway point. The breakin and breakaway points are shown in fig 4.28.1.

[Also the value of K for s = –2.82 is less than the value of K for s = –1.18, therefore when root locus travel from s = –2.82 to –1.18, the value of K increases]

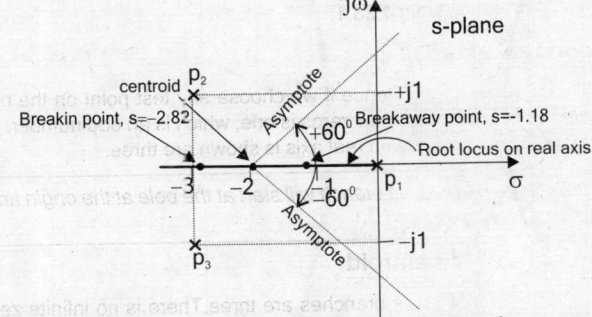

Fig 4.28.1 : *Figure showing the asymptotes, root locus on real axis and location of poles, zeros, centroid, breakin and breakaway points.*

Step 5: To find the angle of departure

Consider the complex pole p_2 shown in fig 4.28.2. Draw vectors from all other poles to the pole p_2 as shown in fig 4.28.2. Let the angle of these vectors be θ_1 and θ_2.

Here, $\theta_1 = 180° - \tan^{-1}(1/3) = 161.6°$

$\theta_2 = 90°$

Angle of departure from the complex pole p_2 $\Big\}$ $= 180° - (\theta_1 + \theta_2)$

$$= 180° - (161.6° + 90°) = -71.6° \approx -72°$$

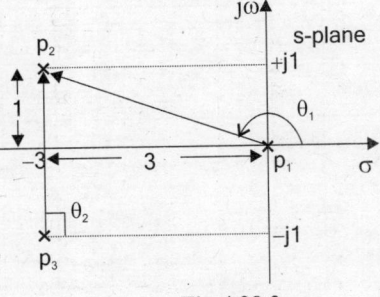

Fig 4.28.2

The angle of departure at complex pole p_3 is negative of the angle of departure at complex pole p_2.

$\therefore$ Angle of departure at pole p_3 = +72

Mark the angles of departure at complex poles using protractor.

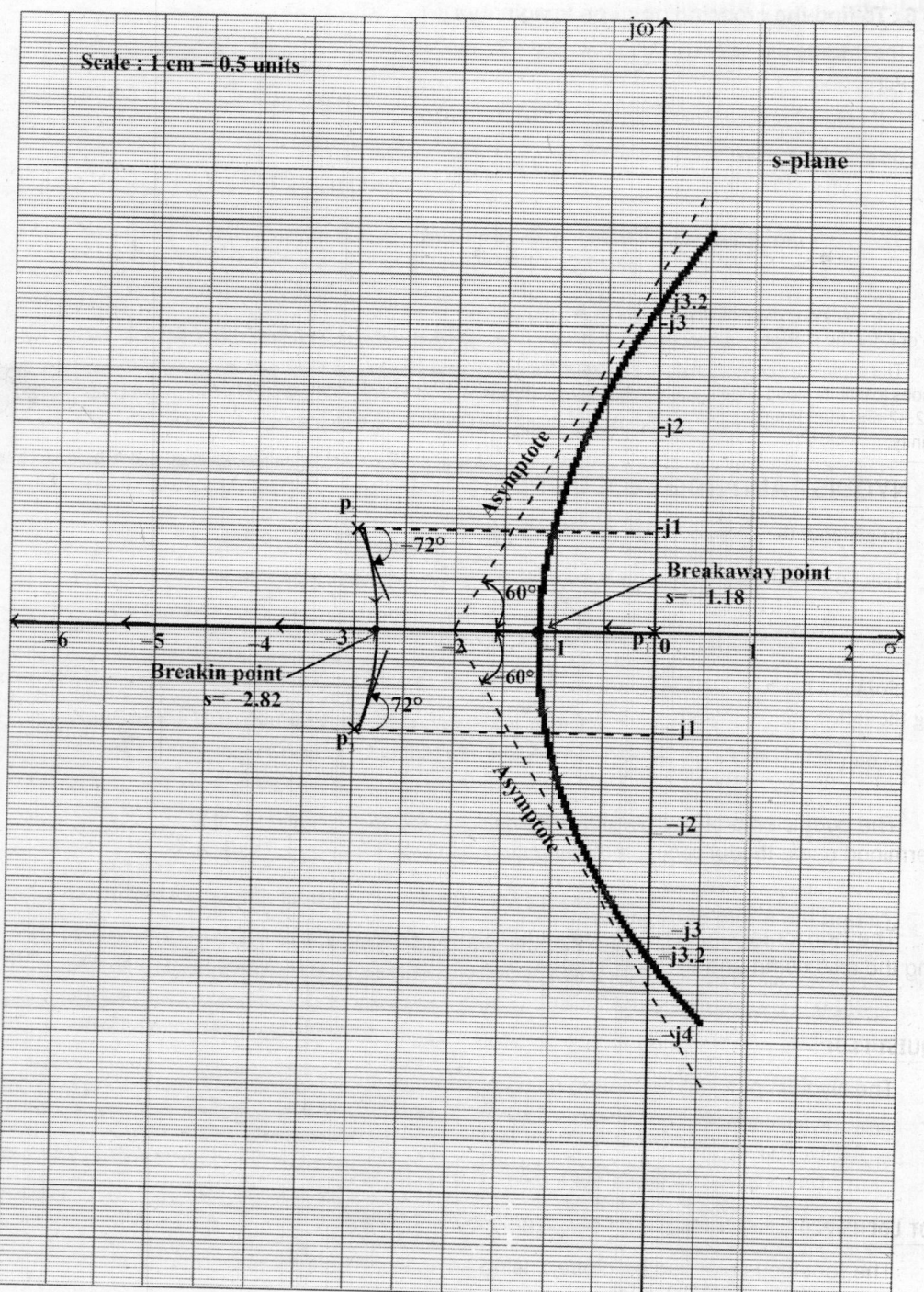

Fig 4.28.3 : *Root locus sketch of,* $1 + G(s) = 1 + \dfrac{K}{s(s^2 + 6s + 10)}$.

Step 6 : To find the crossing point on imaginary axis

The characteristic equation is given by, $s(s^2 + 6s + 10)K = 0$ $\Rightarrow$ $s^3 + 6s^2 + 10s + K = 0$

Put $s = j\omega$.

$$(j\omega)^3 + 6(j\omega)^2 + 10(j\omega) + K = 0 \quad \Rightarrow \quad -j\omega^3 - 6\omega^2 + j10\omega + K = 0$$

On equating imaginary part to zero we get,

$$-\omega^3 + 10\omega = 0$$
$$\omega^3 = 10\omega$$
$$\omega^2 = 10$$
$$\omega = \pm\sqrt{10} = \pm 3.16 \approx \pm 3.2$$

On equating real part to zero we get,

$$-6\omega^2 + K = 0$$
$$K = 6\omega^2$$
$$= 6 \times 10 = 60$$

The root locus crosses imaginary axis at $\pm j3.2$ and the gain K corresponding to this point is 60. This is the limiting value of K for the stability of the system.

The complete root locus sketch is shown in fig 4.28.3. The root locus has three branches. One branch starts at s = 0 and goes to infinity along negative real axis. The other two root locus branches starts at $s = -3 \pm j1$ and enter the real axis at s = -2.82 and then breakaway from real axis at s = -1.18. Finally they travel parallel to asymptotes to meet the zeros at infinity.

4. 9 NYQUIST AND ROOT LOCUS PLOTS USING MATLAB

In general, the open loop transfer function of a system is denoted as G(s).

Let, G(s) be a rational function of "s", as shown below.

$$G(s) = \frac{b_0 s^M + b_1 s^{M-1} + b_2 s^{M-2} + \ldots\ldots + b_{M-1}s + b_M}{a_0 s^N + a_1 s^{N-1} + a_2 s^{N-2} + \ldots\ldots + a_{N-1}s + a_N}$$

For drawing Nyquist and root locus plots, the transfer function G(s) is declared as a function of s using the following commands.

```
s=tf('s');
Gs=(b0*s^M+b1*s^(M-1)+...+bM)/(a0*s^N+a1*s^(N-1)+...+aN);
```

The coefficients of numerator and denominator polynomials of the transfer function are determined using the following command.

```
[num_cof den_cof]=tfdata(Gs);
```

The horizontal and vertical axes range for the Nyquist and root locus plots can be specified using the axis command as shown below.

```
axis([x_start x_end y_start y_end]);
```

NYQUIST PLOT

The Nyquist plot can be plotted using any one of the following commands.

```
nyquist(Gs);
nyquist(Gs,'k');
nyquist(num_cof, den_cof);
```

ROOT LOCUS PLOT

The root locus plot can be plotted using any one of the following commands.

```
rlocus(Gs);
rlocus(Gs,'k');
rlocus(num_cof, den_cof);
```

PROGRAM 4.1

Write a MATLAB program to draw the Nyquist plot of the system governed by the following open loop transfer function.

G(s) = 240/s(s+2)(s+10).

```
%program to plot Nyquist plot
clear all
clc
s=tf('s');
disp('The given transfer function is');
Gs=240/(s*(s+2)*(s+10))

nyquist(Gs,'k');
axis([-4 0.5 -2 2]); grid;
```

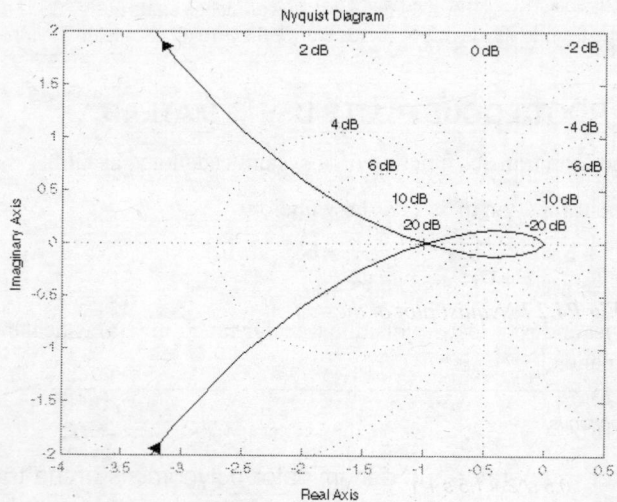

Fig P4.1 : Nyquist plot of the system given in problem 4.1.

OUTPUT

The given transfer function is,

Transfer function:

$$\frac{240}{s^3 + 12 s^2 + 20 s}$$

The Nyquist plot of program 4.1 is shown in fig P4.1.

PROGRAM 4.2

Write a MATLAB program to draw the Nyquist plot of the system governed by the following open loop transfer function.

G(s) = (1+0.5s)(1+s)/(1+10s)(s-1)

```
%program to plot Nyquist plot

clear all
clc
```

```
s=tf('s');
disp('The given transfer function is,')
Gs=((1+0.5*s)*(1+s))/((1+10*s)*(s-1))

nyquist(Gs,'k');
axis([-1.2 0.2 -1 1]);
grid;
```

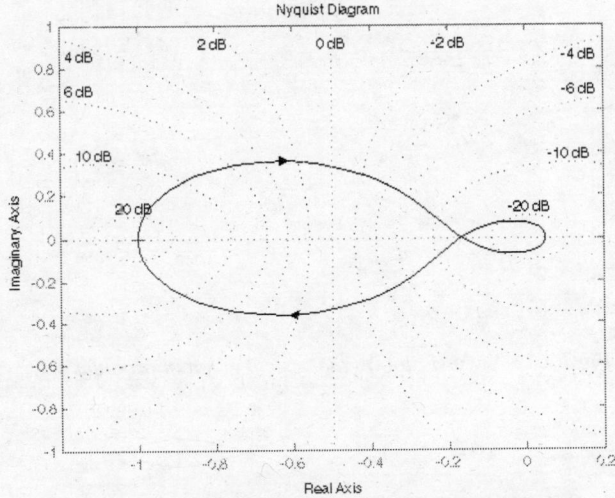

Fig P4.2 : *Nyquist plot of the system given in problem 4.2.*

OUTPUT

The given transfer function is,

Transfer function:

$$\frac{0.5\ s^2 + 1.5\ s + 1}{10\ s^2 - 9\ s - 1}$$

The Nyquist plot of program 4.2 is shown in fig P4.2.

PROGRAM 4.3

Write a MATLAB program to draw the Nyquist plot of the system governed by the following open loop transfer function.

$$G(s) = (s+2)/(s+1)(s-1)$$

```
%program to plot Nyquist plot
clear all
clc
s=tf('s');
disp('The given transfer function is,');
Gs=(s+2)/((s+1)*(s-1))

nyquist(Gs,'k');
axis([-2.5 0.2 -1 1]);
grid;
```

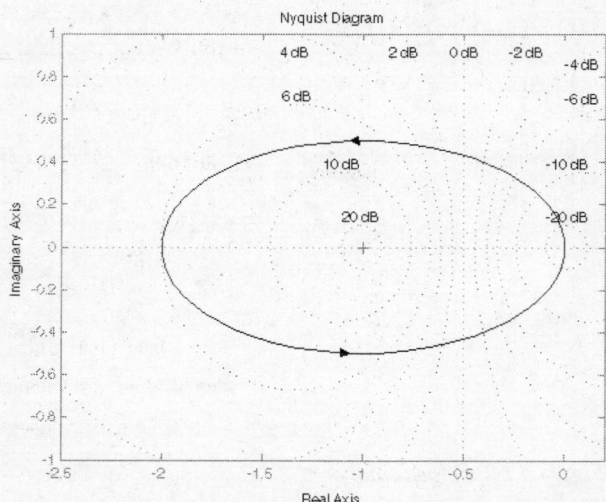

Fig P4.3 : *Nyquist plot of the system given in problem 4.3.*

OUTPUT

The given transfer function is,

Transfer function:

$$\frac{s + 2}{s^2 - 1}$$

The Nyquist plot of program 4.3 is shown in fig P4.3.

PROGRAM 4.4

Write a MATLAB program to draw the root locus plot of the unity feedback system governed by the following open loop transfer function.

$$G(s) = 1/s(s^2+4s+13)$$

```
%program to plot root locus
clear all
clc
s=tf('s');
disp('The given transfer function is,');
Gs=1/(s*(s^2+4*s+13))
rlocus(Gs,'k');
axis([-3 2 -6 6]);              %specify x and y axis limits
sgrid([0.5,0.707],[90.5,1,2]);  %specify the s-grid lines to draw
```

OUTPUT

The given open loop transfer function G(S) is,

Transfer function:

$$\frac{1}{s^3 + 4\,s^2 + 13\,s}$$

The root locus plot of program 4.4 is shown in fig P4.4.

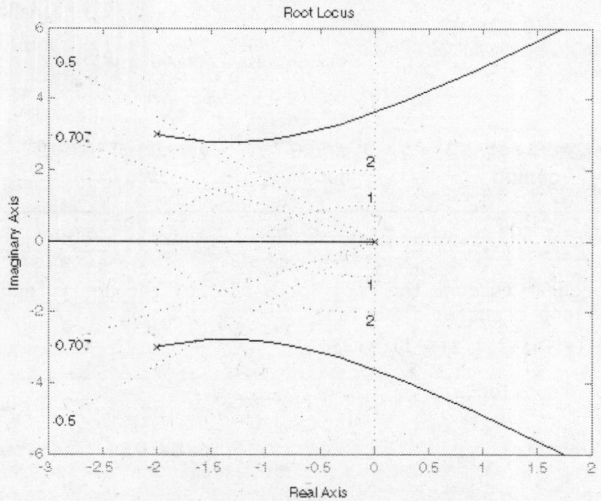

Fig P4.4 : *Root locus plot of the system given in problem 4.4.*

PROGRAM 4.5

Write a MATLAB program to draw the root locus plot of the unity feedback system governed by the following open loop transfer function.

$$G(s) = 1/s(s-4)(s^2+4s+20)$$

```
%program to plot root locus
clear all
clc
s=tf('s');
disp('The given transfer function is,');
Gs=1/(s*(s+4)*(s^2+4*s+20))
rlocus(Gs,'k'); axis([-8 4 -6 6]); sgrid;
```

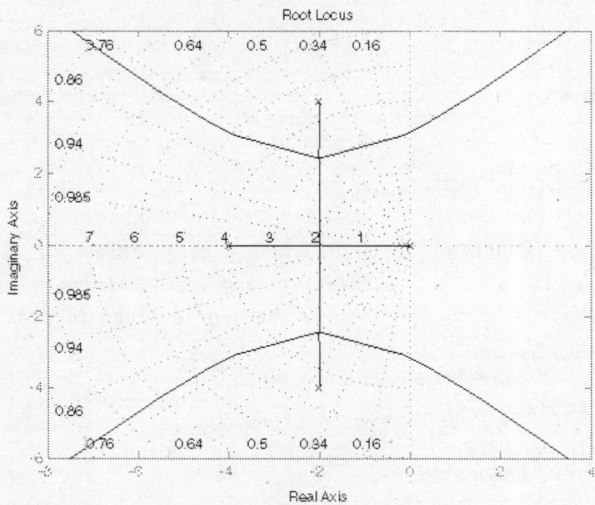

Fig P4.5 : *Root locus plot of the system given in problem 4.5.*

OUTPUT

The given open loop transfer function G(S) is,

Transfer function:

$$\frac{1}{s^4 + 8 s^3 + 36 s^2 + 80 s}$$

The root locus plot of program 4.5 is shown in fig P4.5.

PROGRAM 4.6

Write a MATLAB program to draw the root locus plot of the unity feedback system governed by the following open loop transfer function.

$$G(s) = (s^2 + 6s + 25)/s(s+1)(s+2)$$

```
%program to plot root locus
clear all
clc
s=tf('s');
disp('The given transfer function is,');
Gs=(s^2+6*s+25)/(s*(s+1)*(s+2))

rlocus(Gs,'k');
axis([-6 2 -6 6]); sgrid;
```

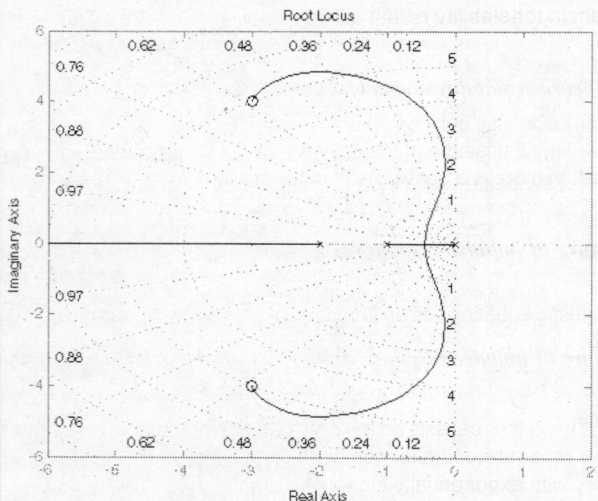

Fig P4.6 : *Root locus plot of the system given in problem 4.6.*

OUTPUT

The given open loop transfer function G(S) is,

Transfer function:

$$\frac{s^2 + 6 s + 25}{s^3 + 3 s^2 + 2 s}$$

The root locus plot of program 4.6 is shown in fig P4.6.

4. 10 SHORT - ANSWERS QUESTIONS

Q4.1   **Define BIBO stability.**

A linear relaxed system is said to have BIBO stability if every bounded (finite) input results in a bounded (finite) output.

Q4.2   **What is impulse response?**

The impulse response of a system is the response of a system for impulse input and it is given by inverse Laplace transform of the system transfer function.

Q4.3   **What is the requirement for BIBO stability?**

The requirement for BIBO stability is that, $\int_{0}^{\infty} m(t)\, dt \angle \infty$,

where m(t) is impulse response of the system.

Q4.4   **What is characteristic equation?**

The denominator polynomial of C(s)/R(s) is the characteristic equation of the system.

Q4.5   **How the roots of characteristic equation are related to stability?**

If the roots of characteristic equation has positive real part then the impulse response of the system is not bounded (the impulse response will be infinite as $t \rightarrow \infty$). Hence the system will be unstable. If the roots have negative real part then the impulse response is bounded (the impulse response becomes 0 as $t \rightarrow \infty$). Hence the system will be stable.

Q4.6   **What is the necessary condition for stability?**

The necessary condition for stability is that all the coefficients of the characteristic polynomial must be positive.

Q4.7   **What is the relation between stability and coefficient of characteristic polynomial?**

If the coefficients of characteristic polynomial are negative or zero, then some of roots lie on right half of s-plane. Hence the system is unstable. If the coefficients of characteristic polynomial are positive and if no coefficient is zero then there is a possibility of the system to be stable provided all the roots are lying on left half of s-plane.

Q4.8   **What will be the nature of impulse response when the roots of characteristic equation are lying on imaginary axis?**

If the roots of characteristic equation lies on imaginary axis the nature of impulse response is oscillatory.

Q4.9   **What will be the nature of impulse response if the roots of characteristic equation are lying on right half of s-plane?**

When the roots are lying on the real axis on the right half of s-plane, then the response is exponentially increasing. When the roots are complex conjugate and lying on the right half of s-plane, then the response is oscillatory with exponentially increasing amplitude.

Q4.10   **What is the principle of argument?**

The principle of argument states that let F(s) be an analytic function and if an arbitrary closed contour in the clockwise direction is chosen in the s-plane so that F(s) is analytic at every point of the contour. Then the corresponding F(s)-plane contour mapped in the F(s)-plane will encircle the origin, N times in the anticlockwise direction, where N is the difference between number of poles ,P and zeros Z of F(s) that are enclosed by the chosen closed contour in the s-plane. (i.e., N = P -Z).

Q4.11   **What is the necessary and sufficient condition for stability?**

The necessary and sufficient condition for stability is that all of the elements in the first column of the routh array should be positive.

Q4.12   *What is routh stability criterion?*

Routh criterion states that the necessary and sufficient condition for stability is that all of the elements in the first column of the routh array be positive. If this condition is not met, the system is unstable and the number of sign changes in the elements of the first column of routh array corresponds to the number of roots of characteristic equation in the right half of the s-plane.

Q4.13   *What is auxiliary polynomial?*

In the construction of routh array a row of all zero indicates the existence of an even polynomial as a factor of the given characteristic equation. In an even polynomial the exponents of s are even integers or zero only. This even polynomial factor is called auxiliary polynomial. The coefficients of auxiliary polynomial are given by the elements of the row just above the row of all zeros.

Q4.14   *What is quadrantal symmetry?*

The symmetry of roots with respect to both real and imaginary axis is called quadrantal symmetry.

Q4.15   *In routh array what conclusion you can make when there is a row of all zeros?*

All zero row in routh array indicates the existence of an even polynomial as a factor of the given characteristic equation. The even polynomial may have roots on imaginary axis.

Q4.16   *What is limitedly stable system ?*

For a bounded input signal, if the output has constant amplitude oscillations then the system may be stable or unstable under some limited constraints. Such a system is called limitedly stable.

Q4.17   *What is Nyquist stability criterion?*

If G(s)H(s)-contour in the G(s)H(s)-plane corresponding to Nyquist contour in s-plane encircles the point −1+j0 in the anti-clockwise direction as many times as the number of right half s-plane poles of G(s)H(s). Then the closed loop system is stable.

Q4.18   *What is root locus?*

The path taken by a root of characteristic equation when open loop gain K is varied from 0 to ∞ is called root locus.

Q4.19   *What is magnitude criterion?*

The magnitude condition states that s=s_a will be a point on root locus if for that value of s magnitude of G(s)H(s) is equal to 1, (i.e. |G(s)H(s)| = 1) .

Let, $G(s)H(s) = \dfrac{K(s+z_1)(s+z_2)(s+z_3).....}{(s+p_1)(s+p_2)(s+p_3)......}$

   ∴ For s = s_a be a point in root locus,

$$|G(s)H(s)| = \dfrac{K|s_a+z_1||s_a+z_2||s_a+z_3|......}{|s_a+p_1||s_a+p_2||s_a+p_3|.....} = 1$$

$$\left[\text{or } |G(s)H(s)| = K\,\dfrac{\text{Product of length of vectors from open loop zeros to the point } s_a}{\text{Product of length of vectors from open loop poles to the point } s_a} = 1 \right]$$

Q4.20   *What is angle criterion?*

The angle criterion states that s = s_a will be a point on root locus if for that value of s the argument or phase of G(s)H(s) is equal to an odd multiple of 180°, [i.e., ∠G(s)H(s) = ±180° (2q+1)].

Let, $G(s)H(s) = \dfrac{K(s+z_1)(s+z_2)(s+z_3).....}{(s+p_1)(s+p_2)(s+p_3)......}$

∴ For s = s_a be a point on root locus,

∠G(s)H(s) = ∠$(s_a + z_1)$ + ∠$(s_a + z_2)$ + ∠$(s_a + z_3)$ .... −∠$(s_a + p_1)$ −∠$(s_a + p_2)$..... = ±180°(2q + 1)

$$\left[\text{or} \left(\begin{matrix} \text{Sum of angles} \\ \text{of vectors from zeros} \\ \text{to the point } s = s_a \end{matrix} \right) - \left(\begin{matrix} \text{Sum of angles} \\ \text{of vectors from poles} \\ \text{to the point } s = s_a \end{matrix} \right) = \pm 180^\circ (2q + 1) \right]$$

Q4.21. *How will you find the gain K at a point on root locus?*

The gain K at a point $s = s_a$ on root locus is given by,

$$K = \frac{\text{Product of length of vector from open loop poles to the point } s_a}{\text{Product of length of vector from open loop zeros to the point } s_a}$$

Q4.22 *How will you find root locus on real axis?*

To find the root locus on real axis, choose a test point on real axis. If the total number of poles and zeros on the real axis to the right of this test point is odd number, then the test point lies on the root locus. If it is even then the test point does not lie on the root locus.

Q4.23 *What are asymptotes? How will you find the angle of asymptotes?*

Asymptotes are straight lines which are parallel to root locus going to infinity and meet the root locus at infinity.

$$\text{Angles of asymptotes} = \frac{\pm 180^\circ (2q + 1)}{n - m} \; ; \; q = 0, 1, 2, \ldots (n - m)$$

Q4.24 *What is centroid? How the centroid is calculated?*

The meeting point of asymptotes with real axis is called centroid. The centroid is given by,

$$\text{Centroid} = \frac{\text{Sum of poles} - \text{Sum of zeros}}{n - m}$$

Q4.25 *What are breakaway and breakin point? How to determine them?*

At breakaway point the root locus breaks from the real axis to enter into the complex plane. At breakin point the root locus enters the real axis from the complex plane.

To find the breakaway or breakin points, form an equation for K from the characteristic equation, and differentiate the equation of K with respect to s. Then find the roots of equation dK/ds = 0. The roots of dK/ds = 0 are breakaway or breakin points, provided for this value of root, the gain K should be positive and real.

Q4.26 *How to find the crossing points of root locus in imaginary axis.*

Method (i)   :   By Routh hurwitz criterion.

Method (ii)   :   By letting $s = j\omega$ in the characteristic equation and separate the real and imaginary parts. These two equations are equated to zero. Solve the two equations for ω and K. The value of ω gives the point where the root locus crosses imaginary axis and the value of K is the gain corresponding to the crossing point.

Q4.27 *What is dominant pole?*

The dominant pole is a pair of complex conjugate pole which decides transient response of the system. In higher order systems the dominant poles are very close to origin and all other poles of the system are widely separated and so they have less effect on transient response of the system.

Q4.28 *How will you fix dominant pole on root locus and find the gain K corresponding to the dominant pole?*

The dominant poles are given by roots of a quadratic factor, $s^2 + 2\zeta\omega_n s + \omega_n^2 = 0$.

$$\therefore \; s = \frac{-2\zeta\omega_n \pm \sqrt{4\zeta^2\omega_n^2 - 4\omega_n^2}}{2} = -\zeta\omega_n \pm j\omega_n \sqrt{1 - \zeta^2}$$

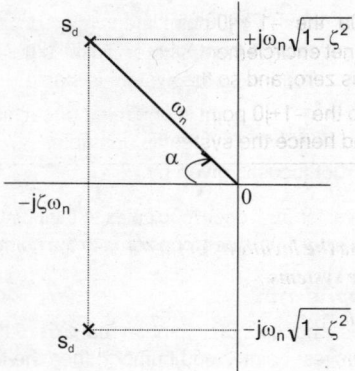

Fig Q4.28

The dominant pole can be plotted on the s-plane as shown in fig Q4.28.

In the right angle triangle OAP,

$$\cos\alpha = \frac{\zeta\omega_n}{\omega_n} = \zeta \qquad \therefore \; \alpha = \cos^{-1}\zeta$$

To fix a dominant pole on root locus draw a line at an angle of $\cos^{-1}\zeta$ with respect to negative real axis. The meeting point of this line with root locus will give the location of dominant pole. The value of K corresponding to dominant pole can be obtained from magnitude condition.

$$K = \frac{\text{Product of length of vectors from open loop poles to dominant pole}}{\text{Product of length of vectors from open loop zeros to dominant pole}}$$

Q4.29   *For the system represented by the following characteristic equation say whether the necessary condition for stability is satisfied or not.*

(i) $s^4 + 3s^3 + 4s^2 + 5s + 10 = 0$

(ii) $s^6 - 2s^5 + s^3 + s^2 + s + 6 = 0$

(iii) $s^3 - 8s^2 + 7s + 6 = 0$

(iv) $s^5 + 4s^4 - 5s^3 - 4s^2 + 2s + 1 = 0$

In equation (i) All the coefficients are positive and so the necessary condition for stability is satisfied.

In equation (ii), (iii) and (iv) some of the coefficients are negative and some of the coefficients are missing. Hence the necessary condition for stability is not satisfied.

Q4.30   *Check the stability of the system whose Nyquist plot are shown in figQ 4.30 a & b, if there is no open loop poles on right half of s-plane.*

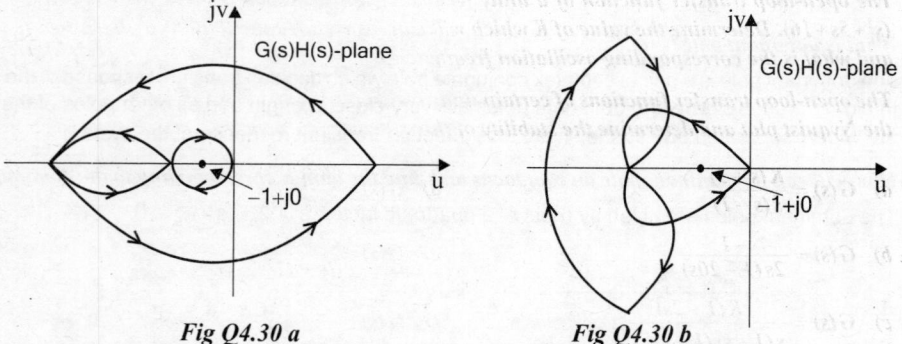

Fig Q4.30 a   *Fig Q4.30 b*

In contour shown in fig Q 4.30a the $-1 + j0$ point is encircled once in clockwise direction and once in anticlockwise direction. Hence net encirclement is zero. Since no poles are lying on right half of s-plane and net encirclement of $-1+j0$ is zero, and so the system is stable.

In contour shown in fig Q4.30b the $-1+j0$ point is encircled once in anticlockwise direction but there is no pole on right half hence and hence the system is unstable.

4.11   EXERCISES

E.4.1   *Using routh criterion determine the locations of the roots of the following characteristic equations and comment on the stability of the systems.*

a) $2s^5 + 2s^4 + 5s^3 + 5s^2 + 3s = 0$

d) $s^5 + s^4 + 4s^3 + 24s^2 + 3s + 63 = 0$

b) $3s^4 + 10s^3 + 5s^2 + 5s + 3 = 0$

e) $s^5 + 2s^4 + 24s^3 + 48s^2 - 25s - 50 = 0$

c) $2s^6 + 4s^5 + s^4 - 32s^3 + 51s^2 + 3s + 15 = 0$

b) $s^6 + 3s^5 + 3s^4 + 9s^3 + 8s^2 + 6s + 4 = 0$

E.4.2   *The characteristic equations for certain feedback control systems are given below. In each case, determine the range of values of K, for which the system is stable.*

a) $s^4 + 3s^3 + 3s^2 + 3s + K = 0$

c) $s^5 + s^4 + s^2 + s + K = 0$

b) $s^5 + s^4 + Ks^3 + s^2 + s - 1 = 0$

d) $s^4 + s^3 + 3Ks^2 + (K + 2)s + 4 = 0$

e) $s^4 + s^3 + 3(K + 1)s^2 + (7K + 5)s + (4K + 7) = 0$

E.4.3   *Open-loop transfer functions of certain unity feedback systems are given below. In each case determine the location of closed loop poles in the s-plane, using routh criterion. Comment on the stability of closed loop system.*

a) $G(s) = \dfrac{200(1 + s)}{s(1 + 0.1s)(1 + 0.2s)(1 + 0.5s)}$

b) $G(s) = \dfrac{(s + 1)}{s(s - 1)(s^2 + 4s + 16)}$

c) $G(s) = \dfrac{10}{(s + 2)(s + 4)(s^2 + 6s + 25)}$

d) $G(s) = \dfrac{2.5}{s(s + 5)(0.1s + 1)}$

E.4.4   *Open-loop transfer functions of certain unity feedback systems are given below. In each case determine the range of values of K for which the system is stable.*

a) $G(s) = \dfrac{K(s + 13)}{s(s + 3)(s + 7)}$

b) $G(s) = \dfrac{K(s + 1)}{s(s - 1)(s + 6)}$

c) $G(s) = \dfrac{K(s + 2)}{s(s + 5)(s^2 + 2s + 5)}$

d) $G(s) = \dfrac{K(s - 1)}{s(s + 2)}$

E.4.5   *The open-loop transfer function of a unity feedback control system is given by $G(s) = K(s + 2)/s(s - 2)$ $(s^2 + 5s + 16)$. Determine the value of K which will cause sustained oscillations in the closed-loop system and what is the corresponding oscillation frequencies?*

E.4.6   *The open-loop transfer functions of certain unity feedback system are given below. In each case, sketch the Nyquist plot and determine the stability of the system.*

a) $G(s) = \dfrac{K(s + 3)}{s(s - 1)}$

d) $G(s) = \dfrac{K(s + 3)}{s(s - 1)}$

b) $G(s) = \dfrac{-1}{2s(1 - 20s)}$

e) $G(s) = \dfrac{K}{(s + 1)(s + 1.5)(s + 2)}$

c) $G(s) = \dfrac{K(1 + 2s)}{s(1 + s)(1 + s + s^2)}$

f) $G(s) = \dfrac{K}{s(s + 1)(s^2 + 2s + 2)}$

E.4.7   *Determine the phase margin and gain margin of the system with following transfer functions.*

   a) $G(s) = \dfrac{2}{(s+1)^2}$ b) $G(s) = \dfrac{20}{s(s+1)(s^2+2s+2)}$

E.4.8   *The open-loop transfer function of a unity feedback system is given by, G(s)= K/s(1+0.5s)(1+s).*

   a) *Determine the value of K so that the gain margin of the system is 6 db.*

   b) *Determine the value of K so that the phase margin of the system is 30°.*

E.4.9   *The open-loop transfer functions of certain unity feedback systems are given below. Sketch the root locus of each system.*

   a) $G(s) = \dfrac{K(s+2)}{(s+3)^2(s^2+2s+17)}$ e) $G(s) = \dfrac{K(s+2)}{(s+3)^2(s^2+2s+17)}$

   b) $G(s) = \dfrac{K}{s(s+3)^2(s^2+2s+2)}$ f) $G(s) = \dfrac{K}{s(s^2+8s+20)}$

   c) $G(s) = \dfrac{K(s^2+2s+10)}{s(s+2)(s+4)}$ g) $G(s) = \dfrac{K}{s(s+2)(s^2+2s+2)}$

   d) $G(s) = \dfrac{K(s+1)}{s^2(s+12)}$ h) $G(s) = \dfrac{K(s^2+1)}{s(s+2)}$

E.4.10   *A unity feedback system has an open-loop transfer function, G(s) = K/s(s²+8s+32). Sketch the root locus and determine* ⋯ *closed loop poles with ζ=0.5. Determine the value of K at this point.*

E.4.11   *Draw the root locus plot for a unity feedback system having forward path transfer function,*

   $G(s) = K/s(s+1)(s+5).$

   a) *Determine the value of K which gives continuous oscillations and the frequency of oscillation.*

   b) *Determine the value of K corresponding to a dominant closed loop pole with damping ratio,*

   $\zeta = 0.7.$

CHAPTER 5

LINEAR SYSTEM DESIGN

5.1 INTRODUCTION TO DESIGN USING COMPENSATORS

The control systems are designed to perform specific tasks. The requirements of a control system are usually specified as performance specifications. The specifications are generally related to accuracy, relative stability and speed of response.

In time domain, the transient state performance specifications are given in terms of rise time, maximum overshoot, settling time and/or damping ratio. In frequency domain, the transient state performance specifications are given in terms of phase margin, gain margin, resonant peak and/or bandwidth. The steady state requirements are given in terms of error constants.

When performance specifications are given for single input, single output linear time invariant systems, then the system can be designed by using root locus or frequency response plots. To be more precise, when time domain specifications are given, root locus technique is employed in designing the system. If frequency domain specifications are given, frequency response plots like Bode plots are used in designing the system.

The first step in design is the adjustment of gain to meet the desired specifications. In practical systems, adjustment of gain alone will not be sufficient to meet the given specifications. In many cases, increasing the gain may result in poor stability or instability. In such cases, it is necessary to introduce additional devices or components in the system to alter the behaviour and to meet the desired specifications.

Such a redesign or addition of a suitable device is called compensation. A device inserted into the system for the purpose of satisfying the specifications is called compensator. The compensators basically introduce pole and/or zero in open loop transfer function to modify the performance of the system.

The design problem may be stated as follows : *When a set of specifications are given for a system, then a suitable compensator should be designed so that the overall system will meet the given specification.*

The compensation schemes used for feedback control system is either series compensation or parallel (feedback) compensation.

In series compensation, the compensator with transfer function, $G_c(s)$ is placed in series with plant. In feedback compensation, the signal from some element is feedback to the input and a compensator with transfer function $G_c(s)$ is placed in the resulting inner feedback path. The series and parallel compensation schemes are shown in fig 5.1 and 5.2 respectively.

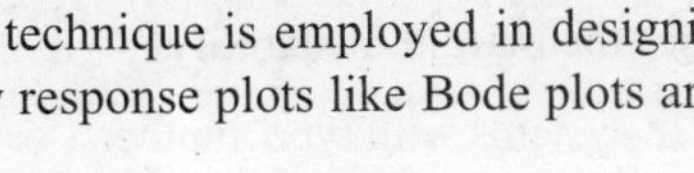

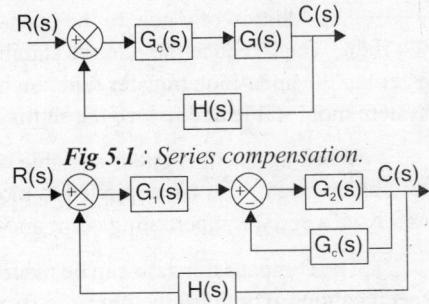

Fig 5.1 : *Series compensation.*

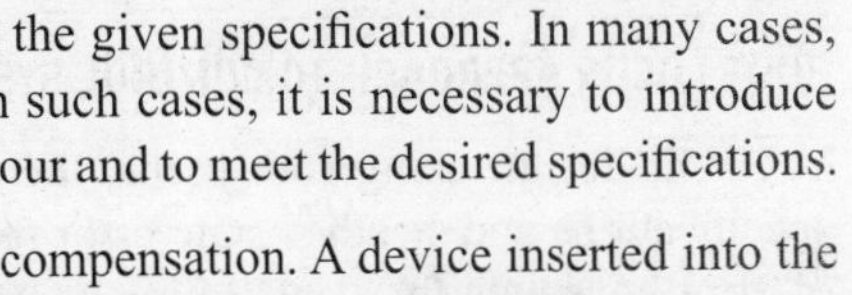

Fig 5.2 : *Feedback/parallel compensation.*

The choice between series compensation and parallel compensation depends on

1. *Nature of signals in the system.*
2. *Power levels at various points.*
3. *Components available.*
4. *Designer's experience..*
5. *Economic considerations and so on.*

The compensator may be electrical, mechanical, hydraulic, pneumatic or other type of device or network. Usually, an electric network or electronic device serves as compensator in many control systems. The different types of electrical or electronic compensators used are ***Lag compensator, Lead compensator*** *and* ***Lag-lead compensator.***

In control systems, compensation is required in the following situations,

1. *When the system is absolutely unstable, then compensation is required to stabilize the system and also to meet the desired performance.*
2. *When the system is stable, compensation is provided to obtain the desired performance.*

The systems with type number 2 and above are usually absolutely unstable systems. Hence for systems with type number 2 and above, lead compensation is required, because the lead compensator increases the margin of stability.

In systems with type number 1 or 0, stability is achieved by adjusting the gain. In such cases any of the three compensators-lag, lead and lag-lead may be used to obtain the desired performance. The particular choice of compensator is based on the factors that are discussed in the following sections.

ROOT LOCUS APPROACH TO CONTROL SYSTEM DESIGN

In design using root locus, the desired behaviour is specified in terms of transient response specifications and steady - state error requirement. The steady - state error is usually specified in terms of error constants for standard inputs, while the transient response requirement is specified in terms of peak overshoot, settling time, rise time, etc., for a step input. The transient response specifications can be translated into desired locations for a pair of dominant closed loop poles.

In order to meet the desired specifications, the root loci are reshaped so that they pass through the points where the dominant closed loop poles are located. The root loci are reshaped by introducing a compensator. The compensator will add a pole and/or a zero in the open loop transfer function of the system.

The addition of a pole to the open-loop transfer function has the effect of pulling the root locus to the right, which reduce the relative stability of the system and increase the settling time. The addition of a zero to the open-loop transfer function has the effect of pulling the root locus to the left which make the system more stable and reduce the settling time.

When a system is either unstable or stable but has undesirable transient response characteristics a lead compensator can be employed to modify the root locus. The transfer function of lead compensator will have a zero (compensating zero) and a pole (compensating pole).

The compensator zero can be placed on the real axis by trial-and-error to satisfy transient response specifications. The introduction of zero will amplify high frequency noise which is eliminated by the compensating pole. The compensating pole is located on real axis such that it makes negligible effect on the root locus in the region where the two dominant closed loop poles are located.

If the pole is located far away from zero then it will not be effective in suppressing the noise. If the pole is too close to zero then it will not allow the zero to do its job. In order to avoid this conflict, the pole is located at 3 to 10 times the value of zero location.

The lag compensator is employed when a stable system has satisfactory transient response characteristics but unsatisfactory steady state characteristics, i.e., error requirement. The transfer function of lag compensator will have a zero (compensating zero) and a pole (compensating pole).

In order to preserve the transient response characteristics the compensating pole and zero should have negligible effect on shape of root locus. This is achieved by placing the compensating pole and zero very close to each other. If the pole and zero are located close to the origin then the error constant will increase which will reduce the steady state error.

The lag-lead compensator is employed when both the transient and steady-state characteristics are not satisfactory. The lead compensation will improve the transient response and lag compensation will reduce the steady state error.

The advantage in design using root locus technique is that the information about closed loop transient response and frequency response are directly obtained from the pole-zero configuration of the system in the s-plane.

FREQUENCY RESPONSE APPROACH TO CONTROL SYSTEM DESIGN

The objective of frequency domain design is to reshape the frequency-response characteristics so that the desired specifications are satisfied. The frequency domain design can be carried out using Nyquist plot, Bode plot or Nichols chart. But Bode plot are popularly used for design because they are easier to draw and modify.

In design using bode plots the desired performance specifications are given in terms of frequency domain specifications and steady-state error requirement.

The stability requirement is specified in terms of phase margin and resonant peak. The transient response requirements are specified in terms of gain crossover frequency, bandwidth and resonant frequency. The error requirement is specified in terms of static error constants (K_p, K_v or K_a).

Note : *In case the transient response specifications are given in time domain, we can translate them into frequency domain specifications using the following formulae.*

Phase margin, $\gamma = \tan^{-1}\dfrac{2\zeta}{[\sqrt{4\zeta^4+1}-2\zeta^2]^{\frac{1}{2}}} \approx 100\zeta$	Resonant peak, $M_r = 1/[2\zeta\sqrt{1-\zeta^2}]$
Gain crossover frequency, $\omega_{gc} = \omega_n[\sqrt{4\zeta^4+1}-2\zeta^2]^{\frac{1}{2}}$	Resonant frequency, $\omega_r = \omega_n\sqrt{1-2\zeta^2}$
Bandwidth, $\omega_b = \omega_n[(1-2\zeta^2)+\sqrt{4\zeta^4-4\zeta^2+2}]^{\frac{1}{2}}$	

The low frequency region of bode plot provides information regarding the steady state performance and high frequency region provides information regarding the transient-state performance. The medium frequency (or mid-frequency) range provides information regarding relative stability. Therefore, the low frequency region of the bode plot is reshaped by lag compensation to improve steady state performance. The high frequency region of the bode plot is reshaped by lead compensation to improve transient-state performance.

When the system requires improvement in both steady-state and transient state, lag-lead compensation can be employed to alter both the low and high frequency regions of bode plot.

The primary function of lead compensator is to reshape the frequency response curve to provide sufficient phase lead angle to offset the excessive phase lag associated with the components of the plant. The primary function of lag compensator is to provide attenuation in the high frequency region to achieve sufficient phase margin.

The advantage in frequency-domain design is that the effects of disturbances, sensor noise and plant uncertainties are relatively easy to visualize and assess in frequency domain. Another advantage of using frequency response is the ease with which experimental information can be used for design purposes.

A disadvantage of frequency-response design is that it gives us information on closed-loop system's transient response indirectly, while the root locus design gives this information directly.

5.2 LAG COMPENSATOR

A compensator having the characteristics of a lag network is called a lag compensator. If a sinusoidal signal is applied to a lag network, then in steady state the output will have a phase lag with respect to input.

Lag compensation results in a large improvement in steady state performance but results in slower response due to reduced bandwidth. The attenuation due to the lag compensator will shift the gain crossover frequency to a lower frequency point where the phase margin is acceptable. Thus, the lag compensator will reduce the bandwidth of the system and will result in slower transient response.

Lag compensator is essentially a low pass filter and so high frequency noise signals are attenuated. If the pole introduced by the compensator is not cancelled by a zero in the system, then lag compensator increases the order of the system by one.

S-PLANE REPRESENTATION OF LAG COMPENSATOR

The lag compensator has a pole at $s = -1/\beta T$ and a zero at $s = -1/T$. The pole-zero plot of lag compensator is shown in fig 5.3. Here, $\beta > 1$, so the zero is located to the left of the pole on the negative real axis. The general form of lag compensator transfer function is given by equation (5.1).

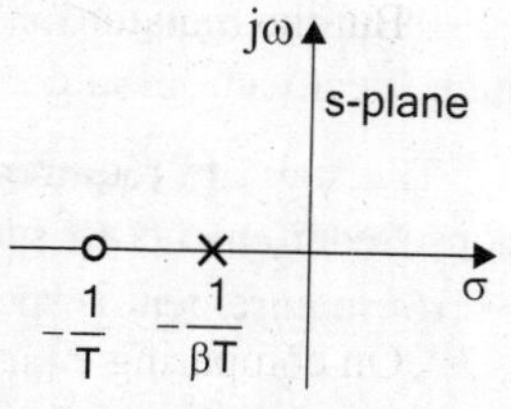

Fig 5.3 : *Pole-zero plot of lag compensator.*

Transfer function of lag compensator, $G_c(s) = \dfrac{s + z_c}{s + p_c} = \dfrac{s + \dfrac{1}{T}}{s + \dfrac{1}{\beta T}}$ .....(5.1)

where, $T > 0$ and $b > 1$

The zero of lag compensator, $z_c = \dfrac{1}{T}$ .....(5.2)

The pole of lag compensator, $p_c = \dfrac{1}{\beta T} = \dfrac{1}{\beta} z_c$ .....(5.3)

From equation(5.2) we get, $T = \dfrac{1}{z_c}$ .....(5.4)

From equation (5.3) we get, $\beta = \dfrac{z_c}{p_c}$ .....(5.5)

REALIZATION OF LAG COMPENSATOR USING ELECTRICAL NETWORK

The lag compensator can be realised by the R-C network shown in fig 5.4.

Let, $E_i(s)$ = Input voltage

$E_o(s)$ = Output voltage

In the network shown in fig 5.4, the input voltage is applied to the series combination of R_1, R_2 and C. The output voltage is obtained across series combination of R_2 and C.

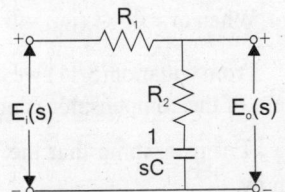

Fig 5.4 : Electrical lag compensator.

By voltage division rule,

$$E_o(s) = E_i(s) \frac{(R_2 + 1/sC)}{(R_1 + R_2 + 1/sC)} = E_i(s) \frac{(sCR_2 + 1)/sC}{[sC(R_1 + R_2) + 1]/sC} = E_i \frac{(sCR_2 + 1)}{[sC(R_1 + R_2) + 1]}$$

The transfer function of the electrical network is the ratio of output voltage to input voltage,

$$\left.\begin{array}{l}\text{Transfer function} \\ \text{of electrical network}\end{array}\right\} \frac{E_o(s)}{E_i(s)} = \frac{CR_2\left(s + \dfrac{1}{CR_2}\right)}{C(R_1 + R_2)\left[s + \dfrac{1}{C(R_1 + R_2)}\right]}$$

$$= \frac{\left(s + \dfrac{1}{R_2 C}\right)}{\left(\dfrac{R_1 + R_2}{R_2}\right)\left[s + \dfrac{1}{((R_1 + R_2)/R_2)R_2 C}\right]} \qquad \text{.....(5.6)}$$

But the transfer function of lag compensator is given by,

$$G_c(s) = \frac{\left(s + \dfrac{1}{T}\right)}{\left(s + \dfrac{1}{3T}\right)} \qquad \text{.....(5.7)}$$

On comparing equations (5.6) and (5.7) we get,

$$\frac{E_o(s)}{E_i(s)} = \frac{1}{\beta} \frac{\left(s + \dfrac{1}{T}\right)}{\left(s + \dfrac{1}{\beta T}\right)} \qquad \text{.....(5.8)}$$

where, $T = R_2 C$ and $\beta = (R_1 + R_2)/R_2$

The transfer function of RC network as given by equation(5.8) is similar to the general form with an attenuation of $1/\beta$ (since $\beta > 1$, $(1/\beta) < 1$). If the attenuation is not required then an amplifier with gain β can be connected in cascade with RC network to nullify the attenuation.

FREQUENCY RESPONSE OF LAG COMPENSATOR

Consider the general form of lag compensator,

$$G_c(s) = \frac{\left(s + \dfrac{1}{T}\right)}{\left(s + \dfrac{1}{\beta T}\right)} = \frac{(sT + 1)/T}{(s\beta T + 1)/\beta T} = \beta \frac{(1 + sT)}{(1 + s\beta T)} \qquad \text{.....(5.9)}$$

Sinusoidal transfer function of lag compensator is obtained by letting $s = j\omega$ in equation (5.9).

$$\therefore\ G_c(j\omega) = \frac{\beta(1+j\omega T)}{(1+j\omega\beta T)} \qquad \qquad \dots\dots(5.10)$$

When $\omega = 0$, $G_c(j\omega) = \beta$ $\qquad\qquad\qquad\qquad\qquad\qquad\qquad\qquad \dots\dots(5.11)$

From equation(5.11) we can say that the lag compensator provides a dc gain of β (here $\beta > 1$). If the dc gain of the compensator is not desirable then it can be eliminated by a suitable attenuation.

Let us assume that the gain β is eliminated by a suitable attenuation network. Now, $G_c(j\omega)$ is given by,

$$G_c(j\omega) = \frac{1+j\omega T}{1+j\omega\beta T} = \frac{\sqrt{1+(\omega T)^2}\ \angle \tan^{-1}\omega T}{\sqrt{1+(\omega\beta T)^2}\ \angle \tan^{-1}\omega\beta T} \qquad \dots\dots(5.12)$$

The sinusoidal transfer function shown in equation(5.12) has two corner frequencies and they are denoted as ω_{c1} and ω_{c2}.

Here, $\omega_{c1} = \dfrac{1}{\beta T}$ and $\omega_{c2} = \dfrac{1}{T}$

Since, $\beta T > T$, $\omega_{c1} < \omega_{c2}$

Let, $A = |G_c(j\omega)|$ in db $= 20\log\dfrac{\sqrt{1+(\omega T)^2}}{\sqrt{1+(\omega\beta T)^2}}$ $\qquad\qquad \dots\dots(5.13)$

At very low frequencies i.e., upto ω_{c1}, $\omega T \ll 1$ and $\omega\beta T \ll 1$.

$$\therefore\ A \approx 20\log 1 = 0$$

In the frequency range from ω_{c1} to ω_{c2}, $\omega T \ll 1$ and $\omega\beta T \gg 1$.

$$\therefore\ A \approx 20\log\frac{1}{\sqrt{(\omega\beta T)^2}} = 20\log\frac{1}{\omega\beta T}$$

At very high frequencies i.e., after ω_{c2}, $\omega T \ll 1$ and $\omega\beta T \gg 1$.

$$\therefore\ A \approx 20\log\frac{\sqrt{(\omega T)^2}}{\sqrt{(\omega\beta T)^2}} = 20\log\frac{1}{\beta}$$

The approximate magnitude plot of lag compensator is shown in fig 5.5. The magnitude plot of Bode plot of $G_c(j\omega)$ is a straight line through 0 db upto ω_{c1}, then it has a slope of -20 db/dec upto ω_{c2} and after ω_{c2} it is a straight line with a constant gain of $20\log(1/\beta)$.

Let, $\phi = \angle G_c(j\omega)$

$$\therefore\ \phi = \tan^{-1}\omega T - \tan^{-1}\omega\beta T$$

As $\omega \to 0$, $\phi \to 0$

As $\omega \to \infty$, $\phi \to 0$

As ω is varied from 0 to ∞, the phase angle decreases from 0 to a negative maximum value of ϕ_m at $\omega = \omega_m$, then increases from this maximum value to 0. The phase plot of lag compensator is shown in fig 5.5. It can be shown that the frequency at which maximum phase lag occurs is the geometric mean of the two corner frequencies.

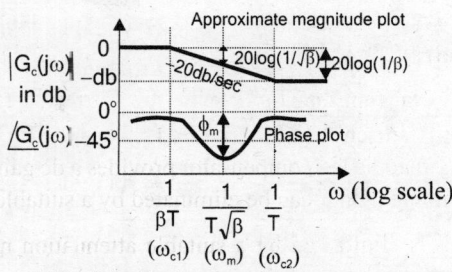

Fig 5.5 : Bode plot of lag compensator.

Frequency of maximum phase lag, $\omega_m = \sqrt{\omega_{c1}\omega_{c2}} = \sqrt{\dfrac{1}{\beta T} \cdot \dfrac{1}{T}} = \dfrac{1}{T\sqrt{\beta}}$

From the bode plot of lag compensator, we observe that lag compensator has a dc gain of unity while it offers a high frequency gain of $(1/\beta)$ [In decibels, it is 20 log $(1/\beta)$]. It means that the high frequency noise is attenuated in passing through the network and so the signal to noise ratio is improved. A typical choice of $\beta = 10$.

DETERMINATION OF ω_m and ϕ_m

The frequency ω_m can be determined by differentiating ϕ with respect to ω and equating $d\phi/d\omega$ to zero as shown below.

From equation (5.12) we get,

Phase of $G_c(j\omega)$, $\qquad \phi = \angle G_c(j\omega) = \tan^{-1}\omega T - \tan^{-1}\omega\beta T$ .....(5.14)

On differentiating the equation (5.14) we get, $\boxed{\dfrac{d}{d\theta}(\tan^{-1}\theta) = \dfrac{1}{1+\theta^2}}$

$$\frac{d\phi}{d\omega} = \frac{1}{1+(\omega T)^2}\,T - \frac{1}{1+(\omega\beta T)^2}\,\beta T \qquad\qquad .....(5.15)$$

When $\omega = \omega_m$, $d\phi/d\omega = 0$

Hence, replace ω by ω_m in equation (5.15) and equate to zero.

$$\therefore \frac{1}{1+(\omega_m T)^2}\,T - \frac{1}{1+(\omega_m\beta T)^2}\,\beta T = 0 \quad\Rightarrow\quad \frac{T}{1+(\omega_m T)^2} = \frac{\beta T}{1+(\omega_m\beta T)^2}$$

On cross multiplication we get,

$$1+(\omega_m\beta T)^2 = \beta[1+(\omega_m T)^2] \quad\Rightarrow\quad (\omega_m\beta T)^2 - \beta(\omega_m T)^2 = \beta - 1$$

$$\therefore \beta(\omega_m T)^2(\beta-1) = (\beta-1) \quad\Rightarrow\quad \omega_m^2 = \frac{1}{T^2\beta}$$

$$\therefore \omega_m = \frac{1}{T\sqrt{\beta}}$$

Frequency corresponding to maximum phase lag, $\therefore \omega_m = \dfrac{1}{T\sqrt{\beta}}$ .....(5.16)

The maximum phase angle ϕ_m can be calculated from the knowledge of β and viceversa. The relations between ϕ_m and β are derived below.

From equation (5.14) we get, $G_c(j\omega) = \phi = \tan^{-1}\omega T - \tan^{-1}\omega\beta T$

On taking tan on either side we get,

$$\tan\phi = \tan[\tan^{-1}\omega T - \tan^{-1}\omega\beta T]$$

$$\boxed{\tan(A-B) = \frac{\tan A - \tan B}{1 + \tan A \tan B}}$$

$$= \frac{\tan T(\tan^{-1}\omega T) - \tan(\tan^{-1}\omega\beta T)}{1 + \tan(\tan^{-1}\omega T).\tan(\tan^{-1}\omega\beta T)} = \frac{\omega T - \omega\beta}{1 + \omega^2 T^2.\beta} = \frac{\omega T(1-\beta)}{1 + \beta(\omega T)^2}$$

$$\boxed{\omega_m = \frac{1}{T\sqrt{\beta}}}$$

At $\omega = \omega_m$, $\phi = \phi_m$, $\therefore$ $\tan\phi_m = \dfrac{\omega_m T(1-\beta)}{1 + \beta(\omega_m T)^2} = \dfrac{\dfrac{1}{T\sqrt{\beta}}T(1-\beta)}{1 + \beta.\dfrac{1}{T^2\beta^2}T^2} = \dfrac{(1-\beta)/\sqrt{\beta}}{1+1} = \dfrac{(1-\beta)}{2\sqrt{\beta}}$

$$\therefore \text{ Maximum lag angle, } \boldsymbol{\phi_m = \tan^{-1}\left[\frac{1-\beta}{2\sqrt{\beta}}\right]} \qquad\qquad .....(5.17)$$

To find the value of β from ϕ_m

From the equation (5.17), it is evident that $(1-\beta)$ and $2\sqrt{\beta}$ are the two sides of right angled triangle. Hence construct a right angle triangle as shown in fig 5.6.

With reference to figure 5.6, $\sin\phi_m = \dfrac{1-\beta}{1+\beta}$

$\sin\phi_m (1+\beta) = (1-\beta)$

$\sin\phi_m + \beta\sin\phi_m = 1-\beta$

$\beta\sin\phi_m + \beta = 1 - \sin\phi_m$

$\beta(\sin\phi_m + 1) = 1 - \sin\phi_m$

$$\therefore \ \boldsymbol{\beta = \frac{1-\sin\phi_m}{1+\sin\phi_m}}$$

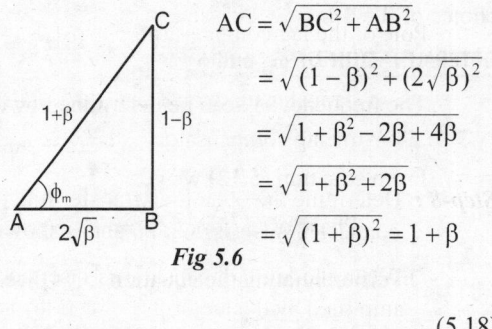

$AC = \sqrt{BC^2 + AB^2}$

$= \sqrt{(1-\beta)^2 + (2\sqrt{\beta})^2}$

$= \sqrt{1+\beta^2 - 2\beta + 4\beta}$

$= \sqrt{1+\beta^2 + 2\beta}$

$= \sqrt{(1+\beta)^2} = 1+\beta$

Fig 5.6

$$.....(5.18)$$

PROCEDURE FOR THE DESIGN OF LAG COMPENSATOR USING BODE PLOT

The following steps may be followed to design a lag compensator using bode plot and to be connected in series with transfer function of uncompensated system, G(s).

Step-1 : Choose the value of K in uncompensated system to meet the steady state error requirement.

Step-2 : Sketch the bode plot of uncompensated system. *[Refer Chapter-3 for the procedure to sketch bode plot].*

Step-3 : Determine the phase margin of the uncompensated system from the bode plot. If the phase margin does not satisfy the requirement then lag compensation is required.

Step-4 : Choose a suitable value for the phase margin of the compensated system.

Let γ_d = Desired phase margin as given in specifications.

γ_n = Phase margin of compensated system.

Now, $\gamma_n = \gamma_d + \in$

where $\in$ = Additional phase lag to compensate for shift in gain crossover frequency.

Choose an initial value of $\in = 5°$.

Step-5 : Determine the new gain crossover frequency, ω_{gcn}. The new ω_{gcn} is the frequency corresponding to a phase margin of γ_n on the bode plot of uncompensated system.

Let, ϕ_{gcn} = Phase of $G(j\omega)$ at new gain crossover frequency, ω_{gcn}

Now, $\gamma_n = 180° + \phi_{gcn}$ (or) $\phi_{gcn} = \gamma_n - 180°$

The new gain crossover frequency, ω_{gcn} is given by the frequency at which the phase of $G(j\omega)$ is ϕ_{gcn}.

Step-6 : Determine the parameter, β of the compensator. The value of β is given by the magnitude of $G(j\omega)$ at new gain crossover frequency, ω_{gcn}. Find the db gain (A_{gcn}) at new gain crossover frequency, ω_{gcn}.

$$\text{Now, } A_{gcn} = 20\log\beta \quad (\text{or}) \quad \frac{A_{gcn}}{20} = \log\beta, \quad \therefore \beta = 10^{A_{gcn}/20}$$

Step-7 : Determine the transfer function of lag compensator.

Place the zero of the compensator arbitrarily at $1/10^{th}$ of the new gain crossover frequency, ω_{gcn}.

$$\therefore \text{ Zero of the lag compensator, } z_c = \frac{1}{T} = \frac{\omega_{gcn}}{10}$$

$$\text{Now, } T = \frac{10}{\omega_{gcn}}$$

Pole of the lag compensator, $p_c = 1/\beta T$

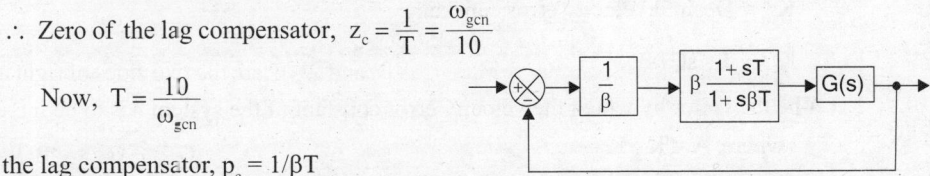

$$\left.\begin{array}{c}\text{Transfer function}\\\text{of lag compensator}\end{array}\right\} G_c(s) = \frac{s + \dfrac{1}{T}}{s + \dfrac{1}{\beta T}} = \beta\left(\frac{1 + sT}{1 + s\beta T}\right)$$

Fig 5.7 : Block diagram of lag compensated system.

Step-8 : Determine the open loop transfer function of compensated system. The lag compensator is connected in series with plant as shown in fig 5.7.

When the lag compensator is inserted in series with plant, the open loop gain of the system is amplified by the factor β ($\beta>1$). If the gain produced is not required then attenuator with gain $1/\beta$ can be introduced in series with the lag compensator to nullify the gain produced by lag compensator.

The open loop transfer function of the compensated system,

$$G_o(s) = \frac{1}{\beta}.G_c(s).G(s) = \frac{1}{\beta}.\beta\frac{(1+sT)}{(1+s\beta T)}.G(s) = \frac{(1+sT)}{(1+s\beta T)}.G(s)$$

Step-9 : Determine the actual phase margin of compensated system. Calculate the actual phase angle of the compensated system using the compensated transfer function at new gain crossover frequency, ω_{gcn}.

Let, ϕ_{gco} = Phase of $G_o(j\omega)$ at $\omega = \omega_{gcn}$

Actual phase margin of the compensated system, $\gamma_o = 180° + \phi_{gco}$

If the actual phase margin satisfies the given specification then the design is accepted. Otherwise repeat the procedure from step 4 to 9 by taking $\in$ as 5° more than previous design.

PROCEDURE FOR DESIGN OF LAG COMPENSATOR USING ROOT LOCUS

The following steps may be followed to design a lag compensator using root locus and to be connected in series with the transfer function of uncompensated system.

Step-1 : Draw the root locus of uncompensated system. *[Refer chapter-4 for the procedure to construct root locus].*

Step-2 : Determine the dominant pole, s_d. Draw a straight line through the origin with an angle $\cos^{-1}\zeta$ with respect to negative real axis. The intersection point of the straight line with root locus gives the dominant pole, s_d.

Step-3 : Determine the open loop gain of the uncompensated system at $s = s_d$. Let this gain be K. The open loop gain K at $s = s_d$ on root locus is given by,

$$K = \frac{\text{Product of vector lengths from } s_d \text{ to open loop poles}}{\text{Product of vector lengths from } s_d \text{ to open loop poles}} \quad \text{(vector length measured to scale)}$$

Step-4 : Calculate the parameter, β of the compensator.

Let, K_{vu} = Velocity error constant of uncompensated system.

   K_{vd} = Desired velocity error constant.

   $K_{vu} = \underset{s \to 0}{Lt}\ sG(s)$

Let A be the factor by which the velocity error constant of the system has to be increased,

   where, $A = K_{vd}/K_{vu}$

Choose β such that it is 10 to 20% greater than A.

   $\therefore \beta = (6.1 \text{ to } 6.2) \times A$.

Step-5 : Determine the transfer function of lag compensator. The zero of the lag compensator (1/T) is chosen to be 10% of the second pole of uncompensated system.

   $\therefore$ Zero of the compensator, $z_c = (-1/T) = 0.1 \times (\text{second pole of } G(s))$

   Now, $T = 1/[-0.1 \times (\text{second pole of } G(s))]$

Pole of the lag compensator, $p_c = -1/\beta T$.

$$\left.\begin{array}{l}\text{Transfer function} \\ \text{of lag compensator}\end{array}\right\}\ G_c(s) = \frac{s + \dfrac{1}{T}}{s + \dfrac{1}{\beta T}} = \beta\frac{(1 + sT)}{(1 + s\beta T)}$$

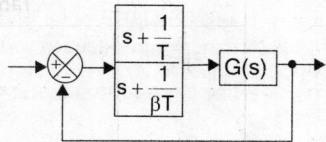

Fig 5.8 : Block diagram of lag compensated system.

Step-6 : Determine the open loop transfer function of the compensated system. The lag compensator is connected in series with the plant as shown in fig 5.8

$$\left.\begin{array}{l}\text{Open loop transfer function} \\ \text{of compensated system}\end{array}\right\}\ G_o(s) = G_c(s).G(s) = \frac{\left(s + \dfrac{1}{T}\right)}{\left(s + \dfrac{1}{\beta T}\right)}.G(s)$$

Step-7 : Check whether the compensated system satisfies the steady state error requirement. If it is satisfied, then the design is accepted otherwise repeat the design by modifying the locations of poles and zeros of the compensator.

EXAMPLE 5.1

A unity feedback system has an open loop transfer function, $G(s) = K/s(1+2s)$. Design a suitable lag compensator so that phase margin is 40° and the steady state error for ramp input is less than or equal to 0.2.

SOLUTION

Step-1 :   Calculation of gain, K.

Given that, $e_{ss} \le 0.2$ for ramp input. Let $e_{ss} = 0.2$

We know that, $e_{ss} = 1/K_v$ for ramp input.

$\therefore$ Velocity error constant, $K_v = \dfrac{1}{e_{ss}} = \dfrac{1}{0.2} = 5$.

By definition of velocity error constant, $K_v = \underset{s \to 0}{Lt}\, sG(s)H(s)$

Since the system is unity feedback system, $H(s) = 1$.

$\therefore K_v = \underset{s \to 0}{Lt}\, s\, G(s) = \underset{s \to 0}{Lt}\, s\dfrac{K}{s(1+2s)} = K \qquad \therefore K = 5$

Step-2 : Bode plot of uncompensated system.

Given that, $G(s) = 5/s(1+2s)$

Let $s = j\omega, \quad \therefore G(j\omega) = 5/j\omega(1+j2\omega)$.

MAGNITUDE PLOT

The corner frequency is, $\omega_c = 1/2 = 0.5$ rad/sec

The various terms of $G(j\omega)$ are listed in table-1. Also the table shows the slope contributed by each term and the change in slope at the corner frequency.

TABLE-1

Term	Corner frequency rad/sec	Slope db/dec	Change in slope db/dec
$\dfrac{5}{j\omega}$	–	–20	–
$\dfrac{1}{1+j2\omega}$	$\omega_c = \dfrac{1}{2} = 0.5$	–20	$-20-20 = -40$

Choose a low frequency ω_l such that $\omega_l < \omega_c$ and choose a high frequency ω_h such that $\omega_h > \omega_c$.

Let $\omega_l = 0.1$ rad/sec and $\omega_h = 10$ rad/sec

Let $A = |G(j\omega)|$ in db

At $\omega = \omega_l$, $\quad A = 20\log\left|\dfrac{5}{j\omega}\right| = 20\log\dfrac{5}{0.1} = 34$ db

At $\omega = \omega_c$, $\quad A = 20\log\left|\dfrac{5}{j\omega}\right| = 20\log\dfrac{5}{0.5} = 20$ db

At $\omega = \omega_h$, $\quad A = \left[\text{slope from } \omega_c \text{ to } \omega_h \times \log\dfrac{\omega_h}{\omega_c}\right] + A_{(at\ \omega=\omega_c)} = -40 \times \log\dfrac{10}{0.5} + 20 = -32$ db

Let the points a, b and c be the points corresponding to frequencies ω_l, ω_c and ω_h respectively on the magnitude plot. In a semilog graph sheet choose appropriate scales and fix the points a, b and c. Join the points by straight lines and mark the slope on respective region. Magnitude plot is shown in fig 5.1.1.

PHASE PLOT

The phase angle of $G(j\omega)$ as a function of ω is given by, $\phi = \angle G(j\omega) = -90° - \tan^{-1} 2\omega$

The phase angle of $G(j\omega)$ are calculated for various values of ω and listed in table-2.

<u>TABLE-2</u>

ω rad/sec	0.1	0.5	6.0	5	10
ϕ deg	−101	−135	−153	−174	−177

On the same semilog sheet take another y-axis, choose appropriate scale and draw phase plot as shown in fig 5.1.1.

Step-3 : Determination of phase margin of uncompensated system.

Let, ϕ_{gc} = Phase of $G(j\omega)$ at gain crossover frequency (ω_{gc}).

and γ = Phase margin of uncompensated system.

From the bode plot of uncompensated system we get, $\phi_{gc} = -162°$.

Now, $\gamma = 180° + \phi_{gc} = 180° - 162° = 18°$

The system requires a phase margin of 40°, but the available phase margin is 18° and so lag compensation should be employed to improve the phase margin.

Step-4 : Choose a suitable value for the phase margin of compensated system.

The desired phase margin, $\gamma_d = 40°$.

$\therefore$ Phase margin of compensated system, $\gamma_n = \gamma_d + \in$

Let initial choice of $\in = 5°$

$\therefore \gamma_n = \gamma_d + \in = 40° + 5° = 45°$

Step 5 : Determine new gain crossover frequency.

Let ω_{gcn} = New gain crossover frequency and ϕ_{gcn} = Phase of $G(j\omega)$ at ω_{gcn}

Now, $\gamma_n = 180° + \phi_{gcn}$

$\therefore \phi_{gcn} = \gamma_n - 180° = 45° - 180° = -135°$

From the bode plot we found that, the frequency corresponding to a phase of -135° is 0.5 rad/sec.

$\therefore$ New gain crossover frequency, ω_{gcn} = 0.5 rad/sec.

Step-6 : Determine the parameter, β

From the bode plot we found that, the db magnitude at ω_{gcn} is 20 db.

$\therefore |G(j\omega)|$ in db at ($\omega = \omega_{gcn}$) = A_{gcn} = 20 db

Also, $A_{gcn} = 20 \log \beta$; $\therefore \beta = 10^{A_{gcn}/20} = 10^{20/20} = 10.$

Step-7 : Determine the transfer function of lag compensator.

The zero of the compensator is placed at a frequency one-tenth of ω_{gcn}.

$\therefore$ Zero of the lag compensator, $z_c = \dfrac{1}{T} = \dfrac{\omega_{gcn}}{10}$

Now, $T = \dfrac{10}{\omega_{gcn}} = \dfrac{10}{0.5} = 20$

Pole of the lag compensator, $p_c = \dfrac{1}{\beta T} = \dfrac{1}{10 \times 20} = \dfrac{1}{200} = 0.005$

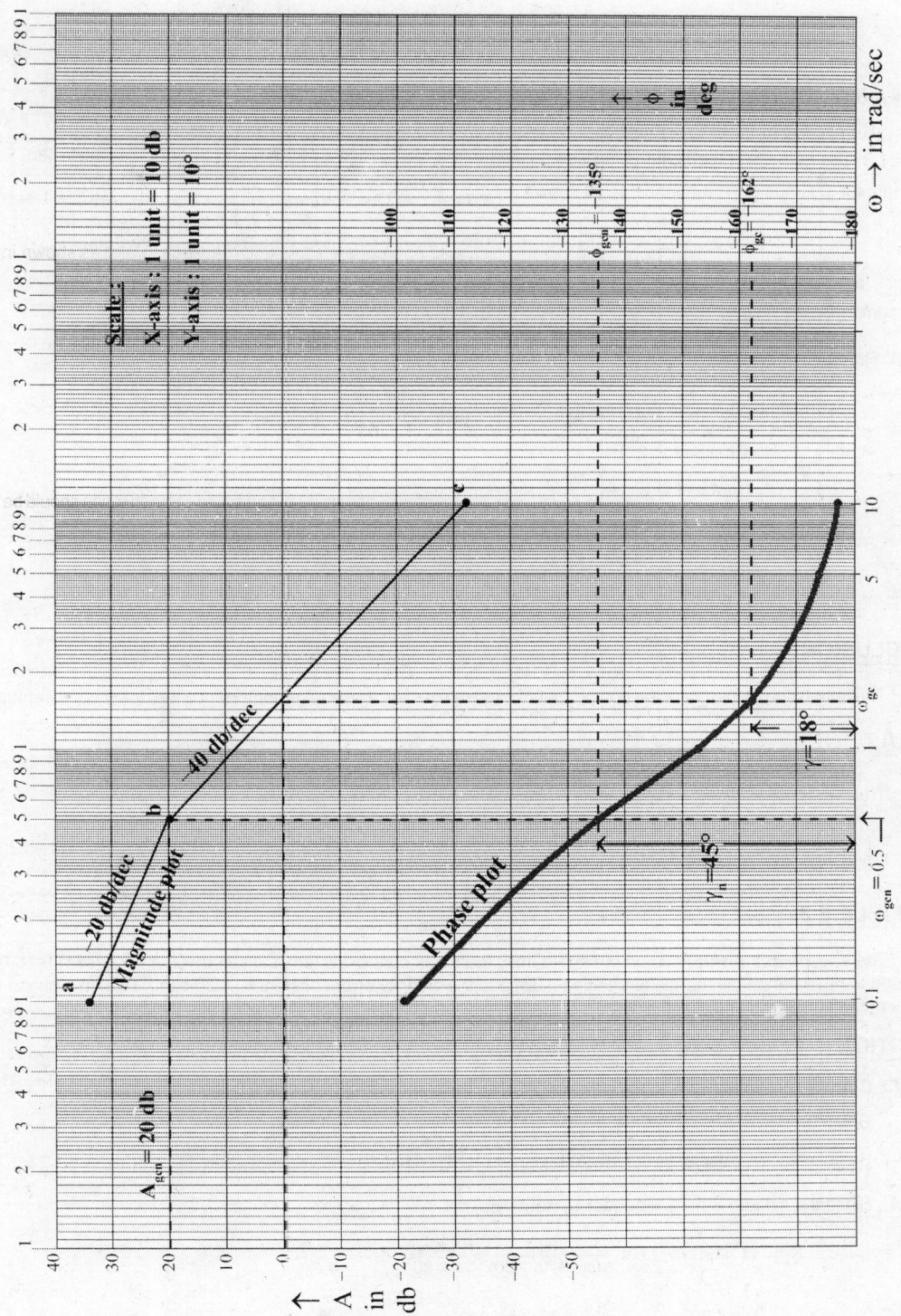

Fig 5.1.1 : Bode plot of $G(j\omega) = 5/j\omega(1+j2\omega)$.

Transfer function of lag compensator, $G_c(s) = \dfrac{s + \dfrac{1}{T}}{s + \dfrac{1}{\beta T}} = \beta \dfrac{1 + sT}{1 + s\beta T} = 10 \dfrac{(1 + 20s)}{(1 + 200s)}$

Step-8 : Determine the open loop transfer function of compensated system.

The block diagram of the compensated system is shown in fig 5.1.2. The gain of compensator is nullified by introducing an attenuator in series with compensator, as shown in fig 5.1.2

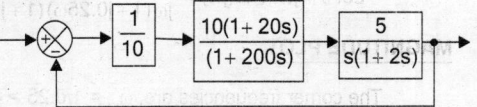

Fig 5.1.2 : *Block diagram of lag compensated system.*

Open loop transfer function of compensated system $\Bigg\}$ $G_o(s) = \dfrac{1}{10} \times \dfrac{10(1 + 20s)}{(1 + 200s)} \times \dfrac{5}{s(1 + 2s)} = \dfrac{5(1 + 20s)}{s(1 + 200s)(1 + 2s)}$

Step-9 : Determine the actual phase margin of compensated system.

On substituting $s = j\omega$ in $G_o(s)$ we get, $G_o(j\omega) = \dfrac{5(1 + j20\omega)}{j\omega(1 + j200\omega)(1 + j2\omega)}$

Let, ϕ_o = Phase of $G_o(j\omega)$

ϕ_{gco} = Phase of $G_o(j\omega)$ at $\omega = \omega_{gcn}$

ϕ_o = $\tan^{-1} 20\omega - 90° - \tan^{-1} 200\omega - \tan^{-1} 2\omega$

At $\omega = \omega_{gcn}$, $\phi_o = \phi_{gco} = \tan^{-1} 20\,\omega_{gcn} - 90° - \tan^{-1} 200\omega_{gcn} - \tan^{-1} 2\,\omega_{gcn}$

$\therefore \phi_{gco} = \tan^{-1}(20 \times 0.5) - 90° - \tan^{-1}(200 \times 0.5) - \tan^{-1}(2 \times 0.5) = -140°$.

CONCLUSION

The actual phase margin of the compensated system satisfies the requirement. Hence the design is acceptable.

RESULT

Transfer function of lag compensator, $G_c(s) = \dfrac{10(1 + 20s)}{(1 + 200s)} = \dfrac{(s + 0.05)}{(s + 0.005)}$

Open loop transfer function of compensated system, $G_o(s) = \dfrac{5(1 + 20s)}{s(1 + 200s)(1 + 2s)}$

EXAMPLE 5.2

The open loop transfer function of certain unity feedback control system is given by G(s) = K/s(s+4) (s+80). It is desired to have the phase margin to be atleast 33° and the velocity error constant K_v = 30 sec⁻¹. Design a phase lag series compensator.

SOLUTION

Step-1 : Calculation of gain, K

Given that, $K_v = 30 \text{ sec}^{-1}$

By definition of velocity error constant, $K_v = \underset{s \to 0}{Lt}\ s\,G(s)\,H(s)$

Since the system is unity feedback system, H(s) = 1.

$\therefore K_v = \underset{s \to 0}{Lt}\ sG(s) = \underset{s \to 0}{Lt}\ s\,\dfrac{K}{s(s + 4)(s + 80)} = \dfrac{K}{4 \times 80}$

i.e., $\dfrac{K}{4 \times 80} = 30$, $\therefore K = 30 \times 80 \times 4 = 9600$

Step-2 : Bode plot of uncompensated system

$$G(s) = \frac{9600}{s(s+4)(s+80)} = \frac{9600/4 \times 80}{s\left(1+\dfrac{s}{4}\right)\left(1+\dfrac{s}{80}\right)} = \frac{30}{s(1+0.25s)(1+0.0125s)}$$

Let $s = j\omega$, $\therefore G(j\omega) = \dfrac{30}{j\omega\,(1+j0.25\omega)(1+j0.0125\omega)}$

MAGNITUDE PLOT

The corner frequencies are, $\omega_{c_1} = 1/0.25 = 4$ rad/sec and $\omega_{c_2} = 1/0.0125 = 80$ rad/sec. The various terms of $G(j\omega)$ are listed in table-1. Also the table shows the slope contributed by each term and the change in slope at the corner frequency.

TABLE-1

Term	Corner frequency rad/sec	Slope db/dec	Change in slope db/dec
$\dfrac{30}{j\omega}$	—	-20	—
$\dfrac{1}{1+j0.25\omega}$	$\omega_{c1} = \dfrac{1}{0.25} = 4$	-20	$-20-20 = -40$
$\dfrac{1}{1+j0.0125\omega}$	$\omega_{c2} = \dfrac{1}{0.0125} = 80$	-20	$-40-20 = -60$

Choose a low frequency ω_l such that $\omega_l < \omega_{c1}$ and choose a high frequency ω_h such that $\omega_h > \omega_{c2}$.

Let $\omega_l = 1$ rad/sec and $\omega_h = 100$ rad/sec.

Let $A = |G(j\omega)|$ in db

At $\omega = \omega_l$, $\quad A = 20\log\left|\dfrac{30}{j\omega}\right| = 20\log\dfrac{30}{1} = 29.5\,db \approx 30\,db$

At $\omega = \omega_{c1}$, $\quad A = 20\log\left|\dfrac{30}{j\omega}\right| = 20\log\dfrac{30}{4} = 17.5\,db \approx 18\,db$

At $\omega = \omega_{c2}$, $\quad A = \left[\text{slope from } \omega_{c1} \text{ to } \omega_{c2} \times \log\dfrac{\omega_{c2}}{\omega_{c1}}\right] + A_{(at\,\omega=\omega_{c1})}$

$$= -40\log\dfrac{80}{4} + 18 = -34\,db$$

At $\omega = \omega_h$, $\quad A = \left[\text{slope from } \omega_{c2} \text{ to } \omega_h \times \log\dfrac{\omega_h}{\omega_{c2}}\right] + A_{(at\,\omega=\omega_{c2})}$

$$= -60\log\dfrac{100}{80} + (-34) = -40\,db$$

Let the points a, b, c and d be the points corresponding to frequencies ω_l, ω_{c1}, ω_{c2} and ω_h respectively on the magnitude plot. In a semilog graph sheet choose appropriate scales and fix the points a, b, c and d. Join the points by straight lines and mark the slope on the respective region. The magnitude plot is shown in fig 5.2.1.

PHASE PLOT

The phase angle of $G(j\omega)$ as a function of ω is given by,

$$\phi = \angle G(j\omega) = -90° - \tan^{-1}0.25\omega - \tan^{-1}0.0125\omega$$

The phase angle of $G(j\omega)$ are calculated for various values of ω and listed in table-2.

TABLE-2

ω rad/sec	1	4	10	50	80	100
ϕ deg	−104	−138	−164 ≈ −165	−207 ≈ −208	−222	−229 ≈ −230

On the same semilog sheet take another y-axis, choose appropriate scale and draw phase plot as shown in fig 5.2.1.

Step-3 : Determination of phase margin of uncompensated system.

Let, ϕ_{gc} = Phase of $G(j\omega)$ at gain crossover frequency (ω_{gc}).

γ = Phase margin of uncompensated system.

From the bode plot of uncompensated system we found that, ϕ_{gc} = −168°.

Now, $\gamma = 180° + \phi_{gc} = 180° - 168° = 12°$

The system requires a phase margin of atleast 33°, but the available phase margin is 12° and so lag compensation should be employed to improve the phase margin.

Step-4 : Choose a suitable value for the phase margin of compensated system.

The desired phase margin, $\gamma_d = 33°$.

$\therefore$ Phase margin of compensated system, $\gamma_n = \gamma_d + \in$

Let initial choice of $\in = 5°$; $\therefore \gamma_n = \gamma_d + \in = 33° + 5° = 38°$

Step 5 : Determine new gain crossover frequency.

Let ω_{gcn} = New gain crossover frequency and ϕ_{gcn} = Phase of $G(j\omega)$ at ω_{gcn}

Now, $\gamma_n = 180° + \phi_{gcn}$

$\therefore \phi_{gcn} = \gamma_n - 180° = 38° - 180° = -142°$

From the bode plot we found that, the frequency corresponding to a phase of −142° is 4.7 rad/sec.

$\therefore$ New gain crossover frequency, ω_{gcn} = 4.7 rad/sec.

Step-6 : Determine the parameter, β

From the bode plot we found that, the db magnitude at ω_{gcn} as 16 db.

$\therefore |G(j\omega)|$ in db at ($\omega = \omega_{gcn}$) = A_{gcn} = 16 db

Also, $A_{gcn} = 20 \log \beta$; $\therefore \beta = 10^{A_{gcn}/20} = 10^{16/20} = 6.3$.

Step-7 : Determine the transfer function of lag compensator.

The zero of the compensator is placed at a frequency one-tenth of ω_{gcn}.

$\therefore$ Zero of the lag compensator, $z_c = \dfrac{1}{T} = \dfrac{\omega_{gcn}}{10}$

Now, $T = \dfrac{10}{\omega_{gcn}} = \dfrac{10}{4.7} = 2.13$

Pole of the lag compensator, $p_c = \dfrac{1}{\beta T} = \dfrac{1}{6.3 \times 2.13} = \dfrac{1}{13.419}$

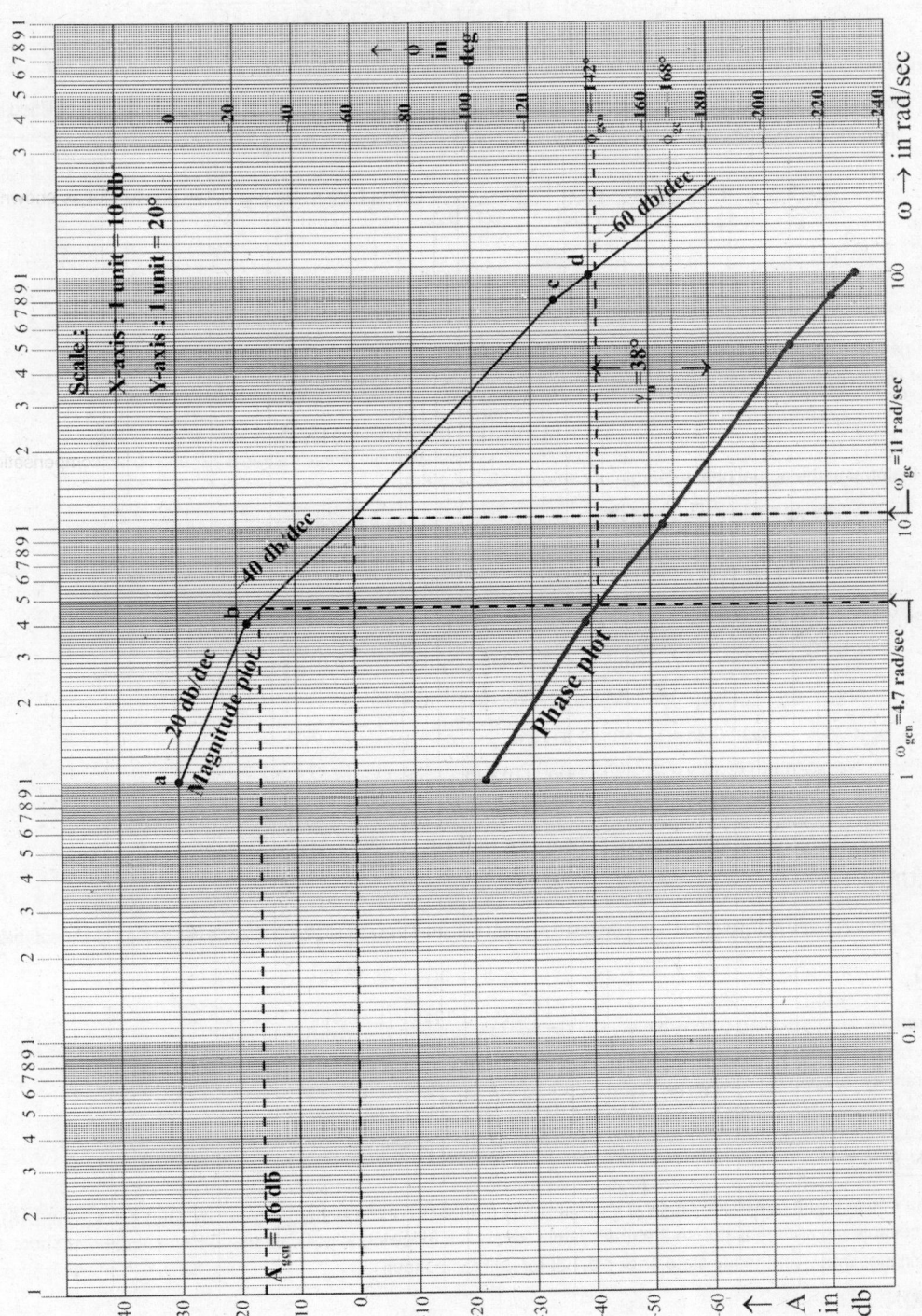

Fig 5.2.1 : Bode plot of $G(j\omega) = 30/j\omega(1+j0.25\omega)(1+j0.0125\omega)$.

Transfer function
of lag compensator
$\Bigg\}$ $G_c(s) = \dfrac{s + \dfrac{1}{T}}{s + \dfrac{1}{\beta T}} = \beta\, \dfrac{1 + sT}{1 + s\beta T} = 6.3\, \dfrac{(1 + 2.13s)}{(1 + 13.419s)}$

Step-8 : Determine the open loop transfer function of the compensated system.

The block diagram of the compensated system is shown in fig 5.2.2. The gain of the compensator is nullified by introducing an attenuator in series with the compensator, as shown in fig 5.2.2.

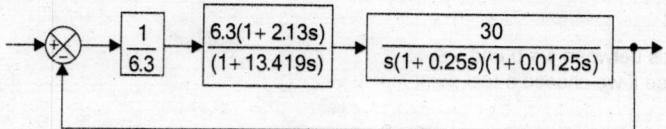

Fig 5.2.2 : *Block diagram of lag compensated system.*

Open loop transfer function
of compensated system
$\Bigg\}$ $G_o(s) = \dfrac{1}{6.3} \times \dfrac{6.3(1 + 2.13s)}{(1 + 13.419s)} \times \dfrac{30}{s(1 + 0.25s)(1 + 0.0125s)}$

$$= \dfrac{30(1 + 2.13s)}{s(1 + 13.419s)(1 + 0.25s)(1 + 0.0125s)}$$

Step-9 : Determine the actual phase margin of compensated system.

On substituting $s = j\omega$ in $G_o(s)$ we get,

$$G_o(j\omega) = \dfrac{30(1 + j2.13\omega)}{j\omega(1 + j13.419\omega)(1 + j0.25\omega)(1 + j0.0125\omega)}$$

Let ϕ_o = Phase of $G_o(j\omega)$ and ϕ_{gco} = Phase of $G_o(j\omega)$ at $\omega = \omega_{gcn}$

$\phi_o = \tan^{-1} 2.13\omega - 90° - \tan^{-1} 13.419\omega - \tan^{-1} 0.25\omega - \tan^{-1} 0.0125\omega$

At $\omega = \omega_{gcn}$, $\phi_o = \phi_{gco} = \tan^{-1} 2.13\,\omega_{gcn} - 90° - \tan^{-1} 13.419\omega_{gcn}$

$\qquad - \tan^{-1} 0.25\omega_{gcn} - \tan^{-1} 0.0125\,\omega_{gcn}$

$\therefore \phi_{gco} = \tan^{-1}(2.13 \times 4.7) - 90° - \tan^{-1}(13.419 \times 4.7)$

$\qquad - \tan^{-1}(0.25 \times 4.7) - \tan^{-1}(0.0125 \times 4.7) = -147°.$

Actual phase margin of compensated system, $\gamma_o = 180° + \phi_{gco} = 180° - 147° = 33°$

CONCLUSION

The actual phase margin of the compensated system satisfies the requirement. Hence the design is acceptable.

RESULT

Transfer function of lag compensator, $G_c(s) = \dfrac{6.3(1 + 2.13s)}{(1 + 13.419s)} = \dfrac{(s + 0.469)}{(s + 0.074)}$

Transfer function of
lag compensated system
$\Bigg\}$ $G_o(s) = \dfrac{30(1 + 2.13s)}{s(1 + 13.419s)(1 + 0.25s)(1 + 0.0125s)}$

EXAMPLE 5.3

The forward path transfer function of a certain unity feedback control system is given by $G(s) = K/s\,(s+2)(s+8)$. Design a suitable lag compensator so that the system meets the following specifications. (i) Percentage overshoot ≤ 16% for unit step input, (ii) Steady state error ≤ 0.125 for unit ramp input.

SOLUTION

Step-1 : Sketch the root locus of uncompensated system.

To find poles of open loop system

Given that, $G(s) = K / s(s + 2)(s + 8)$.

The poles of open loop transfer function are the roots of the equation, $s(s+2)(s+8) = 0$.

∴ The poles are lying at $s = 0, -2, -8$.

Let us denote poles by, p_1, p_2 and p_3. Here, $p_1 = 0$, $p_2 = -2$ and $p_3 = -8$.

To find root locus on real axis

The segments of real axis between $s = 0$ and $s = -2$ and the segment of real axis between $s = -8$ and $s = -\infty$ will be part of root locus. Because if we choose a test point in this segment then to the right of this point we have odd number of real poles and zeros.

To find angles of asymptotes and centroid

Since there are three poles, the number of root locus branches are three. There is no finite zero and so all the three root locus branches will meet the zeros at infinity. Hence the number of asymptotes required is three.

$$\text{Angles of asymptotes} = \frac{\pm 180°(2q+1)}{n-m} \; ; \quad q = 0, 1, 2, \dots n - m.$$

Here, $n = 3$ and $m = 0$. ∴ $q = 0, 1, 2, 3$.

When, $q = 0$, Angles $= \frac{\pm 180°}{3} = \pm 60°$

When, $q = 1$, Angles $= \frac{\pm 180°(2 + 1)}{3} = \pm 180°$

∴ Angles of asymptotes are, $+60°, -60°$ and $\pm 180°$

$$\text{Centroid} = \frac{\text{sum of poles} - \text{sum of zeros}}{n - m} = \frac{0 - 2 - 8}{3} = -3.33$$

To find breakaway point

Closed loop transfer function, $\dfrac{C(s)}{R(s)} = \dfrac{G(s)}{1+G(s)} = \dfrac{\dfrac{K}{s(s+2)(s+8)}}{1+\dfrac{K}{s(s+2)(s+8)}} = \dfrac{K}{s(s+2)(s+8)+K}$

The characteristic equation is, $s(s+2)(s+8) + K = 0$.

∴ $K = -s(s + 2)(s + 8) = -(s^3 + 10s^2 + 16s)$

On differentiating K with respect to s we get,

$dK/ds = -(3s^2 + 20s + 16)$

Put $dK/ds = 0$, $3s^2 + 20s + 16 = 0$

$$s = \frac{-20 \pm \sqrt{20^2 - 4 \times 3 \times 16}}{2 \times 3} = -0.9 \text{ or } -5.7$$

When $s = -0.9$, $K = -(s^3 + 10s^2 + 16s) = -((-0.9)^3 + 10(-0.9)^2 + 16(-0.9)) = 7$

When $s = -5.7$, $K = -(s^3 + 10s^2 + 16s) = -((-5.7)^3 + 10(-5.7)^2 + 16(-5.7)) = -48$

For $s = -0.9$, the value of K is positive and real and so it is actual breakaway point.

To find crossing point on imaginary axis

The characteristic equation is, $s(s+2)(s+8) + K = 0$.

∴ $s^3 + 10s^2 + 16s + K = 0$

Put s = jω.

$$(j\omega)^3 + 10(j\omega)^2 + 16(j\omega) + K = 0 \qquad \Rightarrow \quad -j\omega^3 - 10\omega^2 + j16\omega + K = 0$$

On equating imaginary part to zero we get,

$$-j\omega^3 + j16\omega = 0 \quad \Rightarrow \quad -j\omega^3 = -j16\omega \quad \Rightarrow \quad \omega^2 = 16$$

$$\therefore \ \omega = \pm\sqrt{16} = \pm 4$$

Hence the root locus crosses the imaginary axis at +j4 and −j4. The complete root locus sketch is shown in fig 5.3.1.

Step-2 : Determine the dominant pole s_d.

Given that, %M_p = 16%

We know that, % $M_p = e^{-\zeta\pi/\sqrt{1-\zeta^2}}$; $\therefore \ e^{-\zeta\pi/\sqrt{1-\zeta^2}} = 0.16$

On taking natural log we get, $\dfrac{-\zeta\pi}{\sqrt{1-\zeta^2}} = ln\,0.16 = -1.83$

On squaring we get, $\dfrac{\zeta^2\pi^2}{1-\zeta^2} = 3.3489$

On cross multiplication we get,

$$\zeta^2\pi^2 = 3.3489(1-\zeta^2) \quad \Rightarrow \quad \zeta^2\pi^2 + 3.3489\zeta^2 = 3.3489 \quad \Rightarrow \quad \zeta^2(\pi^2 + 3.3489) = 3.3489$$

$$\therefore \zeta = \sqrt{\dfrac{3.3489}{\pi^2 + 3.3489}} = 0.5$$

$$\therefore \ \cos^{-1}\zeta = \cos^{-1}0.5 = 60°$$

Draw a straight line at an angle of 60° with respect to real axis as shown in fig 5.3.1. The meeting point of this line with root locus is the dominant pole, s_d.

From fig 5.3.1. we get, s_d = −0.75 ± j6.35 (Dominant poles occur as conjugate poles).

Step 3 : To find gain K, at s = s_d

$$K = \frac{\text{Product of vector lengths from open loop poles to } s_d}{\text{Product of vector lengths from open loop zeros to } s_d}$$

Product of vector lengths from poles = $l_1 \times l_2 \times l_3$

From root locus plot of fig 5.3.1. we get, l_1 = 6.5 ; l_2 = 6.8 and l_3 = 7.35

Note : *Vector lengths are measured to scale.*

Since there is no finite zero, the product of vector lengths from zeros is unity.

$$\therefore K = l_1 \times l_2 \times l_3 = 6.5 \times 6.8 \times 7.35 = 19.845 \approx 20$$

Step 4 : To find parameter, β

Given that, G(s) = K/s(s + 2)(s + 8) = 20/s(s + 2)(s + 8)

Velocity error constant of uncompensated system $\Big\}$ $K_{vu} = \underset{s\to0}{Lt}\ sG(s) = \underset{s\to0}{Lt}\ s\,\dfrac{20}{s(s+2)(s+8)} = \dfrac{20}{2\times8} = 1.25$

It is given that steady state error, $e_{ss} \le 0.125$ for ramp input.

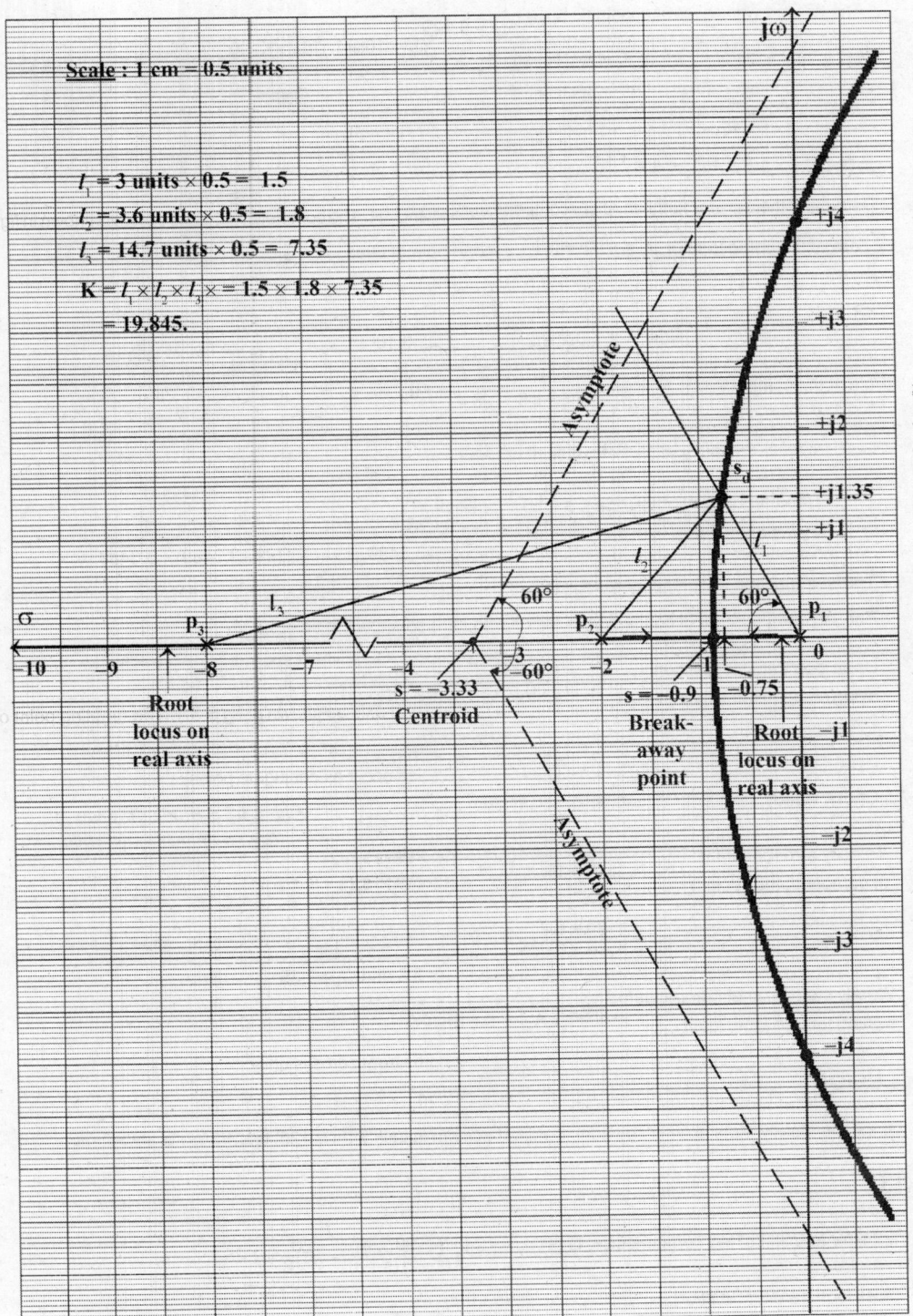

Fig 5.3.1. : Root locus sketch of $1+G(s) = 1 + K/s(s+2)(s+8)$.

Scale : 1 cm = 0.5 units

l_1 = 3 units × 0.5 = 1.5
l_2 = 3.6 units × 0.5 = 1.8
l_3 = 14.7 units × 0.5 = 7.35

$K = l_1 × l_2 × l_3 = 1.5 × 1.8 × 7.35$
= 19.845.

∴ The desired velocity error constant, $K_{vd} = \dfrac{1}{e_{ss}} = \dfrac{1}{0.125} = 8$

Let A be the factor by which K_v is increased.

∴ $A = \dfrac{K_{vd}}{K_{vu}} = \dfrac{8}{1.25} = 6.4$

Let, $\beta = 1.2 \times A = 1.2 \times 6.4 = 7.68$

Step 5 : Determine the transfer function of lag compensator.

Zero of the compensator, $z_c = -1/T = 0.1 \times$ second pole of G(s)

$$= 0.1 \times (-2) = -0.2$$

Now, $T = 1/0.2 = 5$

Pole of the compensator, $P_c = \dfrac{-1}{\beta T} = \dfrac{-1}{7.68 \times 5} = \dfrac{-1}{38.4} = -0.026$

∴ $\dfrac{1}{\beta T} = 0.026$ and $\beta T = 38.4$

$$\left.\begin{array}{l}\text{Transfer function of}\\\text{lag compensator}\end{array}\right\} G_c(s) = \dfrac{s + 1/T}{s + 1/\beta T} = \dfrac{(s + 0.2)}{(s + 0.026)}$$

Step-6 : Transfer function of compensated system.

The lag compensator is connected in series with G(s) as shown in fig 5.3.2.

$$\left.\begin{array}{l}\text{Open loop transfer function}\\\text{of lag compensated system}\end{array}\right\} G_o(s) = \dfrac{(s + 0.2)}{(s + 0.026)} \times \dfrac{20}{s(s + 2)(s + 8)}$$

$$= \dfrac{20(s + 0.2)}{s(s + 2)(s + 8)(s + 0.026)}$$

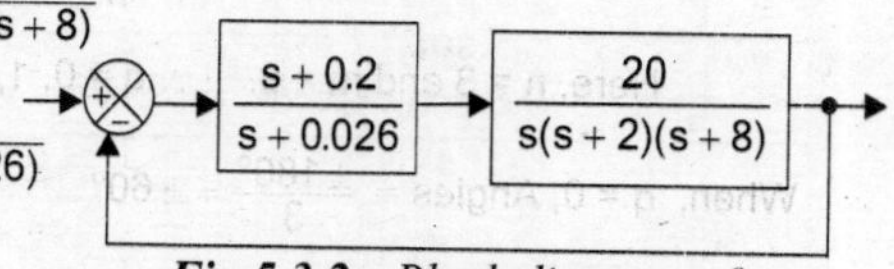

Step-7 : Check steady state error of compensated system

Fig 5.3.2 : *Block diagram of lag compensated system.*

$$\left.\begin{array}{l}\text{Velocity error constant}\\\text{of compensated system}\end{array}\right\} K_{vc} = \operatorname*{Lt}_{s \to 0} sG_o(s) = \operatorname*{Lt}_{s \to 0} s\, \dfrac{20(s + 0.2)}{s(s + 2)(s + 8)(s + 0.026)}$$

$$= \dfrac{20 \times 0.2}{2 \times 8 \times 0.026} = 9.615$$

$$\left.\begin{array}{l}\text{Steady state error of compensated}\\\text{system for unit ramp input}\end{array}\right\} e_{ss} = \dfrac{1}{K_{vc}} = \dfrac{1}{9.615} = 0.104$$

CONCLUSION

Since the steady state error of the compensated system is less than 0.125, the design is acceptable.

RESULT

Transfer function of lag compensator, $G_c(s) = (s + 0.2)/(s + 0.026)$.

$$\left.\begin{array}{l}\text{Open loop transfer function}\\\text{of lag compensated system}\end{array}\right\} G_o(s) = \dfrac{20(s + 0.2)}{s(s + 2)(s + 8)(s + 0.026)}$$

EXAMPLE 5.4

The controlled plant of a unity feedback system is $G(s) = K/s(s +10)^2$. It is specified that velocity error constant of the system be equal to 20, while the damping ratio of the dominant roots be 0.707. Design a suitable cascade compensation scheme to meet the specifications.

SOLUTION

Step-1 : Sketch the root locus of uncompensated system

To find poles of open loop system

Given that, $G(s) = K/s(s + 10)^2$

The poles of open loop transfer function are the roots of the equation $s(s + 10)^2 = 0$.

$\therefore$ The poles are lying at $s = 0, -10, -10$.

Let us denote poles by, p_1, p_2 and p_3. Here, $p_1 = 0$, $p_2 = -10$ and $p_3 = -10$.

To find root locus on real axis

The entire negative real axis will be a part of root locus. Because if we choose a test point on negative real axis then to the right of this point we have odd number of real poles and zeros.

To find angles of asymptotes and centroid

Since there are three poles, the number of root locus branches are three. There is no finite zero and so all the three root locus branches will meet the zeros at infinity. Hence the number of asymptotes required is three.

$$\text{Angles of asymptotes} = \frac{\pm 180°(2q + 1)}{n - m} \; ; \quad q = 0, 1, 2, \dots\dots n - m.$$

Here, $n = 3$ and $m = 0$. $\therefore q = 0, 1, 2, 3$.

When, $q = 0$, Angles $= \dfrac{\pm 180°}{3} = \pm 60°$

When, $q = 1$, Angles $= \dfrac{\pm 180°(2 + 1)}{3} = \pm 180°$

$\therefore$ Angles of asymptotes are, $+ 60°, - 60°$ and $\pm 180°$

$$\text{Centroid} = \frac{\text{sum of poles} - \text{sum of zeros}}{n - m} = \frac{0 - 10 - 10}{3} = \frac{-20}{3} = -6.6$$

To find breakaway point

Closed loop transfer function, $\dfrac{C(s)}{R(s)} = \dfrac{G(s)}{1 + G(s)} = \dfrac{\dfrac{K}{s(s+10)^2}}{1 + \dfrac{K}{s(s+10)^2}} = \dfrac{K}{s(s+10)^2 + K}$

The characteristic equation is, $s(s + 10)^2 + K = 0$

$\therefore K = -s(s + 10)^2 = -(s^3 + 20s^2 + 100s)$

On differentiating K with respect to s we get, $dK/ds = -(3s^2 + 40s + 100)$.

Put $dK/ds = 0$, $3s^2 + 40s + 100 = 0$

$$s = \frac{-40 \pm \sqrt{40^2 - 4 \times 3 \times 100}}{2 \times 3} = \frac{-40 \pm 20}{6} = -3.33, -10$$

When $s = -3.33$, $K = -[(-3.33)^3 + 20 (-3.33)^2 + 100(-3.33)] = 148.14$

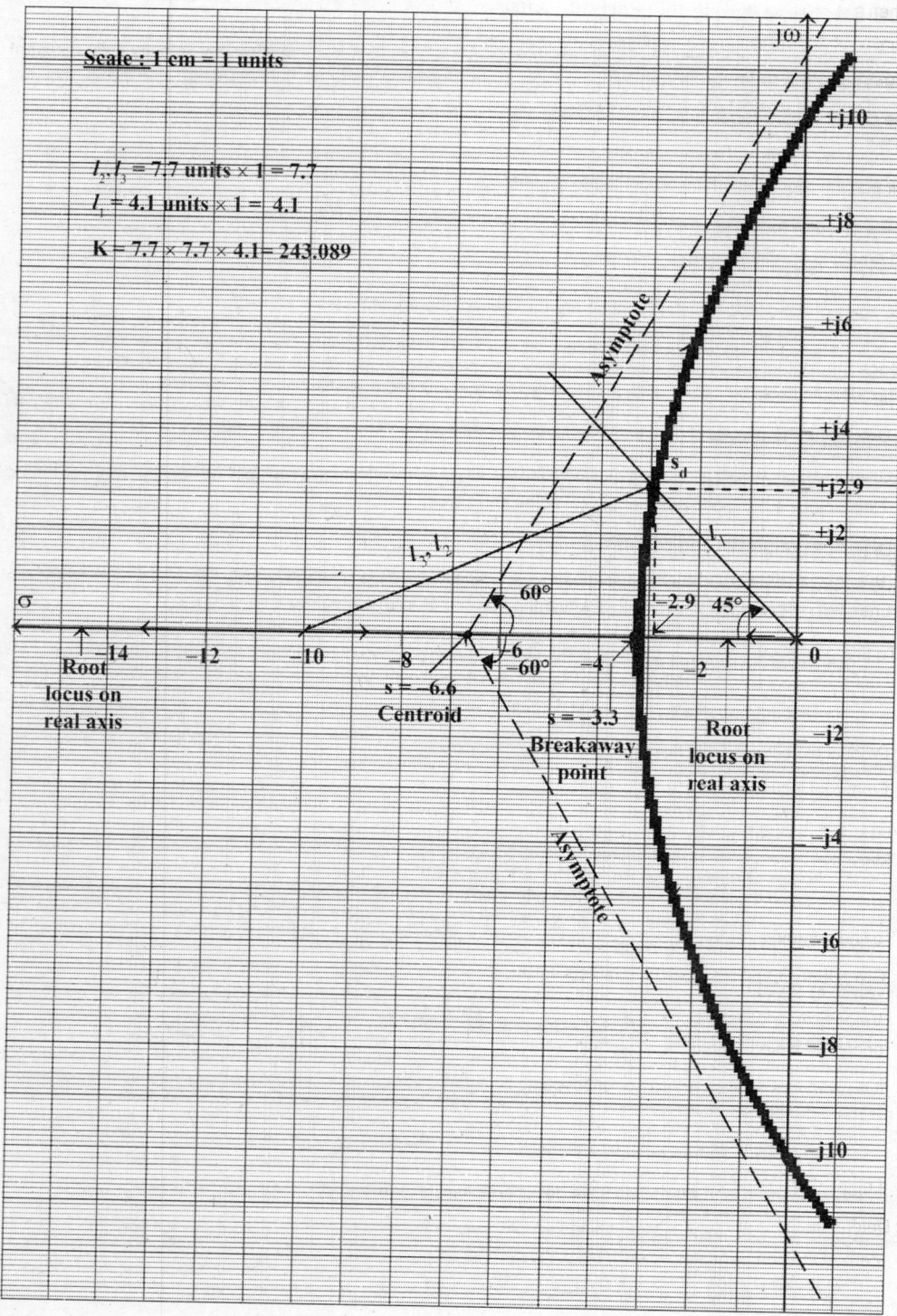

Scale : 1 cm = 1 units

$l_2, l_3 = 7.7$ units $\times 1 = 7.7$
$l_1 = 4.1$ units $\times 1 = 4.1$

$K = 7.7 \times 7.7 \times 4.1 = 243.089$

Fig 5.4.1. : Root locus sketch of $1 + G(s) = 1 + K/s(s+10)^2$.

When s = –10,   K = –[(–10)^3 + 20 (–10)^2 + 100(–10)] = 0

For s = –3.33, the value of K is positive and real and so it is actual breakaway point.

To find crossing point of imaginary axis

The characteristic equation is,   s(s+10)^2+ K = 0

$$\therefore s^3 + 20s^2 + 100s + K = 0$$

Put s = jω.

$$\therefore (j\omega)^3 + 20(j\omega)^2 + 100(j\omega) + K = 0 \quad \Rightarrow \quad -j\omega^3 - 20\omega^2 + j100\omega + K = 0$$

On equating imaginary part to zero we get,

$$-j\omega^3 + j100\omega = 0 \quad \Rightarrow \quad -j\omega^3 = -j100\omega \quad \Rightarrow \quad \omega^2 = 100$$

$$\therefore \quad \omega = \pm\sqrt{100} = \pm 10$$

Hence the root locus crosses the imaginary axis at +j10 and –j10. The complete root locus sketch is shown in fig 5.4.1.

Step-2 : Determine the dominant pole s$_d$.

Given that, ζ = 0.707

$$\therefore \cos^{-1}\zeta = \cos^{-1}0.707 = 45.008 \approx 45°$$

Draw a straight line at an angle of 45° with respect to real axis as shown in fig 5.4.1. The meeting point of this line with root locus is the dominant pole, s$_d$.

From fig 5.4.1. we get, s$_d$ = –2.9 ± j2.9 (Dominant poles occur as conjugate poles).

Step-3 : To find gain K, at s = s$_d$

$$K = \frac{\text{Product of vector length from open loop poles to } s_d}{\text{Product of vector length from open loop zeros to } s_d}$$

Product of vector lengths from poles = $l_1 \times l_2 \times l_3$

From root locus plot of fig 5.4.1. we get,

$$l_1 = 4.1 \; ; \; l_2 = l_3 = 7.7$$

| **Note :** *Lengths should be measured to scale.* |

Since there is no finite zero, the product of vector lengths from zeros is unity.

$$K = l_1 \times l_2 \times l_3 = 4.1 \times 7.7 \times 7.7 = 243.089 \approx 240$$

Step-4 : To find parameter, β

Given that, G(s) = K/s(s – 10)^2 = 240/s(s + 10)^2

Velocity error constant of uncompensated system $\left. \right\}$ $K_{vu} = \underset{s \to 0}{Lt}\, s\, G(s) = \underset{s \to 0}{Lt}\, s\, \dfrac{240}{s(s+10)^2} = \dfrac{240}{10^2} = 2.4$

It is given that desired velocity error constant, K$_{vd}$ should be 20. Let A be the factor by which K$_v$ is increased.

$$\therefore \; A = \frac{K_{vd}}{K_{vu}} = \frac{20}{2.4} = 8.33$$

Let, β = 1.2 × A = 1.2 × 8.33 = 9.996 ≈ 10

Step-5 : Determine the transfer function of lag compensator

Zero of the compensator, z$_c$ = –1/T = 0.1 × second pole of G(s) = 0.1 × (–10) = –1.0

$$\therefore \; 1/T = 1.0 \text{ and } \quad T = 1.0$$

Pole of the compensator, $p_c = \dfrac{-1}{\beta T} = \dfrac{-1}{10 \times 1} = -0.1$

$$\therefore \frac{1}{\beta T} = 0.1 \quad \text{and} \quad \beta T = 10$$

Transfer function of lag compensator $\left.\begin{array}{l}\end{array}\right\}$ $G_c(s) = \dfrac{s + \dfrac{1}{T}}{s + \dfrac{1}{\beta T}} = \dfrac{(s+1)}{(s+0.1)}$

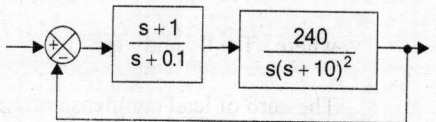

Step-6 : Determine the open loop transfer function of compensated system.

The block diagram of lag compensated system is shown in fig 5.4.2 and in this, the lag compensator is connected in series with G(s).

Fig 5.4.2 : Block diagram of lag compensated system.

Open loop transfer function of compensated system $\left.\begin{array}{l}\end{array}\right\}$ $G_o(s) = \dfrac{(s+1)}{(s+0.1)} \times \dfrac{240}{s(s+10)^2} = \dfrac{240(s+1)}{s(s+10)^2(s+0.1)}$

Step-7 : Check K_v of compensated system

Velocity error constant of compensated system $\left.\begin{array}{l}\end{array}\right\}$ $K_{vc} = \underset{s \to 0}{Lt}\, sG_o(s)$

$$= \underset{s \to 0}{Lt}\, s \frac{240(s+1)}{s(s+10)^2(s+0.1)} = \frac{240}{10^2 \times 0.1} = 24$$

CONCLUSION

Since the velocity error constant of the compensated system is greater than the desired value, the design is accepted.

RESULT

Transfer function of lag compensator $\left.\begin{array}{l}\end{array}\right\}$ $G_c(s) = \dfrac{(s+1)}{(s+0.1)}$

Open loop transfer function of lag compensated system $\left.\begin{array}{l}\end{array}\right\}$ $G_o(s) = \dfrac{240(s+1)}{s(s+10)^2(s+0.1)}$

5.3 LEAD COMPENSATOR

A compensator having the characteristics of a lead network is called a lead compensator. If a sinusoidal signal is applied to the lead network, then in steady state the output will have a phase lead with respect to the input.

The lead compensation increases the bandwidth, which improves the speed of response and also reduces the amount of overshoot. Lead compensation appreciably improves the transient response, whereas there is a small change in steady state accuracy. Generally, lead compensation is provided to make an unstable system as a stable system.

A lead compensator is basically a high pass filter and so it amplifies high frequency noise signals. If the pole introduced by the compensator is not cancelled by a zero in the system, then lead compensation increases the order of the system by one.

S-PLANE REPRESENTATION OF LEAD COMPENSATOR

The lead compensator has a zero at $s = -1/T$ and a pole at $s = -1/\alpha T$. The pole-zero plot of lead compensator is shown in fig 5.9. Here, $\alpha < 1$, so the zero is closer to the origin than the pole. The general form of lead compensator transfer function is given by equation (5.19),

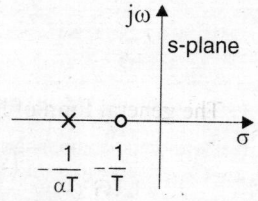

Fig 5.9 : Pole-zero plot of lead compensator.

$$G_c(s) = \frac{s + z_c}{s + p_c} = \frac{\left(s + \dfrac{1}{T}\right)}{\left(s + \dfrac{1}{\alpha T}\right)} \quad\quad\quad .....(5.19)$$

where, $T > 0$ and $\alpha < 1$

The zero of lead compensator, $z_c = \dfrac{1}{T}$ .....(5.20)

The pole of lead compensator, $p_c = \dfrac{1}{\alpha T}$ .....(5.21)

From equation (5.20) we get, $T = \dfrac{1}{z_c}$ ......(5.22)

From equation (5.21) we get, $\alpha = \dfrac{z_c}{p_c}$ .....(5.23)

REALIZATION OF LEAD COMPENSATOR USING ELECTRICAL NETWORK

The lead compensator can be realised by the RC network shown in fig 5.10.

Let, $E_i(s) =$ Input voltage, and $E_0(s) =$ Output voltage

In the network shown in fig 5.10, the input voltage is applied to the series combination of $(R_1 \| C)$ and R_2. The output voltage is obtained across R_2.

By voltage division rule,

Output voltage, $E_0(s) = E_i(s) \dfrac{R_2}{R_2 + \dfrac{\left(R_1 + \dfrac{1}{sC}\right)}{\left(R_1 + \dfrac{1}{sC}\right)}}$

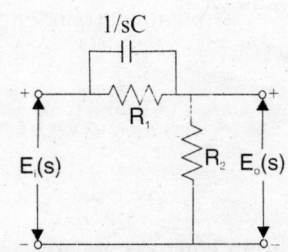

Fig 5.10 : Electrical lead compensator.

$$E_0(s) = E_i(s) . \frac{R_2}{R_2 + \dfrac{R_1}{(R_1 Cs + 1)}} = E_i(s) \frac{R_2}{\dfrac{R_2(R_1 Cs + 1) + R_1}{R_1 Cs + 1}}$$

The transfer function of the electrical network is the ratio of output voltage to input voltage.

Transfer function of electrical network $\left.\right\}$ $\dfrac{E_0(s)}{E_i(s)} = \dfrac{R_2(R_1 Cs + 1)}{[R_1 R_2 Cs + R_2 + R_1]} = \dfrac{R_1 C R_2 \left[s + \dfrac{1}{R_1 C}\right]}{R_1 C R_2 \left[s + \dfrac{(R_1 + R_2)}{R_1 C R_2}\right]}$

$$= \frac{\left[s + \dfrac{1}{R_1 C}\right]}{\left[s + \left(\dfrac{1}{R_2 /(R_1 + R_2)}\right) \dfrac{1}{R_1 C}\right]} \quad\quad .....(5.24)$$

The general form of lead compensator transfer function is,

$$G_c(s) = \frac{\left(s + \dfrac{1}{T}\right)}{\left(s + \dfrac{1}{\alpha T}\right)} \quad\quad\quad .....(5.25)$$

On comparing equations (5.24) and (5.25) we get,

$$\frac{E_o(s)}{E_i(s)} = \frac{s + \dfrac{1}{T}}{\left(s + \dfrac{1}{\alpha T}\right)} \qquad\qquad .....(5.26)$$

where, $T = R_1C$ and $\alpha = \dfrac{R_2}{R_1 + R_2}$

The transfer function of the RC network is similar to the general form of transfer function of lead compensator.

FREQUENCY RESPONSE OF LEAD COMPENSATOR

Consider the general form of lead compensator,

$$G_c(s) = \frac{s + \dfrac{1}{T}}{\left(s + \dfrac{1}{\alpha}\right)} = \frac{(1 + sT)/T}{(s\alpha T + 1)/\alpha T} = \alpha \,\frac{(1 + sT)}{(1 + \alpha sT)} \qquad .....(5.27)$$

The sinusoidal transfer function of lead compensator is obtained by letting $s = j\omega$ in equation (5.27).

$$\therefore \ G_c(j\omega) = \alpha \,\frac{(1 + j\omega T)}{(1 + j\omega\alpha T)} \qquad\qquad .....(5.28)$$

When $\omega = 0$, $G_c(j\omega) = \alpha$ .....(5.29)

From equation (5.29) we can say that the lead compensator provides an attenuation of α (Here $\alpha < 1$). If the attenuation of the compensator is not desirable then it can be eliminated by a suitable amplifier.

Let us assume that the attenuation α is eliminated by a suitable amplifier network. Now, $G_c(j\omega)$ is given by,

$$G_c(j\omega) = \frac{(1 + j\omega T)}{(1 + j\omega\alpha T)} = \frac{\sqrt{1 + (\omega T)^2} \ \angle \tan^{-1}\omega T}{\sqrt{1 + (\omega\alpha T)^2} \ \angle \tan^{-1}\omega\alpha T} \qquad .....(5.30)$$

Sinusoidal transfer function shown in equation (5.30) has two corner frequencies ω_{c1} and ω_{c2}.

Here, $\omega_{c1} = \dfrac{1}{T}$ and $\omega_{c2} = \dfrac{1}{\alpha T}$.   Since, $T > \alpha T$, $\omega_{c1} < \omega_{c2}$.

Let $A = |G_c(j\omega)|$ in db $= 20 \log \dfrac{\sqrt{1 + (\omega T)^2}}{\sqrt{1 + (\omega\alpha T)^2}}$ .....(5.31)

At very low frequencies i.e., upto ω_{c1}, $\omega T \ll 1$ and $\omega\alpha T \ll 1$

$$\therefore \ A \approx 20 \log 1 = 0$$

In the frequency range from ω_{c1} to ω_{c2}, $\omega T \gg 1$ and $\omega\alpha T \ll 1$

$$\therefore \ A \approx 20 \log \sqrt{(\omega T)^2} = 20 \log(\omega T)$$

At very high frequencies i e., after ω_{c2}, $\omega T \gg 1$ and $\omega\alpha T \gg 1$

$$\therefore A \approx 20 \log \frac{\sqrt{(\omega T)^2}}{\sqrt{(\omega\alpha T)^2}} = 20 \log \frac{1}{\alpha}$$

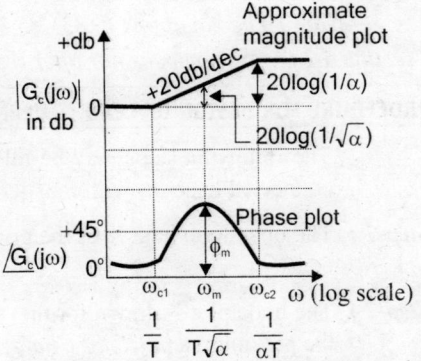

The approximate magnitude plot of lead compensator is shown in fig 5.11. The magnitude plot of Bode plot of $G_c(j\omega)$ is a straight line through 0 db upto ω_{c1}, then it has a slope of +20 db/decade upto ω_{c2} and after ω_{c2} it is a straight line with a constant gain of 20 log (1/α).

Let, $\phi = \angle G_c(j\omega)$.

$$\therefore \phi = \tan^{-1}\omega T - \tan^{-1}\omega\alpha T$$

As $\omega \to 0$, $\phi \to 0$

As $\omega \to \infty$, $\phi \to 0$

Fig 5.11 : *Bode plot of lead compensator.*

As ω is varied from 0 to ∞, the phase angle increases from 0 to a maximum value of ϕ_m at $\omega = \omega_m$, then decreases from this maximum value to 0.

It can be shown that the frequency at which maximum phase lead occurs is the geometric mean of the two corner frequencies,

Frequency of maximum phase lead, $\omega_m = \sqrt{\omega_{c1}.\omega_{c2}} = \sqrt{\dfrac{1}{T} \cdot \dfrac{1}{\alpha T}} = \dfrac{1}{T\sqrt{\alpha}}$

The choice of α is governed by the inherent noise in control systems. From the Bode plot of the lead network, we observe that the high frequency noise signals are amplified by a factor $1/\alpha$, while the low frequency control signals undergo unit amplification. Thus the signal/noise ratio at the output of the lead compensator is poorer than at its input. To prevent the signal/noise ratio at the output from deteriorating excessively, it is recommended that the value of α should not be less than 0.07. A typical choice of $\alpha = 0.6$. Also it is advisable to provide two cascaded lead networks when ϕ_m required (i.e., phase lead required) is more than 60°.

Determination of ω_m, ϕ_m and α

The frequency ω_m can be determined by differentiating ϕ with respect to ω and equating $d\phi/d\omega$ to zero.

From equation (5.32) we get,

Phase of $G_c(j\omega)$, $\phi = \angle G_c(j\omega) = \tan^{-1}\omega T - \tan^{-1}\alpha\omega T$ .....(5.32)

On differentiating the equation (5.32) with respect to ω and equating $d\phi/d\omega$ to zero, we get the frequency corresponding to maximum phase lead as, $\omega_m = 1/T\sqrt{\alpha}$.

$$\therefore \text{Frequency corresponding to maximum phase lead, } \omega_m = \frac{1}{T\sqrt{\alpha}} \qquad .....(5.33)$$

Also we can express ϕ_m in terms of α and α in terms of ϕ_m as shown below.

$$\phi_m = \tan^{-1}\left(\frac{1-\alpha}{2\sqrt{\alpha}}\right) \qquad\qquad .....(5.34)$$

$$\alpha = \frac{1 - \sin \phi_m}{1 + \sin \phi_m} \qquad \qquad .....(5.35)$$

> **Note :** *The equations (5.33), (5.34) and (5.35) can be derived by a similar analysis shown in section 5.2 for lag compensator after replacing β by α.*

PROCEDURE FOR DESIGN OF LEAD COMPENSATOR USING BODE PLOT

The following steps may be followed to design a lead compensator using bode plot and to be connected in series with transfer function of uncompensated system, G(s).

Step-1 : The open loop gain K of the given system is determined to satisfy the requirement of the error constant.

Step-2 : The bode plot is drawn for the uncompensated system using the value of K, determined from the previous step. *[Refer Chapter-3 for the procedure to sketch bode plot].*

Step-3 : The phase margin of the uncompensated system is determined from the bode plot.

Step-4 : Determine the amount of phase angle to be contributed by the lead network by using the formula given below,

$$\phi_m = \gamma_d - \gamma + \epsilon$$

where,

ϕ_m = Maximum phase lead angle of the lead compensator

γ_d = Desired phase margin

γ = Phase margin of the uncompensated system

ϵ = Additional phase lead to compensate for shift in gain crossover frequency

Choose an initial choice of as ϵ 5°

> **Note :** *If ϕ_m is more than 60° then realize the compensator as cascade of two lead compensator with each compensator contributing half of the required angle.*

Step-5 : Determine the transfer function of lead compensator

Calculate α using the equation, $\alpha = \dfrac{1 - \sin \phi_m}{1 + \sin \phi_m}$

From the bode plot, determine the frequency at which the magnitude of G(jω) is $-20 \log 1/\sqrt{\alpha}$ db. This frequency is ω_m.

Calculate T from the relation, $\omega_m = \dfrac{1}{T\sqrt{\alpha}} \qquad \therefore T = \dfrac{1}{\omega_m \sqrt{\alpha}}$

$$\left. \begin{array}{l} \text{Transfer function of} \\ \text{lead compensator} \end{array} \right\} G_c(s) = \frac{s + \dfrac{1}{T}}{s + \dfrac{1}{\alpha T}} = \frac{\alpha(1 + sT)}{(1 + \alpha sT)}$$

Step-6 : Determine the open loop transfer function of compensated system.

The lag compensator is connected in series with G(s) as shown in fig 5.12. When the lead network is inserted in series with the plant, the open loop gain of the system is attenuated by the factor α (α<1), so an amplifier with the gain of 1/α has to be introduced in series with the compensator to nullify the attenuation caused by the lead compensator.

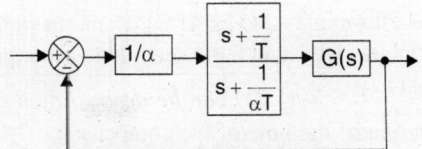

Fig 5.12 : *Block diagram*
of lead compensated system.

Open loop transfer function ⎫
of the overall system   ⎬ $G_o(s) = \dfrac{1}{\alpha} \times \dfrac{s + \dfrac{1}{T}}{s + \dfrac{1}{\alpha T}} \times G(s)$

$$= \frac{1}{\alpha} \times \frac{\alpha(1+sT)}{(1+s\alpha T)} \times G(s) = \frac{(1+sT)}{(1+s\alpha T)} \times G(s)$$

Step-7 :   Verify the design.

Finally the Bode plot of the compensated system is drawn and verify whether it satisfies the given specifications. If the phase margin of the compensated system is less than the required phase margin then repeat step 4 to 10 by taking $\in$ as 5^0 more than the previous design.

PROCEDURE FOR DESIGN OF LEAD COMPENSATOR USING ROOT LOCUS

The following steps may be followed to design a lead compensator using root locus and to be connected in series with transfer function of uncompensated system, G(s).

Step-1 :   Determine the dominant pole, s_d from the given specifications.

Dominant pole, $s_d = -\zeta\omega_n \pm j\omega_n \sqrt{1 - \zeta^2}$

> **Note :** *If ζ alone is specified and ω_n is not available, then draw the root locus and from the root locus find the dominant pole. Refer example 5.8.*

Step-2 :   Mark the poles and zeros of open loop transfer function and the dominant pole on the s-plane. Let the dominant pole be point P.

Step-3 :   Determine the angle to be contributed by lead network.

Let, ϕ = Angle to be contributed by lead network to make the point P as a point on root locus.

Draw vectors from all open loop poles and zeros to point P. Measure the angle contributed by the vectors. *[For procedure to find angle contribution by vectors refer root locus in Chapter-4].*

Now, $\phi = \begin{pmatrix} \text{Sum of angles} \\ \text{contributed by poles} \\ \text{of uncompensated system} \end{pmatrix} - \begin{pmatrix} \text{Sum of angles} \\ \text{contributed by zeros} \\ \text{of uncompensated system} \end{pmatrix} \pm n\,180°$

where n is an odd integer so that n180° is nearest to the difference between angles contributed by poles and zeros.

> **Note :** *If the angle to be contributed is more than 60° then realise the compensator as cascade of two lead compensators with each compensator contributing half of the required angle.*

Step-4 :   Determine the pole and zero of the lead compensator.

Let point O be the origin of s-plane and point P be the dominant pole. Draw straight lines OP and AP such that AP is parallel to x-axis as shown in fig 5.13.

Draw a line PC so as to bisect the angle APO [∠APO] where the point C is on the real axis. With line PC as reference, draw angles BPC and CPD such that each equal to $\phi/2$. Here the points B and D are located on the real axis.

Now the point B is the location of the pole of the compensator $(-1/\alpha T)$ and the point D is the location of the zero of the compensator $(-1/T)$. From the values of point D and B compute T and α.

Step-5 : Determine the transfer function of lead compensator

$$\left.\begin{array}{l}\text{Transfer function of} \\ \text{lead compensator}\end{array}\right\} G_c(s) = \frac{\left(s + \dfrac{1}{T}\right)}{\left(s + \dfrac{1}{\alpha T}\right)}$$

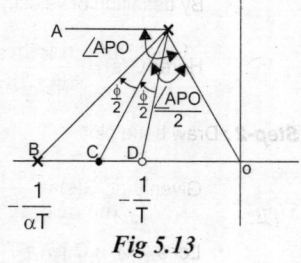

$$= \alpha \frac{(1 + sT)}{(1 + s\alpha T)}$$

Fig 5.13

Step-6 : Determine open loop transfer function of lead compensated system.

The lead compensator is connected in series with the plant as shown in fig 5.14.

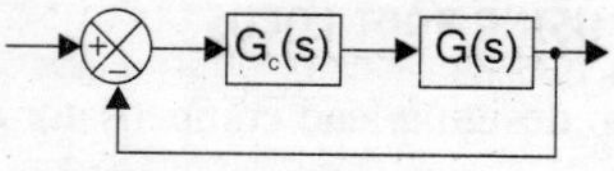

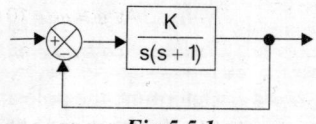

Fig 5.14 : Block diagram of lead compensated system.

Open loop transfer function of compensated system, $G_0(s) = G_c(s) \, G(s)$.

The open loop gain K is given by the value of gain of $s = s_d$. The value of gain, K is determined from pole-zero plot of lead compensated system and by using the magnitude condition given below.

$$K = \frac{\text{Product of vector lengths from all poles to } s = s_d}{\text{Product of vector lengths from all zeros to } s = s_d}$$

> **Note :** *The length of vectors should be measured to scale. For details of magnitude condition, refer root locus in chapter-4.*

Step-7 : Check whether the compensated system satisfies the error requirement. If the error requirement is satisfied then the design is accepted. Otherwise repeat the design by altering the location of poles and zeros by trial and error, without changing the value of ϕ.

> **Note :** *If the open loop gain K is specified in the problem, then take the gain at $s = s_d$ as K_1. Find a parameter, A where $A = K_1/K$. Now introduce an amplifier with gain, A in cascade with compensator to account for reduction in gain due to attenuation by parameter, α. Now, $G_0(s) = A\, G_c(s)\, G(s)$.*

EXAMPLE 5.5

Design a phase lead compensator for the system shown in fig 5.5.1 to satisfy the following specifications. (i) The phase margin of the system $\geq 45°$. (ii) Steady state error for a unit ramp input $\leq 1/15$. (iii) The gain crossover frequency of the system must be less than 7.5 rad/sec.

Fig 5.5.1

SOLUTION

Step-1 : Determine K .

Given that, steady state error, $e_{ss} \leq 1/15$ for unit ramp input

When the input is unit ramp, $e_{ss} = 1/K_v = 1/15.$ $\therefore K_v = 15$

By definition of velocity error constant we get, $K_v = \underset{s \to 0}{Lt}\, s\, G(s)H(s)$

Here, $G(s) = \dfrac{K}{s(s+1)}$ and $H(s) = 1$ $\therefore K_v = \underset{s \to 0}{Lt}\, s\, \dfrac{K}{s(s+1)} = K$

Step-2 : Draw bode plot.

Given that, $G(s) = \dfrac{K}{s(s+1)} = \dfrac{15}{s(s+1)}$

Let $s = j\omega$, $\therefore G(j\omega) = \dfrac{15}{j\omega(1+j\omega)}$

MAGNITUDE PLOT

The corner frequency is, $\omega_{c1} = 1$ rad/sec.

The various terms of $G(j\omega)$ are listed in table-1. Also the table shows the slope contributed by each term and the change in slope at the corner frequency.

TABLE-1

Term	Corner frequency rac/sec	Slope db/dec	Change in slope db/dec
$\dfrac{15}{j\omega}$	—	-20	—
$\dfrac{1}{1+j\omega}$	$\omega_{c1} = 1$	-20	$-20 + (-20) = -40$

Choose a low frequency ω_l such that $\omega_l < \omega_{c1}$ and choose a high frequency ω_h such that $\omega_h > \omega_{c1}$.

Let, $\omega_l = 0.1$ rad/sec and $\omega_h = 10$ rad/sec.

Let, $A = |G(j\omega)|$ in db

Let us calculate A at ω_l, ω_{c1} and ω_h.

At $\omega = \omega_l = 0.1\,\text{rad/sec},$ $A = 20\log\left|\dfrac{15}{j\omega}\right| = 20\log\dfrac{15}{0.1} = 43.5\,\text{db} \approx 44\,\text{db}$

At $\omega = \omega_{c1} = 1\,\text{rad/sec},$ $A = 20\log\left|\dfrac{15}{j\omega}\right| = 20\log\dfrac{15}{1} = 23.5\,\text{db} \approx 24\,\text{db}$

At $\omega = \omega_h = 10\,\text{rad/sec},$ $A = \left[\text{slope from } \omega_{c1} \text{ to } \omega_h \times \log\dfrac{\omega_h}{\omega_{c1}}\right] + A_{(\text{at } \omega = \omega_{c1})}$

$$= -40 \times \log\dfrac{10}{1} + 24 = -16\,\text{db}$$

Let the points a, b and c be the points corresponding to frequencies ω_l, ω_{c1}, and ω_h respectively on the magnitude plot. In a semilog graph sheet choose appropriate scales and fix the points a, b and c. Join the points by straight lines and mark the slope on respective region. Magnitude plot is shown in fig 5.5.2.

PHASE PLOT

The phase angle of $G(j\omega)$ as a function of ω is given by,

$$\phi = \angle G(j\omega) = 90° - \tan^{-1}\omega$$

The phase angle of $G(j\omega)$ are calculated for various values of ω and listed in table-2.

TABLE-2

ω rad/sec	0.1	0.5	1	2	5	10
ϕ deg	−96	−117	−135	−153	−169	−174

On the same semilog sheet take another y-axis, choose appropriate scale and draw phase plot as shown in fig 5.5.2.

Step-3 : Determine the phase margin of uncompensated system.

Let, ϕ_{gc} = Phase of $G(j\omega)$ at gain crossover frequency (ω_{gc}).

γ = Phase margin of uncompensated system.

From the bode plot of uncompensated system we get, $\phi_{gc} = -167°$.

Now, $\gamma = 180° + \phi_{gc} = 180° - 167° = 13°$

The system requires a phase margin of 45°, but the available phase margin is 13° and so lead compensation should be employed to improve the phase margin.

Step-4 : Find ϕ_m

The desired phase margin, $\gamma_d \geq 45°$

Let additional phase lead required, $\epsilon = 5°$

Maximum lead angle, $\phi_m = \gamma_d - \gamma + \epsilon = 45° - 13° + 5° = 37°$

Step-5 : Determine the transfer function of lead compensator.

$$\alpha = \frac{1 - \sin\phi_m}{1 + \sin\phi_m} = \frac{1 - \sin 37°}{1 + \sin 37°} = 0.2486 \approx 0.25$$

The db magnitude corresponding to $\left.\right\}$ $\omega_m = -20\log\frac{1}{\sqrt{\alpha}} = -20\log\frac{1}{\sqrt{0.25}} = -6\,db$

From the bode plot of uncompensated system the frequency ω_m corresponding to a db gain of -6 db is found to be 5.6 rad/sec.

$$\therefore \omega_m = 5.6 \text{ rad/sec.}$$

Now, $T = \dfrac{1}{\omega_m\sqrt{\alpha}} = \dfrac{1}{5.6\sqrt{0.25}} = 0.357 \approx 0.36$

Transfer function of the lead compensator $\left.\right\}$ $G_c(s) = \dfrac{s + \dfrac{1}{T}}{s + \dfrac{1}{\alpha T}} = \alpha\,\dfrac{(1 + sT)}{(1 + s\alpha T)} = 0.25\,\dfrac{(1 + 0.36s)}{(1 + 0.09s)}$

Step-6 : Open loop transfer function of compensated system.

The block diagram of the lead compensated system is shown in fig 5.5.3.

The compensator will provide an attenuation of α. To compensate for that, an amplifier of gain $1/\alpha$ is introduced in series with compensator.

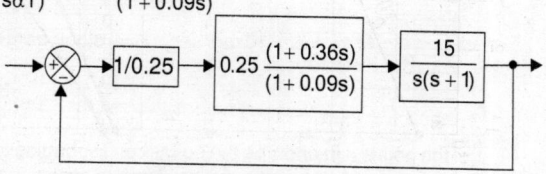

Fig 5.5.3 : *Block diagram of lead compensated system.*

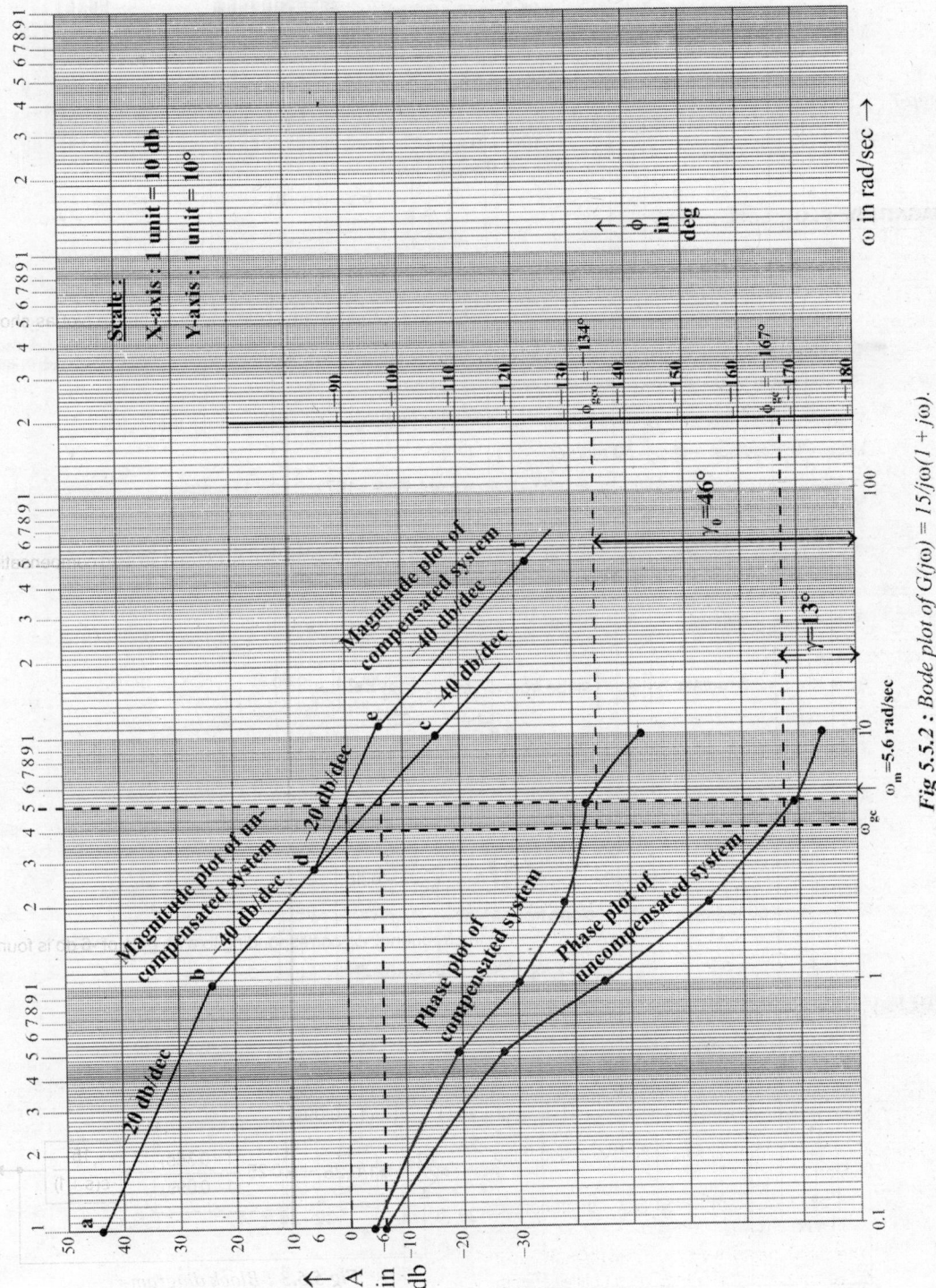

Fig 5.5.2 : Bode plot of G(jω) = 15/jω(1 + jω).

Open loop transfer function of compensated system $\bigg]$ $G_o(s) = \dfrac{1}{0.25} \times \dfrac{0.25(1+0.36s)}{(1+0.09s)} \times \dfrac{15}{s(s+1)}$

$$= \dfrac{15(1+0.36s)}{s(1+0.09s)(1+s)}$$

Step-7 : Draw the bode plot of compensated system to verify the design.

Put $s = j\omega$ in $G_o(s)$, $\quad \therefore \; G_o(j\omega) = \dfrac{15(1+j0.36\omega)}{j\omega(1+j0.09\omega)(1+j\omega)}$

MAGNITUDE PLOT

The corner frequencies are ω_{c1}, ω_{c2} and ω_{c3}.

$\omega_{c1} = \dfrac{1}{1} = 1$ rad/sec ; $\quad \omega_{c2} = \dfrac{1}{0.36} = 2.8$ rad/sec ; $\quad \omega_{c3} = \dfrac{1}{0.09} = 11.1$ rad/sec

The various terms of $G_o(j\omega)$ are listed in table-3. Also the table shows the slope contributed by each term and the change in slope at the corner frequency.

Choose a low frequency ω_l such that $\omega_l < \omega_{c1}$ and choose a high frequency ω_h such that $\omega_h > \omega_{c3}$.

Let $\omega_l = 0.1$ rad/sec and $\omega_h = 50$ rad/sec

Let $A_o = |G_o(j\omega)|$ in db

At $\omega = \omega_l = 0.1$ rad/sec, $\quad A_o = 20 \log \left| \dfrac{15}{j\omega} \right| = 20 \log \dfrac{15}{0.1} = 43.5$ db ≈ 44 db

At $\omega = \omega_{c1} = 1$ rad/sec, $\quad A_o = 20 \log \left| \dfrac{15}{j\omega} \right| = 20 \log \dfrac{15}{1} = 23.5$ db ≈ 24 db

At $\omega = \omega_{c2} = 2.8$ rad/sec, $\quad A_o = \left[\text{slope from } \omega_{c1} \text{ to } \omega_{c2} \times \log \dfrac{\omega_{c2}}{\omega_{c1}} \right] + \left(\begin{array}{c} \text{gain at} \\ \omega = \omega_{c1} \end{array} \right)$

$$= -40 \times \log \dfrac{2.8}{1} + 24 = 6 \text{ db}$$

At $\omega = \omega_{c3} = 11.1$ rad/sec, $\quad A_o = \left[\text{slope from } \omega_{c2} \text{ to } \omega_{c3} \times \log \dfrac{\omega_{c3}}{\omega_{c2}} \right] + \left(\begin{array}{c} \text{gain at} \\ \omega = \omega_{c2} \end{array} \right)$

$$= -20 \times \log \dfrac{11.1}{2.8} + 6 = -6 \text{ db}$$

At $\omega = \omega_{ch} = 50$ rad/sec, $\quad A_o = \left[\text{slope from } \omega_{c3} \text{ to } \omega_h \times \log \dfrac{\omega_h}{\omega_{c3}} \right] + \left(\begin{array}{c} \text{gain at} \\ \omega = \omega_{c3} \end{array} \right)$

$$= -40 \times \log \dfrac{50}{11.1} + (-6) = -32 \text{ db}$$

TABLE-3

Term	Corner frequency rad/sec	Slope db/dec	Change in slope db/dec
$\dfrac{15}{j\omega}$	—	−20	—
$\dfrac{1}{1+j\omega}$	$\omega_{c1} = 1$	−20	−20 −20 = −40

Term	Corner frequency rad/sec	Slope	Change in slope db/dec db/dec
$1 + j0.36\omega$	$\omega_{c2} = \dfrac{1}{0.36} = 2.8$	$+20$	$-40 + 20 = -20$
$\dfrac{1}{1 + j0.09\omega}$	$\omega_{c3} = \dfrac{1}{0.09} = 11.1$	-20	$-20 - 20 = -40$

Let the points a, b, d, e and f be the points corresponding to frequencies ω_l, ω_{c1}, ω_{c2}, ω_{c3} and ω_h respectively on the magnitude plot of compensated system. The magnitude plot of compensated system is drawn on the same semilog graph sheet by using the same scales as shown in fig 5.5.2.

PHASE PLOT

The phase angle of $G_0(j\omega)$ as a function of ω is given by,

$$\phi_0 = \angle G_0(j\omega) = \tan^{-1}0.36\omega - 90° - \tan^{-1}0.09\omega - \tan^{-1}\omega.$$

The phase angle of $G_0(j\omega)$ are calculated for various values of ω and listed in table- 4.

TABLE-4

ω rad/sec	0.1	0.5	1	2	5	10
ϕ_0 deg	-94	-109	-120	-128	-132	-142

In the same semilog sheet and by using the same scales, the phase plot of compensated system is sketched as shown in fig 5.5.2.

Let, ϕ_{gco} = Phase of $G_0(j\omega)$ at new gain crossover frequency.

γ_0 = Phase margin of compensated system.

From the bode plot of compensated system we get, $\phi_{gco} = -134°$.

Now, $\gamma_0 = 180° + \phi_{gco} = 180° - 134° = 46°$

CONCLUSION

The phase margin of the compensated system is satisfactory. Hence the design is acceptable.

RESULT

The transfer function of lead compensator, $G_c(s) = \dfrac{0.25(1 + 0.36s)}{(1 + 0.09s)} = \dfrac{(s + 2.78)}{(s + 11.11)}$

Open loop transfer function of lead compensated system $\left.\right\}$ $G_0(s) = \dfrac{15(1 + 0.36s)}{s(1 + 0.09s)(1 + s)}$

EXAMPLE 5.6

Design a lead compensator for a unity feedback system with open loop transfer function, $G(s) = K/s\,(s+1)(s+5)$ to satisfy the following specifications (i) Velocity error constant, $K_v \geq 50$ (ii) Phase margin is $\geq 20°$.

SOLUTION

Step-1 :   Determine K

Given that, $K_v \geq 50$,   Let $K_v = 50$

By definition of velocity error constant, K_v we get,

$$K_v = \underset{s \to 0}{Lt}\ s\,G(s) = \underset{s \to 0}{Lt}\ s\,\frac{K}{s(s+1)(s+5)} = \frac{K}{5}$$

$$\therefore\ K = 5 \times K_v = 5 \times 50 = 250$$

Step-2 : Draw bode plot

Given that, $G(s) = \dfrac{K}{s(s+1)(s+5)} = \dfrac{250}{s(s+1)(s+5)}$

$$= \frac{250}{s(1+s) \times 5 \times (1+s/5)} = \frac{50}{s(1+s)(1+0.2s)}$$

Let $s = j\omega$, $\therefore\ G(j\omega) = \dfrac{50}{j\omega(1+j\omega)(1+j0.2\omega)}$

MAGNITUDE PLOT

The corner frequency are, $\omega_{c1} = 1$ rad/sec and $\omega_{c2} = 1/0.2 = 5$ rad/sec.

The various terms of $G(j\omega)$ are listed in table-1. Also the table shows the slope contributed by each term and the change in slope at the corner frequencies.

TABLE-1

Term	Corner frequency rad/sec	Slope db/dec	Change in slope db/dec
$\dfrac{50}{j\omega}$	—	-20	—
$\dfrac{1}{1+j\omega}$	$\omega_{c1} = 1$	-20	$-20 + (-20) = -40$
$\dfrac{1}{1+j0.2\omega}$	$\omega_{c2} = \dfrac{1}{0.2} = 5$	-20	$-40 - 20 = -60$

Choose a low frequency ω_l such that $\omega_l < \omega_{c1}$ and choose a high frequency ω_h such that $\omega_h > \omega_{c2}$.

Let $\omega_l = 0.5$ rad/sec and $\omega_h = 10$ rad/sec

Let $A = |G(j\omega)|$ in db

At $\omega = \omega_l$, $\qquad A = 20 \log\left|\dfrac{50}{j\omega}\right| = 20 \log \dfrac{50}{0.5} = 40$ db

At $\omega = \omega_{c1}$, $\qquad A = 20 \log\left|\dfrac{50}{j\omega}\right| = 20 \log \dfrac{50}{1} = 34$ db

At $\omega = \omega_{c2}$, $\qquad A = \left[\text{slope from } \omega_{c1} \text{ to } \omega_{c2} \times \log \dfrac{\omega_{c2}}{\omega_{c1}}\right] + A_{(\text{at }\omega=\omega_{c1})}$

$$= -40 \times \log \frac{5}{1} + 34 = 6 \text{ db}$$

At $\omega = \omega_h$, $\qquad A = \left[\text{slope from } \omega_{c2} \text{ to } \omega_h \times \log \dfrac{\omega_h}{\omega_{c2}}\right] + A_{(\text{at }\omega=\omega_{c2})}$

$$= -60 \times \log \frac{10}{5} + 6 = -12 \text{ db}$$

Let the points a, b, c and c be the points corresponding to frequencies ω_l, ω_{c1}, ω_{c2} and ω_h respectively on the magnitude plot. In a semilog graph sheet choose appropriate scales and fix the points a, b, c and d. Join the points by straight lines and mark the slope on the respective region. The magnitude plot is shown in fig 5.6.2.

PHASE PLOT

The phase angle of $G(j\omega)$ as a function of ω is given by,

$$\phi = \angle G(j\omega) = -90° - \tan^{-1}\omega - \tan^{-1}0.2\omega$$

The phase angle of $G(j\omega)$ are calculated for various values of ω and listed in table-2.

TABLE-2

ω rad/sec	0.1	0.5	1.0	5	10
ϕ deg	$-96°$	-122	-146	-214	-238

On the same semilog sheet take another y-axis, choose appropriate scale and draw phase plot as shown in fig 5.6.2.

Step-3 : Determine the phase margin

Let, ϕ_{gc} = Phase of $G(j\omega)$ at gain crossover frequency.

γ = Phase margin of uncompensated system.

From the bode plot of uncompensated system we get, $\phi_{gc} = -224°$.

Now, $\gamma = 180° + \phi_{gc} = 180° - 224° = -44°$

The phase margin of the system is negative and so the system is unstable. Hence lead compensation is required to make the system stable and to have a phase margin of 20°.

Step-4 : Find ϕ_m

The desired phase margin, $\gamma_d \geq 20°$

Let additional phase lead required, $\epsilon = 5°$

Maximum lead angle, $o_m = \gamma_d - \gamma + \epsilon = 20° - (-44°) + 5° = 69°$

Since the lead angle required is greater than 60°, we have to realise the lead compensator as cascade of two lead compensators with each compensator providing half of the required phase lead angle.

$$\therefore \phi_m = \frac{69°}{2} = 34.5°$$

Step-5 : Determine the transfer function of lead compensator

$$\alpha = \frac{1 - \sin\phi_m}{1 + \sin\phi_m} = \frac{1 - \sin 34.5°}{1 + \sin 34.5°} = 0.28$$

The db magnitude corresponding to $\left.\right\}$ $\omega_m = -20\log = -20\log\frac{1}{\sqrt{0.28}} = -5.5\,db$

From the bode plot of uncompensated system the frequency, ω_m corresponding to a db gain of -5.5 db is found to be 7.8 rad/sec.

$$\therefore \omega_m = 7.8 \text{ rad/sec.}$$

Now, $T = \dfrac{1}{\omega_m\sqrt{\alpha}} = \dfrac{1}{7.8\sqrt{0.28}} = 0.24$

Transfer function of
the lead compensator $\left.\begin{array}{c}\\\\\end{array}\right\}$ $G_c(s) = \dfrac{\left(s + \dfrac{1}{T}\right)^2}{\left(s + \dfrac{1}{\alpha T}\right)^2} = \alpha^2 \dfrac{(1+sT)^2}{(1+s\alpha T)^2}$

$$= (0.28)^2 \frac{(1+0.24s)^2}{(1+0.28 \times 0.24s)^2} = 0.0784 \frac{(1+0.24s)^2}{(1+0.067s)^2}$$

Step-6 : Open loop transfer function of compensated system.

The block diagram of the compensated system
is shown in fig 5.6.1.

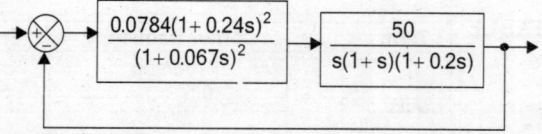

The attenuation provided by the compensator
can be retained to reduce the large value of
open loop gain, so that the unstable system
can be easily brought to stable region.

Fig 5.6.1 : Block diagram of lead compensated system.

Let $G_0(s)$ be open loop transfer function of compensated system.

$$G_0(s) = \frac{0.0784(1+0.24s)^2}{(1+0.067s)^2} \times \frac{50}{s(1+s)(1+0.2s)} = \frac{4(1+0.24s)^2}{s(1+s)(1+0.2s)(1+0.067s)^2}$$

Step-7 : Draw the bode plot of compensated system to verify the design.

Put, $s = j\omega$ in $G_0(s)$, $\therefore G_0(j\omega) = \dfrac{4(1+j0.24\omega)^2}{j\omega(1+j\omega)(1+j0.2\omega)(1+j0.067\omega)^2}$

MAGNITUDE PLOT

The corner frequencies are ω_{c1}, ω_{c2}, ω_{c3} and ω_{c4}.

$$\omega_{c1} = 1 \; ; \; \omega_{c2} = \frac{1}{0.24} = 4.2 \; ; \; \omega_{c3} = \frac{1}{0.2} = 5 \; ; \; \omega_{c4} = \frac{1}{0.067} = 15$$

The various terms of $G_0(j\omega)$ are listed in table-3. Also the table shows the slope contributed by each term and
the change in slope at the corner frequency.

Choose a low frequency ω_l such that $\omega_l < \omega_{c1}$ and choose a high frequency ω_h such that $\omega_h > \omega_{c4}$.

Let $\omega_l = 0.5$ rad/sec and $\omega_h = 30$ rad/sec

Let $A_0 = |G(j\omega)|$ in db

At $\omega = \omega_l$, $A_0 = 20\log\left|\dfrac{4}{j\omega}\right| = 20\log\dfrac{4}{0.5} = 18$ db

At $\omega = \omega_{cl}$, $A_0 = 20\log\left|\dfrac{4}{j\omega}\right| = 20\log\dfrac{4}{1} = 12$ db

At $\omega = \omega_{c2}$, $A_0 = \left[\text{slope from } \omega_{c1} \text{ to } \omega_{c2} \times \log\dfrac{\omega_{c2}}{\omega_{c1}}\right] + A_0 \text{ at } (\omega = \omega_{c1})$

$$= -40 \times \log\frac{4.2}{1} + 12 = -13 \text{ db}$$

At $\omega = \omega_{c3}$, $A_0 = \left[\text{slope from } \omega_{c2} \text{ to } \omega_{c3} \times \log\dfrac{\omega_{c3}}{\omega_{c2}}\right] + A_0 \text{ at } (\omega = \omega_{c2})$

$$= -40 \times \log\frac{4.2}{1} + 12 = -13 \text{ db}$$

At $\omega = \omega_{c4}$, $A_0 = \left[\text{slope from } \omega_{c3} \text{ to } \omega_{c4} \times \log \frac{\omega_{c4}}{\omega_{c3}} \right] + A_0 \text{ at } (\omega = \omega_{c3})$

$$= -20 \times \log \frac{15}{5} + (-13) = -22.5 \, db \approx -23 \, db$$

At $\omega = \omega_h$, $A_0 = \left[\text{slope from } \omega_{c4} \text{ to } \omega_h \times \log \frac{\omega_h}{\omega_{c4}} \right] + A_0 \text{ at } (\omega = \omega_{c4})$

$$= -60 \times \log \frac{30}{15} + (-23) = -41 \, db$$

TABLE-3

Term	Corner frequency rad/sec	Slope db/dec	Change in slope db/dec
$4/j\omega$	–	– 20	–
$\dfrac{1}{1 + j\omega}$	$\omega_{c1} = 1$	– 20	$-20 - 20 = -40$
$(1 + j0.24\omega)^2$	$\omega_{c2} = \dfrac{1}{0.24} = 4.2$	+ 40	$-40 + 40 = 0$
$\dfrac{1}{1 + j0.2\omega}$	$\omega_{c3} = \dfrac{1}{0.2} = 5$	– 20	$0 - 20 = -20$
$\dfrac{1}{(1 + j0.067\omega)^2}$	$\omega_{c4} = \dfrac{1}{0.067} = 15$	– 40	$-20 - 40 = -60$

Let the points e, f, g, h, i and j be the points corresponding to frequencies ω_l, ω_{c1}, ω_{c2}, ω_{c3}, ω_{c4} and ω_h respectively on the magnitude plot of compensated system. The magnitude plot of compensated system is drawn on the same semilog graph sheet by using the same scales as shown in fig 5.6.2.

PHASE PLOT

The phase angle of $G_c(j\omega)$ as a function of ω is given by,

$$\phi_0 = \angle G_0(j\omega) = 2 \tan^{-1} 0.24\omega - 90° - \tan^{-1}\omega - \tan^{-1} 0.2\omega - 2 \tan^{-1} 0.067\omega.$$

The phase angle of $G_c(j\omega)$ are calculated for various values of ω and listed in table - 4.

TABLE-4

ω rad/sec	0.1	0.5	6.0	2.0	5	10	15
$\angle G_0(j\omega)$ deg	–94	–112	–127	–139 ≈ -140	–150	–171	–189 ≈ -188

In the same semilog sheet and by using the same scales, the phase plot of compensated system is sketched as shown in fig 5.6.2.

Let, ϕ_{gc0} = Phase of $G_0(j\omega)$ at new gain crossover frequency (ω_{gcn}).

and γ_0 = Phase margin of compensated system.

From the bode plot of compensated system we get, $\phi_{gc0} = -140°$.

Now, $\gamma_0 = 180° + \phi_{gc0} = 180° - 140° = 40°$

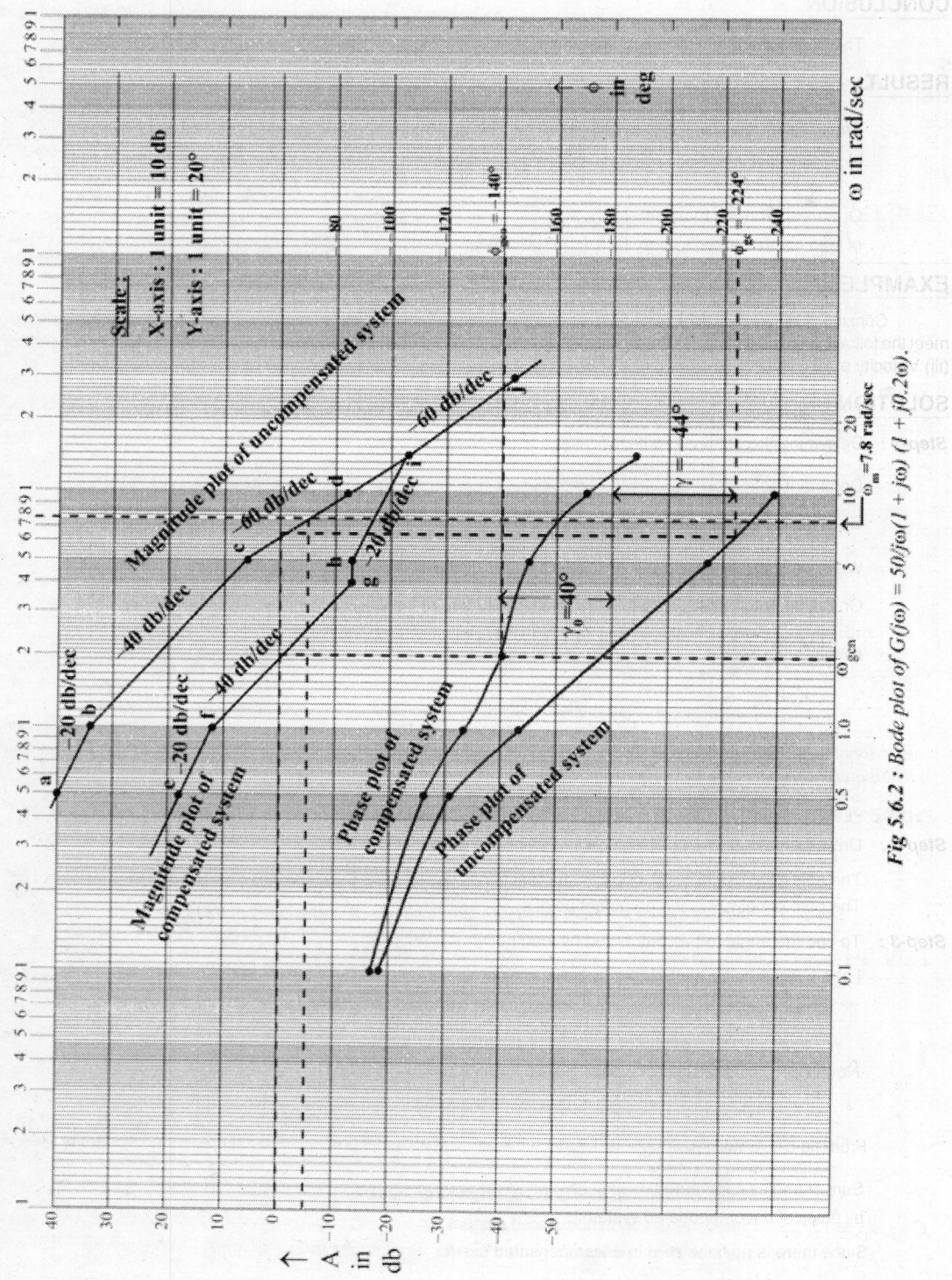

Fig 5.6.2 : Bode plot of G(jω) = 50/jω(1 + jω) (1 + j0.2ω).

CONCLUSION

The phase margin of the compensated system is satisfactory. Hence the design is acceptable.

RESULT

The transfer function of lead compensator $\Big\}$ $G_c(s) = \dfrac{0.0784(1+0.24s)^2}{(1+0.067s)^2} = \dfrac{(s+4.17)^2}{(s+14.92)^2}$

Open loop transfer funct on of lead compensated system $\Big\}$ $G_0(s) = \dfrac{4(1+0.24s)^2}{s(1+s)(1+0.2s)(1+0.067s)^2}$

EXAMPLE 5.7

Consider a unity feedback system with open loop transfer function, $G(s) = K/s(s+8)$. Design a lead compensator to meet the following specifications. (i) Percentage peak overshoot = 9.5%. (ii) Natural frequency of oscillation, $\omega_n = 12$ rad/sec. (iii) Velocity error constant, $K_v \geq 10$.

SOLUTION

Step-1 : Determine the dominant pole, s_d.

Dominant pole, $s_d = -\zeta\omega_n \pm j\omega_n\sqrt{1-\zeta^2}$.

Given that, $\omega_n = 12$ rad/sec and $\%M_p = 9.5\%$.

We know that, $\% M_p = e^{-\zeta\pi/\sqrt{1-\zeta^2}} \times 100$ $\therefore\ e^{-\zeta\pi/\sqrt{1-\zeta^2}} = 0.095$

On taking natural log we get, $-\zeta\pi/\sqrt{1-\zeta^2} = \ln(0.095)$

On squaring we get, $\dfrac{\zeta^2\pi^2}{1-\zeta^2} = (\ln 0.095)^2 = 5.54$.

$\therefore\ \zeta^2\pi^2 = 5.54 - 5.54\zeta^2$ $\Rightarrow$ $\zeta^2\pi^2 + 5.54\zeta^2 = 5.54$ $\Rightarrow$ $\zeta^2(\pi^2 + 5.54) = 5.54$

$\therefore\ \zeta = \sqrt{\dfrac{5.54}{\pi^2 + 5.54}} = 0.6$

$\therefore\ s_d = -(0.6 \times 12) \pm j12 \times \sqrt{1-0.6^2} = -7.2 \pm j9.6$

Step-2 : Draw the pole-zero plot

The pole-zero plot of open loop transfer function is shown fig 5.7.1. Poles are represented by the symbol "x". The pole at point P is the dominant pole, s_d.

Step-3 : To find the angle to be contributed by lead network.

Let ϕ = Angle to be cont ibuted by lead network to make point, P as a point on root locus.

Now, $\phi = \left(\begin{array}{c}\text{Sum of angles}\\ \text{contributed by poles}\\ \text{of uncompensated system}\end{array}\right) - \left(\begin{array}{c}\text{Sum of angles}\\ \text{contributed by zeros}\\ \text{of uncompensated system}\end{array}\right) \pm n180°$

From fig 5.7.1, we get,

Sum of angles contributed by poles of uncompensated system $\Big\}$ $= \theta_1 + \theta_2 = 127° + 85° = 212°$

Since there is no finite zero in uncompensated system, there is no angle contribution by zeros.

$\therefore\ \phi = 212° \pm n180°$

Since there is no finite zerro in uncompensated system, there is no angle contribution by zeros.

$$\therefore \phi = 212° \pm n180°$$

Let $n = 1$, $\therefore \phi = 212° - 180° = 32°$.

Step-4 : To find the pole and zero of the compensator

Draw a line AP parallel to x-axis as shown in fig 5.7.1. The bisector PC is drawn to bisect the angle APO. The angles CPD and BPC are constructed as shown in fig 5.7.1. Here $\angle CPD = \angle BPC = \phi/2 = 32°/2 = 16°$.

From fig 5.7.1.

Pole of the compensator, $p_c = -16.25$

Zero of the compensator, $z_c = -9.1$

We know that, $z_c = -1/T$ $\therefore T = 1/9.1 = 0.11$

We know that, $p_c = -1/\alpha T$ $\therefore \alpha T = 1/16.25$ (or) $\alpha = 1/(T \times 16.25) = 0.56$.

Step-5 : Determine the transfer function of lead compensator

Transfer function of lead compensator $\Big\}$ $G_c(s) = \dfrac{\left(s + \dfrac{1}{T}\right)}{\left(s + \dfrac{1}{\alpha T}\right)} \dfrac{(s+9.1)}{(s+16.25)}$

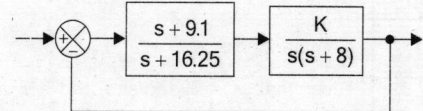

Step-6 : Determine the open loop transfer function of lead compensated system.

The block diagram of lead compensated system is shown in fig 5.7.2.

Fig 5.7.2 : *Block diagram of lead compensated system.*

Open loop transfer function of lead compensated system $\Big\}$ $G_0(s) = \dfrac{(s+9.1)}{(s+16.25)} \times \dfrac{K}{s(s+8)} = \dfrac{K(s+9.1)}{s(s+8)9s+16.25)}$

Here the value of K is given by the value of gain at the dominant pole, s_d on the root locus. From magnitude condition K is given by,

$$K = \frac{\text{Product of vector lengths from all poles to } s = s_d}{\text{Product of vector lengths from all zeros to } s = s_d}$$

From fig 5.7.1, we get,

$$K = \frac{l_1 \times l_2 \times l_4}{l_3} = \frac{12.15 \times 9.2 \times 13.3}{9.9} = 150$$

$$\therefore G_0(s) = \frac{150(s+9.1)}{s(s+8)(s+16.25)}$$

Step-7 : Check for error requirement

For the compensated system, the velocity error constant is given by,

$$K_v = \underset{s \to 0}{\text{Lt}} \, s\, G_0(s) = \underset{s \to 0}{\text{Lt}} \, s\, \frac{150(s+9.1)}{s(s+8)(s+16.25)} = \frac{150 \times 9.1}{8 \times 16.25} = 10.5$$

CONCLUSION

Since the velocity error constant of the compensated system, satisfies the requirement, the design is acceptable.

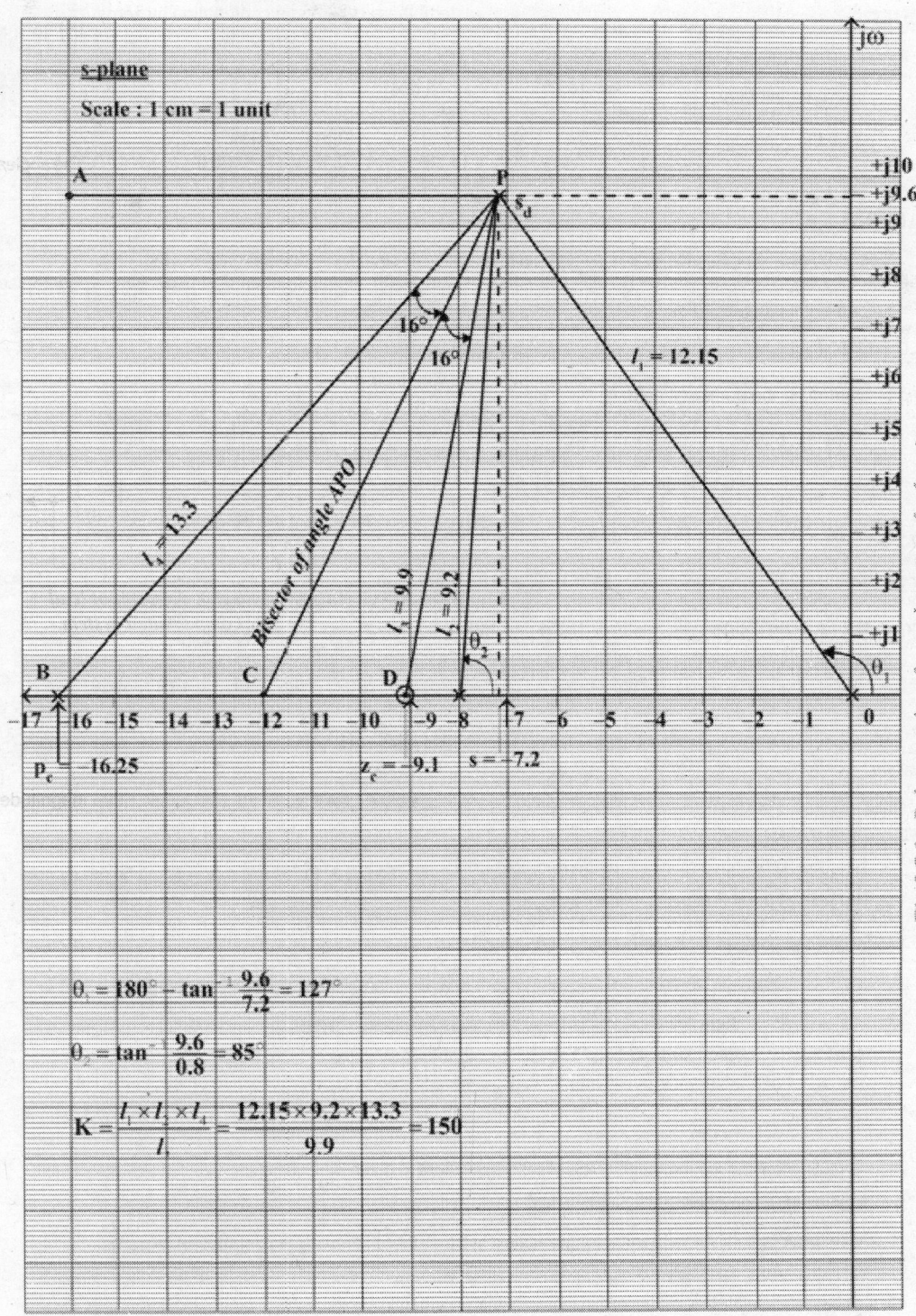

s-plane

Scale : 1 cm = 1 unit

$l_2 = 13.3$

Bisector of angle APO

$l_4 = 9.9$

$l_3 = 9.2$

$l_1 = 12.15$

θ_2

θ_1

A   P s_d

16°

16°

B

C

D

$p_c = 16.25$

$z_c = -9.1$ $s = -7.2$

+j10
+j9.6
+j9
+j8
+j7
+j6
+j5
+j4
+j3
+j2
+j1

jω

-17   -16   -15   -14   -13   -12   -11   -10   -9   -8   -7   -6   -5   -4   -3   -2   -1   0

$$\theta_1 = 180° - \tan^{-1}\frac{9.6}{7.2} = 127°$$

$$\theta_2 = \tan^{-1}\frac{9.6}{0.8} = 85°$$

$$K = \frac{l_1 \times l_3 \times l_2}{l_4} = \frac{12.15 \times 9.2 \times 13.3}{9.9} = 150$$

Fig 5.7.1 : Pole zero plot of open loop transfer function.

RESULT

Transfer function of lead compensator $\Big\}$ $G_c(s) = \dfrac{(s+9.1)}{(s+16.25)} = 0.56 \dfrac{(1+0.11s)}{(1+0.06s)}$

Transfer function of lead compensated system $\Big\}$ $G_0(s) = \dfrac{150(s+9.1)}{s(s+8)(s+16.25)} = \dfrac{10.5(1+0.11s)}{s(1+0.125s)(1+0.06s)}$

EXAMPLE 5.8

Design a lead compensator for a unity feedback system with open loop transfer function $G(s) = K/s\,(s+4)\,(s+7)$ to meet the following specifications. (i) % Peak overshoot = 12.63%. (ii) Natural frequency of oscillation, ω_n = 8 rad/sec. (iii) Velocity error constant, $K_v \geq 2.5$.

SOLUTION

Step-1 : Determine the dominant pole, s_d.

Dominant pole, $s_d = -\zeta\omega_n \pm j\omega_n\sqrt{1-\zeta^2}$

Given that, ω_n = 8 rad/sec and %M_p = 12.63%.

% $M_p = e^{-\zeta\pi/\sqrt{1-\zeta^2}} \times 100$; $\therefore e^{-\zeta\pi/\sqrt{1-\zeta^2}} = 0.1263$

On taking natural log we get, $-\zeta\pi/\sqrt{1-\zeta^2} = \ln(0.1263)$

On squaring we get, $\dfrac{\zeta^2\pi^2}{1-\zeta^2} = (\ln 0.1263)^2 = 4.28$

$\therefore \zeta^2\pi^2 = 4.28 - 4.28\zeta^2 \quad \Rightarrow \quad \zeta^2\pi^2 + 4.28\zeta^2 = 4.28 \quad \Rightarrow \quad \zeta^2(\pi^2 + 4.28) = 4.28$

$\therefore \zeta = \sqrt{\dfrac{4.28}{\pi^2 + 4.28}} = 0.55$

$\therefore s_d = -(0.55 \times 8) \pm j8 \times \sqrt{1 - 0.55^2} = -4.4 \pm j6.68 = -4.4 \pm j6.7$

Step-2 : Draw the pole-zero plot

The pole-zero plot of open loop transfer function is shown fig 5.8.1. Poles are represented by the symbol "x". The pole at point P is the dominant pole, s_d.

Step-3 : To find the angle to be contributed by lead network.

Let ϕ = Angle to be contributed by lead network to make point, P as a point on root locus.

Now, $\phi = \left(\begin{array}{c} \text{Sum of angles} \\ \text{contributed by poles} \\ \text{of uncompensated system} \end{array} \right) - \left(\begin{array}{c} \text{Sum of angles} \\ \text{contributed by zeros} \\ \text{of uncompensated system} \end{array} \right) \pm n180°$

From fig 5.8.1, we get,

$\left. \begin{array}{c} \text{Sum of angles contributed} \\ \text{by poles of uncompensated system} \end{array} \right\} = \theta_1 + \theta_2 + \theta_3 = 123° + 93° + 69° = 285°$

Since there is no finite zero in uncompensated system, there is no angle contribution by zeros.

$\therefore \phi = 285° \pm n180°$

Let n = 1,

$\therefore \phi = 285° - 180° = 105°$.

Since the angle contribution is more than 60°, the lead compensator is realised as cascade of two compensators with each compensator, contributing half of the required angle.

$$\therefore \phi = 105/2 = 52.5° \approx 52°$$

Step-4 : To find the poles and zeros of the compensator

Draw a line AP parallel to x-axis as shown in fig 5.8.6. The bisector PC is drawn to bisect the angle APO. The angles CPD and BPC are constructed as shown in fig 5.8.6. Here $\angle CPD = \angle BPC = \phi/2 = 52°/2 = 26°$.

From fig 5.8.1.

Pole of the compensator, $p_c = -13.55$

Zero of the compensator, $z_c = -4.65$

We know that, $z_c = -1/T$; $\qquad \therefore T = 1/4.65 = 0.215$

We know that, $p_c = -1/\alpha T$; $\qquad \therefore \alpha T = 1/13.55$ (or) $\alpha = 1/T \times 13.55 = 0.343$.

Step-5 : Determine the transfer function of lead compensator

Transfer function of lead compensator $\left. \right\}$ $G_c(s) = \dfrac{\left(s + \dfrac{1}{T}\right)^2}{\left(s + \dfrac{1}{\alpha T}\right)^2} = \dfrac{(s + 4.65)^2}{(s + 13.55)^2}$

Step-6 : Determine the open loop transfer function of lead compensated system.

Block diagram of lead compensated system is shown in fig 5.8.2.

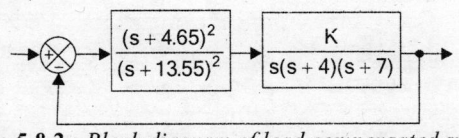

Fig 5.8.2 : *Block diagram of lead compensated system.*

Open loop transfer function of lead compensated system $\left. \right\}$ $G_0(s) = \dfrac{(s + 4.65)^2}{(s + 13.55)^2} \times \dfrac{K}{s(s + 4)(s + 7)}$

Here the value of K is given by the value of gain at the dominant pole, s_d on the root locus. From magnitude condition K is given by,

$$K = \dfrac{\text{Product of vector lengths from all poles to } s = s_d}{\text{Product of vector lengths from all zeros to } s = s_d}$$

From fig 5.8.1, we get,

$$K = \dfrac{l_1 \times l_2 \times l_4 \times l_5^2}{l_3^2} = \dfrac{3.1 \times 6.8 \times 7.3 \times 11.5^2}{6.75^2} = 1167$$

$$\therefore G_0(s) = \dfrac{1167(s + 4.65)^2}{s(s + 4)(s + 7)(s + 13.55)^2}$$

> **Note :** Compensator pole and zero are double pole and zero and so length l_3 and l_5 are squared.

Step-7 : Check for error requirement

For the compensated system, the velocity error constant is given by,

$$K_v = \underset{s \to 0}{Lt}\, sG_0(s) = \underset{s \to 0}{Lt}\, s \dfrac{1167(s + 4.65)^2}{s(s + 4)(s + 7)(s + 13.55)^2} = \dfrac{1167 \times 4.65^2}{4 \times 7 \times 13.55^2} = 4.91$$

CONCLUSION

Since the velocity error constant of the compensated system, satisfies the requirement, the design is acceptable.

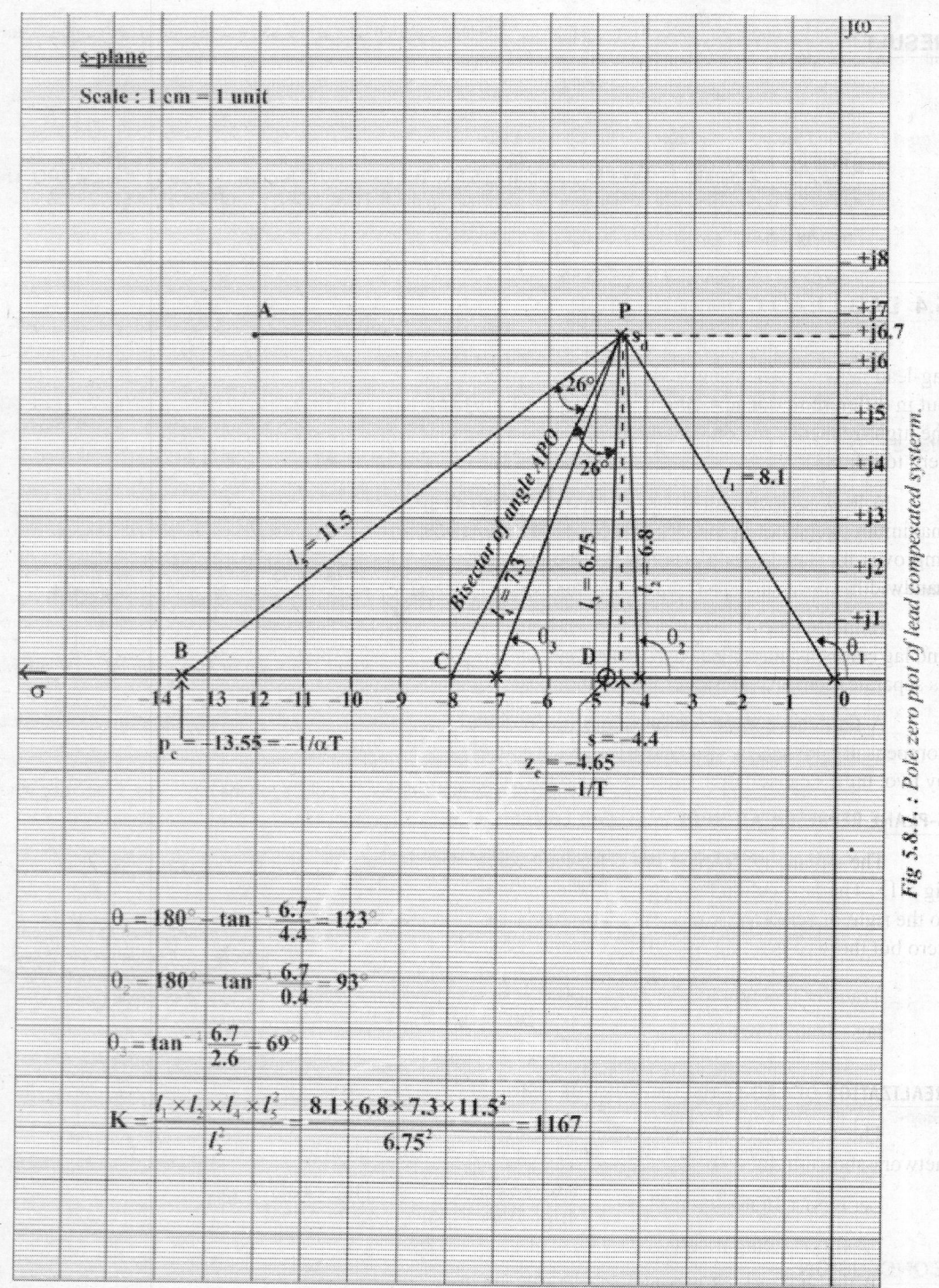

Fig 5.8.1. : Pole zero plot of lead compesated systerm.

RESULT

Transfer function of lead compensator $\left.\right\}$ $G_c(s) = \dfrac{(s+4.65)^2}{(s+13.55)^2} = 0.1178\,\dfrac{(1+0.215s)^2}{(1+0.0738s)^2}$

Transfer function of lead compensated system $\left.\right\}$ $G_0(s) = \dfrac{1167(s+4.65)^2}{s(s+4)(s+7)(s+13.55)^2}$

$$= \frac{4.91(1+0.215s)^2}{s(1+0.25s)(s+0.143s)(1+0.0738s)^2}$$

5.4 LAG-LEAD COMPENSATOR

A compensator having the characteristics of lag-lead network is called lag-lead compensator. In a lag-lead network when sinusoidal signal is applied, both phase lag and phase lead occurs in the output, but in different frequency regions. Phase lag occurs in the low frequency region and phase lead occurs in the high frequency region (i.e) the phase angle varies from lag to lead as the frequency is increased from zero to infinity.

A lead compensator basically increases bandwidth and speeds up the response and decreases the maximum overshoot in the step response. Lag compensation increases the low frequency gain and thus improves the steady state accuracy of the system, but reduces the speed of responses due to reduced bandwidth.

If improvements in both transient and steady state response are desired, then both a lead compensator and lag compensator may be used simultaneously, rather than introducing both a lead and lag compensator as separate elements. However it is economical to use a single lag-lead compensator.

A lag-lead compensation combines the advantages of lag and lead compensations. Lag-lead compensator possess two poles and two zeros and so such a compensation increases the order of the system by two, unless cancellation of poles and zeros occurs in the compensated system.

S-PLANE REPRESENTATION OF LAG-LEAD COMPENSATOR

The s-plane representation of lag-lead compensator is shown in fig 5.15. The lag section has one real pole and one real zero with pole to the right of zero. The lead section also has one real pole and one real zero but the zero is to the right of the pole.

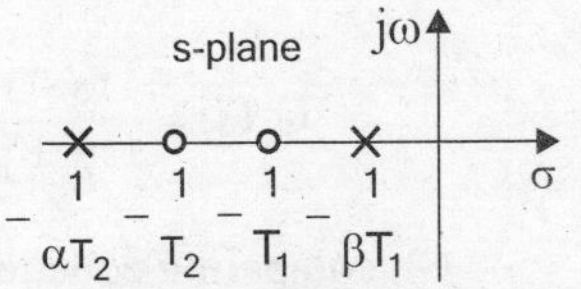

Fig 5.15 : *Pole-zero plot of lag-lead compensator.*

Transfer function of lag – lead compensator $\left.\right\}$ $G_c(s) = \underbrace{\dfrac{(s+1/T_1)}{(s+1/\beta T_1)}}_{\text{lag section}}\,\underbrace{\dfrac{(s+1/T_2)}{(s+1/\alpha T_1)}}_{\text{lead section}}$

.....(5.36)

REALIZATION OF LAG-LEAD COMPENSATOR USING ELECTRICAL NETWORK

The lag-lead compensator can be realized by the R-C network shown in fig 5.16.

Let $E_i(s)$ = Input voltage

$E_0(s)$ = Output voltage.

In the network shown in fig 5.16, the input voltage is applied to the series combination of $R_1 \parallel C_1$, R_2 and C_2. The output voltage, is obtained across series combination of R_2 and C_2. By voltage division rule,

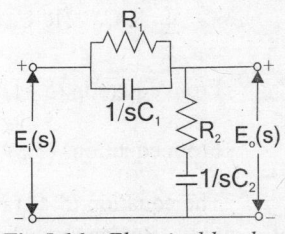

Fig 5.16 : *Electrical lag-lead compensator.*

$$E_o(s) = E_i(s) \frac{R_2 + \dfrac{1}{sC_2}}{\left(R_1 \,||\,^{\prime} \dfrac{1}{sC_1}\right) + R_2 + \dfrac{1}{sC_2}}$$

$$\therefore \frac{E_o(s)}{E_i(s)} = \frac{\dfrac{sR_2C_2 + 1}{sC_2}}{R_1 \dfrac{\dfrac{1}{sC_1}}{R_1 + \dfrac{1}{sC_1}} + \dfrac{sR_2C_2 + 1}{sC_2}} = \frac{\dfrac{sR_2C_2 + 1}{sC_2}}{\dfrac{R_1}{sR_1C_1 + 1} + \dfrac{sR_2C_2 + 1}{sC_2}}$$

$$\therefore \frac{E_o(s)}{E_i(s)} = \frac{\dfrac{sR_2C_2 + 1}{sC_2}}{\dfrac{sR_1C_2 + (sR_1C_1 + 1)(sR_2C_2 + 1)}{(sR_1C_1 + 1)sC_2}} = \frac{(sR_1C_1 + 1)(sR_2C_2 + 1)}{sR_1C_2 + (sR_1C_1 + 1)(sR_2C_2 + 1)}$$

$$= \frac{R_1C_1R_2C_2\left(s + \dfrac{1}{R_1C_1}\right)\left(s + \dfrac{1}{R_2C_2}\right)}{sR_1C_2 + R_1C_1R_2C_2\left(s + \dfrac{1}{R_1C_1}\right)\left(s + \dfrac{1}{R_2C_2}\right)}$$

On dividing the numerator and denominator by $R_1C_1R_2C_2$ we get,

$$\frac{E_o(s)}{E_i(s)} = \frac{\left(s + \dfrac{1}{R_1C_1}\right)\left(s + \dfrac{1}{R_2C_2}\right)}{\dfrac{s}{R_2C_1} + \left(s + \dfrac{1}{R_1C_1}\right)\left(s + \dfrac{1}{R_2C_2}\right)} = \frac{\left(s + \dfrac{1}{R_1C_1}\right)\left(s + \dfrac{1}{R_2C_2}\right)}{s^2 + s\left(\dfrac{1}{R_1C_1} + \dfrac{1}{R_2C_2} + \dfrac{1}{R_2C_1}\right) + \dfrac{1}{R_1R_2C_1C_2}} \quad .....(5.37)$$

The transfer function of lag-lead compensator is given by,

$$G_c(s) = \frac{(s + 1/T_1)(s + 1/T_2)}{(s + 1/\beta T_1)(s + 1/\alpha T_2)} = \frac{(s + 1/T_1)(s + 1/T_2)}{s^2 + s((1/\beta T_1) + (1/\alpha T_2)) + 1/\alpha\beta T_1 T_2} \quad .....(5.38)$$

On comparing equations (5.37) and (5.38) we get,

$$T_1 = R_1 C_1 \qquad\qquad\qquad\qquad\qquad\qquad\qquad\qquad\qquad\qquad .....(5.39)$$

$$T_2 = R_2 C_2 \qquad\qquad\qquad\qquad\qquad\qquad\qquad\qquad\qquad\qquad .....(5.40)$$

$$R_1 R_2 C_1 C_2 = \alpha\beta\, T_1 T_2 \qquad\qquad\qquad\qquad\qquad\qquad\qquad\qquad .....(5.41)$$

$$\frac{1}{R_1C_1} + \frac{1}{R_2C_2} + \frac{1}{R_2C_1} = \frac{1}{\beta T_1} + \frac{1}{\alpha T_2} \qquad\qquad\qquad\qquad .....(5.42)$$

From equation (5.41) we get, $\qquad \alpha\beta = \dfrac{R_1 R_2 C_1 C_2}{T_1 T_2} \qquad\qquad\qquad .....(5.43)$

From equations (5.39), (5.40) and (5.43), we can say that, $\alpha\beta = 1$ $\qquad .....(5.44)$

The equation (5.44) implies that a single lag-lead network does not allow an independent choice of α and β. (But separate lag and lead network will allow independent choice of α and β). Hence in the transfer function of electrical lag-lead compensator replace α by $1/\beta$ as shown below.

$$\therefore \frac{E_o(s)}{E_i(s)} = G_c(s) = \frac{(s + 1/T_1)(s + 1/T_2)}{(s + 1/\beta T_1)(s + \beta/T_2)} \qquad .....(5.45)$$

where $\beta > 1$, $T_1 = R_1 C_1$, $T_2 = R_2 C_2$ and $\dfrac{1}{R_1 C_1} + \dfrac{1}{R_2 C_2} + \dfrac{1}{R_2 C_1} = \dfrac{1}{\beta T_1} + \dfrac{\beta}{T_2}$

FREQUENCY RESPONSE OF LAG-LEAD COMPENSATOR

Consider the transfer function of lag-lead compensator.

$$G_c(s) = \frac{(s + 1/T_1)(s + 1/T_2)}{(s + 1/\beta T_1)(s + 1/\alpha T_2)} = \alpha\beta \frac{(1 + sT_1)(1 + sT_2)}{(1 + s\beta T_1)(1 + s\alpha T_2)} \qquad .....(5.46)$$

The sinusoidal transfer function of lag-lead compensator is obtained by letting $s = j\omega$ in equation (5.46).

$$\therefore G_c(j\omega) = \alpha\beta \frac{(1 + j\omega T_1)(1 + j\omega T_2)}{(1 + j\omega\beta T_1)(1 + j\omega\alpha T_2)} \qquad .....(5.47)$$

For a single lag-lead compensator, $\alpha\beta = 1$. Hence from equation (5.47) we can say that the lag-lead compensator provides a dc gain of unity.

$$\therefore G_c(j\omega) = \frac{(1 + j\omega T_1)(1 + j\omega T_2)}{(1 + j\omega\beta T_1)(1 + j\omega\alpha T_2)} \qquad .....(5.48)$$

The sinusoidal transfer function shown in equation (5.48) has four corner frequencies and they are ω_{c1}, ω_{c2}, ω_{c3} and ω_{c4}, where $\omega_{c1} < \omega_{c2} < \omega_{c3} < \omega_{c4}$.

Here $\omega_{c1} = \dfrac{1}{\beta T_1}$; $\omega_{c2} = \dfrac{1}{T_1}$; $\omega_{c3} = \dfrac{1}{T_2}$ annd $\omega_{c4} = \dfrac{1}{\alpha T_2}$

By an analysis similar to that of lag compensator the bode plot of lag-lead compensator is sketched as shown in fig 5.17.

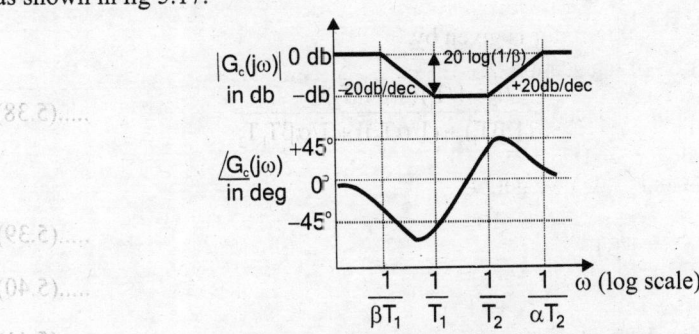

Fig 5.17 : Bode plot of lag-lead compensator.

PROCEDURE FOR DESIGN OF LAG-LEAD COMPENSATOR USING BODE PLOT

The lag-lead compensator is employed only when a large error constant and a large bandwidth are required. First design a lag section and then take $\alpha = 1/\beta$ and design a lead section. The step by step procedure for the design of lag-lead compensator is given below.

Step-1 : Determine the open loop gain K of the uncompensated system to satisfy the specified error requirement.

Step-2 : Draw the bode plot of uncompensated system.

Step-3 : From the bode plot determine the phase margin of the uncompensated system.

Let, ϕ_{gc} = Phase of $G(j\omega)$ at gain crossover frequency.

γ = Phase margin of uncompensated system.

Now, $\gamma = 180° + \phi_{gc}$

If the phase margin is not satisfactory then compensation is required.

Step-4 : Choose a new phase margin

Let, γ_d = Desired phase margin

Now, new phase margin, $\gamma_n = \gamma_d + \in$

Choose an initial value of $\in = 5°$.

Step-5 : From the bode plot, determine the new gain crossover frequency, which is the frequency corresponding to a phase margin of γ_n.

Let, ω_{gcn} = New gain crossover frequency

ϕ_{gcn} = Phase of $G(j\omega)$ at ω_{gcn}

γ_n = $180° + \phi_{gcn}$ (or) $\phi_{gcn} = \gamma_n - 180°$

In the phase plot of uncompensated system, the frequency corresponding to a phase of ϕ_{gcn} is the new gain crossover frequency ω_{gcn}.

Choose the gain crossover frequency of the lag compensator, ω_{gcl}, somewhat greater than ω_{gcn} (i.e., choose ω_{gcl} such that $\omega_{gcl} > \omega_{gcn}$).

Step-6 : Calculate β of lag compensator.

Let, $A_{gcl} = |G(j\omega)|$ in db at $\omega = \omega_{gcl}$

From the bode plot find A_{gcl}

Now, $A_{gcl} = 20 \log \beta$ (or) $\beta = 10^{(A_{gc}/20)}$

Step-7 : Determine the transfer function of lag section

The zero of the lag compensator is placed at a frequency one-tenth of ω_{gcl}.

$\therefore$ Zero of lag compensator, $z_{cl} = 1/T_1 = \omega_{gcl}/10$

Now, $T_1 = 10/\omega_{gcl}$

Pole of lag compensator, $p_{cl} = 1/\beta T_1$

$\left. \begin{array}{l} \text{Transfer function} \\ \text{of lag section} \end{array} \right\}$ $G_1(s) = \dfrac{(s + 1/T_1)}{(s + 1/\beta T_1)} = \beta \dfrac{(1 + sT_1)}{(1 + s\beta T_1)}$

Step-8 : Determine the transfer function of lead section

Take, $\alpha = 1/\beta$

From the bode plot find ω_m which is the frequency at which the db gain is $-20\log(1/\sqrt{\alpha})$

Now $T_2 = \dfrac{1}{\omega_m \sqrt{\alpha}}$

$\left. \begin{array}{l} \text{Transfer function} \\ \text{of lead section} \end{array} \right\}$ $G_2(s) = \dfrac{(s + 1/T_2)}{(s + 1/\alpha T_2)} = \alpha \dfrac{(1 + sT_2)}{(1 + s\alpha T_2)}$

Step-9 :   Determine the transfer function of lag-lead compensator.

Transfer function of lag-lead compensator, $G_c(s) = G_1(s) \times G_2(s) = \beta \dfrac{(1+sT_1)}{(1+s\beta T_1)} \times \alpha \dfrac{(1+sT_2)}{(1+s\alpha T_2)}$

Since $\alpha = \dfrac{1}{\beta}$,

$$G_c(s) = \dfrac{(1+sT_1)(1+sT_2)}{(1+s\beta T_1)(1+s\alpha T_2)}$$

Step-10 :   Determine the open loop transfer function of compensated system.

The lag-lead compensator is connected in series with $G(s)$ as shown in fig 5.18.

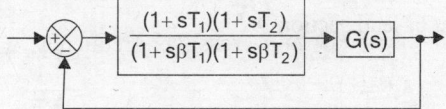

Fig 5.18 : *Block diagram of lag-lead compensated system.*

$\left.\begin{array}{l}\text{Open loop transfer function}\\ \text{of compensated system}\end{array}\right\}$ $G_0(s) = \dfrac{(1+sT_1)(1+sT_2)}{(1+s\beta T_1)(1+s\alpha T_2)} \times G(s)$

Step-11 :   Draw the bode plot of compensated system and verify whether the specifications are satisfied or not. If the specifications are not satisfied then choose another choice of α such that, $\alpha < 1/\beta$ and repeat the steps 8 to 11.

PROCEDURE FOR DESIGN OF LAG-LEAD COMPENSATOR USING ROOT LOCUS

The lag-lead compensation is employed to improve both the transient and steady state responses of a system. First design a lead section to realize the required ζ and ω_n for the dominant closed loop poles. Then determine the error constant of lead compensated system. If it is satisfactory then only lead compensation will meet the requirement. If the error constant has to be increased then design a lag section. The step-by-step procedure for the design of lag-lead compensator is given below.

Step-1 :   Determine the dominant pole, s_d

$$s_d = -\zeta\omega_n \pm \omega_n \sqrt{\zeta^2 - 1}$$

where, ζ = Damping ratio ; ω_n = Natural frequency of oscillation, rad/sec.

Step-2 :   Mark the poles and zeros of open loop transfer function and the dominant pole on the s-plane. Let the dominant pole be point P.

Step-3 :   Find the angle to be contributed by lead network to make the point P as a point on root locus.

Let, ϕ = Angle to be contributed by lead network to make point, P as a point on root locus.

Draw vectors from all open loop poles and zeros to point P. Measure the angle contributed by the vectors. [For the procedure to find angle contribution by vectors refer root locus in Chapter-5].

Now, $\phi = \left(\begin{array}{c}\text{Sum of angles}\\ \text{contributed by poles}\\ \text{of uncompensated system}\end{array}\right) - \left(\begin{array}{c}\text{Sum of angles}\\ \text{contributed by zeros}\\ \text{of uncompensated system}\end{array}\right) \pm n180°$

where n is an odd integer, so that n180° is nearest to the difference between angles contributed by poles and zeros.

Step-4 : Determine the pole and zero of the lead section.

Let point O be the origin of s-plane and point P be the dominant pole. Draw straight lines OP and AP such that AP is parallel to x-axis as shown in fig.5.19. Draw a line PC so as to bisect the angle APO [$\angle$APO] where the point C is on the real axis. With line PC as reference, draw angles BPC and CPD such that each equal to $\phi/2$. Here the points B and D are located on the real axis.

Now the point B is the location of the pole of the compensator $(-1/\alpha T_2)$ and the point D is the location of the zero of the compensator $(-1/T_2)$. Compute T_2 and α from the values of point D and B.

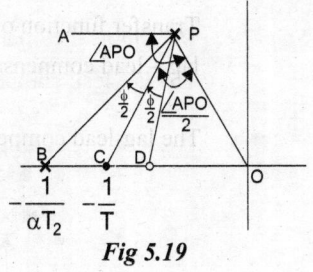

Step-5 : Determine the transfer function of lead section.

$$\left.\begin{matrix}\text{Transfer function}\\ \text{of lead section}\end{matrix}\right\} G_2(s) = \frac{\left(s + \dfrac{1}{T_2}\right)}{\left(s + \dfrac{1}{\alpha T_2}\right)}$$

Fig 5.19

Step-6 : Determine the open loop gain, K.

The open loop gain K is the value of gain at $s = s_d$. The value of gain, K is determined from pole-zero plot of lead compensated system and by using the magnitude condition given below.

$$K = \frac{\text{Product of vector lengths from all poles to } s = s_d}{\text{Product of vector lengths from all zeros to } s = s_d}$$

Note : The length of vectors should be measured to scale. For details of magnitude condition refer root locus in Chapter-4.

$$\left.\begin{matrix}\text{Open loop transfer function}\\ \text{of lead compensated system}\end{matrix}\right\} G_{02}(s) = G_2(s) \times G(s)$$

Step-7 : Determine the velocity error constant of lead compensated system.

$$\left.\begin{matrix}\text{Velocity error constant of}\\ \text{lead compensated system}\end{matrix}\right\} K_{v2} = \underset{s \to 0}{Lt} \, s \, G_{02}(s)$$

If K_{v2} satisfies the requirement then only lead compensation is sufficient but if K_{v2} is less than the desired value then provide lag compensation.

Step-8 : Determine the parameter, β of lag section.

Let, K_{vd} = Desired velocity error constant ; A = Factor by which K_v is increased.

Now, $A = K_{vd}/K_{v2}$. Select β, such that $\beta > A$. [i.e., $\beta = (1.1 \text{ to } 1.2) \times A$]

Step-9 : Determine the transfer function of lag section. Choose the zero of lag section as 10% of the second pole of uncompensated system.

$\therefore$ Zero of lag section, $z_{c1} = 0.1 \times$ second pole of G(s)

Also, $z_{c1} = \dfrac{-1}{T_1}$; $\therefore T_1 = \dfrac{-1}{z_{c1}}$; $\therefore$ Pole of lag section, $p_{c1} = \dfrac{-1}{\beta T_1}$

Transfer function of lag section, $G_1(s) = \dfrac{\left(s + \dfrac{1}{T_1}\right)}{\left(s + \dfrac{1}{\beta T_1}\right)}$

Step-10 : Determine the transfer function of lag-lead compensator and compensated system

$$\left.\begin{array}{l}\text{Transfer function of}\\[4pt]\text{lag – lead compensator}\end{array}\right\} G_c(s) = G_1(s) \times G_2(s) = \frac{\left(s + \dfrac{1}{T_1}\right)\left(s + \dfrac{1}{T_2}\right)}{\left(s + \dfrac{1}{\beta T_1}\right)\left(s + \dfrac{1}{\alpha T_2}\right)}$$

The lag-lead compensator is connected in series with G(s) as shown in fig 5.20.

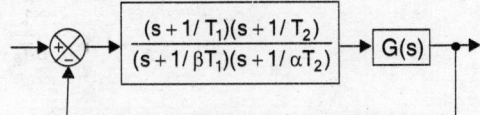

Fig 5.20 : *Block diagram of lag-lead compensated system*

$$\left.\begin{array}{l}\text{Open loop transfer function of}\\[4pt]\text{lag – lead compensated system}\end{array}\right\} G_0(s) = \frac{\left(s + \dfrac{1}{T_1}\right)\left(s + \dfrac{1}{T_2}\right)}{\left(s + \dfrac{1}{\beta T_1}\right)\left(s + \dfrac{1}{\alpha T_2}\right)} \times G(s)$$

Step-11 : Check the velocity error constant of compensated system, if it is satisfactory then the design is accepted, otherwise repeat the design by modifying the locations of poles and zeros of the compensator.

EXAMPLE 5.9

Consider the unity feedback system whose open loop transfer function is G(s) = K/s(s + 3) (s + 6). Design a lag-lead compensator to meet the following specifications. (i) Velocity error constant, K_v = 80. (ii) Phase margin, $\gamma \geq 35°$.

SOLUTION

Step-1 : Determine K

For unity feedback system,

Velocity error constant, $K_v = \underset{s \to 0}{Lt}\, sG(s)$

Given that, K_v = 80.

$\therefore \underset{s \to 0}{Lt}\, s\,G(s) = \underset{s \to 0}{Lt}\, s\,\dfrac{K}{s(s+3)(s+6)} = 80 \quad \dfrac{K}{3 \times 6} = 80 \quad K = 80 \times 3 \times 6 = 1440$

$\therefore G(s) = \dfrac{1440}{s(s+3)(s+6)} = \dfrac{1440}{s \times 3(1 + s/3) \times 6(1 + s/6)} = \dfrac{80}{s(1 + 0.33s)(1 + 0.167s)}$

Step-2 : Bode plot of uncompensated system.

In G(s), put s = jω

$\therefore G(j\omega) = \dfrac{80}{j\omega(1 + j0.33\omega)(1 + j0.167\omega)}$

MAGNITUDE PLOT

The corner frequencies are ω_{c1} and ω_{c2}.

Here $\omega_{c1} = 1/0.33 = 3$ rad/sec and $\omega_{c2} = 1/0.167 = 6$ rad/sec.

The various terms of $G(j\omega)$ are listed in table-1. Also the table shows the slope contributed by each term and the change in slope at the corner frequency.

TABLE-1

Term	Corner frequency rad/sec	Slope db/dec	Change in slope db/dec
$\dfrac{80}{j\omega}$	—	−20	—
$\dfrac{1}{1+j0.33\omega}$	$\omega_{c1} = \dfrac{1}{0.33} = 3$	−20	$-20 - 20 = -40$
$\dfrac{1}{1+j0.167\omega}$	$\omega_{c2} = \dfrac{1}{0.167} = 6$	−20	$-40 - 20 = -60$

Choose a low frequency ω_l such that $\omega_l < \omega_{c1}$ and choose a high frequency ω_h such that $\omega_h > \omega_{c2}$

Let $\omega_l = 0.5$ rad/sec and $\omega_h = 20$ rad/sec.

Let $A = |G(j\omega)|$ in db

At $\omega = \omega_l$, $\quad A = 20\log\dfrac{80}{\omega} = 20\log\dfrac{80}{0.5} = 44$ db

At $\omega = \omega_{c1}$, $\quad A = 20\log\dfrac{80}{\omega} = 20\log\dfrac{80}{3} = 28.5$ db ≈ 28 db

At $\omega = \omega_{c2}$, $\quad A = \left[\text{slope from } \omega_{c1} \text{ to } \omega_{c2} \times \log\dfrac{\omega_{c2}}{\omega_{c1}}\right] + A \text{ at } (\omega = \omega_{c1})$

$\qquad\qquad = -40 \times \log\dfrac{6}{3} + 28 = 16$ db

At $\omega = \omega_h$, $\quad A = \left[\text{slope from } \omega_{c2} \text{ to } \omega_h \times \log\dfrac{\omega_h}{\omega_{c2}}\right] + A \text{ at } (\omega = \omega_{c2})$

$\qquad\qquad = -60 \times \log\dfrac{20}{6} + 16 = -15$ db

Let the points a, b, c and d be the points corresponding to frequencies ω_l, ω_{c1}, ω_{c2} and ω_h respectively on the magnitude plot. In a semilog graph sheet choose appropriate scales and fix the points a, b, c and d. Join the points by straight lines and mark the slope on the respective region. The magnitude plot is shown in fig 5.9.2.

PHASE PLOT

The phase angle of $G(j\omega)$ as a function of ω is given by

$\phi = \angle G(j\omega) = -90° - \tan^{-1} 0.33\omega - \tan^{-1} 0.167\omega.$

The phase angle of $G(j\omega)$ are calculated for various values of ω and listed in table-2.

TABLE-2

ω rad/sec	0.5	1.0	3.0	6	10	20
$\angle G(j\omega)$ deg	−104	−118	−161 ≈ -160	−198	−222	−244.7 ≈ -244

On the same semilog sheet take another y-axis, choose appropriate scale and draw phase plot as shown in fig 5.9.2.

Step-3 : Find phase margin of uncompensated system.

Let, ϕ_{gc} = Phase of G'(jω) at gain crossover frequency

γ = Phase margin of uncompensated system.

From the bode plot of uncompensated system we get, $\phi_{gc} = -226°$

Now, $\gamma = 180° + \phi_{gc} = 180° - 226° = -46°$

Step-4 : Choose a new phase margin

The desired phase margin, $\gamma_d = 35°$

The phase margin of compensated system, $\gamma_n = \gamma_d + \epsilon$

Let initial choice of $\epsilon = 5°$

$\therefore \gamma_n = \gamma_d + \epsilon = 35° + 5° = 40°$.

Step-5 : Determine new gain crossover frequency

Let, ω_{gcn} = New gain crossover frequency and ϕ_{gcn} = Phase of G(jω) at ω_{gcn}

Now, $\gamma_n = 180° + \phi_{gcn}$, $\therefore \phi_{gcn} = \gamma_n - 180° = 40° - 180° = -140°$

From the bode plot we found that the frequency corresponding to a phase of −140° is 1.8 rad/sec.

Let, ω_{gcl} = Gain crossover frequency of lag compensator.

Choose ω_{gcl} such that, $\omega_{gcl} > \omega_{gcn}$. Let $\omega_{gcl} = 4$ rad/sec.

Step-6 : Calculate β of lag compensator

From the bode plot we found that the db magnitude at ω_{gcl} is 23 db.

$\therefore |G(j\omega)|$ in db at $(\omega = \omega_{gcl}) = A_{gcl} = 23$ db.

Also, $A_{gcl} = 20 \log \beta$; $\therefore \beta = 10^{A_{gcl}/20} = 10^{23/20} = 14$

Step-7 : Determine the transfer function of lag section.

The zero of the lag compensator is placed at a frequency one-tenth of ω_{gcl}.

$\therefore$ Zero of lag compensator, $z_{c1} = \dfrac{1}{T_1} = \dfrac{\omega_{gcl}}{10}$

Now, $T_1 = \dfrac{10}{\omega_{gcl}} = \dfrac{10}{4} = 2.5$

Pole of lag compensator, $p_{c1} = \dfrac{1}{\beta T_1} = \dfrac{1}{14 \times 2.5} = \dfrac{1}{35}$

Transfer function of lag section $\Big\}$ $G_1(s) = \beta \dfrac{(1+sT_1)}{(1+s\beta T_1)} = 14 \dfrac{(1+2.5s)}{(1+35s)}$

Step-8 :   Determine the transfer function of lead section.

Let $\alpha = 1/\beta$; $\therefore \alpha = 1/14 = 0.07$

$\left.\begin{array}{l}\text{The db gain (magnitude)}\\ \text{corresponding to } \omega_m\end{array}\right\} = -20 \log \dfrac{1}{\sqrt{\alpha}} = -20 \log \dfrac{1}{\sqrt{0.07}} = -11.5 \text{ db} \approx -12 \text{ db}$

From the bode plot of uncompensated system the frequency ω_m corresponding to a db pair of -12 db is found to be 17 rad/sec.

$\therefore \omega_m = 17 \text{ rad/sec}$

$\therefore T_2 = \dfrac{1}{\omega_m \sqrt{\alpha}} = \dfrac{1}{17\sqrt{0.07}} = 0.22$

$\left.\begin{array}{l}\text{Transfer function}\\ \text{of lead section}\end{array}\right\} G_2(s) = \alpha \dfrac{(1 + sT_2)}{(1 + s\alpha T_2)} = 0.07 \dfrac{(1 + 0.22s)}{(1 + 0.0154s)}$

Step-9 :   Determine the transfer function of lag-lead compensator.

$\left.\begin{array}{l}\text{Transfer function}\\ \text{of lag – lead compensator}\end{array}\right\} G_c(s) = G_1(s) \times G_2(s) = 14 \dfrac{(1 + 2.5s)}{(1 + 35s)} \times 0.07 \dfrac{(1 + 0.22s)}{(1 + 0.0154s)}$

$$= \dfrac{(1 + 2.5s)(1 + 0.22s)}{(1 + 35s)(1 + 0.0154s)}$$

Step-10 :   Determine open loop transfer function of compensated system.

The lag-lead compensator is connected in series with G(s) as shown in fig 5.9.1.

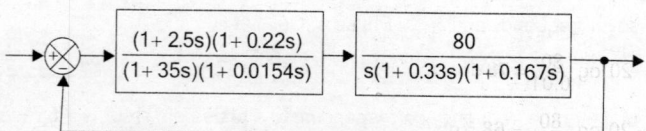

Fig 5.9.1 : *Block diagram of lag-lead compensated system.*

$\left.\begin{array}{l}\text{Open loop transfer}\\ \text{function of}\\ \text{compensated system}\end{array}\right\} G_0(s) = \dfrac{80(1 + 2.5s)(1 + 0.22s)}{s(1 + 35s)(1 + 0.0154s)(1 + 0.33s)(1 + 0.167s)}$

Step-11 :   Bode plot of compensated system.

Put $s = j\omega$ in $G_0(s)$

$\therefore G_0(j\omega) = \dfrac{80(1 + j2.5\omega)(1 + j0.22\omega)}{j\omega(1 + j35\omega)(1 + j0.0154\omega)(1 + j0.33\omega)(1 + j0.167\omega)}$

MAGNITUDE PLOT

There are six corner frequencies, which are given below.

$\omega_{c1} = \dfrac{1}{35} = 0.03 \text{ rad/sec}$; $\omega_{c2} = \dfrac{1}{2.5} = 0.4 \text{ rad/sec}$; $\omega_{c3} = \dfrac{1}{0.33} = 3 \text{ rad/sec}$;

$\omega_{c4} = \dfrac{1}{0.22} = 4.5 \text{ rad/sec}$; $\omega_{c5} = \dfrac{1}{0.167} = 6 \text{ rad/sec}$; $\omega_{c6} = \dfrac{1}{0.0154} = 65 \text{ rad/sec}$

The various terms of $G_0(j\omega)$ are listed in table-3. Also the table shows the slope contributed by each term and the change in slope at the corner frequency.

TABLE-3

Term	Corner frequency rad/sec	Slope db/dec	Change in slope db/dec
$\dfrac{80}{j\omega}$	—	−20	—
$\dfrac{1}{1+j35\omega}$	$\omega_{c1} = \dfrac{1}{35} = 0.03$	−20	−20 − 20 = − 40
$1+j2.5\omega$	$\omega_{c2} = \dfrac{1}{2.5} = 0.4$	+20	− 40 + 20 = − 20
$\dfrac{1}{1+j0.33\omega}$	$\omega_{c3} = \dfrac{1}{0.33} = 3$	−20	−20 − 20 = − 40
$1+j0.22\omega$	$\omega_{c4} = \dfrac{1}{0.22} = 4.5$	+20	− 40 + 20 = − 20
$\dfrac{1}{1+j0.167\omega}$	$\omega_{c5} = \dfrac{1}{0.167} = 6$	−20	20 − 20 = − 40
$\dfrac{1}{1+j0.0154\omega}$	$\omega_{c6} = \dfrac{1}{0.0154} = 65$	−20	− 40 − 20 = − 60

Choose a low frequency ω_l such that $\omega_l < \omega_{c1}$ and choose a high frequency ω_h such that $\omega_h > \omega_{c6}$.

Let $\omega_l = 0.01$ rad/sec and $\omega_h = 80$ rad/sec

Let $A_0 = |G_0(j\omega)|$ in db.

At $\omega = \omega_1$, $\qquad A_0 = 20\log\dfrac{80}{0.01} = 78$ db

At $\omega = \omega_{c1}$, $\qquad A_0 = 20\log\dfrac{80}{0.03} = 68.5$ db ≈ 68 db

At $\omega = \omega_{c2}$, $\qquad A_0 = -40\times\log\dfrac{0.4}{0.03} + 68 = 23$ db

At $\omega = \omega_{c3}$, $\qquad A_0 = -20\times\log\dfrac{3}{0.4} + 23 = 5$ db

At $\omega = \omega_{c4}$, $\qquad A_0 = -40\times\log\dfrac{4.5}{3} + 5 = -2$ db

At $\omega = \omega_{c5}$, $\qquad A_0 = -20\times\log\dfrac{6}{4.5} + (-2) = -4$ db

At $\omega = \omega_{c6}$, $\qquad A_0 = -40\times\log\dfrac{65}{6} + (-4) = -45$ db

At $\omega = \omega_h$, $\qquad A_0 = -60\times\log\dfrac{80}{65} + (-45) = -50$ db

Using the values of A_0 at various frequencies the magnitude plot of compensated system is drawn as shown in fig 5.9.2.

PHASE PLOT

The phase angle of $G_0(j\omega)$ as a function of ω is given by

$$\phi_0 = \angle G_0(j\omega) = \tan^{-1}2.5\omega + \tan^{-1}0.22\omega - 90° - \tan^{-1}35\omega - \tan^{-1}0.0154\omega - \tan^{-1}0.33\omega - \tan^{-1}0.167\omega.$$

The phase angle of $G_0(j\omega)$ are calculated for various values of ω and listed in table-4.

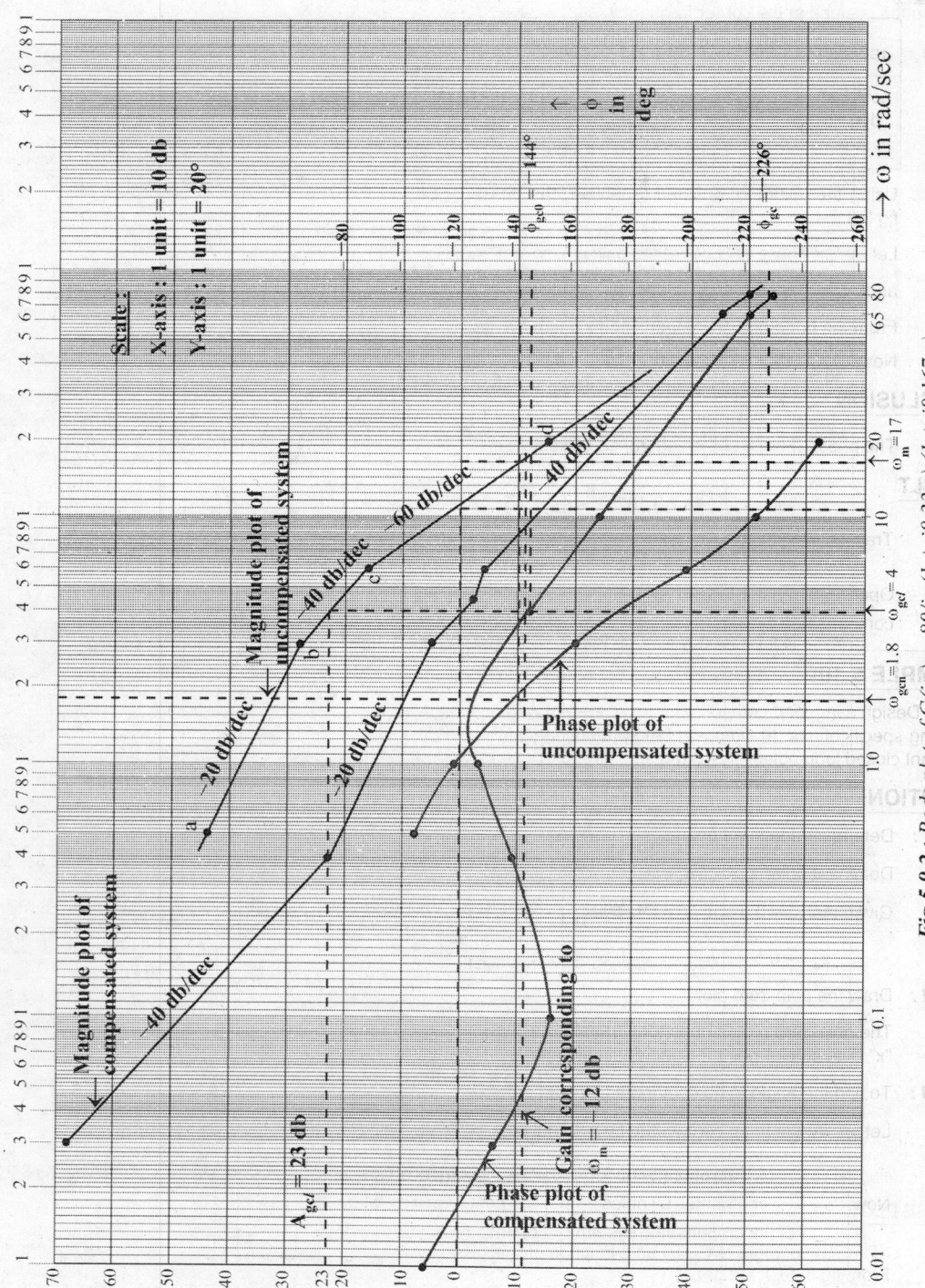

Fig 5.9.2 : Bode plot of G(jω) = 80/jω(1 + j0.33ω) (1 + j0.167ω).

TABLE-4

ω rad/sec	0.01	0.03	0.1	0.4	1	4	10	65	80
$\angle G_0(j\omega)$ deg	–108	–132	–152	–138	–126	–144	–168	–221 $\approx$ –220	–228

Using the values of ϕ_0 listed in table-4, the phase plot of compensated system is sketched as shown in fig 5.9.2.

Let ϕ_{gco} = Phase of $G_0(j\omega)$ at the gain crossover frequency of compensated system.

and γ_0 = Phase margin of compensated system.

From the bode plot of compensated system, we get, $\phi_{gco} = -144°$

Now, $\gamma_0 = 180° + \phi_{gco} = 180° - 144° = 36°$

CONCLUSION

The phase margin of the compensated system is satisfactory. Hence the design is acceptable.

RESULT

Transfer function of lag – lead compensator, $G_c(s) = \dfrac{(1+2.5s)(1+0.22s)}{(1+35s)(1+0.0154s)}$

Open loop transfer function of compensated system $\Big\}$ $G_0(s) = \dfrac{80(1+2.5s)(1+0.22s)}{s(1+35s)(1+0.0154s)(1+0.33s)(1+0.167s)}$

EXAMPLE 5.10

Design a lag-lead compensator for a system with open loop transfer function G(s) = K/s (s + 0.5) to satisfy the following specifications. (i) Damping ratio of dominant closed-loop poles, ζ = 0.5.(ii) Undamped natural frequency of dominant closed loop poles, ω_n = 5 rad/sec. (iii) Velocity error constant, K_v = 80 sec^{-1}.

SOLUTION

Step-1 : Determine dominant pole, s_d.

Dominant pole, $s_d = -\zeta\omega_n \pm j\omega_n\sqrt{1-\zeta^2}$

Given that, ζ = 0.5 and ω_n = 5 rad/sec.

$\therefore s_d = -0.5 \times 5 \pm j5 \times \sqrt{1-0.5^2} = -2.5 \pm j4.3$

Step-2 : Draw the pole-zero plot

The pole-zero plot of open loop transfer function is shown fig 5.10.1. Poles are represented by the symbol "x". The pole at point P is the dominant pole, s_d.

Step-3 : To find the angle to be contributed by lead network.

Let ϕ = Angle to be contributed by lead network to make point, P as a point on root locus.

Now, $\phi = \left(\begin{array}{c}\text{Sum of angles}\\\text{contributed by poles}\\\text{of uncompensated system}\end{array}\right) - \left(\begin{array}{c}\text{Sum of angles}\\\text{contributed by zeros}\\\text{of uncompensated system}\end{array}\right) \pm n180°$

From fig 5.10.1, we get,

Sum of angles contributed by poles of uncompensated system $\Big\} = \theta_1 + \theta_2 = 120° + 115° = 235°$

Since there is no finite zero in uncompensated system, there is no angle contribution by zeros.

$$\therefore \phi = 235° \pm n180°$$

Let $n = 1$, $\therefore \phi = 235° - 180° = 55°$.

Step-4 :   To find the pole and zero of the lead section

Draw a line AP parallel to x-axis as shown in fig 5.10.1. The bisector PC is drawn to bisect the angle APO. The angles CPD and BPC are constructed as shown in fig 5.10.1. Here $\angle CPD = \angle BPC = \phi/2 = 55°/2 = 27.5° \approx 27°$.

From fig 5.10.1., Pole of the lead section, $p_{c2} = -10$

Zero of the lead section, $z_{c2} = -2.65$.

We know that, $z_{c2} = -1/T_2$ $\therefore T_2 = 1/2.65 = 0.377$

We know that, $p_{c2} = -1/\alpha T_2$ $\therefore \alpha T_2 = 1/10$ (or) $\alpha = 1/(T_2 \times 10) = 0.265$.

Step-5 :   Transfer function of lead compensator

Transfer function of lead section, $G_2(s) = \dfrac{(s + 1/T_2)}{(s + 1/\alpha T_2)} = \dfrac{(s + 2.65)}{(s + 10)}$

Step-6 :   To find gain, K

Open loop transfer function of lead compensated system $\Bigg\}$ $G_{02}(s) = G_2(s) \times G(s) = \dfrac{(s + 2.65)}{(s + 10)} \times \dfrac{K}{s(s + 0.5)} = \dfrac{K(s + 2.65)}{s(s + 0.5)(s + 10)}$

Here the value of K is given by the value of gain at the dominant pole s_d on the root locus. From magnitude condition K is given by,

$$K = \frac{\text{Product of vector lengths from all poles to } s = s_d}{\text{Product of vector lengths from all zeros to } s = s_d}$$

From fig 5.10.1, we get,

$$K = \frac{l_1 \times l_2 \times l_4}{l_3} = \frac{5 \times 4.75 \times 7.75}{4.3} = 42.8$$

$$G_{02}(s) = \frac{42.8(s + 2.65)}{s(s + 0.5)(s + 10)}$$

Step-7 :   To find velocity error constant of lead compensated system.

Let, K_{v2} = Velocity error constant of lead compensated system.

$$\therefore K_{v2} = \underset{s \to 0}{Lt} \; s. G_{02}(s) = \underset{s \to 0}{Lt} \; s \; \frac{42.8(s + 2.65)}{s(s + 0.5)(s + 10)} = \frac{42.8 \times 2.65}{0.5 \times 10} = 22.684$$

Step-8 :   To find the parameter, β

Let K_{vd} = Desired velocity error constant

A = The factor by which K_v is increased

Now, $A = K_{vd}/K_{v2} = 80/22.684 = 3.5267$

Select β, such that $\beta > A$. Let $\beta = 4$.

Step-9 :   To find the transfer function of lag section.

Let zero of lag section, $z_{c1} = 0.1 \times$ second pole of $G(s) = 0.1 \times (-0.5) = -0.05$

Also, $z_{c1} = \dfrac{-1}{T_1}$; $\therefore T_1 = \dfrac{1}{0.05} = 20$

Pole of lag section, $p_{c1} = \dfrac{-1}{\beta T_1} = \dfrac{-1}{4 \times 20} = \dfrac{-1}{80} = -0.0125$

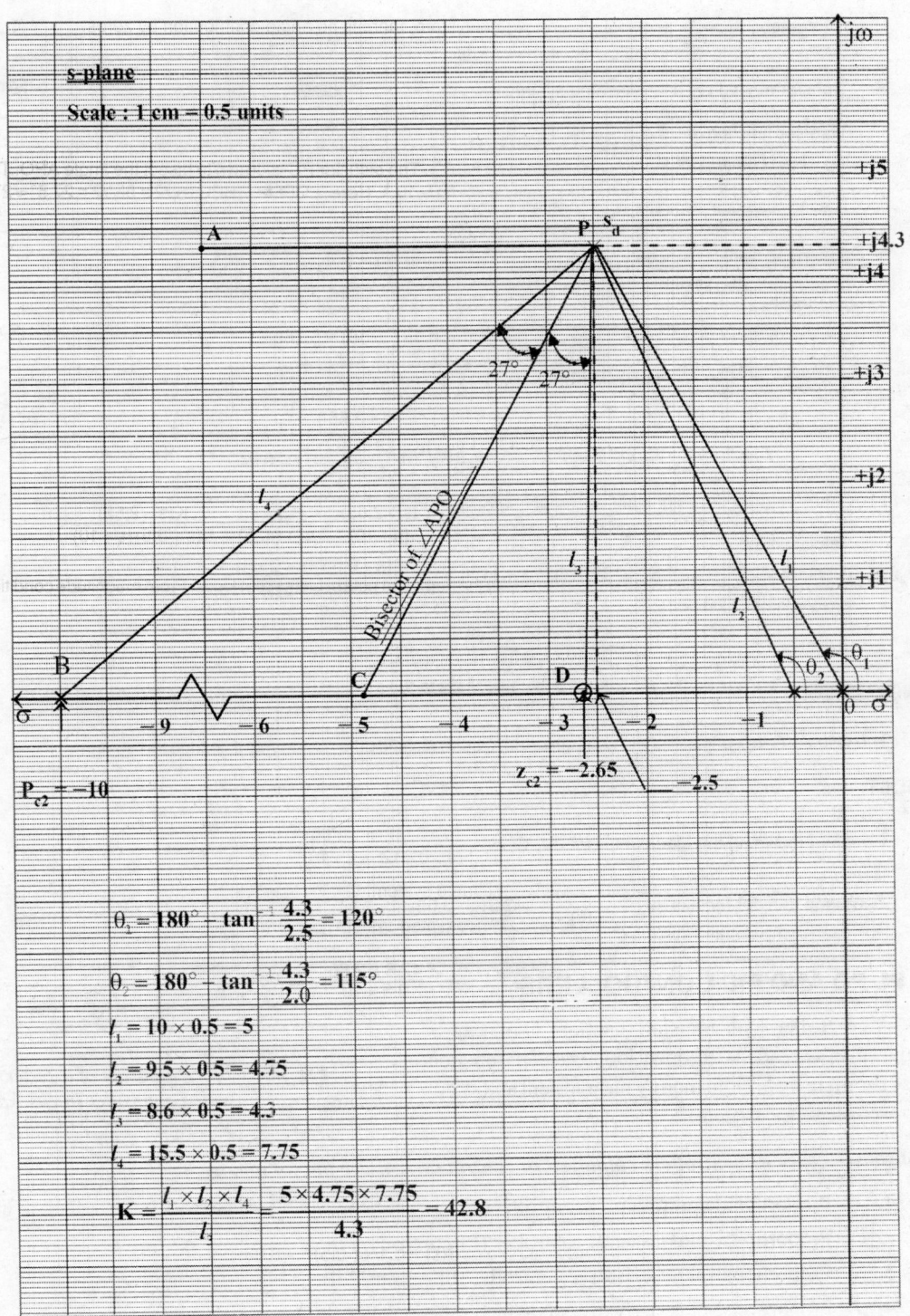

Fig 5.10.1. : Pole zero plot of lead compensated system.

s-plane

Scale : 1 cm = 0.5 units

$\theta_1 = 180° - \tan^{-1} \dfrac{4.3}{2.5} = 120°$

$\theta_2 = 180° - \tan^{-1} \dfrac{4.3}{2.0} = 115°$

$l_1 = 10 \times 0.5 = 5$

$l_2 = 9.5 \times 0.5 = 4.75$

$l_3 = 8.6 \times 0.5 = 4.3$

$l_4 = 15.5 \times 0.5 = 7.75$

$K = \dfrac{l_1 \times l_2 \times l_4}{l_3} = \dfrac{5 \times 4.75 \times 7.75}{4.3} = 42.8$

Transfer function of lag section, $G_1(s) = \dfrac{(s + 1/T_1)}{(s + 1/\beta T_1)} = \dfrac{(s + 0.05)}{(s + 0.0125)}$

Step-10 : Transfer function of compensated system

Transfer function of lag – lead compensator $\Big\}$ $G_c(s) = G_1(s) \times G_2(s) = \dfrac{(s + 0.05)(s + 2.65)}{(s + 0.0125)(s + 10)}$

The lag-lead compensator is connected in series with G(s) as shown in fig 5.10.2.

Open loop transfer function of compensated system $\Big\}$ $G_0(s) = \dfrac{42.8(s + 0.05)(s + 2.65)}{s(s + 0.0125)(s + 0.5)(s + 10)}$

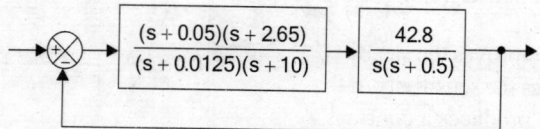

Fig 5.10.2 : Block diagram of lag-lead compensated system.

Step-11 : Check velocity error constant of compensated system.

Velocity error constant of compensated system $\Big\}$ $K_{vc} = \underset{s \to 0}{\text{Lt}} \, s \, G_0(s) = \underset{s \to 0}{\text{Lt}} \, s \, \dfrac{42.8(s + 0.05)(s + 2.65)}{s(s + 0.0125)(s + 0.5)(s + 10)}$

$$= \dfrac{42.8 \times 0.05 \times 2.65}{0.0125 \times 0.5 \times 10} = 90.7$$

CONCLUSION

The velocity error constant of the compensated system satisfies the requirement. Hence the design is accepted.

RESULT

Transfer function of lag-lead compensator, $G_c(s) = \dfrac{(s + 0.05)(s + 2.65)}{(s + 0.0125)(s + 10)}$

Open loop transfer function of lag – lead compensated system $\Big\}$ $G_0(s) = \dfrac{42.8(s + 0.05)(s + 2.65)}{s(s + 0.0125)(s + 0.5)(s + 10)}$

5.5 PI, PD AND PID CONTROLLERS

A controller with transfer function, $G_c(s)$ can be introduced in cascade with open loop transfer function, G(s) as shown in fig 5.21, to modify the transient and steady state response of the system.

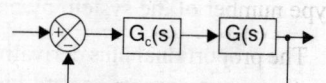

Fig 5.21 : Block diagram of a system with a controller in cascade.

The different types of controllers employed in control system are the following

1. Proportional controller (P-controller)
2. Proportional-plus-integral controller (PI-controller)
3. Proportional-plus-derivative controller (PD-controller)
4. Proportional-plus-derivative-plus-integral controller (PID-controller)

The proportional controller is a device that produces an output signal, u(t), which is proportional to the input signal, e(t).

In P-controller, $u(t) \propto e(t)$

$$\therefore u(t) = K_p e(t) \qquad .....(5.49)$$

where, K_p = Proportional gain or constant.

On taking laplace transform of equation (5.49) we get,

$$U(s) = K_p E(s)$$

Transfer function of P-controller, $\mathbf{G_c(s)} = \dfrac{\mathbf{U(s)}}{\mathbf{E(s)}} = \mathbf{K_p}$ .....(5.50)

The proportional controller improves the steady state tracking accuracy, disturbance signal rejection and relative stability. It also decreases the sensitivity of the system to parameter variations. The proportional control is not used alone because it produces a constant steady state error.

The proportional plus integral controller (PI-controller) is a device that produces an output signal, u(t) consisting of two terms-one proportional to input signal, e(t) and the other proportional to the integral of input signal, e(t).

In PI-controller, $u(t) \propto \left[e(t) + \int e(t)\,dt \right]$ .....(5.51)

$$\therefore u(t) = K_p e(t) + K_i \int e(t)\,dt$$

where, K_p = Proportional gain

K_i = Integral constant or gain.

On taking Laplace transform of equation (5.51) with zero initial conditions we get,

$$U(s) = K_p E(s) + K \frac{E(s)}{s} = E(s)\left(K_p + \frac{K_i}{s} \right)$$

Transfer function of PI controller, $\mathbf{G_c(s)} = \dfrac{\mathbf{U(s)}}{\mathbf{E(s)}} = \mathbf{K_p} + \dfrac{\mathbf{K_i}}{\mathbf{s}}$ .....(5.52)

The PI controller reduces the steady state error. The introduction of PI controller increases the order and type number of the system by one.

The proportional plus derivative controller is a device that produces an output signal, u(t), consisting of two terms-one proportional to input signal, e(t) and the other proportional to the derivative of input signal, e(t).

In PD controller, $u(t) \propto \left[e(t) + \dfrac{d}{dt} e(t) \right]$

$$\therefore u(t) = K_p e(t) + K_d \frac{d}{dt} e(t) \qquad .....(5.53)$$

where, K_p = Proportional gain

K_d = Derivative constant or gain

On taking Laplace transform of equation (5.53) with zero initial conditions we get,

$$U(s) = K_p E(s) + K_d s E(s) = E(s) [K_p + K_d s]$$

Transfer function of PD controller, $G_c(s) = \dfrac{U(s)}{E(s)} = K_p + K_d s$ .....(5.54)

The PD controller increases the damping of the system which results in reducing the peak overshoot.

The PID controller is a device which produces an output signal, u(t) consisting of three terms-one proportional to input signal, e(t), another one proportional to integral of input signal, e(t) and the third one proportional to derivative of input signal, e(t).

In PID controller, $u(t) \alpha \left[e(t) + \int e(t)\, dt + \dfrac{d}{dt} e(t) \right]$

$\therefore\ u(t) = K_p e(t) + K_i \int e(t)\, dt + K_d \dfrac{d}{dt} e(t)$ .....(5.55)

On taking laplace transform of equation (5.55) we get,

$$U(s) = K_p E(s) + K_i \frac{E(s)}{s} + K_d\, sE(s) = E(s)\left[K_p + \frac{K_i}{s} + K_d s \right]$$

Transfer function of PID-controller, $G_c(s) = \dfrac{U(s)}{E(s)} = K_p + \dfrac{K_i}{s} + K_d s$ .....(5.56)

The PID controller have the combined effect of all the three control actions-i.e., proportional, integral and derivative control actions. Hence the introduction of PID controller stabilises the gain, reduces the steady state error and peak overshoot of the system.

DESIGN OF PI, PD AND PID CONTROLLERS IN FREQUENCY DOMAIN

In frequency domain PI, PD and PID controllers can be designed to satisfy a specified gain margin and error constant.

Let ω_1 = Gain crossover frequency of the system with cascade PI/PD/PID controller.

        A_1 = $|G(j\omega)|$ at ω_1

        ϕ_1 = Phase of $G(j\omega)$ at ω_1

        γ_u = Phase margin of uncompensated system at ω_1

        γ_d = Desired phase margin at ω_1

        θ = Angle to be contributed by PI/PD/PID controller at gain crossover frequency, ω_1 to achieve a phase margin of γ_d.

By definition of phase margin, $\gamma_u = 180° + \phi_1$

The desired phase margin, γ_d is given by, $\gamma_d = \theta + \gamma_u$

    $\therefore \theta = \gamma_d - \gamma_u$ .....(5.57)

Transfer function of PID controller } $G_c(s) = K_p + \dfrac{K_i}{s} + K_d s = \dfrac{K_d s^2 + K_p s + K_i}{s}$

On replacing s by $j\omega$ in $G_c(s)$ we get,

$$G_c(j\omega) = \frac{K_d(j\omega)^2 + K_p(j\omega) + K_i}{j\omega} = \frac{(-K_d \omega^2 + K_i) + jK_p \omega}{j\omega}$$ .....(5.58)

Let, $A_c = |G_c(j\omega)|$ at $\omega = \omega_1$

θ = Phase of $G_c(j\omega)$ at $\omega = \omega_6$.

$\therefore$ At $\omega = \omega_1$, $G_c(j\omega) = G_c(j\omega_1) = |G_c(j\omega_1)| \angle G_c(j\omega_1) = A_c \angle \theta$ .....(5.59)

The magnitude of open loop transfer function at gain crossover frequency is unity.

$\therefore$ At $\omega = \omega_1$, $|G_c(j\omega_1).G(j\omega_1)| = 1$

$\therefore$ $A_c.A_1 = 1$ (or) $A_c = \dfrac{1}{A_1}$ .....(5.60)

From equation (5.59) and (5.60) we can write,

$$G_c(j\omega_1) = \dfrac{1}{A_1} \angle \theta$$ .....(5.61)

From equation (5.58) at $\omega = \omega_1$ we get,

$$G_c(j\omega_1) = \dfrac{(-K_d\omega_1^2 + K_i) + jK_p\omega_1}{j\omega_1}$$ .....(5.62)

On equating, equations (5.61) and (5.62) we get,

$$\dfrac{-K_d\omega_1^2 + K_i + jK_p\omega_1}{j\omega_1} = \dfrac{1}{A_1} \angle \theta \quad \Rightarrow \quad -K_d\omega_1^2 + K_i + jK_p\omega_1 = j\omega_1 \dfrac{1}{A_1} \angle \theta$$

Here, $j\omega_1 = \omega_1 \angle 90°$

$\therefore$ $-K_d\omega_1^2 + K_i + jK_p\omega_1 = \omega_1 \angle 90° \dfrac{1}{A_1} \angle \theta \quad \Rightarrow \quad -K_d\omega_1^2 + K_i + jK_p\omega_1 = \dfrac{\omega_1}{A_1} \angle (90° + \theta)$

$\therefore$ $-K_d\omega_1^2 + K_i + jK_p\omega_1 = \dfrac{\omega_1}{A_1} \cos(90° + \theta) + j\dfrac{\omega_1}{A_1}\sin(90° + \theta)$

$$-K_d\omega_1^2 + K_i + jK_p\omega_1 = -\dfrac{\omega_1}{A_1}\sin\theta + j\dfrac{\omega_1}{A_1}\cos\theta$$ .....(5.63)

On equating real parts of equation (5.63) we get,

$$-K_d\omega_1^2 + K_i = -\dfrac{\omega_1}{A_1}\sin\theta \quad \Rightarrow \quad K_d + \dfrac{K_i}{-\omega_1^2} = \dfrac{-\omega_1\sin\theta}{-\omega_1^2 A_1}$$

$$\therefore \boldsymbol{K_d = \dfrac{\sin\theta}{\omega_1 A_1} + \dfrac{K_i}{\omega_1^2}}$$ .....(5.64)

On equating imaginary parts of equation (5.63) we get,

$$K_p\omega_1 = \dfrac{\omega_1}{A_1}\cos\theta$$

$$\therefore \boldsymbol{K_p = \dfrac{\cos\theta}{A_1}}$$ .....(5.65)

The equations (5.64) and (5.65) are used to design cascade PD or PI or PID controller. For designing a PD controller put $K_i = 0$, in equation (5.64). Hence the constants K_p and K_d of PD controller are given by the following two equations,

$$\boldsymbol{K_d = \dfrac{\sin\theta}{\omega_1 A_1}} \quad \text{and} \quad \boldsymbol{K_p = \dfrac{\cos\theta}{A_1}}$$

For designing a PI controller put $K_d = 0$, in equation (5.64). Hence the constants K_p and K_i of PI controller are given by the following two equations,

$$K_i = \frac{-\omega_1 \sin\theta}{A_1} \quad \text{and} \quad K_p = \frac{\cos\theta}{A_1}$$

For designing a PID controller, determine the constant K_i to satisfy the specified error constant. Then calculate K_d and K_p from equations (5.64) and (5.65).

PROCEDURE FOR DESIGN OF PD/PI/PID CONTROLLER IN FREQUENCY DOMAIN

The following procedure can be followed to design a PD/PI/PID controller when the given specifications are desired phase margin, γ_d at a gain crossover frequency, ω_6.

Step-1 : Determine the magnitude and phase of uncompensated open loop sinusoidal transfer function [i.e., $G(j\omega)$]

Let, $A_1 = |G(j\omega)|$ at $\omega = \omega_1$ and $\phi_1 = \angle G(j\omega)$ at $\omega = \omega_1$

Step-2 : Determine the phase margin of uncompensated system and the angle to be contributed by the controller to achieve the desired phase margin.

Let, γ_u = Phase margin of uncompensated system.

γ_d = Desired phase margin at ω_1

θ = Phase angle of the controller at $\omega = \omega_1$

Now, $\gamma_u = 180° + \phi_1$; $\theta = \gamma_d - \gamma_u$

Step-3 : Determine the transfer function of the controller.

a) PD controller

Derivative constant, $K_d = \dfrac{\sin\theta}{\omega_1 A_1}$

Proportional constant, $K_p = \dfrac{\cos\theta}{A_1}$

Transfer function of PD controller $\Big\}$ $G_c(s) = (K_p + K_d s) = K_p\left(1 + \dfrac{K_d}{K_p}s\right)$

b) PI controller

Integral constant, $K_i = \dfrac{-\omega_1 \sin\theta}{\omega_1 A_1}$

Proportional constant, $K_p = \dfrac{\cos\theta}{A_1}$

Transfer function of PI controller $\Big\}$ $G_c(s) = \left(K_p + \dfrac{K_i}{s}\right) = \dfrac{K_i\left(1 + \dfrac{K_p}{K_i}s\right)}{s}$

c) PID controller

Transfer function of PID controller $\Big\}$ $G_c(s) = \left(K_p + K_d s + \dfrac{K_i}{s}\right) = \dfrac{K_d\left(s^2 + \dfrac{K_p}{K_d}s + \dfrac{K_i}{K_d}\right)}{s}$

Evaluate K_i such that the compensated system satisfies the error requirement. For example if the compensated system is type-1 system then, $K_v = \underset{s \to 0}{\text{Lt}}\ sG_c(s)G(s)$ will give the value of K_i.

Derivative constant, $K_d = \dfrac{\sin\theta}{\omega_1 A_1} + \dfrac{K_i}{\omega_1^2}$;   Proportional constant, $K_p = \dfrac{\cos\theta}{A_1}$

Step-4 : Determine the open loop transfer function of compensated system.

The transfer function of the controller is placed in cascade with G(s) as shown in fig 5.22.

Open loop transfer function of compensated system, $G_0(s) = G_c(s) \times G(s)$

Step-5 : Verify the design by calculating phase margin of compensated system.

Let $A_0 = |G_0(j\omega)|$ at $\omega = \omega_1$

$\phi_0 = \angle G_0(j\omega)$ at $\omega = \omega_1$

$\gamma_0 = $ Phase margin of compensated system.

Now, $\gamma_0 = 180° + \phi_0$

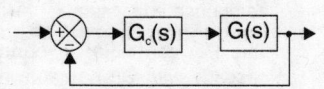

Fig 5.22 : Block diagram of a system with cascade controller.

It can be observed that $A_0 \approx 1$ and γ_0 satisfies the specifications.

EXAMPLE 5.11

Consider a unity feedback system with open loop transfer function, G(s)=5/s(s+0.5)(s+1). Design a PD controller so that the phase margin of the system is 30° at a frequency of 1.2 rad/sec.

SOLUTION

Step-1 : To find magnitude and phase of $G(j\omega_1)$

Given that, $G(s) = \dfrac{5}{s(s+0.5)(s+1)} = \dfrac{5}{s \times 0.5\left(1+\dfrac{s}{0.5}\right)(1+s)} = \dfrac{10}{s(1+2s)(1+s)}$

Put, $s = j\omega$ in G(s)

$\therefore G(j\omega) = \dfrac{10}{j\omega(1+j2\omega)(1+j\omega)} = \dfrac{10}{\omega\angle 90° \sqrt{1+4\omega^2}\angle\tan^{-1}2\omega\sqrt{1+\omega^2}\angle\tan^{-1}\omega}$

$|G(j\omega)| = \dfrac{10}{\omega\sqrt{1+4\omega^2}\sqrt{1+\omega^2}}$

$\angle G(j\omega) = -90° - \tan^{-1}2\omega - \tan^{-1}\omega$

The gain crossover frequency of compensated system, $\omega_1 = 1.2$ rad/sec.

Let, $A_1 = |G(j\omega)|$ at $\omega = \omega_1$; $\phi_1 = \angle G(j\omega)$ at $\omega = \omega_1$

$\therefore A_1 = \dfrac{10}{1.2 \times \sqrt{1+4\times 1.2^2} \times \sqrt{1+1.2^2}} = 2.052$

$\phi_1 = -90° - \tan^{-1}(21.2) - \tan^{-1}(1.2) = -207.5°$

Step-2 : To find γ_u and θ

Let $\gamma_u = $ Phase margin of uncompensated system

$\gamma_d = $ Desired phase margin of compensated system.

$\theta = $ Phase of $G_c(j\omega)$ at $\omega = \omega_1$

Now, $\gamma_u = 180° + \phi_1 = 180° + (-207.5°) = -27.5°$

$\theta = \gamma_d - \gamma_u = 30° - (-27.5°) = 57.5°$

Step-3 :   To find transfer function of PD-controller.

Derivative constant, $K_d = \dfrac{\sin\theta}{\omega_1 A_1} = \dfrac{\sin 57.5°}{1.2 \times 2.052} = 0.343$

Proportional constant, $K_p = \dfrac{\cos\theta}{A_1} = \dfrac{\cos 57.5°}{2.052} = 0.262$

Transfer function of PD controller $\Big\}$ $G_c(s) = (K_p + K_d s) = K_p\Big(1 + \dfrac{K_d}{K_p}s\Big) = 0.262\Big(1 + \dfrac{0.343}{0.262}s\Big) = 0.262(1 + 1.3s)$

Step-4 :   To find open loop transfer function of compensated system.

The PD-controller is connected in cascade with G(s) as shown in figure 5.11.1

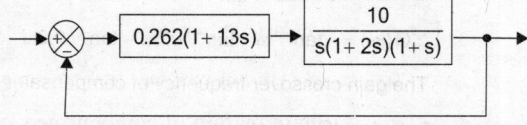

Fig 5.11.1 : Block diagram of compensated system.

Open loop transfer function of compensated system $\Big\}$ $G_0(s) = G_c(s) \times G(s) = 0.262(1 + 1.3s) \times \dfrac{10}{s(1 + 2s)(1 + s)}$

$$= \dfrac{2.62(1 + 1.3s)}{s(1 + 2s)(1 + s)}$$

Step-5 :   To verify the design

Put $s = j\omega$ in $G_0(s)$,

$\therefore$ $G_0(j\omega) = \dfrac{2.62(1 + j1.3\omega)}{j\omega(1 + j2\omega)(1 + j\omega)} = \dfrac{2.62\sqrt{1 + 1.69\omega^2}\angle\tan^{-1}1.3\omega}{\omega\angle 90°\sqrt{1 + 4\omega^2}\angle\tan^{-1}2\omega\sqrt{1 + \omega^2}\angle\tan^{-1}\omega}$

Let $A_0 = |G_0(j\omega)|$ and $\phi_0 = G_0(j\omega)$

$\therefore$ $A_0 = \dfrac{2.62\sqrt{1 + 1.69\omega^2}}{\omega\sqrt{1 + 4\omega^2}\sqrt{1 + \omega^2}}$

$\phi_0 = \tan^{-1}1.3\omega - 90° - \tan^{-1}2\omega - \tan^{-1}\omega$

At $\omega = \omega_1$, $A_0 = A_{01} = \dfrac{2.62\sqrt{1 + 1.69 \times 1.2^2}}{1.2 \times \sqrt{1 + 4 \times 1.2^2} \times \sqrt{1 + 1.2^2}} \approx 1$

At $\omega = \omega_1$, $\phi_0 = \phi_{01} = \tan^{-1}(1.3 \times 1.2) - 90° - \tan^{-1}(2 \times 1.2) - \tan^{-1}1.2 = -150°$

Phase margin of compensated system, $\gamma_0 = 180° + \phi_{01} = 180° - 150° = 30°$

CONCLUSION

The phase margin of the compensated system is satisfactory. Hence the design is acceptable.

RESULT

1. Transfer function of PD controller, $G_c(s) = (0.262 + 0.343s) = 0.262(1 + 1.3s)$

2. Open loop transfer function of compensated system, $G_0(s) = \dfrac{2.62(1 + 1.3s)}{s(1 + 2s)(1 + s)}$

EXAMPLE 5.12

Consider a unity feedback system with open loop transfer function, $G(s) = \dfrac{100}{(s + 1)(s + 2)(s + 5)}$ .Design a PI controller, so that the phase margin of the system is 60° at a frequency of 0.5 rad/sec.

SOLUTION

Step-1 :   To find magnitude and phase of $G(j\omega_1)$

Given that, $G(s) = \dfrac{100}{(s + 1)(s + 2)(s + 5)} = \dfrac{100}{(s + 1) \times 2 \times \big(1 + \frac{s}{2}\big) \times 5 \times \big(1 + \frac{s}{5}\big)} = \dfrac{10}{(s + 1)(1 + 0.5s)(1 + 0.2s)}$

Put, $s = j\omega$ in $G(s)$

$$\therefore \ G(j\omega) = \frac{10}{(1 + j\omega)(1 + j0.5\omega)(1 + j0.2\omega)}$$

$$= \frac{10}{\sqrt{1 + \omega^2} \angle \tan^{-1}\omega \sqrt{1 + 0.25\omega^2} \angle \tan^{-1}0.5\omega \sqrt{1 + 0.04\omega^2} \angle \tan^{-1}0.2\omega}$$

$$|G(j\omega)| = \frac{10}{\sqrt{1 + \omega^2} \angle \tan^{-1}\omega \sqrt{1 + 0.25\omega^2} \angle \tan^{-1}0.5\omega \sqrt{1 + 0.04\omega^2} \angle \tan^{-1}0.2\omega}$$

$$\angle G(j\omega) = -\tan^{-1}\omega - \tan^{-1}0.5\omega - \tan^{-1}0.2\omega$$

The gain crossover frequency of compensated system, $\omega_1 = 0.5$ rad/sec.

Let $\ A_1 = |G(j\omega)|$ at $\omega = \omega_1$

$\quad \phi_1 = \angle G(j\omega)$ at $\omega = \omega_1$

$$\therefore \ A_1 = \frac{10}{\sqrt{1 + 0.5^2} \times \sqrt{1 + 0.25 \times 0.5^2} \times \sqrt{1 + 0.04 \times 0.5^2}} = 8.63$$

$$\phi_1 = -\tan^{-1}0.5 - \tan^{-1}(0.5 \times 0.5) - \tan^{-1}(0.2 \times 0.5) = -46°$$

Step-2 : To find γ_u and θ

Let $\ \gamma_u$ = Phase margin of uncompensated system.

$\quad \gamma_d$ = Desired phase margin of uncompensated system.

$\quad \theta$ = Phase of $G_c(j\omega)$ at $\omega = \omega_1$

Now, $\gamma_u = 180° + \phi_1 = 180° - 46° = 134°$

$$\theta = \gamma_d - \gamma_u = 60° - 134° = -74°$$

Step-3 : To find transfer function of PI-controller.

Integral constant, $K_i = \dfrac{-\omega_1 \sin\theta}{A_1} = \dfrac{-0.5 \times \sin(-74°)}{8.63} = 0.056$

Proportional constant, $K_p = \dfrac{\cos\theta}{A_1} = \dfrac{\cos(-74°)}{8.63} = 0.032$

Transfer function of PI controller $\Bigg\} \ G_c(s) = \left(K_p + \dfrac{K_i}{s} \right) = \left(0.032 + \dfrac{0.056}{s} \right)$

$$= \frac{0.032s + 0.056}{s} = \frac{0.056\left(\dfrac{0.032}{0.056}s + 1 \right)}{s} = \frac{0.056(1 + 0.57s)}{s}$$

Step-4 : To find open loop transfer function of compensated system.

The PI-controller is connected in cascade with $G(s)$ as shown in figure 5.12.1

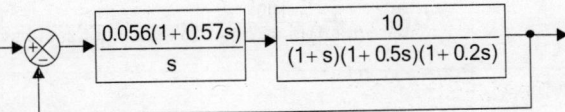

Fig 5.12.1 : *Block diagram of compensated system.*

Open loop transfer function of compensated system $\Bigg\} \ G_0(s) = G_c(s) \times G(s) = \dfrac{0.056(1 + 0.57s)}{s} \times \dfrac{10}{(1 + s)(1 + 0.5s)(1 + 0.2s)}$

$$= \frac{0.56(1 + 0.57s)}{s(1 + s)(1 + 0.5s)(1 + 0.2s)}$$

Step-5 :   To verify the design

Put $s = j\omega$ in $G_0(s)$,

$$\therefore\ G_0(j\omega) = \frac{0.56(1+j0.57\omega)}{(j\omega)(1+j\omega)(1+j0.5\omega)(1+j0.2\omega)}$$

$$\therefore\ G_0(j\omega) = \frac{0.56\sqrt{1+(0.57\omega)^2}\angle\tan^{-1}0.57\omega}{\omega\angle 90°\sqrt{1+\omega^2}\angle\tan^{-1}\omega\sqrt{1+(0.5\omega)^2}\angle\tan^{-1}0.5\omega\sqrt{1+(0.2\omega)^2}\angle\tan^{-1}0.2\omega}$$

Let $A_0 = |G_0(j\omega)|$ and $\phi_0 = \angle G_0(j\omega)$

$$\therefore\ A_0 = \frac{0.56\sqrt{1+0.3249\omega^2}}{\omega\sqrt{1+\omega^2}\sqrt{1+0.25\omega^2}\sqrt{1+0.04\omega^2}}$$

$$\phi_0 = \tan^{-1}0.57\omega - 90° - \tan^{-1}\omega - \tan^{-1}0.5\omega - \tan^{-1}0.2\omega$$

At $\omega = \omega_1$, $A_0 = A_{01} = \dfrac{0.56\times\sqrt{1+0.3249\times 0.5^2}}{0.5\sqrt{1+0.5^2}\sqrt{1+0.25\times 0.5^2}\sqrt{1+0.04\times 0.5^2}} \approx 1$

At $\omega = \omega_1$, $\phi_0 = \phi_{01} = \tan^{-1}(0.57\times 0.5) - 90° - \tan^{-1}0.5 - \tan^{-1}(0.5\times 0.5) - \tan^{-1}(0.2\times 0.5) = -120°$

Phase margin of compensated system, $\gamma_0 = 180° + \phi_{01} = 180° - 120° = 60°$

CONCLUSION

The phase margin of the compensated system, meets the given specification. Hence the design is acceptable.

RESULT

1. Transfer function of PI controller $\Big\}$ $G_c(s) = \left(0.032 + \dfrac{0.056}{s}\right) = \dfrac{0.056(1+0.57s)}{s}$

2. Open loop transfer function of compensated system $\Big\}$ $G_0(s) = \dfrac{0.56(1+0.57s)}{s(1+s)(1+0.5s)(1+0.2s)}$

EXAMPLE 5.13

Consider a unity feedback system with open loop transfer function, $G(s) = \dfrac{100}{(s+1)(s+2)(s+10)}$. Design a PID controller, so that the phase margin of the system is 45° at a frequency of 4 rad/sec and the steady state error for unit ramp input is 0.6.

SOLUTION

Step-1 :   To find magnitude and phase of $G(j\omega_1)$

$$G(s) = \frac{100}{(s+1)(s+2)(s+10)} = \frac{100}{(1+s)\times 2\times\left(1+\frac{s}{2}\right)\times 10\times\left(1+\frac{s}{10}\right)} = \frac{5}{(1+s)(1+0.5s)(1+0.1s)}$$

Put, $s = j\omega$ in $G(s)$

$$\therefore\ G(j\omega) = \frac{5}{(1+j\omega)(1+j0.5\omega)(1+j0.1\omega)}$$

$$= \frac{5}{\sqrt{1+\omega^2}\angle\tan^{-1}\omega\sqrt{1+0.25\omega^2}\angle\tan^{-1}0.5\omega\sqrt{1+0.01\omega^2}\angle\tan^{-1}0.1\omega}$$

$$|G(j\omega)| = \frac{5}{\sqrt{1+\omega^2}\sqrt{1+0.25\omega^2}\sqrt{1+0.01\omega^2}}$$

$$\angle G(j\omega) = -\tan^{-1}\omega - \tan^{-1}0.5\omega - \tan^{-1}0.1\omega$$

The gain crossover frequency of compensated system, ω_1 = 4 rad/sec.

Let, A_1 = $|G(j\omega)|$ at $\omega = \omega_1$

ϕ_1 = $\angle G(j\omega)$ at $\omega = \omega_1$

$\therefore A_1 = \dfrac{5}{\sqrt{1+4^2} \times \sqrt{1+0.25 \times 4^2} \times \sqrt{1+0.01 \times 4^2}} = 0.5$

$\phi_1 = -\tan^{-1}4 - \tan^{-1}(0.5 \times 4) - \tan^{-1}(0.1 \times 4) = -161°$

Step-2 : To find γ_u and θ

Let γ_u = Phase margin of uncompensated system

γ_d = Desired phase margin of compensated system.

θ = Phase of $G_c(\omega)$ at $\omega = \omega_1$

Now, $\gamma_u = 180° + \phi_1 = 180° -161° = 19°$

$\theta = \gamma_d - \gamma_u = 45° - 19° = 26°$

Step-3 : To find transfer function of PID-controller.

Given that, steady state error, e_{ss} = 0.1 for unit ramp input.

$\therefore$ Velocity error constant, $K_v = \dfrac{1}{e_{ss}} = \dfrac{1}{0.1} = 10$

The velocity error constant of compensated system is given by,

$K_v = \underset{s \to 0}{Lt}\ sG_c(s)\,G(s)$

Here $G_c(s) = K_p + K_d s + \dfrac{K_i}{s} = \dfrac{K_d s^2 + K_p s + K_i}{s}$

and $G(s) = \dfrac{5}{(1+s)(1+0.5s)(1+0.1s)}$

$\therefore K_v = \underset{s \to 0}{Lt}\ s\ \dfrac{(K_d s^2 + K_p s + K_i)}{s} \times \dfrac{5}{(1+s)(1+0.5s)(1+0.1s)} = 10$

$\therefore 5K_i = 10 \qquad \Rightarrow \qquad K_i = \dfrac{10}{5} = 2$

Derivative constant, $K_d = \dfrac{\sin\theta}{\omega_1 A_1} + \dfrac{K_i}{\omega_1^2} = \dfrac{\sin(26)°}{4 \times 0.5} + \dfrac{2}{4^2} = 0.344$

Proportional constant, $K_p = \dfrac{\cos\theta}{A_1} = \dfrac{\cos 26°}{0.5} = 1.8$

Transfer function of PID controller $\Big\}$ $G_c(s) = \left(K_p + K_d s + \dfrac{K_i}{s}\right) = \left(1.8 + 0.344s + \dfrac{2}{s}\right) = \dfrac{0.344s^2 + 1.8s + 2}{s}$

$\qquad = \dfrac{0.344\left(s^2 + \dfrac{1.8}{0.344}s + \dfrac{2}{0.344}\right)}{s} = \dfrac{0.344\,(s^2 + 5.23s + 5.81)}{s}$

Step-4 : To find open loop transfer function of compensated system

PID-controller is connected in cascade with G(s) as shown in figure 5.13.1.

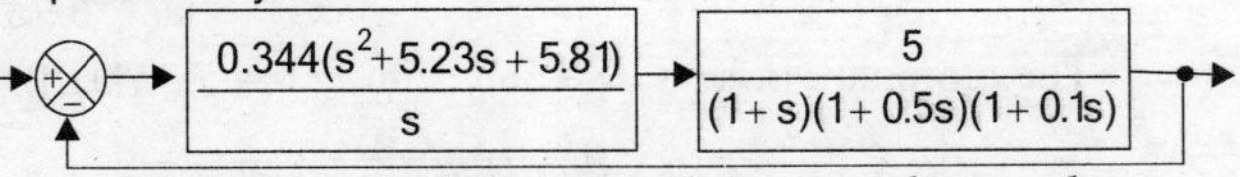

Fig 5.13.1 : *Block diagram of system with cascade PID controller.*

Open loop transfer function of compensated system

$$G_0(s) = G_c(s) \times G(s) = \frac{0.344\,(s^2 + 5.23s + 5.81)}{s} \times \frac{5}{(1+s)(1+0.5s)(1+0.1s)}$$

$$= \frac{1.72\,(s^2 + 5.23s + 5.81)}{s(1+s)(1+0.5s)(1+0.1s)}$$

Step-5 : To verify the design

Put s = jω in $G_0(s)$,

$$\therefore\ G_0(j\omega) = \frac{1.72\,(-\omega^2 + j5.23\omega + 5.81)}{j\omega\,(1+j\omega)(1+j0.5\omega)(1+j0.1\omega)}$$

$$= \frac{1.72\sqrt{(5.81 - \omega^2)^2 + (5.23\omega)^2}\ \angle \tan^{-1}\dfrac{5.23}{5.81 - \omega^2}}{\omega\angle 90°\ \sqrt{1+\omega^2}\ \angle \tan^{-1}\omega\ \sqrt{1+(0.5\omega)^2}\ \angle \tan^{-1}0.5\omega\ \sqrt{1+(0.1\omega)^2}\ \angle \tan^{-1}0.1\omega}$$

Let $A_0 = |\,G_0(j\omega)\,|$ and $\phi_0 = \angle G_0(j\omega)$

$$\therefore\ A_0 = \frac{1.72\sqrt{(5.81 - \omega^2)^2 + (5.23\omega)^2}}{\omega\sqrt{1+\omega^2}\ \sqrt{1+(0.5\omega)^2}\ \sqrt{1+(0.1\omega)^2}}$$

$$\phi_0 = \tan^{-1}\frac{5.23}{5.81 - \omega^2} - 90° - \tan^{-1}\omega - \tan^{-1}0.5\omega - \tan^{-1}0.1\omega,\ \text{for } \omega < \sqrt{5.81}$$

$$= 180° + \tan^{-1}\frac{5.23\omega}{5.81 - \omega^2} - 90° - \tan^{-1}\omega - \tan^{-1}0.5\omega - \tan^{-1}0.1\omega,\ \text{for } \omega > \sqrt{5.81}$$

At $\omega = \omega_1$, $A_0 = A_{01} = \dfrac{1.72\sqrt{(5.81 - 4^2)^2 + (5.23 \times 4)^2}}{4 \times \sqrt{1+4^2} \times \sqrt{(1+(0.5 \times 4)^2} \times \sqrt{1+(0.1\times 4)^2}} = 1$

At $\omega = \omega_1$, $\phi_0 = \phi_{01} = 180° + \tan^{-1}\dfrac{5.23 \times 4}{5.81 - 4^2} - 90° - \tan^{-1}4 - \tan^{-1}(0.5 \times 4) - \tan^{-1}(0.1 \times 4) = -135°$

Phase margin of compensated system, $\gamma_0 = 180° + \phi_{01} = 180° - 135° = 45°$

CONCLUSION

The phase margin of the compensated system, meets the given specification. Hence the design is acceptable.

RESULT

1. Transfer function of PID controller
$$G_c(s) = \left(1.8 + 0.344s + \frac{2}{s}\right) = \frac{0.344\,(s^2 + 5.23s + 5.81)}{s}$$

2. Open loop transfer function of compensated system
$$G_0(s) = \frac{1.72\,(s^2 + 5.23s + 5.81)}{s(1+s)(1+0.5s)(1+0.1s)}$$

DESIGN OF PD, PI AND PID CONTROLLER USING ROOT LOCUS TECHNIQUE

In this technique a controller can be introduced in cascade with open loop system so that, it can have a pair of dominant closed loop poles, to satisfy a specified time domain specifications ζ and ω_n.

The dominant pole, $s_d = -\zeta\omega_n \pm j\omega_n\sqrt{1-\zeta^2} = D\angle\beta$ .....(5.66)

where, $D = |s_d| = \sqrt{(\zeta\omega_n)^2 + \omega_n^2(1-\zeta^2)}$ and $\beta = \angle s_d = \tan^{-1}\dfrac{\omega_n\sqrt{1-\zeta^2}}{-\zeta\omega_n}$

ζ = Damping ratio and ω_n = Natural frequency of oscillation, rad/sec.

Let, $G(s)$ = Open loop transfer function of the system.

$G_c(s)$ = Transfer function of PID controller.

$\therefore$ Open loop transfer function of compensated system = $G_c(s)\,G(s)$ .....(5.67)

For the dominant pole, s_d to be a point on root locus, the magnitude condition says that,

$$G_c(s_d)G(s_d) + 1 = 0 \qquad\qquad .....(5.68)$$

$$\therefore G_c(s_d)G(s_d) = -1 \qquad\qquad .....(5.69)$$

Let $G(s_d) = A_d \angle \phi_d$

where $A_d = |G(s_d)|$ and $\phi_d = \angle G(s_d)$ .....(5.71)

Also, $-1 = 1 \angle 180°$

$$\therefore\; G_c(s_d) = \frac{-1}{G(s_c)} = \frac{1\angle 180°}{A_d \angle \phi_d} = \frac{1}{A_d}\angle(180° - \phi_d)$$

The transfer function of PID controller is given by,

$$G_c(s) = K_p + \frac{K_i}{s} + K_d s = \frac{K_p s + K_i + K_d s^2}{s} = \frac{K_p s^2 + K_p s + K_i}{s} \qquad .....(5.72)$$

Replace s by s_d in equation (5.72).

$$\therefore\; G_c(s_d) = \frac{K_d s_d^2 + K_p s_d + K_i}{s_d} \qquad\qquad .....(5.73)$$

From equation (5.66) we know that $s_d = D\angle\beta$.

$$s_d = D\angle\beta = D\cos\beta + jD\sin\beta \qquad\qquad .....(5.74)$$

$$s_d^2 = (D\angle\beta)^2 = D^2\angle 2\beta = D^2\cos 2\beta + jD^2\sin 2\beta \qquad\qquad .....(5.75)$$

On substituting for s_d and s_d^2 from equations (5.74) and (5.75) in equation (5.73) we get,

$$G_c(s_d) = \frac{K_d[D^2\cos 2\beta + jD^2\sin 2\beta] + K_p[D\cos\beta + jD\sin\beta] + K_i}{D\angle\beta} \qquad .....(5.76)$$

On equating equations (5.71) and (5.76) we get,

$$\frac{1}{A_d}\angle(180° - \phi_d) = \frac{K_d[D^2\cos 2\beta + jD^2\sin 2\beta] + K_p[D\cos\beta + jD\sin\beta] + K_i}{D\angle\beta}$$

$$\frac{1}{A_d}\angle(180° - \phi_d)D\angle\beta = K_d D^2\cos 2\beta + K_p D\cos\beta + K_i + j(K_d D^2\sin 2\beta + K_p D\sin\beta)$$

$$\frac{D}{A_d}(\angle 180° - (\phi_d - \beta)) = K_d D^2\cos 2\beta + K_p D\cos\beta + K_i + j(K_d D^2\sin 2\beta + K_p D\sin\beta)$$

$$\frac{D}{A_d}\cos(180° - (\phi_d - \beta)) + j\frac{D}{A_d}\sin(180 - (\phi_d - \beta)) = K_d D^2\cos 2\beta + K_p D\cos\beta + K_i$$
$$+ j(K_d D^2\sin 2\beta + K_p D\sin\beta)$$

$$-\frac{D}{A_d}\cos(\phi_d - \beta) + j\frac{D}{A_d}\sin(\phi_d - \beta) = K_d D^2\cos 2\beta + K_p D\cos\beta \qquad .....(5.77)$$
$$+ K_i + j(K_d D^2\sin 2\beta + K_p D\sin\beta)$$

On equating real parts of equation (5.77) we get,

$$-\frac{D}{A_d}\cos(\phi_d - \beta) = K_d D^2 \cos 2\beta + K_p D \cos \beta + K_i \qquad .....(5.78)$$

On equating imaginary parts of equation (5.77) we get,

$$\frac{D}{A_d}\sin(\phi_d - \beta) = K_d D^2 \sin 2\beta + K_p D \sin \beta$$

$$\therefore \; K_p D \sin \beta = \frac{D}{A_d}\sin(\phi_d - \beta) - K_d D^2 \, 2 \sin \beta \cos \beta \qquad \boxed{\because \; \sin 2\beta = 2 \sin \beta \cos \beta}$$

$$\therefore \; K_p = \frac{\sin(\phi_d - \beta)}{A_d \sin \beta} - 2 K_d D \cos \beta \qquad .....(5.79)$$

On substituting for K_p from equation (5.79) in equation (5.78) we get,

$$-\frac{D}{A_d}\cos(\phi_d - \beta) = K_d D^2 \cos 2\beta + \left[\frac{\sin(\phi_d - \beta)}{A_d \sin \beta} - 2 K_d D \cos \beta\right] D \cos \beta + K_i$$

$$-\frac{D}{A_d}\cos(\phi_d - \beta) = K_d D^2 \cos 2\beta + \frac{D}{A_d}\frac{\cos \beta}{\sin \beta}\sin(\phi_d - \beta) - \frac{D}{A_d}\frac{\cos \beta \sin(\phi_d - \beta)}{\sin \beta} - K_i$$

$$\therefore \; K_d D^2 (\cos 2\beta - 2\cos^2 \beta) = -\frac{D}{A_d}\cos(\phi_d - \beta) - \frac{D}{A_d}\frac{\cos \beta \sin(\phi_d - \beta)}{\sin \beta} - K_i$$

Put, $\cos 2\beta = 1 - 2\sin^2\beta$

$$K_d D^2 (1 - 2\sin^2\beta - 2\cos^2\beta) = -\frac{D}{A_d}\frac{\sin \beta \cos(\phi_d - \beta) + \cos \beta \sin(\phi_d - \beta)}{\sin \beta} - K_i$$

$$K_d D^2 (1 - 2(\sin^2\beta + \cos^2\beta)) = -\frac{D}{A_d}\frac{\sin(\beta + \phi_d - \beta)}{\sin \beta} - K_i$$

$$\therefore -K_d D^2 = -\frac{D}{A_d}\frac{\sin \phi}{\sin \beta} - K_i \qquad \boxed{\begin{array}{l}\textbf{\textit{Note :}} \; sin^2\theta + cos^2\theta = 1 \\ sin(A + B) = sinAsinB + cosAsinB\end{array}}$$

$$\therefore \; \mathbf{K_d = \frac{\sin \phi}{D A_d \sin \beta} + \frac{K_i}{D^2}} \qquad .....(5.80)$$

On substituting for K_d from equation (5.80) in equation (5.79) we get,

$$K_p = \frac{\sin(\phi_d - \beta)}{A_d \sin \beta} - 2\left[\frac{\sin \phi_d}{D A_d \sin \beta} + \frac{K_i}{D^2}\right] D \cos \beta$$

$$K_p = \frac{\sin \phi_d \cos \beta - \cos \phi_d \sin \beta}{A_d \sin \beta} - \frac{2D \cos \beta \sin \phi_d}{D A_d \sin \beta} - \frac{2K_i D \cos \beta}{D^2}$$

$$= \frac{\sin \phi_d \cos \beta - \cos \phi_d \sin \beta - 2\cos \beta \sin \phi_d}{A_d \sin \beta} - \frac{2K_i \cos \beta}{D}$$

$$= \frac{-(\sin \beta \cos \phi_d + \cos \beta \sin \phi_d)}{A_d \sin \beta} - \frac{2K_i \cos \beta}{D} \qquad \boxed{\sin(A + B) = \sin A \cos B + \cos A \sin B}$$

$$\therefore \; \mathbf{K_p = \frac{-\sin(\beta + \phi_d)}{A_d \sin \beta} - \frac{2K_i \cos \beta}{D}} \qquad .....(5.81)$$

The equations (5.80) and (5.81) are used to design a PID controller. Here the parameter K_i is first determined such that the compensated system satisfies the specified error requirement. Then the parameters K_d and K_p are calculated using equations (5.80) and (5.81) respectively.

For designing a PD controller put $K_i = 0$ in equations (5.80) and (5.81). Hence the design equations are,

$$K_d = \frac{\sin\phi_d}{DA_d \sin\beta} \quad \text{and} \quad K_p = \frac{-\sin(\beta + \phi_d)}{A_d \sin\beta}$$

For designing a PI controller, put $K_d = 0$ in equation (5.80). Hence the design equations are,

$$K_i = \frac{-D \sin\phi_d}{A_d \sin\beta} \quad \text{and} \quad K_p = \frac{-\sin(\beta + \phi_d)}{A_d \sin\beta} - \frac{2K_i \cos\beta}{D}$$

PROCEDURE FOR DESIGN OF PD/ PI/PID CONTROLLER

The following procedure can be followed to design a PD/PI/PID controller when the specifications are damping ratio, ζ and natural frequency of oscillation, ω_n.

Step-1 : Determine the dominant pole, s_d and calculate its magnitude and phase.

Dominant pole, $s_d = -\zeta\omega_n \pm j\omega_n \sqrt{1 - \zeta^2}$

Let $D = |s_d|$ and $\beta = \angle s_d$

By considering the dominant pole at $-\zeta\omega_n + j\omega_n \sqrt{1 - \zeta^2}$ we get,

$$D = \sqrt{\zeta^2\omega_n^2 + \omega_n^2(1 - \zeta^2)} \quad \text{and} \quad \beta = \tan^{-1}\frac{\sqrt{1 - \zeta^2}}{-\zeta}$$

Step-2 : Determine the magnitude and phase of G(s) at $s = s_d$

Let $A_d = |G(s)|$ at $s = s_d$ (i.e., $A_d = |G(s_d)|$)

and, $\phi_d = \angle G(s)$ at $s = s_d$ (i.e., $\phi_d = \angle G(s_d)$)

Step-3 : Determine the transfer function of PD/PI/PID controller.

a. **PD controller**

Derivative constant, $K_d = \dfrac{\sin\phi_d}{DA_d \sin\beta}$; Proportional constant, $K_p = \dfrac{-\sin(\beta + \phi_d)}{A_d \sin\beta}$

Transfer function of PD controller $\Bigg\}$ $G_c(s) = K_p + K_d s = K_d\left(s + \dfrac{K_p}{K_d}\right)$

b. **PI controller**

Integral constant, $K_i = \dfrac{-D \sin\phi_d}{A_d \sin\beta}$

Proportional constant, $K_p = \dfrac{-\sin(\beta + \phi_d)}{A_d \sin\beta} - \dfrac{2K_i \cos\beta}{D}$

Transfer function of PI controller $\Bigg\}$ $G_c(s) = K_p + \dfrac{K_i}{s} = \dfrac{K_p s + K_i}{s} = \dfrac{K_p(s + (K_i/K_p))}{s}$

c. **PID controller**

Transfer function of PID controller $\Bigg\}$ $G_c(s) = K_p + \dfrac{K_i}{s} + K_d s = \dfrac{K_p s + K_i + K_d s^2}{s} = \dfrac{K_d\left(s^2 + \dfrac{K_p}{K_d}s + \dfrac{K_i}{K_d}\right)}{s}$

Determine K_i from the specified error constant, such that the compensated system meets the error requirement.

For example, if the system is type-0 system and velocity error constant, K_v is specified then K_i is obtained by evaluating the following expression.

$$K_v = \underset{s \to 0}{Lt} \, sG_c(s) \, G(s)$$

Calculate the parameter K_d and K_p using the following equations

Proportional constant, $K_p = \dfrac{-\sin(\beta + \phi_d)}{A_d \sin\beta} - \dfrac{2K_i \cos\beta}{D}$

Derivative constant, $K_d = \dfrac{\sin\phi_d}{DA_d \sin\beta} + \dfrac{K_i}{D^2}$

Step-4 : Verify the design

Open loop transfer function of compensated system, $G_0(s) = G_c(s)G(s)$.

The design is accepted if the root locus of compensated system pass through the dominant pole, s_d. This can be verified from the magnitude condition, which states that the point $s = s_d$ will be a point on root locus if $1 + G_0(s_d) = 0$, where $G_0(s_d)$ is the value of $G_0(s)$ at $s = s_d$. It can be shown that $G_0(s_d) = -1$

EXAMPLE 5.14

Consider a unity feedback system with open loop transfer function, $G(s) = 20/s\,(s+2)\,(s+4)$. Design a PD controller so that the closed loop has a damping ratio of 0.8 and natural frequency of oscillation as 2 rad/sec.

SOLUTION

Step-1 :   To find dominant pole, s_d

Dominant pole, $s_d = -\zeta\omega_n \pm j\omega_n\sqrt{1-\zeta^2}$

Given that, $\zeta = 0.8$ and $\omega_n = 2$ rad/sec. $\therefore s_d = -0.8 \times 2 \pm j2 \times \sqrt{1-0.8^2} = -1.6 \pm j1.2$

Let $D = |s_d|$ and $\beta = \angle s_d$

By considering dominant pole at $(-1.6 + j1.2)$ we get,

$s_d = -1.6 + j1.2 = 2\angle143°$ $\qquad \therefore D = 2$ and $\beta = 143°$

Step-2 :   To find magnitude and phase of $G(s)$ at $s = s_d$.

Let $A_d = |G(s_d)|$ and $\phi_d = \angle G(s_d)$

Given that, $G(s) = 20/s\,(s+2)\,(s+4)$

Note : Use polar to rectangular conversion.

$G(s_d) = \dfrac{20}{s_d(s_d+2)(s_d+4)} = \dfrac{20}{(-1.6+j1.2)(-1.6+j1.2+2)(-1.6+j1.2+4)}$

$= \dfrac{20}{(-1.6+j1.2)(0.4+j1.2)(2.4+j1.2)} = \dfrac{20}{2\angle143° \times 1.26\angle71° \times 2.68\angle26°}$

$= \dfrac{20}{2 \times 1.26 \times 2.68} \angle(-143° - 71° - 26°) = 2.96\angle-240°$

$\therefore A_d = 2.96$ and $\phi_d = -240°$

Step-3 : Determine the transfer function of PD controller

Derivative constant, $K_d = \dfrac{\sin\phi_d}{DA_d\sin\beta} = \dfrac{\sin(-240°)}{2\times 2.96\times\sin(143°)} = 0.243$

Proportional constant, $K_p = \dfrac{-\sin(\beta+\phi_d)}{A_d\sin\beta} = \dfrac{-\sin(143°-240°)}{2.96\sin(143°)} = 0.557$

Transfer function of PD controller $\Big\}$ $G_c(s) = K_p + K_p s = (0.557 + 0.243s) = 0.243\left(\dfrac{0.557}{0.243} + s\right) = 0.243\,(2.292 + s)$

Step-4 : Verify the design

Open loop transfer function of compensated system $\Big\}$ $G_0(s) = G_c(s)G(s) = 0.243\,(s+2.292)\dfrac{20}{s(s+2)(s+4)} = \dfrac{4.86\,(s+2.292)}{s(s+2)(s+4)}$

$G_c(s_d) = 0.243\,(s_d + 2.292) = 0.243(-1.6 + j1.2 + 2.292)$

$\qquad = 0.243\,(0.692 + j1.2) = 0.243 \times (1.385\angle 60°) = 0.337\angle 60°$

$G_0(s_d) = G_c(s_d)G(s_d) = 0.337\angle 60° \times 2.96\angle -240°$

$\qquad\qquad = 1\angle -180° = -1$

Since $G_0(s_d) = -1$, the root locus will pass through s_d and so the design is accepted.

RESULT

Transfer function of PD controller $\Big\}$ $G_c(s) = (0.557 + 0.243s) = 0.243\,(s + 2.292)$

Open loop transfer function of compensated system $\Big\}$ $G_0(s) = \dfrac{4.86\,(s+2.292)}{s(s+2)(s+4)}$

EXAMPLE 5.15

Consider a unity feedback system with open loop transfer function, $G(s) = 4/(s+1)(s+5)$. Design a PI controller so that the closed loop has a damping ratio of 0.9 and natural frequency of oscillation as 2.5 rad /sec.

SOLUTION

Step-1 : To find dominant pole, s_d

Dominant pole, $s_d = -\zeta\omega_n \pm j\omega_n\sqrt{1-\zeta^2}$

Given that, $\zeta = 0.9$ and $\omega_n = 2.5$ rad/sec.

$\qquad \therefore s_d = -0.9\times 2.5 \pm j2.5\times\sqrt{1-0.9^2} = -2.25 \pm j1.09$

> *Note : Use polar to rectangular conversion.*

Let $D = |s_d|$ and $\beta = \angle s_d$

By considering the dominant pole at $(-2.25 + j1.09)$ we get,

$\qquad s_d = -2.25 + j1.09 = 2.5\angle 154°$

$\qquad \therefore D = 2.5$ and $\beta = 154°$

Step-2 : To find magnitude and phase of $G(s)$ at $s = s_d$.

Let $A_d = |G(s_d)|$ and $\phi_d = \angle G(s_d)$

Given that, $G(s) = 4/(s+1)(s+5)$

$G(s_d) = \dfrac{4}{(s_d+1)(s_d+5)} = \dfrac{4}{(-2.25+j1.09+1)(-2.25+j1.09+5)}$

$\qquad = \dfrac{4}{(-1.25+j1.09)(2.75+j1.09)} = \dfrac{4}{(1.66\angle 139°)(2.96\angle 21°)}$

$\qquad = \dfrac{4}{1.66\times 2.96}\angle(-139°-21°) = 0.81\angle -160°$

> *Note : Use polar to rectangular conversion.*

$\qquad \therefore A_d = 0.81$ and $\phi_d = -160°$

Step-3 :   Determine the transfer function of PI controller

Integral constant, $K_i = \dfrac{-D \sin \phi_d}{A_d \sin \beta} = \dfrac{-2.5 \sin(-160°)}{0.81 \sin(154°)} = 2.4$

Proportional constant, $K_p = \dfrac{-\sin(\beta + \phi_d)}{A_d \sin \beta} - \dfrac{2K_i \cos \beta}{D} = \dfrac{-\sin(154° - 160°)}{0.81 \sin(154°)} - \dfrac{2 \times 2.4 \cos(154°)}{2.5} = 2.02$

Transfer function of PI controller $\Big\}$ $G_c(s) = K_p + \dfrac{K_i}{s} = 2.02 + \dfrac{2.4}{s} = \dfrac{2.02s + 2.4}{s} = \dfrac{2.02\left(s + \dfrac{2.4}{2.02}\right)}{s} = \dfrac{2.02(s + 1.19)}{s}$

Step-4 :   Verify the design

Open loop transfer function of compensated system $\Big\}$ $G_0(s) = G_c(s)\,G(s) = \dfrac{2.02(s + 1.19)}{s} \times \dfrac{4}{(s+1)(s+5)} = \dfrac{8.08(s + 1.19)}{s(s+1)(s+5)}$

$G_c(s_d) = \dfrac{2.02(s_d + 1.19)}{s_d} = \dfrac{2.02(-2.25 + j1.09 + 1.9)}{-2.25 + j1.09}$

$= \dfrac{2.02(-1.06 + j1.09)}{-2.25 + j1.09} = \dfrac{2.02 \times 1.52 \angle 134°}{2.5 \angle 154°} = 1.228 \angle -20°$

$\therefore\ G_0(s_d) = G_c(s_d)\,G(s_d) = 1.228 \angle -20° \times 0.81 \angle -160° = 1 \angle -180° = -1$

Since $G_0(s_d) = -1$, the root locus will pass through s_d and so the design is accepted.

RESULT

Transfer function of PI controller $\Big\}$ $G_c(s) = \left(2.02 + \dfrac{2.4}{s}\right) = \dfrac{2.02(s + 1.19)}{s}$

Open loop transfer function of compensated system $\Big\}$ $G_0(s) = \dfrac{8.08(s + 1.19)}{s(s+1)(s+5)}$

EXAMPLE 5.16

Consider a unity feedback system with open loop transfer function, $G(s) = 75/(s+1)(s+3)(s+8)$. Design a PID controller to satisfy the following specifications. (a) The steady state error for unit ramp input should be less than 0.08. (b) Damping ratio = 0.8. (c) Natural frequency of oscillation = 2.5 rad/sec.

SOLUTION

Step-1 :   To find the dominant pole, s_d

Dominant pole, $s_d = -\zeta \omega_n \pm j \omega_n \sqrt{1 - \zeta^2}$

Given that, $\zeta = 0.8$ and $\omega_n = 2.5$ rad/sec.

$\therefore\ s_d = -0.8 \times 2.5 \pm j2.5 \times \sqrt{1 - 0.8^2} = -2 \pm j1.5$

Let $D = |s_d|$ and $\beta = \angle s_d$

By considering the dominant pole at $(-2 + j1.5)$ we get,

$s_d = -2 + j1.5 = 2.5 \angle 143°$

$\therefore\ D = 2.5$ and $\beta = 143°$

Note : Use polar to rectangular conversion.

Step-2 :   To find magnitude and phase of $G(s)$ at $s = s_d$.

Let $A_d = |G(s_d)|$ and $\phi_d = \angle G(s_d)$

Given that, $G(s) = 75/(s+1)(s+3)(s+8)$

$$G(s_d) = \frac{75}{(s_d+1)(s_d+3)(s_d+8)} = \frac{75}{(-2+j1.5+1)(-2+j1.5+3)(-2+j1.5+8)}$$

$$= \frac{75}{(-1+j1.5)(1+j1.5)(6+j1.5)} = \frac{75}{(1.8\angle 124°)(1.8\angle 56°)(6.18\angle 14°)}$$

$$= \frac{75}{1.8 \times 1.8 \times 6.18} \angle(-124° - 56° - 14°) = 3.75\angle - 194°$$

$\therefore A_d = 3.75$ and $\phi_d = -194°$

Step-3 : Determine the transfer function of PID controller

Given that, $e_{ss} \leq 0.08$ for unit ramp input.

$\therefore$ Velocity error constant, $K_v \geq \dfrac{1}{e_{ss}} = \dfrac{1}{0.08} = 12.5$

Transfer function of PID controller $\Big\}$ $G_c(s) = \left(K_p + \dfrac{K_i}{s} + K_d s\right) = \left(\dfrac{K_d s^2 + K_p s + K_i}{s}\right)$

Open loop transfer function of compensated system $\Big\}$ $G_0(s) = G_c(s)\,G(s)$

$$= \frac{(K_d s^2 + K_p s + K_i)}{s} \times \frac{75}{(s+1)(s+3)(s+8)} = \frac{75(K_d s^2 + K_p s + K_i)}{s(s+1)(s+3)(s+8)}$$

Velocity error constant of compensated system $\Big\}$ $= \underset{s \to 0}{\mathrm{Lt}}\, sG_0(s) = \underset{s \to 0}{\mathrm{Lt}}\, s\,\dfrac{75(K_d s^2 + K_p s + K_i)}{s(s+1)(s+3)(s+8)} = \dfrac{75K_i}{3 \times 8} = 3.125 K_i$

But the velocity error constant of compensated system should be greater than or equal to 12.5.

$$3.125\, K_i = 12.5 \quad \Rightarrow \quad K_i = \frac{12.5}{3.125} = 4$$

Derivative constant, $K_d = \dfrac{\sin\phi_d}{DA_d \sin\beta} + \dfrac{K_i}{D^2} = \dfrac{\sin(-194°)}{2.5 \times 3.75 \times \sin(143°)} + \dfrac{4}{2.5^2} = 0.68$

Proportional constant, $K_p = \dfrac{-\sin(\beta + \phi_d)}{A_d \sin\beta} - \dfrac{2K_i \cos\beta}{D} = \dfrac{-\sin(143° - 194°)}{3.75 \times \sin(143°)} - \dfrac{2 \times 4 \times \cos(143°)}{2.5} = 2.9$

Transfer function of PID controller $\Big\}$ $G_c(s) = K_p + \dfrac{K_i}{s} + K_d s = 2.9 + \dfrac{4}{s} + 0.68s = \dfrac{0.68s^2 + 2.9s + 4}{s}$

$$= \frac{0.68\left(s^2 + \dfrac{2.9}{0.68}s + \dfrac{4}{0.68}\right)}{s} = \frac{0.68(s^2 + 4.26s + 5.88)}{s}$$

Step-4 : Verify the design

Open loop transfer function of compensated system $\Big\}$ $G_0(s) = G_c(s)\,G(s) = \dfrac{0.68(s^2 + 4.26s + 5.88)}{s} \times \dfrac{75}{(s+1)(s+3)(s+8)}$

$$= \frac{51(s^2 + 4.26s + 5.88)}{s(s+1)(s+3)(s+8)}$$

$$G_c(s_d) = \frac{0.68(s_d^2 + 4.26s_d + 5.88)}{s_d}$$

$$s_d = 2.5\angle 143° = -2 + j1.5$$

$$s_d^2 = (2.5\angle 143°)^2 = 2.5^2 \angle 2 \times 143° = 6.25\angle 286° = 1.72 - j6$$

$$G_c(s_d) = \frac{0.68(1.72 - j6 + 4.26(-2 + j1.5) + 5.88)}{-2 + j1.5} = \frac{0.68(-0.92 + j0.39)}{-2 + j1.5} = \frac{0.68 \times 1\angle 157°}{2.5\angle 143°} = 0.272\angle 14°$$

$$\therefore \ G_0(s_d) = G_c(s_d)G(s_d) = 0.272\angle 14° \times 3.75\angle - 194° = 1\angle - 180° = -1$$

Since $G_0(s_d) = -1$, the root locus will pass through s_d and so the design is accepted.

RESULT

Transfer function of PID controller $\Bigg\}$ $G_c(s) = \left(2.9 + \dfrac{4}{s} + 0.68s\right) = \dfrac{0.68(s^2 + 4.26s + 5.88)}{s}$

Open loop transfer function of compensated system $\Bigg\}$ $G_0(s) = \dfrac{51(s^2 + 4.26s + 5.88)}{s(s + 1)(s + 3)(s + 8)}$

5.6 FEEDBACK COMPENSATION

In feedback compensation a compensating device is placed in an internal feedback path around one or more components of the forward path. It is also called minor loop feedback compensation.

The decision to use a feedback compensation is sometimes a matter of convenience, sometimes a matter of necessity and for some problems it can be shown that feedback scheme will offer better performance. For example, use of velocity feedback for damping is very effective and quite popular. The feedback compensation scheme may be necessary in one of the following situations.

1. Feedback compensation is provided when derivative action is needed on controlled variable in order to avoid the derivative kick on set point changes.

2. In non-electrical systems, when suitable cascade device is not available, feedback compensation is employed.

3. In systems subjected to frequent load disturbances, feedback compensation may be preferred as it provides greater stiffness against load disturbances.

Besides all the factors discussed above, the available components and designer's experience and preferences influence the choice between a cascade and a feedback compensation scheme. A commonly used arrangement for feedback compensation is shown in fig 5.23.

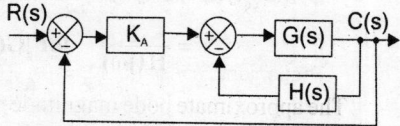

Fig 5.23 : Feedback compensation scheme.

The most popular schemes employ rate feedback or tachometer feedback in which, $H(s) = K_t s$. The disadvantage in the rate feedback is that the system velocity error constant, K_v is reduced. This undesirable effect can be eliminated by reducing the feedback signal in the region of low frequencies. Hence a high pass filter of the type shown in fig 5.24 can be connected in cascade with rate device as shown in fig 5.25.

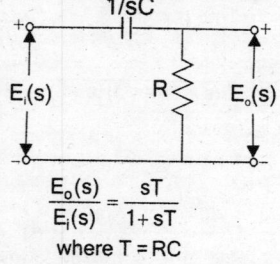

$$\frac{E_0(s)}{E_i(s)} = \frac{sT}{1 + sT}$$

where T = RC

Fig 5.24 : High pass filter.

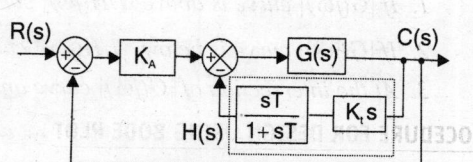

Fig 5.25 : Feedback compensation with high-pass filter in the feedback path.

DESIGN OF FEEDBACK COMPENSATION SCHEME USING BODE PLOT

The minor loop feedback in fig 5.23 can be eliminated and the forward path transfer function, G(s) can be replaced by an equivalent forward path transfer function, $G_{eq}(s)$ as shown in fig 5.26.

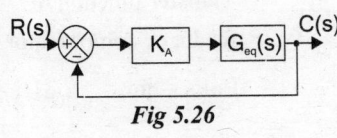

Fig 5.26

Equivalent transfer function, $G_{eq} = \dfrac{G(s)}{1 + G(s)H(s)}$ .....(5.82)

$\left.\begin{array}{l}\text{Open loop transfer function}\\\text{of compensated system}\end{array}\right\}$ $G_0(s) = K_A G_{eq}(s)$ .....(5.83)

The transfer function, $G_{eq}(s)$ and the parameter K_t and K_A can be designed in frequency domain using bode plot.

Put s = jω in $G_{eq}(s)$,   ∴ $G_{eq}(j\omega) = \dfrac{G(j\omega)}{1 + G(j\omega)H(j\omega)}$ .....(5.84)

In general there will be a range of frequencies for which $|G(j\omega)|$ is large and some other range for which it is small. This is also true for H(jω) and the ranges of frequencies will probably overlap but will not be identical.

In some frequency range $|G(j\omega) H(j\omega)| \gg 1$ and in this range the equation(5.84) can be written as

$$G_{eq}(j\omega) \approx \frac{G(j\omega)}{1 + G(j\omega)H(j\omega)} = \frac{1}{H(j\omega)}$$ .....(5.85)

In some frequency range $|G(j\omega) H(j\omega)| \ll 1$ and in this range the equation(5.84) can be written as,

$$G_{eq} \approx G(j\omega)$$ .....(5.86)

From equations (5.85) and (5.86) we can conclude that,

$$\mathbf{G_{eq}(j\omega) = G(j\omega)} \quad \textbf{for } \mathbf{|G(j\omega)H(j\omega)| \leq 1}$$

$$= \frac{1}{\mathbf{H(j\omega)}} \quad \textbf{for } \mathbf{|G(j\omega)H(j\omega)| \geq 1}$$ .....(5.87)

The approximate bode magnitude plot of G(jω) and 1/H(jω) are drawn and by graphically interpreting equation(5.87), the $G_{eq}(j\omega)$ can be determined.

The db magnitude of G(jω)H(jω) can be expressed as,

$$20\log |G(j\omega)H(j\omega)| = 20\log |G(j\omega)| - 20 \log |1/H(j\omega)|$$ .....(5.88)

From equation (5.88) and bode magnitude plots of G(jω) &1/H(jω) we can obtain following conclusions.

1. *If $|G(j\omega)|$ curve is above $|1/H(j\omega)|$ curve, then $|G(j\omega)H(j\omega)| > 1$ and $G_{eq}(j\omega) \approx |1/H(j\omega)|$*

2. *If $|G(j\omega)|$ curve is below $|1/H(j\omega)|$ curve, then $|G(j\omega)H(j\omega)| < 1$ and $G_{eq}(j\omega) \approx |G(j\omega)|$*

3. *At the intersection of $|G(j\omega)|$ curve and $|1/H(j\omega)|$ curve, $|G(j\omega)H(j\omega)| = 1$.*

PROCEDURE FOR DESIGN USING BODE PLOT

Step-1 : Draw the magnitude plot of G(jω) on semilog graph sheet.

Step-2 : Draw the magnitude plot of 1/H(jω) on the same graph sheet and using the same scales. The feedback compensator consists of a rate device and a high pass filter in cascade.

Transfer function of feedback compensator $\left.\right\}$ $H(s) = K_t s \dfrac{sT}{1+sT} = \dfrac{K_t T s^2}{1+sT}$

Put $s = j\omega$, $\quad \therefore$ $H(j\omega) = K_t T \dfrac{(j\omega)^2}{1+j\omega T}$

Now, $\dfrac{1}{H(j\omega)} = \dfrac{1}{K_t T} \dfrac{1+j\omega T}{(j\omega)^2}$

By taking an initial value of $K_t T = 1$ and assuming suitable value for T, draw the bode plot of $1/H(j\omega)$. (Typical values of T is in the range of 1 to 2 sec).

Step-3 : Determine the desired gain crossover frequency of feedback compensated system.

Let, ω_{gc} = Desired gain crossover frequency.

γ_d = Desired (or specified) phase margin.

Choose a new value of phase margin, γ_n such that, $\gamma_n = \gamma_d + \epsilon$

Let initial choice of ϵ be 10° to 25°

Now, $\gamma_n = 180° + \phi_{hgc}$

where, $\phi_{hgc} = \angle(1/H(j\omega))$ at $\omega = \omega_{gc}$

From the transfer function of $1/H(j\omega)$ we get, $\phi_{hgc} = \tan^{-1}\omega_{gc} T - 180°$

$\therefore \gamma_n = 180° + \tan^{-1}\omega_{gc} T - 180° \quad \Rightarrow \quad \omega_{gc} T = \tan\gamma_n \quad \Rightarrow \quad \omega_{gc} = \dfrac{\tan\gamma_n}{T}$

Step-4 : Determine the parameter K_t

Now shift $|1/H(j\omega)|$ curve vertically up or down such that it crosses $|G(j\omega)|$ curve at a frequency atleast 4 times greater than ω_{gc}. Let the $|1/H(j\omega)|$ curve be shifted by $\pm x$ db ("+" for upward shift and "−" for downward shift).

Now, $20\log\dfrac{1}{K_t T} = \pm x \quad \Rightarrow \quad \log\dfrac{1}{K_t T} = \pm\dfrac{x}{20} \quad \Rightarrow \quad \dfrac{1}{K_t T} = 10^{\pm x/20} \quad \Rightarrow \quad K_t = \dfrac{1}{T \times 10^{\pm x/20}}$

Step-5 : Determine the transfer function, $G_{eq}(j\omega)$ from the bode plots of $G(j\omega)$ and $1/H(j\omega)$.

A typical magnitude plot of $G(j\omega)$ and $1/H(j\omega)$ with $K_t T = 1$ are shown in fig 5.27. The magnitude plot of $1/H(j\omega)$ is shifted to cross $|G(j\omega)|$ curve as mentioned in step-4. Let the shifted plot of $1/H(j\omega)$ intersects, the $|G(j\omega)|$ curve as shown in fig 5.28 and the frequencies corresponding to crossing points be ω_{c1} and ω_{c3}.

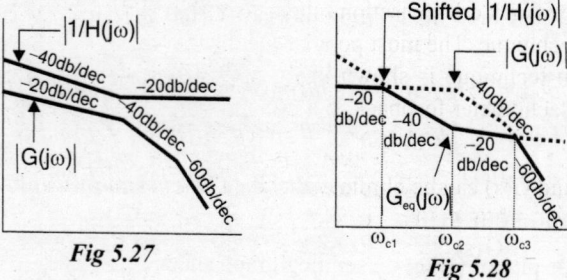

Fig 5.27   *Fig 5.28*

In the frequency range, ω_{c3} to ∞, the magnitude plot of $G_{eq}(j\omega)$ is given by portion of $|G(j\omega)|$ curve in the range, ω_{c3} to ∞.

In the frequency range, 0 to ω_{c1}, the magnitude plot of $G_{eq}(j\omega)$ is given by portion of $|G(j\omega)|$ curve in the range, 0 to ω_{c1}.

In the frequency range, ω_{c1} to ω_{c3}, the magnitude plot of $G_{eq}(j\omega)$ is given by portion of shifted $|1/H(j\omega)|$ curve in the range, ω_{c1} to ω_{c3}.

The bode magnitude plot of $G_{eq}(j\omega)$ is shown as a bold curve in fig 5.28.

From the bode magnitude plot of $G_{eq}(j\omega)$ the transfer function $G_{eq}(j\omega)$ can be determined. The example plot shown in fig 5.28 can be interpreted as follows,

1. *The slope in the starting portion of the curve is −20 db/dec and it is due to an integral factor K/jω where K is given by G(jω) when ω = 0.*

2. *At ω = ω_{c1}, the slope changes from −20 db/dec to −40 db/dec. This is due to a first order factor 1/(1+jωT$_1$), where T$_1$ = 1/ω_{c1}.*

3. *At ω = ω_{c2}, the slope changes from −40 db/dec to −20 db/dec. This is due to a first order factor (1+jωT$_2$), where T$_2$ = 1/ω_{c2}.*

4. *At ω = ω_{c3}, the slope changes from −20 db/dec to −60 db/dec. This is due to a first order factor 1/(1+jωT$_3$)^2, where T$_3$ = 1/ω_{c3}. Hence transfer function, G$_{eq}$(jω) can be written as,*

$$G_{eq}(j\omega) = \frac{K(1+j\omega T_2)}{j\omega(1+j\omega T_1)(1+j\omega T_3)^2} \quad \text{Put, } j\omega = s, \quad \therefore \ G_{eq}(s) = \frac{K(1+sT_2)}{s(1+sT_1)(1+sT_3)^2}$$

Step-6 : Determine the parameter, K_A.

From the bode magnitude plot of $G_{eq}(j\omega)$ find the vertical shift to make ω_{gc} as the gain crossover frequency of $G_{eq}(j\omega)$, (i.e., at ω = ω_{gc}, $|G_{eq}(j\omega)|$ in db = 0). Let the vertical shift be $\pm$ x db (+ for upward shift and − for downward shift).

$$\text{Now, } 20\log K_A = \pm x \implies \log K_A = \frac{\pm x}{20} \implies K_A = 10^{\pm x/20}$$

Step-7 : Verify the design.

Determine the velocity error constant, K_v of the feedback compensated system.

$$\text{where, } K_v = \underset{s \to 0}{\text{Lt}} \ s \, K_A G_{eq}(s)$$

If the velocity error constant is satisfactory then the design is acceptable. Otherwise repeat the design with another suitable value of T or $\in$.

DESIGN OF FEEDBACK COMPENSATION SCHEME USING ROOT LOCUS TECHNIQUE

In time domain the feedback compensation scheme can be designed using root locus technique. The most popular feedback scheme used in this design technique is shown in fig 5.29. It employs a rate device or a tachometer feedback in a minor loop around G(s).

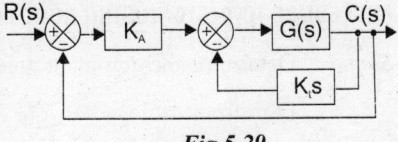

Fig 5.29

The minor loop around G(s) can be eliminated and an equivalent forward transfer function can be obtained as shown below.

$$G_{eq}(s) = \frac{G(s)}{1 + G(s)K_t s} \qquad \qquad .....(5.89)$$

$$\text{Let } G(s) = \frac{K}{s(s + 1/T_1)(s + 1/T_2)} \qquad \qquad .....(5.90)$$

From equation(5.89) and (5.90) we can write,

$$G_{eq}(s) = \dfrac{\dfrac{K}{s(s+1/T_1)(s+1/T_2)}}{1+\dfrac{K}{s(s+1/T_1)(s+1/T_2)}K_t s} = \dfrac{K}{s(s+1/T_1)(s+1/T_2)+K K_t s} \qquad .....(5.91)$$

Closed loop transfer function, $\qquad \dfrac{C(s)}{R(s)} = \dfrac{K_A G_{eq}(s)}{1+K_A G_{eq}(s)} \qquad .....(5.92)$

On substituting for $G_{eq}(s)$ from equation(5.91) in equation(5.92) we get,

$$\dfrac{C(s)}{R(s)} = \dfrac{K_A \dfrac{K}{s(s+1/T_1)(s+1/T_2)+K K_t s}}{1+K_A \dfrac{K}{s(s+1/T_1)(s+1/T_2)+K K_t s}} = \dfrac{K_A K}{s(s+1/T_1)(s+1/T_2)+K K_t s+K_A K} \qquad ....(5.93)$$

The denominator polynomial of closed loop transfer function, $C(s)/R(s)$ is the characteristic equation. Hence from equation(5.93) the characteristic equation is,

$$s(s+1/T_1)(s+1/T_2)+K K_t s+K_A K = 0 \qquad .....(5.94)$$

The equation(5.94) can be rearranged in the form of $1+G(s)H(s) = 0$ as shown below.

On dividing equation(5.94) by $s(s+1/T_1)(s+1/T_2)$, we get,

$$1+\dfrac{K K_t (s+K_A/K_t)}{s(s+1/T_1)(s+1/T_2)} = 0 \qquad .....(5.95)$$

From equation(5.95) we get a new loop transfer function $[G(s) H(s)]_{new}$,

$$\textbf{where, } |G(s)H(s)|_{new} = \dfrac{K K_t (s+K_A/K_t)}{s(s+1/T_1)(s+1/T_2)} \qquad .....(5.96)$$

From equation(5.96) we can conclude that the effect of feedback is to introduce a zero in the loop transfer function. Hence the design procedure includes a search for a suitable location for zero of the compensator.

In this design, the root locus is made to pass through a pair of dominant poles so that the system has desired transient response. The dominant poles are determined from the time domain specifications. The zero as defined by equation(5.96) is located on the real axis such that the angle condition is satisfied at the dominant pole, s_d.

PROCEDURE TO DESIGN USING ROOT LOCUS TECHNIQUE

Step-1 : Determine the dominant pole, s_d.

Dominant pole, $s_d = -\zeta\omega_n + j\omega_n\sqrt{1-\zeta^2}$

where, ζ = Damping ratio

ω_n = Natural frequency of oscillation, rad/sec

Step-2 : Draw the pole-zero plot of $G(s)$

In an ordinary graph sheet, take suitable scales and mark the poles by "x" and zeros by "o". Also mark the dominant pole, s_d. Let point P be the dominant pole.

Step-3 : Determine the new loop transfer function.

Choose a compensation scheme as shown in fig 5.29. Determine the closed loop transfer function $C(s)/R(s)$. The new loop transfer function is obtained by reconstructing the characteristic equation of $C(s)/R(s)$ in the form $1+[G(s)H(s)]_{new}$. In the new loop transfer function, there will be an additional term, $(s+K_A/K_t)$ which contributes a zero, $z_c = -K_A/K_t$.

Step-4 : Locate the zero introduced by the feedback compensator

Let, $\phi =$ Angle contributed by zero of compensator to make point, P as a point on root locus.

Now, $\phi = \begin{pmatrix} \text{Sum of angles contributed} \\ \text{by poles of } G(s) \end{pmatrix} - \begin{pmatrix} \text{Sum of angles contributed} \\ \text{by zeros of } G(s) \end{pmatrix} \pm n180°$

where, n is an odd integer.

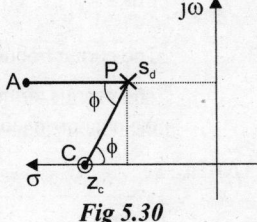

With line AP as reference draw a line PC such that $\angle APC = \phi$, as shown in fig 5.30. This line will intersect the real axis at point C. The value corresponding to point C is the zero of the compensator, z_c.

Step-5 : Determine the parameter K_A and K_t.

From $[G(s)H(s)]_{new}$ we get,

$z_c = -K_A/K_t$

Fig 5.30

Choose a suitable value for K_t and calculate K_A.

Step-6 : Determine the gain K of G(s)

Let, $[G(s)H(s)]_{new} = \dfrac{K\,K_t(s+K_A/K_t)}{s(s+1/T_1)(s+1/T_2)}$

The value of $K\,K_t$ can be obtained from the magnitude condition.

$\therefore\ K\,K_t = \dfrac{\text{Product of vector lengths from poles to } s = s_d}{\text{Product of vector lengths from zeros to } s = s_d}$

Step-7 : Verify the design.

Check the velocity error constant, K_v of the new loop transfer function.

> **Note :** *Vector lengths should be measured to scale.*

$K_v = \underset{s \to 0}{\text{Lt}}\ s[G(s)H(s)]_{new}$

If the velocity error constant is satisfactory then the design is accepted otherwise this procedure calls for modifications in time domain specifications.

EXAMPLE 5.17

Design a feedback compensation scheme for a unity feedback system with open loop transfer function, $G(s) = 3/s(s+1)$ to satisfy the following specifications. (1) Phase margin of system should be atleast 45°. (2) Velocity error constant, $K_v \geq 20$.

SOLUTION

The feedback compensation scheme is shown in fig 5.17.1. The feedback transfer function, H(s) consists of a rate device in cascade with a high pass filter. The minor-loop feedback can be eliminated and replaced by an equivalent transfer function, $G_{eq}(s)$ as shown in fig 5.17.2.

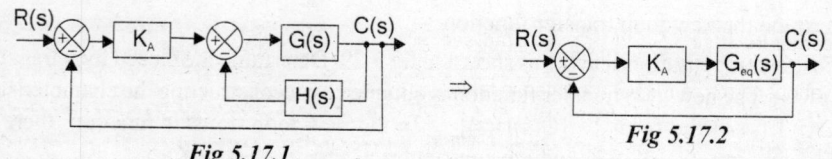

Fig 5.17.1 ⟹ Fig 5.17.2

$$H(s) = K_t s \frac{sT}{1+sT} = \frac{K_t T s^2}{(1+sT)} \quad \text{and} \quad G_{eq}(s) = \frac{G(s)}{1+G(s)H(s)}$$

Step-1 : To draw the magnitude plot of $G(j\omega)$.

Given that, $G(s) = \dfrac{3}{s(s+1)}$

Put $s = j\omega$, $\therefore G(j\omega) = \dfrac{3}{j\omega(1+j\omega)}$

The corner frequency is, $\omega_c = 1$ rad/sec.

The various terms of $G(j\omega)$ are listed in table-1. Also the table shows the slope contributed by each term and the change in slope at the corner frequency.

TABLE-1

Term	Corner frequency rad/sec	Slope db/dec	Change in slope db/dec
$\dfrac{3}{j\omega}$	—	−20	—
$\dfrac{1}{1+j\omega}$	$\omega_c = 1$	−20	−20 −20 = − 40

Choose a low frequency, ω_l such that $\omega_l < \omega_{c1}$ and choose a high frequency, ω_h such that $\omega_h > \omega_c$.

Let $\omega_l = 0.5$ rad/sec and $\omega_h = 10$ rad/sec.

Let $A = |G(j\omega)|$ in db

At $\omega = \omega_l$, $A = 20 \log \dfrac{3}{\omega} = 20 \log \dfrac{3}{0.5} = 15.5 \approx 16$ db

At $\omega = \omega_c$, $A = 20 \log \dfrac{3}{\omega} = 20 \log \dfrac{3}{1} = 9.5$ db ≈ 10 db

At $\omega = \omega_h$, $A = \left[\text{slope from } \omega_c \text{ to } \omega_h \times \log \dfrac{\omega_h}{\omega_c} \right] + A \text{ at } (\omega = \omega_c) = -40 \times \log \dfrac{10}{1} + 10 = -30$ db

Let the points a, b and c be the points corresponding to frequencies ω_l, ω_c and ω_h respectively on the magnitude plot. In a semilog graph sheet choose appropriate scales and fix the points a, b and c. Join the points by straight lines and mark the slope on the respective region. The magnitude plot is shown in fig 5.17.3.

Step-2 : Draw the magnitude plot of $1/H(j\omega)$.

Transfer function of feedback compensator $\Big\}$ $H(s) = \dfrac{K_t T s^2}{1+sT}$

Put $s = j\omega$, $\therefore H(j\omega) = \dfrac{K_t T (j\omega)^2}{(1+j\omega T)}$; Now, $\dfrac{1}{H(j\omega)} = \dfrac{1}{K_t T} \cdot \dfrac{1+j\omega T}{(j\omega)^2}$

Let $K_t T = 1$, $\therefore \dfrac{1}{H(j\omega)} = \dfrac{1+j\omega T}{(j\omega)^2}$; Let $T = 1.25$ sec, $\therefore \dfrac{1}{H(j\omega)} = \dfrac{1+j1.25\omega}{(j\omega)^2}$

The corner frequency, $\omega_{ch} = \dfrac{1}{1.25} = 0.8$ rad/sec

The various terms of $1/H(j\omega)$ are listed in table-2.

TABLE-2

Term	Corner frequency rad/sec	Slope db/dec	Change in slope db/dec
$\dfrac{1}{(j\omega)^2}$	–	– 40	–
$1 + j1.25\omega$	$\omega_{ch} = \dfrac{1}{1.25} = 0.8$	+ 20	– 40 + 20 = – 20

Choose a low frequency, ω_{lh} such that $\omega_{lh} < \omega_{ch}$ and choose a high frequency, ω_{hh} such that $\omega_{hh} > \omega_{ch}$.

Let $\omega_{lh} = 0.1$ rad/sec and $\omega_{hh} = 2.0$ rad/sec.

Let $H = |1/H(j\omega)|$ in db

At $\omega = \omega_{lh}$, $H = 20\log\dfrac{1}{\omega^2} = 20\log\dfrac{1}{0.1^2} = 40$ db

At $\omega = \omega_{ch}$, $H = 20\log\dfrac{1}{\omega^2} = 20\log\dfrac{1}{0.8^2} = 4$ db

At $\omega = \omega_{hh}$, $H = \left[\text{slope from } \omega_{ch} \text{ to } \omega_{hh} \times \log\dfrac{\omega_{hh}}{\omega_{ch}}\right] + H \text{ at } (\omega = \omega_{ch}) = -20 \times \log\dfrac{2}{0.8} + 4 = -4$ db

Let the points d, e and f be the points corresponding to frequencies ω_{lh}, ω_{ch} and ω_{hh} respectively on the magnitude plot. In the same semilog graph sheet and using the same scales draw the magnitude plot as shown in fig 5.17.3.

Step-3 :   Determine the gain cross-over frequency.

   l et ω_{gc} = Desired gain crossover frequency

   γ_d = Desired phase margin

Choose a new value of phase margin, γ_n such that, $\gamma_n = \gamma_d + \epsilon$.

Let $\epsilon = 25°$, $\therefore \gamma_n = \gamma_d + \epsilon = 45° + 25° = 70°$.

Let $\phi_{hgc} = \angle (1/H(j\omega))$ at $\omega = \omega_{gc}$

From the transfer function of $1/H(j\omega)$ we get,

$$\phi_{hgc} = \tan^{-1}\omega_{gc}T - 180° = \tan^{-1}1.25\omega_{gc} - 180°$$

Here, $\gamma_n = 180° + \phi_{hgc}$

$\therefore 70° = 180° + \tan^{-1}1.25\omega_{gc} - 180°$ $\Rightarrow$ $1.25\omega_{gc} = \tan 70°$ $\Rightarrow$ $\omega_{gc} = \dfrac{\tan 70°}{1.25} = 2.2$ rad/sec

Step-4 :   Determine the parameter, K_t

The $|1/H(j\omega)|$ curve is shifted down to cross $|G(j\omega)|$ curve at $\omega = 10$ rad/sec. (here $4\,\omega_{gc} = 4 \times 2.2 = 8.8$, hence frequency corresponding to crossing point is chosen to be 10 rad/sec).

From the bode plots we found that $|1/H(j\omega)|$ curve is shifted by 12 db downwards.

$\therefore 20\log\dfrac{1}{K_t T} = -12$ db $\Rightarrow$ $\log\dfrac{1}{K_t T} = \dfrac{-12}{20}$ $\Rightarrow$ $\dfrac{1}{K_t T} = 10^{-12/20} = 0.2512$

$\therefore K_t = \dfrac{1}{T \times 0.2512} = \dfrac{1}{1.25 \times 0.2512} = 3.185$

Step-5 : Determine the new transfer function, $G_{eq}(j\omega)$

From the bode plots it is observed that the shifted $|1/H(j\omega)|$ curve crosses the $|G(j\omega)|$ curve at $\omega = \omega_{c1} = 0.08$ rad/sec and $\omega = \omega_{c3} = 10$ rad/sec.

In the frequency range, $0 < \omega < \omega_{c1}$, $\quad |G_{eq}(j\omega)| = |G(j\omega)|$

In the frequency range, $\omega_{c1} < \omega < \omega_{c2}$, $\quad |G_{eq}(j\omega)| = |1/H(j\omega)|$ (Shifted curve)

In the frequency range, $\omega_{c2} < \omega < \infty$, $\quad |G_{eq}(j\omega)| = |G(j\omega)|$

The bode magnitude plot of $G_{eq}(j\omega)$ is shown as a bold curve in fig 5.17.3.

The bode magnitude plot of $G_{eq}(j\omega)$ is interpreted as follows,

1. The slope in the starting portion of the curve is -20 db/dec and it is due to an integral factor $K/j\omega$, where K is given by $G(j\omega)$ when $\omega = 0$,

   $|G(j\omega)|_{\omega = 0} = 3$, $\qquad \therefore K = 3$

2. At $\omega = \omega_{c1}$, the slope changes from -20 db/dec to -40 db/dec. This is due to a first order factor $1/(1+j\omega T_1)$ where, $T_1 = 1/\omega_{c1} = 1/0.08 = 12.5$ sec.

3. At $\omega = \omega_{c2}$, the slope changes from -40 db/dec to -20 db/dec. This is due to a first order factor $(1+j\omega T_2)$ where, $T_2 = 1/\omega_{c2} = 1/0.8 = 1.25$ sec.

4. At $\omega = \omega_{c3}$, the slope changes from -20 db/dec to -40 db/dec. This is due to a first order factor, $1/(1+j\omega T_3)$ where, $T_3 = 1/\omega_{c3} = 1/10 = 0.1$ sec.

Hence the transfer function, $G_{eq}(j\omega)$ can be written as,

$$G_{eq}(j\omega) = \frac{K(1+j\omega T_2)}{j\omega(1+j\omega T_1)(1+j\omega T_3)} = \frac{3(1+j1.25\omega)}{j\omega(1+j12.5\omega)(1+j0.1\omega)}$$

Put $j\omega = s$, $\quad \therefore G_{eq}(s) = \dfrac{3(1+1.25s)}{s(1+12.5s)(1+0.1s)}$

Step-6 : Determine the parameter, K_A

From the bode plot it is observed that $|G_{eq}(j\omega)|$ curve should be shifted 17 db upwards to make $\omega_{gc} = 2.2$ rad/sec as the gain crossover frequency of $G_{eq}(j\omega)$ (i.e., to make $|G_{eq}(j\omega)|$ in db = 0 at $\omega = \omega_{gc} = 2.2$ rad/sec).

Now, $20 \log K_A = 17$ db $\quad \Rightarrow \quad \log K_A = \dfrac{17}{20} \quad \Rightarrow \quad K_A = 10^{17/20} = 7.08$

Step-7 : Verify the design

$\left.\begin{array}{l}\text{Open loop transfer function}\\\text{of feedback compensated system}\end{array}\right\} = K_A G_{eq}(s) = 7.08 \times \dfrac{3(1+1.25s)}{s(1+12.5s)(1+0.1s)} = \dfrac{21.24(1+1.25s)}{s(1+12.5s)(1+0.1s)}$

Velocity error constant, $K_v = \underset{s\to 0}{Lt}\, s\, K_A\, G_{eq}(s) = \underset{s\to 0}{Lt}\, s\, \dfrac{21.24(1+1.25s)}{s(1+12.5s)(1+0.1s)} = 21.24$

CONCLUSION

Since the velocity error constant of the compensated system satisfies the requirement, the design is accepted.

RESULT

$K_A = 7.08$, $\qquad K_t = 3.185 \qquad$ and $\qquad T = 6.25$ sec.

$\left.\begin{array}{l}\text{Open loop transfer function}\\\text{of feedback compensated system}\end{array}\right\} = \dfrac{21.24(1+1.25s)}{s(1+12.5s)(1+0.1s)}$

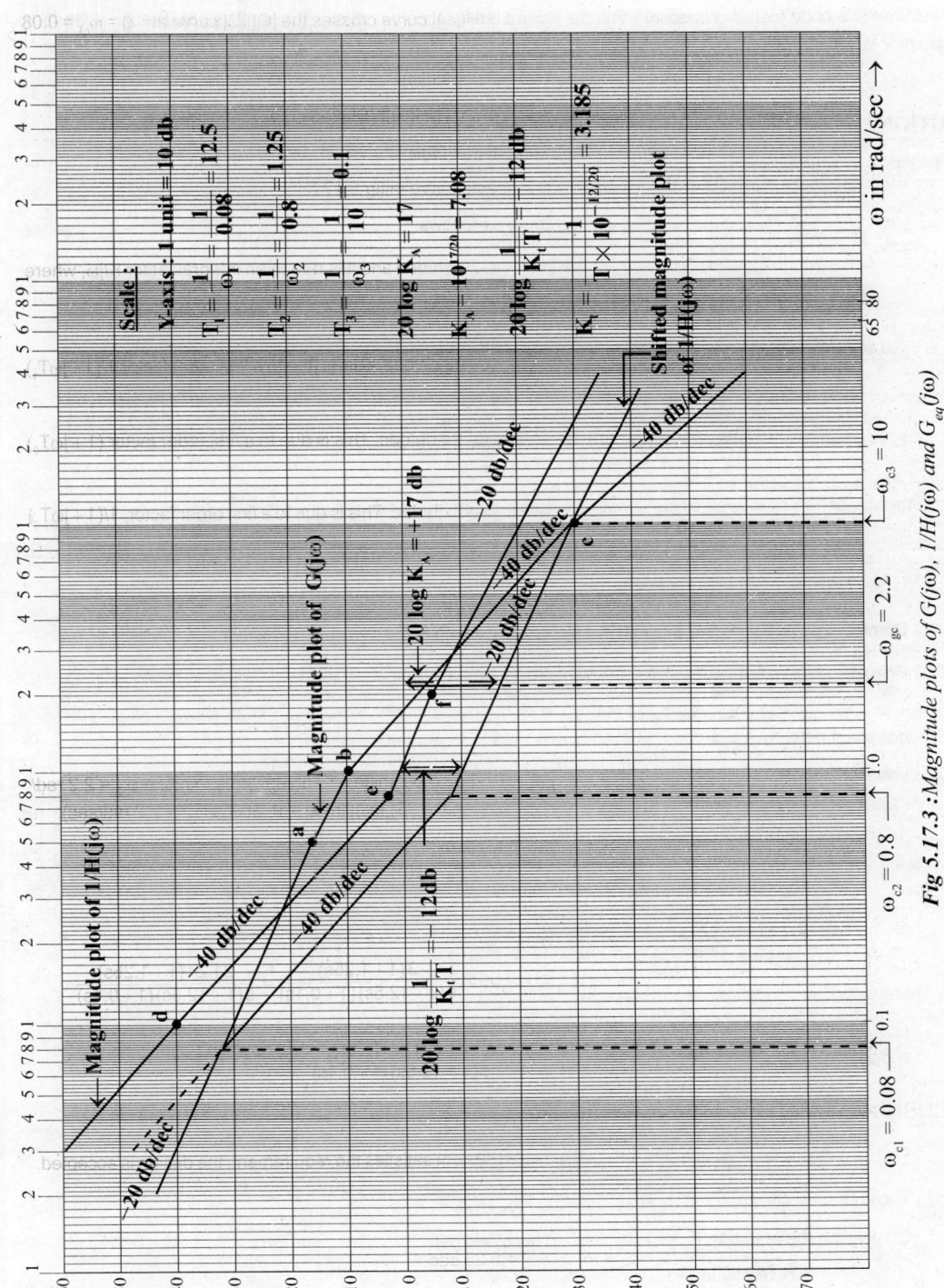

Fig 5.17.3 : Magnitude plots of G(jω), 1/H(jω) and G_eq(jω)

Scale :

Y-axis : 1 unit = 10 db

$T_1 = \dfrac{1}{\omega_1} = \dfrac{1}{0.08} = 12.5$

$T_2 = \dfrac{1}{\omega_2} = \dfrac{1}{0.8} = 1.25$

$T_3 = \dfrac{1}{\omega_3} = \dfrac{1}{10} = 0.1$

$20 \log K_A = 17$

$K_A = 10^{17/20} = 7.08$

$20 \log \dfrac{1}{K_t T} = -12 \, db$

$K_t = \dfrac{1}{T \times 10^{-12/20}} = 3.185$

EXAMPLE 5.18

Consider a unity feedback system with open loop transfer function, $G(s) = K/s(s+2)(s+5)$. Design a feedback compensator to satisfy the following specifications.

(a) Maximum overshoot, $M_p \leq 12\%$, (b) Settling time, $t_s \leq 3$ sec for 2% error and (c) Velocity error constant, $K_v \geq 4$.

SOLUTION

Step-1 : To find dominant pole, s_d

Given that $M_p \leq 12\%$

Let $M_p = 10\%$, $\quad \therefore M_p = e^{-\zeta\pi/\sqrt{1-\zeta^2}} \times 100 = 10\% \quad \Rightarrow \quad e^{-\zeta\pi/\sqrt{1-\zeta^2}} = 0.1$

On taking natural log we get, $-\zeta\pi/\sqrt{1-\zeta^2} = \ln(0.1) = -2.3$

On squaring we get, $\dfrac{\zeta^2\pi^2}{1-\zeta^2} = 5.29$

$\therefore \zeta^2\pi^2 = 5.29 - \zeta^2 5.29 \quad \Rightarrow \quad \zeta^2\pi^2 + 5.29\,\zeta^2 = 5.29 \quad \Rightarrow \quad \zeta = \sqrt{\dfrac{5.29}{\pi^2 + 5.29}} = 5.29 \approx 0.6$

Given that, $t_s \leq 3$ sec for 2% error. Let, $t_s = 3$ sec for 2% error.

Hence for 2% error, $e^{-\zeta\omega_n t_s} = \dfrac{2}{100}$

On taking natural log, we get,

$-\zeta\omega_n t_s = \ln(0.02) = -4 \quad \Rightarrow \quad \omega_n = \dfrac{-4}{-\zeta t_s} = \dfrac{4}{0.6 \times 3} = 2.22$ rad/sec

Dominant pole, $s_d = -\zeta\omega_n + j\omega_n\sqrt{1-\zeta^2} = -(0.6 \times 2.22) + j2.22\sqrt{1-0.6^2} = -1.3 + j1.8$

Step-2 : Pole-zero plot of $G(s)$.

The poles of $G(s)$ are located at $s = 0$, -2 and -5. The poles are marked by "x" as shown in fig 5.18.1. The dominant pole, s_d is also marked in fig 5.18.1. and it is denoted by point P.

Step-3 : Determine the new transfer function.

Let us choose a feedback compensation scheme as shown in fig 5.18.2.

Given that, $G(s) = \dfrac{K}{s(s+2)(s+5)}$

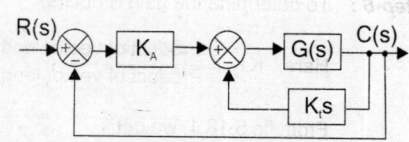

Fig 5.18.2

$$G_{eq}(s) = \dfrac{G(s)}{1+G(s)K_t s} = \dfrac{\dfrac{K}{s(s+2)(s+5)}}{1+\dfrac{K}{s(s+2)(s+5)}K_t s} = \dfrac{K}{s(s+2)(s+5)+KK_t s}$$

Closed loop
transfer function $\left.\right\}$ $\dfrac{C(s)}{R(s)} = \dfrac{K_A G_{eq}(s)}{1+K_A G_{eq}(s)}$

$$= \dfrac{K_A \dfrac{K}{s(s+2)(s+5)+KK_t s}}{1+K_A \dfrac{K}{s(s+2)(s+5)+KK_t s}} = \dfrac{KK_A}{s(s+2)(s+5)+KK_t s+KK_A}$$

The characteristic equation is, $s(s+2)(s+5)+KK_t s + KK_A = 0$ .....(5.18.1)

On dividing the equation (5.18.1) by $s(s+2)(s+5)$ we get,

$$1 + \dfrac{KK_t(s+K_A/K_t)}{s(s+2)(s+5)} = 0$$ .....(5.18.2)

The equation (5.18.2) is in the form of $1 + G(s)H(s) = 0$

$$\therefore [G(s)H(s)]_{new} = \frac{K K_t (s + K_A / K_t)}{s(s+2)(s+5)}$$ .....(5.18.3)

From the equation (5.18.3) we get,

Zero introduced by the feedback compensator, $z_c = -K_A / K_t$ .....(5.18.4)

Step-4 : To locate the zero of the compensator

Let ϕ = Angle contributed by the zero, z_c to make point P as a point on root locus.

Now, $\phi = \left(\begin{array}{c} \text{Sum of angles contributed by} \\ \text{poles of uncompensated system} \end{array} \right) \pm 180°$

From pole-zero plot of fig 5 18.1, we get,

$\left. \begin{array}{c} \text{Sum of angles contributed} \\ \text{by poles of uncompensated system} \end{array} \right\} = \theta_1 + \theta_2 + \theta_3 = 126° + 69° + 26° = 221°$

Let $n = 1$, $\therefore \phi = 221° - 180° = 41°$.

With line AP as reference draw a line PC such that $\angle APC = \phi$ as shown in fig 5.18.1. This line intersects the real axis at point C.

From the pole-zero plot of fig 5.18.1, we get,

Zero of the compensator, $z_c = -3.35$ .....(5.18.5)

Step-5 : To find K_A and K_t

On equating equations (5.18.4) and (5.18.5) we get, $K_A/K_t = 3.35$

Let $K_t = 1.2$, $\therefore K_A = K_t \times 3.35 = 1.2 \times 3.35 = 4.02$

Step-6 : To determine the gain K of G(s)

Here, $K K_t = \dfrac{\text{Product of vector lengths from poles to } s = s_d}{\text{Product of vector lengths from zeros to } s = s_d}$

From fig 5.18.1, we get,

$$K K_t = \frac{l_1 \times l_2 \times l_4}{l_3} = \frac{2.25 \times 1.95 \times 4.1}{2.75} = 6.62 \qquad \therefore K = \frac{6.62}{K_t} = \frac{6.62}{1.2} = 5.5$$

Step-7 : To verify the design

Velocity error constant, $K_v = \underset{s \to 0}{Lt} \, s [G(s)H(s)]_{new}$

$$= \underset{s \to 0}{Lt} \, s \frac{K K_t (s + K_A / K_t)}{s(s+1)(s+5)} = \frac{K K_A}{1 \times 5} = \frac{5.5 \times 4.02}{5} = 4.422$$

CONCLUSION

Since the velocity error constant of the compensated system is satisfactory, the design is accepted.

RESULT

$K = 5.5$; $K_t = 6.2$; $K_A = 4.02$

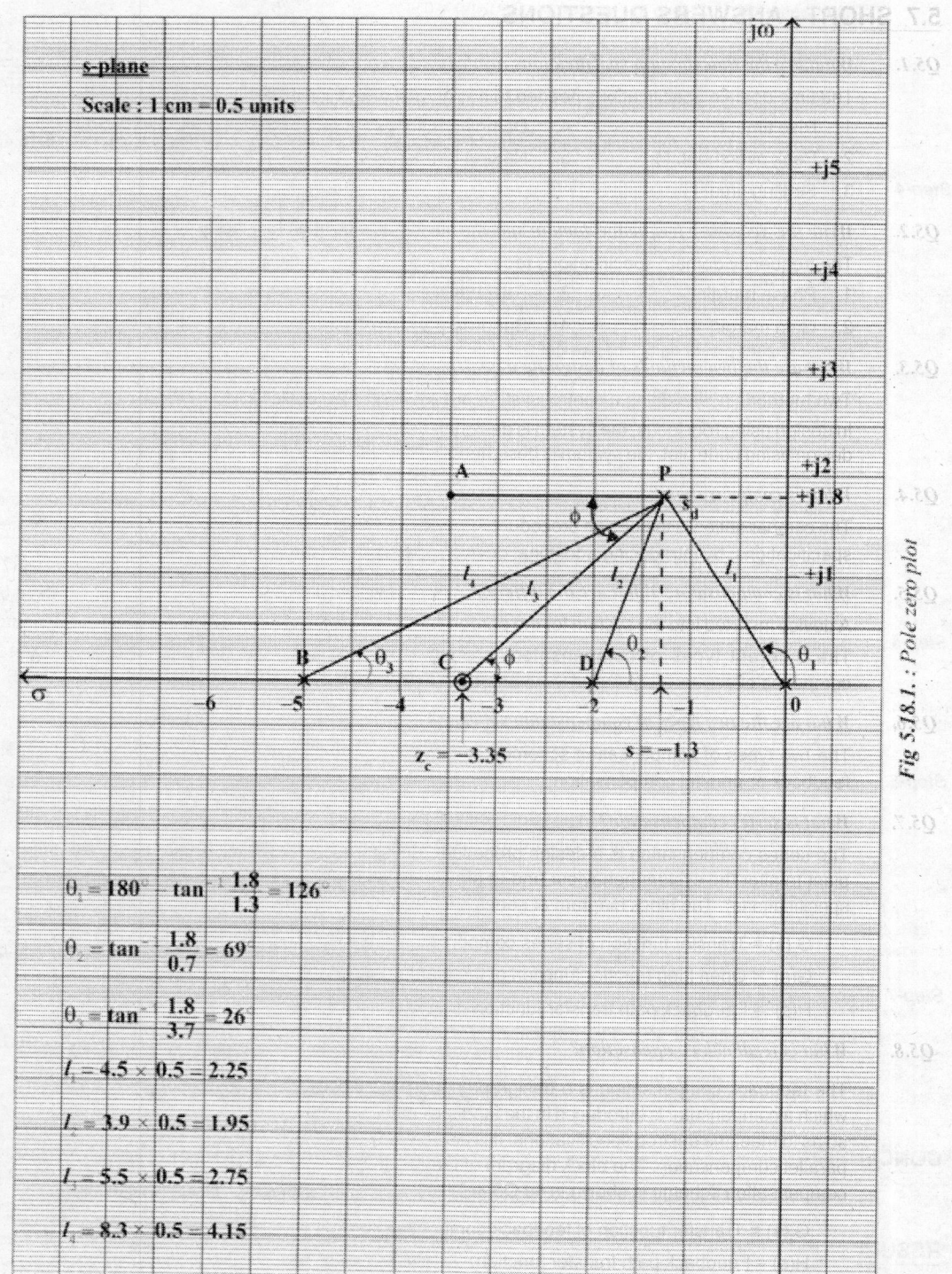

Fig 5.18.1. : Pole zero plot

$\theta_1 = 180° - \tan^{-1} \dfrac{1.8}{1.3} = 126°$

$\theta_2 = \tan^{-1} \dfrac{1.8}{0.7} = 69°$

$\theta_3 = \tan^{-1} \dfrac{1.8}{3.7} = 26°$

$l_1 = 4.5 \times 0.5 = 2.25$

$l_2 = 3.9 \times 0.5 = 1.95$

$l_3 = 5.5 \times 0.5 = 2.75$

$l_4 = 8.3 \times 0.5 = 4.15$

5.7 SHORT - ANSWERS QUESTIONS

Q5.1. *What are the time domain specifications needed to design a control system?*

The time domain specifications needed to design a control system are

1. Rise time, t_r
2. Peak overshoot, M_p
3. Settling time, t_s
4. Damping ratio, ζ
5. Natural frequency of oscillation, ω_n

Q5.2. *Write the necessary frequency domain specifications for design of a control system.*

The frequency domain specifications required to design a control system are,

1. Phase margin
2. Gain margin
3. Resonant peak
4. Bandwidth

Q5.3. *What are the two methods of designing a control system?*

Two methods of designing a control system are design using root locus and design using bode plot.

In design using root locus, the system is designed to satisfy the specified time domain specifications. In design using bode plot, the system is designed to satisfy the specified frequency domain specifications.

Q5.4. *What is compensation?*

The compensation is the design procedure in which the system behaviour is altered to meet the desired specifications, by introducing additional device called compensator.

Q5.5. *What is compensator? What are the different types of compensator?*

A device inserted into the system for the purpose of satisfying the specifications is called compensator. The different types of compensators are lag compensator, lead compensator and lag-lead compensator.

Q5.6. *What are the two types of compensation schemes?*

The two types of compensation schemes employed in control system are series compensation and feedback or parallel compensation.

Q5.7. *What is series compensation?*

The series compensation is a design procedure in which a compensator is introduced in series with plant to alter the system behaviour and to provide satisfactory performance (i.e., to meet the desired specifications). The block diagram of series compensation scheme is shown in fig Q5.7.

$G_c(s)$ = Transfer function of series compensator

G(s) = Open loop transfer function of the plant.

H(s) = Feedback path transfer function.

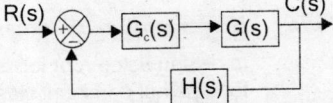

Fig Q5.7 : Series compensation.

Q5.8. *What is feedback compensation?*

The feedback compensation is a design procedure in which a compensator is introduced in the feedback path so as to meet the desired specifications. It is also called parallel compensation. The block diagram of feedback compensation scheme is shown in fig Q5.8.

$G_c(s)$ = Transfer function of feedback compensator

H(s) = Feedback path transfer function.

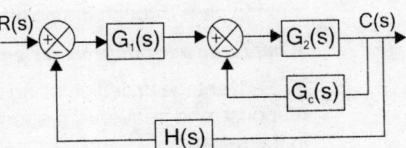

Fig Q5.8 : Feedback compensation.

$G_1(s), G_2(s)$ = Open loop transfer function of the components of the plant.

Q5.9.   *What are the factors to be considered for choosing series or shunt/feedback compensation?*

The choice between series, shunt or feedback compensation depends on the following

1. Nature of signals in the system.
2. Power levels at various points.
3. Components available.
4. Designer's experience.
5. Economic considerations.

Q5.10. *When lag/lead/lag-lead compensation is employed?*

Lag compensation is employed for a stable system for improvement in steady state performance.

Lead compensation is employed for stable/unstable system for improvement in transient-state performance.

Lag-lead compensation is employed for stable/unstable system for improvement in both steady-state and transient state performance.

Q5.11. *Why compensation is necessary in feedback control system?*

In feedback control systems compensation is required in the following situations.

1. When the system is absolutely unstable, then compensation is required to stabilize the system and also to meet the desired performance.

2. When the system is stable, compensation is provided to obtain the desired performance.

Q5.12. *Discuss the effect of adding a pole to open loop transfer function of a system.*

The addition of a pole to open loop transfer function of a system will reduce the steady-state error. The closer the pole to origin lesser will be the steady-state error. Thus the steady-state performance of the system is improved.

Also the addition of pole will increase the order of the system, which inturn makes the system less stable than the original system.

Q5.13. *Discuss the effect of adding a zero to open loop transfer function of a system.*

The addition of a zero to open loop transfer function of a system will improve the transient response. The addition of zero reduces the rise time. If the zero is introduced close to origin then the peak overshoot will be larger. If the zero is introduced far away from the origin in the left half of s-plane then the effect of zero on the transient response will be negligible.

Q5.14. *How root loci are modified when a zero is added to open loop transfer function?*

The addition of a zero to open loop transfer function will pull the root locus to the left which make the system more stable and reduce the settling time.

Q5.15. *How the root loci are modified when a pole is added to open loop transfer function of the system?*

The addition of a pole to the open-loop transfer function has the effect of pulling the root locus to the right, which reduce the relative stability of the system and increase the settling time.

Q5.16. *How control system design is carried using root locus?*

In design using root locus the transient response specifications are translated into desired locations for a pair of dominant closed loop poles.

In order to satisfy the performance specifications, the root loci should pass through these points. Hence a compensator is introduced in open loop transfer function which will reshape the root loci and force them to pass through the points where the dominant closed loop poles are located.

Q5.17. *What is the advantage in design using root locus.*

The advantage in design using root locus technique is that the information about closed loop transient response and frequency response are directly obtained from the pole-zero configuration of the system in the s-plane.

Q5.18. *What are the informations that can be obtained from frequency response plots?*

The low frequency region of frequency response plot provides information regarding the steady-state performance and high frequency region provides information regarding the transient performance of the system. The mid-frequency region provides information regarding relative stability.

Q5.19. *What are the advantages and disadvantages in frequency domain design.*

The advantages of frequency domain design are the following.

1.   The effect of disturbances, sensor noise and plant uncertainities are easy to visualize and asses in frequency domain.

2.   The experimental information can be used for design purposes.

The disadvantage of frequency response design is that it gives the information on closed loop system's transient response indirectly.

Q5.20. *What is lag compensation?*

The lag compensation is a design procedure in which a lag compensator is introduced in the system so as to meet the desired specifications.

Q5.21. *What is lag compensator? Give an example.*

A compensator having the characteristics of a lag network is called lag compensator. If a sinusoidal signal is applied to a lag compensator, then in steady state the output will have a phase lag with respect to input. An electrical lag compensator can be realised by a R-C network. The R-C network shown in fig Q5.21 is an example of electrical lag compensator.

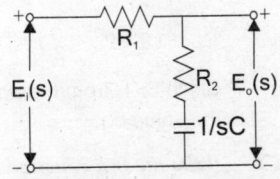

Fig Q5.21 : Lag compensator.

Q5.22. *Write the transfer function of lag compensator and draw its pole-zero plot.*

$$\text{Transfer function of} \atop \text{lag compensator} \} \; g_c(s) = \frac{s + \dfrac{1}{T}}{s + \dfrac{1}{\beta T}}$$

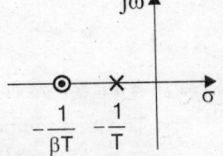

The lag compensator has a pole at s = –1/βT and a zero at s = –1/T. Since β > 1 and T > 0, the pole of lag compensator is nearer to origin. The pole-zero plot of lag compensator is shown in fig Q5.22.

Fig Q5.22 : Pole-zero plot of lag compensator.

Q5.23. *What are the characteristics of lag compensation? When lag compensation is employed?*

The lag compensation improves the steady state performance, reduce the bandwidth and increases the rise time. (The increase in rise time results in slower transient response). If the pole introduced by the compensator is not cancelled by a zero in the system then the lag compensator increases the order of the system by one.

When a system is stable and does not satisfy the steady-state performance specifications then lag compensation can be employed so that the system is redesigned to satisfy the steady-state requirements.

Q5.24. *Draw the bode plot of lag compensator.*

Let, $G_c(j\omega)$ = Sinusoidal transfer function of lag compensator.

The approximate magnitude plot and phase plot at $G_c(j\omega)$ are shown in fig Q5.24.

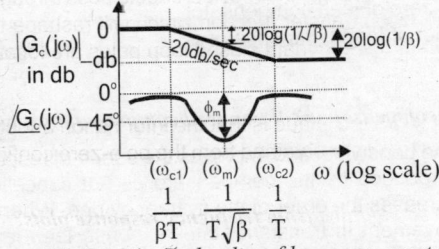

Fig Q5.24 : Bode plot of lag compensator.

Q5.25. *When maximum phase lag occurs in lag compensator? Give the expressions for maximum lag angle and the corresponding frequency.*

The maximum phase lag occurs at the geometric mean of two corner frequencies of the lag compensator.

Maximum phase lag angle, $\phi_m = \tan^{-1}\dfrac{(1-\beta)}{2\sqrt{\beta}}$

Frequency corresponding to maximum phase lag angle $\left.\right\}$ $\omega_m = \sqrt{\omega_{c1}\omega_{c2}} = \sqrt{\dfrac{1}{T}\dfrac{1}{\beta T}} = \dfrac{1}{T\sqrt{\beta}}$

Q5.26. *What is the relation between ϕ_m and β in lag compensator?*

In lag compensator the ϕ_m and β are related by the expressions,

$$\phi_m = \tan^{-1}\left(\dfrac{1-\beta}{2\sqrt{\beta}}\right) \quad \text{or} \quad \beta = \dfrac{1-\sin\phi_m}{1+\sin\phi_m}$$

Since $\beta>1$, from the above expressions we can conclude that, larger the value of β the larger will be the value of ϕ_m.

Q5.27. *Write the two equations that relates β and ϕ_m of lag compensator.*

The following two equations relates ϕ_m and β of lag compensator.

$$\phi_m = \tan^{-1}\dfrac{1-\beta}{2\sqrt{\beta}} \quad ; \quad \beta = \dfrac{1-\sin\phi_m}{1+\sin\phi_m}$$

Q5.28. *What is lead compensation?*

The lead compensation is a design procedure in which a lead compensator is introduced in the system so as to meet the desired specifications.

Q5.29. *What is lead compensator? Given an example.*

A compensator having the characteristic of a lead network is called a lead compensator. If a sinusoidal signal is applied to a lead compensator, then in steady state the output will have a phase lead with respect to input. An electrical lead compensator can be realised by a R-C network. The R-C network shown in fig Q5.29. is an example of electrical lead compensator.

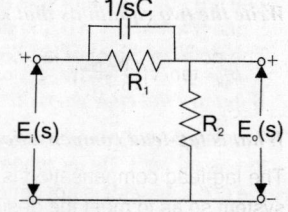

Fig Q5.29 : Electrical lead compensator.

Q5.30. *Write the transfer function of lead compensator and draw its pole-zero plot.*

Transfer functio of lead compensator, $g_c(s) = \dfrac{s+\dfrac{1}{T}}{s+\dfrac{1}{aT}}$

The lead compensator has a pole at $s = -1/\alpha T$ and a zero at $s = -1/T$. Since $\alpha < 1$ and $T > 0$, the zero of lead compensator is nearer to origin. The pole-zero plot of lead compensator is shown in fig Q5.30.

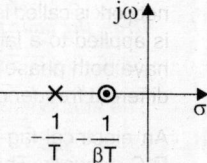

Fig Q5.30 : Pole-zero plot of lead compensator.

Q5.31. *What are the characteristics of lead compensation? When lead compensation is employed?*

The lead compensation increases the bandwidth and improves the speed of response. It also reduces the peak overshoot. If the pole introduced by the compensator is not cancelled by a zero in the system, then lead compensation increases the order of the system by one. When the given system is stable/unstable and requires improvement in transient state response then lead compensation is employed.

Q5.32. **Draw the bode plot of lead compensator.**

Let, $G_c(j\omega)$ = Sinusoidal transfer function of lead compensator. The approximate magnitude plot and phase plot of $G_c(j\omega)$ are shown in fig Q5.32.

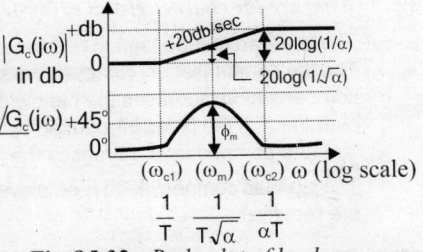

Fig Q5.32 : *Bode plot of lead compensator.*

Q5.33. **What is the relation between ϕ_m and α in lead compensator?**

In lead compensator the ϕ_m and α are related by the expression,

$$\phi_m = \tan^{-1}\left(\frac{1-\alpha}{2\sqrt{\alpha}}\right) \quad \text{or} \quad \alpha = \frac{1-\sin\phi_m}{1+\sin\phi_m}$$

Since $\alpha < 1$ from the above expressions we can conclude that, smaller the value of α the larger will be the value of ϕ_m.

Q5.34. **When maximum phase lead occurs in lead compensator? Give the expressions for maximum lead angle and the corresponding frequency.**

The maximum phase lead occurs at the geometric mean of two corner frequencies of the lead compensator.

Maximum phase lead angle, $\phi_m = \tan^{-1}\dfrac{(1-\alpha)}{2\sqrt{\alpha}}$

Frequency corresponding to maximum phase lead angle $\left.\right\}$ $\omega_m = \sqrt{\omega_{c1}\omega_{c2}} = \sqrt{\dfrac{1}{T}\dfrac{1}{\alpha T}} = \dfrac{1}{T\sqrt{\alpha}}$

Q5.35. **Write the two equations that relates α and ϕ_m of lead compensator**

$$\phi_m = \tan^{-1}\left(\frac{1-\alpha}{2\sqrt{\alpha}}\right) \quad ; \quad \alpha = \frac{1-\sin\phi_m}{1+\sin\phi_m}$$

Q5.36. **What is lag-lead compensation?**

The lag-lead compensation is a design procedure in which a lag-lead compensator is introduced in the system so as to meet the desired specifications.

Q5.37. **What is lag-lead compensator? Given an example.**

A compensator having the characteristics of lag-lead network is called lag-lead compensator. If a sinusoidal signal is applied to a lag-lead compensator then the output will have both phase lag and ead with respect to input, but in different frequency regions.

An electrical lag-lead compensator can be realised by a R-C network. The R-C network shown in fig Q5.37 is an example of electrical lag-lead compensator.

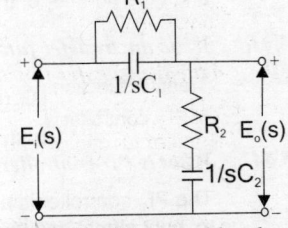

Fig 5.37 : *Electrical lag-lead compensator.*

Q5.38. **Write the transfer function of lag-lead compensator and draw its pole-zero plot.**

Transfer function of lag – lead compensator $\left.\right\}$ $G_c(s) = \underbrace{\dfrac{(s+1/T_1)}{(s+1/\beta T_1)}}_{\text{lag section}} \underbrace{\dfrac{(s+1/T_2)}{(s+1/\alpha T_2)}}_{\text{lag section}}$

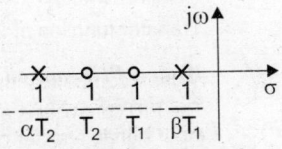

Fig 5.38 : *Pole-zero plot of lag-lead compensator*

Q5.39. *What are the characteristics of lag-lead compensation? When lag-lead compensation is employed?*

The lag-lead compensation has the characteristics of both lag compensation and lead compensation. The lag compensation improves the steady state performance and decreases the bandwidth. The lead compensation increases the bandwidth and improves the speed of response. It also reduces the peak overshoot. If the poles introduced by the compensator is not cancelled by zeros in the system then the lag-lead compensator increases the order of the system by two.

The lag-lead compensation is employed when improvements in both steady-state and transient response are required.

Q5.40. *Draw the bode plot of lag-lead compensator.*

Let, $G_c(j\omega)$ = Sinusoidal transfer function of lag-lead compensator.

The approximate magnitude plot and phase plot of $G_c(j\omega)$ are shown in fig Q5.40.

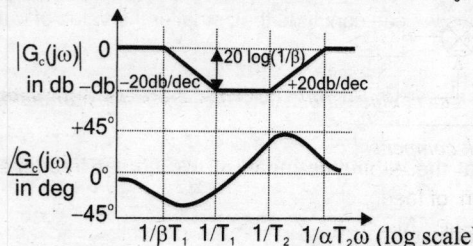

Fig 5.40 : Bode plot of lag-lead compensator.

Q5.41. *What is P-controller and what are its characteristics?*

The proportional controller is a device that produces an output signal which is proportional to the input signal.

The proportional controller improves the steady state tracking accuracy, disturbance signal rejection and relative stability. It also decreases the sensitivity of the system to parameter variations.

Q5.42. *What is PI-controller and what are its effect on system performance?*

The PI-controller is a device that produces an output signal consisting of two terms-one proportional to input signal and the other proportional to the integral of input signal.

The introduction of PI-controller in the system reduces the steady state error and increases the order and type number of the system by one.

Q5.43. *Write the transfer function of PI-controller.*

Transfer function of PI – controller $\Bigg\}$ $G_c(s) = K_p + \dfrac{K_i}{s} = \dfrac{K_p(s + K_i/K_p)}{s}$

Q5.44. *What is PD-controller and what are its effect on system performance?*

The PD-controller is a device that produces an output signal consisting of two terms-one proportional to input signal and the other proportional to the derivative of input signal.

The PD-controller increases the damping of the system which results in reducing the peak overshoot.

Q5.45. *Write the transfer function of PD-controller.*

Transfer function of PD-controller, $G_c(s) = K_p + K_d s = K_d(s + Kp/Kd)$

Q5.46. *What is PID controller and what are its effect on system performance?*

The PID controller is a device which produces an output signal consisting of three terms-one proportional to input signal, another one proportional to integral of input signal and the third one proportional to derivative of input signal.

The PID controller stabilises the gain, reduces the steady state error and peak overshoot of the system.

Q5.47. *Write the transfer function of PID controller.*

$\left.\begin{array}{l}\text{Transfer function}\\\text{of PID – controller}\end{array}\right\}$ $G_c(s) = K_p + \dfrac{K_i}{s} + K_d s = \dfrac{K_d s^2 + K_p s + K_i}{s}$

Q5.48. *What is feedback compensation?*

The feedback compensation is a design procedure in which a compensator is placed in an internal feedback path around one or more components of the forward path so as to meet the desired specifications.

Q5.49. *Draw the block diagram of a feedback compensation scheme.*

The block diagram of a popular feedback scheme employed in control systems is shown in fig Q5.49.

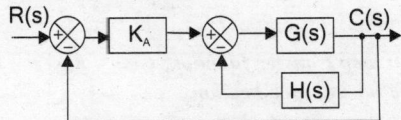

Fig Q5.49 : *Feedback compensation scheme.*

Here, H(s) = Transfer function of feedback compensator

K_A = A parameter to adjust the velocity error constant of the system.

Q5.50. *What is the disadvantage in rate feedback and how it is eliminated?*

The disadvantage in the rate feedback is that the system velocity error constant, K_v is reduced. This undesirable effect can be eliminated by reducing the feedback signal in the low frequencies by introducing a high pass filter in cascade with rate device as shown in fig Q5.50.

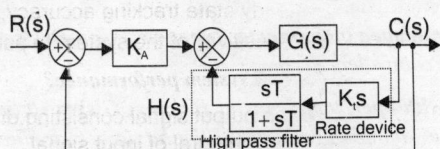

Fig Q5.50 : *Feedback compensation with high-pass filter in cascade with rate device.*

5.8 EXERCISES

E5.1 *Design a phase lag network for a system having $G(s) = K/s(1+0.2s)^2$ to have a phase margin of 30°.*

E5.2 *The open loop transfer function of a certain unity feedback control system is given by $G(s) = K/s(s+1)$. It is desired to have the velocity error constant, $K_v = 10$ and the phase margin to be atleast 60°. Design a phase lag series compensator.*

E5.3 *The controlled plant of a unity feedback system is $G(s) = K/s(s+5)$. It is specified that velocity error constant of the system be equal to 15, while the damping ratio is 0.6 and velocity error is less than 0.25 rad per unit ramp input. Design a suitable lag compensator.*

E5.4 *Consider the unity feedback system with an open loop transfer function $G(s) = K/s(s+2)^2$. Design a suitable lead compensator so that phase margin is atleast 50° and velocity error constant is 20 s^{-6}.*

E5.5 *The open loop transfer function of certain unity feedback control system is given by $G(s) = K/s(0.1s+1)$ $(0.2s+1)$. It is desired to have the phase margin to be atleast 30°. Design a suitable phase lead series compensator.*

E5.6 *A unity feedback system with open loop transfer function $G(s) = K/s^2 (s+1.5)$ is to be lead compensated to satisfy the following specifications.*

    (i)   Damping ratio = 0.45

    (ii)   Undamped natural frequency, $\omega_n = 2.2$ rad/sec

    (ii)   Velocity error constant, $K_v = 30$.

E5.7 *Consider a unity feedback control system whose forward transfer function is $G(s) = K / s \, (s + 2) \, (s + 8)$. Design a lag-lead compensator so that $K_v = 80 \ s^{-1}$ and dominant closed loop poles are located at $-2 \pm j2\sqrt{3}$.*

E5.8 *The open loop transfer function of uncompensated system is $G(s) = K/s \, (s +1)(s + 4)$. Design a lag-lead compensator to meet the following specifications.*

    (i)   Velocity error constant ≥ 5

    (ii)   Damping ratio = 0.4.

E5.9 *Consider a unity feedback system with open loop transfer function, $G(s) = K/s \, (2s+1) \, (0.5s+1)$. Design a suitable lag-lead compensator to meet the following specifications.*

    (i) $K_v = 30$ *(ii)   Phase margin ≥ 50*

E5.10 *Consider a unity feedback system with open loop transfer function, $G(s)=1/s(s+1)$. Design a PD controller so that the phase margin of the system is $30°$ at a frequency of 2 rad/sec.*

E5.11 *A unity feedback system has an open loop transfer function as $G(s) = 50/(s+3) \, (s +1)$. Design a PI controller so that phase margin of the system is $35°$ at a frequency of 1.2 rad/sec.*

E5.12 *Consider a unity feedback system with open loop transfer function, $G(s)=20/(s+0.5) \, (s+2) \, (s+4)$. Design a PID controller so that the phase margin of the system is $30°$ at a frequency of 2 rad/sec and steady state error for unit ramp input is 0.1.*

E5.13 *Consider a unity feedback system with open loop transfer function $G(s) = 10/s(s + 4)$. The dominant poles are $-2 \pm j\sqrt{6}$. Design a suitable PD controller.*

E5.14 *Consider a unity feedback system with open loop transfer function $G(s) = 10/(s +1) \, (s +2)$. Design a PI controller so that the closed loop has damping ratio of 0.707 and natural frequency of oscillation as 1.2 rad/sec.*

E5.15 *Consider a unity feedback system with $G(s) = 50/(s+2) \, (s+10)$. Design a PID controller to satisfy the following specifications.*

    (i) $K_v \geq 2$

    (ii)   Damping ratio = 0.6

    (iii)   Natural frequency of oscillations = 2 rad/sec.

E5.16 *Design a feedback compensation scheme for a unity feedback system with open loop transfer function $G(s) = 1/s^2 \, (s + 5)$ to satisfy the following specifications. (i) phase margin of the system should be atleast $50°$. (ii) Velocity error constant, $K_v \geq 20$.*

E5.17 *Consider a unity feedback system with open loop transfer function, $G(s) = K/s \, (s+1) \, (s+4)$. Design a feedback compensator to satisfy the following specifications.*

    (a)   Maximum overshoot, $M_p \leq 12\%$

    (b)   Settling time, $t_s = 10s$

    (c)   Velocity error constant, $K_v = 10$.

CHAPTER 6

STATE SPACE ANALYSIS

6.1 INTRODUCTION

The state variable approach is a powerful tool/technique for the analysis and design of control systems. The analysis and design of the following systems can be carried using state space method.

1. Linear system

2. Non-linear system

3. Time invariant system

4. Time varying system

5. Multiple input and multiple output system.

The state space analysis is a modern approach and also easier for analysis using digital computers. The conventional (or old) methods of analysis employs the transfer function of the system. The drawbacks in the transfer function model and analysis are,

1. Transfer function is defined under zero initial conditions.

2. Transfer function is applicable to linear time invariant systems.

3. Transfer function analysis is restricted to single input and single output systems.

4. Does not provide information regarding the internal state of the system.

The state variable analysis can be applied for any type of systems. The analysis can be carried with initial conditions and can be carried on multiple input and multiple output systems. In this method of analysis, it is not necessary that the state variables represent physical quantities of the system, but variables that do not represent physical quantities and those that are neither measurable nor observable may be chosen as state variables.

6.2 STATE SPACE FORMULATION

The **state** of a dynamic system is a minimal set of variables (known as state variables) such that the knowledge of these variables at $t = t_0$ together with the knowledge of the inputs for $t \geq t_0$, completely determines the behaviour of the system for $t > t_0$. (or) A set of variables which describes the system at any time instant are called **state variables**.

In the state variable formulation of a system, in general, a system consists of m-inputs, p-outputs and n-state variables. The state space representation of the system may be visualized as shown in fig 6.1.

Let,   State variables $=$ $x_1(t), x_2(t), x_3(t), \ldots\ldots\ldots\ldots x_n(t)$

       Input variables $=$ $u_1(t), u_2(t), u_3(t), \ldots\ldots\ldots\ldots u_m(t)$

       Output variables $=$ $y_1(t), y_2(t), y_3(t), \ldots\ldots\ldots\ldots y_p(t)$

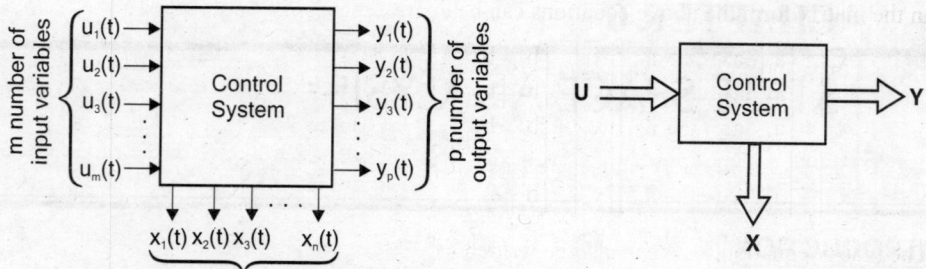

n number of state variables

Fig 6.1 : State space representation of system.

The different variables may be represented by the vectors (column matrix) as shown below.

Input vector $\mathbf{U}(t) = \begin{bmatrix} u_1(t) \\ u_2(t) \\ \vdots \\ u_m(t) \end{bmatrix}$; Output vector $\mathbf{Y}(t) = \begin{bmatrix} y_1(t) \\ y_2(t) \\ \vdots \\ y_m(t) \end{bmatrix}$; State variable vector $\mathbf{X}(t) = \begin{bmatrix} x_1(t) \\ x_2(t) \\ \vdots \\ x_m(t) \end{bmatrix}$

STATE EQUATIONS

The state variable representation can be arranged in the form of *n* number of first order differential equations as shown below.

$$\frac{dx_1}{dt} = \dot{x}_1 = f_1(x_1, x_2, \ldots\ldots x_n \quad ; \quad u_1, u_2 \ldots\ldots u_m)$$

$$\frac{dx_2}{dt} = \dot{x}_2 = f_2(x_1, x_2, \ldots\ldots x_n \quad ; \quad u_1, u_2 \ldots\ldots u_m)$$

$$\vdots$$

$$\frac{dx_n}{dt} = \dot{x}_n = f_n(x_1, x_2, \ldots\ldots x_n \quad ; \quad u_1, u_2 \ldots\ldots u_m) \qquad \ldots\ldots(6.1)$$

The n numbers of differential equations may be written in vector notation as,

$$\dot{\mathbf{X}}(t) = f(\mathbf{X}(t), \mathbf{U}(t)) \qquad \ldots\ldots(6.2)$$

The set of all possible values which the input vector $\mathbf{U}(t)$ can have (assume) at time t forms the input space of the system. Similarly, the set of all possible values which the output vector $\mathbf{Y}(t)$ can assume at time t forms the output space of the system and the set of all possible values which the state vector $\mathbf{X}(t)$ can assume at time t forms the state space of the system.

6.3 STATE MODEL OF LINEAR SYSTEM

The state model of a system consists of the state equation and output equation. The state equation of a system is a function of state variables and inputs as defined by equation (6.2). For linear time invariant systems the first derivatives of state variables can be expressed as a linear combination of state variables and inputs.

$$\dot{x}_1 = a_{11}x_1 + a_{12}x_2 + \ldots\ldots + a_{1n}x_n + b_{11}u_1 + b_{12}u_2 + \ldots\ + b_{1m}u_m$$

$$\dot{x}_2 = a_{21}x_1 + a_{22}x_2 + \ldots\ldots + a_{2n}x_n + b_{21}u_1 + b_{22}u_2 + \ldots\ + b_{2m}u_m$$

$$\vdots$$

$$\dot{x}_n = a_{n1}x_1 + a_{n2}x_2 + \ldots\ldots + a_{nn}x_n + b_{n1}u_1 + b_{n2}u_2 + \ldots\ + b_{nm}u_m \qquad \ldots\ldots(6.3)$$

where, the coefficients a_{ij} and b_{ij} are constants.

In the matrix form the above equations can be expressed as,

$$\begin{bmatrix} \dot{x}_1 \\ \dot{x}_2 \\ \dot{x}_3 \\ \vdots \\ \dot{x}_n \end{bmatrix} = \begin{bmatrix} a_{11} & a_{12} & \cdots & a_{1n} \\ a_{21} & a_{22} & \cdots & a_{2n} \\ a_{31} & a_{32} & \cdots & a_{3n} \\ \vdots & \vdots & \vdots & \vdots \\ a_{n1} & a_{n2} & \cdots & a_{nn} \end{bmatrix} \begin{bmatrix} x_1 \\ x_2 \\ x_3 \\ \vdots \\ x_n \end{bmatrix} + \begin{bmatrix} b_{11} & b_{12} & \cdots & b_{1n} \\ b_{21} & b_{22} & \cdots & b_{2n} \\ b_{31} & b_{32} & \cdots & b_{3n} \\ \vdots & \vdots & \vdots & \vdots \\ b_{n1} & b_{n2} & \cdots & b_{nm} \end{bmatrix} \begin{bmatrix} u_1 \\ u_2 \\ u_3 \\ \vdots \\ u_m \end{bmatrix} \qquad \ldots (6.4)$$

The matrix equation (6.4) can also be written as, $\dot{X}(t) = A\,X(t) + B\,U(t)$ .....(6.5)

where, $X(t)$ = State vector of order $(n \times 1)$

$U(t)$ = Input vector of order $(m \times 1)$

A = System matrix of order $(n \times n)$

B = Input matrix of order $(n \times m)$.

> **Note :** *For convenience the input, output and state variables are denoted as u_1, u_2, ... , y_1, y_2, ... and x_1, x_2, ... ; but actually they are functions of time, t.*

The equation $\dot{X}(t) = A\,X(t) + B\,U(t)$ is called the state equation of Linear Time Invariant (LTI) system.

The output at any time are functions of state variables and inputs.

$\therefore$ Output vector, $Y(t) = f(X(t), U(t))$ .....(6.6)

Hence the output variables can be expressed as a linear combination of state variables and inputs.

$$y_1 = c_{11}x_1 + c_{12}x_2 + \ldots + c_{1n}x_n + d_{11}u_1 + d_{12}u_2 + \ldots + d_{1m}u_m$$
$$y_2 = c_{21}x_1 + c_{22}x_2 + \ldots + c_{2n}x_n + d_{21}u_1 + d_{22}u_2 + \ldots + d_{2m}u_m$$
$$\vdots$$
$$y_p = c_{p1}x_1 + c_{p2}x_2 + \ldots + c_{pn}x_n + d_{p1}u_1 + d_{p2}u_2 + \ldots + d_{pm}u_m \qquad \ldots (6.7)$$

where the coefficients c_{ij} and d_{ij} are constants.

In the matrix form the above equations can be expressed as,

$$\begin{bmatrix} y_1 \\ y_2 \\ y_3 \\ \vdots \\ y_p \end{bmatrix} = \begin{bmatrix} c_{11} & c_{12} & \cdots & c_{1n} \\ c_{21} & c_{22} & \cdots & c_{2n} \\ c_{31} & c_{32} & \cdots & c_{3n} \\ \vdots & \vdots & \vdots & \vdots \\ c_{p1} & c_{p2} & \cdots & c_{pn} \end{bmatrix} \begin{bmatrix} x_1 \\ x_2 \\ x_3 \\ \vdots \\ x_n \end{bmatrix} + \begin{bmatrix} d_{11} & d_{12} & \cdots & d_{1m} \\ d_{21} & d_{22} & \cdots & d_{2m} \\ d_{31} & d_{32} & \cdots & d_{3m} \\ \vdots & \vdots & \vdots & \vdots \\ d_{p1} & d_{p2} & \cdots & d_{pm} \end{bmatrix} \begin{bmatrix} u_1 \\ u_2 \\ u_3 \\ \vdots \\ u_m \end{bmatrix} \qquad \ldots (6.8)$$

The matrix equation (6.8) can also be written as, $Y(t) = C\,X(t) + D\,U(t)$ .....(6.9)

where, $X(t)$ = State vector of order $(n \times 1)$

$U(t)$ = Input vector of order $(m \times 1)$

$Y(t)$ = Output vector of order $(p \times 1)$

C = Output matrix of order $(p \times n)$

D = Transmission matrix of order $(p \times m)$

The equation $Y(t) = C\,X(t) + D\,U(t)$ is called the output equation of Linear Time Invariant (LTI) system.

The state model of a system consists of state equation and output equation. (or) The state equation and output equation together called as state model of the system. Hence the state model of a linear time invariant system (LTI) system is given by the following equations.

$$\dot{X}(t) = A\,X(t) + B\,U(t) \qquad \text{................. State equation.}$$

$$Y(t) = C\,X(t) + D\,U(t) \qquad \text{................. Output equation.}$$

6.4 STATE DIAGRAM

The pictorial representation of the state model of the system is called *State diagram.* The State diagram of the system can be either in Block Diagram form or in Signal flow graph form.

The state diagram describes the relationships among the state variables and provides physical interpretations of the state variables. The time domain state diagram may be obtained directly from the differential equation governing the system and this diagram can be used for simulation of the system in analog computers.

The s-domain state diagram can be obtained from the transfer function of the system. The state diagram provides a direct relation between time domain and s-domain. [i.e., the time domain equations can be directly obtained from the s-domain state diagram].

The state diagram (Block diagram and signal flow graph) of a state model is constructued using three basic elements *Scalar, Adder* and *Integrator*.

Scalar : The scalar is used to multiply a signal by a constant. The input signal x(t) is multiplied by the scalar **a** to give the output, **a x(t)**.

Adder : The adder is used to add two or more signals. The output of the adder is the sum of incoming signals.

Integrator : The integrator is used to integrate the signal. They are used to integrate the derivatives of state variables to get the state variables. The initial conditions of the state variable can be added by using an adder after integrator.

The time domain and s-domain elements of block diagram are shown in table-6.1. The time domain and s-domain elements of signal flow graph are shown in table-6.2.

TABLE-6.1 : Elements of Block Diagram

Element	Time domain	s-domain
Scalar	$x(t) \rightarrow \boxed{a} \rightarrow ax(t)$	$X(s) \rightarrow \boxed{a} \rightarrow aX(s)$
Adder	$x_1(t),\ x_2(t) \rightarrow \otimes \rightarrow x_1(t)+x_2(t)$	$X_1(s),\ X_2(s) \rightarrow \otimes \rightarrow X_1(s)+X_2(s)$
Integrator	$\dot{x}(t) \rightarrow \boxed{\int} \rightarrow \otimes \rightarrow \int_0^t \dot{x}(t)dt + x(0)$, x(0)	$X(s) \rightarrow \boxed{1/s} \rightarrow \otimes \rightarrow \dfrac{X(s)}{s} + \dfrac{X(0)}{s}$, X(0)/s

Table-6.2 : Elements of Signal Flow Graph

Element	Time domain	s-domain
Scalar	$x(t)$ $\xrightarrow{a}$ $a\,x(t)$	$X(s)$ $\xrightarrow{a}$ $a\,X(s)$
Adder	$x_1(t)$, $x_2(t)$ with gains 1, 1 → $x_1(t) + x_2(t)$	$X_1(s)$, $X_2(s)$ with gains 1, 1 → $X_1(s) + X_2(s)$
Integrator	$\dot{x}(t) \xrightarrow{\int dt}$ with $x(0)$, gain 1 → $x(t)+x(0)$	$X(s) \xrightarrow{\frac{1}{s}}$ 1, $\frac{X(0)}{s}$ → $\frac{X(s)}{s} + \frac{X(0)}{s}$

The state model of linear time invariant System is given by the equations.

$$\dot{X}(t) = A\,X(t) + B\,U(t) \qquad \text{................ State equation.}$$

$$Y(t) = C\,X(t) + D\,U(t) \qquad \text{................ Output equation.}$$

The time domain block diagram representation of the state model is shown in fig 6.2. and the time domain signal flow graph representation of the system is shown in fig 6.3.

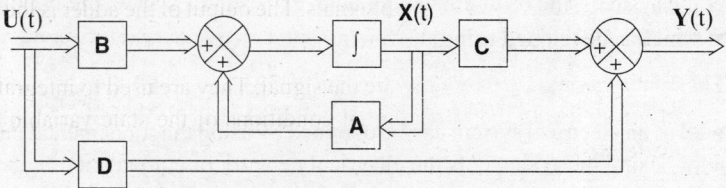

Fig 6.2 : Block diagram of state model.

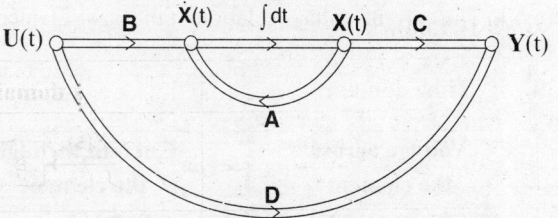

Fig 6.3 : Signal flow graph of state model.

CONSTRUCTION OF TIME DOMAIN STATE DIAGRAM

In state space modelling, n-numbers of first order differential equations are formed for an nth order system. In order to integrate n-numbers of first derivatives, the state diagram requires n-numbers of integrators. Therefore the first step in constructing the state diagram is to draw n-numbers of integrators. Mark the input to the integrators as first derivatives of state variables and so the output of the integrators are state variables. [If initial conditions are given, then they can be added at the output of integrators using adders].

In each state equation, the first derivative of state variable is expressed as a function of state variables and inputs. Therefore from the knowledge of a state equation, the state variables and inputs are multiplied by appropriate scalars and then added to get the first derivative of a state variable. Now, the first derivative of the state variable is given as input to the corresponding integrator. Similarly the input of all other integrators are obtained by considering the state equations one by one.

Each output equation is a function of state variables and inputs. Therefore from the knowledge of an output equation, the state variables and inputs are multiplied by appropriate scalars and then added to get an output. Similar procedure is followed to generate all other outputs.

6.5 STATE - SPACE REPRESENTATION USING PHYSICAL VARIABLES

In state-space modelling of systems, the choice of state variables is arbitrary. One of the possible choice of state variables is the physical variables. The physical variables of electrical systems are current or voltage in the R, L and C elements. The physical variables of mechanical systems are displacement, velocity and acceleration. The advantages of choosing the physical variables (or quantities) of the system as state variables are the following,

1. *The state variables can be utilized for the purpose of feedback.*
2. *The implementation of design with state variable feedback becomes straight forward.*
3. *The solution of state equation gives time variation of variables which have direct relevance to the physical system.*

The drawback in choosing the physical quantities as state variables is that the solution of state equation may become a difficult task.

In state space modelling using physical variables, the state equations are obtained from the differential equations governing the system. The differential equations governing a system are obtained from a basic model of the system which is developed using the fundamental elements of the system.

ELECTRICAL SYSTEM

The basic model of an electrical system can be obtained by using the fundamental elements Resistor, Capacitor and Inductor. Using these elements the electrical network or equivalent circuit of the system is drawn. Then the differential equations governing the electrical systems can be formed by writing Kirchoff 's Current Law equations by choosing various nodes in the network or Kirchoff's Voltage Law by choosing various closed path in the network. The current-voltage relation of the basic elements R, L and C are given in table 6.3.

TABLE-6.3

Element	Voltage across the element	Current through the element
$i(t)$ R $+$ ⟋⟍⟋⟍ $-$ $v(t)$	$v(t) = Ri(t)$	$i(t) = \dfrac{v(t)}{R}$
$i(t)$ L $+$ ⌇⌇⌇⌇ $-$ $v(t)$	$v(t) = L \dfrac{d}{dt} i(t)$	$i(t) = \dfrac{1}{L} \int v(t)\, dt$
$i(t)$ C $+$ ╫ $-$ $v(t)$	$v(t) = \dfrac{1}{C} \int i(t)\, dt$	$i(t) = C \dfrac{dv(t)}{dt}$

A minimal number of state variables are chosen for obtaining the state model of the system. The best choice of state variables in electrical system are currents and voltages in energy storage elements. The energy storage elements are inductance and capacitance. The physical variables in the differential equations are replaced by state variables and the equations are rearranged as first order differential equations. These set of first order equations constitutes the state equation of the system.

The inputs to the system are exciting voltage sources or current sources. The outputs in electrical system are usually voltages or currents in energy dissipating elements. The resistance is energy dissipating element in electrical network. In general the output variables can be any voltage or current in the network.

MECHANICAL TRANSLATIONAL SYSTEM

The basic model of mechanical translational system can be obtained by using three basic elements mass, spring and dash-pot. When a force is applied to a mechanical translational system, it is opposed by opposing forces due to mass, friction and elasticity of the system. The forces acting on a body are governed by Newton's second law of motion.

The differential equations governing the system are obtained by writing force balance equations at various nodes in the system. A node is a meeting point of elements. The table-6.4 shows the force balance equations of idealized elements.

List of symbols used in mechanical translational system are

$$
\begin{aligned}
y &= \text{Displacement, m} \\
v &= dy/dt = \text{Velocity, m/sec} \\
a &= dv/dt = d^2y/dt^2 = \text{Acceleration, m/sec}^2 \\
f &= \text{Applied force, N (Newton)} \\
f_m &= \text{Opposing force offered by mass of the body, N} \\
f_k &= \text{Opposing force offered by the elasticity of the body (spring), N} \\
f_b &= \text{Opposing force offered by the friction of the body (dash-pot), N} \\
M &= \text{Mass, Kg} \\
K &= \text{Stiffness of spring, N/m} \\
B &= \text{Viscous friction coefficient, N/(m/sec).}
\end{aligned}
$$

Guidelines to form the state model of mechanical translational systems

1. For each node in the system one differential equation can be framed by equating the sum of applied forces to the sum of opposing forces. Generally, the nodes are mass elements of the system, but in some cases the nodes may be without mass element.

2. Assign a displacement to each node and draw a free body diagram for each node. The free body diagram is obtained by drawing each mass of node separately and then marking all the forces acting on it.

3. In the free body diagram, the opposing forces due to mass, spring and dash-pot are always act in a direction opposite to applied force. The displacement, velocity and acceleration will be in the direction of applied force or in the direction opposite to that of opposing force.

4. For each free body diagram write one differential equation by equating the sum of applied forces to the sum of opposing forces.

5. Choose a minimum number of state variables. The choice of state variables are displacement, velocity or acceleration.

Table 6.4 : Force Balance Equations of Idealized Elements

Elements	Force balance equations
$f \rightarrow \boxed{M} \quad \mapsto y$ (fixed)	$f = f_m = M \dfrac{d^2y}{dt^2}$
$f \rightarrow \dashv\vdash B \quad \mapsto y$ (fixed)	$f = f_b = B \dfrac{dy}{dt}$
$f \rightarrow \dashv\vdash B \quad \mapsto y_1 \quad \mapsto y_2$	$f = f_b = B \dfrac{d}{dt}(y_1 - y_2)$
$f \rightarrow \mathrm{\infty\infty} K \quad \mapsto y$ (fixed)	$f = f_k = Ky$
$f \rightarrow \mathrm{\infty\infty} K \quad \mapsto y_1 \quad \mapsto y_2$	$f = f_k = K(y_1 - y_2)$

6.   The physical variables in differential equations are replaced by state variables and the equations are rearranged as first order differential equations. These set of first order equations constitute the state equation of the system.

7.   The inputs are the applied forces and the outputs are the displacement, velocity or acceleration of the desired nodes.

MECHANICAL ROTATIONAL SYSTEM

The basic model of mechanical rotational system can be obtained by using three basic elements moment of inertia of mass, rotational dash-pot and rotational spring. When a torque is applied to a mechanical rotational system, it is opposed by opposing torques due to moment of inertia, friction and elasticity of the system. The torque acting on a body are governed by Newton's second law of motion.

The differential equations governing the system are obtained by writing torque balance equations at various nodes in the system. A node is a meeting point of elements. The table-6.5 shows the torque balance equations of the idealized elements.

List of symbols used in mechanical rotational system

θ	=	Angular displacement, rad
$d\theta/dt$	=	Angular velocity, rad/sec
$d^2\theta/dt^2$	=	Angular acceleration, rad/sec^2
T	=	Applied torque, N-m
J	=	Moment of inertia, Kg-m^2/rad
B	=	Rotational frictional coefficient, N-m/(rad/sec)
K	=	Stiffness of the spring, N-m/rad

Guidelines to form the state model of mechanical rotational systems

1. For each node in the system one differential equation can be framed by equating the sum of applied torques to the sum of opposing torques. Generally the nodes are mass elements but in some cases the nodes may be without mass element.

2. Assign an angular displacement to each node and draw a free body diagram for each node. The free body diagram is obtained by drawing each node separately and then drawing all the torques acting on it.

3. In the free body diagram, the opposing torques due to moment of inertia, spring and dash-pot are always act in a direction opposite to applied force. The angular displacement, velocity and acceleration will be in the direction of applied torque or in the direction opposite to that of opposing torque.

TABLE-6.5 : TORQUE BALANCE EQUATIONS OF IDEALIZED ELEMENTS

Elements	Torque balance equations
J, T, θ	$T = T_j = J \dfrac{d^2\theta}{dt^2}$
T, θ, B	$T = T_b = B \dfrac{d\theta}{dt}$
T, θ₁, B, θ₂	$T = T_b = B \dfrac{d}{dt}(\theta_1 - \theta_2)$
T, θ, K	$T_k = K\theta$
T, θ₁, K, θ₂	$T = T_k = K(\theta_1 - \theta_2)$

4. For each free body diagram write one differential equation by equating the sum of applied torques to the sum of opposing torques.

5. Choose a minimum number of state variables. The choice of state variables are angular displacement, velocity or acceleration.

6. The physical variables in differential equations are replaced by state variables and the equations are rearranged as first order differential equations. These set of first order equations constitute the state equation of the system.

7. The inputs are the applied torques and the outputs are the angular displacement, velocity or acceleration of the desired nodes.

EXAMPLE 6.1

Obtain the state model of the electrical network shown in fig 1 by choosing minimal number of state variables.

SOLUTION

Let us choose the current through the inductances i_1, i_2 and voltage across the capacitor v_c as state variables. The assumed directions of currents and polarity of the voltage are shown in fig 2.

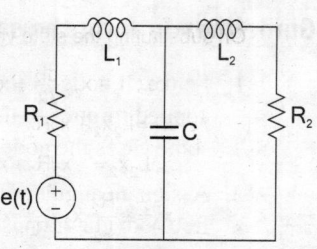

Fig 1.

> **Note :** *The best choice of state variables in electrical network are currents and voltages in energy storage elements.*

Let the three state variables x_1, x_2 and x_3 be related to physical quantities as shown below.

$$x_1 = i_1 = \text{Current through } L_1$$

$$x_2 = i_2 = \text{Current through } L_2$$

$$x_3 = v_c = \text{Voltage across capacitor.}$$

At node A, by Kirchoff's Current Law (refer fig 3.),

$$i_1 + i_2 + C\frac{dv_c}{dt} = 0 \qquad .....(6.1.1)$$

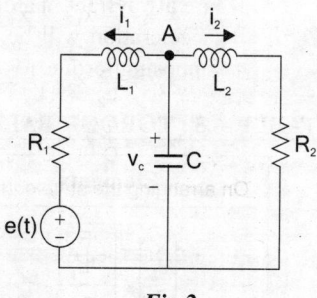

Fig 2.

On substituting the state variables for physical variables in equation (6.1.1) we get,

$$\text{(i.e., } i_1 = x_1, \ i_2 = x_2 \text{ and } \frac{dv_c}{dt} = \dot{x}_3 \text{)}$$

$$x_1 + x_2 + C\dot{x}_3 = 0$$

$$C\dot{x}_3 = -x_1 - x_2$$

$$\dot{x}_3 = -\frac{1}{C}x_1 - \frac{1}{C}x_2 \qquad .....(6.1.2)$$

By Kirchoff's Voltage Law in the closed path shown in fig 4 we get,

$$e(t) + i_1 R_1 + L_1\frac{di_1}{dt} = v_c \qquad .....(6.1.3)$$

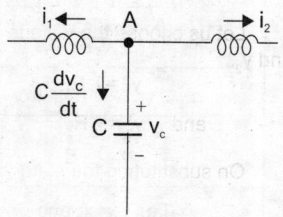

Fig 3.

On substituting the state variables for physical variables in equation (6.1.3) we get,

$$\text{(i.e., } i_1 = x_1, \ \frac{di_1}{dt} = \dot{x}_1 \text{ and } v_c = x_3\text{)}$$

$$e(t) + x_1 R_1 + L_1 \dot{x}_1 = x_3$$

Also, let $u(t) = e(t) = $ input to the system

$$\therefore \ u + x_1 R_1 + L_1\dot{x}_1 = x_3$$

$$L_1\dot{x}_1 = x_3 - x_1 R_1 - u$$

$$\dot{x}_1 = -\frac{R_1}{L_1}x_1 + \frac{1}{L_1}x_3 - \frac{1}{L_1}u \qquad .....(6.1.4)$$

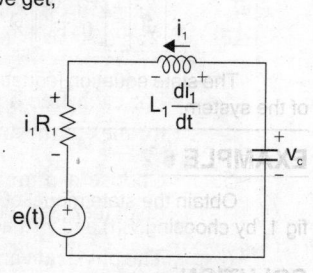

Fig 4.

By Kirchoff's Voltage Law in the closed path shown in fig 5 we get,

$$v_c = L_2\frac{di_2}{dt} + i_2 R_2 \qquad ......(6.1.5)$$

On substituting the state variables for physical variables in equation (6.1.5) we get,

(i.e., $i_2 = x_2$, $\dfrac{di_2}{dt} = \dot{x}_2$ and $v_c = x_3$)

$$x_3 = L_2\dot{x}_2 + x_2 R_2$$

$$\therefore\ L_2\dot{x}_2 = -x_2 R_2 + x_3$$

$$\dot{x}_2 = -\frac{R_2}{L_2}x_2 + \frac{1}{L_2}x_3 \qquad\qquad .....(6.1.6)$$

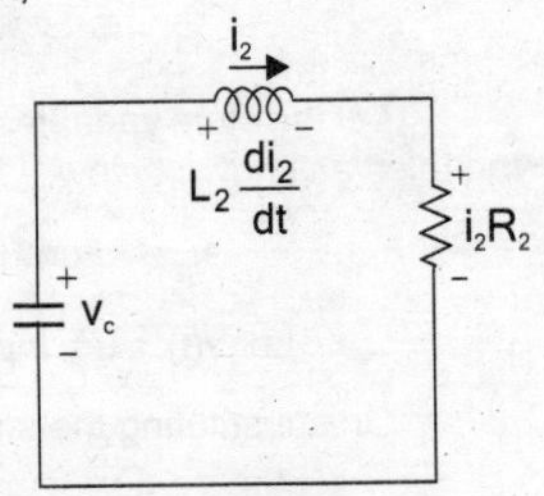

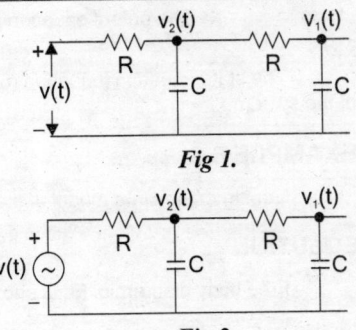

Fig 5.

The equations (6.1.2), (6.1.4) and (6.1.6) are the state equations of the system. Hence the state equations of the system are,

$$\dot{x}_1 = -\frac{R_1}{L_1}x_1 + \frac{1}{L_1}x_3 - \frac{1}{L_1}u$$

$$\dot{x}_2 = -\frac{R_2}{L_2}x_2 + \frac{1}{L_2}x_3$$

$$\dot{x}_3 = -\frac{1}{C}x_1 - \frac{1}{C}x_2$$

On arranging the state equations in the matrix form we get,8

$$\begin{bmatrix} \dot{x}_1 \\ \dot{x}_2 \\ \dot{x}_3 \end{bmatrix} = \begin{bmatrix} -\dfrac{R_1}{L_1} & 0 & \dfrac{1}{L_1} \\ 0 & -\dfrac{R_2}{L_2} & \dfrac{1}{L_2} \\ -\dfrac{1}{C} & -\dfrac{1}{C} & 0 \end{bmatrix} \begin{bmatrix} x_1 \\ x_2 \\ x_3 \end{bmatrix} + \begin{bmatrix} -\dfrac{1}{L_1} \\ 0 \\ 0 \end{bmatrix} [u] \qquad .....(6.1.7)$$

Let us choose the voltage across the resistances as output variables and the output variables are denoted by y_1 and y_2.

$$\therefore\ y_1 = i_1 R_1 \qquad\qquad .....(6.1.8)$$

$$\text{and}\quad y_2 = i_2 R_2 \qquad\qquad .....(6.1.9)$$

On substituting the state variables in equations (6.1.8) and (6.1.9) we get,

(i.e., $i_1 = x_1$ and $i_2 = x_2$)

$$y_1 = x_1 R_1 \qquad ; \qquad y_2 = x_2 R_2$$

On arranging the above equations in the matrix form we get

$$\begin{bmatrix} y_1 \\ y_2 \end{bmatrix} = \begin{bmatrix} R_1 & 0 \\ 0 & R_2 \end{bmatrix} \begin{bmatrix} x_1 \\ x_2 \end{bmatrix} \qquad\qquad .....(6.1.10)$$

The state equation [equation (6.1.7)] and output equation [equation (6.1.10)] together constitute the state model of the system.

EXAMPLE 6.2

Obtain the state model of the electrical network shown in fig 1. by choosing $v_1(t)$ and $v_2(t)$ as state variables.

SOLUTION

Connect a voltage source at the input as shown in fig 2.

Convert the voltage source to current source as shown in fig 3.

At node 1, by Kirchoff's Current Law we can write (refer fig 4)

$$\frac{v_1 - v_2}{R} + C\frac{dv_1}{dt} = 0 \qquad .....(6.2.1)$$

At node 2, by Kirchoff's Current Law, we can write (Refer fig 5)

Fig 1.

Fig 2.

$$\frac{v_2 - v_1}{R} + \frac{v_2}{R} + C\frac{dv_2}{dt} = \frac{v(t)}{R} \qquad .....(6.2.2)$$

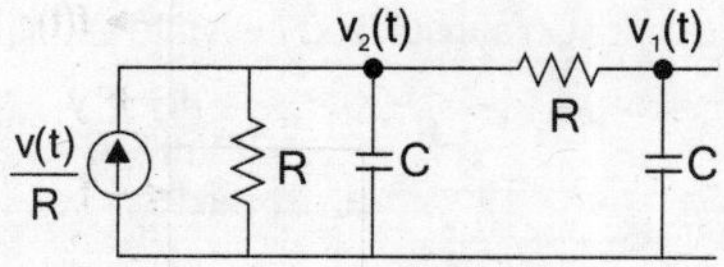

Let the state variables be x_1 and x_2 and they are related to physical variables as shown below.

$$v_1 = x_1 \text{ and } v_2 = x_2$$

Let $v(t) = u = $ input.

Fig 3.

On substituting the state variables in equations (6.2.1) and (6.2.2) we get,

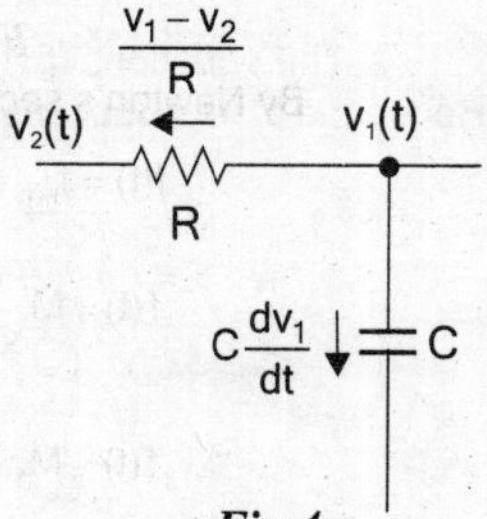

$$\frac{x_1 - x_2}{R} + C\frac{dx_1}{dt} = 0 \qquad .....(6.2.3)$$

$$\frac{x_2 - x_1}{R} + \frac{x_2}{R} + C\frac{dx_2}{dt} = \frac{u}{R} \qquad .....(6.2.4)$$

From equation (6.2.3) we get, $\frac{x_1}{R} - \frac{x_2}{R} + C\dot{x}_1 = 0$

$$\therefore C\dot{x}_1 = -\frac{x_1}{R} + \frac{x_2}{R}$$

$$\dot{x}_1 = -\frac{1}{RC}x_1 + \frac{1}{RC}x_2 \qquad ......(6.2.5)$$

Fig 4.

From equation (6.2.4) we get, $\frac{x_2}{R} - \frac{x_1}{R} + \frac{x_2}{R} + C\dot{x}_2 = \frac{u}{R}$

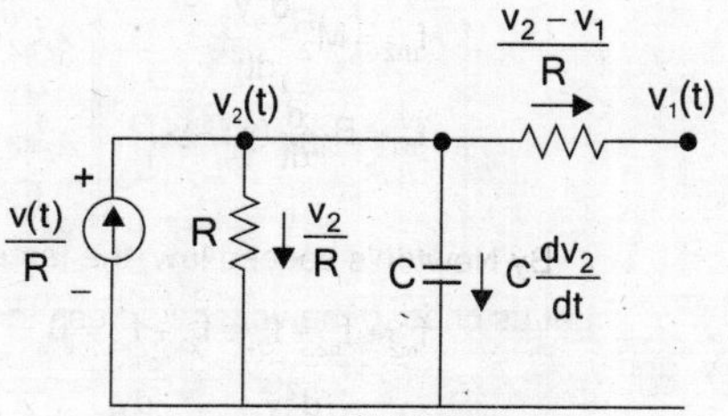

$$\therefore C\dot{x}_2 = \frac{x_1}{R} - \frac{x_2}{R} - \frac{x_2}{R} + \frac{u}{R}$$

$$\dot{x}_2 = \frac{1}{RC}x_1 - \frac{2}{RC}x_2 + \frac{1}{RC}u \qquad ......(6.2.6)$$

The equation (6.2.5) and (6.2.6) are state equations of the system. Hence the state equation of the system are

$$\dot{x}_1 = -\frac{1}{RC}x_1 + \frac{1}{RC}x_2$$

$$\dot{x}_2 = \frac{1}{RC}x_1 - \frac{2}{RC}x_2 + \frac{1}{RC}u$$

Fig 5.

On arranging the state equations in the matrix form,

$$\begin{bmatrix} \dot{x}_1 \\ \dot{x}_2 \end{bmatrix} = \begin{bmatrix} -\dfrac{1}{RC} & \dfrac{1}{RC} \\ \dfrac{1}{RC} & \dfrac{-2}{RC} \end{bmatrix} \begin{bmatrix} x_1 \\ x_2 \end{bmatrix} + \begin{bmatrix} 0 \\ \dfrac{1}{RC} \end{bmatrix} [u] \qquad .....(6.2.7)$$

The output, $y = v_1(t) = x_1$

$$\therefore \text{ The output equation is } y = [1 \quad 0] \begin{bmatrix} x_1 \\ x_2 \end{bmatrix} \qquad .....(6.2.8)$$

The state equation [equation (6.2.7)] and output equation [equation (6.2.8)] together constitute the state model of the system.

EXAMPLE 6.3

Construct the state model of mechanical system shown in fig 1.

SOLUTION

Free body diagram of M_1 is shown in fig 2.

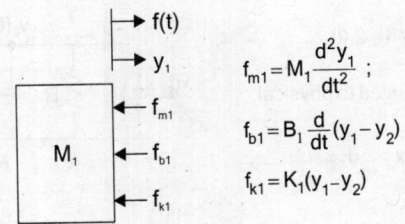

Fig 2.

By Newton's second law, the force balance equation at node M_1 is,

$$f(t) = f_{m1} + f_{b1} + f_{k1}$$

$$f(t) = M_1 \frac{d^2y_1}{dt^2} + B_1 \frac{c}{dt}(y_1 - y_2) + K_1(y_1 - y_2)$$

$$f(t) = M_1 \frac{d^2y_1}{dt^2} + B_1 \frac{dy_1}{dt} - B_1 \frac{dy_2}{dt} + K_1 y_1 - K_1 y_2 \qquad .....(6.3.1)$$

Free body diagram of M_2 is shown in fig 3

$$f_{m2} = M_2 \frac{d^2y_2}{dt^2} \quad ; \quad f_{b2} = B_2 \frac{dy_2}{dt}$$

$$f_{b1} = B_1 \frac{d}{dt}(y_2 - y_1) \quad ; \quad f_{k2} = K_2 y_2$$

$$f_{k1} = K_1(y_2 - y_1)$$

By Newton's second law, the force balance equation at node M_2 is,

$$f_{m2} + f_{b2} + f_{b1} + f_{k2} + f_{k} = 0$$

$$\therefore M_2 \frac{d^2y_2}{dt^2} + B_2 \frac{dy_2}{dt} + B_1 \frac{d}{dt}(y_2 - y_1) + K_2 y_2 + K_1(y_2 - y_1) = 0$$

$$M_2 \frac{d^2y_2}{dt^2} + B_2 \frac{dy_2}{dt} + B_1 \frac{dy_2}{dt} - B_1 \frac{dy_1}{dt} + K_2 y_2 + K_1 y_2 - K_1 y_1 = 0 \qquad .....(6.3.2)$$

Let us choose four state variables x_1, x_2, x_3 and x_4. Also, let the input $f(t) = u$. The state variables are related to physical variables as follows

$$x_1 = y_1 \quad ; \quad x_2 = y_2 \quad ; \quad x_3 = \frac{dy_1}{dt} \quad ; \quad x_4 = \frac{dy_2}{dt} \quad ; \quad \dot{x}_3 = \frac{d^2y_1}{dt^2} \quad ; \quad \dot{x}_4 = \frac{d^2y_2}{dt^2}$$

On substituting $y_1 = x_1$; $y_2 = x_2$; $\frac{dy_1}{dt} = \dot{x}_3$; $\frac{dy_2}{dt} = x_4$; $\frac{d^2y_1}{dt^2} = \dot{x}_3$ and $f(t) = u$ in equation (6.3.1) we get,

$$u = M_1 \dot{x}_3 + B_1 x_3 - B_1 x_4 + K_1 x_1 - K_1 x_2$$

$$M_1 \dot{x}_3 = -B_1 x_3 + B_1 x_4 - K_1 x_1 + K_1 x_2 + u$$

$$\therefore \dot{x}_3 = -\frac{K_1}{M_1} x_1 + \frac{K_1}{M_1} x_2 - \frac{B_1}{M_1} x_3 + \frac{B_1}{M_1} x_4 + \frac{1}{M_1} u \qquad .....(6.3.3)$$

On substituting $y_1 = x_1$; $y_2 = x_2$; $\frac{dy_1}{dt} = x_3$; $\frac{dy_2}{dt} = x_4$ and $\frac{d^2y_2}{dt} = \dot{x}_4$ in equation (6.3.2) we get,

$$M_2 \dot{x}_4 + B_2 x_4 + B_1 x_4 - B_1 x_3 + K_1 x_2 + K_2 x_2 - K_1 x_1 = 0$$

$$\therefore M_2 \dot{x}_4 = -B_2 x_4 - B_1 x_4 + B_1 x_3 - K_2 x_2 - K_1 x_2 + K_1 x_1 = 0$$

$$= -(B_2 + B_1) x_4 + B_1 x_3 - (K_2 + K_1) x_2 + K_1 x_1$$

$$\therefore \dot{x}_4 = \frac{K_1}{M_2} x_1 - \frac{(K_1 + K_2)}{M_2} x_2 + \frac{B_1}{M_2} x_3 - \frac{(B_1 + B_2)}{M_2} x_4 \qquad .....(6.3.4)$$

The state variable $x_1 = y_1$.

On differentiating $x_1 = y_1$ with respectt to t we get, $\dfrac{dx_1}{dt} = \dfrac{dy_1}{dt}$

Let $\dfrac{dx_1}{dt} = \dot{x}_1$ and $\dfrac{dy_1}{dt} = x_3$; $\therefore \dot{x}_1 = x_3$

The state variable, $x_2 = y_2$.

On differentiating $x_2 = y_2$ with respect to t we get, $\dfrac{dx_2}{dt} = \dfrac{dy_2}{dt}$

Let $\dfrac{dx_2}{dt} = \dot{x}_2$ and $\dfrac{dy_2}{dt} = x_4$; $\therefore \dot{x}_2 = x_4$ ....(6.3.6)

The equations (6.3.3) to (6.3.6) are state equations of the mechanical system. Hence the state equations of the mechanical system are,

$$\dot{x}_1 = x_3$$

$$\dot{x}_2 = x_4$$

$$\dot{x}_3 = -\frac{K_1}{M_1}x_1 + \frac{K_1}{M_1}x_2 - \frac{B_1}{M_1}x_3 + \frac{B_1}{M_1}x_4 + \frac{1}{M_1}u$$

$$\dot{x}_4 = \frac{K_1}{M_2}x_1 - \frac{(K_1+K_2)}{M_2}x_2 + \frac{B_1}{M_2}x_3 - \frac{(B_1+B_2)}{M_2}x_4$$

On arranging the state equations in the matrix form, we get,[u]

$$\begin{bmatrix} \dot{x}_1 \\ \dot{x}_2 \\ \dot{x}_3 \\ \dot{x}_4 \end{bmatrix} = \begin{bmatrix} 0 & 0 & 1 & 0 \\ 0 & 0 & 0 & 1 \\ -\frac{K_1}{M_1} & \frac{K_1}{M_1} & -\frac{B_1}{M_1} & \frac{B_1}{M_1} \\ \frac{K_1}{M_2} & -\frac{(K_1+K_2)}{M_2} & \frac{B_1}{M_2} & \frac{(B_1+B_2)}{M_2} \end{bmatrix} \begin{bmatrix} x_1 \\ x_2 \\ x_3 \\ x_4 \end{bmatrix} + \begin{bmatrix} 0 \\ 0 \\ \frac{1}{M_1} \\ 0 \end{bmatrix} [u]$$ .....(6.3.7)

Let the displacements y_1 and y_2 be the outputs of the system.

$\therefore$ $y_1 = x_1$ and $y_2 = x_2$.

The output equation in matrix form is given by,

$$\begin{bmatrix} y_1 \\ y_2 \end{bmatrix} = \begin{bmatrix} 1 & 0 & 0 & 0 \\ 0 & 1 & 0 & 0 \end{bmatrix} \begin{bmatrix} x_1 \\ x_2 \\ x_3 \\ x_4 \end{bmatrix}$$ .....(6.3.8)

The state equation [equation (6.3.7)] and the output equation [equation (6.3.8)] together called state model of the system.

EXAMPLE 6.4

Obtain the state model of the mechanical system shown in fig 1 by choosing a minimum of three state variables.

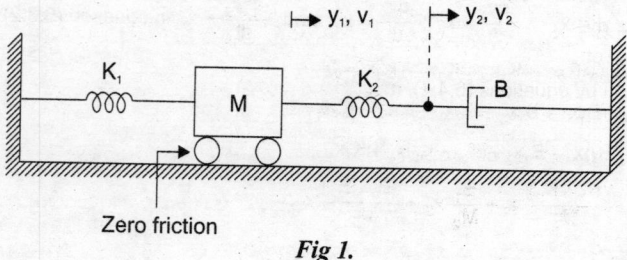

Fig 1.

SOLUTION

Let the three state variables be x_1, x_2 and x_3 and they are related to physical variables as shown below.

$$x_1 = y_1 \; ; \qquad x_2 = y_2 \; ; \quad x_3 = \frac{dy_1}{dt} = v_1$$

Free body diagram of mass M is shown in fig 2.

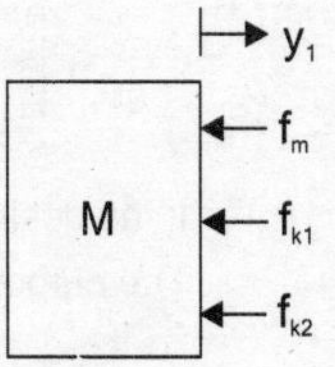

$$f_m = M \frac{d^2y_1}{dt^2} \; ; \quad f_{k1} = K_1 y_1 \; ; \quad f_{k2} = K_2(y_1 - y_2)$$

By Newton's second law, the force balance equation at node M is,

$$f_m + f_{k1} + f_{k2} = 0$$

Fig 2.

$$M \frac{d^2y_1}{dt^2} + K_1 y_1 + K_2(y_1 - y_2) = 0$$

$$M \frac{d^2y_1}{dt^2} + K_1 y_1 + K_2 y_1 - K_2 y_2 = 0 \qquad \qquad .....(6.4.1)$$

Put $\dfrac{d^2y_1}{dt^2} = \dot{x}_3$; $y_1 = x_1$, $y_2 = x_2$ in equation(6.4.1)

$$M\dot{x}_3 + K_1 x_1 + K_2 x_1 - K_2 x_2 = 0$$

$$M\dot{x}_3 + (K_1 + K_2) x_1 - K_2 x_2 = 0$$

$$\dot{x}_3 = -\frac{K_1 + K_2}{M} + \frac{K_2}{M} x_2 \qquad \qquad .....(6.4.2)$$

The free body diagram of node 2 (meeting point of K_2 and B) is shown in fig 3.

$$f_b = B \frac{dy_2}{dt} \; ; \quad f_{k2} = K_2(y_2 - y_1)$$

Writing force balance equation at the meeting point of K_2 and B we get,

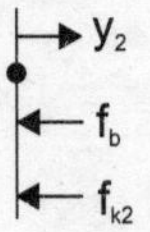

$$f_b + f_{k2} = 0$$

$$B \frac{dy_2}{dt} + K_2(y_2 - y_1) = 0$$

Fig 3.

$$\therefore \frac{dy_2}{dt} = \frac{K_2}{B} y_1 - \frac{K_2}{B} y_2$$

Put $\dfrac{dy_2}{dt} = \dot{x}_2$, $y_1 = x_1$ and $y_2 = x_2$

$$\therefore \dot{x}_2 = \frac{K_2}{B} x_1 - \frac{K_2}{B} x_2 \qquad \qquad .....(6.4.3)$$

The state variable, $x_1 = y_1$. On differentiating this expression with respect to t we get.

$$\frac{dx_1}{dt} = \frac{dy_1}{dt}$$

Let $\dfrac{dx_1}{dt} = \dot{x}_1$ and $\dfrac{dy_1}{dt} = x_3$; $\therefore \dot{x}_1 = x_3$ .....(6.4.4)

The state equations are given by equations (6.4.4), (6.4.3) and (6.4.2)

$$\dot{x}_1 = x_3$$

$$\dot{x}_2 = \frac{K_2}{B} x_1 - \frac{K_2}{B} x_2$$

$$\dot{x}_3 = -\frac{K_1 + K_2}{M} x_1 + \frac{K_2}{M} x_2$$

On arranging the state equations in the matrix form,

$$\begin{bmatrix} \dot{x}_1 \\ \dot{x}_2 \\ \dot{x}_3 \end{bmatrix} = \begin{bmatrix} 0 & 0 & 1 \\ \dfrac{K_2}{B} & -\dfrac{K_2}{B} & 0 \\ -\dfrac{K_1+K_2}{M} & \dfrac{K_2}{B} & 0 \end{bmatrix} \begin{bmatrix} x_1 \\ x_2 \\ x_3 \end{bmatrix} \qquad .....(6.4.5)$$

If the desired outputs are y_1 and y_2, then $y_1 = x_1$ and $y_2 = x_2$

The output equation in the matrix form is given by

$$\begin{bmatrix} y_1 \\ y_2 \end{bmatrix} = \begin{bmatrix} 1 & 0 \\ 0 & 1 \end{bmatrix} \begin{bmatrix} x_1 \\ x_2 \end{bmatrix} \qquad .....(6.4.6)$$

The state equation [equation(6.4.5)] and the output equation [equation(6.4.6)] together constitute the state model of the system.

EXAMPLE 6.5

Determine the state model of armature controlled dc motor.

SOLUTION

The speed of DC motor is directly proportional to armature voltage and inversely proportional to flux. In armature controlled DC motor the desired speed is obtained by varying the armature voltage. This speed control system is an electro-mechanical control system. The electrical system consists of the armature and the field circuit but for analysis purpose, only the armature circuit is considered because the field is excited by a constant voltage. The mechanical system consist of the rotating part of the motor and load connected to the shaft of the motor. The armature controlled DC motor speed control system is shown in fig 1.

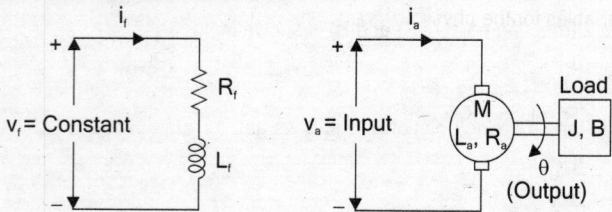

Fig 1 : Armature controlled DC motor.

Let R_a = Armature resistance, Ω

L_a = Armature inductance, H

i_a = Armature current, A

v_a = Armature voltage, V

e_b = Back emf, V

K_t = Torque constant, N-m/A

T = Torque developed by motor, N-m

θ = Angular displacement of shaft, rad

ω = $d\theta/dt$ = Angular velocity of the shaft, rad/sec

J = Moment of inertia of motor and load, Kg-m²/rad

B = Frictional coefficient of motor and load, N-m/(rad/sec)

K_b = Back emf constant, V/(rad/sec).

The equivalent circuit of armature is shown in fig 2

By Kirchoff's Voltage Law, we can write

$$i_a R_a + L_a \frac{di_a}{dt} + e_b = v_a \qquad .....(6.5.1)$$

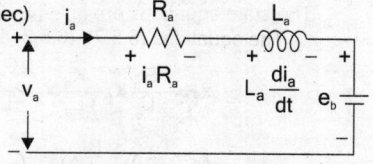

Fig 2 : Equivalent circuit of armature.

Torque of DC motor is proportional to the product of flux and current. Since flux is constant in this system, the torque is proportional to i_a alone.

$$T \propto i_a$$

$\therefore$ Torque, $T = K_t \, i_a$ .....(6.5.2)

The mechanical system of the motor is shown in fig 3. The differential equation governing the mechanical system of motor is given by,

$$J \frac{d^2\theta}{dt^2} + B \frac{d\theta}{dt} = T \qquad\qquad .....(6.5.3)$$

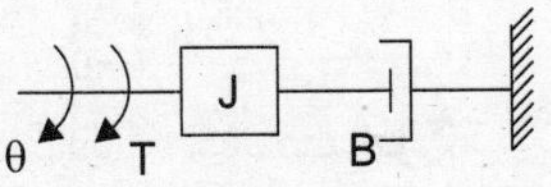

Fig 3.

The back emf of DC machine is proportional to speed (angular velocity) of shaft

$$\therefore e_b \propto \frac{d\theta}{dt} \quad ; \qquad \text{Back emf, } e_b = K_b \frac{d\theta}{dt} \qquad\qquad .....(6.5.4)$$

From equation (6.5.1) and (6.5.4) we get,

$$i_a R_a + L_a \frac{di_a}{dt} + K_b \frac{d\theta}{dt} = v_a \qquad\qquad .....(6.5.5)$$

From equation (6.5.2) and (6.5.3) we get,

$$J \frac{d^2\theta}{dt^2} + B \frac{d\theta}{dt} = K_t i_a \qquad\qquad .....(6.5.6)$$

The equations (6.5.5) and (6.5.6) are the differential equations governing the armature controlled dc motor.

Let us choose i_a, ω and θ as state variables to model the armature controlled dc motor. The physical variables i_a, ω and θ are related to the general notation of state variables x_1, x_2 and x_3 as shown below

$$x_1 = i_a \quad ; \quad x_2 = \omega = d\theta/dt \text{ and } x_3 = \theta$$

The input to the motor is the armature voltage, v_a and let $v_a = u$, where u is the general notation for input variable.

On substituting the state variables for the physical variables in equation (6.5.5) we get,

$$x_1 R_a + L_a \frac{dx_1}{dt} + K_b x_2 = u$$

Let $\dfrac{dx_1}{dt} = \dot{x}_1$, $\quad \therefore x_1 R_a + L_a \dot{x}_1 + K_b x_2 = u$

$$\dot{x}_1 = -\frac{R_a}{L_a} x_1 - \frac{K_b}{L_a} x_2 + \frac{1}{L_a} u \qquad\qquad .....(6.5.7)$$

On substituting the state variables for physical variables in equation (6.5.6) we get,

$$J \frac{d^2 x_3}{dt^2} + B \frac{dx_3}{dt} = K \cdot x_1$$

Let $\dfrac{d^3 x_3}{dt^2} = \dot{x}_2$ and $\dfrac{dx_3}{dt} = x_2$, $\quad \therefore J \dot{x}_2 + B x_2 = K_1 x_1$

$$\dot{x}_2 = \frac{K_1}{J} x_1 - \frac{B}{J} x_2 \qquad\qquad .....(6.5.8)$$

The state variable $x_3 = \theta$. On differentiating $x_3 = \theta$ with respect to t we get,

$$\frac{dx_3}{dt} = \frac{d\theta}{dt}$$

Put $\dfrac{dx_3}{dt} = \dot{x}_3$ and $\dfrac{d\theta}{dt} = x_2$

$$\therefore \dot{x}_3 = x_2 \qquad\qquad .....(6.5.9)$$

The equations (6.5.7), (6.5.8) and (6.5.9) are the state equations of the system.

$$\dot{x}_1 = -\frac{R_a}{L_a} x_1 - \frac{K_b}{L_a} x_2 + \frac{1}{L_a} u$$

$$\dot{x}_2 = \frac{K_1}{J} x_1 - \frac{B}{J} x_2$$

$$\dot{x}_3 = x_2$$

On arranging the state equations in the matrix form,

$$\begin{bmatrix} \dot{x}_1 \\ \dot{x}_2 \\ \dot{x}_3 \end{bmatrix} = \begin{bmatrix} -\dfrac{R_a}{L_a} & -\dfrac{K_b}{L_a} & 0 \\ \dfrac{K_1}{J} & -\dfrac{B}{J} & 0 \\ 0 & 1 & 0 \end{bmatrix} \begin{bmatrix} x_1 \\ x_2 \\ x_3 \end{bmatrix} + \begin{bmatrix} \dfrac{1}{L_a} \\ 0 \\ 0 \end{bmatrix} [u] \qquad \ldots (6.5.10)$$

Let the desired outputs be i_a, ω and θ. Let us equate the desired output quantities to standard notation y_1, y_2 and y_3 as shown below.

$$y_1 = i_a \; ; \qquad y_2 = \omega = d\theta/dt \; ; \quad y_3 = \theta$$

On relating the outputs to state variables we get,

$$y_1 = x_1 \qquad ; \quad y_2 = x_2 \qquad ; \quad y_3 = x_3$$

∴ The output equation in the matrix form is

$$\begin{bmatrix} y_1 \\ y_2 \\ y_3 \end{bmatrix} = \begin{bmatrix} 1 & 0 & 0 \\ 0 & 1 & 0 \\ 0 & 0 & 1 \end{bmatrix} \begin{bmatrix} x_1 \\ x_2 \\ x_3 \end{bmatrix} \qquad \ldots (6.5.11)$$

The state equation [equation (6.5.10)] and the output equation [equation (6.5.11)] together constitute the state model of the armature controlled dc motor.

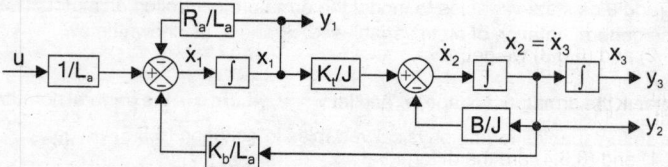

Fig 4 : Block diagram representation of the state model of armature controlled dc motor.

EXAMPLE 6.6

Determine the state model of field controlled dc motor.

SOLUTION

The speed of a DC motor is directly proportional to armature voltage and inversely proportional to flux. In field controlled DC motor the armature voltage is kept constant and the speed is varied by varying the flux of the machine. Since flux is directly proportional to field current, the flux is varied by varying field current. The speed control system is an electromechanical control system. The electrical system consists of armature and field circuit but for analysis purpose, only field circuit is considered because the armature is excited by a constant voltage. The mechanical system consists of the rotating part of the motor and the load connected to the shaft of the motor. The field controlled DC motor speed control system is shown in fig 1.

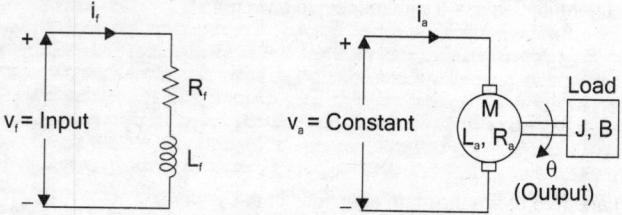

Fig 1 : Field controlled DC motor.

Let R_f =   Field resistance, Ω

       L_f =   Field inductance, H

       i_f =   Field current, A

v_f =   Field voltage, V

θ =   Angular displacement of the motor shaft, rad

ω = $d\theta/dt$ = Angular velocity of the motor shaft, rad/sec

T   =   Torque developed by motor, N-m

K_{tf} =   Torque constant, N-m/A

J   =   Moment of inertia of rotor and load, Kg-m^2/rad

B   =   Frictional coefficient of rotor and load, N-m/(rad/sec).

The equivalent circuit of field is shown in fig 2

By Kirchoff's voltage law, we can write

$$R_f i_f + L_f \frac{di_f}{dt} = v_f \qquad .....(6.6.1)$$

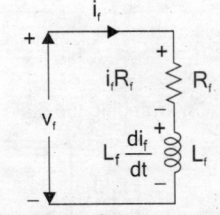

Fig 2 : *Equivalent circuit of field.*

The torque of DC motor is proportional to product of flux and armature current. Since armature current is constant in this system, the torque is proportional to flux alone, but flux is proportional to field current.

$$\therefore T \propto i_f \ ; \ \text{Torque, } T = K_{tf} i_f \qquad .....(6.6.2)$$

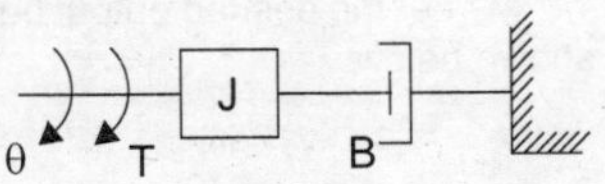

The mechanical system of the motor is shown in fig 3. The differential equation governing the mechanical system of the motor is given by,

$$J \frac{d^2\theta}{dt^2} + B \frac{d\theta}{dt} = T \qquad .....(6.6.3)$$

Fig 3.

From equation (6.6.2) and (6.6.3) we get,

$$J \frac{d^2\theta}{dt^2} + B \frac{d\theta}{dt} = K_{tf} i_f \qquad .....(6.6.4)$$

The equations (6.6.1) and (6.6.4) are the differential equations governing the field controlled dc motor.

Let us choose i_f, ω and θ as state variable to model the field controlled dc motor. The physical variables i_f, ω and θ are related to the general notation of state variables x_1, x_2 and x_3 as shown below.

$$x_1 = i_f \ ; \quad x_2 = \omega = d\theta\backslash dt \ ; \quad x_3 = \theta$$

The input to the system is the field voltage v_f. Let $v_f = u$, where u is the general notation for input.

On substituting the state variables and input variable for the physical variables in equation(6.6.1) we get,

$$R_f x_1 + L_f \frac{dx_1}{dt} = u$$

Let $\frac{dx_1}{dt} = \dot{x}_1, \quad \therefore R_f x_1 + L_f \dot{x}_1 = u$

$$\dot{x}_1 = -\frac{R_f}{L_f} x_1 + \frac{1}{L_f} u \qquad .....(6.6.5)$$

On substituting the state variables for the physical variables in equation (6.6.4) we get,

$$J \frac{d^2 x_3}{dt^2} + B \frac{dx_3}{dt} = K_{tf} x_1$$

Let $\frac{d^2 x_3}{dt^2} = \dot{x}_2$ and $\frac{dx_3}{dt} = x_2, \quad \therefore J\dot{x}_2 + Bx_2 = K_{tf}x_1$

$$\dot{x}_2 = \frac{K_{tf}}{J} x_1 - \frac{B}{J} x_2 \qquad .....(6.6.6)$$

The state variable $x_3 = \theta$. On differentiating $x_3 = \theta$ with respect to t we get,

$$\frac{dx_3}{dt} = \frac{d\theta}{dt}$$

Put $\frac{dx_3}{dt} = \dot{x}_3$ and $\frac{d\theta}{dt} = x_2 \qquad \therefore \dot{x}_3 = x_2 \qquad .....(6.6.7)$

The equations (6.6.5), (6.6.6) and (6.6.7) are the state equations of the system.

$$\dot{x}_1 = -\frac{R_f}{L_f} x_1 + \frac{1}{L_f} u$$

$$\dot{x}_2 = \frac{K_{tf}}{J} x_1 - \frac{B}{J} x_2$$

$$\dot{x}_3 = x_2$$

On arranging the state equations in the matrix form,

$$\begin{bmatrix} \dot{x}_1 \\ \dot{x}_2 \\ \dot{x}_3 \end{bmatrix} = \begin{bmatrix} -\dfrac{R_f}{L_f} & 0 & 0 \\ \dfrac{K_{tf}}{J} & -\dfrac{B}{J} & 0 \\ 0 & 1 & 0 \end{bmatrix} \begin{bmatrix} x_1 \\ x_2 \\ x_3 \end{bmatrix} + \begin{bmatrix} \dfrac{1}{L_f} \\ 0 \\ 0 \end{bmatrix} [u] \qquad \qquad .....(6.6.8)$$

Let the desired output be ω and θ. Let us equate the desired output quantities to standard notation y_1 and y_2 as shown below.

$$y_1 = \omega \quad ; \quad y_2 = \theta$$

On relating the outputs to state variable we get,

$$y_1 = x_2 \quad ; \quad y_2 = x_3$$

The output equation in the matrix form is

$$\begin{bmatrix} y_1 \\ y_2 \end{bmatrix} = \begin{bmatrix} 0 & 1 & 0 \\ 0 & 0 & 1 \end{bmatrix} \begin{bmatrix} x_1 \\ x_2 \\ x_3 \end{bmatrix} \qquad \qquad .....(6.6.9)$$

The state equation [equation (6.6.8)] and the output equation [equation (6.6.9)] together constitute the state model of the system.

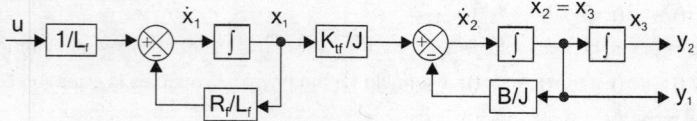

Fig 4 : *Block diagram representation of the state model field controlled dc motor.*

6.6 STATE SPACE REPRESENTATION USING PHASE VARIABLES

The phase variables are defined as those particular state variables which are obtained from one of the system variables and its derivatives. Usually the variable used in the system output and the remaining state variables are the derivatives of the output. The state model using phase variables can be easily determined if the system model is already known in the differential equation or transfer function form. There are three methods of modelling a system using phase variables and they are explained in the following sections.

Method 1

Consider the following nth order linear differential equation relating the output y(t) to the input u(t) of a system,

$$\overset{n.}{y} + a_1 \overset{(n-1).}{y} + a_2 \overset{(n-2).}{y} + \ldots\ldots\ldots + a_{n-2}\ddot{y} + a_{n-1}\dot{y} + a_n y = b\,u \qquad \qquad .....(6.10)$$

By choosing the output **y** and their derivatives as state variables, we get,

$$x_1 = y$$
$$x_2 = \dot{y}$$
$$x_3 = \ddot{y}$$
$$\vdots$$
$$x_n = \overset{(n-1)\cdot}{y} \quad ; \quad \therefore \dot{x}_n = \overset{n\cdot}{y}$$

On substituting the state variables in the differential equation governing the system [equation (6.10)], we get,

$$\dot{x}_n + a_1 x_n + a_2 x_{n-1} + \dots\dots + a_{n-2} x_3 + a_{n-1} x_2 + a_n x_1 = b\,u$$

$$\therefore \dot{x}_n = -a_n x_1 - a_{n-1} x_2 - a_{n-2} x_3 - \dots\dots - a_2 x_{n-1} - a_1 x_n + b\,u$$

The state equations of the system are

$$\dot{x}_1 = x_2$$
$$\dot{x}_2 = x_3$$
$$\vdots$$
$$\dot{x}_{n-1} = x_n$$
$$\dot{x}_n = -a_n x_1 - a_{n-1} x_2 - a_{n-1} x_3 - \dots\dots - a_2 x_{n-1} - a_1 x_n + b\,u$$

On arranging the above equations in the matrix form we get,

$$
\begin{bmatrix} \dot{x}_1 \\ \dot{x}_2 \\ \dot{x}_3 \\ \vdots \\ \dot{x}_{n-1} \\ \dot{x}_n \end{bmatrix} =
\begin{bmatrix}
0 & 1 & 0 & 0 & \cdots & 0 \\
0 & 0 & 1 & 0 & \cdots & 0 \\
0 & 0 & 0 & 1 & \cdots & 0 \\
& \vdots & & & & \\
0 & 0 & 0 & 0 & \cdots & 1 \\
-a_n & -a_{n-1} & -a_{n-2} & -a_{n-3} & \cdots & -a_1
\end{bmatrix}
\begin{bmatrix} x_1 \\ x_2 \\ x_3 \\ \vdots \\ x_{n-1} \\ x_n \end{bmatrix} +
\begin{bmatrix} 0 \\ 0 \\ 0 \\ \vdots \\ 0 \\ b \end{bmatrix} [u]
\qquad \dots (6.11)
$$

or $\dot{X} = A\,X + B\,U$

Here the matrix **A** (system matrix) has a very special form. It has all 1's in the upper off-diagonal, its last row is comprised of the negative of the coefficients of the original differential equation and all other elements are zero. This form of matrix **A** is known as *Bush form* (or) *Companion form.*

Also note that **B** matrix has the speciality that all its elements except the last element are zero. The output being $y = x_1$, the output equation is given by,

$$
y = \begin{bmatrix} 1 & 0 & 0 & \cdots\cdots & 0 \end{bmatrix}
\begin{bmatrix} x_1 \\ x_2 \\ x_3 \\ \vdots \\ x_n \end{bmatrix}
\qquad \dots (6.12)
$$

(or) $Y = CX$

The advantage in using phase variables for state space modelling is that the system state model can be written directly by inspection from the differential equation governing the system.

Method 2

Consider the following n^{th} order differential equation governing the output y(t) to the input u(t) of a system.

$$\overset{n.}{y} + a_1 \overset{(n-1).}{y} + \ldots\ldots + a_{n-1}\dot{y} + a_n y = b_0 \overset{m.}{u} + b_1 \overset{(m-1).}{u} + \ldots\ldots + b_{m-1}\dot{u} + b_m u \qquad \ldots\ldots(6.13)$$

let $n = m = 3$

$$\therefore \dddot{y} + a_1\ddot{y} + a_2\dot{y} + a_3 y = b_0\dddot{u} + b_1\ddot{u} + b_2\dot{u} + b_3 u \qquad \ldots\ldots(6.14)$$

On taking Laplace transform of equation (6.14) with zero initial conditions we get,

$$s^3 Y(s) + a_1 s^2 Y(s) + a_2 s Y(s) + a_3 Y(s) = b_0 s^3 U(s) + b_1 s^2 U(s) + b_2 U(s) + b_3 U(s)$$

$$(s^3 + a_1 s^2 + a_2 s + a_3)Y(s) = (b_0 s^3 + b_1 s^2 + b_2 s + b_3)U(s)$$

$$\therefore \frac{Y(s)}{U(s)} = \frac{b_0 s^3 + b_1 s^2 + b_2 s + b_3}{s^3 + a_1 s^2 + a_2 s + a_3}$$

$$= \frac{s^3\left(b_0 + \dfrac{b_1}{s} + \dfrac{b_2}{s^2} + \dfrac{b_3}{s^3}\right)}{s^3\left(1 + \dfrac{a_1}{s} + \dfrac{a_2}{s^2} + \dfrac{a_3}{s^3}\right)} = \frac{b_0 + \dfrac{b_1}{s} + \dfrac{b_2}{s^2} + \dfrac{b_3}{s^3}}{1 - \left(-\dfrac{a_1}{s} - \dfrac{a_2}{s^2} - \dfrac{a_3}{s^3}\right)} \qquad \ldots\ldots(6.15)$$

From the Mason's gain formula, the transfer function of the system is given by,

$$T(s) = \frac{1}{\Delta} \sum_K P_K \Delta_K \qquad \ldots\ldots(6.16)$$

where, P_K = path gain of Kth forward path.

$\Delta = 1 -$ (sum of loop gain of all individual loops)

$\qquad +$ (sum of gain products of all possible combinations of two non-touching loops) $- \cdots\cdots$

$\Delta_K = \Delta$ for that part of the graph which is not touching K^{th} forward path.

The transfer function of a system with four forward paths and with three feedback loops (touching each other) is given by,

$$T(s) = \frac{P_1 + P_2 + P_3 + P_4}{1 - (P_{11} + P_{12} + P_{13})} \qquad \ldots\ldots(6.17)$$

On comparing equation (6.15) and (6.17) we get,

$$P_1 = b_0 \quad ; \quad P_2 = \frac{b_1}{s} \quad ; \quad P_3 = \frac{b_2}{s^2} \quad \text{and} \quad P_4 = \frac{b_3}{s^3}$$

$$P_{11} = -\frac{a_1}{s} \quad ; \quad P_{12} = -\frac{a_2}{s^2} \quad \text{and} \quad P_{13} = -\frac{a_3}{s^3}$$

Hence for the system represented by the transfer function as that of equation (6.15), a signal flow graph can be constructed as shown in the fig 6.4. The signal flow is constructed such that all $\Delta_K = 1$ and all loops are touching loops.

Let us assign state variables at the output of each integrator in the signal flow graph. Hence at the input of each integrator, the first derivative of the state variable will be available. The state equations are formed by summing all the incoming signals to the nodes, whose values correspond to first derivative of state variables.

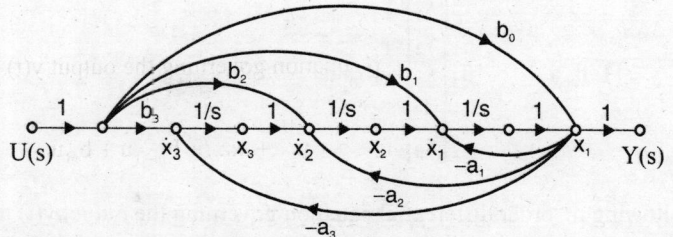

Fig 6.4 : Signal flow graph of the system represented by the equation 6.15.

By summing up the incoming signals to node $\dot{x}_1$ we get, *(Refer fig 6.4a)*

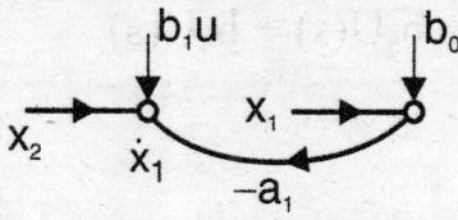

$$\dot{x}_1 = -a_1(x_1 + b_0 u) + x_2 + b_1 u \qquad .....(6.18)$$

$$\therefore \ \dot{x}_1 = -a_1 x_1 + x_2 + (b_1 - a_1 b_0)u$$

Fig 6.4a.

By summing up the incoming signals to node we get, *(Refer fig 6.4b)*

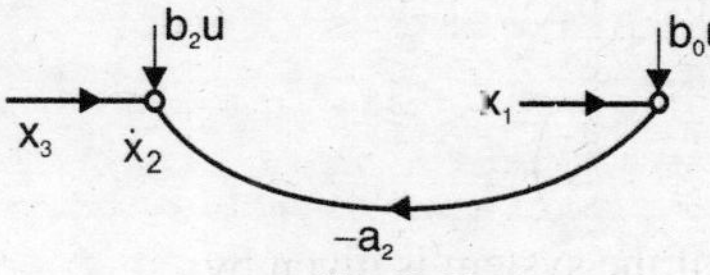

$$\dot{x}_2 = -a_2(x_1 + b_0 u) + x_3 + b_2 u \qquad .....(6.19)$$

$$\therefore \ \dot{x}_2 = -a_2 x_1 + x_3 + (b_2 - a_2 b_0)u$$

Fig 6.4b.

By summing up the incoming signals to node we get, *(Refer fig 6.4c)*

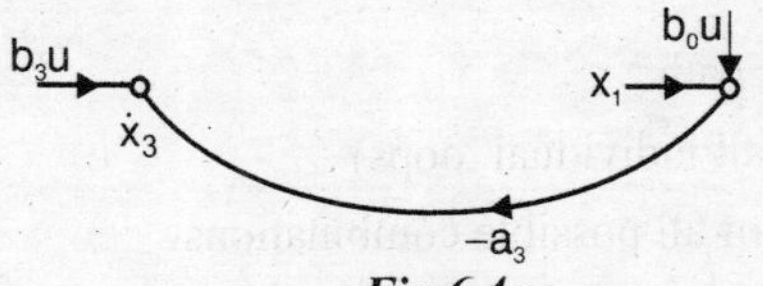

$$\dot{x}_3 = -a_3(x_1 + b_0 u) + b_3 u$$

$$\therefore \ \dot{x}_3 = -a_3 x_1 + (b_3 - a_3 b_0)u \qquad .....(6.20)$$

Fig 6.4c.

The output equation is given by the sum of incoming signals to output node.

$$\therefore \ y = x_1 + b_0 u \qquad .....(6.21)$$

On arranging the state equations and the output equations in the matrix form, we get,

$$\begin{bmatrix} \dot{x}_1 \\ \dot{x}_2 \\ \dot{x}_3 \end{bmatrix} = \begin{bmatrix} -a_1 & 1 & 0 \\ -a_2 & 0 & 1 \\ -a_3 & 0 & 0 \end{bmatrix} \begin{bmatrix} x_1 \\ x_2 \\ x_3 \end{bmatrix} + \begin{bmatrix} b_1 - a_1 b_0 \\ b_2 - a_2 b_0 \\ b_3 - a_3 b_0 \end{bmatrix} [u] \qquad .....(6.22)$$

$$y = \begin{bmatrix} 1 & 0 & 0 \end{bmatrix} \begin{bmatrix} x_1 \\ x_2 \\ x_3 \end{bmatrix} + [b_0]u \qquad .....(6.23)$$

The above results can be generalized for an nth order differential equation, and the general state model for m = n is given below.

$$\begin{bmatrix} \dot{x}_1 \\ \dot{x}_2 \\ \vdots \\ \dot{x}_{n-1} \\ \dot{x}_n \end{bmatrix} = \begin{bmatrix} -a_1 & 1 & 0 & \cdots & 0 \\ -a_2 & 0 & 1 & \cdots & 0 \\ \vdots & \vdots & \vdots & \vdots & \vdots \\ -a_{n-1} & 0 & 0 & \cdots & 1 \\ -a_n & 0 & 0 & \cdots & 0 \end{bmatrix} \begin{bmatrix} x_1 \\ x_2 \\ \vdots \\ x_{n-1} \\ x_n \end{bmatrix} + \begin{bmatrix} b_1 - a_1 b_0 \\ b_2 - a_2 b_0 \\ \vdots \\ b_{n-1} - a_{n-1} b_0 \\ b_n - a_n b_0 \end{bmatrix} [u] \qquad .....(6.24)$$

$$y = \begin{bmatrix} 1 & 0 & 0 & \cdots\cdots & 0 \end{bmatrix} \begin{bmatrix} x_1 \\ x_2 \\ x_3 \\ \vdots \\ x_n \end{bmatrix} + [b_0] u \qquad \qquad .....(6.25)$$

Method 3

Consider the following n^{th} order differential equation governing the output y(t) to the input u(t) of a system.

$$\overset{n.}{y} + a_1 \overset{(n-1).}{y} + \text{............} + a_{n-1}\dot{y} + a_n y = b_0 \overset{m.}{u} + b_1 \overset{(m-1).}{u} + \text{......} + b_{m-1}\dot{u} + b_m u \qquad .....(6.26)$$

let n = m = 3

$$\therefore \ \dddot{y} + a_1 \ddot{y} + a_2 \dot{y} + a_2 y = b_0 \dddot{u} + b_1 \ddot{u} + b_2 \dot{u} + b_3 u \qquad .....(6.27)$$

On taking Laplace transform of equation (6.27) with zero initial conditions, we get

$$s^3 Y(s) + a_1 s^2 Y(s) + a_2 s Y(s) + a_3 Y(s) = b_0 s^3 U(s) + b_1 s^2 U(s) + b_2 s U(s) + b_3 U(s)$$

$$(s^3 + a_1 s^2 + a_2 s + a_3) Y(s) = (b_0 s^3 + b_1 s^2 + b_2 s + b_3) U(s)$$

$$\therefore \ \frac{Y(s)}{U(s)} = \frac{b_0 s^3 + b_1 s^2 + b_2 s + b_3}{s^3 + a_1 s^2 + a_2 s + a_3}$$

Let $\dfrac{Y(s)}{U(s)} = \dfrac{X_1(s)}{U(s)} \cdot \dfrac{Y(s)}{X_1(s)}$

where, $\dfrac{X_1(s)}{U(s)} = \dfrac{1}{s^3 + a_1 s^2 + a_2 s + a_3}$ .....(6.28)

and $\dfrac{Y(s)}{X_1(s)} = b_0 s^3 + b_1 s^2 + b_2 s + b_3$ .....(6.29)

On cross multiplying the equation (6.28) we get,

$$X_1(s)[s^3 + a_1 s^2 + a_2 s + a_3] = U(s)$$

$$s^3 X_1(s) + a_1 s^2 X_1(s) + a_2 s X_1(s) + a_3 X_1(s) = U(s) \qquad .....(6.30)$$

On taking inverse Laplace transform of equation (6.30) we get,

$$\dddot{x}_1 + a_1 \ddot{x}_1 + a_2 \dot{x}_1 + a_3 x_1 = u \qquad .....(6.31)$$

Let the state variable be, x_1, x_2 and x_3

where, $x_2 = \dot{x}_1$

and $x_3 = \ddot{x}_1 = \dot{x}_2$; $\therefore \ \dot{x}_3 = \dddot{x}_1$

On substituting the state variables in equation (6.31) we get,

$$\dot{x}_3 + a_1 x_3 + a_2 x_2 + a_3 x_1 = u$$

$$\therefore \ \dot{x}_3 = - a_3 x_1 - a_2 x_2 - a_1 x_3 + u$$

The state equations are,

$$\dot{x}_1 = x_2$$

$$\dot{x}_2 = x_3$$

$$\dot{x}_3 = - a_3 x_1 - a_2 x_2 - a_1 x_3 + u$$

On cross multiplying the equation(6.29) we get,

$$Y(s) = b_0 s^3 X_1(s) + b_1 s^2 X_1(s) + b_2 s X_1(s) + b_3 X_1(s) \qquad .....(6.32)$$

On taking inverse Laplace transform of equation(6.32), we get,

$$y = b_0 \dddot{x}_1 + b_1 \ddot{x}_1 + b_2 \dot{x}_1 + b_3 x_1 \qquad .....(6.33)$$

On substituting the state variables in equation (6.33) we get,

$$y = b_0 \dot{x}_3 + b_1 x_3 + b_2 x_2 + b_3 x_1 \qquad .....(6.34)$$

Put $\dot{x}_3 = -a_3 x_1 - a_2 x_2 - a_1 x_3 + u$ in equation (6.34)

$$\therefore \ y = b_0(-a_3 x_1 - a_2 x_2 - a_1 x_3 + u) + b_1 x_3 + b_2 x_2 + b_3 x_1$$

$$y = (b_3 - a_3 b_0)x_1 + (b_2 - a_2 b_0)x_2 + (b_1 - a_1 b_0)x_3 + b_0 u \qquad .....(6.35)$$

The equation (6.35) is the output equation.

On arranging the state equations and output equations in the matrix form, we get,

$$\begin{bmatrix} \dot{x}_1 \\ \dot{x}_2 \\ \dot{x}_3 \end{bmatrix} = \begin{bmatrix} 0 & 1 & 0 \\ 0 & 0 & 1 \\ -a_3 & -a_2 & -a_1 \end{bmatrix} \begin{bmatrix} x_1 \\ x_2 \\ x_3 \end{bmatrix} + \begin{bmatrix} 0 \\ 0 \\ 1 \end{bmatrix} [u] \qquad .....(6.36)$$

$$y = [(b_3 - a_3 b_0) \ (b_2 - a_1 b_0)] + \begin{bmatrix} x_1 \\ x_2 \\ x_3 \end{bmatrix} [b_0] \, u \qquad .....(6.37)$$

The above results can be generalized for an nth order differential equation and the general state model for m = n is given below.

$$\begin{bmatrix} \dot{x}_1 \\ \dot{x}_2 \\ \vdots \\ \dot{x}_{n-1} \\ \dot{x}_n \end{bmatrix} = \begin{bmatrix} 0 & 1 & 0 & \cdots & 0 \\ 0 & 0 & 1 & \cdots & 0 \\ \vdots & \vdots & \vdots & & \vdots \\ 0 & 0 & 0 & \cdots & 1 \\ -a_n & -a_{n-1} & -a_{n-2} & \cdots & -a_1 \end{bmatrix} \begin{bmatrix} x_1 \\ x_2 \\ \vdots \\ x_{n-1} \\ x_n \end{bmatrix} + \begin{bmatrix} 0 \\ 0 \\ \vdots \\ 0 \\ 1 \end{bmatrix} [u] \qquad .....(6.38)$$

$$y = [(b_n - a_n b_0) \ (b_{n-1} - a_{n-1} b_0) \cdots (b_2 - a_2 b_0)(b_1 - a_1 b_0)] \begin{bmatrix} x_1 \\ x_2 \\ \vdots \\ x_{n-1} \\ x_n \end{bmatrix} + b_0 u \qquad .....(6.39)$$

Advantages of Phase Variables

The state space model can be directly formed by inspection from the differential equations governing the system. The phase variables provide a link between the transfer function design approach and time-domain design approach.

Disadvantage of Phase Variables

The phase variables are not physical variables of the system and therefore are not available for measurement and control purposes.

EXAMPLE 6.7

Construct a state model for a system characterized by the differential equation,

$$\frac{d^3y}{dt^3} + 6\frac{d^2y}{dt^2} + 11\frac{dy}{dt} + 6y + u = 0$$

Give the block diagram representation of the state model.

SOLUTION

Let us choose y and their derivatives as state variables. The system is governed by third order differential equation and so the number of state variables are three.

The state variables x_1, x_2 and x_3 are related to phase variables as follows

$$x_1 = y$$

$$x_2 = \frac{dy}{dt} = \dot{x}_1$$

$$x_3 = \frac{d^2y}{dt^2} = \dot{x}_2$$

Put $y = x_1$, $\frac{dy}{dt} = x_2$ and $\frac{d^2y}{dt^2} = x_3$ and $\frac{d^3y}{dt^3} = \dot{x}_3$ in the given equation,

$$\therefore \dot{x}_3 + 6x_3 + 11x_2 + 6x_1 + u = 0$$

or $\dot{x}_3 = -6x_1 - 11x_2 - 6x_3 - u$

The state equations are

$$\dot{x}_1 = x_2$$

$$\dot{x}_2 = x_3$$

$$\dot{x}_3 = -6x_1 - 11x_2 - 6x_3 - u$$

On arranging the state equations in the matrix form we get,

$$\begin{bmatrix} \dot{x}_1 \\ \dot{x}_2 \\ \dot{x}_3 \end{bmatrix} = \begin{bmatrix} 0 & 1 & 0 \\ 0 & 0 & 1 \\ -6 & -11 & -6 \end{bmatrix} \begin{bmatrix} x_1 \\ x_2 \\ x_3 \end{bmatrix} + \begin{bmatrix} 0 \\ 0 \\ -1 \end{bmatrix} [u]$$

Here, y = output

But, $y = x_1$

$$\therefore \text{ The output equation is, } y = \begin{bmatrix} 1 & 0 & 0 \end{bmatrix} \begin{bmatrix} x_1 \\ x_2 \\ x_3 \end{bmatrix}$$

The state equation and output equation, constitutes the state model of the system.

The block diagram form of the state diagram of the system is shown in fig 1.

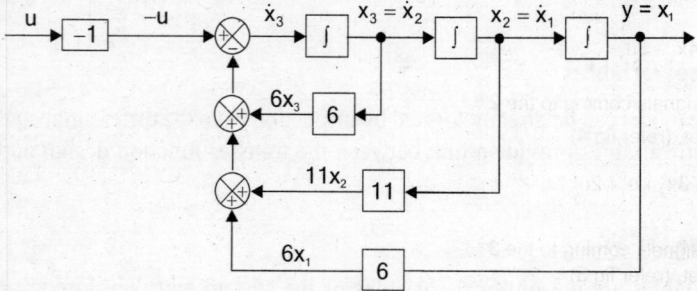

Fig 1 : Block diagram form of state diagram.

EXAMPLE 6.8

The state diagram of a system is shown in fig 1. Assign state variables and obtain the state model of the system.

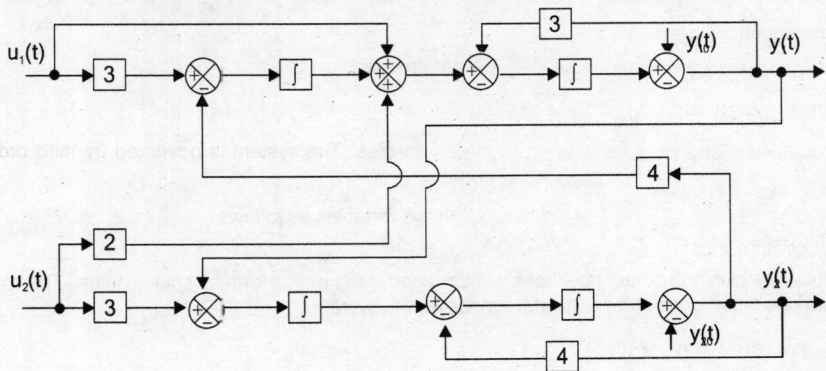

Fig 1.

SOLUTION

Since there are 4-integrators in the state diagram we can assign, 4 state variables. The state variables can be assigned at the output of the integrators as shown in fig 2. Hence at the input of the integrator, the first derivative of the state variable will be available. The state equations are formed by summing all the incoming signals to the input of the integrator and equating to the corresponding first derivative of the state variable.

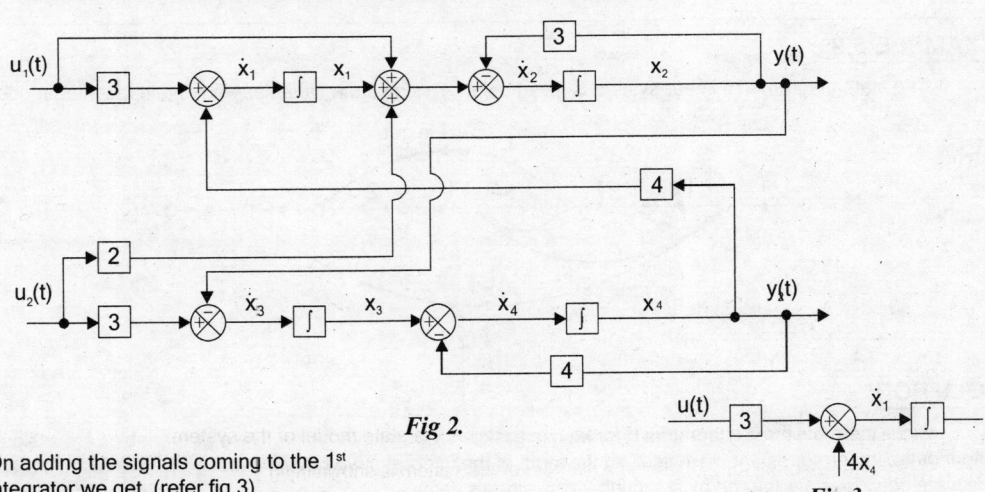

Fig 2.

On adding the signals coming to the 1st integrator we get, (refer fig 3)

$$\dot{x}_1 = -4x_4 + 3u_1$$

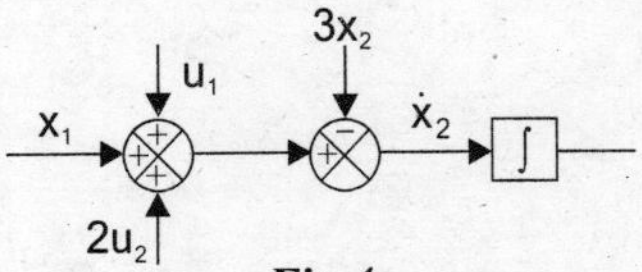

Fig 3.

On adding the signals coming to the 2nd integrator we get, (refer fig 4)

$$\dot{x}_2 = x_1 - 3x_2 + u_1 + 2u_2$$

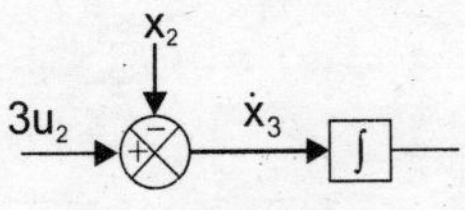

Fig 4.

On adding the signals coming to the 3rd integrator we get, (refer fig 5)

$$\dot{x}_3 = -x_2 + 3u_2$$

Fig 5.

On adding the signals coming to the 4th integrator we get, (refer fig 6)

$$\dot{x}_4 = x_3 - 4x_4$$

The state equations are

$$\dot{x}_1 = -4x_4 + 3u_1$$

$$\dot{x}_2 = x_1 - 3x_2 + u_1 + 2u_2$$

$$\dot{x}_3 = -x_2 + 3u_2$$

$$\dot{x}_4 = x_3 - 4x_4$$

The output equations are, $y_1 = x_2$ and $y_2 = x_4$.

The state equations and output equations are arranged in the matrix form as shown below. The state equations and output equations together constitute the state model of the system.

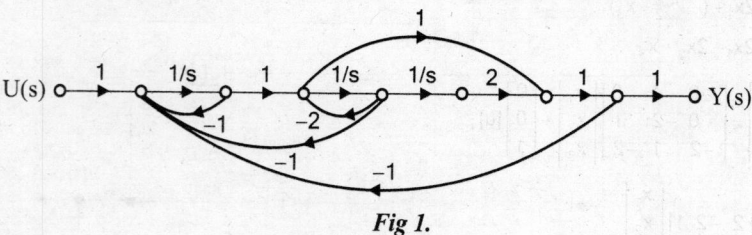

Fig 6.

$$\begin{bmatrix} \dot{x}_1 \\ \dot{x}_2 \\ \dot{x}_3 \\ \dot{x}_4 \end{bmatrix} = \begin{bmatrix} 0 & 0 & 0 & -4 \\ 1 & -3 & 0 & 0 \\ 0 & -1 & 0 & 0 \\ 0 & 0 & 1 & -4 \end{bmatrix} \begin{bmatrix} x_1 \\ x_2 \\ x_3 \\ x_4 \end{bmatrix} + \begin{bmatrix} 3 & 0 \\ 1 & 2 \\ 0 & 3 \\ 0 & 0 \end{bmatrix} \begin{bmatrix} u_1 \\ u_2 \end{bmatrix}$$

$$\begin{bmatrix} y_1 \\ y_2 \end{bmatrix} = \begin{bmatrix} 0 & 1 & 0 & 0 \\ 0 & 0 & 0 & 1 \end{bmatrix} \begin{bmatrix} x_1 \\ x_2 \\ x_3 \\ x_4 \end{bmatrix}$$

EXAMPLE 6.9

The state diagram of a linear system is given below. Assign state variables and obtain the state model.

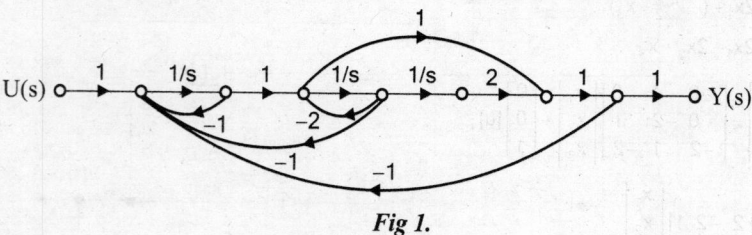

Fig 1.

SOLUTION

Since there are three integrators (1/s) we can assign three state variables. The state variables are assigned at the output of the integrator as shown in fig 2. At the input of the integrator we have the first derivative of the state variable. The state equations are formed by summing all the signals at the input of integrator and equating to the corresponding first derivative of state variable.

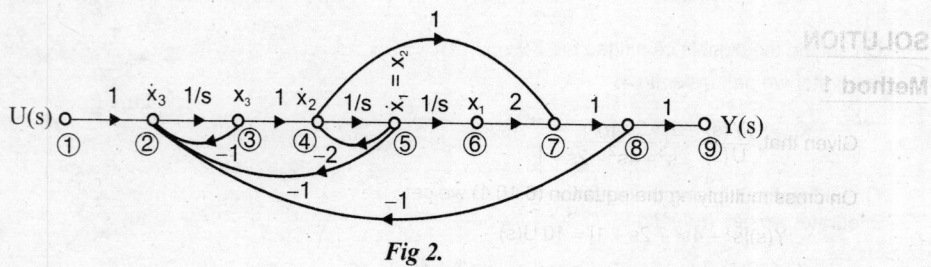

Fig 2.

On adding the signals coming to node-5, we get, (refer fig 3).

$$\dot{x}_1 = x_2$$

On adding the signals coming to node-4, we get, (refer (fig 4)

$$\dot{x}_2 = -2x_2 + x_3$$

On adding the signals coming to node-2, we get, (refer (fig 5)

$$\dot{x}_3 = -(\dot{x}_2 + 2x_1) - x_2 - x_3 + u = -\dot{x}_2 - 2x_1 - x_2 - x_3 + u$$

Put $\dot{x}_2 = -2x_2 + x_3$

$$\therefore \dot{x}_3 = 2x_2 - x_3 - 2x_1 - x_2 - x_3 + u = -2x_1 + x_2 - 2x_3 + u$$

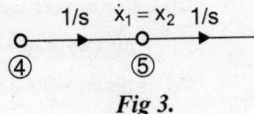

Fig 3.

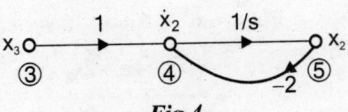

Fig 4.

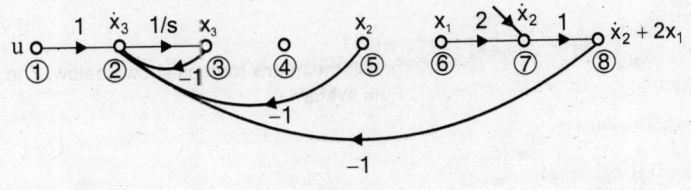

Fig 5.

The state equations are

$$\dot{x}_1 = x_2$$

$$\dot{x}_2 = -2x_2 + x_3$$

$$\dot{x}_3 = -2x_1 + x_2 - 2x_3 + u$$

The output equation is obtained by adding the signals coming to output node (refer fig 6)

$$y = 2x_1 + \dot{x}_2$$

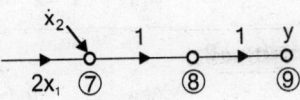

Fig 6.

Put $\dot{x}_2 = -2x_2 + x_3$

$$y = 2x_1 + (-2x_2 + x_3)$$

$$y = 2x_1 - 2x_2 + x_3$$

$$\begin{bmatrix} \dot{x}_1 \\ \dot{x}_2 \\ \dot{x}_3 \end{bmatrix} = \begin{bmatrix} 0 & 1 & 0 \\ 0 & -2 & 1 \\ -2 & 1 & -2 \end{bmatrix} \begin{bmatrix} x_1 \\ x_2 \\ x_3 \end{bmatrix} + \begin{bmatrix} 0 \\ 0 \\ 1 \end{bmatrix} [u]$$

$$y = \begin{bmatrix} 2 & -2 & 1 \end{bmatrix} \begin{bmatrix} x_1 \\ x_2 \\ x_3 \end{bmatrix}$$

EXAMPLE 6.10

Obtain the state model of the system whose transfer function is given as,

$$\frac{Y(s)}{U(s)} = \frac{10}{s^3 + 4s^2 + 2s + 1}$$

SOLUTION

Method 1

Given that, $\dfrac{Y(s)}{U(s)} = \dfrac{10}{s^3 + 4s^2 + 2s + 1}$ .....(6.10.1)

On cross multiplying the equation (6.10.1) we get,

$$Y(s)[s^3 + 4s^2 + 2s + 1] = 10\,U(s)$$

$$s^3 Y(s) + 4s^2 Y(s) + 2s\,Y(s) + Y(s) = 10\,U(s)$$ .....(6.10.2)

On taking inverse Laplace transform of equation (6.10.2) we get,

$$\dddot{y} + 4\ddot{y} + 2\dot{y} + y = 10u \hspace{3cm} .....(6.10.3)$$

Let us define state variables as follows,

$$x_1 = y \hspace{0.3cm} ; \hspace{0.3cm} x_2 = \dot{y} \hspace{0.3cm} ; \hspace{0.3cm} x_3 = \ddot{y}$$

Put $\dddot{y} = \dot{x}_3$; $\ddot{y} = x_3$; $\dot{y} = x_2$ and $y = x_1$ in the equation (6.10.3)

$$\therefore \dot{x}_3 + 4x_3 + 2x_2 + x_1 = 10u$$

or $\dot{x}_3 = -x_1 - 2x_2 - 4x_3 + 10u$

The state equations are

$$\dot{x}_1 = x_2 \hspace{0.3cm} ; \hspace{0.3cm} \dot{x}_2 = x_3 \hspace{0.3cm} ; \hspace{0.3cm} \dot{x}_3 = -x_1 - 2x_2 - 4x_3 + 10u$$

The output equation is $y = x_1$

The state model in the matrix form is,

$$\begin{bmatrix} \dot{x}_1 \\ \dot{x}_2 \\ \dot{x}_3 \end{bmatrix} = \begin{bmatrix} 0 & 1 & 0 \\ 0 & 0 & 1 \\ -1 & -2 & -4 \end{bmatrix} \begin{bmatrix} x_1 \\ x_2 \\ x_3 \end{bmatrix} + \begin{bmatrix} 0 \\ 0 \\ 10 \end{bmatrix} [u] \hspace{1cm} y = \begin{bmatrix} 1 & 0 & 0 \end{bmatrix} \begin{bmatrix} x_1 \\ x_2 \\ x_3 \end{bmatrix}$$

Method 2

$$\frac{Y(s)}{U(s)} = \frac{10}{s^3 + 4s^2 + 2s + 1} = \frac{10}{s^3\left(1 + \dfrac{4}{s} + \dfrac{2}{s^2} + \dfrac{1}{s^3}\right)}$$

$$= \frac{10/s^3}{1 - \left(-\dfrac{4}{s} - \dfrac{2}{s^2} - \dfrac{1}{s^3}\right)}$$

The signal flow graph for the above transfer function can be constructed as shown in fig 1 with a single forward path consisting of three integrators and with path gain $10/s^3$. The graph will have three individual loops with loop gains $-4/s$, $-2/s^2$, and $-1/s^3$.

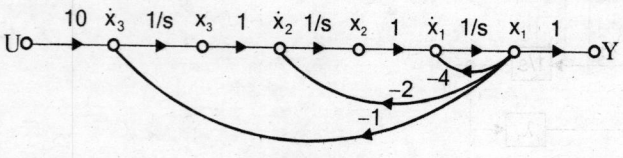

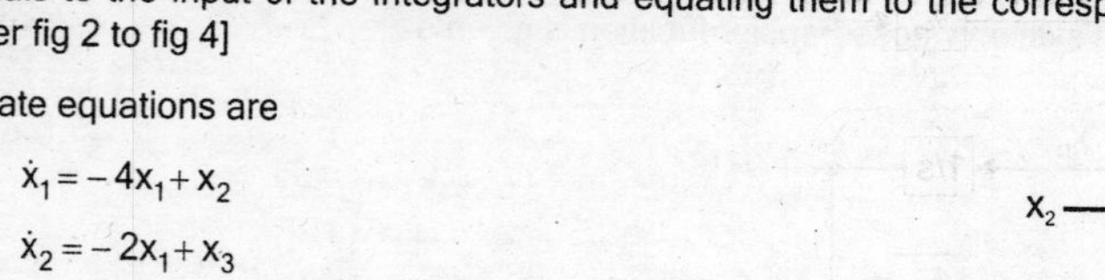

Fig 1.

Assign state variables at the output of the integrator (1/s). The state equations are obtained by summing the incoming signals to the input of the integrators and equating them to the corresponding first derivative of the state variable. [Refer fig 2 to fig 4]

The state equations are

$$\dot{x}_1 = -4x_1 + x_2$$

$$\dot{x}_2 = -2x_1 + x_3$$

$$\dot{x}_3 = -x_1 + 10u$$

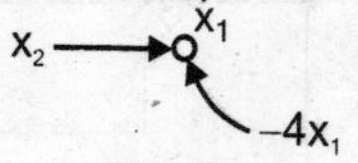

The output equation is, $y = x_1$

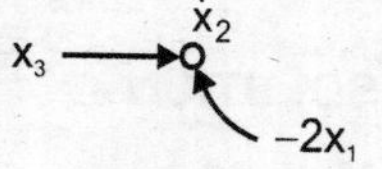

The state model in the matrix form is,

$$\begin{bmatrix} \dot{x}_1 \\ \dot{x}_2 \\ \dot{x}_3 \end{bmatrix} = \begin{bmatrix} -4 & 1 & 0 \\ -2 & 0 & 1 \\ -1 & 0 & 0 \end{bmatrix} \begin{bmatrix} x_1 \\ x_2 \\ x_3 \end{bmatrix} + \begin{bmatrix} 0 \\ 0 \\ 10 \end{bmatrix} [u] \hspace{0.5cm} ; \hspace{0.5cm} y = \begin{bmatrix} 1 & 0 & 0 \end{bmatrix} \begin{bmatrix} x_1 \\ x_2 \\ x_3 \end{bmatrix}$$

Fig 2.

Fig 3.

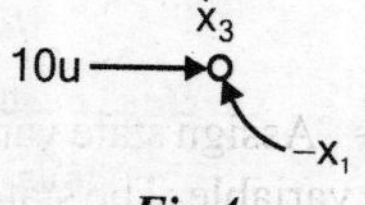

Fig 4.

6.7 STATE SPACE REPRESENTATION USING CANONICAL VARIABLES

In canonical form (or normal form) of state model, the system matrix $\mathbf{A}$ will be a diagonal matrix. The elements on the diagonal are the poles of the transfer function of the system.

By partial fraction expansion, the transfer function $Y(s)/U(s)$ of the nth order system can be expressed as shown in equation (6.40).

$$\frac{Y(s)}{U(s)} = b_0 + \frac{C_1}{s+\lambda_1} + \frac{C_2}{s+\lambda_2} + \cdots\cdots + \frac{C_n}{s+\lambda_n} \qquad \qquad .....(6.40)$$

where $C_1, C_2, C_3, \cdots\cdots C_n$ are residues and $\lambda_1, \lambda_2, \cdots\cdots \lambda_n$ are roots of denominator polynomial (or poles of the system).

The equation (6.40) can be rearranged as shown below.

$$\frac{Y(s)}{U(s)} = b_0 + \frac{C_1}{s\left(1+\dfrac{\lambda_1}{s}\right)} + \frac{C_2}{s\left(1+\dfrac{\lambda_2}{s}\right)} + \cdots\cdots + \frac{C_n}{s\left(1+\dfrac{\lambda_n}{s}\right)}$$

$$= b_0 + \frac{C_1/s}{1+\lambda_1/s} + \frac{C_2/s}{1+\lambda_2/s} + \cdots\cdots + \frac{C_n/s}{1+\lambda_n/s}$$

$$\therefore \; Y(s) = b_0 U(s) + \left[\frac{1/s}{1+(1/s)\times\lambda_1} \times C_1\right] U(s) + \left[\frac{1/s}{1+(1/s)\times\lambda_2} \times C_2\right] U(s) +$$

$$\cdots\cdots + \left[\frac{1/s}{1+(1/s)\times\lambda_n} \times C_n\right] U(s) \qquad \qquad .....(6.41)$$

The equation (6.41) can be represented by a block diagram as shown in fig 6.5

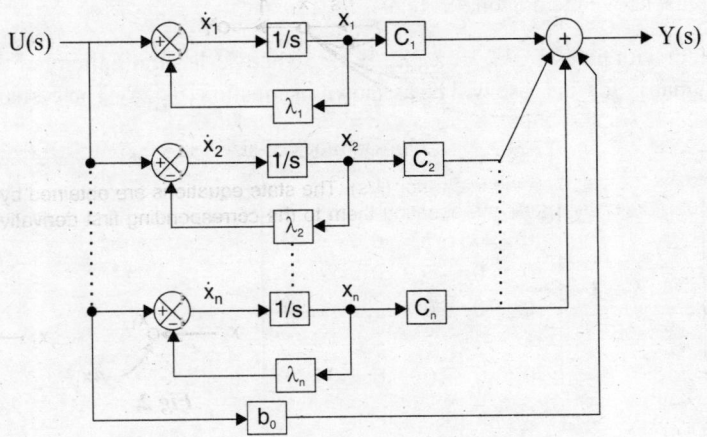

Fig 6.5 : *Block diagram of canonical state model.*

Assign state variables at the output of integrator. The input of the integrator will be first derivative of state variable. The state equations are formed by adding the incoming signals to the integrator and equating to first derivative of state variable. The state equations are,

$$\dot{x}_1 = -\lambda_1 x_1 + u$$

$$\dot{x}_2 = -\lambda_2 x_2 + u$$

$$\vdots$$

$$\dot{x}_n = -\lambda_n x_n + u$$

The output equation is, $y = C_1 x_1 + C_2 x_2 + \ldots\ldots\ldots C_n x_n + b_0 u$

The canonical form of state model in the matrix form is given below

$$\begin{bmatrix} \dot{x}_1 \\ \dot{x}_2 \\ \dot{x}_3 \\ \vdots \\ \dot{x}_n \end{bmatrix} = \begin{bmatrix} -\lambda_1 & 0 & 0 & \cdots & 0 \\ 0 & -\lambda_2 & 0 & \cdots & 0 \\ 0 & 0 & -\lambda_3 & \cdots & 0 \\ \vdots & \vdots & \vdots & & \vdots \\ 0 & 0 & 0 & \cdots & -\lambda_n \end{bmatrix} \begin{bmatrix} x_1 \\ x_2 \\ x_3 \\ \vdots \\ x_n \end{bmatrix} + \begin{bmatrix} 1 \\ 1 \\ 1 \\ \vdots \\ 1 \end{bmatrix} [u]$$

$$y = [C_1 \ C_2 \ C_3 \ \cdots\cdots \ C_n] \begin{bmatrix} x_1 \\ x_2 \\ x_3 \\ \vdots \\ x_n \end{bmatrix} + [b_0] [u] \qquad \ldots\ldots(6.42)$$

The advantage of canonical form is that the state equations are independent of each other. The disadvantage is that the canonical variables are not physical variables and so they are not available for measurement and control.

When a pole of the transfer function has multiplicity, the canonical state model will be in a special form called Jordan canonical form. In this form the system matrix **A** will have a Jordan block of size q × q, correspond to a pole of value λ_i with multiplicity q. In the Jordan block the diagonal element will be the poles and the element just above the diagonal is one.

Consider a system with poles $\lambda_1, \lambda_1, \lambda_1, \lambda_4, \lambda_5, \ldots \lambda_n$, where λ_1 has multiplicity of three. The input matrix (**B**) and system matrix for this case will be as shown in equation (6.42a). The system matrix is also denoted as **J**.

Jordan block of size 3×3

$$B = \begin{bmatrix} 0 \\ 0 \\ 1 \\ 1 \\ \vdots \\ 1 \end{bmatrix} ; \quad A = J = \begin{bmatrix} -\lambda_1 & 1 & 0 & 0 & \cdots & 0 \\ 0 & -\lambda_1 & 1 & 0 & \cdots & 0 \\ 0 & 0 & -\lambda_1 & 0 & \cdots & 0 \\ 0 & 0 & 0 & -\lambda_4 & \cdots & 0 \\ \vdots & \vdots & \vdots & \vdots & & \vdots \\ 0 & 0 & 0 & 0 & \cdots & -\lambda_n \end{bmatrix} \qquad \ldots\ldots(6.42a)$$

The transfer function of the system for this case is given by equation (6.42a) and the block diagram is shown in fig 6.5a.

$$\frac{Y(s)}{U(s)} = b_0 + \frac{C_1}{(s+\lambda_1)^3} + \frac{C_2}{(s+\lambda_1)^2} + \frac{C_3}{s+\lambda_1} + \frac{C_4}{s+\lambda_4} + \ldots\ldots + \frac{C_n}{s+\lambda_n} \qquad \ldots\ldots(6.42b)$$

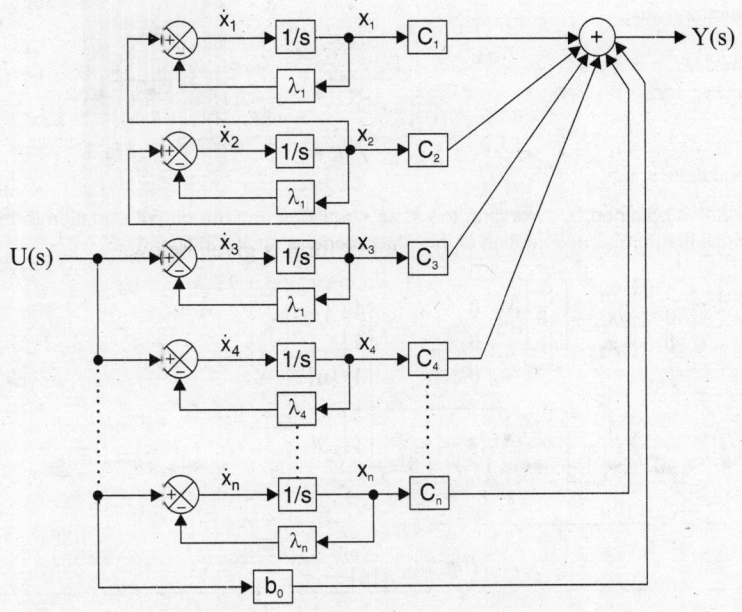

Fig 6.5a : *Block diagram of Jordan canonical state model.*

EXAMPLE 6.11

A feedback system has a closed-loop transfer function

$$\frac{Y(s)}{U(s)} = \frac{10(s+4)}{s(s+1)(s+3)}$$

Construct three different state models for this system and give block diagram representation for each state model.

SOLUTION

Model 1

$$\frac{Y(s)}{U(s)} = \frac{10(s+4)}{s(s+1)(s+3)} = \frac{10s+40}{s(s^2+4s+3)} = \frac{10s+40}{s^3+4s^2+3s} = \frac{10s+40}{s^3\left(1+\dfrac{4}{s}+\dfrac{3}{s^2}\right)} = \frac{\dfrac{10}{s^2}+\dfrac{40}{s^3}}{1-\left(-\dfrac{4}{s}-\dfrac{3}{s^2}\right)}$$

A signal flow graph for the above transfer function can be constructed as shown fig 1, with two forward paths and two individual loops. The forward path gains are $10/s^2$ and $40/s^3$. The loop gains are $-4/s$ and $-3/s^2$.

Assign state variables at the output of integrator as shown in fig 1 and so the input of integrator is first derivative of state variable. The state equations are obtained by summing all the incoming signals to the integrator and equating to the corresponding first derivative of the state variable. [Refer fig 2 to 4].

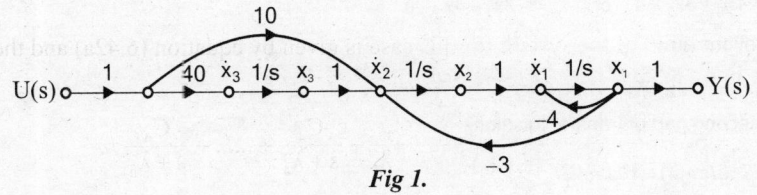

Fig 1.

The state equations are

$$\dot{x}_1 = -4x_1 + x_2$$
$$\dot{x}_2 = -3x_1 + x_3 + 10u$$
$$\dot{x}_3 = 40u$$

The output equation is, $y = x_1$

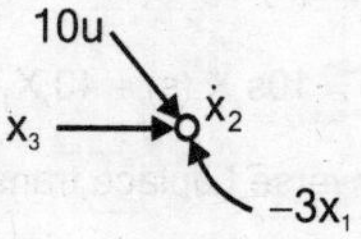

Fig 2.   **Fig 3.**   **Fig 4.**

The state model is obtained by arranging the state equations and the output equation in the matrix form as shown below. The block diagram representation of this state model is shown in fig 6.11.5.

$$\begin{bmatrix} \dot{x}_1 \\ \dot{x}_2 \\ \dot{x}_3 \end{bmatrix} = \begin{bmatrix} -4 & 1 & 0 \\ -3 & 0 & 1 \\ 0 & 0 & 0 \end{bmatrix} \begin{bmatrix} x_1 \\ x_2 \\ x_3 \end{bmatrix} + \begin{bmatrix} 0 \\ 10 \\ 40 \end{bmatrix} [u] \quad ; \quad y = \begin{bmatrix} 1 & 0 & 0 \end{bmatrix} \begin{bmatrix} x_1 \\ x_2 \\ x_3 \end{bmatrix}$$

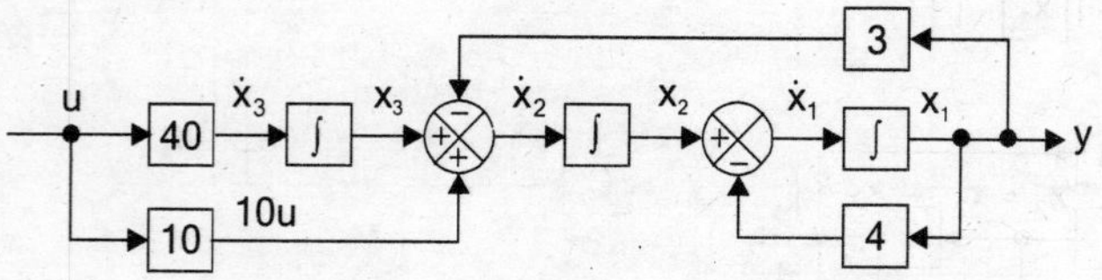

Fig 5.

Model 2

Given that, $\dfrac{Y(s)}{U(s)} = \dfrac{10(s+4)}{s(s+1)(s+3)}$

Let, $\dfrac{Y(s)}{U(s)} = \dfrac{X_1(s)}{U(s)} \cdot \dfrac{Y(s)}{X_1(s)} = \dfrac{10(s+4)}{s(s+1)(s+3)}$

Let, $\dfrac{X_1(s)}{U(s)} = \dfrac{1}{s(s+1)(s+3)}$ and $\dfrac{Y(s)}{X_1(s)} = 10(s+4)$

$$\dfrac{X_1(s)}{U(s)} = \dfrac{1}{s(s+1)(s+3)} = \dfrac{1}{s(s^2+4s+3)} \quad ; \quad \therefore \dfrac{X_1(s)}{U(s)} = \dfrac{1}{s^3+4s^2+3s} \qquad \qquad .....(1)$$

On cross multiplying the equation (1) we get,

$$X_1(s)[s^3 + 4s^2 + 3s] = U(s)$$

$$\therefore s^3 X_1(s) + 4s^2 X_1(s) + 3s X_1(s) = U(s) \qquad \qquad .....(2)$$

On taking inverse Laplace transform of equation (2) we get,

$$\dddot{x}_1 + 4\ddot{x}_1 + 3\dot{x}_1 = u \qquad \qquad .....(3)$$

Let the state variables be x_1, x_2 and x_3 ; where $x_2 = \dot{x}_1$ and $x_3 = \ddot{x}_1$.

Put $\dot{x}_1 = x_2$, $\ddot{x}_1 = x_3$ and $\dddot{x}_1 = \dot{x}_3$ in equation(3)

$$\therefore \dot{x}_3 + 4x_3 + 3x_2 = u \quad (or) \quad \dot{x}_3 = -3x_2 - 4x_3 + u$$

The state equations are

$$\dot{x}_1 = x_2 \quad ; \quad \dot{x}_2 = x_3 \quad ; \quad \dot{x}_3 = -3x_2 - 4x_3 + u$$

Consider the second part of transfer function,

$$\dfrac{Y(s)}{X_1(s)} = 10(s+4) = 10s + 40 \qquad \qquad .....(4)$$

On cross multiplying equation (4) we get,

$$Y(s) = 10s\, X_1(s) + 40\, X_1(s) \qquad\qquad .....(6.11.5)$$

On taking inverse Laplace transform of equation (5) we get,

$$y = 10\dot{x}_1 + 40x_1$$

Put $\dot{x}_1 = x_2$, $\qquad\qquad \therefore\ y = 10x_2 + 40x_1 = 40x_1 + 10x_2$

Here, $y = 40x_1 + 10x_2$ is the output equation. The state model in the matrix form is shown below. The block diagram representation of this state model is shown in fig 6.

$$\begin{bmatrix} \dot{x}_1 \\ \dot{x}_2 \\ \dot{x}_3 \end{bmatrix} = \begin{bmatrix} 0 & 1 & 0 \\ 0 & 0 & 1 \\ 0 & -3 & -4 \end{bmatrix} \begin{bmatrix} x_1 \\ x_2 \\ x_3 \end{bmatrix} + \begin{bmatrix} 0 \\ 0 \\ 1 \end{bmatrix}[u] \quad;\quad y = \begin{bmatrix} 40 & 10 & 0 \end{bmatrix} \begin{bmatrix} x_1 \\ x_2 \\ x_3 \end{bmatrix}$$

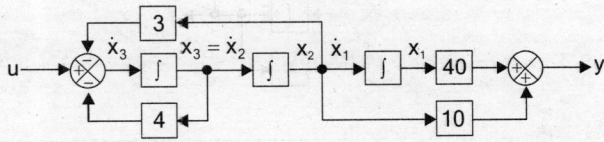

Fig 6.

Model 3

$$\frac{Y(s)}{U(s)} = \frac{10(s+4)}{s(s+1)(s+3)}$$

By partial fraction expansion $Y(s)/U(s)$ can be expressed as,

$$\frac{Y(s)}{U(s)} = \frac{10(s+4)}{s(s+1)(s+3)} = \frac{A}{s} + \frac{B}{s+1} + \frac{C}{s+3}$$

$$A = \frac{10(s+4)}{s(s+1)(s+3)}\bigg|_{s=0} = \frac{10 \times 4}{1 \times 3} = \frac{40}{3}$$

$$B = \frac{10(s+4)}{s(s+3)}\bigg|_{s=-1} = \frac{10(-1+4)}{-1(-1+3)} = \frac{10 \times 3}{-1 \times 2} = -15$$

$$C = \frac{10(s+4)}{s(s+1)}\bigg|_{s=-3} = \frac{10(-3+4)}{-3(-3+1)} = \frac{10 \times 1}{-3 \times (-2)} = \frac{5}{3}$$

$$\frac{Y(s)}{U(s)} = \frac{40/3}{s} - \frac{15}{s+1} + \frac{5/3}{s+3}$$

$$\therefore\ \frac{Y(s)}{U(s)} = \frac{40/3}{s} - \frac{15}{s(1+1/s)} + \frac{5/3}{s(1+3/s)} \qquad\qquad .....(6)$$

The equation (6) can be rearranged as shown below

$$\therefore\ Y(s) = \left[\frac{1}{s} \times \frac{40}{3} \right] U(s) - \left[\frac{1/s}{1 + \frac{1}{s} \times 1} \times 15 \right] U(s) + \left[\frac{1/s}{1 + \frac{1}{s} \times 3} \times \frac{5}{3} \right] U(s) \qquad .....(7)$$

The block diagram of the equation (7) is shown in fig 7.

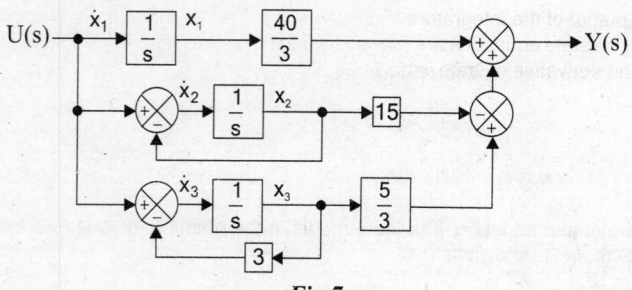

Fig 7.

Assign state variables at the output of the integrator as shown in fig 7. At the input of the integrator, the first derivative of the state variables will be available. The state equations are obtained by adding incoming signals to the integrator and equating to the corresponding first derivative of the state variable.

The state equations are

$$\dot{x}_1 = u$$
$$\dot{x}_2 = -x_2 + u$$
$$\dot{x}_3 = -3x_3 + u$$

The output equation is $y = \dfrac{40}{3}x_1 - 15x_2 + \dfrac{5}{3}x_3$

The state model in the matrix form is shown below. The fig 7 is the block diagram representation of this state model.

$$\begin{bmatrix} \dot{x}_1 \\ \dot{x}_2 \\ \dot{x}_3 \end{bmatrix} = \begin{bmatrix} 0 & 0 & 0 \\ 0 & -1 & 0 \\ 0 & 0 & -3 \end{bmatrix} \begin{bmatrix} x_1 \\ x_2 \\ x_3 \end{bmatrix} + \begin{bmatrix} 1 \\ 1 \\ 1 \end{bmatrix} [u] \quad ; \quad y = \begin{bmatrix} \dfrac{40}{3} & -15 & \dfrac{5}{3} \end{bmatrix} \begin{bmatrix} x_1 \\ x_2 \\ x_3 \end{bmatrix}$$

EXAMPLE 6.12

Determine the canonical state model of the system, whose transfer function is $T(s) = 2(s+5)/[(s+2)(s+3)(s+4)]$.

SOLUTION

By partial fraction expansion,

$$\frac{Y(s)}{U(s)} = \frac{2(s+5)}{(s+2)(s+3)(s+4)} = \frac{A}{s+2} + \frac{B}{s+3} + \frac{C}{s+4}$$

$$A = \frac{2(s+5)}{(s+3)(s+4)}\bigg|_{s=-2} = \frac{2(-2+5)}{(-2+3)(-2+4)} = \frac{2\times 3}{1\times 2} = 3$$

$$B = \frac{2(s+5)}{(s+2)(s+4)}\bigg|_{s=-3} = \frac{2(-3+5)}{(-3+2)(-3+4)} = \frac{2\times 2}{-1\times 1} = -4$$

$$C = \frac{2(s+5)}{(s+2)(s+3)}\bigg|_{s=-4} = \frac{2(-4+5)}{(-4+2)(-4+3)} = \frac{2\times 1}{-2\times(-1)} = 1$$

$$\therefore \frac{Y(s)}{U(s)} = \frac{3}{s+2} - \frac{4}{s+3} + \frac{1}{s+4} \qquad \qquad .....(1)$$

The equation (6.12.1) can be rearranged as shown below

$$\frac{Y(s)}{U(s)} = \frac{3}{s+2} - \frac{4}{s+3} + \frac{1}{s+4}$$

$$\therefore Y(s) = \left[\frac{\dfrac{1}{s}}{1+\dfrac{1}{s}\times 2}\times 3\right]U(s) - \left[\frac{\dfrac{1}{s}}{1+\dfrac{1}{s}\times 3}\times 4\right]U(s) + \left[\frac{\dfrac{1}{s}}{1+\dfrac{1}{s}\times 4}\right]U(s) \qquad ......(2)$$

The equation (6.12.2) can be represented by the block diagram in fig 1.

Assign state variables at the output of the integrators as shown in fig 1. At the input of the integrators we have first derivative of the state variables. The state equations are formed by adding all the incoming signals to the integrator and equating to the corresponding first derivative of state variable.

The state equations are

$$\dot{x}_1 = -2x_1 + u \quad ; \quad \dot{x}_2 = -3x_2 = u \quad ; \quad \dot{x}_3 = -4x_3 + u$$

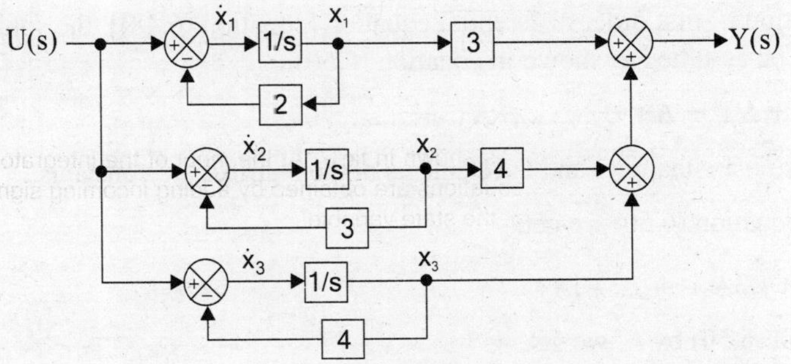

Fig 1.

The output equation is, $y = 3x_1 - 4x_2 + x_3$

The state model in matrix form is given below

$$\begin{bmatrix} \dot{x}_1 \\ \dot{x}_2 \\ \dot{x}_3 \end{bmatrix} = \begin{bmatrix} -2 & 0 & 0 \\ 0 & -3 & 0 \\ 0 & 0 & -4 \end{bmatrix} \begin{bmatrix} x_1 \\ x_2 \\ x_3 \end{bmatrix} [u] \quad ; \quad y = \begin{bmatrix} 3 & -4 & 1 \end{bmatrix} \begin{bmatrix} x_1 \\ x_2 \\ x_3 \end{bmatrix}$$

6.8 SOLUTION OF STATE EQUATIONS

SOLUTION OF HOMOGENEOUS STATE EQUATIONS (Solution of state equations without input or excitation)

Consider a first order differential equation, with initial condition, $x(0) = x_0$.

$$\frac{dx}{dt} = ax \quad ; \quad x(0) = x_0 \qquad \qquad .....(6.43)$$

On rearranging equation (6.43) we get, $\dfrac{dx}{x} = a \, dt$ .....(6.44)

On integrating equation (6.44) we get,

$$\log x = at + C$$

$$\therefore \; x = e^{(at + C)} = e^{at}.e^{C} \qquad \qquad .....(6.45)$$

When $t = 0$, from equation (6.45) we get, $x = x(0) = e^{C}$

Given that, $x(0) = x_0 \; ; \; \therefore e^{C} = x_0$

On substituting the initial condition in equation (6.46), we get the solution of first order differential equation as

$$x = e^{at} x_0 \qquad \qquad .....(6.46)$$

We know that, $\quad e^{x} = \left[1 + x + \dfrac{1}{2!}x^2 + \ldots\ldots + \dfrac{1}{n!}x^n + \ldots\ldots \right] \qquad .....(6.47)$

From equations (6.46) and (6.47) we get,

$$x = e^{at} X_0 = \left(1 + at + \frac{1}{2!} a^2 t^2 + \frac{1}{3!} a^3 t^3 + \cdots + \frac{1}{i!} a^i t^i + \cdots\right) x_0 \qquad \text{.....(6.48)}$$

Consider the state equations without input vector, (i.e., homogeneous state equation)

$$\dot{X}(t) = A\, X(t) \quad ; \quad X(0) = X_0 \qquad \text{.....(6.49)}$$

where, $X(0)$ is the initial condition vector.

By analogy of the solution of first order differential equation [equation (6.48)], the solution of the matrix or vector equation can be assumed as shown in equation (6.50)

$$X(t) = A_0 + A_1 t + A_2 t^2 + A_3 t^3 + \cdots + A_i t^i + \cdots \qquad \text{.....(6.50)}$$

where, $A_0, A_1, A_2, \ldots A_i \ldots$ are matrices and the elements of the matrices are constants.

On differentiating the equation (6.50) we get,

$$\dot{X}(t) = A_1 + 2A_2 t + 3A_3 t^2 + \cdots + iA_i t^{i-1} + \cdots \qquad \text{.....(6.51)}$$

On multiplying equation (6.50) by A, we get,

$$Ax(t) = A\left[A_0 + A_1 t + A_2 t^2 + A_3 t^3 + \cdots + A_i t^i + \cdots\right] \qquad \text{.....(6.52)}$$

From equation (6.49), we know that $\dot{X}(t) = A\, X(t)$. Therefore we can equate the coefficients of equal powers of t in equations (6.51) and (6.52) as shown below.

On equating constants we get,

$$A_1 = A\, A_0$$

On equating coefficients of t we get,

$$2A_2 = A\, A_1$$

$$\therefore\ A_2 = \frac{1}{2} A\, A_1$$

Put $A_1 = A\, A_0$

$$\therefore\ A_2 = \frac{1}{2} A\, A A_0$$

$$A_2 = \frac{1}{2} A^2 A_0$$

On equating coefficients of t^2 we get,

$$3A_3 = A\, A_2$$

$$\therefore\ A_3 = \frac{1}{3} A\, A_2$$

Put $A_2 = \frac{1}{2} A^2 A_0$

$$\therefore\ A_3 = \frac{1}{3} A \times \frac{1}{2} A^2 A_0$$

$$A_3 = \frac{1}{3!} A^3 A_0$$

Similarly, on equating coefficient of t^i we get,

$$A_i = \frac{1}{i!} A^i A_0$$

In the above analysis, the matrices A_1, A_2, A_3, etc., are expressed in terms of A and A_0. Hence replace the matrices $A_1, A_2, A_3 \ldots A_i$ in the assumed solution of $X(t)$ [i.e., equation (6.50)] by the equivalent terms obtained above.

$$\therefore\ X(t) = A_0 + A A_0 t + \frac{1}{2!} A^2 A_0 t^2 + \frac{1}{3!} A^3 A_0 t^2 + \cdots + \frac{1}{i!} A^i A_0 t^i + \cdots$$

$$= \left[1 + At + \frac{1}{2!} A^2 t^2 + \frac{1}{3!} A^3 t^3 + \cdots + \frac{1}{i!} A^i t^i + \cdots\right] A_0 \qquad \text{.....(6.53)}$$

where I is the unit matrix.

It is given that, when $t = 0$, $\quad X(t) = X(0) = X_0 \qquad$ .....(6.54)

From equation (6.53) when t = 0, we get

$$\mathbf{X}(t)\big|_{t=0} = \mathbf{X}(0) = \mathbf{A}_0 \qquad \qquad .....(6.55)$$

From equations (6.54) and (6.55) we get,

$$\mathbf{A}_0 = \mathbf{X}_0 \qquad \qquad .....(6.56)$$

On substituting for $\mathbf{A}_0$ from equation (6.56) in equation (6.53) we get,

$$\mathbf{X}(t) = \left[1 + \mathbf{A}t + \frac{1}{2!}\,\mathbf{A}^2 t^2 + \frac{1}{3!}\,\mathbf{A}^3 t^3 + ... + \frac{1}{i!}\,\mathbf{A}^i t^i + ...\right]\mathbf{X}_0 \qquad .....(6.57)$$

Each of the term inside the brackets is an n × n matrix. Because of the similarity of the entity inside the bracket with a scalar exponential of e^{at}, we call it a matrix exponential, which may be written as,

$$e^{\mathbf{A}t} = 1 + \mathbf{A}t + \frac{1}{2!}\,\mathbf{A}^2 t^2 + \frac{1}{3!}\,\mathbf{A}^3 t^3 + ... + \frac{1}{i!}\,\mathbf{A}^i t^i + ... \qquad .....(6.58)$$

Hence the solution of the state equation is

$$\mathbf{X}(t) = e^{\mathbf{A}t}\,\mathbf{X}_0 \qquad \qquad .....(6.59)$$

The matrix $e^{\mathbf{A}t}$ is called state transition matrix and denoted by $\phi(t)$. From the solution of the state equations it is observed that the initial state $\mathbf{X}_0$ at t = 0, is driven to state $\mathbf{X}(t)$ at time t by state transition matrix.

SOLUTION OF NON-HOMOGENEOUS STATE EQUATIONS (Solution of state equations with input or excitation)

The state equation of nth order system is given by

$$\dot{\mathbf{X}}(t) = \mathbf{A}\,\mathbf{X}(t) + \mathbf{B}\,\mathbf{U}(t) \quad ; \quad \mathbf{X}(0) = \mathbf{X}_0 \qquad .....(6.60)$$

where $\mathbf{X}_0$ is initial condition vector.

The state equation of equation (6.60) can be rearranged as shown below,

$$\dot{\mathbf{X}}(t) - \mathbf{A}\,\mathbf{X}(t) = \mathbf{B}\,\mathbf{U}(t) \qquad \qquad .....(6.61)$$

Premultiply both sides of equation (6.61) by $e^{-\mathbf{A}t}$

$$e^{-\mathbf{A}t}\left[\dot{\mathbf{X}}(t) - \mathbf{A}\,\mathbf{X}(t)\right] = e^{-\mathbf{A}t}\,\mathbf{B}\,\mathbf{U}(t)$$

$$e^{-\mathbf{A}t}\,\dot{\mathbf{X}}(t) + e^{-\mathbf{A}t}(-\mathbf{A})\,\mathbf{X}(t) = e^{-\mathbf{A}t}\,\mathbf{B}\,\mathbf{U}(t) \qquad .....(6.62)$$

Consider the differential of $e^{-\mathbf{A}t}\,\mathbf{X}(t)$

$$\frac{d}{dt}\left(e^{-\mathbf{A}t}\,\mathbf{X}(t)\right) = e^{-\mathbf{A}t}\,\dot{\mathbf{X}}(t) + e^{-\mathbf{A}t}(-\mathbf{A})\,\mathbf{X}(t) \qquad .....(6.63)$$

On comparing equations (6.62) and (6.63) we can write,

$$\frac{d}{dt}\left(e^{-\mathbf{A}t}\,\mathbf{X}(t)\right) = e^{-\mathbf{A}t}\,\mathbf{B}\,\mathbf{U}(t)$$

$$d\left(e^{-\mathbf{A}t}\,\mathbf{X}(t)\right) = e^{-\mathbf{A}t}\,\mathbf{B}\,\mathbf{U}(t)\,dt \qquad .....(6.64)$$

On integrating the equation (6.64) between limits 0 to t we get,

$$e^{-\mathbf{A}t}\,\mathbf{X}(t) = \mathbf{X}_0 + \int_0^t e^{-\mathbf{A}\tau}\,\mathbf{B}\,\mathbf{U}(\tau)\,d\tau \qquad .....(6.65)$$

where $\mathbf{X}_0$ = Initial condition vector = Integral constant

τ = Dummy variable substituted for t.

Premultiply both sides of equation (6.65) by e^{At},

$$e^{At} e^{-At} \mathbf{X}(t) = e^{At} \mathbf{X}_0 + e^{At} \int_0^t e^{-A\tau} \mathbf{B} \, \mathbf{U}(\tau) \, dt$$

$$\mathbf{X}(t) = e^{At} \mathbf{X}_0 + e^{At} \int_0^t e^{-A\tau} \mathbf{B} \, \mathbf{U}(\tau) \, dt \qquad \qquad .....(6.66)$$

The term e^{At} independent of the integral variable τ, and so e^{At} can be brought inside the integral function.

$$\therefore \quad \mathbf{X}(t) = e^{At} \mathbf{X}_0 + \int_0^t e^{At} e^{-A\tau} \mathbf{B} \, \mathbf{U}(\tau) \, dt$$

$$\mathbf{X}(t) = e^{At} \mathbf{X}_0 + \int_0^t e^{A(t-\tau)} \mathbf{B} \, \mathbf{U}(\tau) \, dt \qquad \qquad .....(6.67)$$

The equation (6.67) is the solution of state equation, when the initial conditions are known at $t = 0$. If initial conditions are known at $t = t_0$ then the solution of state equation is given by equation (6.68).

$$\mathbf{X}(t) = e^{A(t-t_0)} \mathbf{X}(t_0) + \int_{t_0}^t e^{A(t-\tau)} \mathbf{B} \, \mathbf{U}(\tau) \, d\tau \qquad \qquad .....(6.68)$$

The state transition matrix e^{At} is denoted by the symbol $\phi(t)$, i.e., $\phi(t) = e^{At}$

Hence, $e^{A(t-t_0)}$ can be expressed as, $e^{A(t-t_0)} = \phi(t - t_0)$ $\qquad \qquad .....(6.69)$

and, $e^{A(t-t)}$ can be expressed as, $e^{A(t-\tau)} = \phi(t - \tau)$ $\qquad \qquad .....(6.70)$

The equations (6.67) and (6.68) can also be expressed as

$$\mathbf{X}(t) = \phi(t) \, \mathbf{X}(0) + \int_0^t \phi(t-\tau) \mathbf{B} \, \mathbf{U}(\tau) \, d\tau \text{ if the initial conditions are known at } t = 0 \qquad .....(6.71)$$

$$\mathbf{X}(t) = \phi(t-t_0) \, \mathbf{X}(t_0) + \int_{t_0}^t \phi(t-\tau) \mathbf{B} \, \mathbf{U}(\tau) \, d\tau \text{ if the initial conditions are known at } t = t_0 \qquad .....(6.72)$$

PROPERTIES OF STATE TRANSITION MATRIX

1. $\phi(0) = e^{A \times 0} = \mathbf{I}$ (unit matrix)

2. $\phi(t) = e^{At} = (e^{-At})^{-1} = [\phi(-t)]^{-1}$

    or $\quad \phi^{-1}(t) = \phi(-t)$

3. $\phi(t_1 + t_2) = e^{A(t_1 + t_2)} = (e^{At_1})(e^{At_2}) = \phi(t_1) \, \phi(t_2)$

COMPUTATION OF STATE TRANSITION MATRIX

The state transition matrix (e^{At}) can be computed by any one of the following two methods.

Method 1 : *Computation of e^{At} using matrix exponential.*

Method 2 : *Computation of e^{At} using Laplace transform.*

Computation of state transition matrix using matrix exponential

In this method, the e^{At} is computed using the matrix exponential of equation (6.58), which is also given below.

$$e^{At} = I + At + \frac{1}{2!} A^2 t^2 + \frac{1}{3!} A^3 t^3 + \dots\dots + \frac{1}{i!} A^i t^i + \dots\dots$$

where, e^{At} = State transition matrix of order n × n

     A = System matrix of order n × n

     I = Unit matrix of order n × n.

The disadvantage in this method is that each term of e^{At} will be an infinite series and the convergence of the infinite series are obtained by trial and error.

Computation of State Transition Matrix by Laplace Transform Method

Consider the state equation without input vector, ...(6.73)

On taking Laplace transform of equation (6.73) we get,

$$s\,X(s) - X(0) = A\,X(s)$$

$$s\,X(s) - A\,X(s) = X(0)$$

$$s\,I\,X(s) - A\,X(s) = X(0), \qquad \text{where, } I \text{ is a unit matrix.}$$

$$(sI{-}A)\,X(s) = X(0)$$

Premultiply both sides by $(sI{-}A)^{-1}$

$$X(s) = (sI{-}A)^{-1}\,X(0)$$

On taking inverse Laplace transform we get,

$$X(t) = \mathcal{L}^{-1}\,[(sI{-}A)^{-}\,X(0)]$$

$$X(t) = \mathcal{L}^{-1}\,[(s\,I - A)^{-1}]\,X(0) \qquad\qquad .....(6.74)$$

On comparing equation (6.74) with the solution of state equation, $X(t) = e^{At}\,X(0)$ we get,

$$e^{At} = \mathcal{L}^{-1}\,[(sI - A)^{-1}\,] \qquad \text{or} \qquad \mathcal{L}[e^{At}] = (sI - A)^{-1} \qquad .....(6.75)$$

We know that, $e^{At} = \phi(t)$,

$$\therefore \; \mathcal{L}[e^{At}] = \mathcal{L}[\phi(t)] = \phi(s) \qquad\qquad .....(6.76)$$

where, $\phi(s) = (sI - A)^{-1}$ and it is called resolvant matrix.

From the system matrix, A the resolvant matrix, $\phi(s)$ can be computed. By taking inverse Laplace transform of resolvant matrix, the state transition matrix is computed, from which the solution of state equation is obtained.

The solution of state equation is given by,

$$X(t) = e^{At}\,X(0)$$

$$\therefore \; X(t) = \mathcal{L}^{-1}\,[\phi(s)]\,X(0) \qquad\qquad .....(6.77)$$

where, $\phi(s) = (sI - A)^{-1}$

Consider the state equation with forcing function (input or excitation),

$$\dot{X} = AX + BU \qquad\qquad .....(6.78)$$

On taking Laplace transform of equation (6.78) we get,

$$s\,\mathbf{X}(s) - \mathbf{X}(0) = \mathbf{A}\,\mathbf{X}(s) + \mathbf{B}\,\mathbf{U}(s)$$

$$s\mathbf{I}\,\mathbf{X}(s) - \mathbf{A}\,\mathbf{X}(s) = \mathbf{X}(0) + \mathbf{B}\,\mathbf{U}(s), \qquad \text{where } \mathbf{I} \text{ is the unit matrix.}$$

$$(s\mathbf{I}-\mathbf{A})\,\mathbf{X}(s) = \mathbf{X}(0) + \mathbf{B}\,\mathbf{U}(s) \qquad\qquad .....(6.79)$$

Premultiply the equation (6.79) by $(s\mathbf{I} - \mathbf{A})^{-1}$

$$\therefore \;\; \mathbf{X}(s) = (s\mathbf{I} - \mathbf{A})^{-1}\,\mathbf{X}(0) + (s\mathbf{I} - \mathbf{A})^{-1}\,\mathbf{B}\,\mathbf{U}(s)$$

$$= \phi(s)\,\mathbf{X}(0) + \phi(s)\,\mathbf{B}\,\mathbf{U}(s) \qquad\qquad .....(6.80)$$

On taking inverse Laplace transform of equation (6.80) we get,

$$\mathbf{X}(t) = \phi(t)\,\mathbf{X}(0) + \mathcal{L}^{-1}\left[\phi(s)\,\mathbf{B}\,\mathbf{U}(s)\right] \qquad\qquad .....(6.81)$$

The equation (6.81) is the solution of state equation with forcing function.

EXAMPLE 6.13

Consider the matrix A. Compute e^{At} by two methods.

$$A = \begin{bmatrix} 0 & 1 \\ -2 & -3 \end{bmatrix}$$

SOLUTION

Method 1

$$e^{At} = \left[I + At + \frac{1}{2!}A^2 t^2 + \frac{1}{3!}A^3 t^3 + \dots\dots \right]$$

$$A = \begin{bmatrix} 0 & 1 \\ -2 & -3 \end{bmatrix}$$

$$A^2 = A.A = \begin{bmatrix} 0 & 1 \\ -2 & -3 \end{bmatrix}\begin{bmatrix} 0 & 1 \\ -2 & -3 \end{bmatrix} = \begin{bmatrix} -2 & -3 \\ 6 & 7 \end{bmatrix}$$

$$A^3 = A^2.A = \begin{bmatrix} -2 & -3 \\ 6 & 7 \end{bmatrix}\begin{bmatrix} 0 & 1 \\ -2 & -3 \end{bmatrix} = \begin{bmatrix} 6 & 7 \\ -14 & -15 \end{bmatrix}$$

$$A^4 = A^3.A = \begin{bmatrix} 6 & 7 \\ -14 & -15 \end{bmatrix}\begin{bmatrix} 0 & 1 \\ -2 & -3 \end{bmatrix} = \begin{bmatrix} -14 & -15 \\ 30 & 31 \end{bmatrix}$$

$$e^{At} = I + At + \frac{1}{2!}A^2 t^2 + \frac{1}{3!}A^3 t^3 + \frac{1}{2!}A^4 t^4 + \dots\dots$$

$$= \begin{bmatrix} 1 & 0 \\ 0 & 1 \end{bmatrix} + \begin{bmatrix} 0 & 1 \\ -2 & -3 \end{bmatrix}t + \frac{1}{2}\begin{bmatrix} -2 & -3 \\ 6 & 7 \end{bmatrix}t^2 + \frac{1}{6}\begin{bmatrix} 6 & 7 \\ -14 & -15 \end{bmatrix}t^3 + \frac{1}{24}\begin{bmatrix} -14 & -15 \\ 30 & 31 \end{bmatrix}t^4 + \dots\dots$$

$$= \begin{bmatrix} 1 & 0 \\ 0 & 1 \end{bmatrix} + \begin{bmatrix} 0 & 1 \\ -2t & -3t \end{bmatrix} + \begin{bmatrix} -t^2 & -\frac{3}{2}t^2 \\ 3t^2 & \frac{7}{2}t^2 \end{bmatrix} + \begin{bmatrix} t^3 & \frac{7}{6}t^3 \\ -\frac{7}{3}t^3 & -\frac{5}{2}t^3 \end{bmatrix} + \begin{bmatrix} -\frac{7}{12}t^4 & \frac{5}{8}t^4 \\ \frac{5}{4}t^4 & \frac{31}{24}t^4 \end{bmatrix}$$

$$= \begin{bmatrix} 1 - t^2 + t^3 - \frac{7}{12}t^4 + \dots\dots & t - \frac{3}{2}t^2 + \frac{7}{6}t^3 - \frac{5}{8}t^4 + \dots\dots \\ -2t + 3t^2 - \frac{7}{3}t^3 + \frac{5}{4}t^4 + \dots\dots & 1 - 3t + \frac{7}{2}t^2 - \frac{5}{2}t^3 + \frac{31}{24}t^4 + \dots\dots \end{bmatrix}$$

The each term in the matrix is an expansion of e^{at}. The convergence of series is obtained by trial and error. Consider the expansion of e^{-t} and e^{-2t}.

$$e^{-t} = 1 - t + \frac{1}{2!}t^2 - \frac{1}{3!}t^3 + \frac{1}{4!}t^4 \dots = 1 - t + \frac{1}{2}t^2 - \frac{1}{6}t^3 + \frac{1}{24}t^4 \dots$$

$$e^{-2t} = 1 - 2t + \frac{1}{2!}2^2t^2 - \frac{1}{3!}2^3t^3 + \frac{1}{4!}2^4t^4 \dots = 1 - 2t + 2t^2 - \frac{4}{3}t^3 + \frac{2}{3}t^4 \dots$$

$$2e^{-t} - e^{-2t} = 2 - 2t + t^2 - \frac{1}{3}t^3 + \frac{1}{12}t^4 \dots - 1 + 2t - 2t^2 + \frac{4}{3}t^3 - \frac{2}{3}t^4 + \dots$$

$$= 1 - t^2 + t^3 - \frac{7}{12}t^4 + \dots$$

$$e^{-t} - e^{-2t} = 1 - t + \frac{1}{2}t^2 - \frac{1}{6}t^3 + \frac{1}{24}t^4 \dots - 1 + 2t - 2t^2 + \frac{4}{3}t^3 - \frac{2}{3}t^4 + \dots$$

$$= t - \frac{3}{2}t^2 + \frac{7}{6}t^3 - \frac{5}{8}t^4 + \dots$$

$$-2e^{-t} + 2e^{-2t} = -2 + 2t - t^2 + \frac{1}{3}t^3 - \frac{1}{12}t^4 + \dots + 2 - 4t + 4t^2 - \frac{8}{3}t^3 + \frac{4}{3}t^4 \dots$$

$$= -2t + 3t^2 - \frac{7}{3}t^3 + \frac{5}{4}t^4 \dots$$

$$-e^{-t} + 2e^{-2t} = -1 + t - \frac{1}{2}t^2 + \frac{1}{6}t^3 - \frac{1}{24}t^4 + \dots + 2 - 4t + 4t^2 - \frac{8}{3}t^3 + \frac{4}{3}t^4 \dots$$

$$= 1 - 3t + \frac{7}{2}t^2 - \frac{5}{2}t^3 + \frac{31}{24}t^4 + \dots$$

$$\therefore e^{At} = \begin{bmatrix} 2e^{-t} - e^{-2t} & e^{-t} - e^{-2t} \\ -2e^{-t} + 2e^{-2t} & -e^{-t} + 2e^{-2t} \end{bmatrix}$$

Method 2

$$A = \begin{bmatrix} 0 & 1 \\ -2 & -3 \end{bmatrix}$$

$$e^{At} = \phi(t) = \mathcal{L}^{-1}[(sI - A)^{-1}]$$

$$sI - A = s\begin{bmatrix} 1 & 0 \\ 0 & 1 \end{bmatrix} - \begin{bmatrix} 0 & 1 \\ -2 & -3 \end{bmatrix} = \begin{bmatrix} s & 0 \\ 0 & s \end{bmatrix} - \begin{bmatrix} 0 & 1 \\ -2 & -3 \end{bmatrix} = \begin{bmatrix} s & -1 \\ 2 & s+3 \end{bmatrix}$$

Let, $\Delta = |sI - A|$ = determinant of $(sI - A)$

$$\therefore \Delta = |sI - A| = \begin{vmatrix} s & -1 \\ 2 & s+3 \end{vmatrix} = s(s+3) + 2 = s^2 + 3s + 2 = (s+2)(s+1)$$

$$\phi(s) = [sI - A]^{-1} = \frac{[\text{Cofactor of } (sI - A)]^T}{\text{Determinant of } (sI - A)} = \frac{[\text{Cofactor of } (sI - A)]^T}{\Delta}$$

$$\therefore \phi(s) = \frac{1}{(s+1)(s+2)}\begin{bmatrix} s+3 & 1 \\ -2 & s \end{bmatrix}$$

$$\phi(s) = \begin{bmatrix} \dfrac{s+3}{(s+1)(s+2)} & \dfrac{1}{(s+1)(s+2)} \\ \dfrac{-2}{(s+1)(s+2)} & \dfrac{s}{(s+1)(s+2)} \end{bmatrix}$$

By partial fraction expansion, $\phi(s)$ can be written as,

$$\phi(s) = \begin{bmatrix} \dfrac{A_1}{s+1} + \dfrac{B_1}{s+2} & \dfrac{A_2}{s+1} + \dfrac{B_2}{s+2} \\ \dfrac{A_3}{s+1} + \dfrac{B_3}{s+2} & \dfrac{A_4}{s+1} + \dfrac{B_4}{s+2} \end{bmatrix}$$

$$\frac{s+3}{(s+1)(s+2)} = \frac{A_1}{s+1} + \frac{B_1}{s+1}$$

$$A_1 = \frac{s+3}{s+2}\bigg|_{s=-1} = 2$$

$$B_1 = \frac{s+3}{s+1}\bigg|_{s=-2} = -1$$

$$\frac{1}{(s+1)(s+2)} = \frac{A_2}{s+1} + \frac{B_2}{s+2}$$

$$A_2 = \frac{1}{s+2}\bigg|_{s=-1} = 1$$

$$B_2 = \frac{1}{s+1}\bigg|_{s=-2} = -1$$

$$\frac{-2}{(s+1)(s+2)} = \frac{A_3}{s+1} + \frac{B_3}{s+2}$$

$$A_3 = \frac{-2}{s+2}\bigg|_{s=-1} = -2$$

$$B_3 = \frac{-2}{s+1}\bigg|_{s=-2} = 2$$

$$\frac{s}{(s+1)(s+2)} = \frac{A_4}{s+1} + \frac{B_4}{s+2}$$

$$A_4 = \frac{s}{s+2}\bigg|_{s=-1} = -1$$

$$B_4 = \frac{s}{s+1}\bigg|_{s=-2} = 2$$

$$\therefore \ \phi(s) = \begin{bmatrix} \dfrac{2}{s+1} - \dfrac{1}{s+2} & \dfrac{1}{s+1} - \dfrac{1}{s+2} \\ \dfrac{-2}{s+1} + \dfrac{2}{s+2} & \dfrac{-1}{s+1} - \dfrac{2}{s+2} \end{bmatrix}$$

On taking inverse Laplace transform of $\phi(s)$ we get $\phi(t)$, where $\phi(t) = e^{At}$

$$\therefore \ e^{At} = \phi(t) = \begin{bmatrix} 2e^{-t} - e^{-2t} & e^{-t} - e^{-2t} \\ -2e^{-t} + 2e^{-2t} & -e^{-t} + 2e^{-2t} \end{bmatrix}$$

It is observed that the results of both the methods are same.

EXAMPLE 6.14

Given that, $A_1 = \begin{bmatrix} \sigma & 0 \\ 0 & \sigma \end{bmatrix}$; $A_2 = \begin{bmatrix} 0 & \omega \\ -\omega & 0 \end{bmatrix}$; $A = \begin{bmatrix} \sigma & \omega \\ -\omega & \sigma \end{bmatrix}$ compute e^{At}.

SOLUTION

Here $A = A_1 + A_2$

$$\therefore \ e^{At} = e^{(A_1 + A_2)t} = e^{A_1 t} e^{A_2 t}$$

$$sI - A_1 = s\begin{bmatrix} 1 & 0 \\ 0 & 1 \end{bmatrix} - \begin{bmatrix} \sigma & 0 \\ 0 & \sigma \end{bmatrix} = \begin{bmatrix} s & 0 \\ 0 & s \end{bmatrix} - \begin{bmatrix} \sigma & 0 \\ 0 & \sigma \end{bmatrix} = \begin{bmatrix} s-\sigma & 0 \\ 0 & s-\sigma \end{bmatrix}$$

$$\Delta_1 = |sI - A_1| = \begin{vmatrix} s-\sigma & 0 \\ 0 & s-\sigma \end{vmatrix} = (s-\sigma)^2$$

$$[sI - A_1]^{-1} = \frac{1}{\Delta_1}\begin{bmatrix} s-\sigma & 0 \\ 0 & s-\sigma \end{bmatrix} = \frac{1}{(s-\sigma)^2}\begin{bmatrix} s-\sigma & 0 \\ 0 & s-\sigma \end{bmatrix} = \begin{bmatrix} \dfrac{1}{s-\sigma} & 0 \\ 0 & \dfrac{1}{s-\sigma} \end{bmatrix}$$

$$e^{A_1 t} = \mathcal{L}^{-1}\left[(sI - A_1)^{-1}\right] = \begin{bmatrix} e^{-\sigma t} & 0 \\ 0 & e^{-\sigma t} \end{bmatrix}$$

$$sI - A_2 = s\begin{bmatrix} 1 & 0 \\ 0 & 1 \end{bmatrix} - \begin{bmatrix} 0 & \omega \\ -\omega & 0 \end{bmatrix} = \begin{bmatrix} s & 0 \\ 0 & s \end{bmatrix} - \begin{bmatrix} 0 & \omega \\ -\omega & 0 \end{bmatrix} = \begin{bmatrix} s & -\omega \\ \omega & s \end{bmatrix}$$

$$\Delta_2 = \left| sI - A_2 \right| = \begin{vmatrix} s & -\omega \\ \omega & s \end{vmatrix} = s^2 + \omega^2$$

$$[sI - A_2]^{-1} = \frac{1}{\Delta_2}\begin{bmatrix} s & \omega \\ -\omega & s \end{bmatrix} = \frac{1}{s^2+\omega^2}\begin{bmatrix} s & \omega \\ -\omega & s \end{bmatrix} = \begin{bmatrix} \dfrac{s}{s^2+\omega^2} & \dfrac{\omega}{s^2+\omega^2} \\ \dfrac{-\omega}{s^2+\omega^2} & \dfrac{s}{s^2+\omega^2} \end{bmatrix}$$

$$e^{A_2 t} = \mathcal{L}^{-1}[(sI - A_2)^{-1}] = \begin{bmatrix} \cos\omega t & \sin\omega t \\ -\sin\omega t & \cos\omega t \end{bmatrix}$$

$$e^{At} = e^{A_1 t} e^{A_2 t} = \begin{bmatrix} e^{-\sigma t} & 0 \\ 0 & e^{-\sigma t} \end{bmatrix}\begin{bmatrix} \cos\omega t & \sin\omega t \\ -\sin\omega t & \cos\omega t \end{bmatrix}$$

$$\therefore\ e^{At} = \begin{bmatrix} e^{-\sigma t}\cos\omega t & e^{-\sigma t}\sin\omega t \\ -e^{-\sigma t}\sin\omega t & e^{-\sigma t}\cos\omega t \end{bmatrix}$$

EXAMPLE 6.15

For a system represented by state equation $\dot{X}(t) = A\,X(t)$

The response is $X(t) = \begin{bmatrix} e^{-2t} \\ -2e^{-2t} \end{bmatrix}$ when $X(0) = \begin{bmatrix} 1 \\ -2 \end{bmatrix}$

and $X(t) = \begin{bmatrix} e^{-t} \\ -e^{-t} \end{bmatrix}$ when $X(0) = \begin{bmatrix} 1 \\ -1 \end{bmatrix}$

Determine the system matrix A and the state transition matrix.

SOLUTION

The solution of state equation is, $X(t) = e^{At} X(0)$ .....(1)

Premultiply the equation (1) by e^{-At}

$e^{-At} X(t) = e^{-At} e^{At} X(0)$

$\therefore\ e^{-At} X(t) = X(0)$ .....(2)

One of the response is $X(t) = \begin{bmatrix} e^{-2t} \\ -2e^{-2t} \end{bmatrix}$ and $X(0) = \begin{bmatrix} 1 \\ -2 \end{bmatrix}$

On substituting the response in equation (2) we get,

$$e^{-At}\begin{bmatrix} e^{-2t} \\ -2e^{-2t} \end{bmatrix} = \begin{bmatrix} 1 \\ -2 \end{bmatrix}$$ .....(3)

Let $e^{-At} = \begin{bmatrix} e_{11} & e_{12} \\ e_{21} & e_{22} \end{bmatrix}$ .....(4)

From equation (3) and (4) we can write

$$\begin{bmatrix} e_{11} & e_{12} \\ e_{21} & e_{22} \end{bmatrix}\begin{bmatrix} e^{-2t} \\ -2e^{-2t} \end{bmatrix} = \begin{bmatrix} 1 \\ -2 \end{bmatrix}$$ .....(5)

On multiplying the equation (5) we get the following two equations,

$e_{11}e^{-2t} - 2e_{12}e^{-2t} = 1$ .....(6)

$e_{21}e^{-2t} - 2e_{22}e^{-2t} = -2$ .....(7)

The second solution of state equation is $X(t) = \begin{bmatrix} e^{-t} \\ -e^{-t} \end{bmatrix}$ and $X(0) = \begin{bmatrix} 1 \\ -1 \end{bmatrix}$

On substituting this solution in equation (6.15.2) we get,

$$e^{-At} \begin{bmatrix} e^{-t} \\ -e^{-t} \end{bmatrix} = \begin{bmatrix} 1 \\ -1 \end{bmatrix} \qquad \qquad .....(8)$$

From equation (6.15.4) and (6.15.8) we can write,

$$\begin{bmatrix} e_{11} & e_{12} \\ e_{21} & e_{22} \end{bmatrix} \begin{bmatrix} e^{-t} \\ -e^{-t} \end{bmatrix} = \begin{bmatrix} 1 \\ -1 \end{bmatrix} \qquad \qquad .....(9)$$

On multiplying the equation (6.15.9) we get the following two equations,

$$e_{11}\, e^{-t} - e_{12}\, e^{-t} = 1 \qquad \qquad .....(10)$$
$$e_{21}\, e^{-t} - e_{22}\, e^{-t} = -1 \qquad \qquad .....(11)$$

Equation (10) $\times\, e^{-t} \Rightarrow e_{11} e^{-2t} - e_{12}\, e^{-2t} = e^{-t}$

Equation (6) $\times\, 1 \quad \Rightarrow e_{11} e^{-2t} - 2e_{12}\, e^{-2t} = 1$

$$\underline{\qquad (-) \qquad (+) \qquad\qquad (-) \qquad\qquad}$$

On subtracting, $\qquad\qquad e_{12} e^{-2t} = e^{-t} - 1 \qquad\qquad .....(12)$

From equation (6.15.12) we get

$$e_{12} = \frac{e^{-t}-1}{e^{-2t}} = \frac{e^{-t}}{e^{-2t}} - \frac{1}{e^{-2t}} = e^{t} - e^{-2t} \qquad\qquad .....(13)$$

From equation(6), $e_{11} = \dfrac{1 + 2e_{12}e^{-2t}}{e^{-2t}}$

Put $e_{12} = e^{t} - e^{2t}$, $\therefore e_{11} = \dfrac{1 + 2(e^{t} - e^{2t})e^{-2t}}{e^{-2t}} = \dfrac{1 + 2e^{-t} - 2}{e^{-2t}} = \dfrac{2e^{-t} - 1}{e^{-2t}}$

$$= \frac{2e^{-t}}{e^{-2t}} - \frac{1}{e^{-2t}} = 2e^{t} - e^{-2t}$$

Equation(11) $\times\, e^{-t} \Rightarrow e_{21} e^{-2t} - e_{22}\, e^{-2t} = -e^{-t}$

Equation(7) $\times\, 1 \quad \Rightarrow e_{21} e^{-2t} - 2e_{22}\, e^{-2t} = -2$

$$\underline{\qquad (-) \qquad (+) \qquad\qquad (+) \qquad\qquad}$$

On subtracting $\qquad\qquad e_{22} e^{-2t} = 2 - e^{-t} \qquad\qquad .....(14)$

From equation (14) we get,

$$e_{22} = \frac{-e^{-t}+2}{e^{-2t}} = \frac{-e^{-t}}{e^{-2t}} + \frac{2}{e^{-2t}} = -e^{t} + 2e^{2t}$$

From equation(11), $e_{21} = \dfrac{-1 + e_{22}e^{-t}}{e^{-t}}$

Put $e_{22} = -e^{t} + 2e^{2t}$, $\therefore e_{21} = \dfrac{-1 + (e^{-t} + 2e^{2t})}{e^{-t}}$

$$e_{21} = \frac{-1 - 1 + 2e^{t}}{e^{-t}} = \frac{-2}{e^{-t}} + \frac{2e^{t}}{e^{-t}} = -2e^{t} + 2e^{2t}$$

$$\therefore\ e^{-At} = \begin{bmatrix} e_{11} & e_{12} \\ e_{21} & e_{22} \end{bmatrix} = \begin{bmatrix} 2e^{t} - e^{2t} & e^{t} - e^{2t} \\ -2e^{t} + 2e^{2t} & -e^{t} + 2e^{2t} \end{bmatrix}$$

$$\therefore\ e^{At} = \begin{bmatrix} 2e^{-t} - e^{-2t} & e^{t} - e^{-2t} \\ -2e^{-t} + 2e^{-2t} & -e^{-t} + 2e^{-2t} \end{bmatrix}$$

e^{At} is the state transition matrix.

We know that, $\mathcal{L}[e^{At}] = \phi(s)$

where $\phi(s) = (sI - A)^{-1}$; $\phi(s)^{-1} = (sI - A)$ or $A = sI - \phi(s)^{-1}$

$$\phi(s) = \mathcal{L}[e^{At}] = \begin{bmatrix} \dfrac{2}{s+1} - \dfrac{1}{s+2} & \dfrac{1}{s+1} - \dfrac{1}{s+2} \\ \dfrac{-2}{s+1} + \dfrac{2}{s+2} & \dfrac{-1}{s+1} + \dfrac{2}{s+2} \end{bmatrix}$$

$$\begin{bmatrix} \dfrac{2(s+2) - (s+1)}{(s+1)(s+2)} & \dfrac{(s+2) - (s+1)}{(s+1)(s+2)} \\ \dfrac{-2(s+2) + 2(s+1)}{(s+1)(s+2)} & \dfrac{-(s+2) + 2(s+1)}{(s+1)(s+2)} \end{bmatrix} = \begin{bmatrix} \dfrac{s+3}{(s+1)(s+2)} & \dfrac{1}{(s+1)(s+2)} \\ \dfrac{-2}{(s+1)(s+2)} & \dfrac{s}{(s+1)(s+2)} \end{bmatrix}$$

Determinant of $\phi(s) = \dfrac{s(s+3) + 2}{(s+1)^2 (s+2)^2} = \dfrac{s^2 + 3s + 2}{(s+1)^2 (s+2)^2}$

$$= \dfrac{(s+1) + (s+2)}{(s+1)^2 (s+2)^2} = \dfrac{1}{(s+1)(s+2)}$$

$$\phi(s)^{-1} = (s+1)(s+2) \begin{bmatrix} \dfrac{s}{(s+1)(s+2)} & \dfrac{-1}{(s+1)(s+2)} \\ \dfrac{2}{(s+1)(s+2)} & \dfrac{s+3}{(s+1)(s+2)} \end{bmatrix} = \begin{bmatrix} s & -1 \\ 2 & s+3 \end{bmatrix}$$

$$A = sI - \phi(s)^{-1} = s \begin{bmatrix} 1 & 0 \\ 0 & 1 \end{bmatrix} - \begin{bmatrix} s & -1 \\ 2 & s+3 \end{bmatrix} = \begin{bmatrix} s & 0 \\ 0 & s \end{bmatrix} - \begin{bmatrix} s & -1 \\ 2 & s+3 \end{bmatrix} = \begin{bmatrix} 0 & 1 \\ -2 & -3 \end{bmatrix}$$

RESULT

$$A = \begin{bmatrix} 0 & 1 \\ -2 & -3 \end{bmatrix} \; ; \; e^{At} = \begin{bmatrix} 2e^{-t} - e^{-2t} & e^{-t} - e^{-2t} \\ -2e^{-t} + 2e^{-2t} & -e^{-t} + 2e^{-2t} \end{bmatrix}$$

EXAMPLE 6.16

The state equation and initial condition vector of an linear time-invariant system are given below. Determine the solution of state equation.

$$\begin{bmatrix} \dot{x}_1 \\ \dot{x}_2 \end{bmatrix} = \begin{bmatrix} 1 & 0 \\ 1 & 1 \end{bmatrix} \begin{bmatrix} x_1 \\ x_2 \end{bmatrix} \; ; \; x_0 = \begin{bmatrix} 1 \\ 0 \end{bmatrix}$$

SOLUTION

Here $A = \begin{bmatrix} 1 & 0 \\ 1 & 1 \end{bmatrix}$; $sI - A = s \begin{bmatrix} 1 & 0 \\ 0 & 1 \end{bmatrix} - \begin{bmatrix} 1 & 0 \\ 1 & 1 \end{bmatrix} = \begin{bmatrix} s & 0 \\ 0 & s \end{bmatrix} - \begin{bmatrix} 1 & 0 \\ 1 & 1 \end{bmatrix} = \begin{bmatrix} s-1 & 0 \\ -1 & s-1 \end{bmatrix}$

$|sI - A| = \begin{vmatrix} s-1 & 0 \\ -1 & s-1 \end{vmatrix} = (s-1)^2 - 0 = (s-1)^2$

$(sI - A)^{-1} = \dfrac{1}{(s-1)^2} \begin{bmatrix} s-1 & 0 \\ 0 & s-1 \end{bmatrix} = \begin{bmatrix} \dfrac{1}{s-1} & 0 \\ \dfrac{1}{(s-1)^2} & \dfrac{1}{s-1} \end{bmatrix}$

$e^{At} = \phi(t) = \mathcal{L}^{-1}[\phi(s)] = \mathcal{L}^{-1}[(sI - A)^{-1}] \begin{bmatrix} e^t & 0 \\ te^t & e^t \end{bmatrix}$

The solution of the state equation is, $X(t) = e^{At} X_0 = \begin{bmatrix} e^t & 0 \\ te^t & e^t \end{bmatrix} \begin{bmatrix} 1 \\ 0 \end{bmatrix} = \begin{bmatrix} e^t \\ te^t \end{bmatrix}$

6.9 TRANSFORMATION OF STATE MODEL

The state model of a system is not unique and it can be formed using physical variables, phase variables or canonical variables. The physical variables are useful from application point of view, because they can be measured and used for control purposes. However, the state model using physical variables is not convenient for investigation of system properties and evaluation of time response. But the canonical state model is most convenient for time domain analysis. In canonical model the system matrix A will be a diagonal matrix. Therefore each component state variable equation is a first order equation and is decoupled from all other component state variable equation.

When a non-diagonal system matrix A has distinct eigenvalues, it can be converted to diagonal matrix by a similarity transformation using modal matrix, M. Due to this the state model is transformed to canonical form. When a non-diagonal system matrix has multiple eigenvalues, it can be converted to Jordan matrix by a similarity transformation using modal matrix, M. Due to this the state model is transformed to Jordan canonical form.

CANONICAL FORM OF STATE MODEL

Consider the state equation of a system, $\dot{X} = AX + BU$, where X is the state variable vector of order $n \times 1$. Let us define a new state variable vector Z, such that $X = MZ$, where M is the Modal matrix or Diagonalization matrix.

The state model of the nth order system is given by,

$$\dot{X} = AX + BU$$

$$Y = CX + DU$$

On substituting $X = MZ$ in the state model of the system, we get

$$\dot{X} = AMZ + BU \qquad .....(6.82)$$

$$Y = CMZ + DU \qquad .....(6.83)$$

Premultiply equation (6.82) by M^{-1}

$$\therefore \ M^{-1}X = M^{-1}AMZ + M^{-1}BU \qquad .....(6.84)$$

The relation governing X and Z is, $X = MZ$. $\qquad .....(6.85)$

On differentiating equation (6.85), we get, $\dot{X} = M\dot{Z}$. $\qquad .....(6.86)$

On premultiplying the equation (6.86) by M^{-1} we get,

$$M^{-1}\dot{X} = \dot{Z} \qquad .....(6.87)$$

From equations (6.84) and (6.87), we get,

$$\dot{Z} = M^{-1}AMZ + M^{-1}BU \qquad .....(6.88)$$

Let, $M^{-1}AM = \Lambda$ (called grammian matrix) $\qquad .....(6.89)$

$$M^{-1}B = \tilde{B} \qquad .....(6.90)$$

$$CM = \tilde{C} \qquad .....(6.91)$$

From equations (6.83) and (6.88) to (6.91) the transformed state model is obtained as shown below.

$$\dot{Z} = AZ + \tilde{B}U \qquad .....(6.92)$$

$$Y = \tilde{C}Z + DU \qquad .....(6.93)$$

In the transformed state model [equations (6.92) and (6.93)] the grammian matrix Λ, will be a diagonal matrix and the transformed state model is called canonical form of state model. The model matrix $\mathbf{M}$ is obtained from Eigenvectors. When the system matrix $\mathbf{A}$ is in the companion or Bush form then the modal matrix is given by a special matrix called vander monde matrix, $\mathbf{V}$, shown below.

i.e., If $\mathbf{A} = \begin{bmatrix} 0 & 1 & 0 & \cdots & 0 \\ 0 & 0 & 1 & \cdots & 0 \\ \vdots & \vdots & \vdots & & \vdots \\ 0 & 0 & 0 & \cdots & 1 \\ -a_n & -a_{n-1} & -a_{n-2} & \cdots & -a_1 \end{bmatrix}$ then $\mathbf{M} = \mathbf{V} = \begin{bmatrix} 1 & 1 & \cdots & \cdots & 1 \\ \lambda_1 & \lambda_1 & \cdots & \cdots & \lambda_1 \\ \lambda_1^2 & \lambda_1^2 & & & \lambda_n^2 \\ \vdots & \vdots & & & \vdots \\ \lambda_1^{n-1} & \lambda_1^{n-1} & \cdots & \cdots & \lambda_1^{n-1} \end{bmatrix}$

> **Note :** *When the state model is available in canonical form then* $\mathbf{M} = \mathbf{M}^{-1} = \mathbf{I}$.

JORDAN CANONICAL FORM OF STATE MODEL

If the eigenvalues has multiplicity then the system matrix cannot be diagonalized. In this case, the transformation as explained in previous section will give a Jordan matrix, $\mathbf{J}$ where $\mathbf{J} = \mathbf{M}^{-1}\mathbf{AM}$. The transformed state model is called Jordan canonical form and it is given by equation (6.94) and (6.95).

$$\dot{\mathbf{Z}} = \mathbf{JZ} + \tilde{\mathbf{B}}\mathbf{U} \qquad \qquad .....(6.94)$$

$$\mathbf{Y} = \tilde{\mathbf{C}}\mathbf{U} + \mathbf{DU} \qquad \qquad .....(6.95)$$

where, $\quad \mathbf{J} = \mathbf{M}^{-1}\mathbf{AM}$

$$\tilde{\mathbf{B}} = \mathbf{M}^{-1}\mathbf{B}$$

$$\tilde{\mathbf{C}} = \mathbf{CM}$$

The Jordan matrix $\mathbf{J}$ will have a Jordan block of size $q \times q$ correspond to a eigenvalue of λ_i with multiplicity q. In the Jordan block the diagonal elements will be eigenvalue and the element just above the diagonal is one.

Consider a system matrix with eigenvalues $\lambda_1, \lambda_1, \lambda_1, \lambda_4, \lambda_5 \ldots\ldots\lambda_n$ where λ_1 has multiplicity of three. The Jordan matrix for this case will be as shown in equation (6.96).

$$\mathbf{J} = \begin{vmatrix} \lambda_1 & 1 & 0 & 0 & \cdots & 0 \\ 0 & \lambda_1 & 1 & 0 & \cdots & 0 \\ 0 & 0 & \lambda_1 & 0 & \cdots & 0 \\ 0 & 0 & 0 & \lambda_4 & \cdots & 0 \\ \vdots & \vdots & \vdots & \vdots & & \vdots \\ 0 & 0 & 0 & 0 & \cdots & \lambda_n \end{vmatrix}$$ — Jordan block of size 3×3 $\qquad .....(6.96)$

If the system matrix $\mathbf{A}$ is in the companion form and has multiple eigen values, then the modal matrix is given by modified vander monde matrix shown in equation (6.97).

Let λ_1 has multiplicity of q and the other eigen values are $\lambda_{q+1}, \lambda_{q+2}, \ldots\ldots\lambda_n$.

$$\mathbf{M} = \mathbf{V} = \begin{bmatrix} 1 & 0 & 0 & \cdots & 1 & \cdots & 1 \\ \lambda_1 & 1 & 0 & \cdots & \lambda_{q+1} & \cdots & \lambda_n \\ \lambda_1^2 & 2\lambda_1 & 1 & \cdots & \lambda_{q+1}^2 & \cdots & \lambda_n^2 \\ \lambda_1^3 & 3\lambda_1^2 & 3\lambda_1 & \cdots & \lambda_{q+1}^3 & \cdots & \lambda_n^3 \\ \vdots & \vdots & \vdots & & \vdots & & \vdots \\ \vdots & \vdots & \vdots & & \vdots & & \vdots \\ \lambda_1^{(n-1)} & \dfrac{d(\lambda_1^{n-1})}{d\lambda_1} & \dfrac{1}{2!}\dfrac{d^2(\lambda_1^{n-1})}{d\lambda_1^2} & \cdots & \lambda_{q+1}^{n-1} & \cdots & \lambda_n^{n-1} \end{bmatrix} \qquad .....(6.97)$$

COMPUTATION OF STATE TRANSITION MATRIX BY CANONICAL TRANSFORMATION

Consider the state equation without input,

$$\dot{\mathbf{X}}(t) = \mathbf{A}\,\mathbf{X}(t) \qquad\qquad .....(6.98)$$

The solution of state equation is

$$\mathbf{X}(t) = e^{\mathbf{A}t}\,\mathbf{X}(0) \qquad\qquad .....(6.99)$$

Let us assume that the system matrix **A** is non-diagonal and its eigen values are distinct. Now the system matrix **A** can be diagonalized by a similarity transformation using modal matrix, **M**. Due to this transformation we get a new state vector $\mathbf{Z}(t)$. The relation governing the old and new state vector is $\mathbf{X}(t) = \mathbf{M}\mathbf{Z}(t)$. Also, the state equation (6.98) modifies to the form shown in equation (6.100).

$$\dot{\mathbf{Z}}(t) = \mathbf{M}^{-1}\mathbf{A}\,\mathbf{M}\,\mathbf{Z}(t) = \Lambda\,\mathbf{Z}(t) \qquad\qquad .....(6.100)$$

where, $\Lambda = \mathbf{M}^{-1}\mathbf{A}\mathbf{M} = \begin{bmatrix} \lambda_1 & 0 & 0 & \cdots & 0 \\ 0 & \lambda_2 & 0 & \cdots & 0 \\ \vdots & \vdots & \vdots & & \vdots \\ 0 & 0 & 0 & \cdots & \lambda_n \end{bmatrix}$

The solution of the modified state equation (equation (6.100)) is given by equation (6.101).

$$\mathbf{Z}(t) = e^{\Lambda t}\,\mathbf{Z}(0) \qquad\qquad .....(6.101)$$

The matrix $e^{\Lambda t}$ can be expressed as an infinite series shown below,

$$e^{\Lambda t} = I + \Lambda t + \frac{1}{2!}\Lambda^2 t^2 + \frac{1}{3!}\Lambda^3 t^3 + \ldots\ldots$$

$$e^{\Lambda t} = \begin{bmatrix} 1 & 0 & \cdots & 0 \\ 0 & 1 & \cdots & 0 \\ \vdots & \vdots & & \vdots \\ 0 & 0 & \cdots & 1 \end{bmatrix} + \begin{bmatrix} \lambda_1 & 0 & \cdots & 0 \\ 0 & \lambda_2 & \cdots & 0 \\ \vdots & \vdots & & \vdots \\ 0 & 0 & \cdots & \lambda_n \end{bmatrix}t + \frac{1}{2!}\begin{bmatrix} \lambda_1^2 & 0 & \cdots & 0 \\ 0 & \lambda_2^2 & \cdots & 0 \\ \vdots & \vdots & & \vdots \\ 0 & 0 & \cdots & \lambda_n^2 \end{bmatrix}t^2 + \frac{1}{3!}\begin{bmatrix} \lambda_1^3 & 0 & \cdots & 0 \\ 0 & \lambda_2^3 & \cdots & 0 \\ \vdots & \vdots & & \vdots \\ 0 & 0 & \cdots & \lambda_n^3 \end{bmatrix}t^3 + \ldots\ldots$$

$$e^{\Lambda t} = \begin{bmatrix} 1 + \lambda_1 t + \frac{1}{2!}\lambda_1^2 t^2 + \frac{1}{3!}\lambda_1^3 t^3 + \ldots & \ldots\ldots\ldots & 0 \\ & 1 + \lambda_2 t + \frac{1}{2!}\lambda_2^2 t^2 + \frac{1}{3!}\lambda_1^3 t^3 + \ldots & \\ 0 & & 1 + \lambda_n t + \frac{1}{2!}\lambda_n^2 t^2 + \frac{1}{3!}\lambda_n^3 t^3 + \end{bmatrix}$$

$$\therefore\ e^{\mathbf{A}t} = \begin{bmatrix} e^{\lambda_1 t} & 0 & \cdots & 0 \\ 0 & e^{\lambda_2 t} & \cdots & 0 \\ \vdots & \vdots & & \vdots \\ 0 & 0 & \cdots & e^{\lambda_n t} \end{bmatrix} \qquad\qquad .....(6.102)$$

We know that, $\mathbf{M}\,\mathbf{Z}(t) = \mathbf{X}(t)$ $\qquad\qquad .....(6.103)$

On premultiplying the equation (6.103) by $\mathbf{M}^{-1}$ we get,

$$\mathbf{Z}(t) = \mathbf{M}^{-1}\,\mathbf{X}(t) \qquad\qquad .....(6.104)$$

From equation (6.104) when t = 0, we get, $\mathbf{Z}(0) = \mathbf{M}^{-1}\mathbf{X}(0)$ .....(6.105)

Using equations (6.104) and (6.105) the equation (6.101) can be written as,

$$\mathbf{M}^{-1}\mathbf{X}(t) = e^{\Lambda t}\mathbf{M}^{-1}\mathbf{X}(0) \qquad .....(6.106)$$

On premultiplying, the equation (6.106) by M we get,

$$\mathbf{X}(t) = \mathbf{M}\, e^{\Lambda t}\, \mathbf{M}^{-1}\mathbf{X}(0) \qquad .....(6.107)$$

On comparing equations (6.99) and (6.107) we get,

The state transition matrix, $e^{At} = \mathbf{M}\, e^{\Lambda t}\, \mathbf{M}^{-1}$ .....(6.108)

When the eigenvalues are distinct the equation (6.108) can be used to compute the state transition matrix e^{At}.

When the eigenvalues have multiplicity the system matrix cannot be diagonalized but can be transformed to Jordan matrix. When one of the eigenvalues λ_1 repeats q times the solution of state equation and state transition matrix are given by equations (6.109) and (6.110) respectively.

$$\mathbf{X}(t) = \mathbf{M}\,\mathbf{Q}(t)\, e^{\Lambda t}\mathbf{M}^{-1}\mathbf{X}(0) \qquad .....(6.109)$$

$$e^{At} = \mathbf{M}\,\mathbf{Q}(t)\, e^{\Lambda t}\,\mathbf{M}^{-1} \qquad .....(6.110)$$

$$\text{where, } \mathbf{Q}(t) = \begin{bmatrix} 1 & t & \frac{1}{2!}t^2 & \cdots & \frac{1}{(q-1)!}t^{q-1} & 0 & \cdots & 0 \\ 0 & 1 & t & \cdots & \frac{1}{(q-2)!}t^{q-2} & 0 & \cdots & 0 \\ 0 & 0 & 1 & \cdots & \frac{1}{(q-3)!}t^{q-3} & 0 & \cdots & 0 \\ \vdots & \vdots & \vdots & & & \vdots & & \vdots \\ 0 & 0 & 0 & \cdots & t & 0 & & 0 \\ 0 & 0 & 0 & \cdots & 1 & 0 & & 0 \\ 0 & 0 & 0 & \cdots & 0 & 1 & & 0 \\ \vdots & \vdots & \vdots & & & \vdots & & \vdots \\ 0 & 0 & 0 & & 0 & 0 & & 1 \end{bmatrix} \qquad .....(6.111)$$

> *Note : In this case the matrix $e^{\Lambda t}$ is same as equation (6.102)*

EXAMPLE 6.17

A linear time invariant system is described by the following state model.

$$\begin{bmatrix} \dot{x}_1 \\ \dot{x}_2 \\ \dot{x}_3 \end{bmatrix} = \begin{bmatrix} 0 & 1 & 0 \\ 0 & 0 & 1 \\ -6 & -11 & -6 \end{bmatrix}\begin{bmatrix} x_1 \\ x_2 \\ x_3 \end{bmatrix} + \begin{bmatrix} 0 \\ 0 \\ 2 \end{bmatrix}[u] \quad \text{and} \quad y = [1\ 0\ 0]\begin{bmatrix} x_1 \\ x_2 \\ x_3 \end{bmatrix}$$

Transform this state model into a canonical state model. Also compute the state transition matrix, e^{At}.

SOLUTION

To find eigenvalues

$$[\lambda I - A] = \lambda\begin{bmatrix} 1 & 0 & 0 \\ 0 & 1 & 0 \\ 0 & 0 & 1 \end{bmatrix} - \begin{bmatrix} 0 & 1 & 0 \\ 0 & 0 & 1 \\ -6 & -11 & -6 \end{bmatrix} = \begin{bmatrix} \lambda & 0 & 0 \\ 0 & \lambda & 0 \\ 0 & 0 & \lambda \end{bmatrix} - \begin{bmatrix} 0 & 1 & 0 \\ 0 & 0 & 1 \\ -6 & -11 & -6 \end{bmatrix} = \begin{bmatrix} \lambda & -1 & 0 \\ 0 & \lambda & -1 \\ 6 & 11 & \lambda+6 \end{bmatrix}$$

The characteristic equation is, $|\lambda I - A| = 0$

$$|\lambda I - A| = \begin{vmatrix} \lambda & -1 & 0 \\ 0 & \lambda & -1 \\ 6 & 11 & \lambda+6 \end{vmatrix} = \lambda[\lambda(\lambda+6)+11]+1\times6 = 0$$

$\therefore \lambda(\lambda^2+6\lambda+11)+6 = 0$

$\lambda^3+6\lambda^2+11\lambda+6 = 0$

$\lambda = -1$ is one of the root of the above equation,

$\therefore (\lambda+1)+(\lambda^2+5\lambda+6) = 0$

$(\lambda+1)(\lambda+2)(\lambda+3) = 0$

The eigen values are $\lambda = -1, -2, -3$; i.e., $\lambda_1 = -1, \lambda_2 = -2$ and $\lambda_3 = -3$

$$\begin{array}{r|rrrr} \lambda = -1 & 1 & 6 & 11 & 6 \\ & & -1 & -5 & -6 \\ \hline & 1 & 5 & 6 & \boxed{0} \end{array}$$

To find modal matrix

The system matrix is given in the companion form and therefore modal matrix, M is given by vander monde matrix, V.

$$M = V = \begin{bmatrix} 1 & 1 & 1 \\ \lambda_1 & \lambda_2 & \lambda_3 \\ \lambda_1^2 & \lambda_2^2 & \lambda_3^2 \end{bmatrix} = \begin{bmatrix} 1 & 1 & 1 \\ -1 & -2 & -3 \\ 1 & 4 & 9 \end{bmatrix}$$

Alternatively modal matrix can be found from eigenvectors. The alternate method is shown below,

$$[\lambda_1 I - A] = (-1)\begin{bmatrix} 1 & 0 & 0 \\ 0 & 1 & 0 \\ 0 & 0 & 1 \end{bmatrix} - \begin{bmatrix} 0 & 1 & 0 \\ 0 & 0 & 1 \\ -6 & -11 & -6 \end{bmatrix} = \begin{bmatrix} -1 & -1 & 0 \\ 0 & -1 & -1 \\ 6 & 11 & 5 \end{bmatrix}$$

Let the cofactors of $[\lambda_1 I - A]$ along Ist row be C_{11}, C_{12} and C_{13}

$$C_{11} = (+1)\begin{vmatrix} -2 & -1 \\ 11 & 4 \end{vmatrix} = 3 \; ; \; C_{12} = (-1)\begin{vmatrix} 0 & -1 \\ 6 & 4 \end{vmatrix} = -6 \; ; \; C_{13} = (+1)\begin{vmatrix} 0 & -2 \\ 6 & 11 \end{vmatrix} = 12$$

$\therefore$ Eigen vector corresponding to $\lambda_1 = m_1 = \begin{bmatrix} C_{11} \\ C_{12} \\ C_{13} \end{bmatrix} = \begin{bmatrix} 6 \\ -6 \\ 6 \end{bmatrix} = \begin{bmatrix} 1 \\ -1 \\ 1 \end{bmatrix}$

Note : The elements of eigenvector can be divided by a constant.

$$[\lambda_2 I - A] = (-2)\begin{bmatrix} 1 & 0 & 0 \\ 0 & 1 & 0 \\ 0 & 0 & 1 \end{bmatrix} - \begin{bmatrix} 0 & 1 & 0 \\ 0 & 0 & 1 \\ -6 & -11 & -6 \end{bmatrix} = \begin{bmatrix} -1 & -1 & 0 \\ 0 & -2 & -1 \\ 6 & 11 & 4 \end{bmatrix}$$

Let the cofactors of $[\lambda_2 I - A]$ along Ist row be C_{11}, C_{12} and C_{13}

$$\therefore C_{11} = (+1)\begin{vmatrix} -2 & -1 \\ 11 & 4 \end{vmatrix} = 3 \; ; \; C_{12} = (-1)\begin{vmatrix} 0 & -1 \\ 6 & 4 \end{vmatrix} = -6 \; ; \; C_{13} = (+1)\begin{vmatrix} 0 & -2 \\ 6 & 11 \end{vmatrix} = 12$$

$\therefore$ Eigen vector corresponding to $\lambda_2 = m_2 = \begin{bmatrix} C_{11} \\ C_{12} \\ C_{13} \end{bmatrix} = \begin{bmatrix} 3 \\ -6 \\ 12 \end{bmatrix} = \begin{bmatrix} 1 \\ -2 \\ 4 \end{bmatrix}$

$$[\lambda_3 I - A] = (-3)\begin{bmatrix} 1 & 0 & 0 \\ 0 & 1 & 0 \\ 0 & 0 & 1 \end{bmatrix} - \begin{bmatrix} 0 & 1 & 0 \\ 0 & 0 & 1 \\ -6 & -11 & -6 \end{bmatrix} = \begin{bmatrix} -3 & -1 & 0 \\ 0 & -3 & -1 \\ 6 & 11 & 3 \end{bmatrix}$$

Let the cofactors of $[\lambda_3 I - A]$ along Ist row be C_{11}, C_{12} and C_{13}

$$C_{11} = (+1)\begin{vmatrix} -3 & -1 \\ 11 & 3 \end{vmatrix} = 2 \; ; \; C_{12} = (-1)\begin{vmatrix} 0 & -1 \\ 6 & 3 \end{vmatrix} = -6 \; ; \; C_{13} = (+1)\begin{vmatrix} 0 & -3 \\ 6 & 11 \end{vmatrix} = 18$$

$\therefore$ Eigen vector corresponding to $\lambda_3 = m_3 = \begin{bmatrix} C_{11} \\ C_{12} \\ C_{13} \end{bmatrix} = \begin{bmatrix} 2 \\ -6 \\ 18 \end{bmatrix} = \begin{bmatrix} 1 \\ -3 \\ 9 \end{bmatrix}$

The modal matrix, $M = [m_1 \; m_2 \; m_3] = \begin{bmatrix} 1 & 1 & 1 \\ -1 & -2 & -3 \\ 1 & 4 & 9 \end{bmatrix}$

It is observed that the modal matrix obtained from vander monde matrix and from eigenvectors are same.

To find M^{-1}

$$M^{-1} = \frac{(\text{Cofactor of } M)^T}{\text{Determinant of } M} = \frac{M_{cof}^T}{\Delta_M}$$

$$\Delta_M = \begin{vmatrix} 1 & 1 & 1 \\ -1 & -2 & -3 \\ 1 & 4 & 9 \end{vmatrix} = 1(-18 + 12) - 1(-9 + 3) + 1(-4 + 2) = -2$$

$$M_{cof}^T = \begin{bmatrix} -6 & 6 & -2 \\ -5 & 8 & -3 \\ -1 & 2 & -1 \end{bmatrix}^T = \begin{bmatrix} -6 & -5 & -1 \\ 6 & 8 & 2 \\ -2 & -3 & -1 \end{bmatrix} \quad \therefore M^{-1} = \frac{1}{-2}\begin{bmatrix} -6 & -5 & -1 \\ 6 & 8 & 2 \\ -2 & -3 & -1 \end{bmatrix}$$

<u>To find canonical form of state model</u>

$$\left.\begin{array}{l} \text{Grammian matrix} \\ \Lambda = M^{-1}AM \end{array}\right\} = \frac{1}{-2}\begin{bmatrix} -6 & -5 & -1 \\ 6 & 8 & 2 \\ -2 & -3 & -1 \end{bmatrix}\begin{bmatrix} 0 & 1 & 0 \\ 0 & 0 & 1 \\ -6 & -11 & -6 \end{bmatrix}\begin{bmatrix} 1 & 1 & 1 \\ -1 & -2 & -3 \\ 1 & 4 & 9 \end{bmatrix}$$

$$= \frac{1}{-2}\begin{bmatrix} 6 & 5 & 1 \\ -12 & -16 & -4 \\ 6 & 9 & 3 \end{bmatrix}\begin{bmatrix} 1 & 1 & 1 \\ -1 & -2 & -3 \\ 1 & 4 & 9 \end{bmatrix} = \begin{bmatrix} -1 & 0 & 0 \\ 0 & -2 & 0 \\ 0 & 0 & -3 \end{bmatrix}$$

$$\tilde{B} = M^{-1}B = -\frac{1}{2}\begin{bmatrix} -6 & -5 & -1 \\ 6 & 8 & 2 \\ -2 & -3 & -1 \end{bmatrix}\begin{bmatrix} 0 \\ 0 \\ 2 \end{bmatrix} = \begin{bmatrix} 1 \\ -2 \\ 1 \end{bmatrix}$$

$$\tilde{C} = CM = [1 \; 0 \; 0]\begin{bmatrix} 1 & 1 & 1 \\ -1 & -2 & -3 \\ 1 & 4 & 9 \end{bmatrix} = [1 \; 1 \; 1]$$

The canonical form of state model is,

$$\dot{Z} = \Lambda Z + \tilde{B}U$$

$$Y = \tilde{C}Z + DU \quad \text{(Here DU is not defined)}$$

$$\begin{bmatrix} \dot{z}_1 \\ \dot{z}_2 \\ \dot{z}_3 \end{bmatrix} = \begin{bmatrix} -1 & 0 & 0 \\ 0 & -2 & 0 \\ 0 & 0 & -3 \end{bmatrix} \begin{bmatrix} z_1 \\ z_2 \\ z_3 \end{bmatrix} + \begin{bmatrix} 1 \\ -2 \\ 1 \end{bmatrix} [u] \quad ; \quad Y = \begin{bmatrix} 1 & 1 & 1 \end{bmatrix} \begin{bmatrix} z_1 \\ z_2 \\ z_3 \end{bmatrix}$$

To compute state transition matrix, e^{At}

The state transition matrix, $e^{At} = M\, e^{\Lambda t}\, M^{-1}$

$$e^{\Lambda t} = \begin{bmatrix} e^{\lambda_1 t} & 0 & 0 \\ 0 & e^{\lambda_2 t} & 0 \\ 0 & 0 & e^{\lambda_3 t} \end{bmatrix} = \begin{bmatrix} e^{-t} & 0 & 0 \\ 0 & e^{-2t} & 0 \\ 0 & 0 & e^{-3t} \end{bmatrix}$$

$$\therefore\ e^{At} = \begin{bmatrix} 1 & 1 & 1 \\ -1 & -2 & -3 \\ 1 & 4 & 9 \end{bmatrix} \begin{bmatrix} e^{-t} & 0 & 0 \\ 0 & e^{-2t} & 0 \\ 0 & 0 & e^{-3t} \end{bmatrix} \left(-\frac{1}{2}\right) \begin{bmatrix} -6 & -5 & -1 \\ 6 & 8 & 2 \\ -2 & -3 & -1 \end{bmatrix}$$

$$= \begin{bmatrix} e^{-t} & e^{-2t} & e^{-3t} \\ -e^{-t} & -2e^{-2t} & -3e^{-3t} \\ e^{-t} & 4e^{-2t} & 9e^{-3t} \end{bmatrix} \begin{bmatrix} 3 & 2.5 & 0.5 \\ -3 & -4 & -1 \\ 1 & 1.5 & 0.5 \end{bmatrix}$$

$$= \begin{bmatrix} 3e^{-t} - 3e^{-2t} + e^{-3t} & 2.5e^{-t} - 4e^{-2t} + 1.5e^{-3t} & 0.5e^{t} - e^{-2t} + 0.5e^{-3t} \\ -3e^{-t} + 6e^{-2t} - 3e^{-3t} & -2.5e^{-t} + 8e^{-2t} - 4.5e^{-3t} & -0.5e^{-t} + 2e^{-2t} - 1.5e^{-3t} \\ 3e^{-t} - 12e^{-2t} + 9e^{-3t} & 2.5e^{-2t} - 16e^{-2t} + 13.5e^{-3t} & 0.5e^{-t} - 4e^{-2t} + 4.5e^{-3t} \end{bmatrix}$$

EXAMPLE 6.18

Convert the following system matrix to canonical form and hence calculate the state transition matrix e^{At}.

$$A = \begin{bmatrix} 4 & 1 & -2 \\ 1 & 0 & 2 \\ 1 & -1 & 3 \end{bmatrix}$$

SOLUTION

To find eigenvalues

$$[\lambda I - A] = \lambda \begin{bmatrix} 1 & 0 & 0 \\ 0 & 1 & 0 \\ 0 & 0 & 1 \end{bmatrix} - \begin{bmatrix} 4 & 1 & -2 \\ 1 & 0 & 2 \\ 1 & -1 & 3 \end{bmatrix} = \begin{bmatrix} \lambda - 4 & -1 & 2 \\ -1 & \lambda & -2 \\ -1 & 1 & \lambda - 3 \end{bmatrix}$$

$$[\lambda I - A] = \begin{bmatrix} \lambda - 4 & -1 & 2 \\ -1 & \lambda & -2 \\ -1 & 1 & \lambda - 3 \end{bmatrix} = (\lambda - 4)[\lambda(\lambda - 3) + 2] + 1(-\lambda + 3 - 2) + 2(-1 + \lambda)$$

$$= (\lambda - 4)(\lambda^2 - 3\lambda + 2) - \lambda + 1 - 2 + 2\lambda = (\lambda - 4)(\lambda - 1)(\lambda - 2) + (\lambda - 1)$$

$$= (\lambda - 1)[(\lambda - 4)(\lambda - 2) + 1] = (\lambda - 1)[\lambda^2 - 6\lambda + 8 + 1]$$

$$= (\lambda - 1)(\lambda^2 - 6\lambda + 9) = (\lambda - 1)(\lambda - 3)^2$$

The eigenvalues are $\lambda_1 = 1$, $\lambda_2 = 3$, $\lambda_3 = 3$. Since one of the eigenvalue has a multiplicity of 2 (repeats two times) the canonical form will be a Jordan canonical form.

To find eigenvectors

$$[\lambda_1 I - A] = (1)\begin{bmatrix} 1 & 0 & 0 \\ 0 & 1 & 0 \\ 0 & 0 & 1 \end{bmatrix} - \begin{bmatrix} 4 & 1 & -2 \\ 1 & 0 & 2 \\ 1 & -1 & 3 \end{bmatrix} = \begin{bmatrix} -3 & -1 & 2 \\ -1 & 1 & -2 \\ -1 & 1 & -2 \end{bmatrix}$$

Let the cofactors of $[\lambda_1 I - A]$ along Ist row be C_{11}, C_{12} and C_{13}.

$$C_{11} = (+1)\begin{vmatrix} 1 & -2 \\ 1 & -2 \end{vmatrix} = 0 \quad ; \quad C_{12} = (-1)\begin{vmatrix} -1 & -2 \\ -1 & -2 \end{vmatrix} = 0 \quad ; \quad C_{13} = (+1)\begin{vmatrix} -1 & 1 \\ -1 & 1 \end{vmatrix} = 0$$

$$\therefore m_1 = \begin{bmatrix} C_{11} \\ C_{12} \\ C_{13} \end{bmatrix} = \begin{bmatrix} 0 \\ 0 \\ 0 \end{bmatrix} \quad \text{The cofactors along 1st row gives null solution.}$$

Let C_{21}, C_{22} and C_{23} be cofactors of $[\lambda_1 I - A]$ along IInd row.

$$C_{21} = (-1)\begin{vmatrix} -1 & 2 \\ 1 & -2 \end{vmatrix} = 0 \; ; \; C_{22} = (+1)\begin{vmatrix} -3 & 2 \\ -1 & -2 \end{vmatrix} = 8 \; ; \; C_{23} = (-1)\begin{vmatrix} -3 & -1 \\ -1 & 1 \end{vmatrix} = 4$$

$$\therefore m_1 = \begin{bmatrix} C_{21} \\ C_{22} \\ C_{23} \end{bmatrix} = \begin{bmatrix} 0 \\ 8 \\ 4 \end{bmatrix}$$

$$[\lambda_2 I - A] = \lambda_2 \begin{bmatrix} 1 & 0 & 0 \\ 0 & 1 & 0 \\ 0 & 0 & 1 \end{bmatrix} - \begin{bmatrix} 4 & 1 & -2 \\ 1 & 0 & 2 \\ 1 & -1 & 3 \end{bmatrix} = \begin{bmatrix} \lambda_2 - 4 & -1 & 2 \\ -1 & \lambda_2 & -2 \\ -1 & 1 & \lambda_2 - 3 \end{bmatrix}$$

The cofactor of $[\lambda_2 I - A]$ along Ist row be C_{11}, C_{12} and C_{13}.

$$C_{11} = (+1)\begin{vmatrix} \lambda_2 & -2 \\ 1 & \lambda_2 - 3 \end{vmatrix} = \lambda_2(\lambda_2 - 3) + 2 = \lambda_2^2 - 3\lambda_2 + 2$$

$$C_{12} = (-1)\begin{vmatrix} -1 & -2 \\ 1 & \lambda_2 - 3 \end{vmatrix} = (-1)[(-1)(\lambda_2 - 3) - 2] = \lambda_2 - 1$$

$$C_{13} = (+1)\begin{vmatrix} -1 & \lambda_2 \\ -1 & 1 \end{vmatrix} = -1 + \lambda_2 = \lambda_2 - 1$$

Let m_2 be the independent eigenvector corresponding to $\lambda_2 = 3$.

$$\text{Now, } m_2 = \begin{bmatrix} C_{11} \\ C_{12} \\ C_{13} \end{bmatrix} = \begin{bmatrix} \lambda_2^2 - 3\lambda_2 + 2 \\ \lambda_2 - 1 \\ \lambda_2 - 1 \end{bmatrix} = \begin{bmatrix} 3^2 - (3 \times 3) + 2 \\ 3 - 1 \\ 3 - 1 \end{bmatrix} = \begin{bmatrix} 2 \\ 2 \\ 2 \end{bmatrix}$$

The eigenvector m_3 is given by,

$$m_3 = \begin{bmatrix} \dfrac{d}{d\lambda_2} C_{11} \\ \dfrac{d}{d\lambda_2} C_{12} \\ \dfrac{d}{d\lambda_2} C_{13} \end{bmatrix} \begin{bmatrix} \dfrac{d}{d\lambda_2}(\lambda_2^2 - 3\lambda_2 + 2) \\ \dfrac{d}{d\lambda_2}(\lambda_2 - 1) \\ \dfrac{d}{d\lambda_2}(\lambda_2 - 1) \end{bmatrix} = \begin{bmatrix} 2\lambda_2 - 3 \\ 1 \\ 1 \end{bmatrix} = \begin{bmatrix} 2 \times 3 - 3 \\ 1 \\ 1 \end{bmatrix} = \begin{bmatrix} 3 \\ 1 \\ 1 \end{bmatrix}$$

To find canonical form of system matrix

The modal matrix is given by, $M = [m_1 \ m_2 \ m_3] = \begin{bmatrix} 0 & 2 & 3 \\ 8 & 2 & 1 \\ 4 & 2 & 1 \end{bmatrix}$

$$M^{-1} = \frac{[\text{Cofactor of M}]^T}{\text{Determinant of M}} = \frac{M_{cof}^T}{\Delta_M} \quad ; \quad \Delta_M = \begin{vmatrix} 0 & 2 & 3 \\ 8 & 2 & 1 \\ 4 & 2 & 1 \end{vmatrix} = -8 + 24 = 16$$

$$M_{cof}^T = \begin{bmatrix} 0 & -4 & 8 \\ 4 & -12 & 8 \\ -4 & 24 & -16 \end{bmatrix}^T = \begin{bmatrix} 0 & 4 & -4 \\ -4 & -12 & 24 \\ 8 & 8 & -16 \end{bmatrix}$$

$$\therefore \ M^{-1} = \frac{1}{\Delta_M} M_{cof}^T = \frac{1}{16} \begin{bmatrix} 0 & 4 & -4 \\ -4 & -12 & 24 \\ 8 & 8 & -16 \end{bmatrix} = \frac{1}{4} \begin{bmatrix} 0 & 1 & -1 \\ -1 & -3 & 6 \\ 2 & 2 & -4 \end{bmatrix}$$

$$\therefore \ M^{-1}AM = \frac{1}{4} \begin{bmatrix} 0 & 1 & -1 \\ -1 & -3 & 6 \\ 2 & 2 & -4 \end{bmatrix} \begin{bmatrix} 4 & 1 & -2 \\ 1 & 0 & 2 \\ 1 & -1 & 3 \end{bmatrix} \begin{bmatrix} 0 & 2 & 3 \\ 8 & 2 & 1 \\ 4 & 2 & 1 \end{bmatrix}$$

$$= \frac{1}{4} \begin{bmatrix} 0 & 1 & -1 \\ -1 & -7 & 14 \\ 6 & 6 & -12 \end{bmatrix} \begin{bmatrix} 0 & 2 & 3 \\ 8 & 2 & 1 \\ 4 & 2 & 1 \end{bmatrix} = \frac{1}{4} \begin{bmatrix} 4 & 0 & 0 \\ 0 & 12 & 4 \\ 0 & 0 & 12 \end{bmatrix} = \begin{bmatrix} 1 & 0 & 0 \\ 0 & 3 & 1 \\ 0 & 0 & 3 \end{bmatrix}$$

$$\therefore \ J = M^{-1}AM = \begin{bmatrix} 1 & 0 & 0 \\ 0 & 3 & 1 \\ 0 & 0 & 3 \end{bmatrix} \longleftarrow \text{Jordan block with multiplicity of 2}$$

To compute state transition matrix, e^{At}

Since the eigen values have a multiplicity of 2, the state transition matrix is given by

The state transition matrix, $e^{At} = M \ Q(t) \ e^{\Lambda t} M^{-1}$

$$Q(t) = \begin{bmatrix} 1 & t & \frac{t^2}{2} \\ 0 & 1 & t \\ 0 & 0 & 1 \end{bmatrix} \quad ; \quad e^{\Lambda t} = \begin{bmatrix} e^{\lambda_1 t} & 0 & 0 \\ 0 & e^{\lambda_2 t} & 0 \\ 0 & 0 & e^{\lambda_3 t} \end{bmatrix} = \begin{bmatrix} e^t & 0 & 0 \\ 0 & e^{3t} & 0 \\ 0 & 0 & e^{3t} \end{bmatrix}$$

$$\therefore \ e^{At} = \begin{bmatrix} 0 & 2 & 3 \\ 8 & 2 & 1 \\ 4 & 2 & 1 \end{bmatrix} \begin{bmatrix} 1 & t & t^2/2 \\ 0 & 1 & t \\ 0 & 0 & 1 \end{bmatrix} \begin{bmatrix} e^t & 0 & 0 \\ 0 & e^{2t} & 0 \\ 0 & 0 & e^{3t} \end{bmatrix} \begin{bmatrix} 0 & 1 & -1 \\ -1 & -3 & 6 \\ 2 & 2 & -4 \end{bmatrix} \times \frac{1}{4}$$

$$= \begin{bmatrix} 0 & 2 & 2t+3 \\ 8 & 8t+2 & 4t^2+2t+1 \\ 4 & 4t+2 & 2t^2+2t+1 \end{bmatrix} \begin{bmatrix} e^t & 0 & 0 \\ 0 & e^{3t} & 0 \\ 0 & 0 & e^{3t} \end{bmatrix} \begin{bmatrix} 0 & 0.25 & -0.25 \\ -0.25 & -0.75 & 1.5 \\ 0.5 & 0.5 & -1 \end{bmatrix}$$

$$= \begin{bmatrix} 0 & 2e^{3t} & (2t+3)e^{3t} \\ 8e^t & (8t+2)e^{3t} & (4t^2+2t+1)e^{3t} \\ 4e^t & (4t+2)e^{3t} & (2t^2+2t+1)e^{3t} \end{bmatrix} \begin{bmatrix} 0 & 0.25 & -0.25 \\ -0.25 & -0.75 & 1.5 \\ 0.5 & 0.5 & -1.0 \end{bmatrix}$$

$$= \begin{bmatrix} -0.5e^{3t}+(t+1.5)e^{3t} & -1.5e^{3t}+(t+1.5)e^{3t} & 3e^{3t}-(2t+3)e^{3t} \\ (-2t-0.5)e^{3t} & 2e^t+(-6t-1.5)e^{3t} & -2e^t+(12t+3)e^{3t} \\ +(2t^2+t+0.5)e^{3t} & +(2t^2+t+0.5)e^{3t} & -(4t^2+2t+1)e^{3t} \\ (-t-0.5)e^{3t} & e^t+(-3t-1.5)e^{3t} & -e^t+(6t+3)e^{3t} \\ +(t^2+t+0.5)e^{3t} & +(t^2+t+0.5)e^{3t} & -(2t^2+2t+1)e^{3t} \end{bmatrix}$$

$$= \begin{bmatrix} (t+1)e^{3t} & te^{3t} & -2t\,e^{3t} \\ (2t^2-t)e^{3t} & 2e^t+(2t^2-5t-1)e^{3t} & -2e^t-(4t^2-10t-2)e^{3t} \\ t^2e^{3t} & e^t+(t^2-2t-1)e^{3t} & -e^t-(2t^2-4t-2)e^{3t} \end{bmatrix}$$

6.10 CONCEPTS OF CONTROLLABILITY AND OBSERVABILITY

CONTROLLABILITY

The controllability verifies the usefulness of a state variable. In the controllability test we can find, whether the state variable can be controlled to achieve the desired output. The choice of state variables is arbitrary while forming the state model. After determining the state model, the controllability of the state variable is verified. If the state variable is not controllable then we have to go for another choice of state variable.

Definition of controllability

A system is said to be completely state controllable if it is possible to transfer the system state from any initial state $X(t_0)$ *to any other desired state* $X(t_d)$ *in specified finite time by a control vector* $U(t)$.

The controllability of a state model can be tested by Kalman's test or Gilbert's test.

Gilbert's method of testing controllability

Case(i) : When the system matrix has distinct eigenvalues

In this case the system matrix can be diagonalized and the state model can be converted to canonical form.

Consider the state model of the system,

$$\dot{X} = AX + BU$$

$$Y = CX + DU$$

The state model can be converted to canonical form by a transformation, $X = MZ$, where M is the modal matrix and Z is the transformed state variable vector.

The transformed state model is given by,

$$\dot{Z} = \Lambda Z + \tilde{B}U$$

$$Y = \tilde{C}Z + DU$$

where, $\Lambda = \mathbf{M}^{-1}\mathbf{A}\mathbf{M}$

$$\tilde{\mathbf{B}} = \mathbf{M}^{-1}\mathbf{B}$$

$$\tilde{\mathbf{C}} = \mathbf{C}\mathbf{M}$$

In this case the necessary and sufficient condition for complete controllability is that, the matrix $\tilde{\mathbf{B}}$ *must have no rows with all zeros. If any row of the matrix* $\tilde{\mathbf{B}}$ *is zero then the corresponding state variable is uncontrollable.*

Case(ii) : When the system matrix has repeated eigenvalues

In this case, the system matrix cannot be diagonalized but can be transformed to Jordan canonical form.

Consider the state model of the system,

$$\dot{\mathbf{X}} = \mathbf{A}\mathbf{X} + \mathbf{B}\mathbf{U}$$

$$\mathbf{Y} = \mathbf{C}\mathbf{X} + \mathbf{D}\mathbf{U}$$

The state model can be transformed to Jordan canonical form by a transformation, $\mathbf{X} = \mathbf{M}\mathbf{Z}$, where $\mathbf{M}$ is modal matrix and $\mathbf{Z}$ is the transformed state variable vector.

The transformed state model is given by,

$$\dot{\mathbf{Z}} = \mathbf{J}\mathbf{Z} + \tilde{\mathbf{B}}\mathbf{U}$$

$$\mathbf{Y} = \tilde{\mathbf{C}}\mathbf{Z} + \mathbf{D}\mathbf{U}$$

where, $\mathbf{J} = \mathbf{M}^{-1}\mathbf{A}\mathbf{M}$

$$\tilde{\mathbf{B}} = \mathbf{M}^{-1}\mathbf{B}$$

$$\tilde{\mathbf{C}} = \mathbf{C}\mathbf{M}$$

In this case, the system is completely controllable if the elements of any row $\tilde{\mathbf{B}}$ *of that correspond to the last row of each Jordan block are not all zero and the rows corresponding to other state variables must not have all zeros.*

KALMAN'S METHOD OF TESTING CONTROLLABILITY

Consider a system with state equation, . For this system, a composite matrix, $\mathbf{Q}_C$ can be formed such that,

$$\mathbf{Q}_C = [\mathbf{B} \quad \mathbf{A}\mathbf{B} \quad \mathbf{A}^2\mathbf{B} \quad .... \quad \mathbf{A}^{n-1}\mathbf{B}] \qquad \qquad .....(6.112)$$

where, n is the order of the system (n is also equal to number of state variables).

In this case the system is completely state controllable if the rank of the composite matrix, $\mathbf{Q}_C$ *is n.*

The rank of the matrix is n, if the determinant of ($n \times n$) composite matrix $\mathbf{Q}_C$ is non-zero. i.e, if $|\mathbf{Q}_C| \neq 0$, then rank of $\mathbf{Q}_C = n$ and the system is completely state controllable.

The advantage in kalman's test is that the calculations are simpler. But the disadvantage in kalman's test is that, we can't find the state variable which is uncontrollable. But in Gilbert's method we can find the uncontrollable state variable which is the state variable corresponding to the row of $\tilde{\mathbf{B}}$ which has all zeros.

Condition for complete state controllability in the s-plane

A necessary and sufficient condition for complete state controllability is that no cancellation of poles and zeros occurs in the transfer function of the system. If cancellation occurs then the system cannot be controlled in the direction of the cancelled mode.

OBSERVABILITY

In observability test we can find whether the state variable is observable or measurable. The concept of observability is useful in solving the problem of reconstructing unmeasurable state variables from measurable ones in the minimum possible length of time. In state feedback control the estimation of unmeasurable state variables is essential in order to construct the control signals.

Definition of observability

A system is said to be completely observable if every state $X(t)$ *can be completely identified by measurements of the output* $Y(t)$ *over a finite time interval.*

The observability of a system can be tested by either Gilbert's method or Kalman's method.

Gilbert's method of testing observability

Consider a state model of nth order system, $\dot{X} = AX + BU$; $Y = CX = DU$

The state model can be transformed to a canonical or Jordan canonical form by a transformation, $X = MZ$, where M is the modal matrix and Z is the transformed state variable vector.

The transformed state model is,

$$\dot{Z} = \Lambda Z + \tilde{B}U \qquad \qquad \dot{Z} = JZ + \tilde{B}U$$
$$\qquad \qquad \qquad (or)$$
$$Y = \tilde{C}Z + DU \qquad \qquad Y = \tilde{C}Z + DU$$

where, $\Lambda = M^{-1}AM$; if eigenvalues are distinct ; $\tilde{B} = M^{-1}B$

$J = M^{-1}AM$; if eigenvalues have multiplicity ; $\tilde{C} = CM$

The necessary and sufficient condition for complete observability is that none of the columns of the matrix $\tilde{C}$ *be zero. If any of the column's of* $\tilde{C}$ *has all zeros then the corresponding state variable is not observable.*

Kalman's Test for observability

Consider a system with state model, $\dot{X} = AX + BU$; $Y = CX + DU$

For this system, a composite matrix, Q_0 can be formed such that,

$$Q_0 = [C^T \quad A^T C^T \quad (A^T)^2 C^T \quad (A^T)^3 C^T \quad .... \quad (A^T)^{n-1}C^T] \qquad .....(6.113)$$

where, n is the order of the system (n is also equal to number of state variables)

In this case, the system is completely observable if the rank of composite matrix, Q_0 *is n.*

The rank of the matrix is n, if the determinant of n × n composite matrix Q_0 is non-zero. The disadvantage in Kalman's test is that, the nonobservable state variables cannot be determined.

Condition for complete observability in the s-plane

The necessary and sufficient condition for complete observability is that no cancellation of poles and zeros occurs in the transfer function. If cancellation occurs, the cancelled mode cannot be observed in the output.

RELATIONSHIPS BETWEEN CONTROLLABILITY, OBSERVABILITY & TRANSFER FUNCTIONS

The concepts of controllability and observability play an important role in the design of control systems in state space. They govern the existence of a complete solution to the control system design problem. The solution to this problem may not exist if the system considered is not controllable.

It is important to note that all physical systems are controllable and observable. However, the mathematical models of these systems may not posses the property of the controllability or observability. Then it is necessary to know the conditions under which a system is controllable and observable and the designer can seek another state model which is controllable and observable.

Duality property

The concepts of controllability and observability are dual concepts and it is propossed by Kalman as principle of duality.

The principle of duality states that a system is completely state controllable if and only if its dual system is completely observable or viceversa. [i.e., if the system is observable then its dual is controllable]. Using the principle of duality, the observability of a given system can be checked by testing the state controllability of its dual or vice-versa.

Consider the system S_1, described by the state model shown below,

$$\dot{X} = AX + BU$$

$$Y = CX$$

Let the dual of system S_1 be denoted as S_2 and the dual system S_2 is described by the following state model.

$$\dot{Z} = A^T Z + C^T V$$

$$N = B^T Z$$

where, Z = State vector of dual system

V = Input vector of dual system

N = Output vector of dual system

For the system S_1 the composite matrix, Q_{C1} for controllability is given by equation (6.114) and the composite matrix, Q_{O1} for observability is given by equation (6.115).

$$Q_{C1} = [B \quad AB \quad A^2B \ldots\ldots\ldots\ldots A^{n-1} B] \qquad \ldots\ldots(6.114)$$

$$Q_{O1} = [C^T \quad A^TC^T \quad (A^T)^2 C^T \ldots\ldots\ldots\ldots (A^T)^{n-1} C^T] \qquad \ldots\ldots(6.115)$$

For the dual system S_2 the composite matrix, Q_{C2} for controllability is given by equation (6.116) and the composite matrix Q_{O2} for observability is given by equation (6.117)

$$Q_{C2} = [C^T \quad A^TC^T \quad (A^T)^2 C^T \ldots\ldots\ldots\ldots (A^T)^{n-1} C^T] \qquad \ldots\ldots(6.116)$$

$$Q_{O2} = [A \quad AB \quad A^2B \ldots\ldots\ldots\ldots A^{n-1} B] \qquad \ldots\ldots(6.117)$$

From equations (6.114) and (6.117) we get $Q_{C1} = Q_{O2}$, hence if the system S_1 is controllable then its dual system S_2 is observable.

From equations (6.115) and (6.116) we get $Q_{O1} = Q_{C2}$, hence if the system S_1 is observable then its dual system S_2 is controllable.

Effect of pole-zero cancellation in transfer function

The concepts of controllability and observability are closely related to the properties of the transfer function. Consider an nth order system with distinct eigenvalues. The transfer function of the system can be expressed as a ratio of two polynomials as shown in equation (6.118).

$$T(s) = \frac{Y(s)}{U(s)} = \frac{b_0 s^m + b_1 s^{m-1} + \ldots\ldots + b_{m-1}s_1 + b_m}{s^n + a_1 s^n + \ldots\ldots + a_{n-1}s + a_n} \quad ; \quad m < n \qquad \ldots\ldots(6.118)$$

$$= \frac{K(s + \beta_1)(s + \beta_2)\ldots\ldots(s + \beta_m)}{(s + \lambda_1)(s + \lambda_2)\ldots\ldots(s + \lambda_n)} \qquad \ldots\ldots(6.119)$$

By partial fraction expansion technique the equation (6.119) can be written as,

$$\frac{Y(s)}{U(s)} = \frac{C_1}{s + \lambda_1} + \frac{C_2}{s + \lambda_2} + \ldots\ldots + \frac{C_i}{s + \lambda_i} + \ldots\ldots + \frac{C_n}{s + \lambda_n} \qquad \ldots\ldots(6.120)$$

where $C_1, C_2, C_3, \ldots\ldots C_n$ are residues.

If the transfer function has identical pair of pole and zero at $\beta_i = \lambda_i$, then $C_i = 0$. The effect of this cancellation on controllability and observability properties depends on the choice of state variables [or depends on the method of forming state model].

In one method of state space modelling using canonical of variables, the $C_i = 0$, will appear in input (control) vector **B** and the the state x_i is uncontrollable. In another method of state space modelling using canonical variables, the $C_i = 0$, will appear in output vector **C** and the state x_i is shielded from observation.

From the above discussion we can conclude that if cancellation of pole-zero occurs in the transfer function of a system, then the system will be either not state controllable or unobservable, depending on how the state variables are defined (or chosen). If the transfer function does not have pole-zero cancellation, the system can always be represented by completely controllable and observable state model.

EXAMPLE 6.19

Write the state equations for the system shown in fig 1 in which x_1, x_2 and x_3 constitute the state vector. Determine whether the system is completely controllable and observable.

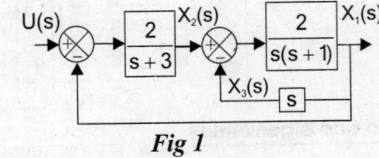

Fig 1

SOLUTION

To find state model

The state equations are obtained by writing equations for the output of each block and then taking inverse Laplace transform.

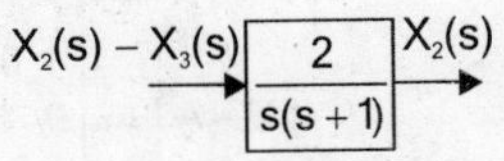

With reference to fig 2 we can write,

$$X_1(s) = [\,X_2(s) - X_3(s)\,]\left[\frac{2}{s(s+1)}\right]$$

$$s(s+1)\,X_1(s) = 2X_2(s) - 2X_3(s)$$

$$s^2 X_1(s) + sX_1(s) = 2X_2(s) - 2X_3(s)$$

On taking inverse Laplace transform

$$\ddot{x}_1 + \dot{x}_1 = 2x_2 - 2x_3$$

With reference to fig 3, we can write,

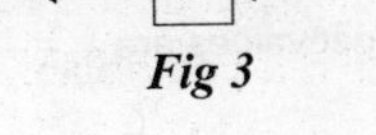

$$X_3(s) = sX_1(s)$$

$X_2(s) - X_3(s)$ $\boxed{\dfrac{2}{s(s+1)}}$ $X_2(s)$

Fig 2

$$\ldots\ldots(6.19.1)$$

$X_3(s)$ $\boxed{s}$ $X_1(s)$

Fig 3

On taking inverse Laplace transform

$$x_3 = \dot{x}_1 \qquad \qquad .....(2)$$

With reference to fig 4 we can write

$$X_2(s) = [U(s) - X_1(s)] \left[\frac{2}{s+3} \right]$$

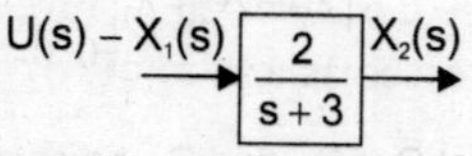

$$X_2(s)\,(s+3) = 2U(s) - 2X_1(s)$$

$$sX_2(s) + 3X_2(s) = 2U(s) - 2X_1(s)$$

Fig 4

On taking inverse Laplace transform

$$\dot{x}_2 + 3x_2 = 2u - 2x_1$$

$$\dot{x}_2 = -2x_1 - 3x_2 + 2u \qquad \qquad .....(3)$$

From equation (2) we get, $\dot{x}_1 = x_3$; $\therefore \ddot{x}_1 = \dot{x}_3$

Put $\dot{x}_1 = x_3$ and $\ddot{x}_1 = \dot{x}_3$ in equation (1)

$$\therefore \ \dot{x}_3 + x_3 = 2x_2 - 2x_3$$

$$\dot{x}_3 = 2x_2 - 2x_3 - x_3$$

$$\dot{x}_3 = 2x_2 - 3x_3 \qquad \qquad .....(4)$$

The state equation are given by equations (2), (3) and (4)

$$\dot{x}_1 = x_3$$

$$\dot{x}_2 = -2x_1 - 3x_2 + 2u$$

$$\dot{x}_3 = 2x_2 - 3x_3$$

The output equation is, $y = x_1$

The state model in the matrix form is,

$$\begin{bmatrix} \dot{x}_1 \\ \dot{x}_2 \\ \dot{x}_3 \end{bmatrix} = \begin{bmatrix} 0 & 0 & 1 \\ -2 & -3 & 0 \\ 0 & 2 & -3 \end{bmatrix} \begin{bmatrix} x_1 \\ x_2 \\ x_3 \end{bmatrix} + \begin{bmatrix} 0 \\ 2 \\ 0 \end{bmatrix} [u] \quad ; \quad y = \begin{bmatrix} 1 & 0 & 0 \end{bmatrix} \begin{bmatrix} x_1 \\ x_2 \\ x_3 \end{bmatrix}$$

To find eigenvalues

Here the system matrix, $A = \begin{bmatrix} 0 & 0 & 1 \\ -2 & -3 & 0 \\ 0 & 2 & -3 \end{bmatrix}$

The characteristic equation is, $|\lambda I - A| = 0$

$$[\lambda I - A] = \lambda \begin{bmatrix} 1 & 0 & 0 \\ 0 & 1 & 0 \\ 0 & 0 & 1 \end{bmatrix} - \begin{bmatrix} 0 & 0 & 1 \\ -2 & -3 & 0 \\ 0 & 2 & -3 \end{bmatrix} = \begin{bmatrix} \lambda & 0 & -1 \\ 2 & \lambda+3 & 0 \\ 0 & -2 & \lambda+3 \end{bmatrix}$$

$$|\lambda I - A| = \begin{vmatrix} \lambda & 0 & -1 \\ 2 & \lambda+3 & 0 \\ 0 & -2 & \lambda+3 \end{vmatrix} = \lambda(\lambda+3)^2 - 1(-4) = \lambda(\lambda^2 + 6\lambda + 9) + 4$$

$$= \lambda^3 + 6\lambda^2 + 9\lambda + 4 = (\lambda+1)(\lambda^2 + 5\lambda + 4)$$

$$= (\lambda+1)(\lambda+1)(\lambda+4) = (\lambda+1)^2(\lambda+4)$$

The eigenvalues are $\lambda_1 = -1$, $\lambda_2 = -1$, and $\lambda_3 = -4$

$\lambda = -1$	1	6	9	4
$\downarrow$		-1	-5	-4
	1	5	4	0

To find eigenvectors

$$[\lambda_1 I - A] = \lambda_1 \begin{bmatrix} 1 & 0 & 0 \\ 0 & 1 & 0 \\ 0 & 0 & 1 \end{bmatrix} - \begin{bmatrix} 0 & 0 & 1 \\ -2 & -3 & 0 \\ 0 & 2 & -3 \end{bmatrix} = \begin{bmatrix} \lambda_1 & 0 & -1 \\ 2 & \lambda_1+3 & 0 \\ 0 & -2 & \lambda_1+3 \end{bmatrix}$$

Let C_{11}, C_{12} and C_{13} be cofactors along Ist row of the matrix $[\lambda_1 I - A]$

$$C_{11} = (+1) \begin{vmatrix} \lambda_1+3 & 0 \\ -2 & \lambda_1+3 \end{vmatrix} = (\lambda_1+3)^2 = \lambda_1^2 + 6\lambda_1 + 9$$

$$C_{12} = (-1) \begin{vmatrix} 2 & 0 \\ 0 & \lambda_1+3 \end{vmatrix} = -(2(\lambda_1+3)) = -2\lambda_1 - 6$$

$$C_{13} = (+1) \begin{vmatrix} 2 & \lambda_1+3 \\ 0 & -2 \end{vmatrix} = -4$$

$$m_1 = \begin{bmatrix} C_{11} \\ C_{12} \\ C_{13} \end{bmatrix} = \begin{bmatrix} \lambda_1^2 + 6\lambda_1 + 9 \\ -2\lambda_1 - 6 \\ -4 \end{bmatrix} = \begin{bmatrix} 1 - 6 + 9 \\ 2 - 6 \\ -4 \end{bmatrix} = \begin{bmatrix} 4 \\ -4 \\ -4 \end{bmatrix}$$

$$m_2 = \begin{bmatrix} \dfrac{d}{d\lambda_1} C_{11} \\ \dfrac{d}{d\lambda_1} C_{12} \\ \dfrac{d}{d\lambda_1} C_{13} \end{bmatrix} = \begin{bmatrix} 2\lambda + 6 \\ -2 \\ 0 \end{bmatrix} = \begin{bmatrix} -2 + 6 \\ -2 \\ 0 \end{bmatrix} = \begin{bmatrix} 4 \\ -2 \\ 0 \end{bmatrix}$$

$$[\lambda_3 I - A] = (-4) \begin{bmatrix} 1 & 0 & 0 \\ 0 & 1 & 0 \\ 0 & 0 & 1 \end{bmatrix} - \begin{bmatrix} 0 & 0 & 1 \\ -2 & -3 & 0 \\ 0 & 2 & -3 \end{bmatrix} = \begin{bmatrix} -4 & 0 & -1 \\ 2 & -1 & 0 \\ 0 & -2 & -1 \end{bmatrix}$$

Let C_{11}, C_{12} and C_{13} be the cofactors along Ist row of the matrix $[\lambda_3 I - A]$.

$$C_{11} = (+1) \begin{vmatrix} -1 & 0 \\ -2 & -1 \end{vmatrix} = 1 \; ; \; C_{12} = (-1) \begin{vmatrix} 2 & 0 \\ 0 & -1 \end{vmatrix} = 2 \; ; \; C_{13} = (+1) \begin{vmatrix} 2 & -1 \\ 0 & -2 \end{vmatrix} = -4$$

$$\therefore \; m_3 = \begin{bmatrix} C_{11} \\ C_{12} \\ C_{13} \end{bmatrix} = \begin{bmatrix} 1 \\ 2 \\ -4 \end{bmatrix}$$

To find canonical form of state model

The modal matrix, M is given by,

$$M = [m_1 \; m_2 \; m_3] \begin{bmatrix} 4 & 4 & 1 \\ -4 & -2 & 2 \\ -4 & 0 & -4 \end{bmatrix}$$

$$M^{-1} = \frac{[\text{Cofactor of M}]^T}{\text{Determinant of M}} = \frac{M_{cof}^T}{D_M}$$

$$\Delta_M = \begin{vmatrix} 4 & 4 & 1 \\ -4 & -2 & 4 \\ -4 & 0 & -4 \end{vmatrix} = 4(8) - 4(24) + 1(-8) = 32 - 96 - 8 = -72$$

$$M_{cof}^T = \begin{bmatrix} 8 & -24 & -8 \\ 16 & -12 & -16 \\ 10 & -12 & 8 \end{bmatrix}^T = \begin{bmatrix} 8 & 16 & 10 \\ -24 & -12 & -12 \\ -8 & -16 & 8 \end{bmatrix}$$

$$M^{-1} = \frac{1}{-72} \begin{bmatrix} 8 & 16 & 10 \\ -24 & -12 & -12 \\ -8 & -16 & 8 \end{bmatrix} = \frac{1}{18} \begin{bmatrix} -2 & -4 & -2.5 \\ 6 & 3 & 3 \\ 2 & 4 & -2 \end{bmatrix}$$

$$J = M^{-1}AM = \frac{1}{18} \begin{bmatrix} -2 & -4 & -2.5 \\ 6 & 3 & 3 \\ 2 & 4 & -2 \end{bmatrix} \begin{bmatrix} 0 & 0 & 1 \\ -2 & -3 & 0 \\ 0 & 2 & -3 \end{bmatrix} \begin{bmatrix} 4 & 4 & 1 \\ -4 & -2 & 2 \\ -4 & 0 & -4 \end{bmatrix}$$

$$= \frac{1}{18} \begin{bmatrix} 8 & 7 & 5.5 \\ -6 & -3 & -3 \\ -8 & -16 & 8 \end{bmatrix} \begin{bmatrix} 4 & 4 & 1 \\ -4 & -2 & 2 \\ -4 & 0 & -4 \end{bmatrix}$$

$$= \frac{1}{18} \begin{bmatrix} -18 & 18 & 0 \\ 0 & -18 & 0 \\ 0 & 0 & -72 \end{bmatrix} = \begin{bmatrix} -1 & 1 & 0 \\ 0 & -1 & 0 \\ 0 & 0 & -4 \end{bmatrix} \quad \text{Jordan Block}$$

$$\tilde{B} = M^{-1}B = \frac{1}{18} \begin{bmatrix} -2 & -4 & -2.5 \\ 6 & 3 & 3 \\ 2 & 4 & -2 \end{bmatrix} \begin{bmatrix} 0 \\ 2 \\ 0 \end{bmatrix} = \begin{bmatrix} -8/18 \\ 6/18 \\ 8/18 \end{bmatrix} = \begin{bmatrix} -4/9 \\ 3/9 \\ 4/9 \end{bmatrix}$$

$$\tilde{C} = CM = \begin{bmatrix} 1 & 0 & 0 \end{bmatrix} \begin{bmatrix} 4 & 4 & 1 \\ -4 & -2 & 2 \\ -4 & 0 & -4 \end{bmatrix} = \begin{bmatrix} 4 & 4 & 1 \end{bmatrix}$$

The Jordan canonical form of state model is shown below,

$$\dot{Z} = JZ + \tilde{B}U \; ; \; Y = \tilde{C}Z + DU \quad \text{(Here DU is not defined)}$$

$$\begin{bmatrix} \dot{z}_1 \\ \dot{z}_2 \\ \dot{z}_3 \end{bmatrix} = \begin{bmatrix} -1 & 1 & 0 \\ 0 & -1 & 0 \\ 0 & 0 & -4 \end{bmatrix} \begin{bmatrix} z_1 \\ z_2 \\ z_3 \end{bmatrix} + \begin{bmatrix} -4/9 \\ 3/9 \\ 4/9 \end{bmatrix} [u] \; ; \; Y = \begin{bmatrix} 4 & 4 & 1 \end{bmatrix} \begin{bmatrix} z_1 \\ z_2 \\ z_3 \end{bmatrix}$$

CONCLUSION

It is observed that the elements of the rows of are not all zeros. Hence the system is completely controllable (or state controllable).

It is observed that the elements of the columns of are not all zeros. Hence the system is completely observable [i.e, all the state variables are observable].

ALTERNATE METHOD

KALMAN'S TEST FOR CONTROLLABILITY

$$A^2 = A.A = \begin{bmatrix} 0 & 0 & 1 \\ -2 & -3 & 0 \\ 0 & 2 & -3 \end{bmatrix} \begin{bmatrix} 0 & 0 & 1 \\ -2 & -3 & 0 \\ 0 & 2 & -3 \end{bmatrix} = \begin{bmatrix} 0 & 2 & -3 \\ 6 & 9 & -2 \\ -4 & -12 & 9 \end{bmatrix}$$

$$A.B = \begin{bmatrix} 0 & 0 & 1 \\ -2 & -3 & 0 \\ 0 & 2 & -3 \end{bmatrix}\begin{bmatrix} 0 \\ 2 \\ 0 \end{bmatrix} = \begin{bmatrix} 0 \\ -6 \\ 4 \end{bmatrix}$$

$$A^2.B = \begin{bmatrix} 0 & 2 & -3 \\ 6 & 9 & -2 \\ -4 & -12 & 9 \end{bmatrix}\begin{bmatrix} 0 \\ 2 \\ 0 \end{bmatrix} = \begin{bmatrix} 4 \\ 18 \\ -24 \end{bmatrix}$$

The composite matrix for controllability, $Q_C = \begin{bmatrix} B & AB & A^2B \end{bmatrix}$

$$= \begin{bmatrix} 0 & 0 & 4 \\ 2 & -6 & 18 \\ 0 & 4 & -24 \end{bmatrix}$$

Determinant of $Q_C = \begin{vmatrix} 0 & 0 & 4 \\ 2 & -6 & 18 \\ 0 & 4 & -24 \end{vmatrix} = 4 \times 8 = 32$; Since $|Q_C| \neq 0$, the rank of $Q_C = 3$

Hence the system is completely state controllable.

KALMAN'S TEST FOR OBSERVABILITY

$$A^T = \begin{bmatrix} 0 & 0 & 1 \\ -2 & -3 & 0 \\ 0 & 2 & -3 \end{bmatrix}^T = \begin{bmatrix} 0 & -2 & 0 \\ 0 & -3 & 2 \\ 1 & 0 & -3 \end{bmatrix}$$

$$C^T = \begin{bmatrix} 1 & 0 & 0 \end{bmatrix}^T = \begin{bmatrix} 1 \\ 0 \\ 0 \end{bmatrix}$$

$$(A^T)^2 = \begin{bmatrix} 0 & -2 & 0 \\ 0 & -3 & 2 \\ 1 & 0 & -3 \end{bmatrix}\begin{bmatrix} 0 & -2 & 0 \\ 0 & -3 & 2 \\ 1 & 0 & -3 \end{bmatrix} = \begin{bmatrix} 0 & 6 & -4 \\ 2 & 9 & -12 \\ -3 & -2 & 9 \end{bmatrix}$$

$$A^T C^T = \begin{bmatrix} 0 & -2 & 0 \\ 0 & -3 & 2 \\ 1 & 0 & -3 \end{bmatrix}\begin{bmatrix} 1 \\ 0 \\ 0 \end{bmatrix} = \begin{bmatrix} 0 \\ 0 \\ 1 \end{bmatrix}$$

$$(A^T)^2 C^T = \begin{bmatrix} 0 & 6 & -4 \\ 2 & 9 & -12 \\ -3 & -2 & 9 \end{bmatrix}\begin{bmatrix} 1 \\ 0 \\ 0 \end{bmatrix} = \begin{bmatrix} 0 \\ 2 \\ -3 \end{bmatrix}$$

The composite matrix for observability $\left.\begin{matrix}\\\\\end{matrix}\right\}$ $Q_0 = \begin{bmatrix} C^T & A^T C^T & (A^T)^2 C^T \end{bmatrix} = \begin{bmatrix} 1 & 0 & 0 \\ 0 & 0 & 2 \\ 0 & 1 & -3 \end{bmatrix}$

Determinant of $Q_0 = \begin{vmatrix} 1 & 0 & 0 \\ 0 & 0 & 2 \\ 0 & 1 & -3 \end{vmatrix} = 1 \times -2 = -2$; Since $|Q_0| \neq 0$, the rank of $Q_0 = 3$

Hence the system is completely observable. (or all the state variables of the system are observable).

6.11 STATE SPACE REPRESENTATION OF DISCRETE TIME SYSTEMS

The state variable analysis techniques of continuous time systems can be extended to the discrete-time system. The discrete form of state space representation is quite analogous to the continuous form.

In the state variable formulation of a discrete time system, in general, a system consists of m-inputs, p-outputs and n-state variables. The state space representation of discrete-time system may be visualized as shown in fig 6.6.

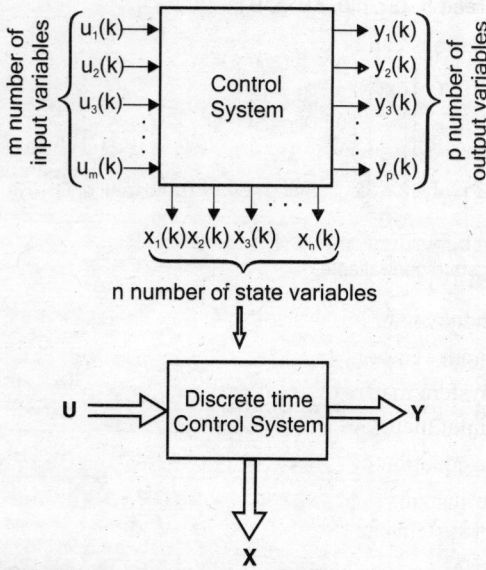

Fig 6.6 : State space representation of discrete time system.

Let,   State variables = $x_1(k)$, $x_2(k)$, $x_3(k)$, ... $x_n(k)$

       Input variables = $u_1(k)$, $u_2(k)$, $u_3(k)$, ... $u_m(k)$

       Output variables = $y_1(k)$, $y_2(k)$, $y_3(k)$, ... $y_p(k)$

The different variables may be represented by the vectors (column matrix) as shown below.

$$\text{Input vector } U(k) = \begin{bmatrix} u_1(k) \\ u_2(k) \\ \vdots \\ u_m(k) \end{bmatrix} \quad ; \quad \text{Output vector } Y(k) = \begin{bmatrix} y_1(k) \\ y_2(k) \\ \vdots \\ y_p(k) \end{bmatrix} \quad ; \quad \text{State variable vector } X(k) = \begin{bmatrix} x_1(k) \\ x_2(k) \\ \vdots \\ x_n(k) \end{bmatrix}$$

> **Note :** *The simplified notation x(k), y(k) and u(k) are used to denote x(kT), y(kT) and u(kT) respectively. Also for convenience the variables are denoted* x_1, x_2, x_3, ... ; y_1, y_2, y_3, ... *and* u_1, u_2, u_3 .....

The state equation of a discrete time system is a set of n-numbers of first order difference equations.

$$x_1(k + 1) = f_1(x_1, x_2, ...x_n \; ; \; u_1, u_2, ...u_m)$$

$$x_2(k + 1) = f_2(x_1, x_2, ...x_n \; ; \; u_1, u_2, ...u_m)$$

$$\vdots$$

$$x_n(k + 1) = f_n(x_1, x_2, ...x_n \; ; \; u_1, u_2, ...u_m)$$

For linear time invariant discrete time systems the above difference equations can be expressed as a linear combination of state variables and inputs.

$$x_1(k+1) = a_{11}x_1 + a_{12}x_2 + \dots + a_{1n}x_n + b_{11}u_1 + b_{12}u_2 + \dots + b_{1m}u_m$$
$$x_2(k+1) = a_{21}x_1 + a_{22}x_2 + \dots + a_{2n}x_n + b_{21}u_1 + b_{22}u_2 + \dots + b_{2m}u_m$$
$$\vdots$$
$$x_n(k+1) = a_{n1}x_1 + a_{n2}x_2 + \dots + a_{nn}x_n + b_{n1}u_1 + b_{n2}u_2 + \dots + b_{nm}u_m$$

where, the coefficients a_{ij} and b_{ij} are constants.

In the matrix form the above equations can be expressed as,

$$\begin{bmatrix} x_1(k+1) \\ x_2(k+1) \\ \vdots \\ x_n(k+1) \end{bmatrix} = \begin{bmatrix} a_{11} & a_{12} & \cdots & a_{1n} \\ a_{21} & a_{22} & \cdots & a_{2n} \\ \vdots & \vdots & & \vdots \\ a_{n1} & a_{n2} & \cdots & a_{nn} \end{bmatrix} \begin{bmatrix} x_1 \\ x_2 \\ \vdots \\ x_n \end{bmatrix} + \begin{bmatrix} b_{11} & b_{12} & \cdots & b_{1m} \\ b_{21} & b_{22} & \cdots & b_{2m} \\ \vdots & \vdots & & \vdots \\ b_{n1} & b_{n2} & \cdots & b_{nm} \end{bmatrix} \begin{bmatrix} u_1 \\ u_2 \\ \vdots \\ u_m \end{bmatrix} \quad \dots(6.121)$$

The matrix equation (6.121) can be written in the vector notation as

$$\mathbf{X}(k+1) = \mathbf{A}\mathbf{X}(k) + \mathbf{B}\mathbf{U}(k) \quad \dots(6.122)$$

where, $\mathbf{X}(k)$ = State vector of order (n × 1)
$\mathbf{U}(k)$ = Input vector of order (m × 1)
$\mathbf{A}$ = System matrix of order (n × n)
$\mathbf{B}$ = Input matrix of order (n × m)

The equation (6.122) is the state equation of (linear time invariant) discrete time system.

The output at any discrete time instant, k are functions of state variables and inputs. Hence the output variables of linear time invariant system can be expressed as a linear combination of state variables and inputs.

$$y_1 = c_{11}x_1 + c_{12}x_2 + \dots + c_{1n}x_n + d_{11}u_1 + d_{12}u_2 + \dots + d_{1m}u_m$$
$$y_2 = c_{21}x_1 + c_{22}x_2 + \dots + c_{2n}x_n + d_{21}u_1 + d_{22}u_2 + \dots + d_{2m}u_m$$
$$\vdots$$
$$y_p = c_{p1}x_1 + c_{p2}x_2 + \dots + c_{pn}x_n + d_{p1}u_1 + d_{p2}u_2 + \dots + d_{pm}u_m$$

where, the coefficients c_{ij} and d_{ij} are constants.

In the matrix form the above equations can be expressed as,

$$\begin{bmatrix} y_1 \\ y_2 \\ \vdots \\ y_p \end{bmatrix} = \begin{bmatrix} c_{11} & c_{12} & \cdots & c_{1n} \\ c_{21} & c_{22} & \cdots & c_{2n} \\ \vdots & \vdots & & \vdots \\ c_{p1} & c_{p2} & \cdots & c_{pn} \end{bmatrix} \begin{bmatrix} x_1 \\ x_2 \\ \vdots \\ x_n \end{bmatrix} + \begin{bmatrix} d_{11} & d_{12} & \cdots & d_{1m} \\ d_{21} & d_{22} & \cdots & d_{2m} \\ \vdots & \vdots & & \vdots \\ d_{p1} & d_{p2} & \cdots & d_{pm} \end{bmatrix} \begin{bmatrix} u_1 \\ u_2 \\ \vdots \\ u_m \end{bmatrix} \quad \dots(6.123)$$

The matrix equation (6.123) can be written in the vector notation as,

$$\mathbf{Y}(k) = \mathbf{C}\mathbf{X}(k) + \mathbf{D}\mathbf{U}(k) \quad \dots(6.124)$$

where, $\mathbf{X}(k)$ = State vector of order (n × 1)
$\mathbf{U}(k)$ = Input vector of order (m × 1)
$\mathbf{Y}(k)$ = Output vector of order (p × 1)
$\mathbf{C}$ = Output matrix of order (p × n)
$\mathbf{D}$ = Transmission matrix of order (p × m)

The equation (6.124) is the output equation of (linear time invariant) discrete time system.

The state equation and output equation are together called as state model of the system. Hence the state model of discrete time system is given by the following equations.

$$X(k+1) = A X(k) + B U(k) \quad \text{........ State equation}$$

$$Y(k) = C X(k) + D U(k) \quad \text{......... Output equation}$$

STATE DIAGRAM OF DISCRETE TIME SYSTEM

The state diagram of discrete time system can be either in block diagram form or signal flow graph form. The three fundamental elements, scalar, adder and unit delay elements are used to construct the state diagram. These basic elements are shown in table-6.6.

Scalar : The scalar is used to multiply a signal by a constant.

Adder : The adder is used to add two or more signals.

Unit delay : The unit delay element will delay the signal passing through it by one sample time.

TABLE-6.6 : Basic Elements of State Diagram of Discrete Time System

Element	Block diagram representation	Signal flow graph representation
Scalar	$x(k) \rightarrow \boxed{a} \rightarrow ax(k)$	$x(k) \xrightarrow{a} a\,x(k)$
Adder	$x_1(k), x_2(k) \rightarrow \oplus \rightarrow x_1(k)+x_2(k)$	$x_1(k), x_2(k) \xrightarrow{1,\,1} x_1(k) + x_2(k)$
Unit delay	$x(k+1) \rightarrow \boxed{z^{-1}} \rightarrow x(k)$	$x(k+1) \xrightarrow{z^{-1}} x(k)$

The block diagram representation of the state model of discrete time system is shown in fig 6.7. and the signal flow graph representation is shown in fig 6.8.

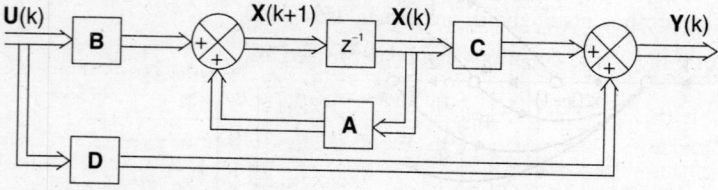

Fig 6.7 : Block diagram representation of discrete time system.

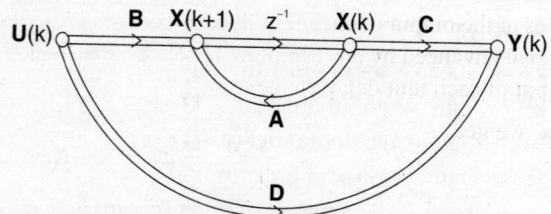

Fig 6.8 : Signal flow graph representation of discrete time system.

PHASE VARIABLE FORM OF STATE MODEL

The discrete time system is governed by n^{th} order difference equation. The general form of nth order difference equation is,

$$y(k) = -\sum_{j=1}^{n} a_j \, y(k-j) + \sum_{j=0}^{m} b_j \, u(k-j) \qquad .....(6.125)$$

where, a_j and b_j are constants for time invariant system.

On expanding the summation of equation (6.125) and rearranging we get,

$$y(k) + a_1 \, y(k-1) + a_2 \, y(k-2) +.........+a_n \, y(k-n) = b_0 u(k) + b_1 u(k-1) + b_2 u(k-2) +......+ b_m u(k-m)$$
$$.....(6.126)$$

On taking $\mathbb{Z}$-transform of equation (6.126) with zero initial conditions we get,

$$y(z) + a_1 \, z^{-1} y(z) + a_2 \, z^{-2} y(z) +.........+ a_n \, z^{-n} y(z) = b_0 U(z) + b_1 \, z^{-1} U(z) + b_2 \, z^{-2} U(z) +......+ b_m z^{-m} U(z)$$

$$Y(z)\left[1 + a_1 \, z^{-1} + a_2 \, z^{-2} +......+ a_n \, z^{-n} \right] = \left[b_0 + b_1 \, z^{-1} + b_2 \, z^{-2} +........+ b_m \, z^{-m} \right] U(z)$$

When $m = n$,

$$\frac{Y(z)}{U(z)} = \frac{b_0 + b_1 z^{-1} + b_2 z^{-2} + ... + b_m z^{-m}}{1 + a_1 z^{-1} + a_2 z^{-2} + ... + a_n z^{-n}} \qquad .....(6.127)$$

The equation (6.127) is the transfer function of the discrete time system. The equation (6.127) can be expressed as shown below.

$$\frac{Y(z)}{U(z)} = \frac{b_0 + b_1 z^{-1} + b_2 z^{-2} + ... + b_m z^{-m}}{1 - (-a_1 z^{-1} - a_2 z^{-2} - ..... - a_n z^{-n})} \qquad .....(6.128)$$

On comparing the equation (6.128) with Mason's gain formula, a signal flow graph can be constructed. Each numerator term of equation (6.128) represent a forward path gain and each denominator term of equation (6.128) represent a individual loop gain. The signal flow graph will not have any non-touching loops. The signal flow graph when n=3 is shown in fig 6.9.

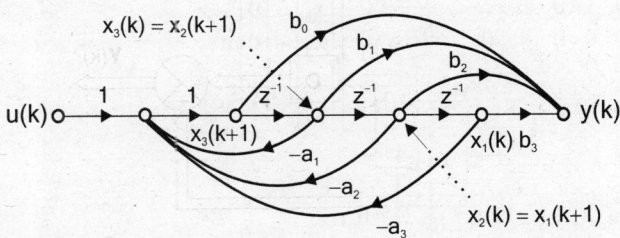

Fig 6.9 : *Signal flow graph representation of discrete time system.*

Let us assign state variables at the output of each unit delay element. Hence the signal at the input of unit delay element will be a signal advanced by one sampling time. The state equations are obtained by summing up the signals at the input of each unit delay element.

In fig 6.9, at node x_2 and x_3 we get,

$$x_1(k+1) = x_2 \qquad .....(6.129)$$

$$x_2(k+1) = x_3 \qquad .....(6.130)$$

By summing up the incoming signals to node $x_3(k+1)$ we get, (refer fig 6.9a)

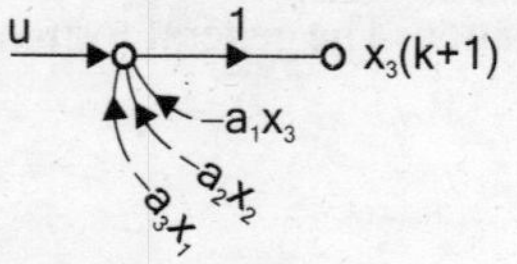

$$x_3(k+1) = -a_3 x_1 - a_2 x_2 - a_1 x_3 + u \qquad .....(6.131)$$

Fig 6.9a.

By summing up the incoming signals to node y we get, (refer fig 6.9b).

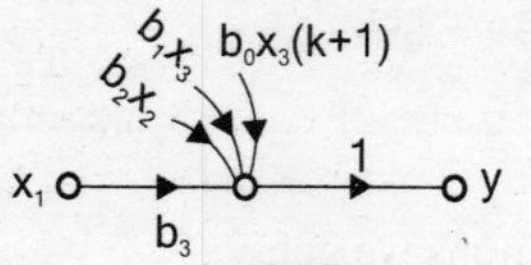

$$y = b_3 x_1 + b_2 x_2 + b_1 x_3 + b_0 x_3(k+1) \qquad .....(6.132)$$

Fig 6.9b.

Substitute for $x_3(k+1)$ from equation (6.131) in equation (6.132).

$$y = b_3 x_1 + b_2 x_2 + b_1 x_3 + b_0 (-a_3 x_1 - a_2 x_2 - a_1 x_3 + u)$$

$$\therefore y = (b_3 - b_0 a_3) x_1 + (b_2 - b_0 a_2) x_2 + (b_1 - b_0 a_1) x_3 + b_0 u \qquad .....(6.133)$$

The equations (6.129), (6.130) and (6.131) are state equations and equation (6.133) is the output equation. On arranging the state equations and output equations in the matrix form we get,

$$\begin{bmatrix} x_1(k+1) \\ x_2(k+1) \\ x_3(k+1) \end{bmatrix} = \begin{bmatrix} 0 & 1 & 0 \\ 0 & 0 & 1 \\ -a_3 & -a_2 & -a_1 \end{bmatrix} \begin{bmatrix} x_1 \\ x_2 \\ x_3 \end{bmatrix} + \begin{bmatrix} 0 \\ 0 \\ 1 \end{bmatrix} u \qquad .....(6.134)$$

$$y = [(b_3 - b_0 a_3)(b_2 - b_0 a_2)(b_1 - b_0 a_1)] \begin{bmatrix} x_1 \\ x_2 \\ x_3 \end{bmatrix} + b_0 u \qquad .....(6.135)$$

The equations (6.134) and (6.135) can be extended to n^{th} order system as shown below.

$$\begin{bmatrix} x_1(k+1) \\ x_2(k+1) \\ x_3(k+1) \\ \vdots \\ x_n(k+1) \end{bmatrix} = \begin{bmatrix} 0 & 1 & 0 & \cdots & 0 \\ 0 & 0 & 1 & \cdots & 0 \\ 0 & 0 & 0 & \cdots & 0 \\ \vdots & \vdots & \vdots & & \vdots \\ -a_n & -a_{n-1} & -a_{n-2} & \cdots & -a_1 \end{bmatrix} \begin{bmatrix} x_1 \\ x_2 \\ x_3 \\ \vdots \\ x_n \end{bmatrix} + \begin{bmatrix} 0 \\ 0 \\ 0 \\ \vdots \\ 1 \end{bmatrix} u \qquad .....(6.136)$$

$$y = [(b_n - b_0 a_n)(b_{n-1} - b_0 a_{n-1})] \begin{bmatrix} x_1 \\ x_1 \\ \vdots \\ x_1 \end{bmatrix} + b_0 u \qquad .....(6.137)$$

The equations (6.136) and (6.137) is the phase variable form of state model of discrete time n^{th} order system.

CANONICAL FORM OF STATE MODEL

The transfer function of equation (6.127) can be expressed as a summation using partial fraction technique as shown below.

$$\frac{Y(z)}{U(z)} = b_0 + \frac{C_1}{z + \lambda_1} + \frac{C_2}{z + \lambda_2} + ... + \frac{C_n}{z + \lambda_n}$$

where C_1, C_2 ,...... C_n are residues and λ_1 , λ_2 , ... λ_n are poles of the system.

$$\frac{Y(z)}{U(z)} = b_0 + \frac{C_1}{z\left(1 + \frac{\lambda_1}{z}\right)} + \frac{C_2}{z\left(1 + \frac{\lambda_2}{z}\right)} + \dots\dots + \frac{C_n}{z\left(1 + \frac{\lambda_n}{z}\right)}$$

$$\frac{Y(z)}{U(z)} = b_0 + \frac{z^{-1}C_1}{1 + z^{-1}\lambda_1} + \frac{z^{-1}C_2}{1 + z^{-1}\lambda_2} + \dots\dots + \frac{z^{-1}C_n}{1 + z^{-1}\lambda_n} \qquad\dots..(6.138)$$

The equation (6.138) can be used to construct the block diagram of state model as shown in fig (6.10).

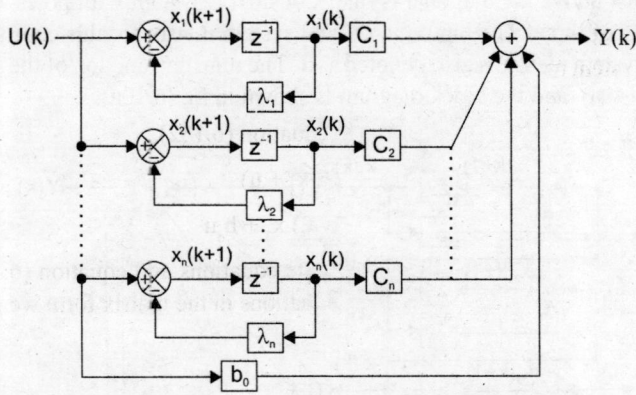

Fig 6.10 : *Block diagram of canonical state model of discrete time system.*

Let us assign the state variables at the output of unit delay elements. Hence the signal at the input of unit delay element will be a signal advanced by one sampling time. The state equations are obtained by summing up the signals at the input of each unit delay element.

The state equations are

$$x_1(k+1) = -\lambda_1 x_1 + u$$

$$x_2(k+1) = -\lambda_2 x_2 + u$$

$$x_3(k+1) = -\lambda_3 x_3 + u$$

$$\vdots \qquad \vdots$$

$$x_n(k+1) = -\lambda_n x_n + u$$

The output equation is

$$y = c_1 x_1 + c_2 x_2 + c_3 x_3 + \dots.. + c_n x_n + b_0 u$$

The state equations and output equation can be expressed in the matrix form as shown below. The equations (6.139) and (6.140) is the canonical form of state model of discrete time n^{th} order system.

$$\begin{bmatrix} x_1(k+1) \\ x_2(k+1) \\ x_3(k+1) \\ \vdots \\ x_n(k+1) \end{bmatrix} = \begin{bmatrix} -\lambda_1 & 0 & 0 & \cdots & 0 \\ 0 & -\lambda_2 & 0 & \cdots & 0 \\ 0 & 0 & -\lambda_3 & \cdots & 0 \\ \vdots & \vdots & \vdots & & \vdots \\ 0 & 0 & 0 & \cdots & -\lambda_n \end{bmatrix} \begin{bmatrix} x_1 \\ x_2 \\ x_3 \\ \vdots \\ x_n \end{bmatrix} + \begin{bmatrix} 1 \\ 1 \\ 1 \\ \vdots \\ 1 \end{bmatrix} u \qquad\dots..(6.139)$$

$$y = \begin{bmatrix} C_1 & C_2 & C_3 & \cdots & C_n \end{bmatrix} \begin{bmatrix} x_1 \\ x_2 \\ x_3 \\ \vdots \\ x_n \end{bmatrix} + b_0 u \qquad \qquad .....(6.140)$$

When a pole of the transfer function has multiplicity then the state model will be in a special form called Jordan canonical form. In this form the system matrix **A** will have a Jordan block of size q×q, correspond to a pole of value λ_i with multiplicity q. In the Jordan block the diagonal element will be the poles and the element just above the diagonal is one. Consider a system with poles $\lambda_1, \lambda_2, \lambda_3, \lambda_4, ... \lambda_n$, where λ_1 has multiplicity of three. The input matrix and system matrix for this case will be as shown in equation (6.139a). The system matrix is also denoted as **J**. The transfer function of the system for this case is given by equation (6.140a) and the block diagram is shown in fig (6.10a).

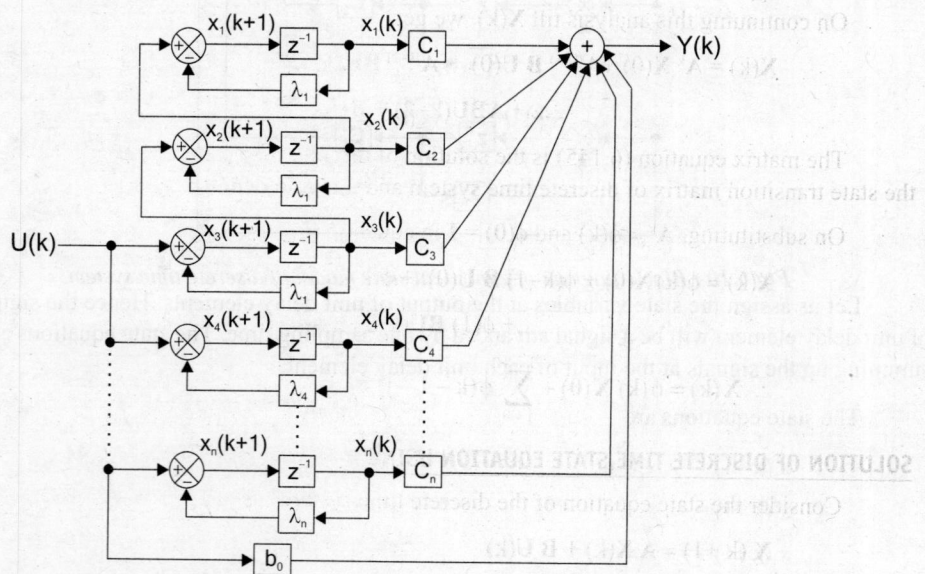

Fig 6.10a : *Block diagram of Jordan canonical state model of discrete time system.*

$$B = \begin{bmatrix} 0 \\ 0 \\ 1 \\ 1 \\ \vdots \\ 1 \end{bmatrix} \quad ; \quad A = J = \begin{bmatrix} -\lambda_1 & 1 & 0 & 0 & \cdots & 0 \\ 0 & -\lambda_1 & 1 & 0 & \cdots & 0 \\ 0 & 0 & -\lambda_1 & 0 & \cdots & 0 \\ 0 & 0 & 0 & -\lambda_4 & \cdots & 0 \\ \vdots & \vdots & \vdots & \vdots & & \vdots \\ 0 & 0 & 0 & 0 & \cdots & -\lambda_n \end{bmatrix} \qquad .....(6.139a)$$

$$\frac{Y(z)}{U(z)} = b_0 + \frac{C_1}{(z+\lambda_1)^3} + \frac{C_2}{(z+\lambda_1)^2} + \frac{C_3}{z+\lambda_1} + \frac{C_4}{z+\lambda_4} + ... + \frac{C_n}{z+\lambda_n} \qquad .....(6.140a)$$

SOLUTION OF DISCRETE TIME STATE EQUATION

The state equation of discrete time system is given by

$$\mathbf{X}(k+1) = \mathbf{A}\,\mathbf{X}(k) + \mathbf{B}\,\mathbf{U}(k) \qquad .....(6.141)$$

When k = 0, the equation (6.141) can be written as

$$\mathbf{X}(1) = \mathbf{A}\,\mathbf{X}(0) + \mathbf{B}\,\mathbf{U}(0) \qquad .....(6.142)$$

When k = 1, the equation (6.141) can be written as

$$\mathbf{X}(2) = \mathbf{A}\,\mathbf{X}(1) + \mathbf{B}\,\mathbf{U}(1) \qquad .....(6.143)$$

On substituting for $\mathbf{X}(1)$ from equation (6.142) in equation (6.143) we get,

$$\mathbf{X}(2) = \mathbf{A}\,[\mathbf{A}\,\mathbf{X}(0) + \mathbf{B}\,\mathbf{U}(0)] + \mathbf{B}\mathbf{U}(1)$$

$$\therefore \mathbf{X}(2) = \mathbf{A}^2\,\mathbf{X}(0) - \mathbf{A}\,\mathbf{B}\,\mathbf{U}(0) + \mathbf{B}\mathbf{U}(1) \qquad .....(6.144)$$

On continuing this analysis till $\mathbf{X}(k)$ we get,

$$\mathbf{X}(k) = \mathbf{A}^k\,\mathbf{X}(0) + \mathbf{A}^{(k-1)}\,\mathbf{B}\,\mathbf{U}(0) + \mathbf{A}^{(k-2)}\,\mathbf{B}\mathbf{U}(1) + ...$$

$$......+ \mathbf{A}\mathbf{B}\mathbf{U}(k-2) + \mathbf{B}\mathbf{U}(k-1) \qquad .....(6.145)$$

The matrix equation (6.145) is the solution of discrete time state equation. The matrix $\mathbf{A}^k$ is called the state transition matrix of discrete time system and it is also denoted by $\phi(k)$.

On substituting, $\mathbf{A}^k = \phi(k)$ and $\phi(0) = \mathbf{I}$ in equation (6.145) we get

$$\mathbf{X}(k) = \phi(k)\,\mathbf{X}(0) + \phi(k-1)\,\mathbf{B}\,\mathbf{U}(0) + \phi(k-2)\,\mathbf{B}\mathbf{U}(1) + ...$$

$$......+ \phi(1)\,\mathbf{B}\mathbf{U}(k-2) + \phi(0)\,\mathbf{B}\mathbf{U}(k-1)$$

$$\therefore \mathbf{X}(k) = \phi(k)\,\mathbf{X}(0) + \sum_{j=0}^{k-1} \phi(k-1-j)\,\mathbf{B}\mathbf{U}(j) \qquad .....(6.146)$$

SOLUTION OF DISCRETE TIME STATE EQUATION USING $\mathbb{Z}$-TRANSFORM

Consider the state equation of the discrete time system

$$\mathbf{X}(k+1) = \mathbf{A}\,\mathbf{X}(k) + \mathbf{B}\,\mathbf{U}(k) \qquad .....(6.147)$$

On taking $\mathbb{Z}$-transform of equation (6.147) we get

$$z\,\mathbf{X}(z) - z\,\mathbf{X}(0) = \mathbf{A}\,\mathbf{X}(z) + \mathbf{B}\,\mathbf{U}(z)$$

$$z\,\mathbf{X}(z) - \mathbf{A}\,\mathbf{X}(z) = z\,\mathbf{X}(0) + \mathbf{B}\,\mathbf{U}(z)$$

$$(z\mathbf{I} - \mathbf{A})\,\mathbf{X}(z) = z\,\mathbf{X}(0) + \mathbf{B}\,\mathbf{U}(z) \qquad .....(6.148)$$

On premultiplying the equation (6.148) by $(z\mathbf{I}-\mathbf{A})^{-1}$ we get

$$\mathbf{X}(z) = (z\mathbf{I} - \mathbf{A})^{-1}\,z\,\mathbf{X}(0) + (z\mathbf{I} - \mathbf{A})^{-1}\,\mathbf{B}\,\mathbf{U}(z) \qquad .....(6.149)$$

On taking inverse $\mathbb{Z}$-transform of equation (6.149) we get $\mathbf{X}(k)$

$$\therefore \mathbf{X}(k) = \mathbb{Z}^{-1}\{\mathbf{X}(z)\} = \mathbb{Z}^{-1}\{(z\mathbf{I} - \mathbf{A})^{-1}\,z\mathbf{X}(0) + (z\mathbf{I} - \mathbf{A})^{-1}\,\mathbf{B}\mathbf{U}(z)\}$$

$$\mathbf{X}(k) = \mathbb{Z}^{-1}\{(z\mathbf{I} - \mathbf{A})^{-1}z\}\mathbf{X}(0) + \mathbb{Z}^{-1}\{(z\mathbf{I} - \mathbf{A})^{-1}\,\mathbf{B}\mathbf{U}(z)\} \qquad .....(6.150)$$

The equation (6.150) is the solution of discrete time state equation.

PROPERTIES OF STATE TRANSITION MATRIX OF DISCRETE TIME SYSTEM

1. $\phi(0) = \mathbf{I}$

2. $\phi^{-1}(k) = \phi(-k)$

3. $\phi(k, k_0) = \phi(k - k_0) = \mathbf{A}^{(k-k_0)}$; where, $k > k_0$

COMPUTATION OF STATE TRANSITION MATRIX

The state transition matrix $\mathbf{A}^k$ can be computed by any one of the following methods.

Method 1 : Computation of $\mathbf{A}^k$ using $\mathbf{Z}$-transform

Method 2 : Computation of $\mathbf{A}^k$ by canonical transformation

Method 3 : Computation of $\mathbf{A}^k$ by Cayley - Hamilton theorem

The computation of $\mathbf{A}^k$ using $\mathbf{Z}$-transform have been dealt in this section.

On comparing equations (6.151) and (6.146) [or (6.145)] we can write,

State transition matrix, $\mathbf{A}^k = \mathbf{Z}^{-1}\{(z\mathbf{I} - \mathbf{A})^{-1}z\}$ .....(6.151)

The equation (6.151) can be used to compute the state transition matrix, $\mathbf{A}^k$.

EXAMPLE 6.20

A discrete-time system has the transfer function

$$\frac{Y(z)}{U(z)} = \frac{4z^3 - 12z^2 + 13z - 7}{(z - 1)^2(z - 2)}$$

Determine the state model of the system in (a) Phase variable form and (b) Jordan canonical form.

SOLUTION

a. Phase variable form of state model

Given that, $\dfrac{Y(z)}{U(z)} = \dfrac{4z^3 - 12z^2 + 13z - 7}{(z - 1)^2(z - 2)}$

$$= \frac{z^3(4 - 12z^{-1} + 13z^2 - 7z^3)}{(z^2 - 2z + 1)(z - 2)} = \frac{z^3(4 - 12z^{-1} + 13z^{-2} - 7z^{-3})}{z^3 - 2z^2 - 2z^2 + 4z + z - 2}$$

$$= \frac{z^3(4 - 12z^{-1} + 13z^2 - 7z^3)}{z^3(1 - 4z^{-1} + 5z^{-2} - 2z^{-3})} = \frac{4 - 12z^{-1} + 13z^{-2} - 7z^{-3}}{1 - (4z^{-1} - 5z^{-2} + 2z^{-3})} \qquad .....(6.20.1)$$

The equation (6.20.1) can be used to construct the signal flow graph of the discrete time system shown in fig 1.

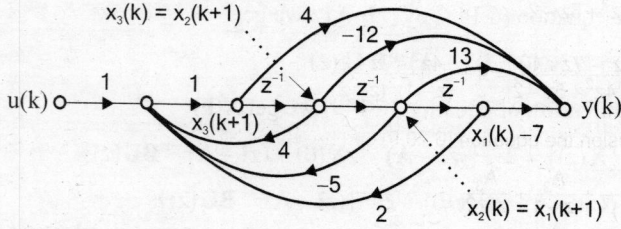

Fig 1 : Signal flow graph for transfer function of equation (6.20.1).

From node x_2 and x_3 we get,

$$x_1(k+1) = x_2 \qquad\qquad .....(6.20.2)$$

$$x_2(k+1) = x_3 \qquad\qquad .....(6.20.3)$$

By summing up the incoming signals to node $x_3(k+1)$ we get, (refer fig 6.20.2)

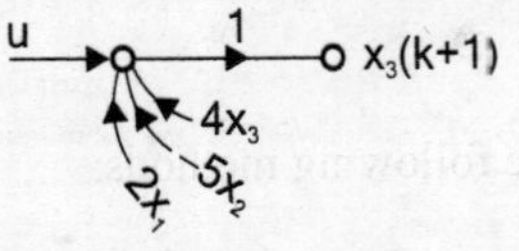

$$x_3(k+1) = 2x_1 - 5x_2 + 4x_3 + u \qquad\qquad .....(6.20.4)$$

Fig 2.

The equations (6.20.2) to (6.20.4) are the state equations. The output equation is obtained by summing up the incoming signals to output node (Refer fig 3)

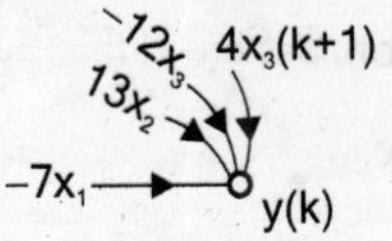

$$y = -7x_1 + 13x_2 - 12x_3 + 4x_3(k+1) \qquad\qquad .....(6.20.5)$$

Fig 3.

On substituting for $x_3(k+1)$ from equation (6.20.4) in equation (6.20.5) we get,

$$y = -7x_1 + 13x_2 - 12x_3 + 4[2x_1 - 5x_2 + 4x_3 + u]$$

$$y = (-7 + 8)x_1 + (13 - 20)x_2 + (-12 + 16)x_3 + 4u$$

$$\therefore\ y = x_1 - 7x_2 + 4x_3 + 4u \qquad\qquad .....(6.20.6)$$

The state equations and output equation can be arranged in the matrix form as shown below.

$$\begin{bmatrix} x_1(k+1) \\ x_2(k+1) \\ x_3(k+1) \end{bmatrix} = \begin{bmatrix} 0 & 1 & 0 \\ 0 & 0 & 1 \\ 2 & -5 & 4 \end{bmatrix} \begin{bmatrix} x_1 \\ x_2 \\ x_3 \end{bmatrix} + \begin{bmatrix} 0 \\ 0 \\ 1 \end{bmatrix} u \qquad\qquad .....(6.20.7)$$

$$y = \begin{bmatrix} 1 & -7 & 4 \end{bmatrix} \begin{bmatrix} x_1 \\ x_2 \\ x_3 \end{bmatrix} + 4u \qquad\qquad .....(6.20.8)$$

The equations (6.20.7) and (6.20.8) is the phase variable form of state model.

b. Jordan Canonical form of state model

Given that, $\dfrac{Y(z)}{U(z)} = \dfrac{4z^3 - 12z^2 + 13z - 7}{(z-1)^2(z-2)}$

$$= \frac{4z^3 - 12z^2 - 13z - 7}{z^3 - 4z^2 + 5z - 2}$$

	4	
$z^3 - 4z^2 + 5z - 2$	$4z^3 - 12z^2 + 13z - 7$	
	$4z^3 - 16z^2 + 20z - 8$	
	$4z^2 - 7z + 1$	

$$= 4 + \frac{4z^2 - 7z + 1}{z^3 - 4z^2 + 5z - 2} = 4 + \frac{4z^2 - 7z + 1}{(z-1)^2(z-2)} \qquad\qquad .....(6.20.9)$$

By partial fraction expansion the equation (6.20.9) can be written as,

$$\frac{Y(z)}{U(z)} = 4 + \frac{A_1}{(z-1)^2} + \frac{A_2}{z-1} + \frac{A_3}{z-2}$$

$$A_1 = (z-1)^2 \frac{4z^2 - 7z + 1}{(z-1)^2(z-2)} \bigg|_{z=1} = \frac{4z^2 - 7z + 1}{z-2} \bigg|_{z=1} = \frac{4-7+1}{1-2} = 2$$

$$A_2 = \frac{d}{dz}\left[(z-1)^2 \frac{4z^2 - 7z + 1}{(z-1)^2(z-2)}\right]\Bigg|_{z=1} = \frac{d}{dz}\left[\frac{4z^2 - 7z + 1}{z-2}\right]\Bigg|_{z=1}$$

$$= \frac{(8z-7)(z-2) - (4z^2 - 7z + 1)}{(z-2)^2}\Bigg|_{z=1} = \frac{(8-7)(1-2) - (4-7+1)}{(1-2)^2} = 1$$

$$A_3 = (z-2) \frac{4z^2 - 7z + 1}{(z-1)^2(z-2)}\Bigg|_{z=2} = \frac{4z^2 - 7z + 1}{(z-1)^2}\Bigg|_{z=2} = \frac{4 \times 2^2 - 7 \times 2 + 1}{(2-1)^2} = 3$$

$$\therefore \frac{Y(z)}{U(z)} = 4 + \frac{2}{(z-1)^2} + \frac{1}{z-1} + \frac{3}{z-2} \qquad\qquad .....(6.20.10)$$

The equation (6.20.10) can be used to construct the block diagram of the discrete time system shown in fig 4.

Let us assign state variable at the output of each unit delay element. The state equations are formed by summing up the incoming signals to the unit delay element.

The state equations are,

$$x_1(k+1) = x_1 + x_2$$
$$x_2(k+1) = x_2 + u$$
$$x_3(k+1) = 2x_3 + u$$

The output equation is,

$$y = 2x_1 + x_2 + 3x_3 + 4u$$

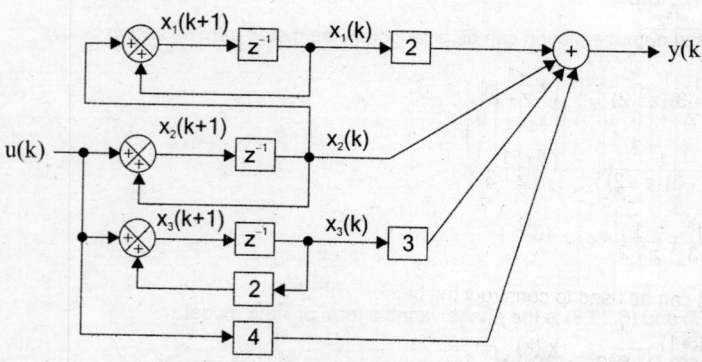

Fig 4 : Block diagram.

On arranging the state equations and the output equation in the matrix form we get,

$$\begin{bmatrix} x_1(k+1) \\ x_2(k+1) \\ x_3(k+1) \end{bmatrix} = \begin{bmatrix} 1 & 1 & 0 \\ 0 & 1 & 0 \\ 0 & 0 & 2 \end{bmatrix}\begin{bmatrix} x_1 \\ x_2 \\ x_3 \end{bmatrix} + \begin{bmatrix} 0 \\ 1 \\ 1 \end{bmatrix}u \qquad\qquad .....(6.20.11)$$

$$y = \begin{bmatrix} 2 & 1 & 3 \end{bmatrix}\begin{bmatrix} x_1 \\ x_2 \\ x_3 \end{bmatrix} + 4u \qquad\qquad .....(6.20.12)$$

The equations (6.20.11) and (6.20.12) constitute Jordan canonical form of state model of the system.

EXAMPLE 6.21

A discrete time system is described by the difference equation,

$$y(k + 2) + 5y(k + 1) + 6\,y(k) = u(k)$$

$$y(0) = y(1) = 0; \quad T = 1 sec.$$

(a)   Determine a state model in canonical form.

(b)   Find the state transition matrix

(c)   For input $u(k) = 1$; $k \geq 1$, find the output $y(k)$.

SOLUTION

(a) To determine the canonical form of state model

Given that, $y(k + 2) + 5y(k + 1) + 6\,y(k) = u(k)$ .....(6.21.1)

and $y(0) = y(1) = 0;$ $T = 1 sec.$

On taking z-transform of equation (6.21.1) with zero initial conditions we get,

$$z^2\,Y(z) + 5z\,Y(z) + 6Y(z) = U(z)$$

$$(z^2 + 5z + 6)\,Y(z) = U(z)$$

$$\therefore \frac{Y(z)}{U(z)} = \frac{1}{z^2 + 5z + 6} = \frac{1}{(z + 3)(z + 2)} \qquad .....(6.21.2)$$

By partial fraction expansion the equation (6.21.2) can be expressed as,

$$\frac{Y(z)}{U(z)} = \frac{A_1}{z + 3} + \frac{A_2}{z + 2}$$

$$A_1 = (z + 3)\,\frac{1}{(z + 3)(z + 2)}\bigg|_{z = -3} = \frac{1}{z + 2}\bigg|_{z = -3} = \frac{1}{-3 + 2} = -1$$

$$A_2 = (z + 2)\,\frac{1}{(z + 3)(z + 2)}\bigg|_{z = -2} = \frac{1}{z + 3}\bigg|_{z = -2} = \frac{1}{-2 + 3} = 1$$

$$\therefore \frac{Y(z)}{U(z)} = \frac{-1}{z + 3} + \frac{1}{z + 2} \qquad .....(6.21.3)$$

The equation (6.21.3) can be used to construct the block diagram shown in fig 1.

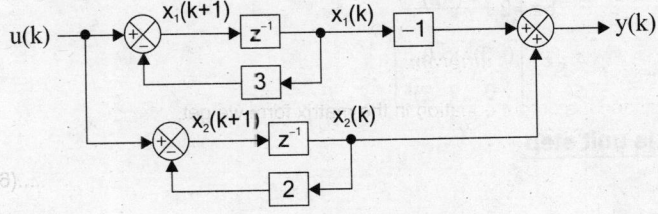

Fig 1 : *Block diagram.*

From the block diagram, the state equations are

$$x_1(k + 1) = -3x_1 + u$$

$$x_2(k + 1) = -2x_2 + u$$

The output equation is

$$y = -x_1 + x_2$$

On arranging the state equations and the output equation in the matrix form, we get,

$$\begin{bmatrix} x_1(k+1) \\ x_2(k+1) \end{bmatrix} = \begin{bmatrix} -3 & 0 \\ 0 & -2 \end{bmatrix} \begin{bmatrix} x_1 \\ x_2 \end{bmatrix} + \begin{bmatrix} 1 \\ 1 \end{bmatrix} u \qquad .....(6.21.4)$$

$$y = \begin{bmatrix} -1 & 1 \end{bmatrix} \begin{bmatrix} x_1 \\ x_2 \end{bmatrix} \qquad .....(6.21.5)$$

The equations (6.21.4) and (6.21.5) constitute the canonical form of state model.

(b) To find state transition matrix A^k

The state transition matrix, $A^k = \mathcal{z}^{-1}\{[zI - A]^{-1}z\}$

From the state equation we get, $A = \begin{bmatrix} -3 & 0 \\ 0 & -2 \end{bmatrix}$

$$\therefore zI - A = z\begin{bmatrix} 1 & 0 \\ 0 & 1 \end{bmatrix} - \begin{bmatrix} -3 & 0 \\ 0 & -2 \end{bmatrix} = \begin{bmatrix} z+3 & 0 \\ 0 & z+2 \end{bmatrix}$$

$$|zI - A| = \begin{vmatrix} z+3 & 0 \\ 0 & z+2 \end{vmatrix} = (z+2)(z+3)$$

$$[zI - A]^{-1} = \frac{1}{|zI - A|}\begin{bmatrix} z+2 & 0 \\ 0 & z+3 \end{bmatrix}^T = \frac{1}{(z+2)(z+3)}\begin{bmatrix} z+2 & 0 \\ 0 & z+3 \end{bmatrix}$$

$$= \begin{bmatrix} \dfrac{1}{z+3} & 0 \\ 0 & \dfrac{1}{z+2} \end{bmatrix}$$

$$A^k = \mathcal{z}^{-1}\{[zI - A]^{-1}z\} = \mathcal{z}^{-1}\begin{bmatrix} \dfrac{z}{z+3} & 0 \\ 0 & \dfrac{z}{z+2} \end{bmatrix} \qquad .....(6.21.6)$$

We know that $\mathcal{z}^{-1}\left\{\dfrac{z}{z-a}\right\} = a^k$

$$\therefore \mathcal{z}^{-1}\left\{\dfrac{z}{z+3}\right\} = (-3)^k \quad \text{and} \quad \mathcal{z}^{-1}\left\{\dfrac{z}{z+2}\right\} = (-2)^k$$

$$\therefore \text{State transition matrix,} \qquad A^k = \begin{bmatrix} (-3)^k & 0 \\ 0 & (-2)^k \end{bmatrix} \qquad .....(6.21.7)$$

(c) To find y(k) when the input is unit step

Consider the state equation,

$$X(k+1) = A X(k) + B U(k) \qquad .....(6.21.8)$$

On taking $\mathcal{z}$-transform of equation(6.21.8) we get

$$z X(z) - z X(0) = AX(z) + BU(z)$$

Let us assume that, $X(0) = 0$

$$\therefore z X(z) - A X(z) = B U(z)$$

$$(zI - A) X(z) = B U(z) \qquad .....(6.21.9)$$

On premultiplying both sides of equation (6.21.9) by $(zI - A)^{-1}$ we get

$$X(z) = (zI - A)^{-1} BU(z) \qquad \qquad .....(6.21.10)$$

Given that, $u(k) = 1, \quad \therefore \; U(z) = Z\{u(k)\} = \dfrac{z}{z-1} \qquad \qquad .....(6.21.11)$

From equation (6.21.6) we get, $[zI - A]^{-1} = \begin{bmatrix} \dfrac{1}{z+3} & 0 \\ 0 & \dfrac{1}{z+2} \end{bmatrix}$

From equation (6.21.4) we get, $B = \begin{bmatrix} 1 \\ 1 \end{bmatrix}$

$$\therefore \; X(z) = \begin{bmatrix} \dfrac{1}{z+3} & 0 \\ 0 & \dfrac{1}{z+2} \end{bmatrix} \begin{bmatrix} 1 \\ 1 \end{bmatrix} \begin{bmatrix} \dfrac{z}{z-1} \end{bmatrix} = \begin{bmatrix} \dfrac{1}{z+3} \\ \dfrac{1}{z+2} \end{bmatrix} \begin{bmatrix} \dfrac{z}{z-1} \end{bmatrix} = \begin{bmatrix} \dfrac{z}{(z-1)(z+3)} \\ \dfrac{z}{(z-1)(z+2)} \end{bmatrix}$$

We know that $X(z) = \begin{bmatrix} x_1(z) \\ x_2(z) \end{bmatrix}$ and $X(k) = \begin{bmatrix} x_1(k) \\ x_2(k) \end{bmatrix}$

Also $X(k) = Z^{-1}\{X(z)\}$

$$\therefore \; X(z) = \begin{bmatrix} X_1(z) \\ X_2(z) \end{bmatrix} = \begin{bmatrix} \dfrac{z}{(z-1)(z+3)} \\ \dfrac{z}{(z-1)(z+2)} \end{bmatrix} \text{ and } X(k) = \begin{bmatrix} x_1(z) \\ x_2(z) \end{bmatrix} = \begin{bmatrix} Z^{-1}\{X_1(z)\} \\ Z^{-1}\{X_2(z)\} \end{bmatrix}$$

$$\therefore \; x_1(k) = Z^{-1}\left\{ \dfrac{z}{(z-1)(z+3)} \right\} \text{ and } x_2(k) = Z^{-1}\left\{ \dfrac{z}{(z-1)(z+2)} \right\}$$

By partial fraction expansion,

$$\dfrac{z}{(z-1)(z+3)} = z\left[\dfrac{1}{(z-1)(z+3)} \right] = z\left[\dfrac{A_1}{z-1} + \dfrac{B_1}{z+3} \right]$$

$$A_1 = \dfrac{z}{(z-1)(z+3)}(z-1)\bigg|_{z=1} = \dfrac{1}{z+3}\bigg|_{z=1} = \dfrac{1}{1+3} = \dfrac{1}{4}$$

$$B_1 = \dfrac{1}{(z-1)(z+3)}(z+3)\bigg|_{z=-3} = \dfrac{1}{z-1}\bigg|_{z=-3} = \dfrac{1}{-3-1} = -\dfrac{1}{4}$$

$$\therefore \; \dfrac{z}{(z-1)(z+3)} = z\left[\dfrac{A_1}{z-1} + \dfrac{B_1}{z+3} \right] = \left[\dfrac{1}{4}\dfrac{z}{z-1} - \dfrac{1}{4}\dfrac{z}{z+3} \right]$$

$$x_1(k) = Z^{-1}\left\{ \dfrac{z}{(z-1)(z+3)} \right\} = Z^{-1}\left[\dfrac{1}{4}\dfrac{z}{z-1} - \dfrac{1}{4}\dfrac{z}{z+3} \right] = \dfrac{1}{4}u(k) - \dfrac{1}{4}(-3)^k$$

By partial fraction expansion,

$$\dfrac{z}{(z-1)(z+2)} = z\left[\dfrac{1}{(z-1)(z+2)} \right] = z\left[\dfrac{A_2}{z-1} + \dfrac{B_2}{z+2} \right]$$

$$A_2 = \dfrac{1}{(z-1)(z+2)}(z-1)\bigg|_{z=1} = \dfrac{1}{z+2}\bigg|_{z=1} = \dfrac{1}{1+2} = \dfrac{1}{3}$$

$$B_2 = \dfrac{1}{(z-1)(z+2)}(z+2)\bigg|_{z=-2} = \dfrac{1}{z-1}\bigg|_{z=-2} = \dfrac{1}{-2-1} = -\dfrac{1}{3}$$

$$\therefore \frac{z}{(z-1)(z+2)} = z\left[\frac{A_2}{z-1} + \frac{B_2}{z+2}\right] = \left[\frac{1}{3}\frac{z}{z-1} - \frac{1}{3}\frac{z}{z+2}\right]$$

$$x_2(k) = z^{-1}\left\{\frac{z}{(z-1)(z+2)}\right\} = z^{-1}\left[\frac{1}{3}\frac{z}{z-1} - \frac{1}{3}\frac{z}{z+2}\right] = \frac{1}{3}u(k) - \frac{1}{3}(-2)^k$$

From equation (6.21.5) we get,

Response or Output, $y(k) = \begin{bmatrix} -1 & 1 \end{bmatrix}\begin{bmatrix} x_1(k) \\ x_2(k) \end{bmatrix}$

$$\therefore \ y(k) = -x_1(k) + x_2(k)$$

$$= -\frac{1}{4}u(k) + \frac{1}{4}(-3)^k + \frac{1}{3}u(k) - \frac{1}{3}(-2)^k$$

$$= \frac{1}{4}(-3)^k - \frac{1}{3}(-2)^k + \frac{-3+4}{12}u(k)$$

$$= \frac{1}{4}(-3)^k - \frac{1}{3}(-2)^k + \frac{1}{12}u(k)$$

6.12 STATE SPACE ANALYSIS USING MATLAB

STATE SPACE ANALYSIS OF CONTINUOUS TIME SYSTEM

In general, the transfer function of a continuous time system is denoted as G(s).

Let, G(s) be a rational function of "s", as shown below.

$$G(s) = \frac{b_0 s^M + b_1 s^{M-1} + b_2 s^{M-2} + \dots + b_{M-1}s + b_M}{a_0 s^N + a_1 s^{N-1} + a_2 s^{N-2} + \dots + a_{N-1}s + a_N}$$

For state space analysis, the transfer function G(s) is declared as a function of "s" using the following commands.

```
s=tf('s');
Gs=(b0*s^M+b1*s^(M-1)+...+bM)/(a0*s^N+a1*s^(N-1)+...+aN);
```

For state space analysis, the coefficients of the numerator and denominator polynomials are declared as two arrays as shown below.

```
NUM_COF = [B0 B1 B2 ........ BM];
DEN_COF = [A0 A1 A2 ........ AN];
```

For state space analysis, the elements of A, B, C, D matrices of state model are declared as arrays as shown below.

```
A = [A11 A12...;   A21 A22...;   A31 A32...;   ........];
B = [B11 B12...;   B21 B22...;   B31 B32...;   ........];
C = [C11 C12...;   C21 C22...;   C31 C32...;   ........];
D = [D11 D12...;   D21 D22...;   D31 D32...;   ........];
```

The state model of a continuous time system can be determined from the transfer function of the system using the following command.

```
[A, B, C, D]=TF2SS(NUM_COF, DEN_COF);
```

The transfer function of a continuous time system can be determined from the state model of the system using the following command.

```
[num_cof, den_cof]=ss2tf(A, B, C, D);
Gs = tf(num_cof, den_cof);
```

The controllability matrix of the state model of a continuous time system can be determined using the following command.

```
Qc = ctrb(A, B);
```

The observability matrix of the state model of a continuous time system can be determined using the following command.

```
Qo = obsv(A, C);
```

STATE SPACE ANALYSIS OF DISCRETE TIME SYSTEM

In general, the transfer function of a discrete time system is denoted as $G(z)$.

Let, $G(z)$ be a rational function of "z", as shown below.

$$G(z) = \frac{b_0 z^M + b_1 z^{M-1} + b_2 z^{M-2} + ... + b_{M-1} z + b_M}{a_0 z^N + a_1 z^{N-1} + a_2 z^{N-2} + ... + a_{N-1} z + a_N}$$

For state space analysis, the coefficients of the numerator and denominator polynomials of discrete time system transfer function are declared as two arrays as shown below.

```
num_cof = [b0 b1 b2 ........ bM];
den_cof = [a0 a1 a2 ........ aN];
```

For state space analysis, the transfer function $G(z)$ is declared as a function of "z" using the following commands.

```
z=tf('z'); Ts = SamplingTime;
Gz=tf(num_cof, den_cof, Ts);
```

For state space analysis, the elements of A, B, C, D matrices of state model are declared as arrays as shown below.

```
A = [a11 a12...;   a21 a22...;   a31 a32...;   .........];
B = [b11 b12...;   b21 b22...;   b31 b32...;   .........];
C = [c11 c12...;   c21 c22...;   c31 c32...;   .........];
D = [d11 d12...;   d21 d22...;   d31 d32...;   .........];
```

The state model of a discrete time system can be determined from the transfer function of the system using the following commands.

```
z=tf('z'); Ts = SamplingTime;
Gz=tf(num_cof, den_cof, Ts);
[A, B, C, D]=tf2ss(num_cof, den_cof);
```

The transfer function of a discrete time system can be determined from the state model of the system using the following commands.

```
syms z; Ts=SamplingTime;
[num_cof, den_cof]=ss2tf(A, B, C, D);
Gz=tf(num_cof, den_cof, Ts);
```

PROGRAM 6.1

Write a MATLAB program to determine the state model of the system governed by the following transfer function.

$$G(s) = 10/(s^3+4s^2+2s+1)$$

```
%Program to determine the state model from the transfer function
clc
clear all
s=tf('s');
disp('The given transfer function is,');
Gs=10/(s^3+4*s^2+2*s+1)

disp('Numerator & denominator coefficients of transfer function are,');
num_cof=[0 0 0 10]
den_cof=[1 4 2 1]

disp('The state model of the given system is,');
[A,B,C,D]=tf2ss(num_cof, den_cof)      %compute matrices A,B,C,D
                                       %of state model
```

OUTPUT

```
The given transfer function is,
Transfer function:
             10
    ---------------------
    s^3 + 4 s^2 + 2 s + 1
Numerator & denominator coefficients of given transfer function are,
num_cof =
        0     0     0    10

den_cof =
        1     4     2     1
The state model of the given system is,
A =
    -4    -2    -1
     1     0     0
     0     1     0
B =
     1
     0
     0
C =
     0     0    10
D =
     0
```

PROGRAM 6.2

Write a MATLAB program to determine the transfer function of the system governed by the following state model.

$$A=\begin{bmatrix} 0 & 1 & 0 \\ 0 & 0 & 1 \\ -1 & -2 & -4 \end{bmatrix} \quad ; \quad B=\begin{bmatrix} 0 \\ 0 \\ 10 \end{bmatrix} \quad ; \quad C=[1\ 0\ 0] \quad ; \quad D=[0]$$

```
%Program to determine transfer function from state model.

clear all
clc
disp('The given state model is,');
A=[0 1 0; 0 0 1; -1 -2 -4]
B=[0; 0; 10]
C=[1 0 0]
D=[0]

disp('Numerator & dencminator coefficients of transfer function are,');
[num_cof, den_cof]=ss2tf(A,B,C,D)
disp('The tranfer function of the system is,');
Gs=tf(num_cof, den_cof)
```

OUTPUT

```
The given state model is,
A =
      0      1      0
      0      0      1
     -1     -2     -4
B =
      0
      0
     10
C =
      1      0      0
D =
      0
Numerator & denominator coefficients of the transfer function are,
num_cof =
          0      0.0000      0.0000     10.0000

den_cof =
     1.0000      4.0000      2.0000      1.0000

The tranfer function cf the system is,
Transfer function:
          6.217e-015 s^2 + 1.776e-015 s + 10
          -----------------------------------
                 s^3 + 4 s^2 + 2 s + 1
```

PROGRAM 6.3

Write a MATLAB program to determine the controllability of the system governed by the following state model.

$$A = \begin{bmatrix} 0 & 0 & 1 \\ -2 & -3 & 0 \\ 0 & 2 & -3 \end{bmatrix} \quad ; \quad B = \begin{bmatrix} 0 \\ 2 \\ 0 \end{bmatrix} \quad ; \quad C = [1\ 0\ 0] \quad ; \quad D = [0]$$

```
%Program to determine controllability of state model of a system
clear all
clc
disp('The given state model is,');
A=[0 0 1;-2 -3 0;0 2 -3]    % Elements of System matrix 'A'
B=[0;2;0]                   % Elements of Input Matrix 'B'
```

```
C=[1 0 0]                        % Elements of Output Matrix 'C'
D=[0]                            % Elements of Transmission Matrix 'D'

disp('The controllability matrix is,');
Qc=ctrb(A,B)                     % evaluating Controllability matrix
disp('Number of uncontrollable states, N_ucs is,');
N_ucs=length(A)-rank(Qc)

disp('Determinant of controllability matrix is,');
Det_Qc = det(Qc)                 % Determinant of Controllability Matrix
rank_Qc=rank(Qc)                 % rank of the system matrix
n=size(A);                       % order of the system matrix

if Det_Qc ~= 0 & n==rank_Qc;
    disp('The system is completely controllable.');
else
    disp('The system is not completely controllable.');
end;
```

OUTPUT

```
The given state model is,
A =
        0       0       1
       -2      -3       0
        0       2      -3
B =
        0
        2
        0
C =
        1       0       0
D =
        0

The controllability matrix is,
Qc =
        0       0       4
        2      -6      18
        0       4     -24

Number of uncontrollable states, N_ucs is,
N_ucs =
        0

Determinant of controllability matrix is,
Det_Qc =
       32

rank_Qc =
        3

The system is completely controllable.
```

PROGRAM 6.4

Write a MATLAB program to determine the observability of the system governed by the following state model.

$$A = \begin{bmatrix} 0 & 0 & 1 \\ -2 & -3 & 0 \\ 0 & 2 & -3 \end{bmatrix} \quad ; \quad B = \begin{bmatrix} 0 \\ 2 \\ 0 \end{bmatrix} \quad ; \quad C = [1\ 0\ 0] \quad ; \quad D = [0]$$

```
%Program to determine observability of state model of a system
clear all
clc
disp('The given state model is,');
A=[0 0 1;-2 -3 0;0 2 -3]        % Elements of System matrix 'A'
B=[0;2;0]                       % Elements of Input Matrix 'B'
C=[1 0 0]                       % Elements of Outout Matrix 'C'
D=[0]                           % Elements of Transmission Matrix 'D'

disp('The observability matrix is,');
Qo = obsv(A,C)                  % evaluating observability matrix
disp('Number of unobservable states, N_uos');
N_uos=length(A)-rank(Qo)

disp('Determinant of observability Matrix is');
Det_Qo = det(Qo)                % Determinant of observability matrix
rank_Qo=rank(Qo)                % rank of the system matrix
n=size(A);                      % order of the system matrix

if Det_Qo ~= 0 & n==rank_Qo;disp('The system is completely observable.');
else
    disp('The system is not completely observable.');
end;
```

OUTPUT

```
The given state model is,
A =
     0     0     1
    -2    -3     0
     0     2    -3
B =
     0
     2
     0
C =
     1     0     0
D =
     0

The observability matrix is,
Qo =
     1     0     0
     0     0     1
     0     2    -3

Number of unobservable states, N_uos
N_uos =
     0

Determinant of observability Matrix is
Det_Qo =
    -2

rank_Qo =
     3

The system is completely observable.
```

PROGRAM 6.5

Write a MATLAB program to determine the state model of the discrete time system governed by the following transfer function.

$$G(z) = (4z^3-12z^2+13z-7)/(z^3-4z^2+5z-2)$$

```
%Program to determine the state model
%of discrete time system from the transfer function
clc
clear all

disp('Numerator & denominator coefficients of transfer function are,');
num_cof=[4 -12 13 -7]
den_cof=[1 -4 5 -2]

z=tf('z');
Ts=0.1;

disp('The transfer function of the given system is,');
Gz=tf(num_cof, den_cof, Ts)

disp('The state model of the given system is,');
[A,B,C,D]=tf2ss(num_cof,den_cof)      %compute matrices A,B,C,D
                                      %of state model
```

OUTPUT

Numerator & denominator coefficients of the given transfer function are,

num_cof =

4	-12	13	-7

den_cof =

1	-4	5	-2

The transfer function of the given system is,

Transfer function:

$$\frac{4\ z^3\ -\ 12\ z^2\ +\ 13\ z\ -\ 7}{z^3\ -\ 4\ z^2\ +\ 5\ z\ -\ 2}$$

Sampling time: 0.1

The state model of the given system is,

A =

4	-5	2
1	0	0
0	1	0

B =

1
0
0

C =

4	-7	1

D =

4

PROGRAM 6.6

Write a MATLAB program to determine the transfer function of the discrete time system governed by the following state model.

$$A = \begin{bmatrix} 0 & 1 & 0 \\ 0 & 0 & 1 \\ 2 & -5 & 4 \end{bmatrix} \quad ; \quad B = \begin{bmatrix} 0 \\ 0 \\ 1 \end{bmatrix} \quad ; \quad C = \begin{bmatrix} 1 & -7 & 4 \end{bmatrix} \quad ; \quad D = [4]$$

```
%Program to determine the transfer function
%of discrete time system from the state model

clear all
clc

disp('The given state model of discrete time system is,');
A=[0 1 0; 0 0 1; 2 -5 4]
B=[0; 0; 1]
C=[1 -7 4]
D=[4]

syms z
Ts=0.1;    %specify sampling time

disp('Numerator & denominator coefficients of transfer function are,');
[num_cof, den_cof]=ss2tf(A,B,C,D)

disp('The tranfer function of the discrete time system is,');
Gz=tf(num_cof, den_co=,Ts)
```

OUTPUT

```
The given state model of discrete time system is,

A =
     0     1     0
     0     0     1
     2    -5     4
B =
     0
     0
     1
C =
     1    -7     4
D =
     4


Numerator & denominator coefficients of the transfer function are,
num_cof =
     4.0000  -12.0000   13.0000   -7.0000

den_cof =
     1.0000   -4.0000    5.0000   -2.0000

The tranfer function of the discrete time system is,
Transfer function:
     4 z^3 - 12 z^2 + 13 z - 7
     -------------------------
     z^3 - 4 z^2 + 5 z - 2
Sampling time: 0.1
```

6.13 SHORT - ANSWERS QUESTIONS

Q6.1   *What are the advantages of state space analysis?*

1. The state space analysis is applicable to any type of systems. They can be used for modelling and analysis of linear & non-linear systems, time invariant & time variant systems and multiple input & multiple output systems.

2. The state space analysis can be performed with initial conditions.

3. The variables used to represent the system can be any variables in the system.

4. Using this analysis the internal states of the system at any time instant can be predicted.

Q6.2   *What are the drawbacks in transfer function model analysis?*

1. Transfer function is defined under zero initial conditions.

2. Transfer function is applicable to linear time invariant systems.

3. Transfer function analysis is restricted to single input and output systems.

4. Does not provides information regarding the internal state of the system.

Q6.3   *What is state and state variable?*

The state is the condition of a system at any time instant, t. A set of variable which describes the state of the system at any time instant are called state variables.

Q6.4   *What is a state vector?*

The state vector is a $(n \times 1)$ column matrix (or vector) whose elements are state variables of the system, (where n is the order of the system). It is denoted by $X(t)$.

Q6.5   *Write the state model of nth order system?*

The state model of a system consists of state equation and output equation. The state model of a nth order system with m-inputs and p-outputs are

$$\dot{X}(t) = AX(t) + BU(t) \qquad .....\text{state equation}$$

$$Y(t) = CX(t) + DU(t) \qquad .....\text{output equation}$$

where $X(t)$ = State vector of order $(n \times 1)$; $U(t)$ = Input vector of order $(m \times 1)$

A = System matrix of order $(n \times n)$; B = Input matrix of order $(n \times m)$

$Y(t)$ = Output vector of order $(p \times 1)$; C = Output matrix of order $(p \times n)$.

D = Transmission matrix of order $(p \times m)$.

Q6.6   *What is state space?*

The set of all possible values which the state vector $X(t)$ can have (or assume) at time t forms the state space of the system.

Q6.7   *What is input and output space?*

The set of all possible values which the input vector $U(t)$ can have (or assume) at time t forms the input space of the system.

The set of all possible values which the output vector $Y(t)$ can have (or assume) at time t forms the output space.

Q6.8   *The state model of a linear time invariant system is given by*

$$\dot{X}(t) = AX(t) + BU(t)$$

$$Y(t) = CX(t) + DU(t)$$

Obtain the expression for transfer function of the system.

Solution : Given that $\dot{X}(t) = AX(t) + BU(t)$ .....(1)

and $Y(t) = CX(t) + DU(t)$ .....(2)

On taking Laplace transform of equation (1) with zero initial conditions we get,

$$s X(s) = A X(s) + B U(s)$$

$$s X(s) - A X(s) = B U(s)$$

$$(sI - A) X(s) = B U(s) \qquad .....(3)$$

On premultiplying equation (3) by $(sI-A)^{-1}$ we get

$$X(s) = (sI - A)^{-1} B U(s) \qquad .....(4)$$

On taking Laplace transform of equation (2) we get,

$$Y(s) = C X (s) + D U(s) \qquad .....(5)$$

Substitute for X(s) from equation (4) in equation (5) we get,

$$Y(s) = C [(sI - A)^{-1} B U(s)] + D U (s)$$

$$= [C (sI - A)^{-1} B + D] U (s)$$

$$\therefore \frac{Y(s)}{U(s)} = C (sI - A)^{-1} B + D \qquad .....(6)$$

The equation (6) is the transfer function of the system.

Q6.9 **What is state diagram?**

The pictorial representation of the state model of the system is called state diagram. The state diagram of the system can be either in block diagram or in signal flow graph form.

Q6.10 **Draw the block diagram representation of state model?**

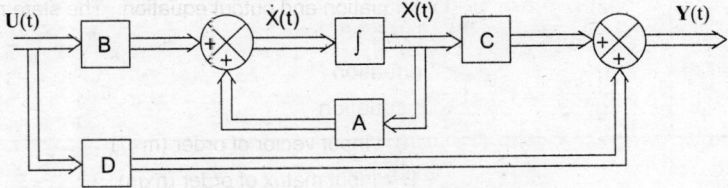

Fig Q6.10 : *Block diagram of state model.*

Q6.11 **Draw the signal flow graph representation of state model?**

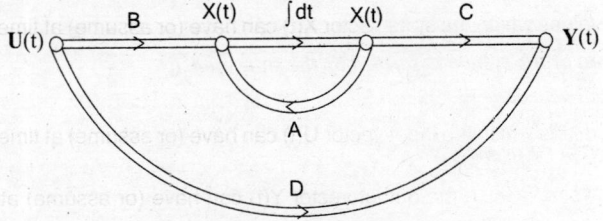

Fig Q6.11 : *Signal flow graph of state model.*

Q6.12 **What are the basic elements used to construct the state diagram?**

The basic elements used to construct the state diagram are Scalar, Adder and Integrator.

Q6.13 **Sketch the basic elements used to construct the block diagram of a state model.**

The basic elements used to construct block diagram of a state model are shown in the following table.

Elements	Time domain	s-domain
Scalar	$x(t)$ → a → $a\,x(t)$	$X(s)$ → a → $a\,X(s)$
Adder	$x_1(t)$ → (+) → $x_1(t)+x_2(t)$, $x_2(t)$	$X_1(s)$ → (+) → $X_1(s)+X_2(s)$, $X_2(s)$
Integrator	$\dot{x}(t)$ → $\int$ → (+) → $\int_0^t \dot{x}(T)dT + x(0)$, $x(0)$	$X(s)$ → $1/s$ → (+) → $\dfrac{X(s)}{s}+\dfrac{X(0)}{s}$, $X(0)/s$

Q6.14 *Sketch the basic elements used to construct the signal flow graph of a state model.*

The basic elements used to construct the signal flow graph of a state model are shown in the following table.

Elements	Time domain	s-domain
Scalar	$x(t)$ —a→ $a\,x(t)$	$X(s)$ —a→ $a\,X(s)$
Adder	$x_1(t)$ —1→, $x_2(t)$ —1→ $x_1(t) + x_2(t)$	$X_1(s)$ —1→, $X_2(s)$ —1→ $X_1(s) + X_2(s)$
Integrator	$\dot{x}(t)$ —$\int dt$→ $x(t)+x(0)$, $x(0)$ —1→	$X(s)$ —$\dfrac{1}{s}$→ $\dfrac{X(s)}{s}+\dfrac{X(0)}{s}$, $\dfrac{X(0)}{s}$ —1→

Q6.15 *Draw the block diagram of the system described by the state model,*

$$\begin{bmatrix} \dot{x}_1 \\ \dot{x}_2 \\ \dot{x}_3 \end{bmatrix} = \begin{bmatrix} 0 & 1 & 0 \\ 0 & 0 & 1 \\ 0 & a_2 & a_3 \end{bmatrix} \begin{bmatrix} x_1 \\ x_2 \\ x_3 \end{bmatrix} + \begin{bmatrix} 0 \\ 0 \\ 1 \end{bmatrix} u \text{ and } y = x_1$$

Solution

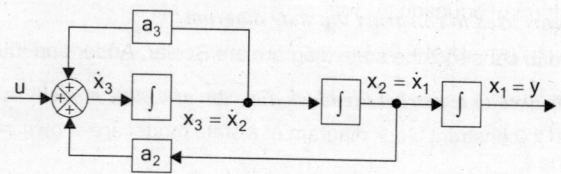

Q6.16   *Draw the signal flow graph of the system described by the state model.*

$$\begin{bmatrix} \dot{x}_1 \\ \dot{x}_2 \\ \dot{x}_3 \end{bmatrix} = \begin{bmatrix} a_1 & a_2 & 0 \\ 1 & 0 & 1 \\ 0 & 1 & 0 \end{bmatrix} \begin{bmatrix} x_1 \\ x_2 \\ x_3 \end{bmatrix} + \begin{bmatrix} 1 \\ 0 \\ 0 \end{bmatrix} u \quad and \quad y = x_3$$

<u>Solution</u>

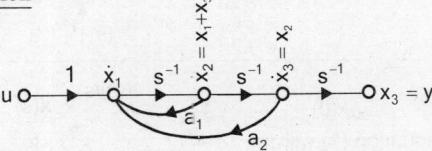

Q6.17   *Determine the state model of the system represented by the block diagram of fig Q6.17.*

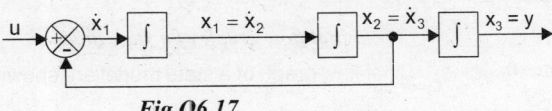

Fig Q6.17

<u>Solution</u>

$$\begin{bmatrix} \dot{x}_1 \\ \dot{x}_2 \\ \dot{x}_3 \end{bmatrix} = \begin{bmatrix} 0 & 1 & 0 \\ 0 & 0 & 1 \\ 0 & -2 & a_1 \end{bmatrix} \begin{bmatrix} x_1 \\ x_2 \\ x_3 \end{bmatrix} + \begin{bmatrix} 1 \\ 0 \\ 0 \end{bmatrix} u \quad and \quad y = x_3$$

Q6.18   *Determine the state model of the system represented by the signal flow graph of fig Q6.18.*

Fig Q6.18

<u>Solution</u>

$$\begin{bmatrix} \dot{x}_1 \\ \dot{x}_2 \\ \dot{x}_3 \end{bmatrix} = \begin{bmatrix} 0 & 1 & 0 \\ 0 & 0 & 1 \\ 0 & -2 & a_1 \end{bmatrix} \begin{bmatrix} x_1 \\ x_2 \\ x_3 \end{bmatrix} + \begin{bmatrix} 0 \\ 0 \\ 1 \end{bmatrix} u \quad and \quad y = x_1$$

Q6.19   *A system is characterized by the differential equation,*

$$\frac{d^2y}{dt^2} + 10\frac{dy}{dt} + 7y - u = 0$$

Determine its transfer function.

<u>Solution</u>

Given that, $\dfrac{d^2y}{dt^2} + 10\dfrac{dy}{dt} + 7y - u = 0$ .....(1)

On taking Laplace transform of equation (1) with zero initial conditions we get,

$s^2\, Y(s) + 10\, s\, Y(s) + 7\, Y(s) - U(s) = 0$

$(s^2 + 10\, s + 7)\, Y(s) = U(s)$

$\therefore \dfrac{Y(s)}{U(s)} = \dfrac{1}{s^2 + 10s + 7}$ .....(2)

The equation (2) is the transfer function of the system.

Q6.20 *The transfer function of a system is given by* $\dfrac{Y(s)}{U(s)} = \dfrac{10}{4s^2 + 2s + 1}$. *Determine the differential equation governing the system.*

Solution

Given that, $\dfrac{Y(s)}{U(s)} = \dfrac{10}{4s^2 + 2s + 1}$

$\therefore [4s^2 + 2s + 1]\, Y(s) = 10\, U(s)$

$4s^2\, Y(s) + 2s\, Y(s) + Y(s) = 10\, U(s)$ .....(1)

On taking inverse Laplace transform of equation (1) we get,

$$4\dfrac{d^2y}{dt^2} + 2\dfrac{dy}{dt} + y = 10u$$

$$4\dfrac{d^2y}{dt^2} + 2\dfrac{dy}{dt} + y - 10u = 0$$ .....(2)

The equation (2) is the differential equation governing the system.

Q6.21 *What are the advantages of state space modelling using physical variable?*

The advantages of choosing the physical variable are the following,

1. The state variable can be utilized for the purpose of feedback
2. The implementation of design with state variable feedback becomes straight forward.
3. The solution of state equation gives time variation of variables which have direct relevance to the physical system.

Q6.22 *What are phase variables?*

The phase variables are defined as those particular state variables which are obtained from one of the system variable and its derivatives. Usually the variable used is the system output and the remaining state variables are then derivatives of the output.

Q6.23 *What is bush form or companion form of state model?*

In bush form or companion form of state model, the system matrix, A has all 1's in the upper off-diagonal and its last row is comprised of the negative of the coefficients of the original differential equation and all other elements are zero. The companion form of state model is shown below

$$
\begin{bmatrix} \dot{x}_1 \\ \dot{x}_2 \\ \dot{x}_3 \\ \vdots \\ \dot{x}_{n-1} \\ \dot{x}_n \end{bmatrix}
=
\begin{bmatrix}
0 & 1 & 0 & 0 & \cdots & 0 \\
0 & 0 & 1 & 0 & \cdots & 0 \\
0 & 0 & 0 & 1 & \cdots & 0 \\
 & & \vdots & & & \\
0 & 0 & 0 & 0 & \cdots & 1 \\
-a_n & -a_{n-1} & -a_{n-1} & -a_{n-1} & \cdots & -a_{n-1}
\end{bmatrix}
\begin{bmatrix} x_1 \\ x_2 \\ x_3 \\ \vdots \\ x_{n-1} \\ x_n \end{bmatrix}
+
\begin{bmatrix} 0 \\ 0 \\ 0 \\ \vdots \\ 0 \\ b \end{bmatrix} [u]
$$

$$
y = \begin{bmatrix} 1 & 0 & 0 & \cdots & 0 \end{bmatrix}
\begin{bmatrix} x_1 \\ x_2 \\ x_3 \\ \vdots \\ x_n \end{bmatrix}
$$

Q6.24 *What are the advantages in choosing phase variables for state space modelling?*

1. Using phase variables the system state model can be written directly by inspection from the differential equation governing the system.

2. The phase variables provides a link between the transfer function design approach and time-domain design approach.

Q6.25 *What is the disadvantage in choosing phase variable for state-space modelling?*

The disadvantage in choosing phase variables is that the phase variables are not physical variables of the system and therefore are not available for measurement and control purposes.

Q6.26 *Write the canonical form of state model of nth order system.*

In canonical form of state model, the system matrix, A will be a diagonal matrix. The canonical form of state model in the matrix form is given below.

$$\begin{bmatrix} \dot{x}_1 \\ \dot{x}_2 \\ \dot{x}_3 \\ \vdots \\ \dot{x}_n \end{bmatrix} = \begin{bmatrix} \lambda_1 & 0 & 0 & \cdots & 0 \\ 0 & \lambda_2 & 0 & \cdots & 0 \\ 0 & 0 & \lambda_3 & \cdots & 0 \\ \vdots & \vdots & \vdots & & \vdots \\ 0 & 0 & 0 & \cdots & \lambda_n \end{bmatrix} \begin{bmatrix} x_1 \\ x_2 \\ x_3 \\ \vdots \\ x_n \end{bmatrix} + \begin{bmatrix} 1 \\ 1 \\ 1 \\ \vdots \\ 1 \end{bmatrix} [u]$$

$$y = \begin{bmatrix} c_1 & c_2 & c_3 & \cdots & c_n \end{bmatrix} \begin{bmatrix} x_1 \\ x_2 \\ x_3 \\ \vdots \\ x_n \end{bmatrix} + [b_0][u]$$

Q6.27 *What is the advantage and the disadvantage in canonical form of state model.*

The advantage of canonical form is that the state equations are independent of each other. The disadvantage is that the canonical variables are not physical variables and so they are not available for measurement and control.

Q6.28 *What is state transition matrix and how it is related to state of a system?*

The matrix exponential e^{At} is called state transition matrix. In the expanded form,

$$e^{At} = I + At + \frac{1}{2!} A^2 t^2 + \frac{1}{3!} A^3 t^3 + \dots + \frac{1}{1!} A^i t^i + \dots$$

The state transition matrix is used to find the state of the system, at any time instant t, from the knowledge of the state at time, t_0.

When the input is zero, $X(t) = e^{At} X(t_0)$

When the input vector is U(t), $X(t) = e^{A(t-t_0)} X(t_0) + \int_{t_0}^{t} e^{A(t-t)} B U(\tau) dt$

where, X(t) = State vector at time, t and $X(t_0)$ = State vector at time, t_0.

Q6.29 *Write the properties of state transition matrix.*

The following are the properties of state transition matrix.

1. $\phi(0) = e^{A \times 0} = I$ (Unit matrix)

2. $\phi(t) = e^{At} = (e^{-At})^{-1} = [\phi(-t)]^{-1}$

3. $\phi(t_1 + t_2) = e^{A(t_1 + t_2)} = e^{At_1} e^{At_2} = \phi(t_1)\phi(t_2) = \phi(t_2)\phi(t_1)$

Q6.30   *Write the solution of homogeneous state equations.*

The solution of homogeneous state equation is, $X(t) = e^{At} X_0$

where, $X(t)$ = State vector at time, t

e^{At} = State transition matrix

and X_0 = Initial condition vector at t = 0.

Q6.31   *Write the solution of non-homogeneous state equations.*

The solution of non-homogeneous state equation is

$$X(t) = e^{A(t-t_0)} X(t_0) + \int_{t_0}^{t} e^{A(t-t)} B U(\tau)\, dt$$

where, $X(t)$ = State vector of time, t ; $X(t_0)$ = Initial condition vector at t = t_0.

B = Input matrix   and $U(t)$ = Input vector

If initial conditions are known at t = 0, then put t_0 = 0.

Q6.32   *What is resolvant matrix?*

The Laplace transform of state transition matrix is called resolvant matrix.

Resolvant matrix, $\phi(s) = \mathcal{L}[\phi(t)] = \mathcal{L}[e^{At}]$

Also, $\phi(s) = [sI - A]^{-1}$

Q6.33   *Write the state model of nth order discrete time system.*

The state model of a system consists of state equation and output equation. The state model of a nth order discrete time system with m-inputs and p-outputs are

$X(k+1) = A X(k) + B U(k)$ ..... state equation

$Y(k) = C X(k) + D U(k)$ ..... output equation

where, $X(k)$ =   State vector of order (n ×1)
              $Y(k)$ =   Output vector of order (p×1)
              $U(k)$ =   Input vector of order (m×1)
              A   =   System matrix of order (n×n)
              B   =   Input matrix of order (n × m)
              C   =   Output matrix of order (p×n)
              D   =   Transmission matrix of order (p×m)

Q6.34   *What are the fundamental elements used to construct the state diagram of discrete time system?*

The fundamental elements used to construct the state diagram of discrete time system are scalar, adder and unit delay element.

Q6.35   *Sketch the basic elements used to construct the block diagram of discrete time system.*

The basic elements used to construct the block diagram of discrete time system are shown below.

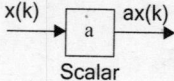   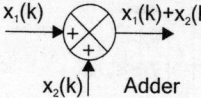   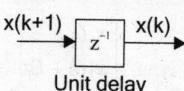

Q6.36   *Draw the basic elements used to construct the signal flow graph of discrete time system.*

The basic elements used to construct the signal flow graph of discrete time system are shown below.

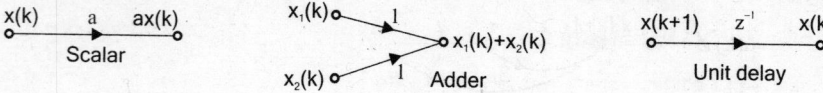

Q6.37   **Draw the block diagram representation of the state model of discrete time system.**

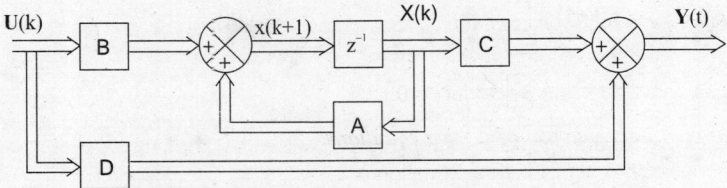

Fig Q6.37 : *Block diagram representation of discrete time system.*

Q6.38   **Draw the block diagram representation of the state model of discrete time system.**

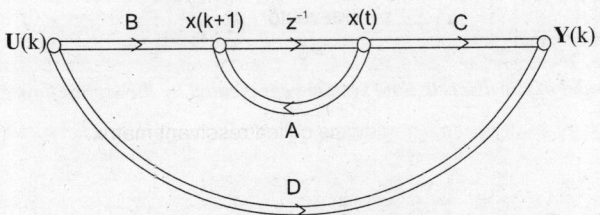

Fig Q6.38 : *Signal flow graph representation of discrete time system.*

Q6.39   **Draw the block diagram of the discrete time system described by the state model,**

$$\begin{bmatrix} x_1(k+1) \\ x_2(k+1) \end{bmatrix} = \begin{bmatrix} 4 & 0 \\ 0 & 2 \end{bmatrix} \begin{bmatrix} x_1(k) \\ x_2(k) \end{bmatrix} + \begin{bmatrix} 1 \\ 1 \end{bmatrix} u \quad ; \quad y(k) = \begin{bmatrix} 3 & 5 \end{bmatrix} \begin{bmatrix} x_1(k) \\ x_2(k) \end{bmatrix} + 10\, u(k)$$

<u>Solution</u>

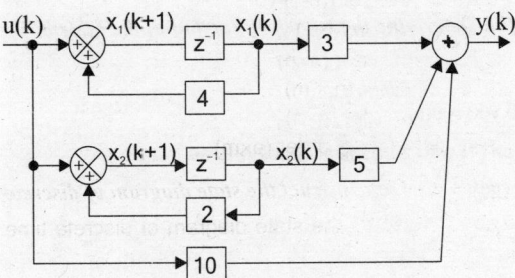

Q6.40   **Draw the signal flow graph of the discrete time system described by the state model.**

$$\begin{bmatrix} x_1(k+1) \\ x_2(k+1) \end{bmatrix} = \begin{bmatrix} 0 & 1 \\ 2 & 3 \end{bmatrix} \begin{bmatrix} x_1(k) \\ x_2(k) \end{bmatrix} + \begin{bmatrix} 0 \\ 1 \end{bmatrix} u \quad ; \quad y(k) = x_1(k)$$

<u>Solution</u>

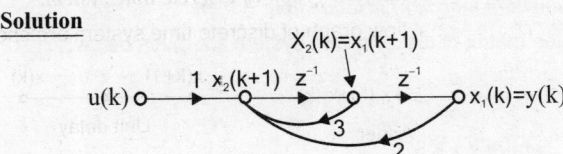

Q6.41   *Determine the state model of the discrete time system represented by the block diagram of fig Q6.41.*

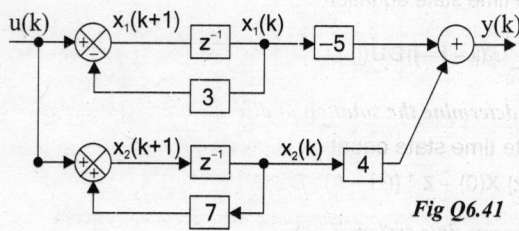

Fig Q6.41

Solution

$$\begin{bmatrix} x_1(k+1) \\ x_2(k+1) \end{bmatrix} = \begin{bmatrix} -3 & 0 \\ 0 & 7 \end{bmatrix} \begin{bmatrix} x_1(k) \\ x_2(k) \end{bmatrix} + \begin{bmatrix} 1 \\ 1 \end{bmatrix} u \quad ; \qquad y(k) = \begin{bmatrix} -5 & 4 \end{bmatrix} \begin{bmatrix} x_1(k) \\ x_2(k) \end{bmatrix}$$

Q6.42   *Determine the state model of the discrete time system represented by the signal flow graph of fig Q6.42.*

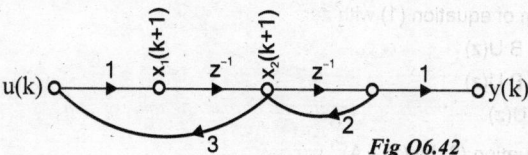

Fig Q6.42

Solution

$$\begin{bmatrix} x_1(k+1) \\ x_2(k+1) \end{bmatrix} = \begin{bmatrix} 0 & 0 \\ 1 & -2 \end{bmatrix} \begin{bmatrix} x_1(k) \\ x_2(k) \end{bmatrix} + \begin{bmatrix} 1 \\ 1 \end{bmatrix} u(k) \quad ; \quad y(k) = x_2(k)$$

Q6.43   *A discrete time system is described by the difference equation,*

$y(k+2) + 3\,y(k+1) + 5\,y(k) = u(k)$. Determine the transfer function of the system.

Solution

Given that, $y(k+2) + 3\,y(k+1) + 5\,y(k) = u(k)$ .....(1)

On taking z-transform of equation (1) with zero initial conditions we get,

$z^2\,Y(z) + 3z\,Y(z) + 6\,Y(z) = U(z)$

$(z^2 + 3z + 6)\,Y(z) = U(z)$

$$\therefore \quad \frac{Y(z)}{U(z)} = \frac{1}{z^2 + 3z + 6}$$ .....(2)

The equation(2) is the transfer function of the system.

Q6.44   *What is state transition matrix of discrete time system ?*

The matrix A^k (or $\phi(k)$) is called the state transition matrix of discrete time system.
It is given by, $A^k = Z^{-1}\{(zI - A)^{-1}z\}$
The state transition matrix is used to find the state of the system, at any discrete time instant k.

Q6.45   *Write the properties of the state transition matrix of discrete time system.*

The properties of the state transition matrix of discrete time system are given below.

1. $\phi(0) = I$
2. $\phi^{-1}(k) = \phi(-k)$
3. $\phi(k, k_0) = \phi(k-k_0) = A^{(k-k_0)} \quad ; \quad$ where $k > k_0$.

Q6.46 *Write the solution of discrete time state equation.*

The solution of discrete time state equation is

$$X(k) = \phi(k)X(0) + \sum_{j=0}^{k-1} \phi(k-1-j)B\,U(j)$$

Q6.47 *Write the expression to determine the solution of discrete time state equation using z-transform.*

The solution of discrete time state equation is given by

$$X(k) = z^{-1}\{(zI-A)^{-1}z\}X(0) + z^{-1}\{(zI-A)^{-1}B\,U(z)\}$$

Q6.48 *The state model of a discrete time system is given by,*

$$X(k+1) = A\,X(k) + B\,U(k)$$
$$Y(k) = C\,X(k) + D\,U(k)$$

Determine its transfer function.

Solution

The state equation is, $X(k+1) = A\,X(k) + B\,U(k)$ .....(1)

On taking z-transform of equation (1) with zero initial conditions we get,

$z\,X(z) = A\,X(z) + B\,U(z)$

$z\,X(z) - A\,X(z) = B\,U(z)$

$(zI - A)\,X(z) = B\,U(z)$ .....(2)

On premultiplying equation (2) by $(zI - A)^{-1}$ we get,

$X(z) = (zI - A)^{-1}\,B\,U(z)$ .....(3)

The output equation is, $Y(k) = C\,X(k) + D\,U(k)$ .....(4)

On taking z-transform of equation (4) we get,

$Y(z) = C\,X(z) + D\,U(z)$ .....(5)

On substituting for $X(z)$ from equation (3) in equation (5) we get,

$Y(z) = C\,(zI - A)^{-1}\,BU(z) + DU(z)$

$Y(z) = [C\,(zI - A)^{-1}\,B + D]\,U(z)$

$$\therefore\ \frac{Y(z)}{U(z)} = C(zI - A)^{-1}B + D$$ .....(6)

The equation (6) is the transform function of discrete time system.

Q6.49 *Define the characteristic equation of a matrix.*

The characteristic equation of a n x n matrix A is the nth degree polynomial of equation, $|\lambda I - A| = 0$, where I is the unit matrix.

Q6.50 *What are eigenvalues and eigenvector?*

A non-zero column vector X is an eigenvector of a square matrix A, if there exists a scalar λ such that $A\,X = \lambda\,X$, then λ is an eigenvalue of A.

Q6.51 *How the eigenvalues are calculated?*

The eigenvalues of a matrix A are the roots of characteristic equation of A. The characteristic equation is $|\lambda I - A| = 0$. Solving the characteristic equation for λ gives the eigenvalues of A.

Q6.52 *Write any two properties of eigenvalues.*

1. A matrix and its transpose have the same eigenvalues.

2. The product of the eigenvalues (counting multiplicities) of the matrix equals the determinant of the matrix.

Q6.53 *How the eigenvectors are calculated, when the eigenvalues are distinct?*

When the eigenvalues are distinct, then there exists only one independent eigenvector corresponding to each eigenvalue. The independent eigenvector corresponding to eigenvalue λ_i may be obtained by taking cofactors of matrix $[\lambda_i I - A]$ along any row.

If C_{k1}, C_{k2}, ... C_{kn} are cofactors of matrix $[\lambda_i I - A]$ along k^{th} row then the eigenvector m_i corresponding to λ_i is given by,

$$m_i = \begin{bmatrix} C_{k1} \\ C_{k2} \\ \vdots \\ C_{kn} \end{bmatrix} \quad ; \quad K = 1 \text{ or } 2 \text{ or } ......n$$

Q6.54 *What is similarity transformation?*

The process of transforming a square matrix A to another similar matrix B by a transformation $P^{-1} AP = B$ is called similarity transformation. The matrix P is called transformation matrix.

Q6.55 *What is meant by diagonalization?*

The process of converting the system matrix A into a diagonal matrix by a similarity transformation using the modal matrix M is called diagonalization.

Q6.56 *What is modal matrix?*

The modal matrix is a matrix used to diagonalize the system matrix. It is also called diagonalization matrix.

If   A   = System matrix

     M   = Modal matrix

and M^{-1} = Inverse of modal matrix, then $M^{-1} AM$ will be a diagonalized system matrix.

> **Note :** Refer appendix for modal matrix

Q6.57 *How the modal matrix is determined?*

The modal matrix M can be formed from eigenvectors. Let m_1, m_2, m_3, ... m_n be the eigenvectors of a nth order system. Now the modal matrix M is obtained by arranging all the eigenvectors columnwise as shown below

Modal matrix, $M = [m_1 \quad m_2 \quad m_3 \quad .... \quad m_n]$

Q6.58 *What is canonical form of state model?*

If the system matrix, A is in the form of diagonal matrix then the state model is called canonical form [In diagonal matrix, we have the eigenvalues on the main diagonal and all other elements are zero].

Q6.59 *What is the transformation used to diagonalize a system matrix?*

The transformation used to diagonalize a system matrix is $X = MZ$

where   X = Original state vector

        Z = Transformed state vector   ;   M = Modal matrix.

Q6.60. *Write the transformed canonical state model of a system.*

The transformed state model of a system is,

$$\dot{Z} = \Lambda Z + \tilde{B}U$$
$$Y = \tilde{C}Z + DU$$

where,   Z = Transformed state vector   ; $\tilde{B} = M^{-1}B$

         M = Modal matrix   ; $\tilde{C} = CM$

         $\Lambda = M^{-1} AM$

Q6.61   *When modal matrix is called vander monde matrix.*

When the system matrix A is in the companion or bush form then the modal matrix is given by a special matrix called vander monde matrix, V, shown below,

$$M = V = \begin{bmatrix} 1 & 1 & \cdots & 1 \\ \lambda_1 & \lambda_2 & \cdots & \lambda_n \\ \lambda_1^2 & \lambda_2^2 & \cdots & \lambda_n^2 \\ \vdots & \vdots & & \vdots \\ \lambda_1^{n-1} & \lambda_2^{n-1} & \cdots & \lambda_n^{n-1} \end{bmatrix}$$

Q6.62   *What is Jordan canonical form?*

When the eigenvalues have multiplicity the system matrix cannot be diagonalized. But the transformation, **X = MZ** will transform the system matrix to a form called Jordan matrix, **J**, where **J** is **M⁻¹ AM**. The transformed state model in this case is called Jordan canonical form.

Q6.63   *What is Jordan matrix?*

The Jordan matrix is the transformed system matrix using the transformation, **X = MZ**. The Jordan matrix J, will have a Jordan block of size q x q correspond to a eigenvalue of λ_1 with multiplicity q. In the Jordan block the diagonal elements will be eigenvalues and elements just above the diagonal is one. A typical Jordan matrix is shown below,

$$J = \begin{bmatrix} \lambda_1 & 1 & 0 & 0 & \cdots & 0 \\ 0 & \lambda_1 & 1 & 0 & \cdots & 0 \\ 0 & 0 & \lambda_1 & 0 & \cdots & 0 \\ 0 & 0 & 0 & \lambda_4 & \cdots & 0 \\ \vdots & \vdots & \vdots & \vdots & & \vdots \\ 0 & 0 & 0 & 0 & \cdots & \lambda_n \end{bmatrix}$$

Q6.64   *How the state transition matrix e^{At} is computed by canonical transformation.*

When the eigenvalues are distinct the following procedure can be used to compute e^{At}.

1.   Compute the eigenvalues $\lambda_1, \lambda_2, \ldots\ldots, \lambda_n$ and eigenvector $m_1, m_2, \ldots\ldots, m_n$ of the system matrix A.
2.   Determine the modal matrix M and M⁻¹. Modal matrix, M = [m_1 m_2 ..... m_n].
3.   The state transition matrix e^{At} is given by, e^{At} = M M⁻¹

$$\text{where,} \quad e^{At} = \begin{bmatrix} e^{\lambda_1 t} & 0 & \cdots & 0 \\ 0 & e^{\lambda_2 t} & \cdots & 0 \\ \vdots & \vdots & & \vdots \\ 0 & 0 & \cdots & e^{\lambda_n t} \end{bmatrix}$$

Q6.65   *Define controllability.*

A system is said to be completely state controllable if it is possible to transfer the system state from any initial state $X(t_0)$ at any other desired state X(t), in specified finite time by a control vector U(t).

Q6.66   *What is the need for controllability test?*

The controllability test is necessary to find the usefulness of a state variable. If the state variables are controllable then by controlling (i.e., varying) the state variables the desired outputs of the system are achieved.

Q6.67   *State the condition for controllability by Gilbert's method.*

Case(i)   : When eigenvalues are distinct

Consider the canonical form of state model shown below which is obtained by using the transformation X = MZ.

$$\dot{Z} = \Lambda Z + \tilde{B}U$$
$$Y = \tilde{C}Z + DU$$

where, $\Lambda = M^{-1} AM$; $\tilde{C} = CM$

$\tilde{B} = M^{-1}B$ and   M = Modal matrix

In this case the necessary and sufficient condition for complete controllability is that, the matrix must have no rows with all zeros. If any row of the matrix is zero then the corresponding state variable is uncontrollable.

Case (ii) : When eigenvalues have multiplicity

In this case the state model can be converted to Jordan canonical form shown below,

$$\dot{Z} = JZ + \tilde{B}U$$
$$Y = \tilde{C}Z + DU$$;   where $J = M^{-1} AM$

In this case the system is completely controllable, if the elements of any row of that correspond to the last row of each Jordan block are not all zero and all rows corresponding to other state variable must not have all zeros

Q6.68   *State the condition for controllability by Kalman's method.*

For a nth order system described by state equation, , we can form a composite matrix, Q_c

where, $Q_c = [B \quad AB \quad A^2 B \quad .... \quad A^{n-1} B]$

In this case the system is completely state controllable if the rank of the composite matrix Q_c is n.

Q6.69   *What is the advantage and the disadvantage in Kalman's test for controllability?*

The advantage in Kalman's test is that the calculations are simpler. But the disadvantage in Kalman's test is that, we can't find the state variable which is uncontrollable. But in Gilbert's method we can find the uncontrollable state variable which is the state variable corresponding to the row of which has all zeros.

Q6.70   *Define observability.*

A system is said to be completely observable if every state X(t) can be completely identified by measurements of the output Y(t) over a finite time interval.

Q6.71   *What is the need for observability test?*

The observability test is necessary to find whether the state variables are measurable or not. If the state variables are measurable then the state of the system can be determined by practical measurements of the state variables.

Q6.72   *State the condition for observability by Gilbert's method.*

Consider the transformed canonical or Jordan canonical form of the state model shown below which is obtained by using the transformation , X = MZ

$$\dot{Z} = \Lambda Z + \tilde{B}U \qquad \qquad \dot{Z} = JZ + \tilde{B}U$$
$$\qquad \qquad \text{or}$$
$$Y = \tilde{C}Z + DU \qquad \qquad Y = \tilde{C}Z + DU$$

where $\tilde{C} = CM$ and M = Modal matrix.

The necessary and sufficient condition for complete observability is that none of the columns of the matrix $\tilde{C}$ be zero. If any of the column is of $\tilde{C}$ has all zeros then the corresponding state variable is not observable.

Q6.73   *State the condition for observability by Kalman's method?*

For a nth order system described by state model,

$$\dot{X} = AX + BU$$
$$Y = CX + DU$$

we can form a composite matrix Q_o, where

$$Q_0 = \left[C^T \quad A^T C^T \quad (A^T)^2 C^T \quad (A^T)^3 C^T \quad \cdots \quad (A^T)^{n-1} C^T \right]$$

In this case the system is completely observable if the rank of composite matrix, Q_o is n.

Q6.74   *What is the advantage and the disadvantage in Kalman's test for observability?*

The advantage in Kalman's test is that the calculations are simpler. The disadvantage in Kalman's test is that the nonobservable state variables cannot be determined.

Q6.75   *State the duality between controllability and observability.*

The concepts of controllability and observability are dual concepts and it is proposed by Kalman as principle of duality.

The principle of duality states that a system is completely state controllable if and only if its dual system is completely observable or viceversa (i.e., if the system is observable then its dual is controllable).

Q6.76   *What is the effect of pole-zero cancellation in transfer function.*

If cancellation of pole-zero occurs in the transfer function of a system, then the system will be either not state controllable or unobservable depending on how the state variables are defined (or chosen).

6.14   EXERCISES

E6.1   *Obtain the state model of the electrical network shown below by choosing minimal number of state variables.*

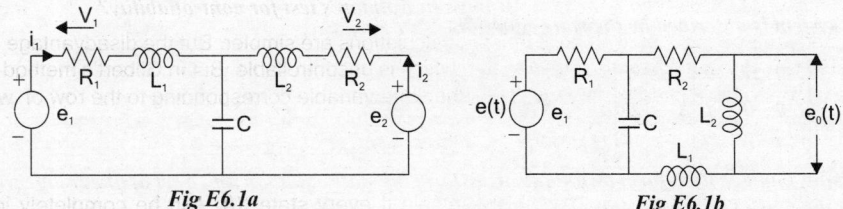

Fig E6.1a   *Fig E6.1b*

E6.2   *Construct the state model of mechanical systems shown in fig E6.2a & E6.2b.*

E6.3   *Obtain the state model of the system whose transfer function is given as*

a) $\dfrac{1}{s^2 + 3s + 2}$

b) $\dfrac{s+3}{s^2 + 3s + 2}$

c) $\dfrac{s^2 + 4s + 3}{s^2 + 9s + 20}$

d) $\dfrac{s^2 + 6s + 8}{(s+3)(s^2 + 2s + 2)}$

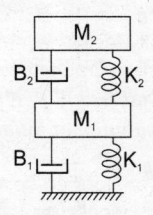

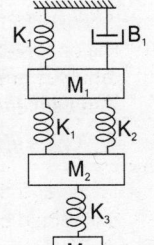

Fig E6.2a

Fig E6.2b

E6.4   *Construct a state model for the systems characterized by the following differential equations.*

a) $\dddot{y} + 3\ddot{y} + 2\dot{y} = \dot{u} + u$

b) $\dddot{y} + 6\ddot{y} + 11\dot{y} + 6y = u$

c) $\ddot{y} + 2\dot{y} + y = \dot{u} + u$

E6.5   *In the state diagrams of systems shown in figE6.5a and figE6.5b. Assign state variables and obtain state model of the system.*

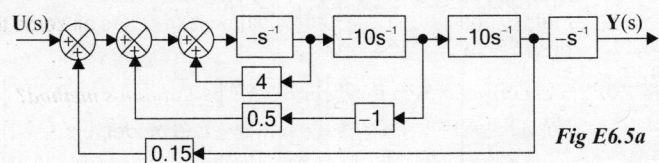

Fig E6.5a

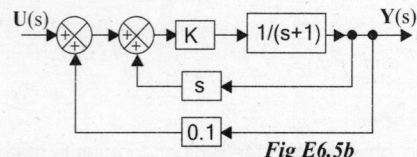

Fig E6.5b

E6.6   *The state diagram of a linear system is shown in fig E6.6. Assign state variables and obtain the state model.*

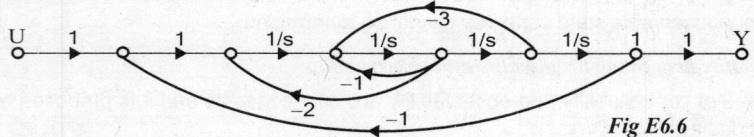

Fig E6.6

E6.7   *For the system matrices given below compute e^{At}*

a) $A = \begin{bmatrix} 0 & 1 \\ 1 & 1 \end{bmatrix}$ b) $A = \begin{bmatrix} -2 & 1 \\ 1 & -2 \end{bmatrix}$ c) $A = \begin{bmatrix} -2 & 0 \\ 0 & -1 \end{bmatrix}$ d) $A = \begin{bmatrix} -1 & 0 \\ 0 & -3 \end{bmatrix}$

E6.8   *The following facts are known about the linear system, $\dot{x}(t) = A\,x(t)$.*

If $x(0) = \begin{bmatrix} 2 \\ -1 \end{bmatrix}$ then $x(t) = \begin{bmatrix} e^{-2t} \\ -4e^{-2t} \end{bmatrix}$; If $x(0) = \begin{bmatrix} 1 \\ -2 \end{bmatrix}$ then $x(t) = \begin{bmatrix} e^{-t} \\ -2e^{-t} \end{bmatrix}$

Find e^{-At} and hence A.

E6.9   *A system is described by the state equation.*

$$\dot{X} = \begin{bmatrix} 0 & 1 & 0 \\ 0 & 0 & 1 \\ -1 & 0 & -3 \end{bmatrix} X + \begin{bmatrix} 0 \\ 0 \\ 1 \end{bmatrix} u \quad ; \quad X(0) = x_0$$

Using Laplace transform technique, transform the state equation into a set of linear algebraic equations in the form $X(s) = \phi(s)\,X(0) + \phi(s)\,B\,U(s)$

E6.10   *Consider the state equation*

$$\begin{bmatrix} \dot{x}_1 \\ \dot{x}_2 \end{bmatrix} = \begin{bmatrix} 0 & 1 \\ -1 & -2 \end{bmatrix} \begin{bmatrix} x_1 \\ x_2 \end{bmatrix}$$

Compute the solution of homogeneous equation, assuming $x_1(2) = 2$.

E6.11   *Compute the solution of following state equations.*

a) $\dot{X} = \begin{bmatrix} 0 & 0 & -2 \\ 0 & 1 & 0 \\ 1 & 0 & 3 \end{bmatrix} X$; $X(0) = \begin{bmatrix} 0 \\ 1 \\ 0 \end{bmatrix}$ b) $\begin{bmatrix} \dot{x}_1 \\ \dot{x}_2 \end{bmatrix} = \begin{bmatrix} 1 & 0 \\ 1 & 1 \end{bmatrix} \begin{bmatrix} x_1 \\ x_2 \end{bmatrix} + \begin{bmatrix} 1 \\ 1 \end{bmatrix} u(t)$ with $X_0 = \begin{bmatrix} 1 \\ 1 \end{bmatrix}$

E6.12   *A linear time invariant system is described by the following state model. Obtain the canonical form of the state model.*

a) $\begin{bmatrix} \dot{x}_1 \\ \dot{x}_2 \\ \dot{x}_3 \end{bmatrix} = \begin{bmatrix} 0 & 0 & -20 \\ 1 & 0 & -24 \\ 0 & 1 & -9 \end{bmatrix} \begin{bmatrix} x_1 \\ x_2 \\ x_3 \end{bmatrix} + \begin{bmatrix} 3 \\ 1 \\ 0 \end{bmatrix} u$

b) $\begin{bmatrix} \dot{x}_1 \\ \dot{x}_2 \\ \dot{x}_3 \end{bmatrix} = \begin{bmatrix} 0 & 1 & 0 \\ -2 & -2 & -1 \\ 0 & 0 & -2 \end{bmatrix} \begin{bmatrix} x_1 \\ x_2 \\ x_3 \end{bmatrix} + \begin{bmatrix} 0 \\ 1 \\ 1 \end{bmatrix} u$ and $y = \begin{bmatrix} 3 & 1 & 0 \end{bmatrix} \begin{bmatrix} x_1 \\ x_2 \\ x_3 \end{bmatrix}$

c) $\begin{bmatrix} \dot{x}_1 \\ \dot{x}_2 \\ \dot{x}_3 \end{bmatrix} = \begin{bmatrix} 1 & 0 & 2 \\ -1 & 1 & 1 \\ 0 & 3 & -1 \end{bmatrix} \begin{bmatrix} x_1 \\ x_2 \\ x_3 \end{bmatrix} + \begin{bmatrix} 1 \\ 1 \\ 0 \end{bmatrix} u$ and $y = \begin{bmatrix} 1 & 0 & 1 \end{bmatrix} \begin{bmatrix} x_1 \\ x_2 \\ x_3 \end{bmatrix}$

d) $\begin{bmatrix} \dot{x}_1 \\ \dot{x}_2 \end{bmatrix} = \begin{bmatrix} 0 & 1 \\ -4 & -5 \end{bmatrix} \begin{bmatrix} x_1 \\ x_2 \end{bmatrix} + \begin{bmatrix} 0 \\ 1 \end{bmatrix} u$ and $y = \begin{bmatrix} \dfrac{1}{3} & -\dfrac{1}{3} \end{bmatrix} \begin{bmatrix} x_1 \\ x_2 \end{bmatrix}$

e) $\begin{bmatrix} \dot{x}_1 \\ \dot{x}_2 \\ \dot{x}_3 \end{bmatrix} = \begin{bmatrix} -1 & 0 & 0 \\ 0 & -2 & 0 \\ 0 & 0 & -3 \end{bmatrix} \begin{bmatrix} x_1 \\ x_2 \\ x_3 \end{bmatrix} + \begin{bmatrix} 1 \\ -2 \\ 0 \end{bmatrix} u$ and $y = \begin{bmatrix} 1 & 1 & 0 \end{bmatrix} \begin{bmatrix} x_1 \\ x_2 \\ x_3 \end{bmatrix}$

f) $\begin{bmatrix} \dot{x}_1 \\ \dot{x}_2 \end{bmatrix} = \begin{bmatrix} -1 & 0 \\ 0 & -3 \end{bmatrix} \begin{bmatrix} x_1 \\ x_2 \end{bmatrix} + \begin{bmatrix} 1 \\ 1 \end{bmatrix} [r]$ and $y = \begin{bmatrix} -1 & -2 \end{bmatrix} \begin{bmatrix} x_1 \\ x_2 \end{bmatrix}$

g) $\begin{bmatrix} \dot{x}_1 \\ \dot{x}_2 \end{bmatrix} = \begin{bmatrix} 0 & 1 \\ -12 & -7 \end{bmatrix} \begin{bmatrix} x_1 \\ x_2 \end{bmatrix} + \begin{bmatrix} 0 \\ 1 \end{bmatrix} [r]$ and $y = \begin{bmatrix} -10 & -4 \end{bmatrix} \begin{bmatrix} x_1 \\ x_2 \end{bmatrix} + [u]$

h) $\begin{bmatrix} \dot{x}_1 \\ \dot{x}_2 \\ \dot{x}_3 \end{bmatrix} = \begin{bmatrix} -2 & 0 & 1 \\ 1 & -3 & 0 \\ 1 & 1 & -1 \end{bmatrix} \begin{bmatrix} x_1 \\ x_2 \\ x_3 \end{bmatrix} + \begin{bmatrix} 1 \\ 0 \\ 1 \end{bmatrix} u$ and $y = \begin{bmatrix} 2 & 1 & -1 \end{bmatrix} \begin{bmatrix} x_1 \\ x_2 \\ x_3 \end{bmatrix}$

i) $\begin{bmatrix} \dot{x}_1 \\ \dot{x}_2 \\ \dot{x}_3 \end{bmatrix} = \begin{bmatrix} 0 & 1 & 0 \\ -2 & -2 & 0 \\ 0 & 0 & -3 \end{bmatrix} \begin{bmatrix} x_1 \\ x_2 \\ x_3 \end{bmatrix} + \begin{bmatrix} 0 \\ 1 \\ 1 \end{bmatrix} u$ and $y = \begin{bmatrix} \dfrac{14}{5} & \dfrac{6}{5} & -\dfrac{1}{5} \end{bmatrix} \begin{bmatrix} x_1 \\ x_2 \\ x_3 \end{bmatrix}$

E6.13 *Determine the state controllability for the systems represented by the following state equations.*

a) $\begin{bmatrix} \dot{x}_1 \\ \dot{x}_2 \end{bmatrix} = \begin{bmatrix} 2 & 1 \\ 0 & -1 \end{bmatrix} \begin{bmatrix} x_1 \\ x_2 \end{bmatrix} + \begin{bmatrix} 1 \\ 0 \end{bmatrix} u$

b) $\begin{bmatrix} \dot{x}_1 \\ \dot{x}_2 \end{bmatrix} = \begin{bmatrix} 0 & 1 \\ -1 & 0 \end{bmatrix} \begin{bmatrix} x_1 \\ x_2 \end{bmatrix} + \begin{bmatrix} 1 \\ 0 \end{bmatrix} u$

c) $\begin{bmatrix} \dot{x}_1 \\ \dot{x}_2 \end{bmatrix} = \begin{bmatrix} 0 & 1 \\ -1 & -2 \end{bmatrix} \begin{bmatrix} x_1 \\ x_2 \end{bmatrix} + \begin{bmatrix} 1 \\ -1 \end{bmatrix} u$

E6.14 *Examine the observability of the systems given below using canonical form.*

a) $\begin{bmatrix} \dot{x}_1 \\ \dot{x}_2 \\ \dot{x}_3 \end{bmatrix} = \begin{bmatrix} 2 & -2 & 3 \\ 1 & 1 & 1 \\ 1 & 3 & -1 \end{bmatrix} \begin{bmatrix} x_1 \\ x_2 \\ x_3 \end{bmatrix} + \begin{bmatrix} 11 \\ 1 \\ -14 \end{bmatrix} u$ and $y = \begin{bmatrix} -3 & 5 & -2 \end{bmatrix} \begin{bmatrix} x_1 \\ x_2 \\ x_3 \end{bmatrix}$

b) $\begin{bmatrix} \dot{x}_1 \\ \dot{x}_2 \\ \dot{x}_3 \end{bmatrix} = \begin{bmatrix} 0 & 1 & 0 \\ 0 & 1 & 1 \\ 0 & -2 & -3 \end{bmatrix} \begin{bmatrix} x_1 \\ x_2 \\ x_3 \end{bmatrix} + \begin{bmatrix} 0 \\ 0 \\ 1 \end{bmatrix} u$ and $y = \begin{bmatrix} 3 & 4 & 1 \end{bmatrix} \begin{bmatrix} x_1 \\ x_2 \\ x_3 \end{bmatrix}$

E6.15 *Convert the following system matrix to canonical form.*

a) $A = \begin{bmatrix} -3 & 0 & 1 \\ 1 & 1 & 0 \\ 2 & -1 & 2 \end{bmatrix}$ b) $A = \begin{bmatrix} 1 & 1 & -1 \\ 0 & 2 & 1 \\ 1 & 2 & 1 \end{bmatrix}$ c) $A = \begin{bmatrix} 1 & 2 & 1 \\ -1 & 0 & 2 \\ 1 & 3 & -1 \end{bmatrix}$

ANSWER FOR EXERCISE PROBLEMS

E6.1a
$$\begin{bmatrix} \dot{x}_1 \\ \dot{x}_2 \\ \dot{x}_3 \end{bmatrix} = \begin{bmatrix} -R_1/L_1 & 0 & -1/L_2 \\ 0 & -R_1/L_1 & -1/L_2 \\ 1/C & 1/C & 0 \end{bmatrix} \begin{bmatrix} x_1 \\ x_2 \\ x_3 \end{bmatrix} + \begin{bmatrix} 1/L_1 & 0 & 0 \\ 0 & 1/L_2 & 0 \\ 0 & 0 & 0 \end{bmatrix} \begin{bmatrix} u_1(t) \\ u_2(t) \\ 0 \end{bmatrix} \; ; \; \begin{bmatrix} y_1 \\ y_2 \\ y_3 \end{bmatrix} = \begin{bmatrix} R_1 & 0 & 0 \\ 0 & R_2 & 0 \\ 0 & 0 & 0 \end{bmatrix} \begin{bmatrix} x_1 \\ x_2 \\ x_3 \end{bmatrix}$$

E6.1b
$$\begin{bmatrix} \dot{x}_1 \\ \dot{x}_2 \\ \dot{x}_3 \end{bmatrix} = \begin{bmatrix} 0 & -1/C & 0 \\ 1/L_1 & -R_2/L_1 & 0 \\ 0 & 0 & 0 \end{bmatrix} \begin{bmatrix} x_1 \\ x_2 \\ x_3 \end{bmatrix} + \begin{bmatrix} 1/CR_1 & 0 & 0 \\ 0 & -1/L_1 & 0 \\ 0 & -1/C_2 & 0 \end{bmatrix} \begin{bmatrix} u_1(t) \\ u_2(t) \\ 0 \end{bmatrix} \; ; \; \begin{bmatrix} y_1 \\ y_2 \end{bmatrix} = \begin{bmatrix} 0 & 0 \\ 0 & R_2 \end{bmatrix} \begin{bmatrix} x_1 \\ x_2 \end{bmatrix} + \begin{bmatrix} 1 & 0 \\ 0 & 0 \end{bmatrix} \begin{bmatrix} u_1 \\ u_2 \end{bmatrix}$$

E6.2a
$$\begin{bmatrix} \dot{x}_1 \\ \dot{x}_2 \\ \dot{x}_3 \\ \dot{x}_4 \end{bmatrix} = \begin{bmatrix} 0 & 0 & 1 & 0 \\ 0 & 0 & 0 & 1 \\ -K_2 & K_2 & -B_2 & B_2 \\ 0 & -K_1 & 0 & -B_1 \end{bmatrix} \begin{bmatrix} x_1 \\ x_2 \\ x_3 \\ x_4 \end{bmatrix} \; ; \; \begin{bmatrix} y_1 \\ y_2 \end{bmatrix} = \begin{bmatrix} 1 & 0 & 0 & 0 \\ 0 & 1 & 0 & 0 \end{bmatrix} \begin{bmatrix} x_1 \\ x_2 \\ x_3 \\ x_4 \end{bmatrix}$$

E6.2b
$$\begin{bmatrix} \dot{x}_1 \\ \dot{x}_2 \\ \dot{x}_3 \\ \dot{x}_4 \\ \dot{x}_5 \\ \dot{x}_6 \end{bmatrix} = \begin{bmatrix} 0 & 0 & 0 & 1 & 1 & 0 \\ 0 & 0 & 0 & 0 & 0 & 0 \\ 0 & 0 & 0 & 0 & 0 & 1 \\ -\dfrac{(K_1+K_2+K_3)}{M_1} & \dfrac{K_2+K_3}{M_1} & 0 & \left(\dfrac{-B_1}{M_1}\right) & 0 & 0 \\ \left(\dfrac{K_2+K_3}{M_2}\right) & -\dfrac{(K_2+K_3+K_4)}{M_2} & \left(\dfrac{-K_4}{M_2}\right) & 0 & 0 & 0 \\ 0 & \left(\dfrac{K_3}{M_2}\right) & \left(\dfrac{-K_3}{M_3}\right) & 0 & 0 & 0 \end{bmatrix} + \begin{bmatrix} 0 \\ 0 \\ 0 \\ 0 \\ 0 \\ 1/M_3 \end{bmatrix} u(t)$$

$$\begin{bmatrix} y_1 \\ y_2 \\ y_3 \end{bmatrix} = \begin{bmatrix} 1 & 0 & 0 & 0 & 0 & 0 \\ 0 & 1 & 0 & 0 & 0 & 0 \\ 0 & 0 & 1 & 0 & 0 & 0 \end{bmatrix} \begin{bmatrix} x_1 \\ x_2 \\ x_3 \\ x_4 \\ x_5 \\ x_6 \end{bmatrix}$$

E6.3a
$$\begin{bmatrix} \dot{x}_1 \\ \dot{x}_2 \end{bmatrix} = \begin{bmatrix} -3 & 1 \\ -2 & 0 \end{bmatrix} \begin{bmatrix} x_1 \\ x_2 \end{bmatrix} + \begin{bmatrix} 0 \\ 1 \end{bmatrix} u \; ; \; Y = \begin{bmatrix} 1 & 0 \end{bmatrix} \begin{bmatrix} x_1 \\ x_2 \end{bmatrix}$$

E6.3b
$$\begin{bmatrix} \dot{x}_1 \\ \dot{x}_2 \end{bmatrix} = \begin{bmatrix} -2 & 0 \\ 0 & -1 \end{bmatrix} \begin{bmatrix} x_1 \\ x_2 \end{bmatrix} + \begin{bmatrix} 1 \\ 1 \end{bmatrix} u \; ; \; Y = \begin{bmatrix} -1 & 2 \end{bmatrix} \begin{bmatrix} x_1 \\ x_2 \end{bmatrix}$$

E6.3c
$$\begin{bmatrix} \dot{x}_1 \\ \dot{x}_2 \end{bmatrix} = \begin{bmatrix} 0 & 1 \\ -20 & -9 \end{bmatrix} \begin{bmatrix} x_1 \\ x_2 \end{bmatrix} + \begin{bmatrix} 0 \\ 1 \end{bmatrix} u \; ; \; Y = \begin{bmatrix} -17 & -5 \end{bmatrix} \begin{bmatrix} x_1 \\ x_2 \end{bmatrix} + u$$

E6.3d
$$\begin{bmatrix} \dot{x}_1 \\ \dot{x}_2 \\ \dot{x}_3 \end{bmatrix} = \begin{bmatrix} -6 & 0 & 0 \\ -8 & 0 & 0 \\ -5 & 0 & 0 \end{bmatrix} \begin{bmatrix} x_1 \\ x_2 \\ x_3 \end{bmatrix} + \begin{bmatrix} 8 \\ 6 \\ 1 \end{bmatrix} u \; ; \; Y = \begin{bmatrix} 1 & 0 & 0 \end{bmatrix} \begin{bmatrix} x_1 \\ x_2 \\ x_3 \end{bmatrix}$$

E6.4a
$$\begin{bmatrix} \dot{x}_1 \\ \dot{x}_2 \\ \dot{x}_3 \end{bmatrix} = \begin{bmatrix} 0 & 1 & 0 \\ 0 & 0 & 1 \\ 0 & -2 & -3 \end{bmatrix} \begin{bmatrix} x_1 \\ x_2 \\ x_3 \end{bmatrix} + \begin{bmatrix} 1 & 0 & 0 \\ 0 & 1 & 0 \\ 0 & 0 & 0 \end{bmatrix} \begin{bmatrix} u_1 \\ u_2 \\ 0 \end{bmatrix}$$

E6.4b
$$\begin{bmatrix} \dot{x}_1 \\ \dot{x}_2 \\ \dot{x}_3 \end{bmatrix} = \begin{bmatrix} 0 & 1 & 0 \\ 0 & 0 & 1 \\ -6 & -11 & 6 \end{bmatrix} + \begin{bmatrix} 0 \\ 0 \\ 1 \end{bmatrix} u \; ; \; Y = \begin{bmatrix} 1 & 0 & 0 \end{bmatrix} \begin{bmatrix} x_1 \\ x_2 \\ x_3 \end{bmatrix}$$

$$Y = \begin{bmatrix} 1 & 0 & 0 \end{bmatrix} \begin{bmatrix} x_1 \\ x_2 \\ x_3 \end{bmatrix}$$

E6.4c $\begin{bmatrix} \dot{x}_1 \\ \dot{x}_2 \end{bmatrix} = \begin{bmatrix} 0 & 1 \\ -1 & -2 \end{bmatrix} \begin{bmatrix} x_1 \\ x_2 \end{bmatrix} + \begin{bmatrix} 0 & 0 \\ 1 & 1 \end{bmatrix} \begin{bmatrix} u_1 \\ u_2 \end{bmatrix}$; $Y = \begin{bmatrix} 1 & 0 & 0 \end{bmatrix} \begin{bmatrix} x_1 \\ x_2 \\ x_3 \end{bmatrix}$

E6.5a $\begin{bmatrix} \dot{x}_1 \\ \dot{x}_2 \\ \dot{x}_3 \\ \dot{x}_4 \end{bmatrix} = \begin{bmatrix} 0 & -10 & 0 & 0 \\ 0 & 0 & -10 & 0 \\ 0 & 0 & 0 & -10 \\ 0 & -0.15 & 0.5 & -4 \end{bmatrix} \begin{bmatrix} x_1 \\ x_2 \\ x_3 \\ x_4 \end{bmatrix} + \begin{bmatrix} 0 \\ 0 \\ 0 \\ 1 \end{bmatrix} u$; $Y = \begin{bmatrix} -1 & 0 & 0 & 0 \end{bmatrix} \begin{bmatrix} x_1 \\ x_2 \\ x_3 \\ x_4 \end{bmatrix}$

E6.5b $\dot{x}_1 = \dfrac{1 - 0.1K}{1 - K} x_1 + \dfrac{K}{1 - K} u(s)$

E6.6 $\begin{bmatrix} \dot{x}_1 \\ \dot{x}_2 \\ \dot{x}_3 \\ \dot{x}_4 \end{bmatrix} = \begin{bmatrix} 0 & 1 & 0 & 0 \\ 0 & 0 & 1 & 0 \\ 0 & -3 & 1 & 1 \\ -1 & 0 & -2 & 0 \end{bmatrix} \begin{bmatrix} x_1 \\ x_2 \\ x_3 \\ x_4 \end{bmatrix} + \begin{bmatrix} 0 \\ 0 \\ 0 \\ 1 \end{bmatrix} u$; $Y = \begin{bmatrix} 1 & 0 & 0 & 0 \end{bmatrix} \begin{bmatrix} x_1 \\ x_2 \\ x_3 \\ x_4 \end{bmatrix}$

E6.7a $e^{At} = \begin{bmatrix} 0.2764e^{1.618t} + 0.7236e^{-0.618t} & 0.4472e^{1.618t} - 0.4472e^{-0.618t} \\ 0.4472e^{1.618t} - 0.4472e^{-0.618t} & 0.7236e^{1.618t} + 0.2764e^{-0.618t} \end{bmatrix}$

E6.7b $e^{At} = 0.5 \begin{bmatrix} e^{-t} + e^{-3t} & e^{-t} - e^{-3t} \\ e^{-t} - e^{-3t} & e^{-t} + e^{-3t} \end{bmatrix}$ **E6.7c** $e^{At} = \begin{bmatrix} e^{-2t} & 0 \\ 0 & e^{-t} \end{bmatrix}$ **E6.7d** $e^{At} = \begin{bmatrix} e^{-t} & 0 \\ 0 & e^{-3t} \end{bmatrix}$

E6.8 $e^{-At} = \begin{bmatrix} 2(e^t - e^{2t}) & \frac{1}{2}e^t - e^{-2t} \\ e^{2t} - 4e^t & \frac{1}{2}e^t - e^{-2t} \end{bmatrix}$

E6.9 $X(s) = \dfrac{1}{s^3 + 3s^2 + 1} \begin{bmatrix} s & -1 & 0 \\ 0 & s & -1 \\ 1 & 0 & s+3 \end{bmatrix} x_0 + \dfrac{1}{s^3 + 3s^2 + 1} \begin{bmatrix} 0 \\ -1 \\ s+3 \end{bmatrix} u(s)$ **E6.10** $e^{At} = \begin{bmatrix} te^t + e^{-t} & -te^{-t} \\ te^{-t} & -te^{-t} + e^{-t} \end{bmatrix}$

E6.11a $X(t) = \begin{bmatrix} 0 \\ e^t \\ 0 \end{bmatrix}$ **E6.11b** $X(t) = \begin{bmatrix} e^t + te^t \\ e^t \end{bmatrix} + \begin{bmatrix} e^t + te^t \\ e^t \end{bmatrix} u(t)$

E6.12   a) $\begin{bmatrix} \dot{z}_1 \\ \dot{z}_2 \\ \dot{z}_3 \end{bmatrix} = \begin{bmatrix} -2 & 1 & 0 \\ 0 & -2 & 0 \\ 0 & 0 & -5 \end{bmatrix} \begin{bmatrix} z_1 \\ z_2 \\ z_3 \end{bmatrix} + \begin{bmatrix} 0.22 \\ 0.33 \\ 0.22 \end{bmatrix} [u]$

     b) $\begin{bmatrix} \dot{z}_1 \\ \dot{z}_2 \\ \dot{z}_3 \end{bmatrix} = \begin{bmatrix} -2 & 0 & 0 \\ 0 & 0.7321 & 0 \\ 0 & 0 & -2.732 \end{bmatrix} \begin{bmatrix} z_1 \\ z_2 \\ z_3 \end{bmatrix} + \begin{bmatrix} 0.5 \\ 0.0528 \\ 0.1969 \end{bmatrix} [u]$; $Y = \begin{bmatrix} -1 & 16.929 & 3.0722 \end{bmatrix} \begin{bmatrix} z_1 \\ z_2 \\ z_3 \end{bmatrix}$

     c) $\begin{bmatrix} \dot{z}_1 \\ \dot{z}_2 \\ \dot{z}_3 \end{bmatrix} = \begin{bmatrix} -1.777 & 0 & 0 \\ 0 & 1.3889 + j1.4174 & 0 \\ 0 & 0 & 1.3889 - j1.4174 \end{bmatrix} \begin{bmatrix} z_1 \\ z_2 \\ z_3 \end{bmatrix} + \begin{bmatrix} 0.8943 + j0.1974 \\ -0.6047 + j0.1506 \\ -0.8481 + j0.6649 \end{bmatrix} [u]$

         $Y = \begin{bmatrix} 1.2799 & (-0.36 - j1.3124) & (2.36 + j3.124) \end{bmatrix} \begin{bmatrix} z_1 \\ z_2 \\ z_3 \end{bmatrix}$

d) $\begin{bmatrix} \dot{z}_1 \\ \dot{z}_2 \end{bmatrix} = \begin{bmatrix} -1 & 0 \\ 0 & -4 \end{bmatrix} \begin{bmatrix} z_1 \\ z_2 \end{bmatrix} + \begin{bmatrix} -1 \\ 1 \end{bmatrix} [u]$; $Y = \begin{bmatrix} \frac{1}{3} & \frac{1}{3} \end{bmatrix} \begin{bmatrix} z_1 \\ z_2 \end{bmatrix}$

e) $\begin{bmatrix} \dot{z}_1 \\ \dot{z}_2 \\ \dot{z}_3 \end{bmatrix} = \begin{bmatrix} -1 & 0 & 0 \\ 0 & -2 & 0 \\ 0 & 0 & -3 \end{bmatrix} \begin{bmatrix} z_1 \\ z_2 \\ z_3 \end{bmatrix} + \begin{bmatrix} 0.5 \\ 2 \\ 0.5 \end{bmatrix} [u]$; $Y = \begin{bmatrix} 2 & -1 & 0 \end{bmatrix} \begin{bmatrix} z_1 \\ z_2 \\ z_3 \end{bmatrix}$

f) $\begin{bmatrix} \dot{z}_1 \\ \dot{z}_2 \end{bmatrix} = \begin{bmatrix} -1 & 0 \\ 0 & -3 \end{bmatrix} \begin{bmatrix} z_1 \\ z_2 \end{bmatrix} + \begin{bmatrix} 0.5 \\ -0.5 \end{bmatrix} [u]$; $Y = \begin{bmatrix} -2 & 4 \end{bmatrix} \begin{bmatrix} z_1 \\ z_2 \end{bmatrix}$

g) $\begin{bmatrix} \dot{z}_1 \\ \dot{z}_2 \end{bmatrix} = \begin{bmatrix} -3 & 0 \\ 0 & -11 \end{bmatrix} \begin{bmatrix} z_1 \\ z_2 \end{bmatrix} + \begin{bmatrix} 1 \\ -1 \end{bmatrix} [u]$; $Y = \begin{bmatrix} 2 & 6 \end{bmatrix} \begin{bmatrix} z_1 \\ z_2 \end{bmatrix} + u$

h) $\begin{bmatrix} \dot{z}_1 \\ \dot{z}_2 \\ \dot{z}_3 \end{bmatrix} = \begin{bmatrix} -2 & 0 & 0 \\ 0 & -0.5858 & 0 \\ 0 & 0 & -3.4142 \end{bmatrix} \begin{bmatrix} z_1 \\ z_2 \\ z_3 \end{bmatrix} + \begin{bmatrix} -0.6667 \\ 0.3738 \\ -1.0405 \end{bmatrix} [u]$

$Y = \begin{bmatrix} -5 & -2.4145 & 0.4142 \end{bmatrix} \begin{bmatrix} z_1 \\ z_2 \\ z_3 \end{bmatrix}$

i) $\begin{bmatrix} \dot{z}_1 \\ \dot{z}_2 \\ \dot{z}_3 \end{bmatrix} = \begin{bmatrix} -3 & 0 & 0 \\ 0 & 1+j & 0 \\ 0 & 0 & -1-j \end{bmatrix} \begin{bmatrix} z_1 \\ z_2 \\ z_3 \end{bmatrix} + \begin{bmatrix} 0.2 \\ -(0.25+j0.25) \\ -0.25+j0.25 \end{bmatrix}$

$Y = \begin{bmatrix} -1\left(\frac{2}{5}+j\frac{14}{5}\right) & \left(\frac{2}{5}-j\frac{14}{5}\right) \end{bmatrix} \begin{bmatrix} z_1 \\ z_2 \\ z_3 \end{bmatrix}$

E6.13

a) Non-controllable

b) Controllable

c) Controllable

E6.14

a) Non-observable

b) Observable

E6.15

a) $\begin{bmatrix} 1.1848 & 0 & 0 \\ 0 & 2.2267 & 0 \\ 0 & 0 & -3.4115 \end{bmatrix}$

b) $\begin{bmatrix} 3 & 0 & 0 \\ 0 & 0.5+j1.6583 & 0 \\ 0 & 0 & 0.5-j1.6583 \end{bmatrix}$

c) $\begin{bmatrix} -1.943 & 0 & 0 \\ 0 & 1.9715+j1.3321 & 0 \\ 0 & 0 & 1.9715-j1.3321 \end{bmatrix}$

CHAPTER 7

SAMPLED DATA CONTROL SYSTEMS

7.1 INTRODUCTION

When the signal or information at any or some points in a system is in the form of discrete pulses, then the system is called discrete data system. In control engineering the discrete data system is popularly known as sampled data system.

The control system becomes a sampled data system in any one of the following situations.

1. *When a digital computer or microprocessor or digital device is employed as a part of the control loop.*

2. *When the control components are used on time sharing basis.*

3. *When the control signals are transmitted by pulse modulation.*

4. *When the output or input of a component in the system is a digital or discrete signal.*

The controllers are provided in control systems to modify the error signal for better control action. If the controllers are constructed using analog elements then they are called analog controllers and their input and output are analog signals, which are continuous functions of time. The analog controllers are complex, costlier and once fabricated it is difficult to alter the controllers.

A digital controller can be employed to implement complex or time shared control functions. [In time shared controller, a single controller will perform more than one function]. The digital controller are simple, versatile, programmable, fast acting and less costlier than analog controllers.

The digital controller can be a special purpose computer (microprocessor based system) or a general purpose computer or it is constructed using non-programmable digital devices. When computer or microprocessor is involved then the controller becomes programmable and its easier to alter the control functions by modifying the program instructions.

A sampled-data control system using digital controller is shown in fig 7.1. The input and output signal in a digital computer will be digital signals, but the error signal (input to the controller) to be modified by the controller and the control signal (output of the controller) to drive the plant are analog in nature. Hence a sampler and an analog-to-digital converter (ADC) are provided at the computer input. A digital to analog converter (DAC) and a hold circuit are provided at the computer output.

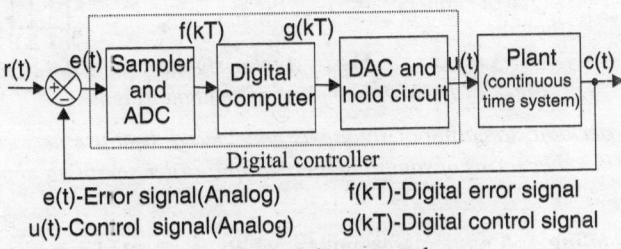

e(t)-Error signal(Analog)   f(kT)-Digital error signal
u(t)-Control signal(Analog)   g(kT)-Digital control signal

Fig 7.1 : *Sampled-data control system.*

The sampler converts the continuous time-error signal into a sequence of pulses and ADC produces a binary code (binary number) for each sample. These codes are the input data to the digital computer which process the binary codes and produces another stream of binary codes as output. The DAC and hold circuit converts the output binary codes to continuous time signal (Analog signal) called control signal. This output control signal is used to drive the plant.

ADVANTAGES OF DIGITAL CONTROLLERS

1. *The digital controllers can perform large and complex computation with any desired degree of accuracy at very high speed. In analog controllers the cost of controllers increases rapidly with the increase in complexity of computation and desired accuracy.*

2. *The digital controllers are easily programmable and so they are more versatile.*

3. *Digital controllers have better resolution.*

ADVANTAGES OF SAMPLED DATA CONTROL SYSTEMS

1. *The sampled data control systems are highly accurate, fast and flexible.*

2. *Use of time sharing concept of digital computer results in economical cost and space.*

3. *Digital transducers used in the system have better resolution.*

4. *The digital components used in the system are less affected by noise, nonlinearities and transmission errors of noisy channel.*

5. *The sampled data system require low power instruments which can be built to have high sensitivity.*

6. *Digital coded signals can be stored, transmitted, retransmitted, detected, analysed or processed as desired.*

7. *The system performance can be modified by compensation techniques.*

7.2 SAMPLING PROCESS

Sampling is the conversion of a continuous-time signal (or analog signal) into a discrete-time signal obtained by taking samples of the continuous time signal (or analog signal) at discrete time instants. Thus if f(t) is the input to the sampler as shown in fig 7.2, the output is f(kT), where T is called the sampling interval or sampling period. The reciprocal of T, i.e., $1/T = F_s$ is called the sampling rate (or samples per second or sampling frequency). This type of sampling is called periodic sampling, since samples are obtained uniformly at intervals of T seconds.

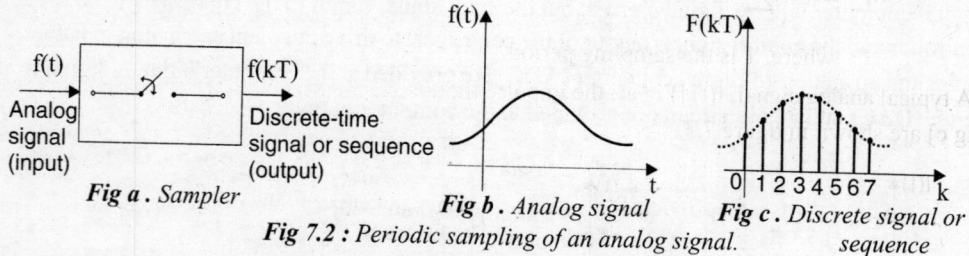

f(t)		f(kT)
Analog signal (input)		Discrete-time signal or sequence (output)

Fig a . Sampler   *Fig b . Analog signal*   *Fig c . Discrete signal or sequence*

Fig 7.2 : *Periodic sampling of an analog signal.*

(In this book only periodic sampling of signals is considered, because periodic sampling is most widely used in practice. The other forms of sampling are multiple-order sampling, multiple-rate sampling and Random sampling).

Multiple-order sampling : A particular sampling pattern is repeated periodically.

     Multiple-rate sampling   : In this method two simultaneous sampling operations with different time periods are carried out on the signal to produce the sampled output.

     Random sampling   : (In this case the sampling instants are random.)

     The sampling frequency F_s (=1/T) must be selected large enough such that the sampling process will not result in any loss of spectral information (i.e., if the spectrum of the analog signal can be recovered from the spectrum of the discrete - time signal, there is no loss of information). A guideline for choosing the sampling frequency is given by sampling theorem.

<u>**SAMPLING THEOREM :**</u>   *A band limited continuous time signal with highest frequency (bandwidth) f_m hertz, can be uniquely recovered from its samples provided that the sampling rate F_s is greater than or equal to $2f_m$ samples per second.*

    From the sampling theorem we can infer that the knowledge of frequency content of a signal is essential while choosing the sampling frequency.

    For processing the sampled signals by digital means, it has to be converted to binary codes and this conversion process is called quantization and coding. The process of converting a discrete time continuous valued signal into a discrete time discrete valued signal is called quantization. In quantization the value of each signal sample is represented by a value selected from a finite set of possible values called quantization levels. The difference between the unquantized sample and the quantized output is called the quantization error. The coding is the process of representing each discrete value by an n-bit binary sequence (or code or number). The process of sampling, quantization and coding are performed by sample/hold circuit and ADC.

7.3 ANALYSIS OF SAMPLING PROCESS IN FREQUENCY DOMAIN

    The sampling process explained in the previous section is equivalent to multiplying the analog signal, f(t) with a impulse train, $\delta_T(t)$ to produce the sampled signal, $f_s(t)$. Let the impulse train consists of pulses of area, Δ. Hence the impulse sampled signal, $f_s(t)$ can be expressed as,

$$f_s(t) = f(t)\, \Delta \delta_T(t) \qquad\qquad .....(7.1)$$

Mathematically, the impulse train, $\delta_T(t)$ can be expressed as,

$$\delta_T(t) = \sum_{k=-\infty}^{+\infty} \delta(t - KT) \qquad\qquad .....(7.2)$$

$$f_s(t) = \Delta f(t) \sum_{k=-\infty}^{+\infty} \delta(t - KT) \qquad\qquad .....(7.3)$$

       where, T is the sampling period.

    A typical analog signal, f(t) [Fig a]; the impulse train, $\delta_T(t)$ [Fig b] and the impulse sampled signal, $f_s(t)$ [Fig c] are shown in figure 7.3.

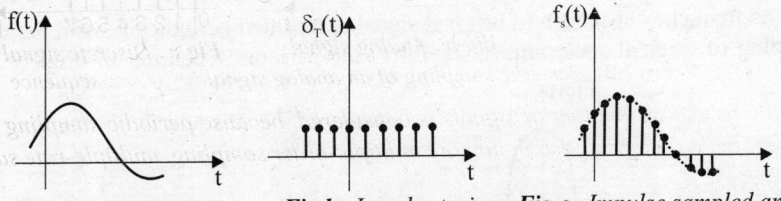

   Fig a . *Analog signal*   **Fig b .** *Impulse train*   **Fig c .** *Impulse sampled analog signal*

               Fig 7.3 : *Impulse sampling of an analog signal.*

The frequency content (frequency response) of a signal can be obtained from the fourier transform of the signal [i.e., Fourier transform converts the time domain signal to frequency domain signal]. Hence the frequency response of the impulse sampled signal can be obtained by taking fourier transform of equation (7.3).

The fourier transform of a single-valued function, f(t) is defined as

$$F\{f(t)\} = F(\omega) = \int_{-\infty}^{+\infty} f(t)\, e^{-j\omega t}\, dt \qquad .....(7.4)$$

On taking fourier transform of $f_s(t)$ using the definition of fourier transform we get,

$$F\{f_s(t)\} = F_s(\omega) = \int_{-\infty}^{+\infty} f_s(t)\, e^{-j\omega t}\, dt$$

$$= \int_{-\infty}^{\infty} \Delta\, f(t) \sum_{k=-\infty}^{+\infty} \delta(t-kT)\, e^{-j\omega t}\, dt \qquad .....(7.5)$$

Mathematically the equation (7.5) represents, the convolution of two signals, f(t) and δ(t – kT). The convolution theorem of fourier transform says that, the convolution of two time domain signals is equivalent to the product of their individual fourier transforms. Therefore, fourier transform of $f_s(t)$ can be expressed as a product of fourier transform of f(t) and δ(t – kT).

$$\therefore\ F_s(\omega) = \frac{\Delta}{2\pi} F\{f(t)\} . F\left\{\sum_{k=-\infty}^{+\infty} \delta(t-kT)\right\} \qquad .....(7.6)$$

Let, $F\{f(t)\} = F(\omega)$

$$.....(7.7)$$

$$F\left\{\sum_{k=-\infty}^{+\infty} \delta(t-kT)\right\} = \omega_s \sum_{k=-\infty}^{+\infty} \delta(\omega-k\omega_s) \qquad .....(7.8)$$

where, $\omega_s = 2\pi/T$ = Sampling frequency in rad/sec.

Using equations (7.7) and (7.8), the equation (7.6) can be written as,

$$F_s(\omega) = \frac{\Delta}{2\pi} \times F(\omega) \times \omega_s \sum_{k=-\infty}^{+\infty} \delta(\omega-k\omega_s) = \frac{\Delta}{2\pi} \frac{2\pi}{T} \sum_{k=-\infty}^{+\infty} F(\omega)\delta(\omega-k\omega_s)$$

Since $F(\omega)\, \delta(\omega-k\omega_s) = F(\omega-k\omega_s)$

$$\boxed{F_s(\omega) = \frac{\Delta}{T} \sum_{k=-\infty}^{+\infty} F(\omega-k\omega_s)} \qquad .....(7.9)$$

The equation (7.9) gives the frequency spectrum of the impulse sampled signal.

Let F(ω) be a band-limited signal with a maximum frequency of ω_m. The frequency spectrum of F(ω) is shown in fig 7.4(a), which is a plot of |F(ω)| Vs ω. The frequency spectrum of impulse sampled signal, i.e, |$F_s(\omega)$| Vs ω, is shown in fig 7.4(b), when $\omega_s > 2\,\omega_m$ and in fig 7.4(c), when $\omega_s < 2\,\omega_m$.

In fig 7.4(b) the frequency spectrum of original signal is repeated periodically with period ω_s and there is no overlapping of original spectrum. In fig 7.4(c) the periodic repetition of original spectrum overlaps.

(a)

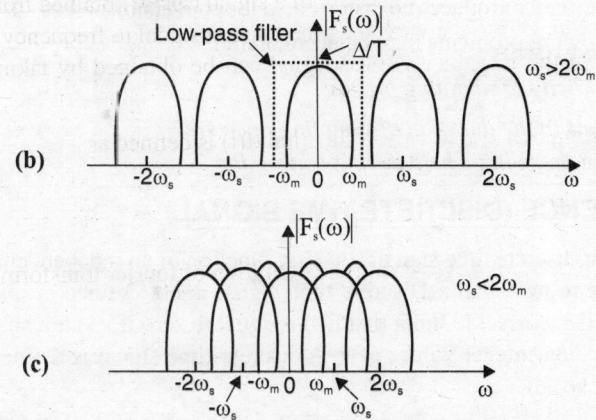

(b)

(c)

Fig 7.4 : *Fourier spectra of input signal and its impulse sampled version.*

From fig 7.4 it is observed that, as long as $\omega_s \geq \omega_m$, the original spectrum is preserved (since there is no overlapping) in the sampled signal and can be extracted from it by low-pass filtering. This fact was proposed as shanon's sampling theorem, which states that the information contained in a signal is fully preserved in the sampled version as long as the sampling frequency is at least twice the maximum frequency in the signal.

7.4 RECONSTRUCTION OF SAMPLED SIGNALS USING HOLD CIRCUITS

The hold circuits are popularly used in the process of analog-to-digital conversion (ADC) and digital-to-analog conversion (DAC). In ADC process the hold circuit is used to hold the sample until the quantization and coding for the current sample is complete.

In DAC process various types of hold circuits are used to convert the discrete time signal to analog signal. The simplest hold circuit is the zero order hold (ZOH). In zero order hold circuits the signal is reconstructed such that the value of reconstructed signal for a sampling period is same as the value of last received sample. The schematic diagram of sampler and zero order hold (ZOH) is shown in fig 7.5. The signal reconstruction by zero order hold (ZOH) circuit is illustrated in fig 7.6.

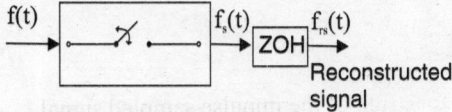

Fig 7.5 : *Sampler and ZOH.*

The high frequencies present in the reconstructed signal are easily filtered out by the various elements of the control system, because the control system is basically a low-pass filter.

In a first-order hold, the last two signal samples (current and previous sample) are used to reconstruct the signal for the current sampling period. Similarly higher order hold circuits can be devised. First or higher-order hold circuits offer no particular advantage over the zero-order hold. In sampled data control systems, the zero-order hold when used in conjunction with a high sampling rate provides a satisfactory performance.

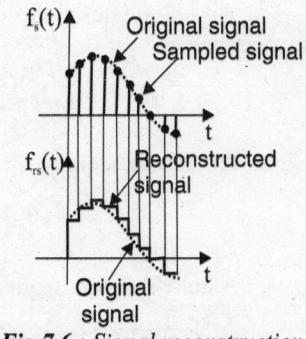

Fig 7.6 : *Signal reconstruction by ZOH.*

An ideal sample/hold circuit introduces no distortion in the conversion process. However, in practical sample/hold circuits the following problems may be encountered.

1. *Errors in the periodicity of sampling process.*
2. *Nonlinear variations in the duration of sampling aperture.*
3. *Droop (changes) in the voltage held during conversion.*

7.5 DISCRETE SEQUENCE (DISCRETE TIME SIGNAL)

A discrete sequence or discrete time signal, f(k), is a function of an independent variable, k, which is an integer. It is important to note that a Discrete time signal is not defined at instants between two successive samples. Also, it is incorrect to think that f(k) is equal to zero if k is not an integer. Simply the signal f(k) is not defined for non-integer values of k. A discrete-time signal is defined for every integer value of k in the range $-\infty < k < \infty$.

Since a digital signal is represented by a set of numbers it is also called a sequence. (i.e., the terms signal and sequence refers the digital or discrete time signal).

METHODS OF REPRESENTING A DISCRETE-TIME SIGNAL OR SEQUENCE

1. Functional representation

$$f(k) = 1 \quad ; \quad k = 1, 3$$

$$4 \quad ; \quad k = 2$$

$$0 \quad ; \quad \text{other } k$$

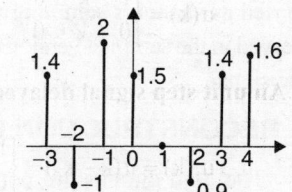

Fig 7.7 : Graphical representation of a discrete-time signal.

2. Graphical representation

The graphical representation of a discrete sequence is shown in fig 7.7.

3. Tabular representation

k	.....	−2	−1	0	1	2	....
f(k)	.....	0	0	0	1	4	....

4. Sequence representation

An infinite duration signal or sequence with the time origin (k = 0) indicated by the symbol ↑ is represented as

$$f(k) = \{......1, 2, 1, 4, 1, 0, 0.....\}$$
$$\uparrow$$

An infinite sequence f(k), which is zero for k < 0, may be represented as

$$f(k) = \{2, 1, 4, 1, 0, 0.....\} \qquad \text{(or)} \qquad f(k) = \{2, 1, 4, 1....\}$$
$$\uparrow \hspace{6cm} \uparrow$$

A finite duration sequence with the time origin (k = 0) indicated by the symbol ↑ is represented as

$$f(k) = \{3, -1, -2, 5, 0, 4, ...\}$$
$$\uparrow$$

A finite duration sequence that satisfies the condition f(k) = 0 for k < 0 may be represented as

$$f(k) = \{2, 1, 4, 1\} \qquad \text{(or)} \qquad f(k) = \{2, 1, 4, 1\}$$
$$\uparrow$$

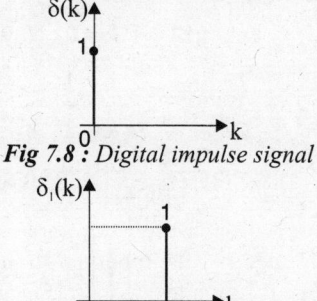

Fig 7.8 : Digital impulse signal

SOME ELEMENTARY DISCRETE-TIME SIGNALS

1. Digital impulse signal or unit sample sequence,

$$\delta(k) = \begin{cases} 1 & ; \ k = 0 \\ 0 & ; \ k \neq 0 \end{cases}$$

An impulse delayed by k_0,

$$\delta_1(k) = \delta(k - k_0) = \begin{cases} 1 & ; \ k = k_0 \\ 0 & ; \ k \neq k_0 \end{cases}$$

Fig 7.9 : Delayed impulse signal

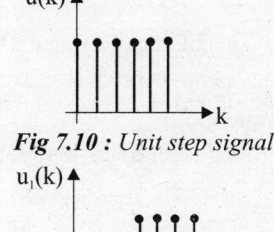

2. Unit step signal

$$u(k) = \begin{cases} 1 & ; \ k \geq 0 \\ 0 & ; \ k < 0 \end{cases}$$

Fig 7.10 : Unit step signal

An unit step signal delayed by k_0

$$u_1(k) = u(k - k_0) = \begin{cases} 1 & ; \ k \geq k_0 \\ 0 & ; \ k < k_0 \end{cases}$$

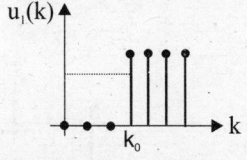

Fig 7.11 : Delayed Unit step signal

The unit step is related to digital impulse by the summation relation

$$u(k) = \sum_{m=0}^{\infty} \delta(k - m)$$

3. Ramp signal

$$u_r(k) = \begin{cases} k & ; \ k \geq 0 \\ 0 & ; \ k < 0 \end{cases}$$

Fig 7.12 : Ramp signal

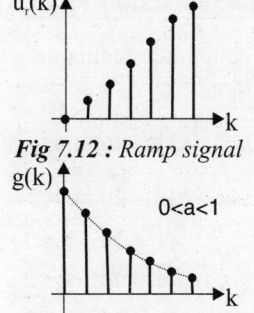

4. Exponential signal

$$g(k) = \begin{cases} a^k & ; \ k \geq 0 \\ 0 & ; \ k < 0 \end{cases}$$

Fig 7.13 : Exponential signal.

MATHEMATICAL OPERATIONS ON DISCRETE TIME SIGNALS

1. Shifting in time

A signal f(k) may be shifted in time by replacing the independent variable k by (k − m), where m is an integer. If m is a positive integer, the time shift results in a delay by m units of time. If m is a negative integer, the time shift results in an advance of the signal by |m| units in time. The delay results in shifting each sample of f(k) to right. The advance results in shifting each sample of f(k) to left.

Example

Let, $f(k) = 1$ $k = 0$ Now, $f_1(k) = f(k+2) = 1$ $k = -2$ and $f_2(k) = f(k-2) = 1$ $k = 2$

$\quad\quad\quad\quad$ 2 $k = 1$ $\quad\quad\quad\quad\quad\quad\quad\quad$ 2 $k = -1$ $\quad\quad\quad\quad\quad\quad\quad\quad$ 2 $k = 3$

$\quad\quad\quad\quad$ 3 $k = 2$ $\quad\quad\quad\quad\quad\quad\quad\quad$ 3 $k = 0$ $\quad\quad\quad\quad\quad\quad\quad\quad$ 3 $k = 4$

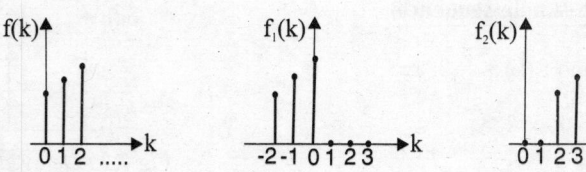

2. Folding or reflection or Transpose

The folding of a signal $f(k)$ is performed by changing the sign of the time base k in the signal $f(k)$. The folding operation produces a signal $f(-k)$ which is mirror image of $f(k)$ with respect to time origin $k = 0$.

Example

Let $f(k) = k$ $-3 \le k \le 3$; $f_1(k) = f(-k) = -k$ $\quad$ $-3 \le k \le 3$

3. Amplitude scaling or scalar multiplication

Amplitude scaling of a signal by a constant A is accomplished by multiplying the value of every signal sample by A.

Let $c(k)$ be amplitude scaled signal of $f(k)$, then $c(k) = Af(k)$

Let $f(k) = 20$ $k = 0$ $\quad\quad$ and $A = 0.1$; $c(k) = 2.0$ $k = 0$

$\quad\quad\quad\quad$ 36 $k = 1$ $\quad\quad\quad\quad\quad\quad\quad\quad\quad\quad\quad\quad$ 3.6 $k = 1$

$\quad\quad\quad\quad$ 40 $k = 2$ $\quad\quad\quad\quad\quad\quad\quad\quad\quad\quad\quad\quad$ 4.0 $k = 2$

$\quad\quad\quad\quad$ −15 $k = 3$ $\quad\quad\quad\quad\quad\quad\quad\quad\quad\quad$ −1.5 $k = 3$

4. Time scaling or down sampling

In a signal, $f(k)$, if k is replaced by μk, where μ is an integer, then it is called time scaling or down sampling.

Example : If $f(k) = a^k$; $k \ge 0$, then $f_1(k) = f(2k) = a^k$ for even values of k

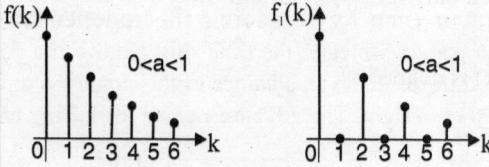

5. Signal (or vector) addition

The sum of two signals $f_1(k)$ and $f_2(k)$ is a signal $c(k)$, whose value at any instant is equal to the sum of the samples of these two signals at that instant.

i.e. $c(k) = f_1(k) + f_2(k)$; $-\infty < k < \infty$.

Example

Let $f_1(k) = \{1, 2, -1, 2\}$ and $f_2(k) = \{-2, 1, 3, 1\}$

$c(k) = f_1(k) + f_2(k) = \{-1, 3, 2, 3\}$

6. Signal (or vector) multiplication

Signal multiplication results in the product of two signals on a sample-by-sample basis. The product of two signals $f_1(k)$ and $f_2(k)$ is a signal $c(k)$, whose value at any instant is equal to the product of the sample of these two signals at that instant. The product is also called modulation.

Example

Let $f_1(k) = \{1, 2, -1, 2\}$ and $f_2(k) = \{-2, 1, 3, 1\}$

$c(k) = f_1(k) . f_2(k) = \{-2, 2, -3, 2\}$

7.6 $\mathcal{Z}$ -TRANSFORM

Transform techniques are an important tool in the analysis of signals and linear time invariant systems. The Laplace transforms are popularly used for analysis of continuous time signals and systems. Similarly $\mathcal{Z}$-transform plays an important role in analysis and representation of linear discrete time systems. The $\mathcal{Z}$-transform provides a method for the analysis of discrete time systems in the frequency domain which is generally more efficient than its time domain analysis.

DEFINITION OF $\mathcal{Z}$-TRANSFORM

Let, $f(k)$ = Discrete time signal or sequence

$F(z) = \mathcal{Z}\{f(k)\} = \mathcal{Z}$-transform of $f(k)$

The $\mathcal{Z}$-transform of a discrete time signal or sequence is defined as the power series

$$F(z) = \sum_{k=-\infty}^{\infty} f(k) \, z^{-k} \qquad \qquad .....(7.10)$$

where, z is a complex variable.

The sequence of equation (7.10) is considered to be two sided and the transform is called two sided $\mathcal{Z}$-transform, since the time index k is defined for both positive and negative values. If the sequence $f(k)$ is one sided sequence, (i.e. $f(k)$ is defined only for positive value of k) then the $\mathcal{Z}$-transform is called one sided $\mathcal{Z}$-transform.

The one sided $\mathcal{Z}$-transform of $f(k)$ is defined as,

$$F(z) = \sum_{k=0}^{\infty} f(k) \, z^{-k}$$

REGION OF CONVERGENCE

Since the $\mathcal{Z}$-transform is an infinite power series, it exists only for those values of z for which the series converges. The region of convergence, (ROC) of F(z) is the set of all values of z, for which F(z) attains a finite value. The ROC of a finite-duration signal is the entire z-plane, except possibly the point z = 0 and z = ∞. These points are excluded, because z^k (when k > 0) becomes unbounded for z = ∞ and z^{-k} (when k < 0) becomes unbounded for z = 0.

The complex variable z can be expressed in the polar form as,

$$z = r\, e^{j\theta} \qquad\qquad .....(7.11)$$

where, r = |z| and θ = $\angle z$

On substituting for z from equation (7.11) in equation (7.10) we get,

$$F(z) = \sum_{k=-\infty}^{\infty} f(k)\, z^{-k} = \sum_{k=-\infty}^{\infty} f(k)\, (re^{j\theta})^{-k} = \sum_{k=-\infty}^{\infty} f(k)\, r^{-k} e^{-j\theta k} \qquad .....(7.12)$$

Now, $|F(z)| = \sum_{k=-\infty}^{\infty} |f(k)\, r^{-k}|$ $\qquad\qquad .....(7.13)$

In the ROC of F(z), $|F(z)| < \infty$.

From equation (7.13) we observe that $|F(z)|$ is finite, if the sequence $f(k)\, r^{-k}$ is absolutely summable.

To find the ROC, the equation (7.13) can be expressed as,

$$|F(z)| = \sum_{k=-\infty}^{\infty} |f(k)\, r^{-k}| = \sum_{k=-\infty}^{\infty} |f(k)\, r^{-k}| + \sum_{k=-\infty}^{\infty} |f(k)\, r^{-k}|$$

$$|F(z)| = \sum_{k=-\infty}^{\infty} |f(-k)\, r^{k}| + \sum_{k=-\infty}^{\infty} \left|\frac{f(k)}{r^{k}}\right| \qquad\qquad .....(7.14)$$

If F(z) converges in some region of the complex plane, both summations in equation (7.14) must be finite.

If the first sum of equation (7.14) converges, there must exist values of r small enough for $f(-k)\, r^k$ to be absolutely summable. Hence the ROC for the first sum consists of all points inside a circle of radius, r_1 as shown in fig 7.14

If the second sum of equation (7.14) converges, there must exist large values of r for which $f(k)\, /\, r^k$ is absolutely summable. Hence the ROC for the second sum consists of all points outside a circle of radius, r_2 as shown in fig 7.15,

Therefore, the ROC of F(z) is the region in between two circles of radius r_1 and r_2 as shown in fig 7.16, where $r_2 < r < r_1$.

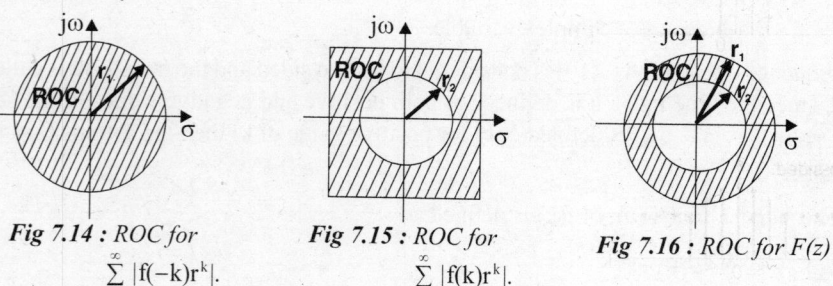

Fig 7.14 : *ROC for* $\sum_{k=1}^{\infty} |f(-k)r^k|.$ **Fig 7.15 :** *ROC for* $\sum_{k=0}^{\infty} |f(k)r^k|.$ **Fig 7.16 :** *ROC for F(z)*

TABLE-7.1 : Characteristic Families of Signals with Their Corresponding ROC

SIGNAL	ROC
Finite-Duration Signals	
Causal (or right sided)	Entire z-plane except z=0
Anti Causal (or left sided)	Entire z-plane except z=∞
Two-sided	Entire z-plane except z=0 and z = ∞
Infinite-Duration Signals	
Causal (or right sided)	$|z| > r_2$
Anti Causal (or left sided)	$|z| < r_1$
Two-sided	$r_2 < |z| < r_1$

TABLE-7.2 : Properties of One-Sided $\mathbb{Z}$-Transform

Note : $F(z) = \mathbb{Z}\{f(k)\}$; $F_1(z) = \mathbb{Z}\{f_1(k)\}$; $F_2(z) = \mathbb{Z}\{f_2(k)\}$

Property	Discrete Sequence	z-transform
Linearity	$a_1 f_1(k) + a_2 f_2(k)$	$a_1 F_1(z) + a_2 F_2(z)$
Shifting, $m \geq 0$	$f(k + m)$	$z^m F(z) - \displaystyle\sum_{i=0}^{m=-1} f(i) \, z^{m-1}$
	$f(k - m)$	$z^{-m} F(z)$
Multiplication by k^m (or differentiation in z-domain)	$k^m f(k)$	$\left(-z \dfrac{d}{dz}\right)^m F(z)$
Scaling in z-domain (or multiplication by a^k)	$a^k f(k)$	$F(a^{-1}z)$
Time reversal	$f(-k)$	$F(z^{-1})$
Conjugation	$f^*(k)$	$F^*(z^*)$
Convolution	$\displaystyle\sum_{m=0}^{k} h(k-m)\, r(m)$	$H(z)\, R(z)$
Initial value	$f(0) = \underset{z \to \infty}{\text{Lt}}\, F(z)$	
Final value	$f(\infty) = \underset{z \to 1}{\text{Lt}}(1-z)^{-1} F(z)$ $= \underset{z \to 1}{\text{Lt}}(z-1) F(z)$ if $F(z)$ is analytic for $|z| > 1$	

TABLE-7.3 : Some Common One Sided Ƶ-transform Pairs

> **Note :** *Two sided sequence can be converted to one sided sequence by multiplying by u(k).*

f(t) ; t ≥ 0	f(k) or f(kT) ; k ≥ 0	F(z)
	$\delta(k)$	1
	$u(k)$ or 1	$z/(z-1)$
	a^k	$z/(z-a)$
	$k\, a^k$	$\dfrac{az}{(z-a)^2}$
	$k^2\, a^k$	$\dfrac{az(z+a)}{(z-a)^3}$
	$(k+1)\, a^k$	$\dfrac{z^2}{(z-a)^2}$
	$\dfrac{(k+1)(k+2)}{2!}\, a^k$	$\dfrac{z^3}{(z-a)^3}$
	$\dfrac{(k+1)(k+2)(k+3)}{3!}\, a^k$	$\dfrac{z^4}{(z-a)^4}$
	$\dfrac{a^k}{k!}$	$e^{az^{-1}}$
t	kT	$\dfrac{Tz}{(z-1)^2}$
t^2	$(kT)^2$	$\dfrac{T^2 z(z+1)}{(z-1)^3}$
e^{-at}	e^{-akT}	$\dfrac{z}{z-e^{-aT}}$
te^{-aT}	kTe^{-akT}	$\dfrac{zTe^{-aT}}{(z-e^{-aT})^2}$
$\sin \omega t$	$\sin \omega kT$	$\dfrac{z\sin \omega T}{z^2 - 2z\cos \omega T + 1}$
$\cos \omega t$	$\cos \omega kT$	$\dfrac{z(z-\cos \omega T)}{z^2 - 2z\cos \omega T + 1}$

GEOMETRIC SERIES

A geometric series is a series in which consecutive elements differ by a constant ratio. Such a series can be written in the form,

$$f(k) = \sum_{k=M_1}^{M_2} C \qquad \qquad .....(7.15)$$

where, C is a constant and M_1 and M_2 are any two numbers.

If C is a complex number, where $|C| < 1$, then by Taylor's series expansion we can write,

$$\frac{1}{1-C} = 1 + C + C^2 + ... = \sum_{k=0}^{\infty} C^k \qquad \qquad .....(7.16)$$

Applying the result in the reverse direction yields the infinite geometric series sum formula

$$\therefore \sum_{k=0}^{\infty} C^k = \frac{1}{1-C} \qquad \qquad .....(7.17)$$

The equation (7.17) is the infinite geometric series sum formula.

We can also compute the sum of a finite number of elements in a geometric series. Let us consider the following sum,

$$1 + C + C^2 + ... + C^{M-1} = \sum_{k=0}^{m-1} C^k \qquad \qquad .....(7.18)$$

The sum of the finite duration sequence in equation (7.18) can be expressed as the difference between the sum of two infinite duration sequences as shown in equation (7.19).

$$\sum_{K=0}^{m-1} C^k = \sum_{K=0}^{\infty} C^k - \sum_{K=M}^{\infty} C^k \qquad \qquad .....(7.19)$$

Now, $\sum_{K=M}^{\infty} C^k = C^M + C^{M+1} + C^{M+2} + ...$

$$= C^M + C^M C + C^M C^2 + ... = C^M (1 + C + C^2 + C^3 + ...)$$

$$= C^M \left(\sum_{K=0}^{\infty} C^k \right) \qquad \qquad .....(7.20)$$

From equations (7.19) and (7.20) we can write,

$$\sum_{k=0}^{M-1} C^k = \sum_{K=0}^{\infty} C^k - C^M \sum_{K=0}^{\infty} C^k = (1 - C^M) \sum_{K=0}^{\infty} C^k$$

$$= (1 - C^M) \left(\frac{1}{1-C} \right) = \frac{1 - C^M}{1 - C} = \frac{C^M - 1}{C - 1} \qquad \qquad .....(7.21)$$

$$\text{when } C = 1, \sum_{K=0}^{M-1} C^k = M \qquad \qquad .....(7.22)$$

The equations (7.21) and (7.22) are finite geometric series sum formula.

> *Note :* *The infinite geometric series sum formula requires that the magnitude of C be strictly less than unity, but the finite geometric series sum formula is valid for any value of C.*

EXAMPLE 7.1

Determine the z-transform and their ROC of the following discrete sequences
(a) f(k) = {3, 2, 5, 7}   (b) f(k) = {2, 4, 5, 7, 3}

↑

SOLUTION

(a)   **Given that, f(k) = {3, 2, 5, 7}**

i.e., $f(0) = 3$; $f(1) = 2$; $f(2) = 5$; $f(3) = 7$

and $f(k) = 0$ for $k < 0$ and for $k > 3$

By the definition of $\mathcal{Z}$-transform,s

$$\mathcal{Z}\{f(k)\} = F(z) = \sum_{k=-\infty}^{\infty} f(k)\, z^{-k}$$

The given sequence is a finite duration sequence, hence the limits of summation can be changed as $k = 0$ to $k = 3$.

$$\therefore F(z) = \sum_{k=0}^{3} f(k)\, z^{-k}$$

On expanding the summation we get,

$$F(z) = f(0)z^0 + f(1)z^{-1} + f(2)z^{-2} + f(3)z^{-3}$$

$$= 3 + 2z^{-1} + 5z^{-2} + 7z^{-3}$$

Here, $F(z)$ is bounded (i.e., finite) except when $z = 0$, therefore the ROC is entire z-plane except $z = 0$.

(b)   **Given that, f(k) = {2, 4, 5, 7, 3}**
                                    ↑

i.e, $f(-2) = 2$; $f(-1) = 4$; $f(0) = 5$; $f(1) = 7$; $f(2) = 3$

and $f(k) = 0$ for $k < -2$ and for $k > 2$.

By the definition of $\mathcal{Z}$-transform,

$$\mathcal{Z}\{f(k)\} = F(z) = \sum_{k=-\infty}^{\infty} f(k)\, z^{-k}$$

The given sequence is a finite duration sequence, hence the limits of summation can be changed as $k = -2$ to $k = 2$.

$$\therefore F(z) = \sum_{k=-2}^{2} f(k)\, z^{-k}$$

On expanding the summation we get,

$$F(z) = f(-2)z^2 + f(-1)z^1 + f(0)z^0 + f(1)\,a^{-1} + f(2)\,z^{-2}$$

$$= 2z^2 + 4z + 5 + 7z^{-1} + 3z^{-2}$$

Here, $F(z)$ is bounded (i.e., finite) except when $z = 0$ and $z = \infty$, therefore the ROC is entire z-plane except $z = 0$ and $z = \infty$.

EXAMPLE 7.2

Determine the $\mathcal{Z}$-transform of the following discrete sequences.

(a) $f(k) = u(k)$ (b) $f(k) = (1/2)^k\, u(k)$ (c) $f(k) = \alpha^k\, u(-k-1)$

SOLUTION

(a)   **Given that, f(k) = u(k)**

$u(k)$ is a discrete unit step sequence, which is defined as

$$u(k) = 1 \text{ for } k \geq 0$$

$$= 0 \text{ for } k < 0$$

By the definition of $\bar{z}$-transform,

$$\bar{z}\{f(k)\} = F(z) = \sum_{k=-\infty}^{\infty} f(k)\, z^{-k}$$

Here, F(z) is an infinite geometric series and it converges if $|z| > 1$ (i.e., $|z^{-1}| < 1$). Using infinite geometric series sum formula we get,

$$F(z) = \frac{1}{1-z^{-1}} = \frac{1}{1-1/z} = \frac{z}{z-1}$$

(b)   **Given that, f(k) = (1/2)^k u(k)**

u(k) is a discrete unit step sequence, which is defined as

$$u(k) = 1 \text{ for } k \geq 0$$
$$= 0 \text{ for } k < 0$$

$$\therefore f(k) = (1/2)^k \text{ for } k \geq 0$$
$$= 0 \qquad \text{for } k < 0$$

By the definition of $\bar{z}$-transform,

$$\bar{z}\{f(k)\} = F(z) = \sum_{k=-\infty}^{\infty} f(k)\, z^{-k}$$

$$= \sum_{k=0}^{\infty} (1/2)^k\, z^{-k} = \sum_{k=0}^{\infty} \left(\frac{1}{2} z^{-1}\right)^k$$

Here, F(z) is an infinite geometric series and it converges if $|z| > 1$ (i.e., $|z^{-1}| < 1$). Using infinite geometric series sum formula we get,

$$F(z) = \frac{1}{1-\frac{1}{2}z^{-1}} = \frac{1}{1-\frac{1}{2}\cdot\frac{1}{z}} = \frac{2z}{2z-1}$$

(c)   **Given that, f(k) = (1/2)^k u(−k−1)**

u(−k−1) is a discrete unit step sequence, which is defined as

$$u(-k-1) = 0 \text{ for } k \geq 0$$
$$= 1 \text{ for } k \leq -1$$

$$\therefore f(k) = 0 \text{ for } k \geq 0$$
$$= \alpha^k \text{ for } k \leq -1$$

By the definition of $\bar{z}$-transform,

$$\bar{z}\{f(k)\} = F(z) = \sum_{k=-\infty}^{\infty} f(k)\, z^{-k}$$

$$= \sum_{k=-\infty}^{-1} (\alpha^k)\, z^{-k} = \sum_{k=1}^{\infty} \alpha^{-k} z^k$$

$$= \sum_{k=1}^{\infty} (\alpha^{-1} z)^k = \sum_{k=0}^{\infty} (\alpha^{-1} z)^k - 1$$

Using infinite geometric series sum formula we get,

$$F(z) = \frac{1}{1-\alpha^{-1}z} - 1 = \frac{1}{1-\frac{z}{\alpha}} - 1 = \frac{\alpha}{\alpha-z} - 1$$

$$= \frac{\alpha-\alpha+z}{\alpha-z} = \frac{z}{\alpha-z}$$

EXAMPLE 7.3

Find the one sided $\bar{z}$-transform of the following discrete sequences.

(a) $f(k) = k\, a^{(k-1)}$ (b) $f(k) = k^2$

SOLUTION

(a) **Given that f(k) = k a$^{(k-1)}$**

The one sided $\bar{z}$-transform of a^k is given by

$$\bar{z}\{a^k\} = \sum_{k=0}^{\infty} a^k z^{-k} = \sum_{k=0}^{\infty} (az^{-1})^k \qquad \qquad .....(7.3.1)$$

Using infinite geometric series sum formula,

$$\bar{z}\{a^k\} = \frac{1}{1-az^{-1}} = \frac{1}{1-a/z} = \frac{z}{z-a} \qquad \qquad .....(7.3.2)$$

From equations (7.3.1) and (7.3.2) we get

$$\sum_{k=0}^{\infty} a^k z^{-k} = \frac{z}{z-a}$$

On expanding the summation in the above equation we get,

$$1 + az^{-1} + a^2 z^{-2} + a^3 z^{-3} + ... = \frac{z}{z-a} \qquad \qquad .....(7.3.3)$$

On differentiating the equation (7.3.3) we get,

$$-az^{-2} - 2a^2 z^{-3} - 3a^3 z^{-4} .... = \frac{(z-a) \times 1 - z \times 1}{(z-a)^2} = \frac{-a}{(z-a)^2} \qquad .....(7.3.4)$$

On multiplying the equation (7.3.4) by -(z/a) we get,

$$z^{-1} + 2az^{-2} + 3a^2 z^{-3} + ... = \frac{z}{(z-a)^2} \qquad \qquad .....(7.3.5)$$

The infinite series on the left hand side of the equation (7.3.5) can be expressed as a summation and the equation (7.3.5) is written as shown below.

$$\sum_{k=1}^{\infty} k\, a^{(k-1)} z^{-k} = \frac{z}{(z-a)^2} \qquad \qquad .....(7.3.6)$$

By definition of $\bar{z}$-transform, the one sided $\bar{z}$-transform of $k\, a^{(k-1)}$ is given by,

$$\bar{z}\{ka^{(k-1)}\} = \sum_{k=0}^{\infty} k\, a^{(k-1)} z^{-k} = \sum_{k=1}^{\infty} k\, a^{(k-1)} z^{-k} \qquad \qquad .....(7.3.7)$$

(Because, $k\, a^{(k-1)} = 0$ when $k = 0$)

On comparing equations (7.3.6) and (7.3.7) we get,

$$\bar{z}\{k\, a^{(k-1)}\} = \frac{z}{(z-a)^2}$$

(b) **Given that, f(k) = k^2**

Let us multiply the given discrete sequence by a discrete unit step sequence,

$$\therefore f(k) = k^2\, u(k)$$

> Note : Multiplying a one sided sequence by u(k) will not alter its value.

By the property of z-transform, we get,

$$\mathcal{Z}\{k^m u(k)\} = \left(-z\frac{d}{dz}\right)^m U(z)$$

where, $U(z) = \mathcal{Z}\{u(k)\} = \dfrac{z}{z-1}$

$$\therefore -z\frac{d}{dz}U(z) = -z\left[\frac{d}{dz}\left(\frac{z}{z-1}\right)\right] = -z\left[\frac{z-1-z}{(z-1)^2}\right] = \frac{z}{(z-1)^2}$$

$$\left(-z\frac{d}{dz}\right)^2 U(z) = -z\frac{d}{dz}\left[-z\frac{d}{dz}U(z)\right]$$

$$= -z\frac{d}{dz}\left(\frac{z}{(z-1)^2}\right) = -z\left(\frac{(z-1)^2 - z\times 2(z-1)}{(z-1)^4}\right)$$

$$= -z\left(\frac{(z-1)(z-1-2z)}{(z-1)^4}\right) = -z\left(\frac{-(z+1)}{(z-1)^3}\right) = \frac{z(z+1)}{(z-1)^3}$$

$$\therefore \mathcal{Z}\{f(k)\} = \mathcal{Z}\{k^2 u(k)\} = \left(-z\frac{d}{dz}\right)^2 U(z) = \frac{z(z+1)}{(z-1)^3}$$

EXAMPLE 7.4

Find the one sided z-transform of the discrete sequences generated by mathematically sampling the following continuous time functions

(a) t^2 (b) $\sin \omega t$ (c) $\cos \omega t$

SOLUTION

(a)   **Given that, f(t) = t²**

The discrete sequence is generated by replacing t by kT, where T is the sampling time period.

$$\therefore f(k) = (kT)^2 = k^2 T^2 = k^2 g(k)$$

where, $g(k) = T^2$

By the definition of one sided z-transform we get,

$$G(z) = \mathcal{Z}\{g(k)\} = \mathcal{Z}\{T^2\} = \sum_{k=0}^{\infty} T^2 z^{-k} = T^2 \sum_{k=0}^{\infty} (z^{-1})^k = T^2\left(\frac{1}{1-z^{-1}}\right) = \frac{T^2 z}{z-1}$$

By the property of z-transform we get,

$$\mathcal{Z}\{f(k)\} = F(z) = \left(-z\frac{d}{dz}\right)^2 G(z) = -z\frac{d}{dz}\left(-z\frac{d}{dz}G(z)\right)$$

$$= -z\frac{d}{dz}\left(-z\frac{d}{dz}\frac{T^2 z}{z-1}\right) = -z\frac{d}{dz}\left(-z\times\frac{(z-1)T^2 - T^2 z}{(z-1)^2}\right)$$

$$= -z\frac{d}{dz}\left(\frac{zT^2}{(z-1)^2}\right) = -z\times\frac{(z-1)^2 T^2 - zT^2\times 2(z-1)}{(z-1)^4}$$

$$= -z\times\frac{(z-1)(zT^2 - T^2 - 2zT^2)}{(z-1)^4} = -z\times\frac{-zT^2 - T^2}{(z-1)^3} = \frac{zT^2(z+1)}{(z-1)^3}$$

(b)   **Given that, f(t) = sin t**

The discrete sequence is generated by replacing t by kT, where T is the sampling time period.

$$\therefore f(k) = \sin(\omega kT)$$

By the definition of one sided z-transform,

$$\mathcal{Z}\{f(k)\} = F(z) = \sum_{k=0}^{\infty} f(k) z^{-k} = \sum_{k=0}^{\infty} \sin \omega kT \times z^{-k}$$

We know that, $\sin \theta = (e^{j\theta} - e^{-\theta})/2j$

$$\therefore F(z) = \sum_{k=0}^{\infty} \frac{e^{j\omega kT} - e^{-j\omega kT}}{2j} z^{-k} = \frac{1}{2j} \sum_{k=0}^{\infty} e^{j\omega kT} z^{-k} - \frac{1}{2j} \sum_{k=0}^{\infty} e^{-j\omega kT} z^{-k}$$

We know that, $\mathcal{Z}\{e^{\pm akT}\} = \sum_{k=0}^{\infty} e^{\pm akT} z^{-k} = \dfrac{z}{z - e^{\pm aT}}$

$$\therefore F(z) = \frac{1}{2j} \frac{z}{z - e^{j\omega T}} - \frac{1}{2j} \frac{z}{z - e^{-j\omega T}}$$

$$= \frac{z(z - e^{-j\omega T}) - z(z - e^{j\omega T})}{2j(z - e^{j\omega T})(z - e^{-j\omega T})} = \frac{z^2 - ze^{-j\omega T} - z^2 + ze^{j\omega T}}{2j(z^2 - ze^{-j\omega T} - ze^{j\omega T} + e^{j\omega T} \cdot e^{-j\omega T})}$$

$$= \frac{z(e^{j\omega T} - e^{-j\omega T})/2j}{z^2 - z(e^{j\omega T} + e^{-j\omega T}) + 1}$$

We know that, $\sin \theta = (e^{j\theta} - e^{-j\theta})/2j$ and $\cos \theta = (e^{j\theta} + e^{-j\theta})/2$

$$\therefore F(z) = \frac{z \sin \omega T}{z^2 - 2z \cos \omega T + 1}$$

(c)   **Given that, f(t) = cos t**

The discrete sequence is generated by replacing t by kT, where T is the sampling time period.

$$\therefore f(k) = \cos(\omega kT)$$

By the definition of one sided $\mathcal{Z}$-transform,

$$\mathcal{Z}\{f(k)\} = F(z) = \sum_{k=0}^{\infty} f(k) z^{-k} = \sum_{k=0}^{\infty} \cos \omega kT \times z^{-k}$$

We know that, $\cos \theta = (e^{j\theta} + e^{-j\theta})/2$

$$\therefore F(z) = \sum_{k=0}^{\infty} \frac{e^{j\omega kT} + e^{-j\omega kT}}{2} z^{-k} = \frac{1}{2} \sum_{k=0}^{\infty} e^{j\omega kT} z^{-k} + \frac{1}{2} \sum_{k=0}^{\infty} e^{j\omega kT} z^{-k}$$

We know that, $\mathcal{Z}\{e^{\pm akT}\} = \sum_{k=0}^{\infty} e^{\pm akT} z^{-k} = \dfrac{z}{z - e^{\pm aT}}$

$$\therefore F(z) = \frac{1}{2} \frac{z}{z - e^{j\omega T}} + \frac{1}{2} \frac{z}{z - e^{-j\omega T}}$$

$$= \frac{z(z - e^{-j\omega T}) + z(z - e^{j\omega T})}{2(z - e^{j\omega T})(z - e^{-j\omega T})} = \frac{z^2 - ze^{-j\omega T} + z^2 - ze^{j\omega T}}{2(z^2 - ze^{-j\omega T} - ze^{j\omega T} + e^{j\omega T} \cdot e^{-j\omega T})}$$

$$= \frac{2z^2 - z(e^{j\omega T} + e^{-j\omega T})}{2[z^2 - z(e^{j\omega T} + e^{-j\omega T}) + 1]} = \frac{z^2 - z(e^{j\omega T} + e^{-j\omega T})/2}{z^2 - z(e^{j\omega T} + e^{-j\omega T} - e^{-j\omega T}) + 1}$$

We know that, $\cos \theta = (e^{j\theta} + e^{-j\theta})/2$

$$\therefore F(z) = \frac{z(z - \cos \alpha T)}{z^2 - 2z \cos \omega T + 1}$$

EXAMPLE 7.5

Find the one sided $\mathcal{Z}$-transform of the discrete sequences generated by mathematically sampling the following continuous time function,

    (a) $e^{-at} \cos \omega t$ (b) $e^{-at} \sin \omega t$

SOLUTION

(a)   **Given that, f(t) = $e^{-at} \cos$ t**

The discrete sequence is generated by replacing t by kT, where T is the sampling time period.

$$\therefore f(k) = e^{-akt} \cos \omega kT$$

By the definition of one sided $\mathcal{Z}$-transform we get,

$$F(z) = \mathcal{Z}\{f(k)\} = \sum_{k=0}^{\infty} e^{-akT} \cos \omega kT \, z^{-k} = \sum_{k=0}^{\infty} e^{-akT} \left(\frac{e^{j\omega kT} + e^{-j\omega kT}}{2} \right) z^{-k}$$

$$= \frac{1}{2} \sum_{k=0}^{\infty} \left(e^{-aT} e^{j\omega T} z^{-1} \right)^k + \frac{1}{2} \sum_{k=0}^{\infty} \left(e^{-aT} e^{-j\omega T} z^{-1} \right)^k$$

From infinite geometric sum series formula we know that, $\displaystyle\sum_{k=0}^{\infty} C^k = \frac{1}{1-C}$

$$\therefore F(z) = \frac{1}{2} \frac{1}{1 - e^{-aT} e^{j\omega T} z^{-1}} + \frac{1}{2} \frac{1}{1 - e^{-aT} e^{-j\omega T} z^{-1}}$$

$$= \frac{1}{2} \frac{1}{1 - e^{j\omega T}/ze^{+aT}} + \frac{1}{2} \frac{1}{1 - e^{-j\omega T}/ze^{+aT}}$$

$$= \frac{1}{2} \left[\frac{ze^{aT}}{ze^{aT} - e^{j\omega T}} + \frac{ze^{aT}}{ze^{aT} - e^{-j\omega T}} \right]$$

$$= \frac{1}{2} \left[\frac{ze^{aT}(ze^{aT} - e^{-j\omega T}) + ze^{aT}(ze^{aT} - e^{-j\omega T})}{(ze^{aT} - e^{-j\omega T})(ze^{aT} - e^{-j\omega T})} \right]$$

$$= \frac{ze^{aT}}{2} \left[\frac{ze^{aT} - e^{-j\omega T} + ze^{aT} - e^{j\omega T}}{(ze^{aT})^2 - ze^{aT}e^{-j\omega T} - ze^{aT}e^{j\omega T} + e^{j\omega}e^{-j\omega T}} \right]$$

$$= \frac{ze^{aT}}{2} \left[\frac{2z e^{aT} - (e^{-j\omega T} + e^{-j\omega T})}{z^2 e^{2aT} - ze^{aT}(e^{j\omega T} + e^{-j\omega T}) + 1} \right]$$

$$= \left[\frac{z e^{aT}(z e^{aT} - \cos \omega T)}{z^2 e^{2aT} - 2z e^{aT} \cos \omega T + 1} \right] \qquad \left(\because \cos \theta = \frac{e^{j\theta} + e^{-j\theta}}{2} \right)$$

(b)   **Given that, $f(t) = e^{-at} \sin t$**

The discrete sequence f(k) is generated by replacing t by kT, where T is the sampling time period.

$$\therefore f(k) = e^{-akt} \sin \omega kT$$

By the definition of one sided $\mathcal{Z}$-transform we get,

$$F(z) = \mathcal{Z}\{f(k)\} = \sum_{k=0}^{\infty} e^{-akT} \sin \omega kT \, z^{-k}$$

$$= \sum_{k=0}^{\infty} e^{-akT} \left(\frac{e^{j\omega kT} - e^{-j\omega kT}}{2j} \right) z^{-k}$$

$$= \frac{1}{2j} \sum_{k=0}^{\infty} \left(e^{-aT} e^{j\omega T} z^{-1} \right)^k - \frac{1}{2j} \sum_{k=0}^{\infty} \left(e^{-aT} e^{-j\omega T} z^{-1} \right)^k$$

From infinite geometric sum series formula we know that, $\displaystyle\sum_{k=0}^{\infty} C^k = \frac{1}{1-C}$

$$\therefore F(z) = \frac{1}{2j} \frac{1}{1 - e^{-aT} e^{j\omega T} z^{-1}} - \frac{1}{2j} \frac{1}{1 - e^{-aT} e^{-j\omega T} z^{-1}}$$

$$= \frac{1}{2j} \frac{1}{1 - e^{j\omega T}/ze^{aT}} - \frac{1}{2j} \frac{1}{1 - e^{-j\omega T}/ze^{aT}}$$

$$= \frac{1}{2j} \frac{ze^{aT}}{ze^{aT} - e^{j\omega T}} - \frac{1}{2j} \frac{ze^{aT}}{ze^{aT} - e^{-j\omega T}}$$

$$= \frac{1}{2j} \left[\frac{ze^{aT}(ze^{aT}-e^{-j\omega T}) - ze^{aT}(ze^{aT}-e^{j\omega T})}{(ze^{aT}-e^{j\omega T})(ze^{aT}-e^{-j\omega T})} \right]$$

$$= \frac{1}{2j} \left[\frac{(ze^{aT})[ze^{aT}-e^{-j\omega T}-ze^{aT}+e^{j\omega T}]}{(ze^{aT})^2 - ze^{aT}e^{-j\omega T} - ze^{aT}e^{j\omega T} + e^{j\omega T}e^{-j\omega T}} \right]$$

$$= \left[\frac{ze^{aT}[e^{j\omega T}-e^{-j\omega T}]/2j}{z^2e^{2aT} - ze^{aT}(e^{j\omega T}+e^{-j\omega T})+1} \right]$$

$$= \frac{ze^{aT}\sin\omega T}{z^2e^{2aT} - 2ze^{aT}\cos\omega T + 1}$$

INVERSE Z-TRANSFORM

The following methods are employed to recover the original discrete sequence from its Z-transform

1. *Direct evaluation by contour integration (or) complex inversion integral.*

2. *Partial fraction expansion.*

3. *Power series expansion.*

The inverse Z-transform by partial fraction expansion method and power series expansion method are presented in this section. The inverse Z-transform by contour integration is beyond the scope of the book.

PARTIAL FRACTION EXPANSION METHOD

Let f(k) = Discrete sequence

and $F(z) = Z\{f(k)\} = Z$-Transform of f(k).

The function F(z) can be expressed as a ratio of two polynomials in z as shown below.

$$F(z) = \frac{b_0 z^m + b_1 z^{m-1} + b_2 z^{m-2} + \ldots\ldots + b_{m-1}z + b_m}{z^n + a_1 z^{n-1} + a_2 z^{n-2} + \ldots\ldots + a_{n-1}z + a_n} \quad ; \quad \text{where } m \le n$$

The function F(z) can be expressed as a series of sum terms by partial fraction expansion technique.

$$\therefore \ F(z) = A_0 + \sum_{i=1}^{n} \frac{A_i}{(z+p_i)} \qquad\qquad \ldots\ldots(7.23)$$

where, A_0 is a constant, $A_1, A_2, \ldots A_n$ are residues and $p_1, p_2, \ldots p_n$ are poles of F(z).

Note : *Sometimes it will be convenient to express F(z)/z as a series of sum terms instead of F(z).*

Once the function F(z) is expressed as a series of sum terms, the inverse Z-transform of F(z) is given by the sum of inverse Z-transform of each term in equation (7.23) [The inverse Z-transform of each term of equation (7.23) can be obtained from standard Z-transform pairs].

The coefficients of the polynomials of F(z) are assumed real and so the roots of the polynomial are real and/or complex conjugate pairs (i.e., complex roots will occur only in conjugate pairs). Hence on factorizing the denominator polynomial we get the following cases. (The roots of the denominator polynomial are poles of F(z)).

Case (i) : *When roots (or poles) are real and distinct*

Case (ii) : *When roots (or poles) have multiplicity*

Case (iii) : *When roots (or poles) are complex conjugate.*

Case (i) *When roots (or poles) are real and distinct*

In this case F(z) can be expressed as,

$$F(z) = \frac{b_0 z^m + b_1 z^{m-1} + b_2 z^{m-2} + \dots + b_{m-1} z + b_m}{(z + p_1)(z + p_2) \dots (z + p_n)}$$

$$= A_0 + \frac{A_1}{(z + p_1)} + \frac{A_2}{(z + p_2)} + \dots + \frac{A_n}{(z + p_n)}$$

where, A_0 is a constant ; $A_1, A_2 \dots A_n$ are residues and $p_1, p_2, \dots p_n$ are poles.

The constant A_0 is present when $m = n$ (i.e., when the order of numerator and denominator polynomial are equal). The value of A_0 is obtained by dividing the numerator polynomial by denominator polynomial.

The residue A_i is evaluated by multiplying both sides of H(z) by $(z + p_i)$ and letting $z = -p_i$.

$$\therefore \; A_i = (z + p_i) F(z) \big|_{z = -p_i}$$

Case (ii) *When roots (or poles) have multiplicity*

Let one of pole has a multiplicity of q. (i.e., repeats q times). In this case F(z) can be expressed as,

$$F(z) = \frac{b_0 z^m + b_1 z^{m-1} + b_2 z^{m-2} + \dots + b_{m-1} z + b_m}{(z + p_1)(z + p_2) \dots (z + p_x)^q \dots (z + p_n)}$$

$$= A_0 + \frac{A_1}{(z + p_1)} + \frac{A_2}{(z + p_2)} + \dots + \frac{A_{x0}}{(z + p_x)^q} + \frac{A_{x1}}{(z + p_x)^{q-1}} + \dots$$

$$+ \dots + \frac{A_{x(q-1)}}{(z + p_x)} + \dots + \frac{A_n}{(z + p_n)}$$

where, $A_{x0}, A_{x1}, \dots A_{x(q-1)}$ are residues of repeated root (or pole), $z = -p_x$.

The constant A_0 and residues of distinct real roots are evaluated as explained in case(i).

The residue A_{xr} of repeated root is obtained as shown below.

$$A_{xr} = \frac{1}{r!} \frac{d^r}{dz^r} \left[(z + p_x)^q F(z) \right] \big|_{z = -p_x} \; ; \quad \text{where } r = 0, 1, 2, \dots (q - 1)$$

Case (iii) *When roots (or poles) are complex conjugate*

Let F(z) has one pair of complex conjugate pole. In this case F(z) can be expressed as,

$$F(z) = \frac{b_0 z^m + b_1 z^{m-1} + b_2 z^{m-2} + \dots + b_{m-1} z + b_m}{(z + p_1)(z + p_2) \dots (z^2 + az + b)(z + p_n)}$$

$$= A_0 + \frac{A_1}{z + p_1} + \frac{A_2}{z + p_2} + \dots + \frac{A_x}{z + \sigma + j\omega} + \frac{A_x^*}{z + \sigma + j\omega} + \dots + \frac{A_n}{z + p_n}$$

The constant A_0 and residues of real and non-repeated roots are evaluated as explained in case(i).
The residue A_x is evaluated as that of case(i) and the residue A_x^* is conjugate of A_x.

POWER SERIES EXPANSION METHOD

Let f(k) = Discrete sequence

and $F(z) = \mathbb{Z}\{f(k)\} = \mathbb{Z}$-transform of f(k).

By the definition of $\mathbb{Z}$-transform we get,

$$F(z) = \sum_{k=-\infty}^{\infty} f(k)\, z^{-k}$$

On expanding the summation we get,

$$F(z) = \ldots\ldots f(-3)z^{-(-3)} + f(-2)z^{-(-2)} + f(-1)z^{-(-1)} + f(0)z^0$$
$$+ f(1)z^{-1} + f(2)z^{-2} + f(3)z^{-3} + \ldots\ldots \qquad\ldots\ldots(7.24)$$

If the given function, F(z) can be expressed as a power series of z by long division then on comparing the coefficients of z with that of equation (7.24), the samples of f(k) are determined. [i.e., the coefficient of z^i is the ith sample f(i) of the sequence f(k)].

Note : *The different method of evaluation of inverse $\mathbb{Z}$-transform of a funcion F(z) will result in different type of mathematical expressions. But on evaluating the expressions for each value of k, we may get a same sequence.*

EXAMPLE 7.6

Determine the inverse $\mathbb{Z}$-transform of the following function,

(a) $F(z) = \dfrac{1}{1 - 1.5z^{-1} + 0.5z^{-2}}$ (b) $F(z) = \dfrac{z^2}{z^2 - z + 0.5}$

(c) $F(z) = \dfrac{1 + z^{-1}}{1 - z^{-1} + 0.5z^{-2}}$ (d) $F(z) = \dfrac{1}{(1 - z^{-1})(1 - z^{-1})^2}$

SOLUTION

(a) **Given that,** $F(z) = \dfrac{1}{1 - 1.5z^{-1} + 0.5z^{-2}}$

$$F(z) = \frac{1}{1 - 1.5z^{-1} + 0.5z^{-2}} = \frac{1}{1 - \dfrac{1.5}{z} + \dfrac{0.5}{z^2}}$$

$$= \frac{z^2}{z^2 - 1.5z + 0.5} = \frac{z^2}{(z-1)(z-0.5)}$$

$$\therefore \frac{F(z)}{z} = \frac{z}{(z-1)(z-0.5)}$$

By partial fraction expansion, F(z)/z can be expressed as

$$\frac{F(z)}{z} = \frac{A_1}{z-1} + \frac{A_2}{z-0.5}$$

$$A_1 = \frac{F(z)}{z}(z-1)\bigg|_{z=1} = \frac{z}{(z-1)(z-0.5)}(z-1)\bigg|_{z=1}$$

$$= \frac{z}{(z-0.5)}\bigg|_{z=1} = \frac{1}{1-0.5} = 2$$

$$A_2 = \frac{F(z)}{z}(z-0.5)\Big|_{z=0.5} = \frac{z}{(z-1)(z-0.5)}(z-0.5)\Big|_{z=0.5}$$

$$= \frac{0.5}{0.5-1} = -1$$

$$\therefore \frac{F(z)}{z} = \frac{2}{z-1} - \frac{1}{z-0.5}$$

$$\therefore F(z) = \frac{2z}{z-1} - \frac{z}{z-0.5}$$

We know that $\mathcal{Z}\{a^k\} = \dfrac{z}{z-a}$ and $\mathcal{Z}\{u(k)\} = \dfrac{z}{z-1}$

On taking inverse $\mathcal{Z}$-transform of F(z) we get,

$$f(k) = 2\,u(k) - (0.5)^k \;\; ; \;\; k \geq 0$$

(Here we consider only one sided $\mathcal{Z}$-transform) .

(b) **Given that,** $F(z) = \dfrac{z^2}{z^2 - z + 0.5}$

$$F(z) = \frac{z^2}{z^2 - z + 0.5} = \frac{z^2}{(z - 0.5 - j0.5)(z - 0.5 + j0.5)}$$

$$\therefore \frac{F(z)}{z} = \frac{z}{(z - 0.5 - j0.5)(z - 0.5 + j0.5)}$$

By partial fraction expansion, we can write,

$$\frac{F(z)}{z} = \frac{A}{z - 0.5 - j0.5} + \frac{A^*}{z - 0.5 + j0.5}$$

$$A = \frac{F(z)}{z}(z - 0.5 - j0.5)\Big|_{z = 0.5 + j0.5}$$

$$= \frac{z}{(z - 0.5 - j0.5)(z - 0.5 + j0.5)}(z - 0.5 - j0.5)\Big|_{z = 0.5 + j0.5}$$

$$= \frac{0.5 + j0.5}{j1} = 0.5 - j0.5$$

$$\therefore A^* = (0.5 - j0.5)^* = 0.5 + j0.5$$

$$\therefore \frac{F(z)}{z} = \frac{0.5 - j0.5}{z - 0.5 - j0.5} + \frac{0.5 + j0.5}{z - 0.5 + j0.5}$$

$$\therefore F(z) = \frac{(0.5 - j0.5)z}{z - (0.5 - j0.5)} + \frac{(0.5 + j0.5)z}{z - (0.5 - j0.5)}$$

We know that $\mathcal{Z}\{a^k\} = \dfrac{z}{z-a}$

On taking inverse $\mathcal{Z}$-transform of F(z) we get,

$$f(k) = (0.5 - j0.5)(0.5 + j0.5)^k + (0.5 + j0.5)(0.5 - j0.5)^k$$

$$= -j\left(\frac{0.5}{-j} + 0.5\right)(0.5 + j0.5)^k + j\left(\frac{0.5}{j} + 0.5\right)(0.5 - j0.5)^k$$

$$= -j(0.5 + j0.5)(0.5 + j0.5)^k + j(0.5 - j0.5)(0.5 - j0.5)^k$$

$$= -j(0.5 + j0.5)^{(k+1)} + j(0.5 - j0.5)^{(k+1)}$$

The roots of the quadratic $z^2 - z + 0.5 = 0$ are

$$z = \frac{1 \pm \sqrt{1 - 4 \times 0.5}}{2}$$

$$= 0.5 \pm j0.5$$

(c)   **Given that,** $F(z) = \dfrac{1 - z^{-1}}{1 - z^{-1} + 0.5z^{-2}}$

$$F(z) = \frac{1 + z^{-1}}{1 + z^{-1} + 0.5z^{-2}} = \frac{1 + 1/z}{1 - \dfrac{1}{z} + \dfrac{0.5}{z^2}}$$

> The roots of the quadratic
> $z^2 - z + 0.5 = 0$ are
> $z = \dfrac{1 \pm \sqrt{1 - 4 \times 0.5}}{2}$
> $= 0.5 \pm j0.5$

$$= \frac{\dfrac{z+1}{z}}{\dfrac{z^2 - z + 0.5}{z^2}} = \frac{z(z+1)}{(z^2 - z + 0.5)} = \frac{z(z+1)}{(z - 0.5 - j0.5)(z - 0.5 + j0.5)}$$

By partial fraction expansion, we can write,

$$\frac{F(z)}{z} = \frac{(z+1)}{(z - 0.5 - j0.5)(z - 0.5 + j0.5)} = \frac{A}{z - 0.5 - j0.5} + \frac{A^*}{z - 0.5 + j0.5}$$

$$A = \frac{F(z)}{z}(z - 0.5 - j0.5)\Big|_{z = 0.5 + j0.5}$$

$$= \frac{(z+1)}{(z - 0.5 - j0.5)(z - 0.5 + j0.5)}(z - 0.5 - j0.5)\Big|_{z = 0.5 + j0.5}$$

$$= \frac{0.5 + j0.5 + 1}{0.5 + j0.5 - 0.5 + j0.5} = \frac{1.5 + j0.5}{j1} = -j1.5 + 0.5 = 0.5 - j1.5$$

$$A^* = (0.5 - j1.5)^* = 0.5 + j1.5$$

$$\therefore \frac{F(z)}{z} = \frac{0.5 - j1.5}{z - 0.5 - j0.5} + \frac{0.5 + j1.5}{z - 0.5 + j0.5}$$

$$F(z) = (0.5 - j1.5)\frac{z}{z - (0.5 + j0.5)} + (0.5 + j1.5)\frac{z}{z - (0.5 - j0.5)}$$

We know that $\bar{z}\{a^k\} = \dfrac{z}{z - a}$

On taking inverse $\bar{z}$-transform of F(z) we get,

$$f(k) = (0.5 - j1.5)(0.5 + j0.5)^k + (0.5 + j1.5)(0.5 - j0.5)^k \ ; \ \text{for } k \geq 0$$

(d)   **Given that,** $F(z) = \dfrac{1}{(1 + z^{-1})(1 - z^{-1})^2}$

$$F(z) = \frac{1}{(1 + z^{-1})(1 + z^{-1})^2} = \frac{1}{\left(1 + \dfrac{1}{z}\right)\left(1 - \dfrac{1}{z}\right)^2}$$

$$= \frac{1}{\dfrac{(z-1)}{z}\left(\dfrac{z-1}{z}\right)^2} = \frac{z^3}{(z+1)(z-1)^2}$$

$$\therefore \frac{F(z)}{z} = \frac{z^2}{(z+1)(z-1)^2}$$

By partial fraction expansion, we can write,

$$\frac{F(z)}{z} = \frac{A_1}{z+1} + \frac{A_2}{(z-1)^2} + \frac{A_3}{z-1}$$

$$A_1 = \frac{F(z)}{z}(z+1)\Big|_{z=-1} = \frac{z^2}{(z+1)(z-1)^2}(z+1)\Big|_{z=-1} = \frac{z^2}{(z-1)^2}\Big|_{z=-1} = \frac{(-1)^2}{(-1-1)^2} = 0.25$$

$$A_2 = \frac{F(z)}{z} (z-1)^2 \Big|_{z=1} = \frac{z^2}{(z+1)(z-1)^2} (z-1)^2 \Big|_{z=1}$$

$$= \frac{z^2}{z+1} \Big|_{z=1} = \frac{1}{1+1} = 0.5$$

$$A_3 = \frac{d}{dz}\left[\frac{F(z)}{z} (z-1)^2 \right]\Big|_{z=1} = \frac{d}{dz}\left[\frac{z^2}{(z+1)(z-1)^2} (z-1)^2 \right]\Big|_{z=1}$$

$$= \frac{d}{dz}\left[\frac{z^2}{z+1} \right]\Big|_{z=1} = \frac{(z+1)2z - z^2}{(z+1)^2}\Big|_{z=1} = \frac{(1+1)\times 2 - 1}{(1+1)^2} = \frac{3}{4} = 0.75$$

$$\therefore \quad \frac{F(z)}{z} = \frac{0.25}{z+1} + \frac{0.5}{(z-1)^2} + \frac{0.75}{z-1}$$

$$F(z) = 0.25 \frac{z}{z+1} + 0.5 \frac{z}{(z-1)^2} + 0.75 \frac{z}{z-1}$$

$$= 0.25 \frac{z}{z-(-1)} + 0.5 \frac{z}{(z-1)^2} + 0.75 \frac{z}{z-1}$$

We know that $Z\{a^k\} = \dfrac{z}{z-a}$; $Z\left\{\dfrac{az}{(z-a)^2}\right\} = ka^k$ and $Z\{u(k)\} = \dfrac{z}{z-1}$

On taking inverse Z-transform of F(z) we get,

$$f(k) = 0.25\,(-1)^k + 0.5k(1)^k + 0.75\,u(k)$$

$$f(k) = 0.25(-1)^k + 0.5k + 0.75\,u(k) \ ; \ \text{for } k \geq 0$$

EXAMPLE 7.7

Determine the inverse Z-transform of the following z-domain functions.

(a) $F(z) = \dfrac{3z^2 + 2z + 1}{z^2 - 3z + 2}$ 　　　　　(b) $F(z) = \dfrac{3z^2 + 2z + 1}{z^2 + 3z + 2}$

(c) $F(z) = \dfrac{z - 0.4}{z^2 + z + 2}$ 　　　　　(d) $F(z) = \dfrac{z - 4}{(z-1)(z-2)^2}$

SOLUTION

(a)　**Given that,** $F(z) = \dfrac{3z^2 + 2z + 1}{z^2 - 3z + 2}$

$$F(z) = \frac{3z^2 + 2z + 1}{z^2 - 3z + 2} = 3 + \frac{11z - 5}{z^2 - 3z + 2}$$

$$= 3 + \frac{11z - 5}{(z-1)(z-2)}$$

By partial fraction expansion we get, $F(z) = 3 + \dfrac{A_1}{z-1} + \dfrac{A_2}{z-2}$

$$A_1 = \frac{11z-5}{(z-1)(z-2)}(z-1)\Big|_{z=1} = \frac{11z-5}{(z-2)}\Big|_{z=1} = \frac{11-5}{1-2} = -6$$

$$A_2 = \frac{11z-5}{(z-1)(z-2)}(z-2)\Big|_{z=2} = \frac{11z-5}{(z-1)}\Big|_{z=2} = \frac{11\times 2 - 5}{2-1} = 17$$

$$
\begin{array}{r}
3 \\
z^2 - 3z + 2\ \overline{\smash{\big)}\ 3z^2 + 2z + 1} \\
\underline{3z^2 - 9z + 6} \\
11z - 5
\end{array}
$$

$$\therefore F(z) = 3 - \frac{6}{z-1} + \frac{17}{z-2}$$

$$= 3 - 6\frac{1}{z}\frac{z}{z-1} + 17\frac{1}{z}\frac{z}{z-2}$$

$$= 3 - 6z^{-1}\frac{z}{z-1} + 17z^{-1}\frac{z}{z-2}$$

We know that, $\mathcal{Z}\{\delta(k)\} = 1$; $\mathcal{Z}\{u(k)\} = \frac{z}{z-1}$ and $\mathcal{Z}\{a^k\} = \frac{z}{z-a}$

By time shifting property we get,

$$\mathcal{Z}\{u(k-1)\} = z^{-1}\frac{z}{z-1} \quad \text{and} \quad \mathcal{Z}\{a^{(k-1)}\} = z^{-1}\frac{z}{z-a}$$

On taking inverse $\mathcal{Z}$-transform of F(z) we get,

$$f(k) = 3\,\delta(k) - 6\,u(k-1) + 17 \times 2^{(k-1)}\,u(k-1) \ ; \ \text{for } k \geq 0$$

Note : The term $2^{(k-1)}$ is multiplied by u(k–1), because this term have samples only after $k \geq 1$.

(b)   **Given that,** $F(z) = \dfrac{3z^2 + 2z + 1}{z^2 + 3z + 2}$

$$F(z) = \frac{3z^2 + 2z + 1}{z^2 + 3z + 2} = 3 - \frac{7z+5}{z^2 + 3z + 2}$$

$$= 3 - \frac{7z+5}{(z+1)(z+2)}$$

$$\begin{array}{r} 3 \\ z^2+3z+2\overline{\smash{\big)}\,3z^2 + 2z + 1} \\ \underline{3z^2 - 9z + 6} \\ -7z - 5 \end{array}$$

By partial fraction expansion we get, $F(z) = 3 - \dfrac{A_1}{z+1} - \dfrac{A_2}{z+2}$

$$A_1 = \frac{7z+5}{(z+1)(z+2)}(z+1)\Big|_{z=-1} = \frac{7z+5}{(z+2)}\Big|_{z=-1} = \frac{7\times(-1)+5}{-1+2} = -2$$

$$A_2 = \frac{7z+5}{(z+1)(z+2)}(z+2)\Big|_{z=-2} = \frac{7z+5}{z+1}\Big|_{z=-2} = \frac{7\times(-2)+5}{-2+1} = 9$$

$$\therefore F(z) = 3 + \frac{2}{z+1} - \frac{9}{z+2}$$

$$= 3 + 2\frac{1}{z}\frac{z}{z-(-1)} - 9\frac{1}{z}\frac{z}{z-(-2)}$$

$$= 3 + 2z^{-1}\frac{z}{z-(-1)} - 9z^{-1}\frac{z}{z-(-2)}$$

We know that, $\mathcal{Z}\{\delta(k)\} = 1$ and $\mathcal{Z}\{a^k\} = \dfrac{z}{z-a}$

By time shifting property,

$$\mathcal{Z}\{a^{(k-1)}\} = z^{-1}\frac{z}{z-a}$$

On taking inverse $\mathcal{Z}$-transform of F(z) we get,

$$f(k) = 3\,\delta(k) + 2(-1)^{(k-1)}\,u(k-1) + 9(-2)^{(k-1)}\,u(k-1) \ ; \ \text{for } k \geq 0$$

Note : The term $2^{(k-1)}$ is multiplied by u(k–1), because these term have samples only after $k \geq 1$.

(c)   **Given that,** $F(z) = \dfrac{z-0.4}{z^2 + z + 2}$

$$F(z) = \frac{z-0.4}{z^2+z+2}$$

$$= \frac{z-0.4}{(z+0.5 - j\sqrt{7}/2)(z+0.5+j\sqrt{7}/2)}$$

The roots of the quadratic $z^2 + z + 2 = 0$ are

$$z = \frac{-1 \pm \sqrt{1 - 4 \times 2}}{2}$$

$$= -0.5 \pm \frac{j\sqrt{7}}{2}$$

By partial fraction expansion we get, $F(z) = \dfrac{A}{z + 0.5 - j\sqrt{7}/2} + \dfrac{A^*}{z + 0.5 + j\sqrt{7}/2}$

$A = \dfrac{z - 0.4}{(z + 0.5 - j\sqrt{7}/2)(z + 0.5 + j\sqrt{7}/2)}(z + 0.5 - j\sqrt{7}/2)\Big|_{z = -0.5 + j\sqrt{7}/2}$

$= \dfrac{z - 0.4}{(z + 0.5 + j\sqrt{7}/2)}\Big|_{z = -0.5 + j\sqrt{7}/2} = \dfrac{-0.5 + j\sqrt{7}/2 - 0.4}{-0.5 + j\sqrt{7}/2 + 0.5 + j\sqrt{7}/2}$

$= \dfrac{-0.9 + j\sqrt{7}/2}{j\sqrt{7}} = \dfrac{-0.9}{j\sqrt{7}} + \dfrac{j\sqrt{7}/2}{j\sqrt{7}} = 0.5 + j\dfrac{0.9}{\sqrt{7}} = 0.5 + j0.34$

$\therefore\ A^* = (0.5 + j0.34)^* = 0.5 - j0.34$

$\therefore\ F(z) = \dfrac{0.5 + j0.34}{z + 0.5 - j\sqrt{7}/2} + \dfrac{0.5 - j0.34}{z + 0.5 + j\sqrt{7}/2}$

$= (0.5 + j0.34)\dfrac{1}{z}\dfrac{z}{z + 0.5 - j\sqrt{7}/2} + (0.5 - j0.34)\dfrac{1}{z}\dfrac{z}{z + 0.5 + j\sqrt{7}/2}$

$= (0.5 + j0.34)z^{-1}\dfrac{z}{z - (-0.5 - j\sqrt{7}/2)} + (0.5 - j0.34)z^{-1}\dfrac{z}{z - (-0.5 - j\sqrt{7}/2)}$

We know that, $\mathcal{z}\{a^k\} = \dfrac{z}{z - a}$

By time shifting property we get, $\mathcal{z}\{a^{(k-1)}\} = z^{-1}\dfrac{z}{z - a}$

On taking inverse $\mathcal{z}$-transform of $F(z)$ we get,

$f(k) = (0.5 + j0.34)(-0.5 + j\sqrt{7}/2)^{(k-1)}u(k-1)$

$\qquad\qquad + (0.5 - j0.34)(-0.5 - j\sqrt{7}/2)^{(k-1)}u(k-1)\ ;\ \text{for } k \geq 0$

> Note : Since the term $a^{(k-1)}$ is valid only for $k \geq 1$, it is multiplied by $u(k-1)$.

(d)   **Given that,** $F(z) = \dfrac{z - 4}{(z - 1)(z - 2)^2}$

By partial fraction expansion we get,

$F(z) = \dfrac{z - 4}{(z - 1)(z - 2)^2} = \dfrac{A_1}{z - 1} + \dfrac{A_2}{(z - 2)^2} + \dfrac{A_3}{(z - 2)}$

$A_1 = \dfrac{z - 4}{(z - 1)(z - 2)^2}(z - 1)\Big|_{z = 1} = \dfrac{z - 4}{(z - 2)^2}\Big|_{z = 1} = \dfrac{1 - 4}{(1 - 2)^2} = -3$

$A_2 = \dfrac{z - 4}{(z - 1)(z - 2)^2}(z - 2)^2\Big|_{z = 2} = \dfrac{z - 4}{z - 1}\Big|_{z = 2} = \dfrac{2 - 4}{2 - 1} = -2$

$A_3 = \dfrac{d}{dz}\left[\dfrac{z - 4}{(z - 1)(z - 2)^2}(z - 2)^2\right]\Big|_{z = 2} = \dfrac{d}{dz}\left[\dfrac{z - 4}{z - 1}\right]\Big|_{z = 2}$

$= \dfrac{(z - 1) - (z - 4)}{(z - 1)^2}\Big|_{z = 2} = \dfrac{3}{(z - 1)^2}\Big|_{z = 2} = \dfrac{3}{(2 - 1)^2} = 3$

$\therefore\ F(z) = \dfrac{-3}{z - 1} - \dfrac{2}{(z - 2)^2} + \dfrac{3}{z - 2} = -3\dfrac{1}{z}\dfrac{z}{z - 1} - \dfrac{1}{z}\dfrac{2z}{(z - 2)^2} + 3\dfrac{1}{z}\dfrac{z}{z - 2}$

$= -3z^{-1}\dfrac{z}{z - 1} - z^{-1}\dfrac{2z}{(z - 2)^2} + 3z^{-1}\dfrac{z}{z - 2}$

We know that, $z\{u(k)\} = \dfrac{z}{z-1}$; $z\{a^k\} = \dfrac{z}{z-a}$ and $z\{ka^k\} = \dfrac{az}{(z-a)^2}$

By time shifting property we get,

$$z\{u(k-1)\} = z^{-1}\dfrac{z}{z-1} \; ; \; z\{a^{(k-1)}\} = z^{-1}\dfrac{z}{z-a} \text{ and } z\{(k-1)a^{(k-1)}\} = z^{-1}\dfrac{az}{(z-a)^2}$$

On taking inverse z-transform of F(z) we get,

$$f(k) = -3u(k-1) - (k-1)2^{(k-1)}u(k-1) + 3 \times 2^{(k-1)}u(k-1)$$

> Note : Since the term $a^{(k-1)}$ is valid only for $k \geq 1$, it is multiplied by u(k–1).

EXAMPLE 7.8

Determine the inverse z-transform of $F(z) = \dfrac{1}{1 - \dfrac{3}{2}z^{-1} + \dfrac{1}{2}z^{-1}}$

when (a) ROC : $|z| > 1.0$ and (b) ROC : $|z| < 0.5$.

SOLUTION

(a) Since the ROC is the exterior of a circle, we expect f(k) to be causal signal. Hence we can express F(z) as a power series expansion in negative powers of z. On dividing the numerator of F(z) by its denominator we get,

$$F(z) = \dfrac{1}{1 - \dfrac{3}{2}z^{-1} + \dfrac{1}{2}z^{-2}} = 1 + \dfrac{3}{2}z^{-1} + \dfrac{7}{4}z^{-2} + \dfrac{15}{8}z^{-3} + \dfrac{31}{16}z^{-4} + \ldots \ldots \qquad \ldots (7.8.1)$$

$$
\begin{array}{r}
1 - \dfrac{3}{2}z^{-1} + \dfrac{7}{4}z^{-2} + \dfrac{15}{8}z^{-3} + \dfrac{31}{16}z^{-4} + \ldots \\[2mm]
\end{array}
$$

$$1 - \dfrac{3}{2}z^{-1} + \dfrac{1}{2}z^{-1} \overline{\Big)\, 1}$$

$$1 - \dfrac{3}{2}z^{-1} + \dfrac{1}{2}z^{-2}$$

$$\dfrac{3}{2}z^{-1} - \dfrac{1}{2}z^{-2}$$

$$\dfrac{3}{2}z^{-1} - \dfrac{9}{4}z^{-2} + \dfrac{3}{4}z^{-3}$$

$$\dfrac{7}{4}z^{-2} - \dfrac{3}{4}z^{-3}$$

$$\dfrac{7}{4}z^{-2} - \dfrac{21}{8}z^{-3} + \dfrac{7}{8}z^{-4}$$

$$\dfrac{15}{8}z^{-3} - \dfrac{7}{8}z^{-4}$$

$$\dfrac{15}{8}z^{-3} - \dfrac{45}{16}z^{-4} + \dfrac{15}{16}z^{-5}$$

$$\dfrac{31}{16}z^{-4} - \dfrac{15}{16}z^{-5}$$

$$\vdots$$

If F(z) is z-transform of f(k) then, by the definition of z-transform we get,

$$F(z) = z\{f(k)\} = \sum_{k=-\infty}^{\infty} f(k)z^{-k}$$

For a causal signal,

$$F(z) = \sum_{k=0}^{\infty} f(k)z^{-k}$$

On expanding the summation we get,

$$F(z) = f(0) + f(1)\,z^{-1} + f(2)z^{-2} + f(3)z^{-3} + f(4)z^{-4} + \ldots\ldots \qquad \ldots (7.8.2)$$

On comparing the two power series of F(z) [i.e., equation (7.8.1) & (7.8.2)], we get,

$$f(0) = 1 \; ; \; f(1) = \frac{3}{2} \; ; \; f(3) = \frac{15}{8} \; ; \; f(4) = \frac{31}{16} \; ; \; \ldots\ldots$$

$$f(k) = \left\{ 1, \frac{3}{2}, \frac{7}{4}, \frac{15}{8}, \frac{31}{16}, \ldots\ldots \right\} \; ; \; \text{for } k \ge 0$$

(b)   Since the ROC is the interior of a circle, we expect f(k) to be anticausal signal. Hence we can express F(z) as a power series expansion in positive powers of z. Therefore, rewrite the denominator polynomial of F(z) in the reverse order and then the numerator, is divided by the denominator as shown below.

$$
\begin{array}{r}
2z^2 + 6z^3 + 14z^4 + 30z^5 + 62z^6 + \ldots\ldots \\
\hline
\tfrac{1}{2}z^{-2} - \tfrac{3}{2}z^{-1} + 1 \, \Big)\, 1 \\
1 - 3z + 2z^2 \\
\hline
3z - 2z^2 \\
3z - 9z^2 + 6z^3 \\
\hline
7z^2 - 6z^3 \\
7z^2 - 21z^3 + 14z^4 \\
\hline
15z^3 - 14z^4 \\
15z^3 - 45z^4 + 30z^5 \\
\hline
31z^4 - 30z^5 \\
\vdots
\end{array}
$$

$$\therefore F(z) = \frac{1}{1 - \frac{3}{2}z^{-1} + \frac{1}{2}z^{-2}} = \frac{1}{\frac{1}{2}z^{-2} - \frac{3}{2}z^{-1} + 1}$$

$$= 2z^2 + 6z^3 + 14z^4 + 30z^5 + 62z^6 \qquad\qquad \ldots (7.8.3)$$

If F(z) is z-transform of f(k) then, by the definition of z-transform we get,

$$F(z) = Z\{f(k)\} = \sum_{k=-\infty}^{\infty} f(k)\, z^{-k}$$

For an anticausal signal, $F(z) = \displaystyle\sum_{k=-\infty}^{0} f(k)\, z^{-k}$

On expanding the summation we get,

$$F(z) = \ldots\ldots f(-6)z^6 + f(-5)z^5 + f(-4)z^4 + f(-3)z^3 + f(-2)z^2 + f(-1)z + f(0) \qquad \ldots (7.8.4)$$

On comparing the two power series of F(z) [i.e., equations (7.8.3) & (7.8.4)] , we get,

$$f(-6) = 62 \; ; \; f(-5) = 30 \; ; \; f(-4) = 14 \; ; \; f(-3) = 6 \; ;$$

$$f(-2) = 2 \quad ; \; f(-1) = 0 \text{ and } f(0) = 0$$

$$\therefore \; f(k) = \{ \ldots\ldots, 62, 30, 14, 6, 2, 0, 0 \}$$
$$\uparrow$$

7.7 LINEAR DISCRETE-TIME SYSTEMS

A discrete-time system is a device or algorithm that operates on a discrete-time signal called the input or excitation, according to some well-defined rule, to produce another discrete-time signal called the output or the response of the system. We can say that the input signal r(k) is transformed by the system into a signal c(k) and expressed as

$$c(k) = H[r(k)]$$

.....(7.25)

where, H denotes the transformation (also called an operator)

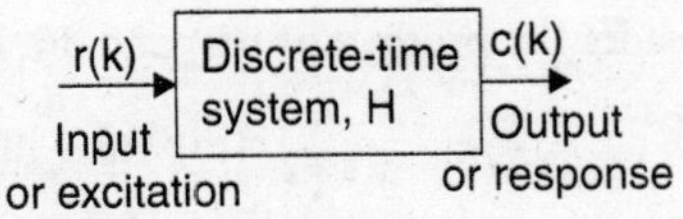

Fig : 7.17

A discrete time system is linear if it obeys the principle of superposition and it is time invariant if its input-output relationship do not change with time.

When the input to a discrete time system is unit impulse, $\delta(k)$ then the output is called impulse response of the system and denoted by h(k)

Fig : 7.18

$$h(k) = H[\delta(k)]$$

.....(7.26)

A linear time invariant discrete time system is characterized by its impulse response h(k) and so the impulse response h(k) is also called weighing sequence.

The input-output description of a discrete-time system consists of mathematical expression or a rule, which explicitly defines the relation between the input and output signals (input-output relationship). It is denoted by

$$r(k) \xrightarrow{\;H\;} c(k)$$

.....(7.27)

The input-output relationship of a linear-time invariant discrete time system, (LDS) can be expressed by N^{th} order constant coefficient difference equation given below.

$$c(k) = -\sum_{m=1}^{N} a_m\, c(k-m) + \sum_{m=0}^{M} b_m\, r(k-m)$$

.....(7.28)

The integer N is called the order of the system and $M \le N$.

Here c(k – m) are past outputs, r(k – m) are past inputs, r(k) is present input and a_k and b_k are constant coefficients.

ANALYSIS OF LINEAR DISCRETE TIME SYSTEM (LDS)

There are two methods of analysing the behaviour or response of a LDS system.

Method 1

The input-output relation of the LDS system is governed by the constant coefficient difference equation of the form shown in equation (7.28). Mathematically the direct solution of equation (7.28) can be obtained to analyse the performance of the system.

Method 2

The given input signal is first decomposed or resolved into a sum of elementary signals. Then using the linearity property of the system, the responses of the system to the elementary signals are added to obtain the total response of the system to the given input signal.

Resolution of discrete time signal (or sequence) into impulses

Let $r(k)$ = Discrete time signal

        $\delta(k)$ = Unit impulse signal

and $\delta(k-m)$ = Delayed unit impulse signal

Consider the product of $r(k)$ and $\delta(k-m)$

$$r(k)\,\delta(k-m) = \begin{cases} r(m)\,\delta(k-m) & ;\ \text{at}\ k = m \\ 0 & ;\ \text{for}\ k \neq m \end{cases} \qquad \text{.....(7.29)}$$

$$\therefore\ r(k)\,\delta(k-m) = \begin{cases} r(m) & ;\ \text{at}\ k = m\ (\because \delta(k-m) = 1\ \text{at}\ k = m) \\ 0 & ;\ \text{for}\ k \neq m \end{cases} \qquad \text{.....(7.30)}$$

The product $r(k)\,\delta(k-m)$ has zero everywhere except at $k = m$. The value of the signal at $k = m$ is the mth sample of the signal $r(k)$ and it is denoted by $r(m)$.

Therefore each multiplication of the signal $r(k)$ by an unit impulse at some delay m, in essence picks out the single value $r(m)$ of the signal $r(k)$ at $k = m$, where the unit impulse is nonzero. Consequently if we repeat this multiplication over all possible delays in the range of, $0 \leq m < \infty$ and sum all the product sequences, the result will be a sequence that is equal to the sequence $r(k)$. Hence $r(k)$ can be expressed as

$$r(k) = \sum_{m=0}^{\infty} r(m)\,\delta(k-m) \qquad \text{.....(7.31)}$$

> **Note :** *Each product $r(k)\,\delta(k-m)$ is an impulse and the summation of impulses gives $r(k)$. Here $r(k)$ is considered as one sided sequence. If $r(k)$ is two sided sequence then the range of m is $-\infty$ to $+\infty$.*

RESPONSE OF LDS SYSTEM FOR ARBITRARY INPUT - THE CONVOLUTION SUM

In a LDS system the response $c(k)$ of the system for arbitrary input $r(k)$ is given by convolution of the input $r(k)$ with the impulse response $h(k)$ of the system. It is expressed as

$$c(k) = r(k) * h(k) \qquad \text{.....(7.32)}$$

      where, the symbol * represents convolution operation.

Proof

Let $c(k)$ be the response of the system H for an input $r(k)$. [Let $r(k)$ be a one sided sequence].

$$\therefore\ c(k) = H[r(k)] \qquad \text{.....(7.33)}$$

The signal $r(k)$ can be expressed as a summation of impulses as,

$$r(k) = \sum_{m=0}^{\infty} r(m)\,\delta(k-m) \qquad \text{.....(7.34)}$$

      where, $\delta(k-m)$ is the delayed unit impulse signal.

From equations (7.33) and (7.34) we get,

$$c(k) = H\left[\sum_{m=0}^{\infty} r(m)\, \delta(k-m) \right] \qquad .....(7.35)$$

The system H is a function of k and not a function of m. Hence by linearity property the equation (7.35) can be written as,

$$c(k) = \sum_{m=0}^{\infty} r(m)\, H[\delta(k-m)] \qquad .....(7.36)$$

Let the response of the LDS system to the unit impulse input $\delta(k)$ be denoted by h(k).

$$\therefore\; h(k) = H[\delta(k)] \qquad .....(7.37)$$

Then by time invariance property the response of the system to the delayed unit impulse input $\delta(k - m)$ is

$$h(k-m) = H[\delta(k-m)] \qquad .....(7.38)$$

Using equation (7.38), the equation (7.36) can be expressed as

$$c(k) = \sum_{m=0}^{\infty} r(m)\, h(k-m) \qquad .....(7.39)$$

The equation of c(k) [equation (7.39)] is called convolution sum. We can say that the input r(k) is convoluted with the impulse response h(k) to yield the output c(k).

$$\therefore\; c(k) = \sum_{m=0}^{\infty} r(m)\, h(k-m) = r(k) * h(k) \qquad .....(7.40)$$

PROPERTIES OF CONVOLUTION

Commutative property   : $r(k) * h(k) = h(k) * r(k)$

Associative property   : $[r(k) * h_1(k)] * h_2(k) = r(k) * [h_1(k) * h_2(k)]$

Distributive property   : $r(k) * [h_1(k) + h_2(k)] = [r(k) * h_1(k)] + [r(k)* h_2(k)]$

7.8 TRANSFER FUNCTION OF LDS SYSTEM (PULSE TRANSFER FUNCTION)

The transfer function of LDS system is given by z-transform of its impulse response. The transfer function of LDS system is also called z-transform function or pulse transfer function.

Let h(k) = Impulse response of a LDS system

Now, $\mathbf{Z}$-transform of $h(k) = \mathbf{Z}\{h(k)\} = H(z)$

∴ **Transfer function of LDS system = H(z)** $\qquad .....(7.41)$

The input-output relationship of a LDS system is governed by a convolution sum of equation (7.40). By taking $\mathbf{Z}$-transform of this convolution sum it can be shown that, H(z) is given by the ratio C(z)/R(z), where C(z) is the $\mathbf{Z}$-transform of output c(k) of LDS system and R(z) is the $\mathbf{Z}$-transform of input r(k) to the LDS system.

Proof

By the definition of one sided $\mathbf{Z}$-transform,

$$C(z) = \mathbf{Z}[c(k)] = \sum_{k=0}^{\infty} c(k)\, z^{-k} \qquad .....(7.42)$$

From equation (7.40), we get, $c(k) = \sum_{m=0}^{\infty} r(m)\, h(k-m)$

On substituting this convolution sum in equation (7.42) we get,

$$C(z) = \sum_{k=0}^{\infty} \left[\sum_{m=0}^{\infty} r(m) h(k-m) \right] z^{-k} \qquad .....(7.43)$$

The order of summation in equation (7.43) can be interchanged. Therefore equation (7.43) can be written as

$$C(z) = \sum_{m=0}^{\infty} r(m) \sum_{k=0}^{\infty} h(k-m) z^{-k} \qquad .....(7.44)$$

Let, p = (k − m), ∴ When k = 0, p = −m
and when k = ∞, p = ∞
Also, k = p + m

On replacing (k − m) by p in equation (7.44) we get

$$C(z) = \sum_{m=0}^{\infty} r(m) \sum_{p=-m}^{\infty} h(p) z^{-(p+m)}$$

$$= \sum_{m=0}^{\infty} r(m) \sum_{p=0}^{\infty} h(p) z^{-p} z^{-m} \qquad (\because h(p) = 0 \text{ ; for } p < 0)$$

$$= \sum_{m=0}^{\infty} r(m) z^{-m} \sum_{p=0}^{\infty} h(p) z^{-p} \qquad .....(7.45)$$

By the definition of one sided $\mathcal{Z}$-transform,

$$\sum_{m=0}^{\infty} r(m) z^{-m} = R(z) \text{ and } \sum_{p=0}^{\infty} h(p) z^{-p} = H(z)$$

Hence equation (7.45) can be written as

$$C(z) = R(z) H(z) \quad \text{(or)} \quad H(z) = \frac{C(z)}{R(z)} \qquad .....(7.46)$$

From equation (7.46) we can conclude that the transfer function of the system is given by the ratio C(z)/R(z).

From the above analysis we can define the transfer function of the LDS system is the ratio of the $\mathcal{Z}$-transform of the output of a system to the $\mathcal{Z}$-transform of the input to the system with zero initial conditions.

Let r(k) = Input of LDS system

and c(k) = Output of a LDS system

Now, $\mathcal{Z}\{r(k)\} = R(z)$ and $\mathcal{Z}\{c(k)\} = C(z)$

∴ Transfer function of LDS system $= \dfrac{C(z)}{R(z)}$ .....(7.47)

The input-output relation of LDS system is governed by the constant coefficient difference equation,

$$c(k) = -\sum_{m=1}^{N} a_m c(k-m) + \sum_{m=0}^{M} b_m r(k-m) \qquad .....(7.48)$$

where, N is the order of the system and M ≤ N.

On taking $\mathcal{Z}$-transform of equation (7.48) we get,

[By time shifting property, $\mathcal{Z}\{c(k-m)\} = z^{-m} C(z)$ and $\mathcal{Z}\{r(k-m)\} = z^{-m} R(z)$]

$$C(z) = -\sum_{m=1}^{N} a_m z^{-m} C(z) + \sum_{m=0}^{M} b_m z^{-m} R(z)$$

$$\therefore C(z) + \sum_{m=1}^{N} a_m z^{-m} C(z) + \sum_{m=0}^{N} b_m z^{-m} R(z) \qquad .....(7.49)$$

On expanding the equation (7.49) with M = N, we get,

$$C(z) + a_1 z^{-1} C(z) + a_2 z^{-2} C(z) + \ldots\ldots + a_N z^{-N} C(z)$$

$$= b_0 R(z) + b_1 z^{-1} R(z) + b_2 z^{-2} R(z) + \ldots\ldots + b_N z^{-N} R(z)$$

$$C(z)\left[1 + a_1 z^{-1} + a_2 z^{-2} + \ldots\ldots + a_N z^{-N}\right] = R(z)\left[b_0 + b_1 z^{-1} + b_2 z^{-2} + \ldots\ldots + b_N z^{-N}\right]$$

$$\therefore \; \frac{C(z)}{R(z)} = \frac{b_0 + b_1 z^{-1} + b_2 z^{-2} + \ldots\ldots + b_N z^{-N}}{1 + a_1 z^{-1} + a_2 z^{-2} + \ldots\ldots + a_N z^{-N}} \qquad\qquad \ldots\ldots(7.50)$$

From the above discussions it is evident that the transfer function of the LDS system can be obtained by taking z-transform of the difference equation governing the system.

EXAMPLE 7.9

The input-output relation of a sampled data system is described by the equation

$$c(k + 2) + 3\, c(k + 1) + 4c(k) = r(k + 1) - r(k)$$

Determine the z-transfer function. Also obtain the weighting sequence of the system.

SOLUTION

Let $R(z) = Z\{r(k)\}$ and $C(z) = Z\{c(k)\}$

By time shifting property, when initial conditions are zero, we get,

$$Z\{c(k + m)\} = z^m\, C(z) \quad \text{and} \quad Z\{r(k + m)\} = z^m\, R(z)$$

Given that, $c(k + 2) + 3\, c(k + 1) + 4c(k) = r(k + 1) - r(k)$

On taking z-transform of the above equation we get,

$$z^2 C(z) + 3z\, C(z) + 4\, C(z) = z\, R(z) - R(z)$$

$$(z^2 + 3z + 4)\, C(z) = (z - 1)\, R(z)$$

$$\therefore \; \frac{C(z)}{R(z)} = \frac{z - 1}{z^2 + 3z + 4}$$

The transfer function of the system, $H(z) = \dfrac{C(z)}{R(z)} = \dfrac{z - 1}{z^2 + 3z + 4}$

The weighing sequence is the impulse response, h(k) of the system. It is given by inverse z-transform of H(z).

$$H(z) = \frac{z - 1}{z^2 + 3z + 4} = \frac{z - 1}{\left(z + \dfrac{3}{2} + j\dfrac{\sqrt{7}}{2}\right)\left(z + \dfrac{3}{2} - j\dfrac{\sqrt{7}}{2}\right)}$$

> The roots of the quadratic $z^2 + 3z + 4 = 0$ are
> $$z = \frac{-3 \pm \sqrt{9 - 4 \times 4}}{2}$$
> $$= -\frac{3}{2} \pm j\frac{\sqrt{7}}{2}$$

By partial fraction technique H(z) can be expressed as

$$H(z) = \frac{A}{z + \dfrac{3}{2} + j\dfrac{\sqrt{7}}{2}} + \frac{A^*}{z + \dfrac{3}{2} - j\dfrac{\sqrt{7}}{2}}$$

$$A = \frac{z - 1}{\left(z + \dfrac{3}{2} + j\dfrac{\sqrt{7}}{2}\right)\left(z + \dfrac{3}{2} - j\dfrac{\sqrt{7}}{2}\right)}\left(z + \dfrac{3}{2} + j\dfrac{\sqrt{7}}{2}\right)\Bigg|_{z = -\frac{3}{2} - j\frac{\sqrt{7}}{2}} = \frac{z - 1}{z + \dfrac{3}{2} - j\dfrac{\sqrt{7}}{2}}\Bigg|_{z = -\frac{3}{2} - j\frac{\sqrt{7}}{2}}$$

$$= \frac{-\dfrac{3}{2} - j\dfrac{\sqrt{7}}{2} - 1}{-\dfrac{3}{2} - j\dfrac{\sqrt{7}}{2} + \dfrac{3}{2} - j\dfrac{\sqrt{7}}{2}} = \frac{-\dfrac{5}{2} - j\dfrac{\sqrt{7}}{2}}{-j\sqrt{7}} = -j\frac{5}{2\sqrt{7}} + \frac{1}{2} = \frac{1}{2} - j\frac{5}{2\sqrt{7}}$$

$$A^* = \left(\frac{1}{2} - j\frac{5}{2\sqrt{7}}\right)^* = \frac{1}{2} + j\frac{5}{2\sqrt{7}}$$

$$H(z) = \frac{\left(\dfrac{1}{2} - j\dfrac{5}{2\sqrt{7}}\right)}{z + \dfrac{3}{2} + j\dfrac{\sqrt{7}}{2}} + \frac{\dfrac{1}{2} + j\dfrac{5}{2\sqrt{7}}}{z + \dfrac{3}{2} - j\dfrac{\sqrt{7}}{2}}$$

$$H(z) = \left(\frac{1}{2} - j\frac{5}{2\sqrt{7}}\right)z^{-1}\frac{z}{z - \left(-\dfrac{3}{2} - j\dfrac{\sqrt{7}}{2}\right)} + \left(\frac{1}{2} + j\frac{5}{2\sqrt{7}}\right)z^{-1}\frac{z}{z - \left(\dfrac{-3}{2} - j\dfrac{\sqrt{7}}{2}\right)}$$

We know that $\mathcal{Z}\{a^k\} = \dfrac{z}{z-a}$

By time shifting property, $\mathcal{Z}\{a^{(k-1)}\} = z^{-1}\dfrac{z}{z-a}$

On taking inverse $\mathcal{Z}$-transform of H(z) we get,

$$h(k) = \left(\frac{1}{2} - j\frac{5}{2\sqrt{7}}\right)\left(-\frac{3}{2} - j\frac{\sqrt{7}}{2}\right)^{(k-1)}u(k-1) + \left(\frac{1}{2} + j\frac{5}{2\sqrt{7}}\right)\left(-\frac{3}{2} + j\frac{\sqrt{7}}{2}\right)^{(k-1)}u(k-1)$$

EXAMPLE 7.10

Solve the difference equation $c(k+2) + 3\,c(k+1) + 2\,c(k) = u(k)$

Given that $c(0) = 1$; $c(1) = -3$; $c(k) = 0$ for $k < 0$

SOLUTION

Let $\mathcal{Z}\{c(k)\} = C(z)$ and $\mathcal{Z}\{u(k)\} = U(z)$

Since u(k) is unit step signal, $U(z) = \dfrac{z}{z-1}$

We know that, if $F(z) = \mathcal{Z}\{f(k)\}$ then

$$\mathcal{Z}\{f(k+m)\} = z^m F(z) - \sum_{i=0}^{m-1} f(i)\, z^{m-i}$$

Given that, $c(k+2) + 3\,c(k+1) + 2\,c(k) = u(k)$

On taking $\mathcal{Z}$-transform of the above equation we get,

$$\mathcal{Z}\{c(k+2)\} + \mathcal{Z}\{3\,c(k+1)\} + \mathcal{Z}\{2\,c(k)\} = \mathcal{Z}\{u(k)\}$$

$$z^2 C(z) - z^2 c(0) - z c(1) + 3\,[z\,C(z) - z\,c(0)] + 2\,C(z) = \frac{z}{z-1}$$

On substituting the initial conditions, $c(0) = 1$ and $c(1) = -3$ we get,

$$z^2 C(z) - z^2 + 3z + 3z\,C(z) - 3z + 2\,C(z) = \frac{z}{z-1}$$

$$(z^2 + 3z + 2)\,C(z) - z^2 = \frac{z}{z-1}$$

$$(z^2 + 3z + 2)\,C(z) = \frac{z}{z-1} + z^2$$

$$(z+1)(z+2)\,C(z) = \frac{z + z^2(z-1)}{(z-1)}$$

$$\therefore\ C(z) = \frac{z[1 + z^2 - z]}{(z-1)(z+1)(z+2)}$$

$$\therefore\ \frac{C(z)}{z} = \frac{z^2 - z + 1}{(z-1)(z+1)(z+2)}$$

By partial fraction expansion technique we can write C(z)/z as,

$$\frac{C(z)}{z} = \frac{A_1}{z-1} + \frac{A_2}{z+1} + \frac{A_3}{z+2}$$

$$A_1 = \frac{z^2 - z + 1}{(z-1)(z+1)(z+2)} (z-1)\bigg|_{z=1} = \frac{z^2 - z + 1}{(z+1)(z+2)}\bigg|_{z=1} = \frac{1 - 1 + 1}{(1+1)(1+2)} = \frac{1}{6}$$

$$A_2 = \frac{z^2 - z + 1}{(z-1)(z+1)(z+2)} (z+1)\bigg|_{z=-1} = \frac{z^2 - z + 1}{(z-1)(z+2)}\bigg|_{z=-1} = \frac{(-1)^2 - (-1) + 1}{(-1-1)(-1+2)} = -\frac{3}{2}$$

$$A_3 = \frac{z^2 - z + 1}{(z-1)(z+1)(z+2)} (z+2)\bigg|_{z=-2} = \frac{z^2 - z + 1}{(z-1)(z+1)}\bigg|_{z=-2} = \frac{(-2)^2 - (-2) + 1}{(-2-1)(-2+1)} = \frac{7}{3}$$

$$\therefore \frac{C(z)}{z} = \frac{1}{6}\frac{1}{z-1} - \frac{3}{2}\frac{1}{z+1} + \frac{7}{3}\frac{1}{z+2}$$

$$C(z) = \frac{1}{6}\frac{1}{z-1} - \frac{3}{2}\frac{1}{z-(-1)} + \frac{7}{3}\frac{z}{z-(-2)}$$

We know that $\mathcal{Z}\{u(k)\} = \frac{z}{z-1}$ and $\mathcal{Z}\{a^k\} = \frac{z}{z-a}$

On taking inverse $\mathcal{Z}$-transform of C(z) we get,

$$c(k) = \frac{1}{6} u(k) - \frac{3}{2} (-1)^k + \frac{7}{3} (-2)^k \; ; \; k \geq 0$$

The above equation of c(k) is the solution of the given difference equation.

7.9 ANALYSIS OF SAMPLER AND ZERO - ORDER HOLD

Consider a pulse sampler with zero-order hold (ZOH) shown in fig 7.19. Let the output of sampler be a pulse train of pulse width Δ. For each input pulse, the ZOH produces a pulse of duration T, where T is the sampling period.

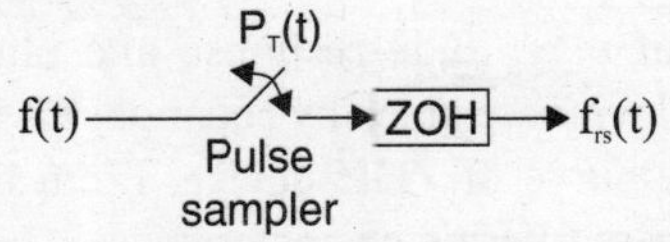

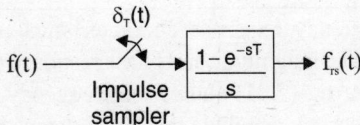

f(t) —→ [Pulse sampler $P_T(t)$] —→ ZOH —→ $f_{rs}(t)$

Fig 7.19 : *Pulse sampler with ZOH.*

f(t) —→ [Impulse sampler $\delta_T(t)$] —→ $\boxed{\dfrac{1 - e^{-sT}}{s}}$ —→ $f_{rs}(t)$

Fig 7.20 : *Equivalent representation of Pulse sampler with ZOH.*

It can be proved that the output of pulse sampler with ZOH can be produced by impulse sampled f(t) when passed through a transfer function,

$$\boxed{G_0(s) = \frac{1 - e^{-sT}}{s}} \quad\quad\quad .....(7.51)$$

Hence the pulse sampler with ZOH can be replaced by an equivalent system consisting of an impulse sampler and a block with transfer function, $(1 - e^{-sT})/s$ as shown in fig 7.20. This equivalent representation offers easier analysis of sampled data control systems.

FREQUENCY RESPONSE CHARACTERISTICS OF ZERO ORDER HOLDING DEVICE

The sinusoidal transfer function of ZOH can be obtained from $G_0(s)$ by replacing s by $j\omega$.

$$\therefore \; G_0(j\omega) = \frac{1 - e^{-j\omega T}}{j\omega} \quad\quad\quad .....(7.52)$$

We know that, $e^{\frac{-j\omega T}{2}} e^{\frac{+j\omega T}{2}} = 1$ $\quad\quad\quad .....(7.53)$

Hence from equation (7.52) and (7.53) we get,

$$G_0(j\omega) = \frac{e^{\frac{-j\omega T}{2}} e^{\frac{j\omega T}{2}} - e^{-j\omega T}}{j\omega} = \frac{e^{\frac{-j\omega T}{2}} e^{\frac{j\omega T}{2}} - e^{\frac{-j\omega T}{2}} e^{\frac{-j\omega T}{2}}}{j\omega}$$

> **Note :** $\sin\theta = \dfrac{e^{j\theta} - e^{-j\theta}}{2j}$

$$= \left(\frac{e^{\frac{j\omega T}{2}} - e^{\frac{-j\omega T}{2}}}{j\omega} \right) e^{\frac{-j\omega T}{2}} = \frac{2}{\omega} \sin \frac{\omega T}{2} e^{\frac{-j\omega T}{2}}$$

$$= T \frac{\sin \dfrac{\omega T}{2}}{\dfrac{\omega T}{2}} e^{\frac{-j\omega T}{2}} \qquad\qquad .....(7.54)$$

We know that, sampling frequency, $\omega_s = \dfrac{2\pi}{T}$

$$\therefore \ T = \frac{2\pi}{\omega_s}$$

On substituting $T = 2\pi/\omega_s$ in equation (7.54) we get,

$$G_0(j\omega) = \frac{2\pi}{\omega_s} \frac{\sin(\pi\omega/\omega_s)}{(\pi\omega/\omega_s)} e^{\frac{-j\pi\omega}{\omega_s}}$$

Magnitude of $G_0(j\omega) = |G_0(j\omega)| = \dfrac{2\pi}{\omega_s} \dfrac{\sin(\pi\omega/\omega_s)}{(\pi\omega/\omega_s)}$ .....(7.55)

Argument (or phase) of $G_0(j\omega) = \angle G_0(j\omega) = \dfrac{-\pi\omega}{\omega_s}$ .....(7.56)

The frequency response characteristics consists of magnitude response and phase response characteristics. The magnitude and phase response of ZOH device are given by equations (7.55) and (7.56) respectively. The fig (7.21) shows the frequency response curve of ZOH device. From the frequency response curve we can conclude that ZOH device has low pass filtering characteristics.

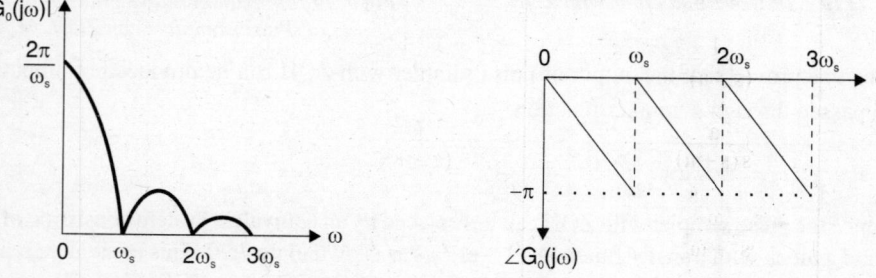

Fig a : *Magnitude response of ZOH device.*   **Fig b :** *Phase response of ZOH device.*

Fig 7.21 : *Frequency response of ZOH device.*

7.10 ANALYSIS OF SYSTEMS WITH IMPULSE SAMPLING

Consider a linear continuous time system fed from an impulse sampler as shown in fig 7.22a. Let H(s) be the transfer function of the system in s-domain. In such a system we are interested in reading the output at sampling instants. This can be achieved by means of a mathematical sampler or read-out sampler.

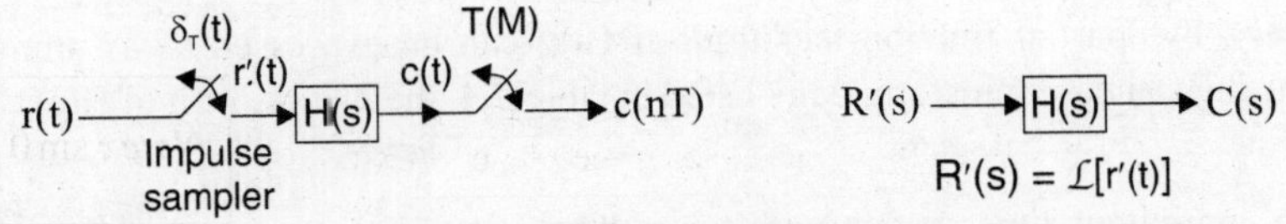

Fig 7.22 : *Linear continuous time system with impulse sampled input.*

For the system shown in fig 7.22b, it can be shown that the z-domain transfer function H(z) can be directly obtained from s-domain transfer function by taking Ƶ-transform of H(s)

i.e., $H(z) = Ƶ\{H(s)\}$ .....(7.57)

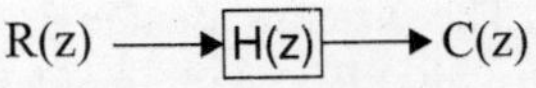

The fig 7.23 shows the Ƶ-transform equivalent of the s-domain system of fig 7.22b.

Fig 7.23 : *The Ƶ-transform equivalent of the system shown in fig 7.22b.*
.....(7.58)

The output in z-domain is given by, $C(z) = H(z) R(z)$

Procedure to find z-transfer function from s-domain transfer function

1. Determine h(t) from H(s), where $h(t) = \mathcal{L}^{-1}[H(s)]$

2. Determine the discrete sequence h(kT) by replacing t by kT in h(t)

3. Take Ƶ-transform of h(kT), which is z-transfer function of the system (i.e., $H(z) = Ƶ\{h(kT)\}$).

TABLE-7.4. Laplace and Ƶ-Transform Pairs

H(s)	H(z)
$\dfrac{1}{s}$	$\dfrac{z}{z-1}$
$\dfrac{1}{s^2}$	$\dfrac{Tz}{(z-1)^2}$
$\dfrac{1}{s^3}$	$\dfrac{T^2 z(z+1)}{2(z-1)^3}$
$\dfrac{1}{s+a}$	$\dfrac{z}{z-e^{-aT}}$
$\dfrac{1}{(s+a)^2}$	$\dfrac{T z e^{-aT}}{(z-e^{-aT})^2}$
$\dfrac{a}{s(s+a)}$	$\dfrac{z(1-e^{-aT})}{(z-1)(z-e^{-aT})}$
$\dfrac{\omega}{s^2+\omega^2}$	$\dfrac{z \sin \omega T}{z^2 - 2z \cos \omega T + 1}$
$\dfrac{s}{s^2+\omega^2}$	$\dfrac{z(z-\cos \omega T)}{z^2 - 2z \cos \omega T + 1}$
$\dfrac{\omega}{(s+a)^2+\omega^2}$	$\dfrac{z e^{-aT} \sin \omega T}{z^2 - 2ze^{-aT} \cos \omega T + e^{-2aT}}$
$\dfrac{s+a}{(s+a)^2+\omega^2}$	$\dfrac{z^2 - ze^{-aT} \cos \omega T}{z^2 - 2ze^{-aT} \cos \omega T + e^{-2aT}}$

Alternatively, by partial fraction technique if H(s) can be expressed as a summation of first order terms then using standard transform pairs listed in table-7.4, the $\mathcal{Z}$-transform of H(s) can be directly obtained.

Consider a continuous time system with transfer function H(s) as shown in fig 7.24a. Let the input r(t) be a continuous time input. To read the continuous output at sampling instants, let us imagine a mathematical sampler at the output stage.

The system shown in fig 7.24a can be equivalently represented by a block of H(s) R(s) with impulse input δ(t) as shown in fig 7.24b. Now the input and so the output does not change by imagining a fictious impulse sampler through which δ(t) is applied to H(s) R(s) as shown in fig 7.24c. For such a system we can prove that

$$C(z) = \mathcal{Z}\{H(s)\,R(s)\} \qquad\qquad .....(7.59)$$

Hence, if C(s) = H(s) R(s) then $C(z) = \mathcal{Z}\{H(s)\,R(s)\} = HR(z)$ .....(7.60)

The function $\mathcal{Z}\{H(s)\,R(s)\}$ is also denoted as HR(z).

When the impulse sampled input is applied to two or more s-domain transfer function in cascade as shown in fig 7.25a, then z-transfer function of the system is given by

$$H(z) = \mathcal{Z}\{H_1(s)\,H_2(s)\} \qquad\qquad .....(7.61)$$

and $\quad C(z) = \mathcal{Z}\{H_1(s)\,H_2(s)\}\,R(z)$ .....(7.62)

$$\text{where, } R(z) = \mathcal{Z}\{R(s)\} \text{ and } R(s) = \mathcal{L}[r(t)]$$

The function $\mathcal{Z}\{H_1(s)\,H_2(s)\}$ is also denoted as $H_1H_2(z)$. The equivalent z-domain system is shown in fig 7.25b.

Consider a system in which impulse sampler is introduced at the input of each block as shown in fig 7.26a.

Now the z-transfer function of the system is given by,

$$H(z) = H_1(z)\,H_2(z) \qquad\qquad .....(7.63)$$

$$\text{where, } H_1(z) = \mathcal{Z}\{H_1(s)\} \text{ and } H_2(z) = \mathcal{Z}\{H_2(s)\}$$

and $\quad C(z) = H_1(z)\,H_2(z)\,R(z)$ .....(7.64)

$$\text{where, } R(z) = \mathcal{Z}\{R(s)\} \text{ and } R(s) = \mathcal{L}[r(t)].$$

The equivalent z-domain system is shown in fig 7.26b.

EXAMPLE 7.11

Determine the z-domain transfer function for the following s-domain transfer functions.

(a) $H(s) = \dfrac{a}{(s+a)^2}$ (b) $H(s) = \dfrac{s}{s^2 + \omega^2}$ (c) $H(s) = \dfrac{a}{s^2 - a^2}$

(d) $H(s) = \dfrac{s+b}{(s+b)^2 + a^2}$ (e) $H(s) = \dfrac{a}{(s+b)^2 + a^2}$

SOLUTION

(a) Given that, $H(s) = \dfrac{a}{(s+a)^2}$

$$h(t) = \mathcal{L}^{-1}[H(s)] = \mathcal{L}^{-1}\left[\dfrac{a}{(s+a)^2}\right] = a\,t\,e^{-at}$$

The discrete sequence h(kT) is obtained by letting t = kT in h(t)

$$\therefore h(kT) = akT\,e^{-akT}$$

z-transfer function, $H(z) = \mathcal{Z}\{h(kT)\}$

$$\therefore H(z) = \mathcal{Z}\{h(kT)\} = \mathcal{Z}\{akT\,e^{-akT}\} = aT \times \mathcal{Z}\{k\,e^{-akT}\}$$

Let $f(k) = e^{-akT}$, $\therefore F(z) = \mathcal{Z}\{f(k)\}$

By the definition of $\mathcal{Z}$-transform,

$$F(z) = \sum_{k=0}^{\infty} f(k)\,z^{-k} = \sum_{k=0}^{\infty} e^{-akT}\,z^{-k} = \sum_{k=0}^{\infty} \left(e^{-aT}z^{-1}\right)^k = \dfrac{1}{1 - e^{-aT}z^{-1}}$$

$$\left(\because \sum_{k=0}^{\infty} C^k = \dfrac{1}{1-C} \text{ if } |C| < 1\right)\left(\begin{array}{c}\text{Infinite geometric}\\ \text{series sum formula}\end{array}\right)$$

$$\therefore F(z) = \dfrac{1}{1 - e^{-aT}/z} = \dfrac{z}{z - e^{-aT}}$$

By the property of $\mathcal{Z}$-transform we get, $\mathcal{Z}\{k\,f(k)\} = -z\dfrac{d}{dz}F(z)$

$$\therefore \mathcal{Z}\{e^{-akT}\} = \mathcal{Z}\{k\,f(k)\} = -z\dfrac{d}{dz}\dfrac{z}{z - e^{-aT}} = -z \times \dfrac{(z - e^{-aT}) - z}{(z - e^{-aT})^2} = \dfrac{z\,e^{-aT}}{(z - e^{-aT})^2}$$

$$\therefore H(z) = aT \times \mathcal{Z}\{k\,e^{-akT}\} = aT \times \dfrac{z\,e^{-aT}}{(z - e^{-aT})^2} = \dfrac{aT\,z\,e^{-aT}}{(z - e^{-aT})^2}$$

(b) **Given that,** $H(s) = \dfrac{s}{s^2 + \omega^2}$

$$h(t) = \mathcal{L}^{-1}[H(s)] = \mathcal{L}^{-1}\left[\dfrac{s}{s^2 + \omega^2}\right] = \cos \omega t$$

The discrete sequence h(kT) is obtained by letting t = kT in h(t)

$$\therefore h(kT) = \cos \omega kT$$

z-transfer function, $H(z) = \mathcal{Z}\{h(kT)\} = \mathcal{Z}\{\cos \omega kT\} = \dfrac{z(z - \cos \omega T)}{z^2 - 2z\cos \omega T + 1}$

[Refer table-7.3 and example 7.4(c)]

(c) **Given that,** $H(s) = \dfrac{a}{s^2 - a^2}$

$$h(t) = \mathcal{L}^{-1}[H(s)] = \mathcal{L}^{-1}\left[\dfrac{a}{s^2 - a^2}\right] = \mathcal{L}^{-1}\left[\dfrac{a}{(s+a)(s-a)}\right]$$

By partial fraction expansion,

$$\dfrac{a}{(s+a)(s-a)} = \dfrac{A_1}{s+a} + \dfrac{A_2}{s-a}$$

$$A_1 = \frac{a}{(s+a)(s-a)}(s+a)\bigg|_{s=-a} = \frac{a}{s-a}\bigg|_{s=-a} = \frac{a}{-a-a} = \frac{-a}{2a} = -\frac{1}{2}$$

$$A_2 = \frac{a}{(s+a)(s-a)}(s-a)\bigg|_{s=a} = \frac{a}{a+a}\bigg|_{s=a} = \frac{a}{a+a} = \frac{a}{2a} = \frac{1}{2}$$

$$\therefore \; h(t) = \mathcal{L}^{-1}\left[-\frac{1}{2}\frac{1}{(s+a)} + \frac{1}{2}\frac{1}{(s-a)}\right] = -\frac{1}{2}e^{-at} + \frac{1}{2}e^{-at}$$

The discrete sequence h(kT) is obtained by letting t = kT in h(t)

$$\therefore \; h(kT) = -\frac{1}{2}e^{-akT} + \frac{1}{2}e^{akT}$$

By the definition of one sided Z-transform,

$$H(z) = \sum_{k=0}^{\infty} h(kT)\, z^{-k} = \sum_{k=0}^{\infty}\left[-\frac{1}{2}e^{-akT} + \frac{1}{2}e^{akT}\right]z^{-k}$$

$$= -\frac{1}{2}\sum_{k=0}^{\infty} e^{-akT}\, z^{-k} + \frac{1}{2}\sum_{k=0}^{\infty} e^{akT}\, z^{-k}$$

$$= -\frac{1}{2}\sum_{k=0}^{\infty} \left(e^{-aT}z^{-1}\right)^{k} + \frac{1}{2}\sum_{k=0}^{\infty} \left(e^{-aT}z^{-1}\right)^{k}$$

From infinite geometric sum series formula we know that,

$$\sum_{k=0}^{\infty} C^{k} = \frac{1}{1-C} \quad ; \quad \text{when} \, |C| < 1$$

$$\therefore \; H(z) = -\frac{1}{2}\frac{1}{1-e^{-aT}z^{-1}} + \frac{1}{2}\frac{1}{1-e^{-aT}z^{-1}} = -\frac{1}{2}\frac{1}{1-e^{-aT}/z} + \frac{1}{2}\frac{1}{1-e^{-aT}/z}$$

$$= -\frac{1}{2}\frac{z}{z-e^{-aT}} + \frac{1}{2}\frac{z}{z-e^{aT}} = \frac{z}{2}\left[\frac{-(z-e^{aT})+z-e^{-aT}}{(z-e^{-aT})(z-e^{aT})}\right]$$

$$= \frac{z}{2}\left[\frac{-z+e^{aT}+z-e^{-aT}}{z^{2}-z\,e^{aT}-ze^{-aT}+e^{-aT}\,e^{aT}}\right] = \frac{z}{2}\left[\frac{e^{aT}-e^{-aT}}{z^{2}-z(e^{aT}+e^{-aT})+1}\right]$$

Since, $\cosh\theta = \dfrac{e^{\theta}+e^{-\theta}}{2}$ and $\sinh\theta = \dfrac{e^{\theta}-e^{-\theta}}{2}$

$$H(z) = \frac{z}{2}\left(\frac{2\sinh aT}{z^{2}-2z\cosh aT+1}\right) = \frac{z\sinh aT}{z^{2}-2z\cosh aT+1}$$

(d) **Given that,** $H(s) = \dfrac{(s+b)}{(s+b)^{2}+a^{2}}$

$$h(t) = \mathcal{L}^{-1}[H(s)] = \mathcal{L}^{-1}\left[\frac{(s+b)}{(s+b)^{2}+a^{2}}\right] = e^{-bt}\cos at$$

The discrete sequence h(kT) is obtained by letting t = kT in h(t)

$$\therefore \; h(kT) = e^{-bkT}\cos akT$$

z-transfer function, $H(z) = Z\{h(kT)\} = Z\{e^{-bkT}\cos akT\}$

From example 7.5(a) we get

$$H(z) = \frac{z\,e^{bT}(z\,e^{bT}-\cos aT)}{z^{2}e^{2bT}-2z\,e^{bT}\cos aT+1}$$

(e) **Given that,** $H(s) = \dfrac{a}{(s+b)^{2}+a^{2}}$

$$h(t) = \mathcal{L}^{-1}[H(s)] = \mathcal{L}^{-1}\left[\frac{a}{(s+b)^{2}+a^{2}}\right] = e^{-bt}\sin at$$

The discrete sequence h(kT) is obtained by letting t = kT in h(t)

$$\therefore h(kT) = e^{-bkT} \sin akT$$

z-transform function, $H(z) = Z\{h(kT)\} = Z\{e^{-bkT} \sin akT\}$

From example 7.5(a) we get, $H(z) = \dfrac{z\, e^{bT} \sin aT}{z^2\, e^{2bT} - 2z\, e^{bT} \cos aT + 1}$

7.11 ANALYSIS OF SAMPLED DATA CONTROL SYSTEMS USING z-TRANSFORM

The analysis of sampled data control systems are performed using the concepts developed in section 7.9 and 7.10. The following points serve as guidelines to determine the output in z-domain and hence the z-transfer function of the sampled data control systems.

1. The pulse sampling is approximated as impulse sampling.

2. The ZOH is replaced by a black with transfer function, $G_0(s) = (1 - e^{-sT})/s$.

3. When the input to a block is impulse sampled signal then the z-transform of the output of the block can be obtained from the z-transform of the input and z-transform of the s-domain transfer function of the block. In determining the output of a block one may come across the following cases.

Case (i) The impulse sampler is located at the input of a block as shown in fig 7.27.

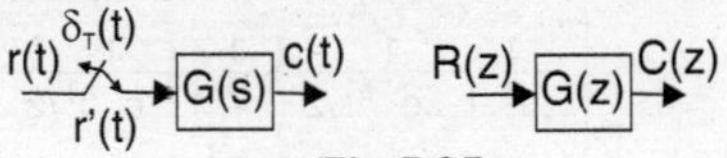

Fig 7.27

In this case, C(z) = G(z) R(z) .....(7.65)

Here, $G(z) = Z\{G(s)\}$; $R(z) = Z\{R'(s)\}$ and $R'(s) = \mathcal{L}[r'(t)]$

Case (ii) The impulse sampler is located at the input of two s-domain cascaded blocks as shown in fig 7.28.

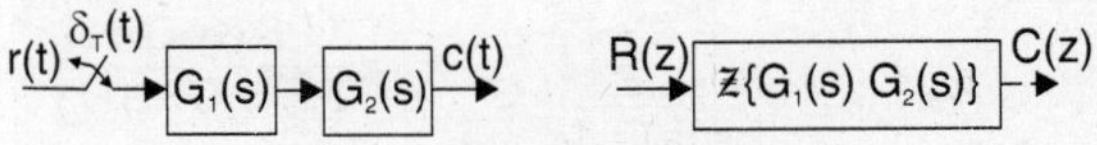

Fig 7.28

In this case, $C(z) = Z\{G_1(s)\, G_2(s)\}\, R(z) = G_1 G_2(z)\, R(z)$ .....(7.66)

Case (iii) The impulse sampler is located at the input of each block as shown in fig 7.29.

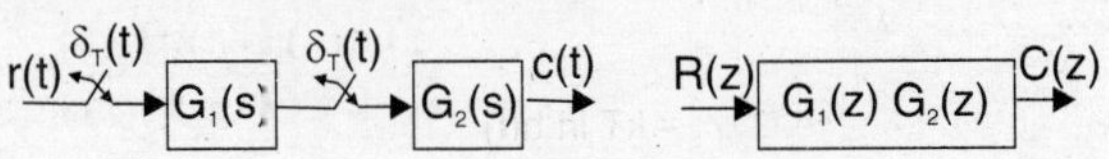

Fig 7.29

In this case, $C(z) = G_1(z)\, G_2(z)\, R(z)$ .....(7.67)

Here, $G_1(z) = Z\{G_1(s)\}$ and $G_2(z) = Z\{G_2(s)\}$

Case(iv) The impulse sampler is located at the input of ZOH in cascade with G(s) as shown in fig 7.30.

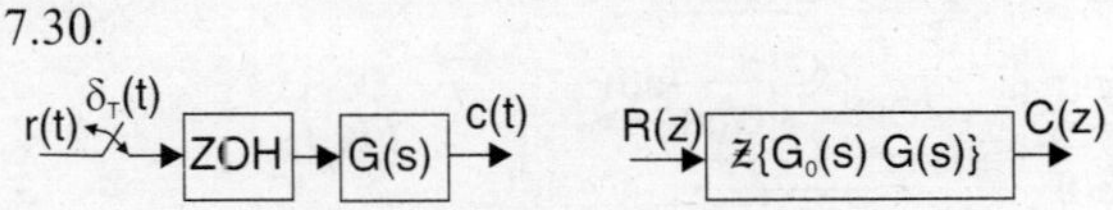

Fig 7.30

In this case, $C(z) = \mathbf{Z}\{G_0(s)\,G(s)\}\,R(z) = (1 - z^{-1})\,\mathbf{Z}\{G(s)/s\}\,R(z)$ .....(7.68)

The table-7.5 shows some configurations of the closed loop sampled data control systems and their corresponding z-domain outputs.

TABLE-7.5

Closed loop sampled data control system	Output in z-domain
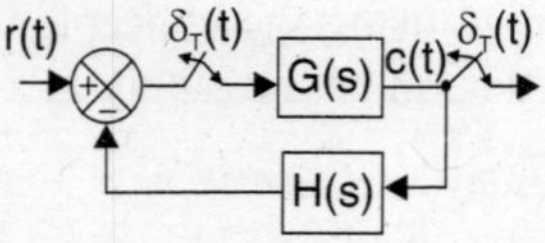	$$C(z) = \dfrac{G(z)\,R(z)}{1 + \mathbf{Z}\{G(s)\,H(s)\}}$$ $$= \dfrac{G(z)\,R(z)}{1 + GH(z)}$$
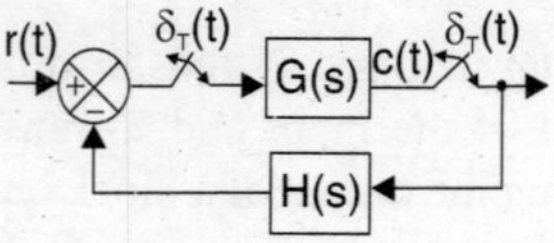	$$C(z) = \dfrac{G(z)\,R(z)}{1 + G(z)\,H(z)}$$
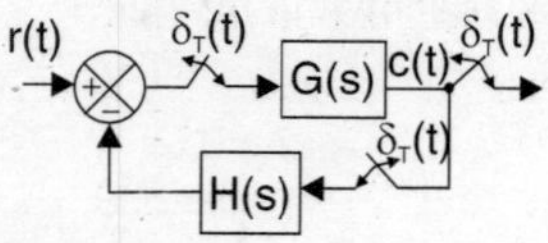	$$C(z) = \dfrac{\mathbf{Z}\{G(s)\,R(s)\}}{1 + \mathbf{Z}\{G(s)\,H(s)\}}$$ $$= \dfrac{GR(z)}{1 + GH(z)}$$
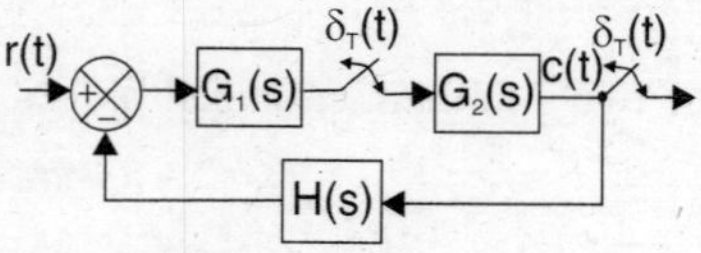	$$C(z) = \dfrac{\mathbf{Z}\{G_1(s)\,R(s)\}\,G_2(z)}{1 + \mathbf{Z}\{G_1(s)\,G_2(s)\,H(s)\}}$$ $$= \dfrac{G_1 R(z)\,G_2(z)}{1 + G_1 G_2\,H(z)}$$
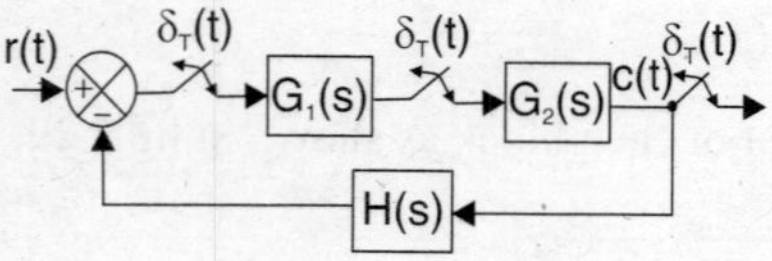	$$C(z) = \dfrac{G_1(z)\,G_2(z)\,R(z)}{1 + G_1(z)\,\mathbf{Z}\{G_2(s)\,H(s)\}}$$ $$= \dfrac{G_1(z)\,G_2(z)\,R(z)}{1 + G_1 G_2\,H(z)}$$

EXAMPLE 7.12

Find C(z)/R(z) for the following closed loop sampled data control systems. Assume all the samplers to be of impulse type.

(a)

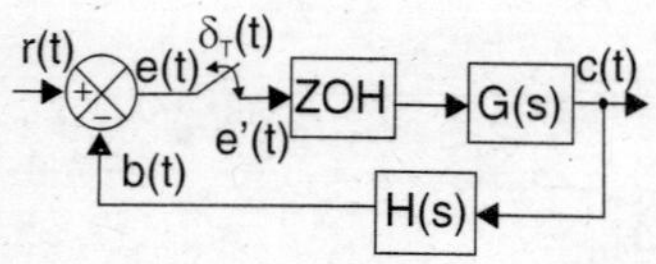

Fig 1a

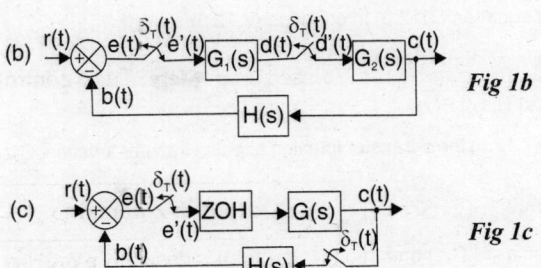

Fig 1b

Fig 1c

SOLUTION

(a) **The ZOH in the system is replaced by $G_0(s)$ as shown in fig 2, where $G_0(s) = (1 - e^{-sT})/s$.**

Let   e(t) = Error signal

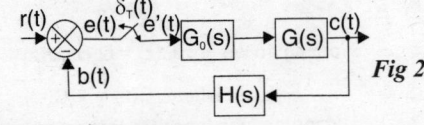

Fig 2

   e'(t) = Impulse sampled error signal

   b(t) = Feedback signal

The input to the cascaded blocks of $G_0(s)$ and G(s) is an impulse sampled signal as shown in fig 2a. It's z-domain equivalent is shown in fig 2b.

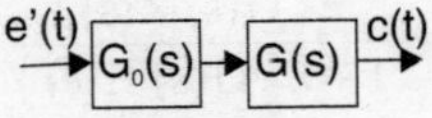

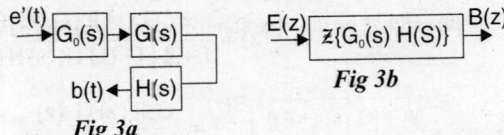

Fig 2a 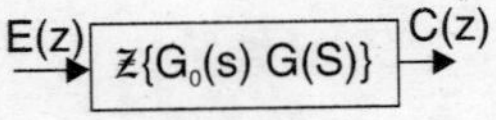   *Fig 2b*

From fig 2b we get, $C(z) = \mathcal{Z}\{G_0(s)\ G(s)\}\ E(z)$ .....(7.12.1)

   Here, $C(z) = \mathcal{Z}\{C(s)\}$; $E(z) = \mathcal{Z}\{E'(s)\}$; $C(s) = \mathcal{L}[c(t)]$ and $E'(s) = \mathcal{L}[e'(t)]$

Fig 3a   *Fig 3b*

The input to the cascaded blocks of $G_0(s)$, G(s) and H(s) is an impulse sampled signal as shown in fig 3a. It's z-domain equivalent is shown in fig 3b.

From fig 3b we get, $B(z) = \mathcal{Z}\{G_0(s)\ G(s)\ H(s)\}\ E(z)$ .....(7.12.2)

   Here, $B(z) = \mathcal{Z}\{B(s)\}$ and $B(s) = \mathcal{L}[b(t)]$

With reference to fig 1.1, at the summing point we get,

   $e(t) = r(t) - b(t)$ .....(7.12.3)

Since e'(t) = e(kT) is an impulse sampled signal, by superposition principle the equation (7.12.3) can be written as,

   $e(kT) = r(kT) - b(kT)$ .....(7.12.4)

      where, e(kT), r(kT) and b(kT) are impulse sampled signals of e(t), r(t) and b(t) respectively.

On taking $\mathcal{Z}$-transform of equation (7.12.4) we get,

   $E(z) = R(z) - B(z)$

   $\therefore R(z) = E(z) + B(z)$ .....(7.12.5)

      where, $R(z) = \mathcal{Z}\{R(s)\}$ and $R(s) = \mathcal{L}[r(t)]$

On substituting for B(z) from equation (7.12.2) in equation (7.12.5) we get,

$$R(z) = E(z) + ƶ\{G_0(s)\ G(s)\ H(s)\}\ E(z)$$

$$= [1 + ƶ\{G_0(s)\ G(s)\ H(s)\}]\ E(z) \qquad \qquad .....(7.12.6)$$

From equations (7.12.1) and (7.12.6) the z-transfer function or pulse transfer function, C(z)/R(z) can be written as,

$$\frac{C(z)}{R(z)} = \frac{ƶ\{G_0(s)\ G(s)\}}{1 + ƶ\{G_0(s)\ H(s)\}} = \frac{G_0G(z)}{1 + G_0GH(z)} \qquad .....(7.12.7)$$

Here, $ƶ\{G_0(s)\ G(s)\}$ is denoted as $G_0G(z)$ and $ƶ\{G_0(s)\ G(s)\ H(s)\}$ is denoted as $G_0GH(z)$.

(b)   **The input to the block $G_2(s)$ is an impulse sampled signal as shown in fig 4a. It's z-domain equivalent is shown in fig 4b.**

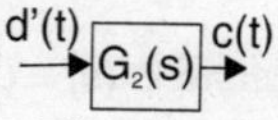

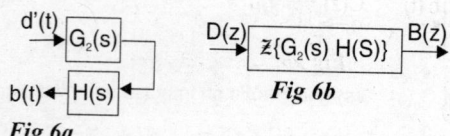

Fig 4a 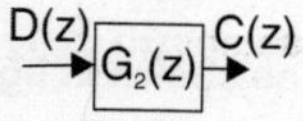   *Fig 4b*

From fig 4b we get, $C(z) = G_2(z)\ D(z)$

$$.....(7.12.8)$$

where, $C(z) = ƶ\{C(s)\}$; $G_2(z) = ƶ\{G_2(s)\}$; $D(z) = ƶ\{D'(s)\}$; $C(s) = \mathcal{L}[c(t)]$ and $D'(s) = \mathcal{L}[d'(t)]$

The input to the block $G_1(s)$ is an impulse sampled signal as shown in fig 5a. It's z-domain equivalent is shown in fig 5b.

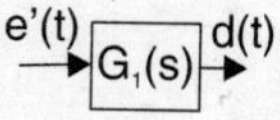

Fig 5a 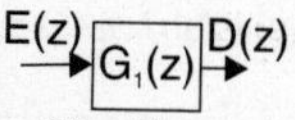   *Fig 5b*

From fig 5b we get, $D(z) = G_1(z)\ E(z)$

$$.....(7.12.9)$$

From equations (7.12.8) and (7.12.9) we get,

$$C(z) = G_2(z)\ G_1(z)\ E(z) \qquad \qquad .....(7.12.10)$$

where, $G_1(z) = ƶ\{G_1(s)\}$; $E(z) = ƶ\{E'(s)\}$ and $E'(s) = \mathcal{L}[e'(t)]$

The input to the cascaded blocks $G_2(s)$ and $H(s)$ is an impulse sampled signal as shown in fig 6a. It's z-domain equivalent is shown in fig 6b.

Fig 6a   *Fig 6b*

From fig 6b we get,

$$B(z) = ƶ\{G_2(s)\ H(s)\}\ D(z)$$

$$.....(7.12.11)$$

On substituting for D(z) from equation (7.12.9) in equation (7.12.11) we get,

$$B(z) = ƶ\{G_2(s)\ H(s)\}\ G_1(z)\ E(z)$$

$$.....(7.12.12)$$

With reference to fig 1b, at the summing point we get,

$$e(t) = r(t) - b(t)$$

$$.....(7.12.13)$$

Since e'(t) = e(kT) is an impulse sampled signal, by superposition principle the equation (7.12.13) can be written as,

$$e(kT) = r(kT) - b(kT)$$

$$.....(7.12.14)$$

where, e(kT), r(kT) and b(kT) are impulse sampled signals of e(t), r(t) and b(t) respectively.

On taking ƶ-transform of equation (7.12.14) we get,

$$E(z) = R(z) - B(z)$$

$$\therefore R(z) = E(z) + B(z) \qquad\qquad .....(7.12.15)$$

On substituting for B(z) from equation (7.12.12) in equation (7.12.15) we get,

$$R(z) = E(z) + \mathcal{Z}\{G_2(s)\;H(s)\}\,G_1(z)\,E(z)$$

$$= [1 + \mathcal{Z}\{G_2(s)\,H(s)\}\,G_1(z)]\,E(z) \qquad\qquad .....(7.12.16)$$

From equations (7.12.10) and (7.12.16) the z-transfer function or pulse transfer function C(z)/R(z) can be written as,

$$\frac{C(z)}{R(z)} = \frac{G_1(z)\,G_2(z)}{1 + \mathcal{Z}\{G_2(s)\,H(s)\}\,G_1(z)} = \frac{G_1(z)\,G_2(z)}{1 + G_2H(z)\,G_1(z)} \qquad\qquad .....(7.12.17)$$

Here $\mathcal{Z}\{G_2(s)\,H(s)\}$ is denoted as $G_2H(z)$

(c) **The ZOH in the system is replaced by $G_0(s)$ as shown in fig 7, where $G_0(s) = (1 - e^{-sT})/s$.**

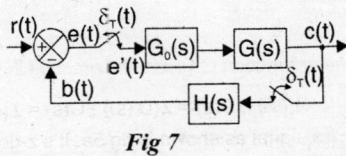

Fig 7

The input to the cascaded blocks of $G_0(s)$ and $G(s)$ is an impulse sampled signal as shown in fig 8a. It's z-domain equivalent is shown in fig 8b.

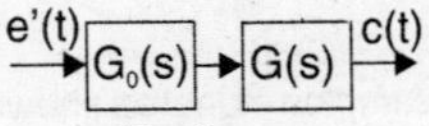

Fig 8a 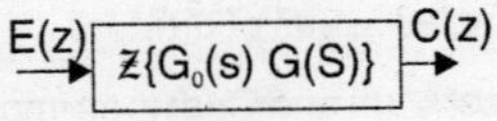   *Fig 8b*

From 8b, we get, $C(z) = \mathcal{Z}\{G_0(s)\,G(s)\}\,E(z)$ .....(7.12.18)

where, $C(z) = \mathcal{Z}\{C(s)\}$; $E(z) = \mathcal{Z}\{E'(s)\}$; $C(s) = \mathcal{L}[c(t)]$ and $E'(s) = \mathcal{L}[e'(t)]$.

The input to the block H(s) is an impulse sampled signal as shown in fig 9a. It's z-domain equivalent is shown in fig 9b.

Fig 9a    *Fig 9b*

From fig 9b, we get, $B(z) = H(z)\,C(z)$ .....(7.12.19)

With reference to fig 7, at the summing point we get,

$$e(t) = r(t) - b(t) \qquad\qquad .....(7.12.20)$$

Since $e'(t) = e(kT)$ is an impulse sampled signal, by principle of superposition the equation (7.12.20) can be written as,

$$e(kT) = r(kT) - b(kT) \qquad\qquad .....(7.12.21)$$

where, e(kT), r(kT) and b(kT) are impulse sampled signals of e(t), r(t) and b(t) respectively.

On taking $\mathcal{Z}$-transform of equation (7.12.21) we get,

$$E(z) = R(z) - B(z) \qquad\qquad .....(7.12.22)$$

On substituting for B(z) from equation (7.12.19) in equation (7.12.22) we get,

$$E(z) = R(z) - H(z)\,C(z) \qquad\qquad .....(7.12.23)$$

On substituting for E(z) from equation (7.12.23) in equation (7.12.18) we get,

$$C(z) = \mathcal{Z}\{G_0(s)\,G(s)\}\,[R(z) - H(z)\,C(z)]$$

$$C(z) = \mathcal{Z}\{G_0(s)\,G(s)\}\,R(z) - \mathcal{Z}\{G_0(s)\,G(s)\}\,H(z)\,C(z)$$

$$C(z) + \mathcal{Z}\{G_0(s)\,G(s)\}\,H(z)\,C(z) = \mathcal{Z}\{G_0(s)\,G(s)\}\,R(z)$$

$$C(z) [1 + \mathbb{z}\{G_0(s) G(s)\} H(z)] = \mathbb{z}\{G_0(s) G(s)\} R(z)$$

$$\therefore \frac{C(z)}{R(z)} = \frac{\mathbb{z}\{G_0(s) G(s)\}}{1 + \mathbb{z}\{G_0(s) G(s)\} H(z)} = \frac{G_0G(z)}{1 + G_0G(z)H(z)} \qquad .....(7.12.24)$$

The equation (7.12.24) is the z-transfer function of the system.

Here $\mathbb{z}\{G_0(s) G(s)\}$ is denoted as $G_0G(z)$.

EXAMPLE 7.13

Find the output C(z) in z-domain for the closed loop sampled data control system shown in fig 1.

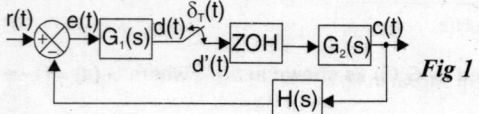

Fig 1

SOLUTION

The ZOH in fig 1 is replaced by a block with transfer function $G_0(s)$ as shown in fig 2 where $G_0(s) = (1 - e^{-sT})/s$.

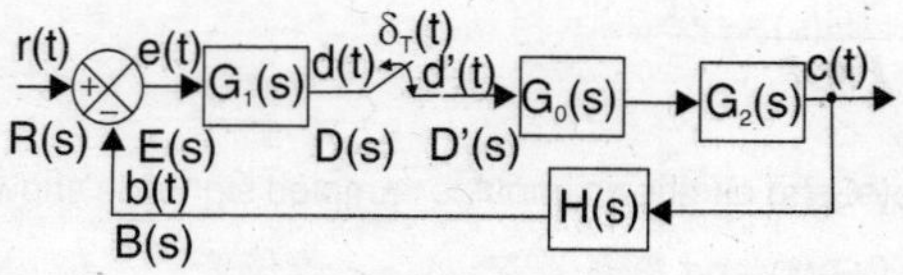

Fig 2

Here, d'(t) = Impulse sampled signal of d(t).

The input to the cascaded blocks of $G_0(s)$ and $G_2(s)$ is an impulse sampled signal as shown in fig 2a. It's z-domain equivalent is shown in fig 2b.

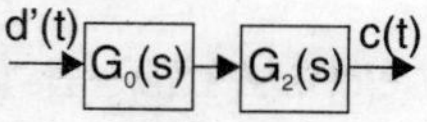

Fig 2a 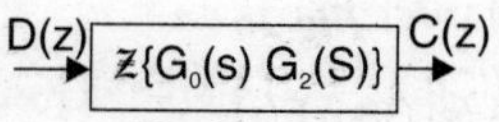   *Fig 2b*

From fig 2b we get, $C(z) = \mathbb{z}\{G_0(s) G_2(s)\} D(z)$ .....(7.13.1)

where, $C(z) = \mathbb{z}\{C(s)\}$; $D(z) = \mathbb{z}\{D'(s)\}$; $C(s) = \mathcal{L}[c(t)]$ and $D'(s) = \mathcal{L}[d'(t)]$.

With reference to fig 2 the following s-domain equations can be obtained.

$$E(s) = R(s) - B(s) \qquad .....(7.13.2)$$
$$D(s) = E(s) G_1(s) \qquad .....(7.13.3)$$
$$B(s) = G_0(s) G_2(s) H(s) D'(s) \qquad .....(7.13.4)$$

On substituting for E(s) from equation (7.13.2) in equation (7.13.3) we get,

$$D(s) = [R(s) - B(s)] G_1(s) = G_1(s) R(s) - G_1(s) B(s) \qquad .....(7.13.5)$$

On substituting for B(s) from equation (7.13.4) in equation (7.13.5) we get,

$$D(s) = G_1(s) R(s) - G_1(s) G_0(s) G_2(s) H(s) D'(s) \qquad .....(7.13.6)$$

On taking z-transform of equation (7.13.6) we get,

$$D(z) = \mathbb{z}\{G_1(s) R(s)\} - \mathbb{z}\{G_1(s) G_0(s) G_2(s) H(s)\} D(z)$$

$$D(z) + \mathbb{z}\{G_0(s) G_1(s) G_2(s) H(s)\} D(z) = \mathbb{z}\{G_1(s) R(s)\}$$

$$D(z) [1 + \mathbb{z}\{G_0(s) G_1(s) G_2(s) H(s)\}] = \mathbb{z}\{G_1(s) R(s)\}$$

$$\therefore D(z) = \frac{\mathbb{z}\{G_1(s) R(s)\}}{1 + \mathbb{z}\{G_0(s) G_1(s) G_2(s) H(s)\}} \qquad .....(7.13.7)$$

> Note : The term $G_0(s)\, G_1(s)\, G_2(s)\, H(s)\, D'(s)$ represents the output of a block with transfer function $G_0(s)\, G_1(s)$ $G_2(s)\, H(s)$ when the input is $D'(s)$.

On substituting for D(z) from equation (7.13.7) in equation (7.13.1) we get,

Output in z-domain, $C(z) = \dfrac{\mathbb{z}\{G_0(s)\, G_2(s)\}\; \mathbb{z}\{G_1(s)\, R(s)\}}{1 + \mathbb{z}\{G_0(s)\, G_1(s)\, G_2(s)\, H(s)\}} = \dfrac{G_0 G_2(z)\, G_1 R(z)}{1 + G_0 G_1 G_2 H(z)}$

where, $\mathbb{z}\{G_0(s)\, G_2(s)\}$ is represented as $G_0 G_2(z)$,

$\mathbb{z}\{G_1(s)\, R(s)\}$ is represented as $G_1 R(z)$ and

$\mathbb{z}\{G_0(s)\, G_1(s)\, G_2(s)\, H(s)\}$ is represented as $G_0 G_1 G_2 H(z)$

EXAMPLE 7.14

For the sampled data control system shown in fig 1, find the response to unit step input, where $G(s) = 1/(s+1)$.

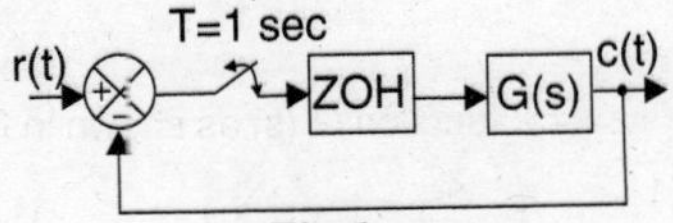

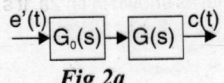

Fig 1

SOLUTION

The ZOH in the system is replaced by $G_0(s)$ as shown in fig 1, where $G_0(s) = (1 - e^{-sT})/s$.

The input to the cascaded blocks of $G_0(s)$ and $G(s)$ is an impulse sampled signal as shown in fig 2a. It's z-domain equivalent is shown in fig 2b.

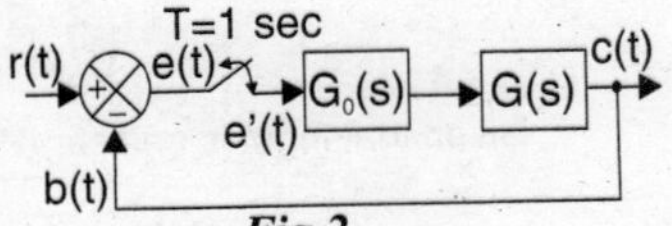

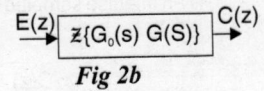

Fig 2

Fig 2a   Fig 2b

From fig 2b we get, $C(z) = \mathbb{z}\{G_0(s)\, G(s)\}\, E(z)$ .....(7.14.1)

With reference to fig 2, at the summing point we get,

$e(t) = r(t) - c(t)$ .....(7.14.2)

Since $e'(t) = e(kT)$ is an impulse sampled signal, the equation (7.14.2) can be written as,

$e(kT) = r(kT) - c(kT)$ .....(7.14.3)

       where, $e(kT)$, $r(kT)$ and $c(kT)$ are impulse sampled signals of $e(t)$, $r(t)$ and $c(t)$ respectively.

On taking z-transform of equation (7.14.3) we get,

$E(z) = R(z) - C(z)$ .....(7.14.4)

On substituting for E(z) from equation (7.14.4) in equation (7.14.1) we get,

$C(z) = \mathbb{z}\{G_0(s)\, G(s)\}\, [R(z) - C(z)]$

$C(z) = \mathbb{z}\{G_0(s)\, G(s)\}\, R(z) - \mathbb{z}\{G_0(s)\, G(s)\}\, C(z)$

$C(z) + \mathbb{z}\{G_0(s)\, G(s)\}\, C(z) = \mathbb{z}\{G_0(s)\, G(s)\}\, R(z)$

$C(z)\, [1 + \mathbb{z}\{G_0(s)\, G(s)\}] = \mathbb{z}\{G_0(s)\, G(s)\}\, R(z)$

$\therefore\; C(z) = \dfrac{\mathbb{z}\{G_0(s)\, G(s)\}\, R(z)}{1 + \mathbb{z}\{G_0(s)\, G(s)\}}$ .....(7.14.5)

We know that, $\mathbb{z}\{G_0(s)\, G(s)\} = (1 + z^{-1})\, \mathbb{z}\left\{\dfrac{G(s)}{s}\right\}$

Here, $G(s) = \dfrac{1}{s+1}$ and $\dfrac{G(s)}{s} = \dfrac{1}{s(s+1)}$

By partial fraction expansion,

25

$$\frac{G(s)}{s} = \frac{1}{s(s+1)} = \frac{A}{s} + \frac{B}{s+1}$$

$$A = \frac{1}{s(s+1)} s \Big|_{s=0} = \frac{1}{s+1}\Big|_{s=0} = 1$$

$$B = \frac{1}{s(s+1)}(s+1)\Big|_{s=-1} = \frac{1}{s}\Big|_{s=-1} = -1$$

$$z\left\{\frac{G(s)}{s}\right\} = z\left\{\frac{1}{s} - \frac{1}{s+1}\right\} = z\left\{\frac{1}{s}\right\} - z\left\{\frac{1}{s+1}\right\}$$

From standard Laplace and z-transform pairs we get,

$$z\left\{\frac{1}{s}\right\} = \frac{z}{z-1} \quad \text{and} \quad z\left\{\frac{1}{s+a}\right\} = \frac{z}{z-e^{-aT}}$$

Here, a = 1 and T = 1

$$\therefore z\left\{\frac{G(s)}{s}\right\} = \frac{z}{z-1} - \frac{z}{z-e^{-T}}$$

Now, $z\{G_0(s)G(s)\} = (1-z^{-1})z\left\{\frac{G(s)}{s}\right\} = (1-z^{-1})\left(\frac{z}{z-1} - \frac{z}{z-e^{-T}}\right)$

$$= \left(1 - \frac{1}{z}\right)\left(\frac{z(z-e^{-1}) - z(z-1)}{(z-1)(z-e^{-1})}\right)$$

$$= \left(\frac{z-1}{z}\right)\left(\frac{z(z-e^{-1}-z+1)}{(z-1)(z-e^{-1})}\right) = \frac{1-e^{-1}}{z-e^{-1}} = \frac{0.632}{z-0.368} \qquad \qquad .....(7.14.6)$$

Given that input is unit step

$$\therefore R(z) = U(z) = \frac{z}{z-1} \qquad \qquad .....(7.14.7)$$

From equations (7.14.5), (7.14.6) and (7.14.7) we get,

$$C(z) = \frac{z\{G_0(s)G(s)\}R(z)}{1 + z\{G_0(s)G(s)\}} = \frac{\left(\dfrac{0.632}{z-0.368}\right)\dfrac{z}{z-1}}{1 + \left(\dfrac{0.632}{z-0.368}\right)}$$

$$= \frac{\dfrac{0.632z}{(z-1)(z-0.368)}}{\dfrac{(z-0.368)+0.632}{(z-0.368)}} = \frac{0.632z}{(z-1)(z-0.368+0.632)} = \frac{0.632z}{(z-1)(z+0.264)}$$

$$\therefore \frac{C(z)}{z} = \frac{0.632}{(z-1)(z+0.264)}$$

By partial fraction expansion,

$$\frac{C(z)}{z} = \frac{A}{z-1} + \frac{B}{z+0.264}$$

$$A = \frac{0.632}{(z-1)(z+0.264)}(z-1)\Big|_{z=1} = \frac{0.632}{z+0.264}\Big|_{z=1} = \frac{0.632}{1+0.264} = 0.5$$

$$B = \frac{0.632}{(z-1)(z+0.264)}(z+0.264)\Big|_{z=-0.264} = \frac{0.632}{z-1}\Big|_{z=-0.264}$$

$$= \frac{0.632}{-0.264-1} = -0.5$$

$$\therefore \frac{C(z)}{z} = \frac{0.5}{z-1} - \frac{0.5}{z+0.264}$$

$$C(z) = 0.5 \frac{z}{z-1} - 0.5 \frac{z}{z-(-0.264)} \qquad .....(7.14.8)$$

We know that,

$$\mathcal{Z}\{1\} = \frac{z}{z-1} \text{ and } \mathcal{Z}\{a^k\} = \frac{z}{z-a}$$

On taking inverse $\mathcal{Z}$-transform of equation (7.14.8) we get,

$$c(k) = 0.5 - 0.5\,(-0.264)^k = 0.5\,[1-(-0.264)^k] \qquad .....(7.14.9)$$

The equation (7.14.9) is the response of given system for unit step input.

7.12 THE z AND s - DOMAIN RELATIONSHIP

Let $r(kT)$ be a discrete sequence which has been obtained by sampling $r(t)$ at a sampling rate of $1/T$. On taking $\mathcal{Z}$-transform of $r(kT)$ we get,

$$\mathcal{Z}\{r(kT)\} = R(z) = \sum_{k=0}^{\infty} r(kT)z^{-k} \qquad .....(7.69)$$

Let, $r(t)$ = Impulse sampled signal of $r(t)$ at the sampling rate of $1/T$.

and $R'(s) = \mathcal{L}[r'(t)]$ = Laplace transform of $r'(t)$.

Now, $\quad r'(t) = \sum_{k=0}^{\infty} r(kT)\,\delta(t-kT) \qquad .....(7.70)$

On taking laplace transform of equation (7.70) we get,

$$R'(s) = \sum_{k=0}^{\infty} r(kT)\,e^{-ksT} \qquad .....(7.71)$$

Let us choose a transformation such that,

$$z = e^{sT} \qquad .....(7.72)$$

$$\therefore \; ln\,z = sT \quad (\text{or}) \quad s = \frac{1}{T}\,ln\,z \qquad .....(7.73)$$

On substituting for s from equation (7.73) in equation (7.71) we get,

$$R'(s) = \sum_{k=0}^{\infty} r(kT)\,e^{\left(-kT.\frac{1}{T}\,lnz\right)}$$

$$= \sum_{k=0}^{\infty} r(kT)\,e^{(ln\,z^{-k})} = \sum_{k=0}^{\infty} r(kT)\,z^{-k} = R(z) \qquad .....(7.74)$$

From equation (7.74) it is obvious that $\mathcal{Z}$-transform of a discrete sequence can be obtained from the Laplace transform of its impulse sampled version, by choosing a transformation, $s = (1/T)\,ln\,z$ (or $z = e^{sT}$).

The transformation, $s = (1/T)\,ln\,z$, maps the s-plane into the z-plane. It can be shown that every section of $j\omega$-axis of length, $N\omega$, maps into the unit circle in the anticlockwise direction where N is an integer and ω_s is the sampling frequency and it can be shown that every strip in the left half s-plane of width ω_s, maps into the interior of the unit circle as shown in fig 7.31.

The above mapping helps in extending the s-plane stability criterion to z-plane. For stability of a system in s-plane the poles of s-domain transfer function should lie on the left half of s-plane. In this transformation the left half of s-plane maps into interior of unit circle. Hence for the stability of the system in z-domain, the poles of the z-transfer function should lie inside the unit circle.

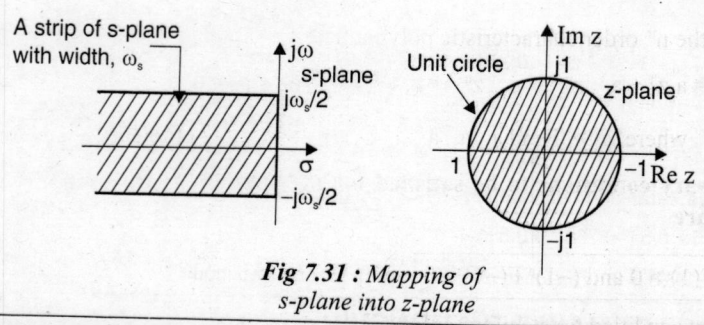

Fig 7.31 : *Mapping of s-plane into z-plane*

7.13 STABILITY ANALYSIS OF SAMPLED DATA CONTROL SYSTEMS

The sampled data control system is stable if all the poles of the z-transfer function of the system lies inside the unit circle in z-plane.

The poles of the transfer function are given by the roots of the characteristic equation. Hence the system stability can be determined from the roots of the characteristic equation.

The z-transfer function of the sampled data control system can be expressed as a ratio of two polynomials in z as shown below.

z-transfer function, $H(z) = \dfrac{C(z)}{R(z)} = A_0 \dfrac{P(z)}{Q(z)}$ .....(7.75)

where, A_0 = constant

$P(z)$ = Numerator polynomial

$Q(z)$ = Denominator polynomial

The characteristic equation is the denominator polynomial of H(z). [i.e., characteristic equation is given by $Q(z) = 0$].

Consider the system shown in fig 7.32. For this system, the z-transfer function is given by,

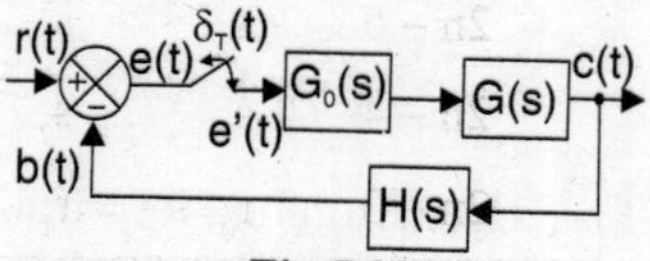

$$H(z) = \frac{C(z)}{R(z)} = \frac{\mathbb{Z}\{G_0(s)\,G(s)\}}{1 + \mathbb{Z}\{G_0(s)\,G(s)\,H(s)\}}$$ .....(7.76)

Fig 7.32

and the characteristic equation is,

$$\boxed{1 + \mathbb{Z}\{G_0(s)\,G(s)\,H(s)\} = 0}$$ .....(7.77)

The following methods are available for the stability analysis of sampled data control systems using the characteristic equation

1.   Jury's stability test

2.   Bilinear transformation

3.   Root locus technique

The Jury's stability test and Bilinear transformation are presented in this book.

JURY'S STABILITY TEST

The Jury's stability test is used to determine whether the roots of the characteristic polynomial lie within a unit circle or not. The Jury's test consists of two parts. One simple test for necessary condition for stability and another test for sufficient condition for stability.

Let F(z) be the nth order characteristic polynomial of a sampled data control system.

$$F(z) = a_n z^n + a_{n-1} z^{n-1} + a_{n-2} z^{n-2} + \ldots + a_2 z^2 + az + a_0 = 0 \qquad \ldots(7.78)$$

where, $a_n > 0$ and $a_0, a_1, a_2, \ldots a_n$ are constant coefficients.

The necessary conditions to be satisfied for the stability of the system with characteristic polynomial, F(z) are

$$\boxed{F(1) > 0 \text{ and } (-1)^n F(-1) > 0} \qquad \ldots(7.79)$$

The sufficient condition for stability can be established through any one of the following two methods.

Method-1 for testing sufficiency

In this method prepare a table as shown below using the coefficients of the characteristic polynomial F(z). The table consists of $(2n - 3)$ rows, where n is the order of the characteristic equation.

Row	z^0	z^1	z^2	......	z^{n-k}	.....	z^{n-2}	z^{n-1}	z^n
1	a_0	a_1	a_2	......	a_{n-k}	.....	a_{n-2}	a_{n-1}	a_n
2	a_n	a_{n-1}	a_{n-2}	......	a_k	.....	a_2	a_1	a_0
3	b_0	b_1	b_2	......	......	.....	b_{n-2}	b_{n-1}	
4	b_{n-1}	b_{n-2}	b_{n-3}	......	......	.....	b_1	b_0	
5	c_0	c_1	c_2	......	......	.....	c_{n-2}		
6	c_{n-2}	c_{n-3}	c_{n-4}	......	......	.....	c_0		
⋮	⋮	⋮							
2n – 5	s_0	s_1	s_2	s_3					
2n – 4	s_3	s_2	s_1	s_0					
2n – 3	r_0	r_1	r_2						

In the above table the elements of row-1 are formed using the coefficients of characteristic polynomial and the row-2 is formed by arranging the elements of row-1 in the reverse order.

The kth element of row-3 is given by, $b_k = \begin{vmatrix} a_0 & a_{n-k} \\ a_n & a_k \end{vmatrix}$

$$\therefore \; b_0 = \begin{vmatrix} a_0 & a_n \\ a_n & \varepsilon_0 \end{vmatrix} \; ; \; b_1 = \begin{vmatrix} a_0 & a_{n-1} \\ a_n & a_1 \end{vmatrix} \; ; \; b_2 = \begin{vmatrix} a_0 & a_{n-2} \\ a_n & a_2 \end{vmatrix} \text{ and so on}$$

The row-4 is formed by arranging the elements of row-3 in reverse order.

The kth element of row-5 is given by, $c_k = \begin{vmatrix} b_0 & b_{n-1-k} \\ b_{n-1} & b_k \end{vmatrix}$

$$\therefore \; c_0 = \begin{vmatrix} b_0 & b_{n-1} \\ b_{n-1} & b_k \end{vmatrix} \; ; \; c_1 = \begin{vmatrix} b_0 & b_{n-2} \\ b_{n-1} & b_1 \end{vmatrix} \; ; \; c_2 = \begin{vmatrix} b_0 & b_{n-3} \\ b_{n-1} & b_2 \end{vmatrix} \text{ and so on}$$

The row-6 is formed by arranging the elements of row-5 in reverse order.

In general we can say that the elements of a row with odd row number are calculated using the elements of the two rows just above the concerned row. The row next to the odd numbered row is formed by arranging the elements of the odd numbered row in reverse order. For calculating the elements of a row the logic explained for b_k and c_k are followed.

The first column elements of the table are used to check the following $(n-1)$ conditions. These $(n-1)$ conditions are the sufficient conditions for stability of the system.

$$\left.\begin{array}{l} |a_0| < |a_n| \\ |b_0| > |b_{n-1}| \\ |c_0| > |c_{n-2}| \\ \vdots \\ |r_0| > |r_2| \end{array}\right\} (n-1) \text{ conditions} \qquad \qquad .....(7.80)$$

If the necessary and sufficient conditions are satisfied then all the poles of the system lies inside the unit circle in z-plane and so the system is stable.

If even one of the condition is not satisfied then the system is unstable.

Method-2 for testing sufficiency

The coefficients of the characteristic polynomial of equation (7.78) are renamed as shown below.

$$F(z) = a_0 z^n + a_1 z^{n-1} + a_2 z^{n-2} + ...+ a_{n-1} z + a_n = 0 \quad ; \quad a_0 > 0 \qquad .....(7.81)$$

Using the coefficients of equation (7.81), construct two square matrices X and Y of order $(n-1)$ as shown below.

$$X = \begin{bmatrix} a_0 & a_1 & a_2 & \cdots & a_{n-2} \\ 0 & a_0 & a_1 & \cdots & a_{n-3} \\ 0 & 0 & a_0 & \cdots & a_{n-4} \\ \vdots & \vdots & \vdots & & \vdots \\ 0 & 0 & 0 & & a_0 \end{bmatrix} \quad ; \quad Y = \begin{bmatrix} a_2 & a_3 & \cdots & a_{n-2} & a_{n-1} & a_n \\ a_3 & a_4 & \cdots & a_{n-1} & a_n & 0 \\ a_4 & a_5 & \cdots & a_n & 0 & 0 \\ \vdots & \vdots & & \vdots & \vdots & \vdots \\ a_n & 0 & \cdots & 0 & 0 & 0 \end{bmatrix}$$

By taking the sum and difference of the two matrices X and Y, construct two more matrices H_1 and H_2.

$$\text{where } H_1 = X + Y \qquad\qquad .....(7.82)$$
$$\text{and } \quad H_2 = X - Y \qquad\qquad .....(7.83)$$

In this method the sufficient condition for stability is that the matrices H_1 and H_2 should be positive innerwise.

If the necessary and sufficient condition for stability are satisfied then the system is stable.

> **Note :** *The necessary conditions for stability are same in both the methods.*

A square matrix A is said to be positive innerwise when all the determinants starting with the centre element and proceeding outwards upto the entire matrix are positive. For example consider the matrix A shown below.

$$A = \begin{bmatrix} a_{11} & a_{12} & a_{13} & a_{14} & a_{15} \\ a_{21} & a_{22} & a_{23} & a_{24} & a_{25} \\ a_{31} & a_{32} & a_{33} & a_{34} & a_{35} \\ a_{41} & a_{42} & a_{43} & a_{44} & a_{45} \\ a_{51} & a_{52} & a_{53} & a_{54} & a_{55} \end{bmatrix}$$

The possible determinants of A starting with centre element are shown in dotted lines and these determinants are also shown below.

$$a_{33} \; ; \; \begin{vmatrix} a_{22} & a_{23} & a_{24} \\ a_{32} & a_{33} & a_{34} \\ a_{42} & a_{43} & a_{44} \end{vmatrix} \; ; \; \begin{vmatrix} a_{11} & a_{12} & a_{13} & a_{14} & a_{15} \\ a_{21} & a_{22} & a_{23} & a_{24} & a_{25} \\ a_{31} & a_{32} & a_{33} & a_{34} & a_{35} \\ a_{41} & a_{42} & a_{43} & a_{44} & a_{45} \\ a_{51} & a_{52} & a_{53} & a_{54} & a_{55} \end{vmatrix}$$

STABILITY ANALYSIS USING BILINEAR TRANSFORMATION

The bilinear transformation maps the interior of unit circle in the z-plane into the left half of the r-plane. In this transformation,

$$r = \frac{z-1}{z+1} \text{ (or) } z = \frac{1+r}{1-r} \qquad .....(7.84)$$

The transformation is performed by substituting, $z = \frac{1+r}{1-r}$ in the characteristic equation of the system.

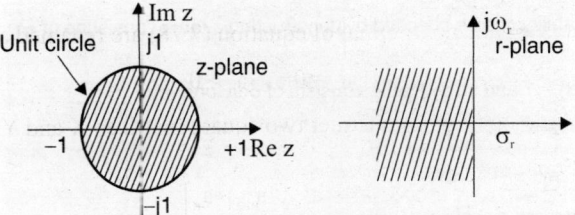

Fig 7.33 : *Mapping of unit circle in z-plane into left half of r-plane*

Consider the characteristic equation of the system in z-plane shown below.

$$a_n z^n + a_{n-1} z^{n-1} + a_{n-2} z^{n-2} + ... + az + a_0 = 0 \; ; \; a_n > 0 \qquad .....(7.85)$$

Using bilinear transformation, the equation (7.85) can be written as,

$$\left(\text{i.e., by substituting } z = \frac{1+r}{1-r} \right)$$

$$a_n \left(\frac{1+r}{1-r} \right)^n + a_{n-1} \left(\frac{1+r}{1-r} \right)^{n-1} + a_{n-2} \left(\frac{1+r}{1-r} \right)^{n-2} + ... + a\left(\frac{1+r}{1-r} \right) + a_0 = 0 \qquad .....(7.86)$$

The equation (7.86) can be simplified and organised into the form

$$\boxed{b_n r^n + b_{n-1} r^{n-1} - .... + b_1 r + b_0 = 0} \qquad .....(7.87)$$

Now the equation (7.87) is the new characteristic equation and for the stability of the system the roots of new characteristic equation should lie on the left half of r-plane. The routh stability criterion can be applied to the new characteristic equation to determine if all its roots lie in the left half of the r-plane.

In this method the necessary condition for stability is that the coefficients of equation (7.87) should be positive. The sufficient condition for stability is that there should not be any sign change in the elements of first column of routh array.

The system is stable if both the necessary and sufficient conditions for stability are satisfied.

EXAMPLE 7.15

Check for stability of the sampled data control systems represented by the following characteristic equation.

(a) $5z^2 - 2z + 2 = 0$

(b) $z^3 - 0.2 z^2 - 0.25 z + 0.05 = 0$

(c) $z^4 - 1.7 z^3 + 1.04 z^2 - 0.268 z + 0.024 = 0$

SOLUTION

(a) **Given that, $F(z) = a_2 z^2 + a_1 z + a_0 = 5z^2 - 2z + 2$**

Check for necessary condition

$$F(z) = 5z^2 - 2z + 2$$

$$F(1) = 5(1)^2 - 2(1) + 2 = 5$$

$$(-1)^n F(-1) = (-1)^2 [5(-1)^2 - 2(-1) + 2] = 9 \quad \text{(Here, n = 2)}$$

Since $F(1) > 0$ and $(-1)^n F(-1) > 0$; the necessary conditions for stability are satisfied.

Check for sufficient condition

The sufficient condition for stability can be checked by constructing a table consisting of $(2n - 3)$ rows as shown below.

Here, n = 2; $\therefore (2n - 3) = 1$ and so the table consists of only one row.

Row	z^0	z^1	z^2
1	a_0	a_1	a_2

Here, $a_0 = 2$; $a_1 = -2$ and $a_2 = 5$

Row	z^0	z^1	z^2
1	2	-2	5

The necessary condition to be satisfied is $|a_0| < |a_2|$

Here, $|a_0| = 2$ and $|a_2| = 5$ and so the condition $|a_0| < |a_2|$ is satisfied.

CONCLUSION

The necessary and sufficient conditions for stability are satisfied. Hence the system is stable.

(b) **METHOD-1 : Stability analysis by Jury's stability test**

Given that $F(z) = a_3 z^3 + a_2 z^2 + a_1 z + a_0 = z^3 - 0.2 z^2 - 0.25 z + 0.05 = 0$

Check for necessary conditon

$$F(z) = z^3 - 0.2 z^2 - 0.25 z + 0.05$$

$$F(1) = (1)^3 - 0.2 (1)^2 - 0.25 (1) + 0.05 = 1 - 0.2 - 0.25 + 0.05 = 0.6$$

$$(-1)^n F(-1) = (-1)^3 [(-1)^3 - 0.2 (-1)^2 - 0.25 (-1) + 0.05] \quad \text{(Here n = 3)}$$

$$= (-1) [-1 - 0.2 + 0.25 + 0.05] = 0.9$$

Since $F(1) > 0$ and $(-1)^n F(-1) > 0$; the necessary conditions for stability are satisfied.

Check for sufficient condition

The sufficient condition for stability can be checked by constructing a table consisting of $(2n - 3)$ rows as shown below.

Here, n = 3, $\therefore (2n - 3) = 3$ and so the table consists of three rows.

Row	z^0	z^1	z^2	z^3
1	a_0	a_1	a_2	a_3
2	a_3	a_2	a_1	a_0
3	b_0	b_1	b_2	

Here, $a_0 = 0.05$; $a_1 = -0.25$; $a_2 = -0.2$ and $a_3 = 1$

$$b_0 = \begin{vmatrix} a_0 & a_3 \\ a_3 & a_0 \end{vmatrix} = \begin{vmatrix} 0.05 & 1 \\ 1 & 0.05 \end{vmatrix} = 0.05^2 - 1 = -0.9975$$

$$b_1 = \begin{vmatrix} a_0 & a_2 \\ a_3 & a_1 \end{vmatrix} = \begin{vmatrix} 0.05 & -0.2 \\ 1 & -0.25 \end{vmatrix} = 0.05 \times (-0.25) - 1 \times (-0.2) = 0.1875$$

$$b_2 = \begin{vmatrix} a_0 & a_1 \\ a_3 & a_2 \end{vmatrix} = \begin{vmatrix} 0.05 & -0.25 \\ 1 & -0.2 \end{vmatrix} = 0.05 \times (-0.2) - 1 \times (-0.25) = 0.24$$

Row	z^0	z^1	z^2	z^3
1	0.05	−0.25	−0.2	1
2	1	−0.2	−0.25	0.05
3	−0.9975	0.1875	0.24	

The necessary condition to be satisfied are, $|a_0| < |a_3|$ and $|b_0| > |b_2|$

$|a_0| < |a_3| \Rightarrow |0.05| < |1|$ - satisfied

$|b_0| > |b_2| \Rightarrow |-0.9975| > |0.25|$ - satisfied

CONCLUSION

The necessary and sufficient conditions for stability are satisfied. Hence the system is stable.

METHOD-2 : Stability analysis by bilinear transformation

Given that, $F(z) = z^3 - 0.2\,z^2 - 0.25\,z + 0.05 = 0$

Let us choose the transformation, $z = \dfrac{1+r}{1-r}$

$$\therefore\; F(r) = \left(\frac{1+r}{1-r}\right)^3 - 0.2\left(\frac{1+r}{1-r}\right)^2 - 0.25\left(\frac{1+r}{1-r}\right) + 0.05 = 0$$

On multiplying throughout by $(1-r)^3$ we get,

$(1 + r)^3 - 0.2\,(1 + r)^2\,(1 - r) - 0.25\,(1 + r)(1 - r)^2 + 0.05(1 - r)^3 = 0$

$(1 + r)\,(1 + r^2 + 2r) - 0.2(1 + r)(1 - r^2) - 0.25(1 - r)(1 - r^2) + 0.05(1 - r)(1 + r^2 - 2r) = 0$

$(1 + r)\,(1 + r^2 + 2r - 0.2 + 0.2r^2) + (1 - r)(-0.25 + 0.25r^2 + 0.05 + 0.05r^2 - 0.1r) = 0$

$(1 + r)(1.2r^2 + 2r + 0.8) + (1 - r)(0.3r^2 - 0.1r - 0.2) = 0$

$(1.2r^2 + 2r + 0.8 + 1.2\,r^3 + 2r^2 + 0.8r) + (0.3\,r^2 - 0.1r - 0.2 - 0.3r^3 + 0.1r^2 + 0.2r) = 0$

$(1.2r^3 + 3.2r^2 + 2.8r + 0.8) + (-0.3r^3 + 0.4r^2 + 0.1r - 0.2) = 0$

$0.9r^3 + 3.6r^2 + 2.9r + 0.6 = 0$

The above equation is the new characteristic equation of the system. The coefficients of the new characteristic equation are positive. Hence the necessary condition for stability is satisfied.

The sufficient condition for stability can be determined by constructing routh array as shown below.

r^3	:	0.9	2.9	...... Row-1
r^2	:	3.6	0.6	...... Row-2
r^1	:	2.75		...... Row-3
r^0	:	0.6		...... Row-4

$r^1 : \dfrac{3.6 \times 2.9 - 0.9 \times 0.6}{3.6}$

$r^1 : 2.75$

$r^0 : \dfrac{2.75 \times 0.6 - 0 \times 3.6}{2.75}$

$r^0 : 0.6$

It is observed that there is no sign change in the elements of first column of routh array. Hence the sufficient condition for stability is satisfied.

CONCLUSION

The necessary and sufficient condition for stability are satisfied. Hence the system is stable.

(c) **Given that,** $F(z) = a_4 z^4 + a_3 z^3 + a_2 z^2 + a_1 z^1 + a_0$

$$= z^4 - 1.7z^3 + 1.04 z^2 - 0.268 z + 0.024$$

Check for necessary condition

$$F(z) = z^4 - 1.7z^3 + 1.04 z^2 - 0.268 z + 0.024$$

$$F(1) = 1 - 1.7 + 1.04 - 0.268 + 0.024 = 0.096$$

$$(-1)^n F(-1) = (-1)^4 [(-1)^4 - 1.7 (-1)^3 + 1.04 (-1)^2 - 0.268 (-1) + 0.024]$$

$$= 1 + 1.7 + 1.04 + 0.268 + 0.024 = 4.032 \quad \text{(Here n = 4)}$$

Here the condition $F(1) > 0$ and $(-1)^n F(-1) > 0$ are satisfied. Therefore the necessary conditions for stability are satisfied.

Check for sufficient condition

The sufficient condition for stability can be checked by constructing a table consisting of $(2n - 3)$ rows as shown below. Here, $n = 4$, $\therefore (2n - 3) = 5$ and so the table consists of five rows.

Row	z^0	z^1	z^2	z^3	z^4
1	a_0	a_1	a_2	a_3	a_4
2	a_4	a_3	a_2	a_1	a_0
3	b_0	b_1	b_2	b_3	
4	b_3	b_2	b_1	b_0	
5	c_0	c_1	c_2		

Here, $a_0 = 0.024$; $a_1 = -0.268$; $a_2 = 1.04$; $a_3 = -1.7$ and $a_4 = 1$

$$b_0 = \begin{vmatrix} a_0 & a_4 \\ a_4 & a_0 \end{vmatrix} = \begin{vmatrix} 0.024 & 1 \\ 1 & 0.024 \end{vmatrix} = 0.024^2 - 1 = -0.9994$$

$$b_1 = \begin{vmatrix} a_0 & a_3 \\ a_4 & a_1 \end{vmatrix} = \begin{vmatrix} 0.024 & -1.7 \\ 1 & -0.268 \end{vmatrix} = 0.024 \times (-0.268) - 1 \times (-1.7) = 1.6936$$

$$b_2 = \begin{vmatrix} a_0 & a_2 \\ a_4 & a_2 \end{vmatrix} = \begin{vmatrix} 0.024 & 1.04 \\ 1 & 1.04 \end{vmatrix} = 0.024 \times 1.04 - 1 \times 1.04 = -1.015$$

$$b_3 = \begin{vmatrix} a_0 & a_1 \\ a_4 & a_3 \end{vmatrix} = \begin{vmatrix} 0.024 & -0.268 \\ 1 & -1.7 \end{vmatrix} = 0.024 \times (-1.7) - 1 \times (-0.268) = 0.2272$$

$$c_0 = \begin{vmatrix} b_0 & b_3 \\ b_3 & b_0 \end{vmatrix} = \begin{vmatrix} -0.9994 & 0.2272 \\ 0.2272 & -0.9994 \end{vmatrix} = 0.9994^2 - 0.2272^2 = 0.9472$$

$$c_1 = \begin{vmatrix} b_0 & b_2 \\ b_3 & b_1 \end{vmatrix} = \begin{vmatrix} -0.9994 & -1.015 \\ 0.2272 & 1.6936 \end{vmatrix} = -0.9994 \times 1.6936 - 0.2272 \times (-1.015) = -1.462$$

$$c_2 = \begin{vmatrix} b_0 & b_1 \\ b_3 & b_2 \end{vmatrix} = \begin{vmatrix} -0.9994 & 1.6936 \\ 0.2272 & -1.015 \end{vmatrix} = -0.9994 \times (-1.015) - 0.2272 \times 1.6936 = 0.6296$$

Row	z^0	z^1	z^2	z^3	z^4
1	0.024	−0.268	1.04	−1.7	1.0
2	1.0	−1.7	1.04	-0.268	0.024
3	−0.9994	1.6936	−1.015	0.2272	
4	0.2272	−1.015	1.6936	−0.9994	
5	0.9472	−1.462	0.6296		

The (n -1) conditions to be satisfied are,

$|a_0| < |a_4|$

$|b_0| > |b_3|$

$|c_0| > |c_2|$

$|a_0| < |a_4| \Rightarrow |0.024| < |1|$ - satisfied

$|b_0| > |b_3| \Rightarrow |-0.9994| > |0.2272|$ - satisfied

$|c_0| > |c_2| \Rightarrow |0.9472| > |0.6296|$ - satisfied

CONCLUSION

The necessary and sufficient conditions for stability are satisfied. Hence the system is stable.

7.14 SHORT - ANSWERS QUESTIONS

Q7.1 *What is sampled data control system?*

When the signal or information at any or some points in a system is in the form of discrete pulses, then the system is called discrete data system or sampled data system.

Q7.2 *When the control system is called sampled data system?*

The control system becomes a sampled data system in any one of the following situations.

1. When a digital computer or microprocessor or digital device is employed as a part of the control loop.

2. When the control components are used on time sharing basis.

3. When the control signals are transmitted by pulse modulation.

4. When the output or input of a component in the system is digital or discrete signal.

Q7.3 *Draw the block diagram of a sampled data control system.*

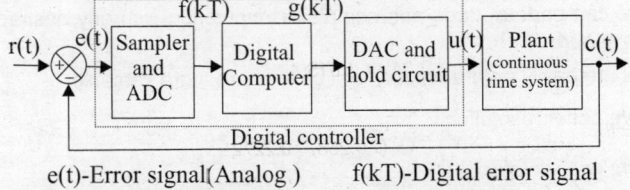

e(t)-Error signal(Analog)   f(kT)-Digital error signal

u(t)-Control signal(Analog)   g(kT)-Digital control signal

Fig Q7.3 : Sampled-data control system

Q7.4 *Distinguish between discrete time systems and continuous time systems.*

The discrete time systems are devices or algorithm that can process (or operate on) discrete-time signals, whereas the continuous time systems are devices that can process (or operator on) analog signals.The input and output signals of discrete-time systems are digital or discrete, but the input and output signals of continuous time systems are analog or continuous time signals.

Q7.5 *Write the advantages and disadvantages of sampled data control systems.*

Advantages of sampled data control system

1. Systems are highly accurate, fast and flexible.

2. Use of time sharing concept of digital computer results in economical cost and space.

3. Digital transducers used in the system have better resolution.

4. The digital components are less affected by noise, non-linearities and transmission errors of noisy channel.

Disadvantages of sampled data control system

1. Conversion of analog signals to discrete-time signals and reconstruction introduce noise and errors in the signal.
2. Additional filters have to be introduced in the system if the components of the system does not have adequate filtering characteristics.

Q7.6   *What is a digital controller?*

A digital controller is a device introduced in the control system to modify the error signal for better control action. The digital controller can be a special purpose computer (microprocessor based system) or a general purpose computer or it is constructed using non-programmable digital devices.

Q7.7   *Compare the analog and digital controller.*

Analog controller	Digital controller
1. Complex	1. Simple
2. Costlier than digital controller	2. Less costlier than analog controller.
3. Slow acting	3. Fast acting
4. Non-programmable	4. Programmable
5. Separate controller should be employed for each control signal	5. A single controller can be used to control more than one signal on time shared basis.

Q7.8   *What are the advantages of digital controllers.*

1. The digital controllers can perform large and complex computation with any desired degree of accuracy at very high speed.
2. The digital controllers are easily programmable and so they are more versatile.
3. Digital controllers have better resolution.

Q7.9   *Explain the terms sampling and sampler.*

Sampling is a process in which the continuous-time signal (or analog signal) is converted into a discrete-time signal by taking samples of the continuous time signal at discrete time instants. Sampler is a device which performs the process of sampling.

Q7.10   *What is periodic sampling?*

The periodic sampling is a sampling process in which the discrete time signal or sequence is obtained by taking samples of continuous time signal periodically or uniformly at intervals of T seconds. Here T is called sampling period and $1/T = F_s$ is called sampling frequency.

Q7.11   *State (shanon's) sampling theorem.*

Sampling theorem states that a bandlimited continuous-time signal with highest frequency f_m, hertz can be uniquely recovered from its samples provided that the sampling rate F_s is greater than or equal to $2f_m$ samples per second.

Q7.12   What is meant by quantization?

The process of converting a discrete-time continuous valued signal into a discrete-time discrete valued signal is called quantization. In quantization the value of each signal sample is represented by a value selected from a finite set of possible values called quantization levels.

Q7.13   What is coding?

The coding is the process of representing each discrete value by a n-bit binary sequence (or code or number).

Q7.14   What are hold circuits?

Hold circuits are devices used to convert discrete time signals to continuous time signals.

Q7.15   What is zero-order hold?

The zero-order hold is a hold circuit in which the signal is reconstructed such that the value of reconstructed signal for a sampling period is same as the value of last received sample. The signal reconstruction by zero order hold is shown in fig Q7.15.

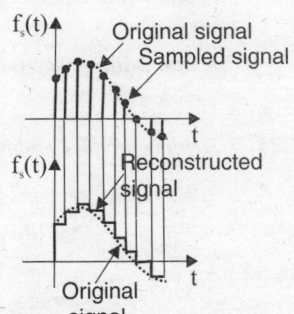

Fig Q7.15 : Signal reconstruction by ZOH

Q7.16   What is first-order hold?

The first-order hold is a hold circuit in which the last two signal samples (current and previous sample) are used to reconstruct the signal for the current sampling period. The reconstructed signal will be a straight line in a sampling period, whose slope is determined by the current sample and previous sample.

Q7.17   Define Acquisition time.

In analog-to-digital conversion process, the Acquisition time is defined as the total time required for obtaining a signal sample and the time for quantizing and coding. It is also called conversion time.

Q3.18   Define Aperture time

The duration of sampling the signal is called aperture time.

Q7.19   Define settling time

In digital-to-analog conversion process the settling time is defined as the time required for the output of the D/A converter to reach and remain within a given fraction of the final value, after application of input code word.

Q7.20   What is "Hold mode droop"?

The changes in signal voltage level in the hold circuits during hold mode (or hold period) is called hold mode droop.

Q7.21   What are the problems encountered in a practical hold circuit?

The problems encountered in practical hold circuit are

1.   Errors in the periodicity of sampling process.

2.   Nonlinear variations in the duration of sampling aperture.

3.   Droop (changes) in the voltage held during conversion.

Q7.22   *How the high frequency noise signals in the reconstructed signal are eliminated?*

The high frequency noise signals (or unwanted signals) introduced by hold circuits in the reconstructed signal are easily filtered out by the various elements of the control system, because the control system is basically a low-pass filter.

Q7.23   *What is discrete sequence?*

A discrete sequence or discrete time signal, f(k) is a function of independent variable, k, which is an integer.

A two sided discrete-time signal f(k) is defined for every integer value of k in the range $-\infty < k < \infty$.

A one sided causal discrete-time signal f(k) is defined for every integer value of k in the range of $0 \le k < \infty$.

A one sided anticausal discrete-time signal f(k) is defined for every integer value of k in the range $-\infty < k \le 0$.

Q7.24   *Define one sided and two-sided Z-transform.*

The Z-transform (two-sided Z-transform) of a discrete sequence, f(k) is defined as the power series,

$$F(z) = Z\{f(k)\} = \sum_{k=0}^{\infty} f(k) z^{-k}$$

where, z is a complex variable.

The notation $Z\{f(k)\}$ is used to denote Z-transform of f(k).

The one sided Z-transform of f(k) is defined as the power series,

$$F(z) = Z\{f(k)\} = \sum_{k=0}^{\infty} f(k) z^{-k}$$

where, z is a complex variable.

Q7.25   *What is region of convergence (ROC)?*

The Z-transform of a discrete sequence is an infinite power series, hence the Z-transform exists only for those values of z for which the series converges. If F(z) is Z-transform of f(k) then the region of convergence (ROC) of F(z) is the set of all values of z, for which F(z) attains a finite value.

Q7.26   *State the final value theorem with regard to Z-transform.*

If f(k) is causal & stable signal and F(z) exists with z = 1 included in the ROC then the final value theorem is given by

$$f(\infty) = \underset{z \to 1}{Lt}\ (1 - z^{-1}) F(z) \qquad ; \quad \text{where } F(z) = Z\{f(k)\}$$

The final value theorem can be applied only if F(z) is analytic for $|z| > 1$.

Q7.27   *State the initial value theorem with regard to Z-transform.*

If f(k) is a causal signal and F(z) exists then the initial value of the signal is given by

$$f(0) = \underset{z \to \infty}{Lt}\ F(z) \qquad ; \quad \text{where } F(z) = Z\{f(k)\}$$

Q7.28   **Define Ƶ-transform of unit step signal.**

The unit step signal, u(k) = 1 for k ≥ 0

The Ƶ-transform at u(k) = $Ƶ\{u(k)\} = \sum_{k=0}^{\infty} z^{-k} = \dfrac{z}{z-1}$

Q7.29   **Find Ƶ-transform of a^k.**

By the definition of z-transform,

$$Ƶ\{a^k\} = \sum_{k=0}^{\infty} a^{-k} z^{-k} = \sum_{k=0}^{\infty} (az^{-1})^k = \dfrac{1}{1-az^{-1}} = \dfrac{1}{1-a/z} = \dfrac{z}{z-a}$$

Q7.30   **Find Ƶ-transform of e^{-akT}.**

By the definition of Ƶ-transform,

$$Ƶ\{e^{-akT}\} = \sum_{k=0}^{\infty} e^{-akT} z^{-k} = \sum_{k=0}^{\infty} (e^{-aT} z^{-1})^k = \dfrac{1}{1-e^{-aT}z^{-1}} = \dfrac{1}{1-e^{-aT}/z} = \dfrac{z}{z-e^{-aT}}$$

Q7.31   **What are the different methods available for inverse Ƶ-transform?**

The inverse Ƶ-transform of a function, F(z) can be obtained by any one of the following methods.

1.   Direct evaluation by contour integration (or) complex inversion integral.
2.   Partial fraction expansion.
3.   Power series expansion.

Q7.32   **What is linear (time-invariant) discrete time system (LDS)?**

A discrete-time system is a device or algorithm that operates on a discrete-time signal called the input or excitation, according to some well-defined rule, to produce another discrete-time signal called the output or the response of the system.

A discrete time system is linear if it obeys the principle of superposition and it is time invariant if its input-output relationship do not change with time.

Q7.33   **What is weighing sequence?**

The impulse response of a linear discrete-time system is called weighing sequence. The impulse response is the output of the system when the input is unit impulse.

Q7.34   **Explain how a discrete-time signal can be expressed as a summation of impulses.**

Multiplication of a discrete signal, r(k) by an unit impulse at some delay m, picks out the value r(m) of the signal r(k) at k = m, where the unit impulse is non-zero. If we repeat this multiplication over all possible delays in the range of 0 ≤ m < ∞ and sum all the product sequences, the result will be a sequence that is equal to the sequence r(k). Hence r(k) can be expressed as

$$r(k) = \sum_{m=0}^{\infty} r(m)\,\delta(k-m)$$

Q7.35   **How the output of a linear discrete-time system (LDS) is related to impulse response?**

The output or response c(k) of a linear discrete time system (LDS) is given by convolution of the input r(k) with the impulse response h(k) of the system. It is expressed as,

c(k) = r(k) * h(k)

where, the symbol * represents convolution operation.

Q7.36   **What is discrete convolution ?**

The convolution of two discrete-time signals (or sequences) is called discrete convolution. The discrete convolution of sequences $f_1(k)$ and $f_2(k)$ is defined as

$$c(k) = \sum_{m=0}^{\infty} f_1(m)\, f_2(k-m)$$

Q7.37   **Write any two properties of discrete convolution.**

The discrete convolution obeys the commutative property and Associative property.

Commutative property : $r(k) * h(k) = h(k) * r(k)$

Associative property   : $[r(k) * h_1(k)] * h_2(k) = r(k) * [h_1(k) * h_2(k)]$

Q7.38   **What is pulse transfer function?**

The transfer function of linear discrete time system is called pulse transfer function or z-transfer function. It is given by the Ƶ-transform of the impulse response of the system. It is also defined as the ratio of Ƶ-transform of output to Ƶ-transform of input of the linear discrete time system.

Pulse transfer function = $H(z) = C(z)/R(z)$

where, $H(z) = Ƶ\{h(k)\}$ = Ƶ-transform of impulse response, $h(k)$

$C(z) = Ƶ$ - transform of output of LDS

$R(z) = Ƶ$-transform of input to LDS.

Q7.39   **What is the equivalent representation of pulse sampler with ZOH?**

The pulse sampler with ZOH shown in fig Q7.40a can be replaced by an equivalent system consisting of an impulse sampler and a block with transfer function, $G_0(s) = (1 - e^{-sT})/s$ as shown in fig Q7.40b.

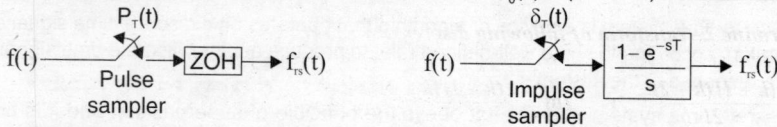

Fig Q7.40 a: *Pulse sampler with ZOH.*   **Fig Q7.40b :***Equivalent representation of Pulse sampler with ZOH.*

Q7.40   **When the z-transfer function of the system can be directly obtained from s-domain transfer function?**

When the input to the system is an impulse sampled signal, the z-transfer function can be directly obtained by taking Ƶ-transform of the s-domain transfer function.

Q7.41   **Give the steps involved in determining the pulse transfer function of G(z) from G(s).**

The following are the steps involved in determing the pulse transfer function G(z) from G(s).
1.   Determine g(t) from G(s), where $g(t) = \mathcal{L}^{-1}[G(s)]$
2.   Determine the discrete sequence g(kT) by replacing t by kT in g(kT).
3.   Take Ƶ-transform of g(kT), which is the required z-transfer function, G(z).

Q7.42   **When a sampled data control system is stable? or what is the stability criterion for sampled data control system?**

The stability criterion for sampled data control system states that the system is stable if all the poles of the z-transfer function of the system lies inside the unit circle in z-plane.

Q7.43   **What is characteristic equation of sampled data system?**

The characteristic equation is the denominator polynomial of the z-transfer function of a sampled data control system.

Q7.44   *What are the necessary conditions to be satisfied for the stability of sampled data control system?*

Let F(z) be the characteristic equation of the system.

Now, the necessary conditions to be satisfied for the stability of the system are,

$$F(1) > 0 \text{ and } (-1)^n F(-1) > 0.$$

Q7.45   *How many rows are formed in Jury's table and what are the sufficient conditions to be checked from this table for stability.*

The Jury's table consists of $(2n - 3)$ rows, where n is the order of the system (and order of the characteristic polynomial).

From the table $(n - 1)$ conditions are checked for ascertaining sufficiency. They are

$$|a_0| < |a_n|$$
$$|b_0| > |b_{n-1}|$$
$$|c_0| > |c_{n-2}|$$
$$\vdots$$
$$|r_0| > |r_2|$$

Q7.46   *What is bilinear transformation.*

The bilinear transformation is a transformation used to map the interior of unit circle in the z-plane into the left half of r-plane. This transformation is achieved by choosing,

$$z = \frac{1+r}{1-r}$$

7.15 EXERCISES

E7.1   *Determine Z̄-transform of following discrete time sequences.*

(a) $\dfrac{(k+1)(k+2)}{2!}$ (b) $\dfrac{(k+1)(k+2)(k+3)}{3!} a^k$ (c) $a^k/k!$ (d) $(kT)^2$

E7.2   *Determine the z-domain transfer function of the following s-domain transfer function.*

(a) $\dfrac{k}{s(s+4)}$ (b) $\dfrac{(a-b)}{(s+a)(s+b)}$ (c) $\dfrac{b}{s(s+b)}$ (d) $\dfrac{s(2s+3)}{(s+1)^2(s+2)}$

E7.3   *Find the inverse Z̄-transform of the following function.*

(a) $\dfrac{5z}{(z-1)(z-2)}$ (b) $\dfrac{5z}{(z-1)^2(z-2)}$ (c) $\dfrac{z(z^2-1)}{(z^2+1)^2}$ (d) $\dfrac{z(z+1)}{(z-1)^3}$

E7.4   *The input output of a sampled data control system is described by the difference equation.*

$$3c(k+2) + 4c(k+1) + c(k) = r(k+2) + 2r(k+1) - 3r(k)$$

where c(0) = 1;   c(1) = −2

Determine the z-transfer function of the system. Also obtain the weighting sequence of the system.

E7.5   *Determine the z-transfer function of two cascaded systems each described by the difference equation*

$$c(k) = 0.5 \, c(k-1) + r(k)$$

E7.6   *Determine the pulse transfer function of a sampled data system shown in fig E7.6.*

E7.7   *Determine the response c(kT) for the system shown in fig E7.7.*

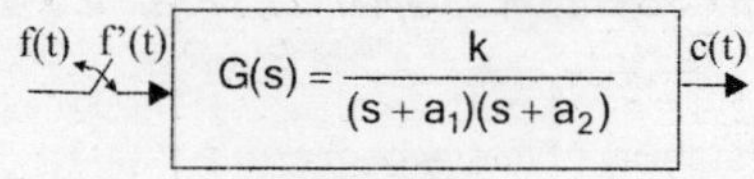

Fig E7.6

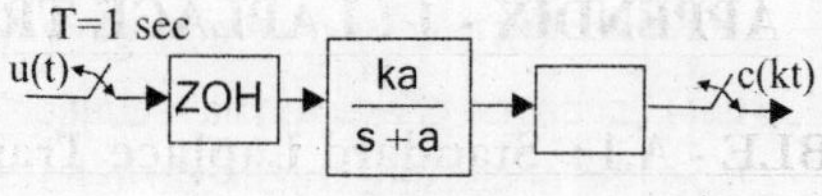

Fig E7.7

E7.8   *Determine the stability of sampled data control systems described by the following characteristic equation.*

   a. $z^3 + 3z^2 - 2.75z + 0.75 = 0$

   b. $z^3 + 4z^2 + 4z + 1 = 0$

   c. $z^4 - 1.4z^3 + 0.4z^2 + 0.08z + 0.002 = 0.$

ANSWER FOR EXERCISE PROBLEMS

E7.1   a) $\dfrac{z^3}{(z-a)^3}$ b) $\dfrac{z^4}{(z-a)^4}$ c) $e^{az^{-1}}$ d) $\dfrac{zT^2(z+1)}{(z-1)^3}$

E7.2   a) $\dfrac{1}{4}\left[\dfrac{z}{(z-1)^2} + \dfrac{ze^{-4T}}{(z-e^{-4T})^2}\right]$ b) $\left[\dfrac{z}{z-e^{-bT}} - \dfrac{z}{z-e^{-4T}}\right]$

   c) $\left[\dfrac{z}{z-1} - \dfrac{z}{z-e^{-bT}}\right]$ d) $\left[\dfrac{2z}{(z-e^{-2T})} - \dfrac{zTe^{-T}}{(z-e^{-T})^2}\right]$

E7.3   a) $5\left[(2)^k - u(k)\right]$ b) $5\left[(2)^k - k(-1)^k - u(k)\right]$ c) $\dfrac{k}{2}\left[(-i)^k + (i)^k\right]$ d) $2k^2a^k + u(k)$

E7.4 $H(z) = \dfrac{4}{(z-2)^2}$

E7.5 $H(z) = \dfrac{6}{z+1} - \dfrac{5.333}{z+0.3333}$; $h(k) = 6(-1)^{k-1} u(k-1) - 5.333(-0.3333)^{k-1} u(k-1)$

E7.6 $\dfrac{k}{a_2-a_1}\left[\dfrac{z}{z-e^{-a_1T}} - \dfrac{z}{z-e^{-a_2T}}\right].$

E7.7 $c(k) = k\left[1 - e^{a(1-k)}\right]$

E7.8   a) Unstable

   b) Unstable

   c) Stable

APPENDIX - 1 : LAPLACE TRANSFORM PAIRS AND PROPERTIES

TABLE - A.1 : Standard Laplace Transform Pairs

S.No.	f(t)	F(s)
1.	Unit impulse, $\delta(t)$	1
2.	Unit step, $u(t)$	$1/s$
3.	t	$1/s^2$
4.	$\dfrac{t^{n-1}}{(n-1)!}$; where, $n = 1,2,3......$	$1/s^n$
5.	e^{-at}	$\dfrac{1}{s+a}$
6.	t^n ; where, $n = 1, 2, 3,....$	$\dfrac{n!}{s^{n+1}}$
7.	$t\,e^{-at}$	$\dfrac{1}{(s+a)^2}$
8.	$\dfrac{1}{(n-1)}\,t^{n-1}\,e^{-at}$; where, $n = 1,2,3....$	$\dfrac{1}{(s+a)^n}$
9.	$t^n\,e^{-at}$; where, $n = 1, 2, 3,.....$	$\dfrac{n!}{(s+a)^{n+1}}$
10.	$\text{Sin }\omega t$	$\dfrac{\omega}{s^2+\omega^2}$
11.	$\text{Cos }\omega t$	$\dfrac{s}{s^2+\omega^2}$
12.	$\text{Sinh }\omega t$	$\dfrac{\omega}{s^2-\omega^2}$
13.	$\text{Cosh }\omega t$	$\dfrac{s}{s^2-\omega^2}$
14.	$e^{-at}\,\text{Sin }\omega t$	$\dfrac{\omega}{(s+a)^2+\omega^2}$
15.	$e^{-at}\,\text{Cos }\omega t$	$\dfrac{s+a}{(s+a)^2+\omega^2}$

TABLE - A1.2 : Properties of Laplace Transforms

Note : $\mathcal{L}\{f(t)\} = F(s);$ $\mathcal{L}\{f_1(t)\} = F_1(s);$ $\mathcal{L}\{f_2(t)\} = F_2(s)$

Property	Time domain signal	s-domain signal
Amplitude scaling	$A\,f(t)$	$A\,F(s)$
Linearity	$a_1\,f_1(t) \pm a_2\,f_2(t)$	$a_1\,F_1(s) \pm a_2\,F_2(s)$
Time differentiation	$\dfrac{d}{dt}f(t)$	$s\,F(s) - f(0)$
	$\dfrac{d^n}{dt^n}f(t)$	$s^n F(s) - \displaystyle\sum_{K=1}^{n} s^{n-K}\, \dfrac{d^{(K-1)}f(t)}{dt^{K-1}}\bigg|_{t=0}$
Time integration	$\displaystyle\int f(t)\,dt$	$\dfrac{F(s)}{s} + \dfrac{\left[\displaystyle\int f(t)\,dt\right]\big|_{t=0}}{s}$
	$\displaystyle\int \ldots\ldots \int f(t)\,(dt)^n$	$\dfrac{F(s)}{s^n} + \displaystyle\sum_{K=1}^{n} \dfrac{1}{s^{n-K+1}}\left[\int \ldots\ldots \int f(t)\,(dt)^K\right]\bigg|_{t=0}$
Frequency shifting	$e^{\pm at}\,f(t)$	$F(s \mp a)$
Time shifting	$f(t \pm \alpha)$	$e^{\pm \alpha s}\,F(s)$
Frequency differentiation	$t\,f(t)$	$-\dfrac{dF(s)}{ds}$
	$t^n\,f(t)$	$(-1)^n\,\dfrac{d^n}{ds^n}\,F(s)$; where, $n = 1, 2, 3, \ldots\ldots$
Frequency integration	$\dfrac{1}{t}\,f(t)$	$\displaystyle\int_s^{\infty} F(s)\,ds$
Time scaling	$f(at)$	$\dfrac{1}{|a|}\,F\!\left(\dfrac{s}{a}\right)$
Periodicity	$f(t + nT)$	$\dfrac{1}{1 - e^{-sT}}\displaystyle\int_0^T f(t)\,e^{-st}\,dt$
Initial value theorem	$\underset{t \to 0}{\text{Lt}}\ f(t) = f(0)$	$\underset{s \to \infty}{\text{Lt}}\ sF(s)$
Final value theorem	$\underset{t \to 0}{\text{Lt}}\ f(t) = f(\infty)$	$\underset{s \to 0}{\text{Lt}}\ sF(s)$
Convolution theorem	$f_1(t) * f_2(t)$ $= \displaystyle\int_{-\infty}^{+\infty} f_1(\lambda)\,f_2(t-\lambda)\,d\lambda$	$F_1(s)F_2(s)$

APPENDIX - 2 : ROOTS OF ALGEBRAIC EQUATION BY LIN'S METHOD

Consider the equation, $s^n + a_{n-1}s^{n-1} + a_{n-2}s^{n-2} + ... + a_2 s^2 + a_1 s + a_0 = 0$

If n is even then the above equation can be expressed as a product of quadratic factors. Each quadratic factor is obtained by assuming a trial divisor and dividing the equation as explained below. The first trial factor uses the three lowest order terms of the original equation. For the above equation, this trial factor is $s^2 + \dfrac{a_1}{a_2}s + \dfrac{a_0}{a_2}$.

The original equation is divided by this first trial factor as follows :

$$
s^2 + \frac{a_1}{a_2}s + \frac{a_0}{a_2} \overline{)
\begin{array}{l}
s^{n-2} + ... \\
\hline
s^n + a_{n-1}s^{n-1} + a_{n-2}s^{n-2} + ...a_2 s^2 + a_1 s_1 + a_0 \\
\hspace{2cm} \vdots \hspace{1.5cm} \vdots \hspace{1.5cm} \vdots \\
\hline
\hspace{2cm} b_2 s^2 + b_1 s_1 + b_0 \\
\hspace{2cm} c_2 s^2 + c_1 s_1 + c_0 \\
\hline
\hspace{2cm} \text{Remainder}
\end{array}}
$$

If the remainder is too large, then the next trial used is $s^2 + \dfrac{b_1}{b_2}s + \dfrac{b_0}{b_2}$.

This procedure is continued until the remainder is negligible. The last trial factor is a quadratic factor of the original equation. The quotient polynomial, which is of order (n-2) contains the remaining factors of the original equation. Lin's method is then applied to the quotient polynomial to obtain the other quadratic factor of the original equation.

We know that complex roots occur in complex conjugate pairs. Therefore, if the highest power of the original equation is odd, there must be atleast one real root. To determine this root, the trial divisor is chosen from the two lowest order terms of the original equation. For the above equation, if n is odd, the first trial divisor is $s + \dfrac{a_0}{a_1}$.

The original equation is divided by this trial divisor and the process outlined above is continued till the remainder is negligible. The last trial divisor gives the real root of the original equation. The remainder equation can be expressed as a product of quadratic factors and they are determined as explained above.

APPENDIX - 3 : MATRIX FUNDAMENTALS

A3.1 DEFINITIONS INVOLVING MATRICES

Matrix : A matrix is an ordered array of elements which may be real numbers, complex numbers, functions or operators. In general the array consists of m rows and n columns. When m = n, the matrix is called square matrix. When n = 1, the matrix is called column matrix or vector. When m = 1, the matrix is called row matrix or vector.

Diagonal matrix : It is a square matrix whose elements other than main diagonal are all zeros.

Unit Matrix : It is a diagonal matrix whose diagonal elements are all equal to unity. The elements other than diagonal are all zeros. It is denoted by **I**.

Transpose : If the rows and columns of an $m \times n$ matrix **A** are interchanged, then the resulting $n \times m$ matrix is called the transpose of **A**. The transpose of **A** is denoted by $\mathbf{A}^T$.

Determinant : A determinant consisting of the elements of a square matrix (in the order given in the matrix) is called the determinant of the matrix.

Symmetric matrix : A square matrix is symmetric if it is equal to its transpose, i.e., $\mathbf{A}^T = \mathbf{A}$. If **A** is a square matrix, then $\mathbf{A} + \mathbf{A}^T$ is a symmetric matrix.

Skew-symmetric matrix : A square matrix is skew-symmetric if it is equal to the negative of its transpose, i.e., $\mathbf{A}^T = -\mathbf{A}$. If **A** is a square matrix then $\mathbf{A} - \mathbf{A}^T$ is a skew symmetric matrix.

Orthogonal Matrix : A matrix **A** is called an orthogonal matrix if it is real and satisfies the relationship $\mathbf{A}^T \mathbf{A} = \mathbf{A} \mathbf{A}^T = \mathbf{I}$.

Minor : If the ith row and j^{th} column of determinant **A** are deleted, the remaining $(n-1)$ rows and columns form a determinant $\mathbf{M}_{ij}$. This determinant is called the minor of the element a_{ij}.

Cofactor : The cofactor C_{ij} of element a_{ij} of the matrix **A** is defined as $C_{ij} = (-1)^{(i+j)} \mathbf{M}_{ij}$, where $\mathbf{M}_{ij}$ is the minor of a_{ij}.

Adjoint matrix : The adjoint matrix of a square matrix **A** is found by replacing each element a_{ij} of matrix **A** by its cofactor C_{ij} and then transposing.

Singular matrix : A square matrix is called singular if its associated determinant is zero. If the determinant of the matrix is nonzero then the matrix is nonsingular.

Rank of matrix : A matrix **A** is said to have a rank r if there exists an $r \times r$ submatrix of **A** which is nonsingular and all other $q \times q$ submatrices are singular, where $q \geq (r+1)$.

Conjugate matrix : The conjugate of a matrix **A** is the matrix in which each element is the complex conjugate of the corresponding element of **A**. The conjugate of **A** is denoted by $\mathbf{A}^*$.

Real matrix : If all the elements of a matrix are real then the matrix is called real matrix. A real matrix is equal to its conjugate.

A3.2 EIGENVALUES AND EIGENVECTORS

A nonzero column vector **X** is an eigenvector of a square matrix **A**, if there exists a scalar λ such that $\mathbf{AX} = \lambda \mathbf{X}$, then λ is a eigenvalue (or characteristic value) of **A**. Eigenvalue may be zero but the corresponding vector may not be a zero vector.

The characteristic equation of n × n matrix **A** is the nth degree polynomial of equation, $|\lambda I - A| = 0$, where **I** is the unit matrix. Solving the characteristic equation for λ gives the eigenvalues of **A**. The eigenvalues may be real, complex or multiples of each other.

Once an eigenvalue is determined it may be substituted into $AX = \lambda X$ and then that equation may be solved for the corresponding eigenvector.

PROPERTIES OF EIGENVALUES AND EIGENVECTORS

1. *The sum of the eigenvalues of a matrix is equal to its trace, which is the sum of the elements on its main diagonal.*

2. *Eigenvectors corresponding to different eigenvalues are linearly independent.*

3. *A matrix is singular if and only if it has a zero eigenvalue.*

4. *If X is an eigenvector of A corresponding to the eigenvalue of λ and A is invertible, then X is an eigenvector of A^{-1} corresponding to its eigenvalue $1/\lambda$.*

5. *If X is an eigenvector of a matrix then KX is also an eigenvector for any nonzero constant K. Here both X and KX correspond to the same eigenvalue.*

6. *A matrix and its transpose have the same eigenvalues.*

7. *The eigenvalues of an upper or lower triangular matrix are the elements on its main diagonal.*

8. *The product of the eigenvalues (counting multiplicities) of the matrix equals the determinant of the matrix.*

9. *If X is an eigenvector of A corresponding to eigenvalue of λ , then X is an eigenvector of A-CI corresponding to the eigenvalue λ- C for any scalar C.*

DETERMINATION OF EIGENVECTORS

Case i : Distinct eigenvalues

If the eigenvalues of A are all distinct, then we have only one independent eigenvector corresponding to any particular eigenvalue λ_i. The eigenvector corresponding to λ_i may be obtained by taking cofactors of matrix $[\lambda_i I - A]$ along any row.

Let, m_i = Eigenvector corresponding to λ_i

Now the eigenvector m_i is given by

$$m_i = \begin{bmatrix} C_{k1} \\ C_{k2} \\ \vdots \\ C_{kn} \end{bmatrix} ; \ k = 1 \text{ or } 2 \text{ or} \dots n \qquad \dots (A3.1)$$

where, $C_{k1}, C_{k2}, \dots C_{kn}$ are cofactors of matrix $[\lambda_i I - A]$ along k^{th} row.

Case ii : Multiple eigenvalues

In this case the eigenvectors corresponding to the distinct eigenvalues are evaluated as mentioned in case (i).

If the matrix has repeated eigenvalues with multiplicity "q", then there exists only one independent eigenvector corresponding to that repeated eigenvalue. If λ_i is a repeated eigenvalue, then the independent vector corresponding to λ_i can be evaluated by taking the cofactor of matrix $[\lambda_i I–A]$ along any row as mentioned in case (1). The remaining (q–1) eigenvectors can be obtained as shown in equation (A3.2).

Let, $m_p = p^{th}$ *eigenvector corresponding to repeated eigenvalue* λ_i.

$$m_p = \begin{bmatrix} \dfrac{1}{p!} \dfrac{d^p}{d\lambda_i^p} C_{k1} \\ \dfrac{1}{p!} \dfrac{d^p}{d\lambda_i^p} C_{k2} \\ \vdots \\ \dfrac{1}{p!} \dfrac{d^p}{d\lambda_i^p} C_{kn} \end{bmatrix} \quad ; \quad p = 1, 2, 3, ...(q-1) \qquad .....(A3.2)$$

where, $C_{k1}, C_{k2}, C_{k3}.....C_{kn}$ *are cofactors of matrix* $[\lambda_i I–A]$ *along* k^{th} *row.*

A3.3 SIMILARITY TRANSFORMATION

The square matrices **A** and **B** are said to be similar if a nonsingular matrix **P** exists such that

$$P^{-1}AP = B \qquad\qquad .....(A3.3)$$

The process of transformation is called similarity transformation and it is a linear transformation. The matrix **P** is called transformation matrix. Also the matrix, **A** can be obtained from **B** by a similarity transformation with a transformation matrix P^{-1},

i.e., $A = P B P^{-1}$ $\qquad\qquad .....(A3.4)$

The similarity transformation can be used for diagonalization of a square matrix. If an $n \times n$ matrix has n linearly independent eigenvectors (i.e., with distinct eigenvalues) then it can be diagonalized by a similarity transformation. If a matrix has multiple eigenvalues then it will not have a complete set of n linearly independent eigenvectors and so it cannot be diagonalized. However such a matrix can be transformed into a Jordan matrix (Jordan canonical form).

The transformation matrix for diagonalization or converting to Jordan form can be obtained from eigenvectors. For a system with n state variables we can find n numbers of eigenvectors $m_1, m_2, m_3, ...,m_n$. The eigenvectors are column vectors of order (n×1). The transformation matrix is obtained by arranging the eigenvectors columnwise as shown in equation (A3.5). This transformation matrix is also called *Modal matrix* and denoted by *M*.

Modal matrix, $M = [\ m_1\ \ m_2\ \ m_3\ .... \ m_n\ ]$ $\qquad\qquad .....(A3.5)$

The similarity transformation will not alter certain properties of the matrix. A property of a matrix is said to be invariant if it is possessed by all similar matrices. The determinant, characteristic equation and trace of a matrix are invariant under a similarity transformation. Since the characteristic equation is invariant the eigenvalues are also invariant under a linear or similarity transformation.

APPENDIX - 4 : Z - TRANSFORM PAIRS AND PROPERTIES

TABLE-A4.1 : Properties of One-Sided Z-Transform

Note : $F(z) = Z\{f(k)\}$; $F_1(z) = Z\{f_1(k)\}$; $F_2(z) = Z\{f_2(k)\}$

Property	Discrete sequence	Z - transform
Linearity	$a_1 f_1(k) + a_2 f_2(k)$	$a_1 F_1(z) + a_2 F_2(z)$
Shifting, $m \geq 0$	$f(k + m)$	$z^m F(z) - \sum_{i=0}^{m=-1} f(i) z^{m-i}$
	$f(k - m)$	$z^{-m} F(z)$
Multiplication by k^m (or differentiation in z-domain)	$k^m f(k)$	$\left(-z \dfrac{d}{dz}\right)^m F(z)$
Scaling in z-domain (or multiplication by a^k)	$a^k f(k)$	$F(a^{-1}z)$
Time reversal	$f(-k)$	$F(z^{-1})$
Conjugation	$f^*(k)$	$F^*(z^*)$
Convolution	$\sum_{m=0}^{k} h(k-m) r(m)$	$H(z) R(z)$
Initial value	$f(0) = \underset{z \to \infty}{Lt}\ F(z)$	
Final value	$f(\infty) = \underset{z \to 1}{Lt}\ (1 - z^{-1}) F(z)$ $= \underset{z \to 1}{Lt}\ (z-1) F(z)$ if $F(z)$ is analytic for $|z| > 1$	

TABLE-A4.2 : Some Common One Sided Ƶ-transform Pairs

> **Note :** *Two sided sequence can be converted to one sided sequence by multiplying by u(k).*

$f(t)$; $t \geq 0$	$f(k)$ or $f(kT)$; $k \geq 0$	$F(z)$
	$\delta(k)$	1
	$u(k)$ or 1	$z/(z-1)$
	a^k	$z/(z-a)$
	$k\,a^k$	$\dfrac{az}{(z-a)^2}$
	$k^2\,a^k$	$\dfrac{az(z+a)}{(z-a)^2}$
	$(k+1)\,a^k$	$\dfrac{z^2}{(z-a)^2}$
	$\dfrac{(k+1)(k+2)}{2!}\,a^k$	$\dfrac{z^3}{(z-a)^3}$
	$\dfrac{(k+1)(k+2)(k+3)}{3!}\,a^k$	$\dfrac{z^4}{(z-a)^4}$
	$\dfrac{a^k}{k!}$	$e^{az^{-1}}$
t	kT	$\dfrac{TZ}{(z-1)^2}$
t^2	$(kT)^2$	$\dfrac{T^2 z(z+1)}{(z-1)^3}$
e^{-at}	e^{-akT}	$\dfrac{z}{z-e^{-aT}}$
te^{-aT}	kTe^{-akT}	$\dfrac{zTe^{-aT}}{(z-e^{-aT})^2}$
$\sin \omega t$	$\sin \omega kT$	$\dfrac{z \sin \omega T}{z^2 - 2z \cos \omega T + 1}$
$\cos \omega t$	$\cos \omega kT$	$\dfrac{z(z - \cos \omega T)}{z^2 - 2z \cos \omega T + 1}$

APPENDIX - 5 : MATLAB COMMANDS

Operators and Special Characters	
+	Plus; addition operator.
−	Minus; subtraction operator.
*	Scalar and matrix multiplication operator.
.*	Array multiplication operator.
∧	Scalar and matrix exponentiation operator.
.∧	Array exponentiation operator.
\	Left-division operator.
/	Right-division operator.
.\	Array left-division operator.
./	Array right-division operator.
:	Colon; generates regularly spaced elements and represents an entire row/column.
( )	Parentheses; encloses function arguments and array indices; overrides precedence.
[ ]	Brackets; enclosures array elements.
.	Decimal point.
...	Ellipsis; line-continuation operator.
,	Comma; separates statements and elements in a row.
;	Semicolon; separates columns and suppresses display.
%	Percent sign; designates a comment and specifies formatting.
_	Quote sign and transpose operator.
._	Nonconjugated transpose operator.
=	Assignment (replacement) operator.

Logical and Relational Operators	
==	Relational operator : equal to.
~=	Relational operator : not equal to.
<	Relational operator : less than.
<=	Relational operator : less than or equal to.
>	Relational operator : greater than.
>=	Relational operator : greater than or equal to.
&	Logical operator : AND.
|	Logical operator : OR.
~	Logical operator : NOT.
xor	Logical operator : EXCLUSIVE OR.

Commands for Managing a Session	
clc	Clears Command window.
clear	Removes variables from memory.
exist	Checks for existence of file or variable.
global	Declares variables to be global.
help	Searches for a help topic.
lookfor	Searches help entries for a keyword.
quit	Stops MATLAB.
who	Lists current variables.
whos	Lists current variables (long display).

Format Codes for fprintf and fscanf	
%s	Format as a string.
%d	Format as an integer.
%f	Format as a floating point value.
%e	Format as a floating point value in scientific notation.
%g	Format in the most compact form : %f or %e.
\n	Insert a new line in the output string.
\t	Insert a tab in the output string.

Array Commands	
cat	Concatenates arrays.
find	Finds indices of nonzero elements.
length	Computes number of elements.
linspace	Creates regularly spaced vector.
logspace	Creates logarithmically spaced vector.
max	Returns largest element.
min	Returns smallest element.
prod	Product of each column.
reshape	Change size
size	Computes array size.
sort	Sorts each column.
sum	Sums each column.

Input/Output Commands	
disp	Displays contents of an array or string.
fscanf	Read formatted data from a file.
format	Controls screen-display format.
fprintf	Performs formatted writes to screen or file.
input	Displays prompts and waits for input.
;	Suppresses screen printing.

Special Matrices	
eye	Creates an identity matrix.
ones	Creates an array of ones.
zeros	Creates an array of zeros.

Program Flow Control	
break	Terminates execution of a loop.
case	Provides alternate execution paths within switch structure.
else	Delineates alternate block of statements.
elseif	Conditionally executes statements.
end	Terminates for, while, and if statements.
error	Displays error messages.
for	Repeats statements a specific number of times
if	Executes statements conditionally.
otherwise	Default part of switch statement.

`return`	Return to the invoking function.
`switch`	Directs program execution by comparing point with case expressions.
`warning`	Display a warning message.
`while`	Repeats statements an indefinite number of times.

Basic xy Plotting Commands	
`axis`	Sets axis limits.
`fplot`	Intelligent plotting of functions.
`grid`	Displays gridlines.
`plot`	Generates xy plot.
`print`	Prints plot or saves plot to a file
`title`	Puts text at top of plot.
`xlabel`	Adds text label to x-axis.
`ylabel`	Adds text label to y-axis.

Plot Enhancement Commands	
`axes`	Creates axes objects.
`gtext`	Enables label placement by mouse.
`hold`	Freezes current plot.
`subplot`	Creates plots in subwindows.
`text`	Places string in figure.

Specialized Plot Commands	
`polar`	Creates polar plot.
`semilogx`	Creates semilog plot (logarithmic abscissa).
`semilogy`	Creates semilog plot (logarithmic ordinate).
`stem`	Creates stem plot.

Exponential and Logarithmic Functions	
`exp(x)`	Exponential; e^x.
`log(x)`	Natural logarithm; $\ln(x)$.
`log10(x)`	Common (base 10) logarithm; $\log(x) = \log_{10}(x)$.
`sqrt(x)`	Square root of x; $\sqrt{x}$.

Complex Functions	
`abs(x)`	Absolute value; $|x|$.
`angle(x)`	Angle of a complex number x.
`conj(x)`	Complex conjugate of x.
`imag(x)`	Imaginary part of a complex number x.
`real(x)`	Real part of a complex number x.

State Space Functions	
`ss2tf`	Computes transfer function from state model.
`tf2ss`	Computes state model from transfer function.

Transform Functions	
`ilaplace`	Returns the inverse Laplace transform.
`iztrans`	Returns the inverse $\mathcal{Z}$-transform.
`laplace`	Returns the Laplace transform.
`ztrans`	Returns the $\mathcal{Z}$-transform.

INDEX

INDEX